Lecture Notes in Artificial Intelligence 5796

Edited by R. Goebel, J. Siekmann, and W. Wahlster

Subseries of Lecture Notes in Computer Science

Ngoc Thanh Nguyen Ryszard Kowalczyk
Shyi-Ming Chen (Eds.)

# Computational Collective Intelligence

## Semantic Web, Social Networks and Multiagent Systems

First International Conference, ICCCI 2009
Wrocław, Poland, October 5-7, 2009
Proceedings

Springer

Series Editors

Randy Goebel, University of Alberta, Edmonton, Canada
Jörg Siekmann, University of Saarland, Saarbrücken, Germany
Wolfgang Wahlster, DFKI and University of Saarland, Saarbrücken, Germany

Volume Editors

Ngoc Thanh Nguyen
Wrocław University of Technology
Institute of Informatics
Str. Janiszweskiego 11/17, 50-370 Wrocław, Poland
E-mail: Ngoc-Thanh.Nguyen@pwr.wroc.pl

Ryszard Kowalczyk
Swinburne University of Technology
Centre for Complex Software Systems and Services
P.O. Box 218, Hawthorn, Victoria 3122, Australia
E-mail: rkowalczyk@ict.swin.edu.au

Shyi-Ming Chen
National Taiwan University of Science and Technology
Department of Computer Science and Information Engineering
G#43, Sec. 4, Keelung Rd., Taipei, 106, Taiwan, R.O.C.
E-mail: smchen@mail.ntust.edu.tw

Library of Congress Control Number: 2009934783

CR Subject Classification (1998): I.2, I.2.11, J.4, K.4.2, H.3.5, D.2, D.3

LNCS Sublibrary: SL 7 – Artificial Intelligence

ISSN 0302-9743
ISBN-10 3-642-04440-9 Springer Berlin Heidelberg New York
ISBN-13 978-3-642-04440-3 Springer Berlin Heidelberg New York



springer.com


Printed in Germany

Typesetting: Camera-ready by author, data conversion by Scientific Publishing Services, Chennai, India
Printed on acid-free paper SPIN: 12753499 06/3180 5 4 3 2 1 0

# Preface

Computational collective intelligence (CCI) is most often understood as a subfield of artificial intelligence (AI) dealing with soft computing methods that enable group decisions to be made or knowledge to be processed among autonomous units acting in distributed environments. The needs for CCI techniques and tools have grown significantly recently as many information systems work in distributed environments and use distributed resources. Web-based systems, social networks and multi-agent systems very often need these tools for working out consistent knowledge states, resolving conflicts and making decisions. Therefore, CCI is of great importance for today's and future distributed systems.

Methodological, theoretical and practical aspects of computational collective intelligence, such as group decision making, collective action coordination, and knowledge integration, are considered as the form of intelligence that emerges from the collaboration and competition of many individuals (artificial and/or natural). The application of multiple computational intelligence technologies such as fuzzy systems, evolutionary computation, neural systems, consensus theory, etc., can support human and other collective intelligence and create new forms of CCI in natural and/or artificial systems. Three subfields in the application of computational intelligence technologies to support various forms of collective intelligence are gaining special attention but they are not the only ones: Semantic Web (as an advanced tool increasing collective intelligence), social network analysis (as the field targeted to the emergence of new forms of CCI), and multiagent systems (as a computational and modeling paradigm especially tailored to capture the nature of CCI emergence in populations of autonomous individuals).

The aim of this conference series (International Conference on Computational Collective Intelligence - ICCCI) is to provide an internationally respected forum for scientific research in the computer-based methods of collective intelligence and their applications in (but not limited to) such fields as the Semantic Web, social networks and multiagent systems.

This volume of the LNCS/LNAI series contains the proceedings of the first event in the ICCCI series (ICCCI 2009) which was held in Wroclaw, Poland, during October 5–7, 2009. The conference was organized by Wroclaw University of Technology (Poland) in cooperation with Swinburne University of Technology (Australia) and National Taiwan University of Science and Technology (Taiwan).

The conference attracted a large number of scientists and practitioners who submitted their papers for four main tracks covering the methodology and applications of computational collective intelligence and three special sessions on specific topics within the field. Each paper was reviewed by two to four members of the International Program Committee. Many of them were reviewed using the double-blind mode. From the submissions for ICCCI 2009 coming from more than 25 countries throughout the world, only 71 papers were selected to be published in the proceedings.

The Program Committee defined the following main topics as related to CCI:

- Semantic Web: semantic annotation of Web data resources; Web Services (service description, discovery, composition); ontology management (mediation and reconciliation, creation, evaluation, merging, alignment, evolution, linking); automatic metadata generation; (semi-) automatic ontology creation; Semantic Web inference schemes; reasoning in the Semantic Web; knowledge portals; information discovery and retrieval in the Semantic Web; etc.
- Social Networks: computational technologies in social networks creation and support; advanced groupware and social networks; models for social network emergence and growth; ontology development in social networks; advanced analysis for social networks dynamics; social networks and semantic communication.
- Multiagent Systems: cooperative distributed problem solving; task and resource allocation; mechanism design, auctions, and game theory; modeling other agents and self; multiagent planning; negotiation protocols; multiagent learning; conflict resolution; trust and reputation management; privacy, safety and security; scalability, robustness and dependability; social and organizational structures; verification and validation; novel computing paradigms (autonomic, grid, P2P, ubiquitous computing); brokering and matchmaking; agent-oriented software engineering, including implementation languages and frameworks; mobile agents; performance, scalability, robustness, and dependability; verification and validation; E-business agents; pervasive computing; privacy, safety, and security.

We would like to thank the invited speakers – Roman Słowiński (Poland), Pierre Lévy (Canada), and Piotr Jędrzejowicz (Poland) – for their interesting and informative talks of world-class standard.

Special thanks go to the Organizing Chair (Radoslaw Katarzyniak) for his efforts in the organizational work. Thanks are due to the Program Committee and the board of reviewers, essential for reviewing the papers to ensure their high quality. We thank the members of the Local Organizing Committee, Publicity Chairs and Special Sessions Chairs. We acknowledge with gratitude the efforts of the Foundation for Development of Wroclaw University of Technology for coordinating the organization of the conference. We extend cordial thanks to the Institute of Informatics and the Faculty of Computer Science of Wroclaw University of Technology for the supports with the administration and network services. Finally, we thank the authors, presenters and delegates for their valuable contribution to this successful event.

Thanks are also due to the many other experts who contributed to making the event a success.

We hope that ICCCI 2009 has significantly contributed to the fulfillment of academic excellence and will lead to even greater successes of ICCCI events in the future.

October 2009

Ngoc Thanh Nguyen
Ryszard Kowalczyk
Shyi-Ming Chen

# ICCCI 2009 Conference Organization

## General Chair

Ngoc Thanh Nguyen
Wroclaw University of Technology, Poland

## Program Chairs

Ryszard Kowalczyk
Swinburne University of Technology, Australia

Shyi-Ming Chen
National Taiwan University of Science and Technology, Taiwan

## Organizing Chair

Radosław Katarzyniak
Wroclaw University of Technology, Poland

## Special Session Chairs

Tokuro Matsuo
Yamagata University, Japan

Janusz Sobecki
Wroclaw University of Technology, Poland

## Publicity Chairs

Jason J. Jung
Yeungnam University, South Korea

Maciej Kiewra
Wroclaw University of Technology, Poland

## Organizing Committee

Maciej Kiewra
Adrianna Kozierkiewicz-Hetmańska
Anna Kozłowska
Wojciech Lorkiewicz

Marcin Maleszka
Bernadetta Mianowska
Marcin Pietranik
Grzegorz Popek
Grzegorz Skorupa
Zbigniew Telec

## Keynote Speakers

Roman Słowiński
Poznań University of Technology, Poland

Pierre Lévy
University of Ottawa, Canada

Piotr Jędrzejowicz
Gdynia Maritime University, Poland

## Special Sessions

1. Nature-Inspired Collective Intelligence NICI 2009
   Heitor Silvério Lopes, Federal University of Technology, Brazil
   Dariusz Król, Wroclaw University of Technology, Poland

2. Dynamics of Real-World Social Networks
   Krzysztof Juszczyszyn, Wrocław University of Technology, Poland
   Jason J. Jung, Yeungnam University, Korea
   Katarzyna Musiał, Wrocław University of Technology, Poland

3. Web Systems Analysis WebSys 2009
   Kazimierz Choroś, Wroclaw University of Technology, Poland

## International Program Committee

| | |
|---|---|
| Stanisław Ambroszkiewicz | Polish Academy of Sciences, Poland |
| Youcef Baghdadi | Sultan Qaboos University, Oman |
| Djamal Benslimane | Lyon 1 University, France |
| Jamal Bentahar | Concordia University, Canada |
| Peter Braun | The Agent Factory GmbH, Germany |
| Paul A. Buhler | College of Charleston, USA |
| Key-Sun Choi | KAIST, Korea |
| Oscar Cordón | European Centre for Soft Computing, Asturias, Spain |
| Jiangbo Dang | Siemens Corporate Research, USA |
| Manuel Núñez García | Universidad Complutense de Madrid, Spain |
| Mauro Gaspari | University of Bologna, Italy |
| Daniela Godoy | Unicen University, Argentina |
| Dominic Greenwood | Whitestein Technologies, Switzerland |

| | |
|---|---|
| Slimane Hammoudi | ESEO, France |
| Tzung-Pei Hong | National University of Kaohsiung, Taiwan |
| Wen-Lian Hsu | Academia Sinica, Taiwan |
| Jingshan Huang | University of South Carolina, USA |
| Do-Sam Hwang | Yeungnam University, Korea |
| Gordan Jezic | University of Zagreb, Croatia |
| Kang-Hyun Jo | Ulsan University, Korea |
| Jerzy Józefczyk | Wrocław University of Technology, Poland |
| Jason J. Jung | Yeungnam University, Korea |
| Yau-Hwang Kuo | National Cheng Kung University, Taiwan |
| Janusz Kacprzyk | Polish Academy of Sciences, Poland |
| Andrzej Kasprzak | Wroclaw University of Technology, Poland |
| Józef Korbicz | University of Zielona Góra, Poland |
| Halina Kwaśnicka | Wrocław University of Technology, Poland |
| Margaret Lyell | Intelligent Automation, USA |
| Huey-Ming Lee | Chinese Culture University, Taiwan |
| Janusz Marecki | IBM T.J. Watson Research Center, USA |
| Ngoc Thanh Nguyen | Wrocław University of Technology, Poland |
| Leo Obrst | The MITRE Corporation, USA |
| Zenon Okraszewski | Wroclaw University of Technology, Poland |
| Tarkko Oksala | Helsinki University of Technology, Finland |
| Julian A. Padget | University of Bath, UK |
| Jeng-Shyang Pan | National Kaohsiung University of Applied Sciences, Taiwan |
| Terry Payne | University of Southampton, UK |
| Giovanna Petrone | University of Turin, Italy |
| Debbie Richards | Macquarie University, Australia |
| Marwan Sabbouh | The MITRE Corporation, USA |
| Francisco García Sánchez | University of Murcia, Spain |
| Quan Z. Sheng | University of Adelaide, Australia |
| Andrzej Skowron | University of Warsaw, Poland |
| Jie Tang | Tsinghua University, China |
| Rainer Unland | University of Duisburg-Essen, Germany |
| Bao Vo | Swinburne University, Australia |
| Sławomir Zadrożny | Polish Academy of Sciences, Poland |

## Program Committees of Special Sessions

### Nature-Inspired Collective Intelligence NICI 2009

| | |
|---|---|
| Ajith Abraham | Norwegian Univ. of Science and Tech, Norway |
| Costin Badica | University of Craiova, Romania |
| Frantisek Capkovic | Slovak Academy of Sciences, Slovakia |
| Bogdan Gabrys | Bournemouth University, UK |
| Małgorzata Kotulska | Wroclaw University of Technology, Poland |
| Mauricio Kugler | Nagoya Institute of Technology, Japan |
| Ana Carolina Lorena | Federal University of the ABC Region, Brazil |
| James J. Lu | Emory University, USA |
| Klaus Meyer-Wegener | University of Erlangen and Nuremberg, Germany |
| Jean-Christophe Nebel | Kingston University, UK |
| Mariusz Nowostawski | University of Otago, New Zealand |
| Witold Pedrycz | University of Alberta, Canada |
| Olga Vitek | Purdue University, USA |

### Dynamics of Real-World Social Networks

| | |
|---|---|
| John G. Breslin | DERI Galway, Ireland |
| Subhasish Dasgupta | George Washington University, USA |
| Paul Davidsson | Blekinge Institute of Technology, Sweden |
| Christo Dichev | Winston Salem State University, USA |
| Bogdan Gabryś | Bournemouth University, UK |
| Tudor Groza | DERI Galway, Ireland |
| Jason J. Jung | Yeungnam University, Korea |
| Krzysztof Juszczyszyn | Wrocław University of Technology, Poland |
| Irwin K. King | The Chinese University of Hong Kong, China |
| Grzegorz Kołaczek | Wrocław University of Technology, Poland |
| Jun Liu | University of Ulster, UK |
| Luis Martínez López | University of Jaén, Spain |
| Katarzyna Musiał | Wrocław University of Technology, Poland |
| Antonio F. Gómez-Skarmeta | Murcia University, Spain |
| Heiner Stuckenschmidt | University of Mannheim, Germany |
| Edward Szczerbicki | University of Newcastle, Australia |

### Web Systems Analysis WebSys 2009

| | |
|---|---|
| Mohamed Hassoun | ENSSIB, Villeurbanne, France |
| Akira Ishikawa | Aoyama Gakuin University, Tokyo, Japan |
| Andreas Jacobsson | Malmö University, Sweden |
| Tarkko Oksala | Helsinki University of Technology, Finland |
| Jakub Piskorski | Joint Research Centre of the European Commission, Italy |
| Andrzej Siemiński | Wrocław University of Technology, Poland |
| Aleksander Zgrzywa | Wrocław University of Technology, Poland |

# Table of Contents

## Keynote Speeches

Rough Set Approach to Knowledge Discovery about Preferences ....... 1
*Roman Słowiński*

Toward a Self-referential Collective Intelligence: Some Philosophical Background of the IEML Research Program ........................ 22
*Pierre Lévy*

A-Teams and Their Applications ................................ 36
*Piotr Jędrzejowicz*

## Collective Decision Making

Local Search Algorithms for Core Checking in Hedonic Coalition Games ............................................................ 51
*Helena Keinänen*

Information Foraging Theory as a Form of Collective Intelligence for Social Search ................................................ 63
*Longo Luca, Barrett Stephen, and Dondio Pierpaolo*

SAM: Semantic Argumentation Based Model for Collaborative Knowledge Creation and Sharing System ............................ 75
*Krissada Maleewong, Chutiporn Anutariya, and Vilas Wuwongse*

A Conception for Modification of Learning Scenario in an Intelligent E-learning System .............................................. 87
*Adrianna Kozierkiewicz-Hetmańska*

Firefly Algorithm for Continuous Constrained Optimization Tasks ...... 97
*Szymon Łukasik and Sławomir Żak*

Distributed Data Mining Methodology with Classification Model Example ........................................................ 107
*Marcin Gorawski and Ewa Płuciennik-Psota*

A Token-Based Mutual Exclusion Approach to Improve Collaboration in Distributed Environments .................................. 118
*Mauricio Paletta and Pilar Herrero*

Discovering Medical Knowledge from Data in Patients' Files ........... 128
*Magdalena Szymkowiak and Beata Jankowska*

Towards an Increase of Collective Intelligence within Organizations Using Trust and Reputation Models ..... 140
*Emil Scarlat and Iulia Maries*

A New Ant Colony Optimization Algorithm with an Escape Mechanism for Scheduling Problems ..... 152
*Tsai-Duan Lin, Chuin-Chieh Hsu, Da-Ren Chen, and Sheng-Yung Chiu*

Recognizing Team Formations in Multiagent Systems: Applications in Robotic Soccer ..... 163
*Huberto Ayanegui-Santiago*

Semi-traces and Their Application in Concurrency Control Problem ..... 174
*Hoang Chi Thanh*

Design of the Directory Facilitator Supporting Fault-Tolerance in Multi-OSGi Agent System ..... 183
*Sang-Hwan Ryu, Seung-Hyun Lee, Kyung-Soo Jang, Ho-Jin Shin, and Dong-Ryeol Shin*

## Multiagent Systems

Agent-Based Provisioning of Group-Oriented Non-linear Telecommunication Services ..... 193
*Vedran Podobnik, Ana Petric, Krunoslav Trzec, Vedran Galetic, and Gordan Jezic*

A Multi-agent Model of Deceit and Trust in Intercultural Trade ..... 205
*Gert Jan Hofstede, Catholijn M. Jonker, and Tim Verwaart*

Implementation of Epistemic Operators for Model Checking Multi-agent Systems ..... 217
*Marina Bagić Babac and Marijan Kunštić*

Meta-game HOLOS as a Multi-agent Decision-Making Laboratory ..... 229
*Rolisław J. Kolbusz, Romuald Kotowski, and Krzysztof Kasianiuk*

Agent-Based Computational Modeling of Emergent Collective Intelligence ..... 240
*Vivek Kumar Singh, Divya Gautam, Rishi Raj Singh, and Ashok K. Gupta*

Fuzzy Cognitive and Social Negotiation Agent Strategy for Computational Collective Intelligence ..... 252
*Amine Chohra, Kurosh Madani, and Dalel Kanzari*

A Multi-agent Architecture for Multi-robot Surveillance ........ 266
*David Vallejo, Paolo Remagnino, Dorothy N. Monekosso, Luis Jiménez, and Carlos González*

Designing Social Agents with Empathic Understanding ........ 279
*Zulfiqar A. Memon and Jan Treur*

Multi-agent Systems in Pedestrian Dynamics Modeling ........ 294
*Jarosław Wąs and Konrad Kułakowski*

Towards a Model for Extraction of Possible Worlds and Accessibility Relation from Cognitive Agent's Experience ........ 301
*Grzegorz Skorupa and Radosław Katarzyniak*

## Social Networks

On Deriving Tagsonomies: Keyword Relations Coming from Crowd ........ 309
*Michal Barla and Mária Bieliková*

D$^2$ISCO: Distributed Deliberative CBR Systems with jCOLIBRI ........ 321
*Sergio González-Sanz, Juan A. Recio-García, and Belén Díaz-Agudo*

Model of a Collaboration Environment for Knowledge Management in Competence-Based Learning ........ 333
*Różewski Przemysław and Ciszczyk Magdalena*

PlWiki – A Generic Semantic Wiki Architecture ........ 345
*Grzegorz J. Nalepa*

Properties of Bridge Nodes in Social Networks ........ 357
*Katarzyna Musiał and Krzysztof Juszczyszyn*

## Semantic Web

Use of Semantic Principles in a Collaborative System in Order to Support Effective Information Retrieval ........ 365
*František Babič, Karol Furdík, Ján Paralič, Peter Bednár, and Jozef Wagner*

Assessing Semantic Quality of Web Directory Structure ........ 377
*Marko Horvat, Gordan Gledec, and Nikola Bogunović*

Computer Aided Requirements Management ........ 389
*Kamil Karwowski, Witold Wysota, and Jacek Wytrębowicz*

BizKB: A Conceptual Framework for Dynamic Cross-Enterprise Collaboration ........ 401
*Hanh Huu Hoang and Thanh Manh Le*

A Simple Parallel Reasoning System for the $\mathcal{ALC}$ Description Logic .... 413
*Adam Meissner*

SemCards: A New Representation for Realizing the Semantic Web ..... 425
*Kristinn R. Thórisson, Nova Spivack, and James M. Wissner*

ExpTime Tableaux for Checking Satisfiability of a Knowledge Base in the Description Logic $\mathcal{ALC}$ .... 437
*Linh Anh Nguyen and Andrzej Szałas*

Semantically Enhanced Intellectual Property Protection System - SEIPro2S .... 449
*Dariusz Ceglarek, Konstanty Haniewicz, and Wojciech Rutkowski*

Semantic Knowledge Representation in Terrorist Threat Analysis for Crisis Management Systems .... 460
*Mariusz Chmielewski, Andrzej Gałka, Piotr Jarema, Kamil Krasowski, and Artur Kosiński*

Consensus Choice for Reconciling Social Collaborations on Semantic Wikis .... 472
*Jason J. Jung and Ngoc Thanh Nguyen*

## Ontology Management

Ontology Mapping Composition for Query Transformation in Distributed Environment .... 481
*Jason J. Jung*

Algebra of Ontology Modules for Semantic Agents .... 492
*Krzysztof Goczyła, Aleksander Waloszek, and Wojciech Waloszek*

Grouping Results of Queries to Ontological Knowledge Bases by Conceptual Clustering .... 504
*Agnieszka Ławrynowicz*

Applying the c.DnS Design Pattern to Obtain an Ontology for Investigation Management System .... 516
*Jolanta Cybulka*

Ontology Applications for Achieving Situation Awareness in Military Decision Support Systems .... 528
*Mariusz Chmielewski*

A Collaborative Ontology-Based User Profiles System .... 540
*Trong Hai Duong, Mohammed Nazim Uddin, Delong Li, and Geun Sik Jo*

Ontology-Based Intelligent Agent for Grid Resource Management ...... 553
*Kyu Cheol Cho, Chang Hyeon Noh, and Jong Sik Lee*

## Special Session: Dynamics of Real-World Social Networks

The Norm Game on a Model Network: A Critical Line ................ 565
*Marcin Rybak, Antoni Dydejczyk, and Krzysztof Kułakowski*

Model for Trust Dynamics in Service Oriented Information Systems .... 573
*Grzegorz Kołaczek*

Collective Prisoner's Dilemma Model of Artificial Society ............. 584
*Marek Chlebuś, Wojciech Kamiński, and Romuald Kotowski*

## Special Session: Nature-Inspired Collective Intelligence

DES Control Synthesis and Cooperation of Agents .................. 596
*František Čapkovič*

Parameter Tuning for the Artificial Bee Colony Algorithm ............ 608
*Bahriye Akay and Dervis Karaboga*

A Modified Ant-Based Approach to Edge Detection ................. 620
*Doğan Aydın*

A Hybrid Evolutionary Approach for the Protein Classification Problem ........................................................ 629
*Denise F. Tsunoda, Heitor S. Lopes, and Alex A. Freitas*

A Family of GEP-Induced Ensemble Classifiers ...................... 641
*Joanna Jędrzejowicz and Piotr Jędrzejowicz*

Handling Dynamic Networks Using Ant Colony Optimization on a Distributed Architecture ........................................ 653
*Sorin Ilie and Costin Badica*

Modelling Shortest Path Search Techniques by Colonies of Cooperating Agents ............................................................ 665
*Dariusz Król and Łukasz Popiela*

Natural Scene Retrieval Based on Graph Semantic Similarity for Adaptive Scene Classification ....................................... 676
*Nuraini Jamil and Sanggil Kang*

## Special Session: Web Systems Analysis

A Hybrid Architecture for E-Procurement ........ 685
*Giner Alor-Hernandez, Alberto Aguilar-Lasserre, Ulises Juarez-Martinez, Ruben Posada-Gomez, Guillermo Cortes-Robles, Mario Alberto Garcia Martinez, Juan Miguel Gomez, Myriam Mencke, and Alejandro Rodriguez Gonzalez*

Localization by Wireless Technologies for Managing of Large Scale Data Artifacts on Mobile Devices ........ 697
*Ondrej Krejcar*

Avoiding Threats Using Multi Agent System Planning for Web Based Systems ........ 709
*Punam Bedi, Vandana Gandotra, Archana Singhal, Vandita Vats, and Neha Mishra*

Using WordNet to Measure the Similarity of Link Texts ........ 720
*Andrzej Siemiński*

Mining Frequent Purchase Behavior Patterns for Commercial Websites ........ 732
*Li-Fu Hsu, Chuin-Chieh Hsu, and Yi-Chen Ku*

Block Map Technique for the Usability Evaluation of a Website ........ 743
*Kazimierz Choros and Monika Muskala*

Global Distribution of HTTP Requests Using the Fuzzy-Neural Decision-Making Mechanism ........ 752
*Leszek Borzemski, Anna Zatwarnicka, and Krzysztof Zatwarnicki*

Deterministic Processing of WWW Pages by the Web Service ........ 764
*Krzysztof Zatwarnicki*

## Special Session: Collective Intelligence for Economic Data Analysis

Comparative Analysis of Regression Tree Models for Premises Valuation Using Statistica Data Miner ........ 776
*Tadeusz Lasota, Piotr Sachnowski, and Bogdan Trawiński*

Electronic Trading on Electricity Markets within a Multi-agent Framework ........ 788
*Mariusz Kaleta, Piotr Pałka, Eugeniusz Toczyłowski, and Tomasz Traczyk*

Comparative Analysis of Premises Valuation Models Using KEEL, RapidMiner, and WEKA ........ 800
*Magdalena Graczyk, Tadeusz Lasota, and Bogdan Trawiński*

A Multi-agent System to Assist with Real Estate Appraisals Using Bagging Ensembles ........ 813
*Tadeusz Lasota, Zbigniew Telec, Bogdan Trawiński, and Krzysztof Trawiński*

Reputation Tracking Procurement Auctions ........ 825
*Ana Petric and Gordan Jezic*

Comparative Analysis of Evolutionary Fuzzy Models for Premises Valuation Using KEEL ........ 838
*Marek Krzystanek, Tadeusz Lasota, and Bogdan Trawiński*

Hybrid Repayment Prediction for Debt Portfolio ........ 850
*Tomasz Kajdanowicz and Przemysław Kazienko*

**Author Index** ........ 859

# Rough Set Approach to Knowledge Discovery about Preferences

Roman Słowiński

Institute of Computing Science, Poznań University of Technology,
60-965 Poznań, and Systems Research Institute,
Polish Academy of Sciences, 01-447 Warsaw, Poland
roman.slowinski@cs.put.poznan.pl

**Abstract.** It is commonly acknowledged that a rational decision maker acts with respect to his/her value system so as to make the best decision. Confrontation of the value system of the decision maker with characteristics of possible decisions (objects) results in expression of preferences of the decision maker on the set of possible decisions. In order to support the decision maker, one must identify his/her preferences and recommend the most-preferred decision concerning either classification, or choice, or ranking. In this paper, we review multiple attribute and multiple criteria decision problems, as well as preference discovery from data describing some past decisions of the decision maker. The considered preference model has the form of a set of "*if..., then...*" decision rules induced from the data. To structure the data prior to induction, we use the Dominance-based Rough Set Approach (DRSA). DRSA is a methodology for reasoning about ordinal data, which extends the classical rough set approach by handling background knowledge about ordinal evaluations of objects and about monotonic relationships between these evaluations. The paper starts with an introduction to preference modeling in multiple attribute and multiple criteria decision problems, then presents the principles of DRSA, together with a didactic example, and concludes with a summary of characteristic features of DRSA in the context of preference modeling.

## 1 Introduction to Preference Modeling in Multiple Attribute and Multiple Criteria Decision Aiding

The aim of scientific decision aiding is to give the Decision Maker (DM) a recommendation concerning a set of *objects* (also called alternatives, solutions, acts, actions, options, possible decisions, ...) evaluated from multiple points of view considered relevant for the problem at hand and called *attributes* (also called features, variables, criteria, ...).

For example, a decision can concern:

1) diagnosis of pathologies for a set of patients, where patients are objects of the decision, and symptoms and results of medical tests are the attributes,
2) assignment of enterprises to classes of risk, where enterprises are objects of the decision, and financial ratio indices and other economic indicators,

N.T. Nguyen, R. Kowalczyk, and S.-M. Chen (Eds.): ICCCI 2009, LNAI 5796, pp. 1–21, 2009.

such as the market structure, the technology used by the enterprise and the quality of management, are the attributes,
3) selection of a car to be bought from among a given set of cars, where cars are objects of the decision, and maximum speed, acceleration, price, fuel consumption, comfort, color and so on, are the attributes,
4) ordering of students applying for a scholarship, where students are objects of the decision, and scores in different disciplines are the attributes.

The following three main categories of decision problems are typically distinguished [41]:

- *classification*, when the decision aims at assigning objects to predefined classes,
- *choice*, when the decision aims at selecting the best objects,
- *ranking*, when the decision aims at ordering objects from the best to the worst.

Looking at the above examples, one can say that 1) and 2) are classification problems, 3) is a choice problem and 4) is a ranking problem.

The above categorization can be refined by distinguishing three kinds of classification problems:

- *taxonomy*, when the value sets of attributes and the predefined classes are not preference ordered,
- *ordinal classification*, when the value sets of attributes are preference ordered [15] while the predefined classes are not, and
- *ordinal classification with monotonicity constraints* (also known as *multiple criteria sorting*), when both the value sets of attributes and the predefined classes are preference ordered [15].

The monotonicity constraints imposed on ordinal classification require that an improvement of an object's evaluation on any attribute should not deteriorate its class assignment. In the above examples, 1) is a taxonomy problem and 2) is an ordinal classification problem with monotonicity constraints.

An important step in Multiple Attribute Decision Aiding concerns construction or selection of attributes describing the objects. They are build on, or selected from among, elementary features of the objects. The aim is to set up a *consistent set of attributes* which makes the pairwise comparison of all objects in the considered set meaningful. In other words, the consistent set of attributes should permit a meaningful distinction of objects, i.e., objects which are indiscernible with respect to a given set of attributes should be considered indifferent; if this was not the case, the given set of attributes would not be consistent.

Very often, the description of objects by attributes is not neutral with respect to *preferences* of the DM, and then there is a need of taking into account that for the DM some values of attributes are more (or less) preferred than others. In such cases, in Multiple Attribute Decision Aiding, the value sets of these attributes are translated into a monotonic preference scale, which may be ordinal or cardinal [43]. Attributes with monotonic preference scales are called *criteria*.

When there is no relationship between value sets of attributes and DM's preferences, then, in order to distinguish such attributes from criteria, one calls them *regular attributes*. For example, in a decision regarding the selection of a car, its price is a criterion because, obviously, a low price is better than a high price. On the other hand, the color of a car is not a criterion but a regular attribute, because red is not intrinsically better than green. One can imagine, however, that also the color of a car could become a criterion if, for example, a DM would consider red better than green.

Let us formalize the concept of criterion a bit more.

Let $x$ denote an object belonging to a universe of discourse $X$. A *criterion* is a real-valued function $g_j$ defined on $X$, reflecting a worth of objects from a DM's point of view, such that in order to compare any two objects $x, x' \in X$ from this point of view, it is sufficient to compare two values: $g_j(x)$ and $g_j(x')$. Without loss of generality, $g_j : X \rightarrow \Re$ for each $j = 1, \dots, n$, and, for all objects $x, x' \in X$, $g_j(x) \geq g_j(x')$ means that "$x$ is at least as good as $x'$ with respect to criterion $g_j$", which is denoted by $x \succeq_j x'$. Therefore, it is supposed that $\succeq_j$ is a complete preorder, i.e., a strongly complete and transitive binary relation defined on $X$ on the basis of evaluations $g_j(\cdot)$.

All points of view being relevant for a decision problem at hand form a *consistent set of criteria* $G = \{g_1, g_2, \dots, g_n\}$. Comparing to the consistency condition of a set of attributes mentioned above, the consistency of the set of criteria involves one condition more, called condition of monotonicity. This condition requires that if for $x, x' \in X$, object $x$ is weakly preferred to object $x'$ (denoted by $x \succeq x'$), then for another object $x'' \in X$, such that $g_j(x'') \geq g_j(x)$, for $j = 1, \dots, n$, object $x''$ should also be weakly preferred to object $x'$ ($x'' \succeq x'$) [42].

For a given finite set of objects $A = \{x_1, x_2, \dots, x_P\} \subseteq X$, and a set of criteria $G$, the only objective information that comes out from comparison of these objects on multiple criteria is a *dominance relation* $\rhd$ in set $A$. Object $x_k$ dominates object $x_h$, which is denoted by $x_k \rhd x_h$, if and only if $g_j(x_k) \geq g_j(x_h)$ for each $j = 1, \dots, n$, and there exists at least one criterion $g_i$, $i \in \{1, \dots, n\}$, such that $g_i(x_k) > g_i(x_h)$. Object $x_k$ is *non-dominated* in set $A$ (Pareto-optimal) if and only if there is no other object $x_h \in A$ dominating $x_k$. Therefore, the dominance relation $\rhd$ is a partial preorder, i.e., a reflexive and transitive binary relation defined on $A$ on the basis of evaluations $g_j(\cdot)$, $j = 1, \dots, n$.

To simplify notation, we will identify $g_j(x)$ with $x_j$, whether it concerns the $j$-th criterion or regular attribute ($j = 1, \dots, n$). Thus, any object $x = (x_1, x_2, \dots, x_n) \in X \subseteq \Re^n$.

Apart from trivial cases, the dominance relation $\rhd$ is rather poor and leaves many objects incomparable – these are all non-dominated objects in set $A$. In order to enrich the dominance relation and make the objects in $A$ more comparable, one needs additional information about value system of the DM, called *preference information*. This information permits to build a more or less explicit model of DM's preferences, called *preference model*. The preference model relates the decision to evaluations of the objects on the considered criteria. In other words, the preference model aggregates evaluations of objects on multiple criteria. It is inducing a preference structure in set $A$. A proper exploitation of this structure leads then to a *recommendation* in terms of sorting, or choice, or ranking of objects from set $A$.

It follows from above that the preference information and the preference model are two crucial components of both Multiple Criteria Decision Aiding. The many methods existing in both fields differ just by these two components. Below, with respect to these two components, we review some recent trends in Multiple Attribute Decision Aiding.

As to the preference information, it depends on the adopted methodology: prices and interest rates for cost-benefit analysis, cost coefficients in objectives and technological coefficients in constraints for mathematical programming, a training set of decision examples for neural networks and machine learning, substitution rates for a value function of Multi-Attribute Utility Theory, pairwise comparisons of objects in terms of intensity of preference for the Analytic Hierarchy Process, attribute weights and several thresholds for ELECTRE methods, and so on (see the state-of-the-art survey [5]). This information has to be provided by the DM, possibly assisted by an analyst.

Very often this information is not easily definable. For example, this is the case of the price of many immaterial goods and of the interest rates in cost-benefit analysis, or the case of the coefficients of objectives and constraints in mathematical programming models. Even if the required information is easily definable, like a training set of decision examples for neural networks, it is often processed in a way which is not clear for the DM, such that (s)he cannot see what are the exact relations between the provided information and the final recommendation. Consequently, very often the decision aiding method is perceived by the DM as a *black box* whose result has to be accepted because the analyst's authority guarantees that the result is "right". In this context, the aspiration of the DM to find good reasons to make decision is frustrated and rises the need for a more transparent methodology in which the relation between the original information and the final recommendation is clearly shown. Such a transparent methodology searched for has been called *glass box* [26]. Its typical representative is using a training set of decision examples as the input preference information.

The decision examples may either by provided by the DM on a set of real or hypothetical objects, or may come from observation of DM's past decisions. Such an approach follows the paradigm of inductive learning used in artificial intelligence [36], or robust ordinal regression becoming popular in operational research [30]. It is also concordant with the principle of posterior rationality postulated by March [35] since it emphasizes the discovery of DM's intentions as an interpretation of actions rather than as a priori position. This paradigm has been used to construct various preference models from decision examples, e.g., the general additive utility functions [29,6], the outranking relations [37,30], the monotonic decision trees [11], and the set of "*if* ..., *then* ..." decision rules [23].

Of particular interest is the last model based on decision rules – it has been introduced to decision analysis by Greco, Matarazzo and Słowiński [16,19,46]. A popular saying attributed to Slovic is that "people make decisions and then search for rules that justify their choices". The rules explain the preferential attitude of the DM and enable understanding of the reasons of his/her past decisions. The recognition of the rules by the DM [34] justifies their use for decision support. So, the preference model in the form of rules derived from decision examples fulfills both explanation and recommendation goals of decision aiding.

For example, in case of a medical diagnosis problem, the decision rule approach requires as input information a set of examples of previous diagnoses, from which some diagnostic rules are induced, such as "if there is symptom $\alpha$ and the test result is $\beta$, then there is pathology $\gamma$". Each one of such rules is directly related to examples of diagnoses in the input information, where there is symptom $\alpha$, test result $\beta$ and pathology $\gamma$.

The decision rules setting up the preference model have a special syntax which involves partial evaluation profiles and dominance relations on these profiles. The traditional preference models, which are the utility function and the outranking relation, can be represented in terms of equivalent decision rules. The clarity of the rule representation of preferences enables one to see the limits of these aggregation functions. Several studies [20,21,45] presented an axiomatic characterization of all three kinds of preference models in terms of conjoint measurement theory and in terms of a set of decision rules. The given axioms of "cancelation property" type are the weakest possible. In comparison to other studies on the characterization of preference models, these axioms do not require any preliminary assumptions about the scales of preferences of criteria. A side-result of these investigations is that the decision rule preference model is the most general among all known models.

Preference information given in terms of decision examples is often inconsistent. For example, objects with the same description by the set of attributes may be assigned to different classes. This explains the interest in *rough set theory* proposed by Pawlak [38]. Rough set theory permits to *structure the data* set such that decision classes are represented by pairs of ordinary sets called *lower* and *upper approximations*. The differences between upper and lower approximations are called boundary sets, and their cardinalities indicate to what degree the data set is inconsistent. Moreover, rough set theory provides useful information about the role of particular attributes and their subsets in the approximation of decision classes. Induction of decision rules from data structured in this way permits to obtain certain or approximate decision rules [39,44]. For example, in the above diagnostic context, cases where the presence of different pathologies is associated with the presence of the same symptoms and test results are inconsistent, and thus they are placed in the boundaries of the classes of pathologies; decision rules supported by these examples are approximate.

As the classical definition of rough sets is based on indiscernibility relation in the set of objects, it can handle only one kind of inconsistency of decision examples – the one related to indiscernibility of objects belonging to different decision classes. While this is sufficient for classification of taxonomy type, the classical rough set approach fails in case of ordinal classification with monotonicity constraints, where the value sets of attributes, as well as decision classes, are preference ordered. In this case, decision examples may be inconsistent in the sense of violation of the dominance principle which requires that an object $\mathbf{x}$ dominating object $\mathbf{x}'$ on all considered criteria (i.e., $\mathbf{x}$ having evaluations at least as good as $\mathbf{x}'$ on all considered criteria) should also dominate $\mathbf{x}'$ on the decision (i.e., $\mathbf{x}$ should be assigned to at least as good decision class as $\mathbf{x}'$). To deal with this kind of inconsistency, Greco, Matarazzo and Słowiński generalized the classical rough set approach, so as to take into account preference orders and monotonic relationships between evaluations on criteria and assignment to decision classes. This generalization, called Dominance-based Rough Set Approach (DRSA), has been adapted to a large variety of decision problems [16,19,46,7,26,28].

Moreover, DRSA has been adapted to handle granular (fuzzy) information [27], and incomplete information [17,2].

The usefulness of DRSA goes beyond the frame of Multiple Attribute Decision Aiding. This is because the type of monotonic relationships handled by DRSA is also meaningful for problems where preferences are not considered but a kind of monotonicity relating ordered attribute values is meaningful for the analysis of data at hand. Indeed, monotonicity concerns, in general, mutual trends existing between different variables, like distance and gravity in physics, or inflation rate and interest rate in economics. Whenever a relationship between different aspects of a phenomenon is discovered, this relationship can be represented by a monotonicity with respect to some specific measures or perception of the considered aspects, e.g., "the colder the weather, the higher the energy consumption" or "the more a tomato is red, the more it is ripe". The qualifiers, like "cold weather", "high energy consumption", "red" and "ripe", may be expressed either in terms of some measurement units, or in terms of degrees of membership to fuzzy sets representing these concepts.

One should mention, moreover, statistical approaches to processing preference information given in terms of decision examples, adequate for situations where the number of decision examples is very large. Statistical approach to learning preference models from decision examples is today one of the main topics in Machine Learning. This interest is motivated by new challenging applications related to Internet, in particular, recommender systems and information retrieval. In the first, the task is to recommend to the user a new item (like movie or book) that fits her/his preferences. The recommendation is computed on the base of the learning information describing the past behavior of the user. In the latter, the task is to sort (or rank) the documents retrieved by the search engine according to the user's preferences. There are several algorithms that are tailored for these kinds of problems. The most popular are based on *rank loss minimization*. These include variants of support vector machines [32] and boosting [9]. One should also note that there exist several other learning approaches in which preferences are modeled [8,10,40,48]. Moreover, an interesting work has been done in the field of ordinal classification with monotonicity constraints [1,3,4,33].

## 2 Fundamentals of Dominance-Based Rough Set Approach (DRSA)

This section, borrowed from [47], presents fundamentals of the Dominance-based Rough Set approach (DRSA) (for a more complete presentation see, for example, [16,19,23,46]).

Information about objects is represented in the form of an information table. The rows of the table are labelled by objects, whereas columns are labelled by attributes and entries of the table are attribute-values. Formally, an information table (system) is the 4-tuple $\boldsymbol{S} =< U, Q, V, \phi >$, where $U$ is a finite set of objects, $Q$ is a finite set of attributes, $V = \bigcup_{q \in Q} V_q$ and $V_q$ is the value set of the attribute $q$, and $\phi : U \times Q \rightarrow V_q$ is a total function such that $\phi(x, q) \in V_q$ for every $q \in Q$, $x \in U$, called an information function [38]. The set $Q$ is, in general, divided into set $C$ of condition attributes and set $D$ of decision attributes.

Condition attributes with value sets ordered according to decreasing or increasing preference of a decision maker are called *criteria*. For criterion $q \in Q$, $\succeq_q$ is a *weak preference* relation on $U$ such that $x \succeq_q y$ means "$x$ is at least as good as $y$ with respect to criterion $q$". It is supposed that $\succeq_q$ is a complete preorder, i.e. a strongly complete and transitive binary relation, defined on $U$ on the basis of evaluations $\phi(\cdot,q)$. Without loss of generality, the preference is supposed to increase with the value of $\phi(\cdot,q)$ for every criterion $q \in C$, such that for all $x, y \in U$, $x \succeq_q y$ if and only if $\phi(x, q) \geq \phi(y, q)$.

Furthermore, it is supposed that the set of decision attributes $D$ is a singleton $\{d\}$. Values of decision attribute $d$ make a partition of $U$ into a finite number of decision classes, $\boldsymbol{Cl}=\{Cl_t, t = 1, ..., n\}$, such that each $x \in U$ belongs to one and only one class $Cl_t \in \boldsymbol{Cl}$. It is supposed that the classes are preference-ordered, i.e. for all $r$,$s \in \{1, ..., n\}$, such that $r > s$, the objects from $Cl_r$ are preferred to the objects from $Cl_s$. More formally, if $\succeq$ is a *comprehensive weak preference relation* on $U$, i.e. if for all $x$,$y \in U$, $x \succeq y$ means "$x$ is comprehensively at least as good as $y$", it is supposed: $[x \in Cl_r, y \in Cl_s, r > s] \Rightarrow [x \succeq y$ and *not* $y \succeq x]$. The above assumptions are typical for consideration of *ordinal classification problems* (also called *multiple criteria sorting problems*).

The sets to be approximated are called *upward union* and *downward union* of classes, respectively:

$$Cl_t^{\geq} = \bigcup_{s \geq t} Cl_s, \quad Cl_t^{\leq} = \bigcup_{s \leq t} Cl_s, \quad t = 1, \ldots, n.$$

The statement $x \in Cl_t^{\geq}$ means "$x$ belongs to at least class $Cl_t$", while $x \in Cl_t^{\leq}$ means "$x$ belongs to at most class $Cl_t$". Let us remark that $Cl_1^{\geq} = Cl_n^{\leq} = U$, $Cl_n^{\geq} = Cl_n$ and $Cl_1^{\leq} = Cl_1$. Furthermore, for $t = 2, \ldots, n$,

$$Cl_{t-1}^{\leq} = U - Cl_t^{\geq} \text{ and } Cl_t^{\geq} = U - Cl_{t-1}^{\leq}.$$

The key idea of the rough set approach is representation (approximation) of knowledge generated by decision attributes, using "*granules of knowledge*" generated by condition attributes.

In DRSA, where condition attributes are criteria and decision classes are preference ordered, the knowledge to be represented is a collection of upward and downward unions of classes, and the "granules of knowledge" are sets of objects defined using a dominance relation.

$x$ *dominates* $y$ with respect to $P \subseteq C$ (shortly, $x$ *P-dominates* $y$), denoted by $xD_Py$, if for every criterion $q \in P$, $\phi(x,q) \geq \phi(y,q)$. The relation of $P$-dominance is reflexive and transitive, i.e., it is a partial preorder.

Given a set of criteria $P \subseteq C$ and $x \in U$, the "granules of knowledge" used for approximation in DRSA are:

- a set of objects dominating $x$, called *P-dominating set*, $D_P^+(x) = \{y \in U : yD_Px\}$,
- a set of objects dominated by $x$, called *P-dominated set*, $D_P^-(x) = \{y \in U : xD_Py\}$.

Remark that the "granules of knowledge" defined above have the form of upward (positive) and downward (negative) *dominance cones* in the evaluation space.

Let us recall that the *dominance principle* (or Pareto principle) requires that an object $x$ dominating object $y$ on all considered criteria (i.e., $x$ having evaluations at least as good as $y$ on all considered criteria) should also dominate $y$ on the decision (i.e., $x$ should be assigned to at least as good decision class as $y$). This principle is the only objective principle that is widely agreed upon in the multiple criteria comparisons of objects.

Given $P \subseteq C$, the inclusion of an object $x \in U$ to the upward union of classes $Cl_t^{\geq}$ $(t = 2, \ldots, n)$ is *inconsistent with the dominance principle* if one of the following conditions holds:

- $x$ belongs to class $Cl_t$ or better, but it is $P$-dominated by an object $y$ belonging to a class worse than $Cl_t$, i.e., $x \in Cl_t^{\geq}$ but $D_P^+(x) \cap Cl_{t-1}^{\leq} \neq \emptyset$,
- $x$ belongs to a worse class than $Cl_t$ but it $P$-dominates an object $y$ belonging to class $Cl_t$ or better, i.e., $x \notin Cl_t^{\geq}$ but $D_P^-(x) \cap Cl_t^{\geq} \neq \emptyset$.

If, given a set of criteria $P \subseteq C$, the inclusion of $x \in U$ to $Cl_t^{\geq}$, where $t = 2, \ldots, n$, is inconsistent with the dominance principle, then $x$ belongs to $Cl_t^{\geq}$ *with some ambiguity*. Thus, $x$ belongs to $Cl_t^{\geq}$ *without any ambiguity* with respect to $P \subseteq C$, if $x \in Cl_t^{\geq}$ and there is no inconsistency with the dominance principle. This means that all objects $P$-dominating $x$ belong to $Cl_t^{\geq}$, i.e., $D_P^+(x) \subseteq Cl_t^{\geq}$.

Furthermore, $x$ *possibly belongs to* $Cl_t^{\geq}$ with respect to $P \subseteq C$ if one of the following conditions holds:

- according to decision attribute $d$, object $x$ belongs to $Cl_t^{\geq}$,
- according to decision attribute $d$, object $x$ does not belong to $Cl_t^{\geq}$, but it is inconsistent in the sense of the dominance principle with an object $y$ belonging to $Cl_t^{\geq}$.

In terms of ambiguity, $x$ possibly belongs to $Cl_t^{\geq}$ with respect to $P \subseteq C$, if $x$ belongs to $Cl_t^{\geq}$ with or without any ambiguity. Due to the reflexivity of the dominance relation $D_P$, the above conditions can be summarized as follows: $x$ *possibly belongs* to class $Cl_t$ or better, with respect to $P \subseteq C$, if among the objects $P$-dominated by $x$ there is an object $y$ belonging to class $Cl_t$ or better, i.e., $D_P^-(x) \cap Cl_t^{\geq} \neq \emptyset$.

The $P$-*lower approximation* of $Cl_t^{\geq}$, denoted by $\underline{P}(Cl_t^{\geq})$, and the $P$-*upper approximation* of $Cl_t^{\geq}$, denoted by $\overline{P}(Cl_t^{\geq})$, are defined as follows ($t = 1, \ldots, n$):

$$\underline{P}(Cl_t^{\geq}) = \{x \in U : D_P^+(x) \subseteq Cl_t^{\geq}\},$$
$$\overline{P}(Cl_t^{\geq}) = \{x \in U : D_P^-(x) \cap Cl_t^{\geq} \neq \emptyset\}.$$

Analogously, one can define the $P$-*lower approximation* and the $P$-*upper approximation* of $Cl_t^{\leq}$ as follows ($t = 1, \ldots, n$):

$$\underline{P}(Cl_t^{\leq}) = \{x \in U : D_P^-(x) \subseteq Cl_t^{\leq}\},$$
$$\overline{P}(Cl_t^{\leq}) = \{x \in U : D_P^+(x) \cap Cl_t^{\leq} \neq \emptyset\}.$$

The $P$-lower and $P$-upper approximations so defined satisfy the following *inclusion property* for each $t \in \{1, \ldots, n\}$ and for all $P \subseteq C$:

$$\underline{P}(Cl_t^{\geq}) \subseteq Cl_t^{\geq} \subseteq \overline{P}(Cl_t^{\geq}), \quad \underline{P}(Cl_t^{\leq}) \subseteq Cl_t^{\leq} \subseteq \overline{P}(Cl_t^{\leq}).$$

The $P$-lower and $P$-upper approximations of $Cl_t^{\geq}$ and $Cl_t^{\leq}$ have an important *complementarity property*, according to which,

$$\underline{P}(Cl_t^{\geq}) = U - \overline{P}(Cl_{t-1}^{\leq}) \text{ and } \overline{P}(Cl_t^{\geq}) = U - \underline{P}(Cl_{t-1}^{\leq}),\ t = 2, \ldots, n,$$

$$\underline{P}(Cl_t^{\leq}) = U - \overline{P}(Cl_{t+1}^{\geq}) \text{ and } \overline{P}(Cl_t^{\leq}) = U - \underline{P}(Cl_{t+1}^{\geq}),\ t = 1, \ldots, n-1.$$

The $P$-*boundaries* of $Cl_t^{\geq}$ and $Cl_t^{\leq}$, denoted by $Bn_P(Cl_t^{\geq})$ and $Bn_P(Cl_t^{\leq})$ respectively, are defined as follows ($t = 1, \ldots, n$):

$$Bn_P(Cl_t^{\geq}) = \overline{P}(Cl_t^{\geq}) - \underline{P}(Cl_t^{\geq}), \quad Bn_P(Cl_t^{\leq}) = \overline{P}(Cl_t^{\leq}) - \underline{P}(Cl_t^{\leq}).$$

Due to complementarity property, $Bn_P(Cl_t^{\geq}) = Bn_P(Cl_{t-1}^{\leq})$, for $t = 2, \ldots, n$.

For every $P \subseteq C$, the *quality of approximation* of the ordinal classification $\boldsymbol{Cl}$ by a set of criteria $P$ is defined as the ratio of the number of objects $P$-consistent with the dominance principle and the number of all the objects in $U$. Since the $P$-consistent objects are those which do not belong to any $P$-boundary $Bn_P(Cl_t^{\geq})$, $t = 2, \ldots, n$, or $Bn_P(Cl_t^{\leq})$, $t = 1, \ldots, n-1$, the quality of approximation of the ordinal classification $\boldsymbol{Cl}$ by a set of criteria $P$, can be written as

$$\gamma_P(\boldsymbol{Cl}) = \frac{\left| U - \left( \bigcup_{t=2,\ldots,n} Bn_P(Cl_t^{\geq}) \right) \right|}{|U|} = \frac{\left| U - \left( \bigcup_{t=1,\ldots,n-1} Bn_P(Cl_t^{\leq}) \right) \right|}{|U|}.$$

$\gamma_P(\boldsymbol{Cl})$ can be seen as a degree of consistency of the objects from $U$, where $P$ is the set of criteria and $\boldsymbol{Cl}$ is the considered ordinal classification.

Each minimal (with respect to inclusion) subset $P \subseteq C$, such that $\gamma_P(\boldsymbol{Cl}) = \gamma_C(\boldsymbol{Cl})$, is called a *reduct* of $\boldsymbol{Cl}$, and is denoted by $RED_{\boldsymbol{Cl}}$. Let us remark that for a given set $U$ one can have more than one reduct. The intersection of all reducts is called the *core*, and is denoted by $CORE_{\boldsymbol{Cl}}$. Criteria in $CORE_{\boldsymbol{Cl}}$ cannot be removed from consideration without deteriorating the quality of approximation. This means that, in set $C$, there are three categories of criteria:

- *indispensable* criteria included in the core,
- *exchangeable* criteria included in some reducts, but not in the core,
- *redundant* criteria, neither indispensable nor exchangeable, and thus not included in any reduct.

The dominance-based rough approximations of upward and downward unions of classes can serve to induce "*if* ..., *then* ..." decision rules. It is meaningful to consider the following five types of decision rules:

1) Certain $D_{\geq}$-decision rules:
   *if* $x_{q1} \succeq_{q1} r_{q1}$ *and* $x_{q2} \succeq_{q2} r_{q2}$ *and* ... $x_{q_p} \succeq_{q_p} r_{q_p}$,
   *then* $x$ *certainly belongs to* $Cl_t^{\geq}$,
   where,
   for each $w_q, z_q \in X_q$, "$w_q \succeq_q z_q$" means "$w_q$ is <u>at least</u> as good as $z_q$", and $P = \{q_1, \ldots, q_p\} \subseteq C$, $(r_{q_1}, \ldots, r_{q_p}) \in V_{q_1} \times \ldots \times V_{q_p}$, $t \in \{2, \ldots, n\}$.

2) Possible $D_{\geq}$-decision rules:
   *if* $x_{q1} \succeq_{q1} r_{q1}$ *and* $x_{q2} \succeq_{q2} r_{q2}$ *and* ... $x_{q_p} \succeq_{q_p} r_{q_p}$,
   *then* $x$ *possibly belongs to* $Cl_t^{\geq}$,
   where
   $P = \{q_1, \ldots, q_p\} \subseteq C$, $(r_{q_1}, \ldots, r_{q_p}) \in V_{q_1} \times \ldots \times V_{q_p}$, $t \in \{2, \ldots, n\}$.

3) Certain $D_{\leq}$-decision rules:
   *if* $x_{q1} \preceq_{q1} r_{q1}$ *and* $x_{q2} \preceq_{q2} r_{q2}$ *and* ... $x_{q_p} \preceq_{q_p} r_{q_p}$,
   *then* $x$ *certainly belongs to* $Cl_t^{\leq}$,
   where,
   for each $w_q, z_q \in X_q$, "$w_q \preceq_q z_q$" means "$w_q$ is <u>at most</u> as good as $z_q$", and $P = \{q_1, \ldots, q_p\} \subseteq C$, $(r_{q_1}, \ldots, r_{q_p}) \in V_{q_1} \times \ldots \times V_{q_p}$, $t \in \{1, \ldots, n-1\}$.

4) Possible $D_{\leq}$-decision rules:
   *if* $x_{q1} \preceq_{q1} r_{q1}$ *and* $x_{q2} \preceq_{q2} r_{q2}$ *and* ... $x_{q_p} \preceq_{q_p} r_{q_p}$,
   *then* $x$ *possibly belongs to* $Cl_t^{\leq}$,
   where
   $P = \{q_1, \ldots, q_p\} \subseteq C$, $(r_{q_1}, \ldots, r_{q_p}) \in V_{q_1} \times \ldots \times V_{q_p}$, $t \in \{1, \ldots, n-1\}$.

5) Approximate $D_{\geq\leq}$-decision rules:
   *if* $x_{q1} \succeq_{q1} r_{q1}$ *and* ... $x_{q_k} \succeq_{q_k} r_{q_k}$ *and*
   $x_{q_{k+1}} \preceq_{q_{k+1}} r_{q_{k+1}}$ *and* ... $x_{q_p} \preceq_{q_p} r_{q_p}$, *then* $x \in Cl_s^{\geq} \cap Cl_t^{\leq}$,
   where

$O' = \{q_1, \ldots, q_k\} \subseteq C$, $O'' = \{q_{k+1}, \ldots, q_p\} \subseteq C$, $P = O' \cup O''$, $O'$ and $O''$ not necessarily disjoint, $(r_{q_1}, \ldots, r_{q_p}) \in V_{q_1} \times \ldots \times V_{q_p}$, and $s, t \in \{1, \ldots, n\}$, such that $s < t$.

The rules of type 1) and 3) represent certain knowledge extracted from the decision table, while the rules of type 2) and 4) represent possible knowledge. Rules of type 5) represent doubtful knowledge.

**Example illustrating DRSA in the context of ordinal classification**

This subsection presents a didactic example which illustrates the main concepts of DRSA [47]. Let us consider the following ordinal classification problem. Students of a college must obtain an overall evaluation on the basis of their achievements in Mathematics, Physics and Literature. The three subjects are clearly criteria (condition attributes) and the comprehensive evaluation is a decision attribute. For simplicity, the value sets of the criteria and of the decision attribute are the same, and they are composed of three values: bad, medium and good. The preference order of these values is obvious. Thus, there are three preference ordered decision classes, so the problem belongs to the category of ordinal classification. In order to build a preference model of the jury, DRSA is used to analyze a set of exemplary evaluations of students provided by the jury. These examples of ordinal classification constitute an input preference information presented as decision table in Table 1.

Note that the dominance principle obviously applies to the examples of ordinal classification, since an improvement of a student's score on one of three criteria, with other scores unchanged, should not worsen the student's overall evaluation, but rather improve it.

Observe that student $S1$ has not worse evaluations than student $S2$ on all the considered criteria, however, the overall evaluation of $S1$ is worse than the overall evaluation of $S2$. This contradicts the dominance principle, so the two examples of ordinal classification, and only those, are inconsistent. One can expect that the quality of approximation of the ordinal classification represented by examples in Table 1 will be equal to 0.75.

One can observe that in result of reducing the set of considered criteria, i.e., the set of considered subjects, some new inconsistencies can occur. For example, removing from Table 1 the evaluation on Literature, one obtains Table 2, where $S1$ is inconsistent not only with $S2$, but also with $S3$ and $S5$. In fact, student $S1$ has not worse evaluations than students $S2$, $S3$ and $S5$ on all the considered criteria (Mathematics and Physics), however, the overall evaluation of $S1$ is worse than the overall evaluation of $S2$, $S3$ and $S5$.

Observe, moreover, that removing from Table 1 the evaluations on Mathematics, one obtains Table 3, where no new inconsistency occurs, comparing to Table 1.

Similarly, after removing from Table 1 the evaluations on Physics, one obtains Table 4, where no new inconsistencies occur, comparing to Table 1.

The fact that no new inconsistency occurs when Mathematics or Physics is removed, means that the subsets of criteria {Physics, Literature} or {Mathematics,

**Table 1.** Exemplary evaluations of students (examples of ordinal classification)

| Student | Mathematics | Physics | Literature | Overall Evaluation |
|---|---|---|---|---|
| $S1$ | good | medium | bad | bad |
| $S2$ | medium | medium | bad | medium |
| $S3$ | medium | medium | medium | medium |
| $S4$ | good | good | medium | good |
| $S5$ | good | medium | good | good |
| $S6$ | good | good | good | good |
| $S7$ | bad | bad | bad | bad |
| $S8$ | bad | bad | medium | bad |

**Table 2.** Exemplary evaluations of students excluding Literature

| Student | Mathematics | Physics | Overall Evaluation |
|---|---|---|---|
| $S1$ | good | medium | bad |
| $S2$ | medium | medium | medium |
| $S3$ | medium | medium | medium |
| $S4$ | good | good | good |
| $S5$ | good | medium | good |
| $S6$ | good | good | good |
| $S7$ | bad | bad | bad |
| $S8$ | bad | bad | bad |

Literature} contain sufficient information to represent the overall evaluation of students with the same quality of approximation as using the complete set of three criteria. This is not the case, however, for the subset {Mathematics, Physics}. Observe, moreover, that subsets {Physics, Literature} and {Mathematics, Literature} are minimal, because no other criterion can be removed without new inconsistencies occur. Thus, {Physics, Literature} and {Mathematics, Literature} are the reducts of the complete set of criteria {Mathematics, Physics, Literature}. Since Literature is the only criterion which cannot be removed from any reduct without introducing new inconsistencies, it constitutes the core, i.e., the set of indispensable criteria. The core is, of course, the intersection of all reducts, i.e., in our example:

$$\{\text{Literature}\} = \{\text{Physics, Literature}\} \cap \{\text{Mathematics, Literature}\}.$$

In order to illustrate in a simple way the concept of rough approximation, let us limit our analysis to the reduct {Mathematics, Literature}. Let us consider student $S4$. His positive dominance cone $D^+_{\{Mathematics,Literature\}}(S4)$ is composed of all the students having evaluations not worse than him on Mathematics and Literature, i.e., of all the students dominating him with respect to Mathematics and Literature. Thus,

$$D^+_{\{Mathematics,Literature\}}(S4) = \{S4, S5, S6\}.$$

**Table 3.** Exemplary evaluations of students excluding Mathematics

| Student | Physics | Literature | Overall Evaluation |
|---|---|---|---|
| $S1$ | medium | bad | bad |
| $S2$ | medium | bad | medium |
| $S3$ | medium | medium | medium |
| $S4$ | good | medium | good |
| $S5$ | medium | good | good |
| $S6$ | good | good | good |
| $S7$ | bad | bad | bad |
| $S8$ | bad | medium | bad |

**Table 4.** Exemplary evaluations of students excluding Physics

| Student | Mathematics | Literature | Overall Evaluation |
|---|---|---|---|
| $S1$ | good | bad | bad |
| $S2$ | medium | bad | medium |
| $S3$ | medium | medium | medium |
| $S4$ | good | medium | good |
| $S5$ | good | good | good |
| $S6$ | good | good | good |
| $S7$ | bad | bad | bad |
| $S8$ | bad | medium | bad |

On the other hand, the negative dominance cone of student $S4$, $D^{-}_{\{Mathematics,Literature\}}(S4)$, is composed of all the students having evaluations not better than him on Mathematics and Literature, i.e., of all the students dominated by him with respect to Mathematics and Literature. Thus,

$$D^{-}_{\{Mathematics,Literature\}}(S4) = \{S1, S2, S3, S4, S7, S8\}.$$

Similar dominance cones can be obtained for all the students from Table 6. For example, for $S2$, the dominance cones are

$$D^{+}_{\{Mathematics,Literature\}}(S2) = \{S1, S2, S3, S4, S5, S6\}$$

and

$$D^{-}_{\{Mathematics,Literature\}}(S2) = \{S2, S7\}.$$

The rough approximations can be calculated using dominance cones. Let us consider, for example, the lower approximation of the set of students having a "good" overall evaluation $\underline{P}(Cl^{\geq}_{good})$, with $P$={Mathematics, Literature}. Notice that $\underline{P}(Cl^{\geq}_{good}) = \{S4, S5, S6\}$, because positive dominance cones of students $S4$, $S5$ and $S6$ are all included in the set of students with an overall evaluation "good". In other words, this means that there is no student dominating $S4$ or $S5$ or $S6$ while having an overall evaluation worse than "good". From the viewpoint of decision making, this means that, taking into account the available information about evaluation of students on Mathematics and Literature, the fact that student $y$ dominates $S4$ or $S5$ or $S6$ is a *sufficient* condition to conclude that $y$ is a "good" student.

The upper approximation of the set of students with a "good" overall evaluation is $\overline{P}(Cl_{good}^{\geq}) = \{S4, S5, S6\}$, because negative dominance cones of students $S4$, $S5$ and $S6$ have a nonempty intersection with the set of students having a "good" overall evaluation. In other words, this means that for each one of the students $S4$, $S5$ and $S6$, there is at least one student dominated by him with an overall evaluation "good". From the point of view of decision making, this means that, taking into account the available information about evaluation of students on Mathematics and Literature, the fact that student $y$ dominates $S4$ or $S5$ or $S6$ is a *possible* condition to conclude that $y$ is a "good" student.

Let us observe that for the set of criteria $P$={Mathematics, Literature}, the lower and upper approximations of the set of "good" students are the same. This means that examples of ordinal classification concerning this decision class are all consistent. This is not the case, however, for the examples concerning the union of decision classes "at least medium". For this upward union the rough approximations are $\underline{P}(Cl_{medium}^{\geq}) = \{S3, S4, S5, S6\}$ and $\overline{P}(Cl_{medium}^{\geq}) = \{S1, S2, S3, S4, S5, S6\}$. The difference between $\overline{P}(Cl_{medium}^{\geq})$ and $\underline{P}(Cl_{medium}^{\geq})$, i.e., the boundary $Bn_P(Cl_{medium}^{\geq}) = \{S1, S2\}$, is composed of students with inconsistent overall evaluations, which has already been noticed above. From the viewpoint of decision making, this means that, taking into account the available information about evaluation of students on Mathematics and Literature, the fact that student $y$ is dominated by $S1$ and dominates $S2$ is a condition to conclude that $y$ can obtain an overall evaluation "at least medium" with some doubts.

Until now, rough approximations of only upward unions of decision classes have been considered. It is interesting, however, to calculate also rough approximations of downward unions of decision classes. Let us consider first the lower approximation of the set of students having "at most medium" overall evaluation $\underline{P}(Cl_{medium}^{\leq})$. Observe that $\underline{P}(Cl_{medium}^{\leq}) = \{S1, S2, S3, S7, S8\}$, because the negative dominance cones of students $S1, S2, S3, S7$, and $S8$ are all included in the set of students with overall evaluation "at most medium". In other words, this means that there is no student dominated by $S1$ or $S2$ or $S3$ or $S7$ or $S8$ while having an overall evaluation better than "medium". From the viewpoint of decision making, this means that, taking into account the available information about evaluation of students on Mathematics and Literature, the fact that student $y$ is dominated by $S1$ or $S2$ or $S3$ or $S7$ or $S8$ is a *sufficient* condition to conclude that $y$ is an "at most medium" student.

The upper approximation of the set of students with an "at most medium" overall evaluation is $\overline{P}(Cl_{medium}^{\leq}) = \{S1, S2, S3, S7, S8\}$, because the positive dominance cones of students $S1, S2, S3, S7$, and $S8$ have a nonempty intersection with the set of students having an "at most medium" overall evaluation. In other words, this means that for each one of the students $S1, S2, S3, S7$, and $S8$, there is at least one student dominating him with an overall evaluation "at most medium". From the viewpoint of decision making, this means that, taking into account the available information about evaluation of students on Mathematics and Literature, the fact that student $y$ is dominated by $S1$ or $S2$ or $S3$ or $S7$ or $S8$ is a *possible* condition to conclude that $y$ is an "at most medium" student.

Finally, the lower and upper approximations of the set of students having a "bad" overall evaluation are $\underline{P}(Cl_{bad}^{\leq}) = \{S7, S8\}$ and $\overline{P}(Cl_{bad}^{\leq}) = \{S1, S2, S7, S8\}$. The difference between $\overline{P}(Cl_{bad}^{\leq})$ and $\underline{P}(Cl_{bad}^{\leq})$, i.e., the boundary $Bn_P(Cl_{bad}^{\leq}) = \{S1, S2\}$ is composed of students with inconsistent overall evaluations, which has already been noticed above. From the viewpoint of decision making, this means that, taking into account the available information about evaluation of students on Mathematics and Literature, the fact that student $y$ is dominated by $S1$ and dominates $S2$ is a condition to conclude that $y$ can obtain an overall evaluation "bad" with some doubts. Observe, moreover, that $Bn_P(Cl_{medium}^{\geq}) = Bn_P(Cl_{bad}^{\leq}) = \{S1, S2\}$.

Given the above rough approximations with respect to the set of criteria $P$={Mathematics, Literature}, one can induce a set of decision rules representing the preferences of the jury. The idea is that evaluation profiles of students belonging to the lower approximations can serve as a base for some certain rules, while evaluation profiles of students belonging to the boundaries can serve as a base for some approximate rules. The following decision rules have been induced (between parentheses there are id's of students supporting the corresponding rule; the student being a rule base is underlined):

**Rule 1)** *if* the evaluation on Mathematics is (at least) good, and the evaluation on Literature is at least medium, *then* the overall evaluation is (at least) good, $\{\underline{S4}, S5, S6\}$,

**Rule 2)** *if* the evaluation on Mathematics is at least medium, and the evaluation on Literature is at least medium, *then* the overall evaluation is at least medium, $\{\underline{S3}, S4, S5, S6\}$,

**Rule 3)** *if* the evaluation on Mathematics is at least medium, and the evaluation on Literature is (at most) bad, *then* the overall evaluation is bad or medium, $\{\underline{S1}, \underline{S2}\}$,

**Rule 4)** *if* the evaluation on Mathematics is at most medium, *then* the overall evaluation is at most medium, $\{\underline{S2}, \underline{S3}, S7, S8\}$,

**Rule 5)** *if* the evaluation on Literature is (at most) bad, *then* the overall evaluation is at most medium, $\{\underline{S1}, \underline{S2}, \underline{S7}\}$,

**Rule 6)** *if* the evaluation on Mathematics is (at most) bad, *then* the overall evaluation is (at most) bad, $\{\underline{S7}, \underline{S8}\}$.

Notice that rules 1)–2), 4)–6) are certain, while rule 3) is an approximate one. These rules represent knowledge discovered from the available information. In the current context, the knowledge is interpreted as a preference model of the jury. A characteristic feature of the syntax of decision rules representing preferences is the use of expressions "at least" or "at most" a value; in case of extreme values ("good" and "bad"), these expressions are put in parentheses because there is no value above "good" and below "bad".

Even if one can represent all the knowledge using only one reduct of the set of criteria (as it is the case using $P$={Mathematics, Literature}), when considering a larger set of criteria than a reduct, one can obtain a more synthetic representation of knowledge, i.e., the number of decision rules or the number of elementary conditions, or both of them, can get smaller. For example, considering the set of all three criteria, {Mathematics, Physics, Literature}, one can induce a set of decision rules composed of the above rules 1), 2), 3) and 6), plus the following :

**Rule 7)** *if* the evaluation on Physics is at most medium, and the evaluation on Literature is at most medium, *then* the overall evaluation is at most medium, $\{S1, S2, \underline{S3}, S7, S8\}$.

Thus, a complete set of decision rules induced from Table 1 is composed of 5 instead of 6 rules.

Once accepted by the jury, these rules represent its preference model. Assuming that rules 1)–7) in our example represent the preference model of the jury, it can be used to evaluate new students. For example, student $S9$ who is "medium" in Mathematics and Physics and "good" in Literature, would be evaluated as "medium" because his profile matches the premise of rule 2), having as consequence an overall evaluation at least "medium". The overall evaluation of $S9$ cannot be "good", because his profile does not match any rule having as consequence an overall evaluation "good" (in the considered example, the only rule of this type is rule 1), whose premise is not matched by the profile of $S9$).

## 3 Conclusions

Dominance-based rough set theory is a mathematical tool for reasoning about ordinal data describing preferences of a decision maker about objects of his/her decision. It handles inconsistencies by structuring this description into lower and upper approximations, corresponding to certain and possible knowledge about the preferences. Induction algorithms running on these approximations discover, in turn, certain and possible decision rules that facilitate an understanding of the decision maker's preferences, and enable a recommendation concordant with these preferences.

The original version of the rough set approach, based on indiscernibility or tolerance relation deal with data describing problems of taxonomy-type classification, i.e., problems where neither the attributes describing the objects, nor the classes to which the objects are assigned, are ordered. On the other hand, multiple criteria decision aiding deals with problems where descriptions (evaluations) of objects by means of attributes (criteria), as well as decisions in classification, choice and ranking problems, are ordered. Moreover, in data describing multiple criteria decision making, there exist a monotonic relationship between conditions and decisions, like "the bigger the house, the more expensive it is". The generalization of the rough set approach and of the induction algorithms about problems in which order properties and monotonic relationships are important gave birth to the Dominance-based Rough Set Approach (DRSA) which made a breakthrough in scientific decision aiding.

The main features of DRSA are the following:

- preference information necessary to deal with any multiple criteria decision problem, or with decision under uncertainty, is asked to the DM just in terms of exemplary decisions,
- the rough set analysis of preference information supplies some useful elements of knowledge about the decision situation; these are: the relevance of attributes or criteria, the minimal subsets of attributes or criteria (reducts) conveying the relevant knowledge contained in the exemplary decisions, the set of indispensable attributes or criteria (core),
- DRSA can deal with preference information concerning taxonomy-type classification, ordinal classification, choice, ranking, multiobjective optimization and decision under uncertainty,
- the preference model induced from preference information structured by DRSA is expressed in a natural and comprehensible language of "*if..., then...*" decision rules,
- suitable procedures have been proposed to exploit the results of application of the decision rule preference model on a set of objects or pairs of objects in order to workout a recommendation,
- no prior discretization of quantitative condition attributes or criteria is necessary,
- heterogeneous information (qualitative and quantitative, ordered and non-ordered, nominal and ordinal, quantitative and numerical non-quantitative scales of preferences) can be processed within DRSA,
- the proposed methodology fulfils some desirable properties for both rough set approach (the approximated sets include lower approximation and are included in upper approximation, and the complementarity property is satisfied), and for multiple criteria decision analysis (the decision rule preference model is formally equivalent to the non-additive, non-transitive and non-complete conjoint measurement model, and to a more general model for preferences defined on all kinds of scales),
- the decision rule preference model resulting from DRSA is more general than all existing models of conjoint measurement, due to its capacity of handling inconsistent preferences (a new model of conjoint measurement is formally equivalent to the decision rule preference model handling inconsistencies),
- the decision rule preference model fulfils the postulate of transparency and interpretability of preference models in decision support; each decision rule can be clearly identified with those parts of the preference information (decision examples) which support the rule; the rules inform the DM in a quasi-natural language about the relationships between conditions and decisions; in this way, the rules permit traceability of the decision support process and give understandable justifications for the decision to be made,

– the proposed methodology is based on elementary concepts and mathematical tools (sets and set operations, binary relations), without recourse to any algebraic or analytical structures.

## Acknowledgment

The author wishes to acknowledge financial support from the Ministry of Science and Higher Education, grant N N519 314435.

## References

1. Dembczyński, K., Greco, S., Kotłowski, W., Słowiński, R.: Statistical model for rough set approach to multicriteria classification. In: Kok, J.N., Koronacki, J., Lopez de Mantaras, R., Matwin, S., Mladenič, D., Skowron, A. (eds.) PKDD 2007. LNCS (LNAI), vol. 4702, pp. 164–175. Springer, Heidelberg (2007)
2. Dembczyński, K., Greco, S., Słowiński, R.: Rough set approach to multiple criteria classication with imprecise evaluations and assignments. European Journal of Operational Research 198, 626–636 (2009)
3. Dembczyński, K., Kotłowski, W., Słowiński, R.: Ordinal classification with decision rules. In: Raś, Z.W., Tsumoto, S., Zighed, D.A. (eds.) MCD 2007. LNCS (LNAI), vol. 4944, pp. 169–181. Springer, Heidelberg (2008)
4. Dembczyński, K., Kotłowski, W., Słowiński, R.: Ensemble of decision rules for ordinal classification with monotonicity constraints. In: Wang, G., Li, T., Grzymala-Busse, J.W., Miao, D., Skowron, A., Yao, Y. (eds.) RSKT 2008. LNCS (LNAI), vol. 5009, pp. 260–267. Springer, Heidelberg (2008)
5. Figueira, J., Greco, S., Ehrgott, M. (eds.): Multiple Criteria Decision Analysis: State of the Art Surveys. Springer, Berlin (2005)
6. Figueira, J., Greco, S., Słowiński, R.: Building a set of additive value functions representing a reference preorder and intensities of preference: GRIP method. European Journal of Operational Research 195, 460–486 (2009)
7. Fortemps, Ph., Greco, S., Słowiński, R.: Multicriteria decision support using rules that represent rough-graded preference relations. European Journal of Operational Research 188, 206–223 (2008)
8. Frank, E., Hall, M.: A simple approach to ordinal classification. In: Flach, P.A., De Raedt, L. (eds.) ECML 2001. LNCS (LNAI), vol. 2167, pp. 145–156. Springer, Heidelberg (2001)
9. Freund, Y., Iyer, R., Schapire, R.E., Singer, Y.: An efficient boosting algorithm for combining preferences. Journal of Machine Learning Research 6, 933–969 (2003)
10. Fürnkranz, J., Hüllermeier, E.: Pairwise preference learning and ranking. In: Lavrač, N., Gamberger, D., Todorovski, L., Blockeel, H. (eds.) ECML 2003. LNCS (LNAI), vol. 2837, pp. 145–156. Springer, Heidelberg (2003)
11. Giove, S., Greco, S., Matarazzo, B., Słowiński, R.: Variable consistency monotonic decision trees. In: Alpigini, J.J., Peters, J.F., Skowron, A., Zhong, N. (eds.) RSCTC 2002. LNCS (LNAI), vol. 2475, pp. 247–254. Springer, Heidelberg (2002)
12. Greco, S., Inuiguchi, M., Słowiński, R.: Dominance-based rough set approach using possibility and necessity measures. In: Alpigini, J.J., Peters, J.F., Skowron, A., Zhong, N. (eds.) RSCTC 2002. LNCS (LNAI), vol. 2475, pp. 85–92. Springer, Heidelberg (2002)

13. Greco, S., Inuiguchi, M., Słowiński, R.: A new proposal for rough fuzzy approximations and decision rule representation. In: Peters, J.F., Skowron, A., Dubois, D., Grzymała-Busse, J.W., Inuiguchi, M., Polkowski, L. (eds.) Transactions on Rough Sets II. LNCS, vol. 3135, pp. 319–342. Springer, Heidelberg (2004)
14. Greco, S., Inuiguchi, M., Słowiński, R.: Fuzzy rough sets and multiple-premise gradual decision rules. International Journal of Approximate Reasoning 41, 179–211 (2006)
15. Greco, S., Matarazzo, B., Słowiński, R.: A new rough set approach to evaluation of bankruptcy risk. In: Zopounidis, C. (ed.) Operational Tools in the Management of Financial Risks, pp. 121–136. Kluwer Academic Publishers, Dordrecht (1998)
16. Greco, S., Matarazzo, B., Słowiński, R.: The use of rough sets and fuzzy sets in MCDM. In: Gal, T., Stewart, T., Hanne, T. (eds.) Advances in Multiple Criteria Decision Making, ch. 14, pp. 14.1–14.59. Kluwer Academic Publishers, Boston (1999)
17. Greco, S., Matarazzo, B., Słowiński, R.: Dealing with missing data in rough set analysis of multi-attribute and multi-criteria decision problems. In: Zanakis, S.H., Doukidis, G., Zopounidis, C. (eds.) Decision Making: Recent Developments and Worldwide Applications, pp. 295–316. Kluwer, Dordrecht (2000)
18. Greco, S., Matarazzo, B., Słowiński, R.: A fuzzy extension of the rough set approach to multicriteria and multiattribute sorting. In: Fodor, J., De Baets, B., Perny, P. (eds.) Preferences and Decisions under Incomplete Information, pp. 131–154. Physica-Verlag, Heidelberg (2000)
19. Greco, S., Matarazzo, B., Słowiński, R.: Rough sets theory for multicriteria decision analysis. European Journal of Operational Research 129, 1–47 (2001)
20. Greco, S., Matarazzo, B., Słowiński, R.: Preference representation by means of conjoint measurement & decision rule model. In: Bouyssou, D., Jacquet-Lagrèze, E., Perny, P., Słowiński, R., Vanderpooten, D., Vincke, P. (eds.) Aiding Decisions with Multiple Criteria–Essays in Honor of Bernard Roy, pp. 263–313. Kluwer, Dordrecht (2002)
21. Greco, S., Matarazzo, B., Słowiński, R.: Axiomatic characterization of a general utility function and its particular cases in terms of conjoint measurement and rough-set decision rules. European Journal of Operational Research 158, 271–292 (2004)
22. Greco, S., Matarazzo, B., Słowiński, R.: Dominance-Based Rough Set Approach to Knowledge Discovery (I) - General Perspective (II) - Extensions and Applications. In: Zhong, N., Liu, J. (eds.) Intelligent Technologies for Information Analysis, ch. 20 & 21, pp. 513–612. Springer, Berlin (2004)
23. Greco, S., Matarazzo, B., Słowiński, R.: Decision rule approach. In: Figueira, J., Greco, S., Ehrgott, M. (eds.) Multiple Criteria Decision Analysis: State of the Art Surveys, ch. 13, pp. 507–563. Springer, Berlin (2005)
24. Greco, S., Matarazzo, B., Słowiński, R.: Dominance-based rough set approach to case-based reasoning. In: Torra, V., Narukawa, Y., Valls, A., Domingo-Ferrer, J. (eds.) MDAI 2006. LNCS (LNAI), vol. 3885, pp. 7–18. Springer, Heidelberg (2006)
25. Greco, S., Matarazzo, B., Słowiński, R.: Dominance-based rough set approach as a proper way of handling graduality in rough set theory. In: Peters, J.F., Skowron, A., Marek, V.W., Orłowska, E., Słowiński, R., Ziarko, W.P. (eds.) Transactions on Rough Sets VII. LNCS, vol. 4400, pp. 36–52. Springer, Heidelberg (2007)
26. Greco, S., Matarazzo, B., Słowiński, R.: Dominance-based Rough Set Approach to Interactive Multiobjective Optimization. In: Branke, J., Deb, K., Miettinen, K., Słowiński, R. (eds.) Multiobjective Optimization: Interactive and Evolutionary Approaches, ch. 5, pp. 121–156. Springer, Berlin (2008)

27. Greco, S., Matarazzo, B., Słowiński, R.: Granular Computing for Reasoning About Ordered Data: the Dominance-based Rough Set Approach. In: Pedrycz, W., Skowron, A., Kreinovich, V. (eds.) Handbook of Granular Computing, pp. 347–373. Wiley, Chichester (2008)
28. Greco, S., Matarazzo, B., Słowiński, R.: Dominance-based rough set approach to decision under uncertainty and time preference. Annals of Operations Research (2009), DOI:10.1007/s10479-009-0566-8
29. Greco, S., Mousseau, V., Słowiński, R.: Ordinal regression revisited: multiple criteria ranking using a set of additive value functions. European Journal of Operational Research 191, 415–435 (2008)
30. Greco, S., Słowiński, R., Figueira, J., Mousseau, V.: Robust ordinal regression. In: Ehrgott, M., Figueira, J., Greco, S. (eds.) New Advances in Multiple Criteria Decision Analysis. Springer, Berlin (2009)
31. Greco, S., Pawlak, Z., Słowiński, R.: Can Bayesian confirmation measures be useful for rough set decision rules? Engineering Applications of Artificial Intelligence 17, 345–361 (2004)
32. Herbrich, R., Graepel, T., Obermayer, K.: Regression models for ordinal data: a machine learning approach, Technical report TR-99/03, TU Berlin (1999)
33. Kotłowski, W., Dembczyński, K., Greco, S., Słowiński, R.: Stochastic Dominance-based Rough Set Model for Ordinal Classification. Information Sciences 178, 4019–4037 (2008)
34. Langley, P., Simon, H.A.: Fielded applications of machine learning. In: Michalski, R.S., Bratko, I., Kubat, M. (eds.) Machine Learning and Data Mining, pp. 113–129. Wiley, New York (1998)
35. March, J.G.: Bounded rationality, ambiguity, and the engineering of choice. In: Bell, D.E., Raiffa, H., Tversky, A. (eds.) Decision Making, Descriptive, Normative and Prescriptive Interactions, pp. 33–58. Cambridge University Press, New York (1988)
36. Michalski, R.S.: A Theory and Methodology of Inductive Learning. In: Michalski, R.S., Carbonell, J.G., Mitchell, T.M. (eds.) Machine Learning: An Artificial Intelligence Approach, pp. 83–129. Tioga Publishing, Palo Alto (1983)
37. Mousseau, V., Słowiński, R.: Inferring an ELECTRE TRI model from assignment examples. Journal of Global Optimization 12, 157–174 (1998)
38. Pawlak, Z.: Rough Sets: Theoretical Aspects of Reasoning about Data. Kluwer, Dordrecht (1991)
39. Pawlak, Z., Słowiński, R.: Rough set approach to multi-attribute decision analysis. European Journal of Operational Research 72, 443–459 (1994)
40. Rennie, J., Srebro, N.: Loss functions for preference levels: regression with discrete ordered labels. In: Proc. of the IJCAI-2005 Multidisciplinary Workshop on Advances in Preference Handling (2005)
41. Roy, B.: Multicriteria Methodology for Decision Aiding. Kluwer Academic Publishers, Dordrecht (1996)
42. Roy, B., Bouyssou, D.: Aide Multicritère à la Décision: Méthodes et Cas. Economica, Paris (1993)
43. Shoemaker, P.J.H.: The expected utility model: its variants, purposes, evidence and limitations. Journal of Economic Literature 20, 529–562 (1982)
44. Słowiński, R.: Rough set learning of preferential attitude in multi-criteria decision making. In: Komorowski, J., Raś, Z.W. (eds.) ISMIS 1993. LNCS, vol. 689, pp. 642–651. Springer, Heidelberg (1993)

45. Słowiński, R., Greco, S., Matarazzo, B.: Axiomatization of utility, outranking and decision-rule preference models for multiple-criteria classification problems under partial inconsistency with the dominance principle. Control and Cybernetics 31, 1005–1035 (2002)
46. Słowiński, R., Greco, S., Matarazzo, B.: Rough set based decision support. In: Burke, E.K., Kendall, G. (eds.) Search Methodologies: Introductory Tutorials in Optimization and Decision Support Techniques, ch. 16, pp. 475–527. Springer, New York (2005)
47. Słowiński, R., Greco, S., Matarazzo, B.: Rough Sets in Decision Making. In: Meyers, R. (ed.) Encyclopedia of Complexity and Systems Science. Springer, New York (2009)
48. Srebro, N., Rennie, J., Jaakkola, T.: Maximum margin matrix factorizations. In: Advances in Neural Information Processing Systems 17, pp. 1329–1336. MIT Press, Cambridge (2005)

# Toward a Self-referential Collective Intelligence Some Philosophical Background of the IEML Research Program

Pierre Lévy

Fellow of the Royal Society of Canada
Canada Research Chair in Collective Intelligence
University of Ottawa
pierre.levy@uottawa.ca

**Abstract.** The IEML research program promotes a radical innovation in the notation and processing of semantics. IEML (Information Economy MetaLanguage) is a regular language that provides new methods for semantic interoperability, semantic navigation, collective categorization and self-referential collective intelligence. This research program is compatible with the major standards of the Web of data and is in tune with the current trends in social computing.

The paper explains the philosophical relevance of this new language, including the role of media and symbolic systems in human cognition, the hope of a scientific revolution in humanities and the perspective of a leap in human collective intelligence thank to its possible self-reference in the mirror of cyberspace.

**Keywords:** collective intelligence, IEML, semantic space, semantic tagging, semantic interoperability, metalanguage, Web of data, philosophy.

## 1 Introduction: Collective Intelligence as a New Field

Collective Intelligence (CI) is the capacity of human collectives to engage in intellectual cooperation in order to create, innovate and invent [42, 69, 72]. It can be applied at any scale, from work teams to huge networks or even to our whole species. Collective Intelligence is a determining factor in competitiveness, creativity and human development in a knowledge based economy, or in an *information economy.* The more our society depends on the creative management of knowledge, the more this capacity becomes of fundamental importance [3]. There is a growing feeling that there exists a strong correlation between communities collective intelligence and the degree of their human development. CI can be seen as a driving force of human development and within this conceptual framework, conversely, human development provides CI with an environment for growth [47]. As digital technologies give us more and more powerful tools to augment personal and collective cognitive processes, it becomes essential to understand how the collective intelligence processes can be multiplied by digital networks [45]. The topic of the augmentation of CI through digital networks is

N.T. Nguyen, R. Kowalczyk, and S.-M. Chen (Eds.): ICCCI 2009, LNAI 5796, pp. 22–35, 2009.

an emerging research area [32], as the extensive body of literature of knowledge management shows [59, 77, 57, 13], and as the Web 2.0 or social computing attests [75, 71, 62, 66, 51].

Douglas Engelbart can be considered as the main founder of this field [17]. I have been involved in its development since the end of the 1980s with an invention [40], several publications [38, 39, 41, 42, 43, 44, 45, 46] and with the creation in 2002 of the first academic research center exclusively devoted to this subject (the Canada Research Chair in Collective Intelligence, or CRC-IC, at the University of Ottawa). The main result of my work at the CRC-IC is *the foundation of a new scientific research program*, based on the development and technical exploitation of an artificial language called IEML (Information Economy MetaLanguage). IEML is supposed to be developed and used collaboratively for the purpose of augmenting human collective intelligence by technologies. A summary of IEML has been presented in a recent paper [49][1] and I'm currently writing a book on the subject. Due to a lack of space, this text cannot explain in detail the syntax and semantics of IEML. I will limit myself here to an attempt to explicitate the questions and problems to which the IEML research program is supposed to answer. Inevitably, this effort will lead me to evoke the historical, philosophical and epistemological background of IEML.

## 2 The Role of Media and Symbolic Systems in Cognition

There is no doubt that human cognition is grounded in a brain structure and neural activity that is biologically determined[10, 18].[2] Nevertheless, in the recent decades, an impressive body of research has been devoted to the subject of intellectual technologies and symbolic tools [30, 22, 60, 23, 5, 26, 78, 8, 53] The main idea behind this interdisciplinary research is that culture-driven collective memory apparati, communication media and symbolic systems all play a prominent role in shaping personal and social cognitive abilities [28].

The invention of writing has allowed the development of systematically organized knowledge (lists, tables, archives, accountancy, complex hermeneutical procedures) beyond the lore of oral cultures, organized around myth, narratives and rituals [22, 60, 23, 26]. The invention of phonetic, or alphabetic, writing systems limited to more or less thirty signs (as opposed to systems requirering thousands of ideographic signs or mixed systems) led to the social extension of writing and reading abilities and fostered the development of abstract conceptual thinking [29, 56, 52]. The invention of the indo-arabic numerals including notation by position and the zero, made arithmetic simpler and easier mainly by allowing the use of uniform algorithms [33]. Making a multiplication using roman numerals

[1] This text re-uses some parts of "From Social Computing to Self-Referencial Collective Intelligence: the IEML Research Program" [49]but adds original epistemologic and philosophical perspectives.

[2] Note that I don't say that human cognition is *determined by* neural activity, but that it is *grounded in* neural activity.

instead of using indo-arabic numerals is to understand the importance of notations and symbolic systems in the performance of cognitive tasks. In addition to supporting a wide dissemination of information and knowledge, the invention of the printing press led indirectly to the development of several standard scientific notation systems including accurate maps with geometric projections of parallels and meridians, biological classification systems, chemical and mathematical notations [19]. Print technology also fostered the development and progressive formalization of linguistic studies [1] and the creation of metadata systems for the organization of libraries and archives [70]. Note that the development of new symbolic systems did not show up immediately after the invention of the printing press: it took several generations to assimilate and exploit the cognitive possibilities opened by this new medium. In general, cultural evolution follows technological evolution. By analogy, we can easily imagine that the full symbolic exploitation of the new communication and processing environment provided by computer networks has not been achieved.

Even if these historical remarks may seem remote from the central point of this paper, they are indeed very close. They suggest that major advances in human cognition are related to inventions of media and symbolic systems. IEML is precisely a symbolic system designed to exploit the cognitive possibilities - and particularly the avenues for collective intelligence - opened up by the new digital medium.

## 3 Successive Addressing Layers in the Evolution of Cyberspace

In order to understand the germane nature of the IEML research program, it helps to view it in the context of a progressive build-up and usage of the digital memory at the service of human collective intelligence. As illustrated in Figure 1, I see cyberspace being constructed from addressing layers, each depending on the one directly below it. In Figure 1, for each layer, the underlined text describes the type of interconnection, the second line states the addressing system itself, the text in italic suggests the main applications or technical tools developed on this layer and the last lines evoke the principal socio-cultural effects.

The first layer concerns the information bits contained and addressed in a computer's memory. The key point here is that the data and the programs are addressed - in principle - by the same storage system. This design has already been anticipated in the early works of Turing (in the concept of the *universal* Turing machine, the programs and the data are recorded on the same abstract tape)[73], as well as in the first plans of the EDVAC of von Neumann [74, 14]. One of the main purposes of operating systems is to manage the memory of the computers at the bit level.

The second layer, called the Internet layer, identifies the servers in digital networks. The Internet Protocol is an addressing system - now universal - that allows the communication between computers or between digital processing devices. Technically, the Internet was introduced at the end of the sixties, but its

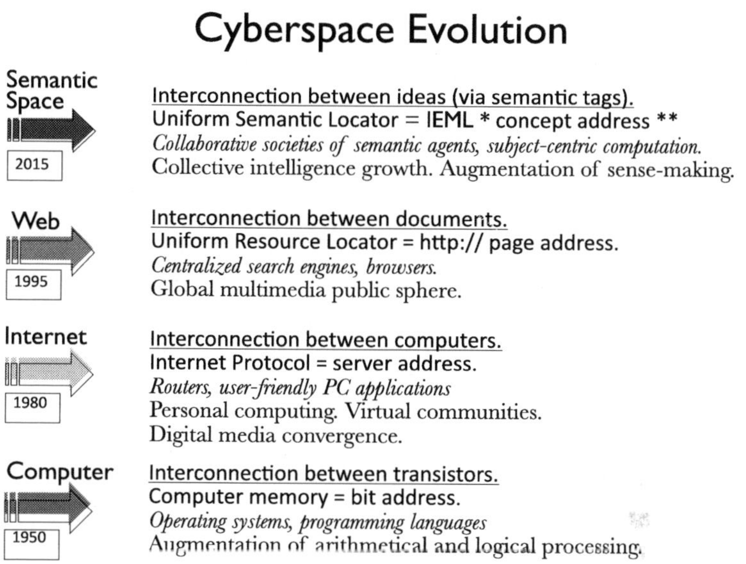


**Fig. 1.** Semantic Space Addressing Layer in the Evolution of Cyberspace

growth and wide public adoption dates back to the 1980s, in parallel with the progressive adoption of personal computers by the public. The Internet and the PC allowed networks of users and institutions to participate in cyberspace by contributing information and by being able to navigate in it.

The third layer, the Web layer, is made of the addresses of the "pages" of hyperdocuments and, by the same token, allows the identification of hyperlinks between those pages [4]. Even if the principle of HTTP was invented in 1991, the Web became a public phenomenon around 1995 thanks to the availability of browsers and numerous websites. The direct consequence of this last layer is the emergence of a new global multimedia public sphere. The recent growth in social computing and social tagging, experimentations in large-scale automatic natural language processing[21], as well as continuous research and development in the "web of data" (or the so-called semantic web)[27, 20], is indeed part of the Web layer but is quick to point towards the next addressing layer, related to semantic computing.

The objective of the IEML research program is to establish this new addressing layer, where significations, or subjects, have addresses ("subject-centric computing" [58]). As suggested in Figure 1, in semantic space, USLs (Uniform Semantic Locators) are used to uniquely address distinct significations and IEML (Information Economy MetaLanguage) is the finite regular language that I have invented for the notation of USLs that are the elements, or "points" of the semantic space. I will sketch now how IEML semantic space will help solving two important issues: the semantic interoperability problem and the problem of self-reference of digital-based collective intelligence.

# 4 The Semantic Interoperability Problem

## 4.1 Collective Categorization in Cyberspace

In cyberspace, for the first time in human history, our species is growing a universally interconnected common memory where ubiquitous data can be accessed and transformed by automatic symbol manipulators. Since the Web only became public around 1995, this new medium is less than one generation old and we are just beginning its techno-cultural exploration. Prior to the Web, there already were intellectual technologies tapping into digital computation power, like spreadsheets, multimedia interactive simulations or hypermedia. But my hypothesis is that the main developments into the full symbolic and cognitive exploitation of the global digital memory are still to come. No generation before ours has been confronted with the challenge of organizing and exploiting an inexhaustible amount of shared data including the various cultural productions of diverse past and present communities. As shown by recent R&D activity addressing the improvement of social tagging [16], or the construction of a "Web of data" organized by ontologies[20, 27], the augmentation of our collective categorization power is a key issue in the development of new symbolic systems fitting the digital medium. It can be argued that categorization is indeed at the core of cognitive processes, and particularly in the case of those human cognitive processes that are driven by (cultural) symbolic systems [37]. Database design and management are becoming the main scientific activity [6] as well as the essence of digital art [54]. The problem of useful data categorization is also at the core of collective intelligence management in companies and businesses. The participation in several social networks with hundred of contacts and the access to data and metadata through global sharing systems like Twitter (hashtags), Delicious, Twine, YouTube or Flickr, make the issue of categorization in personal and collective cloud management urgent [68].

My main hypothesis is that natural languages, as well as notation systems invented before the 21st century, are not appropriate for the current and future scale of the social categorization process. They are not fit to exploit the new interconnected global digital memory and its unprecedented power. Natural languages are in accord with human brain processing, they were not designed to be automatically manipulated. Old notation and writing systems correspond to heavy and slow physical storage and retrieval processes, and not to ubiquitous high speed automatic computing. Ontologies get around the difficulty of categorizing the Web by building rigid logical relationships between opaque chains of symbols (URIs) and strings of characters borrowed from natural languages. But there are several disconnected ontologies. Search engines and advanced folksonomic design bypass the problem of accurate semantic representation of data by operating a statistical analysis on strings of characters originally meant to represent sounds (not meaning). Trying to synchronize and optimize such diverse and massive categorization processes in cyberspace by relying on natural languages and old notation systems is like trying to find powerful algorithms for the manipulation of roman numerals or numbers written in alphabetic notation instead of looking for a better symbolic system. The IEML research program suggests that we should adopt a symbolic system for the notation and manipulation of concepts designed from the very beginning for massively distributed social computation in a technically interconnected global memory. As a support for automatic computing, this symbolic system must be a regular language [11].

### 4.2 Complementarity between IEML and the Web of Data

All IEML expressions are built from a syntactically regular combination of six symbols, called its *primitives*. In IEML a *sequence* is a succession of $3^\lambda$ single primitives, where $\lambda = (0, 1, 2, 3, 4, 5)$. $\lambda$ is called the *layer* of a sequence. For each layer, the sequences have respectively a length of 1, 3, 9, 27, 81 and 243 primitives.

From a syntactic point of view, any IEML expression is nothing else than a *set of sequences*. As there is a distinct semantic for each distinct sequence, there is also a distinct semantic for each distinct set of sequences. In general, *the meaning of a set of sequences corresponds to the union of the meaning of the sequences of this set.* The main result is that any algebraic operation that can be made on sets in general can also be made on semantics (significations) once they are expressed in IEML. This is why IEML semantic space (i.e. the set of all sets of IEML sequences) is a *group of transformations* and can be considered as a *transparent semantic addressing system.*

The IEML research program shares an important goal with the Web of data: it aims at decompartmentalizing online information. But its approach is not to standardize data formats since this is already done by the W3C and other standardization institutions. There are indeed *other obstacles* to information exchange and human collective intelligence in cyberspace *than diverse data formats*: diverse ontologies reflecting different contexts and area of practice, diverse classification systems, diverse folksonomies emerging from social tagging in various social media [25] and, last but not least, multiple natural languages. The ambition of the IEML research program is not to impose some unique super-ontology (note that IEML can be used in the context of ontologies with very different hierarchies of concepts) but rather to provide a pivotal language (endowed with computable semantics) into which at least the higher classes of any ontology and classification system as well as the most popular tags of user-produced folksonomies can be translated.

Contrarily to OWL, and like any natural language, IEML includes verbs, nouns, adjectives, adverbs, conjugations and all kinds of inflections. But by constrast with natural languages, the meaning of all texts in IEML (whatever there complexity) can be automatically reconstructed from the dictionary and grammar of IEML.

As the IEML dictionary can provide diverse natural languages descriptors of IEML terms (basically sequences of layers 0, 1, 2, and 3), IEML can also be used to establish automatic correspondances between tags expressed in different natural languages: once expressed in IEML, a complex concept can be automatically translated to any natural language supported by the IEML dictionary.

According to my perspective, and by contrast with the parallelism between syntax and semantics that is the main feature of IEML, one of the limitations of the current Web of data is the *opacity by design*[3] of its ultimate universal addressing system: the URIs. As an abstract addressing system, the IEML semantic space is in principle independent of the URI address space just as it is independent of any physical or telecommunication addressing system. However, from a synergistic perspective, IEML can bring to the system of URIs a general semantic interconnection and a full group of transformations on semantics. Any IEML expression can indeed be represented by a URI in a straightforward manner, simply by taking any prefix and then adding a URI-encoded syntactic representation of the IEML expression to it. Thus, such IEML-URIs can be directly used as concepts in RDF or OWL and it can offer an alternative grounding to the entities of the Web of data, mapping URIs to such IEML-URIs. Therefore the power of IEML can be leveraged by the existing standards of the Web of data. Symmetrically, the expressive and algebraic properties of IEML can leverage the current Web of data by providing it with a novel grounding that can make it more "semantic".

## 5 Toward a Technoscience of Self-referential Collective Intelligence

As a growing proportion of human interaction, communication and memory use the medium of cyberspace, it becomes in principle feasible to ground interdisciplinary research in social sciences and humanities in a common body of digital data. But as different disciplines and even different schools in the same disciplines have incompatible theoretical frameworks, the new opportunities offered by the extension of cyberspace for the study of human or cultural phenomena are not yet fully exploited.

Between the 16th and 20th centuries, the natural sciences acquired a unique and infinite physical space [35], equipped with a system of universal coordinates and units of measurement. The observational instruments in the natural sciences today are very elaborate in their engineering, and undergo constant progress. The symbolic and conceptual instruments of natural sciences is highly formalized, logically coherent, and largely shared within the scientific community. Mathematicians have their sets, relations, numbers, and functions. Physicists have their mass, energy, and particles. Chemists have their elements, molecules and reactions. Biologists have their biomolecules, DNA, and their intracellular and intercellular pathways of exchange. The theories may abound and diverge, but the language, just like the system of coordinates and measures, remains common to them all, enabling dialogue, controlled testing and an articulated accumulation

[3] See:http://www.w3.org/TR/webarch/#uri-opacity

of discoveries. In terms of knowledge management, we can say that natural sciences have been successful in making a significant portion of their knowledge explicit, so that it could be shared and thus offering mutual enrichment.

By contrast, the humanities and social sciences do not share a cultural space that is unique, infinite, coordinated and measurable. The disciplines are fragmented. Within those disciplines, conflicts between paradigms often limit fruitful dialogue. It is sometimes even difficult to agree on the nature of the disagreements. The observational instruments are not well developed in terms of engineering. Except in certain highly formalized sub-disciplines, the calculability, predictive capacity, and testability of the theories are weak. The main consequences of this situation is that the greater part of the considerable knowledge accumulated by the community of researchers in the humanities and social sciences remains "implicit". That is to say, in terms of knowledge management, the knowledge and expertise accumulated by the humanities are difficult to share in contexts that differ from the initial environment in which they emerged. And yet, the resolution for the difficult problems confronting contemporary humanity demand the effective collaboration of all cultural sciences.

The notion of collective intelligence is a good candidate for the federation of humanistic disciplines and social sciences in a joined effort to make the most of this new universal digital memory. There are at least three reasons to support this claim. First, collective intelligence is the basic engine of human development. Secondly, a comprehensive model of CI can involve any dimension of human society and culture in a scalable, dynamic and holistic way. Thirdly, the objects of social and humanistic sciences are also personal and collective subjects, or expressions of subjects. So, a collective intelligence theoretical framework is a good fit to tackle the subjective, intersubjective and hermeneutical nature of the objects of human sciences, especially if it is constructed from the beginning to describe distributed, fractal and self-referential social cognition.

CI processes should be observed scientifically, and the results of this observation should help various communities to improve dynamically and collaboratively their own collective intelligence. My claim is that this goal cannot be met in the absence of a common semantic coordinate system. This system must be created in a such way as to permit various communities, networks and institutions to express their dynamic collective intelligence activity without forcing them to conform to a unique mould, but allowing them comparisons and exchanges. In this perspective, IEML semantic space can be considered as a conventional geometric framework for the observation of the (practically) infinite abstract universe where collective intelligence processes occur and intertwine. As the printing press, along with new instruments of measurement and observation of the physical world, were the basis of the revolution in natural sciences, I think that the digital medium, along with new instruments of measurement and observation of the information economy in cyberspace, will be the basis of a revolution in cultural sciences. This is why the IEML research program should bring together computer scientists and scholars from humanistic disciplines.

# 6 Conclusion: A Simple Representation of Collective Intelligence

## 6.1 IEML Primitives

IEML is made of a small number of building blocks, six primitive symbols, that are combined and recombined in a regular way in successive layers of composition. The detail of this syntactic combination process will not be explained here. I will just expound the meaning of the six primitive symbols in order to prepare the reading of Figure 2, that shows the simplest possible model of collective intelligence in IEML.

First, in any expression of signification we must be able to distinguish between something (whatever it may be) and nothing, because some syntactic roles must sometimes be left unplayed. Therefore, we need a symbol for blank or empty meaning, just as there are symbols for silence in musical notations and a zero symbol in arithmetic. So the first semantic dialectic of IEML will be between emptiness, noted `E` and fullness.

Now, we must make a distinction within fullness. The terms used in natural language expressions are traditionally divided between verbs and nouns, and a similar distinction can be made (in cognitive terms) between entities and processes. This distinction is very basic in human cognition; it exists in all natural languages. Moreover, the verb / noun distinction is not only useful at the lexical level but also at the propositional level: there are verb phrases and noun phrases. So this distinction is embedded in the deep structure of IEML. That's why fullness is organised by a dialectic of verbishness and nounishness.

If we try to synthesize the dialectic of process, we can say that it is very often organized around a movement or a circulation of energy between two poles. Various cultural traditions have named these two poles differently, but the bipolarity itself is fairly universal: yin and yang, negative and positive, intelligible and perceptible, transcendence and immanence, mind and body, type and token, etc. These two poles are represented in IEML by `U` (virtual) and `A` (actual).[4] The IEML dictionary will take advantage of the virtual / actual dialectic to represent every semantic structure that can be organized by a bipolar symmetry (male and female, singular and plural, and so on).

What are the inner distinctions of the nounish realm? As verbishness shelters bi-polarities, nounishness covers ternarities. The main ternarity lies in the structure of semiotic representation which needs three interdependent poles.

The first pole may be called the *signifier*, for example the sound or the visual representation of the noun "car".

The second pole is the *signified for an interpreter*: the notion of car. We need a (generally human) mind for the interpretation of a linguistic signifier. There is no "notion" outside the context of a living cognitive system able to manipulate symbols. For example, English signifiers do not mean anything for ants or for the sun. Despite the fact that we cannot have access to any signified without any

[4] See [43] for a philosophical account of the virtual / actual dialectic.

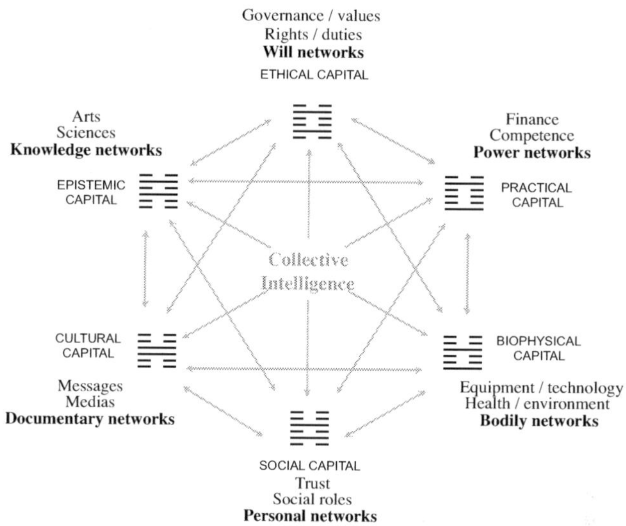


**Fig. 2.** A Model of Collective Intelligence in the Service of Human Development

representation of it, signifieds exist. A signified is of course something abstract, a concept, that is different from the perceptible signifier. As an illustration, we can say that "car", in English, has the *same* meaning (or signified) as "voiture" in French. If there can be two (or more) signifiers for one single signified, signified and signifier must be different entities.

The third pole is the *referent*, the actual thing that is referred to by the signifier: real cars in general, or this particular car, that have weight, a price on the market and can be driven.

The tradition of ternary semiotics begins with Aristotle[5] and is developed by medieval philosophers (*vox*, *conceptus*, *res*) from various linguistic and religious backgrounds [64, 61, 15]. It is pursued by classical philosophers or grammarians, refined by C. S. Peirce [63] and contemporay linguists.

This ternary dialectic[6] is expressed in IEML by three primitive symbols: sign (or signifier) noted `S`, being (or signified for an interpreter) noted `B` and thing (or referent), noted `T`. In general, `S` will be used to connote symbols, documents, language, and representations; `B` to connote people, personality, interiority and community and finally `T` to connote efficiency, power, technology, reality and the material world. We can also use `S`, `B` and `T` in IEML to mark triadic symmetries

[5] See the very beginning of "On Interpretation" [2].

[6] Etymologically, the word "dialectic" comes from the Latin *dialectica* (art of reasoning with method) deriving from the Greek *dia-legein* (through-saying) that means "reasoning" and that is related to the notion of dialogue. So "dialectic" does not derive from any Greek or Latin root meaning "two" or "duality". Therefore a ternary dialectic is indeed possible.

that cannot be derived from the initial semiotic ternarity like: future, present and past for tenses; or first, second and third persons for verbs and pronouns inflections.

Every semantic structure that can be organised by a ternary symmetry will take advantage of the S/B /T dialectic to be represented in IEML. For example: proposition (S), judgement (B) and state of things (T) in logic; legislative (S), judiciary (B) and executive (T) in politics; price (S), property (B) and utility (T) in economy; teachings (S), community (B) and ultimate reality (T) in religion, etc.

### 6.2 A Simple Model of Collective Intelligence in the Service of Human Development

The hexagrams of Figure 2 represent the six poles of collective intelligence and should be read in the following manner. Starting from the top, each line of a hexagram symbolizes a "semantic primitive": E, U, A, S, B, T. Full lines mean that the corresponding primitives are "ON" and broken lines that the corresponding primitives are "OFF". The diagram highlight the symmetry between two dialectics. Vertically, the virtual/actual binary dialectic juxtaposes and joins the two complementary triples: know-want-can/documents-persons-body. Horizontally, the ternary dialectic sign/being/thing juxtaposes and joins the three complementary tuples: know-documents/want-persons/can-bodies.

I have adopted here the network or actor-network theory[7, 36] that is broadly used in human and social sciences, leading to the integration of mathematical tools of graph theory[24, 9, 76]. This diagram shows essentially that *a sustainable collective intelligence implies a continuous exchange of resources between the six kinds of human capitals.* How can we develop methods to measure the value of the six kinds of human capitals and their flow of exchange? How can we extract these measurements from the analysis (preferably automatic) of data and online transactions of a given community or of a network of communities? This is the kind of problems that are posed by the IEML research program.

Human collective intelligence in cyberspace is still in its infancy. In my judgement, the main obstacle that prevents human collective intelligence from crossing the next cognitive threshold is the current absence of systematic self-awareness. Ideally, collective intelligence self-awareness can be reached from various semantic perspectives, at various time-scales and for various communities both large and small. The ethical drive behind the IEML research program is the hypothesis according to which crossing the threshold of *reflexive* collective intelligence will lead to a blooming of human development (economic prosperity, health, education, transmission of cultural heritages, human rights, control of ecosystems, peace, security, research, innovation...). According to this hypothesis, the highest goal of the IEML research program is to provide a symbolic framework for the making of digital tools that can help human collective intelligence observe its own activity in cyberspace, and therefore improve human development.

## References


[1] Auroux, S.: La Révolution technologique de la grammatisation. Mardaga, Liège (1994)

[2] Aristotle: On Interpretation. Translated by Edghill, E. M., `http://classics.mit.edu/Aristotle/interpretation.1.1.html`

[3] Yochai, B.: The Wealth of Networks: How Social Production Transforms Markets and Freedom. Yale UP, New Haven (2006)

[4] Tim, B.-L.: Weaving the Web. Harper, San Francisco (1999)

[5] Jean, B.: L'Ecriture la raison et les Dieux. Gallimard, Paris (1987)

[6] Geoffrey, B.: Memory Practices in the Sciences. MIT Press, Cambridge (2005)

[7] Michel, C. (dir.): La Science et ses réseaux. Genèse et circulation des faits scientifiques. La Découverte, Paris (1989)

[8] Mary, C.: The Craft of Thought, Meditation, Rhetoric and the Making of Images, 400, 1200. Cambridge UP, New York (2000)

[9] Manuel, C.: The information Age. Blackwell, Cambridge (1996, 1997, 1998)

[10] Jean-Pierre, C.: Neuronal Man: The Biology of Mind. Pantheon Books, NY (1985)

[11] Noam, C., Schützenberger, M.P.: The algebraic theory of context free languages. In: Braffort, P., Hirschberg, D. (eds.) Computer Programming and Formal Languages, pp. 118–161. North Holland, Amsterdam (1963)

[12] Henri, B.: Matière et mémoire. Alcan, Paris (1908); 1st edn. (1896)

[13] Kimiz, D.: Knowledge Management in Theory and Practice. Elsevier, NY (2005)

[14] Martin, D.: Engines of Logic: Mathematicians and the Origin of the Computer. Norton & Company, NY (2000)

[15] De Libera, A.: La Querelle des universaux, De Platon à la fin du Moyen-Age. Seuil, Paris (1996)

[16] Christo, D., Xu, J., Dicheva, D., Zhang, J.: A Study on Community Formation in Collaborative Tagging Systems. In: Proceedings of the IEEE/WIC/ACM International Conference on Web Intelligence, Sydney (2008)

[17] Douglas, E.: Augmenting Human Intellect, Technical Report. Stanford Research Institute (1962)

[18] Gerard, E.: Neural Darwinism. Basic Books, NY (1987)

[19] Elisabeth, E.: The Printing Revolution in Early Modern Europe. Cambridge University Press, Cambridge (1983)

[20] Feigenbaum, L., et al.: The semantic Web in Action. Scientific American, pp. 90–97 (December 2007)

[21] Christiane, F. (ed.): Preface by George Miller, Wordnet, An Electronic Lexical Database. MIT Press, Cambridge (1998)

[22] Jack, G.: The Domestication of the Savage Mind. Cambridge UP, New York (1977)

[23] Jack, G.: The Logic of Writing and the Organization of Society. Cambridge UP, New York (1987)

[24] Mark, G.: The Strength of Weak Ties. American Journal of Sociology 78, 1360–1380 (1973)

[25] Harry, H., Valentin, R., Shepherd, H.: The Complex Dynamics of Collaborative Tagging. In: Proceedings of the16th International WWW Conference, pp. 211–220 (2007)

[26] Havelock, E.A.: The Muse Learns to Write, Reflections on Orality and Litteracy from Antiquity to the Present. Yale UP, New Haven (1988)

[27] Jim, H., Dean, A.: Semantic Web for the Working Ontologist. Morgan Kaufmann, Burlington (2008)

[28] Hutchins, E.: Cognition in the Wild. MIT Press, Cambridge (1995)
[29] Harold, I.: Empire and Communication. Toronto UP (1950)
[30] Julian, J.: The Origin of Consciousness in the Breakdown of the Bicameral Mind. Houghton Mifflin, Boston (1976)
[31] Gottlob, F.: Geach, P., Black, M. (eds.): Philosophical Writings of Gottlob Frege. Blackwell, Oxford (1952)
[32] Epaminondas, K.: On the Notion of Collective Intelligence: Opportunity or Challenge? In: Inter. Journal on Organisational and Collective Intelligence (IJOCI), vol. 1(1). Idea Group Publishing, USA (2009)
[33] Robert, K.: The Nothing That Is, a Natural History of Zero. Oxford UP (1999)
[34] Khare, R.: Microformats: the next (small) thing on the semantic Web? IEEE Internet Computing 10(1), 68–75 (2006)
[35] Alexandre, K.: From the Closed World to the Infinite Universe. Harper, NY (1958)
[36] Bruno, L.: Science in Action. Harvard UP, Cambridge (1987)
[37] Claude, L.-S.: La pensée sauvage. Plon, Paris (1962); The Savage Mind. Chicago UP (1966)
[38] Pierre, L.: La Machine Univers. Création, cognition et culture informatique. La Découverte, Paris (1987)
[39] Pierre, L.: Les Technologies de l'intelligence. L'avenir de la pensée à l'ère informatique. La Découverte, Paris (1990)
[40] Pierre, L., Michel, A. (preface of Michel Serres): Les Arbres de connaissances. La Découverte, Paris (1992)
[41] Pierre, L.: De la programmation considérée comme un des beaux-arts. La Découverte, Paris (1992)
[42] Pierre, L.: L'Intelligence collective. Pour une anthropologie du cyberespace. La Découverte, Paris (1994); Collective Intelligence: Mankind's Emerging World in Cyberspace. Perseus Books, Cambridge (1997)
[43] Pierre, L.: Qu'est-ce que le virtuel? La Découverte, Paris (1998); Becoming Virtual. Reality in the Digital Age. Plenum Trade, NY (1998)
[44] Pierre, L.: Cyberculture. Odile Jacob, Paris (1997); Cyberculture. University of Minnesota Press (2001)
[45] Pierre, L.: World Philosophie: le marché, le cyberespace, la conscience. Odile Jacob, Paris (2000)
[46] Pierre, L.: Cyberdémocratie: Essai de philosophie politique. Odile Jacob, Paris (2002)
[47] Pierre, L.: Société du savoir et développement humain. In: Le Canada et la société des savoirs, sous la direction de Patrick Imbert, CR Université d'Ottawa: Canada, enjeux sociaux et culturels dans une société du savoir, pp. 115–175 (2007)
[48] Pierre, L.: Algebraic Structure of IEML Semantic Space. CI Lab Technical Report (2009), http://www.ieml.org/spip.php?article152
[49] Pierre, L.: From Social Computing to Reflexive Collective Intelligence: the IEML Research Program. In: Kapetanios, E., Koutrika, G. (eds.) Information Sciences, Special Issue on Collective Intelligence. Elsevier, Amsterdam (to be published, 2009)
[50] L'espace sémantique
[51] Charlene, L., Josh, B.: Groundswell. Winning in a World Transformed by Social Technologies. Harvard Business Press, Cambridge (2008)
[52] Logan Robert, K.: The Alphabet Effect. Hampton Press, Cresskill (2004); 1st edn. Wm. Morrow, New York (1986)
[53] Logan Robert, K.: The Extended Mind: The Emergence of Language, the Human Mind and Culture. U. of Toronto Press (2007)

[54] Lev, M.: The Language of New Media. MIT Press, Cambridge (2001)
[55] Marshall, M.: The Gutemberg Galaxy. University of Toronto Press (1962)
[56] Marshall, M.: Understanding Media: The Extensions of Man. New American Library, NY (1964)
[57] Morey, D., Maybury, M., Thuraisingham, B. (eds.): Knowledge Management, Classic and Contemporary Works. MIT Press, Cambridge (2000)
[58] Steven, N.: A perspective on the Quest for Global Knowledge Interchange. In: Jack, P., Sam, H. (eds.) Foreword by Douglas Engelbart, XML Topic Maps. Creating and Using Topic Maps for the Web, pp. 31–50. Addison Wesley, Reading (2002)
[59] Nonaka, I., Takeuchi, H.: The Knowledge-creating Company: How Japanese Companies Create the Dynamics of Innovation. Oxford UP, NY (1995)
[60] Walter, O.: Orality and Litteracy, The Technologizing of the Word. Methuen, London (1982)
[61] Claude, P.: Les Mots, les concepts et les choses, la sémantique de Guillaume d' Occam et le nominalisme d' aujourd' hui. Bellarmin-Vrin, Paris-St Laurent (Qc) (1992)
[62] Corina, P.: An Empirical Analysis of the Creation, Use and Adoption of Social Computing Applications. European Commission, Joint Research Centre and Institute for Prospective Technological Studies (2008)
[63] Charles, P.: The Essential Peirce: Selected Philosophical Writings. Indiana UP (1998)
[64] François, R.: La triade sémiotique, le trivium et la sémantique linguistique. Nouveaux actes sémiotiques 9, 54 (1990)
[65] Ferdinand, S.: Cours de linguistique générale. Payot, Lausanne-Paris (1916)
[66] Clay, S.: Here Comes Everybody: The Power of Organizing Without Organizations. Penguin (2008)
[67] Gene, S.: Tagging: People-powered Metadata for the Social Web. Peachpit New Riders, Berkeley (2007)
[68] Nova, S.: `http://novaspivack.typepad.com`
[69] James, S.: The Wisdom of the Crowds. Random House, London (2004)
[70] Elaine, S.: The Intellectual Foundation of Information Organization. MIT Press, Cambridge (2000)
[71] Don, T., Williams, A.D.: Wikinomics, How Mass Collaboration Changes Everything. Portfolio (2007)
[72] Mark, T. (ed.): Collective Intelligence: Creating a Prosperous World at Peace. EIN Press, Oakton (2008)
[73] Allan, T.: On Computable Numbers, with an Application to the Entscheidungsproblem. Proc. London Math. Soc. 42(2), 230–265 (1936)
[74] Von Neumann, J.: First Draft of a Report on the EDVAC, Contract No. W-670-ORD-492, Moore School of Electrical Engineering, Univ. of Penn., Philadelphia, pp. 383–392 (June 30, 1945); Reproduced (partly) in Randell, Brian, Origins of Digital Computers: Selected Papers, pp. 383-392. Springer, Heidelberg (1982)
[75] David, W.: Everything Is Miscellaneous: The Power of the New Digital Disorder. Henri Holt and Cie, USA (2007)
[76] Barry, W.: Computer Networks as Social Networks. Science 293, 2031–2034 (2001)
[77] Etienne, W.: Communities of Practice: Learning, Meaning, and Identity. Cambridge UP, New York (1998)
[78] Frances, Y.: The Art of Memory. University of Chicago Press, Chicago (1974)
[79] Olivier, Z.: Le Management de l'intelligence collective, M2 edns., Paris (2004)

# A-Teams and Their Applications

Piotr Jędrzejowicz

Chair of Information Systems, Gdynia Maritime University,
Morska 83, 81-225 Gdynia, Poland
pj@am.gdynia.pl

## 1 Introduction

The techniques used to solve difficult combinatorial optimization problems have evolved from constructive algorithms to local search techniques, and finally to population-based algorithms. Population-based methods have become very popular. They provide good solutions since any constructive method can be used to generate the initial population, and any local search technique can be used to improve each solution in the population. But population-based methods have the additional advantage of being able to combine good solutions in order to get possibly better ones. The basic idea behind this way of doing is that good solutions often share parts with optimal solutions.

Population based methods are optimization techniques inspired by natural evolution processes. They handle a population of individuals that evolves with the help of information exchange procedures. Each individual may also evolve independently. Periods of cooperation alternate with periods of self-adaptation. Among best-known population-based method are evolutionary algorithms.

Since the publication of Goldberg seminal work [30] different classes of evolutionary algorithms have been developed including genetic algorithms, genetic programming, evolution strategies, differential evolution, cultural evolution, coevolution and population learning algorithms. Not much later studies of the social behavior of organisms have resulted in development of swarm intelligence systems including ant colony optimization and particle swarm optimization (see [46]). Technological advances enabled in the recent years development of various parallel and distributed versions of the population based methods. At the same time, as a result of convergence of many technologies within computer science such as object-oriented programming, distributed computing and artificial life, the agent technology has emerged. An agent is understood here as any piece of software that is designed to use intelligence to automatically carry out an assigned task, mainly retrieving and delivering information.

Tweedale and co-authors [58] outline an abridged history of agents as a guide for the reader to understand the trends and directions of future agent design. This description includes how agent technologies have developed using increasingly sophisticated techniques. It also indicates the transition of formal programming languages into object-oriented programming and how this transition facilitated a corresponding shift from scripted agents (bots) to agent-oriented

N.T. Nguyen, R. Kowalczyk, and S.-M. Chen (Eds.): ICCCI 2009, LNAI 5796, pp. 36–50, 2009.

designs which is best exemplified by multiple agent systems (MAS). A MAS tries to solve complex problems with entities called agents, using their collaborative and autonomous properties [45]. Jennings et al. [37] list the following MAS properties:

- Each agent has partial information or limited capabilities
- There is no global system control
- Data in a MAS are decentralized
- Computation is asynchronous
- Different agents could be heterogeneous, for example, with respect to knowledge representation, reasoning model, solution evaluation criteria, goal, architecture or algorithm.

Paradigms of the population-based methods and multiple agent systems have been during mid nineties integrated within the concept of the asynchronous team of agents (A-Team). In this paper some recent developments in the field of implementation and application of A-Teams are reviewed and analysed. The paper includes a short description of the A-Team concept and its comparison with the related methods and techniques followed by the review of the A-team implementations and applications.

## 2 A-Team Concept

### 2.1 A-Team Definition and Architecture

A-Team is a multi agent architecture, which has been proposed in several papers of S.N. Talukdar and co-authors [52], [54], [56], [57]. It has been claimed that the A-Team framework enables users to easily combine disparate problem solving strategies, each in the form of an agent, and enables these agents to cooperate to evolve diverse and high quality solutions [50].

According to Talukdar [54] an asynchronous team is a collection of software agents that cooperate to solve a problem by dynamically evolving a population of solutions. As Rachlin et al. [50] observed agents cooperate by sharing access to populations of candidate solutions. Each agent works to create, modify or remove solutions from a population. The quality of the solutions gradually evolves over time as improved solutions are added and poor solutions are removed. Cooperation between agents emerges as one agent works on the solutions produced by another. Within an A-Team, agents are autonomous and asynchronous. Each agent encapsulates a particular problem-solving method along with the methods to decide when to work, what to work on and how often to work.

A-Team architecture could be classified as a software multi-agent system that is used to create software assistant agents [44]. According to [2] an asynchronous team (A-Team) is a network of agents (workers) and memories (repositories for the results of work). The paper claims that it is possible to design A-Teams to be effective in solving difficult computational problems. The main design issues are structure of the network and the complement of agents. The discussed paper describes a procedure by which good structures can be learned from experience.

A formal definition of an A-Team was offered in [56]. An A-Team is a set of autonomous agents and a set of memories, interconnected to form a strongly cyclic network, that is, a network in which every agent is in a closed loop. As a consequence, an A-Team can be visualized as a directed hypergraph, called a data flow. Each node represents a complex of overlapping memories. Each arc represents an autonomous agent. Results or trial-solutions accumulate in the memories (just as they do in blackboards) to form populations (like those in genetic algorithms). These populations are time varying. The ground principal of asynchronous teams rests on combining algorithms, which alone could be inept for the task, into effective problem-solving organizations [56]. Talukdar [57] proposed the grammar to provide a means for constructing all asynchronous teams that might be used in solving a given instance of a family of off-line problems. In other words, the grammar constructively defines the space that must be searched if an asynchronous team that is good at solving the given problem-instance is to be found. The primitives of the grammar are:

- Sharable memories, each dedicated to a member of the family-of-problems, and designed to contain a population of trial-solutions to its problem.
- Operators for modifying trial-solutions.
- Selectors for picking trial-solutions.
- Schedulers for determining when selectors and operators are to work.

According to Rachlin et al. [50] the representation of the problem and the solution within an A-Team are problem specific but uniform throughout. It is further suggested that agents come in three flavors: constructors, improvers and destructors. Constructors create initial solutions and add them to the population. Improvers select one or more existing solutions from the population and produce new solutions that are added to the population. Finally, destroyers keep the size of the population of solutions in check. Their main function is to delete clearly sub-optimal or redundant solutions, while keeping promising solutions. This prevents improver agents from wasting effort by working on solutions that are going nowhere.

Following Correa et al. [17] an execution of an A-Team can be described by a set of events. Each event is composed by the following elements: the time at which the event occurs; the input data, if any; the state of the shared memory prior to the occurrence of the event; the state of the shared memory after the occurrence of the event; and the output data, if any. According to this formalism, there is a restricted number of types of events occurring during an A-Team execution:

- The initialization of global variables and the shared memory.
- The reading operation.
- The execution of a heuristic.
- The writing operation.

A-Team architecture offers several advantages. Its main advantage is ability to produce good quality solutions of difficult optimization problems. Rachlin et al. [50] mention modularity, suitability for distributed environments and robustness.

It is also clear that A-Teams demonstrate properties of the collective intelligence system since the collective of agents can produce better solutions than individual members of such collective.

A-Teams are based on a starlike topology. According to Zhu [62] in a starlike topology, the activities of the agents are coordinated or administered by some supervisory (or facilitator) agents designated in the assembly. Only agents that have connections built and specified to the coordinator can interact with each other. An advantage of star-like topology is its loosely enforced control and coordination. Though control and coordination limits the boundary of cooperation the agents can reach, it is desirable when efficiency of cooperation is a main issue that needs to be ensured. The use of facilitators in open agent architecture offers both advantages and weaknesses with respect to scalability and fault tolerance [16]. For example, on the plus side, the grouping of a facilitator with a collection of client agents provides a natural building block from which to construct larger systems. On the minus side, there is the potential for a facilitator to become a communication bottleneck, or a critical point of failure.

## 2.2 Related Architectures

It should be noted that several related agent-based architectures have been proposed in parallel to emergence of A-Teams. One of them is the blackboard architecture enabling cooperation among agents, called knowledge sources, by sharing access to a common memory and allowing agents to work on different parts of the problem [18]. The key difference is that A-Teams do not have a central scheduler responsible for sequencing agent invocations. The blackboard architecture contains three principal components:

- A central, globally accessible, hierarchically organized database called the blackboard that contains the results of applying problem-solving knowledge.
- A set of knowledge sources representing the available problem solving knowledge.
- A scheduler implementing problem solving strategies by analysing the state of the solution and the available knowledge.

Another related concepts are architectures of evolutionary (EMAS) and co-evolutionary (CoEMAS) systems. The main idea of evolutionary multi-agent system is the modeling of evolution process in MAS [39]. In opposition to classical evolutionary algorithms, in EMAS there is no centralized algorithm which manipulates the whole population. All agents are independent and make their own decisions, particularly these concerning reproduction and death. Selection in EMAS is based on a non-renewable resources, which are possessed by agents. Every activity costs some resource and may be realized provided that the agents resource level is high enough. Resources can be gained only from the environment or other agents, and the rule is that better fit agents are given more resources than less fit ones.

Co-evolutionary multi-agent systems (CoEMAS) [47], [48], [27] allow co-evolution of several species of agents. CoEMAS can be applied, for example, to multi-objective optimization and multi-modal function optimization. In Co-EMAS several (usually two) different species co-evolve. One of them represents solutions. The goal of the second is to cooperate (or compete) with the first one in order to force the population of solutions to locate Pareto frontier or proportionally populate and stably maintain niches in multi-modal domain.

The concept of A-Team shares also some similarities with the adaptive memory programming (AMP) [29], [51]. AMP refers to a class of metaheuristics which use a memory to store information collected during the search process. This information is used to bias the solution construction and the future search. After the memory has been initialized new provisory solutions using data stored in the memory is generated and improved with a local search. The improved solution is used to update the memory using knowledge brought by the solution. The procedure iterates until a stopping criterion has been met.

Yang and Zhang [60] proposed a comprehensive framework for distributed problem solving (DPS). In particular, in one of their models (DPS2) tasks are allocated to multiple agents that produce feature solutions that are synthesized into the final solution. This framework advocates use of multiple agents for tackling a single task, thus relying on the use of pluralistic models. Within similar lines Zhang proposed a way to synthesize final solutions in systems where different agents use different inexact reasoning models to solve a problem [61].

Vahidov and Fazlollahi [59] proposed a framework for a pluralistic multiagent decision support system (MADSS). The distinguishing feature of the proposed approach is its organization around human decision making process. The framework builds upon the decision support pyramid with agents organized into teams according to the phases of the problem solving model. The MADSS framework incorporates pluralistic agents that have diverse sets of views and values in approaching the problem and informs the decision maker about this diversity.

Some close similarities to the A-Team architecture characterize the Decision Making Library (Agent-DMSL) [28] a highly customizable decision making library built using Java 1.4 programming language.

Special kinds of artificial agents are those created by analogy with social insects. Social insects (bees, wasps, ants, and termites) have lived on Earth for millions of years. Their behavior in nature is, first and foremost, characterized by autonomy and distributed functioning and self-organizing. In the last couple of years, the researchers started studying the behavior of social insects in an attempt to use the swarm intelligence concept in order to develop various artificial systems. Social insect colonies teach us that very simple organisms can form systems capable of performing highly complex tasks by dynamically interacting with each other. Communication between individual insects in a colony of social insects has been well recognized. The examples of such interactive behavior are bee dancing during the food procurement, ants pheromone secretion, and performance of specific acts which signal the other insects to start performing the same actions. These communication systems between individual insects contribute to the formation of the "collective intelligence" of the social insect colonies. The term "Swarm intelligence", denoting this "collective intelligence" has come into use [10], [11], [12].

# 3 A-Team Implementations

## 3.1 First Generation

A-Teams belong to two broader classes of systems: cooperative multiagent and peer-to-peer (P2P) systems. In cooperative multiagent systems agents work toward achieving some common goals, whereas self-interested agents have distinct goals but may still interact to advance their own goals [41]. In P2P systems a large number of autonomous computing nodes (the peers) pool together their resources and rely on each other for data and services [40].

The reported implementations of the A-Team concept include two broad classes of systems: dedicated A-Teams and platforms, environments or shells used as tools for constructing specialized A-Team solutions. Dedicated (or specialized) A-Teams are usually not flexible and can be used to solving particular problem types only. Among example A-Teams of such type one can mention the OPTIMA system for the general component insertion optimization problem [49] or A-Team with a collaboration protocol based on a conditional measure of agent effectiveness designed for flow optimization of railroad traffic [13]. Majority of applications listed in this section have been using a specialized A-teams.

Among platforms and environments used to implement A-Team concept some well known include IBM A-Team written in C++ with own configuration language [50]. Some implementations of A-Team were based on universal tools like Matlab ([56] and [14]). Some other were written using algorithmic languages like, for example the parallel A-Team of [17] written in C and run under PVM operating system.

The above discussed platforms and environments belong to the first generation of A-Team tools. They are either not portable or have limited portability, they also have none or limited scalability. Agents are not in conformity with the FIPA (The Foundation of Intelligent Psychical Agents) standards and there are no interoperability nor Internet accessibility. Migration of agents is either impossible or limited to a single software platform.

## 3.2 Next Generation

Next generation A-Teams started to appear roughly after the year 2004. Major differences in comparison with the first generation solutions can be found with respect to accessibility, scalability and portability of the next generation solutions. However, there exists still two major classes of solutions. The first includes specialized A-teams designed to solve instances of particular problems. Their architecture is problem-specific and not flexible. The second class covers middleware platforms allowing for an easy implementation of the A-Teams ready to solve instances of arbitrary problems.

Good example of the problem specific A-Team is the implementation proposed by Meneses et al. [42] dedicated to solving probe selection problems. Kernel of the system is a multithreaded preprocessing program written in C++. A dialog-based framework was used to allow for rapid development of the program.

Classes from the Microsoft Foundation Class (MFC) library were utilized for storage of probes, clones and solutions. Individual probes were stored as CStrings. CStringArrays were used to store input clones, probes and probe solutions, and a CObArray was used for Memory M1. A random-number generator was used to randomly select the agent to be called at each iteration.

The middleware platforms supporting implementation of A-Teams, is represented by the JADE-Based A-Team environment (JABAT). Its different versions were proposed in [33], [3], [6] and [26]. The JABAT middleware was built with the use of JADE (Java Agent Development Framework), a software framework proposed by TILAB [9] for the development and run-time execution of peer-to-peer applications. JADE is based on the agents paradigm in compliance with the FIPA specifications and provides a comprehensive set of system services and agents necessary to distributed peer-to peer applications in the fixed or mobile environment. It includes both the libraries required to develop application agents and the run-time environment that provides basic services and must be running on the device before agents can be activated [9]. JADE allows each agent to dynamically discover other agents and to communicate with them according to the peer-to-peer paradigm.

JABAT complies with the requirements of the next generation A-Teams which are portable, scalable and in conformity with the FIPA standards. It has been also extended to become a fully Internet-accessible solution [7], with the ability to not only to access the system in order to solve own combinatorial problems, but also to add own computer to the system as to increase the resources used for solving the problems. It can be also run within grid environment [26].

JABAT produces solutions to optimization problems using a set of optimizing agents, each representing an improvement algorithm. The process of solving a single task (i.e. the problem instance) consists of several steps. At first the initial population of solutions is generated. Individuals forming the initial population are, at the following computation stages, improved by independently acting agents, thus increasing chances for reaching the global optimum. Finally, when the stopping criterion is met, the best solution in the population is taken as the result. How the above steps are carried out is determined by the "working strategy". There may be different working strategies defined in the system, each of them specifying how the initial population of solutions is created, how to choose solutions which are forwarded to the optimizing agents for improvement, how to merge the improved solutions returned by the optimizing agents with the whole population and when to stop searching for better solutions. Main functionality of JABAT can be summarized as follows:

- The system can solve instances of several different problems in parallel.
- The user, having a list of all algorithms implemented for the given problem may choose how many and which of them should be used.
- The optimization process can be carried out on many computers. The user can easily add or delete a computer from the system. In both cases JABAT will adapt to the changes, commanding the optimizing agents working within

the system to migrate. JABAT may also clone some already working agents and migrate the clones, thus increasing the number of concurrently operating agents.
- The system is fed in the batch mode - consecutive problems may be stored and solved later, when the system assesses that there is enough resources to undertake new searches.

To perform the above two classes of agents are used. The first class includes OptiAgents, which are implementations of the improvement algorithms. The second class includes SolutionManagers, which are agents responsible for maintenance and updating of individuals in the common memory. All agents act in parallel. Each OptiAgent is representing a single improvement algorithm (simulated annealing, tabu search, genetic algorithm, local search heuristics etc.). An OptiAgent has two basic behaviors defined. The first is sending around messages on readiness for action including the required number of individuals (solutions). The second is activated upon receiving a message from some SolutionManager containing the problem instance description and the required number of individuals. This behavior involves improving fitness of individuals and resending the improved ones to the sender. A SolutionManager is brought to life for each problem instance. Its behavior involves sending individuals to OptiAgents and updating the common memory.

Main assumption behind the proposed solution is its independence from a problem definition and solution algorithms. Hence, main classes Task and Solution upon which agents act, have been defined at a rather general level. Interfaces of both classes include function ontology(), which returns JADEs ontology designed for classes Task and Solution, respectively. Ontology in JADE is a class enabling definition of the vocabulary and semantics for the content of message exchange between agents. More precisely, an ontology defines how the class is transformed into the text message exchanged between agents and how the text message is used to construct the class (here either Task or Solution). The interface of the main class Task is composed of the following three functions: Task() - class constructor, Solution createSolution() - function generating an initial solution which can be either randomly drawn or empty, TaskOntology ontology() - function returning task ontology. The interface of the main class Solution is composed of the following functions: Solution() - class constructor, Solution(Task t) - constructor producing solution to the given task, SolutionOntology ontology() - function returning solution ontology, Object clone() - function producing copy of the object, Boolean equals(Solution s) - function returning the result of comparison between two solutions, void evaluate() - procedure evaluating the fitness of a solution (stored as a value of the variable fitness). To obtain a solution of the particular problem instance the following actions should be carried out:

- Defining own classes MTask and MSolution inherited from Task and Solution, respectively. In these new classes constructors and other functions need to be over-ridden to assure compatibility between actions and the problem instance requirements.

- Defining own ontologies MTaskOntology and MSolutionOntology inherited from TaskOntology and SolutionOntology, respectively. Both are responsible for translating classes MTask and MSolution into text messages. Messages are more complex than a single task or solution but to produce them the outcome of MTaskOntology and MSolutionOntology class functions are used.
- Defining auxiliary classes and functions as, for example, the Compare function, which could be used by the SolutionManager to compare and sort thus far obtained solutions.

## 4 Example Applications

Already at the early stages of the concept development A-Teams have been successfully applied to a host of problems, ranging from control of electric networks [55], through traveling salesman problems to constraint satisfaction [31]. A-Teams have proven to be successful in addressing hard combinatorial optimization problems where no dominant algorithm exists. Some reported applications of the first generation A-Teams include:

- TSP problem [53].
- Collision avoidance in robotics [38].
- Diagnosis of faults in power systems [15].
- Planning and scheduling in manufacturing [43].
- Flow optimization of railroad traffic [13].
- Bicriteria combinatorial optimization problem and its applications to the design of communication networks for distributed controllers [14].
- Automatic insertion of electronic components [49].
- Parallel and asynchronous approach to give near-optimal solutions to the non-fixed point-to-point connection problem [17].
- Job-shop scheduling [1].

Among the reported implementations of the next generation A-Teams there are two broad application areas, which, in fact remain closely related. The first include A-Teams used to solve computationally hard combinatorial optimization problems. The second area is machine learning and data mining.

In the area of combinatorial optimization one should note the approach to probe selection problems proposed by Meneses et al. [42]. The proposed A-Team can find near-optimal solutions to probe selection problems. The A-Team was implemented and tested on both real and semi-random data with probes of both fixed and varying lengths. The computational experiments showed that the approach is extremely effective and can find results comparable to the best known in a very short computational time. Another advantage is that the A-Team is able to find near-optimal probe sets containing probes of both fixed and varying lengths. The approach is dynamic in the sense that new heuristic agents can be added at any time without requiring any significant changes to the existing code. This allows for continued improvement of the method with the seamless incorporation of new heuristic algorithms.

Several successful implementations of the JABAT based A-Teams have been reported in the field of scheduling. In [34] the flowshop and job-shop scheduling problems are solved by the A-Team implementation of the population learning algorithm. Computational experiment results confirm a very high quality of the obtained solutions. Good quality results in a very competitive time were obtained by the team of agents carrying out several local search procedures with a view to solve instances of the resource constrained project scheduling problem [35]. Jedrzejowicz and Ratajczak-Ropel [36] proposed also an A-Team approach to solving instances of the resource constrained project scheduling problem with minimal and maximal time lags. It was shown that the results obtained by the proposed A-Team architecture are comparable to the best reported in the literature while computation times are much shorter even for more complex problem instances.

Another classical combinatorial optimization problem tackled successfully by the A-Team technique is the vehicle routing. JABAT-based implementation of the A-Team solving the vehicle routing problem (VRP) was proposed in [5]. Although results obtained by the A-Team did not beat those produced by best tabu search algorithms, the advantage of parallelization and use of the distributed environment was visible from the computation time point of view. In [5] the structure and composition of the set of agents solving instances of the VRP were also investigated. The multi agent approach seems especially suitable to deal with the dynamic version of the vehicle routing problem. Barbucha and Jedrzejowicz [4] proposed a modification of the JABAT middleware to support solving the dynamic VRP through allowing to simulate different strategies with respect to simulating allocation of dynamically arriving orders to available vehicles. The idea was further developed in [9] where the dedicated platform MAS-DVRP was developed using the JABAT middleware and incorporating an asynchronous team of agents. The platform was used to carry-out computational experiments aiming at investigating how the degree of dynamism influences system performance. The proposed approach extends the range of algorithms for solving DVRP and offers good quality solutions to the static VRP instances allowing, at the same time, to observe how transition to dynamic version can influence quality of service offered to customers.

A-Teams have proven equally useful and effective in the field of machine learning and data mining. Czarnowski and Jedrzejowicz in [19] proposed a specialized A-Team applied to train the feed-forward artificial neural networks. The approach makes use of the idea which has been conceived for the population learning algorithm [32]. An effective search for solution of the computationally hard problems requires a cocktail of methods applied to a population of solutions, with more advanced procedures being applied only to a more promising population members. The proposed PLANN-Team includes a set of agents and a shared memory of the blackboard type. This memory is supervised by a supervisory agent whose tasks include taking care of communication within the PLANN-Team. The PLANN-Team employs 5 kinds of the search agent: standard mutation, local search, non-uniform mutation gradient mutation and gradient adjustment operator. Computational experiment confirmed that the tool produces high quality neural networks in a reasonable time.

Middleware platform JABAT has been used to implement several A-Teams solving computationally hard optimization problems typical for the field of machine learning. In [21] the A-Team implementation solving data reduction problem was proposed. The approach aimed at reducing the original dataset used for classification in two dimensions including selection of reference instances and removal of irrelevant attributes. Several agents representing different local-search based strategies were employed in parallel to achieve a synergetic effect. It has been shown that a set of simple local-search based agents performing the data reduction task can improve classification accuracy of a classifier. Another data mining application of the A-Team technique was described in [20] where the team of asynchronous agents selects reference vectors under multiple objectives generating Pareto-optimal reference vector sets.

One of the important problems in machine learning and data mining is data reduction. It is well known that in order to avoid an excessive storage and time complexity and to improve generalization accuracy by reducing noise and overfitting it is often advisable to reduce original training set by selecting the most representative information. In [23] the agent-based population learning algorithm reducing simultaneously number of reference vectors and features was implemented using JABAT middleware. In case of several well known benchmark datasets the approach proved competitive to alternative techniques. The success of the approach resulted in proposing an integration of the data pre-processing with the data mining stage [24]. Such an integration is supported by the software shell which allows for an easy implementation of A-Teams carrying data reduction and classification tasks as a single process based on the population learning paradigm.

Agent-based solutions were also proposed to deal with several distributed data mining problems. Czarnowski and Jedrzejowicz [22] proposed to deal with learning classifiers from distributed data sources through applying the A-Team concept implemented using the JABAT middleware. A-Teams are used to select prototypes at the local level, that is, at distributed data sources. The approach was extended in [25] where A-Teams select instances and features, and collaborate to assure a homogenous set of prototypes at the global level.

## 5 Conclusions

Software agents are knowledgeable, autonomous, situated and interactive software entities. Agents' interactions are of special importance when a group of agents interact with each other to solve a problem that is beyond the capability and knowledge of each individual. Efficiency, performance and overall quality of the multiagent applications depend mainly on how the agents interact with each other effectively. There is a convincing evidence that the next generation A-Teams can be effectively used to obtain solutions to difficult optimization problems in various areas. Contemporary A-Teams are composed of autonomous entities that can communicate and collaborate in order to obtain solutions to optimization problems. Available platforms hide all complexity of the distributed architecture and offer sets of predefined objects available to users who can focus on the logic of the A-Team applications and effectiveness of the optimization procedures rather than on middleware issues. Available solutions can be easily moved to new environments. A-Team technology is ready to offer functional, scalable, flexible, efficient, robust and stable A-Team architectures. A-Team architecture can be considered as a convenient tool to implement population-based solutions, parallel metaheuristics and hybrid state-of-the-art algorithms.

## References

1. Aydin, M.E., Fogarty, T.C.: Teams of asynchronous agents for job-shop scheduling problems: an experimental study. Journal of Intelligent Manufacturing 15(4), 455–462 (2004)
2. Baerentzen, L., Avila, P., Talukdar, S.: Learning Network Designs for Asynchronous Teams. In: MAAMAW, pp. 177–196 (1997)
3. Barbucha, D., Czarnowski, I., Jedrzejowicz, P., Ratajczak, E., Wierzbowska, I.: JADE-Based A-Team as a Tool for Implementing Population-Based Algorithms. In: Chen, Y., Abraham, A. (eds.) Intelligent Systems Design and Applications, IDSA, Jinan Shandong China, pp. 144–149. IEEE, Los Alamitos (2006)
4. Barbucha, D., Jedrzejowicz, P.: Multi-agent platform for solving the dynamic vehicle routing problem. In: Proc. 11th Int. IEEE Conference on Intelligent Transportation Systems, Beijing, pp. 517–522 (2008)
5. Barbucha, D., Jedrzejowicz, P.: An experimental investigation of the synergetic effect of multiple agents working together in the A-Team. System Science 34(2), 55–62 (2008)
6. Barbucha, D., Czarnowski, I., Jedrzejowicz, P., Ratajczak-Ropel, E., Wierzbowska, I.: An Implementation of the JADE-base A-Team Environment. International Transactions on Systems Science and Applications 3(4), 319–328 (2008)
7. Barbucha, D., Czarnowski, I., Jedrzejowicz, P., Ratajczak-Ropel, E., Wierzbowska, I.: Web accessible A-team middleware. In: Bubak, M., van Albada, G.D., Dongarra, J., Sloot, P.M.A. (eds.) ICCS 2008, Part III. LNCS, vol. 5103, pp. 624–633. Springer, Heidelberg (2008)
8. Barbucha, D.: Simulating Activities of the Transportation Company through Multi-Agent System Solving the Dynamic Vehicle Routing Problem. In: Hakansson, A., et al. (eds.) KES AMSTA 2009. LNCS (LNAI), vol. 5559, pp. 773–782. Springer, Heidelberg (2009)
9. Bellifemine, F., Caire, G., Poggi, A., Rimassa, G.: JADE. A White Paper, Exp. 3(3), 6–20 (2003)
10. Beni, G.: The concept of cellular robotic system. In: Proceedings of the 1988 IEEE International Symposium on Intelligent Control, pp. 57–62. IEEE Computer Society Press, Los Alamitos (1988)
11. Beni, G., Wang, J.: Swarm intelligence. In: Proceedings of the Seventh Annual Meeting of the Robotics Society of Japan, pp. 425–428. RSJ Press, Tokyo (1989)
12. Beni, G.S.: Hackwood: Stationary waves in cyclic swarms. In: Proceedings of the 1992 International Symposium on Intelligent Control, pp. 234–242. IEEE Computer Society Press, Los Alamitos (1992)

13. Blum, J., Eskandarian, A.: Enhancing Intelligent Agent Collaboration for Flow Optimization of Railroad Traffic. Transportation Research 36(10), 919–930 (2002)
14. Camponogara, E., Talukdar, S.: Designing communication networks for distributed control agents. European Journal of Operational Research 153, 544–563 (2004)
15. Chen, C.L., Talukda, S.N.: Causal Nets for Fault Diagnosis. In: 4th International Conference on Expert Systems Application to Power Systems, Melbourne, Australia, pp. 65–70 (1993)
16. Cheyer, A., Martin, D.: The open agent architecture. Journal of Autonomous Agents and Multi-Agent Systems 4(1), 1438 (2001)
17. Correa, R., Gomes, F.C., Oliveira, C., Pardalos, P.M.: A parallel implementation of an asynchronous team to the point-to-point connection problem. Parallel Computing 29, 447–466 (2003)
18. Craig, I.: Blackboard systems. Alex Publishing Co., Norwood (1995)
19. Czarnowski, I., Jedrzejowicz, P.: An agent-based approach to ANN training. Knowledge-Based Systems 19, 304–308 (2006)
20. Czarnowski, I., Jedrzejowicz, P.: An agent-based approach to the multiple-objective selection of reference vectors. In: Perner, P. (ed.) MLDM 2007. LNCS (LNAI), vol. 4571, pp. 117–130. Springer, Heidelberg (2007)
21. Czarnowski, I., Jedrzejowicz, P.: An Agent-Based Algorithm for Data Reduction. In: Bramer, M., et al. (eds.) Research and Development in Intelligent Systems XXIV, pp. 351–356. Springer, Heidelberg (2007)
22. Czarnowski, I., Jedrzejowicz, P., Wierzbowska, I.: An A-team approach to learning classifiers from distributed data sources. In: Nguyen, N.T., Jo, G.-S., Howlett, R.J., Jain, L.C. (eds.) KES-AMSTA 2008. LNCS (LNAI), vol. 4953, pp. 536–546. Springer, Heidelberg (2008)
23. Czarnowski, I., Jedrzejowicz, P.: Data reduction algorithm for machine learning and data mining. In: Nguyen, N.T., Borzemski, L., Grzech, A., Ali, M. (eds.) IEA/AIE 2008. LNCS (LNAI), vol. 5027, pp. 276–285. Springer, Heidelberg (2008)
24. Czarnowski, I., Jedrzejowicz, P.: A framework for adaptive and integrated classification. In: Rutkowski, L., Tadeusiewicz, R., Zadeh, L.A., Zurada, J.M. (eds.) ICAISC 2008. LNCS (LNAI), vol. 5097, pp. 522–532. Springer, Heidelberg (2008)
25. Czarnowski, I.: Distributed Data Reduction through Agent Collaboration. In: Hakansson, A., et al. (eds.) KES AMSTA 2009. LNCS (LNAI), vol. 5559, pp. 724–733. Springer, Heidelberg (2009)
26. Czarnowski, I., Jedrzejowicz, P., Wierzbowska, I.: A-Team Middleware on a Cluster. In: Hakansson, A., et al. (eds.) KES AMSTA 2009. LNCS (LNAI), vol. 5559, pp. 764–772. Springer, Heidelberg (2009)
27. Dreżewski, R.: A model of co-evolution in multi-agent system. In: Mařík, V., Müller, J.P., Pěchouček, M. (eds.) CEEMAS 2003. LNCS (LNAI), vol. 2691, pp. 314–323. Springer, Heidelberg (2003)
28. Far, B.H.: Modeling and implementation of software agents decision making. In: Proceedings of the Third IEEE International Conference on Cognitive Informatics, pp. 258–267 (2004)
29. Glover, F.: Tabu search and adaptive memory programming advances, applications and challenges. In: Barr, R.S., et al. (eds.) Advances in Metaheuristics, Optimization and Stochastic Modelling Technologies, pp. 1–75. Kluwer Academic Publishers, Boston (1997)
30. Goldberg, D.E.: Genetic Algorithms in Search, Optimization, and Machine Learning. Addison-Wesley, Reading (1989)

31. Gorti, S., Humair, R.D., Sriram, S.N., Talukdar, S.: Solving constraint satisfaction problems using A-Teams. Artificial Intelligence for Engineering Design, Analysis and Manufacturing 10, 1–19 (1996)
32. Jedrzejowicz, P.: Population learning algorithm and its applications. In: Olariu, S., Zomaya, A.Y. (eds.) Handbook of Bioinspired Algorithms and Applications, pp. 555–588. Chapman & Hall, Boca Raton (2006)
33. Jedrzejowicz, P., Wierzbowska, I.: JADE-based A-team environment. In: Alexandrov, V.N., van Albada, G.D., Sloot, P.M.A., Dongarra, J. (eds.) ICCS 2006. LNCS, vol. 3993, pp. 719–726. Springer, Heidelberg (2006)
34. Jedrzejowicz, J., Jedrzejowicz, P.: An investigation of agent-based hybrid approach to solve flowshop and job-shop scheduling problems. In: Apolloni, B., Howlett, R.J., Jain, L. (eds.) KES 2007, Part I. LNCS (LNAI), vol. 4692, pp. 172–179. Springer, Heidelberg (2007)
35. Jedrzejowicz, P., Ratajczak-Ropel, E.: Agent-based approach to solving the resource constrained project scheduling problem. In: Beliczynski, B., Dzielinski, A., Iwanowski, M., Ribeiro, B. (eds.) ICANNGA 2007. LNCS, vol. 4431, pp. 480–487. Springer, Heidelberg (2007)
36. Jederzejowicz, P., Ratajczak-Ropel, E.: Solving the RCPSP/max Problem by the Team of Agents. In: Hakansson, A., et al. (eds.) KES AMSTA 2009. LNCS (LNAI), vol. 5559, pp. 734–743. Springer, Heidelberg (2009)
37. Jennings, N., Sycara, K., Wooldridge, M.: A roadmap of agent research and development. Autonomous Agent and Multi-Agent Systems 1, 7–38 (1998)
38. Kao, J.H., Hemmerle, J.S., Prinz, F.B.: Collision avoidance using asynchronous teams. In: 1996 IEEE International Conference on Robotics and Automation, vol. 2, pp. 1093–1100 (1996)
39. Kisiel-Dorohinicki, M.: Agent-oriented model of simulated evolution. In: Grosky, W.I., Plášil, F. (eds.) SOFSEM 2002. LNCS, vol. 2540, pp. 253–261. Springer, Heidelberg (2002)
40. Koubarakis, M.: Multi-agent systems and peer-to-peer computing: Methods, systems, and challenges. In: Klusch, M., Omicini, A., Ossowski, S., Laamanen, H. (eds.) CIA 2003. LNCS (LNAI), vol. 2782, pp. 46–61. Springer, Heidelberg (2003)
41. Lesser, R.: Cooperative Multiagent Systems: A Personal View of the State of the Art. IEEE Transactions on Knowledge and Data Engineering 11(1), 133–142 (1999)
42. Meneses, C.N., Pardalos, P.M., Ragle, M.: Asynchronous Teams for probe selection problems. Discrete Optimization 5, 74–87 (2008)
43. Murthy, S., Rachlin, J., Akkiraju, R., Wu, F.: Agent-Based Cooperative Scheduling. In: Charniak, E.C. (ed.) Constraints and Agents, AAAI-1997, pp. 112–117 (1997)
44. Müller, J.P.: The right agent (Architecture) to do the right thing. In: Rao, A.S., Singh, M.P., Müller, J.P. (eds.) ATAL 1998. LNCS (LNAI), vol. 1555, pp. 211–225. Springer, Heidelberg (1999)
45. Liau, C.J.: Belief, information acquisition, and trust in multi-agent systems - A modal logic formulation. Artificial Intelligence 149(1), 31–60 (2003)
46. Olariu, S., Zomaya, A.Y. (eds.): Handbook of Bioinspired Algorithms and Applications. Chapman & Hall, Boca Raton (2006)
47. Paredis, J.: Coevolutionary computation. Artificial Life 2(4), 355–375 (1995)
48. Potter, M.A., De Jong, K.A.: Cooperative coevolution: An architecture for evolving coadapted subcomponents. Evolutionary Computation 8(1), 1–29 (2000)
49. Rabak, C.S., Sichman, J.S.: Using A-Teams to optimize automatic insertion of electronic components. Advanced Engineering Informatics 17, 95–106 (2003)

50. Rachlin, J., Goodwin, R., Murthy, S., Akkiraju, R., Wu, F., Kumaran, S., Das, R.: A-teams: An agent architecture for optimization and decision-support. In: Rao, A.S., Singh, M.P., Müller, J.P. (eds.) ATAL 1998. LNCS (LNAI), vol. 1555, pp. 261–276. Springer, Heidelberg (1999)
51. Taillard, E.D., Gambardella, L.M., Gendreau, M., Potvin, J.-Y.: Adaptive memory programming: A unified view of metaheuristics. European Journal of Operational Research 135, 1–16 (2001)
52. Talukdar, S.N., Pyo, S.S., Giras, T.: Asynchronous Procedures for Parallel Processing. IEEE Trans. on PAS 102(11), 3652–3659 (1983)
53. Talukdar, S.N., de Souza, P.S.W.: Scale Efficient Organizations. In: Proc. 1992 IEEE International Conference on Systems, Man, and Cybernetics, Chicago, Il, pp. 1458–1463 (1992)
54. Talukdar, S.N., de Souza, P., Murthy, S.: Organizations for Computer-Based Agents. Engineering Intelligent Systems 1(2), 56–69 (1993)
55. Talukdar, S.N., Ramesh, V.C.: A muti-agent technique for contingency constrained optimal power flows. IEEE Transactions on Power Systems 9(2), 855–861 (1994)
56. Talukdar, S., Baerentzen, L., Gove, A., de Souza, P.: Asynchronous teams: cooperation schemes for autonomous agents. Journal of Heuristics 4(4), 295–332 (1998)
57. Talukdar, S.N.: Collaboration rules for autonomous software agents. Decision Support Systems 24, 269–278 (1999)
58. Tweedale, J., Ichalkaranje, N., Sioutis, C., Jarvis, B., Consoli, A., Phillips-Wren, G.E.: Innovations in multi-agent systems. Journal of Network and Computer Applications 30(3), 1089–1115 (2007)
59. Vahidov, R., Fazlollahi, B.: Pluralistic multi-agent decision support system: a framework and an empirical test. Information & Management 41, 883–898 (2004)
60. Yang, H., Zhang, C.: Definition and application of a comprehensive framework for distributed problem solving. In: Distributed Artificial Intelligence, Architecture and Modeling, First Australian Workshop on DAI, Canberra, pp. 1–16 (1995)
61. Zhang, C.: Cooperation under uncertainty in distributed expert systems. Artificial Intelligence 56, 21–69 (1992)
62. Zhu, Q.: Topologies of agents interactions in knowledge intensive multi-agent systems for networked information services. Advanced Engineering Informatics 20, 31–45 (2006)

# Local Search Algorithms for Core Checking in Hedonic Coalition Games

Helena Keinänen

Helsinki University of Technology, Faculty of Information and Natural Sciences

**Abstract.** Hedonic games have emerged as an important tool in economics and show promise as a useful formalism to model multi-agent coalition formation in AI as well as group formation in social networks. We consider a $coNP$-complete problem of core membership checking in hedonic coalition formation games. No previous algorithms to tackle the problem have been presented. In this work, we overcome this by developing two stochastic local search algorithms for core membership checking in hedonic games. We demonstrate the usefulness of the algorithms by showing experimentally that they find solutions efficiently, particularly for large agent societies.

**Keywords:** multi-agent systems, game theory, core.

## 1 Introduction

In recent years, coalition formation has been the focus of extensive research in multi-agent systems. Coalition formation has also been a main research area in various social, political and economical studies concerning voting, trade and environmental as well as legislative issues. Whenever distinct agents group together to form coalitions, one of the most important questions is to determine stability of the resulting disjoint coalitions. Essentially, this amounts to checking that no group of agents has an incentive to break away from their coalitions, given the individual preferences of all the agents.

Hedonic coalitional games provide a suitable framework for modeling these issues: a stable division of agents into coalitions essentially corresponds to a coalition structure residing in core (a set of all stable partitions of agents into disjoint coalitions) of a hedonic game. Moreover, in hedonic coalitional games the preference an agent assigns to a coalition depends solely on the members of her own coalition and not on the composition of the other coalitions. For this reason, hedonic games are a suitable model for many real-life environments where agents behave rather self-interestedly instead of cooperatively while pursuing their own goals.

Hence, the problem of core membership checking in hedonic games is so important that developing special purpose algorithms for the problem is worthwhile. It is recently shown that the core membership checking in hedonic games is a $coNP$-complete problem [1]. Yet no algorithms have been presented in the literature to tackle the problem. In this paper we thus contribute to the literature

N.T. Nguyen, R. Kowalczyk, and S.-M. Chen (Eds.): ICCCI 2009, LNAI 5796, pp. 51–62, 2009.

by developing two novel algorithms to check core membership in hedonic games. Both of them are stochastic local search algorithms but are based on a very different heuristics. Through extensive numerical experiments, we show that our core membership checking algorithms are practically efficient on hedonic games. Our experiments show that we can easily check core membership on multi-agent societies that are orders of magnitudes larger than reported up till now, involving agents beyond tens up to 5000.

Nevertheless, much previous research effort has been directed at related questions. Recently, the computability of core related questions has been studied in coalitional games [2,3,4,5]. Games with hedonic preferences were introduced in [6]. For hedonic games it was shown that deciding the non-emptiness of the core is $NP$-complete [7]. Hedonic coalitional games have been studied mostly in the context of economics with the main focus on models and stability concepts (see, e.g.,[8,9,10,11]). Also, hedonic games can be used to model multi-agent coordination and group formation in social networks [12]. Stochastic local search algorithms are often considered as the state-of-the-art for finding high quality solutions to various combinatorial problems but they have been only occasionally applied to combinatorial problems arising in game theoretic settings.

## 2 Preliminaries

We study coalition formation in hedonic coalitional games [9,10,1]. Consider a finite set of agents $N = \{1, 2, \ldots, n\}$. Coalition $S$ is a nonempty subset of $N$. We denote the set of all coalitions which contain agent $i \in N$ by $\mathcal{A}^i = \{S \subseteq N \mid i \in S\}$. A coalition structure $CS$ is a partition of $N$ into mutually disjoint coalitions in which, for all $S, S' \subseteq N$, we have $S \cap S' = \emptyset$, $S \neq S'$ and $\bigcup_{S \in CS} = N$. Let $CS(i)$ be the coalition in $CS$ which contains agent $i$, and let $\mathcal{C}(N)$ be the collection of all coalition structures.

In hedonic coalition formation, each agent provides a preference order on those coalitions they may belong to. More precisely, for each agent $i \in N$, we define a preference relation $\succeq_i$ over $\mathcal{A}^i$. The preferences of each agent $i \in N$ over $\mathcal{C}(N)$ are completely determined by the preference relation of $i$, such that for all coalition structures $CS$,$CS' \in \mathcal{C}(N)$, $i$ weakly prefers $CS$ to $CS'$ if and only if $CS(i) \succeq_i CS'(i)$.

Similarly to [9,10,1] we assume the agents' preferences $\succeq = (\succeq_1, \succeq_2, \ldots, \succeq_n)$ are additive, meaning that for each agent $i \in N$, there exists valuation function $v_i : N \to \mathbb{R}$ characteristic $S \succeq_i S'$ such that for all $S, S' \in \mathcal{A}^i$, $S \succeq_i S'$ if and only if

$$\sum_{j \in S} v_i(j) \geq \sum_{j \in S'} v_i(j). \tag{1}$$

We will denote $\sum_{j \in S} v_i(j)$ by $v_i(S)$ for every $i \in N$ and for every $S \in \mathcal{A}^i$.

A hedonic coalitional game is denoted by $\Gamma = \langle N, \succeq \rangle$ where $N = \{1, 2, \ldots, n\}$ is a finite set of agents and $\succeq = (\succeq_1, \succeq_2, \ldots, \succeq_n)$ is a preference order.

In hedonic games with additive preferences, agents' coalition partners can be either friends, enemies or neutral partners. We call agent $j$ a friend (enemy) of agent $i$ if and only if $v_i(j) > 0$ ($v_i(j) < 0$). If $v_i(j) = 0$ agent $i$ considers agent $j$ as a neutral partner. The additive preference order $\succeq$ is based on aversion to enemies, if for each $i \in N$, $v_i(\cdot) \in [-n, 1]$, with $v_i(i) = 1$.

We call a coalition structure $CS$ stable, if for all $S' \subseteq N$, and for all $i \in S'$, $CS(i) \succeq_i S'$. A core of $\Gamma$ is the set of all stable coalition structures of $\Gamma$. We call coalition $S'$ a blocking coalition, if for all $i \in S'$, $S' \succ_i CS(i)$. Additive preferences provides a concise representation to store a hedonic game in computers' memory as this requires only a space quadratic in the number of agents $|N|$. Additive preferences do not guarantee the non-emptiness of the core but the core is always non-empty for preferences based on aversion to enemies [1].

We are concerned with the following problem of checking the core membership in hedonic games which is known to be a $coNP$-complete problem [1].

**Definition 1 (Core Membership Checking Problem).** *Given a coalition structure $CS \in \mathcal{C}(N)$ and hedonic game $\Gamma = (N, \succeq)$, the core membership checking problem is to decide whether $CS$ is in the core of $\Gamma$, and to give a blocking coalition if one exists.*

## 3 Algorithms for Core Membership Checking

We design two distinct stochastic local search algorithms for the core membership checking problem in hedonic games. These algorithms are essentially based on iterative improvement technique but both of them use a very different stochastic heuristics to escape local minima. Given a coalition structure and agents' preferences, we aim to check whether the coalition structure is stable by searching for a blocking coalition from the set of all possible coalitions. One can view this problem as an combinatorial minimization task as follows.

The algorithms search for a blocking coalition from the set of all coalitions by working on coalitions and selecting thereof agents who will be either removed or added to current coalition $S$. The objective is to minimize the number of agents in the coalition $S$ who prefer the given coalition structure $CS$ over $S$. The number of agents in $S$ which prefer coalition structure $CS$ over $S$ is denoted by $Goodness(S)$. If $Goodness(S)$ converges to 0, then a blocking coalition is found and the coalition structure $CS$ is not in the core.

In other words, both algorithms search the space of all coalitions by selecting in each step an agent $i$ in $N$, and proceed to modify the current coalition $S$ by removing from or adding $i$ in $S$, such that $Goodness(S)$ decreases.

Consequently, for both algorithms we use a neighbourhood which maps coalitions to the set of their neighbour coalitions such that the set of neighbours $S'$ of coalition $S$ are defined as $\{S' \in 2^N \setminus \emptyset \mid S' = S \setminus \{i\}$ or $S' = S \cup \{i\}$ where $i \in N\}$. Note that this neighbourhood provides the algorithms with a search landscape where solutions may be reached by at most $|N|$ search steps through the landscape.

### 3.1 Adapted Metropolis Search

We first introduce a simple stochastic local search algorithm obtained by adapting the heuristic of a basic Metropolis dynamics [13]. We call the resulting algorithm outlined below Adapted Metropolis Search (AMS). AMS is parameterised

**Algorithm 1.** Adapted Metropolis Search (AMS)

```
Input: iterationLimit, p, CS, Gamma

determine initial coalition S;
iterations = 0;
while iterations < iterationLimit  do
  if for all i in S, S >_i CS(i) then return S, else:
    S' = S;
    pick a random agent i in N;
    if i is in S then remove i from S', else:
      add i in S'
    if Goodness(S') <= Goodness(S) then S = S', else:
      S = S' with probability p^(Goodness(S')-Goodness(S))
  iterations = iterations+1;
```

by an integer iterationLimit, a real valued parameter $p \in [0, 1]$, input coalition structure $CS$ and the agents' preferences of the corresponding hedonic game $\Gamma$. The AMS begins with some initial coalition $S$. At each iteration it chooses a random neighbouring coalition $S'$ which aims to improve the value of the objective function $Goodness(S)$. Here, $p$ determines the probability of accepting a candidate coalition $S'$ that would lead to worsening of the objective function $Goodness(S)$. The probability to accept worsening solutions depends both on $p$ and on the decline in the quality of the objective function.

Notably, if the value of $p^{(Goodness(S')-Goodness(S))}$ is considerably small, then the algorithm uses essentially a so-called first improvement neighbour selection strategy which always chooses the first seen improving neighbour coalition.

### 3.2 Randomised Iterative Improvement

In this subsection we introduce a more complicated algorithm which is essentially based on randomised iterative improvement (RII) heuristic. Our RII algorithm can be summarized as follows. The above RII algorithm is parameterised by an integer iterationLimit, a real valued noise parameter $p \in [0, 1]$, input coalition structure $CS$ and the agents' preferences of the corresponding hedonic game $\Gamma$. In addition, the algorithm uses as temporary storages a set Improves as well as integer variables min and bestseen.

The RII algorithm starts with an initial candidate coalition $S$, and then performs greedy and random moves with a fixed probability, guided by a dedicated heuristic. The functionality of the RII is that, at each iteration, the agent to be added/removed from coalition $S$ is determined by which agent selection yields a decrease in the objective function $Goodness(S)$. If no decrease is possible, then

**Algorithm 2.** Randomised Iterative Improvement (RII)

```
Input: iterationLimit, p, CS, Gamma

determine initial coalition S;
iterations = 0;
while iterations < iterationLimit  do
  if for all i in S, S >_i CS(i) then return S, else:
  Improves = {};
  for all agents i in N
    S' = S;
    if i is in S then remove i from S', else:
      add i in S'
    if Goodness(S') <= Goodness(S) then add i in Improves
  if Improves is non-empty, then
    pick a random agent i in Improves;
    if i is in S then remove i from S, else:
      add i in S;
  else:
    with probability p:
      pick a random agent i in N;
      if i is in S then remove i from S, else:
      add i in S
    with probability (1-p):
      min = 0;
      bestseen = |N|+1;
      for all agents i in N
        S' = S;
        if i is in S then remove i from S', else:
        add i in S'
        if Goodness(S') < bestseen then
          min = i;
          bestseen = Goodness(S');
      if min is in S then remove min from S, else:
      add min in S;
  iterations = iterations+1;
```

either random coalition neighbour is chosen or a neighbour coalition yielding the smallest increase in $Goodness(S)$.

An essential behaviour of the RII is that a worsening neighbour coalition $S'$ of coalition $S$ is accepted only if a neighbour improving $Goodness(S)$ does not exist. A random worsening neighbour coalition is chosen according to the noise parameter $p$ such that with probability $(1-p)$ the RII algorithm selects a neighbour with the lowest increase in the value of the objective function $Goodness$. Notably, the iterations of the RII algorithm require only time $O(n)$, with $n = |N|$ the number of agents in input game $\Gamma$, which makes the algorithm practically efficient.

## 4 Experimental Setup

We have implemented in the C programming language [14] the AMS and the RII algorithms proposed in this paper as well as several problem instance generators for additive hedonic games[1]. To the best of our knowledge, the AMS and the RII

[1] The implementations reported in this paper are publicly available via the URL http://www.tkk.fi/~brummerk/coreexp09/

are the only algorithms in the literature to check the core membership in hedonic games. Thus, there is no other algorithms that can be used for a numerical comparison in this setting.

To evaluate the AMS and the RII algorithms, we use two classes of randomly generated instances of hedonic coalition formation games. One class is generic hedonic games with additive preferences. Another class of benchmark instances is hedonic games with aversion to enemies preferences.

Problems are generated as follows.

1. A random integer matrix $values$ of size $n \times n$ (with $n = |N|$)is produced to represent the agents' additive preferences. The $i$-th line of matrix $values$ represents agent's $i \in N$ payoffs for the other agents inclusion in the same coalition with $i$. For every agent $i \in N$, and for every coalition $S \in \mathcal{A}^i$ containing $i$, $i$'s value $v_i(S)$ for coalition $S$ is $\Sigma_{j \in S} values[i, j]$.
2. We iterate over each agent's preference list as follows. For agent $i \in N$ and another agent $j \in N \setminus \{i\}$ in $i$'s preference list, we generate a random number $-n \leq k \leq n$ in generic case, and a random number $k \in [-n, 1]$ in aversion to enemies case.
3. The diagonal values for the preference matrix are generated as follows. For every agent $i \in N$, the value of $i$ in the preference list of $i$ is a random number $-n \leq k \leq n$ in the generic additive case, and constant 1 in the case of aversion to enemies preferences.
4. A random coalition structure $CS$ is generated by assigning each agent integer $1 \leq l \leq n$ uniformly at random which represents the coalition where the agent resides in the coalition structure.

To get a clear and detailed picture of the performance of the algorithmics, we have instrumented the implemented source codes in order to measure operation counts and run-lengths that are needed to solve the problem instances. Essentially, for both AMS and RII heuristics the operations counted are the number of coalitions needed to find a blocking coalition. All of the experiments were run on a 1.60GHz Mobile Intel Celeron CPU running Linux Red Hat 7.3 with sufficient main memory, and we used over two weeks of CPU time to obtain the following results.

## 5 Results and Discussion

### 5.1 Noise Parameter Determination

Our first initial experiment aims to find optimal values of the noise parameter $p$ for both algorithms. For this purpose, we run the algorithms on a representative hedonic game instance with generic additive preferences and with the size of 150 agents. Both algorithms are run 10 times on this instance, starting the search from a random initial coalition. For the AMS heuristic we use the values $0.1, 0.3, 0.5, 0.7$ of $p$, and for the RII the values $0.2, 0.4, 0.6, 0.8$ of $p$. For both algorithms, the run-time limit is set to 50000 seen coalitions in this experiment.

Figures 1 and 2 show the results, where the run-lengths (minima and maxima of 10 runs) are given as a function of the number of agents who prefer the given coalition structure $CS$ over the currently checked coalition. A blocking coalition is found as soon as the number of such agents declines to 0. As can be seen from

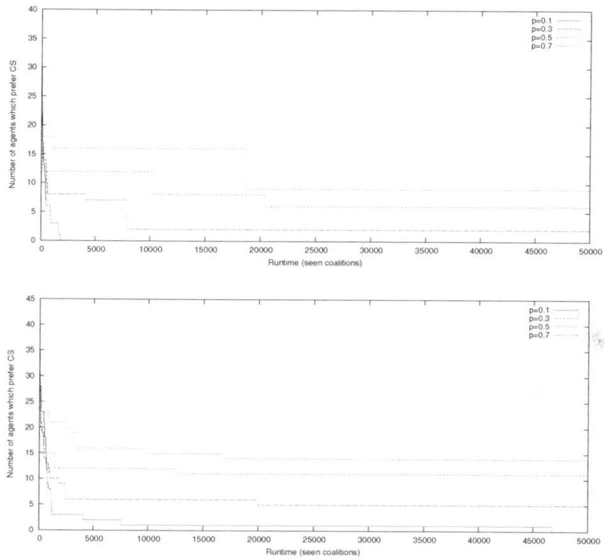


**Fig. 1.** Run-lengths for the AMS with different values of $p$, minima (top) and maxima (down) of 10 runs on an instance with 150 agents

the results in Fig. 1, the AMS with the noise parameter value $p = 0.1$ finds a blocking coalition almost immediately, but for the greater values of $p$ than 0.1 the AMS times out without finding a blocking coalition. In this example, the AMS algorithm works most effectively when the probability to take worsening steps is close to 0. In contrast to AMS, the RII algorithm does not time out so often in this example. As shown in Fig. 2, the RII algorithm works effectively when the noise parameter $p$ lies between 0.4 and 0.6. This experiment shows that the noise parameter $p$ has clearly effect on the performances of both algorithms, and thus must always be tuned to obtain optimal run-times.

### 5.2 Initial Starting Coalitions

The starting point of both algorithms AMS and RII is to determine an initial coalition to begin with the iteration. In our second series of experiments, we aim to find out what is the most beneficial way of determining initial coalitions for both algorithms. As starting points for the algorithms we use either any coalition selected uniformly at random or a singleton coalition consisting of 1 agent selected at random. To compare the run-lengths of the algorithms with

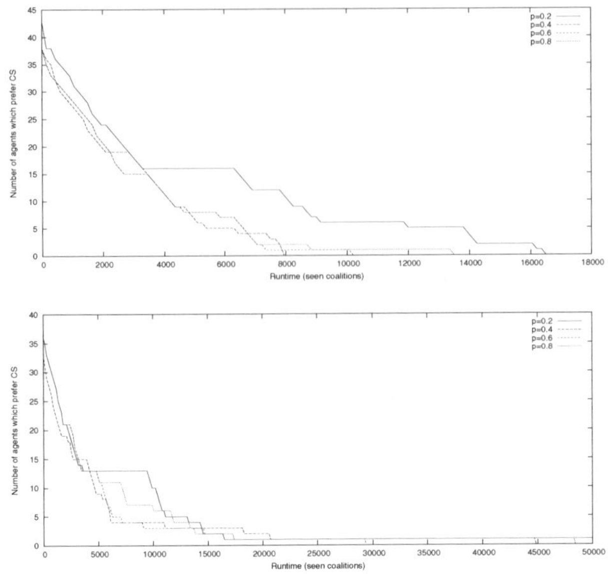


**Fig. 2.** Run-lengths for the RII with different values of $p$, minima (top) and maxima (down) of 10 runs on an instance with 150 agents

the two different starting coalitions, we investigate the execution times to find a blocking coalition on 1000 randomly generated 150-agent game instances. The run-time limits are set to 100000 seen coalitions, and the algorithms are run with approximately optimal parameters, the AMS with $p = 0.1$ and the RII with $p = 0.5$. Figure 3 shows the correlation between different algorithm variants on game instances with generic additive preferences. Figure 4 shows similar results on instances with aversion to enemies preferences. The results clearly show that the AMS with initial singleton coalition outperforms the AMS with a random initial coalition. The same observation holds for the RII, too. Consequently, choosing a singleton coalition as a starting point speeds up considerably both algorithms.

We conjecture that a blocking coalition is most likely to be found quickly with the AMS and RII algorithms, if the search for a blocking coalition starts from a very small coalition and a noise parameter $p$ is used which keeps the checked coalitions rather small. This is due to the definition of the blocking coalition which requires that every coalition member of a blocking coalition has to prefer the coalition over the given coalition structure. Any small coalition involves naturally only a few agents who need to comply with this requirement.

### 5.3 Runtime and Robustness Comparisons

To systematically compare the AMS algorithm with the RII algorithm on large agent societies, we conducted further experiments on problems with sizes 1000/

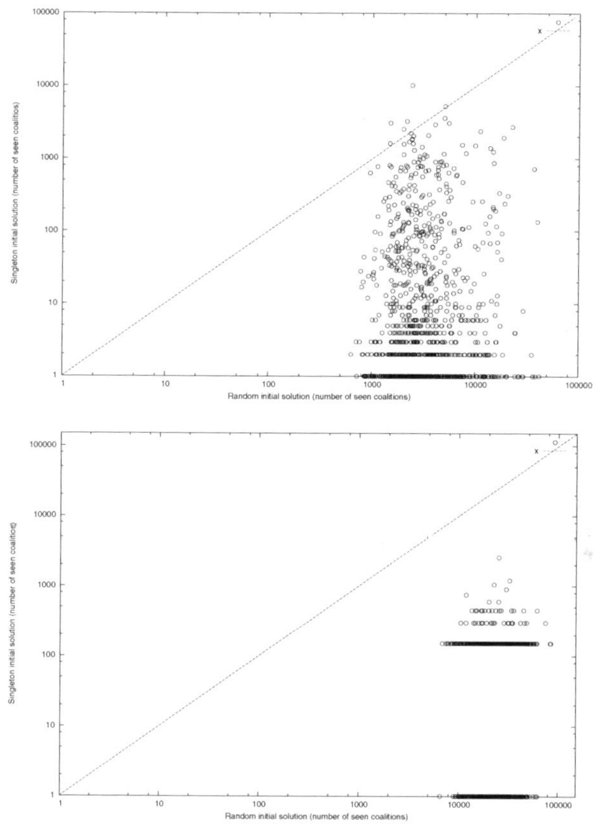


**Fig. 3.** AMS run-times to find a blocking coalition (top) and RII run-times to find a blocking coalition (down), 1000 instances with generic additive preferences, singleton coalition vs. random coalition as a starting point

2000/ 3000/ 4000/ 5000 agents, 100 instances per size for both random and aversion to enemies preferences. The run-time limit for both algorithms is set to 100000 coalitions. For both algorithms with a singleton initial coalition and both preference profile classes, Table 1 summarises run-time characteristics achieved in these tests. For each problem size ($|N|$ number of agents), we show the minimum, median and maximum run-lengths for both algorithms, on 100 instances, each instance run 11 times per algorithm. For both algorithms, minimum run-times are very small considering the large sizes of the agent societies and astronomical sizes of the underlying search spaces ($2^{|N|}$). Notice that median run-times are significantly lower for the AMS than for the RII because the latter always checks the whole neighbourhood. As shown in Table 1, in the case of games with generic additive preferences the AMS often times out already for 1000 agent games while the RII always finds a blocking coalition. Hence, the RII algorithm is more robust on these problems.

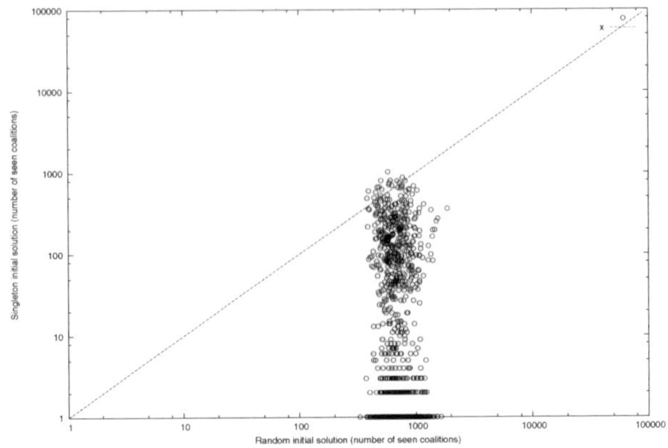


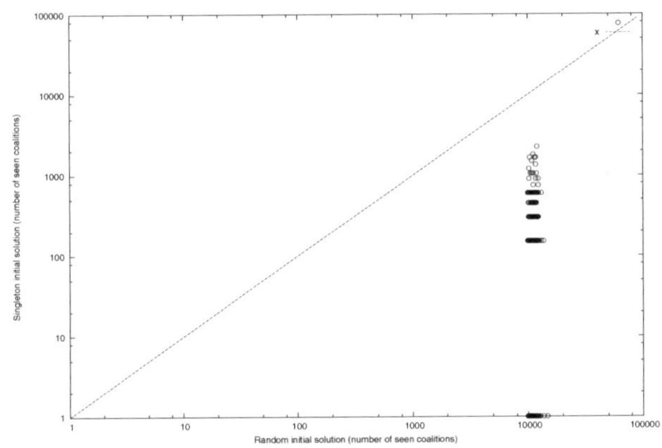


**Fig. 4.** AMS run-times to find a blocking coalition (top) and RII run-times to find a blocking coalition (down), 1000 instances with aversion to enemies preferences, singleton coalition vs. random coalition as a starting point

**Table 1.** Statistics for AMS and RII run-lengths, 11 runs on 100 instances per size (number of agents), - stands for time-out with limit 100000 seen coalitions

| Size | Aversion to Enemies preferences | | | | | | Generic preferences | | | | | |
|---|---|---|---|---|---|---|---|---|---|---|---|---|
| | AMS | | | RII | | | AMS | | | RII | | |
| | min | med | max | min | med | max | min | med | max | min | med | max |
| 1000 | 1 | 4 | 4932 | 1 | 1001 | 26001 | 1 | 3 | - | 1 | 1001 | 19001 |
| 2000 | 1 | 3 | 10430 | 1 | 2001 | 26001 | 1 | 3 | - | 1 | 2001 | 10001 |
| 3000 | 1 | 27 | 22074 | 1 | 3001 | 42001 | 1 | 4 | - | 1 | 3001 | 39001 |
| 4000 | 1 | 4 | 25302 | 1 | 4001 | 72001 | 1 | 3 | - | 1 | 4001 | 36001 |
| 5000 | 1 | 4 | 41322 | 1 | 5001 | 50001 | 1 | 4 | - | 1 | 5001 | - |

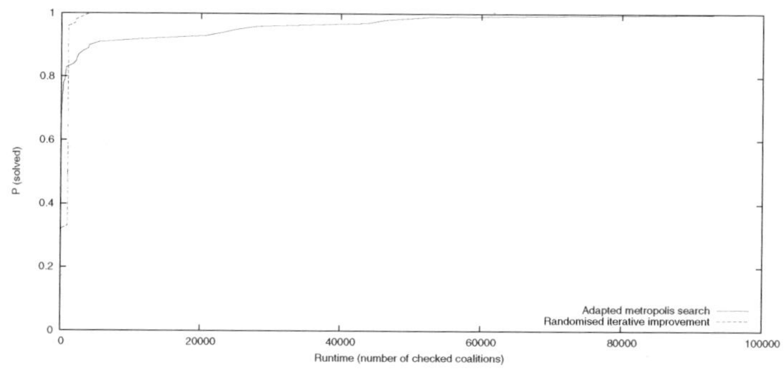


**Fig. 5.** Probability to find a blocking coalition, 100 runs on a hard instance with 1000 agents and generic additive preferences

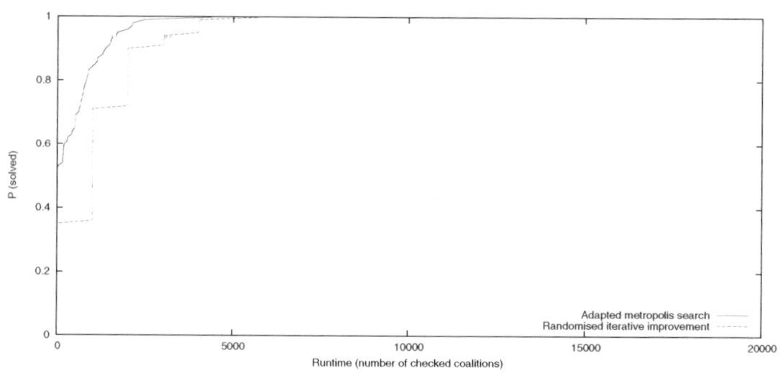


**Fig. 6.** Probability to find a blocking coalition, 100 runs on a hard instance with 1000 agents and aversion to enemies preferences

Finally, we have studied variation of run-lengths in terms of probability to find a blocking coalition using the AMS and the RII algorithms. For this purpose we run 100 times both algorithms on two hard 1000-agent hedonic game instances, one instance with generic additive preference profiles the other with aversion to enemies preferences. Figures 5 and 6 show the probabilities to find a blocking coalition over running times, the AMS is run with $p = 0.6$ and the RII with $p = 0.005$ which are approximately optimal for these particular instances. As can be seen from these plots, run-length variation is quite small for 100 runs of the RII on 1 instance. We observe that in games with generic additive preferences the variation of running times is significantly greater for the AMS than for the RII algorithm, as illustrated in Fig. 5. One may thus prefer to use the RII instead of the AMS in checking the core membership in additive hedonic games.

## 6 Conclusions

We have presented two algorithms for the core membership checking problem in hedonic games. The algorithms are based on different stochastic local search heuristics which aim to find rapidly an agent coalition which serves as a counter-example for the stability of a given coalition structure in a hedonic game instance. We show through extensive practical experiments how the presented algorithms can be used most effectively to solve hard core membership checking problems. In future work, we plan to devise special purpose algorithms to generate from a given hedonic game a coalition structure residing in the core.

## References

1. Sung, S.C., Dimitrov, D.: On core membership testing for hedonic coalition formation games. Operations Research Letters 35, 155–158 (2007)
2. Conitzer, V., Sandholm, T.: Complexity of constructing solutions in the core based on synergies among coalitions. Artificial Intelligence 170, 607–619 (2006)
3. Malizia, E., Palopoli, L., Scarcello, F.: Infeasibility certificates and the complexity of the core in coalitional games. In: IJCAI, pp. 1402–1407 (2007)
4. Elkind, E., Goldberg, L.A., Goldberg, P.W., Wooldridge, M.: Computational complexity of weighted threshold games. In: Proc. of the 22nd National Conf. on Artificial Intelligence, pp. 718–723. AAAI Press, Menlo Park (2007)
5. Elkind, E., Chalkiadakis, G., Jennings, N.R.: Coalition structures in weighted voting games. In: Proc. of the 18th European Conf. on Artificial Intelligence, pp. 393–397. IOS Press, Amsterdam (2008)
6. Drèze, J.H., Greenberg, J.: Hedonic optimality and stability. Econometrica 4, 987–1003 (1980)
7. Ballester, C.: Np-completeness in hedonic games. Games and Economic Behavior 49, 1–30 (2004)
8. Alcalde, J., Romero-Medina, A.: Coalition formation and stability. Social Choice and Welfare 27, 365–375 (2006)
9. Banerjee, S., Konishi, H., Sonmez, T.: Core in a simple coalition formation game. Social Choice and Welfare 18, 135–153 (2001)
10. Bogomolnaia, A., Jackson, M.O.: The stability of hedonic coalition structures. Games and Economic Behavior 38, 201–230 (2002)
11. Iehlé, V.: The core partition of hedonic games. Mathematical Social Sciences 54, 176–185 (2007)
12. Elkind, E., Wooldridge, M.: Hedonic coalition nets. In: Proc. of 8th Int. Conf. on Autonomous Agents and Multiagent Systems, AAMAS 2009 (2009)
13. Metropolis, N., Rosenbluth, A., Rosenbluth, M., Teller, A., Teller, E.: Equations of state calculations by fast computing machines. J. Chem. Phys. 21, 1087–1092 (1953)
14. Kernighan, B.W., Ritchie, D.M.: The C Programming Language. Prentice Hall, Englewood Cliffs (1988)

# Information Foraging Theory as a Form of Collective Intelligence for Social Search

Longo Luca, Barrett Stephen, and Dondio Pierpaolo

Department of Computer Science and Statistics - Trinity College Dublin
{llongo,stephen.barrett,dondiop}@cs.tcd.ie

**Abstract.** The World Wide Web is growing in size and with the proliferation of large-scale collaborative computing environments *Social search* has become increasingly important. The focal point of this recent field is to assign relevance and trustworthiness to web-pages by taking into account the reader's perspective rather than web-masters' point of view. Current web-searching technologies tend to rely on explicit human recommendations, in part because it is hard to obtain user' feedback however these methods are hard to scale. Implicit feedback techniques are a potentially useful alternative. The challenge is in producing implicit web-rankings by reasoning over users' activity during a web-search but without recourse to explicit human intervention. This paper focuses on a novel *Social Search* formal model based on *Information Foraging Theory*, showing a different way to implicitly judge web entities by considering effort expended by users in viewing them. 100 university students were recruited to explicitly evaluate the usefulness of 12 thematic web-sites and an experiment was performed implicitly gathering their web-browsing activity. Correlation indexes were adopted and encouraging results where obtained suggesting the existence of a considerable relationship between explicit feedback and implicit derived judgements. Furthermore, a comparison of the results obtained and the results provided by Google was performed. The proposed nature-inspired approach shows that, by considering the same searching query, *Social search* to be more effective than the Google Page-Rank Algorithm. This evidence supports the presentation of a novel general schema for a *Social search* engine generating implicit web-rankings by taking into account the Collective Intelligence emerged from users by reasoning on their behaviour.

## 1 Introduction

Recently, the concept of *Social search* has been acquiring importance in the World Wide Web as large-scale collaborative computing computations have become feasible. The main advantage of such systems is that the value of web-pages is determined by considering the reader's perspective, rather than merely the perspective of page authors. This approach takes many forms, from the simplest based on sharing bookmarks, to more sophisticated approaches that combine human intelligence with computational paradigms. The *Social search* approach contrasts with that of the leading searching engines, such as Google,

N.T. Nguyen, R. Kowalczyk, and S.-M. Chen (Eds.): ICCCI 2009, LNAI 5796, pp. 63–74, 2009.

whose Page-Rank algorithm relies on the link structure of the Web to find the most authoritative pages. However, a key open challenge in designing *Social search* systems is to identify human values in the Web automatically. A particular practical problem for any potential solution is that users tend to be resistant to explicit and invasive techniques for information gathering, so it is not easy to generate strong recommendations. In contrast, Implicit feedback techniques gather data indirectly to produce implicit web-rankings automatically deduced by reasoning on the activity performed by users over web-pages during a search. Reading time, bookmarking, scrolling, cut-paste, form filling, saving pictures are all considered relevant implicit sources of user preferences [7].

In this paper we propose a novel approach to collaborative *Social search* that takes into account users' activity during Internet browsing sessions, capturing browsing activity indicators such as, for instance, scrolling, clicking, cut & paste. Such activity embodies implicit 'human judgement' where each web-page has been viewed and endorsed by one or more people concluding that it is relevant and worthy of being shared among the community. Appropriate computational analysis allow the achievement of useful page ranking. In particular, in this study, a novel Computational Social Search model, based on Information Foraging theory is presented. This new approach shows how it is possible to evaluate the trustworthiness of a Web page by taking into account the effort performed by users upon Web pages to achieve the goal of finding particular information. The key benefit of systems supporting this kind of collective intelligence is that the *Social search* engine operates concurrently over continuously updated user activity and so it is well positioned to display stronger results synchronised with even rapidly changing of web-content. A further benefit of the approach is a reduction of the impact of link spam since the link structure of web-pages is less important.

The specific contributions of this paper are organised as follows. In section 2 we present the background of *Social Search* and *Information Foraging Theory*, Collective Intelligence and *Computational Trust*; in section 3 we underline the hypothesis; we describe our experiment along with the software tool used in section 4; in 5 we present our formal model and we comment on the results in section 6 by a comparison with Google Page-Rank. We conclude in section 7, presenting future work and open issues related to this research.

## 2 Related Work

*Social search* is a type of web-search technique that infers the relevance of web-search results by considering the opinions of end-users as to the value of web content. Several computational methods have been conceived combining human intelligence with computer paradigms [8] [9] and implicit feedback techniques appear to be useful candidates to automatically evaluate web-entities that users show interest in. Feedback need not to be explicit. When users surf the Internet they interact with a browser so they generate a data set regarding actions such as clicking, scrolling, submitting forms, cutting and pasting text and so on.

This data set presents an indicative pattern of behaviour for the evaluation of web-pages. In [7] three sources of implicit feedback are taken into account: reading time per document, scrolling and interaction, focused on the hypothesis that users will spend more time, scroll more often and interact more with those documents they find relevant. In [10] the authors focused on tasks such as classifying the user with regard to computer usage proficiency or making a detailed assessment of how long it took users to fill in fields of a form. For this, an HTTP proxy was developed, that collect data about mouse movements, keyboard input and more. Similarly in the work of Velayathan and Yamada [11], an unobtrusively framework has been proposed to gather logs of users' behaviour. These logs are then analysed in order to extract effective rules to evaluate web-pages using machine-learning techniques.

The theories of *Information Foraging* [2] are based on the ecological theories of foraging behaviour. The *Optimal Foraging theory* [1] explains the foraging behaviour of organisms in response to the environment in which they live. Foraging theory considers the foraging behaviour of animals by taking into account the payoff that animal obtains from different foraging options. *Information Foraging Theory* seeks to explain and predict how people will best adapt themselves for their information environments and how information environments can thus be best adapted to people. The human propensity to gather and use information to adapt to everyday problems in the world is a core concept in *human psychology* that has been largely ignored in cognitive studies. Humans have not only personal memory, but also external technology, as the Word Wide Web, for storing information. This extended information store thus enables us to perform knowledge based activity, but it is dependent on efficient information retrieval. Information foraging strategies can be adapted to identify the right knowledge, at the right time in order to take useful action or to make appropriate decisions.

The effort spent by users to take an action represents the central aspect of this paper. Information foraging theory assumes that people prefer information-seeking strategies that yield more useful information per unit cost of interaction. Users tend to arrange their environments to optimise this rate of gain, hence they prefer technology designs that improve returns on information foraging. Social Search is premised on the assumption that users are able to generate their own content, building up a data infrastructure with contributions not merely quantitative but in fact qualitative. Individuals thus provide judgements about entities, by using specific graded relevance system such as numbers, letters, descriptions. There are two main ways to provide judgements [6]: explicitly and implicitly. In the former, users can provide feedback using a specific metric. The most important example are eBay and Amazon where buyers and sellers can rate transactions using a graded system. Implicit judgements are inferred from users' behaviour performing specific actions.

There are several definitions of *Collective Intelligence* and a lot of fields of application, but here we focus on the new field of *Social Search*. The goal of the study presented here is to determine the mechanism for the generation of Implicit Feedback correlated with Explicit human judgements as to web-page quality and value. To test this we monitor users' behaviour while surfing the Internet and with appropriate reasoning technique and analysis of the cost of each user activity such as scrolling, bookmarking, printing, filling form, cut and paste, we aim to predict users' interest and to build up a useful and trustworthy rank among web-pages. Our approach differs from the strategy proposed by Pirolli et. al. [3] in their SNIF-ACT model where various backtracking mechanisms are adopted to study how users make navigation choices when browsing over many web-pages until they either give up or find what they were seeking. Similarly, our *Effort formal model* differs from to two other recent models of Web navigation: MESA [4] and COLIDES [5] that simulate the flow of users through the tree structures of linked Web pages.

The relevance of Trust and Reputation in human societies is indisputably recognised. These concepts have been acquiring a great relevance in the area of Distributed Artificial Intelligence as multi-agent systems become feasible. A trust-based decision is a multi-stage process on a specific domain. This process starts identifying and selecting pieces of trust evidence (here browser's events and typically domain specific), conducting an analysis over the application involved. Subsequently, Trust values are produced performing a Trust computation over the evidence set. Both the evidence-extraction and Trust computation steps are informed by a notion of 'trust' in the Trust model and the final Trust decision is taken by considering the computed values along with exogenous factor such as disposition or risk assessments. The proliferation of collaborative environments, for example those based on Wiki, represent good examples in which *Computational Trust* paradigms are applied in order to evaluate the trustworthiness of virtual identities [13]. In this paper we use the concept of Trust to indicate the trustworthiness of a Web-page after the application of our formal model.

## 3 Hypothesis

This work is focused on a comparison between explicit human judgements and implicit derived feedback considering an Internet web-page. We hypothesise that in the context of web-page media, by applying our *formal model* based on *Information Foraging theory*, explicit human judgements are correlated with the corresponding implicit derived feedback. If the answer is positive (i.e., a correlation between them is considerable) it is then possible to build up a collaborative environment achieving good predictions in an implicit and non-invasive way. We seek a method of reasoning over users' behaviour inspired by the *Information Foraging Theory*. We adopt *Computational Trust* and *Reasoning* methodologies to generate a set of ranked results of the most valuable content, as determined by users, which is by implication valuable to other similar users. We refer to this kind of collaborative determiner of value as *Implicit Collaboration* to distinguish from the classic, *Explicit Collaboration*, where users expressly provide judgements and feedback. Furthermore, our hypothesis is focused on the comparison between the rank proposed by Google (Google Vector), the one obtained by our formal model (Implicit vector) and the rank with the average of explicit judgements (Explicit vector). If the Euclidean distance between the Google and the Explicit vector is equal or greater than the distance between the Implicit and the Explicit vector, we can claim that our approach behave as good as Google or even better. This evidence would support our formal model demonstrating how a useful rank can be achieved by using an automatic approach based on reasoning on human behaviour and no more on the link structure of the Web.

## 4 The Experiment

To evaluate a web-page automatically, we have developed a client-side logging tool that unobtrusively gathers logs of users' behaviour through the user's natural interaction while surfing the Internet. Our solution runs transparently without affecting user's browsing experience. We do not provide further technical details about the implementation of the logger because here our goal is to study the correlation between implicit and explicit judgements.[1] The events ($E_i$) logged are presented in the following list: $E_1$: bookmark; $E_2$: print; $E_3$: save as (page); $E_4$: download; $E_5$: submit (text form filling); $E_6$: save as (picture), $E_7$: Im not alive (inactivity period); $E_8$: cut & paste (clipboard using); $E_9$: scroll; $E_{10}$: find in page, $E_{11}$: focused time. These events form a *'user-behaviour pattern'* containing the occurrences generated by a user while surfing a particular domain. We refer at this as $BP_n^d[E_i]$, i.e. a vector containing all the occurrences of events $E_i$ generated by the user $n$ on the domain $d$.

We conducted experiments in order to investigate the ability of our approach to gather logs of users' behaviour. 100 university students with different social background were recruited to participate in this study. The average of ages was between 19 and 35 years old and students have been selected from different schools as Computer Science, Psychology, Arts, Languages, Medicine. The 30% of them usually spend more than six hours per day surfing the Internet, 40% use the World Wide Web just two hours per day and the rest spend just a couple of hours per week browsing web-pages. We asked each of them to organise a trip to Morocco, some weeks long, providing a context-dependent pre-defined list of web-sites, as proposed in the table 1 from which it is possible to collect information. The list has been created selecting 12 URLs from the Google' s list obtained typing in the Google search box the keyword 'morocco' in the second week of October 2008. The selection of the thematic URLs has been made picking up URLs not just from the first page of the Google' s result. Furthermore, the selection of URLs has been made by considering the content and the structure of each web-page to discriminate useful and not useful web-documents. In particular, the authors expected to see whether, for instance, URLs number H, J and K will be automatically considered not relevant by the formal model proposed in this paper.

[1] Collecting implicit judgements raises privacy/anonymity concerns and ethical questions. Considering these issues we designed our solution in such a way that users have to explicitly confirm to use our solution by installing our tool.

**Table 1.** List of experiment's URLs and their position in the Google's Rank

| U.No. | Http link | Google Pos. | Description |
|---|---|---|---|
| A | en.wikipedia.org/wiki/Morocco | 01 | General info |
| B | www.morocco.com/ | 02 | A lot of travelling info |
| C | www.visitmorocco.com | 05 | Cities info |
| D | www.geographia.com/morocco/ | 11 | History & geographic info |
| E | www.morocco-travel.com/ | 13 | Several travelling info |
| F | wikitravel.org/en/Morocco | 14 | General info |
| G | www.magicmorocco.com/ | 19 | Cities info |
| H | rabat.usembassy.gov/ | 26 | Rabat USA embassy |
| I | www.lonelyplanet.com /maps/africa/morocco/ | 29 | Specialised travelling site |
| J | allafrica.com/morocco/ | 39 | Bad structure & useless |
| K | www.worldstatesmen.org/ Morocco.htm | 57 | Plenty of statistics |
| L | www.naturallymorocco.co.uk/ | 67 | Some info |

During the experiment the participants had to naturally interact with the browser, collecting useful data in order to recover this information in the future. Finally, we ask each of them to explicitly provide a judgement of the usefulness of each web-site listed by using a common scale from 1 to 10 (1 means not useful & 10 means very useful). In this experiment we assume that users do not act maliciously, such that the data contained in the log files is the proper representation of the real actions performed by them while surfing the Internet. We assume also that participants do not change their behaviour in order to alter generated logs data. Furthermore, we take into account an *uncertainty* degree both in the judgements provided by users and in the *Computational Trust* paradigms adopted to compute the final value of the usefulness of each web-site. At the end of the experiment a set of noisy information was obtained for each user, containing his/her activity while surfing web-sites. Since volunteers can jump from one web-document to another one of the list, gathered logs need to be filtered and aggregated in order to produce a well defined set of data, i.e. the *'user-behaviour pattern'*. For these reasons, we developed a function named 'filter/aggregator' that analyses data filtering all the URLs not in the list and that aggregates them by grouping per web-page. This module produces a list of 12 *'user-behaviour patterns'* containing the occurrences of events, one for each Internet domain in the list. At this stage, by applying our *Computational Trust* model based on *Information Foraging Theory*, we can generate a unique value from a given pattern indicating the usefulness of a given web-site for a specific user.

## 5 Computational Trust Model Based on Information Foraging Theory

We defined a *Computational Trust* model that extracts the occurrences of events in the *'user-behaviour pattern'* and compute a real value. Since 'scrolling' events are more frequent than 'save as' events or 'cut & paste' events should require more effort than 'click' events, a hierarchy is needed to discriminate their importance. The importance or strength of each event $E_i$ is the complement of the probability $P^d$ of that event to occur in the domain $d$ and we refer as *justification of effort* $(GE^d)$:

$$\forall i \neq (7 \mid 11)\ P^d(E_i) = \sum_{n=1}^{z} BP_n^d[E_i] \cdot \left( \sum_{n=1}^{z} \sum_{x=1}^{y} BP_n^d[E_x] \right)^{-1}, \ x \neq (7 \mid 11)$$

$$GE^d(E_i) = 1 - P^d(E_i)$$

where the left side of the first formula represents the total probability of the event of type $i$ while the right side indicates the sum of the occurrences of all the events $y$ (in this work $y = 11$) generated by all the $z$ users who have previously visited the domain $d$, excluding the event 7 and 11, i.e. the 'inactivity time' and the 'focused time'. In order to monitor inactivity time period, we conceived the *'Im Not Alive'* event, based on a reasonable threshold $(Th)$ of 10 seconds. In other words, if a user does not generate any event for more than 10 seconds, the logger generates an *'Im Not Alive'* event. By considering the focused time $(BP^d[11])$ spent on a domain $d$ by an user and subtracting the occurrences of the *'Im Not Alive'* event $(BP^d[7])$, multiplying by the threshold $(Th)$, it is possible to monitor the activity period. For example, if the focused time is 55 seconds for a web-site $d$, and there are 2 *'Im Not Alive'* events related to it, the activity period spent on the web-site $d$ is $55 - (2 \times 10) = 35$ seconds. By considering the above interpretations, the time-activity function $(Act)$ is formalised as:

$$Act_n^d(BP_n^d[E_{11}],\ BP_n^d[E_7]) = BP_n^d[E_{11}] - (BP_n^d[E_7]\ \cdot\ Th)$$

where $BP_n^d[11]$ is the focus time, $BP_n^d[E_7]$ is the number of occurrences of the *'Im not Alive'* event as gathered by the monitor in seconds, generated by the user $n$ on the domain $d$ and $Th$ is the time threshold defined before (10 seconds). Now we are able to compute the effort done by a user $n$ on a domain $d$: the effort is the total activity per time unit $(E_n^d)$. In particular, the total effort performed by a user $n$ on a domain $d$ is the sum of all the events generated by him/her by surfing the web-domain $d$, multiplied by the *justification of effort*, for each event, over the *activity time*. $y$ refers to all the possible events in the *'user-behaviour pattern'*. Formally:

$$E_n^d = \left(Act_n^d(BP_n^d[E_{11}], BP_n^d[E_7])\right)^{-1} \cdot \sum_{i=1}^{y} BP_n^d[E_i] \cdot GE(E_i)\ ,\ i \neq (7 \mid 11)$$

At this stage we are able to compute the average effort $(AE^d)$ of all the users $z$ who previously have surfed the domain $d$. Formally:

$$AE^d = \frac{1}{z} \sum_{n=1}^{z} (E_n^d)$$

Finally, we can compute the ranking of web-sites by taking into account the average effort on each domain. We adopt a sigmoid function to assign almost null importance to those websites that have not been endorsed/viewed by several web users. The more a web site $d$ is visited, the more the average effort spent on it is emphasised, so its trustworthiness increases underlying the real common users' behaviour on it. Formally the ranked value of a domain $d$ is modelled as:

$$RV^d = AE^d \cdot \frac{1 - e^{-kz^d}}{1 + e^{-kz^d}}$$

where the first part represents the average effort spent on the domain $d$ and the second part is the sigmoid function that assigns 0 to an input of 0 and grows in the number of users $z^d$ who visited the domain $d$. The constant $k$ is a value in [0..1] that supports the threshold of minimum users to start to be confident of the emerged common behaviour. The more this constant tends to 0 the higher the threshold of minimum user is. A value of 1 indicates that with 6 users or more, the sigmoid function returns value in the range $[0.99, 1]$ so it grows very slowly. The constant may be learned with unsupervised techniques but in this experiment is set to $k = 1$. The ranking algorithm of our Computational Trust model based on *Information Foraging Theory* is summarised in the following:

```
For each web-sites d {
    For each event i :
        GE^d(E_i)            Justification of effort of event i
    For each user n :
        E_n^d                  Effort by user n on domain d
    AE^d                           Average efforts on domain d
    RV^d                           Ranking Value of domain d
}
Order web-sites by RV^d in ascending way
```

## 6 Results and Discussions

The 12 values produced by using our schema and the 12 judgements provided by each participant are statistically correlated to test the hypothesis. In this work we adopted the *Pearson's correlation coefficient* that measures the strength of the linear dependence between two variables. One interesting aspect of using the Pearson score is that it corrects for *grade inflation*: if one critic is inclined to give higher scores than another, there can still be perfect correlation if the difference between their scores is consistent. If the correlation value obtained by considering the implicit values and the explicit judgements, for a given web-site and a given volunteer, tends to 1, a linear equation describes the relationship positively with the implicit value increasing with the explicit value. A score of $-1$ shows the inverted relationship between the two values and a value tending to 0 shows that they are independent. In order to test the hypothesis behind this research, the experiment has to produce high correlation values. If the majority of these values tend to 1, our hyphotesis is confirmed and we can sustain there exists a considerable relationship between implicit feedback and explicit human judgements as captured by our model. On the other hand, if our solution produces low correlation values, i.e., values tending to 0 or $-1$, we are

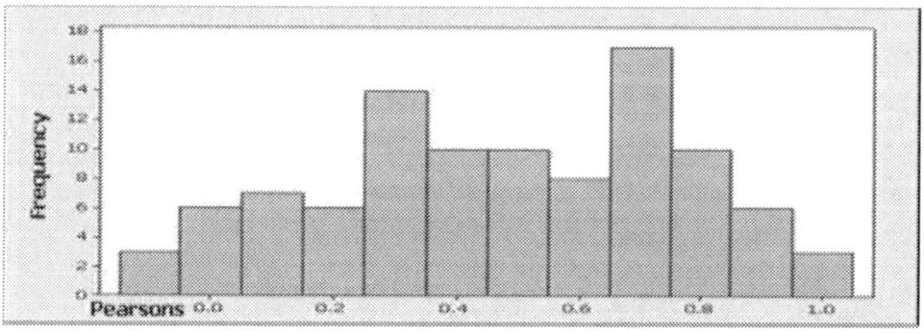


**Fig. 1.** Pearson's Correlation results

not able to satisfy the hypothesis so further studies and different methods have to be considered. The ranked set of Pearson's correlation values obtained from the experiment is depicted in figure 1. Participants with a coefficient close to 1 are the ones who support our hypothesis and their explicit behaviour can be almost perfectly approximated by our Computational Trust model. On the contrary, values close to 0 underline the fact that our model did not succeed in automatically approximate explicit judgements in implicit feedback. Part of the justification can be found in the intrinsic degree of uncertainty in the explicit evaluation provided by user. We are now able to use this evidence to support the hypothesis and examine how, with implicit feedback, an implicit degree of collaboration can be reached automatically. In particular, by considering users' behaviour and by using the schema proposed in this research, it is possible to create a ranked list of web-sites, thus to predict users' interests and tendencies. At the top of this list there will be those web-sites with the higher average of trustworthiness values and at the bottom those ones in which users' behaviour had not huge influence. The ranked list produced is: B, D, C, L, E, F, A, I, G, H, J, K. The web-sites B, D contain useful and detailed information targeted to organise a trip to Morocco as required by the task proposed to the participants. The common behaviour emerged from participants underlines the usefulness of these web sites. URLs C, L, E, F, A, I, G are not so helpful in providing good information about Morocco as the previous ones: this fact is reflected by the moderate users' average effort computed by our model. Web-sites H, I, J are not appropriate for the context at all as users spent very few energy in performing the task. For instance, H relates to the Morocco USA's Embassy web-site: despite the fact it contains key-words related to Morocco, it is totally unrelated and useless to achieve travel information. Finally, the url K belongs to a web-agency offering trip packages already assembled. The achieved results are encouraging as they reflects what the authors expected to notice: the URLs H, J, K in the last positions of the computed ranking, as they do not give any advantage in performing the task. One interesting point in adopting such an approach is that, by enclosing the ranking value to the URLs, more accurate justifications of their position can be achieved. For instance, it is clear that URLs H, J, K, other than being at the bottom of the ranking, are far away from those URLs appropriate for the experiment context. Furthermore, web pages with a very low chance

**Table 2.** Position of each URLs (U.No.) in each evaluation method: our model (Implicit Vector), the Explicit feedback provided by users (Explicit Vector) and the Google Page-Rank (Google Vector)

| U.No. | Effort Avg | Implicit V. | Explicit V. | Google V. |
|---|---|---|---|---|
| A | 0.029 | 7 | 5 | 1 |
| B | 0.054 | 1 | 4 | 2 |
| C | 0.038 | 3 | 6 | 3 |
| D | 0.049 | 2 | 9 | 4 |
| E | 0.031 | 5 | 1 | 5 |
| F | 0.030 | 6 | 2 | 6 |
| G | 0.023 | 9 | 8 | 7 |
| H | 0.012 | 10 | 11 | 8 |
| I | 0.025 | 8 | 7 | 9 |
| J | 0.010 | 11 | 12 | 10 |
| K | 0.006 | 12 | 10 | 11 |
| L | 0.032 | 4 | 3 | 12 |

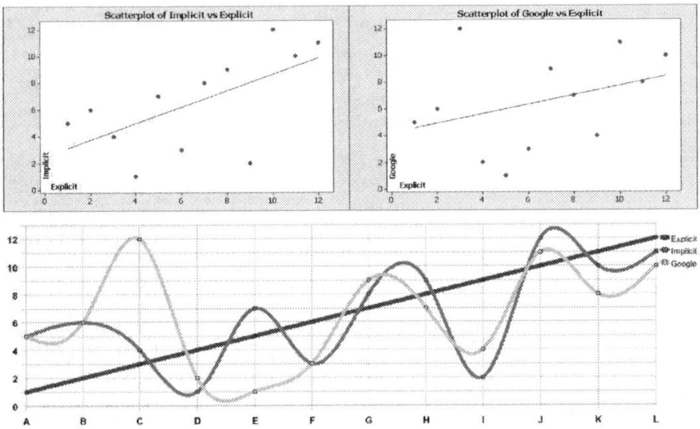


**Fig. 2.** Comparison graph

to be visited, as the url L in position 67 of the Google Rank, can carry useful information as computed by our model.

By taking into account the list proposed in the table 1, ordered by the position of each web-site selected in the Google' s List (table 2 - Google V.) returned typing the search query 'Morocco', and by considering the list (table 2 - Implicit V.) obtained by applying our *Computational Trust* model, it is possible to study the distance of these two vectors from the explicit participants' judgements list (table 2 - Explicit V.) as showed in the figure 2. The scatterplots show both the comparisons: explicit judgements are more correlated with the implicit derived feedback (left), in fact, points rely close to the straight line than the scatterplot in which the Google' s list and the Explicit' s list are compared (right). We adopted the Kendall tau rank correlation coefficient which is a non-parametic statistics used to measure the strength of correspondence between two rankings, i.e. the degree of association of the cross tabulations. The result are encouraging as the Kendall tau coefficient emerged by considering the Google Vector and the Explicit Vector, i.e. $\tau = 0.272$, is less than the coefficient computed by taking into account the Implicit Vector and the Explicit Vector which is $\tau = 0.454$. Furthermore the Euclidean distance of both the Google vector and the Implicit vector compared with the Explicit vector are respectively 13.64 and 10.58 (bottom graph). In C, E, G, K and L cases the implicit obtained feedback are closer to the straight line, i.e., the ones provided explicitly by users: in these cases our strategy succeeded. Url A, B, F and H have the same distance both in the Google and Implicit vectors: in this case we are not able to claim whether our model is effective but at least we can say it behaves as the Google' s one. Eventually, in cases D, I and J, our approach differs from the Google approach which is more appropriate to approximate explicit human judgements. Nevertheless, our formal model produces expressive results in the first half part of the Implicit' s ranking where the more appropriate URLs for the context are expected to be. In fact, the distance of both the lists to the straight line (Explicit vector) are respectively 11.92 (Google Vector) and 7.42 (Implicit vector) for the half part, 6.63 (Google vector) and 7.55 (Implicit vector) for the second half part.

## 7 Conclusions and Open Issues

In this study we performed a comparison between explicit context -dependent human judgements and implicit feedback as computed over data capturing user browsing activity. The experiment included 12 thematic URLs explicitly evaluated by 100 students with different social background and age, providing a degree of usefulness. During browsing sessions we automatically monitored their activity to study their behaviour. A *'user behaviour-pattern'*, containing the occurrences of events, was extracted for each user and web-site and then evaluated by using a *Computational Trust* model based on *Information Foraging Theory.* Even if a small and simple set of information was considered in the *users' behaviour pattern*, the Pearson's coefficient helped us to prove the existence of a encouraging correlation between explicit and implicit judgements. Furthermore, we made a comparison between the explicit judgements provided by user respectively with the ones provided by the Google' s algorithm and our formal model based on users' effort. The Kendall tau coefficient and the Euclidean distance among the vectors prove that our strategy is, in our experiment, more effective than the Google' s Page-Rank to predict user tendencies and interests. This result is promising but supplementary experiments are needed in different contexts and further data has to be collected from more users and more web-pages to try to generalise the ability of our strategy. The authors believe their approach represents a start point to predict users' interests and tendencies building up a third-generation *Social search* engine based on *Implicit Collaboration.* The first-generation ranking engines judged the relevance and quality of a page considering its content. Second-generation ranking metrics started exploiting the link structure of the Web [14]. Our approach try to further improve the ranking metrics by reasoning on users' behaviour while consuming activity over web-pages. Future works will be focused on understanding how good degree of trustworthiness of web-pages can be derived by considering users' behaviour in malicious environments. New reasoning techniques have to be investigated in depth to better evaluate digital entities and to extend our model not just to recommender systems. An important issue will be focused on understanding how to semantically connect searching queries to relevant sets of URLs generated by our *Social search* schema. Privacy issues must be taken also into account in order to guarantee the anonymity of web-users in providing implicit feedback.

## References

1. Stephens, D.W., Krebs, J.R.: Foraging Theory, Princeton, NJ (1986)
2. Pirolli, P.: Information Foraging Theory. Adaptive Interaction with Information. Oxford University Press, Oxford (2007)
3. Pirolli, P., Fu, W.: SNIF-ACT: A Model of Information Foraging on the World Wide Web. In: Brusilovsky, P., Corbett, A.T., de Rosis, F. (eds.) UM 2003. LNCS, vol. 2702, pp. 45–54. Springer, Heidelberg (2003)
4. Miller, C.S., Remington, R.W.: Modeling Information Navigation: implications for Information Architecture. In: HCI (2004)
5. Kitajima, M., Blackmon, M.H., Polson, P.G.: Cognitive Architecture for Website Design and Usability evaluation: Comprehension and Information Scent in Performing by Exploration. In: HCI, Las Vegas (2005)
6. Longo, L., Barrett, S., Dondio, P.: Toward Social Search: from Explicit to Implicit Collaboration to Predict Users' Interests. In: WEBIST 2009 (2009)
7. Kelly, D., et al.: Reading Time, Scrolling and Interaction: exploring Implicit Sources of User Preferences for Relevance Feedback During Interactive Information Retrieval. In: SIGIR 2001, New Orleans, USA (2001)
8. Agichtein, E., Brill, E., Dumais, S.: Improving Web Search Ranking by Incorporating User Behavior Information. In: SIGIR 2006, Seattle, USA (2006)
9. Agichtein, E., Zheng, Z.: Identifying Best Bet Web Search Results by Mining Past User Behavior. In: Kdd 2006, Philadelphia, Pennsylvaia, USA (2006)
10. Atterer, R., et al.: Knowing the User's Every Move - User Activity Tracking for Website Usability Evaluation and Implicit Interaction. In: WWW 2006, Edinburgh, May 23-26 (2006)
11. Velayathan, G., Yamada, S.: Behavior-based Web Page Evaluation. In: WWW 2007, Banff, Alberta, Canada, May 8-12 (2007)
12. Morita, M., Shinoda, Y.: Information Filtering Based on User Behavior analysis and Best Match Text Retrieval. In: 17th ACM SIGIR (1996)
13. Longo, L., Dondio, P., Barrett, S.: Temporal Factors to evaluate trustworthiness of virtual identities. In: IEEE, SECURECOMM 2007, France (2007)
14. Kleinberg, J.: Authoritative sources in a hyperlinked environment. Journal of the ACM 46(5), 604–632 (1999)

# SAM: Semantic Argumentation Based Model for Collaborative Knowledge Creation and Sharing System

Krissada Maleewong[1], Chutiporn Anutariya[1], and Vilas Wuwongse[2]

[1] School of Technology, Shinawatra University,
99 Moo 10 Bangtoey, Samkhok, Pathumthani 12160, Thailand
{krissada,chutiporn}@shinawatra.ac.th
[2] School of Engineering and Technology, Asian Institue of Technology,
P.O. Box 4, Klong Luang, Pathumthani 12120, Thailand
vilasw@ait.ac.th

**Abstract.** Although several knowledge creation and sharing systems available to date have been successful at motivating their members to contribute in content creation and sharing, they still lack formal model to systematically support community deliberation, resolve conflicts, and ensure the quality of knowledge as well as to semantically trace, search and retrieve desired knowledge. This paper aims to fill this need by analyzing the system's key requirements and properties, and then develops *Semantic Argumentation based Model* (*SAM*), which can fulfill the identified requirements and serve as a solid foundation for such a system. SAM harnesses collective intelligence by encouraging multiple members to collaboratively express ideas or solutions regarding a complex issue, and to submit arguments which support or oppose other members' ideas. In principle, an idea accepted or supported by most members is considered as a potential solution to solve the issue. Hence, to allow sophisticated argumentation analysis and to automatically drive the community towards a consensus, *SAM Schema* is constructed for formally and semantically capturing and describing the community deliberation. In addition, SAM employs ontology-based and semantic lexical-based reasoning mechanisms to enhance semantic searching and browsing as well as to facilitate users editing.

**Keywords:** Collaborative Knowledge Creation and Sharing, Argumentation, Collective Intelligence, Semantic Web.

## 1 Introduction

Collaborative knowledge creation and sharing on the Web are continuing to grow with increasing number of contributors and covering dispersing range of disciplines. Today's social Webs, such as wikis, blogs, discussion forums, and social networking sites, are successful at encouraging people to participate actively in content creation and knowledge sharing through a variety of shared interests, and leading to the unanticipated explosion of innovative ideas. However, the current state of the collaborative knowledge creation and sharing systems appears to be less successful at enabling

N.T. Nguyen, R. Kowalczyk, and S.-M. Chen (Eds.): ICCCI 2009, LNAI 5796, pp. 75–86, 2009.

community deliberation and collaboration, as well as enhancing human-machine synergy. Thus, a novel approach to emerging *collective intelligence* is demanded.

By allowing users to easily create and edit articles, Wikipedia has grown to be the world's largest free encyclopedia. Its mechanism, however, introduces edit war and vandalism problems [1, 2], which then leads to an untrustworthiness problem. Moreover, its history pages representing information in reverse chronological order (most recent first) also confuse its users when searching for an evolution of a specific knowledge. In order to enable machine-interpretable contents, Semantic Wikipedia [3] allows users to semantically annotate wiki pages which improve information access and enable knowledge exchange across applications. However, the edit war and vandalism problems remain unsolved.

To effectively capture the deliberation and collaboration, many recent, active researches in knowledge creation and sharing have applied argumentation technologies in various domains as follows. *Compendium* [4] is a knowledge management environment, developed based on *graphical IBIS system* (*gIBIS*) [5] to support group deliberations. *Collaboratorium* [6] proposes a collaborative framework by integrating *IBIS model* [7], *Walton's argumentation schemes* [8], and *Toulmin's argument scheme* [9] to capture discussions as well-structured networks of issues, positions, and arguments. However, they do not have rich formal semantics to handle sophisticated automated processing of argumentative statements such as checking conflicting arguments, ensuring quality of arguments, and querying semantic information. With a focus on ontology engineering domain, *DILIGENT* [10] and *HCOME* [11] develop ontology engineering environments, by applying IBIS model to capture the discussion and allow members to construct an ontology in a collaborative manner.

Although the above studies demonstrate that argumentation theory has been developed in various disciplines of knowledge management for many years, they still lack an effective mechanism for resolving conflicts and determining group agreement. This paper, therefore, aims to fill this need by developing *Semantic Argumentation based Model* (*SAM*), which enables systematic and dynamic knowledge creation and sharing process in a community. The model formalizes *SAM Schema* (*SAMS*) to structurally represent complex argumentation and to allow the semantic description of community deliberation, which can also serve as the community knowledge-base. In order to harness the collective intelligence, several sophisticated automated processing and reasoning mechanism across the collective knowledge are enabled.

The organization of this paper is as follows: Section 2 identifies important requirements for a collaborative knowledge creation and sharing system. Section 3 presents *Semantic Argumentation based Model* (*SAM*) which can fulfill the identified requirements and properties. Section 4 demonstrates an example use case of the developed model. Section 5 evaluates and compares SAM with the current approaches. Section 6 draws conclusions and future research direction.

## 2 Requirements of a Collaborative Knowledge Creation and Sharing System

This section identifies important requirements and properties of a collaborative knowledge creation and sharing system.

*1) Enhancing collaborative knowledge creation and sharing*: In order to foster and facilitate social interactions, a collaborative knowledge management system should encourage many users to collectively create and share their knowledge. The more users contributing in editing contents, checking facts, and examining the writing, the more reliable the published knowledge. Therefore, the system should support collaborative knowledge creation and sharing activities in the systematic and dynamic process, and promotes community deliberation where face-to-face communication is limited.

*2) Simplicity*: In order to encourage multiple users to participate in community deliberation, the system should follow a simple yet effective process with intuitive workflow for motivating members to easily create and share their knowledge. The users-contributed contents should also be intuitively presented or visualized to all members for their clarification and further contribution.

*3) Assuring quality of knowledge*: Collective knowledge created by many users with different skills and interests should be analyzed to assure the quality and reliability of published knowledge based on the following issues.

- *Enabling consensual contents*: The published knowledge should involve the agreement of all or most members in a community, and maintain high quality and reliability of its contents.
- *Ensuring consistent information*: Since the same information often occurs and changes during collaborative process, the system should ensure that all published knowledge about a particular concept, issue or problem is consistent.
- *Detecting conflicting arguments*: All arguments submitted by a member should be consistent. For example, a member can make several arguments to a particular idea based on his/her knowledge, all of which must consistently support or oppose the idea. Arguments made by a member that both support and oppose a certain idea are thus considered to be conflicting arguments, and should then be detected and resolved.

*4) Traceability*: The system should provide the ability to trace and verify the evolution of the community deliberation; hence allowing members to easily follow and structurally query the information about every step in the collaborative knowledge creation and sharing process. For example, a new member may want to trace and justify all the related ideas and arguments involved in creating a particular knowledge, thus facilitating him/her to further contribute.

*5) Enabling automated mediation*: On account of a large community with many contributors having different interests, skills, viewpoints and justifications, their proposed ideas and solutions could be diverse and conflicting. Mediating those disagreeing ideas to achieve a final solution is time consuming and requires a lot of efforts. In addition, such a laborious task when directed by a human moderator can be biased. Therefore, the system should mechanize this mediation task or provide supporting information to assist a human moderator or expert in decision making and achieving the community agreement, which can also speed up the overall knowledge creation process.

*6) Supporting semantic search*: In order to provide precise search and emerge new knowledge, a semantic search should be enabled to derive implicit information, discover similar concepts, and present semantic relations among all relevant results across a community knowledge-base. Moreover, the system should allow users to easily browse the large amount of community knowledge based on their interests.

*7) Interoperability*: Collective knowledge should be formalized using an open, widely-accepted knowledge representation, which allows semantic data interchange across diverse systems for community sharing, reusing as well as local adapting.

# 3 SAM: Semantic Argumentation Based Model

This section develops a *Semantic Argumentation based Model*, namely *SAM*, for supporting collaborative knowledge creation and sharing systems, and achieving the identified requirements and properties as described in Section 2.

## 3.1 Architecture and Process

Figure 1 illustrates the architecture of SAM, which consists of the following five functionalities: (*i*) *Creation and Sharing*, (*ii*) *Analysis and Reasoning*, (*iii*) *Storage*, (*iv*) *Retrieval*, and (*v*) *Reuse*.

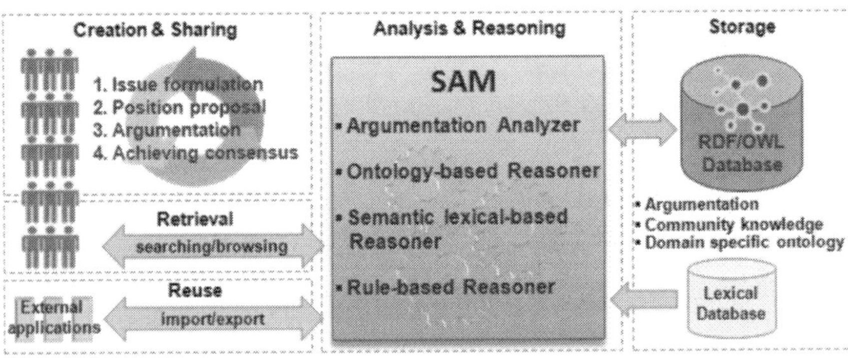


**Fig. 1.** Semantic Argumentation based Model's Architecture

**Creation and sharing:** This function allows multiple users to collaboratively create and share their knowledge in a collective improvement mechanism which is a systematic and dynamic process, and consists of the following steps:

*1) Issue formulation*: To create new knowledge regarding the complex problems and controversial challenges, a community confronts with many *issues/sub-issues* to be resolved by collaborative thinking. Members can raise issues based on their interests for community deliberation.

*2) Position proposal*: For each issue, members are encouraged to express *positions* as alternative solutions, which are represented to all members for their judgments and feedbacks. New members are motivated for critical thinking in order to propose solutions better than the existing ones.

*3) Argumentation*: Each member can submit *arguments* to support or oppose the positions based on his/her clarification. A position supported by many arguments is considered as a potential position to solve the issue. On the other hand, a position made against by many arguments requires a revision in light of the given arguments.

*4) Achieving consensus*: This process aims to elaborate collective knowledge and drive a community towards a consensus using a *collective improvement mechanism* which is similar to *RT Delphi method* [12]. It enhances the speed and efficiency of the model by providing real time information for community deliberation. However, the complete agreement in a large-scale community deliberation is not possible. Therefore, *soft consensus* [13] is employed to evaluate group agreement, and hence allowing the *consensus level* to be identified by a specific threshold. When a position reaches the predefined consensus level, such position is said to be *acceptable* and considered as a potential solution to solve the issue; otherwise the community is encouraged for more deliberation in position proposal and argumentation activities. In addition, the combination of several related potential positions is necessary for resolving a complex problem. To determine an *acceptance level* of each position and discover the potential positions accepted by most members in the community, the deliberation is analyzed by the *argumentation analyzer* (to be described in details next). The result of the analysis is then used to automatically drive the consensus achieving process. However, it is important to note that in certain scenarios and applications, fully automatic achievement of the community agreement may not be possible, and the selection of the final solution relies on an expert's decision. In this case, the results of the argumentation analysis can help support human experts in decision making, mediating the agreement, and achieving a consensus.

**Analysis and reasoning:** This function obtains the community deliberation encoding in RDF/OWL metadata format for sophisticated analysis and productive reasoning across collective knowledge using several components described as follows:

- *Argumentation Analyzer* analyzes community deliberation for assuring the quality of published knowledge and assisting a human moderator or expert to achieve the community agreement. It evaluates an *acceptance level* of each position based on contents and group preference using four measures: a *degree of argument*, a *degree of individual preference*, a *degree of position*, and a *degree of individual's expertise*. The previous work [14] formally defines this four measures and their calculation, while this paper aims to present their intuitive meanings and practical usages as follows:

  *i) Degree of argument* ($\delta_A(a)$) specifies an expressive level of an argument *a* with a possible value within the range [0, 1]. The higher the degree of an argument is, the stronger support or object to the position the argument does. Intuitively, it is evaluated by considering the interrelated components of an argument (i.e., data, backing, qualifier and rebuttal(s)) together with the degree of expertise of the member who makes the argument. The provision of supporting data and backing information helps increase the degree of the argument, while their reliabilities can be approved by community investigation. In addition, an argument submitted by an expert with a high confidence level is considered a high quality one. On the other hand, an argument made against by many rebuttals is considered to have poor quality, hence resulting in a low degree of argument.

  *ii) Degree of individual preference* ($\delta_I(i,p)$) denotes the preference degree of an individual *i* on a position *p* with a possible value within the range [-1, 1]. That is, if the degree is greater than 0, then the individual *i* positively supports the position *p*; the higher the value, the stronger the support. On the other hand, if the value is less than 0, then *i* negatively opposes *p* or *i* disagrees with the position *p*. If the value is equal to 0, then *i* has no preference on *p*. This degree is calculated by taking the average of the degree of arguments submitted by a particular individual *i* on a specific position *p*.

*iii)Degree of position* ($\delta_P(p)$) determines an *acceptance level* of a position *p* and has a value within the range [-1, 1]. Intuitively, it identifies how much the whole community supports or opposes such position *p*. Thus, if the degree reaches a pre-defined threshold, *p* is said to be acceptable by the community and is concerned as a potential position for resolving an issue. In contrast, if the degree has a negative value close to -1, *p* is said to be rejected by the community. The degree is calculated as the *group preference* on *p* by aggregating the degrees of all individual preferences on *p*.

*iv)Degree of individual's expertise* ($\delta_E(i)$) is the capability for resolving an issue of an individual *i*, and is determined by the contributions, continuity, and accuracy that an individual participates in proposing positions and making arguments. The initial value of such degree for all individuals is equal to 1 and is dynamically updated on the basis of the individual knowledge and judgments. The more contributions, longer continuity, and higher accuracy the individual has, the higher degree of individual's expertise is.

▪ *Ontology-based Reasoner* performs semantic reasoning on the community knowledge-base and the system's argumentation database which are represented in RDF/OWL machine-processable format. It enhances knowledge retrieval by supporting ontology-based semantic search and browsing. Therefore, users are encouraged to express their ideas and knowledge in terms of nodes and interconnected links using a predefined schema and controlled semantic relations for generating structural concept networks and formal metadata described in ontology representation. Moreover, the community is motivated to collaboratively conceptualize a domain-specific ontology for classifying users-contributed contents.

▪ *Semantic Lexical-based Reasoner* can help enhance a search performance by adopting a lexical database such as *WordNet* to determine an appropriate semantic relation (such as synonym, antonym, hypernym) of a query term and a term or concept contained in a knowledge-base, hence enabling a user to retrieve more precise search results. Moreover, it also facilitates content creating and editing by deriving and suggesting relevant concepts for their reusing as well as maintaining the consistency of the contents. For example, assume that the community has already created and published knowledge concerning the *Web*'s definition as a hypertext system that operates over the Internet. Later on if another user wants to publish a definition of *the World Wide Web* and is unaware of such existing content about the Web, he/she will then be notified accordingly. Hence, the user can decide to reuse or extend it, if it matches his/her idea; otherwise he/she can make an opposing argument in case of disagreement. On the other hand, the user can define a new *Web* definition if it denotes another concept (e.g. a mesh built by a spider).

▪ *Rule-based Reasoner* enables the formalization of axioms, rules and constraints for governing the knowledge creation and argumentation process. For instance, to avoid conflicting arguments in community deliberation, the following rule can be defined: *all arguments submitted by a member i about a position p must consistently support or oppose such position p*; in other words, the rule constrains that every member must be consistent with his/her argumentation, and is not allowed to have both supporting and opposing arguments for the same position.

**Storage:** In order to allow machine-processable semantics of the users-contributed contents and to enable semantic data interchange, the argumentation, community knowledge, and domain specific ontology are encoded in RDF/OWL metadata format and stored in an RDF/OWL database. The argumentation generated during the community deliberation is stored in order to keep track of the deliberation and enable traceability function, while domain specific ontologies and users-contributed contents are collected in order to elaborate collective knowledge and maintained it as a community knowledge-base.

**Retrieval:** This function allows a user to search and retrieve desired knowledge by formulating a corresponding query. It then semantically evaluates such query by employing the reasoning services supported by *ontology-based reasoner* and *semantic lexical-based reasoner*. The reasoners can derive implicit information, and thus can return more precise and useful results to the user. Moreover, this function allows users-contributed contents to be classified into predefined hierarchical concepts for semantically browsing and displaying of the well-structured knowledge.

**Reuse:** The model provides knowledge reuse function for exchanging information across different platforms. A domain specific ontology or external knowledge can be imported for community sharing and reusing, while community knowledge described and captured in SAMS can be exported for reusing by external applications. Moreover, different users typically have different objectives or requirements. They can export collective knowledge from the community knowledge-base for local adaption or reuse based on their own needs.

### 3.2 *SAMS*: SAM Schema

In order to enable sophisticated analysis, complicated computations, and productive reasoning, as well as discover most-widely accepted positions, *SAM Schema* (*SAMS*) is developed for semantically capturing and describing community deliberation. It formalizes the structure of complex argumentation in a more systematic manner by imposing the burden of proof upon the individuals' judgments. Figure 2 depicts *SAMS* developed by extending *Collective Argumentation Ontology* (*ColaOnto*) [14], which integrates and extends the two widely-accepted argumentation approaches: *Issue-Based Information Systems* (*IBIS*) *model* [7] and *Toulmin's argument scheme* [9].

In *SAMS*, an **issue** is a question, problem, or concern given by an *individual* with a particular *degreeOfExpertise* regarding the domain of interest. An issue can be specialized into specific sub-issues or consist of associated issues. This structure clarifies the domain of interest clearer and hence making it easier for members to understand and contribute. A **position** is a solution posted by an individual in response to a particular issue. A position can consist of several sub-positions, and can raise a new issue for further discussion. For each position, *data* represents a fact that supports the position. Moreover, a position can be extended by a new position for elaborating more details. A position extended by many positions is considered as a potential position to resolve the issue and increased its *degreeOfPosition*. An **argument** is a statement which supports or opposes a position. For each argument, *data* represents a fact, while *backing* is an evidence or reference for the argument. Moreover, a *qualifier* expresses a degree of confidence concerning the posted argument. An argument can be made against by *rebuttals*. This relevant information can be used to evaluate *degreeOfArgument*. For each position and argument, an individual can describe its supported *data* in either textual or structural format. In order to structurally and meaningfully express a position and argument, several widely-accepted knowledge representations, such as ontology, concept map, or mind map, are suggested. However, the ontology representation is recommended for enabling semantic expression and enhancing effective reasoning.

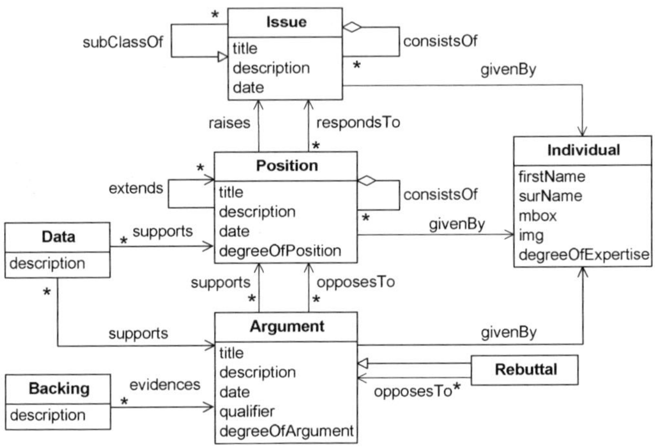


**Fig. 2.** SAMS: SAM Schema4

## 4 Example Use Case

This section conducts a practical example in order to demonstrate the ability and effectiveness of SAM for facilitating collaborative knowledge creation and sharing. Assume that a community is encouraged to develop knowledge about "Environmental problems" in the collaborative process as described follows:

*1) Issue formalization.* As shown in Figure 3, a community formalizes Environmental problems and depicts that Environment problems is a super class of Pollution, Global warming, and Species extinction issues, while the Species extinction issue consists of Causes of extinction and How to preserve endangered species? issues.

*2) Position proposal.* Figure 4 shows Animal testing and Climate change positions responding to Causes of extinction issue. Animal testing position describes its supported data in textual format, while Climate change position expresses its supported data in ontology format. Moreover, a new position, Example of extinct species, extends Climate change position.

The Climate change and Example of extinct species positions described in ontology format allow *ontology-based reasoner* to semantically reason across the

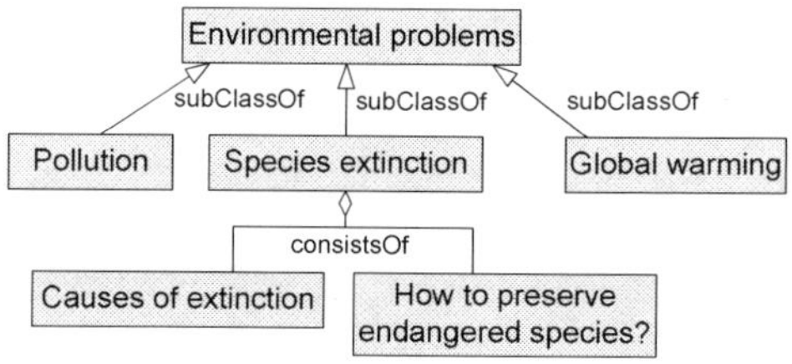


**Fig. 3.** Environmental problems depicted in SAMS and described in RDF format

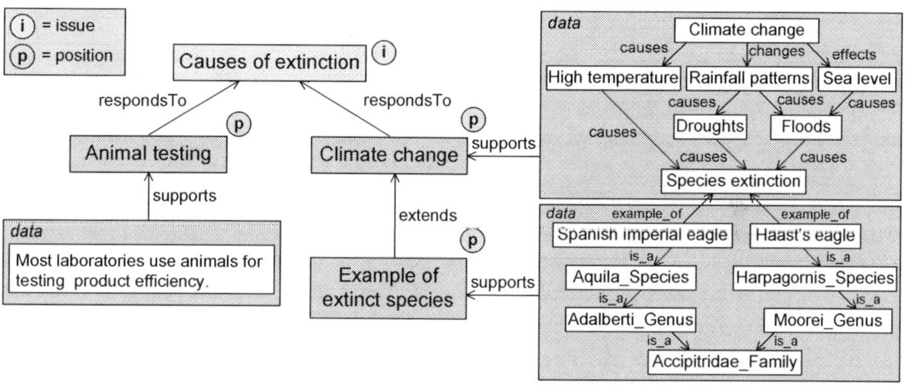


**Fig. 4.** Positions proposed to the causes of extinction issue

contributed contents. For example, a user might search for all extinct *eagles* in *Accipitridae* family, skipping *Moorei* genus. Based on a predefined semantic relation, is_a, which is a transitive property, the reasoner can derive that Spanish imperial eagle is a kind of Accipitridae_Family and is an example of Species Extinction, while Haast's eagle is a Moorei_Genus and hence filtered. Such query is not possible in non-semantic search and requires extensive maintenance without reasoning support. Moreover, by incorporating *semantic lexical-based reasoner*, another user who searches for all extinct *birds* caused by climate change will be able to retrieve all those extinct *eagles* because the reasoner can yield that *eagle* is a hyponym of (kind-of) *bird*.

*3) Argumentation*. As shown in Figure 5, Animal testing position is supported by argument a1 and opposed by argument a2. The argument a1 provides supported data without backing and is opposed by rebuttal r1 which then leads to a revision of the position. The proposer is allowed to revise or withdraw the position, but not allowed to make against his/her proposed position which causes conflicting arguments. On the other hand, Climate change position is supported by arguments a3, a4, and a5, all of which have both supported data and backing information. Therefore, with these strongly supporting arguments, Climate change position is considered as a potential position.

Considering the traceability function, one can observe that by semantically capturing the community deliberation using SAMS, a user can easily and quickly trace through the argumentation visualized in a graphical view. Thus, the user is able to

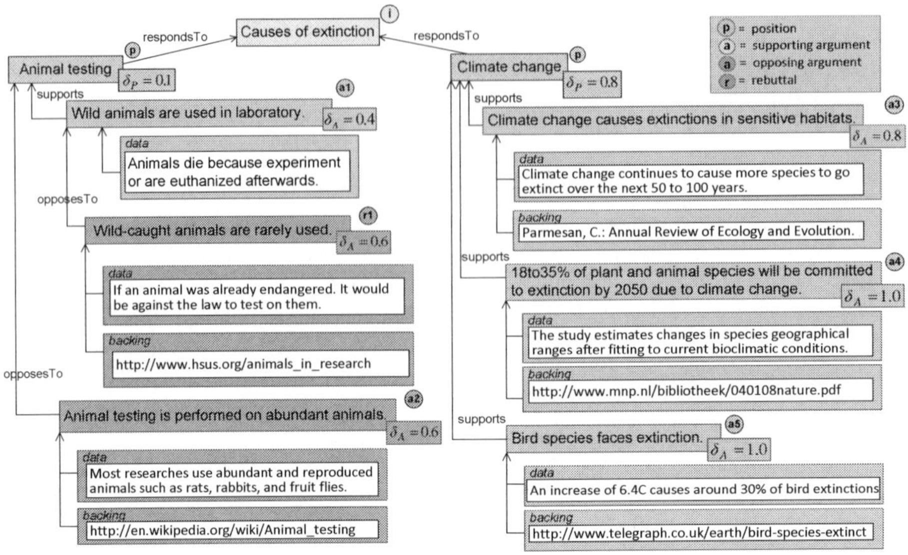


**Fig. 5.** Arguments submitted to animal testing and climate change positions

find all supporting arguments of Climate change position as well as to verify the reasons stating that Animal testing does not cause Species extinction.

*4) Achieving consensus.* Argumentation analyzer examines and calculates degree of arguments, degree of individuals' preference, and degree of positions, while degree of individuals' expertise is initiated to 1 as depicted in Figure 5. Intuitively, Animal testing position is supported by a small degree of argument a1 due to the lack of its backing information and the attacking rebuttal r1, while it is strongly opposed by the degree of argument a2, which provides both supported data and backing information. On the other hand, Climate change position is strongly supported by argument a3, a4 and a5 with high degree of the arguments evaluated based on their supported data and backing information. Moreover, it is extended by Example of extinct species position which implies its reliability and then increases its degree of position.

Assume that the community identified a consensus level at 0.8, which means that the community automatically accepts a position which has the dominance over "*most*" positions based on the consensus. Since, the degree of Climate change position equals to 0.8 and reaches to the consensus level, it is *acceptable* and considered as a potential position for resolving the issue based on the consensus. In addition, the degree of expertise of the individuals who proposes or supports Climate change position will be increased, while that of the individuals who propose or support Animal testing position remains unchanged.

## 5 Model Evaluation and Discussion

This section evaluates *SAM* by comparing it with the existing systems based on the key requirements as demonstrated in Table 1. The table shows that all systems can *enhance collaborative knowledge creation and sharing* by employing different underlying techniques. Most of them except Semantic Wikipedia ensure the *simplicity* of their process by employing certain argumentation theories. Semantic Wikipedia, on the other hand, uses wiki mechanism which has several aforementioned problems. In order to *assure quality of knowledge*, only SAM provides automated argumentation analysis to *enable consensual contents*. It also offers ontology-based and semantic lexical-based term suggestion for *ensuring consistent information*. In addition, DILIGENT and SAM can *detect conflicting arguments* during community deliberation.

Obviously, all systems that are designed based on argumentation theory can support *traceability* function, which is rather limited in a reverse-order history page of Semantic Wikipedia. To mediate group opinions, SAM *enables automated mediation* by applying RT Delphi method and employing soft consensus-based mechanism. On the other hand, Collaboratorium applies manual voting system, while DILIGENT relies on expert judgment for making a decision. Furthermore, Semantic Wikipedia and SAM enhance *semantic search* by employing intelligence reasoners.

**Table 1.** The comparison of SAM with the existing systems

| Requirements | Semantic Wikipedia | Compendium | Collaboratorium | DILIGENT | HCOME | SAM |
|---|---|---|---|---|---|---|
| 1.Enhancing collaborative creation and sharing | √ | √ | √ | domain specific | domain specific | √ |
| 2.Simplicity | wiki mechanism | argumentation process | | | | |
| 3.Assuring quality of knowledge | | | | | | |
| - Enabling consensual contents | – | – | – | expert judgment | community judgment | automated argumentation analysis |
| - Ensuring consistent information | template system | frequency-based term suggestion | – | – | manually mapping | ontology&semantic lexical-based term suggestion |
| - Detecting conflicting arguments | – | – | – | √ | – | √ |
| 4.Traceability | history page | √ | √ | √ | √ | √ |
| 5.Enabling automated mediation | – | – | – | – | – | RT Delphi+soft consensus-based |
| 6.Semantic search | ontology-based | tag-based | – | – | – | ontology& lexical-based |
| 7.Interoperability<br>- Data interchange support | RDF/OWL | XML | – | RDF/OWL | N/A | RDF/OWL |

## 6 Conclusion

The proposed model, *SAM*, facilitates collaborative knowledge creation and sharing by encouraging its members to share their ideas by proposing a position concerning a complex issue, and then submitting their supporting or opposing arguments about other members' ideas. It employs a soft consensus technique to determine the group decision, and considers the position accepted by most members as a potential solution to solve the issue. Thus, to semantically capture the community deliberation, which can further enable sophisticated argumentation analysis, complex reasoning, determining the most widely-accepted positions as well as data interchange, SAM schema (SAMS) is formalized. Based on such semantic argumentation knowledge-base, SAM's argumentation analyzer can dynamically analyze the deliberation and evaluate an acceptance level of each position using its four underlying measures: a *degree of argument*, a *degree of individual preference*, a *degree of position*, and a *degree of individual's expertise*; and hence automatically driving the community toward a consensus. Moreover, SAM constructs the ontology-based and semantic lexical-based reasoners to enhance semantic knowledge retrieval as well as to facilitate semantic content editing. Its rule-based reasoner is also created for governing the knowledge creation and argumentation process which allows definition of rules and constraints to detect and resolve conflicting arguments.

SAM's prototype system development and a detailed user-based evaluation for indicating the practicality and effectiveness in real-world scenarios are underway. Moreover, feedback information should be automatically generated by classifying individuals' knowledge into groups and then responding to a particular user for his/her clarification and revision. This can also support a community to obtain potential positions with a higher consensus level.

## References

1. Viegas, F., Wattenberg, M., Kriss, J., van Ham, F.: Talk before you type: Coordination in Wikipedia. In: 40th HICSS, p. 78a. IEEE press, CA (2007)
2. Brandes, U., Kenis, P., Lerner, J., Raaij, D.: Network Analysis of Collaboration Structure in Wikipedia. In: 18th Int. Conf. on WWW, pp. 731–740. ACM press, NY (2009)
3. Krotzsch, M., Vrandecic, D., Volkel, M., Haller, H., et al.: Semantic Wikipedia. J. Web Semantics 5, 251–261 (2007)
4. Buckingham Shum, S., Selvin, A., Sierhuis, M., Conklin, J., et al.: Hypermedia Support for Argumentation-Based Rationale: 15 Years on from gIBIS and QOC. In: Dutoit, A.H. (ed.) Rationale Management in Software Engineering, pp. 111–132. Springer, Heidelberg (2006)
5. Conklin, J., Begeman, M.L.: gibis: a hypertext tool for exploratory policy discussion. In: Proc. of the 1988 ACM conference on Computer-supported cooperative work (1988)
6. Iandoli, L., Klein, M., Zollo, G.: Can We Exploit Collective Intelligence for Collaborative Deliberation? The Case of the Climate Change Collaboratorium. In: MIT Sloan School of Management, Working Paper, 4675-08 (2008)
7. Rittel, H., Kunz, W.: Issue as elements of information systems, Institute of Urban and Regional Development, University of California, Working Paper 131 (1970)
8. Walton, D.: Argumentation Schemes for Presumptive Reasoning. Erlbaum, Mahwah (1996)
9. Toulmin, S.: The Uses of Argument. University Press, Cambridge (1958)
10. Tempich, C., Simperl, E., Luczak, M., Studer, R., et al.: Argumentation-based ontology engineering. IEEE Intelligent Systems 22(6), 52–59 (2007)
11. Kotis, K., Vouros, G.: Human-centered ontology engineering: The HCOME Methodology. Knowledge and Information Systems 10(1), 109–131 (2005)
12. Gordon, T., Pease, A.: RT Delphi: An Efficient, "Round-Less" Almost Real Time Delphi Method. Technological Forecasting and Social Change 73, 321–333 (2006)
13. Kacprzyk, J., Fedrizzi, M.: A "soft" measure of consensus in the setting of partial (fuzzy) preferences. European Journal of Operations Research 34, 316–325 (1988)
14. Maleewong, K., Anutariya, C., Wuwongse, V.: A Collective Intelligence Approach to Collaborative Knowledge Creation. In: 4th International Conference Semantics, Knowledge, and Grid (SKG 2008), pp. 64–70. IEEE press, DC (2008)

# A Conception for Modification of Learning Scenario in an Intelligent E-learning System*

Adrianna Kozierkiewicz-Hetmańska

Institute of Informatics, Wroclaw University of Technology, Poland
adrianna.kozierkiewicz@pwr.wroc.pl

**Abstract.** The main purpose of an intelligent E-learning system is to guarantee an effective learning and offer the optimal learning path for each student. Learning path should be suitable for student's preferences, abilities, interests, learning styles and especially for his current knowledge. Therefore, if a student has a problem with passing a test it is a signal for the system that the offered learning path is not adequate for this user. System should modify learning scenario based on collected data. In this paper new knowledge structure is proposed. For the defined knowledge structure definitions of a learning scenario and a conception for modification of the learning scenario during a learning process are presented.

## 1 Introduction

Education is a part of living - this theme was firstly taken up in 1929 by Basil Yeaxlee. It means that learning is not confined to childhood and classroom. Lifelong learning is closely related to E-learning. The need of constant learning to improve qualifications, bring skills up-to-date and retrain for a new job require a new form of education. People want to be guaranteed an effective learning in a short time. Therefore, intelligent E-learning systems should be able to adapt the learning material for student's preferences, learning styles, interests, abilities etc., to create learning environment preferable by student, to offer the best learning path for an individual, to offer adaptive testing, to analyze student's mistakes, to modify learning path during learning process, to identify the worst and the best students and to offer them suitable learning materials.

In real life people, before passing exams, attend lectures, read notes or books. If they fail a test, they try to change learning methods. They buy new books, lend colleagues' notes or find private teachers. In E-learning systems if students have problems with learning material they should be offered relearning, changing learning scenario etc. Systems should adapt the learning methods to current student's characteristics. It is a very important task to create an intelligent and personalized learning system. Actually, none of the following systems consist of methods modifying the learning scenario during learning process. Some of them are able to choose next units based on students' knowledge levels or predict suitable presentation methods.

* This research was financially supported by European Union- European Social Found and by Human Capital National Cohesion Strategy under the grant No. II/33/2009.

N.T. Nguyen, R. Kowalczyk, and S.-M. Chen (Eds.): ICCCI 2009, LNAI 5796, pp. 87–96, 2009.

In 1995-1996 first intelligent tutoring web-based systems appeared. One of them is ELM-ART which is used to help people learn LISP programming language. ELM-ART is a personalized textbook and is able to build the most relevant learning path for every learner. Student is provided a special "next" link that is connected to the most suitable educational material. Next units are determined based on student's current level of knowledge. System also provides special links in the form of coloured bullets which allow learner to navigate between units of knowledge. The colour of this bullet informs the student about educational status of next pages [6].

In INSPIRE Lesson Generation Module plans the learning paths and chooses one of 3 levels (remember, use, find) to present outcome concepts based on student's current knowledge level. Presentation Module decides on the appearance of the knowledge modules based on the learning style of the student. Learners with different learning styles view different presentations of the educational material. In INSPIRE student's current level of knowledge is estimated based on adaptive assessment algorithm which uses Item Response Theory [7], [8].

In ActiveMath Course Generator Module is responsible for constructing courses. This module processes goal concepts the user wants to learn, the scenario, student's knowledge level, action history, capabilities and pedagogical rules to determine a learning path. First, system chooses the concept which needs to be learned by the student. Next, examples, exercises and tests are added. In the third step pedagogical rules determine when and which items should be presented and in which order. Finally, the pages are ordered and put together in a proper hierarchy. System ActiveMath does not offer changing of the educational material during learning process [9].

Interesting methods are used in system EDUCE. EDUCE uses predictive engine to build and customise a learning path. System offers four different types of learning material such as verbal/linguistic, visual/spatial, logical/mathematical and musical/rhythmic. The student can choose from viewing only one, all of them or repeatedly viewing some of materials. EDUCE tries to predict at the start of a learning unit which type of material student would prefer. Naïve Bayes algorithm is used for prediction purposes in this system [10].

In paper [11] intelligent tutoring system based on Bayesian approach is presented. In this system Pedagogical Module is responsible for choosing learning material. Pedagogical Module creates a set of suggestions for what the student should learn next. The role of this module is limited to choosing only one of 4 actions: show a new topic, deepen a topic, shallow scan or review previous topic. This decision is made based on Bayesian Network.

In [1] and [12] algorithms based on consensus theory determining opening learning scenario are proposed. This works did not solve the problem of modifying learning scenario.

This work is devoted to modification of learning scenario during a learning process. If students have problems with passing a test it becomes a sign for a system that opening learning scenario could be unsuitable for this learner. System, using information about current student's characteristic such as results of tests, difficulty of lessons, time of learning, the level of satisfaction and data collected during functioning of the system such as the average score for each lesson, the average learning time of each lesson and the average difficulty of each lesson. In this paper a new knowledge structure and a learning scenario are defined. For new knowledge structure original conception of modification of learning scenario during a learning process is proposed.

In the next chapter the idea of intelligent E-learning system with auxiliary definitions are presented. This part of work explains how this system works. Section 3 contains conception of modification learning scenario. Finally, conclusion and future works are described.

## 2 Idea of an Intelligent E-learning System

Intelligent E-learning system should especially adapt to student's needs and preferences. This condition could be fulfilled if assumption that similar students will learn in the same or a very similar way is applied [1].

Learner preferences, which are used for personalization, are stored in a user profile which is created during the registration process. A student's profile is defined as "*representations of some characteristics and attitudes of the learner, which are useful for achieving the adequate and individualized interaction established between computational environments and students*" [3]. It contains two types of data: user data and usage data [2]. User data consists of demographic data (such as login, name, age, sex, educational level, IQ), learning style (related to perception, receiving, processing and understanding of information by a student), abilities (verbal comprehension, word fluency, computational ability, spatial visualization, associative memory, perceptual speed, reasoning) personal character traits (such as concentration, motivation, ambition, self-esteem, level of anxiety, locus of control, open mind, impetuosity, perfectionism) and interests (humanistic science, formal science, the natural science, economics and law, technical science, business and administration, sport and tourism, artistic science, management and organization, education). The usage data contains of information about completed lessons such as: $t_i$ -time spent on reading lesson $l_i$, $score_i$-results of tests lesson $l_i$, $r_i$- rate (it is subjective evaluation of version of lesson $l_i$), $d_i$- difficulty of lesson (number of test's failure refers to lesson $l_i$) and also opening and final scenario, where $i \in \{1,...,q\}$ [4].

After registration process student is classified to a set of similar learners based on his user data. It is assumed that criterion of classification of learner is based on some of user data attributes selected by experts.

Before student begins learning the system proposes the most adequate educational material in the best order. The method of choice of educational material is closely connected with knowledge structure. In described system knowledge is represented in two levels: concepts (multimedia lessons and relations between them) and instances (other forms of lessons for example textual, graphical, interactive etc.). The graphic representation of defined knowledge structure is presented in Figure 1.

Let $L = \{l_1,...,l_q\}$ be the finite set of lessons. Each lesson $l_i$, is related to their version $v_{ik}$ for $k \in \{1,...,m\}$, $i \in \{1,...,q\}$, $V = \bigcup_{i=1,...,q} \bigcup_{k=1,...,m} v_{ik}$ . $Rc$ is called a set of linear relations between lessons. Each of such relations defines the order in which the lessons should be presented to a student, because some lessons should be learned before others. A binary relation $\alpha \in Rc$ is called linear if relation is reflexive, transitive, antisymmetric and total.

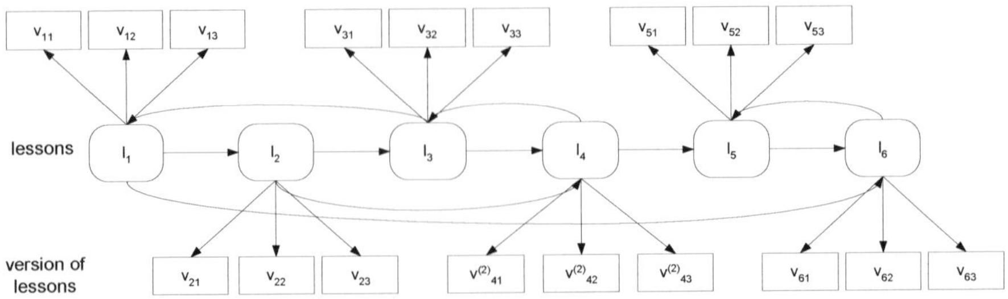


**Fig. 1.** The ontology of knowledge structure

**Definition 1.** *The ontology of knowledge structure is defined by the following triple*

$$(L, V, Rc) \tag{2.1}$$

In the system, except for data stored in users' profiles, also data related to lessons are stored. After each lesson learner has to solve a test. A test consists of assumed number of questions. During learning process we collect the following data:

$$AS_g = [as_{x,j}]_{\substack{x=0,\dots,q,\\ j=1,\dots,q}} \tag{2.2}$$

where: $as_{x,j}$ – 100% minus average the score for lesson $l_j$ which was learnt after lesson $l_x$, $x \in \{0,\dots,q\}$, $j \in \{1,\dots,q\}$

$$AS = [asc_{x,j}]_{\substack{x=0,\dots,q,\\ j=1,\dots,q}} \tag{2.3}$$

where: $asc_{x,j} = \frac{\sum_{y=1}^{G} as_{x,j}}{G}$, for $x \in \{0,\dots,q\}$, $j \in \{1,\dots,q\}$ , $g \in \{1,\dots,G\}$ , G- number of students' class.

$$AD_g = [ad_{x,j}]_{\substack{x=0,\dots,q,\\ j=1,\dots,q}} \tag{2.4}$$

where: $ad_{x,j}$ - difficulty degree of lesson $l_j$, represented by a number of test failures referred to lesson $l_j$ (learnt after lesson $l_x$) divided by a number of all tests taken by students who were learnt lesson $l_j$ after lesson $l_x$, $x \in \{0,\dots,q\}$, $j \in \{1,\dots,q\}$

$$AD = [adc_{x,j}]_{\substack{x=0,\dots,q,\\ j=1,\dots,q}} \tag{2.5}$$

where: $adc_{x,j} = \frac{\sum_{y=1}^{G} ad_{x,j}}{G}$, for $x \in \{0,\dots,q\}$, $j \in \{1,\dots,q\}$ , $g \in \{1,\dots,G\}$

$$AT_g = [at_{x,j}]_{\substack{x=0,\ldots,q,\\ j=1,\ldots,q}} \tag{2.6}$$

where: $at_{x,j}$ – average time of learning lesson $l_j$ which was learnt after lesson $l_x$, $x \in \{0,\ldots,q\}$, $j \in \{1,\ldots,q\}$

$$AT = [atc_{x,j}]_{\substack{x=0,\ldots,q,\\ j=1,\ldots,q}} \tag{2.7}$$

where: $atc_{x,j} = \dfrac{\sum_{y=1}^{G} at_{x,j}}{G}$ for $x \in \{0,\ldots,q\}$, $j \in \{1,\ldots,q\}$ , $g \in \{1,\ldots,G\}$

Let us transform ontology of knowledge structure to a directed and weighted graph.

**Definition 2.** *Knowledge structure is a directed and weighted graph*

$$Gr = (\{L \cup l_o\}, Rc, W) \tag{2.7}$$

*where:* $L = \{l_1, l_2, \ldots, lq\}$ *- set of nodes represents lessons,* $l_i = \{v_{i,1}, v_{i,2}, \ldots, v_{i,m}\}$ *- set of version of lessons corresponding to lesson* $l_i$*,* $l_0$*-starting node, Rc - set of edges represents relationships among lessons,* $W = [w_{x,j}]_{\substack{x=0,\ldots,q\\ j=1,\ldots,q}}$ *- weight matrix, where* $w_{x,j}$ *could equal* $as_{x,j}$ , $asc_{x,j}$ , $ad_{x,j}$ , $adc_{x,j}$ , $at_{x,j}$ or $atc_{x,j}$ , for $i, j \in \{1,\ldots,q\}$ , $x \in \{0,\ldots,q\}$ , $g \in \{1,\ldots,G\}$.

Figure 2 presents obtained graph.

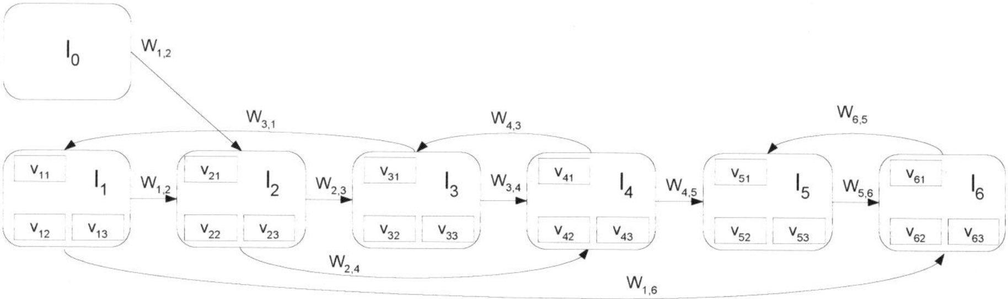


**Fig. 2.** The knowledge structure as a graph

For proposed knowledge structure, learning scenario is given as follows:

**Definition 3.** *A learning scenario s is defined by Hamiltonian path in graph Gr:*

$$s = < n_1, n_2, \ldots, n_q > \tag{2.8}$$

*that is a sequence of nodes* $\{l_1, l_2,..., lq\}$ *represented by exactly one element from each set* $l_{i.}$*, where* $n_i$ *is version of lesson referring to exactly one lesson* $l_{\rho_i}$ *and s does not contain any other versions of lessons referring to the lesson* $l_{\rho_i}$*, for* $i, \rho_i \in \{1,...,q\}$*,* $\rho_i$ *- number of lesson referring to a version of lesson* $n_i$*,*

**Example 1**

For knowledge structure presented in Figure 1 some scenarios are defined as follows:

$s_1 =< v_{21}, v_{43}, v_{31}, v_{12}, v_{61}, v_{51} >$;

$s_2 =< v_{23}, v_{43}, v_{33}, v_{13}, v_{63}, v_{53} >$;

$s_3 =< v_{11}, v_{23}, v_{31}, v_{42}, v_{51}, v_{62} >$

**Example 2**

Figure 3 presents knowledge structure for a computer science course. Let $v_{i,1}$ be a textual version of lesson, $v_{i,2}$ a graphical version of lesson and $v_{i,3}$ an interactive version of lesson for $i \in \{1,...,q\}$. It is assumed that weight $w_{i,j}$ is $ad_{i,j}$, for example if students learnt about graphical editor after operating systems the probability of failures equals 11%, but if students learnt about graphical editor after Microsoft® Word word processor the probability of failures equals 13,4%. If student is offered a scenario $s_1 =< v_{21}, v_{43}, v_{31}, v_{12}, v_{61}, v_{51} >$ it means that student will be taught operating systems first (and is proposed textual version of the lesson), then about Microsoft® Word (interactive version of the lesson), then about graphic editor (textual version of the lesson), about computer network (graphical version of the lesson), about Microsoft® PowerPoint (textual version of the lesson) and finally about Microsoft® Excel (textual version of the lesson).

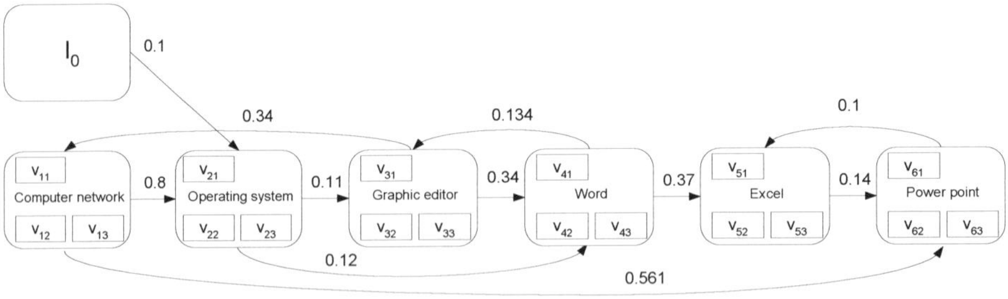


Knowledge structure for a computer science course

Based on assumptions that similar students will learn in the same or a very similar way the opening learning scenario is determined by using the nearest neighbor algorithm. Student is offered a learning scenario which was successfully finished by a student the most similar to the new one. For comparing student's profile Hamming distance is assumed.

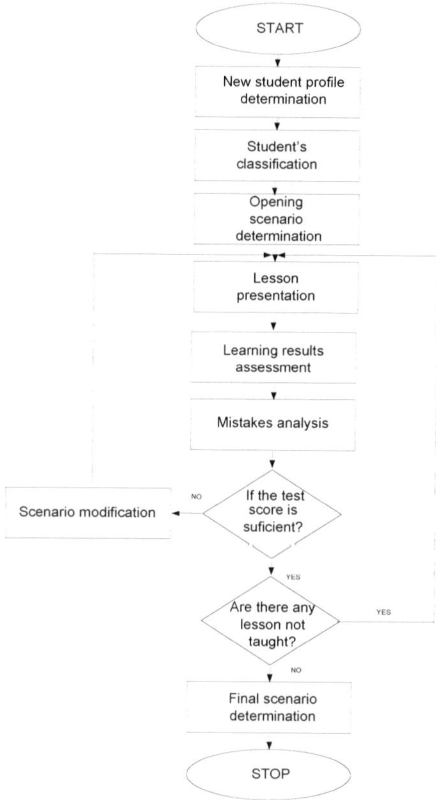


**Fig. 4.** The idea of an intelligent E-learning system [15]

The student begins to learn. System presents first lesson from indicated opening learning scenario. After each lesson student has to pass a test. Tests are adaptive [13] which means that the questions are selected intelligently to fit the student's level of knowledge. This solution minimizes learner's stress and gives more accurate evaluation in a shorter time. If the test score is sufficient student can be presented with the next lesson. Otherwise system decides to change presentation methods and suggests relearning. The conception of modifying the learning scenario during learning process is described in the next section. All changes are stored in the user profile.

Course is finished if all lessons from learning scenario are taught. Then, finally, successful scenario is added to the user profile of the learner. The idea of this system is presented in Figure 4.

## 3 Conception for Modification of Learning Scenario

If a student has a problem with passing a test it is a sign for the system that opening learning scenario is not suitable for this student. System should offer modification of the learning scenario. Modification of a learning scenario is conducted in three steps. If a student has a problem with passing a test for the first time he is offered repetition of the same lesson but in a different version. After another failure system changes lessons' order based on data of students who belongs to the same class. The third step is the final chance. Student could provide false information about himself so it might have happened that he was classified to the improper group. If this possibility is taken into account, in the last step user is offered modification of lessons' order based on all collected data. The graphical representation of this idea is presented in Figure 5.

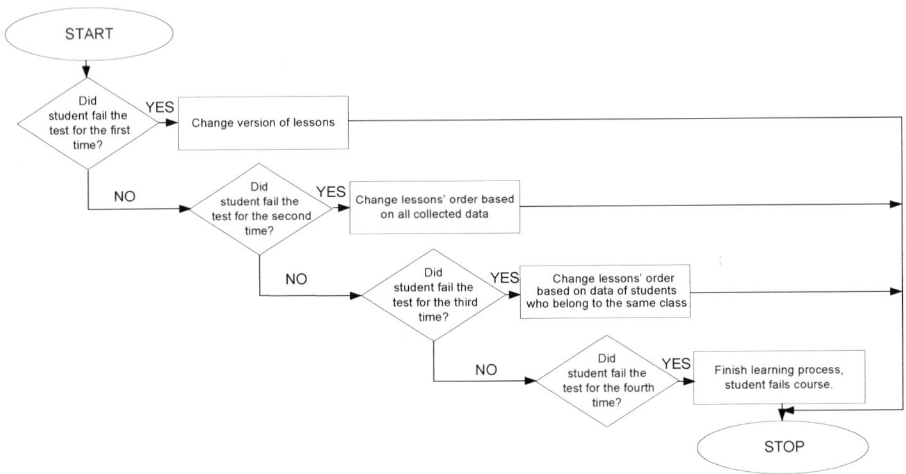


**Fig 5.** Conception of modification of learning scenario in case of a test failure

Problem of modification of learning scenario could be decomposed to 3 smaller problems:

**Problem 1**

*For given data:* $t_i$, $score_i$, $r_i$, $d_i$ *find the version of lesson such that following condition is satisfied:* $\arg\max_{v_{i,k}} P(v_{i,k})\prod P(t_i, score_i, r_i, d_i \mid v_{i,k})$ *for* $k \in \{1,\ldots,m\}$, $i \in \{1,\ldots,q\}$

**Problem 2**

*Let W equal* $AS_g$. *For given W and fixed* $g \in \{1,\ldots,G\}$ *find order* $\alpha^*$ *such that following sum* $\sum_{j=0}^{q} as_{x,j}$ *is minimal for* $x \in \{0,\ldots,q\}$. *For two orders* $\alpha_1^*$ *and* $\alpha_2^*$ *which sums* $\sum_{j=0}^{q} as_{x,j}$ *are equal, choose order such that* $\sum_{j=0}^{q} at_{x,j}$ *is minimal,* $x \in \{0,\ldots,q\}$

**Problem 3**

*Let W equal AD. For given W find order α* such that following sum* $\sum_{j=1}^{q} adc_{x,j}$ *is minimal for* $x \in \{0,...,q\}$. *For two orders* $\alpha_1$* *and* $\alpha_2$* *which sums* $\sum_{j=0}^{q} adc_{x,j}$ *are equal, choose order such that* $\sum_{j=0}^{q} at_{x,j}$ *is minimal.*

Problem 1 can be solved by using Naïve Bayesian Classification method. Problem 2 and 3 can be boiled down to finding the shortest Hamiltonian path. For problem 2 the learning path which was learnt the most effective by students (student achieved the best score) is found. For problem 3 the easiest learning path (students failed the tests the rarest) is found. The shortest Hamiltonian path problem is a NP hard problem and it could be solved by using brute force method or heuristic algorithms.

**Example 3**
For graph presented in Figure 2 there exists one Hamiltonian path: $l_0, l_2, l_4, l_3, l_1, l_6, l_5$

## 4 Conclusions and Further Work

Systems for distance education gain more and more popularity. However, users have strict requirements. They want to be offered the best learning path suitable for their preferences and current knowledge in each step of the learning process. Thus, it is a very important task to determine an opening learning scenario and modify it during the learning process.

In this paper idea of an intelligent E-learning system is proposed. The learner profile, the knowledge structure, the learning scenario and the method of determination of a learning scenario are described. For defined knowledge structure conception of modification of learning scenario during learning process is outlined.

In future work algorithms for solving problems described in section 3 will be worked out. All methods will be verified by experimental tests and proof of correctness will be conducted. Finally, an E-learning system implementation incorporating designed methods is planned.

## References

1. Kukla, E., Nguyen, N.T., Daniłowicz, C., Sobecki, J., Lenar, M.: A model conception for optimal scenario determination in an intelligent learning system. ITSE - International Journal of Interactive Technology and Smart Education 1(3), 171–184 (2004)
2. Kobsa, A., Koenemann, J., Pohl, W.: Personalized Hypermedia Presentation Techniques for Improving Online Customer Relationships. The Knowledge Engineering Review 16(2), 111–155 (2001)
3. Paiva, A., Self, J., Hartley, R.: Externalising Learner Models. In: Proceedings of AIED 1995. AACE Publication (1995)

4. Kozierkiewicz, A.: Content and structure of learner profile in an intelligent E-learning system. In: Nguyen, N.T., Kolaczek, G., Gabrys, B. (eds.) Knowledge Processing and Reasoning for Information Society, EXIT Warsaw (2008)
5. Kozierkiewicz, A.: Determination of Opening Learning Scenarios in Intelligent Tutoring Systems. In: Zgrzywa, A., Choroś, K., Siemiński, A. (eds.) New trend in multimedia nad network information systems. IOS Press, Amsterdam (2008)
6. Brusilovsky, P., Schwarz, E., Weber, G.: ELM-ART: An intelligent tutoring system on World Wide Web. In: Lesgold, A.M., Frasson, C., Gauthier, G. (eds.) ITS 1996. LNCS, vol. 1086, pp. 261–269. Springer, Heidelberg (1996)
7. Gouli, E., Kornilakis, H., Papanikolaou, K., Grigoriadou, M.: Adaptive Assessment Improving Interaction in an Educational Hypermedia System. In: Proceedings of the PanHellenic Conference with International Participation in Human-Computer Interaction, pp. 217–222
8. Grigoriadou, M., Papanikolaou, K., Kornilakis, H., Magoulas, G.: INSPIRE: An iNtelligent system for personalized instruction in a remote environment. In: Reich, S., Tzagarakis, M.M., De Bra, P.M.E. (eds.) AH-WS 2001, SC 2001, and OHS 2001. LNCS, vol. 2266, pp. 215–225. Springer, Heidelberg (2002)
9. Melis, E., Andres, E., Budenbender, J., Frischauf, A., Goguadze, G., Libbrecht, P., Pollet, M., Ullrich, C.: ActiveMath: A Generic and Adaptive Web-Based Learning Environment. International Journal of Artificial Intelligence in Education (2001)
10. Kelly, D., Tangney, B.: Predicting Learning Characteristics in a Multiple Intelligence Based Tutoring System. Intelligent Tutoring Systems, 678–688 (2004)
11. Gamboa, H., Fred, A.: Designing Intelligent Tutoring Systems: a Bayesian Approach. In: Proc. of 3rd International Conference on Enterprise Information Systems, ICEIS 2001 (2001)
12. Nguyen, N.T.: Advanced Methods for Inconsistent Knowledge Management. Springer, New York (2008)
13. Papanastasiou Elena, C.: Computer-adaptive Testing in Science Education. In: Proc. of the 6th Int. Conf. on Computer Based Learning in Science, Cyprus, pp. 965–971 (2003)

# Firefly Algorithm for Continuous Constrained Optimization Tasks

Szymon Łukasik and Sławomir Żak

Systems Research Institute, Polish Academy of Sciences
slukasik@ibspan.waw.pl,
slzak@ibspan.waw.pl

**Abstract.** The paper provides an insight into the improved novel metaheuristics of the Firefly Algorithm for constrained continuous optimization tasks. The presented technique is inspired by social behavior of fireflies and the phenomenon of bioluminescent communication. The first part of the paper is devoted to the detailed description of the existing algorithm. Then some suggestions for extending the simple scheme of the technique under consideration are presented. Subsequent sections concentrate on the performed experimental parameter studies and a comparison with existing Particle Swarm Optimization strategy based on existing benchmark instances. Finally some concluding remarks on possible algorithm extensions are given, as well as some properties of the presented approach and comments on its performance in the constrained continuous optimization tasks.

**Keywords:** firefly algorithm, constrained continuous optimization, swarm intelligence, metaheuristics.

## 1 Introduction

Fireflies, also called lighting bugs, are one of the most special and fascinating creatures in nature. These nocturnal luminous insects of the beetle family *Lampyridae* (order *Coleoptera*), inhabit mainly tropical and temperate regions, and their population is estimated at around 1900 species [1]. They are capable of producing light thanks to special photogenic organs situated very close to the body surface behind a window of translucent cuticle [2]. Bioluminescent signals are known to serve as elements of courtship rituals, methods of prey attraction, social orientation or as a warning signal to predators (in case of immature firefly forms commonly referred to as glowworms). The phenomenon of firefly glowing is an area of continuous research considering both its biochemical [3] and social aspects [4].

Mechanisms of firefly communication via luminescent flashes and their synchronization has been imitated effectively in various techniques of wireless networks design [5], dynamic market pricing [6] and mobile robotics [7]. Firefly Algorithm developed recently by Xin-She Yang at Cambridge University and presented in the Chapter 8 of monograph [8] follows this approach. This swarm

N.T. Nguyen, R. Kowalczyk, and S.-M. Chen (Eds.): ICCCI 2009, LNAI 5796, pp. 97–106, 2009.

intelligence optimization technique is based on the assumption that solution of an optimization problem can be perceived as agent (firefly) which "glows" proportionally to its quality in a considered problem setting. Consequently each brighter firefly attracts its partners (regardless of their sex), which makes the search space being explored more efficiently. Similar nature inspired metaheuristics include: Particle Swarm Optimization (PSO) [9] or Artificial Bee Colony optimization technique (ABC) [10].

This paper is devoted to the detailed study of Firefly Algorithm (FA), its experimental evaluation and possible improvements. It is organized as follows. In the next Section a comprehensive review of the existing FA scheme based on monograph [8] is given as well as some proposals for its extension. The subsequent part of the paper contains the results of parameter studies and some guidelines for its proper assignment. Next the comparison with Particle Swarm Optimization technique is performed. Ultimately, the final part of the paper presents some concluding remarks and suggestions for future work in the subject.

## 2 Firefly Algorithm in Practice

### 2.1 FA Scheme

Let us consider continuous constrained optimization problem where the task is to minimize cost function $f(x)$ for $x \in S \subset \mathbb{R}^n$ i.e. find $x^*$ such as:

$$f(x^*) = \min_{x \in S} f(x) \; . \tag{1}$$

Assume that there exists a swarm of $m$ agents (fireflies) solving above-mentioned problem iteratively and $x_i$ represents a solution for a firefly $i$ in algorithm's iteration $k$, whereas $f(x_i)$ denotes its cost. Initially all fireflies are dislocated in $S$ (randomly or employing some deterministic strategy). Each firefly has its distinctive attractiveness $\beta$ which implies how strong it attracts other members of the swarm. As a firefly attractiveness one should select any monotonically decreasing function of the distance $r_j = d(x_i, x_j)$ to the chosen firefly $j$, e.g. the exponential function:

$$\beta = \beta_0 e^{-\gamma r_j} \tag{2}$$

where $\beta_0$ and $\gamma$ are predetermined algorithm parameters: maximum attractiveness value and absorption coefficient, respectively [8]. Furthermore every member of the swarm is characterized by its light intensity $I_i$ which can be directly expressed as a inverse of a cost function $f(x_i)$. To effectively explore considered search space $S$ it is assumed that each firefly $i$ is changing its position iteratively taking into account two factors: attractiveness of other swarm members with higher light intensity i.e. $I_j > I_i, \forall j = 1, ...m, j \neq i$ which is varying across distance and a fixed random step vector $u_i$. It should be noted as well that if no brighter firefly can be found only such randomized step is being used [8].

To summarize, when taking into consideration all above statements, the algorithm scheme established in [8] can be presented in the following pseudo-code form:

Firefly Algorithm for Constrained Optimization

**Input:**
$f(z)$, $z = [z_1, z_2, ..., z_n]^T$ {*cost function*}
$S = [a_k, b_k], \forall k = 1, ..., n$ {*given constraints*}
$m, \beta_0, \gamma, \min u_i, \max u_i$ {*algorithm's parameters*}
**Output:**
$x_{i^{min}}$ {*obtained minimum location*}

```
begin
   for i=1 to m do
      x_i ← Generate_Initial_Solution ()
   end
   repeat
      i^min ← arg min_i f(x_i)
      x_{i^min} ← arg min_{x_i} f(x_i)
      for i=1 to m do
         for j=1 to m do
            if f(x_j) < f(x_i) then                {move firefly i towards j}
               r_j ← Calculate_Distance (x_i,x_j)
               β ← β_0 e^{-γ r_j}                  {obtain attractiveness}
               u_i ← Generate_Random_Vector (min u_i, max u_i)
               for k=1 to n do
                  x_{i,k} ← (1 − β)x_{i,k} + βx_{j,k} + u_{i,k}
               end
            end
         end
      end
      u_{i^min} ← Generate_Random_Vector (min u_i, max u_i)
      for k=1 to n do
         x_{i^min,k} ← x_{i^min,k} + u_{i^min,k}   {best firefly should move randomly}
      end
   until stop condition true
end
```

In the next part of the paper some technical details of the algorithm will be considered. A closer look will be taken at such important issues as the sensitivity of the parameters, their influences on the convergence rate of the algorithm, the potential improvement, and further development. For more detailed theoretical considerations, MATLAB code and convincing two dimensional demonstrations of the algorithm performance one could refer to the pioneer publication already mentioned [8].

### 2.2 Technical Details

The algorithm presented here makes use of a synergic local search. Each member of the swarm explores the problem space taking into account results obtained by others, still applying its own randomized moves as well. The influence of other solutions is controlled by value of attractiveness (2). It can be adjusted by modifying two parameters: its maximum value $\beta_0$ and an absorption coefficient $\gamma$.

The first parameter describes attractiveness at $r_j = 0$ i.e. when two fireflies are found at the same point of search space $S$. In general $\beta_0 \in [0, 1]$ should be used and two limiting cases can be defined: when $\beta_0 = 0$, that is only non-cooperative distributed random search is applied and when $\beta_0 = 1$ which is equivalent to the scheme of cooperative local search with the brightest firefly strongly determining other fireflies positions, especially in its neighborhood [8].

On the other hand, the value of $\gamma$ determines the variation of attractiveness with increasing distance from communicated firefly. Using $\gamma = 0$ corresponds to no variation or constant attractiveness and conversely setting $\gamma \to \infty$ results in attractiveness being close to zero which again is equivalent to the complete random search. In general $\gamma \in [0, 10]$ could be suggested [8]. It is more convenient, however, to derive $\gamma$ value specifically for the considered problem. Such customized absorption coefficient should be based on the "characteristic length" of the optimized search space. It is proposed here to use:

$$\gamma = \frac{\gamma_0}{r_{max}} \tag{3}$$

or:

$$\gamma = \frac{\gamma_0}{r_{max}^2} \tag{4}$$

wheras $\gamma_0 \in [0, 1]$ and:

$$r_{max} = \max d(x_i, x_j), \forall x_i, x_j \in S\ . \tag{5}$$

Efficiency of both techniques introduced here will be experimentally evaluated in the next Section.

Finally one has to set random step size i.e. its lower and upper bounds ($\min u_i$, $\max u_i$) and define the method of its generation. In [8] it was proposed to use $\min u_i = -0.5\alpha$ and $\max u_i = 0.5\alpha$, with $\alpha \in [0, 1]$ being algorithm's parameter. In consequence $u_i$ for each search space dimension $k$ is supposed to be generated according to:

$$u_{i,k} = \alpha(rand - \frac{1}{2})\ . \tag{6}$$

with $rand \sim U(0, 1)$ - a random number obtained from the uniform distribution. Here it is suggested to use alternative approach i.e. to define random vector as a fraction of firefly distance to search space boundaries:

$$u_{i,k} = \begin{cases} \alpha\, rand_2(b_k - x_{i,k}) & if\ \operatorname{sgn}(rand_1 - 0.5) < 0 \\ -\alpha\, rand_2(x_{i,k} - a_k) & if\ \operatorname{sgn}(rand_1 - 0.5) \geq 0 \end{cases} \tag{7}$$

with two uniform random numbers $rand_1$, $rand_2$ obtained similarly as above.

In the end it could be noted that computational complexity of the algorithm under consideration is $O(m^2)$, so using larger population size leads to substantial increase in calculation time. It can, however, bring significant benefits in terms of algorithm's performance, especially when some deterministic technique of initial swarm displacement is being employed. In the paper, simple random dislocation of fireflies in $S$, instead of such strategy, is being assumed.

## 3 Numerical Experiments and Parameter Studies

The performance of the presented technique was verified experimentally using its MATLAB implementation and a set of 14 benchmark problems (for a detailed list please refer to the Appendix). All tests were conducted for a fixed number of algorithm iterations $l$ and repeated in 100 independent trials with different random number generator seeds. As problems are characterized by different scales on the cost function it was more convenient to use ranking of different algorithm's variants instead of direct analysis of quality indexes $|f_{min} - f(x_{i^{min}})|$. It means that each problem was considered separately with tested configurations being ranked by their performance. Then the final comparison was carried out using medians of obtained ranks. Due to space limitations only most representative results are presented in the paper. The full set of simulation data can be found on the first author's web site (`http://www.ibspan.waw.pl/~slukasik`)

### 3.1 Population Size

Firstly, the influence of swarm size on the algorithm efficiency was analyzed. For such purpose a fixed number of cost function evaluations (2000) and FA variants with $m = \{6, 8, 10, ..., 200\}$ were assumed and suitably decreasing number of iterations $l$ were compared. To make such a comparison more representative variants characterized by significant rounding errors of $m\,l$ product i.e. $||m\,l| - m\,l| > 20$ were rejected. All tests were conducted for $\beta_0 = 1$, $\alpha = 0.01$ and fixed $\gamma = 1$. Although general remarks on optimal number of fireflies cannot be made, two tendencies can be observed. For difficult optimization tasks, such as instances no 2, 3, 4, 8, 9, 13, 14 it is always a better option to use a maximum number of fireflies. It is an observation based on medians of ranks for those problems depicted on Fig. 1. Still, there exists a set of relatively easy optimization problems like 1, 5, 6, 7, 10, 11, 12, 13 where optimal number of fireflies could be found, e.g. the problem of Sphere function minimization is solved most effectively by a set of 28 fireflies as shown on Fig. 2.

The aforementioned remarks should be accompanied by an additional look on associated calculation time which, as noted before, increases significantly with size of the swarm. Taking it into account it is advisable to use reasonable population of 40-50 fireflies and refrain from applying such rule only for more complicated optimization tasks. It is worth noting that similar remarks have been made in [9] with reference to the Particle Swarm Optimization technique.

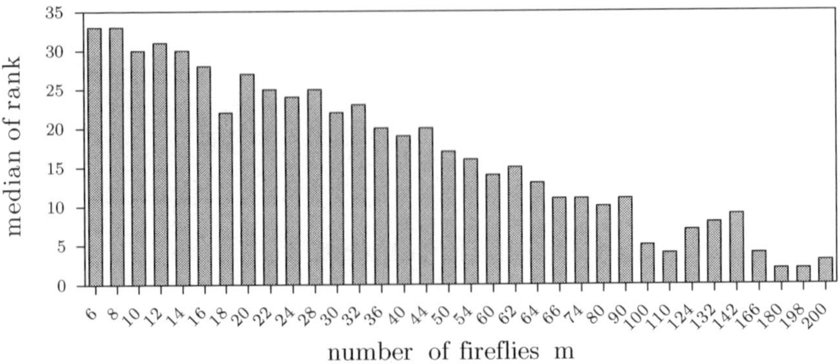


**Fig. 1.** Median of performance ranks for varying population size (problems no 2, 3, 4, 8, 9, 13, 14

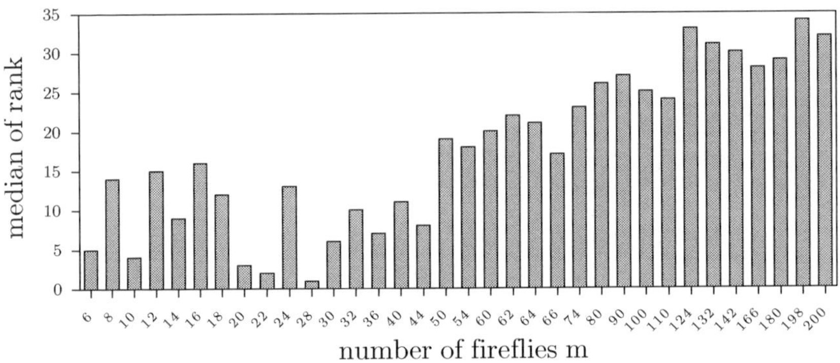


**Fig. 2.** Median of performance ranks for varying population size (problem: 12)

### 3.2 Maximum of Attractiveness Function

In the second series of computational experiments the influence of $\beta_0$ value on the algorithm performance was studied. Testing runs were conducted for $\beta_0 = \{0, 0.1, 0.2, ..., 1.0\}$ with other parameters fixed i.e. $m = 40, l = 250, \alpha = 0.01$ and $\gamma = 1$. Again, each of the tested configurations was ranked by its performance in the considered problem and median of such rank is reported (see Fig. 3).

It is observed that the best option is to use maximum attractiveness value $\beta_0 = 1$ which implies the strongest dependence of fireflies' positions on their brighter neighbors location.

### 3.3 Absorption Coefficient and Random Step Size

Finally changes in the algorithm's performance with varying absorption coefficient $\gamma$ and random step size $\alpha$ were under investigation. Maximum attractiveness $\beta_0 = 1$ was used, with population size $m = 40$ and iteration number $l = 250$.

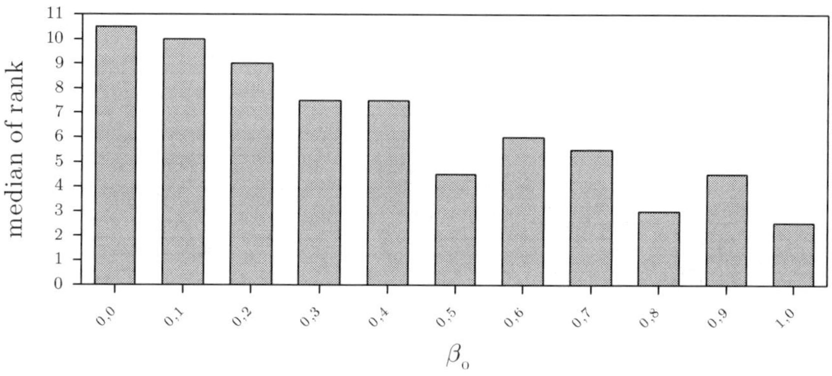


**Fig. 3.** Median of performance ranks with varying maximum of attractiveness function

Firefly Algorithm variants with $\alpha = \{0.001, 0.01, 0.1\}$ and $\gamma = \{0.1, 1.0, 10.0\}$ were tested. Additionally two problem-related techniques of obtaining absorption coefficient (Eq. (3) and (4)) were considered (with $\gamma_0 = \{0.1, 0.2, ..., 1.0\}$), so the overall number of examined configurations reached 75.

The obtained results indicate that for the examined optimization problems variants of the algorithm with $\alpha = 0.01$ are the best in terms of performance. Furthermore it could be advisable to use adaptable absorption coefficient according to (3) with $\gamma_0 = 0.8$ as this configuration achieved best results in the course of executed test runs. Although proposed technique of $\gamma$ adaptation in individual cases often performs worse than fixed $\gamma$ values it has an advantage to be automatic and "tailored" to the considered problem.

## 4 Comparison with Particle Swarm Optimization

Particle Swarm Optimization is a swarm-based technique introduced by Kennedy and Eberhart [11]. It has been intensively developed recently with research studies resulting in numerous interesting both theoretical and practical contributions [9]. The Particle Swarm Optimizer was studied in continuous optimization context in [12], with suggested variants and recommended parameter values being explicitly given.

Experiments reported here involved a performance comparison of Firefly Algorithm with such advanced PSO algorithm defined with constriction factor and the best parameters set suggested in conclusion of [12]. Both algorithm were executed with the same population size $m = 40$, iteration number $l = 250$ and the test was repeated 100 times for its results to be representative. The obtained results are presented in Tab. 1. It contains algorithms' performance indices given as average difference between a result obtained by both techniques $f(x_{i^{min}})$ and actual minimum of cost function $f(x^*)$ with standard deviations given for reference. For the Firefly Algorithm results obtained by the best configuration

**Table 1.** Performance comparison of Firefly Algorithm and Particle Swarm Optimization technique

| Problem | $avg.\vert f(x^*) - f(x_{imin})\vert \pm std.dev.\vert f(x^*) - f(x_{imin})\vert$ PSO | FA($\gamma_0 = 0.8$) | FA(best) |
|---|---|---|---|
| 1 | 5.75E-21 ± 5.10E-02 | 2.35E-04 ± 3.76E-04 | 1.01E-04 ± 1.78E-04 |
| 2 | 4.98E+02 ± 1.59E+02 | 6.59E+02 ± 2.25E+02 | 4.14E+02 ± 2.98E+02 |
| 3 | 0.00E+00 ± 0.00E+00 | 1.42E-01 ± 3.48E-01 | 3.05E-02 ± 1.71E-01 |
| 4 | 3.04E+01 ± 1.09E+01 | 1.63E+01 ± 5.78E+00 | 2.78E+00 ± 5.39E-01 |
| 5 | 4.63E-02 ± 2.59E-02 | 1.56E-01 ± 4.56E-02 | 1.48E-01 ± 4.17E-02 |
| 6 | 2.31E-01 ± 6.17E-01 | 1.10E+00 ± 6.97E-01 | 5.66E-01 ± 3.36E-01 |
| 7 | 1.16E-17 ± 6.71E-17 | 7.90E-05 ± 6.82E-05 | 4.14E-05 ± 2.17E-05 |
| 8 | 2.15E-06 ± 6.51E-15 | 2.80E-02 ± 7.78E-02 | 2.21E-03 ± 1.11E-03 |
| 9 | 5.84E-02 ± 5.99E-02 | 2.18E-01 ± 1.67E-01 | 2.18E-02 ± 2.33E-02 |
| 10 | 2.86E+00 ± 3.52E+00 | 3.09E+00 ± 3.72E+00 | 1.69E-01 ± 1.08E+00 |
| 11 | 4.00E-18 ± 9.15E-18 | 1.63E-06 ± 1.60E-06 | 1.44E-06 ± 1.47E-06 |
| 12 | 7.31E-22 ± 1.61E-21 | 1.59E-06 ± 1.73E-06 | 6.17E-07 ± 6.23E-07 |
| 13 | 4.61E+00 ± 1.09E-01 | 6.46E-03 ± 6.46E-02 | 2.75E-06 ± 5.24E-06 |
| 14 | 3.63E-03 ± 6.76E-03 | 2.59E-02 ± 3.96E-02 | 1.00E-02 ± 1.28E-02 |

selected in Section 3.3 are presented, as well as the best result obtained for each test problem by one of the 75 algorithm variants considered in the same Section.

It is noticeable that Firefly Algorithm is outperformed repeatedly by Particle Swarm Optimizer (PSO was found to perform better for 11 benchmark instances out of 14 being used). It is also found to be less stable in terms of standard deviation. It is important to observe though that the advantage of PSO is vanishing significantly (to 8 instances for which PSO performed better) when one relates it to the best configuration of firefly inspired heuristic algorithm. Consequently, such comparison should be repeated after further steps in the development of the algorithm being described here will be made. Some possible improvements contributing to the Firefly Algorithm performance progress will be presented in the final Section of the paper.

## 5 Conclusion

Firefly Algorithm described here could be considered as an unconventional swarm-based heuristic algorithm for constrained optimization tasks. The algorithm constitutes a population-based iterative procedure with numerous agents (perceived as fireflies) concurrently solving a considered optimization problem. Agents communicate with each other via bioluminescent glowing which enables them to explore cost function space more effectively than in standard distributed random search.

Most heuristic algorithms face the problem of inconclusive parameters settings. As shown in the paper, coherent suggestions considering population size and maximum of absorption coefficient could be derived for the Firefly Algorithm. Still the algorithm could benefit from additional research in the adaptive establishment of absorption coefficient and random step size. Furthermore some additional features like decreasing random step size and more sophisticated procedure of initial solution generation could bring further improvements in the algorithm performance. The algorithm could be hybridized together with other heuristic local search based technique like Adaptive Simulated Annealing [13]. Firefly communication scheme should be exploited then on the higher level of the optimization procedure.

**Acknowledgments.** Authors would like to express their gratitude to Dr Xin-She Yang of Cambridge University for his suggestions and comments on the initial manuscript.

## References

1. Encyclopædia Britannica: Firefly. In: Encyclopædia Britannica. Ultimate Reference Suite. Encyclopædia Britannica, Chicago (2009)
2. Babu, B.G., Kannan, M.: Lightning bugs. Resonance 7(9), 49–55 (2002)
3. Fraga, H.: Firefly luminescence: A historical perspective and recent developments. Journal of Photochemical & Photobiological Sciences 7, 146–158 (2008)
4. Lewis, S., Cratsley, C.: Flash signal evolution, mate choice, and predation in fireflies. Annual Review of Entomology 53, 293–321 (2008)
5. Leidenfrost, R., Elmenreich, W.: Establishing wireless time-triggered communication using a firefly clock synchronization approach. In: Proceedings of the 2008 International Workshop on Intelligent Solutions in Embedded Systems, pp. 1–18 (2008)
6. Jumadinova, J., Dasgupta, P.: Firefly-inspired synchronization for improved dynamic pricing in online markets. In: Proceedings of the 2008 Second IEEE International Conference on Self-Adaptive and Self-Organizing Systems, pp. 403–412 (2008)
7. Krishnanand, K., Ghose, D.: Glowworm swarm based optimization algorithm for multimodal functions with collective robotics applications. Multiagent and Grid Systems 2(3), 209–222 (2006)
8. Yang, X.S.: Nature-Inspired Metaheuristic Algorithms. Luniver Press (2008)
9. Eberhart, R.C., Shi, Y.: Computational Intelligence: Concepts to Implementations. Morgan Kaufmann, San Francisco (2007)
10. Karaboga, D., Basturk, B.: A powerful and efficient algorithm for numerical function optimization: artificial bee colony (ABC) algorithm. Journal of Global Optimization 39(3), 459–471 (2007)
11. Kennedy, J., Eberhart, R.: Particle swarm optimization. In: IEEE International Conference on Neural Networks, 1995. Proceedings, vol. 4, pp. 1942–1948 (1995)
12. Schutte, J.F., Groenwold, A.A.: A study of global optimization using particle swarms. Journal of Global Optimization 31(1), 93–108 (2005)
13. Ingber, L.: Adaptive simulated annealing (ASA): lessons learned. Control & Cybernetics 25(1), 33–55 (1996)
14. Himmelblau, D.M.: Applied Nonlinear Programming. McGraw-Hill, New York (1972)
15. Schwefel, H.P.: Numerical Optimization of Computer Models. John Wiley & Sons, Inc., Chichester (1981)

16. Easom, E.: A survey of global optimization techniques. Master's thesis, University of Louisville (1990)
17. Mühlenbein, H., Schomisch, D., Born, J.: The Parallel Genetic Algorithm as Function Optimizer. Parallel Computing 17(6-7), 619–632 (1991)
18. Griewank, A.: Generalized descent for global optimization. Journal of Optimization Theory and Applications 34, 11–39 (1981)
19. Rosenbrock, H.H.: State-Space and Multivariable Theory. Thomas Nelson & Sons Ltd. (1970)
20. Neumaier, A.: Permutation function, http://www.mat.univie.ac.at/~neum/glopt/my_problems.html
21. Törn, A., Žilinskas, A.: Global Optimization. Springer, Heidelberg (1989)
22. Shekel, J.: Test functions for multimodal search techniques. In: Proceedings of the 5th Princeton Conference on Infomration Science and Systems, pp. 354–359 (1971)
23. Jansson, C., Knüppel, O.: Numerical results for a self-validating global optimization method. Technical Report 94.1, Technical University of Hamburg-Harburg (1994)
24. Bilchev, G., Parmee, I.: Inductive search. In: Proceedings of IEEE International Conference on Evolutionary Computation, pp. 832–836 (1996)
25. Michalewicz, Z.: Genetic Algorithms + Data Structures = Evolution Programs. Springer, Heidelberg (1998)
26. Neumaier, A.: Powersum function, http://www.mat.univie.ac.at/~neum/glopt/my_problems.html

## Appendix: Benchmark Instances

| No. | Name | $n$ | $S$ | $f(x^*)$ | Remarks |
|---|---|---|---|---|---|
| 1 | Himmelblau [14] | 2 | (-6,6) | 0 | four identical local minima |
| 2 | Schwefel [15] | 10 | (-500,500) | 0 | several local minima |
| 3 | Easom [16] | 2 | (-100,100) | -1 | a singleton maximum in a horizontal valley |
| 4 | Rastrigin [17] | 20 | (-5.12,5.12) | 0 | highly multimodal and difficult |
| 5 | Griewank [18] | 5 | (-600,600) | 0 | several local minima |
| 6 | Rosenbrock [19] | 4 | (-2.048,2.048) | 0 | long curved only slightly decreasing valley, unimodal |
| 7 | Permutation [20] | 2 | (-2,2) | 0 | function parameter $\beta$=10 |
| 8 | Hartman3 [21] | 3 | (0,1) | -3.862 | 4 local minima |
| 9 | Hartman6 [21] | 6 | (0,1) | -3.322 | 6 local minima |
| 10 | Shekel [22] | 4 | (0,10) | -10.536 | 10 local minima |
| 11 | Levy 10 [23] | 5 | (-10,10) | 0 | $10^5$ local minima |
| 12 | Sphere [24] | 3 | (-5.12,5.12) | 0 | unimodal |
| 13 | Michalewicz [25] | 5 | (0,$\pi$) | -4.687 | $n!$ local minima |
| 14 | Powersum [26] | 4 | (0,2) | 0 | singular minimum among very flat valleys |

# Distributed Data Mining Methodology with Classification Model Example

Marcin Gorawski and Ewa Płuciennik-Psota

Silesian University of Technology, Institute of Computer Science, Akademicka 16, 44-100 Gliwice, Poland
{Marcin.Gorawski,Ewa.Plucienník}@polsl.pl

**Abstract.** Distributed computing and data mining are two elements essential for many commercial and scientific organizations. Data mining is a time and hardware resources consuming process of building analytical models of data. Distribution is often a part of organizations' structure. Authors propose methodology of distributed data mining by combining local analytical models (build in parallel in nodes of a distributed computer system) into a global one without necessity to construct distributed version of data mining algorithm. Different combining strategies are proposed and their verification method as well. Proposed solutions were tested with data sets coming from UCI Machine Learning Repository.

**Keywords:** Distributed data mining, data analysis, data models, analytical SQL.

## 1 Introduction

Distributed computing is nowadays almost ubiquitous. So is data mining – time and hardware resources consuming process of building data analytical models. More and more organizations operate globally so distributed data processing becomes not only a method of improving performance but a necessity. Distributed data mining can be realized in a few ways [1]. Most popular are meta-learning [2] and distributed versions of local data mining algorithms [3]. We propose methodology of combining local analytical models (build in parallel in nodes of distributed computer system) into a global one without necessity of constructing distributed version of data mining algorithm and with compact global model form. Basic assumptions for proposed solution are (i) a complete horizontal data fragmentation and (ii) a model form understood by a human being. Building global data model consists of two stages. In the first one local models are built in a parallel manner. Second one consists of combining these models into a global data picture. All steps of combining methodology are presented with classification model example.

An analysis of a huge amount of information is feasible only if information systems are used. First, information needs to be accumulated and stored in a persistent structure enabling effective data access and management. The most popular storage way is relational database. Regarding the above mentioned circumstances we have also proposed an enhancement of SQL for data mining of a distributed data structure.

N.T. Nguyen, R. Kowalczyk, and S.-M. Chen (Eds.): ICCCI 2009, LNAI 5796, pp. 107–117, 2009.

# 2 Distributed Data Mining Methodology

Proposed methodology of building distributed data mining models can be divided into three steps:

- Choosing and implementing a data mining algorithm.
- Selecting a data mining model quality measure.
- Working out combining strategy and its quality measure.

In case of a classification quality evaluation can be done by testing local and global models. Verifying combining strategy quality (in other words quality of a global model created with a given strategy, in case of classification model accuracy) is a very important step because it allows to answer a basic question: Are created global models good enough to meet analyst expectations?

## 2.1 Distributed Data Mining Scheme

Fig. 1 shows proposed distributed data mining schema. In the first stage local models are built in parallel using some data mining algorithm (local algorithm). Then local models are combined into a global data picture. Before using any combining strategy it needs to be verified.

We propose two kinds of verifying values. First one is a quality of model built using all data with local algorithm (control model). It is assumed that model built that way is the most accurate [4] so it can be stated that a global model with approximate

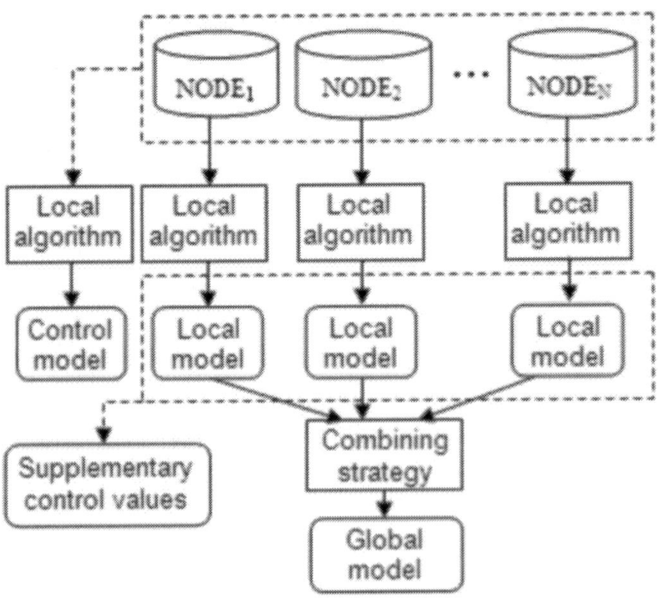


**Fig. 1.** Distributed data mining scheme

accuracy is very good. In other words this is the value we aspire. On the other hand we need to pinpoint some additional quality values which can help, among others to set the lower limit of the quality. We call these values supplementary control values (SCV). As a main SVC we have chosen average quality of local models. In case of classification additionally we also consider quality of global model built as a meta-classifier based on local model voting.

## 2.2 Combining Strategies

We have assumed data model in form of a rule set. For n-dimensional analysis space with $A_1 \ldots A_n$ attributes taking values from sets, adequately $a_1, \ldots, a_n$ and target attribute C taking values from set c, rule has a form: if P then Cl.

$$\begin{gathered} P\colon A_1 \in (\text{sub}(a_1)) \text{ and } A_2 \in (\text{sub}(a_2)) \text{ and } \ldots \text{and } A_n \in (\text{sub}(a_n)),\ Cl\colon C = c^i, \\ \text{where sub(a) denotes a subset of set a, and } c^i \text{ denotes an element of set C,} \\ P\colon \text{sub}(a_i) - \text{denotes a part of premise of rule P for attrib } A_i\colon A_i \in \text{sub}(a_i). \end{gathered} \tag{1}$$

We have defined three types of rules which can appear in a global classification model. The first ones are conflicting rules which coverage can have common parts and conclusions are different. The second ones are subrules – rules which coverage is a subset of coverage of another rule with the same conclusion. The third ones are friendly rules – rules which coverage can have common parts with rules with the same conclusion Rule's conclusion is a value of a class to which object that meets premise condition will be classified. These kinds of rules have to be taken into consideration during combining local models' process and applying global model. Below we present a brief description of our combining strategies. Full algorithms and definitions can be found in [5].

The first stage of combining local classification models is a creation of a set including all rules from local models – basic rule set. The second stage consists of iterative applying of conflicting rules and subrules finding operators on rules set. If conflicting rules are found, one of the conflict eliminating strategies should be applied. The stopping process condition is an unchangeable number of rules in consecutive iterations.

### 2.2.1 Bag Strategy S_1

Every rule has a one vote. During testing or applying model objects are assigned to the class with the highest number of votes (rule's voting). The main disadvantage of this strategy is a relatively large number of rules – it is a sum of number of local model rules (subrules are not absorbed). However, the strategy does not require any rule processing.

### 2.2.2 Absorbing Strategy S_2

Rules which have superior rules are eliminated. Superior rule absorbs its subrules. Rule's strength is appropriately updated – it equals one plus number of absorbed subrules. Rule's strength decides about voting results.

### 2.2.3. Conflicting Rules Separation Strategies

This type of strategies requires some conflicting rules separation methods. Rules separation consists of a separation (moving or removing common parts) of sets of values for one or more attributes in rules premises. Within the confines of rules separation strategies the following variants were created:

- S_31 – removing common parts for all attributes from a rule in which attributes have more numerous value's sets,
- S_32 – removing common parts from both rules,
- S_33 – removing common part for the least informative attribute,
- S_34 – removing common part for the most informative attribute,
- S_35 – removing common part for an attribute with the least numerous common part,
- S_36 – removing common part for an attribute with the most numerous common part.

Attribute's informativity is evaluated with mutual information for an attribute and a target value (class) $I(A_i, C)$ [6].

During rules separation process also another approach can be applied. Common parts are removed from both conflicting rules and a new rule's premise is created from these parts. In this case we have to deal with the situation where for every pair of conflicting rules a new rule can be created. Of course there is another issue, namely the evaluation of a conclusion for the new rule. The way of finding a solution for that issue implies consecutive combining strategies. Value of a new rule's conclusion can be set with value of the conclusion for the rule with most similar premise from basic rule set. We have defined rules premises similarity measure on the basis of Jaccard coefficient:

$$sim(P_1, P_2) = \frac{\sum_{i=1}^{n} w_i \dfrac{|P_1{:}sub(a_i) \cap P_2{:}sub(a_i)|}{\dfrac{|P_1{:}sub(a_i)| + |P_2{:}sub(a_i)|}{2}}}{n} \quad (2)$$

Evaluation of similarity of a new rule's premise and another rule's premise can be done as follows:

- S_41 – unweighted similarity: measure (2) with $w_i = 1$,
- S_421 – weighted similarity: measure (2) with $w_i = 1/|a_i|$ ,
- S_422 – weighted similarity: measure (2) with $w_i = I(A_i, C)$.

$w_i$ denotes a given attribute's weight, for $w_i$ = 1similarity measure takes values from 0 to1.

Fig. 2 shows three rules $R_1$, $R_2$, $R_3$ with premises defined using two attributes $A_1$ (with $a_1$ set containing 7 elements) and $A_2$ (with $a_2$ set containing 3 elements). Common parts of rules $R_2$ and $R_3$ with rule $R_1$ are denoted with crossing lines.

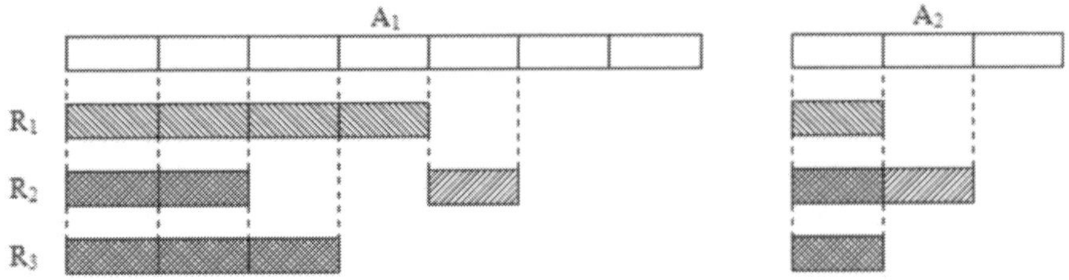


**Fig. 2.** Rules similarity example

The unweighted rules similarities are computed as follows:

$$sim(R_1, R_2) = \frac{\frac{\frac{2}{7}}{2} + \frac{\frac{1}{3}}{2}}{2} = 0{,}619 \qquad sim(R_1, R_3) = \frac{\frac{\frac{3}{7}}{2} + \frac{\frac{1}{2}}{2}}{2} = 0{,}928$$

The weighted rules similarities with $w_i = 1/|a_i|$ are following:

$$sim_w(R_1, R_2) = \frac{\frac{\frac{1}{7}*\frac{2}{7}}{2} + \frac{\frac{1}{3}*\frac{1}{3}}{2}}{2} = 0{,}207 \qquad sim_w(R_1, R_3) = \frac{\frac{\frac{1}{7}*\frac{3}{7}}{2} + \frac{\frac{1}{3}*\frac{1}{2}}{2}}{2} = 0{,}228$$

In case of a weighted similarity the difference between both similarities is smaller. There are two main reasons for this. First, weighted similarity measure has smaller values than unweighted one (weights are fractions). Second, set $a_1$ has much more elements than set $a_2$ so attribute $A_1$ has lower influence on final similarity measure value. Similarity with $w_i = I(A_i, C)$ can be evaluated only for a specific training set.

A new rule conclusion can be also evaluated on the basis of some global statistics, for example a class distribution in a set of objects (from local training sets) belonging to the new rule's coverage. In this strategy (S_43) a conclusion is determined based on most frequent class from the class distribution.

### 2.2.4 Conflicting Rules Separation Example

Rules separation methods are presented with example conflicting rules generated during cars set classification (test sets are described in section 3). We have two conflicting rules to be separated with strategy group S_3*. Common parts of rules premises are marked out with bold font:

- Rule A: if BUYING in (high, **low**, **med**, vhigh) and DOORS in (**2**, 3) and LUG_BOOT in (big, **med**, small) and MAINT in (**low**) and PERSONS in (**more**) and SAFETY in (**med**) then class = unacc.
- Rule B: if BUYING in (**low**, **med**) and DOORS in (**2**) and LUG_BOOT in (**med**) and MAINT In (**low**) and PERSONS in (**more**) and SAFETY in (**med**) then class = acc.

BUYING and LUG_BOOT attributes were added to a rule A premise to obtain a rule complete form. Rule has a complete form if all classification model attributes (analysis dimensions) are presented in a rule's premise. If some rule is not completed we need to add missing attributes to the rule's premise. This process does not influence the rule's coverage because all possible attributes values are included in the rule's premise (for example, LUG_BOOT has three possible values: 'big', 'med' and 'small'). Attributes list ordered by attributes informativity is as follows: DOORS 0,007; LUG_BOOT 0,0315; MAINT 0,0957; BUYING 0,1378; PERSONS 0 ,2145; SAFETY 0 ,256.

Rules premises after applying S_31 separation:

- Rule A: if BUYING in (high, vhigh) and DOORS in (3) and LUG_BOOT in (big, small) and MAINT in (**low**) and PERSONS in (**more**) and SAFETY in (**med**).
- Rule B: if BUYING in (**low, med**) and DOORS in (**2**) and LUG_BOOT in (**med**) and MAINT in (**low**) and PERSONS in (**more**) and SAFETY in (**med**).

It is unfeasible to separate premises for attributes MAINT, PERSONS, SAFETY because they are identical. You cannot remove common parts of premises for two rules if for one of them (or both) removing common part results in empty set, for example: {med} – {med} = ∅ for attribute BUYING in rule B. For S_32 strategy result is the same as for S_31 strategy.

Rules premises after separation with S_33 strategy:

- Rule A: if BUYING in (high, **low**, **med**, vhigh) and DOORS in (3) and LUG_BOOT in (big, **med**, small) and MAINT in (**low**) and PERSONS in (**more**) and SAFETY in (**med**).
- Rule B: if BUYING in (**low, med**) and DOORS in (**2**) and LUG_BOOT in (**med**) and MAINT In (**low**) and PERSONS in (**more**) and SAFETY in (**med**).

Common part (element 2) for attribute with the smallest value of attribute informativity (DOORS) was removed from the rule A premise.

Rules premises after S_34 separation:

- Rule A: if BUYING in (high, vhigh) and DOORS in (**2**,3) and LUG_BOOT in (big, **med**, small) and MAINT in (**low**) and PERSONS in (**more**) and SAFETY in (**med**).
- Rule B: if BUYING in (**low, med**) and DOORS in (**2**) and LUG_BOOT in (**med**) and MAINT In (**low**) and PERSONS in (**more**) and SAFETY in (**med**).

Attributes with the biggest value of informativity are: SAFETY, PERSONS, BUYING, etc. For the first two rules separation is unfeasible. Separation is conducted with BUING attribute – from rule A premise we remove common part: {'med', 'low'} set.

Rules premises after applying S_35 strategy:

- Rule A: if BUYING in (high, **low**, **med**, vhigh) and DOORS in (3) and LUG_BOOT in (big, **med**, small) and MAINT in (**low**) and PERSONS in (**more**) and SAFETY in (**med**).
- Rule B: if BUYING in (**low, med**) and DOORS in (**2**) and LUG_BOOT in (**med**) and MAINT In (**low**) and PERSONS in (**more**) and SAFETY in (**med**).

For all attributes except BUYING common parts in rules premises have the same amount of elements. Removing is possible for DOORS or LUG_BOOT attributes.

Rules premises after applying S_36 strategy:

- Rule A: if BUYING in (high, vhigh) and DOORS in (**2**,3) and LUG_BOOT in (big, **med**, small) and MAINT in (**low**) and PERSONS in (**more**) and SAFETY in (**med**).

- Rule B: if BUYING in (**low**, **med**) and DOORS in (**2**) and LUG_BOOT in (**med**) and MAINT In (**low**) and PERSONS in (**more**) and SAFETY in (**med**).

Attribute BUYING has the most numerous common part.

For strategies S_4* group we take into consideration two rules C and D:

- Rule C: if BUYING in (**med**) and DOORS in (**2**, 3, 4, 5more) and LUG_BOOT in (**small**) and MAINT in (**low**) and PERSONS in (**more**) and SAFETY in (**high**) then class = good.
- Rule D if BUYING in (**med**) and DOORS in (**2**) and LUG_BOOT in (**small**) and MAINT in (**low**, med, high, vhigh) and PERSONS in (**more**) and SAFETY in (**high**) then class = unacc .

Above presented rules are separated by removing common parts from premises (for all attributes for which it is feasible). We obtain two rules C' and D' and a new rule E premise:

- Rule C': if BUYING in (**med**) and DOORS in (3, 4, 5more) and LUG_BOOT in (**small**) and MAINT in (low) and PERSONS in (**more**) and SAFETY in (**high**) then class = good.
- Rule D': if BUYING in (**med**) and DOORS in (2) and LUG_BOOT in (**small**) and MAINT in (med, high, vhigh) and PERSONS in (**more**) and SAFETY in (**high**) then class = unacc.
- Rule E premise (common parts of premises of rules C and D): if BUYING in (med) and DOORS in (2) and LUG_BOOT in (small) and MAINT in (low) and PERSONS in (more) and SAFETY in (high).

Rule E conclusion is set using strategies from S_4* group:

- S_41: then class = unacc,
- S_421: then class = unacc,
- S_422: then class = good,
- S_43: then class = unacc.

## 2.3 EwSQL (Explore with SQL)

SQL enhancement enabling data mining are still not standardized. In 2001 the SQL/MM standard has emerged [7]. This standard defines issues concerning data mining using object features of databases (defining object types and methods) and XML as a data structure definition language. The standard does not define any new SQL constructions for data mining. Such constructions arise within the confines of scientific researches, for example DMQL [8], MSQL [9], Mine Rule [10], [11], MineSQL [12]. Aside from SQL enhancements solutions based on XML have arisen: MQL [13] and KDDML [14]. Some data mining algorithms (clustering [15] and classification [16]) have been implemented with standard SQL constructions. However none of the mentioned above solutions is adapted to a distributed data mining. We have proposed SQL enhancement for distributed data mining EwSQL (Explore with SQL) which allows creating and running data mining queries in a distributed environment using described above strategies. Detailed syntax and data mining query examples can be found in [17].

# 3 Tests

During tests we have used three sets from the UCI Machine Learning Repository [18]. The first one (adult) contains USA census data from 1994 (two valued class, record number: 48 842). Second one's (votes) content encompasses congressional voting results in USA Congress in 1984 (two valued class, record number: 435). Third one (cars) contains data about some car features like price, security level, etc. (four valued class, record number: 1728). Every set was divided into a test set and a training set (used for a global control model). Then training set was divided into three sets with similar sizes. These sets were used for training local models combined then into a global model. During tests local classification models, global control models, global models based on voting and global combined models were built. All created models were then tested with the test sets. Local models were built using the ID3 algorithm, which does not require any parameters [19].

## 3.1 Tests Results

Table 1 contains best accuracy results for combining strategies and voting models in percentage of control model and average local models accuracy.

**Table 1.** Comparison of combining strategies accuracy with voting models

| Set | Max percentage of control model accuracy | | Max percentage of average local models accuracy | |
|---|---|---|---|---|
| | Voting | Strategies | Voting | Strategies |
| adult | 100,30 | 100,08 | 100,74 | 100,55 |
| votes | 100,77 | 100,77 | 106,27 | 106,27 |
| cars | 96,36 | 98,86 | 110,62 | 115,19 |

**Table 2.** Global models accuracy with respect to the control and local models accuracy

| Strategy/ set | Accuracy percentage with respect to control model | | | local models' avg. accuracy | | |
|---|---|---|---|---|---|---|
| | adult | votes | cars | adult | votes | cars |
| S_1 | 100.05 | 100.77 | 86.42 | 100.39 | 106.27 | 93.29 |
| S_2 | 99.27 | 86.81 | 103.88 | 99.61 | 91.55 | 112.14 |
| S_31 | 99.13 | 96.01 | 99.17 | 99.47 | 101.25 | 107.21 |
| S_32 | 99.20 | 96.01 | 98.79 | 99.54 | 101.25 | 106.85 |
| S_33 | 99.15 | 90.22 | 98.27 | 99.49 | 95.13 | 106.17 |
| S_34 | 99.18 | 94.18 | 95.46 | 99.51 | 99.32 | 103.13 |
| S_35 | 99.09 | 91.47 | 86.42 | 99.43 | 96.46 | 93.29 |
| S_36 | 99.21 | 89.15 | 86.42 | 99.55 | 94.01 | 93.29 |
| S_41 | 98.96 | 89.92 | 96.87 | 99.30 | 94.82 | 104.77 |
| S_43 | 99.64 | 99.22 | 94.04 | 99.98 | 104.62 | 101.70 |
| S_421 | 98.94 | 89.92 | 96.87 | 99.27 | 94.82 | 104.77 |
| S_422 | 98.92 | 90.70 | 97.51 | 99.26 | 95.64 | 105.45 |

Values greater than 100 denote then a given model accuracy is better than a control model accuracy (values equal to 100 mean the same accuracy for both models). As we can see best global model built using proposed strategies have accuracy comparable or better than voting models with respect to the control model. The average accuracy of local models with respect to the control model amounted to 99,66% for adult set, 94,83% for votes set and 76,88% for cars set.

For the adult set only strategy S_1 have reached accuracy slightly greater than the control model and the average accuracy of local models. However, the accuracy of remain strategies was about 99% of the control model. Moreover, it has to be pointed out that strategies were better in classification if it comes to class value '>50K', which was represented by about 24% examples in the training set. In case of the votes set also only S_1 strategy was more accurate then control model, but strategies S_31, S_32, and S_43 had accuracy greater than the average from local models. Remain strategies' accuracies oscillated in range of 86% - 99% of control model accuracy. Some strategies were better in classification of object with class value 'republican' (about 38% examples), other in classification of object with class value 'democrat' (about 62% examples). For the cars set the best strategy was S_2 with almost 104% accuracy of control model. Most of the strategies had accuracy greater than local models average. Remain strategies' accuracies range from 86% to 99% of control model accuracy. Strategies S_31 and S_32 were two times better in classification of objects with class 'vgood' and 1,5 times better for objects with class 'good' than control model. Both class values were represented by less than 4% of examples. In case of cars set the best strategy was S_2. For all strategies accuracy for class 'unacceptable' were better than for the control model.

Comparing accuracies of voting models and models built with described above strategies in relation to the control models and average local models accuracies it can be stated that:

- on the average, combining strategies for all sets had a slightly worse accuracy than voting models except for the cars set,
- maximal accuracies for strategies oscillate around voting model accuracy,
- maximal accuracies for strategies in relation to the control model were slightly worse than the voting model accuracy for adult set, in case of remaining sets they were equal or better,
- global models' accuracies range from 86% to 101% of the control model accuracy.

If it comes to the acceleration of model building result were very encouraging. Distributed processing allowed reaching more than three-fold acceleration for the adult set and two-fold for the cars set in comparing of control model building time. In case of the votes set distributed model building last 60% of time needed for control model building.

## 4 Summary

In this article we have presented new methods for creating distributed classification models. Distributed data mining scheme with global models quality verification was described. Proposed solution combines distributed data mining with simplicity of SQL queries. The EwSQL allows simple and clear definition of data mining models features.

Presented solutions were implemented using Oracle DBMS, PL/SQL and Java and tested using official test sets. Accuracy of created global models was compared with accuracy of control models, voting models and average accuracy of local models. We also examined time needed for building and combining models. Obtained results allow stating that proposed solution is very promising and improves model build efficiency without significant loss of accuracy. Of course proposed strategies required further examination especially for conflicting rules modifications. More tests with different nodes number and naturally distributed data are needed.

## References

1. Chan, P., Prodromidis, A., Stolfo, G.: Meta-learning in distributed data mining systems: Issues and approaches. In: Advances of Distributed Data Mining. AAAI Press, Menlo Park (2000)
2. Guo, Y., Reuger, S.M., Sutiwaraphun, J., Forbes-Millot, J.: Meta-learning for parallel data mining. In: Proceedings of the 7th Parallel Computing Workshop (1997)
3. Caragea, D., Silvescu, A., Honavar, V.: Invited Paper. A Framework for Learning from Distributed Data Using Sufficient Statistics and its Application to Learning Decision Trees. International Journal of Hybrid Intelligent Systems 1(2), 80–89 (2004)
4. Grossman, R., Turinsky, A.: A Framework for Finding Distributed Data Mining Strategies That Are Intermediate Between Centralized Strategies and In-Place Strategies. In: Proceedings of Workshop on Distributed and Parallel Knowledge Discovery at KDD-2000, pp. 1–7 (2000)
5. Gorawski, M., Pluciennik, E.: Analytical Models Combining Methodology with Classification Model Example. In: First International Conference on Information Technology, Gdansk (2008), `http://ieeexplore.ieee.org/xpl/freeabs_all.jsp?arnumber=4621623`, ISBN: 978-1-4244-2244-9
6. Pluim, J.P.W., Maintz, J.B.A., Viergever, M.A.: Mutual-information-based registration of medical images: a survey. IEEE Transactions on Medical Imaging 22(8), 986–1004 (2003)
7. International Organization for Standardization (ISO). Information Technology, Database Language, SQL Multimedia and Application Packages, Part 6: Data Mining Draft Standard No. ISO/IEC 13249-6 (2003)
8. Han, J., Fu, Y., Wang, W., Koperski, K., Zaiane, O.: DMQL: A Data Mining Query Language for Relational Database. In: Proc. of the SIGMOD 1996 Workshop on Research Issues on Data Mining and Knowledge Discovery, Montreal, Canada, pp. 27–34 (1996)
9. Imieliński, T., Virmani, A.: MSQL: A Query Language for Database Mining. Data Mining and Knowledge Discovery 3(4), 373–408 (1999)
10. Meo, R., Psaila, G., Ceri, S.: A New SQL-like Operator for Mining Association Rules. In: Proc. 22nd VLDB Conference, Bombaj, India, pp. 122–133 (1996)
11. Meo, R., Psaila, G., Ceri, S.: An Extention to SQL for Mining Association Rules. Data Mining and Knowledge Discovery 2(2), 195–224 (1998)
12. Morzy, T., Zakrzewicz, M.: SQL-like language for database mining. In: Proc. of the First East-European, Symposium on Advances in Databases and Information Systems - ADBIS, St. Petersburg, vol. 1, pp. 311–317(1997)

13. Baglioni, M., Turini, F.: MQL: An algebraic query language for knowledge discovery. In: Cappelli, A., Turini, F. (eds.) AI*IA 2003. LNCS, vol. 2829, pp. 225–236. Springer, Heidelberg (2003)
14. Romei, A., Ruggieri, S., Turini, F.: KDDML: a middleware language and system for knowledge discovery in databases. Data & Knowledge Engineering 57(2), 179–220 (2006)
15. Cereghini, P., Ordonez, C.: SQLEM: Fast Clustering in SQL using the EM Algorithm. In: SIGMOD Conference, pp. 559–570 (2000)
16. Dunemann, O., Sattler, K.: SQL Database Primitives for Decision Tree Classifiers. n: Proc. of the 10th ACM CIKM Int. Conf. on Information and Knowledge Management, pp. 379–386 (2001)
17. Gorawski, M., Pluciennik, E.: Distributed Data Mining by Means of SQL Enhancement. In: Meersman, R., Tari, Z., Herrero, P. (eds.) OTM-WS 2008. LNCS, vol. 5333, pp. 34–35. Springer, Heidelberg (2008)
18. Asuncion, A., Newman, D.J.: UCI Machine Learning Repository. Irvine, CA: University of California, School of Information and Computer Science (2007), `http://www.ics.uci.edu/~mlearn/MLRepository.html`
19. Quinlan, R.: Induction of Decision Trees. Machine Learning 1, 81–106 (1986)

# A Token-Based Mutual Exclusion Approach to Improve Collaboration in Distributed Environments

Mauricio Paletta[1] and Pilar Herrero[2]

[1] Centro de Investigación en Informática y Tecnología de la Computación (CITEC)
Universidad Nacional Experimental de Guayana (UNEG)
Av. Atlántico, Ciudad Guayana, 8050, Venezuela
mpaletta@uneg.edu.ve

[2] Facultad de Informática, Universidad Politécnica de Madrid (UPM)
Campus de Montegancedo S/N; 28660; Madrid, Spain
pherrero@fi.upm.es

**Abstract.** This paper presents a token-based mutual exclusion algorithm designed to be used in Collaborative Distributed Environments by means of multi-agent architecture. This research makes use of a variation of the token-based Naimi-Tréhel's algorithm by considering only three messages in total as well as reducing the needed information sharing. This ensures to improve the effectiveness of agents' communication resources and therefore to improve collaboration in these environments.

**Keywords:** Collaboration, distributed environment, mutual exclusion, multi-agent system.

## 1 Introduction

Collaborative Distributed Environments (CDEs) are those in which multiple users, usually agents, in remote locations participate in shared activities aiming to achieve a common goal. Usually, part of these share activities have to be doing by only one agent at any time by defining a Critical Section (CS). Therefore, a Mutual Exclusion Mechanism (MEM) is required. Aiming to reduce communication resources among agents in the Multi-Agent System (MAS), the coordination algorithm used for the MEM has to be efficient and effective regardless of the number of nodes in the CDE.

However, state of the art MEM best algorithms have a logarithmic message complexity ($O(log(N))$ regarding the number of nodes $N$ as well as to the requirements to store path histories over their lifetime. This paper presents a new MEM for agents' coordination in MAS. This MEM needs only 3 messages regardless of $N$ as well as reducing the needed information sharing. This proposal is used in a MAS-based collaborative process whose details can be reviewed at [1], [2].

The remainder of this paper is organized as follows. Section 2 briefly describes MEM background including Naimi-Tréhel's algorithm as well as some related work associated to this algorithm. The MEM approach presented in this paper is showed in

N.T. Nguyen, R. Kowalczyk, and S.-M. Chen (Eds.): ICCCI 2009, LNAI 5796, pp. 118–127, 2009.

Section 3. Section 4 presents the performance evaluation results doing to the method. Finally, the last section includes the conclusions and outgoing future research related to this work.

## 2 Mutual Exclusion Algorithm Background and Related Work

A distributed MEM ensures that exactly one process can execute the CS at any given time (safety property) and that all CS requests will eventually be satisfied (liveness property) [3]. They can be divided into two groups: permission-based [4], [5], [6] and token-based [7], [8], [9]. Permission-based algorithms are established on the principle that a node should only enter the CS after having received permission from all the other nodes.

In the token-based algorithms a unique system-wide token is shared among all nodes and its possession gives a node the exclusive right to enter the CS. These algorithms usually have an average message cost lower than the ones that permission-based have [3]. They are focused on scalability, hence they minimize communication, sometimes at the expense of over utility [10].

According to Gommands et al [11], a token provides a flexible mechanism that allows the right to access a light-path to be associated with a request from an application. It can be defined as "a shared abstract permission that is presented as part of an access request in each domain". Moreover, token-based methods have been shown to be effective in large-scale team coordination tasks, including the task of information sharing [12].

In the particular case of the Naimi-Tréhel's algorithm [8], it is a token-based algorithm which keeps two data-structures: 1) a logical dynamic tree in which the root is always the last node that will get the token among the current requesting ones; 2) a distributed queue which holds CS requests that have not been satisfied yet.

Initially in this method, the root of the tree is the token holder, elected among all nodes. Each node $i$ into the tree keeps the local variable *last* which points to the last probable owner of the token. Each node $k$ into the queue keeps the variable *next* which points to the next node to whom the token will be granted after $k$ leaves the CS. When a node $i$ wants to enter the CS, it sends a request to its *last*. Node $i$ then sets its *last* to itself and waits for the token. *Last*, then becomes the new root of the tree. Upon receiving $i$'s token request message, node $j$ can take one of the following actions depending on its state: 1) if $j$ is not the root of the tree, it forwards the request to its *last* and then updates its *last* to $i$; 2) if $j$ is the root, it updates its *last* to $i$ and if it holds an idle token, it sends the token to $i$; however, if $j$ holds the token but is in the CS or is waiting for the token, it just sets its *next* to $i$. After executing the CS itself, $j$ will send the token to its *next*.

Regarding to related works associated with the Naimi-Tréhel's algorithm, author in [13] presents an extension to this algorithm in which a token request is associated with a priority and the algorithm satisfies first the requests with the higher priority. A similar strategy is adopted in [14] in which authors treat intra-cluster requests before inter-cluster ones.

Several authors have proposed hierarchical approaches for combining different mutual exclusion algorithms. Detailed information on this topic can be consulted in [3] (Section 5).

On the other hand, most of the existing work on multi-agent coordination (an extensively studied area of MAS) does not scale grade well very large teams. Distributed constraint-based algorithms [15], [16] have high communication requirements that get dramatically worse as the team size in increased. Combinatorial auctions [17] have an exponential number of possible combinations of bids, and frequently use centralized auctioneers that can become severe bottlenecks. Swarminspired approaches [18] have been used for large-scale coordination, but the behavior can be inefficient.

Next section presents the token-based approach proposed in this paper starting with a briefly description of the MAS-based collaboration process where this mechanism is been used.

## 3 The Token-Based Approach

This section presents the token-based mutual exclusion algorithm proposed in the paper. On one hand, the section contains a briefly explanation of the collaborative process by means of MAS architecture where the algorithm is used. On the other hand details of the proposed mechanism are presented.

### 3.1 The Collaborative Process

The collaborative process used for this research [1], [2] is based on the concept of awareness by using the key awareness concepts originally proposed by Herrero et al in [19], [20]. It has a CDE containing a set of $n$ nodes $N_i$ ($1 \leq i \leq n$) and $r$ items or resources $R_j$ ($1 \leq j \leq r$). These resources can be shared as a collaborative mechanism among different nodes (such as power, disk space, data and/or applications, and so on). It has:

1) $N_i$.*Focus*($R_j$): It can be interpreted as the subset of the space (environment/ medium) on which the agent in $N_i$ has focused his attention aiming for interaction/collaboration with, according to the resource (collaboration item) $R_j$.
2) $N_i$.*NimbusState*($R_j$): Indicates the current grade of collaboration that $N_i$ can give over $R_j$. It could have three possible values: *Null*, *Medium* or *Maximum*. If the current grade of collaboration $N_i$ that is given about $R_j$ is not high, and this node could collaborate more over this resource, then $N_i$.*NimbusState*($R_j$) will get the *Maximum* value. If the current grade of collaboration $N_i$ that is given about $R_j$ is high but $N_i$ could collaborate more over this resource, then $N_i$.*NimbusState*($R_j$) would be *Medium*. Finally, if $N_i$ is overloaded, its *NimbusState*($R_j$) would be *Null*.
3) $N_i$.*NimbusSpace*($R_j$): It Represents the subset of the space (CDE) where $N_i$ is currently collaborating with $R_j$. It will determine those nodes that could be taken into account for new collaborative necessities.
4) $R_j$.*AwareInt*($N_a$, $N_b$): This concept quantifies the degree of collaboration over $R_j$ between a pair of nodes $N_a$ and $N_b$. It is manipulated via *Focus* and *Nimbus*

(*State* and *Space*), requiring a negotiation process which is part of the collaboration model where the token-based mutual exclusion strategy presented in this paper is used. Following the awareness classification introduced by Greenhalgh [21], it could be *Full*, *Peripheral* or *Null*.

5) $N_i$.*RequiredTask*($R_1,\ldots,R_p$): Indicates that $N_i$ requires collaboration with all $R_j$ ($1 \le j \le p$) with no ordered sequence of resources.
6) $N_i$.*TaskResolution*($R_j$): Determines the score to collaborate $R_j$ in $N_i$. It is represented with a value within [0, 1]. The closer the value is to 0 the hardest it will be for $N_i$ to collaborate the $R_j$ necessity. The higher the value (closer to 1) is the complete willingness to collaborate.

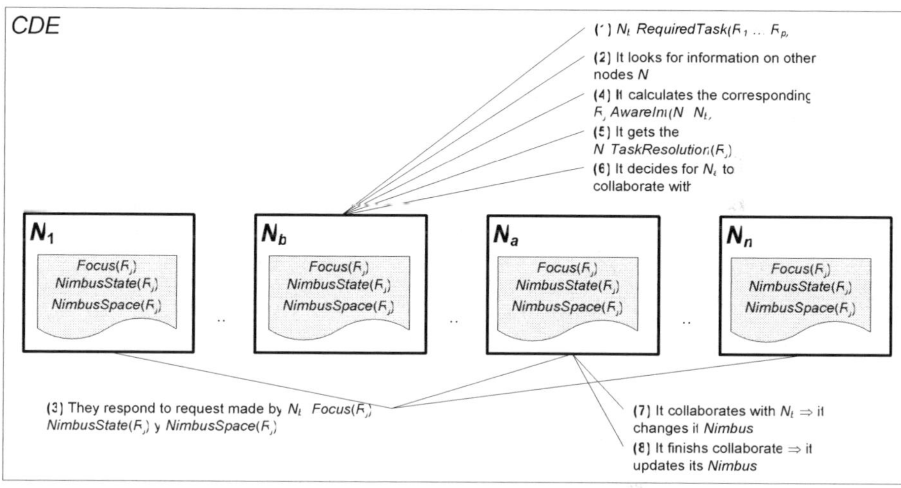


**Fig. 1.** MAS-based collaborative process

By using the concepts previously mentioned, the MAS-based collaborative process in the CDE follows these steps (see Fig. 1):

1) $N_b$ has the necessity to collaborate making use of the resources $R_1,\ldots,R_p$, so that, it generates a $N_b$.*RequiredTask*($R_1,\ldots,R_p$).
2) $N_b$ looks for the CDE current conditions (*Nimbus*/*Focus* related to the other nodes), given by $N_i$.*Focus*($R_j$), $N_i$.*NimbusState*($R_j$), and $N_i$.*NimbusSpace*($R_j$) $\forall i$, $1 \le i \le n$ and $\forall j$, $1 \le j \le r$. This information is used for deciding the most suitable node to collaborate related with any resource $R_j$.
3) Nodes in CDE respond to request for information made by $N_b$. This is done through the exchange of messages between agents by using a protocol defined in this MAS-based collaborative process [1].
4) The current awareness values given by $R_j$.*AwareInt*($N_i$, $N_b$) are obtained based on CDE current conditions.
5) $N_b$ gets the collaboration scores $N_i$.*TaskResolution*($R_j$).

6) For each resource $R_j$ ($1 \le j \le p$) included in $N_b$.*RequiredTask*($R_1,\ldots,R_p$), $N_b$ selects the node $N_a$ whose $N_a$.*TaskResolution*($R_j$) is the more suitable (greatest score). Then, $N_a$ will be the node in which $N_b$ should collaborate on resource $R_j$.
7) Once $N_a$ receives a request for cooperation, it updates its *Nimbus* (given by $N_a$.*NimbusState*($R_j$) and $N_a$.*NimbusSpace*($R_j$)) to maintain the CDE current conditions for future collaborations.
8) Once $N_a$ has finished collaborating with $N_b$ it must update its *Nimbus* (*State* and *Space*).

The nature of a node initiating a collaborative process to answer a *RequiredTask*($R_1,\ldots,R_p$), provokes a change in the conditions of the collaboration levels of the environmental nodes involved in the process. Since this information is required by the process of taking action, the levels of collaboration between the nodes turn into a CS.

### 3.2 The Mutual Exclusion Algorithm

The mutual exclusion strategy used in this MAS-based collaborative model previously explained is a variation of the Naimi-Tréhel's token-based algorithm.

In this approach, the token travels with a queue *Q* which has the nodes that require the exclusive use of the CS and haven't been able to satisfy that need. The mechanism works as follows (see diagram in Fig. 2):

- The first agent comes in to work in the CDE assuming that it has the token. Although by this time *Q* is empty.
- When an agent is required to collaborate according to one or more resources, the following scenarios are presented:
  - The corresponding node has the token, and it is not currently running the collaborative process. The node will then initiate the process.
  - The corresponding node has the token and it is currently collaborating. The id of the agent is added at the end of *Q*. This allows for a reasonable use of the CS and prevents the token to be held by a node for a long period of time.
  - The corresponding node is not in possession of the token. The agent will: 1) accumulate the resources for the future execution of the collaborative process; and 2) send to the network the message INFORM_IF indicating its need to use the CS.

• The token travels between nodes through the INFORM_REF message, being *Q* a parameter of this message. The first node in *Q* is supposed to be the one receiving the message. If this is true, then this entry is eliminated from *Q* and the required collaborative process starts.

• If a node receives the token (INFORM-REF message) and, it hasn't made any collaborative requirement or the agent did the requirements but it is not interested in collaborating any more, then it sends the token to the agent indicated as the first in *Q*. If *Q* is empty then the agent keeps the token until an INFORM-IF is received.

• When an agent holding the token receives an INFORM_IF, the corresponding agent's id associated to this message is placed at the end of *Q* and a corresponding acknowledgement is sent. This is done by sending back the message AGREE to it. If the agent receiving the message is not currently working in any collaborative process then the token is sent back (by using an INFORM-REF) to the agent that sent the INFORM_IF message.

• If the agent is holding the token, $Q$ is empty, and the node must finish its execution among the CDE, then the agent selects between one of the rest of the nodes of the environment by using a random uniform distribution. If a node is selected, the agent that holds the token will send it to the agent associated to the selected node.

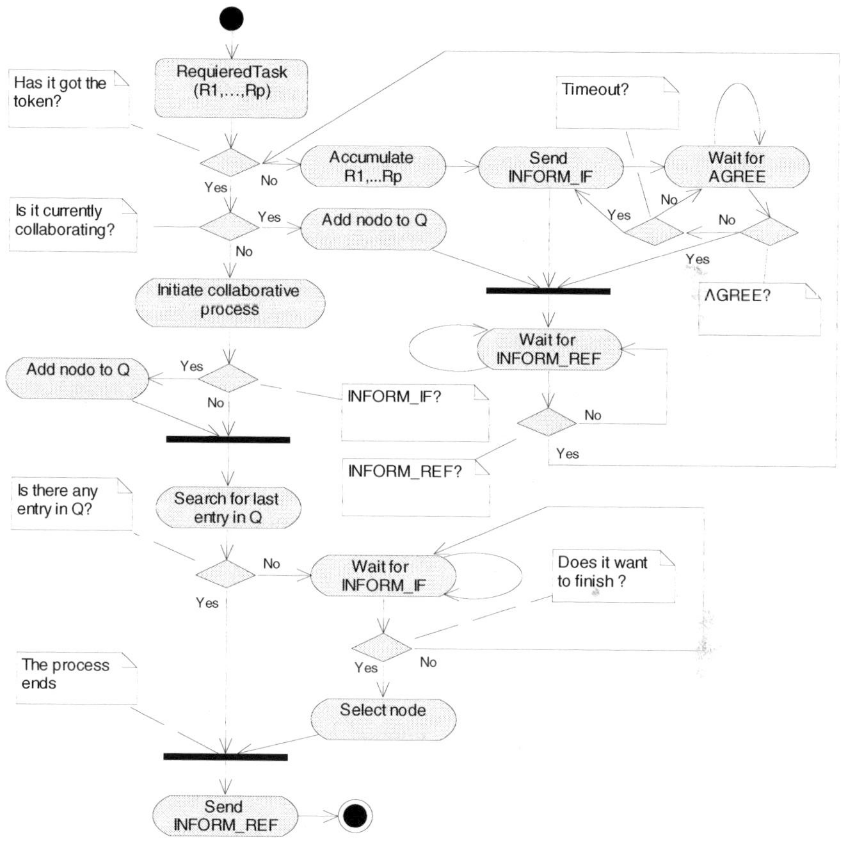


**Fig. 2.** Activity diagram of the mutual exclusion mechanism

In order to reduce the exchange of messages between agents INFORM_IF is created to be a UDP-based message [22]. Because this is a message with no guarantee of reaching the final destination, the confirmation message AGREE is needed. Once an agent sends the INFORM_IF message, it initiates a process of time-waiting for the AGREE message to come. If a timeout condition is reached the agent can now send the INFORM_IF again. INFORM_REF and AGREE are TCP-based messages instead UDP-based message. Therefore, these are the only three messages needed for synchronizing this mutual exclusion mechanism. Fig. 3 shows this protocol.

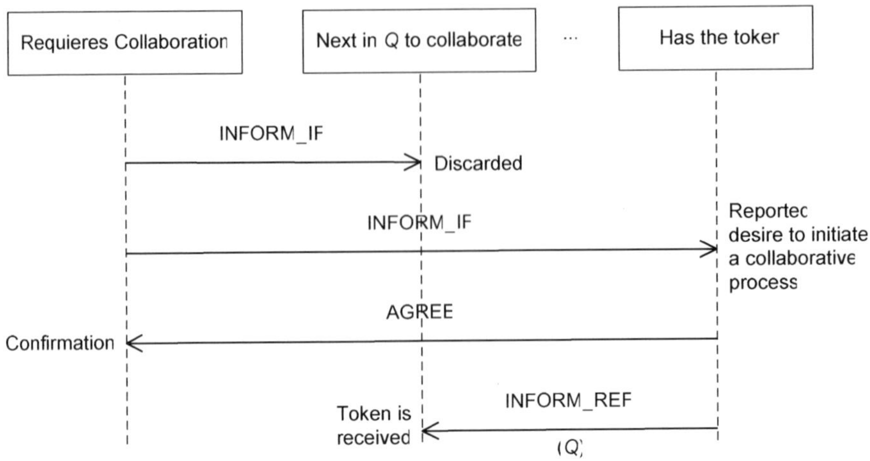


**Fig. 3.** Inter-agents message exchange for mutual exclusion mechanism

## 4 Evaluation

A Multi-Agent System (MAS) used for Collaborative Distributed Environments (CDEs) has been created to evaluate the Mutual Exclusion Mechanism (MEM) presented in this paper. This MAS-based platform has been implemented in JADE [23], [24]. The evaluation of the MEM was conducted with different scenarios by changing the quantity of nodes (agents) from 3 to 9 varying in 1 node for each case. This evaluation was done aiming to rate the MEM capability for managing growth in the environment conditions. Nodes were connected in a TCP/IP-based LAN (Local Area Network). It assumes that each node can directly communicate with any other.

For each scenario the experiment was conducted by creating 50 randomly CDE conditions, i.e. defining $N_a.Focus(R_j)$, $N_a.NimbusState(R_j)$ and $N_a.NimbusSpace(R_j)$ for each $N_a$ ($1 \leq a \leq n$) as well as defining a $N_b.RequiredTask(R_1,\ldots,R_p)$ also in a randomly way. For any CDE condition and each 250 millisecond occurring in 5 minutes, each node decides to enter to the CS with 33% using a uniform distribution. There were a total of 16,398 requests.

The results of the simulation indicate that the mechanism works with only an inter-agent exchange of 3 messages, as expected, and the mutual exclusion of the CS was achieved in 100%. Timeout for receiving AGREE confirmation, (set to 1 second in all the experiments) was reached only a couple of times on the 9 nodes scenario. So that, only in 2 of the 16,398 experimental conditions was necessary to use 4 instead of 3 messages. Therefore, to respond and process 16,398 requests for CS, the MEM required a total of 49,196 messages among the agents varying the number of nodes between 3 and 9. Same experiments with the Naimi-Tréhel's method require approximately 873,450 messages i.e., 1,675% more messages than with the method proposed in this paper. Fig. 4 shows the graphical comparison of the average messages sent by using the proposal method explained in this paper and the Naimi-Tréhel's method.

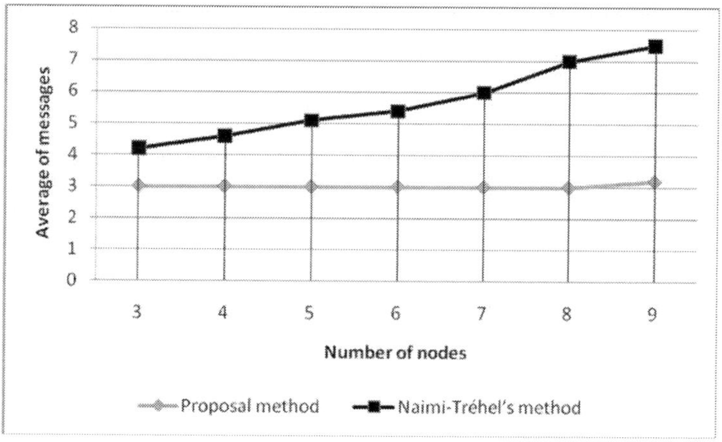


**Fig. 4.** Comparing average messages between the proposal method and the Naimi-Tréhel's method

## 5 Conclusion and Future Work

This paper presents a new mutual exclusion mechanism used for a multi-agent based system in collaborative distributed environments. As far as we know, differences between the proposed algorithm and other similar state of the art algorithms, including the Naimi-Tréhel's method, are the following:

1) The use of UDP-based message to reduce significantly the p2p messages, especially for large number of nodes in the environment.
2) The number of messages to synchronize the mutual exclusion does not depend on the number of nodes; it is fixed to 3.
3) The information that travels with the token it is represented by only one queue that requires little storage space.

Although this method has not yet been tested in a real CDE, it has been designed to be suitable for real environments. In fact the experimentation and validation carried out to present demonstrate that this MEM could be extended to real scenarios in CDE with no problems.

We are currently working on testing this method in real CDE. We are also working in using this strategy in collaborative grid environments. Previous results on this research can be consulted in [25].

## References

1. Paletta, M., Herrero, P.: Awareness-based Learning Model to Improve Cooperation in Collaborative Distributed Environments. In: Håkansson, A., et al. (eds.) Proc. 3rd International KES Symposium on Agents and Multi-agents Systems – Technologies and Applications (KES-AMSTA 2009). LNCS (LNAI), vol. 5559, pp. 793–802. Springer, Heidelberg (2009)

2. Paletta, M., Herrero, P.: Foreseeing Cooperation Behaviors in Collaborative Grid Environments. In: Proc. 7th International Conference on Practical Applications of Agents and Multi-Agent Systems (PAAMS 2009), Advances in Soft Computing, vol. 55, pp. 120–129. Springer, Heidelberg (2009)
3. Sopena, J., Arantes, L.B., Legond-Aubry, F., Sens, P.: The impact of clustering on token-based mutual exclusion algorithms. In: Luque, E., Margalef, T., Benítez, D. (eds.) Euro-Par 2008. LNCS, vol. 5168, pp. 565–575. Springer, Heidelberg (2008)
4. Lamport, L.: Time, clocks, and the ordering of events in a distributed system. Communications of the ACM 21(7), 558–564 (1978)
5. Maekawa, M.: A $\sqrt{N}$ algorithm for mutual exclusion in decentralized systems. ACM Transactions on Computer Systems 3(2), 145–159 (1985)
6. Ricart, G., Agrawala, A.: An optimal algorithm for mutual exclusion in computer networks. Communications of the ACM 24(1), 9–17 (1981)
7. Suzuki, I., Kasami, T.: A distributed mutual exclusion algorithm. ACM Transactions on Computer Systems 3(4), 344–349 (1985)
8. Naimi, M., Trehel, M., Arnold, A.: A log (N) distributed mutual exclusion algorithm based on path reversal. Journal of Parallel and Distributed Computing 34(1), 1–13 (1996)
9. Raymond, K.: A tree-based algorithm for distributed mutual exclusion. ACM Transactions on Computer Systems 7(1), 61–77 (1989)
10. Xu, Y., Scerri, P., Sycara, K., Lewis, M.: Comparing market and token-based coordination. In: Proc. 5th International Joint Conference on Autonomous Agents and Multiagent Systems, pp. 1113–1115. ACM, New York (2006)
11. Gommans, L., Xu, L., Demchenko, Y., Wan, A., Cristea, M., Meijer, R., Laat, C.: Multi-domain lightpath authorization, using tokens. Future Generation Computer Systems 25, 153–160 (2009)
12. Xu, Y., Scerri, P., Yu, B., Okamoto, S., Lewis, M., Sycara, K.: An integrated token-based algorithm for scalable coordination. In: Proc. 4th International Conference on Autonomous Agents and Multiagent Systems, pp. 407–414. ACM, New York (2005)
13. Mueller, F.: Prioritized token-based mutual exclusion for distributed systems. In: Proc. International Parallel Processing Symposium, pp. 791–795 (1998)
14. Bertier, M., Arantes, L., Sens, P.: Distributed mutual exclusion algorithms for grid applications: A hierarchical approach. Journal of Parallel and Distributed Computing 66, 128–144 (2006)
15. Mailler, R., Lesser, V.: Solving distributed constraint optimization problems using cooperative mediation. In: Proc. 3rd International Joint Conference on Autonomous Agents and Multiagent Systems, pp. 438–445. ACM, New York (2004)
16. Modi, P., Shen, W., Tambe, M., Yokoo, M.: An asynchronous complete method for distributed constraint optimization. In: Proc. 2nd International Joint Conference on Autonomous Agents and Multiagent Systems, pp. 161–168. ACM, New York (2003)
17. Hunsberger, L., Grosz, B.: A combinatorial auction for collaborative planning. In: Proc. 4th International Conference on Multi-Agent Systems, pp. 151–158. IEEE Computer Society, Los Alamitos (2000)
18. Cicirello, V., Smith, S.: Wasp nests for self-configurable factories. In: Proc. 5th International Conference on Autonomous Agents, pp. 473–480. ACM, New York (2001)
19. Herrero, P., Bosque, J.L., Pérez, M.S.: An Agents-Based Cooperative Awareness Model to Cover Load Balancing Delivery in Grid Environments. In: Meersman, R., Tari, Z., Herrero, P. (eds.) OTM-WS 2007, Part I. LNCS, vol. 4805, pp. 64–74. Springer, Heidelberg (2007)

20. Herrero, P., Bosque, J.L., Salvadores, M., S. Pérez, M.: Managing Dynamic Virtual Organizations to get Effective Cooperation in Collaborative Grid Environments. In: Meersman, R., Tari, Z. (eds.) OTM 2007, Part II. LNCS, vol. 4804, pp. 1435–1452. Springer, Heidelberg (2007)
21. Greenhalgh, C.: Large Scale Collaborative Virtual Environments. Doctoral Thesis, University of Nottingham (1997)
22. Postel, J.: User Datagram Protocol. USC/Information Sciences Institute, RFC 768 (August. 1980)
23. Bellifemine, F., Poggi, A., Rimassa, G.: JADE – A FIPA-compliant agent framework. In: Proc. International Conference on Practical Applications of Agents and Multi-Agent Systems (PAAM 1999), Telecom Italia internal technical report, pp. 97–108 (1999)
24. Chmiel, K., Gawinecki, M., Kaczmarek, P., Szymczak, M., Paprzycki, M.: Testing the Efficiency of JADE Agent Platform. In: Proc. 3rd International Symposium on Parallel and Distributed Computing (ISPDC), vol. 13(2), pp. 49–57. IEEE Computer Society, Los Alamitos (2005)
25. Paletta, M., Herrero, P.: Learning Cooperation in Collaborative Grid Environments to Improve Cover Load Balancing Delivery. In: Proc. IEEE/WIC/ACM Joint Conferences on Web Intelligence and Intelligent Agent Technology, E3496, pp. 399–402. IEEE Computer Society, Los Alamitos (2008)

# Discovering Medical Knowledge from Data in Patients' Files*

Magdalena Szymkowiak[1] and Beata Jankowska[2]

[1] Poznań University of Technology, Institute of Mathematics, ul. Piotrowo 3a, 60-965 Poznań, Poland
mszymkow@amu.edu.pl
[2] Institute of Control and Information Engineering, Pl. Sklodowskiej Curie 5, 60-965 Poznań, Poland
beata.jankowska@put.poznan.pl

**Abstract.** An Individual Patient's File (IPF) includes information about a patient and the course of his illness over many years. During this period, the doctor, taking permanent care of the patient, puts various data and observations into his IPF. In the paper we propose a kind of probabilistic and statistical analysis of the data stored in collections of IPFs. The collections can be made accessible by means of a medical Semantic Web. The proposed analysis requires the data to be expressed in the form of Bernoulli variables. As a result, it gives us medical hypotheses, with their premises-facts, conclusions and reliabilities. They can be presented in the form of production rules, used in medical Rule-Based Systems for automatic reasoning.

**Keywords:** individual patient's file, Bernoulli variable, conditional probability, statistical analysis, medical rule-based system.

## 1 Introduction

Due to the development of medical sciences, it is insufficient for a medicine doctor to use only his own, even if wide, experience while making diagnostic and therapeutic decisions. These decisions should also depend on empirical knowledge of other specialists, gained during their long-standing observation of many patients. For each patient, the observation results are stored in his Individual Patient's File (IPF). Unfortunately, the way of recording data in IPF is not subjected to any universal norms. The doctors operate with different categories of medical concepts, and with different relationships between these categories; summarizing – they use different medical ontologies. That is why, the records in IPFs differ one from the other in syntax and – first of all – in semantics.

In order to have a possibility of interpreting data stored in IPFs, it is first necessary to link them to proper ontologies [1]. Next, in order to have a possibility of merging them or processing in any way, the two following conditions must be satisfied: the availability of some reference medical ontology, and knowledge of partial mappings between the reference ontology and the source ontologies used in IPFs. It is equivalent to having: a

* The research has been partially supported by PUT under grant PB-43–066/09 BW and by the Polish Ministry of Science and Higher Education under grant N516 369536.

N.T. Nguyen, R. Kowalczyk, and S.-M. Chen (Eds.): ICCCI 2009, LNAI 5796, pp. 128–139, 2009.

medical Semantic Web based on Collections of Individual Patients Files (CIPFs), and metadata giving the interpretation to the data stored in particular IPFs.

The IPF is being built throughout the years for each patient covered by the health care. For example, in Poland, on the strength of Act on the Forms and Range of Medical Documentation, in the IPF there should be [2]:

- general data (some pieces of personal information about a patient);
- the data including essential information about patient's health and a detailed course of patient's diseases, recorded during his visits at specialists since the beginning of the patient's treatment;
- the data including information about all the health services provided for the patient.

Having the medical Semantic Web means having a broad access to CIPFs. A thorough analysis of huge data sets recorded in the CIPFs gives us a possibility to formulate reliable medical hypotheses concerning as well diagnostics as therapy and medical prognosis. The obtained hypotheses could be useful in medical practice, thus, in the paper, we point at some probabilistic and statistical tools that can facilitate designing of those hypotheses.

Let us remark that the data stored in CIPFs can have a quantitative or qualitative form. Besides, they are expressed by means of various individuals: keywords, numbers, chars, strings or logical values. Meanwhile, a probabilistic analysis is possible only in a case of data written in the same format. That is why, we suggest transforming all the data to the homogenous form of zero-one sequences, it means – sequences of Bernoulli variables. This requirement, although substantial and often difficult to satisfy in practice, results in efficient designing hypotheses and determining their importance, one in comparison to the other.

The obtained hypotheses can be expressed in the form of production rules, that compose the knowledge base of a medical Rule-Based System (RBS) aiding medical diagnosis and treatment.

## 2 Knowledge Base of Medical Rule-Based System

A medicine doctor acquires medical knowledge during theoretical lectures, while reading professional textbooks and journals, or studying medical cases. He also derives it from experience – both own and other specialists'. The doctor uses the knowledge in his everyday practice, while taking care of patients, making the treatment decisions and reporting them in the IPFs. If assume an ideal situation in which all the doctor's decisions are fully consistent with his medical knowledge, the IPFs can be regarded as a perfect image of this knowledge. Then, a methodological analysis of data from numerous IPFs could enable discovering global medical knowledge.

Let us consider the case of an expert in pulmonology, with 20 years period of service in a university clinic. The pulmonologist is interested in asthma disease [3], in particular – in bronchial asthma accompanied by diabetes. During his practise, he has recorded data of several hundred patients that have undergone treatment for this disease. As a result of his long-term experience, the doctor came to the conclusion that a routine treatment, consisting in administering intravenous glucocorticosteriods (GCS) to such patients, can give a side effect of coma. The doctor wants to verify the treatment hypothesis, so he acquaints statisticians with the contents of owned IPFs, holding all sensitive data in confidence. He asks them to check the implications between conjunctions of premises-facts among which there is one about administering of GCS to patients with asthma accompanied by diabetes, and a fixed conclusion about the side effect of coma. When the opportunity occurs, the statisticians can discover all dependencies between the data stored in the CIPFs that can be recognised as strong.

Having a medical Semantic Web means an access to data recorded by a great number of doctors (among them – pulmonologists) in CIPFs. A thorough probabilistic analysis of the data can give a result in a form of general medical hypotheses, corresponding to treatment decisions made with a reference to various diseases (among them – pulmonary diseases, also – bronchial asthma accompanied by diabetes). Such the hypotheses, expressed in a form of production rules [4], constitute knowledge bases of specialized medical RBSs.

## 2.1 Production Rule – Its Schema and Reliability

We assume that each hypothesis can be expressed in the form of production rule r which consists of a series of premises-facts $f_1, f_2, \ldots, f_v$ (specifying the rule's constraints) and conclusion c (formulating the rule's result):

```
r: it happens g-PM :
     if     (f1) and
            (f2) and
             ...  and
            (fv)
      then (c)  PM-c
```

**Fig. 1.** Schema of a production rule

Basing on the data stored in accessible CIPFs, apart from discovering and formulating premises-facts and conclusions of production rules, we will determine for each of them – using tools of probabilistic and statistical analysis – a probability mass of conclusion PM-c (reflecting the conclusion reliability), and a global probability mass g-PM (reflecting the reliability of the rule r as a whole). The factor g-PM shows the importance of the rule in comparison to other rules from the knowledge base of the medical RBS.

We will assume that for any patient each premise-fact $f_h$ for $h=1,2,\ldots,v$ is given a probability mass PM-$f_h$ that reflects its reliability. Then the final probability p of the conclusion c of the rule r is being calculated by the formula [5]:

$$p = (g - PM) \cdot \min\{(PM - f_h) : f_h \text{ is a premise} - \text{fact from the rule } r\} \cdot (PM - c) \ . \qquad (1)$$

## 2.2 Types of Rules in Medical RBS

The production rules in the knowledge base of a medical RBS, relate mostly to the diagnostic and therapeutic decisions and prognoses of the treatment of a particular disease.

The diagnostic rules correspond to the hypotheses of making specified diagnostic decisions. Some of them relate to indirect decisions (implementation of additional testing, commission of additional consultation), and other ones correspond to final decisions (proper identification of a particular disease). For a specified patient, the diagnostic decisions depend on the patient's characteristics recorded in his IPF in a form of various data (general data, data including information about the intensity of worrying symptoms, data including the results of essential diagnostic tests).

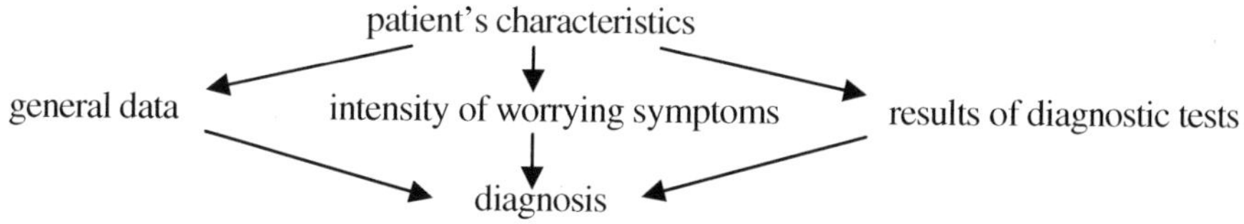


**Fig. 2.** Diagram of a diagnostic rule

The therapeutic rules correspond to the hypotheses of making therapeutic decisions (choosing the appropriate set and dosage of drugs, recommending a surgery, ordering an appropriate diet). For a specified patient, the therapeutic decisions usually depend on the patient's characteristics and the diagnosis that was made before.

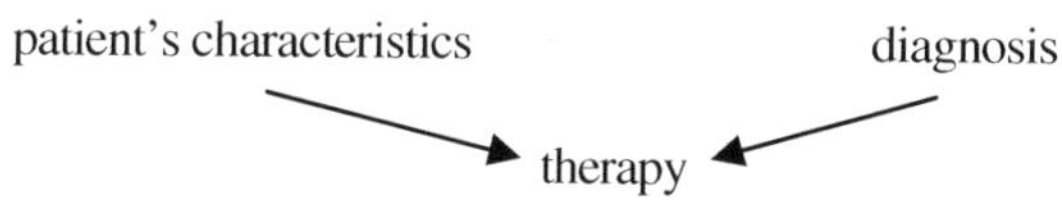


**Fig. 3.** Diagram of a therapeutic rule

Prognostic rules correspond to the hypotheses about the results of therapy (recession of burdensome symptoms, appearance of side effects, improvement of test results, necessity of hospitalization). They often concern the effects that are remote in time. For a specified patient, the treatment effects depend in general on the patient's characteristics, the diagnosis that was made, and the therapy that was used before.

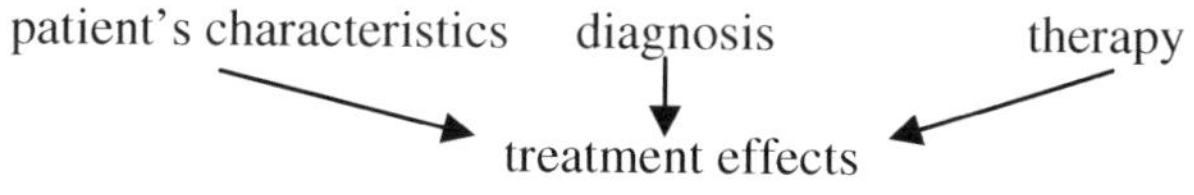


**Fig. 4.** Diagram of a prognostic rule

The diagnostics and therapeutic decisions made and used for each patient, as well as the treatment's effects, should be recorded, without a delay, in the patient's IPF to expand the patient's characteristics.

All diagnostic, therapeutic and prognostic rules, designed on the basis of data from CIPFs that we have an access to, will belong to the knowledge base of the medical RBS intended to aid the treatment of a particular disease.

### 2.3 Rules and Chains of Reasoning

The core of reasonings performed in an RBS [6] consists of successive firing of rules that are active. In a system with uncertainty, only that rule can be recognised as active in which each premise-fact $f_h$ for h=1,2,...,v is satisfied with a probability mass PM-$f_h$ not smaller than some appointed threshold value $\tau$:

$$\underset{h=1,2,\ldots,v}{\forall} \mathrm{PM} - f_h \geq \tau \, . \tag{2}$$

From among all active rules (so called agenda) only the one which is designated by the algorithm of agenda conflict resolution is fired at each step of reasoning. Most inference engines consider the rule's priority (represented by means of the factor g-PM) as the most important criterion of choosing rules from the agenda. Granting priorities to the rules is a task of a great importance – the priorities can influence the final results of reasonings performed in the RBS. An extensive discussion on designing the factor g-PM is placed in the section 6.

Let us notice that a final result of reasoning can be obtained in one step, after having applied the right production rule (e.g. a hypothetical diagnosis, made on the basis of general data, information about intensity of worrying symptoms, and results of essential diagnostic tests), or more frequently, in a number of steps, after having applied a long chain of reasoning. The hypothetical diagnosis, put into the knowledge base as a result of firing a diagnostic rule, can become the premise-fact sufficient for firing a therapeutic rule. Similarly, the conclusions of some diagnostic and therapeutic rules can become premises-facts activating prognostic rules. The premise-fact obtained in such a way is put into the knowledge base with a granted probability, equal to the probability p of the conclusion which becomes this premise-fact. The probability p, calculated by the formula (1), depends directly on the factor PM-c. Therefore and because of our earlier assumption (2) about the probability of the premise-fact in an active rule, also the factor PM-c must be designed carefully, with the awareness of its influence on the chains and the final results of reasonings (see section 5).

### 2.4 Exemplary Rules of Medical RBS

To illustrate the form and meaning of knowledge we are going to obtain from data stored in CIPFs, let us give a simple example. Let us consider the case of a 8-year-old boy with suspicion of moderate asthma exacerbation [3]. A routine treatment consists in administering intravenous glucocorticosteriods (GCS) to such patients. However, the boy suffers not only from moderate bronchial asthma but also from diabetes. In order to describe its present state of health (patient's characteristics), we must introduce a series of premises-facts into the knowledge base. Each of them is given a probability mass PM-f that reflects its reliability. The premises-facts could be as follows (Fig.5):

```
f1: (Age 8)    PM-f 1.0
f2: (Moderate_asthma)     PM-f 1.0
f3: (Wheezing)     PM-f 0.8
f4: (Difficulty_in_breath)     PM-f 0.9
f5: (Pulse 140)     PM-f   1.0
f6: (Diabetes)     PM-f 1.0
```

**Fig. 5.** Exemplary premises-facts from the knowledge base of a medical RBS

Then, in order to make for the boy an initial diagnosis and also provide some hypotheses about an efficient therapy and health prognosis, production rules to the point of bronchial asthma accompanied by diabetes will be needed. These are the rules we will try to obtain from data stored in CIPFs of asthmatic patients. They could be as follows:

```
r1: it happens g-PM 0.95 :
  if     (Asthma) and
         (Difficulty_in_breath) and
         (Pulse 120+)
  then   (Asthma_exacerbation)   PM-c 0.9

r2: it happens g-PM 0.98 :
  if     (Age_range <4 ; 18>) and
         (Moderate_asthma)and
         (Asthma_exacerbation)
  then   (Intravenous_GCS)   PM-c 0.97

r3: it happens g-PM 0.65 :
  if     (Age_range   <4 ; 18>) and
         (Asthma_exacerbation)   and
         (Diabetes) and
         (Intravenous_GCS)
  then   (Results_side_effect_coma)    PM-c   0.32
```

**Fig. 6.** Exemplary rules from the knowledge base of a medical RBS

After having fired the above rules, first $r_1$, $r_2$ and next $r_3$, we obtain the following conclusions about the 8-year-old boy with suspicion of moderate asthma exacerbation:

$c_1$: diagnosis of asthma exacerbation,

$c_2$: suggestion of administering intravenous GCS,

$c_3$: prognosis of a side effect of coma in case of administering intravenous GCS to him.

The final probabilities of the conclusions, calculated by means of the formula (1), will be as follows:

$$\begin{aligned} p_1 &= 0.95 \cdot \min\{1.0, 0.9, 1.0\} \cdot 0.9 = 0.77; \\ p_2 &= 0.98 \cdot \min\{1.0, 1.0, 0.77\} \cdot 0.97 = 0.73; \\ p_3 &= 0.65 \cdot \min\{1.0, 0.77, 1.0, 0.73\} \cdot 0.32 = 0.15. \end{aligned} \tag{3}$$

## 3 Probabilistic Tools for Production Rules Designing

Let r be the rule containing v premises-facts $f_h$ for h=1,2,…,v and the conclusion c. Let $B_h$ denotes the event that the premise-fact $f_h$ is satisfied. Then $B = \bigcap_{h=1}^{v} B_h$ means satisfying the conjunction of all premises-facts of the rule r. Let C denotes the event that the conclusion c is satisfied. Then:

$$P(C/B) = \frac{P(C \cap B)}{P(B)} = \frac{\frac{n(C \cap B)}{n(\Omega)}}{\frac{n(B)}{n(\Omega)}} = \frac{n(C \cap B)}{n(B)}. \tag{4}$$

It means that the conditional probability of C, given the event B, is equal to the ratio of the number of patients who fulfil both the conclusion and the conjunction of the premises-facts to the number of patients who fulfil the conjunction of the premises-facts of the rule r.

The formula (4) can be presented in a different way, where C' denotes the event that the conclusion c is not satisfied:

$$P(C/B) = \frac{n(B \cap C)}{n(B \cap C) + n(B \cap C')}. \tag{5}$$

The whole situation can be described by the following Fig.7., where Ω stands for the set of all patients examined:

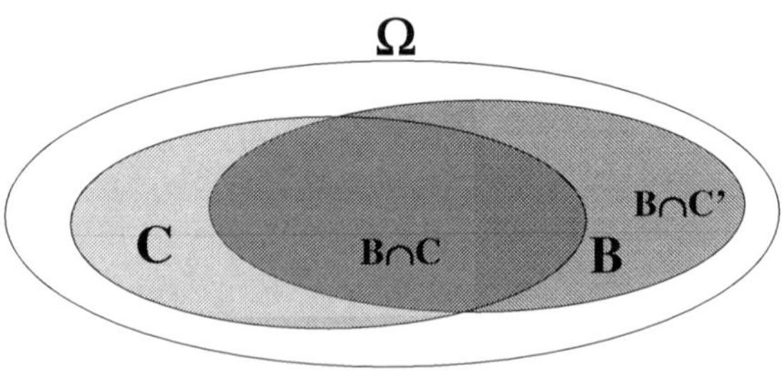


**Fig. 7.** Interdependence between the events B and C described above

The formula (5) proves that the smaller the number of patients who fulfill the conjunction of premises-facts and – simultaneously – do not fulfill the conclusion, the higher the value of conditional probability P(C/B) (in other words – the premises-facts should be specific for the patients fulfilling the conclusion).

$$P(C/B) \to 1 \text{ if } n(B \cap C') \to 0. \tag{6}$$

## 4 Designing Rules Based on Data Stored in CIPFs

Let us assume that each data in the patient's IPF has a form of Bernoulli variable X (it can take only two values, namely 0 and 1). For qualitative data, we determine the condition whose satisfaction or dissatisfaction decides whether the value of the variable is equal to 1 or 0, respectively. For quantitative data, we determine an interval that usually conforms to some medical standards, next we check if the value of the data satisfies or dissatisfies the requirements of the standards (whether it belongs or does not belong to the indicated interval), and finally we set the variable to 1 or 0, respectively.

Sometimes the transformation of data into a form of Bernoulli variables is needed. For example, in order to dichotomize the data "age", we have to decompose it into at least six levels: "paediatric", "adolescent", "young", "adult", "old", and "very old". Such the decomposition influences an essential increase in the number of variables. From the other side, it allows to formulate hypotheses more precisely.

Next, let us assume that CIPFs for a set $D$ of w patients $d_i$ (for i=1,2,..., n) is accessible. Let $X$ be a set of m Bernoulli variables $X_j$ (for j=1,2,...,m) representing data of each patient from the set $D$. We assume that for each patient from $D$, the values of all the variables from $X$ are determined.

Now, the CIPFs can be represented by means of a zero-one n-by-m matrix Z. In this matrix, the rows correspond to the patients from the set $D$ and the columns correspond to the variables from the set $X$, and each element of the matrix Z is as follows:

$$z_{ij} = \begin{cases} 0 & \text{if the value of variable } X_j \text{ relating to the patient } d_i \text{ is equal } 0 \\ 1 & \text{in opp. case.} \end{cases} \qquad (7)$$

In further considerations, one of the variables, for instance the variable $X_k$, will be taken as a representative of the conclusion c, and other variables, for instance the variables $X_1,X_2,\ldots,X_v$ will be taken as representatives of the premises-facts $f_1,f_2,\ldots,f_v$ of the rule r.

Within the set $D$, we can build a subset $B$ of those patients who fulfil the event B – the conjunction of all the premises-facts of the rule r. It can be done according to the formula:

$$d_i \in B \Leftrightarrow \underset{h=1,\ldots,v}{\forall} z_{ih} = 1 \Leftrightarrow \prod_{h=1}^{v} z_{ih} = 1\,. \qquad (8)$$

We claim that this set is not empty, which means that there is at least one patient who fulfils all the premises-facts of the rule.

Similarly, within the set $D$, we can build a subset $C$ of those patients who fulfil the event C – the conclusion c of the rule r. Then, $B \cap C$ stands for a subset of $D$ comprising all those patients who fulfill both the conjunction of all the premises-facts $f_1,f_2,\ldots,f_v$ and the conclusion c. The following exists:

$$d_i \in B \cap C \Leftrightarrow (\forall_{h=1,\dots,v} z_{ih} = 1 \wedge z_{ik} = 1) \Leftrightarrow z_{ik} \cdot \prod_{h=1}^{v} z_{ih} = 1. \tag{9}$$

We will wonder now, which of the variables are the best to be selected as the representatives of the conclusion and the premises-facts of the rule. First, we will propose two methods of selecting a variable $X_k$ to be the conclusion of the rule. The first one consists in consulting a doctor, who can point at the variables corresponding to clinically essential outcomes of clinical trials (in case of an expert in pulmonology mentioned in the section 2, it can be a conclusion about the side effect of coma). In turn, the second method consists in choosing the variable at random, and checking the relations between this randomly selected variable $X_k$ and the remaining variables from the set $X$. The second method allows us to discover all interesting (often unexpected) dependencies between the data stored in the CIPFs. The dependencies can be finally expressed in a form of general medical rules.

Next, we have to decide which variables are the best to be selected as representatives of the premises-facts of the rule. According to the formula (6), if we want to obtain a rule of a great importance, then we should select those variables $X_1, X_2, \dots, X_v$ as its premises-facts that are characteristic of the patients with $X_k = 1$ (only very few patients fulfil the conjunction of premises-facts corresponding to $X_1, X_2, \dots, X_v$ and, simultaneously, do not fulfil the conclusion corresponding to $X_k$). It means, that the following constraint concerning the matrix Z should be satisfied:

$$|B \cap C'| = \sum_{d_i \in D} \left( (1 - z_{ik}) \prod_{h=1}^{v} z_{ih} \right) \to 0. \tag{10}$$

## 5 Determining the Reliability Factor of Conclusion PM-c

For h=1,2,…,v, let $X_h$ be the variable corresponding to premise-fact $f_h$ of the rule r. The variable $X_h$ is represented by h-th column of the matrix Z. Let $X_k$ be the variable corresponding to conclusion c of the rule r. The variable $X_k$ is represented by k-th column of the matrix Z. We can determine the size of the set $B$ (the number of patients who fulfil the conjunction of all the premises-facts $f_1, f_2, \dots, f_v$ of the rule r) and the size of the set $B \cap C$ (the number of patients who fulfil both the conjunction of all the premises-facts $f_1, f_2, \dots, f_v$ and the conclusion c of the rule r):

$$N = |B| = \sum_{d_i \in D} \prod_{h=1}^{v} z_{ih} \qquad L = |B \cap C| = \sum_{d_i \in D} (z_{ik} \cdot \prod_{h=1}^{v} z_{ih}). \tag{11}$$

For determining the value of PM-c, the following formula is being proposed [5]:

$$(\text{PM} - \text{c}) = \frac{\hat{R}}{(1 + \frac{s}{2})} (1 - \alpha), \tag{12}$$

where $\hat{R}$ being a calculated risk of the rule's conclusion is, under consideration that the set $D$ of the patients as a random sample, the point estimate [7] of the probability P(C/B) defined by the formula (4). Then the risk $\hat{R}$ can be given as the ratio:

$$\hat{R} = \frac{L}{N} = \frac{|B \cap C|}{|B|}, \tag{13}$$

Next, the value s from the formula (12) is the span of a Confidence Interval CI including – with the probability $1-\alpha$ – an estimated risk R of the rule's conclusion for the whole population of patients fulfilling these premises-facts. The Confidence Interval CI [8] for the estimated risk R can be calculated from the inequality:

$$P\left(\hat{R} + u_{\frac{\alpha}{2}}\sqrt{\frac{\hat{R}(1-\hat{R})}{N}} < R < \hat{R} + u_{1-\frac{\alpha}{2}}\sqrt{\frac{\hat{R}(1-\hat{R})}{N}}\right) = 1-\alpha, \tag{14}$$

in which the symbols: $u_{\frac{\alpha}{2}}, u_{1-\frac{\alpha}{2}}$ stand for critical values of normal distribution N(0;1) of respective orders.

In order to make the PM-c maximal, the span s of CI should be minimal. It takes place when N – the size of the set $B$ – is large, which means when a great number of patients fulfils the conjunction of the premises-facts. Moreover, since $B \subset D$, it can take place only if the size of the set $D$ of patients, whose IPFs we have the access to, is large enough. This request can be satisfied thanks to the usage of our medical Semantic Web.

## 6 Determining the Global Reliability Factor of Rule g-PM

In turn, our proposal of estimating the global probability g-PM is based on the observation that the rule's importance depends on statistical parameters of evidence, some of which have special meaning. Namely, we will take into account three such parameters $s_1$, $s_2$ and $s_3$, of the semantics given below, and we will propose to calculate g-PM as a product of the form:

$$g-PM = s_1 \cdot s_2 \cdot s_3 \tag{15}$$

The value of $s_1$ depends on the number of patients (N=|$B$|) fulfilling the rule's premises-facts. We consider N to be large enough if it is not less than 100 (the border value of 100 is taken from the statistical considerations on the interval CI [9]). So we obtain:

$$s_1 = \begin{cases} \frac{N}{100} & \text{if } N < 100 \\ 1 & \text{if } N \geq 100 \end{cases} \tag{16}$$

The value of $s_2$ is functionally dependent upon the number of "significant" rule's premises-facts. We can consider the premises-facts on account of their direct influence on the on the number N – the size of the set $B$. We will say the premise-fact is a significant one (influences directly on the number N) if and only if its removal from the rule results in increasing the number N by at least one. We propose $s_2$ for each rule containing one significant premise-fact to have minimum value of ½, and for each rule containing at least five significant premises-facts – to have maximum value of 1. In other intermediate situations, $s_2$ takes discrete values from the interval (½ ; 1). Let us denote the number of significant premises-facts by z. Then we have:

$$s_2 = \begin{cases} \frac{1}{2} & \text{if} \quad z = 1 \\ 1 - \left(\frac{1}{z}\right)^4 & \text{if} \quad 2 \le z \le 4 \\ 1 & \text{if} \quad z \ge 5 \ . \end{cases} \tag{17}$$

At last, the value of $s_3$ will be calculated from the following formula:

$$s_3 = \frac{1}{2}\hat{R} + \frac{1}{2} \ . \tag{18}$$

As we can see, $s_3$ is linearly dependent on $\hat{R} \in \langle 0;1 \rangle$. We propose the both constants of the function to take the value of ½. As an effect, for $\hat{R}$ =0, $s_3$ takes the minimum value of ½, and for $\hat{R}$ =1 – the maximum value of 1.

Summarizing, we claim the rule to be of the greatest importance (to have the highest g-PM value) if: the number of patients fulfilling the rule's premises-facts is large enough (at the best – not less than 100), the number of significant premises-facts is not less than 5, and the percentage of patients fulfilling the rule's conclusion among the patients fulfilling the rules premises-facts is high (at the best – 100%).

## 7 Conclusions and Future Research

In the paper a new proposal of designing a reliable knowledge base of medical Rule-Based System is given. It is based on probabilistic and statistical analysis of the data stored in CIPFs. In order to implement this method, an access to a great number of CIPFs is necessary. Moreover, the data in CIPFs should be represented in a form of zero-one matrices. The access to numerous CIPFs can be provided by means of a medical Semantic Web. It is the subject of our current studies. In fact, since the last decade most of data in CIPFs are coded by means of standard electronic notations, such as HL7, EHR, and others proposed by national health organizations. If only we manage to define schemes of their translation into the form of zero-one matrices, then the requirements for data will be satisfied.

Having had the data, we can start designing production rules. Our goal is to formulate reliable rules, i.e. rules with high global reliability factors g-PM. To this end, using specialist's advice, from the patients' data in CIPFs we choose one that can be considered as the rule's conclusion. Then, from the remaining patients' data we choose next ones (at least five) that can be considered as the rule's premises-facts. They should satisfy the following conditions:

- a lot of patients fulfil the conjunction of the premises-facts;

- a lot of patients fulfil both the conjunction of the premises-facts and the conclusion, however, few patients fulfil the premises-facts and do not fulfil the conclusion.

In the future, the research on other determinants of g-PM will be done. We suppose that one of them is the premise-fact's "weight", which depends on the number of patients for whom this premise-fact is significant. The second determinant can be the "newness" of the premise-fact, which depends on the up-date of the corresponding patients' data.

From the other hand, also a random choice of patients' data for the rule's conclusion can be advantageous. Examining relations between this data and the data chosen randomly for premises-facts can enable discovering unexpected, even for specialists, medical hypotheses.

The proposed method should be tested on real (or at least synthetic) large-scale data. Such the data are now being obtained in cooperation with specialists from Poznan University of Medical Sciences.

## References

1. Dieng-Kuntzusing, R., et al.: Building and using a medical ontology for knowledge management and cooperative work in a health care network. Computers in Biology and Medicine 36, 871–892 (2006)
2. Act on the Forms and Range of Medical Documentation in Poland (in Polish) (2006), `http://www.abc.com.pl/serwis/du/2006/1819.htm`
3. Global Strategy for Asthma Management and Prevention Global Initiative for Asthma. GINA Report (2008)
4. Agraval, R., Imielinski, T., Swani, A.: Mining association rules between sets of items in large databases. In: Proc. of the ACM SIGMOD conference on Management of Data, Washington, D.C., pp. 207–216 (1993)
5. Jankowska, B., Szymkowiak, M.: How to Acquire and Structuralize Knowledge for Medical Rule-Based Systems? In: Kacprzyk, J. (ed.) Studies in Computational Intelligence, vol. 102, pp. 99–116. Springer, Heidelberg (2008)
6. Ligęza, A.: Logical Foundations for Rule-Based Systems. In: Studies in Computational Intelligence, vol. 11. Springer, Heidelberg (2006)
7. Petrie, A., Sabin, C.: Medical Statistics at a Glance. Blackwell Science Ltd., London (2000)
8. Bobrowski, D., Łybacka, K.: Chosen Methods of Statistical Reasoning (in Polish). WPP, Poznań (1995)
9. Krysicki, W., et al.: Mathematical Statistic (in Polish). PWN, Warszawa (1994)

# Towards an Increase of Collective Intelligence within Organizations Using Trust and Reputation Models

Emil Scarlat and Iulia Maries

University of Economic, Economic Cybernetics Department, 15-17 Dorobanti Avenue, Sector 1, Bucharest, Romania
{emil.scarlat@yahoo.com, iulia.maries@hotmail.com}

**Abstract.** Trust and reputation are fundamental concepts in multi-agent systems, but at the same time are significant to human life. The purpose of this paper is to find a way to enhance collective intelligence within organizations. First, we present some perspectives concerning the concepts of collective intelligence, trust and reputation. Then we suggest four computational models of trust and reputation, describing the main characteristics of each model and based on a cognitive model of trust, it is shown up how trust can increase collective intelligence in an organization. We try to simulate agents' behavior using the preferential attachment hypothesis.

**Keywords:** Collective Intelligence, Trust, Reputation, Preferential Attachment.

## 1 Introduction

Artificial intelligence is moving from the paradigm of an isolated and non-situated intelligence to the paradigm of situated, social and collective intelligence. The paradigm of the intelligent or adaptive agents and multi-agent systems, together with the emergence of the information society technologies are responsible for the increasing interest on trust and reputation mechanisms applied to electronic societies. [10]

Collective intelligence is a shared intelligence that emerges from the collaboration of individuals. The study of collective intelligence is considered a subfield of sociology, communication or behavior, computer science or cybernetics. Collective intelligence explores collective behavior from the level of quarks to the level of bacterial, plant, animal and human societies. The concept of collective intelligence can be extended to relationship intelligence. The use of new informational technologies should be oriented to collective intelligence field for helping people think, develop and implement ideas collectively.

The presence of collective intelligence has been felt for a long time: families, companies and states are groups of individuals that at least sometimes act intelligent. Bee and ant colonies are examples of groups of insects that are finding food sources acting intelligent. Even the human brain could be seen as a collection of individual neurons that collectively act intelligent. In the last few years there have been shown up new examples of collective intelligence: Google and Wikipedia. Google takes the

N.T. Nguyen, R. Kowalczyk, and S.-M. Chen (Eds.): ICCCI 2009, LNAI 5796, pp. 140–151, 2009.

collective knowledge created by millions of people for making websites, using sophisticated algorithms and technologies to answer the questions typed in. Wikipedia uses less sophisticated technologies, but very clever organizational principles and motivational techniques, to get people from all over the world to create a collection of knowledge.

Mathematic, a measure applied is the "collective intelligence quotient", which can be easier formalized than the individual intelligence quotient. Individual intelligence has been evaluated based on external results of behavior during real processes or during IQ tests, while elements of collective intelligence, like displacements, actions of beings or exchange of information, can be observed, measured and evaluated. A formal molecular model of computation and mathematical logic for describing the collective intelligence concept has been proposed [13]. The process, random and distributed, is tested in mathematical logics by social structure. Humans, ant and bacterial colonies, and information are modeled as abstract informational molecules that have expressions of mathematical logic. They are displaced quasi-chaotically due to interactions with their environments. Their interactions in the abstract computational space create an inference process perceived as "collective intelligence". The formal definition of collective intelligence has a set of requirements: informational molecules must emerge in a certain computational space, which can be software agents, ants, humans or social structures; than interactions must emerge between computational space, which consists in the ability to solve specific problems; and the emergence of collective intelligence determines specific inferences, in a probabilistic perspective. Based on above conditions, three basic hypotheses are presented:

- Precedence hypothesis: collective intelligence emerged first, as a result of interacting chemical molecules on Earth.
- Hypothesis of origin: life emerged later, from collectively intelligent actions of stabilization or development.
- Hypothesis on cycles: dependency between life and intelligence (individual or collective) is the result of a development cycle of evolution.

Collective intelligence can be defined as a group ability to solve more problems than its individuals [6]. In order to overcome the individual cognitive limits and the difficulties of coordination, a collective mental map can be used. A collective mental map is represented as an external memory with shared access and it can be formalized as a directed graph. The basic mechanism of collective mental map development consists of averaging the individual preferences, amplifying the weak links through positive feedback and integrating specialized sub-networks through division of labor. The efficiency of mental problem-solving depends on the problem representation in the cognitive system. Problem representation can be determined by a set of problem states, a set of possible actions, and a "fitness" criterion, a preference function for selecting the adequate actions that varies with the specific goals and preferences of the agent. Therefore, a mental map represents a highly selective representation of features relevant to problem-solving and consists of problem states, actions that lead from one state to another, and a preference function for determining the best action at any moment. Increasing problem-solving ability needs two complementary processes: enlarging the map with additional states and actions, and improving the preference function. In this situation, the better collective mental map is, the more easily problems will be solved. Intelligent agents are characterized by the quality of their mental maps, knowledge and understanding of their environment, capacities for action or goals.

Collective intelligence is trying to offer a new perspective to different phenomena, another way of thinking about effectiveness, profitability, teamwork or leadership.

The paper is organized as follows: section 2 discusses the definitions and typologies of trust and reputation, than shows a selection of four computational trust and reputation models and their main characteristics. Section 3 describes a cognitive model of trust that can increase collective intelligence within organizations. Simulations of agents' behavior are presented in section 4. Section 5 shows up conclusions and directions for future work.

## 2 Trust and Reputation

In this section we will present a selection of computational trust and reputation models and describe their main characteristics, but first we need to clarify the notions of trust and reputation. This field is quite recent, but in the last years there have been proposed interesting models with direct implementation in different domains.

### 2.1 Trust

Trust is important to human society due to its social component. The concept of trust has different meanings, but Gambetta's point of view is the most significant:

"*... trust (or, symmetrically, distrust) is a particular level of the subjective probability with which an agent assesses that another agent or group of agents will perform a particular action, both before he can monitor such action (or independently of his capacity ever to be able to monitor it) and in a context in which it affects his own action*". [5]

There are significant characteristics of trust mentioned in the above definition:

- Trust is subjective
- Trust is affected by the actions that cannot be monitor
- The level of trust is dependent on how our actions are affected by the other agent's actions.

In a socio-cognitive perspective [3], trust represents an explicit reason-based and conscious form. While trust means different things, the concept can be seen as:

- A mental attitude towards another agent, a disposition
- A decision to rely upon another agent, an intention to delegate and trust
- A behavior, for example the intentional act of trust and the relation between the trustier and the trustee.

The above concepts imply multiple sets of cognitive elements involved in the trustier mind.

**Typologies of trust**

In a social manner, there have been identified three types of trust:

- interpersonal trust (the direct trust that an agent has in another agent)
- impersonal trust (the trust within a system that is perceived through different properties)
- dispositional trust (the general trusting attitude)

## 2.2 Reputation

An agent behavior can be induced by other agents that cooperate, determining a reputation mechanism. The simplest definition of reputation can be the opinion others have of us. Otherwise, reputation represents a perception that an agent has of another agent's intentions or an expectation about an agent's behavior.

Reputation can be defined as "expectation about an agent's behavior based on information about or observations of its past behavior" [1]. This definition considers reputational information based on agent's personal experiences.

## 2.3 Computational Trust and Reputation Models

In the last years new approaches have been proposed, with direct implementation in different domains, in order to determine the level of trust.

Marsh has introduced a computational trust model in the distributed artificial intelligence [9]. An artificial agent can absorb trust and than he can make trust-based decisions. This model proposes a representation of trust as a continuous variable over the range [-1, +1). There are differentiated three types of trust: basic trust (calculated from all agent's experiences), general trust (the trust on another agent without taking into account a specific situation) and situational trust (the trust on another agent taking into account a specific situation). There are proposed three statistical methods to estimate general trust, each determining a different type of agent: the maximum method leads to an optimistic agent (takes the maximum trust value from the experiences he has), the minimum method leads to a pessimistic agent (takes the minimum trust value from the experiences he has) and the mean method that lead to a realistic agent (takes the mean trust value from the experiences he has). Trust values are used in agents' decision whether to cooperate or not with another agent.

Zacharia has proposed two reputation mechanisms (Sporas and Histos) in online communities based on collaborative ratings that an agent receives from others [14]. Sporas takes into consideration only the recent ratings between agents, and users with very high reputation values have smaller rating changes after updates than users with a low reputation. Histos comes as a reply, taking into consideration both direct information and witness information. The reputation value is subjectively assigned by each individual, so reputation mechanisms could generate social changes in users' behavior. A successful mechanism ensures high prediction rates, robustness against manipulability and cooperation incentives of the online community.

Abdul-Rahman and Hailes have suggested a model that allows agents to decide which other agent's opinion they trust more [1]. In their view trust can be observed from two perspectives: as direct trust or as recommender trust. Direct trust can be represented as one of the values: "very trustworthy", "trustworthy", "untrustworthy" or "very untrustworthy". For each partner, the agent has a panel with the number of past experiences in each category, and trust on a partner is given by the degree corresponding to the maximum value in the panel. The model takes into account only the trust coming from a witness, the recommender trust, which is considered "reputation". This approach could not differentiate agents that are lying from those that are telling the truth, but think different, so the model gives more importance to the information coming from agents with similar point of view.

Sabater and Sierra have proposed a modular trust and reputation model (ReGreT) to e-commerce environment [11]. This model takes into consideration three different types of information sources: direct experiences, information from third party agents and social structures. Trust can be determined combining direct experiences with the reputation model. Direct trust is built from direct interactions, using information perceived by the agent itself, and trust is determined from direct experiences. The reputation model is composed of specialized types of reputation: witness reputation (calculated from the reputation coming from witness), neighborhood reputation (calculated from the information regarding social relations between agents) and system reputation (calculated from roles and general properties). Witness reputation is determined based on information from other agents of the community. Neighborhood reputation is expressed based on social environment of the agent and the relations between the agent and that environment. System reputation is considered as objective features of the agent (for example, agent's role in the society). Those components merge and determine a trust model based on direct knowledge and reputation.

## 3 A Cognitive Model of Trust

Cognitive aspects are the basis for the collective intelligence concept and this is the reason of according more and more importance to the cognitive models of trust.

In a socio-cognitive perspective, trust can be seen as a mental attitude. To argue this perspective, have been suggested three statements [3]:

- Only a cognitive agent can trust another agent; it means that only an agent endowed with goals and beliefs (an agent trust another agent relatively to a goal).
- Trust is a mental state, a complex attitude of an agent towards another agent about a specific action relevant for a goal (the agent that feels trust is a cognitive agent endowed with goals and beliefs, but the agent trusted is not necessarily a cognitive agent).

- Trust is a the mental counter-part of delegation (the agent that feels trust depends on the trusted agent's action, what means that the agent that feels trust is "delegating" actions or goals to the other agent; this is the relation between trust and delegation).

Following the above conditions, delegation is an action, the result of a decision, action or decision that creates a social relation among the two agents that interact and the afferent action. In these conditions, trust is a mental state, a social attitude towards another agent.

Another possibility is to consider trust as a level of risk associated with the cooperation with another agent and in this situation, it estimates how likely the agent is to fulfill its commitments. Trust can be derived both from direct interactions among agents and from reputation, built from information received from third parties. Based on trust and reputation, agents can make more informed decisions about whether to interact with others or not.

Castelfranchi and Falcone have elaborated a cognitive model [4] that consider trust as a bet, implying risks, due to the delegating action, and than presented a quantification of trust degree, used to model the delegate decision (to delegate or not to delegate). In their opinion trust represent a mental state, a relational capital for agents that are trusted in a social network.

The main idea is that trust can be seen from two points of view: trustier view (the agent selects the right partners for achieving its own goals) and trustee's view (the agent is selected from partners to establish with them collaboration or cooperation and take advantage from the accumulated trust).

Trust analysis as relational capital starts from a dependence network with potential partners, in which needs, goals, abilities and resources are distributed among the agents, and than inserts the analysis of what it means for an agent to be trusted. Those conditions represent a form of power and could be used by an agent to achieve his own goals. The analysis presents the difference between relational capital and social capital, individual trust capital and collective trust capital.

In order to achieve each goal, an agent needs actions, plans and resources. Formally, let $A_t = \{A_1, \ldots, A_n\}$ be a set of agents. Each agent $A_i \in A_t$ can have associated:

- a set of goals $G_i = \{g_{i1}, \ldots, g_{iq}\}$
- a set of actions $Act_i = \{\alpha_{i1}, \ldots, \alpha_{iz}\}$
- a set of plans $\Pi = \{p_{i1}, \ldots, p_{is}\}$
- a set of resources $R_i = \{r_{i1}, \ldots, r_{im}\}$.

The sets of actions, plans, resources owned by an agent are used to achieve a set of tasks $(\tau_1, \ldots, \tau_r)$.

Considering the above notes, a Dependence Relationship between two agents $A_i$ and $A_j$ with respect to the goal $g_{ik}$, can be seen as an objective dependence or as a subjective dependence: $(A_i, A_j, g_{ik})$. A Dependence Network is represented by the set of dependence relationships among the agents from $A_t$ set at the moment $t$: $(A_t, t)$.

An agent $A_i$ has an Objective Dependence Relationship with another agent $A_j$ if for achieving at least one of his goals, $g_{ik} \in G_i$, agent $A_i$ needs actions, plans and resources that are owned by $A_j$ and not owned by $A_i$. A Subjective Dependence Relationship represents agent's $A_i$ point of view with respect to its dependence relationships. In this case, the dependence relationship can be modeled on the agent subjective interpretation: $B_jG_i$ is the set of goals of the agent $A_i$ believed by the agent $A_j$, $B_jAct_i$ is the set of actions of the agent $A_i$ believed by the agent $A_j$, $B_j \Pi_i$ is the set of plans of the agent $A_i$ believed by the agent $A_j$, and $B_jR_i$ is the set of resources of the agent $A_i$ believed by the agent $A_j$.

For a Dependence Network ($A_t$, $t$), can be expressed a Potential for Negotiation of an agent $A_i$ about a goal $g_{ik}$ as:

- An Objective Potential for Negotiation:

$$OPN(A_i, g_{ik}) = f(\sum_{j=1}^{n} \frac{1}{1+p_{kj}}) \tag{1}$$

- A Subjective Potential for Negotiation:

$$SPN(A_i, g_{ik}) = \sum_{j=1}^{n} \frac{1}{1+p_{kj}} \tag{2}$$

With: ***f*** a monotonous function, ***n*** the number of agents in the $A_t$ set that have a dependence relation with the agent $A_i$ with respect to the goal $g_{ik}$ ($n$ is the number of direct dependences), and $\boldsymbol{p_{kj}}$ the number of agents in the set $A_t$ set that are competitors with the agent $A_i$ on the same actions, plans or resources owned by the agent $A_j$ with respect to the goal $g_{ik}$ (the agent $A_j$ can not satisfy all the agents at the same moment, so $p$ is the number of indirect dependences).

Trust is also implied in a dependence belief (*BD*). To believe to be dependent means both to believe not to be able to perform an action $\alpha$ to achieve a goal $g$, (*BD1*), and to believe that an agent is able to perform an action $\alpha$ in order to achieve the goal $g$, (*BD2*). The second type of dependence belief, (*BDep2*) is a component of trust because represents a positive evaluation of an agent as being able or competent.

The Subjective Potential for Negotiation of an agent $A_i$ with a goal $g_{ik}$ can be represented based on the dependence belief:

$$SPN(A_i, g_{ik}) = \sum_{j=1}^{n} \frac{B_i(DA_j * DW_j)}{1+p_{kj}} \tag{3}$$

With: $\boldsymbol{D(B_i(A_j))}$ a degree of ability of the agent $A_j$ with respect to the goal $g_{ik}$ believed by the agent $A_i$, $\boldsymbol{D(B_i(A_j))}$ a degree of willingness of the agent $A_j$ with respect to the goal $g_{ik}$ believed by the agent $A_i$, and $\boldsymbol{p_{kj}}$ the number of agents in the set $A_t$ set that are competitors with the agent $A_i$ on the same actions, plans or resources owned by the agent $A_j$ with respect to the goal $g_{ik}$.

The Subjective Trust Capital of an agent $A_i$ about a task $\tau_k$ can be expressed as a function:

$$STC(A_i, \tau_k) = \sum_{j=1}^{n} B_i(B_j DA_i * B_j DW_i) \tag{4}$$

With: ***n*** number of agents that need the task $\tau_k$, $\boldsymbol{D(B_{ij}(A_{jk}))}$ the agent's $A_i$ degree of belief with respect the agent's $A_j$ ability about a task $\tau_k$, believed by the agent $A_j$, $\boldsymbol{D(B_{ij}(W_{jk}))}$ the agent's $A_i$ degree of belief with respect the agent's $A_j$ willingness about a task $\tau_k$, believed by the agent $A_j$.

The cumulated trust capital of an agent $A_i$ about a specific task $\tau_k$ is the sum of the abilities and willingness believed by each dependent agent, for all the agents that need that task in the network dependence. The subjectivity occurs because the network dependence and the abilities and willingness believed are considered from the agent's $A_i$ point of view.

The Degree of Trust of an agent $A_i$ on other agent $A_j$ with respect to the task $\tau_k$ can be described:

$$DT(A_i A_j \tau_k) = B_i DA_j * B_i DW_j, \tag{5}$$

and the Self-Trust of the agent $A_i$ with respect to the task $\tau_k$:

$$ST(A_i, \tau_k) = B_i(DA_i * DW_i). \tag{6}$$

The Subjective Usable Trust Capital of an agent $A_i$ about a task $\tau_k$ can be defined based on the above notes:

$$SUTC(A_i, \tau_k) = \sum_{j=1}^{n} \frac{B_i(B_j DA_i * B_j DW_i)}{1 + p_{kj}} \tag{7}$$

With: $\boldsymbol{p_{kj}}$ the number of agents in the dependence network that can achieve the same task with a trust value comparable with agent's $A_i$ trust value.

It should be taken into consideration that even if dependence relationship between agents in a society are important, there will not exist exchanges if trust is not present to enforce connections.

Trust can be an advantage for the trustee, but there is a disadvantage in treating social capital at individual level (relational capital). Sometimes relational capital could be in conflict with the collective capital (for an individual is better to monopolize trust, but for the community is better to distribute it among individuals).

## 4 NetLogo Simulations

Preferential attachment is frequently used in describing social, biological and technological networks and it represents the mechanism of formation models for such networks.

Social networks are interaction networks, where nodes are agents and links between nodes are interactions between agents. In the evolution of social networks, an important hypothesis is that highly connected nodes increase their connectivity faster than their less connected peers, called preferential attachment. Experiments reveal that the rate at which nodes acquire links depends on the node's degree, offering direct quantitative support for the presence of preferential attachment.

The evolving network models are mostly based on two important hypothesis, growth and preferential attachment. The growth hypothesis sustains that networks continuously expand through new nodes and links between the nodes, and the preferential attachment hypothesis sustains that the rate with which a node with $k$ links acquire a new link is a monotonically increasing function of $k$.

### 4.1 Preferential Attachment in NetLogo

This model shows a way of arising networks, when there are a few hubs that have many connections, while the others have only a few. The model starts with two nodes connected by an edge. A new node is added at each step. A new node picks an existing node to connect to randomly, but there is a tendency observed: a node's chance of being selected is directly proportional to the number of connections it already has. In our simulations, we are going to endow agents with two attributes, "reputation" and "intelligence".

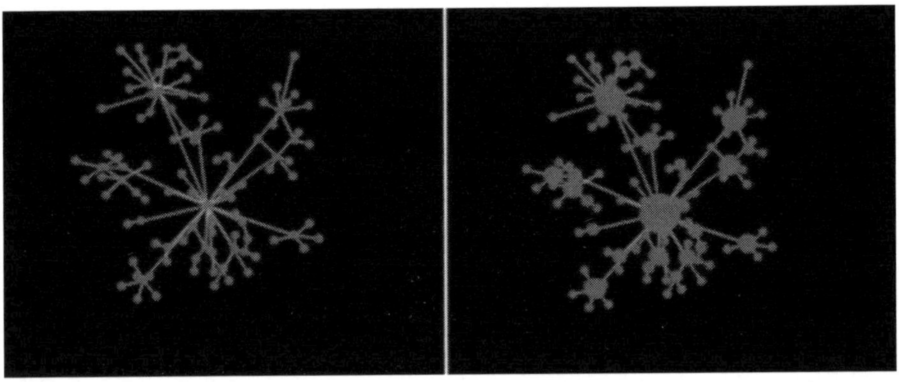

**Fig. 1a), 1b).** There are two nodes that have many connections, while the most of them have only a few (figure 1b) shows resized nodes, for clearance). The popular nodes acquire new links faster than the other ones.

### 4.2 Simulations

The forthcoming simulations will try to show agents' behavior assuming that they are endowed with two attributes, reputation and intelligence. We propose a representation of reputation as a variable over the range [-1, 1] and intelligence as a variable over the range [1, 100]. The simulations are averaged over 56 runs.

The first agent's intelligence is a random value between 1 and 100. Higher the agent intelligence level is, closer to 100 the value of intelligence is. Agents are going to be linked each other only if their intelligence is varying with less than 10. For example, if an agent has intelligence 82, it could have connections only with agents that have intelligence between 72 and 92.

The first agent's reputation is a random value between -1 and 1, usually close by 1. Higher the agent reputation level is, closer to 1 the value of reputation is and the more connections the agent has.

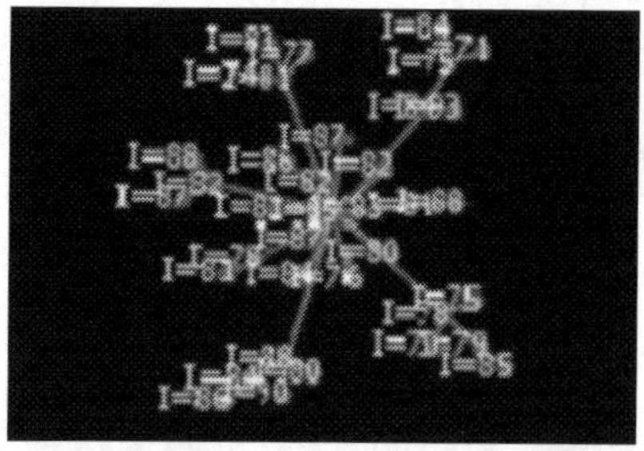

**Fig. 2.** Intelligence values for the agents are presented. The maximum intelligence value is 90 and the minimum intelligence value is 71, but the difference between them is amplified by the growing number of agents.

We can notice that agents make links with those agents that have the closest value of intelligence. For example, if there are two agents, one with intelligence 82 and the other with intelligence 84, the third agent with intelligence 79 will choose to connect to the agent with the closest intelligence, in this case the agent with intelligence 82.

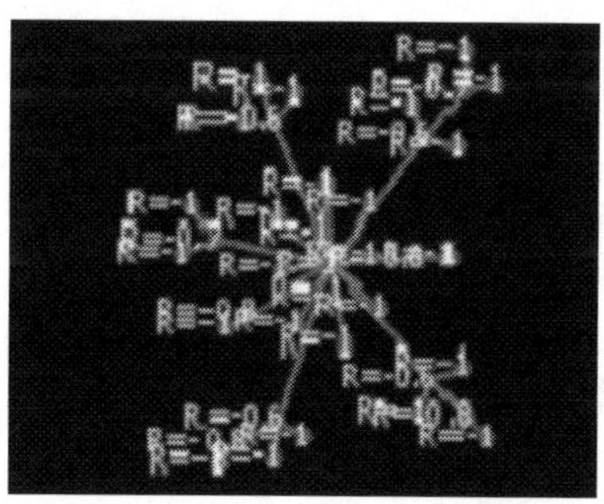

**Fig. 3.** Reputation values are shown. The maximum reputation value is 1 and the minimum reputation value is -1.

Agent's reputation is determined by the number of connections that agent has with other agents. The more connections an agent has, the reputation value is closer to 1. The first agent has frequently reputation value 1 because it is the agent with the highest number of connections.

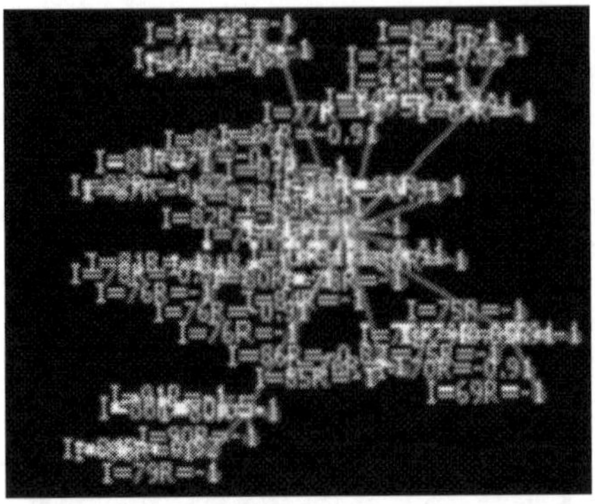

**Fig. 4.** Agents are endowed with the two attributes, reputation and intelligence

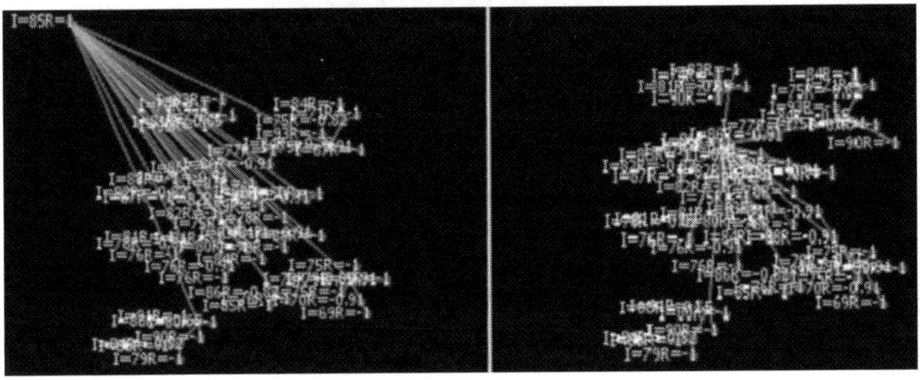

**Fig. 5.** An isolated agent with high level of both reputation and intelligence is drawn back into the crowd by its connections after two or three runs

An agent prefers to connect with another agent that has a high level of reputation, but at the same time with the closest value of intelligence.

## 5 Conclusions

The organizational behavior field is interested in studying organizations as complex social systems. Trust is a decisive source of social capital within social systems. Most of the theories from this field explore individual and collective human behavior within organizations and their central activities try to identify the determinants of intra-organizational cooperation. Managing collective intelligence within an organization implies combining all tools, methods and processes that can lead to connection and cooperation among individual intelligences.

Individual intelligence can not face all the problems in today's world. To successfully deal with problems we need to develop collective intelligence as a global civilization. Collective intelligence can improve competitiveness within organizations in the context of a global market and collective performance has become a critical factor in the organization's development. In this situation creating, developing and sustaining trust among members within teams is the core that leads to performance.

The contribution of this paper is both as a conceptual analysis and practical use of trust in multi-agent systems and social theory. The research is based on the sociological literature, but with emphasis on multi-agent systems. In this paper, the need for effective trust management in open, distributed systems is highlighted.

Our future work will focus on extending the NetLogo simulations by adding new attributes or characteristics to the agents, and furthermore, developing a new model of trust and reputation that would lead to an increase of the collective intelligence in organizations.

**Acknowledgement.** This article is a result of the project "Doctoral Program and Ph.D Students in the education, research and innovation triangle". This project is co-funded by European Social Fund through the Sectorial Operational Programme for Human Resources Development 2007-2013, coordinated by the Academy of Economic Studies, Bucharest.

## References

1. Abdul-Rahman, A., Hailes, S.: Supporting Trust in Virtual Communities. In: Proceedings of the Hawaii International Conference on System Sciences, Maui, Hawaii (2000)
2. Atlee, T.: Co-Intelligence, collective intelligence, and conscious evolution (The Co-Intelligence Institute), Earth Intelligence Network (2008)
3. Castelfranchi, C., Falcone, R.: Principles of Trust for MAS: Cognitive Anatomy, Social Importance, and Quantification. In: Proceedings of the 3rd International Conference on Multi-Agent Systems (1998)
4. Castelfranchi, C., Falcone, R., Marzo, F.: Cognitive Model of Trust as Relational Capital. In: 10th Workshop on Trust, Privacy, Deception and Fraud in Agent Societies, Honolulu, Hawaii (2007)
5. Gambetta, D.: Can We Trust Trust? In: Trust: Making and Breaking Cooperative Relations, ch. 13. University of Oxford, Oxford (2000)
6. Heylighen, F.: Collective Intelligence and its Implementation on the Web: algorithms to develop a collective mental map. Springer, Netherlands (1999)
7. Lévy, P.: Collective Intelligence, A Civilization: Towards a Method of Positive Interpretation. Springer, Heidelberg (2006)
8. Malone, T.W.: What is collective intelligence and what will we do about it (MIT Center for Collective Intelligence) Earth Intelligence Network (2008)
9. Marsh, S.P.: Formalizing Trust as a Computational Concept., PhD Thesis, Department of Computing Science and Mathematics, University of Stirling (1994)
10. Mui, L., Mohtashemi, M., Halberstadt, A.: A Computational Model of Trust and Reputation. In: Proceedings of the 35th Hawaii International Conference on System Sciences (2002)
11. Sabater-Mir, J., Sierra, C.: REGRET: A Reputation Model for Gregarious Societies (2001)
12. Surowiecki, J.: The Wisdom of Crowds. Random House Large Print (2004)
13. Szuba, T., Almulla, M.: Was Collective Intelligence before Life on Earth? In: IPDPS Workshops on Parallel and Distributed Processing (2000)
14. Zacharia, G.: Collaborative Reputation Mechanisms for Online Communities, Master Thesis, Massachusetts Institute of Technology (1999)

# A New Ant Colony Optimization Algorithm with an Escape Mechanism for Scheduling Problems

Tsai-Duan Lin[1], Chuin-Chieh Hsu[2], Da-Ren Chen[1], and Sheng-Yung Chiu[2]

[1] Hwa Hsia Institute of Technology, Department of Information Management, 23554 Taipei County, Taiwan
[2] National Taiwan University of Science and Technology, Department of Information Management, 69042 Taipei, Taiwan
{D9309103,CChsu,M9209004}@mail.ntust.edu.tw, danny@cc.hwh.edu.tw

**Abstract.** Ant colony optimization (ACO) algorithm is an evolutionary technology often used to resolve difficult combinatorial optimization problems, such as single machine scheduling problems, flow shop or job shop scheduling problems, etc. In this study, we propose a new ACO algorithm with an escape mechanism modifying the pheromone updating rules to escape local optimal solutions. The proposed method is used to resolve a single machine total weighted tardiness problem, a flow shop scheduling problem for makespan minimization, and a job shop scheduling problem for makespan minimization. Compared with existing algorithms, the proposed algorithm will resolve the scheduling problems with less artificial ants and obtain better or at least the same, solution quality.

**Keywords:** Ant colony, optimization, escape mechanism, combinatorial, scheduling, makespan.

## 1 Introduction

Most combinatorial optimization scheduling problems are nondeterministic polynomial-time hard (NP-hard) problems; as the scale of the problems becomes bigger, their optimal solutions can hardly be reached within a short time. Consequently, many researchers have introduced various heuristic algorithms, other than an optimization algorithm, to search for approximate solutions [1]. Moreover, various evolutionary algorithms, including genetic algorithms, tabu Search Algorithms, and simulated annealing algorithms, have often been used to solve the scheduling problems [2][3][4][5][6] [7][8]. Although taking more computation time, evolutionary algorithms usually find better solutions than heuristic algorithms do.

Ant colony optimization (ACO) is an evolutionary algorithm that Dorigo [9][10][11][12][13][14][15][16][17][18][19][20][21][22][23] first proposed to model the foraging behavior of real ants for an optimal solution. Since then various kinds of problems have been solved by the ACO algorithm, an efficient one. This study modifies the pheromone updating rule, develops a mechanism to escape the local optimum, and finally reaches a new ACO algorithm with an escape mechanism. The proposed

N.T. Nguyen, R. Kowalczyk, and S.-M. Chen (Eds.): ICCCI 2009, LNAI 5796, pp. 152–162, 2009.

algorithm is, in this study, to solve a single machine total weighted tardiness problem, a flow shop scheduling problem for makespan minimization, and a job shop scheduling problem for makespan minimization and, then, to be tested and examined for its efficiency.

## 2 Background

By depositing pheromone on good solutions, an ACO algorithm can obtain better solutions through iterative search. The general solution-search procedure of ACO is illustrated in Fig. 1.

```
ACO algorithm
   Initialize parameters and pheromone trails;
   While (terminating condition not met) do
      construct ant solutions;
      local search    /*optional*/
      update pheromone trail
   End-While
end
```

**Fig. 1.** The solution-search procedure of the ACO algorithm

### 2.1 Ant Colony System for Resolving $1 \mid \Sigma w_jT_j$ Problem

Dorigo[11] et al. first employed ACO to solve a single machine total weighted tardiness problem, i.e. a problem $1 \mid \Sigma\, w_jT_j$ with $n$ jobs, where each Job j ($j$ =1,…,$n$) possesses data for processing time $p_j$, due date $d_j$, job weight $w_j$, and so on. We are to find the machines to handle the order of the $n$ jobs, i.e. the jobs to be scheduled at each position $p$ ($p$=1,…,$n$); refer to [11] for the evolutional flow. The heuristic value $\eta_{pj}$ denotes the suitability of dispatching Job $j$ at Position $p$; to compute said value, Dorigo [11] uses three dispatch rules, viz. *Earliest Due Date(EDD)*, *Modified Due Date(MDD)*, and *Apparent Urgency(AU)*.

### 2.2 Ant Colony System for Resolving $F_m \mid prmu \mid C_{max}$ Problem

This problem is very similar to the $1 \mid \Sigma\, w_jT_j$ problem, but after ants obtains a complete solution, the algorithm calculates the makespan, not the total tardiness.

To update the global pheromone for $\tau_{pj}$: $\tau_{pj} \leftarrow (1-\sigma)\cdot\tau_{pj} + \Delta\tau_{pj}$, where $\Delta\tau_{pj}$ = $1/C_{max}{}^*$ and $C_{max}{}^*$ is the makespan of the optimal solution obtained up to the current iteration. The heuristic value $\eta_{pj}$, denoting the suitability of dispatching Job $j$ at Position $p$, is computed by Ying [20] who modified Palmer's [24] slope heuristic algorithm. The equations for *slope index* $_j$ for each Job j and $\eta_{pj}$ are listed as follows:

$$slope\ index_j = -\sum_{i=1}^{m}(m-(2i-1))p_{ij} \quad \text{and} \quad \eta_{pj} = slope\ index_j - \min\{slope\ index_j\}+1,$$

where $p_{ij}$ is the processing time of Job j on machine $i$.

### 2.3 Ant Colony System for Resolving $J_m \mid C_{max}$ Problem

With an algorithm that generates active schedules for $J_m \mid C_{max}$ problems [1][25] as a reference, we propose a new ACO algorithm (Algorithm 1) in this study. Since the optimal solution for a scheduling problem must be an active schedule, so must be the scheduling solution rendered by the algorithm; hence, an active-schedule generation will obtain a better solution or has a better chance to obtain an optimal solution. The following proposed algorithm uses a method, in forming scheduling solutions, to render an active schedule:

**Algorithm 1:**

**Step1: Setting parameters**

**Step2: Forming solutions in steps**

For iteration number equal to 1 to Max-Iteration Do

For Ant k equal to 1 to $m_{ant}$ Do

While (jobs are not completely scheduled) Do

1. Calculate the earliest start and finish time of the jobs the precedent jobs of which have been processed; locate the job with the minimum finish time $c^*$; process the job's machine $M^*$ and $M^*$'s current schedulable position $p$;
2. If the processing machine of the jobs whose precedent jobs are all processed already is $M^*$, and the earliest start time < $c^*$, place these jobs $Q_{jM^*}$ onto the candidate list ($L$);
3. Generate a variable $q$ conforming to [0,1] uniform distribution;
   If ($q \le q_0$)
   Job $Q_{jM^*}$ with the maximum value of $\tau_{M^*pj} \cdot \eta^{\beta}_{M^*pj}$ is selected from the candidate list ($L$) and placed at the schedulable position $p$ for machine $M^*$;
   Else
   Calculate the probability of selecting job $Q_{jM^*}$ from the candidate list (L) according to equation (1); select $Q_{jM^*}$ using roulette wheel selection and place it at the schedulable position $p$ of machine M*;

$$P_{O_{j,p}} = \begin{cases} \dfrac{\tau_{M^*pj}\eta^{\beta}_{M^*pj}}{\sum_{candidate\ jobs}\tau_{M^*pj}\eta^{\beta}_{M^*pj}} & \text{, if } Q_{jM^*} \text{ is not scheduled} \\ 0 & \text{, otherwise} \end{cases} \quad (1)$$

4. After the ant at Location $p$ of Machine $M^*$ selects Job $j$, update local pheromone for $\tau_{M^*pj}$ according to equation (2):

$$\tau_{M^*pj} \leftarrow (1-\xi)\cdot\tau_{M^*pj} + \xi\cdot\tau_0 \quad (2)$$

End While Loop

Calculate the makespan according to the solution obtained from each ant; if the solution is better than previous ones, the solution obtained by the present ant is updated as the global optimal solution;

End Ant k Loop

Update global pheromone:

$$\tau_{Mpj} \leftarrow (1-\rho)\cdot\tau_{Mpj} + \Delta\tau_{Mpj} \quad (3)$$

End the iteration number Loop

**Step3: End; output the global optimal solution and makespan.**

## 3 The Methodology

### 3.1 Pheromone Modification Strategy Based on Solution Quality

The solution quality is first determined before proceeding with the addition or subtraction of the pheromone; for example, the updating rule of $1 \mid \Sigma w_jT_j$ problem is as follows:

$\tau_{pj} \leftarrow \tau_{pj} \times$ (The global optimal solution obtained until the last iteration/the solution formed by the present ant)

If $\tau_{pj} > \tau_0$, $\tau_{pj} \leftarrow (1-\xi)\cdot\tau_{pj} + \xi\cdot\tau_0$ is employed to update the local pheromone. The purpose is to render the subsequent ants to have a higher probability to select other solutions, and thus to find better solutions. As to the global pheromone updating, the setting of the original ant colony system shall be maintained.

The iteration accumulation is used to determine if the algorithm stagnates; e.g., for the $1 \mid \Sigma w_jT_j$ problem, we will first determine if the present ant has formed a solution with a smaller total tardiness. If no improvement is rendered, the counter for the non-improving iteration number is added by 1; otherwise, the counter number is reset to 0. In addition, reset Parameter ($q_0[k]$), denoting the development-exploring ratio controlling each Ant $k$, to its initial value so that the ants can re-search for new solutions toward better solutions.

The non-improvement iteration number serves another purpose in pheromone updating: The pheromone $\tau_{pj}$ for an ant selecting Job $j$ at Location p, if [non-improvement iteration number / E] is even and $\tau_{pj} < \tau_0$, $\tau_{pj}$ is set as $\tau_0$. This mechanism can also control the change of search width when the algorithm stagnates. If the algorithm does not stagnate and the non-improvement iteration number is less than 200, then [non-improvement iteration number/E] is 0; if $\tau_{pj} < \tau_0$, then $\tau_{pj}$ is set as $\tau_0$. This step can prevent pheromone diversity from getting too large so as to enhance the algorithm's search width. When stagnation first occurs in the algorithm and the range for the non-improvement iteration number is [200, 400], [non-improvement iteration number / E] is odd; if $\tau_{pj} < \tau_0$, the pheromone will not be adjusted. Meanwhile, the pheromone differences will be greater to prevent ants from searching poorer solutions.

### 3.2 The Exploration-Ratio Change Strategy and the Pheromone Reset Strategy

If the stagnation iteration number parameter ($E$) for activating the escape mechanism is set at 200, the algorithm stagnates when the counter of the non-improvement iteration number is a multiple of 200. Then, the exploration-ratio change strategy and the pheromone reset strategy will be executed.

Taking $1 \mid \Sigma w_jT_j$ problem as an example, the proposed algorithm assumes that ten ants are dispatched at each iteration to form solutions. The parameter controlling the exploration/exploitation ratio is 0.9, and it is reduced, by a fixed value $\alpha = 0.1$, when the algorithm stagnates. When the counter for the non-improvement iteration number is 200, the stagnation occurs for the first time; at this time,

$q_0$[1],…,$q_0$[min{10,200/200}] is reduced by 0.1. If the non-improvement situation persists, the second stagnation occurs when the counter for the non-improvement iteration number turns to 400; at this time, $q_0$[1],…,$q_0$[min{10,200/200}] is further reduced by 0.1. When the optimal solution is improving, the counter of the non-improvement iteration number is reset to 0 and the parameter ($q_0[k]$) controlling the exploration/exploitation ratio for each Ant k is reset to 0.9. Consequently, some ants can have a larger search width, whereas others maintain their original search width. Furthermore, when the algorithm stagnates, all pheromones are reset to the initial pheromone $\tau_0$ to erase ants' previous learning experience.

### 3.3 The ACO Algorithm with Escape Mechanism for $1 \mid \Sigma w_j T_j$ Problem

The first step of the proposed algorithm is to set parameters. In contrast to the ant colony system, each Ant k has a parameter $q_0[k]$ to control the exploration/exploitation ratio when forming solutions. When the algorithm stagnates, the parameter $q_0[k]$ will be changed to alter the ants' search width. When Ant *k* at Location p is to select an unscheduled job *j*, the algorithm will also generate a variable *q* between 0 and 1; if $q \le q_0[k]$ (the value of $q_0[k]$ is between 0 and 1; each $q_0[k]$ is usually set identical at the time the algorithm starts), Job j($j \in s$) with the maximum value of $\tau_{pj} \cdot \eta_{pj}^{\beta}$ is selected from the unscheduled jobs, *s*. On the other hand, equation (1) of Algorithm 2 will be used to calculate the probability of selecting each job; Job j is then selected by roulette wheel selection. On local pheromone updating, we will use $\tau_{pj} \leftarrow (1-\xi) \cdot \tau_{pj} + \xi \cdot \tau_0$ to do the updating if $\tau_{pj} > \tau_0$.

### 3.4 The ACO Algorithm with Escape Mechanism for $F_m \mid prmu \mid C_{max}$ Problem

The solution searching procedure is shown in Algorithm 2; the algorithm needs to calculate the makespan, not the total tardiness. For global pheromone updating, $\Delta\tau_{pj} = 1/C_{max}{}^*$, where $C_{max}{}^*$ is the makespan of the optimal solution obtained until the last iteration. On the other hand, the calculation of the heuristic value will follow that of ant colony system. Local pheromone updating will be executed by $\tau_{pj} \leftarrow (1-\xi) \cdot \tau_{pj} + \xi \cdot \tau_0$ only if $\tau_{pj} > \tau_0$. Furthermore, when the algorithm stagnates, the mechanisms of the exploration-ratio change strategy and the pheromone reset strategy will be also activated.

### 3.5 The ACO Algorithm with Escape Mechanism for $J_m \mid C_{max}$ Problem

Each Ant k has a parameter $q_0[k]$ to control the exploration/exploitation ratio to form solutions. When the algorithm stagnates, each ant's parameter $q_0[k]$ will be changed to alter its search width. Moreover, the ratio of the makespan of the global optimal solution obtained until the last iteration to the makespan of the solution formed by the present ant is taken as the pheromone updating criterion. For local pheromone, it will be updated only if $\tau_{Mpj} > \tau_0$. Also, when [non-improvement iteration number/*E*] is even and $\tau_{Mpj} < \tau_0$, $\tau_{Mpj}$ will be set to be $\tau_0$.

**Algorithm 2**

**Step1: Setting parameters**

**Step2: Forming solutions in steps**

For iteration number =1 to Max-Iteration Do
  For Ant k =1 to $m_{ant}$ Do
    For Location p =1 to $n$ Do
      Generate a variable $q$ conforming to uniform distribution between [0,1];
      If ($q \le q_0[k]$),
        Job j with the maximum $\tau_{pj} \cdot \eta_{pj}^{\beta}$ is selected from the unscheduled jobs;
      Else
        Calculate the probability of Location p selecting Job j according to equation (1) and select Job j using roulette wheel selection;

$$P_{pj} = \begin{cases} \dfrac{\tau_{pj}\eta_{pj}^{\beta}}{\sum_{i \in unscheduled\ jobs} \tau_{pi}\eta_{pi}^{\beta}} & \text{, Job j is unscheduled} \\ 0 & \text{, otherwise} \end{cases} \quad (1)$$

      After the ant at Location p selects Job j, update local pheromone for $\tau_{pj}$ according to equation (2) if $\tau_{pj} > \tau_0$:

$$\tau_{pj} \leftarrow (1-\xi)\cdot\tau_{pj} + \xi\cdot\tau_0 \quad (2)$$

    End Location p Loop
    Calculate the total tardiness according to the solution obtained from each ant; if the solution is better than previous ones, the solution obtained by the present ant is updated as the global optimal solution;
    Execute Pheromone modification strategy based on solution quality;

$\tau_{pj} \leftarrow \tau_{pj} \times$ (the global optimal solution obtained until the last iteration / the solution formed by the present ant) (3)

    If ([non-improvement iteration number/$E$] is an even number and $\tau_{pj} < \tau_0$),
      $\{\tau_{pj} = \tau_0\}$
  End Ant k Loop
  Update global pheromone:

$$\tau_{pj} \leftarrow (1-\rho)\cdot\tau_{pj} + \rho\cdot\Delta\tau_{pj} \quad (4)$$

  If (global optimal solution is not improved), {non-improvement iteration number is added by 1}.
  If (global optimal solution is improved), {non-improvement iteration number is reset to 0; parameter $q_0[1,\ldots,m_{ant}]$ is reset as original value;}.
  If (global optimal solution is a multiple of $E$),
  $\{q_0[1],\ldots,q_0[\min(m_{ant}$, non-improvement iteration number/$E$ )] is reduced by the fixed value of $\alpha$.
  Resetting all pheromones, i.e., $\tau_{pj} \leftarrow \tau_0 (p=1,\ldots,n\ ;\ j=1,\ldots,n)$;}
End iteration number Loop

**Step3: End and output the global optimal solution and makespan**

When the solution forming behaviour of ants falls into stagnation, the algorithm starts to execute the escape mechanism from local optimal solution, including the exploration-ratio change strategy and the pheromone reset strategy.

## 4 Experimental Results and Discussion

This study uses Visual C++ to implement the algorithm program, which is then tested in a computer (Pentium-M, 1.4GHz CPU, 512MB memory) with the bench problems of $1 \mid \Sigma w_j T_j$, $F_m \mid prmu \mid C_{max}$, and $J_m \mid C_{max}$. The aim is to determine if the proposed ACO algorithm with local escape mechanism can dispatch fewer ants but still find solutions better than or at least same as those obtained by a conventional ant colony system.

The benchmark problems for single machine total weighted tardiness provided in OR Library [27] are used for testing. OR Library provides a problem set containing 40 jobs in each of the 125 groups, where every problem has its known optimal solution. Hence, there are 125 groups of problems in the problem set. The benchmark problems provided by Taillard [26] are selected for testing flow shop problem, including 12 problem sets of 20 to 500 jobs and 5 to 20 machines. Each problem set can generate 10 problems; thus the total number of problems is 120.

We adopt the parameter setting scheme introduced by Dorigo [11] for ant colony system: The number of ant ($m_{ant}$) is 20; the local pheromone evaporation rate ($\xi$) and the global pheromone evaporation rate ($\rho$) are both 0.1; the parameter ($\beta$) controlling the significance of heuristic value is 2; the parameter ($q_0$) controlling the exploration/exploitation ratio is set at 0.9, and the initial pheromone is 1/(task number$\times T_{MDD}$), where $T_{MDD}$ is the total tardiness obtained by using MDD dispatch rule for single machine scheduling, and the initial pheromone is 1/(number of task$\times$known optimal solution) for job shop scheduling. Furthermore, the terminating condition for the algorithm is set as the maximum iteration number (3000 iterations). The MDD dispatch rule is used to set the heuristic value for single machine scheduling, but for job shop scheduling is calculated by *slope* heuristics. For the algorithm proposed in this study, the number of ants dispatched ($m_{ant}$) is set at 10; the parameter $q_0[k]$ controlling the exploration/exploitation ratio for each Ant $k$ is set at 0.9; the iteration number parameter ($E$) of stagnation for activating the escape mechanism is set at 200, and the parameter $\alpha$ for reducing the exploration/exploitation ratio is set at a fixed value of 0.1. The rest of the parameters are the same as those of the ant colony system. Table 1 lists the comparison between the two algorithms for the single machine total weighted tardiness problem. Tables 2 and 3 summarize the comparison between ant colony system and the proposed ACO algorithm with escape mechanism for the $F_m \mid prmu \mid C_{max}$ problem.

**Table 1.** Results of the ACO algorithms for $1 \mid \Sigma w_j T_j$ problem

| | Ant colony system | ACO algorithm with escape mechanism |
|---|---|---|
| Iteration number to find optimal solution | 48 | 111 |
| Average differences between average solutions and optimal solutions | 0.020782 | 0.012304 |
| Average time spent for each experiment of each problem | 20.319 | 10.765 |

**Table 2.** Results of ACO algorithms for $F_m$ | *prmu* | $C_{max}$ problem (1)

| Without heuristic value | Ant colony system | ACO algorithm with escape mechanism |
|---|---|---|
| Problem | Mean solutions difference | Mean solutions difference |
| job:20 machine:5 | 0.025815 | 0.020065517 |
| job:20 machine:10 | 0.04501016 | 0.03476405 |
| job:20 machine:20 | 0.03632758 | 0.02988141 |
| job:50 machine:5 | 0.014477171 | 0.012689639 |
| job:50 machine:10 | 0.06429007 | 0.05119182 |
| job:50 machine:20 | 0.08524322 | 0.06716862 |
| job:100 machine:5 | 0.010858361 | 0.009621992 |
| job:100 machine:10 | 0.05328 | 0.046129 |
| job:100 machine:20 | 0.097869 | 0.088479 |
| job:200 machine:10 | 0.046441 | 0.042482 |
| job:200 machine:20 | 0.106244 | 0.100626 |
| job:500 machine:20 | 0.085856 | 0.085075 |

**Table 3.** Results of ACO algorithms for $F_m$ | *prmu* | $C_{max}$ problem (2)

| With heuristic value | Ant colony system | ACO algorithm with escape mechanism |
|---|---|---|
| Problem | Mean solutions difference | Mean solutions difference |
| job:20 machine:5 | 0.016427 | 0.015695 |
| job:20 machine:10 | 0.032548 | 0.030831 |
| job:20 machine:20 | 0.029 | 0.026481 |
| job:50 machine:5 | 0.007248 | 0.006601 |
| job:50 machine:10 | 0.040246 | 0.039406 |
| job:50 machine:20 | 0.06254 | 0.055443 |
| job:100 machine:5 | 0.004834 | 0.004063 |
| job:100 machine:10 | 0.030122 | 0.028358 |
| job:100 machine:20 | 0.072861 | 0.071081 |
| job:200 machine:10 | 0.022088 | 0.020537 |
| job:200 machine:20 | 0.074445 | 0.072901 |
| job:500 machine:20 | 0.051476 | 0.050749 |

For the job shop problem, the column $(n,m)$ in Table 4 denotes that the problem has $n$ jobs and $m$ machines, where each job has to be executed in all the machines, and the processing time is given. The fourth and fifth columns in Table 4 are the solutions of AS and ACS-JSP. Five experiments are conducted for each problem, and an average is calculated for the solutions obtained.

The total iteration number executed is 3000. The number of ants dispatched ($m_{ant}$) in each iterative cycle equals to the number of jobs ($n$) for both ACS-active and ACS-JSP, and that of the proposed algorithm is $n/2$. For other parameters, the local pheromone evaporation rate ($\xi$) and the global pheromone evaporation rate ($\rho$) in ACS-active are both set at 0.1; the parameter ($\beta$) controlling the significance of heuristic value is 1; the parameter ($q_0$) controlling the exploration/exploitation ratio is set at 0.9; and the initial pheromone is 1/(the known optimal solution). The LRT dispatch rule is used to calculate the heuristic value. Also, the parameter $q_0[k]$ controlling the exploration/exploitation ratio for each Ant k is set at 0.9; the iteration number parameter ($E$) of stagnation for activating the escape mechanism is set at 100; and the parameter $\alpha$ for reducing the exploration/exploitation ratio is set at a fixed value of 0.2.

**Table 4.** Results of different ACO algorithms for $J_m \mid C_{max}$ problem

| Problem | $n,m$ | Optimal solution | AS | ACS-JSP | ACS-active | ACO algorithm with escape mechanism |
|---|---|---|---|---|---|---|
| ORB4 | 10,10 | 1005 | 1185 | 1148 | 1060.6 | 1048.2 |
| ORB1 | 10,10 | 1059 | 1405 | 1316 | 1194.4 | 1149 |
| MT10 | 10,10 | 930 | 1195 | 1078 | 1015.4 | 994 |
| LA16 | 10,10 | 945 | 1108 | 1099 | 1012.2 | 988.8 |
| LA17 | 10,10 | 784 | 853 | 896 | 816.6 | 800.8 |
| LA18 | 10,10 | 848 | 943 | 1037 | 928 | 897.8 |
| LA19 | 10,10 | 842 | 988 | 980 | 891.2 | 871.8 |
| LA20 | 10,10 | 902 | 1062 | 1059 | 988.4 | 959 |
| LA21 | 15,10 | 1040 | 1317 | 1215 | 1157.4 | 1142.2 |
| LA22 | 15,10 | 927 | 1187 | 1068 | 1041 | 1006.2 |
| LA23 | 15,10 | 1032 | 1218 | 1199 | 1092 | 1072.8 |
| LA24 | 15,10 | 935 | 1162 | 1167 | 1012 | 1003 |
| LA25 | 15,10 | 977 | 1261 | 1191 | 1056.2 | 1051.4 |

## 5 Conclusion

This study proposes a new ACO algorithm to solve a single machine total weighted tardiness problem, a flow shop scheduling problem for makespan minimization, and a job shop scheduling problem for makespan minimization. In addition to the changes in local pheromone updating, the proposed algorithm adopts a pheromone modification strategy based on solution quality. When the solution forming behavior of ants falls into stagnation, the algorithm starts to execute the escape mechanism from local optimal solutions, including the exploration-ratio change strategy and the pheromone reset strategy.

The results of this study clearly demonstrate the superiority of the proposed algorithm and are summarized as follows: (i) Example validation has confirmed that, under the present parameter setting, the proposed ACO algorithm with escape mechanism can obtain better or at least the same, solution quality with fewer ants in three scheduling problems considered herein, when compared with the solutions obtained by a conventional ant colony optimization system. (ii) For the job shop scheduling problem for makespan minimization, this study uses the flow which generates an active schedule to form scheduling solutions. Example validation has shown that the proposed method performs better than the conventional way of forming scheduling solutions.

Being compared with existing algorithms, the proposed algorithm will resolve the scheduling problems with less artificial ants and obtain better or at least the same, solution quality.

## References


1. Pinedo, M.: Scheduling: theory, algorithm, and systems, 2nd edn. Prentice Hall, Englewood Cliffs (2002)
2. Belarmino, A.-D.: An SA/TS mixture algorithm for the scheduling tardiness problem. European Journal of Operation Research 88, 516–524 (1996)
3. Ben-Daya, M., Al Fawzan, M.: A tabu search approach for the flow shop scheduling problem. European Journal of Operation Research 109, 88–95 (1998)
4. James, R.J.W.: Using tabu search to solve the common due date early/tardy machine scheduling problem. Computers and Operations Research 24, 199–208 (1997)
5. Liaw, C.-F.: A tabu search algorithm for the open shop scheduling problem. Computer and Operations Research 26, 109–126 (1999)
6. Liaw, C.-F.: A hybrid genetic algorithm for the open shop scheduling problem. European Journal of Operation Research 124, 28–42 (2000)
7. Park, B.J., Choi, H.R., Kim, H.S.: A hybrid genetic algorithm for the job shop scheduling problems. Computer and Industrial Engineering 45, 597–613 (2003)
8. Tadahiko, M., Hisao, I., Hideo, T.: Genetic algorithms for flowshop scheduling problems. Computer and Industrial Engineering 30, 1061–1071 (1996)
9. Bauer, A., Bullnheimer, B., Hartl, R.F., Strauss, C.: Minimizing total tardiness on a single machine using ant colony optimization. In: Proceedings of the 1999 Congress on Evolutionary Computation, pp. 1445–1450. IEEE Press, Los Alamitos (1999)
10. Bullnheimer, B., Hartl, R.F.: A New Rank Based Version of the Ant colony system – A Computational Study. Central European Journal of Operations Research 7, 25–38 (1999)
11. den Bensten, M., Stützle, T., Dorigo, M.: Ant Colony Optimization for the Total Weighted Tardiness Problem. In: Deb, K., Rudolph, G., Lutton, E., Merelo, J.J., Schoenauer, M., Schwefel, H.-P., Yao, X. (eds.) PPSN 2000. LNCS, vol. 1917, pp. 611–620. Springer, Heidelberg (2000)
12. Bullnheimer, B., Hartl, R.F., Strauss, C.: An improved Ant colony system algorithm for the Vehicle Routing Problem. Annals of Operations Research 89, 319–328 (1999)
13. Colorni, A., Dorigo, M., Maniezzo, V., Trubian, M.: Ant colony system for job shop scheduling. Belgian Journal of Operations Research 34, 39–53 (1994)
14. Costa, D., Hertz, A.: Ants can color graphs. Journal of the Operational Research Society 48, 295–305 (1997)
15. Gambardella, L.M., Taillard, E., Dorigo, M.: Ant colonies for the quadratic assignment problem. Journal of Operational Research Society 50, 167–176 (1999)
16. Li, Y.-M., Xul: An ant colony optimization heuristic for solving maximum independent set problems. In: Fifth International Conference on Computational Intelligence and Multimedia Applications ICCIMA 2003. Proceedings, September 27-30, pp. 206–211 (2003)
17. Maniezzo, V., Colorni, A.: The ant colony system applied to quadratic assignment problem. IEEE Transactions on Knowledge and Data Engineering 11, 769–784 (1999)
18. Taillard, E., Gambardella, L.M.: Adaptive Memories for the Quadratic Assignment Problem. Technical Report, IDSIA-87-97, IDSIA, Lugano, Switzerland (1997)
19. Vincent, T., Nicolas, M., Fabrice, T., Daniel, L.: An Ant Colony Optimization algorithm to solve a 2-machine bicriteria flowshop scheduling problem. European Journal of Operation Research 142, 250–257 (2002)
20. Ying, K.-C., Liao, C.-J.: An ant colony system for permutation flow-shop sequencing. Computers and Operations Research 31, 791–801 (2004)

21. Jiang, M.-P.: Application of Ant Colony System in Job-shop Scheduling Problem. Master Thesis, Graduate Institute of Industrial Engineering, National Taiwan University of Science and Technology (2003)
22. Hsu, C.-Y.: A Study on Particle Swarm Optimization for Discrete Optimization Problems, Master Thesis, Graduate Institute of Industrial Engineering, National Tsing Hua University (2003)
23. Chen, L.-S.: Ant Colony System for the Single Machine Total Weighted Tardiness Problem with Sequence-dependent Setups. Master Thesis, Graduate Institute of Industrial Engineering, National Taiwan University of Science and Technology (2002)
24. Palmer, D.S.: Sequencing jobs through a multi-stage process in the minimum total time – a quick method of obtaining a near optimum. Operation Research Quarterly 16, 101–107 (1965)
25. Giffler, B., Thompson, G.L.: Algorithm for solving production scheduling problems. Operations Research 8, 487–503 (1960)
26. Taillard, E.: Benchmarks for basic scheduling problems. European Journal of Operation Research 64, 278–285 (1993)
27. Beasley, J.E.: OR-Library, `http://people.brunel.ac.uk/~mastjjb/jeb/info.html`

# Recognizing Team Formations in Multiagent Systems: Applications in Robotic Soccer

Huberto Ayanegui-Santiago

Universidad Autonoma de Tlaxcala
Facultad de Ciencias Basicas, Ingenieria y Tecnologia
Calzada de Apizaquito s/n, 90350,
Apizaco, Tlaxcala, Mexico
hayanegui@gmail.com

**Abstract.** The main purpose of this work is the recognition of soccer team formations by considering a dynamic structural analysis. Traditionally, the recognition of team formations is carried-out without taking into account an expressive representation of relations between players. This kind of approaches are not able to manage the constant changes occurring in the soccer domain, which results in an inefficient way of recognizing formations immerse in dynamic environments. It is presented in this work an efficient model to recognize formations based on a representation that takes into account multiple relations among defender, midfielder and forward players. The proposed model has been tested with different teams in off-line mode showing that it is able to recognize the different main formations used by a team during a match.

**Keywords:** Robotic soccer, Multiagent modeling, Team formations.

## 1 Introduction

The strong dynamic conditions of the soccer game difficult the task of building useful representations able to facilitate the recognition of formation patterns. In addition, if teams play without any strategy, that is, in a chaotic or not organized way, the difficulty of building such representations becomes harder. Nevertheless, teams playing in strategic and organized ways search for respecting predefined structures or patterns [2,4] also known as formations. A formation in a soccer game is a specific structure defining a distribution of players based on their positions within the field of play. A soccer formation is represented by the relation among defenses, midfielders and forwards of a team, for instance, 5:3:2 (five defenders, three mid-fielders, two forwards). The purpose of recognizing a formation of the opponent team is to improve the performance of our team in order to obtain better results [8,9,1].

Our approach is related to the work of Visser and colleagues [10] that recognizes the formation of the opponent team using a neural networks model. This work feeds the observed player positions into a neural network and tries to classify them into a predefined set of formations. If a classification can be done,

N.T. Nguyen, R. Kowalczyk, and S.-M. Chen (Eds.): ICCCI 2009, LNAI 5796, pp. 163–173, 2009.

the appropriate counter-formation is looked up and communicated to the players. The main difference to our approach is that Visser and colleagues do not represents the multiple relations between players.

Riley and Veloso in [7] use a model to identify *home areas* of players to recognize formations. A home area specifies the region of the field in which the agent should generally be. Thus, they propose that identifying home areas, the agents can infer a role in the team (defender, midfielder or forward players). A drawback of this approach is that due to dynamic conditions of the world, the player movements can generate such a wide range extending considerably the home areas, which difficult the task of determining the role of a player.

The ISAAC system [5] is an automated game analysis tool for simulated robotic soccer. It employs a local adjustment approach to suggest small changes (such as *shoot* when closer to the goal) to a team's designer in order to improve the performance. The suggestions are backed up by examples from the games previously analyzed and provided in a useful format easy to examine by the designers. However, to analyze the behavior of a team, ISAAC identifies successful or failure patterns of actions of multiple agents without consider the formation of a team.

It is presented in this work an efficient model to recognize formations based on a representation that takes into account multiple relations among defender, midfielder and forward players. The proposed model has been tested with different teams in off-line mode showing that it is able to recognize the different main formations used by a team during a match.

The test domain for this research is simulated robotic soccer, specifically, the Soccer Server System [6], used in the Robot World Cup Initiative [3], an international AI and robotics research initiative. The system is a rich multi-agent environment including fully distributed team agents in two different teams composed of eleven agents.

The rest of this paper is organized as follows: Section 2 describes the proposed method. Section 3 presents the analysis of results. Section 4 presents a conclusion of the work and identifies future work.

## 2 The Multiple Relation Model

This work is focused on teams that play in strategic and organized ways searching for respecting predefined patterns or formations. In particular, it is emphasized the analysis of formations following an abstract pattern based on a distribution of zones named Defensive (D), Middle (M) and Attack (A), such as in the classic real soccer game. This pattern will be represented as follows: D:M:A.. In this way, a certain number of players are aligned and belong to a specific zone by respecting predefined relations. For practical reasons of analysis the goalkeeper does not form part of a zone.

Due to the dynamic conditions of the soccer game, the players are in constant moving and temporally breaking the alignment with their zone teammate because tactical reasons. It is pointed out in this work that a model is not able to manage the constant changes of the soccer games without a representation able to express explicitly relevant relations between players.

Based on the D:M:A pattern before mentioned and in order to facilitate the analysis in this work, the formations have been classified as: very defensive, defensive, classical, offensive and very offensive as shown in Table 1.

For each pattern of Table 1, a specific structure or template can be defined in order to represent the relation between players. To define a template the following relations are to be considered:

**Table 1.** Formations based on the D:M:A pattern

| Strategy | D | M | A | | D | M | A |
|---|---|---|---|---|---|---|---|
| | 10 | 0 | 0 | | 3 | 2 | 5 |
| | 9 | 1 | 0 | | 3 | 4 | 3 |
| Very | 9 | 0 | 1 | Offensive | 3 | 3 | 4 |
| deffensive | 8 | 2 | 0 | formations | 2 | 8 | 0 |
| strategy | 8 | 0 | 2 | | 2 | 0 | 8 |
| | 8 | 1 | 1 | | 2 | 7 | 1 |
| | 7 | 3 | 0 | | 2 | 1 | 7 |
| | 7 | 0 | 3 | | 2 | 6 | 2 |
| | 7 | 2 | 1 | | 2 | 2 | 6 |
| | 7 | 1 | 2 | | 2 | 5 | 3 |
| Deffensive | 6 | 4 | 0 | | 2 | 3 | 5 |
| formations | 6 | 0 | 4 | | 2 | 4 | 4 |
| | 6 | 3 | 1 | Very | 1 | 9 | 0 |
| | 6 | 1 | 3 | offensive | 1 | 0 | 9 |
| | 6 | 2 | 2 | formations | 1 | 8 | 1 |
| Offensive- | 5 | 5 | 0 | | 1 | 1 | 8 |
| deffensive | 5 | 0 | 5 | | 1 | 7 | 2 |
| formations | 5 | 4 | 1 | | 1 | 2 | 7 |
| | 5 | 1 | 4 | | 1 | 6 | 3 |
| | 5 | 3 | 2 | | 1 | 3 | 6 |
| | 5 | 2 | 3 | | 1 | 5 | 4 |
| Variants | 4 | 6 | 0 | | 1 | 4 | 5 |
| of classicals | 4 | 0 | 6 | | 0 | 0 | 10 |
| formations | 4 | 5 | 1 | | 0 | 10 | 0 |
| | 4 | 1 | 5 | | 0 | 1 | 9 |
| Classical | 4 | 4 | 2 | | 0 | 9 | 1 |
| formations | 4 | 2 | 4 | | 0 | 2 | 8 |
| | 4 | 3 | 3 | | 0 | 8 | 2 |
| Offensive | 3 | 7 | 0 | | 0 | 3 | 7 |
| formations | 3 | 0 | 7 | | 0 | 7 | 3 |
| | 3 | 6 | 1 | | 0 | 4 | 6 |
| | 3 | 1 | 6 | | 0 | 6 | 4 |
| | 3 | 5 | 2 | | 0 | 5 | 5 |

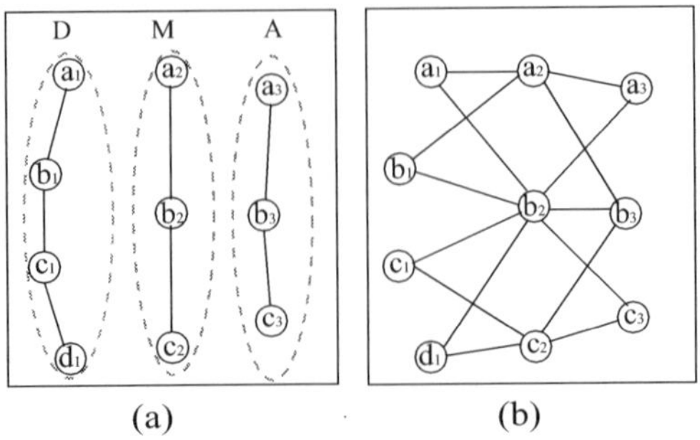


**Fig. 1.** (*a*) *N*eighbor nodes of the same zone are linked. (*b*) Neighbor nodes of neighbor zones are linked. (*c*) Graph obtained from previous process.

- The relation that represents the membership of a player to a zone. A player is a member or belongs to a specific zone if and only if the value of its coordinate x is close to an imaginary line crossing the axis of x in a specific value.
- The relations between neighbor zones. In general, a zone X is neighbor of the zone Y, if and only if there is not an intermediate zone between them. That is, in this case D is a neighbor zone of M and M is neighbor zone of A. But, D is not a neighbor of A.
- The neighborhood relation between players of the same zone. In principle, we can consider the following situation. Let's take the zone D, composed of four players aligned in a vertical line, as shown in Figure 1(a). We can consider that player $a_1$ is neighbor of the player $b_1$ and player $b_1$ is neighbor of player $c_1$, and this of player $d_1$. By transitivity, player $a_1$ should be neighbor of player $d_1$, however, in this work, for practical reasons just players linked by one arc are considered neighbors.
- The neighborhood relation between players that belong to neighbor zones. Once the nodes of same zones have been linked, nodes of neighbor zones based on minimal distances are linked (see Fig. 1(b)). For example, let's specify the neighbors of player $a_1$ of the zone D in Figure 1. Player $a_1$ has player $a_2$ and $c_2$ as neighbors. However, for practical analysis reasons, player $c_2$ is not considered as neighbor of player $a_1$ since in 2D robotic soccer, the ball does not fly, so the interactions between players who are separated considerably, it is null. For example, in Figure 1, players $a_1$ and $c_2$ does not interact between them, however, some players have key roles in a team according to their position, for example in Fig. 1, player $b_2$ interacts with the four neighbors of zone D and with the three neighbors of the attack zone.

Figure 2 shows an example of a 4:3:3 structure according to the relations described previosly.

In order to assign a value to each relation between players, a reference player is placed in the center of the Figure 3(a), thus, it is able to *see* into eight

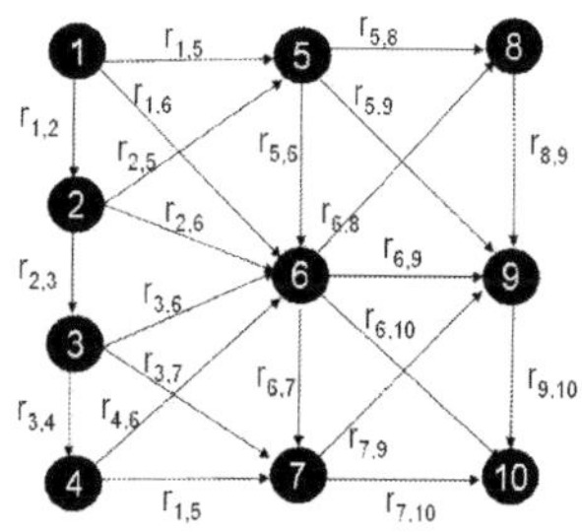


**Fig. 2.** A 4:3:3 formation and their relations

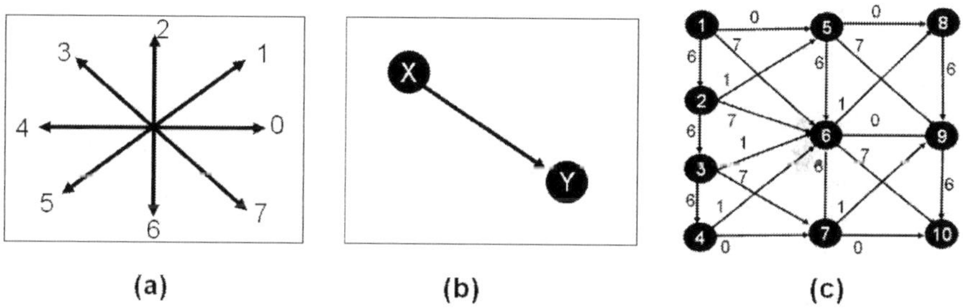


**Fig. 3.** (a) Eight possible orientation values, (b) The agent Y is placed on the orientation 7 with respect to X, (c) Template 4:3:3 with their orientation values

possible orientations. For instance, for the case shown in Figure 3(b), player X see the object Y in the orientation 7. In this way, a relation based on the orientation between players is quantified by the values showed by Figure 3(a). The orientation value 6 between players determines the set of players belonging to a same zone. For instance, there are 3 orientations of value 6 for the defensive zone in Figure 3(c). The orientation values of the template 4:3:3 of the Figure 2 are shown in Figure 3(c).

For instance, template 4:3:3 in Figure 2 can be weighted as is showed in Figure 3(c) where agents from neighbor zones and the same zone are related respectively. The list of formations in Table 1 can be represented in a relational template in a similar way.

A similarity value is a measured that quantifies the similarity between a template and an observed formation in a given cycle. It is considered that bigger is the matching between them more is their similarity and inversely. The measure should be done one-to-one, that is, one player of the template has one and only one correspondent player of the observed formation.

Let $T$ and $O$ be the template and the observation pattern, respectively. The following expression measures the similarity between them:

$$S(T,O) = \sum_{i=1}^{n}(r_{i_T} - r_{i_O}) \quad (1)$$

Where:
$n$: is the number of relations established in the template $T$.
$i$: varies from 1 to $n$.
$r_{i_T}$: is the relation $i$ of the template $T$.
$r_{i_O}$: is the relation $i$ of the observation $O$.

The substraction ($r_{i_T}$ - $r_{i_O}$) of expression (1) refers to a separation between two orientations but not an arithmetic operation. In order to see how the substraction works, let's take as reference the orientations shown in Figure 3(a), and the following example: if $r_{i_T}$ =1 and $r_{i_O}$=3 then $r_{i_T}$ - $r_{i_O}$.= 2. Likewise, If $r_{i_T}$ =0 and $r_{i_O}$=7 then $r_{i_T}$ - $r_{i_O}$=1.

In order to quantify numerically the total similarity between the network representing the template and the observed pattern, they are transformed into a code. Let's take as example of codification the template illustrated in Figure 4(a), which is expressed as follows: 666-66-66-07171710-0710710.
Where:

- 666: represents the relation between agents of the defense zone.
- 66: represents the relation between agents of the middle zone.
- 66: represents the relation between agents of the attack zone.
- 07171710: represents the relations between agents belonging to defensive and middle zones.
- 0710710: represents the relations between agents belonging to middle and offensive zones.

The observed structure is built following the same process used to build the code of the template. In case of the observed structure of Figure 4(b) the codification is: 666-56-66-77171711-0710710. To illustrate how to calculate the similarity between two structures, Figure 4(a) shows a template 4:3:3 and Figure 4(b) is an observed structure obtained in a given cycle.

The code from template is 666-66-66-07171710-0710710 and the code for the observation is 666-56-66-77171711-0710710. Applying Equation (1), Figure 5 shows the result.

The similarity between the template and observed structure is equal to 3 which mean a very good similarity.

Once a current observation has been compared to a set of templates of the Table 1, the more similar template to the current observation is supposed to be the formation played by the team. To determine the set of templates to be compared to the observation structure, it is used a clustering algorithm by considering the following aspects:

- It is taken just the coordinate x of the ten players belonging to the current observed formation;

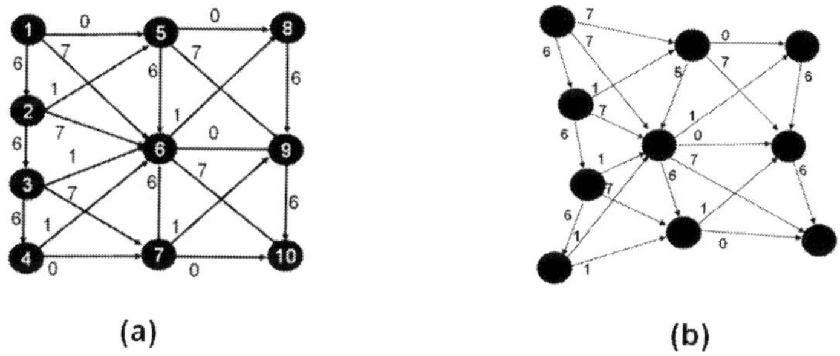


**Fig. 4.** (a) A template with a 4:3:3 formation and (b) an observation

| | | |
|---|---|---|
| Template | = | 666 66 66 07171710 0710710 |
| Observation | = | 666 56 66 77171711 0710710 |
| Distance | = | 1 + 1 + 1 = 3 |

**Fig. 5.** Result of applying Equation (1)

- The agents are grouped as follows: agents with x coordinate lower values form the defensive zone; agents with x coordinate higher values form the attack zone and finally the rest of the agents form the middle zone (values between the lower and the higher values).

We are not able to decide about a formation with only one cycle, it is necessary to consider a sequence of cycles.

## 3 Analysis of Results

In this section, relevant experimental results are obtained from matches of three different soccer teams: TsinghuAeolus 2002 (China), FCPortugal (Portugal) 2002 and RoboSina (Iran) 2005. It has been selected these teams because they illustrate the general application of the proposed work for recognizing different patterns. In the case of TsinghuAeolus, the selected match is TsinghuAeolus vs. Everest. TsinghuAeolus won Everest 7-0 in the final match of the championship 2002.

Figure 6 shows two graphics, each one corresponds to a half time of this match. The two more frequent formations found by the clustering algorithm during the match were 4:3:3 and 4:4:2. The formation 4:3:3 is represented by the darker graphic. Each point of these graphics is composed of a value of similarity in a given cycle. Lower the similarity value bigger the similarity between the observation and the template. Based on this criterion the graphics shows that the structure 4:3:3 is the global formation of the team. As you can see in Figure 6, the team keeps the same formation in both half times of the match.

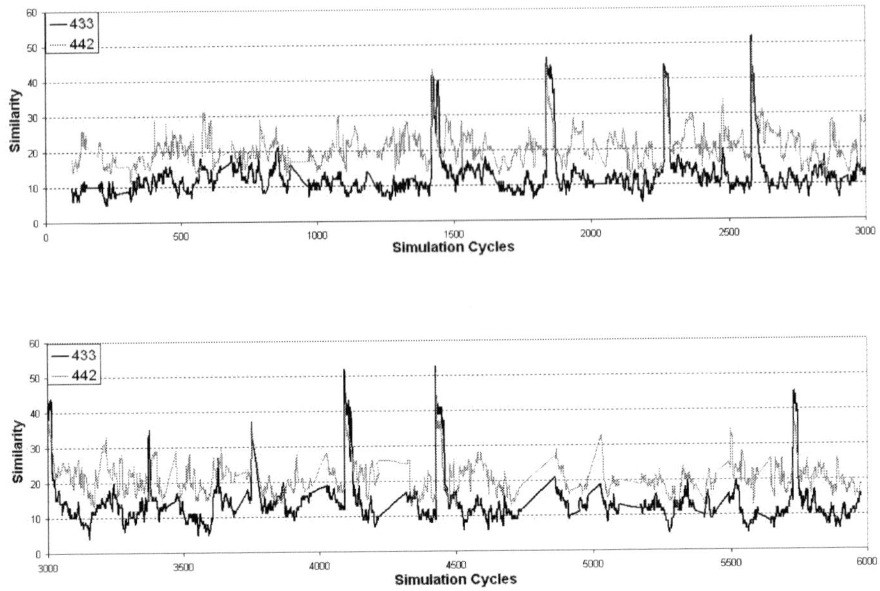


**Fig. 6.** TsinghuAeolus 2002- first and second half time

A very interesting result interpreted from the Figure6 is related with the peaks, which coincide with a transition after scoring a goal of the team TsinghuAeolus.

The second analyzed team is FC Portugal, version 2002. The results illustrated in Figure7 correspond to the match FC Portugal vs UBCDynamo. The two more frequent formations found by the clustering algorithm during the whole match were 4:3:3 and 4:2:4. Figure 7 shows two graphics, each one corresponds to a half time of the whole match. In the first half time, the team started with a formation 4:3:3 and changed to 4:2:4 between cycles 1000 and 2000, just after scoring the second goal. Finally, the team changed to 4:3:3 again. In the second half time, the team started with a formation 4:3:3 and changed to 4:2:4 from the cycle 4000. A possible cause could be that the score was in favor of this team, so the team has possibly decided for a more offensive behavior.

Finally, Figure8 shows the RoboSina team (from Iran) during the last RoboCup championship in the match RoboSina vs, Oxsy. The two more frequent formations found by the clustering algorithm during the match were 4:2:4 and 4:3:3. The formation 4:2:4 is represented by the darker graphic. The structure 4:2:4 is the global formation of the team. As you can see in Figure8, the team keeps the same formation.

It has showed that the proposed model, combined with the clustering algorithm, has been able to detect the global formation of different matches of different teams during different tournaments with very good results. A total of

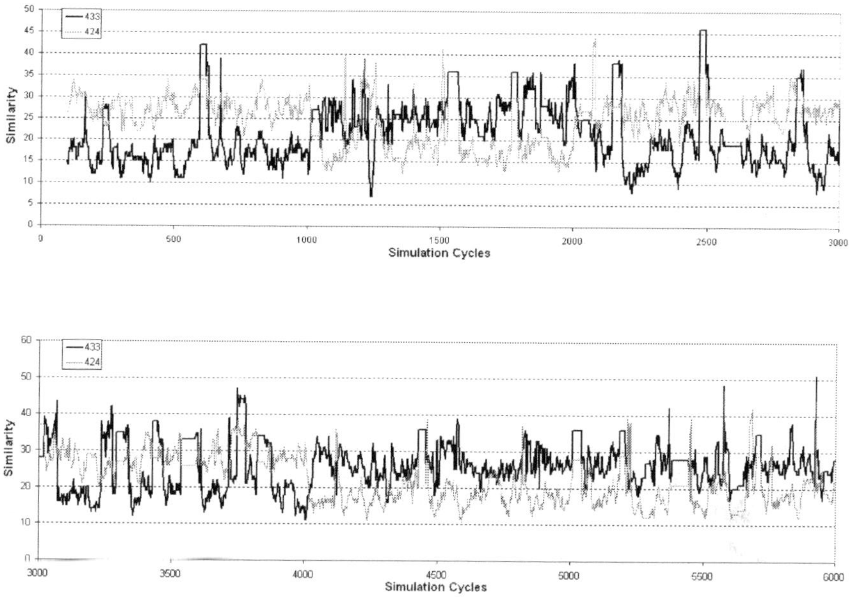


**Fig. 7.** FCPortugal 2002 First and second half time

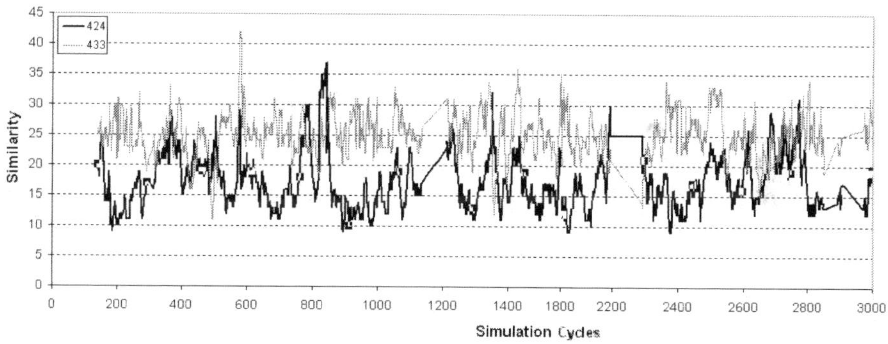


**Fig. 8.** RoboSina 2005 First half time

10 teams and about 30 matches were analyzed. Because the lack of space, it does not present more results of other matches.

To analyze a team during a match two phases are carried-out: in the first phase, a clustering algorithm (k-means) is used to propose the main candidates to be the formations. This algorithm used only data related with the x coordinate of each player to form three different zones and their respective members. This algorithm has an important weakness, it is very sensible to the changes, even if they are not important, of the coordinate x. It has verified in most of the cases that the changes detected in coordinate x do not reveal a real change of formation.

This weakness is due to fact that the information supplied is not enough to detect the fluctuations of a given structure, pointed out that a formation should be considered above all as a structure. On the contrary, the proposed model is provided with a representation model composed of relations between players. These relations build naturally a structure, which can fluctuate due to the positional changes of players during a match, but just relevant changes, and during an important period, are able to break with certain relations and consequently with the given structure representing the formation. The proposed model is able to deal with structures instead of dealing just with positional data.

## 4 Conclusion

This work has been concerned with the pattern recognition of formations in dynamic, complex, competitive and real time domains, such as soccer robotic games. These facts bring about multiple interactions between agents of a team, and make difficult the task of identifying formation patterns. In this work, an efficient model to recognize formations has been presented based on a representation that takes into account multiple relations among players, including: the membership of a player to a zone; the relations between neighbor zones, the neighborhood relation between players of the same zone and the neighborhood relation between players belonging to neighbor zones. Models that do not take into consideration these relations are weak and lack of expressivity whether behavioral patterns belonging to these aspects are to be discovered. In order to cope with this problem, this work has contributed with a multiple relational model which facilitates the discovering formations of a team. It has been shown that the model is able to recognize formations and changes of formations.

As future works, to strength the structural analysis by considering topological changes of the structures or formations can be considered. Another important aspect to improve is to convert the numerical orientations between players in symbolical ones by using a fuzzy approach. This is because the frontiers between orientations values are imprecise. Finally, work in progress is focused on the determination of variants of formations; however, the results obtained so far are not relevant to be shown in this paper. To consider that variants are, in most of the times, the strategic key of the teams to decide the final result of a match.

## References

1. Carmel, D., Markovitch, S.: Incorporating opponent models into adversary search. In: Proceedings of the Thirteenth National Conference on Artificial Intelligence, Portland, OR (1996)
2. Kaminka, G.A., Fidanboylu, M., Chang, A., Veloso, M.M.: Learning the sequential coordinated behavior of teams from observations. In: Kaminka, G.A., Lima, P.U., Rojas, R. (eds.) RoboCup 2002. LNCS (LNAI), vol. 2752, pp. 111–125. Springer, Heidelberg (2003)

3. Kitano, H., Tambe, M., Stone, P., Veloso, M.M., Coradeschi, S., Osawa, E., Matsubara, H., Noda, I., Asada, M.: The robocup synthetic agent challenge 97. In: Kitano, H. (ed.) RoboCup 1997. LNCS, vol. 1395, pp. 62–73. Springer, Heidelberg (1998)
4. Kuhlmann, G., Stone, P., Lallinger, J.: The ut austin villa 2003 champion simulator coach: A machine learning approach. In: Nardi, D., Riedmiller, M., Sammut, C., Santos-Victor, J. (eds.) RoboCup 2004. LNCS (LNAI), vol. 3276, pp. 636–644. Springer, Heidelberg (2005)
5. Nair, R., Tambe, M., Marsella, S., Raines, T.: Automated assistants for analyzing team behaviors. Autonomous Agents and Multi-Agent Systems 8(1), 69–111 (2004)
6. Noda, I., Frank, I.: Investigating the complex with virtual soccer. In: Heudin, J.-C. (ed.) VW 1998. LNCS (LNAI), vol. 1434, pp. 241–253. Springer, Heidelberg (1998)
7. Riley, P., Veloso, M., Kaminka, G.: An empirical study of coaching. In: Distributed Autonomous Robotic Systems, vol. 5, pp. 215–224. Springer, Heidelberg (2002)
8. Steffens, T.: Feature-based declarative opponent-modelling. In: Polani, D., Browning, B., Bonarini, A., Yoshida, K. (eds.) RoboCup 2003. LNCS (LNAI), vol. 3020, pp. 125–136. Springer, Heidelberg (2004)
9. Sycara, K.: Multiagent Systems 10(1):79–93 (1998)
10. Visser, U., Drücker, C., Hübner, S., Schmidt, E., Weland, H.-G.: Recognizing formations in opponent teams. In: Stone, P., Balch, T., Kraetzschmar, G.K. (eds.) RoboCup 2000. LNCS (LNAI), vol. 2019, pp. 391–396. Springer, Heidelberg (2001)

# Semi-traces and Their Application in Concurrency Control Problem

Hoang Chi Thanh

Department of Informatics, Hanoi University of Science, VNUH
334 - Nguyen Trai Rd., Thanh Xuan, Hanoi, Vietnam
{thanhhc@vnu.vn}

**Abstract.** The theory of traces was originated by A. Mazurkiewicz to model non-sequential behaviour of a distributed system. The normal form of a trace shows us an optimal way to execute a process occurred on the system. Nevertheless, the independence of actions in many systems is not static and depends on the history of the systems. To describe this fact, D. Kuske and R. Morin proposed a notion of local independence. Basing on the local independence, we introduce a notion of semi-trace, analyze its structure and the relationship between traces and semi-traces. Furthermore, we investigate some applications of semi-traces in concurrency control problem.

**Keywords:** trace, local independence, concurrency control, distributed system.

## 1 Introduction

The behaviour of a system shows what the system can do. Some mathematical tools have been proposed for representing bebaviours of systems [6], e.g. pomsets, firing sequences, trace languages, event structures, etc. ... Theory of traces was originated by A. Mazurkiewicz to describe nonsequential behaviours of distributed systems [1,3,4-6]. The main idea is based on an independence relation between symbols in an alphabet. The symbols represent actions of a system. Two independent actions can be executed concurrently if they occur adjacently.

Following the theory of traces, the independence of actions is static. It does not depend on the environmental changing. The normal form of a trace points out an optimal way to execute the corresponding process, because it is a shortest sequence of maximal concurrent steps. The authors of [1,4] have constructed some algorithms to find the normal form of a trace directly from its representative. These algorithms become sound tools for transforming sequential processes into concurrent ones. They have been applied for net systems as well [5].

Unfortunately, for many systems, in one situation two actions might be independent while in another situation they cannot be executed concurrently even though they occur adjacently. The theory of traces does not notice the history of a system. In fact, the execution of an action may depend on the sequence of actions already executed. Besides, the independence of a subset of actions is not reflected in the independence

N.T. Nguyen, R. Kowalczyk, and S.-M. Chen (Eds.): ICCCI 2009, LNAI 5796, pp. 174–182, 2009.

relation. To describe this fact, D. Kuske and R. Morin proposed a notion of local independence relation [2]. The notion becomes more appropriate for many systems. Basing on the local independence relation, we introduce a notion of semi-trace, analyze its structure and the relationship between traces and semi-traces. Using semi-trace to represent concurrent process, we apply it in the concurrency control problem.

The paper consists of 4 parts. After the introduction, part 2 is devoted to the notion of local independence relation and semi-trace. An application of semi-traces in the concurrency control problem is presented in part 3. It is a matching algorithm transforming sequential processes of a distributed system into concurrent ones. The conclusion points out some developing researches.

# 2 Local independence and semi-trace

## 2.1 Trace and Trace Language

Let $A$ be a finite alphabet.

**Definition 1**

1) *An independence relation* (over $A$) is a symmetric and irreflexive binary relation $I$ over $A$, i.e.

$$\forall a, b \in A: \ (a,b) \in I \ \Leftrightarrow \ a \neq b \ \wedge \ (b,a) \in I.$$

2) *A concurrency alphabet* is a pair $\mathcal{A} = (A,I)$, where $A$ is an alphabet and $I$ is an independence relation over $A$.

Let $\mathcal{A} = (A, I)$ be a concurrency alphabet. The relation $\cong_I$ over $A^*$ is defined as follows:

$$\forall u, v \in A^*: \ u \cong_I v \ \Leftrightarrow \ \exists u_1, u_2 \in A^*, \exists (a,b) \in I: \ u = u_1 ab u_2 \ \wedge \ v = u_1 ba u_2.$$

The relation $\equiv_I$ over $A^*$ is defined as the reflexive and transitive closure of the relation $\cong_I$, i.e. $\equiv_I = (\cong_I)^*$. It is called *a trace equivalence relation* over $A^*$. The relation is indeed the least equivalence relation over $A^*$ containing the relation $\cong_I$. Each equivalence class $[x]_\equiv$, where $x \in A^*$, is called *a trace* over $\mathcal{A}$. A set of traces is called *a trace language* over $\mathcal{A}$. Trace languages are often used to represent behaviours of systems whilst a trace does a process. Each string in a trace shows a sequential execution of the corresponding process.

Trace composition corresponds to string concatenation: $[x_1]_\equiv \circ [x_2]_\equiv = [x_1.x_2]_\equiv$. So one can compose some traces into a trace or decompose a given trace into some traces. There are many ways to decompose a trace but the following way gives us the normal form of the trace.

**Theorem 1** [1]. Every trace $t = [u]_\equiv$, $t \neq \Lambda$ has a unique decomposition $t = t_1 \circ t_2 \circ \ldots \circ t_m$, $m \geq 1$ such that:

1) Each $t_i$, $i = 1,2, \ldots, m$, is not empty,

2) $t_i = [u_i]_\equiv$, $i = 1, 2, \ldots, m$ and each symbol in $u_i$ occurs only one time and two different symbols in $u_i$ are independent.

3) $\forall i = 1, 2, \ldots, m\text{-}1$, if $t_i = [u_i]_\equiv$ and $t_{i+1} = [u_{i+1}]_\equiv$ then each symbol in $u_{i+1}$ is not independent of some symbol in $u_i$.

In the normal form of a trace $t = [u_1]_\equiv \circ [u_2]_\equiv \circ \ldots \circ [u_m]_\equiv$, each $u_i$ presents a maximal concurrent step. So the normal form of a trace points out the optimal way to execute a concurrent process corresponding to the trace. Some algorithms for finding the normal form of a trace can be found in [1,4].

As review in the introduction, the theory of traces does not notice history of a system. In fact, the execution of an action may depend on the result of execution of actions already executed. Consider the following example.

**Example 2.** We consider a producer - consumer system. Its alphabet is $A = \{p,c\}$, where $p$ represents a production of one item and $c$ a consumption. The sequential behaviour (language) of the system is:

$$L = \{ u \mid u \in A^*, \forall v \in \text{Prefix}(u) : \#_p(v) \geq \#_c(v) \}.$$

i.e. the sequential behaviour of this system describes all possible sequences for which, at each stage, there are at least as many productions as consumptions.

It is easy to think that the independence relation is $I = \{(p,c), (c,p)\}$. The sequence $ppc$ is a sequential process on this system. Hence, $cpp \in [ppc]_\equiv$ but the sequential process $cpp$ can not be executed as lacking item to consume. Further, the normal form of the trace $[ppc]_\equiv$ is $[pc]_\equiv \circ [p]_\equiv$. The concurrent process corresponding to this normal form can not be executed when the system begins to work or no item in the store.

The reason for the above phenomenon is our confusion about the independence of actions in this system. Two actions $p$ and $c$ are independent only when there is at least one item in the store. The independence of actions is local.

## 2.2 Local Independence Relation and Semi-trace

As mentioned in the introduction, not only pairs of actions are independent but also finite sets of actions. Moreover, the independence of actions depends on the history of a system.

**Definition 3.** Let $A$ be an alphabet.

a) *A local independence relation* over $A$ is a non-empty subset $LI$ of $A^* \times 2^A$.

b) The least equivalence relation $\approx_{LI}$ on $A^*$ induced by $LI$, such that:

1. $\forall u, u' \in A^*, \forall a \in A : u \approx_{LI} u' \Rightarrow u.a \approx_{LI} u'.a$,
2. $\forall (u,q) \in LI, \forall q' \subseteq q, \forall v_1, v_2 \in \text{Lin}(q') : u.v_1 \approx_{LI} u.v_2$,

is called *a semi-trace equivalence relation* on $A^*$.

The first property ensures that results of executing an action after execution of two equivalent sequences of actions must be the same. This requirement often is necessary for many systems. The second property requires the results of sequential executions of actions in each concurrent step are the same.

Each equivalence class of a semi-trace equivalence relation is called *a semi-trace*. A set of semi-traces is called *a semi-trace language*.

In order to recognize semi-traces we have to determine more clearly the structure of a semi-trace equivalence relation.

Let $LI$ be a local independence relation over $A^*$. So in each pair $(u,q) \in LI$, $u$ is a history of the system and $q$ is a set of independent actions. The history ensures all actions in the set $q$ be executed concurrently. The following result points out the structure of a semi-trace equivalence relation.

**Theorem 2.** Let $A$ be an alphabet and $LI$ be a local independence relation over $A^*$. The semi-trace equivalence relation $\approx_{LI}$ is the reflexive and transitive closure of the following 'equal' relation $\sim_{LI}$ :

$\forall u, v \in A^*$ :

$$u \sim_{LI} v \Leftrightarrow \exists (u_1,q) \in LI, \exists \alpha, \beta \in A^*, \exists u_2, u_3 \in \text{Lin}(q') \text{ where } q' \subseteq q, \text{ such that: } u = \alpha.u_1.u_2.\beta \text{ and } v = \alpha.u_1.u_3.\beta.$$

Proof: The relation $\sim_{LI}$ is illustrated as the following Fig. 1:

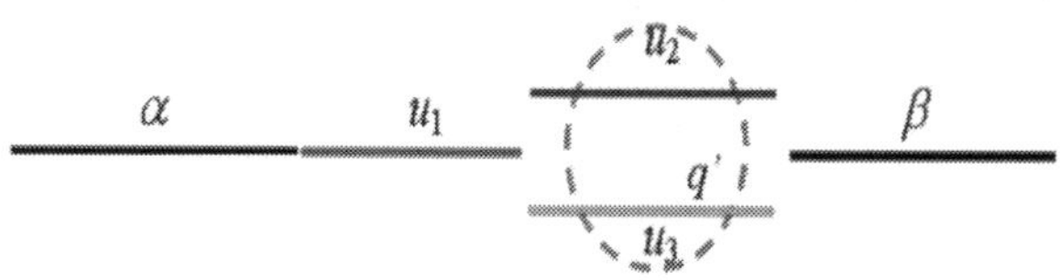


**Fig. 1.** Illustration of the 'equal' relation $\sim_{LI}$

The theorem directly follows from Definition 3 and properties of the reflexive and transitive closure.

**Lemma 3.** Let $t$ be a semi-trace over an alphabet $A$ and let $u$, $v$ be two strings in $t$.

1) $|u| = |v|$ ;
2) $\forall a \in A : \#_a(u) = \#_a(v)$.

Proof: Follows from the definition of the 'equal' relation $\sim_{LI}$ in Theorem 2.

Two semi-trace equivalence strings over an alphabet much have the same length and the same number of occurrences of each symbol. So the results of execution of sequential processes corresponding to these strings are the same.

The local independence relation over that system in Example 2 is:

$$LI = \{(u,\{p\}) \mid u \in A^*, |u| \leq 1\} \cup \{(u,\{p,c\}) \mid u \in A^*, \forall v \in \text{Prefix}(u) : \#_p(v) > \#_c(v)\}.$$

The semi-trace $[ppc]_\approx = \{ppc, pcp\}$. It does not contain the string $cpp$. This is a big distinction between trace and semi-trace.

The relation $\sim_{LI}$ induced by a local independence relation *LI* over $A^*$ defined as in Theorem 2 shows the 'equality' between strings on $A^*$. Two 'equal' strings can be seen as permutations of each other. The relation $\sim_{LI}$ ensures that each subset of an independent set is independent. We can attach these requirements to a local independence relation.

**Definition 4.** A local independence relation *LI* over $A^*$ is *complete* if:

1) $(u,q) \in LI \wedge q' \subseteq q \Rightarrow (u,q') \in LI$ ;
2) $(u,q) \in LI \wedge q' \subseteq q \wedge v \in \mathrm{Lin}(q') \Rightarrow (u.v, q \setminus q') \in LI$ ;
3) $(u,p) \in LI \wedge (u.w_1.v,q) \in LI \Rightarrow (u.w_2.v,q) \in LI$ with $w_1,w_2 \in \mathrm{Lin}(p)$.

We can make a local independence relation be complete. The above definition indicates what pairs to add.

Let denote *CLI* be the local independence relation completed from a local independence relation *LI*. It is easy to see that the local independence relation *LI* is finite if and only if its complete local independence relation *CLI* is finite. A question arises: whether the sets of semi-traces generated by a local independence relation *LI* and by its complete local independence relation *CLI* are the same. This is asserted by the following theorem.

**Theorem 4.** The sets of semi-traces generated by a local independence relation *LI* and by its complete local independence relation *CLI* are the same.

Proof: We show that the relation $\sim_{LI}$ generated by *LI* and the relation $\sim_{CLI}$ generated by *CLI* as in Theorem 2 are identical.

By Definition 4 we get $LI \subseteq CLI$. Thus, $\sim_{LI} \subseteq \sim_{CLI}$.

Assume that $u \sim_{CLI} v$ . Then, $\exists \alpha, \beta \in A^*$, $\exists (u_1,q) \in CLI$ , $\exists u_2, u_3 \in \mathrm{Lin}(q')$ with $q' \subseteq q$ such that: $u = \alpha.u_1.u_2.\beta$ and $v = \alpha.u_1.u_3.\beta$.

Since $(u_1,q) \in CLI$ there exists $(u_1,q'') \in LI$ such that $q \subseteq q''$. So, $q' \subseteq q \subseteq q''$.

Thus: $\exists \alpha, \beta, u_2, u_3 \in A^*$, $\exists (u_1,q'') \in LI$ , $\exists u_2, u_3 \in \mathrm{Lin}(q')$, where $q' \subseteq q''$ such that: $u = \alpha.u_1.u_2.\beta$ and $v = \alpha.u_1.u_3.\beta$. It means: $u \sim_{LI} v$ .

Two relations $\sim_{LI}$ and $\sim_{CLI}$ are identical. The semi-trace equivalence relations over $A^*$ generated by them are the same.

## 2.3 Relationship between Trace and Semi-trace

Let $\mathcal{A} = (A, I)$ be a concurrency alphabet. We define a domain of the independence relation *I* as follows.

**Definition 5.** A subset $B \subseteq A$ is called *a domain* of the relation *I* iff:

1) $\forall a, b \in B : (a,b) \in I$ ,
2) $\forall a \in A \setminus B , \exists b \in B : (a,b) \notin I$ .

An independence relation *I* can be represented by an undirected graph.

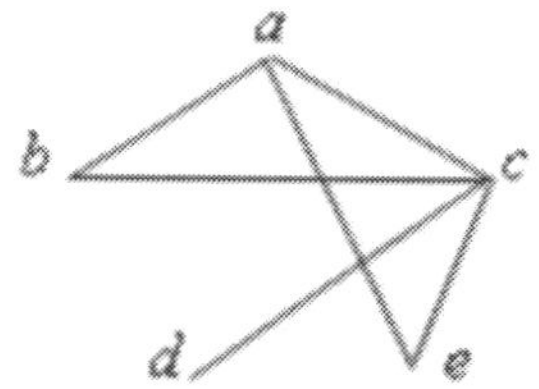


This independence relation has three domains: $\{a,b,c\}$, $\{a,c,e\}$, $\{c,d\}$.

**Fig. 2.** An independence relation

Then, each domain of $I$ is indeed a clique of the graph. Let denote dom($I$) be the set of domains of $I$.

Construct a local independence relation $LI$ over $A^*$ as follows:

$$LI = \{(u,q) \mid u \in A^* \wedge |u| \leq 1, q \in \mathrm{dom}(I)\}.$$

It is easy to show that the trace equivalence relation $\equiv_I$ over $A^*$ is identical to the semi-trace equivalence relation $\approx_{LI}$ over $A^*$. Therefore, in this case, traces and semi-traces are the same. The notion of semi-trace is a real extension of the notion of trace.

# 3 Applying Semi-trace to Concurrency Control Problem

The concurrency control problem was proposed and investigated in [4,5]. The main aim of this problem is to find an optimal way for executing processes occurred in a system. In other word, we have to construct an efficient algorithm for transforming sequential processes into concurrent ones as the following scheme:

**Fig. 3.** Concurrency control scheme for a process

Let $\Sigma$ be a distributed system and let $LI$ be an independence relation over $\Sigma$. In general, the sequential behaviour of the system $\Sigma$ is easy to determine. $L(\Sigma)$ consists of sequential processes occurred in $\Sigma$. Assume that we have constructed the sequential behaviour $L(\Sigma)$ of a system $\Sigma$.

Concurrent steps are being hidden in the sequential processes. We have to discover them. To do so, we use a matching method, that matches sequential processes in $L(\Sigma)$ with pairs of the local independence relation $CLI$ completed from $LI$.

Over some systems, the local independence relation $LI$ is infinite. So is its complete local independence relation $CLI$ (Example 2). The infinitive is indicated by the history in pairs $(u,q)$ of $LI$. Thus, we have to know whether a local independence relation $LI$ is essential or not.

**Definition 6**

1) A pair $(u,q) \in LI$ is called *redundant* if there exists $(u',q) \in LI$ such that $u'$ is a suffix of $u$, i.e. $u = w.u'$ with $w \in A^+$.

2) A local independence relation $LI$ is called *essential* if it does not contain any redundant pair.

Of course, the set of semi-traces generated by a local independence relation and its essential local independence relation are the same. Thus, one can make a local independence relation be finite by ignoring redundant pairs.

We are ready now to construct an efficient algorithm for transforming sequential processes into concurrent ones. The complete local independence relation $CLI$ of $LI$ is stored in computer as follows: Each pair $(u,q) \in CLI$ is stored as a record with four fields: one string field containing the history $u$, one set field containing the step $q$, two integer fields containing the length of the history $u$ and the cardinality of the step $q$. Let denote $k$ the number of pairs. The relation $CLI$ is represented as an array $PAIR[1..k]$ of records sorted by descending of cardinalities of steps. Matching process is performed from right to left. When cutting a subsequence of the input to match, we have to change its right half into a set (in the command 10).

**Algorithm 5**

*Input*: A sequential process in $L(\Sigma)$ stored in a string $t[1..n]$.

*Output*: A concurrent process corresponding to $t$.

```
1 Begin
2   r ← n ;
3   while r ≥ 1 do
4     begin
5       for i ← 1 to k do
6         begin
7           b ← r + 1 - PAIR[i].|q| ;
8           h ← b - PAIR[i].|u| ;
9           OK ← false ;
10      if (t[b..h-1] = PAIR[i].u) ∧ (\overline{t[h..r]} = PAIR[i].q)
11              then
12                begin
13                  write (\overline{t[h..r]}) ;
14                  r ← h - 1 ;
15                  OK ← true ;
16                  break
17                end
18          end ;
19      if not OK then write(t[r]) ;
20      r ← r - 1
21    end
22 End.
```

Note that steps in the output are printed in the order from tail to head. So when executing we proceed in the reverse order.

**Complexity of the algorithm**

Before compairing in the command 7 we have to change the subsequence $t[h..r]$ in to a set $\overline{t[h..r]}$. The set is a subset of actions in Σ. So the complexity of the cycle (5 - 18) is $k.|A|$, where $A$ is the set of actions in Σ. Hence, the total complexity of the algorithm is $n.k.|A|$. The set $A$ and the number $k$ are finite. Thus, the complexity of this algorithm is $O(n)$, i.e. it is linear of the length of input

Back to the producer - consumer system Σ as in Example 2. The system's sequential behaviour is:

$$L(\Sigma) = \{\ u \mid u \in A^*,\ \forall v \in \text{Prefix}(u):\ \#_s(v) \geq \#_b(v)\}.$$

Restricting the local independence relation $LI$ we have the following essential local independence relation:

$$LI' = \{(s,\{s\}), (b,\{s\}), (s,\{s,b\}), (sbs,\{s,b\}), (ssb,\{s,b\})\}.$$

Applying the above algorithm to the sequential process *ssbsbsbbsb* we get the following concurrent process: *s, s,* {s,*b*}, {s,*b*}, {s,*b*}, *b*. The sequential process is executed in 9 steps whilst the corresponding concurrent step is executed in 6 steps.

# 4 Conclusion

Like trace languages, semi-trace languages are a sound tool to represent concurrent behaviours of distributed systems. A local independence relation over a system helps us in representing behaviour of the system more precisely. The algorithm 5 transforms sequential processes occurred on a system into concurrent ones. This algorithm is easy to implement in databases, operating systems, production chain systems … It may be applied to other models of concurrent systems as well.

**Acknowledgment.** This work was supported by the Vietnam National University, Hanoi (Project No. QG-09-01).

# References

1. Aalbersbeg, J.I., Rozenberg, G.: Theory of Traces. Theoretical Computer Science 60, 1–82 (1988)
2. Kuske, D., Morin, R.: Pomsets for Local Trace Languages. In: Palamidessi, C. (ed.) CONCUR 2000. LNCS, vol. 1877, pp. 426–441. Springer, Heidelberg (2000)
3. Mazurkiewicz, A.: Concurrent Program Schemes and Their Interpretations, Aarhus Unvivversity Publication, DAIMI PB-78 (1977)
4. Thanh, H.C.: Algorithms for finding the normal form of traces and synchronous traces. Journal of Computer Science and Cybernetics 17(1), 72–77 (2001)

5. Thanh, H.C.: Transforming sequential processes of a net system into concurrent ones. International Journal of Knowledge-based and Intelligent Engineering Systems 11(6), 391–397 (2007)
6. Thiagarajan, P.S.: Some behavioural aspects of net theory. Theoretical Computer Science 71, 133–153 (1990)

# Design of the Directory Facilitator Supporting Fault-Tolerance in Multi-OSGi Agent System

Sang-Hwan Ryu[1], Seung-Hyun Lee[2], Kyung-Soo Jang[3], Ho-Jin Shin[4], and Dong-Ryeol Shin[2]

[1] Korea Railroad Research Institute, Korea
shryu@krri.re.kr
[2] School of Information & Communication, Sungkyunkwan University, Korea
{lshyun0,drshin}@ece.skku.ac.kr
[3] School of Information & Media, Kyungin Women's College, Korea
ksjang@kic.ac.kr
[4] Dept of. Information Science and Telecommunications, HanShin University, Korea
hjshin@hs.ac.kr

**Abstract.** Multi-agent system and OSGi technologies are essential elements in realizing the upcoming ubiquitous environments. Up until now, different approaches are proposed to support the reliability of multi-agent system level by many researchers. Also, there are mainly focus on fault-tolerance of detection/recovery in the inside system fault or each agent level. We have view on them; will be provide an effective solution of service level for supporting the service reliability of an extended Directory Facilitator (DF) in multi-OSGi agent system. We propose a new, more reliable and extendable DF architecture, and present a detection/recovery algorithm of service fault for safety in railroad system. Through the proposed fault-tolerance architecture, the extended DF of multi-OSGi agent system can improve reliability and interoperability of the service level.

## 1 Introduction

Multi-agent system [1, 3] and Open Service Gateway Initiative (OSGi) [2] technology for ubiquitous environments are becoming increasingly attractive for deployment. OSGi technologies have begun a position as the standardization in the industry and association. Multi-agent environment is autonomous, intelligent, and service-oriented. Also, it can recognize user intention. Specially, OSGi alliance is to specify a Java-based service-delivery platform that allows service-providers, gateway operators, devices and car manufacturers to develop, deploy and manage network-based services in a standardized and cost effective manner. Also, the Foundation for Intelligent Physical Agents (FIPA) [1] establishes multi-agent system standards for the promotion of agent-based technology and interoperability of these standards in systematic. Agents in the FIPA-compliant agent system can provide services to others, and store these services in the DF of the multi-agent system. Users can search for the specific services through the DF, which maintains the services description. The Java Agent

N.T. Nguyen, R. Kowalczyk, and S.-M. Chen (Eds.): ICCCI 2009, LNAI 5796, pp. 183–192, 2009.

Development Framework (JADE) [3] is a popular multi-agent system for supporting function of the DF service. Up until now, different mechanisms for supporting the reliability have been proposed [4]. Reliability as well as fault-tolerance is used to the vital issues for implementation of a multi-agent system [5]. Most of these mechanisms are approaches for detecting and recovering the agent system level or each agent`s fault. However, fault-tolerant mechanism for the DF of the multi-agent system not exists. Therefore, a mechanism focusing on the reliability of agent`s service level and service repository is needed. Also, the advantage of these approaches is not blocked when a failure happens, but can be provided continuously the interoperability and reliability [6]. We try to apply them for safety in our railroad project.

In this paper, our goals are the followings: firstly, we provide the interoperability between OSGi platform and multi-agent system. It means that an important element is the interoperability for service management between heterogeneous platforms. Secondly, the External DF Fault Manager (EDFFM) module is proposed, supporting detection/recovery of DFs' faults. Thirdly, the Service Fault Detection/Recovery Manager (SFDRM) module is proposed for detecting agents` faults. To detect a fault of the DF, the proposed EDFFM module acts as a DF-independent agent in a multi-OSGi agent system. The SFDRM is DF dependent module that monitors service agents and uses our proposed service recovery mechanism. By using the proposed fault-tolerant model, service reliability of the DF can be improved and storage overhead reduced.

The remainder of this paper is organized as follows. Section 2 describes the OSGi platform, the problem background and existing fault-tolerant approaches. Section 3 presents mainly the proposed DF architecture for a service repository supporting fault-tolerance. In Section 4, we introduce the proposed algorithm for detection of agents` fault and recovery of services. Section 5 shows our DF prototype, and evaluation. Finally, conclusion and future work are discussed in Section 6.

## 2 Related Work

In this section, we introduce OSGi and potential DF problems that can be generated. Further, the existing approaches for multi- agent system reliability are introduced.

### 2.1 OSGi

In the OSGi alliances [2], system bundles discuss service lifecycle, management, running methods, as service providers to access local or dynamic networks environments. For example, typical OSGi platform provides as service platform related to all types of networked devices, in home, vehicle, mobile and other environments. For them, OSGi alliance establishes that the committee consists of standard platform and deployment service infrastructure.

The OSGi platform mainly supports the interaction among services through a registry that contains service descriptions published by service providers. Also, OSGi can provide interoperability as an authorized service bundles. It is similar to service management and the DF's service role on the view service management in agent system. But, Interoperability is in agent system. A service description is published it becomes available for other services registered with the framework. The service registry allows service requesters to discover and bind published services. Services in OSGi are Java classes or interfaces that are collected as functional and deployment units referred to as bundles in the OSGi terminology. We are use to OSGi platform as a framework for the service management trough environment called kopflerfish as a service gateway.

## 2.2 Problems of Multi-agent System

We classify the potential problems relevant to the DF in multi-agent systems into three cases. The first case is server's loss. In the case of host`s loss, the DF is also dead. If the host is dead, hosts recovery should be performed to firstly recover the DF.

The second case is the case where a problem is caused by the DF. This case can be generated by an agent`s incorrect message receipt, network problem and so on. Although the DF is dead, service agents don`t know that the DF has experienced failure. So, they operate under the belief that the DF is alive. Similarly, service request agents will attempt to access the DF. In this case, system-wide problems can be generated such as when the problems are the fault of the host.

Finally, a problem could be caused by the agent. An agent fault means that the service agent is unexpectedly dead after service registration in the DF. Most existing DF mechanisms suffer from this deficiency. So, services registered in the DF are continually maintained, even if the service agent is actually dead. In this case, the service request agent succeeds in searching the specific service from the DF, but cannot communicate with the service agent. Therefore, DF architecture supporting service fault-tolerant issues is essential.

## 2.3 Existing Fault-Tolerant Approaches

For multi-agent system`s reliability, the checkpoint concept has been used for a long time [7]. The checkpoint is an active agent`s current state, and is stored in a permanent database. This approach uses the latest checkpoint data stored immediately before an agent fault, and recovers agent failure when an agent experiences a fault. However, frequent store of checkpoint data results in storage flood and communication overhead. So, this mechanism is not appropriate for mobile service environments.

Supervisor and executor agents concepts are used for monitoring the agent`s execution, detecting the suspected fault conditions and reacting to these conditions in [8]. In addition, this is located in an external host, separated by a multi-OSGi agent system, for supervising and monitoring. That is, the external host becomes the main system for fault-tolerance. However, communication overhead is generated because this should periodically observe the multi-OSGi agent system.

Guiyue Jin et al. [9] proposed a fault-tolerant protocol using the check pointing concept and replication scheme in multi-agent system. In addition, this approach uses rear guards and witness agents. The rear guards monitor the execution of actual agents. When failure occurs or they failure is suspected, rear guards transmit the witness agent to the location of the agent. Then, the witness agent executes a series of processes for fault recovery. This approach is most applicable for mobile environments.

The Reliable Agent Layer (RAL) and Fault-Tolerant System Manager (FTSM) have been proposed for multi-agent systems supporting fault-tolerance in [11]. The RAL is a platform-dependent component and is located in the multi-agent system. The RAL mainly stores the agent`s checkpoints. The FTSM is platform-independent component and supports multi-agent system fault-tolerance. This approach supports various multi-agent systems and overcomes many preexisting problems, such agent death, host death, agent unresponsiveness, agent migration faults, certain communication problems, and faults caused by resource unavailability.

The above mentioned approaches propose a fault-tolerant mechanism for improved the service reliability. That is, almost all existing approaches focus on agent system level or agent level fault-tolerance. However, no mechanism is proposed to support the reliability of the service level in the multi- agent system. We are proposed multi-OSGi system environment that includes an extended DF, which stores the service descriptions provided by agents in service level.

## 3 Proposed Architecture

In this paper, we provide to support interoperability by the using of message parsing between the agent and bundle in multi-OSGi agent system. In the FIPA specification [1], as the platform for interoperability recommends agent system based on several resources (e.g. Ontology-resource, Protocol resource, Language resource, and so on). The proposed system for interoperability is designed to integrate the agent and service bundles. Also, we designed the EDFFM, which supports the detection and recovery of DF`s faults, and the SFDRM supporting the detection of an agent`s fault and recovery of the agent`s service. Fig 1 shows the proposed the DF architecture.

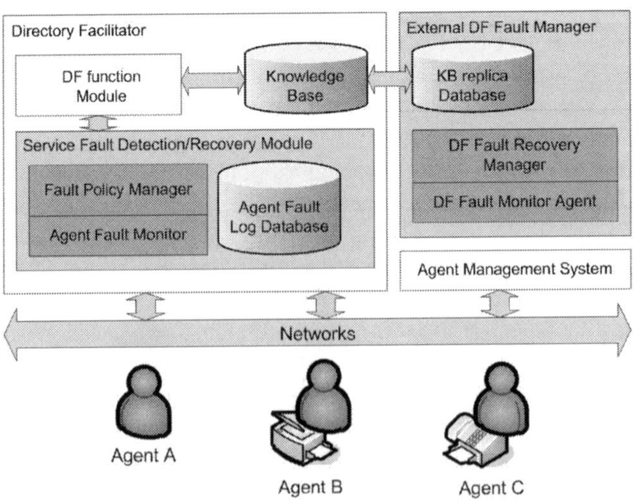


**Fig. 1.** Directory Facilitator supporting Fault-Tolerance

### 3.1 External DF Fault Manager (EDFFM)

The EDFFM is an external application for monitoring the DF`s faults. This can operate independently because it is not located in the DF. In other words, Multi- agent system`s faults have no effect on the EDFFM. This can detect a host or agent system failure. The EDFFM consists of the DF Fault Recovery Manager (DFFRM), DF Fault Monitor Agent (DFFMA) and one permanent database called the Knowledge Base Replica Database (KBRD). The role of each component is as follows:

**DFFRM.** The DFFRM performs DF recovery when DF faults are generated. Faults relevant to the DF are classified into two cases. The first case is the agent`s fault and the second case is the host`s fault. The DFFRM performs the following operations for solving these cases, through the received message:

(1) If the DFFRM receives the *RES (df,loss)* message from the DFFMA, it restarts the DF on the current host. Then, the DFFRM transmits the DF registration message to the Agent Management System (AMS). In addition, it requests all agents in service log data stored in Knowledge Base Replica Database (KBRD) to re-register services.

(2) When the DFFRM receives the message *RES (host,loss)* from the DFFMA, it creates a list of available hosts registered in the AMS. Then, one available host is selected from the list. This executes a new DF module on selected host and requests the AMS deregisters the existing registered DF. The DFFRM requests the AMS to register a new DF on a selected host, after deregistration of the existing system. Finally, the DFFRM requests all agents stored in the KBRD to re-register services with the DF. Fig 2 shows the pseudo code for recovery when a fault occurs in the DF.

```
if (DF is dead at host1) then
    if(host1 is dead) then
    hostlist = available hosts registered in AMS;
    newhost = available host in hostlist;
  else
   newhost = host1;
  end if;
   DFFRM executes new DF on newhost;
   DFFRM requests AMS to deregister DF on host1;
   DFFRM requests AMS to register DF on newhost;
   DFFRM sends request messages for re-registration to agent in KBRD;
end if;
```

**Fig. 2.** Pseudo-code sample for DF recovery

**DFFMA.** The DFFMA monitors the DF`s state. By periodically sending the *REQ (df,alive)* message to the DF, for checking the service repository. This expects the receiving message, *RES (df,alive)*, from the DF. When the DFFMA receives the *RES (df,alive)* from the DF, no action is taken. However, when the DFFMA cannot receive the *RES (df,alive)* message, it suspects DF fault and tests the host where the DF exists. If the host is dead, the *RES (host,loss)* message is generated. In the opposite scenario, when the host reacts to the DFFMA`s test, this generates the *RES (df,loss)* message. The DFFMA transmitted the generated message is transmitted to the DFFRM.

**KBRD.** The KBRD is a permanent storage to store information of service agents and services registered in the DF. It maintains lightweight information of service agents that should again register services with the DF. Figure 3 is the structure of the KBRD. The KBRD includes different parameters such as service agent`s AID, address, provided services, registered time and TTL.

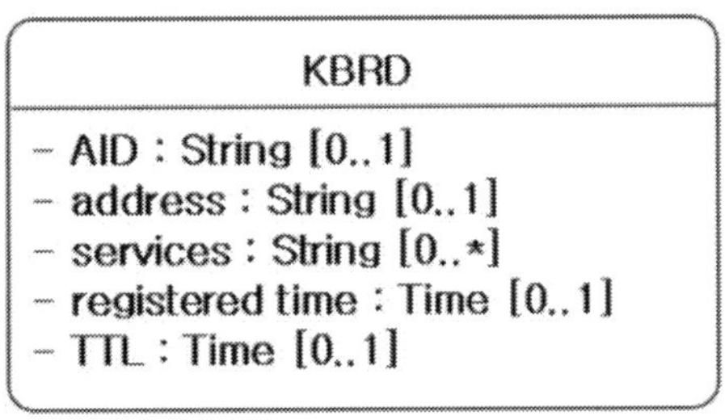


**Fig. 3.** Structure of KBRD

### 3.2 Service Fault Detection/Recovery Module (SFDRM)

The SFDRM is a module for detection/recovery of services and agent`s fault, but the EDFFM is the module for detection/recovery of DF or host`s fault. In this paper, we propose a fault-tolerant mechanism of service repository level and service level in the multi-OSGi agent system. So, the SFDRM does not consider the recovery mechanism of agent faults, but does consider the recovery mechanism of services provided by agents.

The SFDRM considers the detection mechanism of agent`s faults. A detailed service recovery algorithm for service fault detection/recovery is mentioned in Section 4. In contrast to the EDFFM, the SFDRM is a component included in the DF. The SFDRM consists of an Agent Fault Monitor (AFM) for detecting the agent`s fault, and Fault Policy Manager (FPM) for recovering the agent`s service and Agent Fault Log Database (AFLD), which is used maintaining a log of failed agents. Detailed description of each module is provided as follows:

**AFM.** The AFM is a module for monitoring the execution of agents with registered services in the DF. It does not continually monitor all agents, but only agents searched through the DF, by the user agent. In other words, when the SFDRM receives a message for searching the specific service, it searches relevant services in the DF`s

Knowledge Base, and queries whether the owner agent that provides the service is alive. This mechanism can reduce the communication overhead comparison in the existing monitoring mechanism.

**FPM.** The FPM contains policies for service recovery and recovers services using these policies when agent`s fault is detected by the AFM. These policies may be initiated on de-registration of registered service, service`s re-registration after recovery, and search for failed services. The FPM performs processes for fault recovery using these policies.

**AFLD.** The AFLD stores a log of service or agent`s faults. Existing replication scheme stores all agents` information, for recovering service after the agent fault recovery. However, the AFLD maintains only information of the agents that have suffered from a fault. So, the proposed architecture has advantages of reducing communication overhead and increasing storage efficiency. The proposed architecture is suitable for mobile environments, because of these advantages.

## 4 Fault Detection and Recovery Algorithm for Service

In this section, we describe the proposed fault detection and recovery algorithm for service. We assume that the EDFFM checks faults and aliveness of the DF.

First of all, we define the three cases of probable conditions that can be generated from the DF. We describe the proposed mechanism using scenarios of these three cases. Fig 4 shows the algorithm proposed for detecting and recovering the faults. As shown in fig 4, agent A provides and registers service x to the DF. Agent B is the service requester which searches service x.

**Case (a)** is the scenario of normal search process, as shown in case (a) of fig 4. Agent B requests the SFDRM to search the service x. After searching for service x, the SFDRM asks the service agent whether the agent is alive. If agent A is alive, the SFDRM return service x to agent B.

**Case (b)** is when the agent is dead after registering service x. When agent B requests the SFDRM to search the service x, the SFDRM can search service x and ask agent A to identify the existence of agent A. However, agent A does not respond due to a fault. The SFDRM knows that agent B is dead, and deregisters service x`s description in the DF. Finally, it adds service x`s description information to the AFLD. Similarly, agent B fails to search for service x.

**Case (c)** is the scenario that agent A`s service is re-registered in the DF after agent A`s recovery. When agent B searches for service x, the SFDRM first searches service x in the AFLD, because it cannot find service x in the DF. If the SFDRM finds service x in the AFLD, the SFDRM will be certain that agent A is dead. If agent A recovers quickly, the SFDRM can identify the existence of agent A, when service x is requested. Finally, the SFDRM deletes the log data in the AFLD, and re-registers service x in the DF. This service x can be searched by agent B.

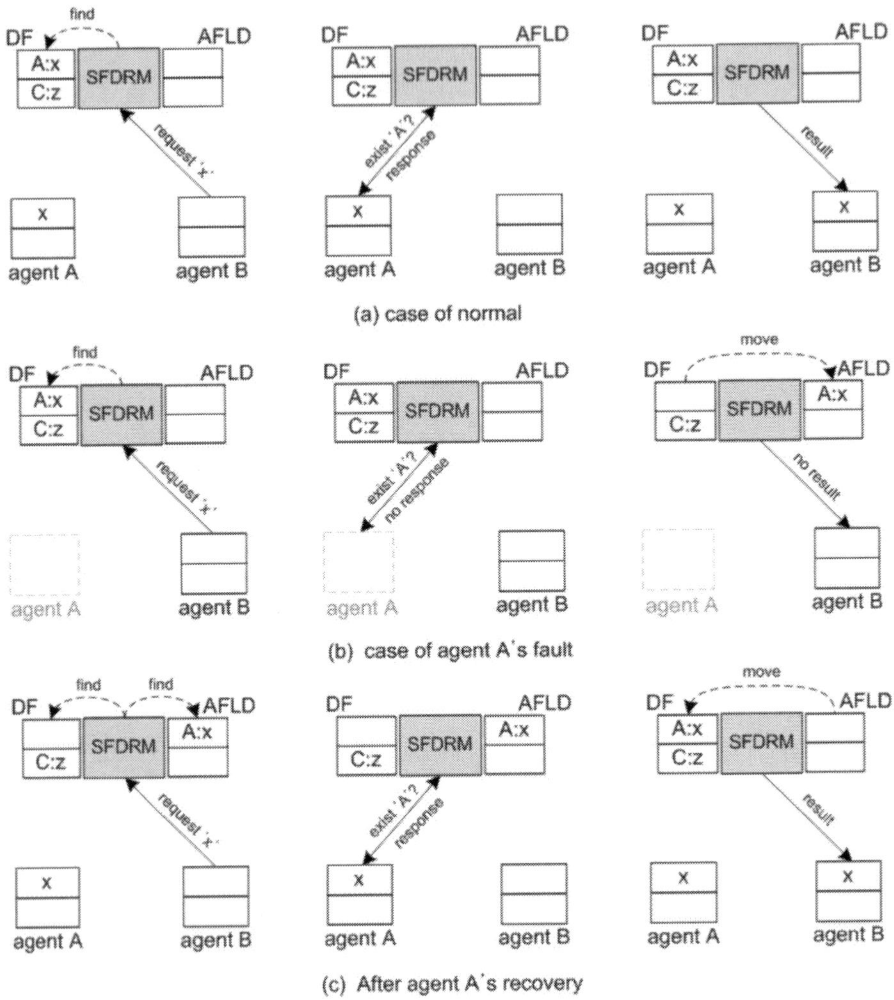


**Fig. 4.** DF's fault detection/recovery algorithm

## 5 Evaluation and Implementation

As described in [8], the reliability can be measured the reliability as a number of each service agent. Therefore, system availability, as the measure of reliability, cannot be applied to them. Especially, a service agent can either successfully complete its tasks or fail to do so. For this reason, the reliability can present the percentage of successful service agent to achieve our goals. Our final goal guarantees the reliability which in fault/recovery algorithm can be supported the interoperability between OSGi platform and multi-agent system. Against successful case, fault case means no guarantee for reliability. It is that calculates the reliability only present the successful service agent which managed registered services in DF agent. As a result, the reliability can be calculated using the following expression that modified from [10]:

$$R = \frac{S}{A} \cdot 100\,[\%] \quad (1)$$

*R:* *R*eliability
*S:* No. of successful Services
*A:* No. of total Agents

Also, our DF prototype is developed using Java 5 and Java Server Page (JSP) through above mentioned architecture. Fig 5 shows the extended DF prototype. To communicate with agents, developed DF uses FIPA-ACL (Agent Communication Language) and FIPA-SL (Semantic Language). To evaluate our DF, we plan to test its performance through comparison among different fault-tolerant mechanisms. All of these mechanisms are similar to ours from the viewpoint of recovery. However, communication and storage overhead are different among them, clearly. Therefore, we will demonstrate the superiority of our approach.

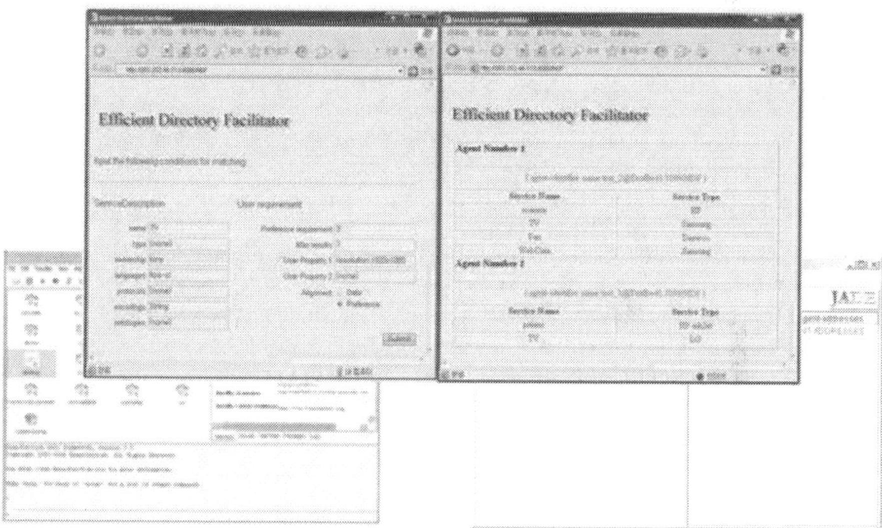


**Fig. 5.** Prototype of Extended DF

## 6 Conclusion and Future Work

In this paper, we proposed an extended Directory Facilitator (DF) architecture for supporting reliability and interoperability in a multi-OSGi agent system. The proposed architecture can detect faults in the DF, store services and recover from faults. The proposed architecture uses a policy-based algorithm, in order to deregister or re-register services registered in the DF, when a service agent experiences a fault. The proposed fault/recovery algorithm can reduce network overhead by communicating only with relevant agents when search is requested, this is in contrast to the existing periodical monitoring mechanism. This solves the problem of DF inefficiency through requesting service re-registration to agents not using the conventional replication scheme. Furthermore, the proposed mechanism uses a lightweight fault log, reflecting it`s suitability to mobile devices with limited resources. Now, we try to apply fault-tolerance as OSGi Agent system for safety in railroad environment.

## Acknowledgment

The research was supported by the "The future rail safety projects" Project of Korea Railroad Research Institute, Republic of Korea.

## Reference

1. FIPA: The Foundation for Intelligent Physical Agents, http://www.fipa.org
2. OSGi Alliance, http://www.osgi.org/Main/HomePage
3. Giovanni Caire: JADE Tutorial (2003), http://jade.tilab.com
4. Bellifemine, F., Bergenti, F., Caire, G., Poggi, A.: JADE-A Java Agent Development Framework. In: Multi-agent Programming. Multiagent Systems, Artificial Societies, and Simulated Organizations, vol. 15, pp. 125–147. Springer, New York (2006)
5. Mohindra, A., Purakayastha, A., Thati, P.: Exploiting non-determinism for reliability of mobile agent systems. In: International Conference on Dependable Systems and Networks, DSN 2000, Proceedings, June 2000, pp. 144–153 (2000)
6. Charlton, P., Bonnefoy, D., Lhuillier, N.: Dealing with interoperability for agent-based services. In: Agents 2001, pp. 236–237 (2001)
7. Lyu, R.M., Wong, Y.T.: A progressive fault tolerant mechanism in mobile agent systems, http://www.cse.cuhk.edu.hk/~lyu/paper_pdf/SCI2003.pdf
8. Patel, R.B., Garg, K.: Fault-tolerant mobile agents computing on open networks, http://www.caip.rutgers.edu/~parashar/AAW-HiPC2003/patel-aaw-hipc-03.pdf
9. Jin, G., Ahn, B., Lee, K.D.: A Fault-Tolerant Protocol for Mobile Agent. In: Laganá, A., Gavrilova, M.L., Kumar, V., Mun, Y., Tan, C.J.K., Gervasi, O. (eds.) ICCSA 2004. LNCS, vol. 3045, pp. 993–1001. Springer, Heidelberg (2004)
10. Tosic, M., Zaslavsky, A.: Reliable Multi-OSGi agent systems with Persistent Publish/Subscribe Messaging. In: Ali, M., Esposito, F. (eds.) IEA/AIE 2005. LNCS (LNAI), vol. 3533, pp. 165–174. Springer, Heidelberg (2005)
11. Tosic, M., Zaslavsky, A.B.: The External Fault-tolerant Layer Supporting Multi-agent Systems from Different Vendors. In: IAT 2005, pp. 435–438 (2005)

# Agent-Based Provisioning of Group-Oriented Non-linear Telecommunication Services

Vedran Podobnik[1], Ana Petric[1], Krunoslav Trzec[2], Vedran Galetic[2], and Gordan Jezic[1]

[1] University of Zagreb, Faculty of Electrical Engineering and Computing, Croatia
[2] Ericsson Nikola Tesla, Croatia
{vedran.podobnik,ana.petric,gordan.jezic}@fer.hr,
{krunoslav.trzec,vedran.galetic}@ericsson.com

**Abstract.** The future of telecommunications is directed towards creating environments aware of user preferences, mobile device capabilities and communication context. Consequently, telecommunication operators have recognized the importance of dynamic formation of user groups according to similar characteristics (i.e., user preferences, user device, communication context, etc.). In this paper, we propose an agent-based solution that supports provisioning of group-oriented telecommunication services by applying ontology-based user profiling as well as semantic clustering techniques. Furthermore, proof-of-concept group-oriented service is presented.

**Keywords:** Intelligent Software Agents, Group-oriented Service, Semantic Clustering, Ontology-based User Profiles.

## 1 Introduction

Actors on telecommunication market [1] are pursuing innovations and launching new value added services in order to increase revenues. This new market demands and technological developments have led to the convergence of different domains (i.e., telecommunications, information technology (IT), the Internet, broadcasting and media) all involved in the telecommunication service provisioning process. The ability to transfer information embodied in different media into digital form to be deployed across multiple technologies is considered to be the most fundamental enabler of convergence [2].

The evolved next-generation network (NGN) [3] should aim at taking changing user demands into account and at creating spontaneous, adaptive services that can be delivered anytime, anywhere, to any device the user prefers. Therefore, the realization of the full potential of convergence will make it necessary for telecommunication operators to deploy dynamic, cooperative and business-aware consistent knowledge layer in the network architecture in order to enable ubiquitous personalized services. Providing such context-aware services transparently to the user is not only challenging from a network point of view, but also imposes severe requirements on the service provisioning.

N.T. Nguyen, R. Kowalczyk, and S.-M. Chen (Eds.): ICCCI 2009, LNAI 5796, pp. 193–204, 2009.

In this paper we focus on a special type of non-linear telecommunication services[1]: *group-oriented services.* We define a *group-oriented service* as a telecommunication service that is provisioned to a group of users with certain similarities (e.g., similar preferences, devices and/or context). The main idea behind group-oriented services is to group users into clusters taking into account users' interests, characteristics of their devices and the context in which they find themselves while requesting a service. To achieve that it is necessary to introduce a rather new approach in the service provisioning process: building implicit social networks of mobile users. Unlike explicit social networks (e.g., Facebook[2], MySpace[3] or LinkedIn[4]), implicit networks are built autonomously based on similarities of user profiles, without the interference of users themselves and in order to provide useful information for telecommunication operators.

Semantic Web technologies are rather novel but very amenable grounding for user clustering [5, 6], while software agents have proven to be very suitable for user profile management [7, 8] and telecommunication processes enhancements [1, 9, 10].

This paper is organized as follows. In Section 2, we present how to use the technologies of the Semantic Web to create an ontology-based profile of telecommunication service user. Section 3 describes semantic clustering of users based on their profiles. In Section 4, a multi-agent system enabling service provisioning in NGN is presented, as well as a proof-of-concept service which demonstrates provisioning of group-oriented services. Section 5 proposes ideas for future research work and concludes the paper.

## 2 User Profiles

The Semantic Web is a vision in which knowledge is organized into conceptual spaces according to meaning, and keyword-based searches are replaced by semantic query answering [11]. Semantic Web languages, such as *Resource Data Framework*[5] *(RDF), RDF Schema*[6] *(RDFS)* and the *Web Ontology Language*[7] *(OWL),* can be used to maintain detailed user profiles. With the help of various query languages, based on *Structured Query Language (SQL)* syntax, it is possible to perform efficient semantic profile matchmaking once the profiles have been created according to a certain standard. Such matchmaking enables us to perform clustering according to true, semantic similarities, rather than keyword matchmaking. Semantic queries are the main means of information retrieval used in current research in this area. Inspiration for a query-based style of reasoning stems directly from the widespread propagation of RDBMS

---

[1] *Non-linear telecommunication services* are interactive services where the user participates in the service provisioning procedure, tailoring the service to his/her preferences, device and/or context [4].

[2] http://www.facebook.com

[3] http://www.myspace.com

[4] http://www.linkedin.com

[5] RDF specifications: *http://www.w3.org/RDF/*

[6] RDFS specifications: *http://www.w3.org/TR/rdf-schema/*

[7] OWL specifications: *http://www.w3.org/TR/owl-features/*

(*Relational Database Management Systems*). Semantic query languages have a number of features in which they differ from SQL queries due to Semantic Web knowledge, which can be either *asserted* (explicitly stated) or *inferred* (implicit), being *network structured,* rather than *relational*. Also, the Semantic Web assumes an OWM (*Open World Model*) in which the failure to derive a fact does not imply the opposite [12], in contrast to *closed world* reasoning where all relations that cannot be found are considered false [13].

In our implementation information was retrieved by means of RDQL (*RDF Data Query Language*) and SeRQL (*Sesame RDF Query Language*) queries. A Sesame [14] repository with OWL support [15] was utilized to store the required knowledge.

## 2.1 Related Work on User Profiling

The W3C is working on the CC/PP[8] (*Composite Capabilities/Preferences Profile*), an RDF-based specification which describes device capabilities and user preferences used to guide the adaptation of content presented to that device. It is structured to allow a client to describe its capabilities by reference to a standard profile, accessible to an origin server or other sender of resource data, and a smaller set of features that are in addition to or different than the standard profile. A set of CC/PP attribute names, permissible values and associated meanings constitute a CC/PP vocabulary.

OMA's (*Open Mobile Alliance*) UAProf[9] (*User Agent Profile*) specification, based on the CC/PP, is concerned with capturing classes of mobile device capabilities which include the hardware and software characteristics of the device. Such information is used for content formatting, but not for content selection purposes. The UAProf specification does not define the structure of user preferences in the profile.

## 2.2 Proposed User Profiles

We extended the CC/PP and UAProf specifications and mapped them to the OWL ontology in order to create the telecommunication service user profile, as shown in List. 1. Opening rows (i.e., rows 1-10) and the closing row (i.e., row 43) must be defined according to XML and RDF specifications, while remaining rows (i.e., rows 11-42) contain information about the user. We distinguish five different types of user information:

- *Rows 11-13* and the *row 42* – Every profile is an instance of a *UserDeviceProfile* (an ontology defining *UserDeviceProfile* is shown in Fig. 1). The profile is described with 20 attributes, as follows;
- *Rows 14-20* – Five attributes defining the user device *hardware*;
- *Rows 21-25* – Three attributes defining the user device *software*;
- *Rows 26-33* – Six attributes defining *user preferences*;
- *Rows 34-41* – Six attributes defining the user *context*;

[8] CC/PP specifications: *http://www.w3.org/Mobile/CCPP/*

[9] UAProf specification: *http://www.openmobilealliance.org/*

```
<?xml version="1.0"?>
<rdf:RDF
    xmlns="http://www.tel.fer.hr/astorm/InfoServiceV2.owl#"
    xmlns:is="http://www.tel.fer.hr/astorm/InfoServiceV2.owl#"
    xmlns:rdf="http://www.w3.org/1999/02/22-rdf-syntax-ns#"
    xmlns:xsd="http://www.w3.org/2001/XMLSchema#"
    xmlns:rdfs="http://www.w3.org/2000/01/rdf-schema#"
    xmlns:owl="http://www.w3.org/2002/07/owl#"
    xml:base="http://www.tel.fer.hr/astorm/User1.owl">

<!-- profile for User1's mobile phone Sony Ericsson K700 -->
    <is:MobilePhoneProfile rdf:ID="SonyEricssonK700">

<!-- hardware -->
    <is:hasAvailableMemory rdf:datatype="http://www.w3.org/2001/XMLSchema#int">18000</is:hasAvailableMemory>
    <is:hasHorizontalScreenResolution rdf:datatype="http://www.w3.org/2001/XMLSchema#int">180</is:hasHorizontalScreenResolution>
    <is:hasVerticalScreenResolution rdf:datatype="http://www.w3.org/2001/XMLSchema#int">230</is:hasVerticalScreenResolution>
    <is:hasScreenBitsPerPixel rdf:datatype="http://www.w3.org/2001/XMLSchema#int">16</is:hasScreenBitsPerPixel>
    <is:hasImei rdf:datatype="http://www.w3.org/2001/XMLSchema#string">35461002-303538-0-34</is:hasImei>

<!-- software -->
    <is:hasOs rdf:resource="http://www.tel.fer.hr/astorm/InfoServiceV2.owl#BasicOs"/>
    <is:hasBrowser rdf:resource="http://www.tel.fer.hr/astorm/InfoServiceV2.owl#SonyEricssonBrowser"/>
    <is:hasJavaVersion rdf:datatype="http://www.w3.org/2001/XMLSchema#int">15</is:hasJavaVersion>

<!-- user preferences -->
    <is:hasPreferredInformationType rdf:resource="http://www.tel.fer.hr/astorm/InfoServiceV2.owl#PlainText"/>
    <is:hasPreferredInformationService rdf:resource="http://www.tel.fer.hr/astorm/InfoServiceV2.owl#CroatiaPoliticsInstance"/>
    <is:hasPreferredLanguage rdf:resource="http://www.tel.fer.hr/astorm/InfoServiceV2.owl#English"/>
    <is:hasPreferredGenre rdf:resource="http://www.tel.fer.hr/astorm/InfoServiceV2.owl#RockMusic"/>
    <is:hasPreferredQoS rdf:resource="http://www.tel.fer.hr/astorm/InfoServiceV2.owl#Silver"/>
    <is:hasPreferredDeliveryType rdf:resource="http://www.tel.fer.hr/astorm/InfoServiceV2.owl#NonStreaming"/>

<!-- context -->
    <is:hasEnvironment rdf:resource="http://www.tel.fer.hr/astorm/InfoServiceV2.owl#InnerSpace"/>
    <is:hasLocation rdf:resource="http://www.tel.fer.hr/astorm/InfoServiceV2.owl#Ina"/>
    <is:hasCoordinatesX rdf:datatype="http://www.w3.org/2001/XMLSchema#float">50.21389</is:hasCoordinatesX>
    <is:hasCoordinatesY rdf:datatype="http://www.w3.org/2001/XMLSchema#float">48.21389</is:hasCoordinatesY>
    <is:atTime rdf:resource="http://www.tel.fer.hr/astorm/InfoServiceV2.owl#Night"/>
    <is:hasSocialActivity rdf:resource="http://www.tel.fer.hr/astorm/InfoServiceV2.owl#WritingPresentation"/>

    </is:MobilePhoneProfile>
</rdf:RDF>
```

**List. 1.** An example of a user profile

Attributes can be classified into one of the following types:

- *Interval*: An interval attribute is defined by a continuous linear scale divided into equal intervals (e.g., in List. 1 *hasAvailableMemory* could be qualified as an interval attribute with integer values);
- *Ordinal* (or *rank*): An ordinal attribute has multiple states that can be ordered in a meaningful sequence. The distance between two states increases when they are further apart in the sequence and intervals between these consecutive states might be different (e.g., in List. 1 *hasPrefferedQoS* could be qualified as an ordinal attribute with values *Bronze, Silver,* and *Gold*);
- *Nominal* (or *categorical*): A nominal attribute takes on multiple states, but these states are not ordered in any way (e.g., in Fig. 1 *hasPrefferedLanguage* is a nominal attribute with values *English*, *Deutsch*, and *Hrvatski*).
- *Binary*: A binary attribute is a nominal attribute that has only two possible states (e.g., *hasPrefferedDeliveryType* can be *streaming* or *nonstreaming*).

## 3 Semantic Clustering

Clustering is a process that results in partitioning a set of objects, which are described by a set of attributes, into clusters. Clustering algorithms rely on distance measures that describe the similarity between objects that need to be grouped. Consequently, objects in resulting clusters are more similar to one another than to those in other clusters. In our case, we compare user profiles and partition users into groups. Such partitioning enables the telecommunication operator to enhance the provisioning of group-oriented services.

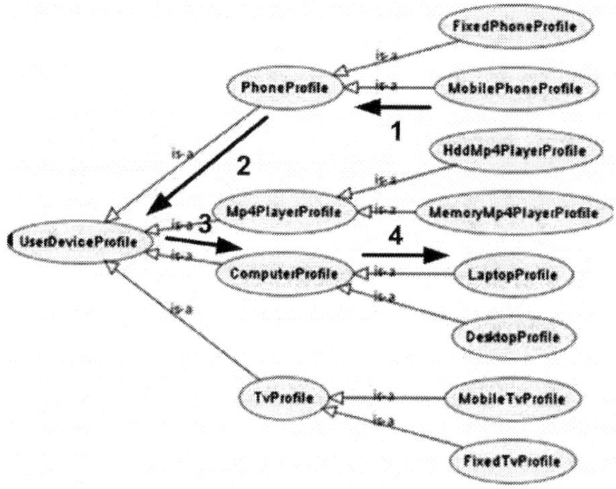


**Fig. 1.** Class hierarchy

In our implementation we do not use standard distance measures to compare profiles, but rather take advantage of a novel approach based on semantic matchmaking. Table 1 shows the comparison of two user profiles. Each attribute in the profile is asserted individually, while the final result is the arithmetic mean of individual attribute scores. The semantic matchmaking procedure is carried out as follows:

- *Position within the class hierarchy*: Each profile is an instance of a certain class from the ontology. Fig. 1 shows how class hierarchy position is transformed into a real number that represents the similarity between two classes, or objects. Greater distance between two classes implies less similarity between classes' instances: we can see that the *MobilePhoneProfile* and the *LaptopProfile* classes are four "steps" away from each other in the hierarchy. The similarity score is calculated by division of 1 by the number steps (in this case 4, so the similarity score equals 0.25);
- *Common attribute types*: When comparing binary and nominal attributes the result is either *0* (if the values are not equal), or *1* (if the values are identical). When comparing ordinal attributes the result is a number between *0* and *1*, depending on the rank of each value. A comparison result of two attributes is a ratio between the smaller and the greater attribute between the two: e.g., when comparing *Silver* and *Gold* levels of *QoS* the similarity score is 0.5;
- *Attributes with object values*: Some attributes' values contain references to other class instances. They can also be compared using the already mentioned approach using the class hierarchy position.

For clustering, we use *k-means* algorithm extended with the *constructive clustering analysis* method that determines the quality of the partition [16]. *K-means* is a partition-based approach: its idea is to segment a set of objects into multiple non-overlapping clusters. A partition-based technique creates an initial partition depending on a specific number of clusters and then attempts to improve the partition iteratively by moving objects between clusters.

**Table 1.** Profile comparison results

| Attribute | Type | Value (profile A) | Value (profile B) | Score |
|---|---|---|---|---|
| **ID** | abstract | Mobile 1 | Laptop 1 | none |
| Class | class | MobilePhoneProfile | LaptopProfile | 0.250 |
| **User preferences** | | | | |
| InformationType | instance | PlainText | Avi | 0.250 |
| InformationService | instance | CroatiaPoliticsInstance | MoviesInstance | 0.142 |
| Language | instance | English | Hrvatski | 0.500 |
| Genre | instance | RockMusic | ThrillerMovie | 0.250 |
| QoS | instance | Silver | Gold | 0.500 |
| DeliveryType | instance | NonStreaming | Streaming | 0.500 |
| **Hardware** | | | | |
| AvailableMemory | integer | 18000 | 1000000 | 0.018 |
| HorizontalResolution | integer | 180 | 1600 | 0.113 |
| VerticalResolution | integer | 230 | 1050 | 0.219 |
| BitsPerPixel | integer | 16 | 32 | 0.500 |
| **Software** | | | | |
| Os | instance | BasicOs | WindowsVista | 0.500 |
| Browser | instance | SonyEricssonBrowser | MozillaFirefox | 0.500 |
| JavaVersion | integer | 15 | 16 | 0.940 |
| **Context** | | | | |
| Environment | instance | InnerSpace | InnerSpace | 1.000 |
| Location | instance | Ina | TrgBanaJelacica | 0.250 |
| CoordinatesX | float | 50,21459 | 50,21779 | 0.990 |
| CoordinatesY | float | 48,21344 | 48,74144 | 0.990 |
| Time | instance | Night | Night | 1.000 |
| SocialActivity | instance | WritingPresentation | CoffeDrinking | 0.250 |
| | | | ***Profile similarity*** | ***0.483*** |

The main task of *constructive clustering* analysis method is finding the optimal number of clusters for a given set of objects. This method uses the *silhouette measure* to determine the optimal number of clusters. To perform *constructive clustering* it is necessary to use one of the non-constructive analysis algorithms (e.g., *k-means*) to perform clustering for a given number of clusters. The *silhouette measure* is calculated for each object in the given set taking into account the resulting partition. It is calculated as follows:

$$s(i) = \begin{cases} 1 - \dfrac{a(i)}{b(i)} & if\ a(i) < b(i) \\ 0 & if\ a(i) = b(i)\ or\ the\ object\ is\ the\ only\ object\ in\ the\ cluster\ A \\ \dfrac{a(i)}{b(i)} - 1 & if\ b(i) < a(i) \end{cases}$$

where *s(i)* is the *silhouette measure* of the object $o_i$, the cluster *A* is the cluster containing the object $o_i$, *a(i)* is the average distance of object $o_i$ to all other objects in the cluster *A*, and *b(i)* is the average distance of the object $o_i$ to all other objects in the cluster *B*, which is the *neighbour* of object $o_i$. The *neighbour* is the cluster closest to the observed object, different from cluster *A*. Usually it is the second best choice for $o_i$. The next step is calculating the *average silhouette width* for the object set, calculated as an average value of *silhouette measures* of all objects for the given partition. When we obtain the *average silhouette width* for each number of clusters *k* between 2 and *n-1* (where *n* denotes the total number of objects to be clustered) the highest one is called the *silhouette coefficient,* and the *k* for which the *silhouette coefficient* is achieved is the optimal number of clusters. List. 2 presents the calculation of *silhouette coefficient* when clustering is done upon 20 user profiles. Firstly, *k-means* algorithm is used for clustering user profiles into 2 to 19 clusters (iterations k=2 to k=19, respectively) and then the *average silhouette width* for every iteration is calculated. One notices that the highest *average silhouette width* is achieved when k=3, so the optimal number of clusters is 3: these clusters are shown in Fig. 2 (approximate coordinates for each profile in a two-dimensional plane is calculated with the procedure called *multidimensional scaling*).

```
Iteration k=2: 0.76560247
Iteration k=3: 0.78083026
Iteration k=4: 0.63344204
Iteration k=5: 0.45255604
Iteration k=6: 0.5378326
Iteration k=7: 0.41642666
Iteration k=8: 0.2297335
Iteration k=9: 0.26506615
Iteration k=10: 0.6089652
Iteration k=11: 0.38601458
Iteration k=12: 0.21313605
Iteration k=13: 0.10924552
Iteration k=14: 0.18178827
Iteration k=15: 0.35817325
Iteration k=16: -5.4418447E-4
Iteration k=17: 0.26543617
Iteration k=18: 0.16281635
Iteration k=19: 0.037484333
...
3 3 3 4 5 3 3 5 3 5 3 3 3 3 4 5 3 3 5 5   0.78083026
...
Optimal number of clusters is 3
```

For *k = 3, average silhouette width* is *0.78083026.*

Partition with the highest *average silhouette width.* Numbers from 3 to 5 represent cluster identifiers, and the last number is the *silhouette coefficient.*

**List. 2.** An example of constructive clustering analysis method

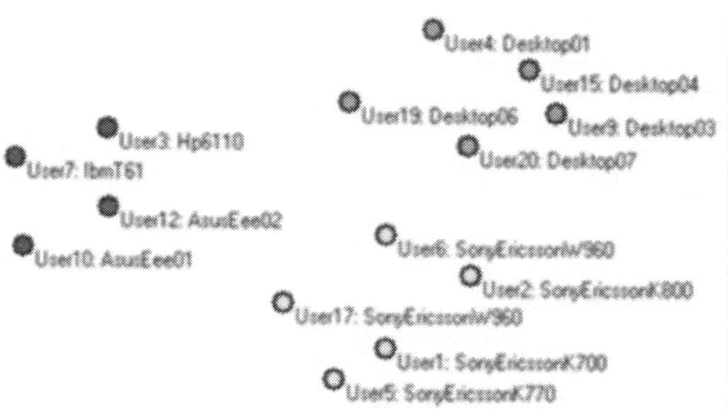


**Fig. 2.** An example of user profile clustering

# 4 Agent-Based Telecommunication Service Provisioning

The Agent-based Service and Telecom Operations Management (A-STORM) multi-agent system is part of the proof-of-concept prototype that deals with agent-based service provisioning in the NGN. This section briefly addresses software agents as a technology used to implement the prototype and the NGN as its environment. The description of the prototype architecture follows and is accompanied by a case study that illustrates a group-oriented telecommunication service.

## 4.1 Multi-agent System for Service Provisioning in NGN

A software agent is a program which autonomously acts on behalf of its principal, while carrying out complex information and communication tasks that have been delegated to it. The concept of software agents appeared in the mid-1990's [17] and resulted in the application of an agent-based computing paradigm in various research domains [9, 18]. A system composed of several software agents, collectively capable of reaching goals that are difficult to achieve by an individual agent or a monolithic system, is called a multi-agent system. The A-STORM multi-agent system was implemented using the Java Agent DEvelopment Framework (JADE[10]). JADE is a software framework used for developing agent-based applications in compliance with the Foundation for Intelligent Physical Agents (FIPA[11]) specifications.

Today we are witnessing the fusion of the Internet and mobile networks into a single, but extremely prominent and globally ubiquitous converged network [19, 20]. One of the fundamental principals in the converged network is the separation of services from transport [21]. This separation represents the horizontal relationship aspect of the network where the transport stratum and the service stratum can be distinguished. The transport stratum encompasses the processes that enable three types of connectivity, namely, user-to-user, user-to-service platform and service platform-to-service platform connectivity. On the other hand, the service stratum comprises processes that enable (advanced) telecom service provisioning assuming that the earlier stated types of connectivity already exists. From the vertical relationship aspect each stratum can have more layers where a data (or user) plane, a control plane and a management plane in each layer can be distinguished.

In our prototype intelligent agents are introduced into the control plane and the management plane of the service stratum. Those agents are in charge of gathering and processing context information that is required for different service provisioning tasks. In particular, the prototype has been developed in order to explore possibilities for implementation of ontology-based user profiling/clustering, context-aware service personalization and rule-based software deployment in the converged network. Fig. 3 shows the architecture of A-STORM multi-agent system used to implement the proof-of-concept prototype.

Business-driven provisioning agents (i.e., the Business Manager Agent, the Provisioning Manager Agent) perform provisioning operations according to business strategies defined in a rule-based system. Service deployment agents (i.e., the Deployment Coordinator Agent, Remote Maintenance Shell agents) provide end-to-end solutions for efficient service deployment by enabling software deployment and maintenance at remote systems. Context management agents (i.e., the Charging Manager Agent, Group Manager Agent, Session Manager Agent, Preference Manager Agent) gather context information from network nodes and devices (e.g., trigger events in SIP/PSTN call model, balance status, device location) and enable user personalization through the execution of context-dependent personal rules. A more detailed description of these agents and their functionalities can be found in [22, 23].

---

[10] http://jade.tilab.com

[11] http://www.fipa.org

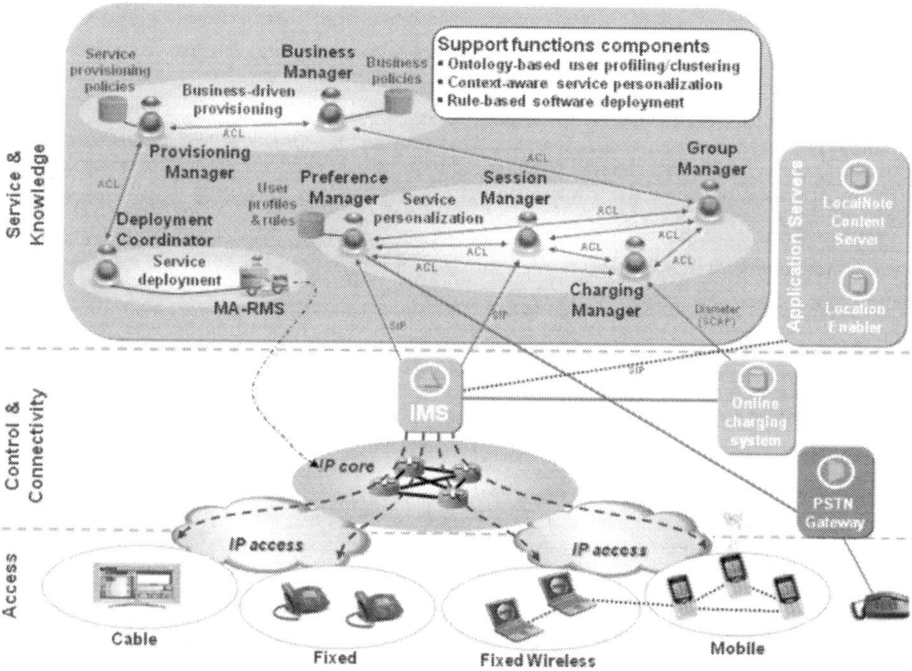


**Fig. 3.** A-STORM proof-of-concept prototype architecture

### 4.2 Proof-of-Concept Group-Oriented Service: Multi-agent-Based Digital Goods Recommender within Implicit Social Networks (MADRIS)

In this section the concept of viral marketing is brought forward. It represents a platform on which regular users may resell digital goods [24]. The proposed combination of Semantic Web and multi-agent technology brings an added value to the concept of viral marketing, which is elaborated through the following case study. The participating entities are described in the previous section and the sequence of actions is accompanied by a sequence diagram in Fig. 4.

If a particular user, e.g., user$_i$, is interested in the *MADRIS* service then his/her Preference Manager Agent PrefMA$_i$ sends a registration request to the Provisioning Manager Agent (ProvMA) (1). (Registration cancellation is also performed in an analogous manner.) The ProvMA provides the Group Manager Agent (GMA) with a list of $m$ user profiles whose users subscribed to the *MADRIS* service (2). The GMA allocates these profiles into $k$ groups according to users' preferences. This is achieved by performing semantic clustering algorithms described in Section 3. The GMA provides the ProvMA with a list of assembled groups and associated profiles (3). In a particular group $G_x$ consisting of $n$ ($n < m$) profiles, the ProvMA informs each of the $n$ PrefMAs of all other PrefMAs within that particular group (4). If the user$_i$ wants to purchase some content, the corresponding PrefMA$_i$ first needs to ascertain that all requirements imposed by the particular content provisioning service are met. Therefore, it consults the ProvMA (5a), which performs matching of the user profile with the service profile [8] and determines whether additional software is necessary for the chosen content to render properly on the user's device. If so, the ProvMA sends the list of required software to the Deployment Coordinator Agent (DCA) (5b), which forwards it, along with deployment procedure specification, to the Remote Maintenance Shell (RMS) agent [25] (5c). The RMS agent deploys and installs the software to the user's mobile device. The PrefMA$_i$ can then download the content so it sends a message to the Business Manager Agent (BMA) requesting to purchase the content and the rights to resell it to $x$ ($x<n$) other users whose profiles belong to its group (6). The BMA sends the requested content to the PrefMA$_i$ and the rights to resell it (7). Being authorised to resell the content, the PrefMA$_i$ has the liberty to sporadically distribute recommendations to any subset of PrefMAs belonging to its group, e.g., PrefMA$_s$..PrefMA$_t$ (8). Should any of PrefMAs that have received the recommendation, say, the PrefMA$_j$, decide to purchase the same content, it informs the BMA about the purchase, along with information on who it received the recommendation from (9).

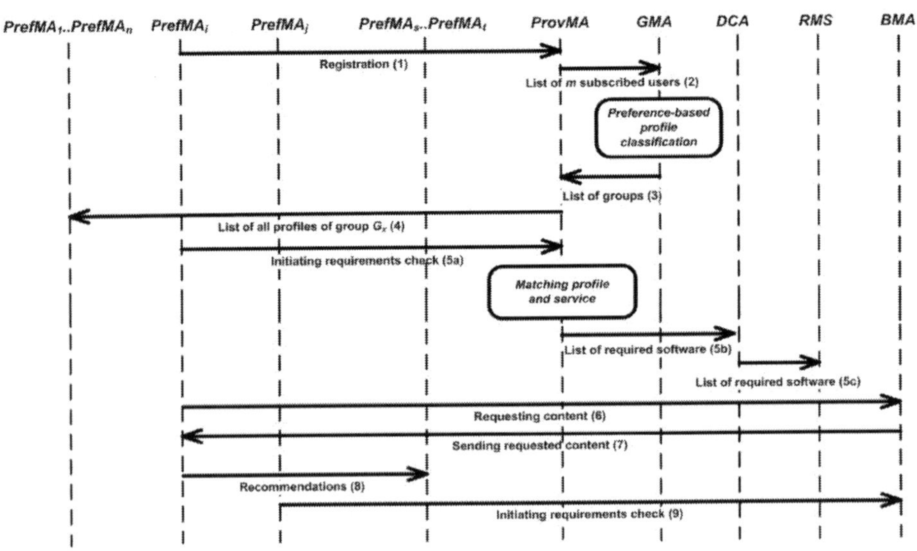


**Fig. 4. The** MADRIS scenario

This approach in marketing digital goods ensures that all involved parties benefit from successful transactions between end-users. Namely, let alone the fact that providing digital goods does not imply almost any production expenses, the service provider need not perform resource-consuming tasks such as advertisement. The advertisement becomes redundant as users occasionally receive recommendations. End-users benefit from this solution as it facilitates choosing from a large number of generally available digital goods, as they are provided with the ability to concentrate on a smaller number of recommended goods. What makes these goods potentially interesting is the fact that recommendations are sent by users from the group assembled according to the similarity in preferences of users whose profiles it encompasses. If a user opts for purchasing a recommended good, not only does the content provider gain income, but the reseller also profits, in accordance with its contract with the provider. In fact, the reseller gains credit for every successful transaction it induces, manifesting by means of reducing the cost of a future purchase. For instance, the user A purchases some digital content from the content provider for a price of $\mathcal{X}$ money units and recommends it to the user B. The user B is offered to buy the recommended content at a lower price (e.g., $0.9 \times \mathcal{X}$ money units). If he/she decides to do so, the content provider also credits the user A by reducing the price of his/her next purchase (e.g., for, $0.2 \times \mathcal{X}$ money units).

## 5 Conclusion and Future Work

In this paper, we describe how context-aware semantic reasoning can affect the course of telecommunication service provisioning and enable agent-based contextual user personalization. The CC/PP and UAProf schemas were mapped to an OWL ontology, while features for user preferences and context support were added. User profiles, describing a number of different mobile devices and user preferences, were created. Furthermore, semantic clustering algorithm was implemented and presented. Finally, a proof-of-concept group-oriented service demonstrating the advantages of combining the Semantic Web and agent technologies was presented.

In many respects, the proposed multi-agent system enables provision of non-linear group-oriented telecommunication services through synergy of both ontology-based knowledge representation and clustering techniques. Therefore, our future research will also be aimed at improving the proposed semantic clustering feature. In particular, we will put an emphasis on implementing Group Manager Agents that are capable of executing both scalable and computationally efficient graph clustering algorithms suitable for dealing with multidimensional data sets such as user profiles.

**Acknowledgements.** This work was carried out within research projects 036-0362027-1639 "Content Delivery and Mobility of Users and Services in New Generation Networks", supported by the Ministry of Science, Education and Sports of the Republic of Croatia, and "Agent-based Service & Telecom Operations Management", supported by Ericsson Nikola Tesla, Croatia.

## References

1. Podobnik, V., Lovrek, I.: Multi-Agent System for Automation of B2C Processes in the Future Internet. In: 27th IEEE Conference on Computer Communications (INFOCOM) Workshops, pp. 1–4. IEEE Press, Phoenix (2008)
2. Ackoff, R.L.: From Data to Wisdom. Journal of Applied System Analysis 16, 3–9 (1989)
3. Knightson, K., Morita, N., Towle, T.: NGN architecture: generic principles, functional architecture, and implementation. IEEE Communications Magazine 43(10), 49–56 (2006)
4. Feijoo, C., et al.: The emergence of IP interactive multimedia services and the evolution of the traditional audiovisual public service regulatory approach. Telematics and Informatics 24(4), 272–284 (2007)
5. Grimnes, G.A., Edwards, P., Preece, A.: Instance based clustering of semantic web resources. In: Bechhofer, S., Hauswirth, M., Hoffmann, J., Koubarakis, M. (eds.) ESWC 2008. LNCS, vol. 5021, pp. 303–317. Springer, Heidelberg (2008)

6. Dietze, S., Gugliotta, A., Domingue, J.: Conceptual Situation Spaces for Semantic Situation-Driven Processes. In: Bechhofer, S., Hauswirth, M., Hoffmann, J., Koubarakis, M. (eds.) ESWC 2008. LNCS, vol. 5021, pp. 599–613. Springer, Heidelberg (2008)
7. Bonnin, J., Lassoued, I., Hamouda, Z.B.: Automatic multi-interface management through profile handling. Mobile Networks and Applications 14(1), 4–17 (2009)
8. Frkovic, F., Podobnik, V., Trzec, K., Jezic, G.: Agent-based user personalization using context-aware semantic reasoning. In: Lovrek, I., Howlett, R.J., Jain, L.C. (eds.) KES 2008, Part I. LNCS (LNAI), vol. 5177, pp. 166–173. Springer, Heidelberg (2008)
9. Trzec, K., Lovrek, I., Mikac, B.: Agent behaviour in double auction electronic market for communication resources. In: Gabrys, B., Howlett, R.J., Jain, L.C. (eds.) KES 2006. LNCS (LNAI), vol. 4251, pp. 318–325. Springer, Heidelberg (2006)
10. Podobnik, V., Petric, A., Trzec, K., Jezic, G.: Software Agents in New Generation Networks: Towards the Automation of Telecom Processes. In: Jain, L.C., Nguyen, N.T. (eds.) Knowledge Processing and Decision Making in Agent-Based Systems, pp. 71–99. Springer, Heidelberg (2009)
11. Antoniou, G., van Harmelen, F.: Semantic Web Primer. MIT Press, Cambridge (2004)
12. Walton, C.: Agency and the Semantic Web. Oxford University Press, New York (2007)
13. Grimm, S., Motik, B.: Closed World Reasoning in the Semantic Web through Epistemic Operators. In: 1st OWL: Experiences and Directions Workshop (OWLED), Galway (2005)
14. Broekstra, J., Kampman, A., van Harmelen, F.: Sesame: A Generic Architecture for Storing and Querying RDF and RDF Schema. In: Horrocks, I., Hendler, J. (eds.) ISWC 2002. LNCS, vol. 2342, pp. 54–68. Springer, Heidelberg (2002)
15. Kiryakov, A., Ognyanov, D., Manov, D.: OWLIM – A pragmatic semantic repository for OWL. In: Dean, M., Guo, Y., Jun, W., Kaschek, R., Krishnaswamy, S., Pan, Z., Sheng, Q.Z. (eds.) WISE 2005 Workshops. LNCS, vol. 3807, pp. 182–192. Springer, Heidelberg (2005)
16. Wei, C.P., Hu, P.J.-H., Kung, L., Tan, J.: E-Health Intelligence: A Multiple-Level Approach for E-Health Data Mining. In: Tan, J. (ed.) E-Health Care Information Systems: An Introduction for Students and Professionals, pp. 330–351. Wiley, San Francisco (2005)
17. Nwana, H.S.: Software Agents: An Overview. Knowledge and Engineering Review 11(3), 205–244 (1996)
18. Petric, A., Podobnik, V., Jezic, G.: The crocodileAgent: Designing a robust trading agent for volatile E-market conditions. In: Nguyen, N.T., Grzech, A., Howlett, R.J., Jain, L.C. (eds.) KES-AMSTA 2007. LNCS (LNAI), vol. 4496, pp. 597–606. Springer, Heidelberg (2007)
19. Berners-Lee, T., Fischetti, M.: Weaving the Web. Harper San Francisco, New York (1999)
20. ITU-T Recommendation Y.2001: General Overview of NGN (2004)
21. ITU-T Recommendation Y.2011: General principles and general reference model for Next Generation Networks (2004)
22. Petric, A., et al.: An Agent Based System for Business-driven Service Provisioning. In: AAAI 2007 Workshop on Configuration, pp. 25–30. AAAI Press, Vancouver (2007)
23. Petric, A., et al.: Agent-Based Support for Context-Aware Provisioning of IMS-Enabled Ubiquitous Services. In: AAMAS 2009 Service-Oriented Computing: Agents, Semantics, and Engineering workshop (SOCASE), pp. 22–28. AAMAS, Budapest (2009)
24. Cattelan, R.G., He, S., Kirovski, D.: Prototyping a novel platform for free-trade of digital content. In: 12th Brazilian Symposium on Multimedia and the Web (WebMedia), pp. 79–88. ACM, Natal (2006)
25. Kusek, M., et al.: Mobile agent based software operation and maintenance. In: 7th International Conference on Telecommunications (ConTEL), pp. 601–608. IEEE, Zagreb (2003)

# A Multi-agent Model of Deceit and Trust in Intercultural Trade

Gert Jan Hofstede[1,2], Catholijn M. Jonker[2], and Tim Verwaart[3]

[1] Wageningen University, Postbus 9109, 6700 HB Wageningen, The Netherlands
gertjan.hofstede@wur.nl
[2] Delft University of Technology, Mekelweg 4, 2628 CD Delft, The Netherlands
c.m.jonker@tudelft.nl
[3] LEI Wageningen UR, Postbus 29703, 2502 LS Den Haag, The Netherlands
tim.verwaart@wur.nl

**Abstract.** Trust is a sine qua non for trade. According to transaction cost economics, a contract always offers some opportunity to defect. In the case of asymmetric product information, where the seller is better informed about product quality than the buyer is, the buyer either has to rely on information provided by the seller or has to check the information by testing the product or tracing the supply chain processes, thus incurring extra transaction cost. An opportunistic seller who assumes the buyer to trust, may deliver a lower quality product than agreed upon. In human decisions to deceive and to show trust or distrust toward business partners, issues like morality, shame, self-esteem, and reputation are involved. These factors depend strongly on trader's cultural background. This paper develops an agent model of deceit and trust and describes a multi-agent simulation where trading agents are differentiated according to Hofstede's dimensions of national culture.

**Keywords:** trust and reputation management, deceit, negotiation, trade partner selection, culture.

## 1 Introduction

A business transaction usually incurs cost on transaction partners, thus reducing the value of the transaction for the party bearing the cost. In transaction cost economics [1] opportunism and the incompleteness of contracts are central issues. Due to bounded rationality, contracts cannot specify solutions for all contingencies that may occur in transactions executed under the contracts. The incompleteness offers contract partners opportunities to defect. As Williamson [1] asserts, not every contract partner will take full advantage of every opportunity to defect. However, it is the uncertainty about a contract partner's opportunism that incurs transaction cost. *Ex ante* and *ex post* types of transaction cost can be distinguished. *Ex ante* are the cost of searching, bargaining, drafting, and safeguarding of contracts. *Ex post* are the cost of monitoring and enforcing task completion. Transaction cost economics is the basis for the process model of trading agents applied in this paper. The process model is depicted in Fig. 1.

N.T. Nguyen, R. Kowalczyk, and S.-M. Chen (Eds.): ICCCI 2009, LNAI 5796, pp. 205–216, 2009.

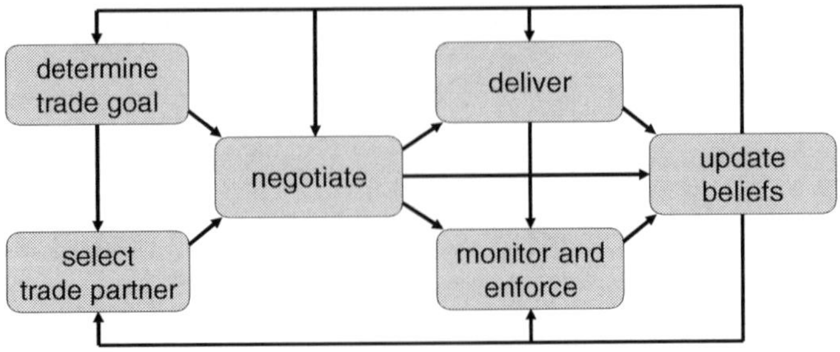


**Fig. 1.** Processes and internal information flows of trading agents

The outcome of successful negotiation is a contract. After that, it comes to delivery. An agent may deliver according to contract, or fail to do so intentionally (opportunism) or unintentionally (by incompetence or a flaw of its quality control system). At the same time, an agent may monitor the other party's delivery and either check if it is according to contract, or trust and accept without checking. Klein Woolthuis et al. [2] studied the relationship between trust and contracts. They concluded that trust can substitute or complement contracts: if trust is high, contracts can either be rather incomplete, because parties do not expect deceit, or more complete but not actively monitored and enforced, as a signal confirming the trusting relationship; if trust is low, a contract can either be rather complete as a safeguard against opportunism, or incomplete because of opportunistic intentions (so, contract incompleteness does not necessarily imply trust).

The trading situation of the simulation presented in this paper is based on the Trust And Tracing game [3]. In this game, players select trade partners and trade imaginary food products that have a value-increasing quality attribute known by the supplier, but invisible to the customer, e.g. "organically grown". The customer can, at the cost of a fee, involve the Tracing Agency to test the actual quality. The Tracing Agency reports the test's outcome to both customer and supplier, and in case of untruthful delivery, punishes the supplier by a fine. Based on experience from negotiation and tracing results, agents update their beliefs about the market, potential partners, and the risks of opportunistic behavior. This paper focuses on the post-contract phase. The models for trust, deceit, and experience-based belief update are described in Section 2.

Human decisions to deceive and to trust are not strictly rational; they are influenced by emotions [4]. As such, they are known to depend on cultural background [5], [6]. G. Hofstede's five dimensions of national cultures [7] are widely used to identify cultural differences. G.J. Hofstede et al. described models for the influence of culture on trade processes, including deceit and trust, for each of the five dimensions separately [8], [9], [10], [11], [12]. However, the differentiation of human behavior cannot be described along a single one of these dimensions. The present paper's goal is to integrate G.J. Hofstede et al.'s individual dimension models, focusing on the decisions whether to deliver truthfully or untruthfully (*deceit*) and whether to trace the delivery or to accept it without tracing (*trust*), and on experience- based belief update about partner's trustworthiness and benevolence (i.e. its inclination to trust). Section 3 presents the model of the influence of culture on deceit and trust. Section 4 describes experimental results from multi-agent simulations. Section 5 concludes the paper with a discussion of the results.

## 2 Modeling Deceit, Trust, and Experience-Based Belief

The simulation model represents the trade process of The Trust And Tracing game [3], where a group of 15-20 participants repeatedly trade commodities of different quality levels for an a priori unknown time. Suppliers are informed about the quality; customers are not informed. Participants are free to select a partner for each transaction, and negotiate about price, quality to be delivered, and conditions of the contract. Customers may (a) avoid deceit by buying low quality or (b) buy high quality and either (b1) accept vulnerability and trust the supplier to deliver according to contract, or (b2) protect themselves by negotiating a guarantee, for instance money back in case deceit would be revealed, or (b3) have the commodity traced in advance (certification). Option (a) is free of cost and risk, and a low price may be negotiated, but the customer has to accept low quality. Options (b1) and (b2) incur risk on the customer (as they offer the seller an opportunity to defect), and additional cost only if the customer decides to monitor (trace) the delivery. The certification option (b3) excludes risk, but always incurs additional cost.

Although trust is also relevant for the processes of partner selection and negotiation, the present paper focuses on the post-contract phase of transactions. It describes the decision whether to deceive or not in the delivery process, the decision whether to trust or to trace in the monitoring and enforcing process, and the update of beliefs resulting from confirmed or violated trust. The remaining part of this section discusses relevant literature from the social sciences and introduces the agent's decision models applied in the simulation.

In experiments using a repetitive ultimatum game with asymmetric information, Boles et al. [4] found that most people do not choose deceptive strategies. However, deceit occurred in their experiments, in particular when stakes were high. So, for deceit to occur, at least two conditions have to be satisfied: motive (substantial advantage for the deceiver) and opportunity (lack of information on the part of the deceived).

As Boles et al. found, the conditions of motive and opportunity are not sufficient for deceit. The decision to deceive depended on interpersonal interactions and the player's satisfaction about the behavior of the other party. They report that "*the bargainers were little like those depicted by rational economic models*" [4] and that "*responders may react emotionally and reject profitable offers in the present when they realize that they have been deceived in the past*" [4].

Role-playing research into cheating on service guarantees by consumers reported by Wirtz and Kum [13] confirms that people are not inclined to seize any opportunity to cheat. Their research also confirms that potential material gain is a condition for cheating, but they found no evidence that people who cheat let their decision depend on the expected amount of payout or the ease of the opportunity to cheat. They report cheating to be related to personality of players (morality, Machiavellianism and self-monitoring). Two factors found to decrease cheating were satisfaction about the deal and the expectation of repeated dealing with the supplier in the future. Wirtz and Kum [13] suggest that a sense of loyalty and trust may reduce cheating. They also refer to Hwang and Burgers [14] that take an economics approach and argue that the high cost of the loss of a trusted partner is an inhibitor of opportunism. Both views indicate that a high-trust relation inhibits deceit.

In the research discussed above, four factors that influence deceit are found: opportunity, expected payout, player's personal traits and values, and player's trust relationship with their counterpart. Steinel and De Dreu [15] conclude on the basis of experiments with the Information Provision Game that, due to greed and maybe to fear of exploitation, individuals are less honest when they experience their counterpart to be competitive rather than cooperative, and that this tendency is stronger for prosocial than for selfish individuals. The importance of the relationship and the behavior of the counterpart is confirmed by Olekalns and Smith [16] who contrast two models of ethical decision making: *fair trade* (my counterpart trusts me, so I will cooperate) and *opportunistic betrayal* (my counterpart trusts me, so I can easily defect). In experiments with Australian undergraduate students they found strong support for *fair trade* as the prevailing model. However, Wirtz and Kum [13] found that individuals scoring high on Machiavellianism in the personality test, were more easily tempted to seize an opportunity to cheat and actually followed what Olekalns and Smith [16] called the *opportunistic betrayal* model.

A general conclusion of the work cited so far in this section is that deceit is less likely to occur when trade partners show trust in each other, even when rational strategies to win the game would suggest cheating. As the purpose of the multi-agent simulation reported in this paper is to represent actual human behavior rather than to apply deception as a strategy to win a game, we cannot employ rational models like the ones proposed by Castelfranchi et al. [17] and Ward and Hexmoor [18].

In the simulation an agent's decision to deceive is modeled as a Bernoulli variable with probability of deceit

$$p(\text{deceit}) = q\,(1\text{-}\,c)\,m_i\,(1\text{-}d_i)\,, \tag{1}$$

where $q$ represents the quality agreed in the current contract ($q$=1 for high quality; $q$=0 for low quality or no opportunity); $c$=1 if certification has been agreed (no opportunity); $c$=0 otherwise; $m_i$ represents the supplier's motive or rationale to deceive customer $i$ ($m_i$ =1 if the supplier expects an extra profit from deceit; $m_i$=0 otherwise, for instance if the customer negotiated a guarantee and the supplier expects the customer to trace the delivery); $d_i$ represents on the interval [0, 1] seller's threshold for deceit toward customer $i$, where $d_i$=1 represents perfect truthfulness. $d_i$ is influenced by seller's personal traits and values (like risk aversion and morality), power and group relations, and seller's estimate of customer's benevolence toward the seller, i.e., seller's trust that the customer will accept deliveries without tracing. Details on $d_i$ and the influence of cultural background are discussed in Section 3.

For the purpose of the simulation, Klein Woolthuis et al.'s [2] narrow definition of trust is adopted. A customer's trust in a particular supplier is defined as the customer's estimate of the probability that the supplier will cooperate and deliver according to contract, even if the supplier has the motive and the opportunity to defect. However, this does not imply that an agent's decision to have a delivery traced can be modeled as a Bernoulli variable with $p(\text{trace})=q(1\text{-}c)(1\text{-}t_j)$ where $q(1\text{-}c)$ represent opportunity as in equation (1) and $t_j$ represents trust in supplier $j$. Additional factors like power and group relationships with the supplier and the agent's cultural background also have their effect on the decision to trace. The effects of relationships and cultural background on the tracing decision are discussed in Section 3.

Trust and distrust develop during social interactions. Visual and auditory contact is relevant to develop trust and detect deceit in human interactions [19]. However, the multi-agent simulation does not support these effects. The only sources of information that can be taken into account are negotiation outcomes and tracing reports, which are relevant in reality as well. Every successful negotiation, resulting in a transaction will strengthen partners' trust in each other. However, customers can decide to trace a delivery and this can have its effects on mutual trust. First, if the result of tracing reveals deceit, the customer's trust in the seller will be reduced. Second, to some extent the fine and the reputational damage resulting from revealed deceit will reinforce the supplier's honesty. However, reinforced honesty will decay to its original level in the course of time. Third, the supplier delivering truthfully may be offended by tracing and the relation may be damaged. For this reason, customers may exercise restraint to trace. Tracing will always reduce the supplier's belief about customer's benevolence. So, the following dynamics have to be modeled:

- development of trust and benevolence belief by successful negotiations;
- for customers: reduction of trust in case of revealed deceit;
- for suppliers: reinforcement of honesty in case of revealed deceit;
- for suppliers: decay of reinforced honesty to a base level;
- for suppliers: reduction of benevolence belief in case of tracing.

Formal models for representing the development of trust were analyzed by Jonker and Treur [20]. They distinguish six types of trust dynamics: blindly positive, blindly negative, slow positive – fast negative, balanced slow, balanced fast, and slow negative – fast positive. The most realistic type of dynamics for trust in trading situations is slow positive – fast negative: it takes a series of positive experiences to develop trust, but trust can be destroyed by a single betrayal (e.g., Boles et al. [4] report that deceit leads to emotional reactions and consequences beyond what is rational; Steinel and De Dreu [15] refer to "*punitive sentiment*" towards deceivers). A consumer's trust in supplier $j$ after the $n$'th experience is updated as follows.

$$\begin{array}{ll} t_{j,n} = t_{j,n-1} + u^{+}(1 - t_{j,n-1}) & \text{if } n^{\text{th}} \text{ experience is positive ,} \\ t_{j,n} = (1-u^{-})\, t_{j,n-1} & \text{if } n^{\text{th}} \text{ experience is negative ,} \\ t_{j,n} = t_{j,n-1} & \text{if } n^{\text{th}} \text{ experience is neither positive nor negative ,} \end{array} \qquad (2)$$

with $0 < u^{+} < u^{-} < 1$ , where $t_{j,n} = 1$ represents complete trust and $t_{j,n} = 0$ represents complete distrust; a successful negotiation counts as a positive experience; a tracing report revealing deceit counts as negative; all other experiences are considered neither negative nor positive with respect to trust.

A supplier's belief $b_{i,n}$ about customer $i$'s benevolence is updated by the same mechanism. Also for the supplier, a successful negotiation counts as a positive experience. However, tracing always counts as a negative experience for a supplier, whether it reveals deceit or not, because it is interpreted as distrust.

An effect of revealed deceit on the supplier's part is that supplier's current honesty $h_k$ (a personal trait, representing the inclination to deliver truthfully) is reinforced to 1, representing maximal honesty. $h_k$ will subsequently decay to a base value $h'$ on each interaction, whether it is successful or not, with a decay factor $f$.

$$h_k = h' + f\,(h_{k-1} - h')\,, \text{ with } 0 < h' < 1 \text{ and } 0 < f < 1\,. \qquad (3)$$

## 3 The Influence of Culture on Deceit and Trust

The preceding section introduced models for deceit, trust and belief update in a process of trade. The roles of deceit and trust are known to be different across cultures [5], [6]. Therefore, a multi-agent simulation of international trade that models the effects of deceit and trust should include the effects of culture. This section proposes an approach to model the effects of culture on the parameters and variables introduced in the previous section (deceit threshold, inclination to trace, and positive and negative trust update factors), based on G. Hofstede's dimensions of culture [7]. First culture and Hofstede's dimensions and their effects on deceit and tracing are discussed; than the agent's decision models are proposed.

Hofstede describes culture as "*the collective programming of the mind that distinguishes the members of one group or category of people from another*" [7], p. 9. This implies that culture is not an attribute of individual people, unlike personality characteristics. It is an attribute of a group that manifests itself through the behaviors of its members. For a trading situation, culture of the trader will manifest itself in four ways. First, culture filters observation. It determines the salience of clues about the acceptability of trade partners and their proposals. Second, culture sets norms for what constitutes an appropriate partner or offer. Third, it sets expectations for the context of the transactions, e.g., the enforceability of regulations and the possible sanctions in case of breach of the rules. Fourth, it sets norms for the kind of action that is appropriate given the other three and, in particular, the difference between the actual situation and the desired situation.

**Table 1.** Hofstede's dimensions of culture [7]

| Dimension | Definition |
|---|---|
| Power Distance | "*The extent to which the less powerful members of institutions and organizations within a country expect and accept that power is distributed unequally*" [7], p. 98 |
| Uncertainty Avoidance | "*The extent to which the members of a culture feel threatened by uncertain or unknown situations*" [7], p. 161 |
| Individualism and Collectivism | "*Individualism stands for a society in which the ties between individuals are loose: Everyone is expected to look after him/herself and her/his immediate family only. Collectivism stands for a society in which people from birth onward are integrated into strong, cohesive in-groups, which throughout people's lifetime continue to protect them in exchange for unquestioning loyalty*" [7], p. 255 |
| Masculinity and Femininity | "*Masculinity stands for a society in which social gender roles are clearly distinct: Men are assumed to be assertive, tough, and focused on material success; women are supposed to be more modest, tender and concerned with the quality of life. Femininity stands for a society in which gender roles overlap: Both men and women are supposed to be modest, tender and concerned with the quality of life.*" [7], p. 297 |
| Long- Versus Short-Term Orientation | "*Long Term Orientation stands for the fostering of virtues oriented towards future rewards, in particular, perseverance and thrift. Its opposite pole, Short Term Orientation, stands for the fostering of virtues related to the past and the present, in particular, respect for tradition, preservation of 'face' and fulfilling social obligations*" [7], p. 359 |

G. Hofstede [7] identified five dimensions to compare national cultures (Table 1). For the dimensions, indices are available for many countries in the world. The indices are usually named as PDI, UAI, IDV, MAS, and LTO. For the multi-agent model, we scale the indices to the interval [0, 1] and refer to the scaled indices as PDI$^*$, UAI$^*$, IDV$^*$, MAS$^*$, and LTO$^*$. E.g., IDV$^*$ refers to the degree of individualism and 1-IDV$^*$ to the degree of collectivism, both in the range [0, 1].

G.J. Hofstede et al. [8], [9], [10], [11], [12], modeled the influence on trade processes of each of the five dimensions separately. However, single dimensions do not fully represent the differentiation of human behavior. A realistic simulation must take the simultaneous effect of all dimensions into account. The purpose of the present paper is to develop a first version of integrated models for deceit, trust and belief update. The remaining part of this section summarizes the effects of individual dimensions as described in [8], [9], [10], [11], [12], and then proposes the integration of the formal specifications reported in those papers.

**Power Distance.** [8] On the dimension of power distance, egalitarian societies are on the one extreme (small power distance), hierarchical societies on the other (large power distance). In hierarchical societies, status and position in the societal hierarchy are the main issue in relations. Trust is only relevant among partners that have equal status. The lower ranked have no choice but to show trust in the higher ranked, whatever belief about their trustworthiness they may have. The higher ranked have no reason to distrust the lower ranked, because they assume that deceit of a higher ranked would not even be considered. With respect to deceit, the higher ranked do not have to fear for repercussions when trading with lower ranked, so the decision, whether to defect or not, merely depends on their morality. The lower ranked on the other hand will not easily consider to defect and will usually cooperate when trading with higher ranked and will only defect if in need.

For egalitarian traders, decisions to deceive and to trust are not influenced by status difference. Trust is equally important in every relation, regardless of partner's status. However, showing distrust may be harmful to relations, so there may be other incentives for benevolent behavior.

**Uncertainty Avoidance.** [9] Uncertainty avoidance must not be confused with risk avoidance. People in uncertainty avoiding societies accept risks they are familiar with, but they fear the unknown. They are willing to take risks in order to reduce uncertainty about things they are not familiar with, or to eliminate them.

Uncertainty avoiding traders fear and distrust strangers. They follow the rules when dealing with familiar relations, but easily deceive strangers. A foreign partner will be distrusted until sufficient evidence for the contrary has been found. Once, in the course of repeated transactions, sufficient evidence for trustworthiness has been found through tracing of deliveries, and partners have become familiar, the uncertainty avoiding may finally come to trust their partners and expect them to follow the rules like they do themselves. After they have come to trust, any unexpected revelation of deceit provokes furious reactions from uncertainty avoiding traders. They will not easily deal again with a partner that abused their trust.

**Individualism and Collectivism.** [10] In individualistic societies, people have a personal identity and are responsible for their personal actions and view a business partner as an individual. In collectivistic societies, a person's identity is primarily given by group memberships (such as extended family, village, and clubs) and relations. People from collectivistic societies feel responsible for their in-group and prefer to trade with their in-group. Serious negotiations with out-group business partners must be preceded by some form of familiarization. In collectivistic societies harmony must be preserved, so the threshold for showing distrust by tracing is high.

In collectivistic societies trust and deceit are based on group memberships and norms. People from collectivistic societies primarily trust in-group members and distrust out-group members. After a long-lasting relation, outsiders may be trusted as in-group members. Deceiving an out-group partner is acceptable if it serves in-group interests. In individualistic societies opportunistic behavior and trust are based on personal interests, personal values, and interpersonal relations.

**Masculinity and Femininity.** [11] On the masculine extreme of the dimension are competitive, performance-oriented societies; on the other are cooperation-oriented societies. A cooperation-oriented trader is interested in the relationship. Building trust is important. In principle, the cooperation-oriented trader does not trace, since in his mind this would constitute ostentation of distrust. If conned, then the cooperation-oriented trader will avoid the conman if possible, or give him one more chance.

Trust is irrelevant in extremely performance-oriented societies. A performance-oriented trader sticks to the contract of the deal, and deceives the trade partner to the limits of the contract without any compunction. As a consequence, the performance-oriented trader sees no problems in dealing again with a trader that conned him in the past: "It's all in the game". The performance-oriented trader always traces the goods after buying, since he expects the possibility of deception. The trader learns from mistakes to make sure that new contracts will not lead to new and uncomfortable surprises on his side.

**Long- Versus Short-Term Orientation.** [12] Traders from long-term oriented societies value their relations. They value a deal not only by the financial pay off, but also by the relational gains. They are inclined to invest in relations by behaving truthfully and by trusting their partners. They value their business relations by the prospect of future business. They have no respect for others that put their relations at stake for some short-term profit. If they turn out to be deceived by a business partner they will not easily forgive the deceiver.

People from short-term oriented cultures find it hard to understand the sacrifice of the long-term oriented. The short-term oriented tend to grab a chance for an easy profit and are willing to put their relations at stake for it, especially if they are in need to fulfil other social obligations, like showing off for family members. They calculate the bottom line of the transaction. Their threshold to deceive or to distrust depends on the value they attach to the relation in their social life. They can understand that a business partner may be tempted to defect if a profitable opportunity occurs, and they have trouble understanding that people from long-term oriented cultures cannot.

**Table 2.** Influence of Hofstede's dimensions of culture and partner characteristics on deceit and trust (+ indicates increasing influence; - indicates decreasing influence; 0 indicates no effect)

| Culture and partner characteristics | Effect on deceit threshold | Effect on inclination to trace | Effect on positive upd. factor | Effect on negative upd. factor |
|---|---|---|---|---|
| Large power distance | | | | |
| - with lower ranked partner | 0 | - | 0 | 0 |
| - with higher ranked p. | + | - | 0 | 0 |
| Small power distance | 0 | 0 | 0 | 0 |
| Uncertainty avoiding | | | | |
| - with familiar partner | 0 | 0 | - | + |
| - with stranger | - | + | - | + |
| Uncertainty tolerant | 0 | 0 | 0 | 0 |
| Individualistic | 0 | 0 | 0 | 0 |
| Collectivistic | | | | |
| - with in-group partner | 0 | - | 0 | + |
| - with out-group partner | - | 0 | 0 | - |
| Masculine (competitive) | - | + | 0 | - |
| Feminine (cooperative) | 0 | - | 0 | 0 |
| Long-term oriented | + | - | 0 | + |
| Short-term oriented | | | | |
| - with respected partner | + | - | 0 | 0 |
| - with other partners | - | 0 | 0 | 0 |

[8], [9], [10], [11], [12] proposed formal models for the influence of individual culture dimensions on trade processes, including effects on deceit threshold, inclination to trace, and positive and negative trust update factors. Table 2 summarizes these effects.

In the multi-agent simulation the decision to deceive is modeled as a random one, with probability of deceit as in equation (1), and $p$(truthful delivery)=1-$p$(deceit). Quality $q$ and certification $c$ are attributes of the contract. If either $q$=0 or $c$=1, there is no opportunity to deceive, so the agent delivers truthfully. If the customer negotiated a guarantee, and the agent is convinced that the partner will trace rather than trust the delivery there is no motive to deceive. The motive depends on value difference between high and low quality $\Delta v$, customer's benevolence $b_j$, fine $r$, and value $v$ to be restituted in case of a guarantee; $g$=1 indicates guarantee, $g$=0 no guarantee):

$$m_i = 1 \text{ if } \Delta v > (1\text{-}b_i)\{r+gv\} \text{ ; otherwise } m_i = 0 . \tag{4}$$

If motive and opportunity are present, the decision is effected by the agent's current honesty $h$ and its belief about the relation with the customer $b_i$. We assume that the deceit threshold toward agent $i$ has max($h$, $b_i$) as a basis, which is modified by cultural effect $e^d_i$ in the direction of 1 if $e^d_i > 0$; in the direction of 0 if $e^d_i < 0$:

$$d_i = \max(h, b_i) + \{1\text{-}\max(h, b_i)\}\ (|e^d_i|+e^d_i)/2 - \max(h, b_i)(\ |e^d_i|\text{-}e^d_i)/2 \ , \tag{5}$$

To model the effect of culture, for both negative and positive modification of the deceit threshold, the index with the maximal effect is selected:

$$e^d_i = \max\{\text{PDI}^*(s_i\text{-}s_j),\ \text{LTO}^*,\ (1\text{-LTO}^*)s_i\} - \max\{\text{UAI}^*D_{ij},\ (1\text{-IDV}^*)D_{ij},\ \text{MAS}^*,\ (1\text{-LTO}^*)(1\text{-}s_i)\}\ . \tag{6}$$

where $s_i$ indicates partner's status or hierarchical position in society and $s_j$ indicates own societal status, both on the interval [0, 1]; $D_{ij}$ represents group distance, i.e. $D_{ij}$=0 represents maximal familiarity; a complete stranger has $D_{ij}$=1.

Similarly, the decision to trace or to trust is modeled, with distrust (1 - $t_j$) as the basis for the decision, with cultural effect $e^t_j$ on the inclination to trace $j$'s deliveries.

$$p(\text{trace}) = q\ (1\text{-}c)\ \{1 - t_j\ -\ t_j\ (|e^t_j| + e^t_j)/2\ -\ (1\text{-}\ t_j)\ (|e^t_j| - e^t_j)/2\ \}\ ; \quad p(\text{trust}) = 1 - p(\text{trace})\ . \tag{7}$$

$$e^t_j = \max(\text{UAI}^*D_{ij},\ \text{MAS}^*) - \max\{\text{PDI}^*|s_i\text{-}s_j|,\ (1\text{-IDV}^*)(1\text{-}D_{ij}),\ 1\text{-MAS}^*,\ \text{LTO}^*,\ (1\text{-LTO}^*)s_j\}\ . \tag{8}$$

The basic values $u^{+\prime}$ and $u^{-\prime}$ of the update factors are modified in a similar way.

$$u^+ = u^{+\prime}\ -\ u^{+\prime}\text{UAI}^*\ . \tag{9}$$

$$u^- = u^{-\prime}\ +\ (1 - u^{-\prime})(|e^{u-}| + e^{u-})/2\ -\ (u^{-\prime} - u^{+\prime})(|e^{u-}| - e^{u-})/2\ ; \tag{10}$$

$$e^{u-} = \max\{\text{UAI}^*,\ (1\text{-IDV}^*),\ \text{LTO}^*\}\ -\ \text{MAS}^*\ . \tag{11}$$

## 4 Example of Results

To test the implementation of the model, simulations were run in an environment where agents could trade repeatedly, approximately 30-40 times per run. For all deals, $q$=1 or $c$=0, forcing the agents to decide on deceit and trust. In all runs, the culture dimensions were set to 0.5, except one dimension, which was set to 0.1 or 0.9 in order to represent a cultural extreme. Agents had labels, visible to all other agents. One represented status (societal rank). Other labels represented group memberships. Taking trust and benevolence developed by rule (2) as a proxy for familiarity, $D_{ij}$ was computed as the minimum of label-based and familiarity-based group distance.

The results presented in Table 3 indicate that culturally differentiated agent behavior at the micro level has impact on macro level statistics. The tendencies are as expected. The tracing rate is high in uncertainty avoiding and masculine societies, indicating low trust. The tracing is effective to reduce cheating. In short-term oriented societies opportunity is given to high-status members. They are not traced and seize the opportunity. In feminine societies honesty is not enforced but the deceit frequency remains low because of the strong inclination to cooperate in these societies.

**Table 3.** Results of simulations in societies with hypothetical cultures. In the hypothetical cultures all scaled cultural dimensions have index 0.5, except one, which has either 0.1 or 0.9.

| Culture and partner characteristics | Percent of transactions traced | Percentage of deceit in traces | Percent of deceit in all transactions | Percentage of deceit discovered |
|---|---|---|---|---|
| Large power dis., mixed ranks | 11 | 7 | 13 | 6 |
| Small power distance | 11 | 12 | 11 | 12 |
| Uncertainty avoiding, ingroup | 59 | 9 | 7 | 77 |
| Uncertainty tolerant | 9 | 21 | 7 | 26 |
| Individualistic | 12 | 7 | 8 | 10 |
| Collectivistic, ingroup partner | 14 | 27 | 14 | 26 |
| Masculine (competitive) | 56 | 14 | 13 | 62 |
| Feminine (cooperative) | 1 | 33 | 13 | 3 |
| Long-term oriented | 6 | 6 | 9 | 4 |
| Short-t. oriented, mixed ranks | 15 | 31 | 25 | 19 |

## 5 Conclusion

Culture is known to have its effects on honesty in trade, and on trust as a mechanism to compensate for the inevitable incompleteness of contracts. Occurrence of deceit, and mechanisms and institutions to reduce it, vary considerably across the world. For research into these mechanisms, multi-agent simulations can be a useful tool.

In intelligent agent research, much attention has been paid to trust. Little research has been published about the simulation of deceit. Publications such as [17] and [18] modeled deceit as a rational strategy to gain advantage in competitive situations. A strictly rational approach of deceit neglects the emotional impact that deceit has, not only on the deceived, but also on the deceivers. Feelings of guilt and shame result from deceiving [5]. The extent to which these feelings prevail is different across cultures [5]. People have emotional thresholds for deceit, that cannot be explained from rational evaluation of cost and benefit, but that are based on morality and cooperative attitudes [4], [13], [15]. Once deceived, people react to an extent that goes beyond rationality [4], especially when they are prosocial rather than selfish [15]. In human decision making a model based on *fair trade* prevails over a model of *opportunistic betrayal* [16]. In addition to psychological factors, rational economic motives can be given for the human inclination to cooperative behavior [14].

This paper contributes by introducing an agent model of deceit and placing it in a cultural context. It takes human deceptive behavior as a point of departure. Building on the work of [8], [9], [10], [11], [12] that modeled single dimensions of culture, this paper proposes an integrated model of culture's effects on deceit and trust. Example results have been generated that verify the implementation and illustrate that cultural effects can be generated. However, for realistic experiments, the model has to be tuned to and calibrated by observations and results of experiments, for instance to simulate effects like the ones reported by Triandis et al. [5] from human experiments on deceit across cultures. That work remains for future research.

## References


1. Williamson, O.E.: Transaction cost economics: how it works; where it is headed. De Economist 146(1), 23–58 (1998)
2. Klein Woolthuis, R., Hillebrand, B., Nooteboom, B.: Trust, Contract and Relationship Development. Organization Studies 26(6), 813–840 (2005)
3. Tykhonov, D., Jonker, C., Meijer, S., Verwaart, T.: Agent-Based Simulation of the Trust and Tracing Game for Supply Chains and Networks. Journal of Artificial Societies and Social Simulation 11(3), 1 (2008), `http://jasss.soc.surrey.ac.uk/11/3/1.html`
4. Boles, T.L., Croson, R.T.A., Murnighan, J.K.: Deception and Retribution in Repeated Ultimatum Bargaining. Organizational Behavior and Human Decision Processes 83(2), 235–259 (2000)
5. Triandis, H.C., et al.: Culture and Deception in Business Negotiations: A Multilevel Analysis. International Journal of Cross Cultural Management 1, 73–90 (2001)
6. Zaheer, S., Zaheer, A.: Trust across borders. Journal of International Business Studies 37, 21–29 (2006)
7. Hofstede, G.: Culture's Consequences, 2nd edn. Sage Publications, Thousand Oaks (2001)
8. Hofstede, G.J., Jonker, C.M., Verwaart, T.: Modeling Power Distance in Trade. In: David, N., Sichmann, J.S. (eds.) MABS 2008. LNCS, vol. 5269, pp. 1–16. Springer, Heidelberg (2009)
9. Hofstede, G.J., Jonker, C.M., Verwaart, T.: Modeling Culture in Trade: Uncertainty Avoidance. In: 2008 Agent-Directed Simulation Symposium (ADSS 2008), Spring Simulation Multiconference 2008, pp. 143–150, SCS, San Diego (2008)
10. Hofstede, G.J., Jonker, C.M., Verwaart, T.: Individualism and Collectivism in Trade Agents. In: Nguyen, N.T., Borzemski, L., Grzech, A., Ali, M. (eds.) IEA/AIE 2008. LNCS (LNAI), vol. 5027, pp. 492–501. Springer, Heidelberg (2008)
11. Hofstede, G.J., Jonker, C.M., Meijer, S., Verwaart, T.: Modeling Trade and Trust across Cultures. In: Stølen, K., Winsborough, W.H., Martinelli, F., Massacci, F. (eds.) iTrust 2006. LNCS, vol. 3986, pp. 120–134. Springer, Heidelberg (2006)
12. Hofstede, G.J., Jonker, C.M., Verwaart, T.: Long-term Orientation in Trade. In: Schredelseker, K., Hauser, F. (eds.) Complexity and Artificial Markets. LNEMS, vol. 614, pp. 107–118. Springer, Heidelberg (2008)
13. Wirtz, J., Kum, D.: Consumer Cheating on Service Guarantees. Journal of the Academy of Marketing Science 32(2), 159–175 (2004)
14. Hwang, P., Burgers, W.P.: Apprehension and Temptation: The Forces against Cooperation. Journal of Conflict Resolution 43(1), 117–130 (1999)
15. Steinel, W., De Dreu, K.W.: Social Motives and Strategic Misrepresentation in Social Decision Making. Journal of Personality and Social Psychology 86, 419–434 (2004)
16. Olekalns, M., Smith, P.L.: Mutually Dependent: Trust, Affect and the Use of Deception in Negotiation. Journal of Business Ethics 85, 347–365 (2009)
17. Castelfranchi, C., Falcone, R., De Rosis, F.: Deceiving in GOLEM: how to strategically pilfer help. In: Castelfranchi, C., Tan, Y.H. (eds.) Trust and Deception in Virtual Societies, pp. 91–110. Kluwer, Dordrecht (2001)
18. Ward, D., Hexmoor, H.: Towards deception in agents. In: AAMAS 2003, Proceedings, pp. 1154–1155. ACM, New York (2003)
19. Burgoon, J.K., Stoner, G.M., Bonito, J.A., Dunbar, N.E.: Trust and Deception in Mediated Communication. In: Proceedings of HICSS 2003 - Track1, p. 44. IEEE, Los Alamitos (2003)
20. Jonker, C.M., Treur, J.: Formal analysis of models for the dynamics of trust based on experiences. In: Garijo, F.J., Boman, M. (eds.) MAAMAW 1999. LNCS, vol. 1647, pp. 120–134. Springer, Heidelberg (1999)

# Implementation of Epistemic Operators for Model Checking Multi-agent Systems

Marina Bagić Babac and Marijan Kunštić

University of Zagreb
Faculty of Electrical Engineering and Computing
Unska 3, HR-10000 Zagreb, Croatia
{marina.bagic,marijan.kunstic}@fer.hr

**Abstract.** The problem of multi-agent system (MAS) specification and verification has been introduced in this paper, Epistemic transition system (ETS) represents an agent as the smallest unit in a multi-agent system, while Epistemic synchronous product (ESP) represents the formal model for a multi-agent system. Therefore, a formal framework for epistemic properties of multi-agent systems has been provided. A special extension of Action computation tree logic with unless operator for epistemic reasoning (ACTLW-ER) is used for MAS model checking. Epistemic operators of ACTLW-ER are implemented by symbolic model checking algorithms using binary decision diagrams.

**Keywords:** Action Computation Tree Logic with Unless Operator, Epistemic Reasoning, Multi-agent Systems.

## 1 Introduction

This paper tries to give one of the solutions to the problem of model checking multi-agent systems (MAS). While the general model checking techniques were studied for various systems [2], we put the emphasis on epistemic multi-agent systems, i.e. the multi-agent systems concerned with the knowledge [3] and knowledge-based logics [5]. Therefore, this paper provides a multi-agent system specification as a formal framework for verification of their dynamic epistemic properties.

The approach of this paper is similar to [10] in the sense of the usage of symbolic model checking with binary decision diagrams (BDDs) to verify the multi-agent system. However, while the approach [10] is based on deontic interpreted system which relies on Kripke structure as a formal specification model for the MAS, this paper gives the formal framework of Epistemic Transition System (ETS) and Epistemic Synchronous Product (ESP).

Another direction of the research in this paper is the extension the Action Computation Tree Logic with Unless Operator logic - ACTLW [6], [7]. This logic finds its fundamentals in Action Computation Tree Logic [4]. However, it serves to verify the generic processes and is not expressive enough for epistemic properties of multi-agent systems. ACTLW was developed for the verification of

N.T. Nguyen, R. Kowalczyk, and S.-M. Chen (Eds.): ICCCI 2009, LNAI 5796, pp. 217–228, 2009.

generic systems based on Labelled Transition System (LTS) [8]. So, in this paper epistemic operators were added to this logic, therefore providing the ACTLW for Epistemic Reasoning (ACTLW-ER). Also, the introduction of colours of actions enriched the approach. Colours are dedicated to each action in order to manipulate over atomic propositions which hold knowledge provided by the agents.

In our previous work [1] we have given some epistemic-temporal operators. Now, for efficiency reasons we use separately temporal operators from epistemic ones such as $\mathbf{K}\ \varphi$ - "an agent knows $\varphi$", and $\mathbf{E}_G\ \varphi$ - "each agent in the group G knows $\varphi$".

In the end, we provide an example of Dining Cryptographers multi-agent system, a security communication protocol from the literature [10].

## 2 Specification of Epistemic Systems

Here we describe the event-based approach to modeling knowledge, one that is typically used in the work on knowledge in game theory and mathematical economics. This approach in [3] uses Aumann structures while we define our own structure in order to extend Aumann structure with data and also retaining some properties of Kripke structure to reason on atomic propositions, i.e. epistemic properties of states. Formally, we define Epistemic Transition System (ETS), combining constructs from LTS [6] and MTS [9] as follows.

**Epistemic Transition System** is a 9-tuple:

$$\mathcal{A} = (\mathcal{S}, \mathcal{S}_0, \mathcal{A}, \delta, \mathcal{K}, \mathcal{C}, f_s, f_a, f_c) \tag{1}$$

where there are:

- $\mathcal{S}$, a set of states where $\mathcal{S}_i \in \mathcal{S}$
- $\mathcal{S}_0$, a set of agent's initial states
- $\mathcal{A}$, a set of actions
- $\delta \subseteq \mathcal{S} \times \mathcal{A} \times \mathcal{S}$, the transition relation
- $\mathcal{K}$, a set of epistemic atoms
- $\mathcal{C}$, a set of actions' colours: $\{\alpha, \kappa, \rho, \epsilon\}$
- $f_a$, a function mapping a set of atoms to each action $f_a : a_i \rightarrow 2^{\mathcal{K}}$
- $f_s$, a function mapping a set of atoms to each state $f_s : S_i \rightarrow 2^{\mathcal{K}}$
- $f_c$, a function mapping an action to its colour $f_c : a_i \rightarrow c \in \mathcal{C}$

We define now a multi-agent system as a collection of ETSs. It is defined as synchronous product of individual ETSs - Epistemic Synchronous Product (ESP).

**Epistemic Synchronous Product** is a 10-tuple:

$$\mathcal{M} = (\mathcal{S}, \mathcal{S}_0, \mathcal{A}, \delta, \mathcal{K}, \mathcal{C}, f_s, f_a, f_c, f_d) \tag{2}$$

where there are:

- $\mathcal{S} \subseteq \mathcal{S}_1 \times \cdots \times \mathcal{S}_n$, a non-empty set of states of MAS;
- $\mathcal{S}_0 \subseteq \mathcal{S}_{0,1} \times \cdots \times \mathcal{S}_{0,n}$, a set of MAS' initial states
- $\mathcal{A} = \mathcal{A}_1 \cup \cdots \cup \mathcal{A}_n$, a finite, non-empty set of actions
- $\delta \subseteq \mathcal{S} \times \mathcal{A} \times \mathcal{S}$, the transition relation;
  - $(S_i, a, S_i') \in \delta_i \wedge \forall j \neq i : a \notin \mathcal{A}_j :$
    $((S_1, ..., S_i, ..., S_n), a, (S_1, ..., S_i', ..., S_n')) \in \delta$
  - $(S_i, a, S_i') \in \delta_i \wedge (S_j, a, S_j') \in \delta_j :$
    $((S_1, ..., S_i, ..., S_j, ..., S_n), a, (S_1, ..., S_i', ..., S_j', ..., S_n')) \in \delta$
- $\mathcal{K}$, a set of (non-)epistemic atoms
- $\mathcal{C}$, a set of actions' colours: $\{\alpha, \kappa, \rho, \epsilon\}$
- $f_c$, a function mapping an action to its colour $f_c : a_i \rightarrow c \in \mathcal{C}$
- $f_a$, a function mapping a set of atoms to each action $f_a : a_i \rightarrow 2^{\mathcal{K}}$
- $f_s$, a function mapping a set of atoms to each state $f_s : \mathcal{S}_i \rightarrow 2^{\mathcal{K}}$
- $f_d$, a function mapping each state to a subset of $\mathcal{K}_1 \times ... \times \mathcal{K}_n$, where $\mathcal{K}_1 = f_{s_i}(S_i)$, a set of atoms of an agent's state.

## 3 Syntax and Semantics of ACTLW for Epistemic Reasoning

Action Computation Tree Logic with Unless Operator for Epistemic Reasoning (ACTLW-ER) syntax and semantics are defined over the ETS and ESP.

If there exists a transition from state $p$ to state $q$, then $q$ is called a successor of $p$. An element $(p, a, q) \in \delta$ is called a *transition* from state $p$ to state $q$ or shortly, an $a$-transition. In order for a transition relation to be a total ternary relation each state of ETS must have at least one successor. A state is called a *deadlocked state* iff there are no transitions from that state.

A sequence of transitions starting and ending in the same state is called a *cycle*. If a path is infinite or ends up in a deadlocked state, it is called an *infinite fullpath* or a *finite fullpath*, respectively. The *empty fullpath* is a finite fullpath with one state and no transitions. The number of transitions in a finite fullpath $\pi$ will be denoted with $|\pi|$.

If $\mathcal{E}$ is an ETS, then a sequence of transitions $(p_0, a_1, p_1), (p_1, a_2, p_2), ...$ where $\forall i \geq 0.\ (p_i, a_{i+1}, p_{i+1}) \in \delta$ is called a *path* $\pi$ in $\mathcal{E}$. Moreover, $p_i$ and $a_i$ are called the $i$-th state and $i$-th action on this path, respectively, and the transition ending in the $i$-th state is called the $i$-th transition on this path. A notation $st(\pi, i)$ and $act(\pi, i)$ are used for identification of particular states and transitions on the path $\pi$ as in [7]:

- $st(\pi, 0)$ is the first state on the path $\pi$,
- $st(\pi, i)$ is a state reached after the $i$-th transition on the path $\pi$ $(i \geq 1)$,
- $act(\pi, i)$ is an action executed during the $i$-th transition on the path $\pi$ $(i \geq 1)$.

A state $p$ where formula $\varphi$ holds is called $\varphi$-state and a transition $(p, a, q)$ where an action formula $\chi$ holds for an action $a$ is called $\chi$-transition. A $\chi$-transition $(p, a, q)$ where formula $\varphi$ holds in state $q$ is called $\{\chi\}\varphi$-transition.

Let $\chi, \varphi$, and $\gamma$ be a *data-action formula*, a *state formula*, and a *path formula*, respectively, iff they meet the following syntactic rules:

$$\chi ::= true \,|\, \alpha \,|\, \alpha(\Delta) \,|\, \kappa \,|\, \kappa(\Delta) \,|\, \rho \,|\, \rho(\Delta) \,|\, \epsilon \,|\, \epsilon(\Delta) \,|\, \tau \,|\, \tau(\Delta) | \neg\chi \,|\, \chi \vee \chi \quad (3)$$

$$\varphi ::= true \,|\, k \,|\, \neg\varphi \,|\, \varphi \wedge \varphi' \,|\, \mathbf{E}\gamma \,|\, \mathbf{A}\gamma \,|\, \mathbf{K}\,\varphi \,|\, \mathbf{E_G}\,\varphi \,|\, \mathbf{C}\varphi \quad (4)$$

$$\gamma ::= \{\chi\}\varphi\,\mathbf{U}\,\{\chi'\}\varphi' \,|\, \{\chi\}\varphi\,\mathbf{W}\,\{\chi'\}\varphi' \quad (5)$$

A data-action formula $\chi$ (3) may be constructed by sending or receiving (! or ?) message to or from port $p$ or may be a Boolean negation ($\neg\chi$) or a Boolean composition (either product or addition) of more than one data-action formulae $\chi_i$. Formal definition of the data-action formula semantics is given in Table 1. The silent action $\tau$ is also included in the set of actions).

A state formula (4) is either Boolean value, or contains a path operators **A** or **E** denoting all or some of the states at a lifecycle path of an agent. Formal definition of the colour-action formula semantics is given in Table 3.

A path formula (5) contains the constructs from the above (actions, states) and temporal operators (**U** and **W**) while the other temporal operators are derived from these ones.

Let $\mathcal{M} = (\mathcal{S}, \mathcal{S}_0, \mathcal{A}, \delta, \mathcal{K}, \mathcal{C}, f_s, f_a, f_c)$ be a multi-agent system. Satisfaction of data-action formula $\chi$ by an action $a \in \mathcal{A}$ (written $a \models \chi$), state formula $\varphi$ by a state $s \in \mathcal{S}$ ($s \models \varphi$), a path formula $\gamma$ by a finite fullpath $\pi$ (written $\pi \models \gamma$), and a path formula $\gamma$ by an infinite fullpath $\sigma$ (written $\sigma \models \gamma$) in a ESP $\mathcal{M}$ is given inductively by the semantic rules given in tables 1, 2, 3 and 4.

**Table 1.** Data-action Semantic Rules of ACTLW for Epistemic Reasoning

| | |
|---|---|
| $a \models true$ | always |
| $a \models \chi$ | iff $a = \chi$ |
| $a \models \chi(\Delta)$ | iff $a = \chi \wedge \Delta = true$ |
| $a \models \tau(\Delta)$ | iff $a = \tau \wedge \Delta = true$ |
| $a \models \neg\chi$ | iff $a \not\models \chi$ |
| $a \models \chi \vee \chi'$ | iff $a \models \chi \vee a \models \chi'$ |

## 4 Symbolic Resolving ACTLW-ER Operators

In a multi-agent system different agents have different knowledge of the world. An agent may need to reason about its own knowledge about the world; it may also need to reason about what other agents know about the world. Reasoning about knowledge refers to the idea that agents in a group take into account not only the facts of the world, but also the knowledge of other agents in the group [3].

**Table 2.** Colours Semantics of ACTLW for Epistemic Reasoning

$$
\begin{array}{lll}
a,\pi \models \kappa(\Delta) & \text{iff} & \exists i \in [1,|\pi|] : act(\pi,i) \models \kappa(\Delta) \wedge st(\pi,i) \models \Delta \wedge \forall j > i \wedge j \leq |\pi| : \\
 & & st(\pi,j) \models \Delta \\
a,\pi \models \epsilon(\Delta) & \text{iff} & \exists i \in [1,|\pi|] : act(\pi,i) \models \epsilon(\Delta) \wedge st(\pi,i) \not\models \Delta \wedge \forall j > i \wedge j \leq |\pi| : \\
 & & st(\pi,j) \not\models \Delta \\
a,\pi \models \alpha(\Delta) & \text{iff} & \exists i \in [1,|\pi|] : act(\pi,i) \models \alpha(\Delta) \wedge st(\pi,i) \models \Delta \\
a,\pi \models \rho(\Delta) & \text{iff} & \exists i \in [1,|\pi|] : act(\pi,i) \models \rho(\Delta) \wedge st(\pi,i) \models true
\end{array}
$$

**Table 3.** State Semantic Rules of ACTLW for Epistemic Reasoning

$$
\begin{array}{ll}
s \models true & \text{always} \\
s \models k & \text{iff } k \in f_s(s) \in \mathcal{K} \\
s \models \neg\varphi & s \not\models \varphi \\
s \models \varphi \wedge \varphi' & s \models \varphi \wedge s \models \varphi' \\
s \models \mathbf{E}\gamma & \text{iff } \exists\pi : s = st(\pi,0) \wedge \pi \models \gamma \\
 & \text{or } \exists\sigma : s = st(\pi,0) \wedge \sigma \models \gamma \\
s \models \mathbf{A}\gamma & \text{iff } \forall\pi \text{ and } \forall\sigma \\
 & \pi : s = st(\pi,0) \\
 & \sigma : s = st(\pi,0) \wedge \sigma \models \gamma \\
s \models \mathbf{K}\varphi & \text{iff } s \models \varphi \\
s \models \mathbf{E_G}\varphi & \text{iff } \forall i \in G : s \models \varphi
\end{array}
$$

**Table 4.** Path Semantic Rules of ACTLW for Epistemic Reasoning

$$
\begin{array}{ll}
\pi \models \varphi\{\chi(\Delta)\}\,\mathbf{U}\,\{\chi'(\Delta')\}\varphi' & \text{iff } st(\pi,0) \models \varphi \wedge \exists i \in [1,|\pi|] : (act(\pi,i) \models \chi'(\Delta') \\
 & \wedge st(\pi,i) \models \varphi') \wedge \forall j \in [1,i-1] : (act(\pi,j) \models \chi(\Delta) \\
 & \wedge st(\pi,j) \models \varphi) \\
 & \\
\sigma \models \varphi\{\chi(\Delta)\}\,\mathbf{U}\,\{\chi'(\Delta')\}\varphi' & \text{iff } st(\sigma,0) \models \varphi \wedge \exists i \in [1,|\sigma|] : (act(\sigma,i) \models \chi'(\Delta') \\
 & \wedge st(\sigma,i) \models \varphi') \wedge \forall j \in [1,i-1] : (act(\sigma,j) \models \chi(\Delta) \\
 & \wedge st(\sigma,j) \models \varphi) \\
 & \\
\pi \models \{\chi\}\varphi\,\mathbf{W}\,\{\chi'\}\varphi' & \text{if } \pi \models \{\chi\}\varphi\,\mathbf{U}\,\{\chi'\}\varphi' \text{ or if} \\
 & \forall i \in [1, len(\pi)] st(\pi,i) \models \varphi \wedge\ act(\pi,i) \models \chi \\
 & \\
\sigma \models \{\chi\}\varphi\,\mathbf{W}\,\{\chi'\}\varphi' & \text{if } \sigma \models \{\chi\}\varphi\,\mathbf{U}\,\{\chi'\}\varphi' \text{ or if} \\
 & \forall i \geq 1 : st(\sigma,i) \models \varphi \wedge\ act(\sigma,i) \models \chi
\end{array}
$$

In this section we reason on semantics of epistemic ACTLW-ER operators. This approach was strongly inspired by [7].

### 4.1 K$\varphi$ Operator

For Kripke structures it is common to express the formula **K**$\varphi$ as "an agent knows $\varphi$", where $\varphi$ is a set of atomic propositions true for that state.

Our approach is event-based so we define agent's knowledge in terms of agent's actions. We say that "an agent knows $\varphi$ if he knows $\varphi$ in his initial state, or during his lifecycle an epistemic $(\chi, \varphi)$-transition occurs, where $\chi$-action is either $\kappa$ or $\kappa(\Delta)$-action."

The definition holds also for a multi-agent system, and is extended for a set of agents according to ESP definition (Def. 2). In order to calculate it we use fixed points (Knaster-Tarski [5]) and calculate a set of $\varphi$-states as the least fixed point of the function.

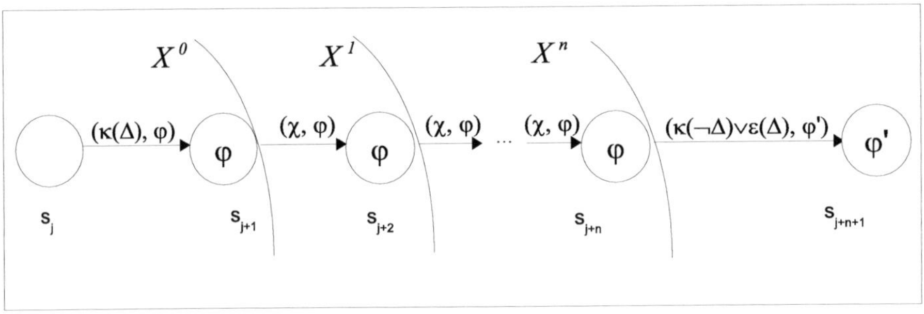


**Fig. 1.** Resolving **K**$\varphi$ Operator without initial knowledge

$$X_{K_i\varphi} = \mathbf{lfp}\, Z.(\{q \in \mathcal{S} \,|\, \exists a \in A_i\, \exists q \in \mathcal{S} \,:\, (q, a, q') \in \delta_{(\kappa(\Delta),\varphi)} \\ \vee \exists a \in A\, \exists q \in Z \,:\, (q, a, q') \in \delta_{\neg(\epsilon(\Delta)\vee\kappa(\neg\Delta),\neg\varphi)}\}) \tag{6}$$

where $X_{K_i\varphi}$ is a set of all states which hold the statement "an agent knows $\varphi$", $Z$ is a function which holds fixed point calculation, i.e. calculates the states holding the statement.

Therefore, we have a following sets of states;

$$X^0_{K_i\varphi} = \{q \in \mathcal{S} \,|\, \exists a \in A_i\, \exists q \in \mathcal{S} \,:\, (q, a, q') \in \delta_{(\kappa(\Delta),\varphi)}\} \tag{7}$$

$$\forall i > 0 : \\ X^i_{K_i\varphi} = X^{i-1}_{K_i\varphi} \bigcup \\ \{q \in \mathcal{S} \,|\, \exists a \in A\, \exists q \in X^{i-1}_{K_i\varphi} \,:\, (q, a, q') \in \delta_{\neg(\epsilon(\Delta)\vee\kappa(\neg\Delta),\neg\varphi)}\} \tag{8}$$

It is also possible for an agent to know $\varphi$ in its initial state. Then, we add $\mathbf{Y}_{K_i\varphi}$ to the Formula 11 where:

$$Y_{K_i\varphi} = \mathbf{lfp}\, Z.(\{q \in \mathcal{S} \,|\, q_i \in S_{i0} \bigcap S_{i\varphi} \\ \vee \exists a \in A\, \exists q \in Z\, :\, (q, a, q') \in \delta_{\neg(\epsilon(\Delta) \vee \kappa(\neg\Delta), \neg\varphi)}\}) \tag{9}$$

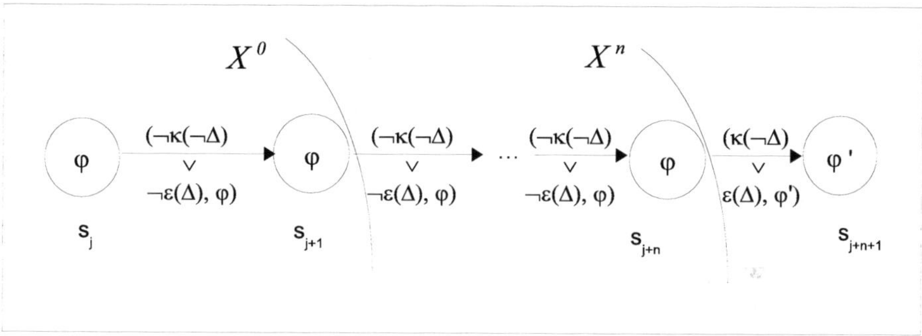


**Fig. 2.** Resolving $\mathbf{K}\varphi$ Operator with initial knowledge

Thus, we have:

$$X'_{K_i\varphi} = X_{K_i\varphi} \bigcup Y_{K_i\varphi} \tag{10}$$

## 4.2 $\mathbf{E}_G\varphi$ Operator

Another epistemic operator $\mathbf{E}_G\varphi$ is read as "every agent in group G knows". According to the previous section we have: $\mathbf{E}_G\varphi = \mathbf{K}_1\varphi + \mathbf{K}_2\varphi + ... + \mathbf{K}_G\varphi$

which is obtained by calculating:
$\mathbf{E}_G\varphi = \bigcap_i \mathrm{X}_{K_i\varphi}$

– without the initial knowledge:

$$\mathbf{E}_G\varphi = \bigcap_i \mathbf{lfp}\, Z.(\{q \in \mathcal{S} \,|\, \exists a \in A_i\, \exists q \in \mathcal{S}\, :\, (q, a, q') \in \delta_{(\kappa(\Delta), \varphi)} \\ \vee \exists a \in A\, \exists q \in Z\, :\, (q, a, q') \in \delta_{\neg(\epsilon(\Delta) \vee \kappa(\neg\Delta), \neg\varphi)}\}) \tag{11}$$

– with the initial knowledge:

$$\mathbf{E}_G\varphi = \bigcap_i (\mathbf{lfp}\, Z.(\{q \in \mathcal{S} \mid \exists a \in A_i\, \exists q \in \mathcal{S} \,:\, (q,a,q') \in \delta_{(\kappa(\Delta),\varphi)}$$
$$\vee \exists a \in A\, \exists q \in Z \,:\, (q,a,q') \in \delta_{\neg(\epsilon(\Delta)\vee\kappa(\neg\Delta),\neg\varphi)}\})$$
$$\bigcup \mathbf{lfp}\, Z.(\{q \in \mathcal{S} \mid q_i \in S_{i0} \bigcap S_{i\varphi}$$
$$\vee \exists a \in A\, \exists q \in Z \,:\, (q,a,q') \in \delta_{\neg(\epsilon(\Delta)\vee\kappa(\neg\Delta),\neg\varphi)}\})) \tag{12}$$

## 5 The Dining Cryptographers Protocol

Three cryptographers are sitting down to dinner at their favorite three-star restaurant. Their waiter informs them that arrangements have been made with the maitre d'hotel for the bill to be paid anonymously. One of the cryptographers might be paying for the dinner, or it might have been NSA (U.S. National Security Agency). The three cryptographers respect each other's right to make an anonymous payment, but they wonder if NSA is paying. They resolve their uncertainty fairly by carrying out the following protocol.

Each cryptographer flips an unbiased coin behind his menu, between him and the cryptographer on his right, so that only the two of them can see the outcome. Each cryptographer then states aloud whether the two coins he can see–the one he flipped and the one his left-hand neighbor flipped–fell on the same side or on different sides. If one of the cryptographers is the payer, he states the opposite of what he sees. An odd number of differences uttered at the table indicates that a cryptographer is paying; an even number indicates that NSA is paying (assuming that the dinner was paid for only once). Yet if a cryptographer is paying, neither of the other two learns anything from the utterances about which cryptographer it is [10].

### 5.1 Specification of Agents in Dining Cryptographers System

Specification of a multi-agent system of dining cryptographers requires at least three agents for the cryptographers and one for the environment. Therefore, we denote each one of them as agent A, B, C and E, respectively. Each Cryptographer Agent has no any prior knowledge before the system startup. Initial knowledge is given to them by the Environment Agent. The Environment Agent supplies each Cryptographer Agent with the necessary information and then lets it communicate to the other agents in the system. To specify the communication between the agents in the system we have explicitly denoted the port's (or channel's) names in order to give precise address of the message destination, i.e. *a-c* stands for a communication point between the Cryptographer Agent A and the Cryptographer Agent C, or *env-c* stands for a communication point between the Environment Agent and the Cryptographer Agent C. The main difference between the three cryptographers agents' specifications are in the epistemic atoms and the addresses of the adjacent agents. The first two actions that each Cryptographer Agent performs are synchronized with the Environment Agent. They learn about their status of paying for the dinner or not and also the status of the coin on their right to notify the adjacent agent on their left.

### 5.2 Model Checking Dining Cryptographers

Model checking Dining Cryptographers protocol requires verification formulae on each of the included agent, specifically on Cryptographers and Environment agents. Due to the symmetry of Cryptographer agents we first reason on properties of only one Cryptographer Agent and Environment Agent, and than on the multi- agent system with four agents in it.

**($F_1$)** If the number of differences in the utterances is even, then Cryptographer Agent knows that non of the cryptographers paid.

**AG** {inform (even)} ($\mathbf{K}_A$ (NOTpaying) $\vee$ $\mathbf{K}_B$ (NOTpaying) $\vee$ $\mathbf{K}_C$ (NOTpaying))

**($F_2$)** All cryptographers know that either one of them or NSA is paying for the dinner.

**AF** {inform (odd) $\vee$ inform (even)} $\mathbf{C}_G$ (paying $\vee$ NOTpaying)

**($F_3$)** If the number of differences in the utterances is odd, then Cryptographer Agent A knows that either Cryptographer Agent B or Cryptographer Agent C paid for the dinner.

**EF** {inform (odd)} $\mathbf{K}_A$ ($\mathbf{K}_B$ (paying) $\vee$ $\mathbf{K}_C$ (paying)

**($F_4$)** There is no path such that Cryptographer Agent A can decide on his statement until he collects the information from other agents.

$\neg$**E** {say (equal) $\vee$ say (NOTequal)}**U** {infoRight (head) $\vee$ infoRight (tail)}

### 5.3 Experimental Results

The simulation run is provided by the EST tool [11], 2nd edition, 5.1 version on 2.00 GHz Intel (R) Core (TM) 2 Duo CPU with 2.00 GB of RAM, running Ubuntu Linux with kernel 5.0.

As the size of the model is defined as $\mid M \mid=\mid S \mid + \mid R \mid$, $S$ is a set of states and $R$ is the temporal relation. Here we define $S$ as the number of all the possible combinations of local states and actions. In this example of Dining Cryptographers there is 14 local states for each Cryptographer Agent and 15 actions for each one; there is 9 local states and 15 actions for the Environment Agent hence $\mid S \mid\approx 1.25{\cdot}10^9$. To define $\mid R \mid$ there must be taken into account that besides the temporal relation, there are also the epistemic relations. Hence, $\mid R \mid$ is the sum of the sizes of temporal and epistemic relations. The approximation is $\mid R \mid$ as $\mid S \mid^2$, hence $\mid M \mid=\mid S \mid + \mid R \mid\approx \mid S \mid^2 \approx 1.56 \cdot 10^{18}$.

The results of BDDs, model checking and time consuming are given in the Tables 7, 8 and 9, respectively.

**Table 5.** Dining Cryptographer Agent A

| | **CRYPTOGRAPHER AGENT A** |
|---|---|
| INIT | = env-a ? infoPaying (NOTpaying); NOT-PAYING |
| | + env-a ? infoPaying (paying); PAYING |
| NOT-PAYING | = env-a ? infoRight (head); |
| | a-b ! tellRight (head); N-RIGHT-HEAD |
| | + env-a ? infoRight (tail); |
| | a-b ! tellRight (tail); N-RIGHT-TAIL |
| N-RIGHT-HEAD | = a-c ? infoLeft (tail); |
| | env-a ! say (equal); H-C-TAIL |
| | + a-c ? infoLeft (head); |
| | env-a ! say (NOTequal); H-C-HEAD |
| H-C-TAIL | = env-a ? info (odd); ! return; INIT |
| | + env-a ? info (even); ! return; INIT |
| H-C-HEAD | = env-a ? info (odd); ! return; INIT |
| | + env-a ? info (even); ! return; INIT |
| N-RIGHT-TAIL | = a-c ? infoLeft (tail); |
| | env-a ! say (NOTequal); T-C-TAIL |
| | + a-c ? infoLeft (head); |
| | env-a ! say (equal); T-C-HEAD |
| T-C-TAIL | = env-a ? info (odd); ! return; INIT |
| | + env-a ? info (even); ! return; INIT |
| T-C-HEAD | = env-a ? info (odd); ! return; INIT |
| | + env-a ? info (even); ! return; INIT |
| PAYING | = env-a ? infoRight (head); |
| | a-b ! tellRight (head); P-RIGHT-HEAD |
| | + env-a ? infoRight (tail); |
| | a-b ! tellRight (tail); P-RIGHT-TAIL |
| P-RIGHT-HEAD | = a-c ? infoLeft (tail); |
| | env-a ! say (equal); H-TAIL |
| | + a-c ? infoLeft (head); |
| | env-a ! say (NOTequal); H-HEAD |
| H-TAIL | = env-a ? info (odd); ! return; INIT |
| | + env-a ? info (even); ! return; INIT |
| H-HEAD | = env-a ? info (odd); ! return; INIT |
| | + env-a ? info (even); ! return; INIT |
| P-RIGHT-TAIL | = a-c ? infoLeft (tail); |
| | env-a ! say (NOTequal); C-TAIL |
| | + a-c ? infoLeft (head); |
| | env-a ! say (equal); C-HEAD |
| C-TAIL | = env-a ? info (odd); ! return; INIT |
| | + env-a ? info (even); ! return; INIT |
| C-HEAD | = env-a ? info (odd); ! return; INIT |
| | + env-a ? info (even); ! return; INIT |

**Table 6.** Environment Agent

| | ENVIRONMENT AGENT |
|---|---|
| INIT | = env-a ! infoPaying (NOTpaying); NOT-PAYING |
| | + env-a ! infoPaying (paying); PAYING |
| NOT-PAYING | = env-a ! infoRight (head); |
| | a-b ? say (NOTequal); N-NOT-EQUAL |
| | + env-a ! infoRight (tail); |
| | a-b ? say (equal); N-EQUAL |
| N-NOT-EQUAL | = env-a ? info (odd); ! return; INIT |
| | + env-a ? info (even); ! return; INIT |
| N-EQUAL | = env-a ? info (odd); ! return; INIT |
| | + env-a ? info (even); ! return; INIT |
| PAYING | = env-a ! infoRight (head); |
| | a-b ? say (equal); P-EQUAL |
| | + env-a ! infoRight (tail); |
| | a-b ? say (NOTequal); P-NOT-EQUAL |
| P-EQUAL | = env-a ? inform (odd); ! return; INIT |
| | + env-a ? inform (even); ! return; INIT |
| P-NOT-EQUAL | = env-a ? inform (odd); ! return; INIT |
| | + env-a ? inform (even); ! return; INIT |

**Table 7.** BDDs for Dining Cryptographers

| $\mid M \mid$ | BDDs nodes |
|---|---|
| $\approx 1.56 \cdot 10^{18}$ | 60 |
| $\approx 6.89 \cdot 10^{22}$ | 78 |
| $\approx 3.04 \cdot 10^{27}$ | 91 |

**Table 8.** Dining Cryptographers: Model Checking Results

| Formula | $\mathbf{F}_1$ | $\mathbf{F}_2$ | $\mathbf{F}_3$ | $\mathbf{F}_4$ |
|---|---|---|---|---|
| Result | FALSE | TRUE | TRUE | TRUE |

**Table 9.** Dining Cryptographers: Time Consuming

| MAS | ESP [$\mu$s] | Verification [$\mu$s] |
|---|---|---|
| 4 agents | 8 | 7 |
| 5 agents | 8 | 7 |
| 6 agents | 9 | 7 |

## 6 Conclusions

This paper has investigated the model checking epistemic properties of multi-agent systems. The specifying model for the agent has been given as the epistemic transition system and the epistemic synchronous product for a multi-agent system. The verification is based on symbolic model checking using binary decision diagrams and special epistemic logic ACTLW-ER - Action Computation Tree Logic with unless operator for Epistemic Reasoning. We have expressed our framework with security communication protocol of Dining Cryptographers. We have given the experimental results due to the memory requirements for building the model for a multi-agent system in BDDs nodes and the average time needed to build and verify the system. The results obtained are encouraging.

## References


1. Bagić, M., Kunštić, M.: Verification of Intelligent Agents with ACTL for Epistemic Reasoning. In: Proceedings of International Conference on Intelligent Agents, Web Technologies and Internet Commerce - IAWTIC 2006, p. 76. IEEE, Los Alamitos (2006)
2. Clarke, E.M., Grumberg, O., Peled, D.A.: Model Checking. MIT Press, Cambridge (2001)
3. Fagin, R., Halpern, J.Y., Moses, Y., Vardi, M.Y.: Reasoning About Knowledge. The MIT Press, Cambridge (2003)
4. Fantecji, A.: Enhancing test coverage by back-tracing model-checker counterexamples. Electronic Notes in Computer Science, vol. 116, pp. 199–211 (2005)
5. Huth, M.R.A., Ryan, M.D.: Logic in Computer Science: Modelling and reasoning about systems. Cambridge University Press, Cambridge (2000)
6. Meolic, R., Kapus, T., Brezocnik, Z.: Verification of concurrent systems using ACTL. In: Hamza, M.H. (ed.) Applied informatics: proceedings of the IASTED international conference AI 2000, Anaheim, Calgary, Zuerich, pp. 663–669. IASTED/ACTA Press, Innsbruck (2000)
7. Meolic, R., Kapus, T., Brezocnik, Z.: An Action Computation Tree Logic With Unless Operator. In: Proceedings of the 1st South-East European workshop on formal methods SEEFM 2003, Thessaloniki, Greece, pp. 100–114 (2003)
8. Milner, R.: Communication and Concurrency, Prentice Hall International Series, Computer Science, UK (1989)
9. Pecheur, C., Raimondi, F.: Symbolic model checking of logics with actions. In: Edelkamp, S., Lomuscio, A. (eds.) MoChArt IV. LNCS (LNAI), vol. 4428, pp. 113–128. Springer, Heidelberg (2007)
10. Raimondi, F., Lomuscio, A.: Automatic verification of deontic and epistemic properties of multi-agent systems by model checking via OBDD's. In: Proceedings of ECAI 2004, pp. 53–57. IOS Press, Amsterdam (2004)
11. University of Maribor, Laboratory of Microcomputer Systems (2009), `http://lms.uni-mb.si/EST/est2ed.html`

# Meta-game HOLOS as a Multi-agent Decision-Making Laboratory

Rolisław J. Kolbusz[1], Romuald Kotowski[2], and Krzysztof Kasianiuk[3]

[1] Krakowskie Przedmieście 43/3, 00-071 Warsaw, Poland
rjkolbusz@poczta.onet.pl
[2] Polish-Japanese Institute of Information Technology, Koszykowa 86, 02-008 Warsaw, Poland
rkotow@pjwstk.edu.pl
[3] Collegium Civitas, Plac Defilad 1, 00-901 Warsaw, Poland
krzysztof.kasianiuk@collegium.edu.pl

**Abstract.** A generalized version of the meta-game HOLOS for strategic decision-making processes is presented and discussed. HOLOS is a practical implementation of the proposed by R.J. Kolbusz [1] New Games' Paradigm (NGP) – an attempt that stems from the fuzzy logic-approach to decision-making processes and makes the latter relatively simple and accessible to players. The proposed complex game system encompasses and exceeds the classic *n*-players game theory formula. HOLOS can be played simultaneously on multiple online arenas. It presupposes that the players' cognitive structures are interdependent with the game structure: players can build the basic rules and agree or disagree on the importance of issues that may occur during the match. They also make the assessment and the review of the game together. The utilities of players and the definition of whether they are in the situation of conflict or cooperation may be multi-dimensional, multi-variable-dependent, and set either before or concluded after the game. An exemplary match will be shown in details.

**Keywords:** meta-game, multi-agent decision-making process, heterothelic game.

## 1 Introduction

The approach towards multi-agent decision-making processes may be seen as a by-product of one's view on the basic nature of the reality [1]. For example, our conclusions on how a particular conflict should be resolved, depends on the introductory assumptions about the nature of the conflict as a whole [2]. The conflict can be resolved differently, if it is concluded that *in fact there is a conflict* between agents, and differently, when it is concluded that *the conflict is in fact* the function of the agents' perceptions of the reality they live in. This in effect is an example of the problem between the *only one reality* (that is partially seen by all agents), and *multiple realities*, and multiple views of these realities. It is also an example of uncountable dilemmas of every decision-maker, as – to be able to make any decision – one needs to be clear of what the decision should be about in the first place.

N.T. Nguyen, R. Kowalczyk, and S.-M. Chen (Eds.): ICCCI 2009, LNAI 5796, pp. 229–239, 2009.

Both types of conclusions determine our multi-agent problem-solving procedures. In the first approach, one of the issues seems to be *how we can detect the clue aspects of the PROBLEM, in order to receive the requested outcome.* In the second approach, more important seems to be the question of *how our VIEWS should meet, due to be able to assess the pre-determined goals.*

In the proposed approach, it is possible to enhance the decision-making capabilities of individual agents and multiple-agent structures (such as organizations and institutions), by involving their processes of reasoning in a complex-game system, which makes individual and collective attitudes toward decision-making processes possible to evolve.

It is our aim to present some of the basic axioms of such a complex system, which are called the Meta-Game HOLOS, and one of its implementations – a Multi-Agent Decision-Making Laboratory (MADML) that is aimed to use as a general framework for experimentation. It is our hypothesis, that HOLOS-based undertakings might give important insight into human-human and human-artificial agents interactions.

The presentation consists of five parts. After introduction (Part 1) the details of the multi-agent decision-making processes and complex games are shortly described, and HOLOS as a Multi-Agent Decision-Making Laboratory is presented (Part 2). In Part 3 basic axioms and conventions of HOLOS are given and discussed. Part 4 discusses two examples of simplified matches as played by several players. The interpretation of the games played within the system is proposed, and some of the questions stemming from such an implementation, and the approach to possible answers are given. In Part 5 concluding remarks and future directions for HOLOS and MADML development are proposed and highlighted.

## 2 Multi-agent Decision-Making Processes and Complex Games

Each problem considered may be deconstructed into a set of alternatives emerging from the unstructured, semi-structured or fully-structured situations [3]. Each decision-making process may be also deconstructed into a series of essential decisions stemming from the particular set of alternatives [4]. A game combines both of the mentioned actions, especially in its heterothelic (utylitary) convention.

Every game is a system consisting of players (agents) and of a structure. No matter, whether we play Go, Poker, Basketball or other games – it is always required that there are

- the entities that interact [5], and
- the structure of action, that emerges from their utilities and expectations or pre-established rules [6].

In addition to this, HOLOS presupposes that the process of game-creation and structure-building is an inherent and clue part of the complex game as a whole. Thus in a heterothelic convention, HOLOS-based games can be utilized as the problem-solving instruments.

A *multi-agent decision-making process* is understood as the process of perceiving, recognizing and resolving problems that are conducted by agents, whose knowledge, capabilities, experience, and attitudes are different and in many ways complementary

[7]. A *game* is understood as everything that is happening and that is being made or done by players in a particular time and space. *Complex game* is every set of games that are inter-dependent, but whose basic goals and schemes of behavior may be conflicting.

The main goal of the MADML is to research the methods of influence on the officers (agents) on different levels of competition to cooperate, to minimize the global and local conflicts and to ask for the cooperation between rivals on all possible areas. It is conducted by the implementation of the creative holistic thinking in the frames of the multi-valued and fuzzy logics. MADML is concentrated on the problems of coexistence, competition and struggle, and finally on the essential cooperation that allows to minimize the scope of the crises, depending on the level of implementation.

## 3 HOLOS – Basic Axioms and Conventions

HOLOS is not a game in its traditional sense. Although many aspects might show some similarities to classical or more recent board games, RPG games, strategic or logic games, it consists of a number of features that make it radically different from the latter. It is rather a tool for complex problem solving that leads to the expansion of the cognitive capabilities of players than a game for fun.

HOLOS and MADML stems from the following assumption – once a problem is described, it can be:

1. deconstructed into the set of basic conventions (see below),
2. played as a game, and
3. discussed and utilized, either in order to
   a. change the *strategy* of the problem resolution or
   b. change the *understanding* of the nature of the problem.

Additionally, HOLOS bases on the conviction, that everything in a game can be a part of the game. Thus it introduces three stages of each meta-game:

1. pre-match (game before the match),
2. match,
3. after-match (game after the match).

In a HOLOS game, players take an active role in all stages of the game. A HOLOS *player* is every agent (self-steering) and *quasi*-agent (externally-steered) being in play, but whose businesses can be (but do not have to be) autonomous and opposite to the businesses of other players. A HOLOS player defines and decides on his own, what is important and/or interesting to him. Players businesses are important not only for him, but they should also be recognized and deduced by other players. It is obvious that such a knowledge can be correlated and influenced in the relations and interactions of the players during the current match and also in the future matches.

During the *first stage* HOLOS player(s) create the time-space of the match(es), and the rules and *conventions* of the game: the goals of the basic and/or complex games, their norms, the rules of assessment, and everything, that – due to the players goals – is with the meta-game connected. They create the conditions and construct the mechanisms for the match, including the systems of observations and – especially for the purposes of MADML – documentation of the game that is to be discussed after the match. Participants establish what in fact they agree that is *important or/and interesting* and what actually forms the base of the elements of the complex game, matches, observations, reports and recordings. During the first stage, players choose what are the pay-offs of all and every moves. They decide and note, which game is the most important one, e.g. they may choose whether they want to win the total match with of the total number of points or one of the component game. This is important especially, when the number of component games is big (starting from three, four, five, six etc.). For example, the strategies can be: to optimize the overall score in all component games, to maximize the scores in some of the component games, to minimize the scores of all other players or of a chosen player, the overall behavior of the players toward each other (e.g. cooperation vs. competition) etc.

This stage is a crucial part of each game. This is time for negotiation. Fundamentals of the constituent matches in a meta-game, its evaluations of observations, reports and recordings can be assumed or presumed, confirmed by the results and verified in the series of matches. Altogether they consist of the businesses of the subjects and/or the time-space of the game, ideas, principles, goals and tasks of the game.

During the *second* stage, players play the match, they have created before. This also means, that they observe and notify what is happening during the match. Every event occurring during the match that has been previously agreed to be *important or/and interesting* is called the *crucial* event. After the movement, the player notifies to the arbiters the successive crucial events of the game generated by his/her action together with the measures of the match; in case of the absence of negative opinions of the rivals it allows the arbiter to confirm the notification and to record it in the proper accounts (of players, constituent matches, total game).

During the *third* stage, evaluations and discussions are directed on the cognitive goals, which involve all aspects of the game that have been actually present in the first two stages. Usually, measures and criteria of evaluations are: collected points, sum of points from all constituent matches (as high as possible) and/or the positions of the players in the all constituent matches (as low as possible); important are also another measures like non-material values, over-countings, etc.

If the meta-game consists of multiple matches, the ending state of the precedent match may be utilized as an introductory state of the succeeding one.

HOLOS involves multiple games in one match or a series of games. Furthermore it can be complicated by adding more games, depending of the number of variables the players wish to introduce to it.

Other important elements of HOLOS are among others:

- introduction of the handicapped and independent quasi-players of the diversified character, importance and meaning, as well as the active subjects, objects and the states of the time-space, together with their businesses and affairs;
- elements of costs and losses of the matches, as well as the consequences of the decisions that may be dependent on the game results and actions;
- coupling the interesting and/or important elements of the constituent games into the total game without losing the individual and independent existence and conclusions;

- application of many criterions used to the evaluations of the constituent games (results and states);
- application of the two or more systems of evaluations, e.g. the sum of the points gained by the players or the positions of the players in the constituent matches.

Introducing the randomness in the match may also be a very important feature as it affects the very basics of the match mechanics. It can be made by simply using the dice (dices) or by the help of the earlier prepared special cards, or by the other means. The match can be changed considerably by simply allowing other ways of making steps (e.g. like the rook or knight in chess). Even with the little experience in playing HOLOS, the players can easily modify the rules to fit the game to their needs.

HOLOS is played on a board (or online equivalent), and the size of the board may be as big, as players wish to. So may be the shapes of the basic parts of it (see below). They can be squares and other geometric shapes, that can be formed into a consistent board.

Each of the conventions mentioned above may be introduced into a HOLOS game as a separate one, and thus constitute a new individual *basic* game, or as a part of a set of conventions, and thus constitute a *complex*-game. Which conventions are to be introduced depends on as many factors as the users may imagine (e.g. time and/or space dependency, crucial event-dependency, random event-dependency, players' will-dependency).

## 4 Example of the Simplified Matches

To be more specific, let us give an example of the very simplified HOLOS game for three players. A description of the preliminary abstract game is presented, conducted in the preparatory stage of the MADML emergence.

Let us assume that the players are the beginners and that they wish to have only the first impression of the system.

### 4.1 The Pre-match

Three players (A, B, C) decided to play three games in one match only. They wanted to spend on the game approximately 1,5 hour. The games were the following:

1. *MOVE*: to make a move from the available list of movements;
2. *REGIONS*: to collect points from the preferred parts of the board;
3. *ODD-EVEN*: to collect points when the sum of digits in the field number was odd or even.

In Fig. 1 the especially prepared board for that match is shown. It is seen that the board consists of 100 fields. Every field on the board had two attributes:

1. the number, and
2. the indication of the geographic position (it can be also identified by the letters and numbers given on the edges of the board, similarly to the chess-board).

The board was divided into four regions: North, South, East, West, so every region consists of 25 fields.

Each player made a choice on two *regions of interest*. After that, players defined their starting field with the one-by-one rule. All decided to start from different corners of the board corresponded with their region of interest.

The second field of interest was the number odd or even, depending on the choice of the player. This information was obtained from the last digits in the number of the field.

| | 0 | 1 | 2 | 3 | 4 | 5 | 6 | 7 | 8 | 9 |
|---|---|---|---|---|---|---|---|---|---|---|
| A | 0<br>NW | 1<br>NW | 2<br>NW | 3<br>NW | 4<br>NW | 5<br>NE | 6<br>NE | 7<br>NE | 8<br>NE | 9<br>NE |
| B | 10<br>NW | 11<br>NW | 12<br>NW | 13<br>NW | 14<br>NW | 15<br>NE | 16<br>NE | 17<br>NE | 18<br>NE | 19<br>NE |
| C | 20<br>NW | 21<br>NW | 22<br>NW | 23<br>NW | 24<br>NW | 25<br>NE | 26<br>NE | 27<br>NE | 28<br>NE | 29<br>NE |
| D | 30<br>NW | 31<br>NW | 32<br>NW | 33<br>NW | 34<br>NW | 35<br>NE | 36<br>NE | 37<br>NE | 38<br>NE | 39<br>NE |
| E | 40<br>NW | 41<br>NW | 42<br>NW | 43<br>NW | 44<br>NW | 45<br>NE | 46<br>NE | 47<br>NE | 48<br>NE | 49<br>NE |
| F | 50<br>SW | 51<br>SW | 52<br>SW | 53<br>SW | 54<br>SW | 55<br>SE | 56<br>SE | 57<br>SE | 58<br>SE | 59<br>SE |
| G | 60<br>SW | 61<br>SW | 62<br>SW | 63<br>SW | 64<br>SW | 65<br>SE | 66<br>SE | 67<br>SE | 68<br>SE | 69<br>SE |
| H | 70<br>SW | 71<br>SW | 72<br>SW | 73<br>SW | 74<br>SW | 75<br>SE | 76<br>SE | 77<br>SE | 78<br>SE | 79<br>SE |
| I | 80<br>SW | 81<br>SW | 82<br>SW | 83<br>SW | 84<br>SW | 85<br>SE | 86<br>SE | 87<br>SE | 88<br>SE | 89<br>SE |
| J | 90<br>SW | 91<br>SW | 92<br>SW | 93<br>SW | 94<br>SW | 95<br>SE | 96<br>SE | 97<br>SE | 98<br>SE | 99<br>SE |

**Fig. 1.** A plain board for a simplified 3-game exemplary match

It was decided that the game will be played without the dice (in order to reduce the scope of randomness of the match), and that every player will obtain the possibility of 25 movements in 8 possible directions, i.e. perpendicular or obliquely, with respect to the borders of the field: 5×1, 5×2, 5×3, 5×4 and 5×5 steps. This was defined as the *range of the step*. When a movement was made, the number of the available movements was getting smaller. The results of the match were to be summarized in Fig. 2.

Every player had his own table of results (see Fig. 3). In the first row (M) the number of the subsequent movement is given and in the second row (P) – the number (position) of the field reached by the player in the respected move. Row (R) shows the range of the movement made.

Every game played in the match had its own row of results. They all were played independently and simultaneously:

1. MOVE. The game was marked one point for one movement, i.e. no movement-no points.
2. REGIONS. The game was marked as follows: if the player reached a certain field, he could receive 0, 2 or 4 points respectively, depending on the description of his regions of interest. Every player received 2 points per region of interest (e.g. it was S and E for Player A). If the field Player A stepped into was marked with two his letters, he obtained 4 points. If the field was marked with other letters (not his), he obtained 0 points.
3. ODD-EVEN. The sum of digits, treated as numbers, creating the number of the field is always the odd or the even number. "0" was agreed to be the even digit. If the number was odd and player was interested in the odd numbers – he got 2 points; in the opposite case he received 0 points. The analogous situation was with the even numbers.

| PLAYER [A, B, C] | | | |
|---|---|---|---|
| Important and/or interesting | Field value | Sum in the game | Final position |
| x | 1 | | |
| SE | 2+2 | | |
| even | 2 | | |

**Fig. 2.** A plain score form for the exemplary match

| Player [A, B, C] | | | | | | | | | | | | | | | | | | | | | | | | | |
|---|---|---|---|---|---|---|---|---|---|---|---|---|---|---|---|---|---|---|---|---|---|---|---|---|---|
| M | 1 | 2 | 3 | 4 | 5 | 6 | 7 | 8 | 9 | 10 | 11 | 12 | 13 | 14 | 15 | 16 | 17 | 18 | 19 | 20 | 21 | 22 | 23 | 24 | 25 |
| P | | | | | | | | | | | | | | | | | | | | | | | | | |
| R | | | | | | | | | | | | | | | | | | | | | | | | | |
| Points in the game | | | | | | | | | | | | | | | | | | | | | | | | | |
| 1 | | | | | | | | | | | | | | | | | | | | | | | | | |
| 2 | | | | | | | | | | | | | | | | | | | | | | | | | |
| 3 | | | | | | | | | | | | | | | | | | | | | | | | | |

**Fig. 3.** A score form for a player taking part in the exemplary match

## 4.2 The Match

The players started from the relevant corners – Player A from the square 99; Player B from the square 9; Player C form the square 0. The players have used the three independent strategies (see Fig. 5). The first player A (SE) made his movements stochastically, without any defined plan or program. The fields he reached were distributed all over the board. Actually, he made the greatest number of movements, but he has won only the first, simplest game and lost two other, more important games. As a final result he was the looser.

The second player B (NE) decided to make short-range movements at the beginning and greater-range steps later, in order to collect points from the neighboring regions.

The third player C (NW) used the "two-region" strategy. He decided to use as many moves as possible on the region of his interest but making use of the long range steps at the beginning of the match.

| Player A | | | | | | | | | | | | | | | | | | | | | | | | | |
|---|---|---|---|---|---|---|---|---|---|---|---|---|---|---|---|---|---|---|---|---|---|---|---|---|---|
| M | 1 | 2 | 3 | 4 | 5 | 6 | 7 | 8 | 9 | 10 | 11 | 12 | 13 | 14 | 15 | 16 | 17 | 18 | 19 | 20 | 21 | 22 | 23 | 24 | 25 |
| P | 99 | 44 | 45 | 85 | 89 | 59 | 79 | 74 | 54 | 64 | 94 | 84 | 95 | 80 | 50 | 5 | 41 | 61 | 65 | 55 | 58 | 98 | | | |
| R | 0 | 5 | 1 | 4 | 4 | 3 | 2 | 5 | 2 | 1 | 3 | 1 | 1 | 5 | 3 | 5 | 4 | 2 | 4 | 1 | 3 | 4 | | | |
| Points in the game | | | | | | | | | | | | | | | | | | | | | | | | | |
| 1 | 1 | 1 | 1 | 1 | 1 | 1 | 1 | 1 | 1 | 1 | 1 | 1 | 1 | 1 | 1 | 1 | 1 | 1 | 1 | 1 | 1 | 1 | | | |
| 2 | 4 | 0 | 2 | 4 | 4 | 4 | 4 | 2 | 2 | 2 | 2 | 4 | 4 | 2 | 2 | 2 | 0 | 2 | 4 | 4 | 4 | 4 | | | |
| 3 | 2 | 2 | 0 | 0 | 0 | 2 | 2 | 0 | 0 | 2 | 0 | 0 | 2 | 2 | 0 | 0 | 0 | 0 | 0 | 2 | 0 | 0 | | | |

| Player B | | | | | | | | | | | | | | | | | | | | | | | | | |
|---|---|---|---|---|---|---|---|---|---|---|---|---|---|---|---|---|---|---|---|---|---|---|---|---|---|
| M | 1 | 2 | 3 | 4 | 5 | 6 | 7 | 8 | 9 | 10 | 11 | 12 | 13 | 14 | 15 | 16 | 17 | 18 | 19 | 20 | 21 | 22 | 23 | 24 | 25 |
| P | 9 | 39 | 49 | 47 | 27 | 29 | 8 | 7 | 16 | 15 | 35 | 38 | 18 | 48 | 78 | 75 | 25 | 69 | 19 | | | | | | |
| R | 0 | 3 | 1 | 2 | 2 | 2 | 1 | 1 | 1 | 1 | 2 | 3 | 2 | 3 | 3 | 3 | 5 | 4 | 5 | | | | | | |
| Points in the game | | | | | | | | | | | | | | | | | | | | | | | | | |
| 1 | 1 | 1 | 1 | 1 | 1 | 1 | 1 | 1 | 1 | 1 | 1 | 1 | 1 | 1 | 1 | 1 | 1 | 1 | 1 | | | | | | |
| 2 | 1 | 1 | 1 | 1 | 1 | 1 | 1 | 1 | 1 | 1 | 1 | 1 | 1 | 1 | 1 | 1 | 1 | 1 | 1 | | | | | | |
| 3 | 4 | 4 | 4 | 4 | 4 | 4 | 4 | 4 | 4 | 4 | 4 | 4 | 4 | 4 | 2 | 2 | 4 | 2 | 4 | | | | | | |

| Player C | | | | | | | | | | | | | | | | | | | | | | | | | |
|---|---|---|---|---|---|---|---|---|---|---|---|---|---|---|---|---|---|---|---|---|---|---|---|---|---|
| M | 1 | 2 | 3 | 4 | 5 | 6 | 7 | 8 | 9 | 10 | 11 | 12 | 13 | 14 | 15 | 16 | 17 | 18 | 19 | 20 | 21 | 22 | 23 | 24 | 25 |
| P | 0 | 4 | 40 | 20 | 24 | 42 | 22 | 52 | 2 | 13 | 63 | 30 | 31 | 1 | 11 | 33 | 3 | 53 | 43 | 10 | 12 | | | | |
| R | 0 | 4 | 4 | 2 | 4 | 2 | 2 | 3 | 5 | 1 | 5 | 3 | 1 | 3 | 1 | 2 | 3 | 5 | 1 | 3 | 2 | | | | |
| Points in the game | | | | | | | | | | | | | | | | | | | | | | | | | |
| 1 | 1 | 1 | 1 | 1 | 1 | 1 | 1 | 1 | 1 | 1 | 1 | 1 | 1 | 1 | 1 | 1 | 1 | 1 | 1 | 1 | 1 | | | | |
| 2 | 4 | 4 | 4 | 4 | 4 | 4 | 4 | 2 | 4 | 4 | 2 | 4 | 4 | 4 | 4 | 4 | 4 | 2 | 4 | 4 | 4 | | | | |
| 3 | 2 | 2 | 2 | 2 | 2 | 2 | 2 | 0 | 2 | 2 | 0 | 0 | 2 | 0 | 2 | 2 | 0 | 0 | 0 | 0 | 0 | | | | |

**Fig. 4.** The filled forms for current results of the exemplary match

The match has been finished, although there have been still many fields un-covered by any of the players. The board after the match is shown below, in Fig. 5. and the final results are collected in Fig. 6.

| | 0 | 1 | 2 | 3 | 4 | 5 | 6 | 7 | 8 | 9 |
|---|---|---|---|---|---|---|---|---|---|---|
| A | 0 NW | 1 NW | 2 NW | 3 NW | 4 NW | 5 NE | 6 NE | 7 NE | 8 NE | 9 NE |
| B | 10 NW | 11 NW | 12 NW | 13 NW | 14 NW | 15 NE | 16 NE | 17 NE | 18 NE | 19 NE |
| C | 20 NW | 21 NW | 22 NW | 23 NW | 24 NW | 25 NE | 26 NE | 27 NE | 28 NE | 29 NE |
| D | 30 NW | 31 NW | 32 NW | 33 NW | 34 NW | 35 NE | 36 NE | 37 NE | 38 NE | 39 NE |
| E | 40 NW | 41 NW | 42 NW | 43 NW | 44 NW | 45 NE | 46 NE | 47 NE | 48 NE | 49 NE |
| F | 50 SW | 51 SW | 52 SW | 53 SW | 54 SW | 55 SE | 56 SE | 57 SE | 58 SE | 59 SE |
| G | 60 SW | 61 SW | 62 SW | 63 SW | 64 SW | 65 SE | 66 SE | 67 SE | 68 SE | 69 SE |
| H | 70 SW | 71 SW | 72 SW | 73 SW | 74 SW | 75 SE | 76 SE | 77 SE | 78 SE | 79 SE |
| I | 80 SW | 81 SW | 82 SW | 83 SW | 84 SW | 85 SE | 86 SE | 87 SE | 88 SE | 89 SE |
| J | 90 SW | 91 SW | 92 SW | 93 SW | 94 SW | 95 SE | 96 SE | 97 SE | 98 SE | 99 SE |

**Fig. 5.** The state of the board after the exemplary match

| PLAYER A | | | |
|---|---|---|---|
| **Important and/or interesting** | **Field value** | **Sum in the game** | **Final position** |
| x | 1 | 22 | 1 |
| SE | 2+2 | 62 | 3 |
| even | 2 | 16 | 3 |
| | | **100** | **3** |

| PLAYER B | | | |
|---|---|---|---|
| **Important and/or interesting** | **Field value** | **Sum in the game** | **Final position** |
| x | 1 | 19 | 3 |
| NE | 2+2 | 70 | 2 |
| odd | 2 | 26 | 1 |
| | | **115** | **2** |

| PLAYER C | | | |
|---|---|---|---|
| **Important and/or interesting** | **Field value** | **Sum in the game** | **Final position** |
| x | 1 | 21 | 2 |
| NW | 2+2 | 78 | 1 |
| even | 2 | 26 | 2 |
| **TOTAL WINNER** | | **123** | **1** |

**Fig. 6.** Numerical results of the exemplary match

### 4.3 After-Match – Remarks on Conflict

The proposed multi-game was very simple and abstract and has been finished within one hour and a half. Nevertheless, some very interesting facts were observed.

The match has not shown any potential for conflict, until Player A stepped on the regions of interest of Player A and than of Player B. The conflict did not emerge, because Players A and B were more interested in exploitation of other parts of the board. It is seen that a smart change of scoring would change the final results dramatically. But this decision has to be made during the pre-match phase of the game.

It has occurred that the fact that one starts the match does not imply that one wins. Although Player A started the game, he defeated, by collecting the least number of points of all players (see Fig. 6).

The presented abstract match was too simplified to show all the capabilities and strengths of HOLOS. Simply changing the rules of ascribing the points to the specific fields of the board could have easily introduce the aspects of cooperation and competition with greater power.

## 5 Concluding Remarks and Future Directions for MADML Development

HOLOS is a complex game system that makes the capabilities and individual characteristics of the agents a part of the game. The utilities of players and the definition of whether they are in the situation of conflict or cooperation may be multi-dimensional, multi-variable-dependent, and set either before or concluded after the game.

It is expected that HOLOS may present a high potential for experimental and quasi-experimental (basic) research – like in MADML – including those on autonomous agents (like people) and quasi-autonomous (AI) agents.

The goal of the Multi-Agent Decision-Making Laboratory is to conduct experimental and quasi-experimental research on decision-making processes and the interrelationship and inter-dependency between the scope of agency and structural factors in the individual and group behavior.

Two types of games will be the subject for future research:

1. abstract games, and
2. story games.

Abstract games are represented by the abstract nominal marks and do not declare the specified contents and values or particular meanings; specific nominal marks of the abstract games give often some associations – they can be introduced by chance or consciously, instrumentally. Thanks to the simplicity of the basic assumptions of the system, also some basic game theory problems may be researched. For instance, a prisoners dilemma, if the goal of the match is to check what are the consequences of acting in a cooperative or non-cooperative manner. Presumable, the first step would be to think of the board construction. The board may be constructed in such a way, that it represents the pay-off of each of two players. The same can be thought for four or more players etc.

Story games are created by adding the stories to the existing abstract games, which enables the verification of the nominal marks in the applied measures; stories are of the various values, they have different ranges and meanings, which on the one hand creates difficulties in the evaluations of the crucial events, but on the other hand allows to be more specific in training of the particular competencies.

## References


1. Kolbusz, R.J.: Gry przyszłości i gry o przyszłość. In: Dombrowska, E. Niedźwiedzka A. (eds.): Twórczość – wyzwanie XXI wieku, pp. 97–109. Impuls, Kraków (2003) (in Polish)
2. Wiener, N.: The Human Use of Human Beings: Cybernetics and Society. Houghton Mifflin, Boston (1950); Wooldridge, M.: An Introduction to Multiagent Systems. John Wiley & Sons Ltd., New York (2002); Sun, R.: Cognition and Multi-Agent Interaction. From Cognitive Modeling to Social Simulation. Rensselaer Polytechnic Institute, New York (2006)
3. Axelrod, R.: The Evolution of Cooperation. Basic Books, New York (1984); Rapoport, A.: Fights, Games and Debates. The University of Michigan Press, Ann Arbor (1992)
4. Ashby, W.R.: An Introduction to Cybernetics. Chapman & Hall, London (1956)
5. von Neumann, J., Morgenstern, O.: The Theory of Games and Economic Behavior. Princeton University Press, Princeton (1944)
6. Mazur, M.: Cybernetyczna teoria układów samodzielnych. Warszawa (1966) (in Polish)
7. Watson, J.: Strategy. An Introduction to Game Theory. University of California, San Diego (2002)
8. Keil, D., Goldin, D.Q.: Indirect interaction in environments for multi-agent systems. In: Weyns, D., Van Dyke Parunak, H., Michel, F. (eds.) E4MAS 2005. LNCS (LNAI), vol. 3830, pp. 68–87. Springer, Heidelberg (2006)

# Agent-Based Computational Modeling of Emergent Collective Intelligence

Vivek Kumar Singh[1], Divya Gautam[1], Rishi Raj Singh[1], and Ashok K. Gupta[2]

[1] Department of Computer Science,
Banaras Hindu University, Varanasi, U.P., India
vivek@bhu.ac.in, divyagautam.bhu@gmail.com, rishi.raj@gmail.com
[2] J. K. Institute of Applied Physics & Technology,
University of Allahabad, Allahabad, U.P., India
akgjkiapt2000@yahoo.com

**Abstract.** Collective Intelligence is a form of intelligence which emerges out of collaboration and coordination of many individual agents. A group of actors performing simple behaviours and interacting with fellow group members & the environment often produce global behaviours which seems intelligent. Understanding the emergence of intelligent collective behaviours in social systems, such as norms & conventions, higher level organizations, collective wisdom and evolution of culture from simple and predictable local interactions; has been an important research question since decades. Agent-based modeling of complex social behaviours by simulating social units as agents and modeling their interactions; provides a new generative approach to understanding the dynamics of emergence of collective intelligence behaviours. In this paper, we have presented an analytical account of nature, form and dynamics of collective intelligence, followed by some of our experimental work on evolution of collective intelligence. The paper concludes with a short discussion of the results and relevant implications for designing systems for achieving desired collective intelligence.

**Keywords:** Collective Intelligence, Agent Based Modeling, Multi-agent Systems, Emergent Phenomenon, Socionics.

## 1 Introduction

Collective Intelligence is defined as the ability of groups to exhibit greater intelligence than its individual members. Human (even insect and animal) groups working together are known to produce various collective behaviours which seem intelligent. Individual actors in social groups generally have a local view, i.e., they interact only with individuals in their neighbourhood. These local and relatively simple interactions can produce emergent global behaviours as seen in case of ant colonies, bird flocks, human settlements, organizations and markets. The foraging behaviour of ants [1], flocking behaviour of birds [2], complex patterns of human settlements [3], forms of organizational behaviour and working of markets [4] are some of the examples of

N.T. Nguyen, R. Kowalczyk, and S.-M. Chen (Eds.): ICCCI 2009, LNAI 5796, pp. 240–251, 2009.

global emergent behaviours, which are also instances of collective intelligence. The collective intelligence of these systems is not a kind of intelligence per se; rather it is a global outcome of collaboration and coordinated interactions of a number of individuals, which cannot be attributed to any individual agent. Understanding how (and possibly why) these complex and intelligent collective behaviours emerge from limited local interactions; have been a primary research question for sociologists & psychologists since decades. Agent Based Modeling seems to provide the right methodology and tool towards this end.

In Agent Based Modeling (ABM) [5], a system is modeled as a set of autonomous agents who can perceive the environment and produce different actions. The action is determined by the behavioural rules which are programmed (also learned) in the agents. The agents represent the actors in the system; environment represents the agent surroundings including other agents; and the behaviour rules model the interaction of the agent with other agents as well as with the environment. ABM can be used to model a number of phenomenons in varied domains like Markets & Economy, Organizations, World Wide Web and Social Systems etc. Many fundamental differences are seen in the dynamics of collective intelligence in simple insects and cognitively more complex human groups. Decentralized groups of insects (ants & bees etc.) show remarkable capability to produce collective intelligence but a similarly organized human group may fail to do so. A number of factors, including the actors (whether simple or cognitively superior), interactions (whether direct or through the environment), structure of group (whether centralized or democratic), affect the nature and form of emergence of collective intelligence. In this paper, we have tried to analyze the nature & form of collective intelligence as a phenomenon & perspective (Section 2), to present a relevant account of use of ABM approach to model and analyze the emergence of collective intelligence in groups in different situations (Section 3) along with our experimental results (Section 4). The paper concludes with a short discussion of experimental outcomes and relevant implications (Section 5).

## 2 Collective Intelligence

Collective Intelligence as a phenomenon is found in a variety of forms since ages. All organizations whether they be companies or sporting teams, families and even countries are all groups of actors doing things individually and at the same time resulting into macroscopic behaviour which seems intelligent. This capability is also seen in insects and animals, which have limited cognitive abilities. Going slightly down the scale one can even take human brain, a collection of individual neurons, working collectively to produce intelligent behaviour, to be an example of this sort.

### 2.1 The Emergent Nature of Collective Intelligence

Emergent behaviour is defined as a behaviour that is not attributed to any individual, but is a global outcome of coordination of individuals. It is action of simple rules combining to produce complex results [6]. Although emergent behaviour comes from individuals, the interactions are what make things difficult to understand. When individuals produce behaviours that affect other individuals and their behaviours, which in turn affect the original individuals, a complex feedback dynamic loop is established, which results into complex behavioural patterns. This complex feedback loop of interactions, limit the possibility of any analysis of emergent phenomenon at any level simpler than that of system as a whole. This indicates that collective intelligence is an emergent phenomenon. It is result of coordinated and collaborative working of many individuals. The individuals need not be extraordinary intelligent. It is the right synergistic combination of a number of individuals that make the group more capable and intelligent than individual members. Collective intelligence as seen in everyday life occurs in two broad forms: *emergence by nature* (working of neural networks, ant and bee colonies, bird flock etc.) and *emergence by design* (in organizations, markets, teams etc.).

## 2.2 Collective Intelligence at Different Cognitive Levels

A number of lower level insects & animals such as ants, bees and birds are known to produce collective behaviours which are often regarded as intelligent. These simple beings have limited vision and highly constrained cognitive capacity; but working as a colony they can produce useful and intelligent global behaviours. The colony is generally organized as a leaderless decentralized group, with every member performing its own duties. The local interactions combine synergistically to produce remarkably valuable global behaviours. However, in case of peer groups of human beings, rather than emergence of useful group level intelligence, a number of factors often inhibit the collective intelligence capability of the group. This raises a new question about the nature and prerequisites for collective intelligence. Do the cognitive superiority and behavioural complexity of human beings play spoilsport? Or if a different organizational topology (with leaders or with meta-norms governing the interactions) is required for collective intelligence in human groups? A deeper and more involved research is necessary to answer these questions. A related aspect of these questions is *the inverse problem* in collective intelligence [7], i.e., how should the group be organized so that it produces desired emergence.

## 2.3 Factors Affecting the Emergence of Collective Intelligence

Surowiecki in his book "The Wisdom of Crowds" [8] distinguished between the madness of crowds vs. wisdom of crowds. He suggests that wisdom of crowds emerges only under right conditions. He lists diversity, independence and decentralization as the key factors necessary for a group to be intelligent. Diversity means that the group should include members with a wide variety of knowledge or abilities. Independence refers to ability of group members to use their own knowledge and abilities without being overly influenced by others. The specific decentralization suggested calls for group members to be aggregated in such a way that finds the right balance between (a) "making individual knowledge globally and collectively useful", and (b) "still allowing it to remain resolutely specific and local". Work in organization theory also suggests factors that can facilitate collective intelligence of groups (particularly in organizational settings). Existence of formal and informal structures, proper modularization of tasks, dense communication structures and incentives for contribution are some of the factors suggested to be useful for innovations & collective intelligence in organizational groups. Researches in social sciences have also suggested some factors affecting collective intelligence, though in a negative manner. The key factors identified to inhibit collective intelligence are biases, selfishness, narrow bandwidth among group members and geographical and cultural boundaries etc [9]. However, there appears to be no universally agreed set of factors facilitating or inhibiting collective intelligence of groups.

## 2.4 Understanding the Dynamics of Collective Intelligence

Though the knowledge of factors affecting collective intelligence can be useful to organize better and more productive groups but the real difficulty is to understand how groups become collectively more intelligent. A detailed analysis of emergence of collective intelligence requires controlled and verifiable experimentation. ABM allows simulated experimentation of working of real world groups. One can form a hypothesis about the dynamics of emergence & factors affecting it; and then test and analyze it by controlled variation of different parameters and observing the results. The fact that the emergent global patterns and system-level behaviours are sometimes entirely of new types, that are not apparent from the behaviours of individual actors, makes a deeper analysis through ABM approach essential. A number of complex social phenomena, where simple local interactions generate intelligent system-level behaviours, have been modeled using ABM approach Some of the representative & relevant work on ABM approach to understand the dynamics of emergent intelligence in social systems can be found in [10], [11] & [12]. Macy & Willer [13] group the work on application of ABM to collective behaviours in two categories: (a) *models of emergent structure* which includes works on cultural differentiation, homophilous clustering, idea diffusion, convergent behaviours and norms; and (b) *models of emergent social order* which include viability of trust, cooperation and collective action in absence of global control. Goldstone & Janssen [14] group the work into following three themes: (a) *patterns and organizations* which include settlement patterns and segregation, human group behaviours and traffic patterns; (b) *social contagion* which include spread of ideas, fashions, cultures and religions; and (c) *cooperation* which include evolution of cooperation, trust and reputations and social norms and conventions. A concrete general theory of the dynamics of emergence of collective intelligence is yet to be worked out.

## 2.5 Collective Intelligence as a Perspective

Szuba in his book "Computational Collective Intelligence" [15] proposes a formal model for the phenomenon of collective intelligence as an unconscious, random, parallel, and distributed computational process, run in mathematical logic by social structure; with prospective applications for optimization of organizations. The findings & suggestions in his work, however, have much broader and deeper inherent implications. Collective Intelligence is not just a phenomenon; it's also a perspective that can be applied to many different kind of phenomenon. To apply the perspective of collective intelligence to a phenomenon, the following elements must be identified: (a) a group of actors; (b) a set of resources available to those actors; (c) a set of actions that the actors take; (d) the collective results of the actions; and (e) a way of evaluating the results. Once these elements are identified, one can analyze how intelligently a given group acted given the resources it had or if they could be connected in some other ways to act more intelligently [16].

Collective intelligence perspective has also inspired a number of computational approaches, which are being applied to solve various optimization problems. While particle swarm algorithms [17], ant colony optimization [18], artificial bee colony [19] draw their inspiration from social behaviours of natural systems; there are also approaches such as computer supported cooperative work (CSCW) techniques, which is another kind of manifestation of collective intelligence. Considerable amount of research has been done in CSCW [20]. Collective Intelligence as a problem solving perspective is applied to problems of combinatorial optimization [21], organizational improvements and creating & extracting knowledge from the World Wide Web [22]. The MIT centre for Collective Intelligence has accelerated the work on collective intelligence by framing the key research question in collective intelligence as "How can people and computers be connected so that collectively they act more intelligently than any individual, group, or computer has ever done before?"[23].

## 3 Applying ABM Approach to Collective Intelligence

The ABM approach helps in a detailed analysis of the *how* and *why* of the emergent collective behaviours. The dynamics of emergence of collective intelligence in swarms is quite different from groups of cognitively superior individuals. While colonies of lower order insects demonstrate the capability to produce intelligent collective behaviours in absence of any leader or direct communication; individuals placed at a higher cognitive level require a different setting for collective intelligence to emerge in the group. The role played by norms, global biases, presence of leaders and the behavioural complexity of individuals in the group, in emergence of collective intelligence (particularly in higher order groups), deserves a deeper study. We devised several agent-based models to try to answer some important questions around these aspects. One of the models also elaborates upon the ability of a collective intelligent group to cope up with obstacles.

### 3.1 Collective Intelligence in Groups with and without Norms

Social norms are standards of behaviour that are based on widely shared beliefs how individual group members ought to behave in a given situation [24]. A norm exists in a given social setting to the extent that individuals usually act in a certain way and are often punished when seen not to be acting in this way [25]. The group members might obey the norm voluntarily if their individual goals are in line with the normatively required behaviour, or might be forced to obey the norm otherwise due to the fear of punishment. A group instead of being beneficial can become a destructive mob if appropriate conditions for collective intelligence are not provided for. Presence of norms in the groups is known to have a positive effect on their behaviour, as they help in streamlining the group's behaviour and avoiding conflicts. A norm can be internal to the group, i.e., in form of some conventions followed by all the members in the group; or it may be external to the group, i.e., in form of external rules applicable to the entire group. To demonstrate the role of norms, we have implemented an extended traffic model [26] to incorporate norms. The experimental setup and results are explained in Section 4. Simulation runs show that moderate traffic can be regulated with internal norms, but as the traffic becomes more complex and dense internal norms fail to maintain smooth flow. External norms, on the other hand, maintain the traffic flow, though with slightly lower average speeds and higher average waiting times. The results are similar to findings in other experimental works on norms [27], [28], [29] & [30].

### 3.2 Observing How a Collective Intelligent Group Reacts to Obstacles

The second aspect we took up for analysis was to see how a group capable of producing self-organizing & collective intelligence behaviours cope up with obstacles introduced to its collective intelligence process. We used Craig Reynolds's "Boid Simulation" [2], one of the earliest & simplest (but important) agent-based models of emergent collective behaviour, as the base point for analysis. The Boid simulation models the interactions of agents (birds) with simple behavioural rules: (a) *Cohesion*: each agent steers toward the average position of nearby flockmates; (b) *Separation*: each agent steers to avoid crowding local flockmates; and (c) *Alignment*: each agent steers toward the average heading of local flockmates. The model is run by initially placing the agents at random locations and then allowing the behavioural rules to operate iteratively. Surprisingly, even with three simple rules applying only locally, leaderless flocks emerge. This demonstrates how limited local interactions can result in interesting global behaviours. Since our primary interest was in reaction to obstacles, we extended the boid model by introducing small obstacles in the flying area and added an additional dimension to the behaviour set of birds, called *obstacle avoidance*. Obstacle avoidance requires the birds to change their alignment whenever they encounter an obstacle in front of them. Considering a flock to be a collective intelligent peer group, it would be proper to expect that the flock would be able to overcome simple obstacles, i.e., the flock should be able to entirely avoid the obstacle, or if the flock breaks due to the obstacle, it should be able to reunite after the obstacle is over. The experimental results, explained in the Section 4, tell that flocks show varied reactions to the obstacle, which depends on factors like how flock encounters the obstacle (whether head on) and the vision of the birds.

### 3.3 Collective Intelligence in Groups with Leaders

Though most of the simulation models of social behaviour and organizations have emphasized the spontaneous emergence of socially coordinated behaviour, groups of real organisms, by contrast, often display social hierarchies with some individuals acting as leader. In complex social hierarchies of human beings, where some individuals may be more influential than others, the question of how does presence of an influential individual (a leader) in the group affects the group's ability to produce collective intelligence is important. A recent research [31] suggests that in a group having sparse direct communication channels, presence of an influential leader is beneficial for collective intelligence. However, with adequate communication (space-independent) added to the group, leaders tends to be detrimental, rather than beneficial. Small leaderless groups outperform large ones in solving collective problems, whereas, larger groups perform better in presence of influential leaders. The emergence of space-independent communication, made possible by modern technology, has made leaders less important in determining a group's ability to solve collective problems. These findings appear to be intuitively correct. As long as groups are small and manageable, they can be highly intelligent, but very big groups find it difficult to act intelligently, they rather become uncontrolled mob.

### 3.4 Collective Intelligence in Complex-Individual Groups

In case of groups with individuals capable of performing complex behaviours, the dynamics of collective intelligence becomes somewhat more difficult to understand. To analyze the complexity involved with higher order individual groups, we simulated two simple behaviour-aggregation settings and compared their results. The first setup is a simple voting model [32], where a number of individuals are placed on a cellular automaton like structure and every individual decides his vote (two-valued) based on the votes by his eight neighbours. All individuals are considered to be doing a neighbourhood aggregation on a single behavioural feature. If initial voting preferences are set randomly, the entire population soon reaches a steady state of almost equal favour for both votes. However, if the individuals in the population are modeled as capable of producing multi-dimensional behaviours, where each behaviour can have multi-valued expressions rather than two valued votes, results are different. Here, when an individual has to decide its own value for a particular behaviour, it polls its neighbourhood for the corresponding behaviour and the value found in majority becomes its own expression for that behaviour [33]. This is done for all the behaviours. We find that even in this case the population reaches to a steady state soon, though the final pattern is relatively more complex and is also very sensitive to initial distribution of behaviour potentials. Introducing a slight disruption in the steady state reached produces entirely new situations. The experimental setup and results are explained in the following section.

## 4 Experimental Results

We have designed the simulations in Net Logo [34], a language highly suitable for modeling and analyzing emergent and self-organizing phenomena. In the first experiment we simulated a traffic grid on a 50 X 50 wrap around agent grid, where there are a number of cars, travelling on a set of single-lane roads with intersections at regular intervals. Every car tries to accelerate and reach higher speed if there are no cars in front of it and is forced to de-accelerate to avoid collision, otherwise. When on intersection a car decides its action according to three different settings, corresponding to the three parameters under study. In the first setting, when on an intersection a car simply waits for the other cars on the intersection to pass to avoid any collision. In the second setting, a car is programmed to always give way to traffic on its left, i.e., when on an intersection a car always looks for any car coming from its left and waits for it to pass before moving ahead. The third setting introduces traffic lights to regulate the traffic. The cars now cross the intersection only if the traffic light is favourable and wait otherwise. First setting corresponds to a group which is largely self organizing, second to a group with internal norm (every car favours the cars coming from its left) and the third to a group with external norm (where traffic lights are the external norms governing the traffic). Several runs of the simulation have been observed with the three settings. A snapshot of the runs in three situations is shown in the Figure 1, 2, 3 & 4. While the self-organizing situation is able to maintain smooth traffic flow for a small number of cars, it results in traffic jams (particularly at intersections) as the number of cars increases. Introduction of internal norm of always giving way to traffic from left allows smooth traffic flow for relatively higher number of cars but it also tends to fail for a larger traffic size. External norms in form of traffic signals, however, are able to maintain smooth traffic, though with reduced average speeds and higher average waiting times for the cars.

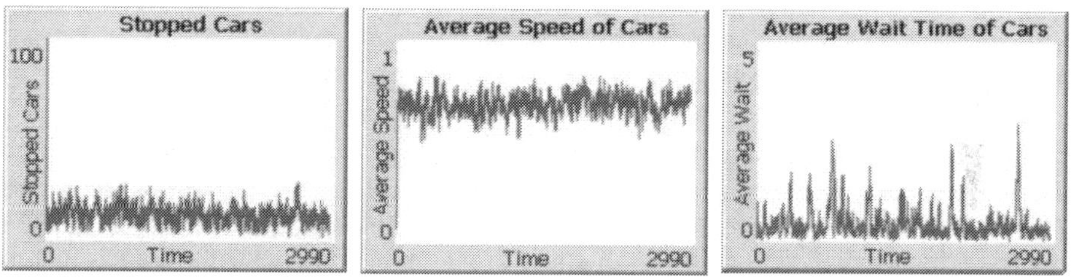


**Fig. 1.** A snapshot of the traffic simulation on a 50 X 50 wrap around agent grid, with number of cars =100 and system is largely self-organizing using only basic traffic rules

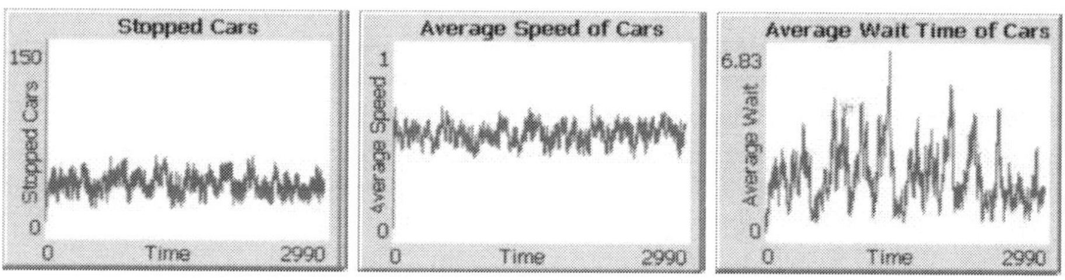


**Fig. 2.** Traffic statistics on a 50 X 50 wrap around agent grid, with number of cars = 150 and in presence of internal norm of always giving way to the traffic coming from the left

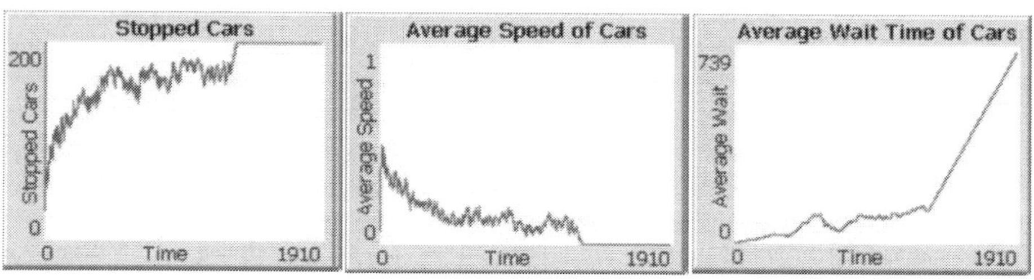


**Fig. 3.** A Snapshot of the traffic model on 50 X 50 wrap around agent grid, with number of cars = 200 and internal norm of giving way to traffic coming from the left

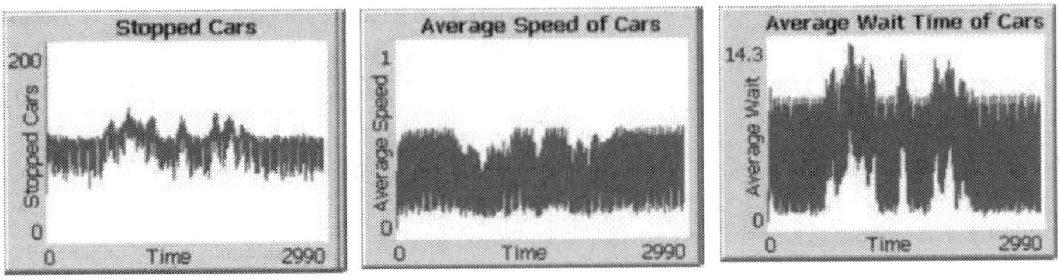


**Fig. 4.** A Snapshot of the traffic model on 50 X 50 wrap around agent grid, with number of cars = 200 and presence of external norm in form of traffic lights

In our second simulation model, we wanted to see how a system (boid model) known for its self-organization and collective intelligence ability, reacts to the obstacle in its way, i.e., whether the bird flock is able to avoid the obstacle completely and still maintain the flock or if the flock gets disrupted in the process. This was done by implementing the boid model on a 30 X 30 wrap around agent grid using Net Logo and introducing a small obstacle (shown as red patch) in the flying space of bird flocks. A bird after seeing an obstacle in front performs *obstacle avoidance* behaviour, which allows it to change its alignment towards left or right in the two dimensional plane. This is coupled with the three basic behaviours of *cohesion*, *separation* and *alignment*. An interesting point to observe is that whether the bird flock remains intact or gets disrupted in the process and in case of disruption whether the disrupted flock is able to reunite into a single flock after the obstacle is over. A snapshot of the simulated runs is shown in Figure 5. When the flock has higher vision (able to see 10 patches ahead), the flock is able to avoid the obstacle with most of the flock intact. However, when the flock's vision is limited to just a few patches ahead, head on collisions do occur. Some of the flock members break apart from the flock but a substantial portion of the flock remains intact. The separated flock members eventually join the original flock or some other flock. When obstacle size is relatively bigger, more number of birds breaks apart from the flock.

In the third simulation experiment of behaviour aggregation, on 30 X 30 wrap around agent grid, with two settings of single two-valued behaviour and multiple multi-valued behaviours (number of behaviours = 5 and each behaviour can take numeric values from 1 to 5), the system reaches to a steady state in both the cases, though the final structure is more complex in the second setting. While simple two-valued behaviour aggregation tends to reach to a steady final state with almost equal size groups of both voted-behaviours; multiple-feature, multiple-valued behaviours result in complex structures which are also sensitive to initial distribution. Introduction of a disruptive noise (by mutating some of the behavioural values of a small number of agents) to the steady state, however, has an interesting effect in the second model. The mutation could be considered as a creative idea or some externally induced influence. Interesting group patterns emerge and with the noise continued at regular intervals, bigger groups often tend to eat-up smaller groups, i.e., the majority behaviour tend to dominate.. A snapshot of the simulation runs of different scenarios is shown in Figure 6.

**Fig. 5.** A snapshot of the obstacle avoidance behaviour of bird flocks. Figure on the left is avoidance behaviour with higher vision; Figures in the middle and on the right, respectively, are snapshot of a low vision flock encountering the obstacle and after the obstacle is over.

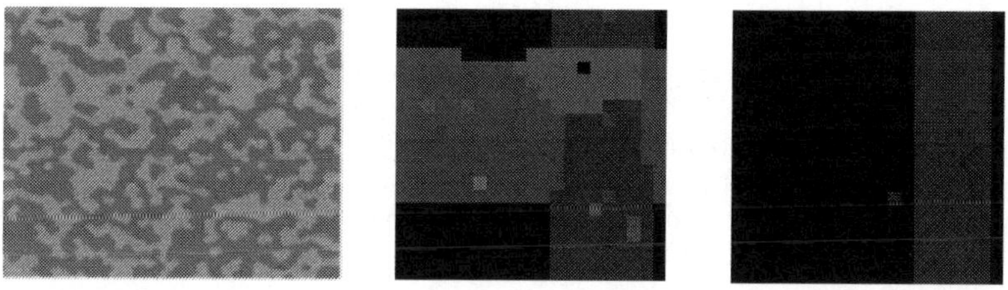

**Fig. 6.** A snapshot of the behaviour aggregation model; leftmost figure is the steady state with single two-valued behaviour, in centre is steady state with multiple multi-valued behaviours and on right is effect of introducing noise to multiple multi-valued behaviours

## 5 Conclusion

In this paper, we have tried to present an analytical account of nature and form of collective intelligence, its different manifestations and factors affecting it. Simulations have been designed to explore the role of norms and behavioural complexity of individuals in the emergence of collective intelligence in groups and the ability of self-organizing collective intelligent groups to overcome obstacles. The role played by norms in groups is highly important. While small groups can by and large self-organize, bigger groups often fail to exploit the group's intelligence in absence of either internal or external norms. As evident in case of traffic simulation, smooth traffic flow can be maintained without any norm only in case of small number of vehicles; as the traffic increases a norm becomes necessary for organized traffic. The relatively local interaction mechanisms known to produce emergent macro-level intelligence, can also work towards groups ability to overcome smaller obstacles as seen in the experiments with the boid simulation. The boid simulation demonstrates that leaderless flocks which are largely self-organized can efficiently overcome smaller obstacles in the way. The affect of behavioural complexity of group members is, however, most difficult to analyze. When individuals in a group are diverse and capable of multiple & complex behaviours, the group is less likely to converge to a collective opinion, rather produces regions of different opinions. However, when small noise is introduced, the majority opinion tends to increase its influence and eventually takes the system to a close to homogeneous opinion. Relative contributions of these factors (norms, leaders, behavioural complexity) vary with the size and complexity of the groups. These findings have implications for designing multi-agent systems as well. One needs to carefully analyze the requirement of internal and external norms while designing a multi-agent system based solution to a situation, particularly in case of a desired self-organization. The behavioural complexity of the agent should also be deeply understood if a macro-level emergent property is expected from the system. Appropriate provisions for communication structure and leader (or coordinator) agents should also be made as and when necessary. Collective intelligence as a phenomenon and perspective is creating a new world of possibilities, including it being the mankind's new emerging world in cyberspace [35].

## References


1. Bonabeau, E., Dorigo, M., Theraulaz, G.: Swarm Intelligence: From Natural to Artificial Systems. Oxford University Press, New York (1999)
2. Reynolds, C.W.: Flocks, Herds and Schools: A Distributed Behavioural Model. Computer Graphics 21, 25–34 (1987)
3. Schelling, T.C.: Dynamic Models of Segregation. Journal of Mathematical Sociology 1, 143–186 (1971)
4. Tesfatison, L.: Introduction to the special issue on Agent-based Computational Economics. Journal of Economic Dynamics and Control (2001)
5. Macal, C.M., North, M.J.: Tutorial on Agent based Modeling and Simulation. In: Kuhl, M.E., Steiger, N.M., Armstrong, F.B., Joines, J.A. (eds.) Proceedings of the 2005 Winter Simulation Conference (2005)
6. Zhengping, L., Cheng, H.S.: A Survey of Emergent Behaviour and its Impacts in Agent-based Systems. In: Proceedings of IEEE International Conference on Industrial Informatics (August 2006)
7. Namatame, A.: Adaptation and Evolution in Complex Systems. World Scientific (2006)
8. Surowiecki, J.: The Wisdom of Crowds. Anchor, New York (2005)
9. Wikipedia, Collective Intelligence (2009), `http://en.wikipedia.org/wiki/Collective_intelligence`
10. Epstein, J.M.: Generative Social Sciences: Studies in Agent-based Computational Modeling. Princeton University Press, Princeton (2007)
11. Axelrod, R.: The Complexity of Cooperation. Princeton University Press, Princeton (1997)
12. Epstein, J.M., Axtell, R.: Growing Artificial Societies: Social science from bottom-up. MIT Press, Cambridge (1996)
13. Macy, M.W., Willer, R.: From factors to actors: computational sociology and agent based modeling. Annual Review of Sociology 28, 143–166 (2002)
14. Goldstone, R.L., Janssen, M.A.: Computational models of collective behaviour. Trends in Cognitive Sciences 9(9) (2005)
15. Szuba, T.M.: Computational Collective Intelligence, 1st edn. Wiley-Interscience, Hoboken (2001)
16. Malone, T.W.: What is Collective Intelligence, Social Text (2005), `http://www.socialtext.net/mit-cci-hci/index.cgi?what_is_collective_intelligence`
17. Banks, A., Vincent, J., Anyakoha, C.: A Review of Particle Swarm Optimization, Part-I: Background & Development. Natural Computing 6, 467–484 (2007)

18. Dorigo, M., Birattari, M., Stutzle, T.: Ant Colony Optimization: Artificial Ants as Computational Intelligence Technique. IEEE Computational Intelligence (November 2006)
19. Karaboga, D., Basturk, B.: On the Performance of Artificial Bee Colony Algorithm. Applied Soft Computing 8, 687–697 (2008)
20. Greif, I. (ed.): Computer Supported Cooperative Work: A Book of Readings. Morgan Kaufmann, San Francisco (1988)
21. Singh, V.K., Gupta, A.K.: Collective Intelligence Based Approaches to Search and Optimization Problems. In: Proceedings of UGC Sponsored National Seminar on Soft Computing Techniques and their Applications at Howrah, India (March 2009)
22. Alag, S.: Collective Intelligence in Action. Manning, New York (2009)
23. Malone, T.W.: What is Collective Intelligence and what we will do about it? (2005), Edited transcript available at MIT Centre for Collective Intelligence, `http://cci.mit.edu`
24. Fehr, E., Fischbacher, U.: Social Norms and Human Cooperation. Trends in Cognitive Science 8(4), 185–190 (2004)
25. Axelrod, R.: An Evolutionary Approach to Norms. The American Political Science Review 80(4), 1095–1111 (1986)
26. Wilensky, U.: NetLogo Traffic Grid Model. models/ TrafficGrid, Center for Connected Learning and Computer-Based Modeling, Northwestern University, Evanston, IL (2003), `http://ccl.northwestern.edu/netlogo/`
27. Axelrod, R.: The Evolution of Cooperation. Basic, New York (1984)
28. Conte, R., Castelfranchi: Understanding the Function of Norms in Social Groups through Simulation. In: Gilbert, N., Conte, R. (eds.) Artificial Societies: The Computer Simulation of Social Life. UCL Press, London (1995)
29. Epstein, J.: Learning to be thoughtless: Social Norms and Individual Computation. Centre on Social and Economic Dynamics Working Papers, No. 6 (2000)
30. Sen, S., Airiau, S.: Emergence of Norms through Social Learning. In: IJCAI-2007, pp. 1507–1512 (2007)
31. Gigliotta, O., Miglino, O., Parisi, D.: Groups of Agents with a Leader. Journal of Artificial Societies and Social Simulation 10(4) (2007)
32. Wilensky, U.: NetLogo Voting Model. Voting, Center for Connected Learning and Computer-Based Modeling, Northwestern University, Evanston, IL (1998), `http://ccl.northwestern.edu/netlogo/models/`
33. Gautam, D., Singh, R.R., Singh, V.K.: Multi-Agent based Models of Social Contagion and Emergent Collective Behaviour. In: Proceedings of the IEEE International Conference on Intelligent Agents and Multi-agent Systems at Chennai, India (July 2009)
34. Wilensky, U.: NetLogo. Centre for Connected Learning and Computer-Based Modeling, Northwestern University, Evanston, IL (1999), `http://ccl.northwestern.edu/netlogo`
35. Levy, P.: Collective Intelligence: Mankind's Emerging World in Cyberspace. Perseus Books, Cambridge (1997)

# Fuzzy Cognitive and Social Negotiation Agent Strategy for Computational Collective Intelligence

Amine Chohra, Kurosh Madani, and Dalel Kanzari

Images, Signals, and Intelligent Systems Laboratory (LISSI / EA 3956),
Paris-East University, Senart Institute of Technology,
Avenue Pierre Point, 77127 Lieusaint, France
{chohra,madani,kanzari}univ-paris12.fr

**Abstract.** Finding the adequate (*win-win* solutions for both parties) negotiation strategy with *incomplete* information for autonomous agents, even in one-to-one negotiation, is a complex problem. Elsewhere, negotiation behaviors, in which the characters such as conciliatory or aggressive define a '*psychological*' aspect of the negotiator personality, play an important role. The aim of this paper to develop a fuzzy cognitive and social negotiation strategy for autonomous agents with incomplete information, where the characters conciliatory, neutral, or aggressive, are suggested to be integrated in negotiation behaviors (inspired from research works aiming to analyze human behavior and those on social negotiation psychology). For this purpose, first, one-to-one bargaining process, in which a buyer agent and a seller agent negotiate over single issue (price), is developed for a time-dependent strategy (based on time-dependent behaviors of Faratin *et al.*) and for a fuzzy cognitive and social strategy. Second, experimental environments and measures, allowing a set of experiments, carried out for different negotiation deadlines of buyer and seller agents, are detailed. Third, experimental results for both time-dependent and fuzzy cognitive and social strategies are presented, analyzed, and compared for different deadlines of agents. The suggested fuzzy cognitive and social strategy allows agents to improve the negotiation process, with regard to the time-dependent one, in terms of agent utilities, round number to reach an agreement, and percentage of agreements.

**Keywords:** Social and cognitive systems, negotiation, behaviors, strategies, adaptation.

## 1 Introduction

Negotiations have received wide attention from the distributed Artificial Intelligence (AI) community as a pervasive mechanism for distributed conflict resolution between intelligent computational agents [1]. In a context where agents must reach agreements (deals) on matters of mutual interest, *negotiation* techniques for reaching agreements (deals) are required. In general, any negotiation settings will have four different components [2]: 1) a negotiation set, the space of possible proposals that agents can make;

N.T. Nguyen, R. Kowalczyk, and S.-M. Chen (Eds.): ICCCI 2009, LNAI 5796, pp. 252 – 265, 2009.

2) a protocol, the legal proposals that agents can make ; 3) a collection of strategies, one for each agent, which determine what proposals agents will make ; 4) an agreement rule that determines the reach agreements (deals) stopping the negotiation. Negotiation usually proceeds in a series of rounds, with every agent making a proposal at every round. The proposals that agents make are defined by their strategy (a way to use the protocol), must be drawn from the negotiation set, and must be legal, as defined by the protocol (which defines possible proposals at different rounds). If agreement is reached, as defined by the agreement rule, then negotiation terminates with agreement deal. These four parameters lead to an extremely rich and complex environment for analysis. Another source of complexity in negotiation is number of agents involved in process, and the way in which these agents interact [2]. First possibility is one-to-one negotiation, second possibility is many-to-one negotiation (which can be treated as a number of concurrent one-to-one negotiations), and third possibility is many-to-many negotiation (hard to handle).

An interesting survey on negotiation models in the AI field is given in [3], [4], [5]. Elsewhere, Lomuscio *et al.* [6] identified the main parameters on which any automated negotiation depends and provided a classification scheme for negotiation models. The environment that a negotiator is situated in greatly impacts the course of negotiation actions. Instead of focusing on analyzing the strategy equilibrium as a function of (the distribution of) valuations and historical information as in game theory, AI researchers are interested in designing flexible and sophisticated negotiation agents in complex environments with incomplete information. Agents have incomplete and uncertain information about each other, and each agent's information (e.g., deadline, utility function, strategy, ...) is its private knowledge.

An important research work has been developed by Faratin *et al.* [7] which devised a negotiation model that defines a range of strategies and behaviors for generating proposals based on time, resource, and behaviors of negotiators. By another way, in the research works developed aiming to analyze and describe the human behavior in [8], twelve categories representing three major parts of the behavior have been defined: the positive socio-emotional part, a neutral task part, and the negative socio-emotional part. In another side, in research works on the social psychology of the negotiation of Rubin and Brown developed in [9], the interpersonal orientation of a person has an influence on his negotiating behavior. It is predominantly concerned with the degree of a person's responsiveness. If he is not responsive, he stands to gain much in negotiating situation due to deliberateness of his behavior. Responsive people are more co-operative and therefore expect positive results. Personality type should therefore be determined first to obtain the best results in negotiation. Thus, negotiation behaviors, in which characters such as conciliatory, neutral, or aggressive define a 'psychological' personality aspect of a negotiator, play an important role.

The aim of this paper to develop a fuzzy cognitive and social negotiation strategy for autonomous agents with incomplete information, where the characters conciliatory, neutral, or aggressive, are suggested to be integrated in negotiation behaviors (inspired from research works aiming to analyze human behavior and those on social negotiation psychology). For this purpose, first, one-to-one bargaining process, in which a buyer agent and a seller agent negotiate over single issue (price), is developed in Sect. 2 for a time-dependent strategy (based on time-dependent behaviors of Faratin *et al.* [7]) and for a fuzzy cognitive and social strategy. Second, experimental environments and measures, allowing a set of experiments carried out for different negotiation deadlines of buyer and seller agents, are detailed in Sect. 3. Third, experimental results for both time-dependent and fuzzy cognitive and social strategies are presented, analyzed, and compared for different deadlines of agents in Sect. 4.

## 2 One to One Negotiation Agent Strategy

In this Section, one-to-one bargaining process, in which a buyer agent and a seller agent negotiate, over single issue (price), is developed for the two strategies.

### 2.1 Negotiation Set

A negotiation set is the space of possible proposals that agents can make. The negotiation set (objects): the range of issues over which an agreement must be reached. Let i represents the negotiating agents, in bargaining bilateral negotiation $i \in \{buyer(b), seller(s)\}$, and j the issues under negotiation, in single issue negotiation j = *price*. The value for issue *price* acceptable by each agent i is $x^i \in [min^i, max^i]$.

### 2.2 Negotiation Protocol

A protocol is the legal proposals that agents can make. The process of negotiation can be represented by rounds, where each round consists of an offer from agent b (buyer) at time $t_1$ and a counter-offer from an agent s (seller) at time $t_2$. Then, a negotiation consists in a sequence of rounds: round1 ($t_1$, $t_2$), round2 ($t_3$, $t_4$), … Thus, for a negotiation between agents b and s, and if agent b starts first, then it should offer in times ($t_1$, $t_3$, $t_5$, …, $t^b_{max}$), and agent s provides counter-offers in ($t_2$, $t_4$, $t_6$, …, $t^s_{max}$), where $t^b_{max}$ and $t^s_{max}$ denote negotiation deadline for agents b and s, respectively. Note that the three different deadline cases are allowed: 1) $t^b_{max} > t^s_{max}$, where considered deadline is $T_{max} = t^s_{max}$ ; 2) $t^b_{max} = t^s_{max}$, where considered deadline is $T_{max} = t^b_{max} = t^s_{max}$ ; 3) $t^b_{max} < t^s_{max}$, where considered deadline is $T_{max} = t^b_{max}$.

For agent b, the proposal to offer or accept is within interval $[min^b, max^b]$, where $max^b$ is the buyer reservation price in negotiation thread, and $min^b$ is the lower bound of a valid offer. Similarly, for agent s, the proposal to offer or accept is within interval $[min^s, max^s]$, where $min^s$ is the seller reservation price and $max^s$ is the upper bound of a valid offer. Initially a negotiator offers most favorable value for himself: agent b starts with $min^b$ and agent s starts with $max^s$. If proposal is not accepted, a negotiator concedes with time proceeding and moves toward other end of the interval.

### 2.3 Negotiation Behaviors

The paces of concession depend on the negotiation behaviors of agent b and agent s which are characterized by negotiation decision functions. For negotiation strategies, time t is one of predominant factors used to decide which value to offer next.

**Time-Dependent Agent Behaviors:** Time-dependent functions are used as negotiation decision functions varying the acceptance value (price) for the offer depending on the remaining negotiation time (an important requirement in negotiation), i.e., depending on t and $t^b_{max}$ for agent b and depending on t and $t^s_{max}$ for agent s. Thus, proposal $x^b[t]$ to be offered by agent b and the one $x^s[t]$ to be offered by agent s at time t, with 0<= t <= $t^i_{max}$ belonging to [0, T - 1], are as follows. The proposal $x^s[t]$ to be offered by agent s at time t, with 0<= t <= $t^s_{max}$ belonging to [0, T - 1], is defined by Eq. (1).

$$x^s[t] = min^s + (1 - \alpha^s(t))\,(max^s - min^s),$$

where $\alpha^s(t)$ are time-dependent functions ensuring that: $0 <= \alpha^s(t) <= 1$, (1)

$\alpha^s(0) = K^s$ (positive constant) and $\alpha^s(t^s_{max}) = 1$.

Such $\alpha^s(t)$ functions can be defined in a wide range according to the way in which $\alpha^s(t)$ is computed (the way they model the concession), e.g., polynomial in Eq.(2).

$$\alpha^s(t) = K^s + (1 - K^s)\left(\frac{min(t, t^s_{max})}{t^s_{max}}\right)^{\frac{1}{\beta}} \quad (2)$$

Indeed, the constant $\beta > 0$ determines the concession pace along time, or convexity degree of the offer curve as a function of the time. By varying $\beta$ a wide range of negotiation behaviors can be characterized: Boulware (B) with $\beta < 1$ and Conceder (C) with $\beta > 1$ [5], and the particular case of Linear (L) with $\beta = 1$.

**Fuzzy Cognitive and Social Agent Behaviors:** The proposal $x^b[t]$ to be offered by agent b at time t, with 0<= t <= $t^b_{max}$ belonging to [0, T - 1], is defined using fuzzy cognitive and social strategy detailed in Sect. 2. 4.

## 2.4 Negotiation Strategies

A collection of strategies, one for each agent, which the role is to determine what proposals agents will make (which behavior should be used at any one instant).

**Time-Dependent Agent Strategy:** During a negotiation *thread* (the sequence of rounds with offers and counter-offers in a two-party negotiation), a negotiation strategy based on time-dependent behaviors defined in [7] consists to define the way in which such behaviors are used. In this paper, each strategy uses individually the behaviors Boulware (B), Linear (L), or Conceder (C) during a negotiation thread.

**Fuzzy Cognitive and Social Agent Strategy:** Suggested fuzzy cognitive and social agent strategy, illustrated by synopsis in Fig. 1, is built of a similarity degree (from ), a Fuzzy System 1 (FS1), an updating of the internal agent behaviors, a difference, a Fuzzy System 2 (FS2), a Fuzzy System 3 (FS3), and an updating of the proposal.

***Similarity Degree.*** Similarity degree S[t], see [10], is computed using the Eq. (3):

$$S[t] = 1 - \frac{\left|x^b[t-1] - x^s[t]\right|}{max\{x^b[t-1], x^s[t]\}} = \frac{min\{x^b[t-1], x^s[t]\}}{max\{x^b[t-1], x^s[t]\}}. \quad (3)$$

***Fuzzy System 1 (FS1).*** For FS1, time deadline t ---> Tmax = [0, 100] is normalized in [0, 1] and character variations is given from t and similarity degree.

*Fuzzy Rule Base 1*: from 2 input variables S[t] and t, 2 output variables giving the character variations ΔAgg[t] and ΔCon[t], and 3 fuzzy membership functions for each variable as defined in Fig. 2, $3^2 = 9$ fuzzy rules are deduced as in Eq. (4).

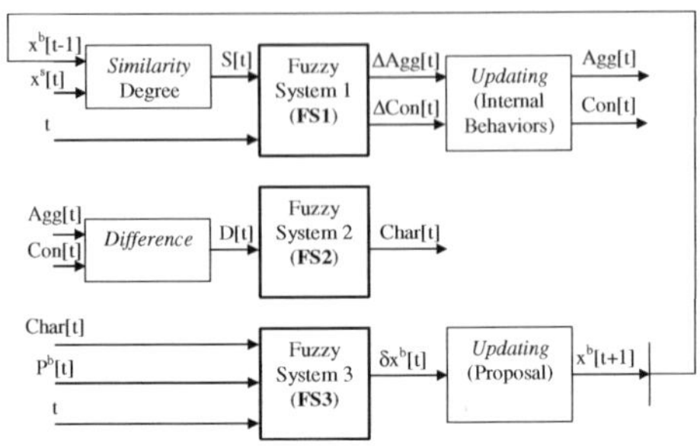


**Fig. 1.** Fuzzy cognitive and social strategy synopsis

The fuzzy linguistic variables, for each fuzzy membership function, are defined as:

- S[t]: Desirable (D), Acceptable (A), and Undesirable (Und),
- t: Favorable (F), Critical (C), and Unfavorable (Unf),
- ΔAgg[t] and ΔCon[t]: Very Negative (VN), Negative (N), Positive (P), and Very Positive (VP).

$$\textit{If } (S^t \textit{ is } D \textit{ and } t \textit{ is } F) \textit{ Then } (\Delta Agg[t] \textit{ is } N \textit{ and } \Delta Con[t] \textit{ is } P), \ldots \tag{4}$$

***Updating (Internal Behaviors).*** Internal behavior updating is achieved by Eq. (5).

$$Agg[t] = Agg[t\text{-}1] + \Delta Agg[t] \textit{ and } Con[t] = Con[t\text{-}1] + \Delta Con[t]. \tag{5}$$

***Difference.*** The difference is achieved using Eq. (6).

$$D[t] = Con[t] - Agg[t]. \tag{6}$$

***Fuzzy System 2 (FS2).*** This system gives new agent character from the difference of internal behavior characters.

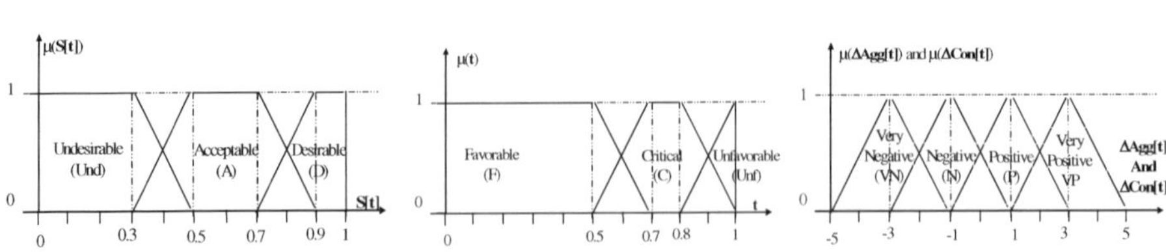


**Fig. 2.** Fuzzy membership functions of S[t], t, and ΔAgg[t] and ACon[t]

*Fuzzy Rule Base 2*: from 1 input variable D[t], 1 output variable Char[t], and 4 fuzzy membership functions, for each variable as defined in Fig. 3, $4^1 = 4$ fuzzy rules are deduced as in Eq. (7). Fuzzy linguistic variables, for each fuzzy membership function, are defined:

- D[t]: Very Negative (VN), Negative (N), Positive (P), and Very Positive (VP),
- Char[t]: Very Aggressive (VAgg), Aggressive (Agg), Conciliatory (Con), and Very Conciliatory (VCon).

$$If\ (D[t]\ is\ VN)\ Then\ (Char[t]\ is\ VAgg),\ \ldots \tag{7}$$

***Fuzzy System 3 (FS3).*** For FS3, profit of agent b is $P^b[t]$ such as defined in Eq. (8) and variation of the proposal is given from new agent character, agent profit, and t.

$$P^b[t] = max^b - x^s[t]. \tag{8}$$

where $max^b$ is the reservation price of agent b and $x^s[t]$ the proposal of agent s.

*Fuzzy Rule Base 3*: from 3 input variables Char[t], $P^b[t]$, t, 1 output variable $\delta x^b[t]$, and 4 fuzzy membership functions for Char[t] defined in Fig. 3, 3 fuzzy membership functions for $P^b[t]$ defined in Fig. 4, 3 fuzzy membership functions for t defined in Fig. 2, and 4 fuzzy membership functions for $\delta x^b[t]$ defined in Fig. 4, $3^2 * 4^1 = 36$ fuzzy rules are deduced in Eq. (9).

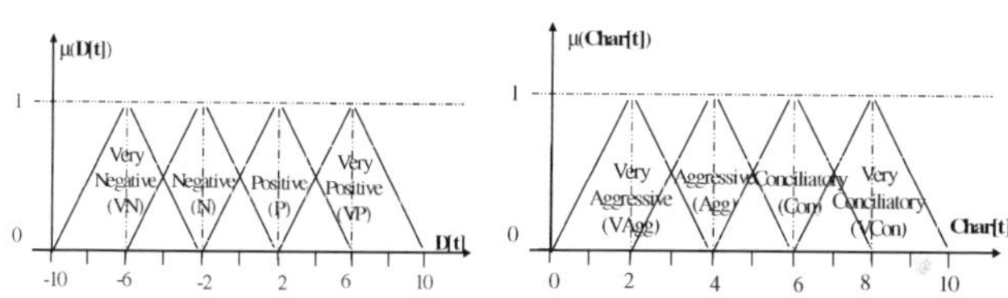


**Fig. 3.** Fuzzy membership functions of D[t] and Char[t]

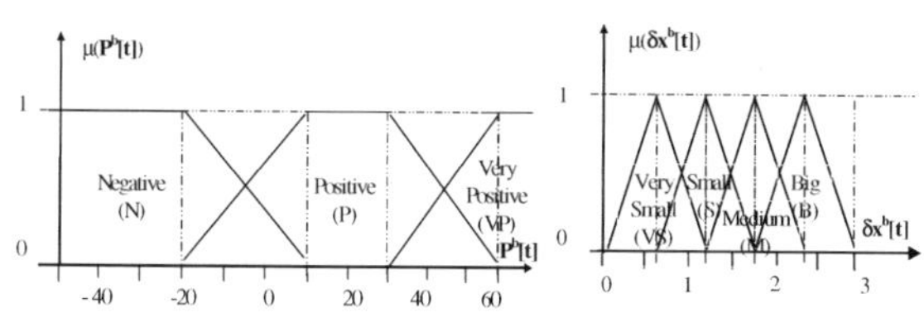


**Fig. 4.** Fuzzy membership functions of $P^b[t]$ and $\delta x^b[t]$

The fuzzy linguistic variables, for each fuzzy membership function, are defined as:

- Char[t]: Very Aggressive (VAgg), Aggressive (Agg), Conciliatory (Con), and Very Conciliatory (VCon),
- $P^b[t]$: Negative (N), Positive (P), and Very Positive (VP),
- t: Favorable (F), Critical (C), and Unfavorable (Unf),
- $\delta x^b[t]$: Very Small (VS), Small (S), Medium (M), Big (B), and null.

$$If\ (Char[t]\ is\ VAgg\ and\ P^b[t]\ is\ N\ and\ t\ is\ F)\ Then\ (\delta x^b[t]\ is\ VS),\ \ldots \tag{9}$$

***Updating (Proposal).*** The new proposal is finally obtained using Eq. (10).

$$x^b[t+1] = x^b[t-1] + \delta x^b[t]. \tag{10}$$

### 2.5 Agreement Rule

An agreement rule is a rule which determines the reach agreements stopping negotiation. Agent b accepts an offer (or a deal) $x^s[t]$ from agent s at time t if it is not worse than the offer he would submit in the next step, i.e., only if the relation given in Eq. (11) is satisfied. Similarly, agent s accepts an offer (or a deal) $x_b[t]$ from agent b at time t only if the relation given in Eq. (11) is satisfied.

$$\begin{cases} x^b(t+1) >= x^s(t) \\ t <= T_{max} \end{cases}, \begin{cases} x^s(t+1) <= x^b(t) \\ t <= T_{max} \end{cases}. \tag{11}$$

## 3 Experiments: Environments and Measures

In this Section, experimental environments and measures are presented and a set of experiments, carried out for different deadlines of agents b and s, are detailed.

### 3.1 Experimental Environments

Environments are defined in bargaining bilateral negotiation between buyer(b) and seller(s), in single issue negotiation j = *price*. The experimental environment is defined by the following variables [ $t^b_{max}$ , $t^s_{max}$ , $T_{max}$, $K^b$, $K^s$, $min^b$, $max^b$, $min^s$, $max^s$].

The negotiation interval (the difference between the minimum and maximum values of agents) for price is defined using two variables: $\theta^i$ (the length of the reservation interval for an agent i) and $\Phi$ (the degree of intersection between the reservation intervals of the two agents, ranging between 0 for full overlap and 0.99 for virtually no overlap). In the experimental environment: $\theta^i$ are randomly selected between the ranges [10, 30] for both agents, and $\Phi$ = 0. The negotiation intervals are then computed, setting $min^b$ = 10, by:

$$min^b = 10,\ max^b = min^b + \theta^b,\ min^s = \theta^b \Phi + min^b,\ \text{and}\ max^s = min^s + \theta^s \tag{12}$$

The analysis and evaluation of negotiation behaviors and strategies developed in [11], indicated that negotiation deadlines significantly influences the performance of the negotiation. From this, the experimental environment is defined from random selection of the round number within [10, 50] which corresponds to a random selection of $T_{max}$ within [20, 100]. The initiator of an offer is randomly chosen because the agent which opens the negotiation fairs better, irrespective of whether agent is a buyer (b) or a seller (s).

**Time-Dependent Agent Behaviors:** Parameter β ranges [12], [13], are defined as:

- $\beta \in [0.01, 0.20]$ for Boulware (B) or *Aggressive (Agg)*,
- $\beta = 1.00$ for Linear (L) or *Neutral (Neu)*,
- $\beta \in [20.00, 40.00]$ for Conceder (C) or *Conciliatory (Con)*.

Then, the constants $K^i$ are chosen as small positive $K^i = 0.1$, for seller agent, in order to not constrain the true behavior of each time-dependent (decision) functions.

**Fuzzy Cognitive and Social Agent Behaviors:** Conciliatory (FCon) and Aggressive (FAgg), where FCon ∈ [0, 10] and FAgg ∈ [0, 10]. Thus, fuzzy cognitive conciliatory, neutral, or aggressive characters are defined as:

- FCon ∈ [0.0, 4.0] and FAgg ∈ [6.0, 10.0] for *Fuzzy Aggressive (FAgg)*,
- FNeu = FCon = FAgg = 5.0 for *Fuzzy Neutral (FNeu)*,
- FCon ∈ [6.0, 10.0] and FAgg ∈ [0.0, 4.0] for *Fuzzy Conciliatory (FCon)*.

### 3.2 Experimental Measures

To produce statistically meaningful results, for each experiment, precise set of environments is sampled from parameters specified in Sect. 3.1 and number of environments used is $N = 200$, in each experiment. This ensures that probability of sampled mean deviating by more than 0.01 from true mean is less than 0.05. To evaluate effectiveness of negotiation behaviors, three measures are considered:

- *Average Intrinsic Utility*, the intrinsic benefit is modeled as the agent's utility for the negotiation's final outcome, in a given environment, independently of the time taken and the resources consumed [7]. This utility, $U^i$, can be calculated for each agent for a *price x* using a linear scoring function, i.e., for a buyer agent b when there is a deal (an agreement) for a *price x* as shown in Eq. (14). If there is no deal in an environment (a particular negotiation), then $U^b = U^s = 0$. Then, the average intrinsic utility for each agent is given by $AU^i$ in Eq. (13), where N is the total number of environments in each experiment, $U^i$ the utility of each agent, for each environment with deal, and $N_D$ is the number of environments with deals.

$$U^b = max^b - x\,,\ and\ U^s = x - max^s\,,\ and\ AU^i = \frac{\sum_{n=1}^{N} U^i[n]}{N_D}. \tag{13}$$

- *Average Round Number*, rounds to reach an agreement (a deal), a lengthy negotiation incurs penalties for resource consumption, thus shrinking the utilities obtained by the negotiators indirectly [14]. The average round number AR is given in Eq. (14), where $R_D$ is the number of rounds, for each environment with deal, and $N_D$ is the number of environments with deals.

- *Percentage of Deals*, the percentage of deals D(%) is obtained from the *Average Deal* AD given in Eq. (14), where $N_D$ is the number of environments with deals. Then, Percentage of Deals D(%) is given in Eq. (14).

$$AR = \frac{\sum_{n=1}^{N} R_D[n]}{N_D}\,,\ and\ AD = \frac{N_D}{N}\,,\ D(\%) = AD\,.\,100\,\%\,. \tag{14}$$

In order to analyze the performance of strategies, two measures are obtained [15], once an agreement is achieved: 1) *utility product*, UP, product, $AU^b.AU^s$, of the utilities is computed as shown in Eq. (15), this measure indicates joint outcome ; 2) *utility difference*, UD, difference, $\left|AU^b - AU^s\right|$, of utilities is computed as shown in Eq. (15), this measure indicates distance between both utilities.

$$UP = AU^b.AU^s \text{ and } UD = \left|AU^b - AU^s\right|. \tag{15}$$

There is an important relation between these product and difference utility measures and a compromise should be taken into account. Even though a high joint outcome is expected (for a relatively *win-win* results), it is also important that the difference between both utilities is low. For this reason, the analysis and evaluation of the obtained results is based not only on the utility product but also on the utility difference (a compromise should then be taken into account).

Also, to analyze performance of strategies, an interesting measure is suggested:

- *average performance*, the Average Performance ($AP^i$) of each agent is an average evaluation strategy measure implying the average negotiation deadline $At^i_{max}$ and three (03) experimental measures, i.e., the average intrinsic utility $AU^i$, the average round number AR, and the average deal AD as shown in Eq. (16).

$$AP^i = \frac{AU^i + (1 - \frac{AR}{At^i_{max}}) + AD}{3}, \text{ with } At^i_{max} = \frac{\sum_{n=1}^{N} t^i_{max}[n]}{N_D}, \tag{16}$$

where $N_D$ is number of environments with deals, and N is total number of environments in each experiment.

## 4 Experimental Results of Negotiation Strategies

In this Section, experimental results of time-dependent and fuzzy cognitive and social negotiation strategies are presented, analyzed, and compared for different deadlines.

### 4.1 TimeDependent(Buyer)-TimeDependent(Seller)

The results presented in the following concern time-dependent strategy for both b and s agents, and they are given for different agent deadlines: $t^b_{max} > t^s_{max}$, $Tmax = t^b_{max} = t^s_{max}$, and $t^b_{max} < t^s_{max}$. These results are obtained, for each deadline, from 9 experiments (Agg-Agg, Agg-Neu, Agg-Con, Neu-Agg, Neu-Neu, Neu-Con, Con-Agg, Con-Neu, and Con-Con). Thus, results shown in Table 1, Table 4, and Table 7 give b and s agent utilities, those in Table 2, Table 5, and Table 8 give average round numbers, percentage of deals, utility products and differences, and those in Table 3, Table 6, and Table 9 give their average performances and absolute difference.

**Table 1.** Results $AU^b$ and $AU^s$: case $t^b_{max} > t^s_{max}$

| Seller | Buyer(Agg) | Buyer(Neu) | Buyer(Con) |
|---|---|---|---|
| Agg | 45.09, 6.27 | 23.86, 25.06 | 2.23, 46.54 |
| Neu | 34.34, 17.62 | 36.60, 18.51 | 6.11, 43.21 |
| Con | 50.23, 0.25 | 49.12, 4.58 | 24.93, 21.48 |

**Table 2.** Results AR, D(%), UP, and UD: case $t^b_{max} > t^s_{max}$

| Seller | Buyer(Agg) | Buyer(Neu) | Buyer(Con) |
|---|---|---|---|
| Agg | 31.000, 2%, 282.71, 38.82 | 22.466, 30%, 597.93, 1.20 | 17.944, 36%, 103.78, 44.31 |
| Neu | 31.200, 17%, 605.07, 16.72 | 15.200, 35%, 677.46, 18.09 | 3.769, 39%, 264.01, 37.10 |
| Con | 24.578, 19%, 12.55, 49.98 | 4.382, 34%, 224.96, 44.54 | 1.000, 39%, 535.49, 3.45 |

**Table 3.** Results $AP^b$, $AP^s$, and ($\left|AP^b - AP^s\right|$): case $t^b_{max} > t^s_{max}$

| Seller | Buyer(Agg) | Buyer(Neu) | Buyer(Con) |
|---|---|---|---|
| Agg | 15.240, 2.171 (13.069) | 8.293, 8.599 (**0.306**) | 1.121, 15.817 (14.696) |
| Neu | 11.706, 6.003 (**5.703**) | 12.586, 6.493 (6.093) | 2.484, 14.835 (12.351) |
| Con | 17.037, 0.275 (16.762) | 16.801, 1.936 (14.865) | 8.769, 7.615 (**1.154**) |

**Table 4.** Results $AU^b$ and $AU^s$: *case* $Tmax = t^b_{max} = t^s_{max}$

| Seller | Buyer(Agg) | Buyer(Neu) | Buyer(Con) |
|---|---|---|---|
| Agg | 22.25, 22.48 | 7.49, 38.36 | 0.93, 47.21 |
| Neu | 40.56, 8.16 | 22.25, 22.26 | 5.50, 46.13 |
| Con | 47.57, 0.94 | 44.47, 5.33 | 25.52, 35.32 |

**Table 5.** Results AR, D(%), UP, and UD: case $Tmax = t^b_{max} = t^s_{max}$

| Seller | Buyer(Agg) | Buyer(Neu) | Buyer(Con) |
|---|---|---|---|
| Agg | 28.906, 32%, 500.18, 0.23 | 25.440, 25%, 287.31, 30.87 | 20.405, 37%, 43.90, 46.28 |
| Neu | 27.592, 20%, 330.96, 32.40 | 16.500, 34%, 495.28, 0.01 | 4.028, 35%, 253.71, 40.63 |
| Con | 21.794, 34%, 44.71, 46.63 | 4.214, 28%, 237.02, 39.14 | 1.000, 28%, 901.36, 9.80 |

**Table 6.** Results $AP^b$, $AP^s$, and ($\left|AP^b - AP^s\right|$): case $Tmax = t^b_{max} = t^s_{max}$

| Seller | Buyer(Agg) | Buyer(Neu) | Buyer(Con) |
|---|---|---|---|
| Agg | 7.696, 7.772 (**0.076**) | 2.772, 13.062 (10.290) | 0.653, 16.079 (15.426) |
| Neu | 13.766, 2.966 (10.800) | 7.771, 7.775 (**0.004**) | 2.260, 15.804 (13.544) |
| Con | 16.182, 0.638 (15.544) | 15.226, 2.179 (13.047) | 8.927, 12.194 (**3.267**) |

**Table 7.** Results $AU^b$ and $AU^s$: case $t^b_{max} < t^s_{max}$

| Seller | Buyer(Agg) | Buyer(Neu) | Buyer(Con) |
|---|---|---|---|
| Agg | 5.03, 55.26 | 2.14, 58.46 | 0.30, 50.02 |
| Neu | 29.93, 22.43 | 16.25, 32.05 | 3.88, 43.40 |
| Con | 52.25, 2.23 | 45.00, 6.47 | 20.11, 34.89 |

**Table 8.** Results AR, D(%), UP, and UD: case $t^b_{max} < t^s_{max}$

| Seller | Buyer(Agg) | Buyer(Neu) | Buyer(Con) |
|---|---|---|---|
| Agg | 30.330, 3%, 277.95, 50.23 | 28.857, 7%, 125.10, 56.32 | 25.500, 18%, 15.00, 49.72 |
| Neu | 13.047, 12%, 671.32, 7.50 | 14.806, 31%, 520.81.15.80 | 4.269, 26%, 168.39, 39.52 |
| Con | 19.185, 27%, 116.51, 50.02 | 3.763, 38%, 291.15, 38.53 | 1.000, 33%, 701.63, 14.78 |

**Table 9.** Results $AP^b$, $AP^s$, and ($\left|AP^b - AP^s\right|$): case $t^b_{max} < t^s_{max}$

| Seller | Buyer(Agg) | Buyer(Neu) | Buyer(Con) |
|---|---|---|---|
| Agg | 1.767, 18.636 (16.869) | 0.829, 19.723 (18.894) | 0.280, 16.960 (16.680) |
| Neu | 10.241, 7.795 (**2.446**) | 5.729, 11.058 (**5.329**) | 1.677, 14.868 (13.191) |
| Con | 17.680, 1.086 (16.594) | 15.428, 2.600 (12.828) | 7.138, 12.069 (**4.931**) |

### 4.2 FuzzyCognitiveandSocial(Buyer)- FuzzyCognitiveandSocial (Seller)

The results presented in the following concern the fuzzy cognitive and social strategy for both buyer(b) and seller(s) agents, and they are given for different deadlines of agents: $t^b_{max} > t^s_{max}$, Tmax = $t^b_{max} = t^s_{max}$, and $t^b_{max} < t^s_{max}$. These results are obtained, for each deadline, from 9 experiments (FAgg-FAgg, FAgg-FNeu, FAgg-FCon, FNeu-FAgg, FNeu-FNeu, FNeu-FCon, FCon-FAgg, FCon-FNeu, and FCon-FCon). Thus, results shown in Table 10, Table 13, and Table 16 give the b and s agent utilities, those in Table 11, Table 14, and Table 17 give average round numbers, percentage of deals, utility products and differences, and those in Table 12, Table 15, and Table 18 give their average performances and absolute difference.

**Table 10.** Results $AU^b$ and $AU^s$: case $t^b_{max} > t^s_{max}$

| Seller | Buyer(FAgg) | Buyer(FNeu) | Buyer(FCon) |
|---|---|---|---|
| FAgg | 8.91, 10.47 | 8.59, 16.56 | 8.19, 20.80 |
| FNeu | 14.64, 8.41 | 12.76, 13.43 | 13.38, 18.75 |
| FCon | 16.68, 6.28 | 16.47, 12.47 | 15.59, 16.86 |

**Table 11.** Results AR, D(%), UP, and UD: case $t^b_{max} > t^s_{max}$

| Seller | Buyer(FAgg) | Buyer(FNeu) | Buyer(FCon) |
|---|---|---|---|
| FAgg | 16.000, 9%, 93.28, 1.56 | 15.666, 12%, 142.25, 7.97 | 14.642, 14%, 170.35, 12.61 |
| FNeu | 14.000, 9%, 123.12, 6.23 | 13.200, 15%, 171.36, 0.67 | 14.000, 19%, 250.87, 5.37 |
| FCon | 11.266, 15%, 104.75, 10.40 | 12.750, 16%, 205.38, 4.00 | 12.833, 6%, 262.84, 1.27 |

**Table 12.** Results $AP^b$, $AP^s$, and ($\left|AP^b - AP^s\right|$): case $t^b_{max} > t^s_{max}$

| Seller | Buyer(FAgg) | Buyer(FNeu) | Buyer(FCon) |
|---|---|---|---|
| FAgg | 3.266, 3.720 (**0.454**) | 3.171, 5.762 (2.591) | 3.048, 7.191 (4.143) |
| FNeu | 5.185, 3.050 (2.135) | 4.580, 4.750 (**0.170**) | 4.798, 6.530 (1.732) |
| FCon | 5.896, 2.382 (3.514) | 5.823, 4.437 (1.386) | 5.496, 5.866 (**0.370**) |

**Table 13.** Results $AU^b$ and $AU^s$: case $Tmax = t^b_{max} = t^s_{max}$

| Seller | Buyer(FAgg) | Buyer(FNeu) | Buyer(FCon) |
|---|---|---|---|
| FAgg | 8.48, 8.33 | 9.79, 16.92 | 8.82, 22.01 |
| FNeu | 13.49, 7.64 | 17.50, 17.18 | 15.96, 21.54 |
| FCon | 21.25, 7.48 | 17.58, 12.65 | 15.44, 15.35 |

**Table 14.** Results AR, D(%), UP, and UD: case $Tmax = t^b_{max} = t^s_{max}$

| Seller | Buyer(FAgg) | Buyer(FNeu) | Buyer(FCon) |
|---|---|---|---|
| FAgg | 12.166, 6%, 70.63, 0.15 | 16.375, 8%, 165.64, 7.13 | 15.500, 8%, 194.12, 13.19 |
| FNeu | 14.000, 6%, 103.06, 5.85 | 18.214, 14%, 300.65, 0.32 | 16.375, 16%, 343.77, 5.58 |
| FCon | 14.750, 16%, 158.95, 13.77 | 13.384, 13%, 222.387, 4.93 | 12.381, 21%, 237.004, 0.09 |

**Table 15.** Results $AP^b$, $AP^s$, and ($\left|AP^b - AP^s\right|$): case $Tmax = t^b_{max} = t^s_{max}$

| Seller | Buyer(FAgg) | Buyer(FNeu) | Buyer(FCon) |
|---|---|---|---|
| FAgg | 3.112, 3.062 (**0.050**) | 3.532, 5.909 (2.377) | 3.213, 7.610 (4.397) |
| FNeu | 4.772, 2.822 (1.950) | 6.112, 6.005 (**0.107**) | 5.615, 7.475 (1.860) |
| FCon | 7.388, 2.798 (4.590) | 6.162, 4.518 (1.644) | 5.481, 5.451 (**0.030**) |

**Table 16.** Results $AU^b$ and $AU^s$: case $t^b_{max} < t^s_{max}$

| Seller | Buyer(FAgg) | Buyer(FNeu) | Buyer(FCon) |
|---|---|---|---|
| FAgg | 7.06, 7.08 | 8.68, 14.83 | 7.91, 18.29 |
| FNeu | 22.67, 11.92 | 14.25, 13.83 | 15.17, 19.44 |
| FCon | 16.14, 6.13 | 16.08, 11.57 | 15.40, 14.51 |

**Table 17.** Results AR, D(%), UP, and UD: case $t^b_{max} < t^s_{max}$

| Seller | Buyer(FAgg) | Buyer(FNeu) | Buyer(FCon) |
|---|---|---|---|
| FAgg | 10.166, 6%, 49.98, 0.02 | 14.400, 5%, 128.72, 6.15 | 13.625, 8%, 144.67, 10.38 |
| FNeu | 5.000, 1%, 270.22, 10.75 | 14.166, 6%, 197.07, 0.42 | 15.500, 10%, 294.90, 4.27 |
| FCon | 11.000, 11%, 98.93, 10.01 | 11.866, 15%, 186.04, 4.51 | 11.833, 18%, 223.45, 0.89 |

**Table 18.** Results $AP^b$, $AP^s$, and ($\left|AP^b - AP^s\right|$): case $t^b_{max} < t^s_{max}$

| Seller | Buyer(FAgg) | Buyer(FNeu) | Buyer(FCon) |
|---|---|---|---|
| FAgg | 2.621, 2.670 (**0.049**) | 3.123, 5.233 (2.110) | 2.883, 6.399 (3.516) |
| FNeu | 7.851, 4.289 (3.562) | 4.985, 4.904 (**0.081**) | 5.294, 6.782 (1.488) |
| FCon | 5.658, 2.367 (3.291) | 5.644, 4.190 (1.454) | 5.428, 5.180 (**0.248**) |

With regard to the time-dependent strategy, the obtained results demonstrate that the suggested fuzzy cognitive and social strategy allows agents to improve the negotiation process in terms of agent utilities, round number to reach an agreement, and percentage of agreements in *almost* all cases:

- $t^b_{max} > t^s_{max}$ with **0.454**, **0.170**, and **0.370** for fuzzy congnitive and social see Table 12 (**5.703**, **0.306**, and **1.154** for time-dependent see Table 3),

- Tmax = $t^b_{max} = t^s_{max}$ with **0.050**, **0.107**, and **0.030** for fuzzy congnitive and social see Table 15 (**0.076**, **0.004**, and **3.267** for time-dependent see Table 6),

- $t^b_{max} < t^s_{max}$ with **0.049**, **0.081**, and **0.248** for fuzzy congnitive and social see Table 18 (**2.446**, **5.329**, and **4.931** for time-dependent see Table 9).

## 5 Conclusion

In this paper, a *fuzzy cognitive* and *social* strategy has been suggested for autonomous agent systems with *incomplete* information intending to find adequate (*win-win* solutions for both parties) strategy, in one-to-one single issue negotiation. With regard to time-dependent strategy, the obtained results demonstrate that suggested fuzzy cognitive and social strategy allows agents to improve negotiation process in terms of agent utilities, round number to reach an agreement, and percentage of agreements.

An interesting alternative for future research could be the learning from interaction in negotiation which is fundamental, from embodied cognitive science and understanding natural intelligence perspectives [16], [17], for understanding human behaviors and developing new solution concepts [18].

## References

1. Rosenschein, J., Zlotkin, G.: Rules of Encounter. MIT Press, Cambridge (1994)
2. Wooldridge, M.: An Introduction to MultiAgent Systems. John Wiley & Sons, England (2002)
3. Jennings, N.R., Faratin, P., Lomuscio, A.R., Parsons, S., Sierra, C., Wooldridge, M.: Automated negotiation: prospects, methods, and challenges. Int. J. of Group Decision and Negotiation 10(2), 199–215 (2001)
4. Gerding, E.H., van Bragt, D., Poutré, J.L.: Scientific Approaches and Techniques for Negotiation: A Game Theoretic and Artificial Intelligence Perspective. CWI, Technical Report, SEN-R0005 (2000)
5. Li, C., Giampapa, J., Sycara, K.: Bilateral negotiation decisions with Uncertain Dynamic outside options. IEEE Trans. on Systems, Man, and Cybernetics, Part C: Special Issue on Game-Theoretic Analysis and Stochastic Simulation of Negotiation Agents 36(1), 1–13 (2006)
6. Lomuscio, A.R., Wooldridge, M., Jennings, N.R.: A classification scheme for negotiation in electronic commerce. Int. J. of Group Decision and Negotiation 12(1), 31–56 (2003)
7. Faratin, P., Sierra, C., Jennings, N.R.: Negotiation decision functions for autonomous agents. International Journal of Robotics and Autonomous Systems 24(3-4), 159–182 (1998)
8. Bales, R.F.: Interaction Process Analysis: A Method for the Study of Small Groups. Addisson Wesley, Cambridge (1950)
9. Rubin, J.Z., Brown, B.R.: The Social Psychology of Bargaining and Negotiation. Academic Presss, New York (1975)

10. Faratin, P., Sierra, C., Jennings, N.R.: Using similarity criteria to make issue trade-offs in automated negotiations. Artificial Intelligence 142, 205–237 (2002)
11. Wang, K.-J., Chou, C.-H.: Evaluating NDF-based negotiation mechanism within an agent-based environment. Robotics and Autonomous Systems 43, 1–27 (2003)
12. Pruitt, D.: Negotiation Behavior. Academic Press, London (1981)
13. Raiffa, H.: The Art and Science of Negotiation. Harvard University Press, Cambridge (1982)
14. Lee, C.-F., Chang, P.-L.: Evaluations of tactics for automated negotiations. Group Decision and Negotiation 17(6), 515–539 (2008)
15. Ros, R., Sierra, C.: A negotiation meta strategy combining trade-off and concession moves. Auton Agent Multi-Agent Sys. 12, 163–181 (2006)
16. Pfeifer, R., Scheier, C.: Understanding Intelligence. MIT press, Cambridge (1999)
17. Chohra, A.: Embodied cognitive science, intelligent behavior control, machine learning, soft computing, and FPGA integration: towards fast, cooperative and adversarial robot team (RoboCup). Technical GMD Report, No. 136, Germany (June 2001) ISSN 1435-2702
18. Zeng, D., Sycara, K.: Benefits of learning in negotiation. In: Proc. of the 14th National Conference on Artificial Intelligence (AAAI-1997), Providence, RI, July 1997, pp. 36–41 (1997)

# A Multi-agent Architecture for Multi-robot Surveillance

David Vallejo[1], Paolo Remagnino[2], Dorothy N. Monekosso[2], Luis Jiménez[1], and Carlos González[1]

[1] Escuela Superior de Informática
David.Vallejo@uclm.es
University of Castilla-La Mancha, Spain
[2] Kingston Roboclub,
Kingston University, United Kingdom

**Abstract.** In this paper we propose a multi-agent architecture that gives support to a set of robots for surveillance tasks, such as environment exploration. We introduce two protocols to coordinate the robots: one of them to carry out role allocation depending on the *mobility* concept and the other one to gather information from the environment. The communication model is based on the use of *communication channels* and a publish-subscribe scheme that guarantees scalability when the number of robots increases. A case study in which the agents that control the robots are trained to recognise predefined patterns and to learn how far they are from the robots and their orientation is presented.

## 1 Introduction

Multi-robot coordination [5] involves the use of mechanisms for robots which may have different skills to cooperate and to carry out global tasks. Multi-robot coordination is directly related to multi-agent coordination [17], in which software agents cooperate by taking into account the actions performed by other agents. Coordination in the robotics field has to deal with relevant issues such as uncertainty management, physical communication issues, and physical robot restrictions in highly dynamic environments. So, a relevant research line in robotics is how to address the coordination among robots sharing a common goal in dynamic environments. To deal with this issue, Multi-Agent Systems (MAS) [16] perfectly fit into this problem, which involves heterogeneous sources of knowledge that cooperate to reach a common aim. On the other hand, a coordination protocol can be used for the robots to interact with one another within the environment in order to make decisions in the right direction. The idea consists in defining a set of rules that robots must follow to achieve partial objectives such as role assignment, task distribution, or even environment exploration.

To face the challenge of deploying multi-robot surveillance systems, we make use of the multi-agent paradigm to control the robots. Specifically, we are interested in how to dynamically establish teams of agents to coordinate them

N.T. Nguyen, R. Kowalczyk, and S.-M. Chen (Eds.): ICCCI 2009, LNAI 5796, pp. 266–278, 2009.

when obtaining information of environment. To do that, we introduce a communication scheme based on event channels and two coordination protocols used to assign roles and gather information. Within this context, an important robot concept is mobility, which represents the measure for a robot to carry out a task depending on its cost at a particular time. To evaluate mobility, we suggest a metric based on Fuzzy Logic [18], which allows us to deal with the uncertainty inherent in relevant issues of robotics, such as distance between positions or robot autonomy. Finally, we also propose the use of a machine learning algorithm for the agents to distinguish patterns and to apply this knowledge when carrying out the environment analysis.

To test the multi-agent architecture, we have deployed a set of robots around a closed environment where there are different objects that contain information. Our goal is to evaluate how the robots interact with one another to study the scenario depending on the number of robots, the number of objects, and the state of each robot in every moment.

The rest of the paper is organised as follows. Section 2 resumes the relevant research lines related with this work. Section 3 introduces the multi-agent architecture, the communication mechanisms, and the coordination protocols developed in this paper. Section 4 describes the platform used to deploy our architecture and the environment built to carry out multi-robot exploration. Finally, Section 5 concludes the paper and discusses future research lines.

## 2 Related Work

### 2.1 Multi-agent Architectures

The Multi-Agent paradigm [16] has been used when designing multi-robot systems and, from a general point of view, in the surveillance domain. In both cases, there are two main issues to take into account: i) the distribution of knowledge, which involves the cooperation of multiple sources of information in order to reach a common aim; and ii) the high number of sensors which requires flexible communication mechanisms to guarantee scalability.

In the case of agent-based surveillance architectures, there exist proposals to deploy agents specialised in classifying objects and detecting behaviours [1], to identify behaviours of people and vehicles [14], to manage and detect traffic in real-time [4], or even to coordinate BDI-based agents in video-surveillance [12]. However, the main limitation of these approaches is scalability when reusing the architecture in different surveillance domains or when dealing with different questions. To face this issue, we propose an architecture based on *reasoning agents*, which request the knowledge needed to deal with a particular problem to another agent named *loader agent*, which is responsible for the ontology management.

On the other hand, there is also a research line that combines multi-agents and robotics since there are several questions that need to be addressed in both areas (such as coordination [5]). Within this context, a recent work was proposed by Posadas et al. [13] in which the authors proposed an agent-based distributed architecture for mobile robot control, paying special attention to the communication framework. Other approaches that use the multi-agent paradigm deal with particular issues such as learning in multi-robot coordination [15] or driving mobile manipulator robots [10].

### 2.2 Role Allocation

Role allocation refers to the process that allows to classify robots in a team depending on certain criteria in order to cooperatively accomplish a goal. For instance, role allocation is a crucial task in the RoboCup[1], where the participants have to design techniques for role allocation and exchange. Within this context, the authors in [7] analyse allocation mechanisms of a number of RoboCup teams. From a general point of view, elements of this problem are a set of robots $R$, a set of roles to allocate $O$, and a function $f$ that evaluate how well a robot $r$ can play the role $o$ (often named utility), that is, $f(r, o, p, E_t)$, where $p$ denotes the priority when choosing between two or more roles and $E_t$ represents the environment state in the moment $t$. There exist different approaches when allocating roles such as those that take into account the environment state [11] [3] or even the use of auctions [6].

### 2.3 Multi-robot Exploration

The problem of exploring an unknown environment is a relevant question in robotics. Within this context, coordinating a number of robots have several advantages over exploration made by a single robot so that robots share the environment information and the exploration can be distributed between robots.

A coordination approach that simultaneously takes into account the cost of reaching a target point and its utility is presented in [2]. In this way, when a robot's goal is to explore a concrete target, the utility of such unexplored area from this target position is reduced. Another interested work was introduced in [8], where an architecture composed of a team of centimetre-scale robots is deployed for environment exploration. Once again, there are contributions from the market economy, such as the proposed in [19].

## 3 Our Approach

### 3.1 Multi-agent Architecture

Figure 1 shows the multi-agent architecture that we propose to deploy knowledge-based distributed systems for surveillance purposes, such as robot exploration. Basically, it consists in a multi-agent system where the agents give support to the surveillance services provided by the system. These services typically involves information retrieval, data storage, surveillance tasks such as behaviour analysis, or even alarm management. The architecture is structured into three different layers:

[1] http://www.robocup.org

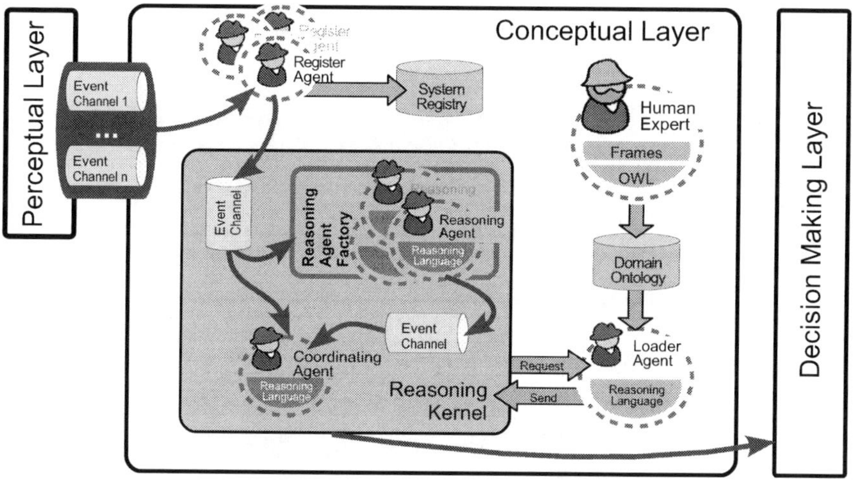


**Fig. 1.** Multi Agent Architecture

1. the **perceptual layer**, which gets information from the environment and it is composed of the sensors used to retrieve data from the monitored environment;
2. the **conceptual layer**, which is composed of agents that perform the analysis of the environment by generating useful knowledge from the perceptual layer feedback;
3. the **decision-making layer**, which manages the process of making the adequate decisions depending on the knowledge obtained from the conceptual layer. A typical action may involve the activation of an alarm if a suspicious static object was detected.

The core of the architecture is the *Reasoning Kernel*. Basically, this component allows to deploy a set of agents that share knowledge about a certain problem. On the one hand, the reasoning agent factory allows to instantiate reasoning agents, which obtain the knowledge needed to complete a task by means of the loader agent. This agent represents the link with the human expert, that is, the person responsible for providing with the knowledge needed to deal with a particular surveillance problem. An instance of this knowledge may be an ontology to represent the knowledge for the normality analysis of moving object trajectories. In the case of the multi-robot problem, a reasoning agent is deployed for each mobile robot. This knowledge may be acquired or learned. Finally, the function of the coordinating agent is to manage the knowledge generated by the reasoning agents.

The key communication component of the architecture is the event channel, which represents a one-way communication channel and allows to deploy

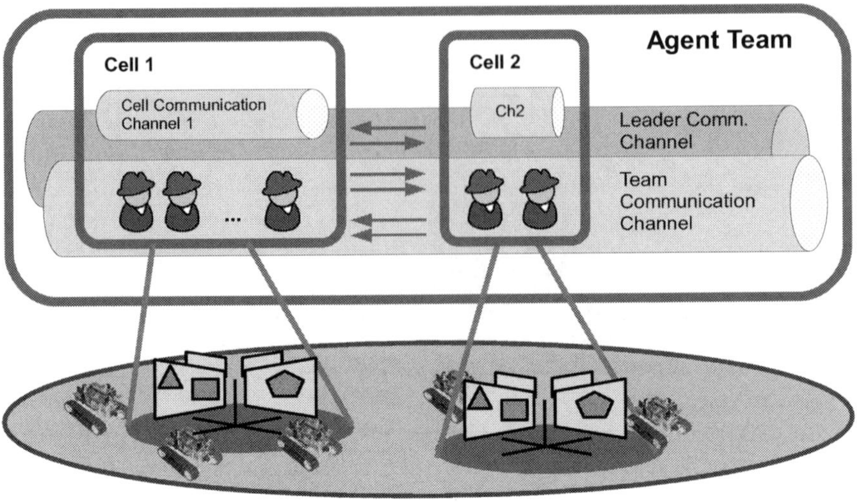


**Fig. 2.** Graphical view of the multi-robot system

content-based communication mechanisms for the agents to share information. The communication model is based on a publish-subscribe scheme to guarantee scalability.

### 3.2 Multi-robot System

The multi-robot system discussed in this paper for surveillance purposes is instantiated from the multi-agent architecture shown in Figure 1. The highest level organisation is an agent team $T$, which is composed of a certain number of cells $C$ ($C \in T$). Each cell is a virtual group of agents so that each agent is responsible for controlling an individual robot. Cells are dynamically set up and may reconfigure their members when required. The main goal of each cell $C_i$ is to obtain as much information as possible from the object $o_i$, where $o_i$ involves a static object to be monitored. Monitoring an object may consists in identifying the object or making use of more advanced techniques to perform such task. Besides, each cell $C_i$ is composed of a number of agents $A_i = \{l_i, E_i\}$, where $l_i$ represents the leader of the cell responsible for analysing $o_i$ and $E_i$ is the set of explorers of the cell $C_i$ (see Figure 2). The leader's primary responsibility is to coordinate the whole cell and its members when monitoring the environment. As previously stated, every agent controls one robot to provide it with the mechanisms needed to cooperatively get information from the environment.

Two main cases are distinguished depending on the number of robots ($n_r$) and the number of objects ($n_o$) to analyse:

- if $n_r \leq n_o$, then each agent will lead the analysis of the objects until completing the study of the whole environment;

- if $n_r > n_o$, then the leaders of the objects will emerge and the other agents, known now as explorers, will associate to a certain cell depending on its mobility to the corresponding object.

The main disadvantage is that it is not possible for cells to have a high number of members. To overcame this problem, leaders use the negotiation protocol described in Section 3.4 to move explorers to balance the environment analysis. It is important to remark that the main purpose of the leader role is related to communication issues but not to establish a strong hierarchy between agents. In fact, leaders are also inherently explorers.

We define the state of an agent $a_j^i$ at instant $t$ as

$$a_j^i(t) =< p_{a_j^i}, \theta, R >, \tag{1}$$

where $a_j^i$ represents the agent $j$ in the cell $C_i$, $p_{a_j^i}$ is the position of agent's $a_j^i$ robot, $\theta$ is the angle that defines the clockwise rotation of agent's $a_j^i$ robot about the axis $Z$, and $R$ specifies the set of roles $r_i \in R$ of agent $a_j^i$.

### 3.3 Communication

Communication among agents is carried out through event channels, which represent one-way communication channels used by the agents to send messages. An event channel provides us with the flexibility needed to deal with the dynamics of our architecture: auction management, set up and dynamic reconfiguration of cells, and communication among the agents of a cell. The communication model of the event channels is based on the publish-subscribe mechanism. The generic communication channel $Ch_g$ is defined as

$$Ch_g =< P, S, L >, \tag{2}$$

where $P = \{p_1, p_2, \ldots, p_n\}$ is the set of publishers of $Ch_g$, $S = \{s_1, s_2, \ldots, s_n\}$ is the set of subscribers of $Ch_g$, and $L = \{l_1, l_2, \ldots, l_n\}$ is the set of links of $Ch_g$ with other event channels. Both publishers and subscribers are the agents that control the physical robots. A link $l_i =< Ch_i, c_i >$ represents a relation of cost $c_i$ between two event channels $Ch_g$ and $Ch_i$ where $Ch_g \neq Ch_i$. The cost $c_i$ is used to filter messages, that is, event channels will forward messages if their costs are lower or equals than the cost link $c_i$. The cost $c_m$ of a message and the cost $c_i$ of a link are estimated depending on the application domain by taking into account the desired behaviour of the parties that exchange messages. To model broadcast messages, links with cost equals to 0 will forward all messages received and messages with cost equals to 0 will be forwarded to all links. From this communication model, we instantiate two specific channels:

- **Team Communication Channel**. The team communication channel $Ch_t$ allows agents within a team to publish and receive messages. The main purpose of this channel is to notify general information concerning the whole team, such as the leader election or the need of analysing a new object. In

several situations, it may be interesting for the leaders to send messages among themselves without notifying the explorers. To deal with this flexibility, we define $Ch_t =< A, A, \{l_l\} >$ where $A$ is the set of agents of the team and $l_l =< Ch_l, c >$, that is, $Ch_l$ is the communication channel for leaders to process private messages of cost $c$. In this way, $Ch_l =< A, l_i \in C_i, \phi > / i = \{1, 2, \ldots, n\}$, where $n$ is the number of cells.
- **Cell Communication Channel.** When a leader $l_i$ rises with the goal of obtaining information from an object $o_i$, it automatically creates a cell communication channel $Ch_i$ which will be used to make easy the coordination between the agents of a cell $C_i$. An explorer $e_j$ interested in joining a cell $C_i$ sends a message to $Ch_t$ with a cost $c$ only the leaders to receive this message ($c$ must be greater than 1). The leader $l_i$ then suggests $e_j$ to join the cell $C_i$ or to join another cell $C_k$ by considering previous negotiations with the leader $l_k$ (see Section 3.4). Cell communication channels are then mainly used for coordinating the analysis of an object.

### 3.4 Coordination Protocols

**A) Role assignment.** As previously introduced in Section 3.2, agents within a cell $C_i$ may play the role of explorers or leaders (which are also explorers). To select the best candidate $l_i$ for analysing an object $o_i$, we employ a simple auction protocol that allows agents to rise as leaders and begin to study the environment. Bids represent how mobile is the robot controlled by a particular agent. In this way, a high mobility value indicates that an agent is well prepared to become a leader of a certain cell and, therefore, of a certain object. Next, we introduce the steps of the auction protocol:

**1. Goal announcement**
The first step consists in getting information about the positions of the objects. We address two different ways to obtain them: i) to use an external camera that has a complete view of the environment or ii) to employ the robots to get partial views. When this process is completed, one of the agents is elected as the auctioneer to announce the objects that have to be analysed. This is done by means of the message *Announcement* that contains the location of the set of objects. Then the auctioneer waits for receiving bids for an adjustable timeout.

**2. Mobility evaluation**
When an agent receives an announcement, it evaluates the mobility of his robot to all objects in order to bid for the best objects for him. Mobility is evaluated by a metric $M =< V, DDV, R >$, where $V = \{v_1, v_2, \ldots, v_n\}$ is the set of variables of $M$, $DDV = \{ddv_1, ddv_2, \ldots, ddv_n\}$ is the set of domain definition for the variables, and $R = \{r_1, r_2, \ldots, r_n\}$ is the set of rules that determine the mobility value. This set of rules takes into account the distance to the object $o_i$ and the current autonomy of the robot to evaluate its mobility. To model these variables and to define the rules that determine the mobility value depending on the physical conditions we use a fuzzy approach [18]. We justify this choice

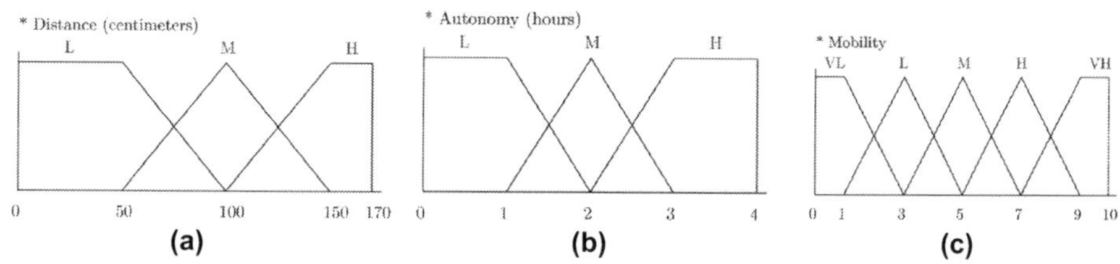


**[0] IF** Distance **IS** L Autonomy IS L **THEN** Mobility **IS** L
**[1] IF** Distance **IS** L Autonomy **IS** M **THEN** Mobility **IS** H
....
**[6] IF** Distance **IS** H Autonomy **IS** L **THEN** Mobility IS VL
**[7] IF** Distance IS H Autonomy **IS** M **THEN** Mobility **IS** L
...

**(d)**

```
<rule name="R1">
  <antecedent varname="Distance" linguisticlabel="L"/>
  <antecedent varname="Autonomy" linguisticlabel="L"/>
  <consequence varname="Mobility" linguisticlabel="L"/>
</rule>
```

**(e)**

**Fig. 3.** Metric for evaluating mobility: **a)** definition of the variable *distance*, **b)** definition of the variable *autonomy*, **c)** definition of the variable *mobility*, **d)** some of the rules of the fuzzy system, **e)** XML definition of the rule *R1* for processing purposes (VL = very low, L = low, M = medium, H = high, VH = very high)

due to the environment uncertainty and the interpretability of the set of rules to define the system. Thus, every $ddv_i = \{l_1, l_2, \ldots, l_n\}$ represents the set of linguistic labels used to define $v_i$. $V$, $DDV$, and some rules of $R$ are defined in Figure 3 according to the platform used to deploy this architecture, as described in Section 4.

**3. Bid submission**
Once the metric has been applied, the agent $a_i$ submits his best bid $b_i$ to the auctioneer by means of the message *Bid*. A bid $b_i =< a_i, o_i, v >$ allows the auctioneer to obtain the information needed to evaluate all the bids. $v$ represents the mobility value of the agent $a_i$ regarding $o_i$.

**4. Bid evaluation and goal assignment**
The auctioneer sorts by mobility the bids received and notifies agents who are now leaders by means of the message *Close*. For instance, if two agents $a_i$ and $a_j$ submit two bids $b_i =< a_i, o, v_i >$ and $b_j =< a_j, o, v_j >$ for the same object $o$ and $v_i > v_j$, then agent $a_i$ will become the leader of the object $o$. In this case, $a_j$ will keep on bidding for becoming the leader of another object.

**5. New iteration**
Each iteration will define one of more leaders. In this point, there may be objects without a leader so that the auction continues until all the objects have a leader or all the robots are leaders. In case of the number of objects $n_o$ be lower than the number of robots $n_r$, then $n_r - n_o$ agents will have to join some of the cells like explorers. Currently, explorers join the cell $C_i$ responsible for analysing the object $o_i$ according to the mobility measure.

**B) Negotiation to gather information.** Once the initial cells $C$ have been formed, the next step is how to distribute the study of an object $o_i$. In this point, each cell $C_i$ constitutes an autonomous entity and uses its internal cell communication channel $Ch_i$. Currently, explorers within a cell follow a simple protocol to surround an object. The method consists in taking the closest position to the object depending on the number of explorers and the mobility of each explorer to the object. These positions are calculated from the object position.

On the other hand, it is possible for the agents to move to another cell to keep on exploring the environment. Currently, the value that conditions this change depends on the number of components of the cell. If this value is greater than a threshold, the leader sends a broadcast message via the leader communication channel $Ch_l$ to ask for a cell with a low number of explorers. If two or more leaders call for this proposal, the explorer will join the cell with the lowest number of explorers.

# 4 Experimental Validation

## 4.1 Experimental Design

**Platform.** We have been working with four Surveyor SRV-1 Blackfin Robots (Figure 4.d) moving around a scenario of 1.2 by 1.3 meter (see Figures 4.e and 4.f). The robot has a digital video camera with resolution from 160x128 to 1280x1024 pixels, a WLAN 802.11b/g interface, and a laser pointer on a quad-motor tracked robotic base. The robot can run C programs or can be remotely controlled by sending predefined commands. The scenario is composed of a number of objects $O$ distributed around the environment. Each object $o_i$ or signpost represents a pattern, as graphically shown in Figure 4.f, which consists of a set of basic geometric shapes of different colours.

**Learning phase.** Before deploying the robots on the environment, there is a learning phase for the agents to be able to recognise objects. Basically, this phase consists on using a neural network which we train with real images (directly taken from the robot camera) by means of the back-propagation method. The input of the neural net is the set of pixels of the image and the output is the distance from the robot to the object, the vision angle, and the identifier of the object to recognise. The goal is that a robot knows how far is an object and the orientation regarding a concrete axis. To carry out this method, a robot takes 36 pictures by varying the distance and the orientation to the object, as graphically shown in Figures 4.d, 4.e and 4.f. To make easy this task, we have built a template to put the robot on different positions of the environment, as exposed in Figure 4.d. Once the neural net has been built and trained, the robots can dynamically use it when obtaining information from the scenario. This process involves taking a picture, feeding on the neural net, and getting the relevant information, that is, the distance, the orientation, and the identifier of the pattern.

## 4.2 Simulations

To evaluate the software architecture proposed and the coordination protocols of Section 3.4, we have tested different scenarios by randomly deploying robots

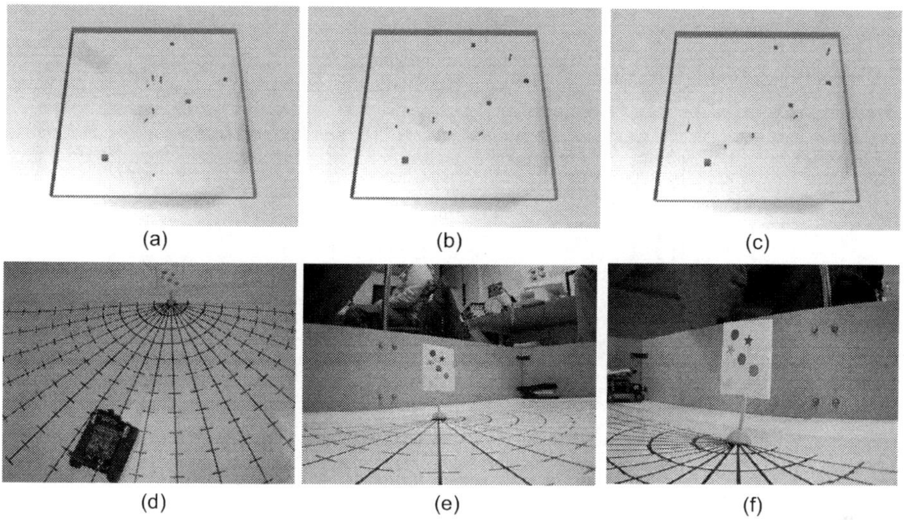


**Fig. 4. (a)(b)(c)** Graphical simulation. **(d)** Learning patterns. **(e)(f)** Image taken from the robot camera.

and objects on a virtual environment. In order to study how the cells are created and how the robots spread around the scenario, we have graphically simulated the robot movements as shown in Figures 4.a, 4.b, and 4.c. On the other hand, Figure 5 resumes the information obtained when running one of these random tests. The values on the auction rounds represent the mobility of a robot when it bids for a specific object. The mobility value for a robot to move towards a signpost is calculated by means of the fuzzy system of Section 3.4 (see Figure 3). The inputs of this rule system are the distance between the robot and the signpost and the autonomy of the robot in that time while the output is the mobility value.

Bold values represent winning bids because they involve the highest value of an auction round according to a particular object. When all the objects have a leader, the remaining robots join to one of the existing cell depending on the mobility value of its respective object, as shown in the last rows of the table.

### 4.3 Implementation Details

Currently, the software used to analyse the environment from the zenithal camera and obtain both robot and object positions is implemented in C++ by using Roborealm[2]. The software developed to move and interact with the robots is also implemented in C++ and represents the public interface employed by agents to control robots. On the other hand, agents are implemented in Python to easily

[2] http://www.roborealm.com/

| | | | | | | | |
|---|---|---|---|---|---|---|---|
| Objects | ID | a | b | c | d | | |
| | Position | [110,35] | [4,78] | [56,87] | [33,110] | | |
| Robots | ID | 1 | 2 | 3 | 4 | 5 | 6 |
| | Position | [66,63] | [54,35] | [102,64] | [17,42] | [40,74] | [120,27] |
| | Autonomy | 2 | 2 | 1 | 1 | 3 | 1 |
| Cells creation | Auction – First Round | | | | | | |
| | Robot/Object | 1 | 2 | 3 | 4 | 5 | 6 |
| | a | | | **3** | | | 3 |
| | b | | | | 3 | **8.66** | |
| | c | **7** | 6.85 | | | | |
| | d | | | | | | |
| | Auction – Second Round | | | | | | |
| | Robot/Object | 1 | 2 | 3 | 4 | 5 | 6 |
| | a | | | | | | |
| | b | | | | | | |
| | c | | | | | | |
| | d | | **5.92** | | 2.36 | | 1.27 |
| | Explorers joining a cell | | | | | | |
| | Robot/Object | 1 | 2 | 3 | 4 | 5 | 6 |
| | a | | | | | | **X** |
| | b | | | | **X** | | |
| | c | | | | | | |
| | d | | | | | | |

**Fig. 5.** Results obtained when simulating one random scenario (the positions are expressed in centimetres and the mobility values range from 0 to 10)

develop fast prototypes to test and debug the system. Besides, the software that manages the neural net and the training process is also implemented in Python.

Since we use different programming languages and different operating systems we have used ZeroC ICE [9], a modern object-oriented middleware to communicate all the components of the systems. This abstraction layer provides us with independence regarding not only programming languages and operating systems but also hardware platforms and communication networks. Because ICE is essentially a multi-threaded middleware, the number of threads of the server that attends the bid submission can be dynamically adapted depending on the number of robots/agents.

## 5 Conclusions

We have presented a multi-agent architecture which can be used to deploy a set of robots with the goal of exploring an environment. The communication model is based on event channels and a publish-subscribe mechanism, which provides the multi-robot system with scalability and flexibility for information exchange. To allocate roles, we have used a simple auction protocol and a metric based on the mobility to determine which is the best robot to lead the study of a certain object. On the other hand, the current information gathering scheme aims at reinforcing the information obtained from objects scattered around the environment. Finally, agents have been trained by using a neural network to recognised predefined patterns and approximate the distance and orientation between the robot camera and a certain object. Both the multi-agent architecture and the coordination protocols have been tested by means of a simulation tool that provides us with graphical feedback in order to debug the whole system.

We are now interested in improving the protocol for the agents to gather information. For instance, if the information of an object is classified as *very important*, then a high number of robots should study such object so that cells composition will adapt according to this need. This fact involves to perform a more complex analysis of the environment on behalf of robots.

## Acknowledgement

This work has been funded by the Regional Government of Castilla-La Mancha under the Research Project PII1C09-0137-6488.

## References

1. Abreu, B., Botelho, L., Cavallaro, A., et al.: Video-based multi-agent traffic surveillance system. In: IEEE Intelligent Vehicles Symposium, pp. 457–462 (2000)
2. Burgard, W., Moors, M., Stachniss, C., Schneider, F.E.: Coordinated multi-robot exploration. IEEE Transactions on Robotics 21(3), 376–386 (2005)
3. Chaimowicz, L., Campos, M.F.M., Kumar, V.: Dynamic role assignment for cooperative robots. In: IEEE International Conference on Robotics and Automation (ICRA 2002), vol. 1, pp. 293–298 (2002)
4. Chen, B., Cheng, H.H., Palen, J.: Integrating mobile agent technology with multi-agent systems for distributed traffic detection and management systems. Transportation Research Part C 17(1), 1–10 (2009)
5. Farinelli, A., Iocchi, L., Nardi, D.: Multirobot systems: a classification focused on coordination. IEEE Transactions on Systems, Man and Cybernetics, Part B 34(5), 2015–2028 (2004)
6. Gerkey, B.P., Mataric, M.J.: Sold!: auction methods for multirobot coordination. IEEE Transactions on Robotics and Automation 18(5), 758–768 (2002)
7. Gerkey, B.P., Mataric, M.: On role allocation in roboCup. In: Polani, D., Browning, B., Bonarini, A., Yoshida, K. (eds.) RoboCup 2003. LNCS (LNAI), vol. 3020, pp. 43–53. Springer, Heidelberg (2004)
8. Grabowski, R., Navarro-Serment, L.E., Paredis, C.J.J., Khosla, P.K.: Heterogeneous Teams of Modular Robots for Mapping and Exploration. Autonomous Robots 8(3), 293–308 (2000)
9. Henning, M.: A New Approach to Object-Oriented Middleware Internet Computing. IEEE 8(1), 66–75 (2004)
10. Hentout, A., Bouzouia, B., Toukal, Z.: Multi-agent Architecture Model for Driving Mobile Manipulator Robots. International Journal of Advanced Robotic Systems 5(3), 257–268 (2008)
11. McMillen, C., Veloso, M.: Distributed, Play-Based Role Assignment for Robot Teams in Dynamic Environments. Distributed Autonomous Robotic Systems, 1–10 (2006)
12. Pavón, J., Gómez-Sanz, J., Fernández-Caballero, A., Valencia-Jiménez, J.J.: Development of intelligent multisensor surveillance systems with agents. Robotics and Autonomous Systems 55(12), 892–903 (2007)

13. Posadas, J.L., Poza, J.L., Simó, J.E., Benet, G., Blanes, F.: Agent-based distributed architecture for mobile robot control. Engineering Applications of Artificial Intelligence 21(6), 805–823 (2008)
14. Remagnino, P., Shihab, A.I., Jones, G.: Distributed intelligence for multi-camera visual surveillance. Pattern recognition 37(4), 675–689 (2004)
15. Wang, Y., de Silva, C.W.: A machine-learning approach to multi-robot coordination. Engineering Applications of Artificial Intelligence 21(3), 470–484 (2008)
16. Weiss, G.: Multiagent Systems: A Modern Approach to Distributed Artificial Intelligence. MIT Press, Cambridge (1999)
17. Woolridge, M., Wooldridge, M.J.: Introduction to multiagent systems. John Wiley & Sons, Inc, New York (2001)
18. Zadeh, L.A.: Fuzzy logic = computing with words. IEEE Transactions on Fuzzy Systems 4(2), 103–111 (1996)
19. Zlot, R., Stentz, A., Dias, M.B., Thayer, S.: Multi-robot exploration controlled by a market economy. In: IEEE International Conference on Robotics and Automation (ICRA 2002), vol. 3, pp. 3016–3023 (2002)

# Designing Social Agents with Empathic Understanding

Zulfiqar A. Memon[1,2] and Jan Treur[1]

[1] Vrije Universiteit Amsterdam, Department of Artificial Intelligence
De Boelelaan 1081, 1081 HV Amsterdam
[2] Sukkur Institute of Business Administration (Sukkur IBA),
Air Port Road Sukkur, Sindh, Pakistan
{zamemon,treur}@few.vu.nl
http://www.few.vu.nl/

**Abstract.** This paper addresses the design of an agent model for a social agent capable of understanding other agents in an empathic way. The model describes how the empathic agent deals with another agent's mental states and the associated feelings, thus not only understanding the other agent's mental state but at the same time feeling the accompanying emotion of the other agent.

## 1 Introduction

For functioning within a social context, one of the important issues is in how far agents have a good understanding of one another. Having understanding of another agent often is related to the notion of mindreading or Theory of Mind (ToM). This is a very wide notion, subsuming various foci of the understanding (such as attention, intention, desire, emotion, or belief states) and various methods for it, for example, based on the Theory Theory perspective or the Simulation Theory perspective as known from the philosophical literature; e.g., (Gärdenfors, 2003; Goldman, 2006).

For humans, one of the deepest and most fundamental forms of mutual understanding is based on the notion of empathy; e.g., (Ickes, 1997; Hoffman, 2000; Preston and Waal, 2002; Decety and Jackson, 2004; Lamm, Batson, and Decety, 2007; Iacoboni, 2005, 2008). Originally (cf. Lipps, 1903) the notion of empathy was named by the German word 'einfühlung' which could be translated as 'feeling into'; e.g., (Preston and Waal, 2002). As this word indicates more explicitly, the notion has a strong relation to feeling: empathic understanding is a form of understanding which includes (but is not limited to) feeling what the other person feels. This paper addresses how an agent can be designed that is able to have empathic understanding of other agents.

A particular challenge here is how to relate understanding of any mental state (such as an attention, belief, desire or intention state) of another agent to a form of understanding which includes feeling the same emotion as the other agent. As one of the points of departure for an approach to address this challenge, Damasio (1999, 2004)'s theory was adopted, describing how any internal state or external stimulus provokes an emotion in the form of a bodily response that is felt via sensing this body state: a body loop. As a variant an as if body loop goes via preparations for a body state directly to a

N.T. Nguyen, R. Kowalczyk, and S.-M. Chen (Eds.): ICCCI 2009, LNAI 5796, pp. 279–293, 2009.

sensory representation of this body state. As another point of departure the Simulation Theory perspective on mindreading (e.g., Goldman, 2006) was adopted, which assumes that mindreading focussing on certain mental states of an observed agent makes use of the same mental states within the observing agent.

In this paper, first the notion of empathy is clarified and positioned (Section 2) and Damasio's theory on the generation of feelings is briefly introduced (Section 3). In Section 4 the designed agent model for empathic understanding is introduced and in Section 5 some simulation results are discussed. Finally, Section 6 is a discussion.

## 2 Empathy and Different Types of Mindreading

Empathy can be considered a specific type of mindreading. In the literature, empathy is described in different manners:

- The ability to put oneself into the mental shoes of another person to understand his or her emotions and feelings (Goldman, 1993)
- A complex form of psychological inference in which observation, memory, knowledge, and reasoning are combined to yield insights into the thoughts and feelings of others (Ickes, 1997)
- An affective response more appropriate to someone else's situation than to one's own (Hoffman, 1982)
- An affective response that stems from the apprehension or comprehension of another's emotional state or condition, and which is similar to what the other person is feeling or would be expected to feel in the given situation (Eisenberg, 2000)

Recurring aspects in such descriptions are on the one hand understanding, having insight in, apprehension or comprehension, and on the other hand feeling the state of the other person. Here the state of the other person may involve emotions felt and/or other mental states of the person. For the sake of simplicity, below notions such as understanding, having insight in, apprehension, comprehension, are indicated as understanding. For example, a person may understand but not feel an emotion felt by another person. These distinctions can be used more generally to obtain a form of classification of different types of mindreading that are possible. More specifically, mindreading can address three types of states of an observed person:

(a) Emotions felt by the person
(b) Other mental states (e.g., attention states, desire, intention, belief states)
(c) Both emotion states and other mental states

Moreover, this not only applies to a person who is observed but also to a person performing the observing. In particular, a person can understand or feel another person's state, or both. Given this, mindreading of another person's state can take three forms:

(1) Feeling the state of another person without understanding it
(2) Understanding the state of another person without feeling it
(3) Both understanding and feeling the state of another person

As the other person's state may involve emotions felt and/or other mental states, the combination of these provides the matrix of possibilities as shown in Table 1.

For example, the possibility indicated as type (1a), 'feeling but not understanding another person's emotion', is a case of emotion contagion as often occurs in the interaction between persons (e.g., Hatfield, Cacioppo, and Rapson, 1994). Here the emotion felt by one person is mirrorred in the emotion felt in the other person. Another specific case is type (2c): 'understanding but not feeling another person's emotions and other mental states'. This is a case that is often assumed to occur in psychopaths who have well-developed skills in mindreading and apply them to their victims thereby serving their own interest, but do not mirror the feelings of their victims (cf. Raine, 1993, pp. 159-165; Blair, 2005). Yet other specific cases are type (2b) which subsumes classicical cases described by the Theory Theory perspective on mindreading (e.g., Goldman, 2006; Bosse, Memon, and Treur, 2007), and type (2a) that subsumes approaches based on dedicated emotion recognition methods, for example, from facial expressions; e.g., (Pantic and Rothkrantz, 2000).

**Table 1.** Different types of mindreading

| ***Observed person*** <br> ***Observing person*** | *Other person's emotions felt* (a) | *Other person's other mental states* (b) | *Other person's emotions felt and other mental states* (c) |
|---|---|---|---|
| *Feeling but not understanding* (1) | Feeling but not understanding another person's emotion | Feeling but not understanding another person's belief, desire, intention, attention, ... | Feeling but not understanding another person's emotions and other mental states |
| *Understanding but not feeling* (2) | Understanding but not feeling another person's emotion | Understanding but not feeling another person's belief, desire, intention, attention, ... | Understanding but not feeling another person's emotions and other mental states |
| *Both understanding and feeling* (3) | Understanding and feeling another person's emotion | Understanding and feeling another person's belief, desire, intention, attention, ... | Understanding and feeling another person's emotions and other mental states |

Some of the descriptions of the notion of empathy (e.g., in the descriptions from (Goldman, 1993; Hoffman, 1982; Eisenberg, 2000) quoted above) concentrate on feelings and mirrorring them, which could be described as being subsumed by type (1a) or (3a). However, other descriptions explicitly involve thoughts as well, of both the observed and observing person (e.g., Ickes, 1997), which makes them subsumed by type (3c). In the current paper this more extended (and challenging) notion of empathy is taken as the aim. Here an extra aspect is that feelings and other mental states are interrelated: usually any mental state of a person that by itself is not an emotion state (for example, a belief, desire, intention or attention state) induces or goes together with a certain emotion state. For example, a belief that something bad is to happen, may relate to feeling fear, or the belief that something good has happened may relate to feeling happiness. Another example of such a relationship is the role of cognitive elements (for example, certain thoughts) in the development, persistence and recurrence of mood disorders such as depressions; e.g., (Ingram, Miranda & Segal, 1998).

## 3 Relating Mental States to Emotions Felt

A question that may arise from the distinctions made in the previous section is whether it is possible to feel a state of another person which by itself is not a feeling, for example, a belief state. An answer to this involves the way in which any mental state in a person induces emotions felt within this person, as described by Damasio (1999, 2004); e.g.:

> 'Even when we somewhat misuse the notion of feeling – as in "I feel I am right about this" or "I feel I cannot agree with you" – we are referring, at least vaguely, to the feeling that accompanies the idea of believing a certain fact or endorsing a certain view. This is because believing and endorsing *cause* a certain emotion to happen. As far as I can fathom, few if any exceptions of any object or event, actually present or recalled from memory, are ever neutral in emotional terms. Through either innate design or by learning, we react to most, perhaps all, objects with emotions, however weak, and subsequent feelings, however feeble.' (Damasio, 2004, p. 93)

From this perspective, if any mental state of an observed person is mirrored within an observing person, by an (assumingly) similar mechanism the associated feeling can also be generated within the observing person. In principle, this can even happen for the case where the observed person has a damaged neural structure causing that this associated feeling is not generated. In this case the observing person can feel the other person's state, whereas the person himself does not feel it. For example, if such a person believes he has won a lottery, he may not feel happiness about it, whereas an observing agent may mirror such a belief and based on that may generate the accompanying feeling of happiness.

In some more detail, emotion generation via a body loop roughly proceeds according to the following causal chain; see Damasio (1999, 2004):

having a mental state → preparation for the induced bodily response →
induced bodily response → sensing the induced bodily response →
sensory representation of the induced bodily response → induced feeling

As a variation, an 'as if body loop' uses a direct causal relation

preparation for the induced bodily response →
sensory representation of the induced bodily response

as a shortcut in the causal chain. In the model used here an essential addition is that the body loop (or as if body loop) is extended to a recursive body loop (or recursive as if body loop) by assuming that the preparation of the bodily response is also affected by the state of feeling the emotion:

feeling → preparation for the bodily response

as an additional causal relation. Such recursiveness is also assumed by Damasio (2004), as he notices that what is felt by sensing is actually a body state which is an internal object, under control of the person:

> 'The brain has a direct means to respond to the object as feelings unfold because the object at the origin is inside the body, rather than external to it. The brain can act directly on the very object it is perceiving. It can do so by modifying the state of the object, or by altering the transmission of signals from it. The object at the origin on the one hand, and the brain map of that object on the other, can influence each other in a sort of reverberative process that is not to be found, for example, in the perception of an external object.' (...)
>
> 'In other words, feelings are not a passive perception or a flash in time, especially not in the case of feelings of joy and sorrow. For a while after an occasion of such feelings begins – for seconds or for minutes – there is a dynamic engagement of the body, almost certainly in a repeated fashion, and a subsequent dynamic variation of the perception. We perceive a series of transitions. We sense an interplay, a give and take.' (Damasio, 2004, pp. 91-92)

Thus the obtained model is based on reciprocal causation relations between emotion felt and body states, as roughly shown in Figure 1.

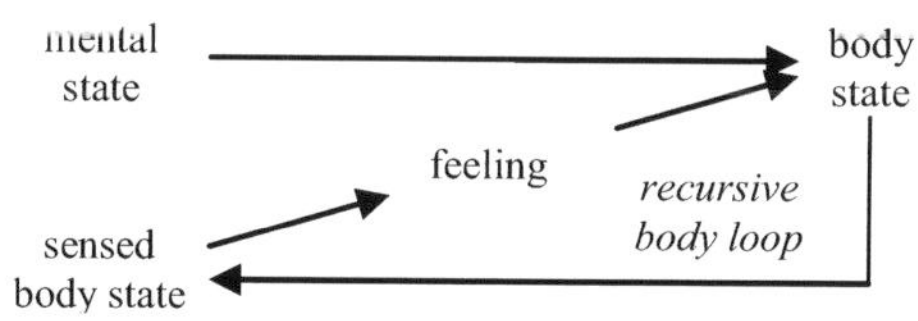


**Fig. 1.** Recursive body loop induced by a given mental state

Both the bodily response and the feeling are assigned a level or gradation, expressed by a number, which is assumed dynamic; for example, the strength of a smile and the extent of happiness. The causal cycle is modelled as a positive feedback loop, triggered by the (any) mental state and converging to a certain level of feeling and body state. Here in each round of the cycle the next body state has a level that is affected by both the mental state and the level of the feeling state, and the next level of the feeling is based on the level of the body state.

## 4 Description of the Designed Agent Model

The design of the mindreading capability for the empathic agent was based on the Simulation Theory perspective; cf. (Goldman, 2006). According to this perspective mindreading is performed by the observing agent by activating the same own mental states as the observed agent, thereby using similar mechanisms as those used by the observed agent. Therefore, a first step is the design of the basic mechanisms to generate a mental state (here en belief state was chosen), and to generate the associated feelings. These basic mechanisms will be used by both agents.

In the description of the detailed agent model the temporal relation a $\twoheadrightarrow$ b denotes that when a state property a occurs, then after a certain time delay (which for each relation instance can be specified as any positive real number), state property b will occur. In this language (called LEADSTO) both logical and numerical calculations can be specified, and a dedicated software environment is available to support specification and simulation; for details see (Bosse, Jonker, Meij & Treur, 2007).

A lottery scenario is used to illustrate the model. Agent A observes both his own lot number and the winning number and creates the corresponding beliefs; as the number in both beliefs is the same, from these the belief that the lottery was won is generated, which leads to an associated feeling of happiness. By communication agent B hears from agent A about the own lot number and the winning number. From this he simulates the process in agent A thus entering an empathic understanding process in which she generates both the belief about the lottery won and the associated feeling. For an overview of the model for agent A, see Figure 2. An overview of the model of agent B is depicted in Figure 3. These pictures also show representations from the detailed specifications explained below.

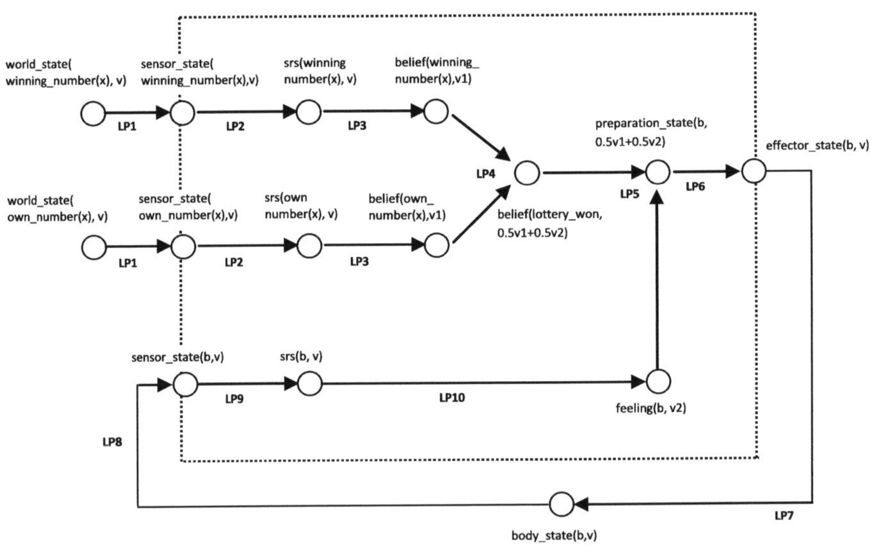


**Fig. 2.** Overview of the agent model for the observed agent A

The detailed specification (both informally and formally) of the agent model for empathic understanding is presented below. Here capitals are used for (assumed universally quantified) variables, e.g. 'B', whereas small letters represents an instance of that variable, e.g. 'b'. All aspects have been formalized numerically by numbers in the interval [0, 1]. First the part is presented that describes the basic mechanisms to generate a belief state (on winning the lottery) and the associated feeling (of happiness). These are used by both agents. The first dynamic property addresses how properties of the world state can be sensed.

**LP1 Sensing a world state**
If world state property *W* occurs of level *V*
then a sensor state for *W* of level *V* will occur.
world_state(W, V) ↠ sensor_state(W, V)

This dynamic property is used by agent A to observe both the own number and the winning number (see Figure 2); to this end the variable W is instantiated by own_number(x) and winning_number(x). Note that communications are also considered world facts; LP1 is used by agent B by instantiating W for communications indicated as communicated_by_to(I, agentA, agentB). From this sensory representations and beliefs are generated according to the next two dynamic properties LP2 and LP3. Note that also for these the variable W is instantiated as before.

**LP2 Generating a sensory representation for a sensed world state**
If a sensor state for world state property *W* with level *V* occurs,
then a sensory representation for *W* with level *V* will occur.
sensor_state(W, V) ↠ srs(W, V)

**LP3 Generating a belief state for a sensory representation**
If a sensory representation for *W* with level *V* occurs,
then a belief for *W* with level *V* will occur.
srs(W, V) ↠ belief(W, V)

Dynamic property LP4 describes how the belief is generated that the lottery was won.

**LP4 Generating a belief on winning the lottery**
If a belief with level *V1* occurs that *X* is the main price winning number of the lottery
and a belief with level *V2* occurs that *X* is the number of the own lot
then a belief with level *0.5V1+0.5V2* will occur that the main price of the lottery was won
belief(winning_number(X), V1) & belief(own_number(X), V2) ↠
belief(lottery_won, 0.5V1+0.5V2)

The emotional response to this belief is the preparation for a specific bodily reaction b, as expressed in dynamic property LP5.

**LP5 From belief that lottery was won and feeling to preparation of a specific body state**
If a belief that the lottery was won with level *V1* occurs and feeling body state b has level *V2*,
then preparation state for body state *b* will occur with level *0.5V1+0.5V2*.
belief(lottery_won, V1) & feeling(b, V2) ↠ preparation_state(b, 0.5V1+0.5V2)

Dynamic properties LP6 to LP10 describe the body loop.

**LP6 From preparation to effector state for body modification**
If preparation state for body state *B* occurs with level *V*,
then the effector state for body state *B* with level *V* will occur.
preparation_state(B, V) ↠ effector_state(B, V)

**LP7 From effector state to modified body**
If the effector state for body state *B* with level *V* occurs,
then the body state *B* with level *V* will occur.
effector_state(B, V) ↠ body_state(B, V)

**LP8 Sensing a body state**
If body state *B* with level *V* occurs,
then this body state *B* with level *V* will be sensed.

body_state(B, V) →→ sensor_state(B, V)

**LP9 Generating a sensory representation of a body state**
If body state *B* with level *V* is sensed,
then a sensory representation for body state *B* with level *V* will occur.

sensor_state(B, V) →→ srs(B, V)

**LP10 From sensory representation of body state to feeling**
If a sensory representation for body state *B* with level *V* occurs,
then *B* is felt with level *V*.

srs(B, V) →→ feeling(B, V)

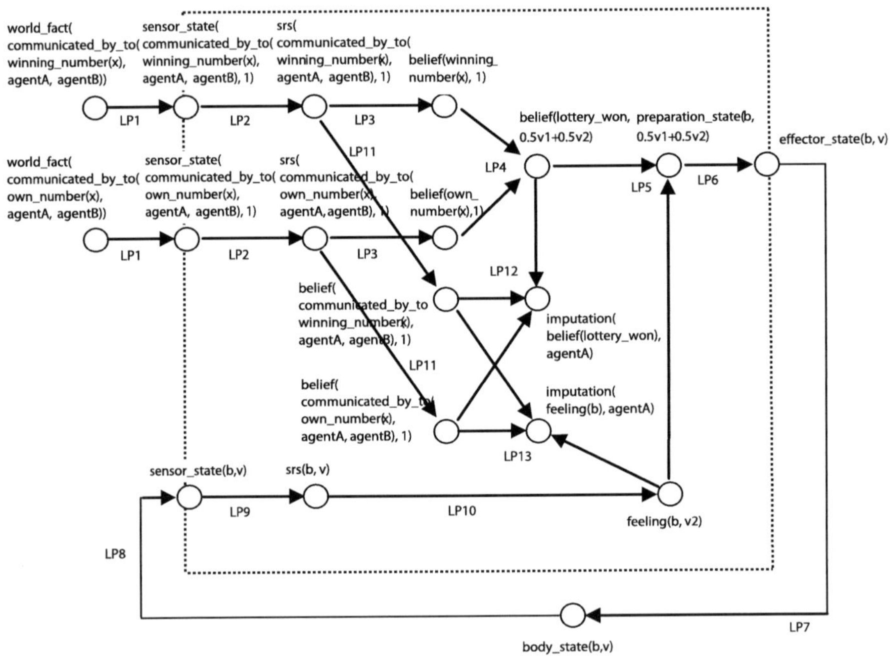


**Fig. 3.** Overview of the agent model for the observing agent B

Above the part of the model was shown that is used by both the observed and observing agent. Next the part of the model is discussed that is particularly involved in the empathic understanding process. This part of the model is used within the observing agent; see Figure 3.

First the communication from the other agent is related to the own beliefs.

**LP11 Affecting own beliefs by communicated information**
If in agent B a sensory representation with level *V* occurs that agent A communicated world fact *W*,
then a belief with level *V* for this world fact will occur.

srs(communicated_by_to (W, agentA, agentB), V) →→ belief(W, V)

Next it is shown how the imputation process takes place for a belief. Here, *th* is a (constant) threshold for imputation. In the simulations shown, th is assumed 0.95 as an example.

**LP12 Imputation of a belief**
If a belief that the lottery was won with level $V1 \geq th$ occurs
and a belief occurs with level $V2 \geq th$ that the own number was communicated by agentA,
and a belief occurs with level $V3 \geq th$ that the winning number was communicated by agentA,
then the belief that the lottery was won will imputed.

```
belief(lottery_won, V1) &
belief(communicated_by_to(own_number(X1), agentA, agentB), V2) &
belief(communicated_by_to(winning_number(X2), agentA, agentB), V3) &
V1≥th & V2≥th & V3≥th
→ imputation(belief(lottery_won), agentA)
```

Finally, feelings are imputed in the following manner.

**LP13 Imputation of a feeling**
If a certain body state *B* is felt, with level $V1 \geq th$
and a belief occurs with level $V2 \geq th$ that the own number was communicated by agentA,
and a belief occurs with level $V3 \geq th$ that the winning number was communicated by agentA,
then feeling *B* will imputed.

```
feeling(B, V1) &
belief(communicated_by_to(own_number(X1), agentA, agentB), V2) &
belief(communicated_by_to(winning_number(X2), agentA, agentB), V3) &
V1≥th & V2≥th & V3≥th
→ imputation(feeling(B), agentA)
```

## 5 Example Simulation Results

Based on the model described in the previous section, a number of simulations have been performed. Some example simulation traces are included in this section as an illustration; see Figure 4 and Figure 5 (here the time delays within the temporal LEADSTO relations were taken 1 time unit). In all of these figures, where time is on the horizontal axis, the upper part shows the time periods, in which the binary logical state properties hold (indicated by the dark lines); for example,

```
world_state(winning_number(X), 1)
belief(lottery_won, 1.0)
imputation(feeling(b), agentA)
```

Below this part, quantitative information for the other state properties values for the different time periods are shown (by the dark lines). For example, in Figure 4, the preparation state for b has value 0.5 at time point 6 which increased to 0.75 at time point 12 and so forth. The graphs show how the recursive body loop approximates a state for feeling with value 1. Notice that in all lower 6 traces i.e. from preparation state to feeling state, the states are activated based on temporal delay between them, as depicted in Figure 2, i.e. preparation state has activation level '0' at time point 0, the successor state effector state has activation level '0' at time point 1 and so on.

Figure 4 shows the simulation for the observed agent based on the basic mechanisms to generate a belief state and to generate the associated feeling as described in the previous section (from LP1 to LP10). As shown in Figure 4 (upper part), the observed agent A notices his own number and the winning number from the world state, shown by the state properties

sensor_state(own_number(X), 1)

and

sensor_state(winning_number(X), 1)

respectively. It then generates the belief that he has won the lottery by comparing the two numbers shown by the state property

belief(lottery_won(X), 1.0)

The lower part of Figure 4 shows the values of the various activation levels over time. Here it is shown that the recursive body loop results in an approximation of convergent activation levels for the states that relate to the feeling and the body state, among others.

Figure 5 shows a simulation trace for the observing agent, depicting the empathic understanding process described in the previous section (in particular using LP11 to LP13, but also using LP1 to LP10 for the underlying basic mechanism). Here it is shown (in the upper part of the Figure 5) that agent A (observed agent) communicates his own number and winning number to the agent B (observing agent), shown by the state properties

sensor_state(communicated_by_to
(own_number(X), agentA, agentB), 1)

and

sensor_state(communicated_by_to
(winning_number(X), agentA, agentB), 1)

respectively. Stepping in the shoes of agent A, then agent B (the observing agent) generates its own beliefs about the lot numbers and about winning lottery belief (which mirror the beliefs of agent A), as shown by the state property

belief(lottery_won, 1.0).

Later agent B imputes this belief (at time point 5) to agent A as shown by state property

imputation(belief(lottery_won), agentA).

As shown in the figure, after generating the associated feeling, agent B also imputes this feeling to agent A, shown by the state property

imputation(feeling(b), agentA)

at time point 35.

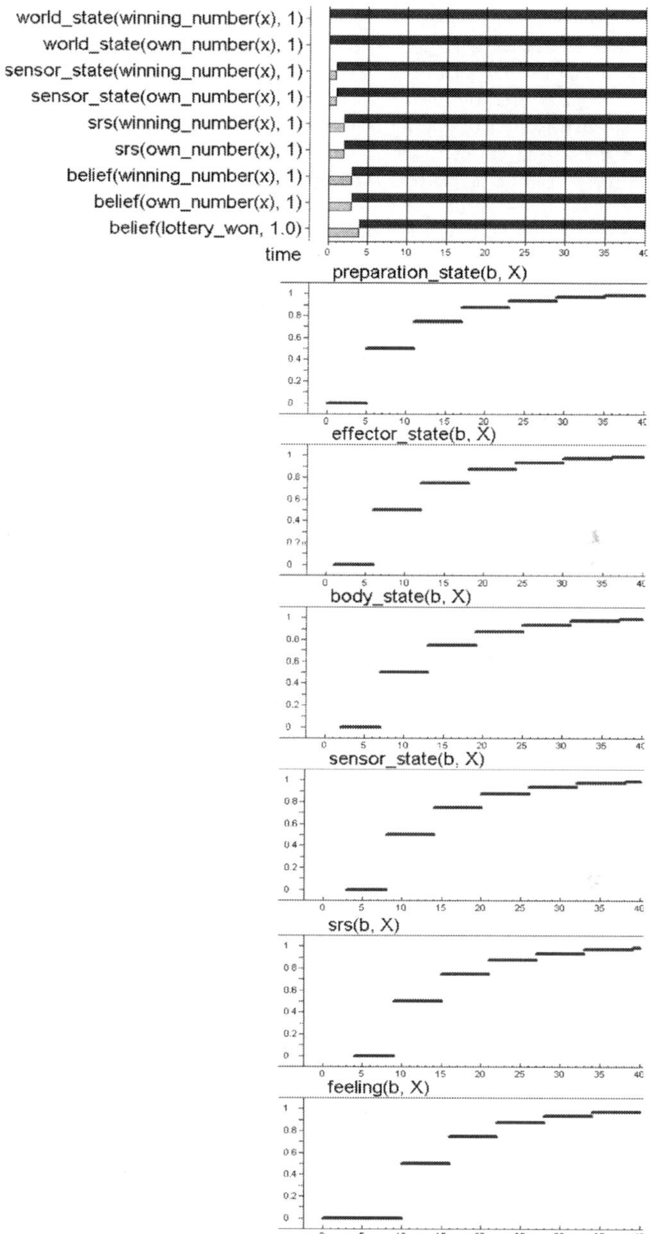


**Fig. 4.** Example simulation trace for the observed agent

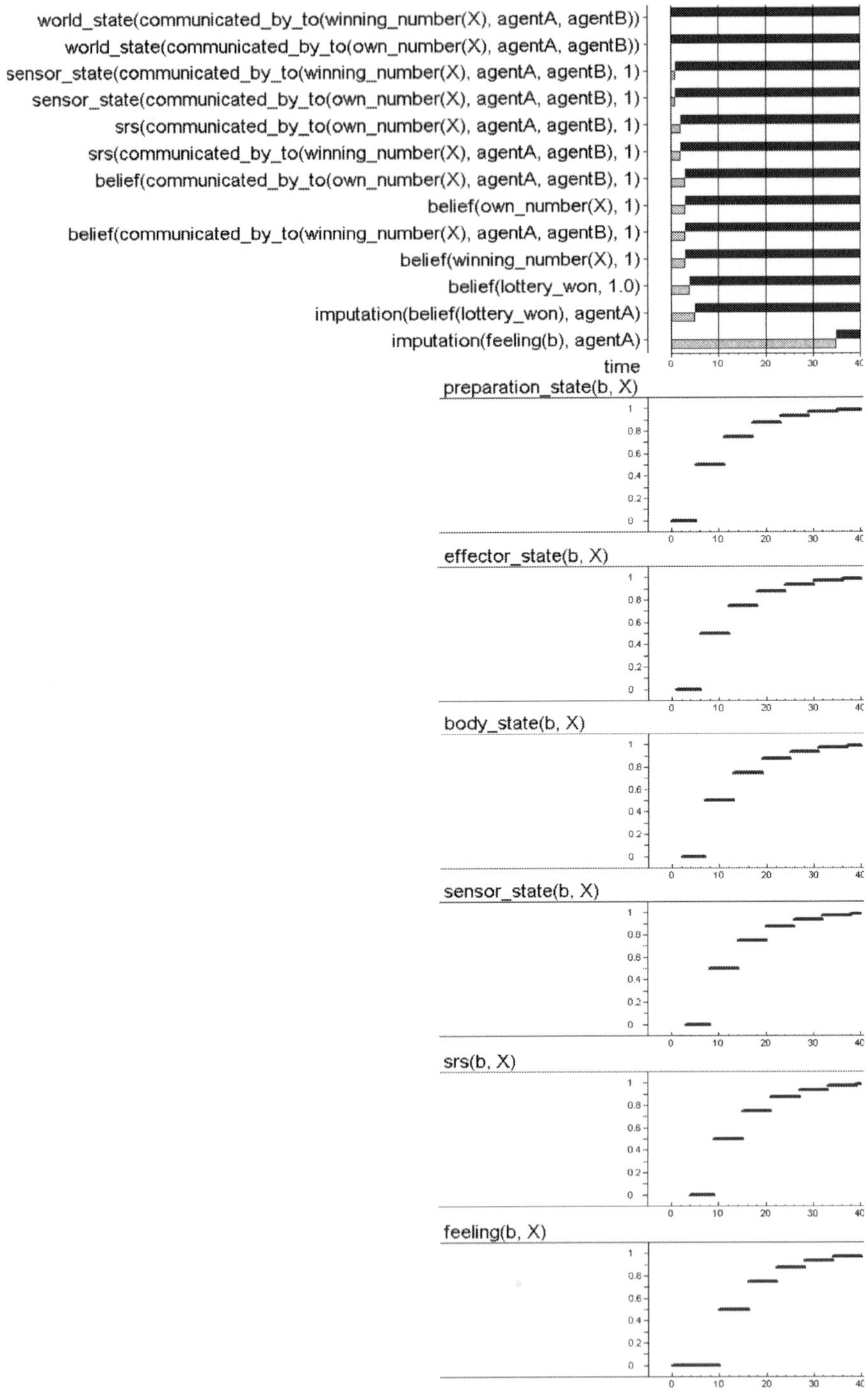


**Fig. 5.** Example simulation trace for the observing agent

## 6 Discussion

For an agent observing another agent, having an empathic understanding of the observed agent is considered a specific type of mindreading. Mindreading as such can focus on certain aspects such as emotion, desire, belief, intention, or attention states (e.g., Gärdenfors, 2003). A characteristic of an empathic response is that the response does not only include that the observing agent understands the mental state of the observed agent, but also feels the corresponding feeling. In this paper the design of an agent model was introduced that is capable of understanding other agents in an empathic way. The model describes how the empathic agent does not only understand another agent's mental state but at the same time feels the accompanying emotion. The proposed model is an extension of an earlier model described in (Bosse, Memon, and Treur, 2009) which does not focus on the more complex issue of empathic understanding, but only on reading another person's emotions. The model proposed in the current paper was based on two main assumptions:

(1) The observing agent performs mindreading using the same mental states as the observed agent
(2) Both agents have a similar mechanism to associate feelings to a given mental state

Concerning assumption (1), to obtain a form of mindreading for which the observing agent generates the same mental state, the Simulation Theory perspective was followed; cf. (Goldman, 2006). According to this perspective mindreading is performed by the observing agent in a simulative manner by activating the same mental states as the observed agent; see also (Hesslow, 2002). This assumption is recently getting more and more support by empirical results, for example, concerning the discovery of the mirror neuron system; e.g., (Rizzolatti and Craighero, 2004; Iacoboni, Molnar-Szakacs, Gallese, Buccino, Mazziotta, and Rizzolatti, 2005; Iacoboni, 2005, 2008; Pineda, 2009; Goldman, 2009).

Concerning assumption (2), to this end a computational model of Damasio (1999, 2004)'s informal theory about the generation of emotion and feeling was exploited. This theory assumes a neural mechanism that involves changes in an agent's sensed body state, triggered by a certain mental state. Assuming that the observed agent and the observing agent indeed have a similar mechanism for this, makes it possible that for a given mental state the observing agent generates the same feeling as the observed agent.

Especially in relation to assumption (2) it can be questioned to which extent the mechanisms to associate feelings to a given mental state are the same for both agents. As it may be considered plausible that basically the mechanisms are similar, it is not difficult to imagine that due to innate and learned individual differences, the empathic reaction may be limited in extent. Indeed, it is often reported that identical twins have a much higher level of mutual empathy than any two persons which are not identical twins. Moreover, it is also often considered that more empathy is shown between two persons when they have had similar experiences in life. Nevertheless, a certain extent of empathy still seems possible between persons which are not genetically identical and have not exactly the same experiences. It is an interesting challenge for future research to develop the introduced model for empathy further by introducing explicit parameters by which such individual differences can be expressed, and for which some notion of extent to which empathy occurs can be defined.

Other models described in the literature usually only address either emotion recognition, or recognition of another type of mental state, or feel another persons feeling; e.g., (Pantic and Rothkrantz, 2000; Goldman, 2006; Bosse, Memon, and Treur, 2007; Hatfield, Cacioppo, and Rapson, 1994). As far as the authors know the model proposed here is unique in the sense that it combines both understanding and feeling of another person's mental states, and takes into account the way in which (other) mental states induce feelings both for the observing and the observed person.

Future work will address a more extensive evaluation and assessment of the model and thereby will explore more variations, for example, of different scenarios with different extents of similarity between the persons, and different values of its parameters such as the threshold value and the weight factors, for example in the generation of the preparation and the belief which were now taken *0.95* and *0.5* respectively.

## References

1. Blair, R.J.R.: Responding to the emotions of others: Dissociating forms of empathy through the study of typical and psychiatric populations. Consciousness and Cognition 14, 698–718 (2005)
2. Bosse, T., Memon, Z.A., Treur, J.: An Adaptive Emotion Reading Model. In: Proc. of the 31st Annual Conference of the Cognitive Science Society, CogSci 2009, pp. 1006–1011. Cognitive Science Society, Austin (2009)
3. Bosse, T., Memon, Z.A., Treur, J.: Emergent Storylines Based on Autonomous Characters with Mindreading Capabilities. In: Lin, T.Y., Bradshaw, J.M., Klusch, M., Zhang, C., Broder, A., Ho, H. (eds.) Proc. of the Sixth Intern. Conf. on Intelligent Agent Technology, IAT 2007, pp. 207–214. IEEE Computer Society Press, Los Alamitos (2007)
4. Damasio, A.: The Feeling of What Happens. Body and Emotion in the Making of Consciousness. Harcourt Brace, New York (1999)
5. Damasio, A.: Looking for Spinoza. Vintage books, London (2004)
6. Decety, J., Jackson, P.L.: The functional architecture of human empathy. Behav. Cogn. Neurosci. Rev. 3, 71–100 (2004)
7. Eisenberg, N.: Emotion, regulation, and moral development. Annu. Rev. Psychol. 51, 665–697 (2000)
8. Gärdenfors, P.: How Homo became sapiens: on the evolution of thinking. Oxford University Press, Oxford (2003)
9. Goldman, A.I.: Simulating Minds: The Philosophy, Psychology, and Neuroscience of Mindreading. Oxford Univ. Press, New York (2006)
10. Goldman, A.I.: Ethics and cognitive science. Ethics 103, 337–360 (1993)
11. Goldman, A.I.: Mirroring, Mindreading, and Simulation. In: Pineda, J.A. (ed.) Mirror Neuron Systems: The Role of Mirroring Processes In Social Cognition. Humana Press Inc. (2009)
12. Hatfield, E., Cacioppo, J.T., Rapson, R.L.: Emotional contagion. Cambridge University Press, New York (1994)
13. Hesslow, G.: Conscious thought as simulation of behavior and perception. Trends Cogn. Sci. 6, 242–247 (2002)

14. Hoffman, M.L.: Empathy and Moral Development. Cambridge University Press, New York (2000)
15. Hoffman, M.L.: Development of prosocial motivation: empathy and guilt. In: Eisenberg, N. (ed.) The Development of Prosocial Behavior, pp. 281–313. Academic Press, New York (1982)
16. Iacoboni, M.: Mirroring People. Farrar, Straus & Giroux, New York (2008)
17. Iacoboni, M.: Understanding others: imitation, language, empathy. In: Hurley, S., Chater, N. (eds.) Perspectives on imitation: from cognitive neuroscience to social science, vol. 1, pp. 77–100. MIT Press, Cambridge (2005)
18. Iacoboni, M., Molnar-Szakacs, I., Gallese, V., Buccino, G., Mazziotta, J.C., Rizzolatti, G.: Grasping the intentions of others with one's own mirror neuron system. PLoS Biol. 3, e79 (2005)
19. Ickes, W.: Empathic Accuracy. The Guilford Press, New York (1997)
20. Ingram, R.E., Miranda, J., Segal, Z.V.: Cognitive vulnerability to depression. Guilford Press, New York (1998)
21. Lamm, C., Batson, C.D., Decety, J.: The neural basis of human empathy – effects of perspective-taking and cognitive appraisal. J. Cogn. Neurosci. 19, 42–58 (2007)
22. Lipps, T.: Einfühlung, innere Nachahmung und Organempfindung. Archiv für die gesamte Psychologie 1, 465–519 (1903)
23. Pantic, M., Rothkrantz, L.J.M.: Expert System for Automatic Analysis of Facial Expressions. Image and Vision Computing Journal 18, 881–905 (2000)
24. Pineda, J.A.: Mirror Neuron Systems: The Role of Mirroring Processes in Social Cognition. Humana Press Inc. (2009)
25. Preston, S.D., de Waal, F.B.M.: Empathy: its ultimate and proximate bases. Behav. Brain Sci. 25, 1–72 (2002)
26. Raine, A.: The Psychopathology of Crime: Criminal Behaviors as a Clinical Disorder. Guilford Publications, New York (1993)
27. Rizzolatti, G., Craighero, L.: The mirror-neuron system. Annu. Rev. Neurosci. 27, 169–192 (2004)

# Multi-agent Systems in Pedestrian Dynamics Modeling

Jarosław Wąs and Konrad Kułakowski

Institute of Automatics
AGH University of Science and Technology
al. Mickiewicza 30, 30-059 Kraków, Poland
jarek@agh.edu.pl, kkulak@agh.edu.pl

**Abstract.** The article presents the use of Multi-agent system (MAS) in pedestrian dynamics modeling. Current trends in pedestrian dynamics are presented with particular focus on a short review of various agent-based models of pedestrian dynamics. The models are briefly discussed and compared.

## 1 Introduction

Increasing interest in crowd behavior modeling is becoming more and more apparent in current years. Architects, civil engineers, safety engineers etc. all need realistic and effective tools for crowd modeling.

There is a close analogy between a crowd of people and a multi-agent system. In both cases the behavior of individuals (e.g. their attitude towards other agents, being cooperative or competitive, etc) has significant impact on the macroscopic view of the system. Because of that, the MAS method seems to be very promising in the area of pedestrian crowd modeling.

The main approaches to pedestrian dynamics can be divided into: macroscopic approach (using hydrodynamical equations [11]) and microscopic approach, where we take into account an individual behavior of particular pedestrians. The classification of microscopic models of pedestrian dynamics has a conventional character and *all of the microscopic models could be interpreted as MAS*! Thus, let us classify microscopic methods of crowd modeling in the most common way:

**Social Forces** method [8] is based on a physical method of Molecular Dynamics. In the Social Forces concept, several forces (physical terms) are essential: first, the term describing the acceleration towards the desired motion velocity; secondly, terms reflecting that a pedestrian keeps a certain distance from other pedestrians and borders; and thirdly, the term modeling attractive effects. In this method, we could observe self-organization of several collective effects of pedestrian behavior[1]. The disadvantage of Social Force lies in rather low computational effectiveness (the method is continuous in time and space)

[1] Althought the model is described by a set of differential equations, it is a microscopic model and can be also interpreted as MAS!

N.T. Nguyen, R. Kowalczyk, and S.-M. Chen (Eds.): ICCCI 2009, LNAI 5796, pp. 294–300, 2009.

**Cellular Automata** (Non-homogeneous) is a method, where space is represented by a lattice and each cell can be free or occupied by a pedestrian. Transition rules are executed in discrete time-steps, which results in pedestrian movement in the model. Generally, movement is realized according to a gradient of potential field generated by all targets [4], [10], [2], [6]. The method can be successfully used for evacuation situations as well as for freeway traffic modeling [16]. Using Cellular Automata for pedestrian dynamics modeling is very effective and easy to understanding. However, there is a limitation: since all individuals are represented homogeneously, more complex models may suffer from lack of flexibility.

**Multi-Agent Systems** is a method in which pedestrians are represented by autonomous, adaptive and decision-making entities (agents). Agents perceive information from environment and they can interact with each other. The method can be used for both: freeway traffic and evacuation modeling. It is a proper and promising method for systems having heterogeneous population with complex and nonlinear interactions among agents. Actually, the most frequently, the MAS method in pedestrian dynamics modeling is *derived from non-homogeneous CA*, by defining some *strategic abilities* of agents for instance target choosing etc.

## 2 Agent-Based Models

In agent-based modeling, a system is modeled as a collection of autonomous decision-making entities called agents. Each agent individually assesses its situation and makes decisions on the basis of a set of rules [3].

Abilities of particular agents can be divided into the following groups 1.

**Table 1.** Agent abilities in pedestrian dynamics

| Type | Level | Characteristic | Example |
|---|---|---|---|
| Operational | Skill-based | Automatic reaction | Avoiding obstacles |
| Tactical | Rule-based | Stereotypic reaction | Lane formation |
| Strategic | Knowledge-based | Decision making | Target choosing |

An important question is about the distinction between the CA and agent-based models. In this paper, it is assumed that as long as operational and tactical abilities of pedestrians are discussed, that CA based model is used. The agent-based model is assumed implicitly every time when strategic abilities of individuals are also important.

Let us take into consideration some recent MAS models dedicated for pedestrian dynamics:

### 2.1 Dijkstra-Timmermans-Jessurun Model (DTJ Model)

DTJ model was developed by Jan Dijkstra, Joran Jessurun and Harry Timmermans [5]. It is now a classical MAS model based on Cellular Automata.

In the model, all pedestrian facilities are modeled as a network of walkway sections (a lattice of cellular automaton) and pedestrian flow is presented as a queuing process (pedestrian occupies one cell on the lattice). Each pedestrian $p_n$ possesses a scenario, which includes for example behavior-, intentions- and activities-properties. A pedestrian $p_n$ is supposed to carry out a set of activities (purchasing a set of goods, becoming involved in window-shopping etc). In the model, the completion of an activity is a key decision point impacting agent activity agenda and time alloted to further (not completed yet) activities. The node of the network where pedestrians may decide to take another route, changing the anticipated duration, is also connected with a decision point [5]. Thus, the pedestrian takes into account not only the shortest route to target, but also *the most attractive* one.

### 2.2 SBBPedis

This model was proposed in [9] for modeling pedestrian traffic in Bern railway station. In this approach, space representation is discrete and agents represent pedestrians.

During generation of an agent, targets, preferred velocity and route are chosen. An agent is also equipped with sensors which make it possible to adapt its route dynamically. The factor requiring route adaptation include static (columns, seats, etc.) and dynamic (other agents) obstacles. Targets (defined in the model as trains) are represented as a set of doors, which can be chosen by an agent.

The model was projected as a very large-scale agent based pedestrian simulation. It was tested with a population exceeding 40 000 agents.

### 2.3 Situated Cellular Agents (SCA)

SCA model is based on Cellular Automata [1]. Specialized representation of the model for crowd modeling (SCA4crowd) was presented in [12]. A space in the model is represented as a non-directed graph of sites, where graph nodes represent possible space locations and graph edges represent possible movement directions. Simulated pedestrians are represented as agents and placed in the nodes of the graph. Pedestrian attitude is represented by the agents internal state.

In the model, the environment has an explicit structure, which is only locally known to agents. Local interaction can occur between adjacent agents to allow coordination of autonomous behaviors (for instance to solve local deadlocks through site-exchange). Distance interactions between agents are possible only indirectly, through the emission of signals (propagation of fields with the possibility of perceiving by other agents). The agent's position in the environment strictly determines perception, interactions and behavior of the particular agent.

### 2.4 Dynamics Navigation

The model was proposed by Teknomo and Millonig [13]. The representation of time and space in the model is discrete. Contrary to CA models and most MAS models (derived from CA), a cell in dynamics navigation can be occupied by any number of pedestrians (the maximum number is defined according to density parameter) 1. All interactions among agents are determined by two functions. On the one hand the first function defines the probability of agent's move towards the given cell (in Moore neighborhood), on the other hand the second function defines relation between speed and density - fundamental diagram (the speed is inversely proportional to the density of pedestrians in cell).

Each agent in the model possesses a set of properties: speed, trajectory, timing, etc. The crucial issue is defining trajectory according to a normalized displacement vector function called a navigation matrix.

**Fig. 1.** Running simulation - static environment (figure from [13])

### 2.5 Social Distances Model

The model was presented by Wąs, Gudowski and Matuszyk in [14] and [15]. It is based on a sociological theory of Social Distances by E.T. Hall (four types of distances-zones among people). The space in the model is represented as a square lattice, while pedestrians are represented as ellipses 2. The center of a cell and centers of an ellipse coincide each other.

The behavior of particular agent depends on an actual state. In a passive state, agents ("observers") are under the influence of other agents in any active states ("intruders"). Depending on their actual state pedestrians proceed according to certain movement algorithms, waiting for an event or action or moving attracted by their aims. The model also includes familiar behaviors (suitable for representing the dynamics of e.g. parents and children or groups of friends).

### 2.6 Alpsim

The Alpsim project was designed as a virtual alpine landscapes planning environment with autonomous agents [7]. In this model every agent is able to move freely, adapt to the environment and make decisions (based on perception and communication with other agents). The area covers 150 square kilometers, and more then a thousand tourists are simulated. Space in the model is represented

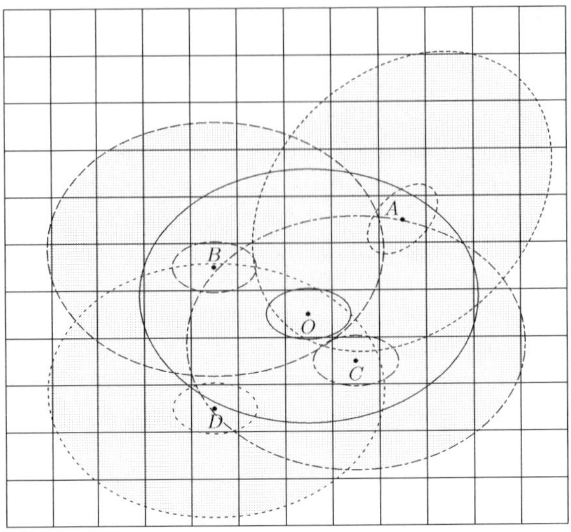


**Fig. 2.** Social distances areas (fig from [14])

as a graph. Agents (tourists) are placed in the nodes of the graph. These agents are given "plans" and they are introduced into the simulation with no "knowledge" of the environment. The agents execute their plans, receiving feedback from the environment as they move throughout the landscape (afterwards the actions are compared to their expectations).

### 2.7 Comparison of Certain Features of Agent-Based Pedestrian Dynamics Models

In this chapter we sum up certain features of MAS models. All the models use discrete representation of space. In table 2, the characteristics of some pedestrians agent-based models are presented.

## 3 Discussion

Multi-agent systems provide a good conceptual platform for complex crowd simulation problems. Thus, usually discrete formalism in the form of "if-then" rules or nonlinear coupling is more useful than continuous representation e.g. differential equations. Another advantage of agent-based simulations is possibility the inclusion learning and adaptation behaviors into the model.

From the perspective of pedestrian dynamics modeling, the MAS systems might be perceived as a convenient simulation method which is easy to implement but difficult to design. The main difficulties originate from the fact that all the interactions among agents, complex and non-linear behaviors, as well as

**Table 2.** Characteristics of chosen agent based-models dedicated for pedestrian dynamics

| Model Feature | DTJ | SBBPedis | Dynamic Navigation | SCA+ | Social Distances | Alpsim |
|---|---|---|---|---|---|---|
| Discrete spatial representation | ✓ | +/− | ✓ | ✓ | ✓ | ✓ |
| Discrete time representation | ✓ | ✓ | ✓ | ✓ | ✓ | ✓ |
| Primary example of environment | Shopping center | Railway station | Building | Building | Tram | Alps |
| Agent 2D representation | square | square | a cell occupancy | a node occupancy | ellipse | a node occupancy |
| Model scale | v.large | large | v.large | large | medium | v.large |
| Collective behavior | low | low | low | high | medium | medium |

learning and adaptation processes have to be taken into account. All of the models have to be precisely and carefully calibrated and validated using qualitative and quantitative methods 3. For evacuation issues the norm ISO TR 13387 regulates the most important aspects of validation. The majority of these aspects (pointed in the norm) is also helpful for validating freeway pedestrian traffic models. It is also self-evident in all causes, expert knowledge is crucial in the validation process.

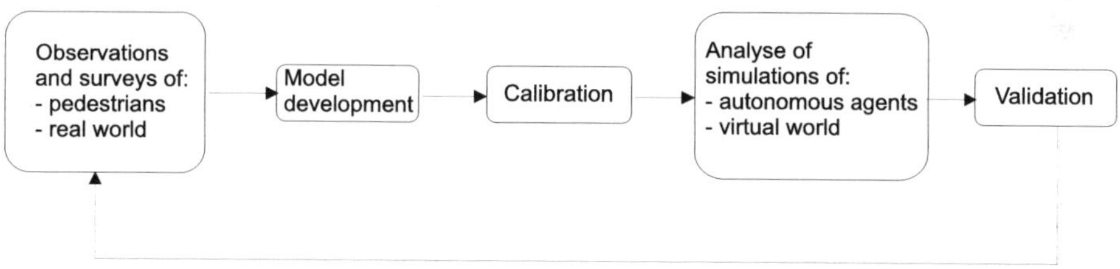


**Fig. 3.** Stages of creating of agent-based models dedicated for pedestrians behavior

The crucial problem nowadays is the efficiency of agent-based methods. It is important to develop realistic but effective simulation. The article shows that non-homogeneous cellular automata offer a very promising perspective for the development of agent-based models as it equips pedestrians with some strategic abilities and some "intelligent" functionalities. In the authors opinion, agent-based modeling based on Cellular Automata will be the most dominant approach in pedestrian dynamics modeling.

## Acknowledgment

This research is financed by the Polish Ministry of Science and Higher Education, Project no: N N516 228735.

## References

1. Bandini, S., Federici, M.L., Vizzari, G.: Situated Cellular Agents Approach to Crowd Modeling and Simulation. Cybernetics and Systems 38(7), 729–753 (2007)
2. Blue V., Adler J.: Cellular Automata Microsimulation for Modeling Bi-Directional Pedestrian Walkways. Transportation Research Journal Part B 35(3), 293-31 (2001)
3. Bonabeau, E.: Agent-based modeling: Methods and techniques for simulating human systems. PNAS 99 (suppl.) (2002)
4. Burstedde, C., Klauck, K., Schadschneider, A., Zittartz, J.: Simulation of pedestrian dynamics using a 2-dimensional cellular automation. Physica A 295(4), 507–525 (2001)
5. Dijkstra J., Timmermans H.J.P., Jessurun J.: A Multi-Agent Cellular Automata System for Visualising Simulated Pedestrian Activity, in: Proceedings of ACRI Theoretical and Practical Issues on Cellular Automata LNCS (2000)
6. Dudek–Dyduch, E., Wąs, J.: Knowledge representation of pedestrian dynamics in crowd: Formalism of cellular automata. In: Rutkowski, L., Tadeusiewicz, R., Zadeh, L.A., Żurada, J.M. (eds.) ICAISC 2006. LNCS (LNAI), vol. 4029, pp. 1101–1110. Springer, Heidelberg (2006)
7. Gloor, C., Stucki, P., Nagel, K.: Hybrid techniques for pedestrian simulations. In: Proceedings of 4th Swiss Transport Research Conference (2004)
8. Helbing, D., Molnar, P.: Social Force Model for Pedestrian Dynamics. Physical Review E 51 (1995)
9. Klügl, F., Rindsfüser, G.: Large-scale Agent-based Pedestrian Simulation, in: Proceedings of 5th German Conference on Multiagent Systems Technologies LNCS (LNAI) Springer (2007)
10. Klüpfel, H.: The simulation of crowds at very large events, In: Traffic and Granular, Flow 2005. Springer, Berlin (2006)
11. Pauls, J.: Movement of People. DiNenno Washington (1995)
12. Redaelli, S.: Cellular Automata and Situated Cellular Agents for Crowd Dynamics Modeling and Simulation, PhD Thesis, University of Milano-Bicocca (2008)
13. Teknomo, K., Millonig, A.: A navigation algorithm for pedestrian simulation in dynamic environments. In: Proceedings of 11th conference on Transport Research (2007)
14. Wąs, J., Gudowski, B., Matuszyk, P.J.: Social distances model of pedestrian dynamics. In: El Yacoubi, S., Chopard, B., Bandini, S. (eds.) ACRI 2006. LNCS, vol. 4173, pp. 492–501. Springer, Heidelberg (2006)
15. Wąs, J.: Multi-agent frame of social distances model. In: Umeo, H., Morishita, S., Nishinari, K., Komatsuzaki, T., Bandini, S. (eds.) ACRI 2008. LNCS, vol. 5191, pp. 567–570. Springer, Heidelberg (2008)
16. Was, J.: Cellular Automata Model of Pedestrian Dynamics for Normal and Evacuation Conditions. In: Proceedings of Intelligent Systems Design and Applications, Wroclaw, pp. 154–159. IEEE CS, Washington (2005)

# Towards a Model for Extraction of Possible Worlds and Accessibility Relation from Cognitive Agent's Experience

Grzegorz Skorupa and Radosław Katarzyniak

Wroclaw University of Technology, Institute of Informatics,
Wybrzeże Wyspiańskiego 27, 50-370 Wrocław, Poland
{grzegorz.skorupa,radoslaw.katarzyniak}@pwr.wroc.pl

**Abstract.** In this paper a problem of finding possible worlds based on agent's partial knowledge gathered from environment observations is defined. Paper marks a direction of research into the ways in which data mining techniques can be used to extract possible worlds and accessibility relation within minds of cognitive agents. The problem is considered in the context of robotics. Problem also connects to recent research trend of merging data mining to agents. Introductory and informal solution is proposed.

**Keywords:** possible worlds, accessibility relation, grounding, cognitive agent.

## 1 Introduction

Since a few years agents and multi-agent systems have become one of the major research directions in computer science. One of the key factors of such popularity are basic assumptions about an agent such as its reactivity, pro-activeness, intelligence and learning abilities [8,14]. Ironically the same features are questioned by critics, because existing methods, related to these features, can't fulfil much of the assumed needs.

If agent systems are to become major breakthrough in computer science and artificial intelligence more general reasoning, predicting and planning methods are required. The need for improvements is crucial especially in agent based robots where agents are considered to play the role of a reasoning mechanism of a robot. Often robots have to be placed in partially known environment and hence must learn its nature in order to meet their design aims. Agents have to be able to discover interesting dependencies in an environment behaviour, adapt to its changes and learn to make proper predictions.

Ability to connect reasoning and discovery mechanisms to agent's experience seems to be crucial in order to ensure correctness while maintaining its generality. Creating such connection between experience and symbols used to express knowledge gives true meaning to the knowledge and is called grounding [6].

In this paper we are addressing *a problem of creating grounded possibilities based only on agent's past interactions with an environment*. In simple words: we are searching for a general method that will allow agent to predict and reasonably back up what situations were, are and will be possible based on what she knows. The searched truly possible situations called accessible possible worlds can be later used as an input

N.T. Nguyen, R. Kowalczyk, and S.-M. Chen (Eds.): ICCCI 2009, LNAI 5796, pp. 301–308, 2009.

for reasoning, prediction and planning. What is strongly stressed in this approach is the autonomy of an agent and encapsulation of possible worlds and possible states. We show that the problem raises many nontrivial questions and doubts even for a very simple environment model.

We are proposing an initial solution that uses data mining methods to discover rules hidden within the environment. Agent is supposed to discover some dependencies among its previous experiences using data mining methods and later obey these dependencies when creating accessible possible worlds. Because we are planning to use data mining, our solution naturally meets the recent trend in joining data mining with agency [5,13], that tends to increase the abilities of agents and allow applicability of agent systems to new domains.

In section 2 an agent's environment is described. Section 3 introduces agent knowledge about the environment. In section 4 a problem of evaluating accessibility of the worlds is defined. Sections 5 and 6 contain a brief description of proposed solution.

## 2 Assumptions about the Environment

The environment agent lives in is assumed to be a tuple:

$$<O,P,T,F>$$

Where $O$ is a fixed, finite and non-empty set of objects. $P=\{P_1, P_2,...,P_K\}$ is a finite and fixed set of binary properties and $T$ is a set of time moments. We assume that $T$ is an ordered set according to $\prec_t$ relation. The intuitive meaning of $t_1 \prec_t t_2$ is that time moment $t_1$ happens before moment $t_2$. For simplicity we shall assume that time moments are consecutive natural numbers, so $T = \mathbb{N}$. The $F$ is a function from properties and time moments to power sets of objects ($F: P \times T \rightarrow 2^O$). $F$ can be interpreted as a world state function that at given time moment assigns objects that currently can be said to have given property.

We define *actual state* at moment $t$ as a:

$$\check{s}(t) =< \check{P}_1(t), \check{P}_2(t), \dots, \check{P}_K(t) >, \text{ where } \check{P}_i(t) = F(P_i, t) \subseteq O \quad (1)$$

If some object $o \in O$ does not belong to $P_i(t)$, it is assumed the object does not have given property. The actual state holds all the information about considered objects properties at some time moment.

The sequence of actual states over time is called *actual world*:

$$\check{S} = (\check{s}(1), \check{s}(2), \dots) \quad (2)$$

The actual world represents all the states of the environment ordered according to time. Both actual worlds and actual states form the real history of environment and describe its real behaviour and objects' properties. For further purposes we shall also need definitions of possible state and possible environment that represent some hypothetical (not necessarily real) history of the environment. The *possible state* is similar to actual state:

$$s(t) =< P_1(t), P_2(t), \dots, P_K(t) >, \text{ where } P_i(t) \subseteq O \quad (3)$$

Let us notice that $P_i(t)$, differently than $\check{P}_i(t)$, does not have to be equal to $F(P_i, t)$. $P_i(t)$ is simply some subset of all objects set $O$.

Similarly to actual world we define *possible world* as some hypothetical history of the environment:

$$S = (s(1), s(2), \dots) \tag{4}$$

By $\mathbb{S}$ we shall denote a set of all possible worlds, hence $S \in \mathbb{S}$ and i.e. $\check{S} \in \mathbb{S}$.

The finite subsequence from time moment 1 until time moment $t$ shall be called as *possible world till moment t* and denoted by:

$$S(t) = (s(1), s(2), \dots, s(t)) \tag{5}$$

This terminology is not coincidentally the same as the one used for modal logics, where possible worlds are used to formally define modalities and their satisfaction relation. We see the possible world as a limit of cognitive activities of the cognitive agent.

## 3 What an Agent Knows

We assume the agent is not omnipotent. She has no ability to observe all the objects and determine their properties at each time moment. Hence the agent does not know what is the actual possible world (eq. (2)). The agent has only partial knowledge about actual states. In other words, there are some objects' properties that agent has not observed, and hence does not know them. The information about some actual state at time moment $t$ shall be called *base profile* at time moment $t$ and defined as a tuple:

$$b(t) = < P_1^+(t), P_1^-(t), P_2^+(t), P_2^-, \dots, P_K^+(t), P_K^+(t) > \tag{6}$$

Each $P_k^+(t)$ and $P_k^-(t)$ is some subset of all objects set $O$. Agent has observed that some object $o \in O$ has property $P_K$ at moment $t$ when $o \in P_k^+(t)$. If $o \in P_k^-(t)$ the agent has observed that object lacks given property. If $o \notin P_k^+(t) \cup P_k^-(t)$ the agent has not observed the state of object's property. It is assumed that that base profiles are autonomously created and encapsulated in the body of cognitive agent.

It is assumed base profile meets following requirements:

$$P_k^+(t) \cup P_k^-(t) \subseteq O\,, \quad k=1,2,\dots,K \tag{7}$$

$$P_k^+(t) \cap P_k^-(t) = \emptyset\,, \quad k=1,2,\dots,K \tag{8}$$

$$P_k^+(t) \subseteq \check{P}_k(t) \quad \text{and} \quad P_k^-(t) \cap \check{P}_k(t) = \emptyset\,, \quad k=1,2,\dots,K \tag{9}$$

Intuitively condition 7 means that agent has only partial knowledge about object properties. Condition 8 means that no object can have some property and lack it at the same time. The condition 9 tells that none of the observed properties is contradictory to actual state – the base profile has no misjudged properties.

The sequence of time ordered base profiles gathered by agent until time $t$ shall be called *knowledge state at time t* and denoted by:

$$B(t) = (b(1), b(2), \ldots, b(t)) \tag{10}$$

When there is no confusion we shall simply use term: knowledge state.

## 4 Possible Worlds Accessibility – The Problem

In modal logics Kripke possible worlds [9] are used to formally define the satisfaction relation of modal sentences of the form "I (agent) know that $\varphi$" where $\varphi$ is the sentence in classical binary logic. Accessibility relation $R \subseteq \mathbb{S} \times \mathbb{S}$ between possible worlds is used to determine the satisfaction of such statements. We say two worlds $S_1$ and $S_2$ are accessible iff $(S_1, S_2) \in R$. In agent systems this relation depends on the knowledge of an agent. Intuitively two worlds are accessible for an agent if, based on what she knows, they are both really possible and not purely hypothetical (impossible). The accessibility relation changes with time due to new observations that agent gathers from the world. For example suppose that an agent does not know the colour of an apple. Hence the apple can be green in one accessible possible world and red in the other. If agent finds out that the apple is red, the world where the apple was assumed to be green is no longer accessible. In terms of agent we can define accessible possible worlds as a set:

$$\overline{\mathbb{S}} = \{S: (\check{S}, S) \in R\} \tag{11}$$

where each accessible possible world is accessible from actual world, based on what agent knows. Further we shall assume that accessibility relation is the equivalence relation[1].

It is usually assumed that accessibility relation is given a priori as an external knowledge [3,7,10] and can't be changed by the agent as in. This implies that this relation has to be somehow created in advance by an expert, when designing an agent. Even for simple environments (as the one assumed above) the relation is extremely complicated because it defines connections between whole worlds that hold information about all the states among all time span. To effectively define the relation one can use some rules that in consequence define what are the possible combinations of properties and objects within a state and among time (between successive states). Still such solution forces the participation of an expert.

The most general problem addressed in this work is: *When it is possible for an agent to autonomously evaluate the accessibility relation based on a knowledge state.*

It is quite clear that such task can't be solved in the most general case because the cognitive agent simply does not possess enough empirical knowledge. The agent has one life and lives in only one (actual) world. The agent can't gather sufficient information from many runs by restarting its existence. If the run of the world is not is not driven by some quite simple rules or disallows their discovery from own experience, one is unable to correctly predict other possible worlds, because only one world is

[1] Assumption about equivalence is common in literature [3, 11].

available. The intuitive fact given above has been formally proven. Statement seems quite clear and we omit the formal proof, we can only mark that the proof can be easily done by a simple example of an environment.

The simplifying idea is not to generate the whole possible worlds that include whole history and future, but to generate possible states based on interactions from the past. The generated possible states from different time moments can be then used to at least partially generate the possible worlds.

In the problem definition we used the term *evaluate* because of reasons given above. By evaluating we mean using some mental process happening within an agent to include real possibilities and exclude merely hypothetical ones. Hence the more specific question is: *What are the required constraints and how to construct the evaluation method in order to be able to prove the convergence with time to the accessible possible states (accessible based on the unknown to the agent accessibility relation).* Please notice it is also not quite clear how the term *convergence* should be defined in order to compile with the common sense and not force to many requirements at the same time. Semi-formally we can say that we are trying to design a method of creating a set of evaluated accessible worlds $\widetilde{\mathbb{S}}(t)$ at some time moment $t$ with the agents knowledge $B(t)$ as input such that:

$$\widetilde{\mathbb{S}}(t) \subseteq \mathbb{S} \text{ that ``converges'' to } \overline{\mathbb{S}} \text{ when } t \to \infty \tag{12}$$

## 5 State of the Art

Similar problems have been tackled in 60's in the context of combinatorial possible worlds introduced by Armstrong [1,2]. Some of the works [4, 12] introduce the need of truthmakers[2]. All of these works are of either philosophical nature or consider very specific cases such as excluding features[3].

In AI there are some works on learning techniques such as Q-Learning that seem to at least partially solve the problem. Modified Q-Learning could be used to approximate the probability of possible states. The basic disadvantage of Q-Learning is that it must know everything about the current state and go many times through all the possible states in order to work correctly.

Usually experts define some logical formulas in order to define accessibility relation. Hence data mining techniques seem to be a good starting point where agent's knowledge can be (after some transformations) treated as mining data. Some of the more interesting techniques are connected to generating associative rules because it seems these can successfully work on partial data and do not treat the state as a whole. The implication like rules also seem to be the most common type of rules considered in data mining because they seem to cover many of the interesting data dependencies. But these algorithms have to be carefully modified in order to fit the addressed

---

[2] Truthmaker is some reasoning that explains why given world is really possible. Intuitively truthmakers can be treated as a different way to define accessibility relation.

[3] Two features exclude when one object can't have them both at the same time. For example ball can't be green and red (all over at the same time moment).

problem. Also some of the rule picking techniques have to be considered to ensure consistency between rules.

## 6 Accessibility through Rules

As mentioned previously the accessibility relation can be defined by some rules describing behavior of the environment. More specifically rules describe which combinations of properties and objects are possible, and which are not.

For example suppose that an agent knows that object $o$ has property $P_1$ at some time point $t$ ($o \in P_1^+(t)$) but does not know the property $P_2$ of the same object (neither $o \in P_2^+(t)$ nor $o \in P_2^-(t)$). If agent knows that there is a rule saying $o \in P_1(t) \rightarrow o \in P_2(t)$ she can say that a world where $o \in P_1(t)$ and $o \in P_2(t)$ is accessible while world where $o \notin P_2(t)$ is not.

If an agent knew the rules, she could effectively generate possible worlds based on knowledge state defined in (10). In this case agent could be treated as an expert system where knowledge state forms input, rules are used to generate different possible worlds (or states for some time moment) that in turn form output. The generated possible worlds must simply satisfy all the rules.

The task becomes harder when the rules are not known in advance. In order to correctly evaluate what are the accessible possible worlds agent has to discover the rules and then create possible worlds (or states). The main advantage of such approach is that accessible possible worlds can be created step by step based on rules that include some local dependencies. Hence the agent can make some preliminary assumptions even when little mining data is available. The main disadvantage is that the approach is very sensitive to complicated dependencies or any types of rules that can't be directly mined from data.

## 7 The Dynamic Accessible Possible States Model

The preliminary model we are proposing is based on rule discovery approach described in previous paragraph. The basic idea is to discover rules using specially suited data mining algorithms and then use these rules to create accessible possible states for different time moments. We would like the rules to be specially suited to the context of what agent currently knows, hence they have to be rediscovered, updated and improved along time flow. Because the knowledge state has some unknown properties the mining data has to be constructed carefully and nontrivially. The unknown properties may result in contradictory rules for which conflicts have to be solved. The figure 1 represents the basic flow of the proposed model, gray boxes and white rounded boxes represent required algorithms and data respectively. Rules are updated each time a new observation (base profile) is made. The observation is added to knowledge state of an agent which in turn is later used by an rules discovery mechanism to update rules set. Later generated rules and knowledge state are used to generate accessible states for time moments of agent's interest.

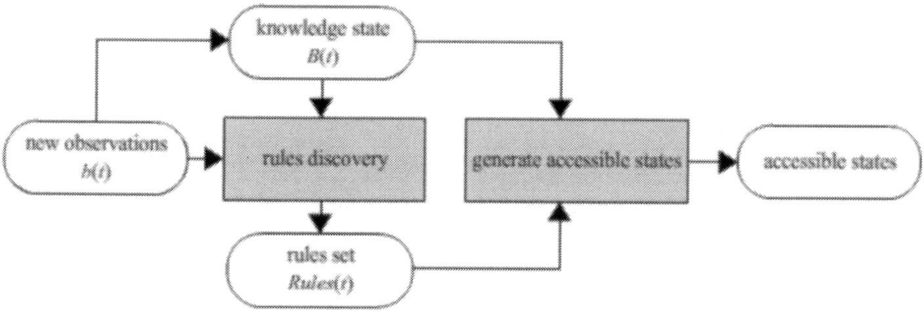


**Fig. 1.** Preliminary possible states evaluating model

Generated accessible states form the agent's ground level mental model representing possible variants of the environment – it's possible history and future. Accessible states can be later used to check the satisfaction of modal statements of the form: "Agent knows that $o \in P_i(t)$". Other possible uses include: prediction, action planning and cost function evaluation.

Some constraints have to be met by the environment for such model to be successful. Basically it must be possible for data mining algorithms to properly discover all the rules, assuming that knowledge state of an agent has enough observations to meet statistical requirements. Each bad or undiscovered rule may result in some accessible possible states being treated as inaccessible or the opposite. The question is: What are the required constraints?

## 8 Conclusions

We have defined a research problem of evaluating the environment behaviour based on a simple environment model, when only partial observations are available to the agent. Some of the key sub-problems have been marked and aspects of the model convergence have been discussed. We have also shown how the problem is located within current research trends.

A preliminary and simple model of constructing accessible states has been proposed. The model is computationally grounded [15] and relies entirely on agent's observations. Hence it can be successfully used in agent systems. The model is grounded and directly connected to observations.

Current learning solutions tend to use global approach treating environment states as a whole and ignoring local dependencies. Such approach leads to very long lasting learning process. The proposed model tries to find local dependencies within states rather than comparing the states globally. We believe such approach will result in quicker learning techniques.

We are planning further research concerning the presented problem and proposed solution in the next years. The following main questions need to be answered:

– When it is possible to successfully evaluate the accessibility relation?

- How to construct a method of evaluating possible worlds an hence the accessibility relation?
- What techniques are most useful when evaluating possible worlds?
- How to define the distance between evaluated possible worlds and accessible worlds?
- How fast can evaluated possible worlds converge to accessible worlds?
- How agent itself could measure its correctness when constructing the accessibility relation?

## References

1. Armstrong, D.: Universals and Scientific Realism, vol. 1-2. Cambridge University Press, Cambridge (1978)
2. Armstrong, D.: A Combinatorial Theory of Possibility. Cambridge University Press, Cambridge (1989)
3. Belardinelli, F., Lomuscio, A.: Quantified epistemic logics for reasoning about knowledge in multi-agent systems. Artifical Intelligence 173, 982–1013 (2009)
4. Bigelow, J.: Real Possibilities. Philosophical Studies 53, 37–64 (1988)
5. Cao, L., Luo, C., Zhang, C.: Agent-mining interaction: An emerging area. In: Gorodetsky, V., Zhang, C., Skormin, V.A., Cao, L. (eds.) AIS-ADM 2007. LNCS (LNAI), vol. 4476, pp. 60–73. Springer, Heidelberg (2007)
6. Harnad, S.: The Symbol Grounding Problem. Physica D 42, 335–346 (1990)
7. Halpern, J., Moses, Y.: A guide to completeness and complexity for modal logics of knowledge and belief. Artificial Intelligence 54, 319–379 (1992)
8. Jennings, N., Sycara, K., Wooldridge, M.: A Roadmap of Agent Research and Development. Autonomous Agents and Multi-Agent Systems 1, 7–38 (1998)
9. Kripke, S.: Semantical Analysis of Modal Logic. Zeitschrift für Mathematische Logic und Grundlagen der Mathematik 9, 67–96 (1963)
10. Levesque, H.: A Logic of Implicit and Explicit Belief. In: Proceedings of the Fourth National Conference on Artificial Intelligence (AAAI-1984), pp. 198–202 (1984)
11. Moore, R.: Reasoning about Knowledge and Action, Technical Note 191, SRI International, AI Center (1980)
12. Schneider, S.: Alien Individuals, Alien Universals, and Armstrong's Combinatorial Theory of Possibility. Southern Journal of Philosophy (2002)
13. Symeonidis, A., Mitkas, P.: Agent Intelligence Through Data Mining. Springer, Heidelberg (2005)
14. Wooldridge, M., Jennings, N.: Intelligent Agents: Theory and Practice. Knowledge Engineering Review 10(2), 115–152 (1995)
15. Wooldridge, M.: Computationally Grounded Theories of Agency. In: Proceedings of ICMAS, International Conference of Multi-Agent Systems, pp. 13–20. IEEE Press, Los Alamitos (2000)

# On Deriving Tagsonomies: Keyword Relations Coming from Crowd

Michal Barla and Mária Bieliková

Institute of Informatics and Software Engineering, Faculty of Informatics and Information Technologies, Slovak University of Technology
Ilkovičova 3, 842 16 Bratislava, Slovakia
{Name.Surname}@fiit.stuba.sk

**Abstract.** Many keyword-based approaches to text classification, information retrieval or even user modeling for adaptive web-based system could benefit from knowledge on relations between various keywords, which gives further possibilities to compare them, evaluate their distance etc. This paper proposes an approach how to determine keyword relations (mainly a parent-child relationship) by leveraging collective wisdom of the masses, present in data of collaborative (social) tagging systems on the Web. The feasibility of our approach is demonstrated on the data coming from the social bookmarking systems delicious and CiteULike.

## 1 Introduction

Phenomenon of the Social Web, with its roots in Web 2.0, is gaining a lot of attention all over the world in both research and practice. We are studying the power and wisdom of masses, when millions of people switched from the passive reading of the content to the active participation in its creation. People are blogging, sharing wikis, connecting themselves in various social applications and above all – tagging almost everything they get into touch: bookmarks, photographs, videos, publications, blogposts, articles etc. People got used to classify items by assigning few simple tags to it and to use those tags for a future retrieval of their favorite items. More, they are often expecting to find a new, yet unseen content by using their own tags. A part of the Web 2.0 success lies in an implicit agreement of masses on a shared (but never explicitly defined) vocabulary used to tag items – folksonomies.

At the same time, vast amount of content requires efficient navigation support, content reorganization or filtering – personalization and adaptation of the web. Its efficiency is dependent on adaptive system's ability to capture and maintain user model. A lot of research was devoted to finding the most suitable, flexible or most generic and all-encompassing user model representation [1,2], however, so far we are not aware of any explicit agreement on an ideal model representation.

The obvious question, when analyzing the success of Web 2.0, is whether an assignments of keywords (tags) to *user* instead of to *pages* (i.e., creation of tag-based user model) could lead to simple, viable and efficient approach to user

N.T. Nguyen, R. Kowalczyk, and S.-M. Chen (Eds.): ICCCI 2009, LNAI 5796, pp. 309–320, 2009.

modeling for adaptive web-based system. The challenge is then to combine the user model with the models of communities coming from the emerging social web and create a solid platform for personalization based on both traditional (e.g., as presented in [3]), and social approaches, such as the one presented in [4].

When considering tag-based user models in a (tag-based) Web 2.0 environment, we are facing the need to be able to compare various tags. We might need to compare user characteristics or even whole user models, represented by tags, to find similarities between users, which could serve for model maintenance as well as for more complex tasks such as online community creation or recommending in a recommender system. We might also need to compare the domain items represented by tags (e.g., a web page) in order to evaluate the explicit or implicit feedback [5] and update the user model appropriately.

In this paper, we propose a method for inferring various relationships between tags, which allows for full-blown usage of the tag-based user model for personalization and adaptation on the Web.

The paper is structured as follows: In section 2 we explain our approach to finding relationships between tags. Section 3 presents data we acquired and results of experiments we performed. In section 4 we summarize the related works, which served as an inspiration for our algorithm for building hierarchies from folksonomies. Finally, we give conclusions.

## 2 Finding Relationships between Tags

Our approach to finding relationships between tags combines three distinct approaches:

1. Deriving of parent-child relationships between tags from a given folksonomy;
2. Determining similarity between tags by applying spreading activation on the top of the folksonomy graph;
3. Interconnecting tags by additional semantic relationships as well as enriching the whole tag corpus by adding external keywords; both coming from the Wordnet lexical database.

### 2.1 Building Hierarchies from Folksonomies

Folksonomy is defined as a hypergraph [6] $H := \langle V,\ E\rangle$, where the set of vertices $V = A \cup T \cup I$ and $A \cap T = \emptyset,\ A \cap I = \emptyset,\ T \cap I = \emptyset$ and the set of ternary edges $E = \{(a,t,i) \mid a \in A,\ t \in T,\ i \in I\}$. A social tagging system can be represented by such a hypergraph with following definitions of the sets A, T and I (we will use them for the rest of the paper as well):

- Actors (users) $A = \{a_1, ..., a_k\}$
- Tags (keywords, concepts) $T = \{t_1, ..., t_l\}$
- Items (objects, instances) $I = \{i_1, ..., i_m\}$

The aforementioned hypergraph can be reduced into three bipartite graphs: a graph holding the associations between actors and tags ($AT$), actors and items ($AI$) and tags and items ($TI$). For example, the $AT$ valued bipartite graph is defined as follows: $AT := \langle A \times T,\ E_{at} \rangle$, $E_{at} = \{(a,\ t) \mid \exists i \in I : (a,\ t,\ i) \in E\}$.

Each such bipartite graph $XY := \langle X \times Y,\ E_{xy} \rangle$ can be furthermore folded into two one-mode graphs $G_X := \langle V_x,\ E_x \rangle$, $G_Y := \langle V_y,\ E_y \rangle$, where $V_x = X$, $V_y = Y$, and $E_x = \{(a,\ b) \mid a,\ b \in X,\ \exists y \in Y : (a,\ y) \in E_{xy},\ (b,\ y) \in E_{xy}\}$, $E_y = \{(a,\ b) \mid a,\ b \in Y,\ \exists x \in X : (x,\ a) \in E_{xy},\ (x,\ b) \in E_{xy}\}$. We can furthermore define a weight of an edge $e_{ab} \in E_i$ in such a one-mode graph as $w(e_{ab}) := |\{k \mid k \in K : (a, k) \in E_{ik},\ (b, k) \in E_{ik}\}|$. In other words, the weight $w(e_{a,b})$ shows the number of times the $a$ and $b$ were linked together in an original bipartite graph.

By folding aforementioned bipartite graphs, we get six different one-mode graphs, each representing different semantic network encoded in the folksonomy. For example, by folding $AT$ graph through tags, we get a social network of actors (users) based on overlapping sets of tags, where the links are between people who have used the same tags with weights showing the number of tags they have used in common. Similarly, we can get a social network based on overlapping sets of items (two people are linked if they have tagged the same item, with weight showing the number of items they have tagged in common). In our work, we were primarily interested in folding $TI$ graph through items, giving us a semantic network of tags.

The proposed approach to creation of hierarchy from the folded folksonomy hypergraph as defined above is based on a rather simple assumptions of set theory:

> *In an ideal situation, the tag $t_a$ is a parent of tag $t_b$ if the set of entities (persons or items) classified under $t_b$ is a subset of the entities under $t_a$.*

In other words, all items classified under narrow tag also appear under the broader tag.

Moreover, since our goal was to produce a reusable hierarchy of tags, which could be mapped to users' interests and use this hierarchy as a basis for reasoning on those interests, we were not interested in tags, which were used only by a small amount of users, even if they were using it quite extensively. We wanted only what "crowd agrees upon" and were filtering-out tags not achieving a certain degree of popularity (i.e., it is not used by at least k% of all users), even if this decision reduced drastically the amount of tags in the resulting hierarchy.

The algorithm 1 shows the basic idea of our approach using a simple pseudo-code. First, we create an artificial *root* of the hierarchy and put it in the set of already processed tags (*ordered tags* in the algorithm). Then, we process all tags from the folksonomy in the following manner:

1. If the tag $t$ does not reach the popularity threshold, we omit it immediately and pass on to the next one.
2. Otherwise, we compute its overlap (intersect) with every tag from the *ordered tags* set, resulting in identifying the tag $t_o$ with a maximum overlap.

3. The parent-child or sibling relationship is established if the ratio of maximum overlap to the overall use of the tag $t$ reaches the pre-defined threshold. The roles of the tags in relationship is determined as follows: if the tag $t_o$ is used significantly more often than $t$, we declare $t$ to be a child of $t_o$ and vice versa. If the usage of both tags is more or less equal, we declare them as siblings.
4. Before the assignment of relationships, we check whether it will not "break" the context in the hierarchy, i.e., we compute an average overlap of all ancestors or children in the branch respectively. If the average overlap falls below the threshold, we create a duplicate of tag $t_o$, assign tag $t$ to it appropriately and make a new branch of it. The creation of the duplicate aims at solving the homonymy problem, where the $t_o$ tag has multiple meanings, depending on a context.

### 2.2 Finding Related Tags by Spreading Activation

Spreading activation is a method for associative retrieval [7] in associative and semantic networks, hence a network data structure consisting of nodes and links modeling relationships between nodes. Searching is done by activating the selected node with an activation energy and spreading this energy through the edges to its *neighbors* so that

$$Energy[neighbor_i] = \frac{Energy[origin]}{|neighbors|}$$

The process runs recursively until convergence. At the end, the nodes' activation levels represent the measure of similarity to the initially chosen node or nodes.

Spreading activation search in the folksonomy finds new relationships, which are not "visible" when considering only set theory assumptions and the algorithm 1. If these relations are added to the already existing parent-child relationships, it allows us to make contextual "jumps" between the tags, which we believe could have an interesting impact for the tag-based user modeling and personalization.

Since the folksonomy does not provide direct links between tags, the spreading activation is performed on either the bipartite graph $TI$ (tags-items) or $TA$ (tags-actors) or on a combination of the two, all depending on the definition what *neighbors* means for the algorithm (i.e., the spreading activation on the $TA$ graph is performed if the *neighbors* of a tag are all actors which used that tag). However, due to the vast amount of connections present in every social tagging system, which leads to enormous time and space complexity of the recursive process, we needed to select a sub-graph of the whole folksonomy. For instance, for the $TI$ bipartite graph, we select the sub-graph by modifying the *neighbors* function of a tag/item so that it will return only popular items/tags, where popularity of a tag or item is defined (following the same principles as in algorithm 1) as a ratio of actors which used that particular tag or tagged that particular item. When spreading through $TA$ graph, we define an actor as popular if he or she used at least $k\%$ of popular tags and tagged at least $l\%$ of popular items.

**Algorithm 1.** Creation of tag hierarchy

```
Data: a folksonomy
Result: Popular tags structured in a hierarchy
create root
ordered tags ← root
for each tag t do
    t.popularity = |t.users| / |overallusers|
    if t.popularity < popularity threshold then
        continue
    end
    t_o ← max( t ∩ tag_x, ∀tag_x ∈ ordered tags )
    if |t_o| / |t| > overlap threshold then
        if |t_o| ≫ |t| then
            if rightcontext?(t_o.ancestors, t) then
                t_o.children ← t
            end
            else
                create a copy t_oc of t_o
                root.children ← t_oc
                t_o.children ← t
            end
        end
        else if |t_o| ≪ |t| then
            if rightcontext?(t_o.children, t) then
                t.parent = t_o.parent
                t.children ← t_o
            end
            else
                create a copy t_oc of t_o
                t.children ← t_oc
                root.children ← t
            end
        end
        else
            t_o.sibling ← t
        end
    end
    else
        root.children ← t
    end
end
```

## 2.3 Applying Knowledge on Keywords from Wordnet

Wordnet [8] (`wordnet.princeton.edu`) is a large lexical database for the English language developed and maintained by Cognitive Science Laboratory at

Princeton University. It groups nouns, verbs, adjectives and adverbs into sets of cognitive synonyms (synsets). Moreover, it provides conceptual-semantic and lexical relations between synsets, i.e., one can browse for hyponyms and hypernyms[1] of a given word, other similar words or even antonyms.

The reason why we did not use Wordnet as our source of keywords and relationships at the first place comes from our plans to leverage our hierarchy for user modeling purposes. Tags acquired from social tagging systems are closer to the user than Wordnet, they are more "webbish". Even more, some relationships which emerge from the social tagging systems could *never* come out of Wordnet (for example, a relationship between words `ie`, `png` and `bugs`, pointing to the well-known problem of Internet Explorer's broken PNG support). Relationships acquired from social tagging systems are like ordinary people used them, not like linguistics have decided them to be.

However, we believe, that Wordnet can still significantly contribute to the quality of the tag-based models built upon the generated tag hierarchy. Information on semantic relationships between words can be used not only to identify particular subtrees, which should be merged (in case of synonymy) or divided (because of ambiguity), but also to add new words (along with their relationships) to the hierarchy, which would raise the probability that we will be able to map user's interest to our keywords.

## 3 Experiment

In order to determine the feasibility of the proposed approach to deriving relationships between tags from folksonomies for the purposes of tag-based user modeling and to determine the optimal setting of the algorithm, we performed several experiments with two different folksonomies. Our main concern was whether the algorithm creates cohesive groups of tags (subtrees) without significant flaws of the context.

### 3.1 Data

We collected a part of delicious bookmarking site (`http://delicious.com`) dataset by periodically polling their RSS feeds. First, we used a 'recent activity' RSS feed to obtain a list of 128 448 unique user login names, next we used this information in user-scoped RSS feeds to obtain all tags and tagged pages for a given login name.

As the second dataset, we took the anonymized folksonomy of users-tags-publications from the CiteULike (`http://citeulike.org`), which is a system for tagging and searching for scholarly papers. Summarization of data we were able to acquire so far is listed in the Table 1.

[1] hypernym – the generic term used to designate a whole class of specific instances. $Y$ is a hypernym of $X$ if $X$ is a (kind of) $Y$

**Table 1.** Overview of the acquired dataset

| | delicious | citeulike |
|---|---|---|
| #usernames: | 128 448 | 44 215 |
| #records: | 2 957 144 | 5 228 356 |
| #processed users: | 2 234 | 44 215 |
| #unique tags: | 220 647 | 294 806 |
| #unique items(pages,publications): | 962 367 | 1 437 245 |

First thing we were interested in was whether these two folksonomies are used in a similar manner. The graph on Fig. 1 depicts a distribution of tags on pages in delicious in a logarithmic scale. We can see that the distribution of tags fits more or less the power-law, with 421 840 pages tagged by one tag only and one page having 171 distinct tags[2].

Similar graph for the publications of the CiteULike dataset is shown on Fig. 2. Again, we see the power law distribution, with 625 658 publications tagged by one tag only and one publication tagged by 1 708 distinct tags[3].

We know already that power law distributions tend to arise in social systems where many people express their preferences among many options. Therefore, by observing the power law in both datasets, we were assured that datasets are valid and contain enough users (which could be an issue especially for our delicious dataset) to perform a crowd-based analysis. However, we found an interesting difference in popularity of tags. In our delicious dataset, we can find many tags which are shared among 5+% of all users whereas in the CiteULike dataset, a tag marked as "most active" on CiteULike website reaches popularity of 1 to 2 percents only. We continue in delicious feeds harvesting in order to determine, whether the overall popularity of tags marked as popular in our current dataset will decrease.

### 3.2 Results

We executed the algorithm 1 (see section 2) several times on both folksonomies with different setups and were observing the resulting trees. The setups of the variables for different runs of the algorithm are summarized in the Table 2. The *floating average overlap threshold* from the table means that the actual threshold is computed "on-the-fly" as a fraction of current parent-child overlap.

The manual inspection of resulting hierarchies proved the viability of our approach, where meaningful relations between keywords were created, which would definitely provide a good basis for a tag-based user modeling. The quality of the result depends highly on the configuration of the algorithm. The rest of this section is devoted to analysis of the impact of algorithm's variables to the created hierarchy.

[2] that page is `http://www.scribd.com/`

[3] the most tagged publication according to the CiteULike linkout database is surprisingly "about:blank"

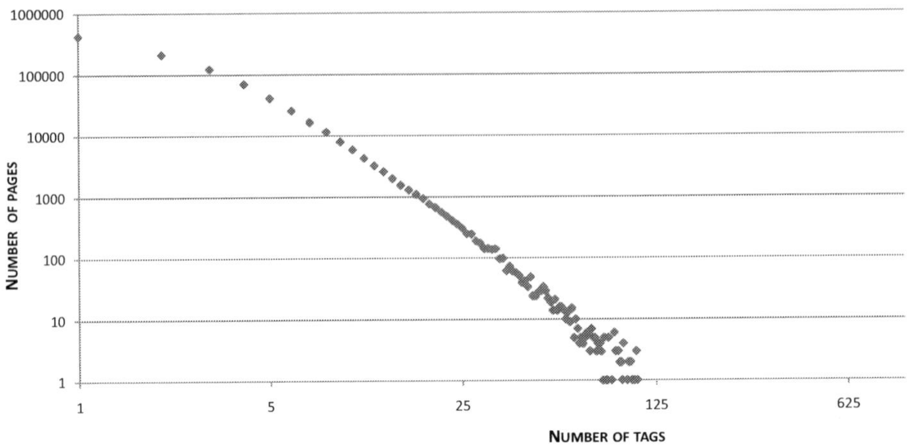


**Fig. 1.** Distribution of tags on pages in delicious dataset

**Table 2.** Overview of algorithm setup

| | deli[1] | deli[2] | deli[3] | cite[1] | cite[2] | cite[3] |
|---|---|---|---|---|---|---|
| popularity threshold: | 5% | 5% | 5% | 1,5% | 0,5% | 0,5% |
| overlap threshold: | 10% | 5% | 5% | 10% | 5% | 5% |
| average overlap (context) threshold: | 6,6% | 3,3% | floating | 6,6% | 3,3% | floating |

We provide summarization of basic attributes of the produced hierarchies in the Table 3, with examples on Fig. 3 and 4 (complete results can be seen at `www.fiit.stuba.sk/~barla/iccci09`). Our taxonomies conform roughly to the criteria on estimating quality of a taxonomy defined in [9], where a high-quality taxonomy should have an average depth of 3 with a maximum depth of 5. On the delicious[1] and delicious[2] datasets, we can observe the impact of the overlap threshold. When set to 10% (delicious[1]), the tagsonomy tends to be more flat with highly consistent subtrees. The 5% setup of delicious[2] led to organizing more tags into subtrees (i.e., less tags on the first level), but showed that the contextual threshold 3,3% was setup weakly (only four tags were duplicated in order to keep up with context). The best results were achieved when contextual threshold was set as floating, according to current parent-child overlap. For instance, tag *currency* on the Fig. 3 which was wrongly assigned in a branch *audio/conversion* was moved into a separate subtree in order to keep the average overlap of tags in the *audio* branch on the higher level.

We were surprised by results coming from the CiteULike folksonomy. The first setup with 10% overlap threshold produced very small hierarchy with only 67 popular tags. It seems that in CiteULike, the crowd did not make an agreement on most appropriate tags for particular publication (i.e., everybody uses his or her own specific tags). One reason for such a difference could be that delicious

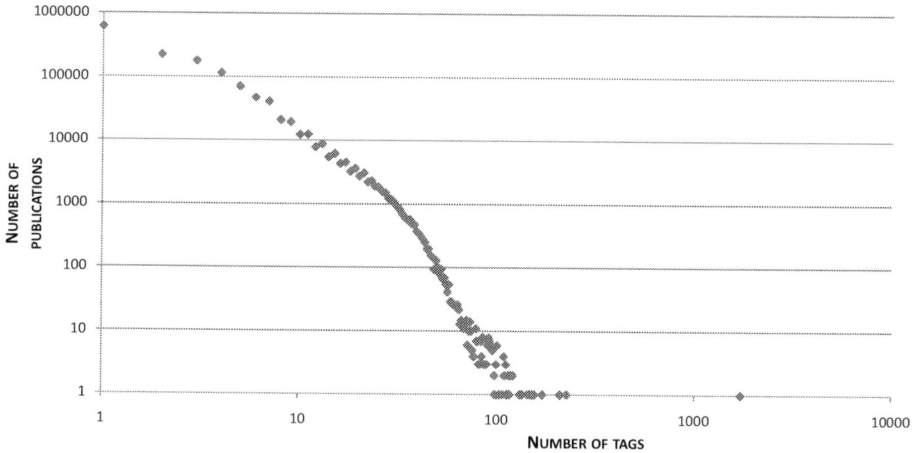


**Fig. 2.** Distribution of tags on publications in CiteULike dataset

**Table 3.** Attributes of the resulting hierarchies

| | deli[1] | deli[2] | deli[3] | cite[1] | cite[2] | cite[3] |
|---|---|---|---|---|---|---|
| #popular tags: | 1085 | 1085 | 1085 | 67 | 505 | 505 |
| #duplicated tags: | 1 | 4 | 94 | 0 | 5 | 25 |
| #first-level tags: | 357 | 172 | 277 | 54 | 83 | 125 |
| #child-less first-lev. tags: | 245 | 84 | 92 | 47 | 38 | 51 |
| average depth: | 1.8194 | 2.2074 | 2.0628 | 1.194 | 2.1743 | 1.9539 |
| maximum depth: | 5 | 5 | 5 | 2 | 5 | 5 |

is pro-actively supporting such an agreement to emerge by recommending tags when users are adding a particular page already present in the system, while CiteULike does not provide such a feature yet. Therefore, we decreased the required popularity to 0,5% for the CiteULike dataset, which resulted in 505 organized tags.

Resulting taxonomies pointed out yet another interesting difference between delicious and CiteULike folksonomies: CiteULike folksonomy contains words considered as English stop-words (and, on, the, for etc.) and (moreover) these stop-words are *popular*. This was something we did not expect at all, that somebody would use stop-words to organize some content. A possible explanation is that CiteULike users tend to post a short sentence as one tag (e.g., "example of a graph analysis"), but the CiteULike system considers each word of a sentence as a tag.

Another phenomenom of the CiteULike dataset is a popular *no-tag* tag, which is assigned automatically by the system if user enters a publication without any tags. Obviously, many people do use CiteULike without taking advantage of tagging system built in it, which was rather surprising finding.

```
3d
  modeling
  blender
  uml
  opengl
```

```
audio
  sound
  voice
  audiobooks
  recording
  sounds
```

```
blog
  blogger
  blogging
    comments
  tumblr
```

```
3d
  model
  modeling
  virtual
  secondlife
  blender
  uml
  opengl
```

```
audio
  conversion
    currency
  convert
  converter
  skype
  voip
  speech
  ...
```

```
blog
  bookmarks
  del.icio.us
  tag
    delicious
  internet
  telephone
  domain
security
    privacy
      rights
      human
  ...
```

**Fig. 3.** Example parts of tagsonomy created from the delicious dataset. The first row contains examples from the deli[1] setup, the second row from the deli[2] setup of the algorithm.

```
social
  trust
  community
  identity
  communities
  virtual
  folksonomy
  tagging
```

```
analysis
  pattern
    matching
  reference
  performance
  recognition
    object
  mass
  distribution
  density
  monitoring
  regression
  estimation
```

```
learning
  machine
  knowledge
    representation
  management
    strategy
  memory
    sleep
  online
  children
  problem
  plasticity
```

**Fig. 4.** Example parts of tagsonomy created from the CiteULike dataset using the cite[2] setup of the algorithm

## 4 Related Works

The raise of the tagging systems naturally provoked increase of interest in analysis of folksonomy data among researchers. We are aware of various approaches in deriving additional knowledge from folksonomies for different purposes.

In [6], Mika presents two approaches to retrieve relationships between tags: concept-mining based on graph clustering algorithms ($\lambda$-set analysis) and set theory assumptions, which are similar to our work, but does not take into account contextual conditions nor popularity of the tags. It seems that it was performing acceptably on a small chosen domain (such as keywords related to semantic web).

Schmitz et al. [10] opted for association rules mining to build-up conceptual structure. Resulting rules have the form *Users assigning the tags from A to some resources often also assign the tags from B to them.* Even if authors did not provide the way how to derive a taxonomy from the mined rules, we can just look on them as on the subsumption relations, which we are deriving from the overlapping sets of tagged resources.

Heymann and Garcia-Molina [11] proposed another approach based on comparing tag vectors and connecting similar tags together. Then, the taxonomy is created according to tags' centrality measure in the created similarity graph.

Schwarzkopf et al. [9] extend both algorithms [10,11] by taking into account a context of a tag, defined similarly to our work. They did not try to filter-out non-popular tags as we do, in order to obtain only a "crowd-agreement" tags, nor they do any further processing in order to enrich furthermore the taxonomy.

Shepitsen et al. [12] propose context-dependent hierarchical agglomerative clustering technique to organize tags into clusters subsequently used for recommendation of resources. As with any other clustering technique, a crucial part is the definition of similarity between items being clustered. Shepitsen et al. used cosine similarity of vectors over the set of tags.

We see several differences of our approach compared to the aforementioned ones. Apart from popularity, configurable bidirectional context checking (to ancestor or to children) and siblings detection, we proposed also an incorporation of other techniques and approaches into one corpus such as spreading activation, which greatly improves the resulting hierarchy and broaden its possible usage.

## 5 Conclusions

In this paper, we have shown a method how wisdom of the masses in the form of social bookmarking folksonomy can be used to create a "tagsonomy" (a taxonomy of tags). We also proposed other techniques with different background such as graph activation search coming from the graph theory and Wordnet's conceptual semantic relationships coming from the cognitive science area, which can contribute and enhance the final taxonomy of tags by adding new "shortcuts" between hierarchically ordered tags.

We performed several experiments with the algorithm on CiteULike and delicious folksonomies, which proved the viability of the approach and pointed out some interesting differences in the two mentioned tagging systems. We have shown that our algorithm for deriving hierarchy from the folksonomy can handle such differences when properly configured. More, the results proved that web 2.0 generated folksonomies can be used, when taking into account tags with a certain level of popularity optionally enriched by Wordnet's synonyms, for user's interest modeling for personalizing and adapting the web (and web 2.0 itself).

In our future work, we plan to evaluate the impact of spreading activation on relationships between tags and compare two different spreading methods (via actors, via items). Our preliminary results showed that the spreading algorithm is able to create meaningful "jumps" in the hierarchy, but is highly dependent on number of links from and to tag (when there are too many links, the energy is divided into neglectable chunks). One way how to solve this issue is to introduce a variable starting energy, depending on a popularity of a given tag.

**Acknowledgment.** This work was partially supported by the Scientific Grant Agency of Slovak Republic, grant No. VG1/0508/09 and the Cultural and Educational Grant Agency of the Slovak Republic, grant No. KEGA 3/5187/07.

## References

1. Heckmann, D., et al.: GUMO – The General User Model Ontology. In: Ardissono, L., Brna, P., Mitrović, A. (eds.) UM 2005. LNCS (LNAI), vol. 3538, pp. 428–432. Springer, Heidelberg (2005)
2. Andrejko, A., Barla, M., Bieliková, M.: Ontology-based User Modeling for Web-based Information Systems. In: Advances in Information Systems Development, pp. 457–468. Springer, Heidelberg (2007)
3. Barla, M., Tvarožek, M., Bieliková, M.: Rule-Based User Characteristics Acquisition from Logs With Semantics for Personalized Web-based Systems. Computing and Informatics (2009) (accepted)
4. Coyle, M., Smyth, B.: (Web Search)shared: Social Aspects of a Collaborative, Community-Based Search Network. In: Nejdl, W., Kay, J., Pu, P., Herder, E. (eds.) AH 2008. LNCS, vol. 5149, pp. 103–112. Springer, Heidelberg (2008)
5. Joachims, T., et al.: Accurately interpreting clickthrough data as implicit feedback. In: SIGIR 2005, pp. 154–161. ACM, New York (2005)
6. Mika, P.: Ontologies are us: A Unified Model of Social Networks and Semantics. J. Web Sem. 5(1), 5–15 (2007)
7. Crestani, F.: Application of Spreading Activation Techniques in Information Retrieval. Artif. Intell. Rev. 11(6), 453–482 (1997)
8. Fellbaum, C.: WordNet: An Electronical Lexical Database. The MIT Press, Cambridge (1998)
9. Schwarzkopf, E., Heckmann, D., Dengler, D., Kröner, A.: Mining the Structure of Tag Spaces for User Modeling. In: Data Mining for User Modeling, Workshop held at UM 2007, pp. 63–75 (2007)
10. Schmitz, C., et al.: Mining Association Rules in Folksonomies. In: Data Science and Classification. Studies in Classification, Data Analysis, and Knowledge Organization, Part VI, pp. 261–270. Springer, Heidelberg (2006)
11. Heymann, P., Garcia-Molina, H.: Collaborative Creation of Communal Hierarchical Taxonomies in Social Tagging Systems. Technical report, Computer Science Department, Stanford University (2006), `http://heymann.stanford.edu/taghierarchy.html` (June 29, 2006)
12. Shepitsen, A., et al.: Personalized Recommendation in Social Tagging Systems using Hierarchical Clustering. In: Recommender Systems 2008, pp. 259–266. ACM, New York (2008)

# D²ISCO: Distributed Deliberative CBR Systems with jCOLIBRI

Sergio González-Sanz, Juan A. Recio-García, and Belén Díaz-Agudo

Department of Software Engineering and Artificial Intelligence,
Universidad Complutense de Madrid, Spain
sergio.gonzalez@ciemat.es, jareciog@fdi.ucm.es, belend@sip.ucm.es

**Abstract.** In this paper we describe D²ISCO: a framework to design and implement deliberative and collaborative Case Based Reasoning (CBR) systems. Using D²ISCO we design and implement distributed CBR systems where each node collaborates, arguments and counterarguments its local results with other nodes to improve the performance of the system's global response. D²ISCO is integrated as a part of jCOLIBRI 2 [1] an established framework in the CBR community. We perform a case study for a collaborative music recommender system and present the results of an experiment of the accuracy of the system results using a fuzzy version of the argumentation system AMAL [2] and a network topology based on a social network.

## 1 Introduction

Case-Based Reasoning (CBR) is one of most successful applied AI technologies of recent years. CBR is based on the intuition that tend to recur. It means that new problems are often similar to previously encountered problems and, therefore, that past solutions may be of use in the current situation [3].

jCOLIBRI 2 [1] is a well established framework in the CBR community that can be used to design different types of CBR systems. However its underlying architecture has taken the conventional so called *single agent, single case base* problem solving approach where one, usually well-maintained, case base functions as the central knowledge resource. Research efforts in the area of *distributed CBR* concentrate on the distribution of resources within CBR architectures and study how it is beneficial in a variety of application contexts. In contrast to single-agent CBR systems, multi-agent systems distribute the case base itself and/or some aspects of the reasoning among several agents. In [4] the research efforts in the area of distributed CBR are categorized using two criteria: (1) how knowledge is organised/managed within the system (i.e. single vs.multiple case bases), and (2) how knowledge is processed by the system (i.e. single vs. multiple processing agents). Much of work in distributed CBR assumes multi-case base architectures involving multiple processing agents differing in their problem solving experiences [5]. The "ensemble effect" [2] shows that a collection of agents with uncorrelated case bases improves the accuracy of any individual. Multiple

N.T. Nguyen, R. Kowalczyk, and S.-M. Chen (Eds.): ICCCI 2009, LNAI 5796, pp. 321–332, 2009.

sources of experience exist when several CBR agents need to coordinate, collaborate, and communicate. Within this purpose AMAL has been proposed as a case-based approach to groups of agents that coordinate, collaborate, and communicate in order to improve their collective and individual decisions by learning from communication and by argumentation over debated outcomes [6].

Our current work, described in this paper, proposes an extension to jCOLIBRI 2 to design deliberative and distributed multiagent CBR systems where the case base itself and/or some aspects of the reasoning process are distributed among several agents. Our current work focuses on distributed *retrieval* processes working on a network of collaborating CBR systems. We deal with aspects such as the topology of the network, the definition of confidence models for the different agents, voting and negotiation techniques between agents to reach a consensus in the final solution. In this paper we consider a simplified type of retrieval-only CBR systems: recommender systems. We perform a case study for a collaborative music recommender system and present the results of an experiment of the accuracy of the system results using a fuzzy version of the argumentation system AMAL and a network topology based on a social network.

In the network of agents every agent should be able to define the trustworthiness regarding the connected agents [7,8]. In the recommender systems arena, in order to provide meaningful results, trust must reflect user similarity to some extent; recommendations only make sense when obtained from like-minded people exhibiting similar taste. Our model is based on having into account the social connections of the collaborative agents, including the level of trust of the agent they collaborate with. We use social trust as the basis for recommender systems [9] [10][11]. Social networks offer an opportunity to get information about the social environment of a given user and associate trustworthiness values to it. If a node receives a query and it cannot give an good answer to it, then it will ask for collaborations with other nodes it has relations or links with. The trust models evolves in time according to the real accuracy of the answers provided by a certain node.

This article is structured as follows: Section 2 introduces D²ISCO, a distributed CBR framework for collective experiences and explains the original AMAL protocol, the fuzzy version proposed in this paper, and how a query is propagated through the nodes of the distributed system. Section 3 describes a case study of a collaborative and Distributed Music Recommender System and runs some experiments to measure the accuracy of the system results using our fuzzy version of the argumentation system AMAL and a network topology based on a social network.

## 2 D²ISCO: Distributed Reasoning for Collective Experiences

D²ISCO[1] uses a protocol rooted in the AMAL argumentation process proposed in [2,6]. Therefore, to understand the features of D²ISCO we need to briefly

[1] D²ISCO: **D**eliberative, **DIS**tributed & **CO**llaborative extension for jCOLIBRI.

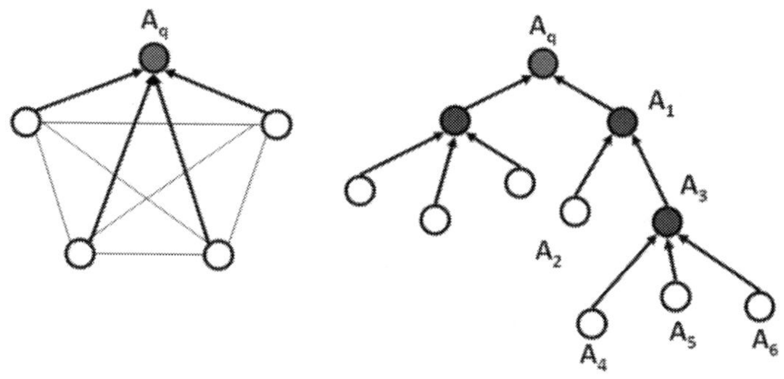


**Fig. 1.** Comparison between AMAL (left) and D²ISCO (right) topologies

depict AMAL and then describe the improvements of our approach. AMAL follows the same mechanism as human committees, first each individual member of a committee exposes his arguments and discuses those of the other members (joint deliberation), and if no consensus is reached, then a voting mechanism is required. The interaction protocol of AMAL allows a group of agents $A_1, \ldots, A_n$ to deliberate about the correct solution of a problem $Q$ by means of an argumentation process. Each of these agents uses a CBR system to find a solution for $Q$ using their own case base and then they start a deliberation process to order the local solutions and find the best of them. If the argumentation process arrives to a consensual solution, the joint deliberation ends; otherwise a weighted vote is used to determine the joint solution [6]. Moreover, AMAL also allows the agents to learn from the counterexamples received from other agents.

The reasoning protocol begins with an agent ($A_q$) issuing a query to the agents that is linked to ($A_1, A_2, ..., A_n$). Each one of these agents retrieves $k$ items from their own case base. Then, an argumentation process consisting on $k$ cycles is performed to defend and discard the proposed items by means of counterexamples. When the process finishes $A_q$ receives at most $k$ trusted items.

The AMAL protocol consists on a series of rounds. In the initial round, each agent states which is its individual local solution for the problem $Q$. Then, at each round an agent can try to rebut the solution or prediction made by any of the other agents giving a counterexample. When an agent receives a counterargument or counterexample, it informs the other agents if it accepts the counterargument (and changes its solution) or not. Moreover, agents have also the opportunity to answer to counterarguments by trying to generate a counterargument to the counterargument.

Next we summarize some of the limitations in the original AMAL approach:

- Regarding the topology, AMAL proposes to link every agent with all the agents in the system. This N to N topology has repercussions in the efficiency of the argumentation process and it is not scalable to real size systems.
- In real scenarios, users –or agents– are organized and linked by means of topologies that are analogous to social networks. Therefore, the N to N topology does not reflect faithfully the relations among users.

- AMAL is based on Description Logic (DLs) case representation and reasoning what implies an additional knowledge representation effort.
- AMAL does not take into account the confidence between agents in the argumentation process.

Our D²ISCO approach for building distributed and deliberative systems improves the AMAL protocol and solves its main drawbacks. The features of our framework are:

- The topology of the systems follows the structure of a social network. This enables to increase easily the number of agents in the system and to incorporate confidence factors in the argumentation process that are obtained from this social network.
- If the social network reflects the preferences of the users, it is also possible to add this information to the argumentation process. This way, social networks have two possible uses: 1) to obtain the confidence among users, and 2) to compute the similarity between users according to its preferences.
- The argumentation process is directed by a lead node/agent $A_q$ that issues the query and organizes the deliberation of its children nodes ($A_c$). This agent is in charge of accepting or rejecting the counterexamples presented by those children agents.
- Our argumentation and case retrieval process is hierarchical. When solving a problem $Q$, the agent that issues the query $A_q$ becomes the root of the whole hierarchy of agents –defined by the structure of the social network. Then, the query is sent to the leafs of the tree and the retrieval follows an inverse direction. The leafs of the tree deliberate with their immediate parent $A_p$ node that organizes the reasoning. When this intermediate deliberation finishes, $A_p$ participates in the deliberation organized by its parent node but this time it takes the role of a children node $A_c$. This behavior is repeated until reaching the root $A_q$. It is important to note that in every intermediate deliberation $A_p$ receives the cases retrieved by its children nodes $A_c$ and incorporates them in its own case base. Figure 1 illustrate the difference between AMAL and D²ISCO topologies. The direction of the arrows represent the forwarding of cases. The left graph shows a typical AMAL net where every agent is linked with every agent. On the right we find the hierarchical topology of D²ISCO. Here we can note how the argumentation process begins with $A_3$ being the organizing agent ($A_c$) and $\{A_4, A_5, A_6\}$ its corresponding children ($A_c$). Afterward, $A_3$ takes part of the deliberation conducted by $A_1$ where $A_c = \{A_2, A_3\}$. Finally, $A_1$ contributes in the final deliberation leaded by $A_q$.
- D²ISCO reasons with the case ratings and it doesn't requires expressive case representations based on DLs. To substitute the reasoning capabilities of DLs, our approach uses a fuzzy decision system. Moreover this fuzzy system takes into account the confidence and similarity between users –obtained from the social network– and the similarity between cases.
- Because we are not using logical formulas to define the counterarguments, our deliberative process applies the concept of *defenses*. Defenses are complete

cases that are highly rated by the agent involved in the deliberation and are offered to trust the arguments presented by the agent.

Once we have described the behavior of our approach, we can focus on its main improvement: the fuzzy decision system. Next section summarizes its main features and illustrates it by means of an example.

### 2.1 Fuzzy Decision System

A key feature of the distributed reasoning protocol described in previous section is the decision system that accepts or rejects the counterexamples. In the original AMAL protocol these arguments are generated using descriptive logic. Our proposal relies in a fuzzy reasoner [12] that allows extending the protocol to cases where a counterexample can not be generated by logic induction.

It is important to note that the agent that leads the argumentation ($A_p$) does not retrieve cases from its own case base, but plays an important role in the argumentation because: (a) defines the confidence in the agents involved in the argumentation process, and (b) decides if a counterexample is accepted o rejected based on its confidence in the involved agents and the goodness value of their items. A counterexample against the case $C_i$ is a case $C_e$ that is rather similar to $C_i$ but it has a low value of goodness. In recommender systems this measure of *goodness* is usually the *rating* of the item given by the corresponding agent/user. A defense against a counterexample $C_e$ is a case $C_d$ that is rather similar to $C_e$ and it has a high value of goodness.

The fuzzy decision system is involved in many steps of the argumentation process to:

- Organize the list of cases returned by an intermediate deliberation. It generates a value $V_t$ measuring the degree of trust for a certain case $C_i$ using: 1) the value of goodness of $C_i$ in its local case base, 2) the similarity between $C_i$ and the current query $Q$, and 3) the compatibility between agent $A_i$ that returns $C_i$ and the agent conducting the deliberation $A_p$.
- Measure the confidence in a counterexample ($V_c$). If an agent $A_i$ presents a counterexample, it includes the a value $V_c$ that indicates its confidence in its argumentation. To obtain this value, the fuzzy system uses: 1) the goodness of $C_e$ in its local case base, 2) the similarity between $C_e$ and $Q$, and 3) the compatibility between agent $A_i$ and $A_p$.
- Decide if the counterexample is accepted ($V_a$). When an agent $A_i$ proposes an counterexample $C_e$ to the conducting agent $A_p$, this last agent has to decide if the counterexample is accepted or rejected based on: 1) the confidence of the counterexample $V_c$, 2) its confidence in $A_i$, and 3) the agent being rebutted. Finally, the counterexample is accepted if it is higher than threshold $\alpha$ defined in the system ($V_a > \alpha$).
- Measure the confidence in a defense ($V_d$). This value is computed by the agent being rebutted when in looks for a defense. It is analogous to $V_c$ and measures its confidence in the defense. The defense will be accepted by $C_q$ if it exceeds another configured threshold $V_d > \beta$.

**Table 1.** Argumentation Example

| | Rating | Artist | Title | Year | Price | Style |
|---|---|---|---|---|---|---|
| Query ($Q$) | 10 | Mike Oldfield | * | * | * | * |
| Case-$A_2$ ($C_2$) | 5,58 | Mike Oldfield | The Millennium Bell | 1999 | 11 | House |
| Case-$A_3$ ($C_3$) | 3,77 | Mike Oldfield | Hergest Ridge | 1974 | 26 | Rock |
| Case-$A_1$ ($C_1$) | 1,82 | Mike Oldfield | Crises | 1983 | 22 | Rock |
| **Round 1 - $A_2$ has the token** | | | | | | |
| $A_2$ counterexample for $C_1$ | 0,5 | Mike Oldfield | Incantations | 1978 | 27 | Rock |
| Accepted because $A_1$ cannot generate a defense | | | | | | |
| Round result: | | | | | | |
| $C_2$ | 5,58 | Mike Oldfield | The Millennium Bell | 1999 | 11 | House |
| $C_3$ | 3,77 | Mike Oldfield | Hergest Ridge | 1974 | 26 | Rock |
| **Round 2 - $A_3$ has the token** | | | | | | |
| $A_3$ counterexample for $C_2$ | 1,36 | Deep Dish | George is on | 2005 | 19 | House |
| $A_2$ defense ($C_e$) | 5,3 | Deep Dish | George is on | 2005 | 19 | House |
| $A_2$ defense is accepted | | | | | | |
| Round result: | | | | | | |
| $C_2$ | 5,58 | Mike Oldfield | The Millennium Bell | 1999 | 11 | House |
| $C_3$ | 3,77 | Mike Oldfield | Hergest Ridge | 1974 | 26 | Rock |
| **Round 3 - $A_1$ has the token** | | | | | | |
| $A_1$ counterExample for $C_2$ | 1,67 | Mike Oldfield | Earth Moving | 1989 | 23 | Pop |
| Accepted because $A_2$ cannot generate a defense | | | | | | |
| Round result: | | | | | | |
| $C_3$ | 3,77 | Mike Oldfield | Hergest Ridge | 1974 | 26 | Rock |

**Example.** To illustrate the behavior of our deliberative recommender, let's use a real example described in Table 1. Here we will not use intermediate nodes for clarity reasons. Also every agent returns only one case. In the example $A_q$ sends the query to $A_1$, $A_2$ and $A_3$, and they answer returning a case in the order shown in the table ($A_2$,$A_3$,$A_1$). This order is used to decide how to propose the examples and counterexamples.

In the first round, $A_2$ begins presenting a counterexample to the case $C_1$ retrieved by $A_1$. $A_q$ decides to accept the counterexample because $A_1$ cannot generate any defense by retrieving another counterexample from its case base. This way, $C_1$ is removed from the initial retrieved set. The following round begins with $A_3$ having the token. Here it presents a counterexample to the case $C_2$ retrieved by $A_2$, but this agent manages to find a defense $D_2$ that is accepted by $A_q$. Therefore, the round finishes without changes in the retrieved set. In the third round, $A_1$ has the token and presents a counterexample for $C_2$ that is accepted. As this case is removed, the only remaining case is $C_3$ that is the solution that $A_q$ obtains for its query.

During each round of the argumentation, the fuzzy system participates several times. To illustrate its behavior let's detail the reasoning process of the first round. It begins when $A_2$ presents a counterexample $C_e$ for $C_1$. The behavior of this subsystem is shown in the first row of Figure 2. On the left we find the inputs and outputs of the system and on the right we have included a representation of the fuzzy formula with ($rat(C_e) = 0.5$). The output value is $V_c = 9$ and it is the

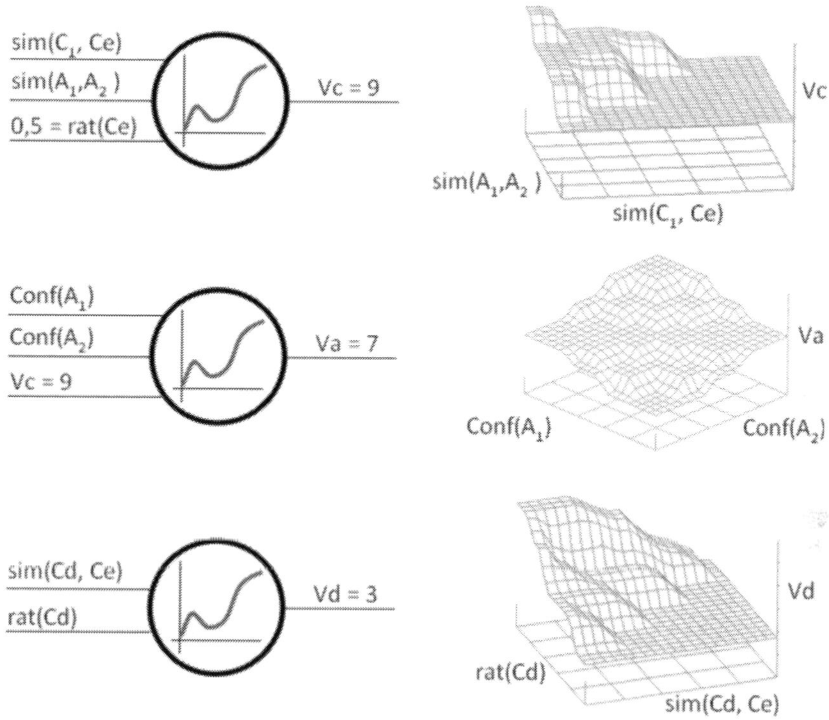


**Fig. 2.** Behavior of the fuzzy decision system

confidence measure sent by $A_2$ to $A_q$. Then, $A_q$ decides if the counterexample is accepted taking into account its confidence in $A_1$ –the case being rebutted–, $A_2$ and $V_c$. This subsystem is shown in the second row of Figure 2 with $V_c$ set to 9. Here, the result is $V_a = 7$ and it exceeds the threshold $\alpha > 6.5$ configured in the system. It means that the counterexample is accepted and sent to $A_1$. Next $A_1$ has the opportunity of presenting a defense. It looks in its case base but it cannot find any case which $V_d > \beta$ that has been set to $\beta = 5$. This step is represented in the third row of Figure 2. Therefore the first round finishes by removing $C_1$ from the retrieval set. If $A_1$ could find a defense, it will be sent –together with $V_d$– to $A_q$ and a reasoning process similar to the one shown in the second row of Figure 2 would happen.

## 2.2 Propagation of the Query through the Distributed System

The propagation of the query through the distributed system has also an important impact in the reasoning process. Here, the topology of the network that links the different agents in the system defines how the argumentation process is performed. Moreover, each link has a confidence value between every pair of agents that is having into account during the reasoning.

Commonly these topologies define a hierarchy of agents where the root node is the agent $A_i$ that issues the query. This way, the query must be distributed through the intermediate nodes to reach every agent in the system. Clearly this hierarchy can have an important influence in the reasoning process by weighting the confidence in the agents according to its distance to the root agent. In our proposal, this influence is reflected on the ranking of the items returned by the agent and the confidence given to counterexamples.

In the fuzzy argumentation system each agent (but $A_i$ that initiates the query) returns as an answer to the query the ordered list resulting of mixing the local results of its reasoning (local similarity search) with the argumentation results with its neighbor agents. This list of results is ordered by the Vt value computed by the fuzzy system.

Another interesting feature is the level of depth of the propagation of the query inside the hierarchy of agents. Usually there is no limit and the query reaches every agent, however, it implies two common problems inherit from any distributed architecture: loops and multiple paths to the same agent. To address these problems our approach uses traces to store the reached agents and to identify the query within the whole system to avoid repeating its processing:

$$id_{query} = < id_{node}, n_{query}, trace >$$

where: $id_{node}$ identifies the node that issues the original query, $n_{query}$ identifies the query inside the distributed system and *trace* is the list of visited agents.

Once described the behavior of our distributed and deliberative recommender, next section presents a case study that proves the benefits of the approach.

## 3 Case Study: Distributed and Collaborative Music Recommender System

In this section we describe a case study in the domain of music recommendation, a classical example of successful recommender applications where there are many users interested on finding and discovering new music that would fulfill their preferences. Moreover, the users of this kind of applications tend to interchange recommendations with other users that have similar preferences. These relationships conform social networks that reflect the similarity in the preferences of the users and allow them to discover new items, and the confidence in the recommendation. This way, we can measure the confidence between two users depending on their corresponding distance in the social network.

The experiments were designed to simulate a real scenario with the highest possible fidelity. As the case base has a catalog of songs, and each user may have a part of this catalog in its internal list of rated items. Every user interacts with its corresponding recommender agent. When a recommender agent receives a query from the user, it forwards the query to the other agents in the system. These agents will use their rated items to recommender songs that fulfill the preferences of the query. Agents are organized according to a social network that ideally reflects the similarity and confidence between users. Our premise is that a real social network

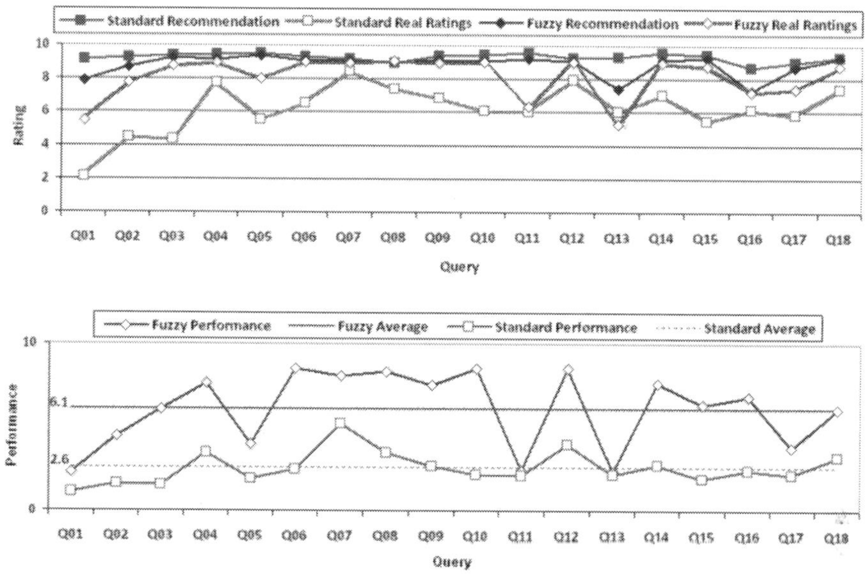


**Fig. 3.** Results using a social network built with Pearson's coefficients

will relate users with similar preferences, but this initial intuition was also tested in our experiments by simulating two different networks: a network that relates users with similar preferences and a random social network. To measure the benefits of the architecture of our recommender we used cross-validation over the rated items of a user, by comparing the recommendations performed by a single agent that manages the whole catalog –this is, the real ratings of the user– and the recommendations of our collaborative approach.

We have simulated a real social network where we have randomly generated the ratings of the users to control the different factors involved in the experiment. The catalog of songs contains 300 real items. Then we used two numbers of users for every experiment: 10 and 20 users. A local product base of 50 songs was assigned to every user in a random way. It means that these local catalogs may overlap. The ratings of the songs were simulated by assigning preference profiles to the users. Then, ratings were generated using probabilistic distributions associated with every preference of the profile. For example, a user that prefers pop songs will rate higher this kind of items according to certain probability.

The most important component of the experiment is the social network that relates users according to their similarity and mutual confidence. Our initial premise was that a real network will reflect this feature, so we decided to generate the network by linking users with similar preferences. To perform this, we have used the Pearson's coefficient. This metric is a common way for measuring the similarity between users in recommender systems. So, we decided to compute this coefficient between every pair of users in the system and create a network link if a pair has a similarity above certain threshold. However, to test the influence of the topology of the network in the recommendation process we also generated another network with random links. Finally, we used these coefficients to define the confidence level between each pair of users/agents. Figure 3 shows some representative results of our case study. These graphs summarize the ratings obtained with our fuzzy approach compared to the standard AMAL protocol. Here we are raising 18 random queries into a network with 50 users linked by means of the Pearson's coefficients. To measure the performance of the system we used a cross-validation process and compared the ratings for the $k$ best recommended items with the real ratings given by the user to these same items. Ideally, a perfect recommender system will return high rated items according to the preferences of the user. To illustrate this measure, the upper graph of Figure 3 shows the ratings of the recommended items together with the real ratings of the user for these items. Note, that the songs retrieved by both systems (fuzzy and standard) can be different for the same query and it implies that the lines showing the average of the real ratings given by the user for these retrieved sets are different too.

As the graph shows, the original AMAL protocol tends to offer high ratings because it simply chooses the best-rated items from every agent. However, this does not reflect the real ratings given to these items by the user, and therefore the two lines *Standard Recommendation* and *Standard Real Ratings* are quite distant. On the other hand, our fuzzy decision system approximates better the preferences of the user, and the differences with the real ratings are lower meanwhile returning high rated items. This is due to the improvement in the negotiation and decision system thanks to the fuzzy reasoning.

It is important to note that during the experiments we obtained no significant differences when modifying the $k$ value. Moreover we found another noticeable feature: the similarity between the recommended items to the query and the similarity of the corresponding real user-rated items to the query was very close. This result allowed us to leave the similarity aside and concentrate our performance measures on the ratings of the items.

Because we are trying to maximize the ratings but minimizing their difference with the real ones, we have defined the following performance measure to test our architecture of recommender systems:

$$performance = \frac{rat_{rec}(r)}{1 + |rat_{rec}(r) - rat_{real}(r)|}$$

where $r$ is the set of $k$ items returned by the system, $rat_{rec}(r)$ is the average of the ratings returned by the recommender for $r$, and $rat_{rec}$ is the average of the real ratings given to $r$ by the user. Figure 3-bottom shows graphically this performance measure for both approaches (fuzzy and original AMAL). Here, a higher value means better performance, and as we can note, the fuzzy approach is always on top. We also include the average of both lines to measure the global performance of the recommendation. These results prove that the fuzzy approach (section 2.1) is significantly better than the original AMAL protocol. Moreover, we have included the performance of the system without any argumentation protocol (neither D²ISCOnor AMAL) by issuing the queries to every agent and selecting the top-rated item. Although these ratings are very high, they do not

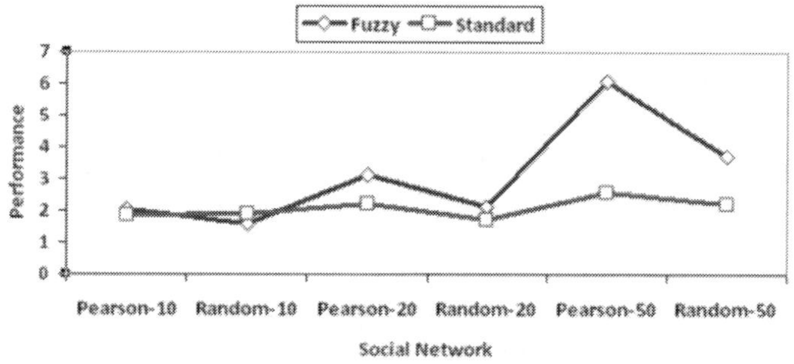


**Fig. 4.** Global results using different social networks

reflect the preferences of the user and the performance of this simple strategy is worse. This fact proves the "ensemble effect" described in [2]: a collection of agents with uncorrelated case bases improves the accuracy of any individual[2].

Finally, we have generated several social networks with different topologies to test the behavior of the architecture of our recommender. Results are shown in Figure 4, where we measure the performance for random and Pearson-based networks with 10, 20 and 50 nodes. Our reasoning process always reports better results with the networks generated using Pearson's coefficients. This fact confirms our premise that social networks will improve the recommendation process. Regarding the fuzzy improvement for the standard AMAL protocol, the Figure shows that the fuzzy approach is better for the networks with 20 and 50 nodes. In the case of the 10-node networks, the fuzzy system is not able to find the minimum similarity relationships between users to perform the decision process correctly. However, this is a coherent and meaningless consequence due to the low number of users in the network.

## 4 Conclusions

This paper describes D²ISCO: a framework to design and implement deliberative and collaborative CBR systems. We have performed a case study for a collaborative music recommender system from a catalog of 300 items and networks of up to 50 nodes where the local catalogs overlap. We have presented the results of an experiment of the accuracy of the system results using a fuzzy version of the argumentation system AMAL and a network topology based on a simulated social network. We have generated several social networks using Pearson's relation between ratings. We measure the performance for random and Pearson-based networks with different number of nodes. Our reasoning process and architecture of recommender always reports better results with the networks generated using Pearson's coefficients. This fact confirms our premise that social

[2] Under some restrictions like the aggregation function (e.g. majority voting) or the retrieval strategy (e.g. content-based k-NN).

networks will improve the recommendation process. Regarding the argumentation process we have shown that the fuzzy improvement of the standard AMAL protocol is better with a bigger number of nodes. Besides, although the original AMAL protocol tends to offer high ratings, the fuzzy decision system approximates better the preferences of the user, and the differences with the real ratings are lower meanwhile returning high rated items. This is due to the improvement in the negotiation and decision system thanks to the fuzzy reasoning.

$D^2$ISCO has been integrated as a part of jCOLIBRI 2 [1] an established framework in the CBR community. As an ongoing work we are designing a set of reusable templates for collaborative and distributed systems, we are proposing a declarative characterization of this kind of systems based on its observable characteristics, like the number of nodes, topology of the network, number of cases of the local case bases, overlapping between the case bases,and the type of argumentation and reasoning processes. We are evaluating the templates in the context of recommender systems. The recommender domain is specially appropriate to work with social networks topologies, because the recommendations can be biased towards the social environment of each node.

## References


1. Díaz-Agudo, B., González-Calero, P.A., Recio-García, J.A., Sánchez-Ruiz-Granados, A.A.: Building cbr systems with jcolibri. Sci. Comput. Program. 69(1-3), 68–75 (2007)
2. Ontañón, S., Plaza, E.: Arguments and counterexamples in case-based joint deliberation. In: Maudet, N., Parsons, S., Rahwan, I. (eds.) ArgMAS 2006. LNCS (LNAI), vol. 4766, pp. 36–53. Springer, Heidelberg (2007)
3. Leake, D.B. (ed.): Case-based reasoning: Experiences, lessons, and future directions. AAAI Press/MIT Press, Menlo Park (1996)
4. Plaza, E., McGinty, L.: Distributed case-based reasoning. Knowledge Eng. Review 20(3), 261–265 (2005)
5. McGinty, L., Smyth, B.: Collaborative case-based reasoning: Applications in personalised route planning. In: Aha, D.W., Watson, I. (eds.) ICCBR 2001. LNCS (LNAI), vol. 2080, pp. 362–376. Springer, Heidelberg (2001)
6. Ontañón, S., Plaza, E.: An argumentation-based framework for deliberation in multi-agent systems. In: Rahwan, I., Parsons, S., Reed, C. (eds.) Argumentation in Multi-Agent Systems. LNCS (LNAI), vol. 4946, pp. 178–196. Springer, Heidelberg (2008)
7. Grandison, T., Sloman, M.: A Survey of Trust in Internet Applications. IEEE Communications Surveys and Tutorials 3(4), 2–16 (2000)
8. Golbeck, J., Hendler, J.A.: Inferring binary trust relationships in web-based social networks. ACM Trans. Internet Techn. 6(4), 497–529 (2006)
9. W., M.D.: Recommending collaboration with social networks: a comparative evaluation. In: CHI 2003: Proceedings of the SIGCHI conference on Human factors in computing systems, pp. 593–600. ACM, New York (2003)
10. Golbeck, J.: Generating predictive movie recommendations from trust in social networks. In: iTrust, pp. 93–104 (2006)
11. Ziegler, C.N., Golbeck, J.: Investigating interactions of trust and interest similarity. Decision Support Systems 43(2), 460–475 (2007)
12. Zimmermann, H.J.: Fuzzy set theory—and its applications, 3rd edn. Kluwer Academic Publishers, Norwell (1996)

# Model of a Collaboration Environment for Knowledge Management in Competence-Based Learning

Różewski Przemysław and Ciszczyk Magdalena

West Pomeranian University of Technology in Szczecin, Faculty of Computer Science and Information Systems, ul. Żołnierska 49, 71-210 Szczecin, Poland
{prozewski,mciszczyk}wi.ps.pl

**Abstract.** In the article the authors propose a model of an educational social agents collaboration between students, teacher, and an e-learning information system (repository). The system will support the competence-based learning process by means of social network creation in order to exchange ontology in the repository environment. The proposed model consists of three integrated sub models: (i) knowledge management model, (ii) student – e-learning information system – teacher interaction organization model (iii) model for quantity assessment of collaboration.

**Keywords:** social networks creation, competence-based learning process, collaboration in learning process.

## 1 Introduction and Problem Statement

The way that educational organizations work has changed radically over the last 10 years with the reason being the new market situation of the educational organization. Nowadays, educational organizations operate as market companies. Globalization, Open and Distance Learning [28], knowledge economy [9] and rapid technology development are origins of this change of situation. The expectations of students and their potential employers regarding the learning outcomes and effectiveness of the learning process are growing. The answers to these expectations are the following ideas: (i) model of a learning process based on competency, (ii) a digital repository with high quality didactical material, (iii) and e-portfolio.

The learning process based on competency [36] is aimed at developing an employee according to the labour markets demand. In the framework of the European initiative – Bologna Process, competence is going to be the tool used in the student evaluation process and at the same time allowing concordance of learning process with the Open and Distance Learning concept [39]. The reason for continuous repository development is resulting from the present situation on the educational market. The students are possessing almost unlimited access to different educational resources, courses and curriculum [16]. In the case of Europe, the European Higher Education Area has been developed and provides a cohesive educational system [17]. In such situation, the quality of the repository has strong influence on the quality of

N.T. Nguyen, R. Kowalczyk, and S.-M. Chen (Eds.): ICCCI 2009, LNAI 5796, pp. 333–344, 2009.

the entire learning process. Additionally, the repository's content – knowledge, has to be constantly updated due to continuous de-actualization of domain knowledge.

The idea is to include students in the repository development process. On one hand, students can expand their knowledge in an active-based learning process, on the other hand, students can record their achievements relying on a market located e-portfolio mechanism.

Student participation in the repository development process and close collaboration with the teacher are the basis for a new context of the educational social network. Such social network changes the student-teacher interaction into student–e-learning information system–teacher interaction. A new e-learning platform and information/knowledge technology development gives opportunity to organize an alternative learning process based on the Internet. The e-learning information system takes over some of the teacher's responsibilities during the student-teacher interaction: (i) the e-learning information system supports traditional learning activities; (ii) the learning-teaching process is relocated to the Internet and takes the form of an e-learning process.

Universities actively support the concept of a social network based on the e-learning technology [7], [34]. The information technology gives possibility to arrange a more efficient student-teacher cooperation. However, such cooperation requires a model of educational social agent collaboration between the students, the teacher, and the e-learning information system (including repository).

In the paper authors consider the following educational situation: teacher has interest in developing the repository in a specific direction; student has interest in his/her competency development and e-portfolio improvement; there is a software environment, which gives possibility to develop and maintain didactical material repository based on the learning object (LO) [46] and appropriate information system [15]; the learning-teaching process is based on competency.

The authors propose a model of an e-learning information system. The system will support the competency-based learning process by means of social network creation in order to exchange ontology in repository environment. The proposed model consists on three integrated sub models: (i) knowledge management model, (ii) student–e-learning information system–teacher interaction organization model (iii) model for quantity assessment of student–e-learning information system–teacher interaction.

## 2 Background

### 2.1 Competence-Based Learning as a New Way of Learning

Research conducted by David C. McClelland proved that the school results and intelligence tests cannot be seen as measures of efficient executing of tasks. He gave an opinion on necessity to define new variables, which he called competences [21].

In the literature a huge number of studies about defining the nature and meaning of the term "competence" can be found [32]. The present theories of human resource management focus the employers assessment on the basis of acquired competences [32], [33]. This emanating from the labor market requirement has become a signal for changes in the learning process. In Europe, this changes in education are especially supported by the Bologna Process [40]. In the frames of the Bologna Process initiative, aimed at creating a European Higher Education Area, one of the goals is to introduce a common Qualification Framework based on competences [1]. In Europe, works to identify and describe the competencies for the learning programs are now in progress [42].

Supporting the concept of competence-based education or competence-based learning causes the search for such recognition of the term of competence, which identifies the learning process with competences. For this assumption the authors propose to use the definition of competence, which concerns the appropriate structure of knowledge [38]. This definition states that competence is the ability to find an effective way of using theoretical knowledge to solve the practical tasks and to verify the found solution. The basis of competence is procedural knowledge joined with appropriate theoretical knowledge. The competent specialist is very knowledgeable about the chosen domain on the practical level. This definition gives the ability to structurize knowledge and on to prepare, on the basis of this structures, appropriate didactic materials.

## 2.2 The Role of the Repository in the Teaching/Learning Process

The etymology of the term "repository" says that it is a place intended for the storage of records and official books with the ability to use them. The present sources signifi cantly extend the nature of functioning of the repositories. Currently they are a place for storing collections of digital documents, shared them in the network with the use of appropriate interface for a selected group of people or with unlimited access [10],[11],[29].

The main task of the repository is to ensure for each user the possibility to read, copy, distribute, print, search or link the full texts of scientific articles and other materials and documents placed in the repositories [11]. This concept is strongly supported by members of the Open Archives Initiative (OAI) [3],[37].

The existing repositories can be used for the purpose of the learning process as a source of outlook shaping, research way development or as a source of didactic materials. By using the repository, there are shared the elements of domain knowledge, mainly in the form of Learning Objects (LOs), which are interpreted as modules of knowledge that arise as a result of the analysis and division of knowledge into a "pieces" [22].

Developing the university repositories is determined by many factors. They become a bridge between the rapid outdating of domain knowledge and the time of reaction to these changes, giving the possibility to quickly adapt and update the didactic materials. A well-developed repository assures good quality of the learning process. In the evaluation of the universities rankings (eg, Webometrics Ranking of World Universities [43]) repositories are becoming the source of new perception regarding the quality of the entire university.

## 2.3 E-Portfolio as a Mechanism for Creating a Personal Development Plan

The first portfolios were prepared in a paper version, but information technology development changed their presentation form into a digital one. The adaptation of e-portfolios to the needs of various scientific fields has caused their use in the learning process, and as a tool to promote the concept of lifelong learning [5].

Nowadays, there is no single generally accepted definition of e-portfolio. However, some authors highlight that the main purpose of e-portfolio is to plan personal development [2], [41], [44]. Reflection is recognized as the main aspect in the process of creating an e-portfolio [1], [6], [8], [30], [44]. It requires analysis of individual activities and possessed experiences. Students try to understand, analyze and formulate the way of personal development. This process increases the awareness of learners and leads to a consciously and intentionally developed learning process [24].

### 2.4 Related Work

According to the scientific studies [19], [20], [31], [35] the stable expansion of e-learning tools and techniques results in creating new conditions and opportunities for realizing the learning process and for cooperation between its actors. Technology has opened a new range of possibilities. The emerging Learning Management Systems (LMS) like Moodle [http://moodle.org/], Blackboard [http://www.blackboard.com/], Oracle e-Leaning [http://ilearning.oracle.com/ilearn/en/learner/jsp/login.jsp] or Caroline [http://www.claroline.net/] are able to support the learning process and transform it from the traditional form into blended learning or distance learning.

Active learning is designed to stimulate the students' motivation and their interest in their own development. As a directed development path planning mechanism e-portfolio can be used. The motivation aspect in the learning process has been discussed in detail from the pedagogical point of view in [23]. Five motivation rules have been defined in this research. They are based on the ARCS model elaborated by Keller and its further extension on the basis of Kuhl, Corno and Zimmerman considerations [12]. This model states that the key elements of the learning process are attention, relevance, confidence, satisfaction, and self-regulatory. It defines conditions, in which students can have a high level of motivation and determination in the learning process. One of the thesis states that motivational gaps regarding a specific situation should be determined and then appropriate strategies should be applied to improve them.

The approach proposed by Kusztina [13] presents the motivation concept in terms of a cooperative game between the learning process participants. The repository mechanism is being introduced for the needs of the conducted game. New knowledge is loaded into the repository as a result of collaboration. Kusztina aims at a formal motivation model, which leads to finding balance between the student's motivation function and the teacher's motivation function.

Including repository in the learning process was also proposed in [25], [26] or [27]. However, there are still problems with defining content organization methods, granulating knowledge into the form of Learning Objects (LOs) and developing methods for repository management, maintenance and development. In [46] a method of preparing the LOs sequence was proposed. However, there is still an open issue the research of costing and resource constrains in the learning process. Regardless of how the mentioned above matters are solved, one problem still remains: conducting the learning process according to the concept of competence-based learning.

# 3 Collaboration Model between Learning Process Participants

## 3.1 The Concept of Collaboration for Knowledge Repository in Competence-Based Learning

In the competence-based learning existing information tools (such as a repository and e-portfolio) can be used and regarded as a final product of the learning process. A repository is used as a direct mechanism in the learning process, while the e-portfolio is an individual property of each student and represents his/her achievements in the learning process.

However, although use of these tools is helpful, it is insufficient for guaranteeing acquisition of competence. For competence-based learning, it is necessary to introduce: new organization of the learning process and collaboration "student-information system-teacher", which considers three levels of learning process design (Fig. 1).

The first level (knowledge level) concerns proper preparation of the learning process from the point of view of the structure of domain knowledge. In accordance with the accepted definition of knowledge, the knowledge structurization process includes preparation of a portion of theoretical and procedural knowledge, and of corresponding test tasks, to verify students' skills/abilities.

The second level (social network level) is the basis for determining conditions for cooperation between the teacher and the students. This cooperation creates a new dimension for social networks, described by appropriate functions of motivation.

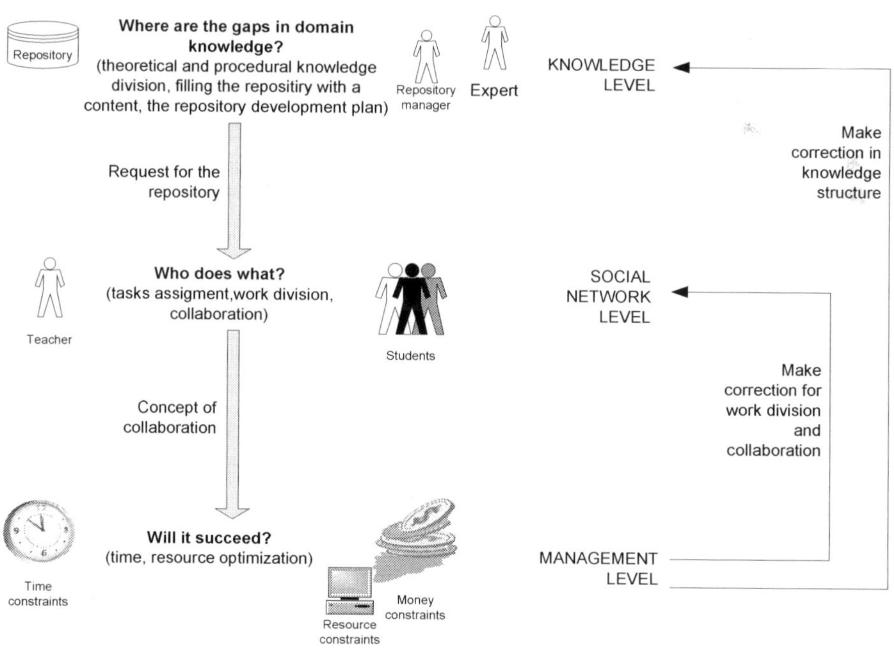


**Fig. 1.** Model of collaboration environment for knowledge management in competence based learning

Checking the conditions of cooperation takes place at the management level. The established cooperation conditions are analyzed from the point of view of the existing constrains (human resource, time resource). The obtained results can be a signal to change the cooperation model. In addition, feedback information is transmitted, concerning further repository development and learning process participants collaboration design.

The combination of these components allows organizing the learning process in accordance with the requirements for competence-based learning, using for this purpose the accepted model of knowledge management, and involving students, who become active and equal participants of the learning process and consciously create their own cognitive process.

### 3.2 Knowledge Level

The knowledge level is based on the heuristic algorithm presented in [14]. The main task of this algorithm is to present the domain knowledge representation in accordance with a specific educational situation. Moreover, the algorithm allows dividing knowledge in accordance with the LOs philosophy, which is based on knowledge modules in digital form, shared in a network, and reusable in a learning process [22]. More details about knowledge division process and LOs sequence preparing can be found in [46].

The methodology used for this purpose should carry out to determine the proportion and depth of domain knowledge in the specified course/subject. It is also necessary to assess the domain knowledge covering completeness by the didactic materials. The point of departure for the model is domain ontology and its division into the theoretical and procedural knowledge excerpts.

The person responsible for the repository content has to ensure a differentiated complexity level for the elaborated topics as well as the decreasing the disproportion between the didactic materials content for the different topics.

The idea is to base the repository development on students' tasks. The task solution prepared by a student can be loaded into the repository when a different method for solving the given problem is used, or when there is a new task solved using a known method or the task and the solution represent a new research area but are in accordance with the goals and scope of the learning process. From the student's point of view the students solutions loaded into the repository can be the basis for e-portfolio and personal learning path development.

### 3.3 Social Network Level

The repository development plan based on ontology is a crucial element to forming cooperation in the learning process. The social network created for the collaboration process consists of teacher, students and repository, which is an activity "accelerated" mechanism in this network (Fig. 2).

The learning process participants' behaviors can be described by appropriate functions of motivation.

The teacher's motivation function aims at influencing the students in a way that will lead to filling the repository with new content. These solutions will be proof of the high competence level acquired by the students. For N – teacher, the teacher's motivation function can be described by a vector $\sigma^N(r_i) = \overline{x}(r_i)$, where $\overline{x}(r_i)$ - resources assigned to solving task $r_i$ (like didactic materials, teacher's time, software, hardware, etc). The teacher's motivation function depends on the complexity of tasks, tasks relevance and other teacher's preferences.

The student's motivation function results from the accepted development plan. For a specified subject, students will decide about their own involvement in the learning process. This strategy depends on many factors: (i) the repository content, (ii) tasks' complexity level, (iii) time needed to solve a task, (iv) potential grade/mark received for the tasks sequent execution. For an actual set of tasks in the repository, the student's motivation function is personal developed for each student (S) and can be described as $\sigma_j^S(r_i) = F(W(s_j), H(r_i), C_j^S(r_i), F^S)$, where $W(s_j)$ - student's base knowledge, $H(r_i)$ - grade/mark, that can be given by a teacher $C_j^S(r_i)$ - costs bear by the student regarding solving the task, $F^S$ - other student's preferences (e.g. student's goals and constrains in the learning process).

The synergy effect and, on its basis, the repository enrichment with new content, requires finding balance between teacher's motivation $\sigma^N$ and student's motivation $\sigma_j^S$.

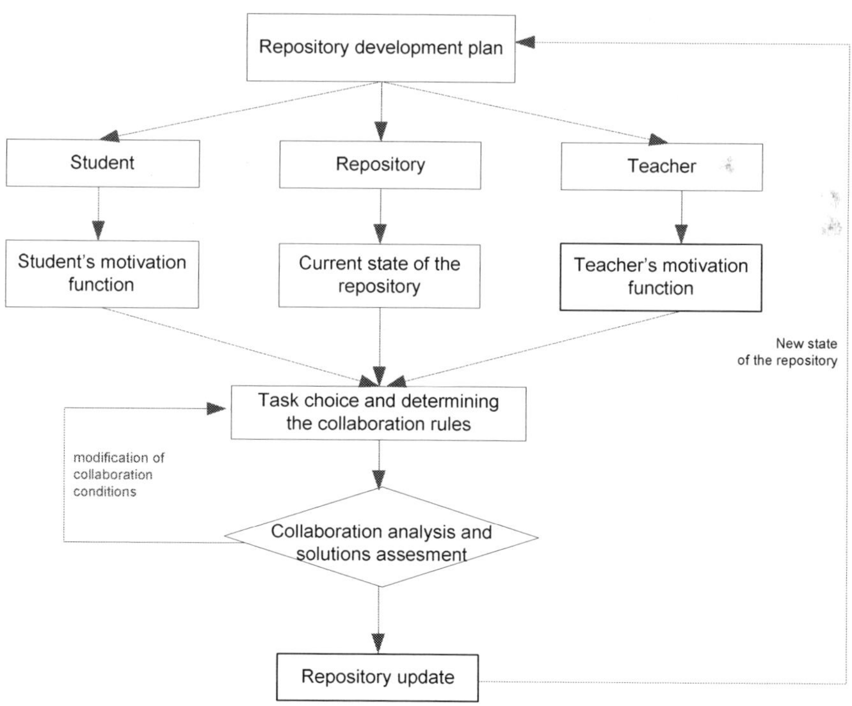


**Fig. 2.** Organization of social networks in the learning process

The form of satisfying the teacher's interests is placing in the repository a properly solved task of a significantly high level of complexity. The form of satisfying the student's interests is minimal summary time costs while obtaining a high grade, which also depends on the complexity level of the task.

Balance between the teacher's and student's motivation function defines the goal function of the task choice: $\Phi(y_j^i) = \alpha\sigma^N + \sigma_j^S = \max_Y, \quad i = 1,2,\ldots,i^*, \ j = 1,2,\ldots j^*$, where α is a weigh coefficient.

The motivation model can be formulated in terms of games theory, as was shown in [18]. However, due to the fact that cooperation between the teacher and the student is conducted with time constraints, financial constrains and resource constraints, simulation tools (e.g. simulation tool Arena) can be used for its analysis.

### 3.4 Management Level

The target state of the repository should contain a set of didactic material units which will completely cover the prepared ontology. The condition of the repository development by prepared by a students' content will be determined by teacher high assessment.

The effectiveness of this type of work with the repository depends on the initial state of the repository, the quantity of students involved in repository development, the state of the desired repository level, meeting the time and resource utilization constraints. The accepted knowledge management and the learning process participants' motivation strategy requires a verification /calculation mechanism.

The analysis of the motivation model assumptions, their adjustment and the effectiveness of the repository development process can be verified through the use of a simulation model. The simulation model lets to analyze the parameters of ill-formalized learning process and proposed cooperation between learning process participants. The assumptions like time between students tasks arrival, time of tasks checking process or delay time became the basis for simulation model elaboration in Arena (Fig. 3). Arena is a simulation tool, developed by Rockwell Software, which allows to model and analyze the real life processes. Some aspects of using Arena for learning process analysis can be found in [45].

Currently the simulation experiment allows analyzing the queue's parameters on a teacher's workstation as well as defining the students' service time by specified input parameters. Fig. 3 presents the results of such simulation. For 55 students, time interval of 6 days, daily time for tasks' examination – 3 hours, expected time for each student – 20 minutes, correction time – 1 day and predicted probability of tasks' evaluation: 70% – exit with promotion without repository development, 15% – placing solution in the repository, 15% – sending back for correction; one can see that a queue on the teacher's workstation occurs. This queue shows that accepted assumptions for the learning process realization should be changed. As it was checked in the model, the time interval can be extended to 8 days to realize the other assumptions.

In addition, simulation allows assessing: limited access to software and hardware resources, maximum size of social networks queue, cost of teacher's work. On the basis of statistical data, the group management strategy can be modified and appropriate adjustments can be made in the repository development plan.

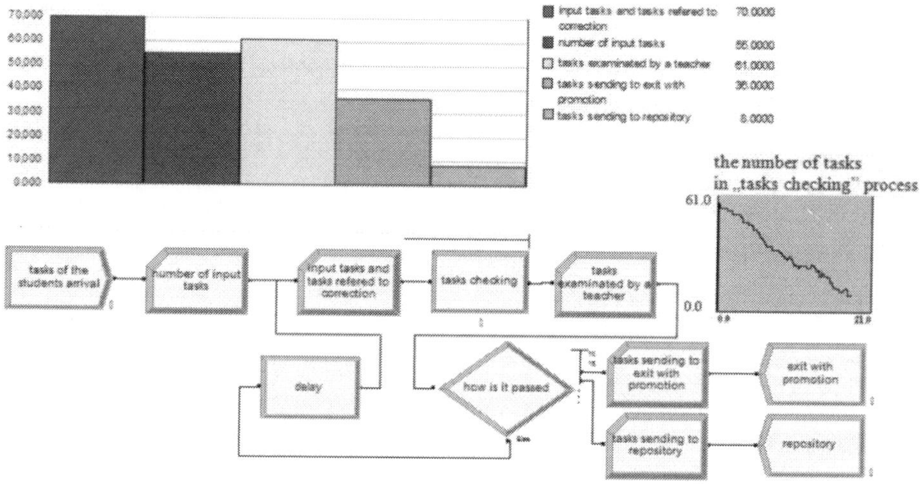


**Fig. 3.** Simulation model of teacher-students collaboration in tasks assessment (Arena software)

## 4 Conclusion

The model of social agent collaboration between students, teacher and an e-learning information system (repository) allows to:

- assessing the degree of the repository fulfillment and its further development plan;
- checking the competence acquired path through the appropriate domain knowledge structurization;
- measuring competence growth through analysis of the domain ontology graph coverage;
- elaborating a model of teacher behavior with successive groups of students;
- modeling and reviewing the group management strategy in accordance with teacher and students preferences.

## References

1. Adorni, G., Coccoli, M.: Implementation of e-Portfolio for Sports Physiotherapy Experts. In: Conference Proceedings ICL 2007 (2007), `http://hal.archives-ouvertes.fr/hal-00197269/en/`
2. Barrett, H.C.: My Online Portfolio Adventure, `http://helenbarrett.com/myportfolio/versions.html#1`
3. Björk, B.-C.: Open access to scientific publications - an analysis of the barriers to change? Information Research 9(2) (2004)
4. Bologna Working Group on Qualifications Frameworks, A Framework for Qualifications of the European Higher Education Area. Report, `http://www.bologna-bergen2005.no/`
5. Cambridge, D.: Layering Networked and Symphonic Selves: A Critical Role for ePortfolios in Employability through Integrative Learning. Campus-Wide Information Systems 24(4), 244–262 (2008)

6. Chang, C.-C.: Enhancing self-perceived effects using Web-based portfolio assessment. Computers in Human Behavior 24, 1753–1771 (2008)
7. Daniel, B., McCalla, G., Schwier, R.: Social Network Analysis techniques: implications for information and knowledge sharing in virtual learning communities. International Journal of Advanced Media and Communication 2(2), 20–34 (2008)
8. e-Portfolio, The Pennsylvania State University, `http://portfolio.psu.edu/reflect_planning`
9. Freeman, L.A.: Information Systems Knowledge: Foundations, Definitions, and Applications. Information Systems Frontiers 3(2), 249–266 (2001)
10. Jones, R.: Giving birth to next generation repositories. International Journal of Information Management 27, 154–158 (2007)
11. Xia, J.: A comparison of Subject and Institutional Repositories in Self Archiving Practices. The Journal of Academic Librarianship 34(6), 489–495 (2008)
12. Keller, J.M.: First principles of motivation to learn and e3-learning. Distance Education 29(2), 175–185 (2008)
13. Kusztina, E.: Concept of an open and distance learning information system. Publisher house of Szczecin University Technology, Szczecin (2006) (in polish)
14. Kusztina, E., Zaikin, O., Ciszczyk, M., Tadeusiewicz, R.: Quality factors for knowledge repository: based on e-Quality project. In: Proceedings of the 14-th Conference EUNIS 2008, Aarhus, Dania (2008), `http://eunis.dk/`
15. Kusztina, E., Zaikin, O., Różewski, P.: On the knowledge repository design and management in E-Learning. In: Lu, J., Ruan, D., Zhang, G. (eds.) E-Service Intelligence: Methodologies, Technologies and applications. Studies in Computational Intelligence, vol. 37, pp. 497–517. Springer, Heidelberg (2007)
16. Kusztina, E., Zaikin, O., Różewski, P., Małachowski, B.: Cost estimation algorithm and decision-making model for curriculum modification in educational organization. European Journal of Operational Research 197(2), 752–763 (2009)
17. Kusztina, E., Zaikine, O., Różewski, P., Tadeusiewicz, R.: Competency framework in Open and Distance Learning. In: Proceedings of the 12th Conference of European University Information Systems EUNIS 2006, Tartu, Estonia, pp. 186–193 (2006)
18. Kusztina, E., Zaikin, O., Tadeusiewicz, R.: Motivation model in Open Distance Learning Systems. Control & Cybernetics (submitted)
19. Martineau, P.: Teaching with Technology. The Education Digest 74(7), 14–18 (2009)
20. Martinez, A., Dimitriadisb, Y., Rubiac, B., Gomez, E., de la Fuentea, P.: Combining qualitative evaluation and social network analysis for the study of classroom social interactions. Computers & Education 41, 353–368 (2003)
21. McClelland, D.: Testing for competence rather than for intelligence. American Psychologist Journal 28, 1–14 (1973)
22. McGreal, R.: A Typology of Learning Object Repositories. In: Adelsberger, H.H., Kinshuk, Pawlowski, J.M., Sampson, D. (eds.) Handbook on Information Technologies for Education and Training, 2nd edn., pp. 5–28. Springer, Heidelberg (2008)
23. Menchaca, M.P., Bekele, T.A.: Learner and instructor identified success factors in distance education. Distance Education 29(3), 231–252 (2008)
24. MOSEP Project 2006-2008, More Self-Esteem with my e-Portfolio, http://www.mosep.org
25. Nachmias, R., Leiba, M., Ram, J.: European virtual N2L University (EVNU) – educational exchange in NANO2LIFE – a network of excellence, `http://www.biu.ac.il/bar-e-learn/eden2007/Leiba_Nachmias_Ram.doc`

26. Paquette, G., Marino, O., Lundgren-Cayrol, K., Leonard, M., de la Teja, I.: Learning Design Repositories – Structure Ontology and Processes, http://dspace.ou.nl/bitstream/1820/765/1/Paper28.pdf
27. Paralic, J., Paralic, M.: Technologies for Support of Knowledge Creation Processes. In: 18th International Conference on Information and Intelligent Systems IIS 2007, pp. 55–61 (2007), http://www.foi.hr/CMS_home/znan_strucni_rad/konferencije/IIS/2007/papers/T01_08.pdf
28. Patru, M., Khvilon, E. (eds.): Open and distance learning: trends, policy and strategy considerations UNESCO, ED.2003/WS/50 (2002)
29. Peters, T.A.: Digital Repositories: Individual, Discipline-based, Institutional, Consortial Or National? The Journal of Academic Librarianship 28(6), 414–417 (2002)
30. Lin, Q.: Preservice teachers' learning experiences of constructing e-portfolios online. Internet and Higher Education 11, 194–200 (2008)
31. Saltz, J.S., Hiltz, S.R., Turoff, M., Passerini, K.: Increasing Participation in Distance Learning Courses. IEEE Internet Computing 11(3), 36–44 (2007)
32. Sampson, D., Fytros, D.: Competence Models in Technology-Enhanced Competence-Based Learning. In: Adelsberger, H.H., Kinshuk, Pawlowski, J.M., Sampson, D. (eds.) Handbook on Information Technologies for Education and Training, 2nd edn., pp. 155–177. Springer, Heidelberg (2008)
33. Sanchez, R.: Understanding competence-based management, identifying and managing five modes of competence. Journal of Business Research 57, 518–532 (2004)
34. Shen, D., Nuankhieo, P., Huang, X., Amelung, C., Laffey, J.: Using Social Network Analysis to Understand Sense of Community in an Online Learning Environment. Journal of Educational Computing Research 39(1), 17–36 (2008)
35. Sims, R.: Rethinking (e)learning: a manifesto for connected generations. Distance Education 29(2), 153–164 (2008)
36. Stefanov, K., Nikolova, N., Ilieva, M., Stefanova, E.: Turning university professors into competent learners. International Journal Emerging Technologies in Learning, iJET 3(S1/2008), 46–52 (2008)
37. Swan, A.: Open Access: What is it and why should we have it?, http://eprints.ecs.soton.ac.uk/13028/
38. Tadeusiewicz, R., Choraś, R.S., Rudowski, R. (eds.): Leksykon haseł związanych z e-nauczaniem. p.16. Wydawnictwo Wyższej Szkoły Humanistyczno-Ekonomicznej, Łódź (2007)
39. Tait, A.: Open and Distance Learning Policy in the European Union 1985-1995. Higher Education Policy 9(3), 221–238 (1996)
40. The Bologna Declaration on the European space for higher education: an explanation, http://ec.europa.eu/education/policies/educ/bologna/bologna.pdf
41. Trestini, M.: The e-portfolio in a professional training scheme: what is at stake and the limits of the scheme, http://dspace.ou.nl/bitstream/1820/1680/14/The-eportfolio-in-a-professional-training-scheme-what-is-at-stake-and-the-limits-of-the-scheme.pdf
42. Tuning Educational Structure. In: Europe Project, http://www.relint.deusto.es/TuningProject/index.htm
43. Webometrics Ranking of World's Universities, http://www.webometrics.info/

44. W. Aalderink: Digital portfolio: tool for flexible learning and teaching in competency focused higher education, In: Proceedings of the 10th Conference EUNIS (2004), `http://www.surfspace.nl/en/Themas/portfolio/Start/Documents/wijnandaalderink-eunis2004.pdf`
45. Zaikin, O.: Queuing modeling of supply chain in intelligent production. Printing & Publishing Lab Faculty of Computer Science and Information Systems Technical University of Szczecin, Szczecin (2002)
46. Zaikine, O., Kushtina, E., Różewski, P.: Model and algorithm of the conceptual scheme formation for knowledge domain in distance learning. European Journal of Operational Research 175(3), 1379–1399 (2006)

# PlWiki – A Generic Semantic Wiki Architecture*

Grzegorz J. Nalepa

Institute of Automatics,
AGH University of Science and Technology,
Al. Mickiewicza 30, 30-059 Kraków, Poland
gjn@agh.edu.pl

**Abstract.** The paper presents a new semantic wiki architecture called *PlWiki*. The most important concept is to provide a strong knowledge representation and reasoning with Horn clauses-based representation. The idea is to use Prolog clauses on the lower level to represent facts and relations, as well as define rules on top of them. On the other hand a higher-level Semantic Web layer using RDF support is provided. This allows for compatibility with Semantic Media Wiki while offering improved representation and reasoning capabilities. Another important idea is provide an extension to already available flexible wiki solution (DokuWiki) instead of modifying existing wiki engine.

## 1 Introduction

Recently the most important development of the Internet concerned not the lower network network layers, but the higher application or service layers related to the Web technology. This is mainly due to the fact, that while the speed and storage capacities of the Web increased by orders of magnitude, its search and processing capabilities remained almost unchanged on the conceptual level.

This phenomena led, almost a decade ago, to the proposal of the *Semantic Web* [1]. In this architecture a number of higher level semantic facilities built on top of the Web would allow not just to *search data* but to *reason with knowledge*. In fact, this was the point where the focus of the Web development moved from *content* (data) to *knowledge* (in a broad sense). A decade later, number of semantic technologies is available and widely used, starting from the data structuring XML, to meta-data annotations with RDF and ontologies with RDFS and OWL (observe the well-known Semantic Web layer cake). While these technologies provided knowledge encoding and representation solutions, the challenge remains to provide an efficient knowledge processing and reasoning with rules on the Web. This is in fact the point, where most of the current Semantic Web research focuses. Recent rule standards from W3C include RIF and SWRL.

Besides knowledge representation and reasoning, a sensible knowledge engineering solution for the Web is another important challenge. While the Semantic Web initiative targets mainly representation aspects, it does not directly address

* The paper is carried out within the AGH UST Project No. 10.10.120.105.

N.T. Nguyen, R. Kowalczyk, and S.-M. Chen (Eds.): ICCCI 2009, LNAI 5796, pp. 345–356, 2009.

the specific problems stemming from the massively parallel and collaborative nature of the Web. Social networks, that provide specific services on top of the Web and the Semantic Web, try to cope with these problems. Recently the technology of the wiki systems has gained importance with respect to the collaborative knowledge acquisition and engineering. The development of *semantic wikis* [2,3,4] allowed to use the Semantic Web methods and tools on top of the existing content-centered wiki solutions.

Existing semantic wikis allow for introduction of semantic information (e.g. meta-data, ontologies) into a wiki. In fact, they often allow to build a wiki around an ontology, which improves their conceptual coherence. Most of the semantic wikis reached a stage where the reasoning capabilities have to be added. This is where some limitations of existing solutions become exposed.

In this paper a new semantic wiki architecture called *PlWiki* is considered. The most important concept is to provide a strong knowledge representation and reasoning with Horn clauses-based representation. But instead of building directly on top of OWL [5] and SWRL [6] a more generic solution is proposed. The idea is to use Prolog [7] clauses to represent facts and define rules on the lower level. On the other hand a higher level Semantic Web layer using RDF and OWL support is provided. Another important idea is not to develop an entirely new wiki engine (e.g. [2]) but to provide an extension to an existing well-established wiki solution. It is argued, that such a generic architecture is both more flexible and efficient. This approach is loosely based on some preliminary ideas earlier considered in [8].

The rest of the paper is organized as follows: In Sect. 2 wikis as web-based knowledge engineering systems are discussed. Then in Sect. 3 the development of semantic and knowledge wikis is considered. The Sect. 4 gives the main motivation of the research. The requirements for the PlWiki system are specified in Sect. 5. The design of the PlWiki plugin for the DokuWiki system is presented in Sect. 6. The use of the plugin is considered on the example introduced in Sect. 7, and the semantic layer is discussed in Sect. 8. Finally a short evaluation of the approach as well as directions for the future work are given in Sect. 9.

## 2 Wikis Technology

Wiki systems appeared in the mid 90s. According to Wikipedia the first system called "wiki" (WikiWikiWeb) was established 15 years ago. Tho goal of these systems was to provide a conceptually simple tool for massively collaborative knowledge sharing and social communication. Wikis were meant to help build certain communities interested in given topics. Clearly some of them grew large and general, such as the Wikipedia.

A wiki system is a community-driven collaboration tool. It allows users to build content in the form of the so-called wiki pages, as well as uploaded media files. Wikipages are plain text documents containing special wiki markup (e.g. for structuring content) thus creating the so-called wikitext. The wikitext is simplistic and human readable, making it a much more accessible tool than

HTML/XML. Pages are identified by a unique keyword (name) and usually grouped within the so-called namespaces. Pages are linked to each other and to external websites creating a hyperwikitext structure.

An important feature of wikis is the integrated version control functionality, very helpful in a collaborative environment. It allows registering all subsequent versions of every page, thus allowing to see introduced differences. All wiki edits may be identified by user names and time stamps, so it is possible to recreate any previous state of the wiki at any given time.

From the technical point of view a wiki has a regular web-based client-server architecture. It is run on the web server and accessed by a regular browser. Wikis introduce a range of access control mechanisms from simple ones, to full-fledged ACL (Access Control Lists) solutions. On the server side wikis require different runtime environment (e.g. PHP/JSP/Python), possibly with a relational database system. A comprehensive comparison of different wiki systems can be found on `http://www.wikimatrix.org`.

One of the most interesting wiki systems for developers is DokuWiki (`http://www.dokuwiki.org`). It is designed to be portable, easy to use and set up. Like number of other solutions DokuWiki is based on PHP. However, it does not require any relational database back-end. It allows for image embedding, and file upload and download. Pages can be arranged into namespaces which act as a tree-like hierarchy similar to directory structure. It provides syntax highlighting for in-page embedded code of programming languages such as: C/C++, Java, XML and others, using GeSHi (`qbnz.com/highlighter`). Furthermore, it supports extensive user authentication and authorization mechanisms including ACL. Its modularized architecture allows the user to extend DokuWiki with plugins which provide additional syntax and functionality. A large number of plugins is available. The templates mechanism provides an easy way to change the presentation layer of the wiki.

All wiki systems provide an abstract representation of the content they store. They all provide standard searching capabilities. However, they lack facilities helping in expressing the semantics of the stored content.[1] This is especially important in the case of collaborative systems, where number of users work together on the content. This is why wikis became one of the main applications and testing areas for the Semantic Web technologies.

## 3 A Challenge of Semantic Wikis

A step in the direction of enriching standard wikis with the semantic information has been performed by the introduction of the so-called *semantic wikis*, such as the IkeWiki [2], OntoWiki [9], SemanticMediaWiki [3], or SweetWiki [4]. In such systems the standard wikitext is extended with semantic annotations. These include relations (represented as RDF triples) and categories (here RDFS is needed). It is possible to query the semantic knowledge, thus providing dynamic

[1] Besides simple tagging mechanisms, that can later be used to create the so-called *folksonomies*.

wiki pages. Ultimately these extension can also allow for building an ontology of the domain to which the content of the wiki is related. This extension introduces not just new content engineering possibilities, but also semantic search and analysis of the content.

However, from the knowledge engineering point of view, expressing semantics is not enough. In fact a knowledge-based system should provide effective knowledge representation and processing methods. In order to extend semantic wikis to knowledge-based systems, ideas to use a problem-solving knowledge have been introduced. An example of such a system is the *KnowWE* semantic wiki [10,11]. In such a system the semantic knowledge is extended with the problem-solving domain-specific knowledge. The system allows for introducing knowledge expressed with decision rules and trees related to the domain ontology. Several semantic wiki systems are available, most of them in the development stage providing demo versions.[2] A recent FP7 project Kiwi (`http://www.kiwi-project.eu`) aims at providing a collaborative knowledge management based on semantic wikis (it is the continuation of IkeWiki effort).

In this paper a generic solution based on the use of Prolog as the language for expressing both the semantics, and the knowledge processing is presented.

## 4 Motivation and Objectives

The semantic wiki technology is a young one. Multiple systems are developed to test new ideas and features. However, number of conceptual challenges remain. Some of the persistent problems are:

- an expressive yet effective (in terms of inference) knowledge representation, allowing for explicitly representing the semantics of the wiki content,
- powerful query and inference facilities, that enable reasoning on top of the gathered knowledge; in fact they are the main factor limiting the practical usability of the knowledge,
- interfaces helping users to encode and use the knowledge they poses,
- integration with the existing technologies that improves the portability and makes the development of the system easier.

All of the existing semantic wiki solutions address these problems in various manners [4]. Some of the most common approaches to cope with these problems include the extensive use of selected Semantic Web technologies to introduce well-founded semantics. This includes the use of RDF for meta-data, as well as RDFS, and possibly OWL, for ontology management, and SPARQL as the query language [3]. Some other [12] introduce extended knowledge representation for problem solving. User interfaces are usually based on simple forms helping to input semantic annotations, as well as editors highlighting the wiki markup [4]. Most of these solutions modify some existing wiki engines, e.g. Media Wiki that powers Wikipedia. In general, they still lack universal rule representation (mostly

[2] See `http://semanticweb.org/wiki/Semantic_Wiki_State_Of_The_Art`

due to the development stage of SWRL). However, it is an ongoing research where an optimal solution is hard to find.

The approach discussed in this paper is different. The basic idea is to allow the use of a logical knowledge representation based on Horn clauses [13] for facts, relations and rules, as well as dynamic queries. This allows not only to represent facts, but also introduce rules for inference. On top of this the Semantic Web layer with RDF may also be provided. The approach is based on the concept of using the Prolog language interface. This also opens up possibilities of powerful querying mechanism, more powerful than SPARQL (while compatibility layer compatible with SPARQL is provided). The solution is developed as an optional extension to an existing modular wiki engine of DokuWiki.

The main objectives of this approach are to enhance both representation and inference features, allow for a complete rule framework in the wiki, and to use a clean integration approach with an existing system. In the next section more detailed system requirements are presented.

## 5 PlWiki Requirements

Starting from the motivation outlined in the previous section, the following main requirements concerning the new wiki solution have been formulated:

1. provide rich knowledge representation, including rules,
2. allow for an efficient and flexible reasoning in the wiki,
3. expressive power equivalent to Horn clauses,
4. provide a Semantic Web layer for the knowledge engineer,
5. extend an existing well-established wiki engine,
6. improve speed and portability for an easy deployment,
7. allow for meta-knowledge suitable for wiki knowledge evaluation.

A decision has been made to build the new wiki with use of the Prolog language [7]. This allows to meet first three requirements since Prolog programs assure both generic and flexible representation with Horn clauses, by providing fact and rule representation. Thus it is possible the represent the domain specific knowledge and reasoning procedures with the same generic representation.

The next requirement is a natural one, considering the existing Semantic Web stack, on which semantic wikis are build. This means the support for meta-data encoding with RDF and ontologies with RDFS (possibly with OWL too). Such a solution allows to extend the wiki using existing ontologies, optionally build with other semantic wikis.

The fifth requirement is a conscious decision based on number of experiences with other semantic wiki engines. There are tens of wiki engines available (see `www.wikimatrix.org`). Most of them are similar w.r.t to main concepts and features. However, there are number of differences when it comes to the wikitext syntax, implementation and runtime environment, as well as extra features. This is why, instead of building yet another wiki engine, or modify an existing one, another solution is proposed. The idea is to use a ready, flexible and extensible wiki engine, that could be optionally extended with knowledge representation and processing capabilities.

The next requirement concerns an easy deployment of the new system. Number of existing solutions impose high requirements on the runtime environment for the wiki, e.g. a database server, a J2EE stack with extra libraries, etc. This makes the installation of such systems harder, limits their portability (i.e. number of hosting solutions do not meet all of these requirements), and lowers their speed (Java-based solutions are often slower than Python or PHP-based ones).

The last requirement concerns the ability to analyze wiki knowledge using procedures specified in the wiki, using the same representation as the wiki contents. This is considered for the future research [14].

In order to meet these requirement a fast, flexible and portable Prolog implementation has been chosen. It provides a rich library stack for the Semantic Web compatibility. In the next section the design of the PlWiki extension for the DokuWiki system built with use of the SWI-Prolog environment is given.

## 6 PlWiki System Design

The main goal of the new knowledge wiki design is to deliver a generic and flexible solution. Instead of modifying an existing wiki engine or implementing a new one, a development of an extension of the DokuWiki system was chosen. To provide a rich knowledge representation and reasoning for the Semantic Web, the SWI-Prolog environment was selected. The basic idea is to build a layered knowledge wiki architecture, where the expressive Prolog representation is used on the lowest knowledge level. This representation is embedded within the wiki text as an optional extension. On top of it number of layers are provided. These include standard meta-data descriptions with RDF and ontologies specification solutions with RDFS and OWL.

The PlWiki stack can be observed in Fig. 1. The stack is based on a simple runtime including the Unix environment with the Unix filesystem, the Apache web server and the PHP stack. Using this runtime the standard DokuWiki installation is run. The PlWiki functionality is implemented with the use of an optional plugin allowing to enrich the wikitext with Prolog clauses, as well run the SWI-Prolog interpreter. It is also possible to extend the wikitext with explicit semantical information encoded with the use of RDF and possibly OWL representation. This layer uses the Semantic Web library provided by SWI-Prolog. An optional decision rule layer is also considered with the use of the HeaRT runtime for the $XTT^2$ framework [15,16].

The main layer interfacing with the DokuWiki engine is presented in Fig. 2. The figure shows the dataflow in the DokuWiki system. DokuWiki provides a flexible plugin system, providing five kinds of plugins (see `www.dokuwiki.org/devel:plugins`):

- *Syntax Plugins*, extending the wikitext syntax,
- *Action Plugins*, redefining selected core wiki operations, (e.g. saving wikipages),

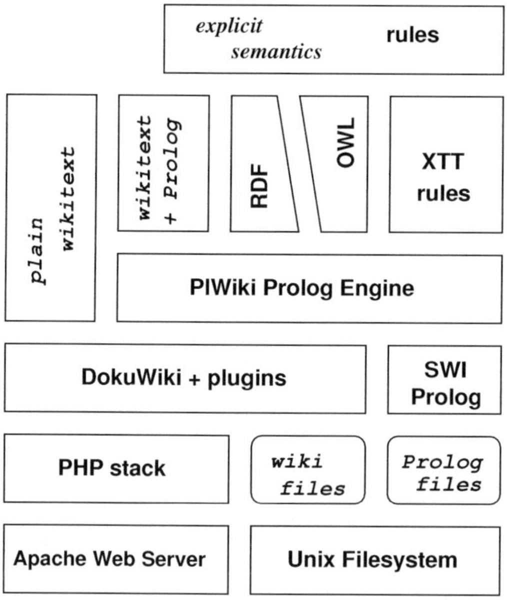


**Fig. 1.** The PlWiki stack

- *Admin Plugins*, providing extra administration functionality,
- *Helper Plugins*, supporting other plugins with generic functions,
- *Renderer Plugins*, allowing to create new export modes (possibly replacing the standard XHTML renderer).

The current version of PlWiki implements both the *Syntax* and *Renderer* functionality. Text-based wikipages are fed to a lexical analyzer (Lexer) which identifies the special wiki markup. The standard DokuWiki markup is extended by a special `<pl>...</pl>` markup that contains Prolog clauses. The stream of tokens is then passed to the Helper that transforms it to special renderer instructions that are parsed by the Parser. The final stage is the Renderer, responsible for creating a client-visible output (e.g. XHTML). In this stage the second part of the plugin is used for running the Prolog interpreter.

The detailed functionality of the PlWiki Syntax Plugin includes parsing the Prolog code embedded in the wikitext, and generating the knowledge base composed of files containing the Prolog code, where each wikipage has a corresponding file in the knowledge base. The PlWiki Renderer plugin is responsible for executing the Prolog interpreter with a given goal, and rendering the results via the standard DokuWiki mechanism.

The PlWiki framework uses the SWI-Prolog environment, licensed under the Lesser GNU Public License (see `www.swi-prolog.org`). It is a mature implementation widely used in research and education as well as for commercial applications. It provides a fast and scalable development environment, including

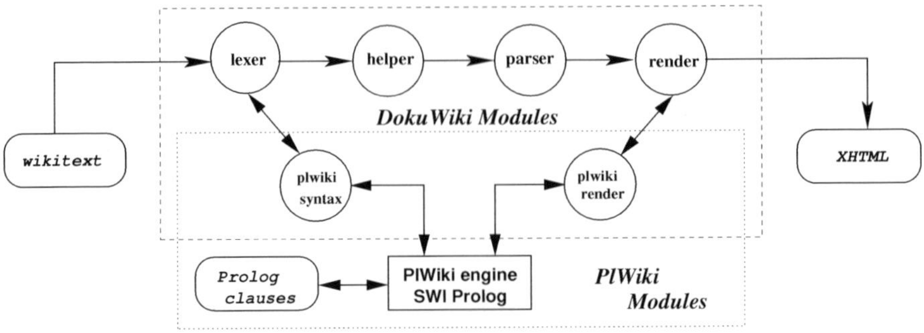


**Fig. 2.** PlWiki plugins

graphics, libraries and interface packages, portable to many platforms, including Unix/Linux platforms, Windows, and MacOS X. SWI-Prolog provides a rich set of libraries, including the *semweb* library for dealing with standards from the W3C standard for the Semantic Web (RDF, RDFS and OWL). This infrastructure is modular, consisting of Prolog packages for reading, querying and storing Semantic Web documents.

One should keep in mind, that the Prolog-based representation is quite close to the natural language. Not only on the semantical level, but to a degree also on the syntactic level. It is possible thanks to the operator definition.

Currently the PlWiki system is under heavy development. The system is being developed by Michał Kotra as part of his MSc Thesis. See `http://home.agh.edu.pl/gjn` for more information.

## 7 Prolog Representation Example

Below basic use examples of the generic Prolog representation are given.

```
<pl>
capital(poland,warsaw).
country(poland).
country(germany).
</pl>
```

This simple statement adds two facts to the knowledge base. The plugin invocation is performed using the predefined syntax. To actually specify the goal (query) for the interpreter the following syntax is used:

```
<pl goal="coutry(X),write(X),nl,fail"></pl>
```

It is possible to combine these two, as follows:

```
<pl goal="country(X),write(X),nl,fail">
country(france).
```

```
country(spain).
</pl>
```

It is possible to specify a given *scope* of the query (in terms of wiki namespaces):

```
<pl goal="country(X),write(X),nl,fail"
    scope="prolog:examples"></pl>
```

A bidirectional interface, allowing to query the wiki contents from the Prolog code is also available, e.g.:

```
<pl goal="consult('lib/plugins/prolog/plwiki.pl'),
          wikiconsult('plwiki/pluginapi'),list."></pl>
```

There are several options how to analyze the wiki knowledge base (that is Prolog files built and extracted from wiki pages). A basic approach is to combine all clauses. More advanced uses allow to select pages (e.g. given namespace) that are to be analyzed.

On top of the basic Prolog syntax, semantic enhancements are possible. These can be easily mapped to Prolog clauses.

## 8 The PlWiki Semantic Layer

Besides the generic Prolog-based knowledge representation features based on pure Prolog clauses, typical semantic wiki features are supported. Semantic Media Wiki (SMW) [3], a standard semantic wiki solution, provides a simple yet flexible mechanism for annotating categories, and properties. In the first version of PlWiki three main features are considered:

- categories definitions as in SMW,
- simple queries from SMW (with SPARQL queries in the future), and
- generic RDF annotations.

To provide a better compatibility with existing solutions parsing of SMW wikitext is provided, with a corresponding Prolog representation available. The wiki user can use the SMW syntax directly in DokuWiki to enter wikitext. The PlWiki plugins transforms the wikitext to Prolog clauses, asserted to the internal knowledge base. In fact these clasues could also be introduced by using the PlWiki `<pl></pl>` tags.

Examples are as follows, with the SMW syntax given first, and the corresponding Prolog representation below.

```
[[Category:Cities]] Warsaw is in Poland.
    wiki_category('Cities,'Warsaw').
Warsaw is [[capital of::Poland]].
    wiki_property(capital_of,subject_page_name,'Poland').
[[attribute::created:=April 22 2009]]
    wiki_attribut(page_uri,created,date(22,april,2009)).
```

The Prolog clauses are asserted to the PlWiki knowledge base by the syntax plugin analyzing the wiki text.

In a similar fashion simple queries are handled. A query for a category or property is simply mapped to a corresponding Prolog query:

```
{{#ask: [[Category:Cities]] [[capital of::Poland}}

wiki_category('Cities',Page),
wiki_property(capital_of,Page,'Poland'),
wiki_out(Page).
```

Plain RDF annotations are also supported. Currently, these are separated from the explicit annotations mentioned above. For compatibility reasons an RDF annotation can be embedded directly in XML serialization, then it is parsed by the corresponding Prolog library, and turned to the internal representation, that can also be used directly. SWI-Prolog's represents RDF triples simply as:

```
rdf(?Subject, ?Predicate, ?Object).
```

So mapping the above example would result in:

```
rdf('Warsaw',capital_of,'Poland').
```

The SWI-Prolog RDF storage in highly optimized. It can be integrated with the provided RDFS and OWL layers, as well as with the *ClioPatria* platform[3] providing also SPARQL queries.

Thanks to the full Prolog engine available in the wiki, the inference options are almost unlimited. Prolog uses backwards chaining with program clauses. However, it is very easy to implement meta-interpreters for forward chaining.

A simple clause finding recently created pages might be as follows:

```
recent_pages(Today,Page) :-
    wiki_attribut(Page,created,date(D,April,2009)),
    I is Today - D,
    I < 7.
```

Compound queries can also be created easily and executed as Prolog predicates.

## 9 Conclusions and Future Development

This paper presents an original idea of implementing a semantic wiki using Prolog for knowledge representation and inference. In the paper a proof-of-concept prototype implementation for the DokuWiki system is described. The system allows for fact, relations and rule representation, using Prolog. It also allows to use Semantic Web languages such as RDF to provide semantic annotations, opening possibilities to implement full support for OWL-based ontologies.

[3] See http://e-culture.multimedian.nl/software/ClioPatria.shtml

In the current version of the prototype it is not possible to check the syntax of the Prolog code, or assist the user in entering it. In the future debugging and syntax highlighting features are planned.

There are obviously some performance issues regarding knowledge processing within a wiki system based on Prolog code interpretation. Extracting knowledge from many pages and processing it could be a time consuming operation. Some smart caching techniques are evaluated. They are based on the caching mechanism present in the DokuWiki system.

An enhanced direct support for wiki markup present in other solutions (e.g. SemanticMediaWiki) is also planned. This should ultimately allow to import ready to use semantic wikis implemented with SMW, and possibly other wikis.

The user interface is an important area for improvement. Besides Prolog editing support, extended interfaces that use semantic forms are also considered. They should also allow ontology edition and visualization.

An important challenge is the rule framework in the wiki. Thanks to Prolog flexibility, it is possible to support number of rule languages built on top the SWI stack, including Semantic Web languages such as SWRL (with Description Logics reasoners integrated), as well as custom solutions. One of the options that is evaluated is the use of $XTT^2$ rule framework [16] that has a transparent Prolog implementation.[4]. On the other hand it provides standalone visual knowledge editors, and formal analysis tools (e.g. for rule verification).

Finally, an important research direction is a knowledge evaluation in the wiki. The distributed knowledge development process in a wiki poses new problems in knowledge engineering when compared to the classic development of monolithic knowledge bases. In case of most of the semantic wikis the focus of the current research is on the knowledge representation, integration and authoring. However, with the growing amount of knowledge contained in semantic wikis, the knowledge quality issues seem to be critical [14]. Using the Prolog interpreter several knowledge evaluation plugins are developed. Thus it is possible to analyze the knowledge stored in the wiki w.r.t. to a number of features [14]. This is an experimental feature that is being developed.

## References

1. Berners-Lee, T., Hendler, J., Lassila, O.: The semantic web: Scientific American. Scientific American (May 2001)
2. Schaffert, S.: Ikewiki: A semantic wiki for collaborative knowledge management. In: WETICE 2006: Proceedings of the 15th IEEE International Workshops on Enabling Technologies: Infrastructure for Collaborative Enterprises, pp. 388–396. IEEE Computer Society, Washington (2006)
3. Krötzsch, M., Vrandecic, D., Völkel, M., Haller, H., Studer, R.: Semantic wikipedia. Web Semantics 5, 251–261 (2007)
4. Buffa, M., Gandon, F., Ereteo, G., Sander, P., Faron, C.: Sweetwiki: A semantic wiki. Web Semantics: Science, Services and Agents on the World Wide Web (in press) (2008)

[4] See HeKatE project website at `http://hekate.ia.agh.edu.pl`

5. McGuinness, D.L., van Harmelen, F.: Owl web ontology language overview, w3c recommendation 10 february 2004. Technical report, W3C (2004)
6. Horrocks, I., Patel-Schneider, P.F., Boley, H., Tabet, S., Grosof, B., Dean, M.: Swrl: A semantic web rule language combining owl and ruleml, w3c member submission May 21, 2004. Technical report, W3C (2004)
7. Bratko, I.: Prolog Programming for Artificial Intelligence, 3rd edn. Addison Wesley, Reading (2000)
8. Nalepa, G.J., Wojnicki, I.: Proposal for a prolog-based knowledge wiki. In: Nalepa, G.J., Baumeister, J. (eds.) 4th Workshop on Knowledge Engineering and Software Engineering (KESE2008) at the 32st German conference on Artificial Intelligence, Kaiserslautern, Germany, September 23, pp. 43–54 (2008)
9. Auer, S., Dietzold, S., Riechert, T.: Ontowiki - a tool for social, semantic collaboration. In: Cruz, I., Decker, S., Allemang, D., Preist, C., Schwabe, D., Mika, P., Uschold, M., Aroyo, L.M. (eds.) ISWC 2006. LNCS, vol. 4273, pp. 736–749. Springer, Heidelberg (2006)
10. Baumeister, J., Reutelshoefer, J., Puppe, F.: Knowwe: community-based knowledge capture with knowledge wikis. In: K-CAP 2007: Proceedings of the 4th international conference on Knowledge capture, pp. 189–190. ACM, New York (2007)
11. Baumeister, J., Puppe, F.: Web-based knowledge engineering using knowledge wikis. In: Proc. of the AAAI 2008 Spring Symposium on Symbiotic Relationships between Semantic Web and Knowledge Engineering, pp. 1–13. Stanford University, USA (2008)
12. Reutelshoefer, J., Baumeister, J., Puppe, F.: Ad-hoc knowledge engineering with semantic knowledge wikis. In: Proc. of SemWiki 2008 - The Wiki Way of Semantics, Workshop co-located with the 5th European Semantic Web Conference, Tenerife, Spain (2008)
13. Ben-Ari, M.: Mathematical Logic for Computer Science. Springer, London (2001)
14. Baumeister, J., Nalepa, G.J.: Verification of distributed knowledge in semantic knowledge wikis. In: FLAIRS-22: Proceedings of the twenty-second international Florida Artificial Intelligence Research Society conference, Sanibel Island, Florida, USA, May 19–21 (2009) (to be published)
15. Nalepa, G.J., Ligęza, A.: A graphical tabular model for rule-based logic programming and verification. Systems Science 31(2), 89–95 (2005)
16. Nalepa, G.J., Ligęza, A.: Xtt+ rule design using the alsv(fd). In: Giurca, A., Analyti, A., Wagner, G. (eds.) ECAI 2008: 18th European Conference on Artificial Intelligence: 2nd East European Workshop on Rule-based applications, RuleApps 2008: Patras, July 22, pp. 11–15. University of Patras, Patras (2008)

# Properties of Bridge Nodes in Social Networks

Katarzyna Musiał and Krzysztof Juszczyszyn

Wroclaw University of Technology, Institute of Computer Science,
Wybrzeze Wyspianskiego 27, Wroclaw, Poland
{katarzyna.musial,krzysztof.juszczyszyn}@pwr.wroc.pl

**Abstract.** The main goal of the paper is to describe the properties of the nodes within a social network that connect the peripheral nodes and peripheral groups with the rest of the network. These nodes are usually called bridging nodes or simply bridges. All the experiments are carried out on the real data, so-called Thurman network. First, the regular cliques, peripheral nodes and peripheral cliques from a network are extracted and then the bridging nodes identified. Afterwards for all nodes their characteristic features, such as social position and degree of nodes, are calculated. Finally, we try to find the correlation between nodes centrality and their degree and the fact if given node is a bridge or not.

**Keywords:** bridging node, regular clique, peripheral node, peripheral clique, social position.

## 1 Introduction

The analysis of the characteristics of the nodes that connect the whole network with the peripheral nodes or peripheral cliques that are loosely connected with the rest of the network is a very interesting research problem. The issue of identifying bridges within a given network is a complex and resource consuming task because it involves an extensive analysis of the groups and cliques existing within a given network [12], [10], [4]. Bridges can be seen as the nodes without which the network will split into two or more subgroups. This concept is similar to the idea of weak ties [3], [6], [1] that also tend to be vital bridges between the two densely knit clumps. In this work bridges are defined as the nodes that (i) connect regular cliques with the peripheral nodes or (ii) connect regular cliques with the peripheral cliques that are loosely connected with the rest of the network. The explanations of such concepts as regular clique, peripheral clique, bridging node and peripheral node as well as the whole method for bridges properties analysis is presented. By defining the properties of the bridges we will be able to identify them without the complex calculations.

## 2 Method for Bridges' Properties Analysis

The main goal of the paper is to identify bridges within the social network and describe their characteristic features. The whole process is presented in the

N.T. Nguyen, R. Kowalczyk, and S.-M. Chen (Eds.): ICCCI 2009, LNAI 5796, pp. 357–364, 2009.

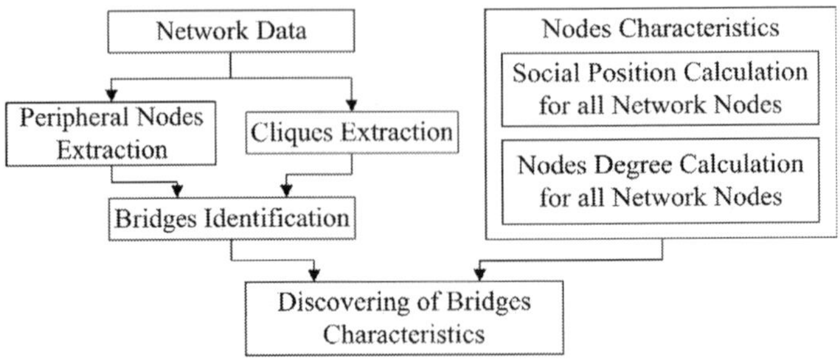


**Fig. 1.** Crucial steps of the proposed method of bridges properties analysis

Figure 1. The method is split into two main groups of activities: (i) the process of bridges identification and (ii) the process of the network nodes properties calculation. Note that the network is represented as a weighted and directed graph.

In the former process, the first step is to extract all cliques existing in a given network. The clique is defined as "a maximal complete subgraph of three or more nodes" [12]. It means that a clique consists of a subset of nodes, all of which are connected to all nodes of the clique. Additionally, in the rest of the network does not exist even one node that is connected to all nodes that belong to the analyzed clique [9], [5]. In the same time, the peripheral nodes and peripheral cliques are identified. The former ones are nodes that do not belong to any clique that exist within an analyzed network. The latter ones are cliques that are loosely connected with the rest of the network, i.e. they do not have any common node with the rest of the extracted cliques. The set of cliques that does not include the peripheral cliques is called a set of regular cliques. Afterwards the bridging nodes can be discovered. Bridges are nodes that belong to the regular clique and connect it with peripheral node or peripheral clique.

The second set of activities is connected with analyzing the characteristic features of the network nodes. Different characteristics from the node perspective can be analyzed, e.g. centrality, prestige, density, social capital [12], [4]. In the presented research, we take into consideration two of them: centrality index named social position [7], [8] as well as the number of incoming, outgoing and mutual relationships of the node [12], [2]. Social position function $SP(x)$ of member $x$ in the social network, respects both the value of node positions of all other network members as well as the level of their activities in relation to $x$ [7], [8]:

$$SP(x) = (1 - \varepsilon) + \varepsilon \cdot \sum_{i=1}^{m_x} (SP(y_i) \cdot C(y_i \to x)) \tag{1}$$

where: $\varepsilon$ – the constant coefficient from the range $[0, 1]$; $y_i$ — $x$s acquaintances, i.e. the members who are in direct relationship to $x$: $C(y_i \to x) > 0$; $C(y_1 \to$

$x$),...,$C(y_m \rightarrow x)$ – the commitment function that denotes the contribution in activity of $y_1$,...,$y_m$ directed to $x$; $m_x$ — the number of $x$'s acquaintances.

The importance of the node in the weighted and directed network, expressed by the social position function, tightly depends on the strength of the relationships that other members of the network maintain with the given node as well as on the social positions of these members. In other words, the member's social position is inherited from others but the level of inheritance depends on the activity of the members directed to this person, i.e. intensity of common interaction, cooperation or communication.

The last stage of the method is to find the features of the bridging nodes that are their specific and individual properties and distinguish them from the other network nodes. If it is possible to find such features, then it will be no need to perform the extensive calculations in order to locate the bridging nodes in the network structure.

## 3 Experiments

The experiments were conducted on the Thurman office social network that is a non–symmetrical network of 15 people who worked in one company. Thurman spent 16 months observing the interactions among employees in the office of a large corporation [11].

The adjacency matrix for the Thurman network is presented in Table 1 where values 1 represent the existence of the connection between two users. In order to obtain the values of commitment function (that are needed to calculate the social position values, see Equation 1) for each individual, value one — from the

**Table 1.** The values of commitment function within the Thurman network

| Member | 12 | 1 | 5 | 2 | 8 | 6 | 3 | 10 | 4 | 9 | 7 | 11 | 13 | 14 | 15 |
|---|---|---|---|---|---|---|---|---|---|---|---|---|---|---|---|
| 12.Emma | – | – | 1 | – | 1 | 1 | – | 1 | – | 1 | – | 1 | 1 | 1 | 1 |
| 1.Ann | – | – | 1 | 1 | 1 | 1 | 1 | – | – | 1 | – | 1 | 1 | – | – |
| 5.Pete | 1 | 1 | – | 1 | 1 | 1 | 1 | 1 | 1 | 1 | 1 | 1 | 1 | 1 | 1 |
| 2.Amy | – | 1 | 1 | – | 1 | 1 | 1 | – | 1 | – | – | – | – | – | – |
| 8.Lisa | 1 | 1 | 1 | 1 | – | 1 | 1 | – | – | 1 | – | – | – | – | – |
| 6.Tina | – | 1 | 1 | 1 | 1 | – | 1 | – | – | – | – | – | – | – | – |
| 3.Katy | – | 1 | 1 | 1 | 1 | 1 | – | – | – | – | – | – | – | – | – |
| 10.Minna | 1 | – | – | 1 | 1 | – | – | – | 1 | – | 1 | – | – | – | – |
| 4.Bill | – | – | – | 1 | – | – | – | 1 | – | – | 1 | – | – | – | – |
| 9.President | 1 | 1 | 1 | 1 | 1 | 1 | 1 | 1 | 1 | – | 1 | 1 | 1 | 1 | 1 |
| 7.Andy | – | – | 1 | – | – | – | – | 1 | 1 | – | – | – | – | – | – |
| 11.Mary | 1 | 1 | – | – | – | – | – | – | – | – | – | – | – | – | – |
| 13.Rose | 1 | 1 | – | – | – | – | – | – | – | – | – | – | – | – | – |
| 14.Mike | 1 | – | – | – | – | – | – | – | – | – | – | – | – | – | – |
| 15. Peg | 1 | – | – | – | – | – | – | – | – | – | – | – | – | – | – |

original matrix Table 1 — is divided by the number of members relationships, e.g. Emma communicates with nine members so her contribution of activity to each of her acquaintances equals $\frac{1}{9}$.

### 3.1 Process of Bridges Extraction

First step of the method is to extract the cliques existing within the network. Note, that the relationships within the Thurman network are weighted and directed. It means that group can be called clique if and only if all the relationships between members are mutual. Four cliques were extracted from the Thurman network (the numbers indicate the ID of the specific user, see Table 1):

1. $C_1 = \{Ann, Amy, Lisa, Katy, Tina, Pete\} = \{1, 2, 8, 3, 6, 5\}$
2. $C_2 = \{Ann, Lisa, Pete, President\} = \{1, 8, 5, 9\}$
3. $C_3 = \{Pete, Emma, Lisa, President\} = \{5, 12, 8, 9\}$
4. $C_4 = \{Bill, Minna, Andy\} = \{4, 10, 7\}$

In order to find the peripheral nodes i.e. these that do not belong to any of the cliques, we apply the formula:

$$PN = V \setminus (C_1 \cup C_2 \cup ... \cup C_n) \tag{2}$$

where: $PN$ — the set of peripheral nodes; $V$ — the set of all nodes in a network; $n$ — number of cliques within a given network

After utilizing the above formula the set $NP$ is obtained:

$$NP = \{Mary, Rose, Mike, Peg\} = \{11, 13, 14, 15\}$$

Next part of the experiments is to identify the peripheral cliques. Let us remind that the peripheral clique ($PC$) is the clique that does not posses even one common node with all other cliques existing within the network. It means that each clique $C_x$ is a $PC$ if and only if the following criterion is met:

$$\sum_{i=1 \wedge i \neq x}^{n} card(C_i \cap C_x) = 0 \tag{3}$$

After the application of the above criterion we obtain:
$C_1 \cap C_2 = \{Ann, Lisa, Pete\} = \{1, 8, 5\}$
$C_1 \cap C_3 = \{Lisa, Pete\} = \{8, 5\}$
$C_1 \cap C_4 = \emptyset$

It does not meet the above criterion, i.e. $\sum_{i=1 \wedge i \neq 1}^{4} card(C_i \cap C_1) \neq 0$ so neither $C_1$ nor $C_2$ is not the peripheral one.
$C_2 \cap C_1 = \{Ann, Lisa, Pete\} = \{1, 8, 5\}$
$C_2 \cap C_3 = \{President, Lisa, Pete\} = \{9, 8, 5\}$
$C_2 \cap C_4 = \emptyset$

This also does not meet the above criterion: $\sum_{i=1 \wedge i \neq 2}^{4} card(C_i \cap C_2) \neq 0$ It means the product of sets is not empty, so $C_2$ is not the peripheral one.
$C_3 \cap C_1 = \{Lisa, Pete\} = \{8, 5\}$
$C_3 \cap C_2 = \{President, Lisa, Pete\} = \{9, 8, 5\}$
$C_3 \cap C_4 = \emptyset$

It does not meet the above criterion, i.e. $\sum_{i=1 \wedge i \neq 3}^{4} card(C_i \cap C_3) \neq 0$ $C_4 \cap C_1 = \emptyset$
$C_4 \cap C_2 = \emptyset$
$C_4 \cap C_3 = \emptyset$
$\sum_{i=1 \wedge i \neq 4}^{4} card(C_i \cap C_4) = 0$, so clique $C_4$ according to Formula 3 is the peripheral clique. From now on $C_4$ clique is a peripheral clique and is excluded from the list of cliques. The groups that remain in the set of cliques are named from now on as regular cliques.

After the identification of regular cliques as well as peripheral nodes and peripheral cliques (see Table 2) the bridges can be uncovered.

**Table 2.** The values of commitment function within the Thurman network

| Regular Cliques | Peripheral Nodes | Peripheral Cliques |
|---|---|---|
| $C_1 = \{1, 2, 8, 3, 6, 5\}$ | | |
| $C_2 = \{1, 8, 5, 9\}$ | $PN = \{11, 13, 14, 15\}$ | $PC = C_4 = \{4, 10, 7\}$ |
| $C_3 = \{5, 12, 8, 9\}$ | | |

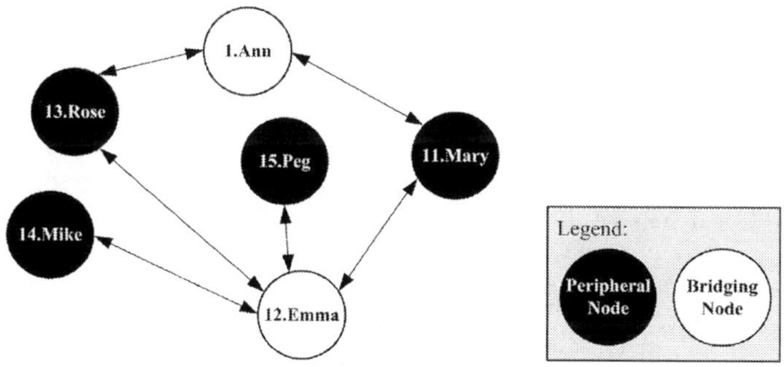


**Fig. 2.** Structure of peripheral nodes in Thurman network and bridges that connect *PN* with the regular cliques

In the first step the bridges that connect the peripheral nodes with the regular cliques are uncovered. In order to perform this, all of the peripheral node's connections are analyzed (Figure 2). It can be easily noticed that Ann and Emma

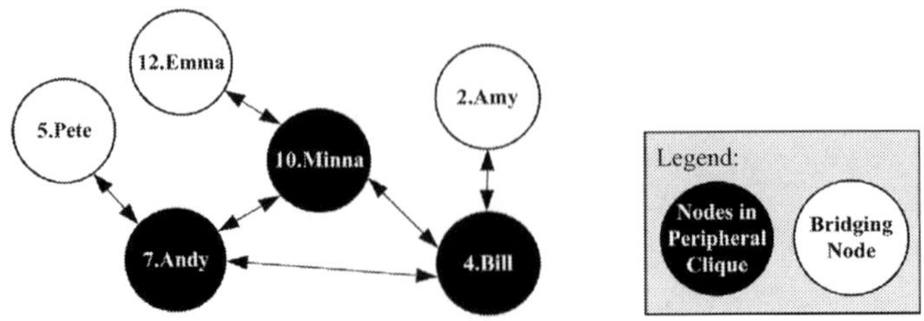


**Fig. 3.** Structure of peripheral clique in Thurman network and bridges that connect *PC* with the regular cliques

are bridges in this case, i.e. a set of bridges $B_1$ equals: $B_1 = \{Ann, Emma\} = \{1, 12\}$. Next, the analysis of peripheral clique and its relationships with the external network (Figure 3) shows that the bridging nodes in this case are $B_2 = \{Emma, Pete, Amy\} = \{12, 5, 2\}$ The final set of bridges $B$ is the sum of the sets $B_1$ and $B_2$, so finally the set of bridges consist of:

$$B = \{Ann, Emma, Pete, Amy\} = \{1, 12, 5, 2\}$$

### 3.2 Properties of Bridges in Social Network

The goal of the next part of the experiments is to investigate the characteristic features of the extracted bridging nodes. For all of the nodes their social position is calculated as well as the number of mutual, incoming and outgoing edges. Before starting this part of the experiments some primary assumptions were made. The initial node positions $NP_0(x) = 1$ are established for every member $x$ in the network. The value of $\varepsilon$ is 0.9 and the stop condition is: no difference in node position values to the precision of 5 digits after the decimal point for all the members in two following iterations, i.e. $\tau = 0.00001$. The calculated values are presented in Table 3. Note, that bridges identified in the first part of experiments bridges posses high social positions and have the biggest number of connections — Emma, Ann, Pete and Amy. Emma has the highest social position and she is the only community member that connects both peripheral nodes (all of them) and the peripheral clique with the regular cliques. Two of the peripheral users (Peg and Mike) communicate only with Emma so she is an crucial node when concerning the cohesion of the whole network. Ann, who has the second highest social position and has 8 mutual connections, binds two peripheral users (Mary and Rose) with the regular cliques. However, these connections are not as crucial as relations between Emma and Peg or Mike, because not only Ann is connected with Mary and Rose. Another person that is connected with Mary and Rose is Emma. Pete, who obtained third highest social position, has also 8 mutual connections and is one of three people that connects the peripheral clique ({Minna, Bill, Andy}) with the regular cliques. Another users who connect the peripheral clique with the regular cliques are Emma and

**Table 3.** The social position values and number of edges in Thurman network

| ID | Member | *SP* Ranking | *SP* | No. of mutual edges | No. of incoming edges | No. of outgoing nodes |
|---|---|---|---|---|---|---|
| 12 | Emma | 1 | 1.90025 | 8 | 0 | 0 |
| 1 | Ann | 2 | 1.56732 | 8 | 0 | 3 |
| 5 | Pete | 3 | 1.48140 | 8 | 0 | 5 |
| 2 | Amy | 4 | 1.38236 | 6 | 2 | 0 |
| 8 | Lisa | 5 | 1.36532 | 7 | 1 | 0 |
| 6 | Tina | 6 | 1.17424 | 5 | 2 | 0 |
| 3 | Katy | 7 | 1.01320 | 5 | 1 | 0 |
| 10 | Minna | 8 | 0.86255 | 3 | 2 | 2 |
| 4 | Bill | 9 | 0.79626 | 3 | 2 | 0 |
| 9 | President | 10 | 0.73712 | 4 | 0 | 9 |
| 7 | Andy | 11 | 0.63676 | 3 | 1 | 0 |
| 11 | Mary | 12 | 0.60897 | 2 | 2 | 0 |
| 13 | Rose | 12 | 0.60897 | 2 | 2 | 0 |
| 14 | Mike | 14 | 0.43264 | 1 | 2 | 0 |
| 15 | Peg | 14 | 0.43264 | 1 | 2 | 0 |

Amy. Amy was identified also as a bridging node and she obtained the fourth position in the *SP* ranking. Additionally, she possesses 6 mutual relationships - 2 relations less than the three other bridging nodes. On the other hand President, who has many connections but low social position (10th in *SP* ranking), is not a bridging node. It means that social position is better measure to identify the bridges than the number of the edges.

## 4 Conclusions

The method of detecting bridging nodes in a directed and weighted social network as well as their properties were investigated. The conducted research reveals that the bridges obtain the highest social position. Two types of bridges can be distinguished: (i) these that connect peripheral nodes with regular cliques and (ii) these that connect peripheral cliques with the regular cliques. Note that the highest social position is obtained by Emma (12) who binds both all of the peripheral nodes and peripheral clique with the regular cliques. Ann (1), Pete (5) and Amy (2) also are the bridging nodes and all of them obtain high social positions: Ann — second, Pete — third and Amy — fourth. This reveals that social position is a good measure that can be used to find the bridging nodes within the network. In other words, there exists the correlation between the social position of the network and the fact if the node is a bridging one. The future work will focus on conducting the research within the complex social networks with big number of nodes and edges. We intend to develop a fast method of identification of the bridge nodes then to use it to track changes in community dynamics (group evolution). The changes in the properties of bridging nodes may be used to track processes of merging, splitting, growth and contraction of cliques in complex social networks.

**Acknowledgment.** The work was supported by The Polish Ministry of Science and Higher Education, grant no. N N516 264935.

## References

1. Burt, R.S.: Structural holes. Harvard University Press, Cambridge (1992)
2. Carrington, P., Scott, J., Wasserman, S.: Models and methods in Social Network Analysis. Cambrige University Press, Cambrige (2005)
3. Granovetter, M.S.: The strength of weak ties: A Network Theory Revisited. Sociological Theory 1, 201–233 (1983)
4. Hanneman, R., Riddle, M.: Introduction to social network methods. Online textbook, `http://faculty.ucr.edu/~hanneman/nettext/` (April 1, 2006)
5. Harary, F., Norman, R.Z., Cartwright, D.: Structural Models: An Introduction to the Theory of Directed Graphs. John Wiley and Sons, New York (1965)
6. Hatala, J.P.: Social Network Analysis in Human Resources Development: A New Methodology. Human Resource Development Review 5(1), 45–71 (2006)
7. Kazienko, P., Musiał, K.: Assessment of Personal Importance Based on Social Networks. In: Gelbukh, A., Kuri Morales, Á.F. (eds.) MICAI 2007. LNCS (LNAI), vol. 4827, pp. 529–539. Springer, Heidelberg (2007)
8. Kazienko, P., Musiał, K., Zgrzywa, A.: Evaluation of Node Position Based on Email Communication. Control and Cybernetics (in press, 2009)
9. Luce, R.D., Perry, A.D.: A method of matrix analysis of group structure. Psychometrica 14, 95–116 (1949)
10. Scott, J.: Social network analysis: A handbook, 2nd edn. Sage, London (2000)
11. Thurman, B.: In the office: Networks and coalitions. Social Networks 2, 47–63 (1979)
12. Wasserman, S., Faust, K.: Social network analysis: Methods and applications. Cambridge University Press, New York (1994)

# Use of Semantic Principles in a Collaborative System in Order to Support Effective Information Retrieval

František Babič, Karol Furdík, Ján Paralič, Peter Bednár, and Jozef Wagner

Centre for Information Technologies, Technical University of Košice,
Letná 9/B, Košice, Slovakia
{frantisek.babic,karol.furdik,jan.paralic,peter.bednar,
jozef.wagner}@tuke.sk

**Abstract.** This paper focuses on information retrieval aspects of a new application in domain of collaborative systems based on utilization of semantic principles for representation of different types of knowledge, collaborative objects and relations between them. Proposed collaborative system (within European IST project called KP-Lab[1]) uses ontologies as common communication framework and exchange format for different types of end-user tools. Theoretical background is provided by innovative theoretical approach called Trialogical learning. Information retrieval in KP-Lab System is supported by designed and implemented text-mining and search services. These two sets of functionalities provide features for management of shared objects, described with content in textual format, as well as with semantic metadata.

**Keywords:** ontologies, semantic metadata, text mining, semantic search, collaborative system.

## 1 Introduction

Domain of collaborative systems became important part of teaching and learning in last years. Important fact is that this approach should not substitute completely face to face seminars and lectures. The main trend is to provide suitable solutions to effective support of traditional learning processes with use of new information or communication technologies in order to investigate the "best practices" and interesting innovative elements and approaches. This approach offers possibility to create application that builds on social relations, on-line or offline communicative channels, sharing different types of objects, awareness support, effective and user-friendly search features, simple and intuitive user environment with possibility to manage video or sound files, etc.

Sharing features in collaborative systems lead to accumulation of large volumes of data in different formats, e.g. doc, pdf, rtf, avi, mpeg, mp3, etc. It is necessary to provide effective support for representation and search methods in these large databases. One possibility is to combine principles of semantic web for data representation and web 2.0 approaches for user friendly access environment.

[1] http://www.kp-lab.org/

N.T. Nguyen, R. Kowalczyk, and S.-M. Chen (Eds.): ICCCI 2009, LNAI 5796, pp. 365–376, 2009.

Presented collaborative system is based just on this idea and provides interesting functionalities for support of collaborative learning or working processes based on underlying Trialogical learning theory (TL). TL provides theoretical framework for modeling of knowledge creation processes in collaborative manner based on suitable technological solution. This approach was one of the main motivation of technological development within KP-Lab project, an integrated IST EU funded project that is contracted for 5years (2006-2011).

This project and comparable solutions are briefly introduced in the following subsections. Result of the implementation work in mentioned project, the KP-Lab system is described in the section 2. Section 3 focuses on our information retrieval functionalities enhanced with semantic technologies. Finally, section 4 concludes the paper with a brief summary.

## 1.1 Related Work

Domain of collaborative systems is nowadays very extensive and includes such fields as Computer-supported Collaborative Learning, Computer-supported Collaborative Work, Virtual learning environment, Collaborative Working Environment, Learning Management System, etc. However, as European Commission states in its report [8]: "the characteristic of current collaborative environments is that they are not integrated and interoperable, that they support mainly point to point and not multipoint conferencing, that they are defined mainly for structured environment providing static artefacts and that they do not support the unstructured orchestration of activities using collaboration aware objects. Finally they focus primarily on peer communication and not flexible team interaction." Results of this report describe a vision and the research topics for further development in the domain of collaborative environment, namely the context-oriented data mapping, support of the shared objects lifecycle, synchronous and asynchronous cross-domain communication/collaboration, etc. Proposed collaborative system KP-Lab is aiming to address these principles, as it is described in the next sections.

KP-Lab System can be compared with different representatives in domain of collaborative systems. We choose for comparison these open source applications as Claroline, FLE3, Moodle, SAKAI; commercial product called BSCW and new solution based on web2.0 principles – Google Apps. These environments were chosen because they have many similar functions and features as KP-Lab System, thus, through the comparison it is possible to bring forward the advantages, benefits and innovations of the proposed approach. Furthermore, all the selected tools are widely in use, therefore, it is essential to be able to show the benefits of new system in order to motivate and convince the future users. Several main advantages of the proposed system can be identified based on performed comparison:

- Multifunctionality – selected open-source solutions don't provide so many features as KP-Lab System, except of commercial BSCW that provides many similar functions but these functions are based on another development approach (using transactional database not semantic repository, processes are modeled with workflows, etc.)

- Orientation on semantic features – shared objects' description consists of two parts (different integrated tools share the same semantic, semantic information can be reused across the tools):
    - metadata (semantic information) is saved into knowledge repository (semantic repository based on ontologies - SWKM)
    - the content is saved into content repository based on Java technologies and access to it is based on G2CR (gateways to content repositories)
- Easily extendable and highly interoperable system, e.g. access to different types of content repositories through G2CR (migration of the data from previous used systems) or import and export of learning objects based on most used standards (IMS, SCORM).

Some examples are provided below:

- Moodle is strongly oriented on the area of integrated modules, while the semantic aspects and capabilities to analyze user's practices are weakly developed.
- Google apps provide a set of tools that are connected through predefined API and user can select any combinations and customize them. Search is provided by Google search engines with possibility to save search results and queries.
- SAKAI emphasizes the phase of development and, like Moodle, can be extended by the modules with new features. SAKAI provides similar functionalities as KP-Lab System, such as shared workspace, job scheduler with calendar, portfolio, discussion and blogs facilities. However, it lacks focus on semantics of the shared objects, its editing, managing, and visualization capabilities.
- BSCW provides advanced functionalities as tagging, communities, templates, and search on different indexing services. Limited support is provided for editing tags or indexes, also collaborative and idea-generation tools are not available.

Each of the collaborative systems includes some search facilities that enable accessing and retrieving the stored information. The searching is usually based on full-text indexing. More advanced retrieval facilities are rarely used – the tag-based search, provided by the BSCW, can be mentioned as an example. The KP-Lab System employs the semantic enhancements of the stored information to provide semantic-based retrieval combined with text-mining methods (cf. section 3).

### 1.2 Trialogical Learning

Trialogical learning [3] refers to the process where participants are collaboratively developing shared objects of activity (such as conceptual artefacts, practices, or products) in a systematic fashion. It concentrates on the interaction through these common objects (or artefacts) of activity, i.e. the interactions between people are mediated by various types of knowledge artefacts, not just among people (as it is in dialogical learning) or within one's mind (as it is in monological learning) – the previous two approaches. This innovative approach provides theoretical framework for knowledge creation processes analogous to some other existing approaches in this domain as Socialisation – Externalisation – Combination - Internalisation (SECI) model, theory of knowledge building, theory of expansive learning or Activity theory.

"Trialogue" in trialogical learning is not about discussion between three persons, it means that individuals (or groups of people) are developing some shared objects of activity within some social or cultural settings, see Fig. 1.

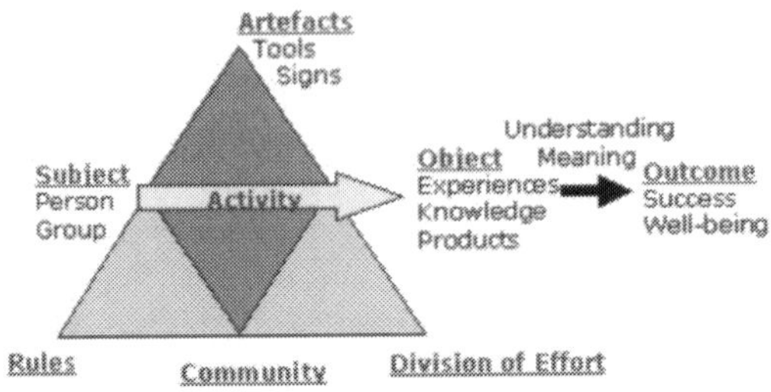


**Fig. 1.** The activity model [7]

The basic concept for trialogical learning is an activity. *Activity* is composed of two elements (subject and object) and mediated by mediators - artefacts. A *subject* is a person or a group engaged in an activity. An *object* (in the sense of "objective") is held by the subject and motivates the activity, giving it a specific direction. The mediation can occur through the use of many different types of *artefacts*, material tools as well as mental tools, including culture, ways of thinking and language. Transforming the object into an *outcome* motivates the existence of the activity.

As a demonstrative example of described approach, we can mention a collaborative creation of an essay. Object is the essay itself; it acts as motivational element of planned activity. Subjects are individual people/participants that will be engaged in evolutionary process, e.g. students and teachers. Artefact or mediator is the environment in which will be the essay created, some type of collaborative document creation tool or wiki engine, etc. *Community* consists of all engaged participants and has some *rules*. These *rules* are defined for single community member or for the whole community as such. *Division of Effort* describes decomposition of all activity; e.g. by means of various process elements (e.g. tasks), which usually have associated responsible member(s) and may have defined also start and end time, inputs and output, etc. After successful evaluation by teacher essay will be an outcome of activity. Every participant can write his/her part or edit, modify the other parts. Every action and every change is saved for monitoring purposes. When some problems emerge, participants can discuss it via a chat or in virtual meetings. Every participant has possibility to make his/her work public or keep it private.

## 2 KP-Lab System

Design and proposed functionalities of KP-Lab System can be seen as results of internal co-evolutionary design process based on long-term discussion with pedagogical partners about their expectations and requirements that cover basic concepts of trialogical learning and their real work. The whole development stressed the usage of semantic principles, effective management of created ontologies, designed and implemented semantic features are available to the users through end-user tools. Architecture of the whole KP-Lab System is based on the platform and collaborative user environment (KP-environment) with integrated end-user applications.

### 2.1 KP-Lab Platform

The KP-Lab platform (see Fig. 3) provides integrated semantic middleware that is based on common semantic framework represented by internal ontology architecture [6]. This set of designed ontologies provides possibility to connect heterogeneous technologies in KP-Lab platform through web services as the common language for the communication and functionalities around shared objects. The core of described architecture is represented by Trialogical Learning Ontology (TLO) that defines core concepts and principles of trialogical learning and provides the common semantics for the data interoperability in whole KP-Lab System [6]. KP-Lab platform is composed of several groups of services and libraries:

- Semantic Knowledge Middleware Services (**SWKM Services** in Fig. 3), providing storage and management services for semantic descriptions (metadata) of the shared objects created by the KP-Lab tools. **Knowledge repository** is implemented within RDFSuite [1] that is based on RDF (Resource Description Framework[2]) standard. This standard enables the creation and exchange of resource metadata as normal Web data. RDFSuite is being developed at FORTH -ICS in Greece and comprises the Validating RDF Parser (VRP), the Schema-Specific Data Base (RSSDB), interpreters for the RDF Query Language (RQL) and RDF Update Language (RUL).
- Content Management Services (**Content Transfer Module - CTM** in Fig. 3) are dedicated to creation and management of regular content (documents in various formats) used in shared objects (content described by metadata), either towards KP-Lab's own content repositories or external content repositories. KP-Lab **Content repository** is implemented through Jackrabbit[3] engine for the compatibility with the JSR-170 standard.
- **Persistence-API (P-API)** is a client library used for managing all Knowledge repository tasks: generating RQL/RUL and persisting/fetching of data from repository. This library [9] provides the generic RDF persistence framework, which allows serialization and deserialization of the Java objects into RDF repositories. It allows developers to focus on the application logic rather than on the RDF language or RQL/RUL, the low level mechanism of storing the metadata in the SWKM.
- **Technical services** cover middleware support services, dedicated to the authorization and identity management, the user management, routing etc.

---

2 http://www.w3.org/RDF/
3 http://jackrabbit.apache.org/

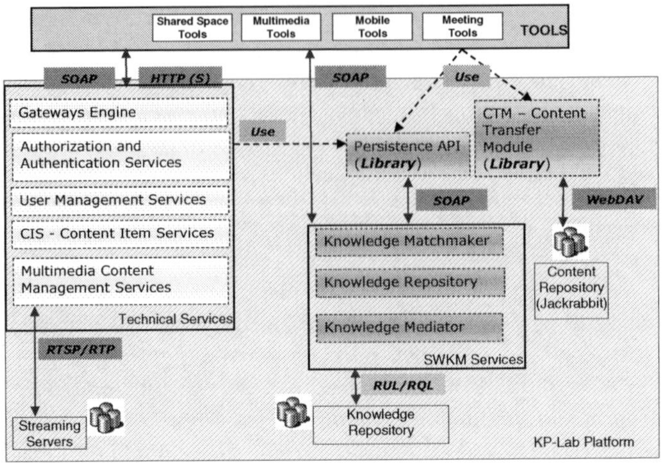


**Fig. 3.** The KP-Lab platform [4]

## 2.2 KP-Environment

KP-environment provides virtual user environment that mediated all user activities and actions within different types of shared objects based on their goals or expectations. The integrated end-user applications have been implemented based on initial analyses, case studies, generic scenarios and requirements identification, to enable representation, realization, analyses and adaptation of existing or innovative knowledge practices in collaborative manner around shared objects, see Fig. 4.

**Shared Space** provides the main features of a learning system aimed at facilitating innovative practices of sharing, creating and working with knowledge in education and workplaces through different types of views. It supports users' collaboration according to different working practices and allows viewing of shared knowledge in flexible manner. It provides a set of tools for knowledge building and process management (parts of knowledge practices can be represented as knowledge processes). The personalised, temporal and faceted views allow users to describe and visualise shared knowledge objects, their associations and state in different arrangements.

**Support tools** provide functionalities that are necessary for effective collaborative work or learning within virtual space, e.g. awareness (on-line or historical based on logs of events), search (free or semantic based) and help (simple help pages, wiki, interactive help based on user's interests or performed actions).

**Common tools** refer to the tightly integrated tools of KPE, which are available inside a shared space for working with shared objects, e.g. possibility to comment or tag (tags based on predefined vocabularies, or own tags) relevant concepts, to create chat or virtual meeting with semi-automatic generation of discussion maps, to re-use interesting concepts based on created templates, possibility to import learning objects in form of packages from other types of collaborative systems, support for personal work within to-do list or calendar.

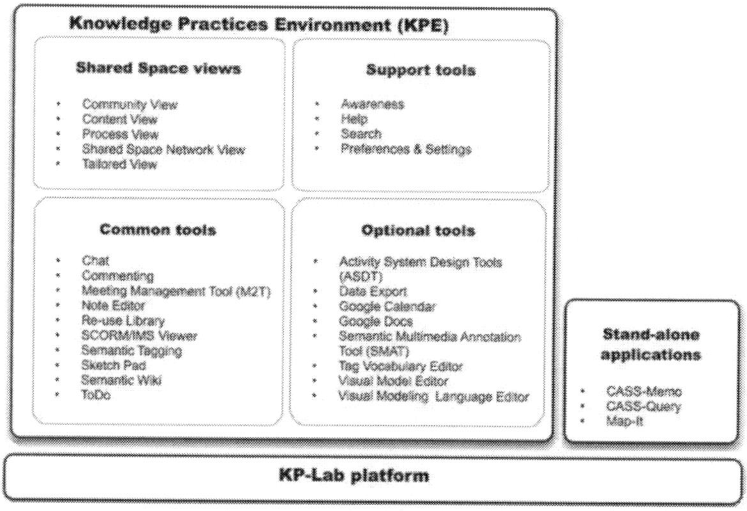


**Fig. 4.** Integrated view on whole KPE architecture [5]

**Optional tools** are loosely integrated applications that can be selected by the user based on its preferences. An optional tool opens directly into the KPE graphical user interface or into a separate browser window. These applications provide interesting functionalities, e.g. possibility to analyze multimedia video clips through tagging features (tags based on predefined vocabularies, or own tags), to export data for research purposes (from three different types of repositories – knowledge, content and awareness), to interact with web2.0 external applications like Google Calendar or Google Docs and last but not least possibility to create and manage visual models and own visual modeling languages.

**Stand-alone applications** are used separately from the KPE due to their implementation as e.g. mobile applications (CASS Memo), or due to the focus on supporting specific pedagogical research methodology (CASS Query).

## 3 Information Retrieval in KP-Lab System

Shared objects in KP-Lab System consist of the content and metadata parts. The content is stored in the Content repository and the metadata in the Knowledge repository (see section 2.1). The content is represented by the documents in different formats, e.g. doc, pdf or rtf, and the metadata are used for more structured description of these documents by title, author, type of document, etc., in the Dublin Core standard[4].

Real usage of KP-Lab System within pilot courses, evaluation cases and other experiments brought a need to store and manipulate large data volumes. It implied a design and implementation of the services for effective information retrieval, employing the access features to Content and Knowledge repositories (P-API or CTM), advanced text-mining methods and semantic search capabilities.

[4] http://dublincore.org/

Various learning or working materials are uploaded into KP-Lab System as shared objects and they are further investigated in a collaborative learning process. These materials can be semantically annotated through predefined vocabularies - taxonomies, concept maps, or domain ontologies, or through free tags – textual descriptions created by users. The ontologies provide a conceptual framework for the semantic annotation by defining a structure of a domain of discourse. In the trialogical learning, the ontology itself is a subject of creation, modification, and evolution in the process of learning as a socially determined and interactive activity [10].

### 3.1 Motivating Scenario

KP-environment enables users (students, teachers, etc.) working on shared objects in one virtual place. It allows the participants of collaborative processes perceiving and handling shared materials, knowledge representations and respective processes in an integrated way, to support a creation of new and innovative knowledge.

As an example, let us assume that a group of students is aiming at creation of documentation resources for a given product. The collection of text-based materials should be organized into a defined process. The collaborative procedure starts with the design of the process elements - tasks accompanied with the relevant resources. Students will use the KP-Lab search functions to retrieve the materials from internal collaboration space or from outside. A query can be formulated on a concrete title of document, its author, type or keywords. These metadata are specified by the semantic tags that accompany each of the documents stored in KP-Lab Shared Space. The external documents can be retrieved by a full-text search and then stored into the Content Repository. The collected materials are then appointed to the process; their context is defined by a set of semantic tags specified by the students. Such a semantic description consists of pre-defined as well as ad-hoc created and customizable semantic tags. Based on text-mining methods, KP-Lab supports this process by providing recommendations of suitable tags. In addition, it allows clustering and categorization of materials into topic groups, enhancements of the semantic tag vocabulary (ontology), and an advanced search that employs the built-in semantic information.

### 3.2 Text-Mining Services

KP-Lab text-mining services have been designed to assist users in creating or updating the semantic descriptions of KP-Lab shared objects [10]. The semi-automatic generation of these descriptions or even of new KP-Lab ontologies relies on the textual information attached to particular objects.

The textual description is analyzed and processed by parsing, part-of-speech tagging, lemmatization, and keywords extraction techniques. The text-mining classification of the shared objects, which is based on a matching of extracted keywords against an existing conceptual model, proposes a set of the most relevant ontology concepts that are suggested for users as suitable for semantic annotation. The classification works in two modes: 1) as a supervised method, employing the classification models created from previously processed training examples, and 2) as a decision-support system employing heuristic rules in the following form:

IF <text contains keywords {$K_1$, $K_2$, ..., $K_i$}>
THEN <shared object should be annotated by concepts {$C_1$, $C_2$, ..., $C_n$}>
(optional: weight=N); N=<0,1>.

In addition, unsupervised text mining techniques such as clustering algorithms are used to find some unseen concepts (or clusters) in the set of analyzed textual resources. These may lead to, e.g., the suggestion to upgrade existing KP-Lab ontologies, as the knowledge of a user group evolves.

Functional requirements for the text-mining services emerged from a discussion on user expectations and the service utilization within end-user applications:

- Creation of a training data set from already annotated shared objects to a predefined set of categories, i.e. concepts of existing domain ontology. The textual descriptions of the objects are pre-processed and transformed into a term-document matrix. The classification service indexes the training data set and stores it into the Mining Object Repository.
- Creation of a classification model, based on the selected algorithm and on a given training data set. Based on the selected implementation platform [2], the kNN (k Nearest Neighbours), SVM (Support Vector Machine), and Perceptron were employed as basic classification algorithms.
- Modification (improvement) of the applied classification model, by changing the texts and/or categories in the training data set, as well as by editing the settings of the algorithm or switching to another algorithm.
- Provision of basic measures for applied classification model, e.g. by means of precision and recall. Support of creation, indexing, and storage of the testing data set(s) that can be used for more exact evaluation of the quality of classification processes.
- Verification and validation of the applied classification model. The model is no longer valid if a portion of training data set (e.g., the term-document matrix or the set of pre-defined categories) was modified. In this case, a re-indexing of the model is needed to make it valid again.
- Classification of a set of unknown (not annotated) objects to the categories used for training. The output of this function is a set of weighted categories (concepts, terms) for each of the classified objects.

Implementation of the classification service (see Fig. 5) is based on the JBowl (Java Bag of Words) library [2], providing a platform for several classification algorithms, tools for processing natural language texts, as well as for some of the clustering techniques [10].

These basic functionalities of proposed text-mining services are extended with some new features to improve the support of semantic tagging and semantic maintenance of the shared objects:

- Assistance in the process of semantic tagging, namely a suggestion of proper tags that semantically match with the textual content of the objects.
- Transformation of free-text tags into the tags predefined in the vocabulary.
- Consistency checking of the semantic tagging. Evaluation of homogeneity, similarities, and differences between the tagging performed by different users.

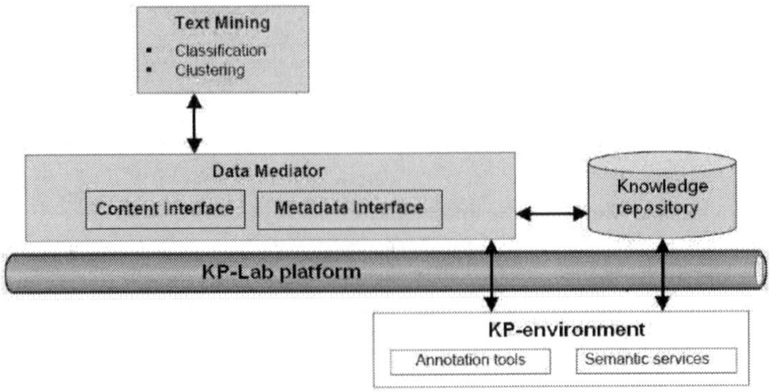


**Fig. 5.** Position of Text-mining services in whole KP-Lab System

- Maintenance of the tag vocabulary. Proposal for adding, modification, or removal of a semantic tag from/to the vocabulary.
- Extension of search results, query expansion with a possibility to tag the search results. Searching for similar objects, based on a textual content, metadata properties, or semantic annotations of a shared object.

### 3.3 KP-Lab Search Services

KP-Lab Search services provide an integrated interface for semantic search and free-term search. In order to effectively evaluate a search query, properties of the objects of activity have to be indexed using the Indexing service. Implementation of the Search services is based on the Solr search server [11], which provides an API

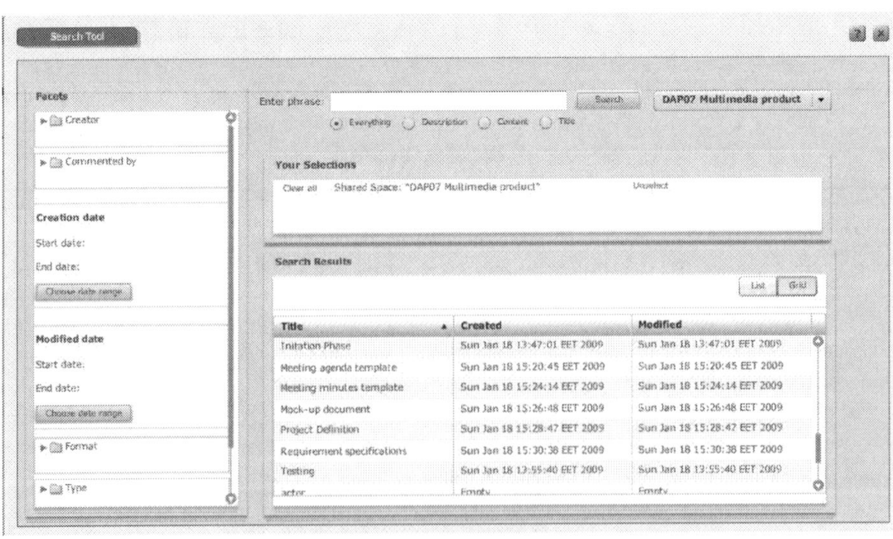


**Fig. 6.** Facet GUI of KP-Lab Search services

for both indexing and faceted searching. Indexing service is integrated with the Persistence API and CTM to simplify maintenance of the index for tool developers. In the case that Persistence API is not used to manage objects of activities, it is possible to invoke Indexing service directly using the Solr HTTP interface.

The Search services enable advanced faceted search (see Fig. 6) based on the metadata and content of shared objects and support user-defined, flexible visualization as well as semantics based classification and clustering of search results.

The users can save their search results as well as save the queries they have made. In next release, Search services will give suggestions by means of saved queries in order to help the user to find the most appropriate way of executing the search.

## 4 Conclusion

Information retrieval in KP-Lab System is represented by designed and implemented text-mining and search services. These two sets of functionalities provide features for management of shared objects, mainly content in textual format. These services are implemented on the middleware layer and end-users can use them within designed facet GUI, integrated part of virtual user environment called KP-environment.

Actual version of the proposed system was published in the beginning of this year and is available on http://2d.mobile.evtek.fi/shared-space/. Centre of information technologies[5], Technical University of Košice Slovakia, is responsible for the design and implementation of Persistence-API, Text-mining and Search services, Knowledge process service, To-Do service, Historical awareness, and partially CTM.

**Acknowledgments.** The work presented in this paper was supported by the following projects: the KP-Lab project, which is supported by European Commission DG INFSO under the IST program, contract No. 27490; the Slovak Research and Development Agency under the contracts No. APVV-0391-06 and No. RPEU-0011-06; the Slovak Grant Agency of Ministry of Education and Academy of Science of the Slovak Republic under grant No. 1/4074/07.

The KP-Lab Integrated Project is sponsored under the 6th EU Framework Programme for Research and Development. The authors are solely responsible for the content of this article. It does not represent the opinion of the KP-Lab consortium or the European Community, and the European Community is not responsible for any use that might be made of data appearing therein.

## References

1. Alexaki, V., Christophides, G., Karvounarakis, D., Plexousakis, K.: Tolle: The ICS-FORTH RDFSuite: Managing Voluminous RDF Description Bases. In: Proc. of the 2nd International Workshop on the Semantic Web (SemWeb 2001), in conjunction with Tenth International World Wide Web Conference (WWW10), Hongkong, pp. 1–13 (2001)
2. Bednar, P., Butka, P., Paralic, J.: Java Library for Support of Text Mining and Retrieval. In: Proceedings of Znalosti 2005, Stará Lesná, Slovakia, pp. 162–169 (2005)

[5] http://web.tuke.sk/fei-cit/index-a.html

3. Hakkarainen, K., Paavola, S.: From monological and dialogical to trialogical approaches to learning. A paper at an international workshop Guided Construction of Knowledge in Classrooms, February 5-8. Hebrew University, Jerusalem (2007)
4. Ionescu, M., et al.: KP-Lab Platform Architecture Dosier. KP-Lab public deliverable D4.2.3 (November 2007)
5. Markkanen, H., et al.: M33 specification of end-user applications. KP-Lab public deliverable D6.6 (December 2008)
6. Markkanen, H., Holi, M.: Ontology-driven knowledge management and interoperability in trialogical learning applications. Article in the KP-Lab book (2008)
7. Nardi, B.A.: Activity Theory and Human-Computer Interaction in Context and Consciousness: Activity Theory and Human Computer Interaction. MIT Press, Cambridge (1996)
8. New Collaborative Working Environments 2020. Report on industry-led FP7 consultations and 3rd Report of the Experts Group on Collaboration@Work. European Commission, Information Society and Media (February 2006)
9. Persistence API, http://kplab.tuke.sk/trac/wiki/kms-persistence-3
10. Smrž, P., Paralič, J., Smatana, P., Furdík, K.: Text Mining Services for Trialogical Learning. In: Proc. of the 6th annual conference Znalosti 2007, Ostrava, Czech Republic, pp. 97-108 (2007) ISBN 978-80-248-1279-3
11. Solr search server specification, http://wiki.apache.org/solr/

# Assessing Semantic Quality of Web Directory Structure

Marko Horvat, Gordan Gledec, and Nikola Bogunović

Faculty of Electrical Engineering and Computing, University of Zagreb
Unska 3, HR-10000 Zagreb, Croatia
{Marko.Horvat2,Gordan.Gledec,Nikola.Bogunovic}@fer.hr

**Abstract.** The administration of a Web directory content and associated structure is a labor intensive task performed by human domain experts. Because of that there always exists a realistic risk of the structure becoming unbalanced, uneven and difficult to use to all except for a few users proficient in a particular Web directory. These problems emphasize the importance of generic and objective measures of Web directories structure quality. In this paper we demonstrate how to formally merge Web directories into the Semantic Web vision. We introduce a set of objective criterions for evaluation of a Web directory's structure quality. Some criteria functions are based on heuristics while others require the application of ontologies.

**Keywords:** Ontology, Ontology Alignment, Semantics, Artificial Intelligence, Semantic Web, Web directory.

## 1 Introduction

The Semantic web vision and related spectrum of technologies have enjoyed rapid development during the last eight years. The initial paper by Tim Berners-Lee [1] introduced the notion of universally described semantics of information and services on the Web. The vision of a Web as a shared common medium for data, information and knowledge exchange, and collaboration, fostered a wealth of development and research. The idea itself was simple but appropriately far reaching. The Semantic web brought the power of managed expressivity provided by ontologies to the World Wide Web (WWW) [2]. Today the research in Semantic web application is not largely focused on the problem of ontologically-based Web directories [3][4][5]. Furthermore, as yet a lot of the effort is unfinished and more systems are in the phase of research and development (R&D) than in everyday production [6].

However, Web directories have simple hierarchical structures which are commonplace and effective for data storage importance and classification. This makes them important applications for data storage or classification, and motivates research in the assessment of their semantic qualities.

The remainder of the paper is organized as follows; the next chapter describes the categories and the structure of Web directories. Mutual associations between the Semantic Web and Web directories, as well as the semantic dimension of categories, are all presented in the third chapter. Quality measures of Web directories are discussed in the fourth chapter. Related publications and our conclusion with outlook for future work are presented at the end of the paper.

N.T. Nguyen, R. Kowalczyk, and S.-M. Chen (Eds.): ICCCI 2009, LNAI 5796, pp. 377–388, 2009.

## 2 Categories and the Structure of Web Directories

In order to explain how Web directories can be positioned within the Semantic web vision it is first necessary to formally define all constituent elements and how they organize to make up a Web directory, and secondly to add semantic annotations to these building elements. A web directory, web catalog or link directory as it is also called, is a structured and hierarchically arranged collection of links to other web sites. Web directories are divided into categories and subcategories with a single top category, often called the root category, or just the root. Each category can have a provisional number of subcategories with each subcategory further subsuming any number of other subcategories, and so on. Furthermore, every category has a unique name and an accompanying Uniform Resource Locator (URL), and can also carry other associated information.

Each category of a Web directory contains a set of links to various sites on the WWW, and a set of links to other categories within the web directory. This basic trait is the most important feature of a Web directory.

Each Web directory has a start page, i.e. a home page, which represents its root category, and every other category of a Web directory has its own adjoined web page. The start page displays subcategories that belong to the root. By following a link to a subcategory, user opens that category's page and browses through its links and subcategories. This process continues until the user finds a link to a web resource that s/he is looking for. In essence, the user can be described as an intelligent agent that traverses the structure of a Web directory looking for specific information.

Since Web directories are always rooted and the order of categories is strictly maintained, it is possible to assign level numbers to categories. The subcategories of the root are the 2$^{nd}$ level categories, and in turn their subcategories are the 3$^{rd}$ level categories, and so on. Maximum level of a Web directory is called depth.

Each category, except the root, has one category above it, which is called its parent. The categories below a certain category (i.e. with a greater level number than the category) are called its children, while categories on the same level as a node are called its siblings. Categories with no children are called terminal categories, and a category with at least one child is sometimes called nonterminal category. Associations between categories are arbitrary, but there must be at least one path between any pair of categories. Disjoint sets of categories are not allowed, as well as parallel links and self-loops. Each nonterminal category must have links to all its children, but can also have links to other categories in the Web directory which are semantically similar, or otherwise analogous to the category.

We will formally designate with $\boldsymbol{C}$ the set of all categories in a Web directory, and $\boldsymbol{R}$ will be the set of all Web resources in a Web directory. One category with unique identification number $n$ is denoted $c_n$. Category has its own characteristic URL *url*. The category $c_n$ must be a member of $\boldsymbol{C}$. $C_n$ is a subset of $\boldsymbol{C}$ that belongs to the category $c_n$, and $R_n$ the subset of $\boldsymbol{R}$ with Web resources that belong to the category $c_n$. In order to be more informative, the categories can also be written as $c_n^l$ with their member level $l$, where $l$ is a natural number smaller than or equal to the depth of a Web directory $L$. Therefore, category is a tuple $c_n = \{n, l, url, C_n, R_n\}$ and can be schematically annotated as in the figure below.

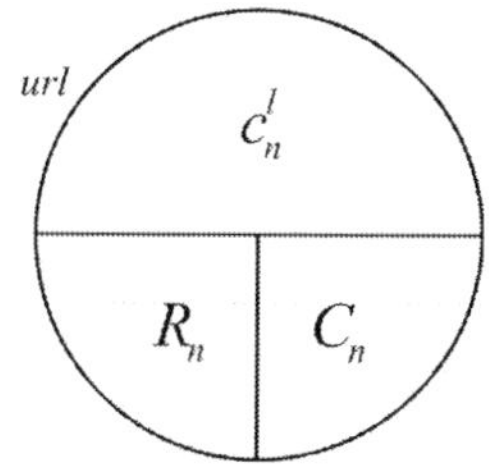


**Fig. 1.** Schematic representation of a single category

We can define a Web directory *wd* to be an element of the WWW. With ***C*** and ***R*** being members of *wd* the algebraic definitions of the elements of a Web directory and their mutual relationships are

$$\begin{gathered} C_n \in c_n \subset \boldsymbol{C} \\ R_n \in c_n \subset \boldsymbol{R} \\ l \in [1, L] \in \boldsymbol{N} \\ C \in wd \subset \text{WWW} \\ R \in wd \subset \text{WWW} \end{gathered} \tag{1}$$

The set of all children categories to $c_n^l$ is $C_n$ while the set of all children one level below is $C_n^{l-1}$ or $C_n^{-1}$, two levels below $C_n^{l-2}$ or $C_n^{-2}$, etc. As can be seen in (1) category is also a Web resource ($c_n \subset \boldsymbol{R}$), as it should be expected since it has unique URL and carries specific information. Furthermore, Web directory itself also becomes a tuple $wd = \{\boldsymbol{C}, \boldsymbol{R}\}$.

Mathematically speaking, Web directories are simple rooted graphs [7]. In this formal respect, categories represent vertexes and connections represent vertices. The path between two vertexes is called the arc, edge or link, and when there is an edge connecting the two vertices, we say that the vertices are adjacent to one another and that the edge is incident on both vertices. The degree of a vertex is equivalent to the number of edges incident on it.

Using the described formalisms, the schema of a simple Web directory with 6 categories distributed in 4 levels, with parent-child associations and two specific links $c_6 \rightarrow c_5$ and $c_3 \rightarrow c_2$ could be depicted with Fig. 2.

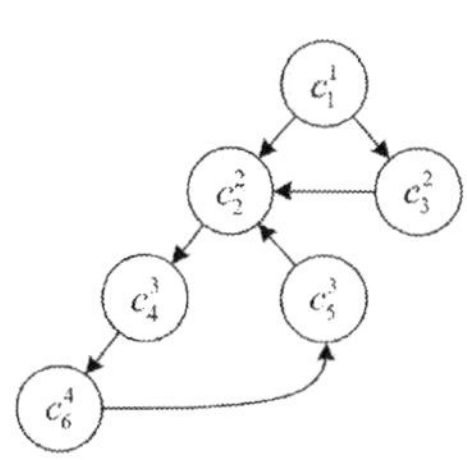


**Fig. 2.** Schematic representation of a Web directory

However, the structure of a Web category (Fig. 3) cannot be described as a tree because more than one path can connect any of its two categories: apart from paths which connect parent/child categories, they can be associated with *ad hoc* cross-links.

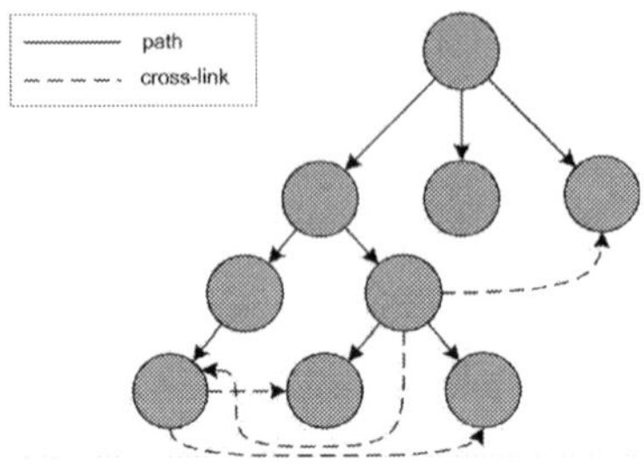


**Fig. 3.** Realistic Web directory with possible multiple paths between two categories

If, for the sake of discussion, all categories of a Web directory except the root had paths only to its children such structure would constitute a rooted tree, as in Fig. 4.

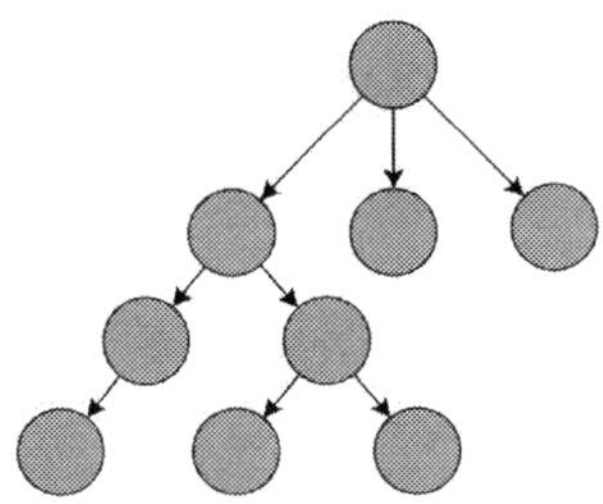

**Fig. 4.** Idealistic Web directory with only one path between any two categories

Sometimes the order of categories appearance is relevant, e.g. the position of links within a category's Web page is prioritized, and in that case we are talking about ordered and rooted simple graphs as an algorithmic definition of a Web directory.

Although the categorization of a Web directory should be defined by a standard and unchanging policy, this is frequently not the case. Web directories often allow site owners to directly submit their site for inclusion, even suggest an appropriate category for the site, and have editors review the submissions. The editors must approve the submission and decide in which category to put the link in. However, rules that influence the editors' decision are not completely objective and are thus difficult to implement unambiguously. Sometimes a site will fall in two or even more categories, or require a new category. Defining a new category is a very sensitive task because it has to adequately represent a number of sites, avoid interfering with domains of other categories, and at the same time the width and depth of the entire directory's structure has to be balanced. A Web directory with elaborate structure at one end and sparse and shallow at the other is confusing for users and difficult to find quality information in. Furthermore, after several sites have been added to a directory it may become apparent that an entirely new categorization could better represent the directory's content. In this case a part of directory's structure or even all of its levels have to be rearranged which is again time and labor consuming task.

## 3 The Semantic Web and Web Directories

At the moment of writing, the resources on the WWW are primarily designed for human and not machine use [8]. To rephrase it, the declarative and procedural knowledge currently offered by various Web sources is shaped in a way that better suites humans and not machines. The vision of the Semantic Web is directly aimed at solving this dichotomy by introducing self-describing documents that carry data and the accompanying metadata together, and thus organize and interconnect available information so it also becomes processable by computer applications [9].

The structure of a Web directory is basically a subjective construction. It depends on human comprehension and the policy taken by the Web directory's administrator, or even on the users that submit sites to the directory. It is important to note that not all Web directories, or even all segments of a Web directory, have the same editorial policy. Clearly, for the sake of a Web directory's informative clarity and usability, the semantic distance between any two categories should be approximately constant, and not dramatically vary from one category to the next. Also, the key for the selection of concepts that represent categories should remain uniform throughout the directory's structure. The only parameters that should be used to judge the quality of a directory are its informative value and usability, to humans and machines equally. In the fifth chapter we will propose several numerical parameters that objectively measure the worth of a directory.

Let's assume that we have at a disposition function *sem* that takes a resource $r_i \in \boldsymbol{R}$ and from its semantic content builds an ontology $o_i \in \boldsymbol{O}$ where $\boldsymbol{R}$ and $\boldsymbol{O}$ are sets of all resources and ontologies, respectively.

$$sem : \boldsymbol{R} \rightarrow \boldsymbol{O} \tag{2}$$

The function *sem* builds an ontology from a resource. In slightly different terms, it creates a solid representation of an abstract property. This property can be described as informal and explicit on the semantic continuum scale [10] and its technical realization is strictly formal. Operations of the function *sem* can be performed by a computer system or a domain expert, in which case we talk about automatic or manual ontology construction, respectively. The necessary mathematical assumption on *sem* is it has well-defined addition and subtraction operators in $\boldsymbol{R}$ and $\boldsymbol{O}$

$$\begin{aligned} &\oplus : \boldsymbol{R} \times \boldsymbol{R} \rightarrow \boldsymbol{R} \\ &\odot : \boldsymbol{R} \times \boldsymbol{R} \rightarrow \boldsymbol{R} \\ &\hat{+} : \boldsymbol{O} \times \boldsymbol{O} \rightarrow \boldsymbol{O} \\ &\hat{-} : \boldsymbol{O} \times \boldsymbol{O} \rightarrow \boldsymbol{O} \end{aligned} \tag{3}$$

This allows application of union operator across these two sets and concatenation of individual resources and ontologies, as well as determining their respective differences

$$\begin{aligned} sem(r_1 \oplus r_2) &= sem(r_1) \hat{+} sem(r_2) \\ sem(r_1 \odot r_2) &= sem(r_1) \hat{-} sem(r_2) \end{aligned} \tag{4}$$

Also, we should define a modulo operator $|\bullet|$ on $\boldsymbol{O}$ as $|\bullet|: \boldsymbol{O} \times \boldsymbol{O} \rightarrow \boldsymbol{O}$.

The *semantic content of a category* can be defined in three ways: *i)* by its Web resources, *ii)* from its subsumed categories, *iii)* as a constant.

By the first definition, semantic content of a category $c_i$ within a Web directory *wd* derives from the semantic content of all its Web resources $r_{ij}$ where $r_{ij} \in R_i \in c_i$ as

$$sem(c_i) = \underset{r_{ij} \in R_i}{\hat{+}} sem(r_{ij}) \tag{5}$$

According to the second definition, the semantic content of $c_i$ can also equal the aggregation of the semantic content of its children categories $c_j \in C_i^{-1} \in c_i$

$$sem(c_i) = \underset{c_j \in C_i^{-1}}{\hat{+}} sem(c_j) \tag{6}$$

Finally, if $c_i$ has no resources $(R_i = \varnothing)$ and subcategories $(C_i = \varnothing)$ it is assumed that the semantic content of $c_i$ is defined by a constant $const_i$ as

$$sem(c_i) = const_i : R_i = \varnothing, C_i = \varnothing \tag{7}$$

Reasoning behind such threefold definition is that the meaning of categories is conformed to the directory's editorial policy. If a category is empty and no resources have been added, it will still have some member semantics attached by the Web directory administrator.

The structure of directory *wd* is *ideal* if for non-empty $\boldsymbol{R}$ and $\boldsymbol{C}$

$$\begin{aligned} &\left| \underset{r_{ij} \in R_i}{\hat{+}} sem(r_{ij}) \hat{-} \underset{c_j \in C_i}{\hat{+}} sem(c_j) \right| = \varnothing \\ &\forall c_i \in wd, R_i \in c_i, C_i \in c_i \end{aligned} \tag{8}$$

That is, the structure of directory *wd* can be considered perfect if and only if for each category $c_i \in wd$ the semantic content of its Web resources $R_i \in c_i$ and subsumed categories $C_i \in c_i$ are equal.

Pragmatically, we can define a neighborhood $\varepsilon$ within $\boldsymbol{O}$ and say that the structure of directory *wd* is *realistically ideal* if

$$\left| \underset{r_{ij}\in R_i}{\hat{+}} sem(r_{ij}) \hat{-} \underset{c_j\in C_i}{\hat{+}} sem\left(c_j\right) \right| \leq \varepsilon \qquad (9)$$
$$\forall c_i \in wd, R_i \in c_i, C_i \in c_i$$

The existence of the function *sem*, with the described properties, is fundamental and indivertible in the ontology-based construction of Web directories.

## 4 Semantic Quality Measures

During or after Web directory's construction it is highly desirable to establish some measures of value of the accomplished process. The criterion functions that will provide these measurements should be objective and universal. Benefits of such measures would be twofold: *i*) they could provide a matching framework between Web directories, and *ii*) they could be used to assess the semantic structure quality of individual Web directories. In other words, by using them structures of any two Web directories could be objectively compared and the criterions could point to potential semantic deficiencies in a directory. Information retrieval in Web directories can be executed either through searching or browsing scenarios. Because of the sheer size of data available on the Web, searching is the dominant retrieval scenario. Several performance measures for evaluation of searching scenarios have already been proposed, such as precision, recall, fall-out and F-measure. However, information seeking by browsing scenarios is interesting in reduced information collections like blogs, RSS feeds, social networks [11], but also individual directory categories. Since information in Web directories may be browsed by intelligent agents as well as human users, the establishment of parameters for objective measurement of Web directory's structure and content is of a significant importance for determining its usability, semantic quality and subsequently other intrinsic characteristics.

We have identified three parameters that can be used to objectively assess the semantic quality of a Web directory. The parameters are:

1. Path ratio
2. Maximum revisit
3. Distance decrease progression

All parameters require observation of user's actions, i.e. browsing pattern of a person or an intelligent agent using the directory. We will assume that the browsing scenario starts at the root category although this is not strictly necessary (nor is often the case in real-world use). The parameters are calculated based on observation of an action of a single user. Each observation represents one browsing session for a specific resource contained within the directory. After the parameters of individual observations are collected they may be statistically processed and aggregated. This data can then cumulatively represent relevant trends and features in the actions of any number of directory's users.

*Path ratio* (PR) is calculated as a proportion between the minimum number of categories between the root and the category with the required Web resource, and the number of categories the user traverses while browsing. Therefore, when browsing for a resource $r$ in a Web directory $wd$ the browse $b(r,wd)$ with the length $|b(r,wd)|$ parameter PR is defined as

$$PR(b) = 1 - \frac{\min|b(r,wd)|}{|b(r,wd)|}, PR(b) \in [0,1\rangle \tag{10}$$

The rationale behind this parameter is that in the case of the optimal, or direct, browse $b^*$ the user will achieve the shortest path between the root and the category with the resource browsed for. In this case $\mathrm{PR}(b')=0$. In a suboptimal, or indirect, browse $b'$ user will traverse at least one category more and $\mathrm{PR}(b')>0$. This is explained in the next figure that illustrates a browsing pattern staring at category $c_1^1$ and ending at $c_9^4$.

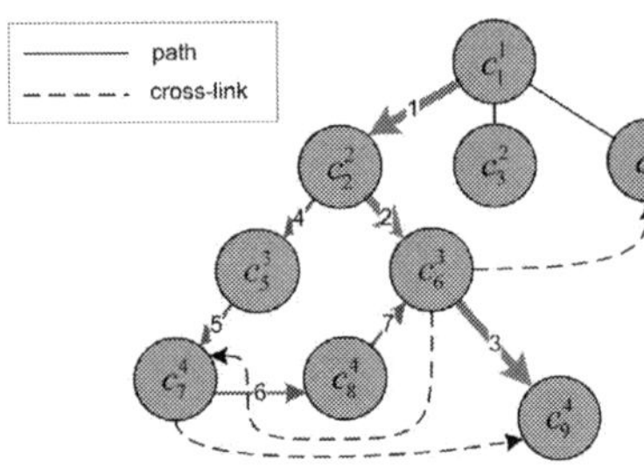


**Fig. 5.** Optimal (direct) and suboptimal (indirect) browse paths in calculating parameter PR

Browse $b_1$ with the path 1→2→3 is optimal because it traces the shortest and the most direct path between the start and the end category so that $\mathrm{PR}(b_1)=0$. While the browse 1→4→5→6→7→3 will also lead to the end resource, it is suboptimal since its length is greater than that of the optimal browse (6 > 3), thus $\mathrm{PR}(b_2) = 1/2$.

*Maximum revisit* (MR) or *maximum category revisit* is a parameter that describes the maximum number of repeated visits to any category while browsing for one resource. Because Web directories are simple rooted graphs with at least one path between any two nodes, there is never a need to visit the same category twice while browsing for a resource. Therefore, MR specifies the level of wander or loitering in a Web directory's structure while browsing (Fig. 9).

The best possible browse $b$ for a resource $r$ in a Web directory $wd$ has

$$\mathrm{MR}(b(r,wd)) = 0 \tag{11}$$

indicating no category revisit, where MR($b$) can be any natural whole number including zero $\mathrm{MR}(b) = 0,1,2,...,n,n+1 \in N \cup \{0\}$.

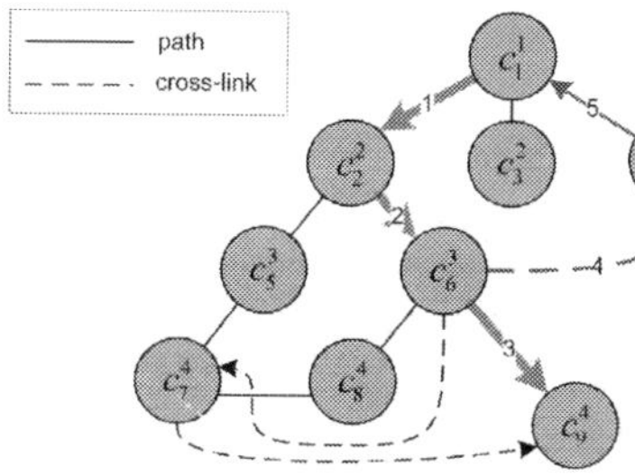


**Fig. 6.** Optimal and suboptimal browsing paths with revisits in calculating parameter MR

In the Fig 6., browse $b_1$ with the path 1→2→3 starting in $c_1^1$ and finishing with $c_9^4$ has $\mathrm{MR}(b_1) = 0$. However, due to the configuration of the directory it is possible to needlessly revisit some or even all categories. This is illustrated in the browse $s_2$ with the path 1→2→4→5→1→2→3 which gives $\mathrm{MR}(b_2) = 1$. Since $\mathrm{MR}(b_1) < \mathrm{MR}(b_2)$ browse $b_1$ is a better then $b_2$.

*Distance decrease progression* (DDP) is an ontology-based parameter. It describes the gradient of semantic convergence toward the resource during one browse. As the user browses categories looking for a particular resource, each category s/he visits should be progressively ontologically closer to the resource. If this is not the case, than either he is loitering or the directory does not have the optimal structure. Parameter $\mathrm{DPP}(s)$ can be defined as a series

$$\mathrm{DPP}(b) = \sum_{i=1}^{n-1} dist(c_i, c_T) - dist(c_{i+1}, c_T) \tag{12}$$

where $c_T$ is the target category containing the resource the user is looking for, $c_i$ is any category being browsed and $n$ is the length, i.e. number of steps, of the browse $b$. It is also necessary to apply a similarity measure $sim : \boldsymbol{C}^2 \rightarrow [0,1]$ between the two categories $c_1, c_2 \in \boldsymbol{C}$ and a distance function $dist(c_1, c_2) = 1 \,/\, sim(c_1, c_2)$ as defined in [12][13]. If the sequence of partial sums $\{s_1, s_2, ..., s_n, s_{n+1}, ...\}$ converges, than the series is also convergent, where

$$s_n = \sum_{k=1}^{m} dist(c_k, c_T) - dist(c_{k+1}, c_T) \tag{13}$$

The search $b'$ is optimal if $\mathrm{DPP}(b')$ converges to 0.

All three parameters described here should be used in conjunction with each other in order to cumulatively describe this important design feature of Web directories.

Node distribution in some Web directories, at a certain level in their structures, does not necessarily have to follow concept semantics partition or this process can be somehow affected and skewed. Examples of this are content division according to date, contributors' names or alphabet, e.g. having node "A" for subnodes with

"Apples" content, "B" for "Bananas", "C" for "Citrus", etc. These nodes would have more in common with a concept "Fruit" that with "Alphabet Letter". Subsequently, mutual semantic distance of such nodes would be great and incompatible with the directory's partition. In order to overcome this problem in calculation of the semantic quality parameters one has to simply ignore semantic value of these nodes at a level $l$ and directly link nodes in levels $l-1$ and $l+1$. By doing this monotone semantic difference between nodes is restored.

Every Web directory should have an easily understandable semantic schema that is reflected in a directory's structure so it becomes self-explanatory which category to browse in order to iteratively and progressively approach the required resource. This issue is closely correlated to the Web usability of directories. However, due to diverse quality of data sources available on the Web it is not easy to construct a directory with an ideal path ratio, maximum revisit and distance decrease progression values. Further planned experiments should provide more information on the everyday applicability of the parameters proposed here.

## 5 Related Work

All previous work regarding coupling of Web directories with ontologies and the Semantic web paradigms have been directed at using Web directories, their data and structure, to extract information from WWW with the goal of document classification and ontology learning. In this paper we presented an exactly opposite approach – using available knowledge to construct a Web directory itself.

The paper by Kavalec [4] which described a mechanism for extraction of information on products and services from the source code of public web pages was especially useful in our work. Papers by Mladenić [5], Li [14] and Brin [15] were also helpful.

We would particularly like to emphasize the work by a research group at FER which introduced ontologies in the search mechanism of the Croatian Web directory, and thus successfully resolved problems of low recall, high recall and low precision and vocabulary mismatch [16]. The Croatian Web directory (http://www.hr/) [17] was founded in February 1994 and its purpose has been to promote and maintain the network information services through the "national WWW homepage" and enable easy navigation in Croatian cyberspace using hierarchically and thematically organized directory of WWW services. At the moment of writing the directory contains 25,185 Web resources listed in 753 categories.

## 6 Conclusion

Web directories are commonplace method for structuring semantically heterogeneous resources. The form of simple rooted graphs is well-suited for information storage and representation in Web and Desktop environments, and in numerous applications ranging from directory trees, bookmarks and generic menus to tables of content. Also, the aspect of social collaboration is very important since networking and the Web enable instant publication and usage of data, ideally within groups of trusted users with the same areas of interest. All this only emphasizes the importance of successful construction, management, information extraction and reuse of all simple rooted graphs data structures and Web directories in particularly.

We would like to advise caution in using publicly available Web directories to learn new ontologies. Structures of Web directories are often biased and influenced by the contributors of resources. Administration of a large directory is an overwhelming task prone to errors. Therefore, it may be better to construct ontologies from smaller directories or from directories with rigid administrative policies. The former directories are more numerous than the latter, but they will also offer less information and in a more specialized area.

The primary goal of this paper was to put forward a series of objective criteria functions for evaluating the quality of Web directories. Two criteria are based in heuristics while the third calls for introduction of ontologies, which is possible only if Web directories are placed in the context of the Semantic web vision.

In the future we would like to apply the presented measures to a real Web directory and through a set of measurements obtain experimental results with relevant statistics about its structure quality. In this we plan to use the Croatian Web directory and its domain "Tourism" as a suitable test category. Also, we shall combine this work with our other efforts to develop ontology-based software for automated construction of Web directories [18]. In this respect the semantic quality measures would be used as control parameters in an iterative process of constructing and refining the Web directory's structure.

**Acknowledgements.** We would wish to thank Siniša Popović at the Laboratory for Interactive Simulation Systems (http://liss.esa.fer.hr/), Faculty of Electrical Engineering and Computing for his kind assistance in defining the algebra used in the paper.

## References

1. Berners-Lee, T., Hendler, J., Lassila, O.: The Semantic Web. Scientific American 284(5), 34–43 (2001)
2. Shadbolt, N., Berners-Lee, T., Hall, W.: The Semantic Web revisited. IEEE Intelligent Systems 21(3), 96–101 (2006)
3. Choi, B.: Making Sense of Search Results by Automatic Web-page Classifications. In: Proc. of WebNet 2001 – World Conference on the WWW and Internet, pp. 184–186 (2001)
4. Kavalec, M., Svátek, V.: Information Extraction and Ontology Learning Guided by Web Directory. In: ECAI Workshop on NLP and ML for ontology engineering, Lyon (2002)
5. Mladenić, D.: Turning Yahoo into an Automatic Web Page Classifier. In: Proc. 13th European Conference on Artificial Intelligence, ECAI 1998, pp. 473–474 (1998)
6. Cardoso, J.: The Semantic Web Vision: Where Are We? IEEE Intelligent Systems 22(5), 84–88 (2007)
7. Sedgewick, R.: Algorithms in C++, 3rd edn. Addison-Wesley, Reading (2001)
8. Horrocks, I.: Ontologies and the Semantic Web. Communications of the ACM 51(12), 58–67 (2008)
9. Antoniou, G., van Harmelen, F.: A Semantic Web Primer. MIT Press, Cambridge (2004)

10. Uschold, M.: Where are the Semantics in the Semantic Web? AI Magazine 24(3), 25–36 (2003)
11. Schraefel, M.C.: Building Knowledge: What's Beyond Keyword Search? IEEE Computer Magazine 42(3), 52–59 (2009)
12. Ehrig, M., Sure, Y.: Ontology mapping - an integrated approach. In: Bussler, C.J., Davies, J., Fensel, D., Studer, R. (eds.) ESWS 2004. LNCS, vol. 3053, pp. 76–91. Springer, Heidelberg (2004)
13. Euzenat, J., Shvaiko, P.: Ontology Matching. Springer, Heidelberg (2007)
14. Li, J., Zhang, L., Yu, Y.: Learning to Generate Semantic Annotation for Domain, Specific Sentences. In: Delugach, H.S., Stumme, G. (eds.) ICCS 2001. LNCS (LNAI), vol. 2120, pp. 44–57. Springer, Heidelberg (2001)
15. Brin, S.: Extracting patterns and relations from the World Wide Web. In: Proc. of the Sixth International Conference on Extending Database Technology (1998)
16. Gledec, G., Matijašević, M., Jurić, D: Improving Search on WWW. HR Web Directory by Introducing Ontologies (2005)
17. Gledec, G., Jurić, J., Matijašević, M., Mikuc, M.: WWW.HR - Experiences with Web-server development and maintenance. In: Proc. of the 18th International Conference on Computers in Telecommunications, 22nd International Convention MIPRO 1999, pp. 93–96 (1999)
18. Horvat, M., Gledec, G., Bogunović, N.: Decision Support Systems in Ontology-Based Construction of Web Directories. Accepted for publication in ICSOFT 2009, Sofia, Bulgaria (2009)

# Computer Aided Requirements Management

Kamil Karwowski, Witold Wysota, and Jacek Wytrębowicz

Institute of Computer Science, Warsaw University of Technology
k.karwowski@elka.pw.edu.pl
w.wysota@ii.pw.edu.pl
j.wytrebowicz@ii.pw.edu.pl

**Abstract.** The paper presents a concept of a requirement management system for ICT projects. The system (further referenced as the CARM framework) makes use of semantic networks by expressing requirements through ontologies. Architecture and details about the system and its use are described. An aspect of using the model-view-controller pattern for designing requirement management systems is also presented.

## 1 Introduction

Computer systems in modern world are the tools that interact with the life of almost every human being. Among them, there are banking systems, booking services, aid and management systems and others. Rapid expansion of their complexity opened a demand for better and better quality of created software. Increasing capacity and throughoutput of computer systems has been causing software development companies to put more stress on producing software of high quality. The quality depends, among others, on the effort spent on requirements engineering.

There are different approaches to managing requirements in ICT projects. Very formal requirement specifications are created for conformance testing of telecommunication equipment, e.g., IPv6 Requirements Catalogue[1]. Yellow sticky notes are used to collect requirements in some extreme programming techniques. There are projects where requirements have to be specified, fixed and formally accepted before designing. There are projects that assume requirement changes during designing. The importance of requirement specification depends on project character, size, cost, and on legacy rules the concerned enterprise comes under.

A requirement itself can have several meanings. It can express an idea, a goal, a necessity and a constraint. IEEE Std 830-1998 defines Recommended Practice for Software Requirements Specifications[1]. Requirements are collected, analyzed, validated, refined, verified, sometimes they are even modeled. Requirement specification is a reference document for testing and future corrections or extensions of the related product.

[1] http://www.ipt.etsi.org/STF295-ph1/home.asp

N.T. Nguyen, R. Kowalczyk, and S.-M. Chen (Eds.): ICCCI 2009, LNAI 5796, pp. 389–400, 2009.

The requirement specification evolves during the project. The document starts to be more and more detailed. In fact, for a big and complex project the specification is composed from several documents. At the first stage of design the project team creates a system (architecture) specification on the base of requirement specification. The created specification of the system can be considered as requirement specification for the next project stage — for implementation. It is desired to keep links between goals definition, functional requirements and implementation requirements. Frequently in real projects these links rest only in the minds of project teams. When time to market is the priority then usually these documents are temporary and have bad quality. Maintenance of requirement documentation needs an effort, time and motivation. It is an obligation to do it in military and safety critical projects. Big enterprises, which foresee future product maintenance and evolution, try to upkeep the requirement documentation. Small enterprises take a risk of reliance on human memory and stable employment. A *Computer Aided Requirements Management* (CARM) system would be a chance to eliminate this risk and to lower the needed effort for requirement maintenance. In many projects we use different requirement representations, which depict selected scope of requirements for different stakeholders of the project. A stakeholder needs a representation which fits his perception and helps performing his task or decision. IEEE Std 1471-2007 defines Recommended Practice for Architectural Description of Software-Intensive Systems[2].

This standard introduces the concept of different views, created for project stakeholders according to the definition of their viewpoints. To avoid inconsistency, the requirement views should be generated from one data set (one specification). However sometimes it could be impossible, because during requirement collection different description techniques can be used (e.g., tabular form and UML diagrams for use cases expression, screenshot examples, PICS[2] tables). Thus a designer has to link the incoming data manually. A Computer Aided Requirements Management (CARM) system would be helpful in this task. Machine-readable links allow for automatic data selection. If the data gathered by CARM has machine readable representation then CARM will be able to generate expected views of requirements.

In the following sections of this paper first we briefly present others existing semantic-based approaches to the problem and then we describe CARM — our environment for managing and processing technical requirements. We show the overall architecture along with details on its internals, present ontologies used and explain their purpose. Finally we present potential uses of the system and show paths for future research.

## 2 Similar Projects

Among the tools aiding project requirements management one can find many examples of classical concepts of requirement engineering (such as requirements dependency matrix) but only a small number of methods are based on semantic

[2] Protocol Implementation Conformance Statement.

networks and ontologies. Most important projects from this field are conducted by the Agile Knowledge Engineering and Semantic Web group[3]. Below we describe three of them, which we find the most worthy of attention.

The first one called OntoWiki[3] is a tool for visualization and editing of semantic knowledge base. The tool aids multiuser initiatives in designing and editing of knowledge bases, with tracking changes and generating different views on instances of data. Not being oriented on requirement management it is a good example of a system, which uses an aspect of different categories of users viewing the same data in different ways. The data is stored as instances of ontologies introduced into the system. The advantage of the platform is the possibility of adding data important from the point of view of a particular user. The data can then be viewed independently of its structure using one of the five predefined views. Unfortunately it is not possible to add new views which greatly limits the use of the system.

OntoWiki is a general solution for exploring custom ontologies. It allows for creating new platforms, which can build upon it. An example of such is SoftWiki[4] — a tool for aiding software development. Its main goal is to make users (shareholders) cooperate in creating software in a process based on requirements. Its main assumptions defined in [5] are:

- multiuser creation, structuralization and management of requirements,
- support for semantic annotations,
- maximizing productivity of developing software,
- storing data in a form readable by both human and machine.

In order to support this ideas, a new requirements ontology has been created — *SoftWiki Ontology for Requirements Engineering* (SWORE[4]), which is the second presented tool. It defines different types of requirements (Goal, Scenario, Functional Requirement, Quality Requirement). This enables the specification of abstract requirements at different levels of granularity[6]. Two classes: Raw Information and Reference Point help to interlink requirements with existing documents or resources. In this case resources can be understood as other domain ontologies.

The third tool we want to present is Software Engineering 2.0 (also called SEOntology — Software Engineering Ontology)[5]. According to [7] SEOntology defines the world's first and only software engineering ontology and a project management ontology in conjunction with a domain ontologies. The main purposes for using ontologies in this project mentioned in [7] are:

- to provide communication and references mechanism for software engineers,
- to provide mechanism for accessing various data,
- to create clearly defined semantic for Web Services,
- to create various semantics for different domains (such as manufacturing, energy and power systems, financial systems).

---

[3] http://aksw.org
[4] http://softwiki.de/SWORE
[5] http://www.seontology.org

The main ontology that describes software engineering contains three sub-models: Domain, Sub Domain and Instance Knowledge. The Domain Knowledge describes software engineering concepts that everyone working on system is agreed upon. The Sub Domain Knowledge contains a specific concepts for particular projects. The Instance Knowledge defines project data.

## 3 Semantic Requirement Management

The idea of CARM has arisen during a project, we have been working on for Telekomunikacja Polska S.A.[8]. In this paper we present the concept of CARM framework. CARM cannot be a closed system. It has to be an open framework, which could be integrated with existing tools already used in a given enterprise. It has to be easily extended to support new requirement representations and new stakeholder views.

We have tried to define the CARM framework in a pragmatic manner. We had a brainstorm with project managers and specialists from telecom R&D. This experience has shown very high expectations for such a framework.

Below we give gathered conclusions. It is expected that CARM will allow to reduce design and deployment cost of new ICT products. CARM should support workers from different departments, which are involved in a project. They have to share a common, consistent project data repository. Requirement specifications should be used not only at design stage, but also during testing, deployment and maintenance of products. A specification should be updated when a defect is discovered or removed or when the product is modified to offer new functionalities. Requirement management support should allow for requirement grouping, hierarchization, prioritizing, versioning, change tracking. It is desired that CARM would be able to discover any requirement inconsistency. Analysts and designers work with different tools they already have and know. They create files in formats defined by these tools. It is expected that CARM will enable to exchange data files with them, and even that CARM will allow to bind requirement specifications with importing data files. The most useful binding is a semantic based one. Efficient inconsistency discovery and any machine aided reasoning about requirements and related documents will be possible only if semantic description of them is given.

The gathered expectations about CARM functionality are very wide. Moreover, they point to functionalities already supported by existing tools, i.e. concurrent versioning systems (CVS, Subversion, Perforce) issue ticket systems (there is a huge number of such solutions), documentation writing (commonly used word processors), report generation (Crystal Reports, OceanGenrap), work-flow management (many systems exist in particular domains). CARM should rather cooperate with them, than to compete with them. We want to concentrate on new values CARM could provide using semantic approach to requirement management.

### 3.1 General Design

The CARM platform will be used in an enterprise, which leads ICT projects. It will be deployed inside the enterprise intranet, what gives desired access for a project team and what gives minimum level of security. CARM should evolve to enable adoptions for particular design environments, and to satisfy future needs, hence SOAapproach is convenient for its development. A web browser is the preferred way of communication between workstations and CARM services. It is expected that a designer can work on his workstations off-line, using a local repository for needed data. The content of local repositories should be synchronized with a central one in similar way, as we do with concurrent versioning systems. A logically central data repository can be implemented in a distributed or redundant way to achieve data security and fast access. Designers, analysts and other people involved in a project, they need different requirement views and they differently process the requirements. CARM will support view management facility. A CARM user will get a view to access requirement documents and data. The view can be considered as an interface to the requirement repository. Such approach suits IEEE 1471 standard. It is worth to notice, that similar approach has been applied in DoDAF[6] and MoDAF[7] recommendations. What differs our approach from those mentioned above, is the use of ontology definitions as a basis for CARM architecture.

### 3.2 Architecture

Logic architecture of CARM platform is shown on figure 1. It consists of three layers:

- *Data Layer* (System Data Area) — a main platform data model. It contains all ontologies available in the system and is responsible for serialization and deserialization of data and for performing basic operations such as retrieving ontology items or modification of the ontologies themselves.
- *Views Layer* — which consists of views that represents various requirement modeling techniques. Each view should relate to proper ontology (or ontologies) for getting data to fill the view.
- *Management Layer* — consisting of modules responsible for platform management (User module — groups, roles, access rights) and additional ontology management (inference/rules mechanism).

Both the *Management Layer* and the *View Layer* can be treated as typical modules built according to preferred practices and solutions. In further subsections we would like to describe the main element of the CARM platform — the *System Data Area*.

For building requirements model as an ontology we define three main assumptions:

[6] Department of Defense Architecture Framework (USA).
[7] Ministry of Defense Architecture Framework (UK).

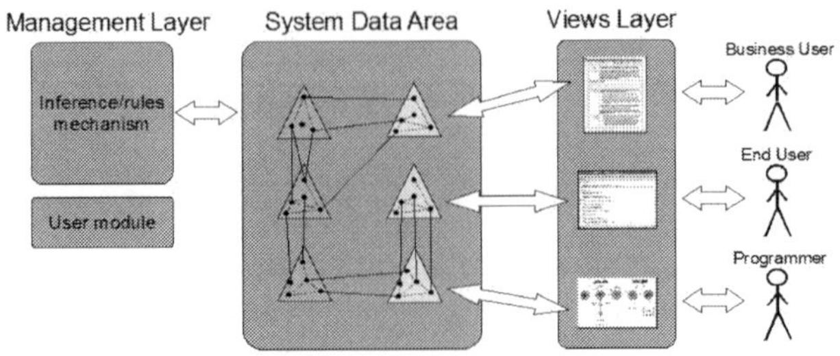


**Fig. 1.** CARM structure

- requirements model should be available for different kinds of users such as: business user, end-user, programmer;
- there should be a possibility to refer external resources such as documentations, code;
- platform meta-data should be constructed in a way that easily allows to add further extensions and modifications.

**Ontology stack.** A solution, that can satisfy the above assumptions, is to create ontologies in a form of a stack. The approach assumes creation of so called *upper ontology*, which is an ontology describing projects that is generic enough to form a basis for other ontologies. This ontology contains general concepts characteristic to the domain of projects we are dealing with. Subsequent layers of the stack (in form of ontologies as well) build upon those from higher layers. The second layer would describe main project domains (computer systems, telecommunications) and the third layer would contain meta-data about specific methodologies. An example of meta-data can be a definition of requirements dependency matrix, which depicts various relations between collected requirements. All three layers of the stack form the framework for adopting requirements (Fig. 3). The concept of an ontology stack is compliant to the foundations of the SUPER project[9].

Our upper ontology — URMOnto (*Upper Requirements Modeling Ontology* — see fig. 2) defines basic concepts such as: project, requirement and stakeholder. It also describes requirements in three more detailed subclasses: goal, task and common requirement; which are base for typical requirement modeling techniques. The goal class helps to create requirements in form of a hierarchy, which describes main project directives. This modeling approach is more suitable for general requirements, not detailed ones. A task is a class characteristic to a given scenario. It assumes that requirements consist of some kinds of steps or sub-tasks arranged in specific order. A *task requirement* is end-user specific (represents scenarios of some functionality) and describes external requirements (mainly communication with others systems and interfaces). The last class, *common requirement* expresses elementary (functional) requirements reflecting what needs to be implemented. Three different subclasses of requirement class focus chiefly on users (they reflect business user, end-user and developer requirements)[10]. Similar approach to requirements classification is presented in SWORE in SoftWiki[6].

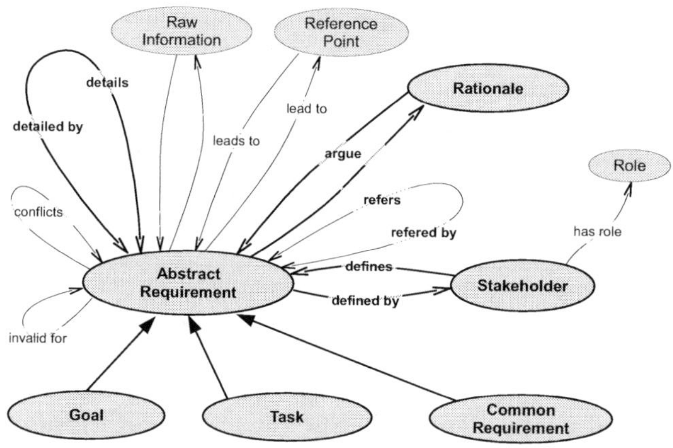


**Fig. 2.** URMOnto

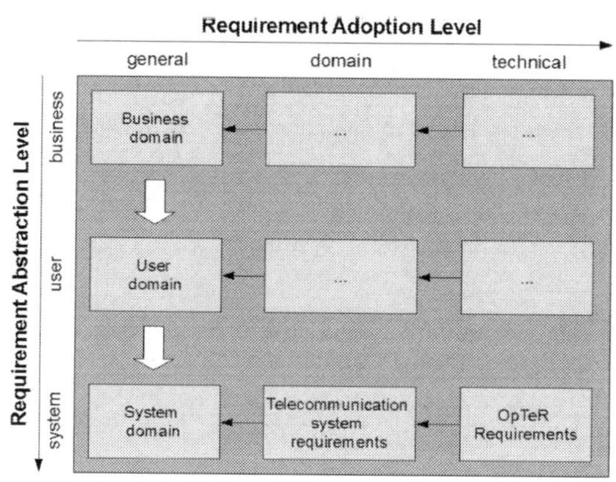


**Fig. 3.** Requirements levels

The second layer of the ontology stack — *Domain Requirements Level* defines different domains of projects. It consists of ontologies that are dictionaries, taxonomies and hierarchies of some concepts. The main purpose of the second layer is to provide consistent and unambiguous dictionary[1]. It defines main concepts that are referenced from requirements.

The final layer consists of ontologies that describe specific modeling methodologies. These ontologies should be based on the concepts defined on the upper layers. An example of a modeling ontology is OpTeR (Operator Technical Requirements) [8] that describes operator requirements in telecommunication area (Fig. 3). We can notice that the most general requirement is defined in the business domain ontology on the general requirements level (upper left corner). On the other hand, the most precise requirement description is defined on the technical level of system abstraction level (bottom right corner).

**Semantic annotations.** When describing requirements a frequent tool used is a reference to an external document or its fragment. It seems convenient to introduce tags (annotations), to make it possible to point them within an ontology description. Some popular data formats like RTF, PDF, ODF or Microsoft Word have support for tagging or commenting documents, which makes it easy to inject semantic annotations to their contents. These file formats that only have commenting support (like all XML-based ones) can be annotated by using qualified (semantics aware) comments.

What is important, it is the meaning of tags. They are not just pins that lets one point an exact place in a text, but they also say what a particular paragraph is about and what semantic value it carries. Such annotations can then be processed by machines that extract sentences and terms from the text. This in turn can be used to show additional information to different users (see next section for details on views). Finally it simply enriches the knowledge database available for the project. Extracting additional requirements or test cases for validating existing requirements is one of the obvious things the mechanism can be used for.

Annotating is a very time consuming and tedious task. The domain of telecommunication projects is a perfect example to demonstrate the problem. Part of the project that gave birth to the idea of CARM was to analyze existing telecom standards and other documents. One of the conclusions was that there were many redundancies and contradictions in those documents, especially if they originate from different organizations (ETSI, ITU-T, ANSI, IEEE). If we wanted to annotate them automatically it might occur that we receive contradicting "facts". Therefore a manual intervention is required to resolve those inconsistencies and it is important to make the framework, which could help with the process as much as possible. We opt for using existing tools for text analysis and data extraction from documents. An example of such a tool is the *Text Onto Miner* framework for semi-automated ontology creation[11].

**Data views.** The task of the Ontology View Layer (Fig. 1) is to enable browsing instances of ontology classes and their attributes. Requirement modeling methodologies are defined as proper ontologies in technical adoption level (Fig. 3). They can be used to retrieve proper data. The retrieval allows to construct views representing adequate modeling methodology (according to selected ontology). However ontologies presented in our approach don't contain information about how data should be displayed (e.g., in form of a table or a graph).

Some requirements modeling techniques are very simple to visualize. For example requirement dependency matrix or the earlier mentioned OpTeR methodology can be presented as typical tables.

On the other hand, there are techniques that have specific visualization elements (e.g., UML). In the CARM platform we defined two solutions for generating views.

First one is an implementation of Model View Controller (MVC) pattern. The model in this approach is an ontology from technical requirements level. By

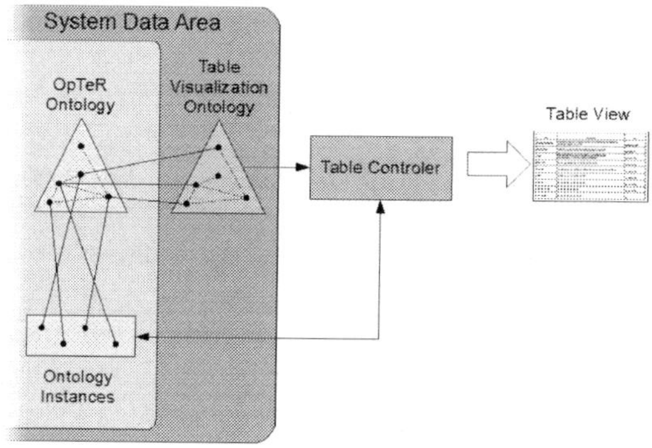


**Fig. 4.** Example of generic view

using a proper controller and view, we can construct any complex view (such as an UML graph). However this solution requires definition of both controller and view for a specific ontology, which further complicates adding new perspectives to the platform needs a lot of coding and any reuse of this code is weak.

The second solution for generating views suits well simple modeling techniques. In this approach there are specific ontologies defined, which describe some types of visualization (e.g., table, graph or list). For each ontology a dedicated controller and view is also defined. In this case, if we want to add a new model to our platform we only define an ontology for that model, which additionally referrs to proper visualization ontology. This helps to links elements representing modeling concepts with elements describing visualization. Figure 4 presents an example of this approach.

### 3.3 Ontologies Used

The basic version of CARM contains four ontologies: upper ontology for requirements and three ontologies from general requirements level. Adding further ontologies (both from domain and technical levels) strongly depends on the project, which we model requirements for. A typical use of CARM platform assumes that technical ontologies with related views and controllers will be constructed as modules and could be reused in other projects. In this approach, when starting a new project with CARM a user can select proper modules (views) from ones already existing and simply add them to the platform. If there is no proper view available, then it should be constructed. Reusing domain ontologies (shown on the second level of figure 3) is similar to what was described for the technical level. In addition to domain level there are already many ontologies, which define dictionaries and taxonomies. We can reuse them. Instead of building a new definition we can adjust an existing one. For example, we can adopt OWL-S[8] — the WebServices ontology, for a concrete ontology of a WebService under design.

[8] http://www.w3.org/Submission/2004/SUBM-OWL-S-20041122

### 3.4 Fields of Use

CARM has been conceived to aid in telecommunication and information projects. In the next few paragraphs we explain some of potential advantages of using the described semantic approach, when dealing with requirements in projects from these domains.

Storing requirements as a semantic network yields many advantages. One of them is the ability to automatically update a specification under design, through reasoning based on already defined (or discovered) requirements. A requirement specification can be extended by data originating from standards or other documents or even from common sense rules applied to gathered knowledge. Wherever exists a large number of documents related to the domain of a project (such as standards, recommendations, acts or manuals) there is a potential benefit of semantically annotating them (as we suggested earlier in section 3.2). This leads directly to discovering knowledge of possible requirements or ways of satisfying them, which can be added to the pool of defined requirements. There are already systems available that allow for ontology creation from natural language documents[11] — using them for this purpose is much advised.

Another advantage comes from centralization of all gathered data (semantic network definitions, ontology instances, specification documents, generated views and reports). Having a system that contains all data related to a project in one place makes it easier to organize, transform and store. The ability to serve data in different formats and different hierarchies or aspects, enables a personalized look at the data, which was described in section 3.2 of this paper. Having different perspectives on the same data means not only that one is able to view it differently. It also means that changes made in one perspective (view) will automagically be transferred to other perspectives allowing other people to benefit from them. This prevents a situation that workers from different departments or teams come to different conclusions, because they are working concurrently on similar sub-domains of a problem, but they create different documents and thus don't share knowledge and new experiences. By deploying CARM the knowledge from one work group "diffuses" to other teams.

There is yet another potential use of a semantic network in requirement engineering. It is possible to create an ontology that is able to associate requirements with services or technical details (such as protocols, devices, etc.). The services or technical elements are used when filling the requirements during implementation or deployment. Basing on the semantic relations and on a simple database, one can receive (through reasoning) suggestions about concrete solutions (algorithms, protocols), or about suppliers that can deliver parts of or even whole components that make up the service in question. It would allow to move some aspects of designing to an automated system, and reduce time and financial effort needed to make a new product available to public.

Finally we have the ability to add to a new project already existing semantic networks and reuse specifications that were created for earlier projects. Collection of ontologies, created in an enterprise, together with related design techniques, together with documentation from earlier projects, are intellectual property of that enterprise. The volume of the data can be huge, e.g., documentation of telecom services managed by an operator. Semantic bindings help in their management and reuse.

### 3.5 CARM Implementation

CARM framework is currently in a prototype stage and is implemented as a multiuser web application. It contains two basic perspectives (both constructed as MVC models): requirements table view and requirements dependency view. For each perspective there is an ontology defined, which is an extension of the URMOnto ontology. We have used the Jena ontology engine and Pellet reasoner for implementation of CARM framework. For presentation layer deployment we have relied on the Apache Tomcat web server and on the Struts 2 presentation framework.

The prototype allows to edit requirements using the table perspective and shows proper changes in the dependency view. Next, it allows to automatically reason about collected requirements. Future efforts on CARM framework are focused on implementing solutions for views generation (for selected simple modeling techniques such as use cases). This will let us to define easily and to add next perspectives to CARM platform.

Experiments performed with the current prototype confirm that semantic requirements description allows for automatic consistency check. Moreover it allows for discovery of facts and to make up the requirement set as a result. However our tests show that ontology processing (mainly the reasoning process with many ontologies involved) takes a lot of time (in the range of seconds), and requires much memory. Moreover the reasoning process generates many sentences but only a part of them is important to a user.

Advantages of the CARM architecture are: that it contains an open base of requirement ontologies and that it allows to add new specialized views of requirements. A project team or an enterprise can build-up the framework following their particular needs and ideas. It is possible to develop the framework in the way which conforms to the IEEE 1471 standard.

## 4 Conclusions

Use of ontologies as a new way to store knowledge is becoming more and more popular. Their use in projects opens new perspectives including advanced methods for inferring and knowledge discovery. Created data representation allows for machine processing without loss of its semantic values. This in turn permits efficient automatic data searching and reasoning. In addition to that, entering new data consistent with a predefined ontology is easy and fast (hence the designer doesn't have to work on a new ontology definition for every project). However we have to realize results of automatic reasoning strongly depend on the volume and quality of earlier collected data. First we have to collect some knowledge in a machine readable form and only then we can take advantage from it.

Unfortunately extended capabilities for describing data also bring more complexity that often requires a significant amount of work to be spent on preparing the semantic models. This in turn draws a conclusion that usefulness of using semantic approach in small projects is very limited. Nevertheless specifications of bigger systems can surely take advantage of such approach by automating some tasks and deriving knowledge from documents written in natural language.

## References

1. IEEE: IEEE recommended practice for software requirements specifications. IEEE Std 830-1998 (1998)
2. IEEE: Systems and software engineering - recommended practice for architectural description of software-intensive systems. ISO/IEC 42010 IEEE Std 1471-2000 First edition 2007-07-15, c1–24 (2007)
3. Auer, S., Dietzold, S., Riechert, T.: OntoWiki - A Tool for Social, Semantic Collaboration. In: Cruz, I., et al. (eds.) ISWC 2006. LNCS, vol. 4273, pp. 736–749. Springer, Heidelberg (2006)
4. Lohmann, S., Heim, P., Auer, S., Dietzold, S., Riechert, T.: Semantifying Requirements Engineering The SoftWiki Approach. In: 4th Intl. Conf. on Semantic Technologies, I-SEMANTICS 2008 (2008)
5. Auer, S., Riechert, T., Fähnrich, K.P.: Softwiki - agiles requirements-engineering für softwareprojekte mit einer großen anzahl verteilter stakeholder. In: Meißner, K., Engelien, M. (eds.) Proc. Workshop Gemeinschaften in Neuen Medien (GeNeMe), Dresden, pp. 97–108. TUDPress, Dresden (2006)
6. Riechert, T., Lauenroth, K., Lehmann, J., Auer, S.: Towards Semantic based Requirements Engineering. In: 7th Intl. Conf. on Knowledge Management (2007)
7. Dillon, T.S., Chang, E., Wongthongtham, P.: Ontology-Based Software Engineering – Software Engineering 2.0. In: Australian Software Engineering Conference, pp. 13–23. IEEE Computer Society, Los Alamitos (2008)
8. Grzegorzewski, M., Karwowski, K., Wysota, W., Wytrębowicz, J.: Metodyka opisu struktur, usług i protokołów hybrydowego środowiska NGN/CSN dla wspomagania działań Operatora. Technical report ordered by R&D Dept. of Telekomunikacja Polska S.A. Serial number: June 26 (501/E/1036/4470) (2008)
9. Szczekocka, E., Frankowski, J., Kupidura, H., Rubach, P., Skorczyk, M.: Project SUPER. R&D Dept. Telekomunikacja Polska S.A. (2007)
10. Gottesdiener, E.: Good practices for developing user requirements. CrossTalk - The Journal of Defense Software Engineering (2008)
11. Gawrysiak, P., Protaziuk, G., Rybiński, H., Delteil, A.: Text onto miner – A semi automated ontology building system. In: An, A., Matwin, S., Raś, Z.W., Ślęzak, D. (eds.) Foundations of Intelligent Systems. LNCS (LNAI), vol. 4994, pp. 563–573. Springer, Heidelberg (2008)

# BizKB: A Conceptual Framework for Dynamic Cross-Enterprise Collaboration

Hanh Huu Hoang and Thanh Manh Le

Hue University
3 Le Loi Street, Hue City, Vietnam
{hhhanh,lmthanh}@hueuni.edu.vn

**Abstract.** Semantics is applied into Business Process Management (BPM) to bridge the gap between the business world and information systems, especially in the context of B2B integration. Current standards such as BPMN, XPDL, BPEL and their combination have not fulfilled the expectation of two communities. The gap is still there: how enterprises can make the cross collaboration each other without 'knowing' their partners; and how a process based on the graphical notation can be fully mapped into the executable process without its semantics. In this paper, we propose a new approach, namely BizKB framework, for the cross-enterprise integration using ontologies and Semantic Web Services technologies in order to realizing business concepts into the executable level using web services.

## 1 Motivation

Bussler [1], Jung et al. [2], and Leyman et al. [3] describe hierarchical structures of business-to-business (B2B) process modeling with web services. The previous researches largely focused on how to design external collaborative processes and execute them automatically. However, relatively little attention has been paid to the following: first, most B2B process modeling methodologies have mainly dealt with automatic execution of collaborative processes, not human interaction. Second, previous methodologies did not effectively support flexible exchange of trading partners and configuration management for dynamic B2B collaboration [4]. In order to ensure a true common understanding about the contents of a communication between business partners, certainty regarding the semantics of the data transferred has to be established. Agreeing upon and deploying a syntax standard is not sufficient [5].

Moreover, the degree of mechanization in BPM is currently very limited. The major obstacle preventing a coherent view on business processes is that the business processes are not accessible to machine reasoning. Additionally, businesses cannot query their process space by logical expressions, e.g. in order to identify activities relevant to comply with regulations [6]. Many research projects addressing ontology engineering and management in the business area overlook the fact that information is never exchanged on a purpose-free basis, but rather, it always travels along with business processes – i.e., material, financial or coordination flows – and serves their specific objectives. The real purpose of ontological engineering methods in the business area is not information integration itself, but the integration of its underlying processes [5].

N.T. Nguyen, R. Kowalczyk, and S.-M. Chen (Eds.): ICCCI 2009, LNAI 5796, pp. 401–412, 2009.

Founded on ontologies Semantic Web technology provides scalable methods and tools for the machine-readable representation of knowledge. Semantic Web Services (SWS) make use of Semantic Web technology to support the automated discovery, substitution, composition, and execution of software components (Web Services). BPM is a natural application for the Semantic Web and SWS technologies, because the latter provide large-scale, standardized knowledge representation techniques for executable artifacts.

The Semantic Web and, in particular, SWS technology offer the promise of integrating applications at the semantic level. By combining them and BPM, and developing one consolidated technology, we want to create horizontal ontologies which describe business processes and vertical telecommunications oriented ontologies to support domain-specific annotation (Fig. 1).

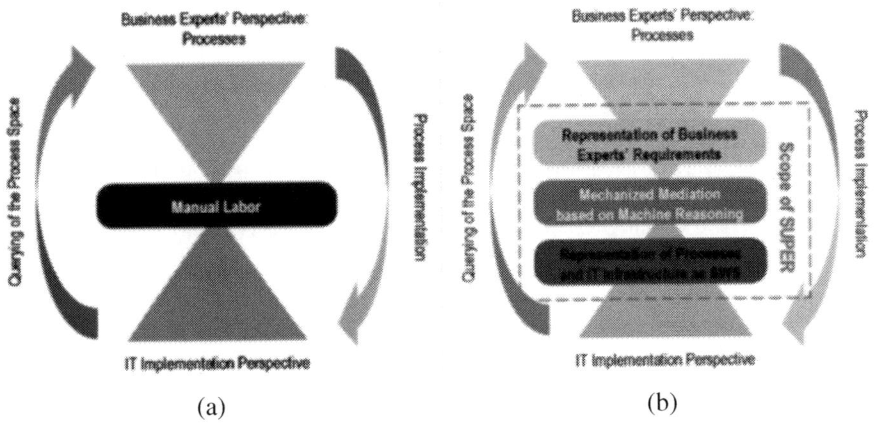


**Fig. 1.** BPM in business and IT perspectives (a) and the semantic-based approach (b) which tends to bridge the gaps using semantic technologies such as ontologies, and Semantic Web Services [6]

Therefore this work aims at providing a semantic-based framework that organizes, shares and uses the knowledge embedded in business processes within existing IT systems and software, and within employees' heads, in order to make companies more adaptive.

Our BizKB approach is a realization forward closer to the Semantic BPM (sBPM) intuition by providing a framework using formal technique to model the BP processes with the domain-oriented perspective. It is also regarded as formal BP repositories for later use of B2B integration or cross-enterprise collaboration, or even for researches of BPM communities. Creating references between the elements of different standards means clarifying their semantics in a given context. Once created, this expert knowledge should be available to other users within the same or even other collaboration contexts. Methods for semantic referencing should preserve the knowledge created [5].

With these motivation and objectives, we structure our paper as follows: Section 2 describes the related work. The need for enterprise collaboration nowadays is discussed in Section 3 respectively. Section 4 presents our motivation on the semantic-based approach for the mentioned issues, and a proposed architecture for the frame-work is presented in Section 5. The paper concludes with remarks and an outlook for the future work in Section 6.

## 2 Related Work

B2B integration and cross-enterprises collaboration have been hot topics in BPM research community recently.

### 2.1 Cross-Enterprise Collaboration

Oh et al [4] tried to tackle difficulties in the B2B integration by layering the integration process into different layers. An integrated process is designed with BPMN[1], divided into the three types of processes, and automatically transformed to XPDL[2] and BPEL[3] for process enactment. Specifically, collaborative processes are mapped to corresponding actual partners through virtual partners. However they failed to define what the virtual partner is; and ignored the semantics in the integration process

In another approach, [7] wanted to solve the integration issues using the approach of formulating collaborative processes in an certain "abstract" level. However they ignore the method to define the level of the abstraction. Furthermore, the approach failed to point out the root of problems in the B2B integration is the semantics of enterprises' business processes.

### 2.2 Semantic Business Process Management Approach

Since the failure of the non-semantic approaches as mentioned above, research efforts have been emerged from the motivation of knowledge management and applying Semantic Web technologies into BPM researches to bring the administrative side and IT side together.

Jenz's BPM Ontology approach [8] argued that the third generation business process management is different in that it provides an integrated view on business processes. According Jenz's, the business oriented view has a counter piece in the form of the IT view, and both must be on an equal footing. The business view can be segmented into three layers: core business ontology layer; industry-specific ontology layer; and organization-specific ontology layer. The IT view is not segmented into layers and is completely organization-specific.

SUPER [6] addresses the ever enduring need of new weaponry in struggle for survival in optimistic business environment where profit margins dramatically drop while competitiveness reaches the new sky high limits. This project answers the two most urgent issues emerging from BPM:

- shift in control of processes from IT professionals to business natives;
- carrying up business process management to a new complexity level.

---

[1] Business Process Modeling Notation (OMG/BPMI), http://www.bpmn.org/
[2] XML Process Definition Language (WfMC), http://www.wfmc.org/xpdl.html
[3] Web Services Business Process Execution Language (OASIS).

The major objective of the SUPER project is to raise BPM to the business level, where it belongs, from the IT level where it mostly resides now [6]. This objective requires that BPM is accessible at the level of semantics of business experts. SUPER's approach has tried to transform existing BPMN and BPEL standards into a semantics-enriched form, respectively called sBPMN (so-called BPMO – Business Process Modeling Ontology) and sBPEL [9, 10] in the attempt to realize their goals.

In the same line, the SemBiz project[4] aims at bridging the gap between the business level perspective and the technical implementation level in Business Process Management (BPM) by semantic descriptions of business processes along with respective tool support. This approach takes emerging frameworks for Semantic Web Services, namely the Web Service Modeling Ontology (WSMO)[5] as a basis for defining an exhaustive semantic description framework for business processes. On basis of this, novel functionalities for BPM on the business level can be supported by inference-based techniques that work on semantic process descriptions.

These three approaches try to bring semantic technologies into their approaches to tackle the challenge in B2B integration. However, SUPER delivers more noticeable deliverables for research communities.

### 2.3 BP Repositories Approach

The goal of the MIT Process Handbook project [11] is to develop rich online libraries for sharing and managing the business knowledge. For example, using the libraries parties can find interesting case studies, best practices, generate innovative ideas about new business possibilities, and develop new computer programs. It is called the Process Handbook – an extensive online knowledge base including entries for over 5000 business activities and a set of software tools for managing this knowledge [11].

The Process Handbook can help people to (1) redesign existing business processes, (2) invent new process, especially those that take advantage of information technology, novel coordination structures, or exception handling approaches, (3) organize and share knowledge about organizational practices, and (4) automatically, or semi-automatically, be used to generate software to support or analyze business processes.

Meanwhile at University of Zurich, an OWLized version of the MIT process handbook has been developed including an ontology and approximately 8000 business processes [12].

## 3 Cross-Enterprise Collaboration and Integration

### 3.1 BPM Standards

BPM standards are primarily grouped into three groups: (1) Graphical (2) Execution and (3) Interchange standards [13] depicted in Fig. 2.

As shown in Fig. 2, graphical standards are currently the highest level of expressing business processes (most natural) while the lowest (most technical) level is the execution standards. While graphical standards diagrammatically express

[4] SemBiz Project, http://www.sembiz.org/

[5] Web Service Modeling Ontology, http://www.wsmo.org/

contemporary business processes, the execution standards aim to automate business processes via computers.

The graphical standards are graph-oriented representations while the execution standards mainly belong to the syntax-based block-oriented representation. Block-oriented languages define a business process' control flow via nesting different elements and attributes of their syntax, while the graph-oriented languages specify control flow by using different kinds of nodes and arcs. Hence, transformations from the graph-oriented graphical standards to the block-oriented execution standards and vice versa are problematic and often experience a loss of information and semantics [14]. This created a need for interchange standards, the middle level of Fig. 2.

**Fig. 2.** BPM Standards Categories

Interchange standards act as the middle layer between the two contrasting modes of representation. Standardization groups (e.g. OMG) which pioneered interchange standards often claim their creations as the missing link between the business analyst and the IT specialist. In our opinion, this is only half the story as there are still many aspects of business process modeling that current standards fail to address such as goals, context, role definitions, and decomposition from high-level goals to automated business processes. In fact, it is more accurate to say that interchange standards are the non-contextual translator between graphical standards and execution standards.

### 3.2 BPM's Missing Pace

Collaborative business processes (CBP) are public business processes [15] that span across enterprises. With more globalised supply chains, there is currently an increasingly popular group of standards which facilitate the business to business (B2B) integration. These standards are technically B2B information exchange standards and they include: the IDEF Family [16], ebXML BPSS [17], RosettaNet PIPs [18], and newly emerging UBL [19].

Though they facilitate businesses' entry into e-commerce instead of the usual fax- and paper-based methods, these standards merely standardize information exchange and do not address the real needs of a dynamic business process collaborations like those discussed in [20]. They focus on the metadata in common for the information exchange between businesses. However, these standards ignore the most importance of the enterprise integration is to understand the meaning of data of each other. Therefore, the interoperability is quite limited.

# 4 BizKB: A Semantics-Based Approach for B2B Integration

## 4.1 CBP Patterns Analysis

In order to prepare the materials for constructing the knowledge base, an analysis will be conducted by a BP analyst to extract the common collaborative patterns from BPs. For example, we have analyzed two processes – Order Management (OM) and

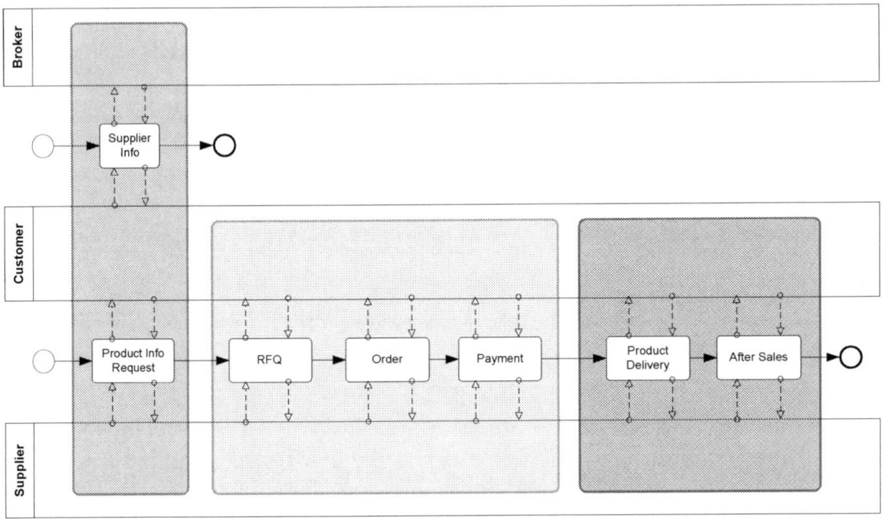


**Fig. 3.** Generic Collaborative OM process

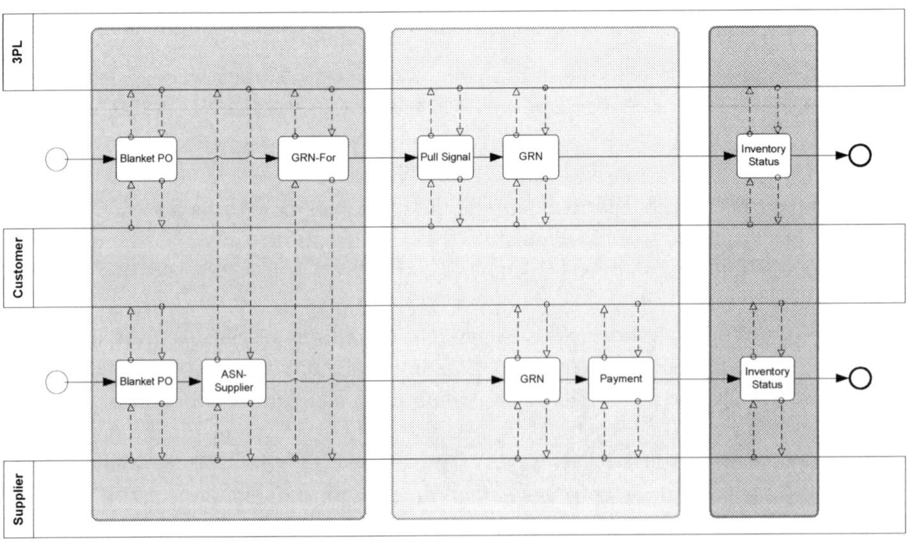


**Fig. 4.** Generic Collaborative VMI process

Vendor Managed Inventory (VMI) – for the collaborative common patterns and find a *three-step* pattern for these processes depicted as follows (Fig. 3 and Fig. 4).

Fig. 3 shows common patterns of the OM process with three parties – Supplier, Customer and Broker – and main activities between them which grouped into three parts: *setup – action – wrap-up*. Similarly, Fig. 4 presents found patterns for generic VMI process which is also described in three steps as same as above.

From these patterns, we identify all detailed level of processes and the rules to decompose from the common patterns into more details in lower levels. All granularity levels and their details will be stored in the BizKB presented later.

## 4.2 Applying Semantics in Formalizing BP Concepts: BPMO-Based Approach

Current approaches to BPM suffer from a lack of automation that would support a smooth transition between the business world and the IT world and do not fulfill the task of the automation of BP integration in the B2B integration context [21]. According to the analysis in the Section 3, the current BPM standards for B2B integration are lack of semantics that cannot coordinate the automated and dynamic B2B integration process which assumes that the partners are unknown. SBPM that is, the combination of Semantic Web and Semantic Web Services technologies with BPM, has been proposed as a solution for overcoming these problems [21].

In our approach, we use BPMO as the modeling notation to represent the CBPs in a semantics-enriched manner. BPMO is based on WSMO which is an ontology language for Semantic Web Services. WSMO is a comprehensive framework for

```
   namespace { _"http://www.ip-super.org/ontologies/BPMO/bpmo-1-4-
20080109#",
        wsmostudio _"http://www.wsmostudio.org#" }

   ontology instanceOntology_Procurement
        nonFunctionalProperties
              wsmostudio#version hasValue "0.7.3"
        endNonFunctionalProperties

        importsOntology
              _"http://www.ip-super.org/ontologies/BPMO/bpmo-1-4-
20080109#"

   instance ReceiveMessageEvent_1223890337875_1300073394 memberOf Re-
ceiveMessageEvent
        hasHomeProcess hasValue Process_1223879899796_1863342747
        hasName hasValue "Receive Message"
        messageFrom hasValue SendMessageEvent_1223890337890_641380045

   instance Process_1223879899796_1863342747 memberOf Process
        hasName hasValue "Broker"
        hasWorkflow hasValue Workflow_1223890337937_995697432
   ....
```

**Fig. 5.** Fragment example of the BPMO-based Procurement process

semantically enabled Service-Oriented Architecture (SOA) technology. It defines semantic description models for four top level notions along with respective reasoning support for managing these: ontologies, Web services, goals, and mediators. WSMO appears to be a suitable extension to BPM in order to overcome the missing paces in current BPM standards for B2B integration [22]. Furthermore, with this modeling methodology, we can tackle the difficulties as follows:

- Ontologies allow to model processes and data as shared conceptual model; their formalization allows semantically enhanced information processing;
- Semantic annotations of Web services allow to precisely detect and automatically executed suitable business functionalities as well as to maintain them;
- Goals allow to specify processes and tasks on the problem layer for which suitable Web services can be detected dynamically at execution time;
- Mediators allow handling potentially occurring mismatches on the data and the process level, therewith enabling semantically stable interchange within and across enterprises if this is not given in advance.

# 5 BizKB Framework

The ultimate goal of the BizKB approach is to build a platform for BP discovery and integration based-on Semantic Web technologies, which supports the process of cross-enterprise collaboration. Many initiatives restrict the range of standards they deal with for political, practical or technical reasons. For companies exposed to different national, industry or enterprise-specific standards – as is practically every business if all of its communications are addressed – this approach is clearly of low practical value. A universally usable methodology will avoid the predefinition of a range of manageable standards [5].

## 5.1 Conceptual Architecture

As depicted in Fig. 6, the overall conceptual architecture of the BizKB framework consists of two main parts: the BizKB and the Process Formulator. The out of framework is the CBP with service profiles of Semantic Web Services attached to CBP.

### 5.1.1 BizKB: The Business Knowledge Base

BizKB is the heart of the framework depicted in Fig. 6 which contains the knowledge of the businesses in the form of BPMO-based collaborative business processes with different levels of the abstraction.

In order to formulate these BPMO-based processes to store in the BizKB, the BP analysts are required as an important human factor of the system. Based on the analysis on the BPs, the found CBP patterns, level of the abstraction and associate business rules are also extracted and realized.

As depicted in Fig. 6, extracted artifacts of BPs are modeled using BPMO according to specific domains and kept in the persistence of BizKB. This repository is considered as the process feeder for the later stage of the CBP pattern discovery and CBPs formulation.

Establishing a complete reference collection as a knowledge base beforehand is very unlikely due to the number of standards, their evolution speed and the cost a complete analysis would create, if it were at all possible. Thus the knowledge base has to be flexible, in the sense that its evolutionary growth is not only possible but also a substantial building criterion. Clearly, an approach that does not start with a fully developed knowledge base shows weaknesses in the starting phase. Due to its initially small knowledge base, references supplied by the system might be erroneous and incomplete. But with the growth of the knowledge base, quality improvement occurs quickly – as many Web 2.0 projects demonstrate [5].

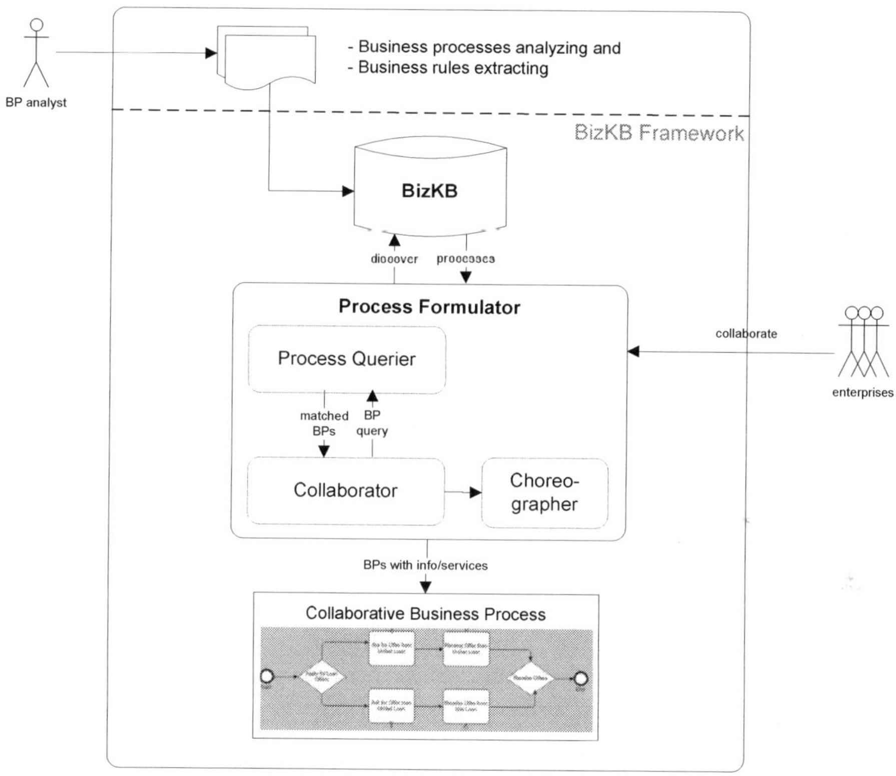


**Fig. 6.** The BizKB conceptual architecture. The framework architecture contains three main parts, the BizKB - BP knowledge base; the Process Formulator component and the Pre-processing stage for business processes analyzing.

### 5.1.2 The Process Formulator

The interactive part of the framework (Fig. 6) is the *Process Formulator* component which consists of two main subparts – *Process Querier* and the *Collaborator*. These parts are interacted by the demanding enterprise to find out the appropriate CBP patterns to form a collaborative business process with the help of the third subpart - *Choreographer*.

The Process Querier helps to find the appropriate process patterns at a certain level of the abstraction. Due to the enterprise's discovery into the BizKB, the detailed level will be matched to the need. For example, in the OM process presented in Fig. 7, one participant wants to identify the process of "Buy" products, however the participant cannot clearly identify parts of the process and related information, the Process Querier can help to indentify the basic patterns, sample processes, and even the generalization levels of the needed process.

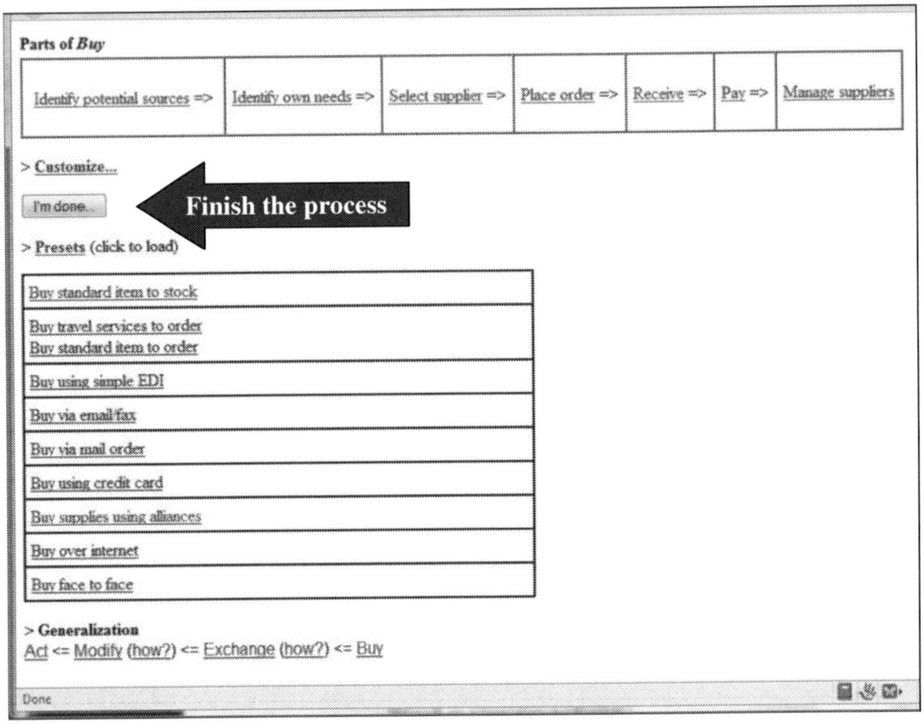


**Fig. 7.** The screen of the Process Querier prototype whose processes are learnt from MIT Process Handbook

After matched processes returned, the Choreographer will coordinate to finalize the output collaborative business process to fulfill the B2B integration demand. The new formed CBP is attached with services profiles for specific Semantic Web Services. This process is serialized using WSMO standard which conforms the unique of the framework's BPMO standard (which is based on WSMO) and benefits from Semantic Web Services advantages.

## 5.2 Mapping Processes from Conceptual Layer to Execution Layer

In order to map the CBP from the conceptual layer to the process for execution, there are phases of services discovery and matching to be carried out to find actual matched partners' services.

In our framework, we do not emphasize the execution phase in which processes are modeled in BPEL – an execution standard using Web Services.

The use of Semantic Web Services within our framework resulted in CBPs provides the appropriate flexibility in this respect. At runtime, *Goals* are be bound to specific Semantic Web Services selected on the basis of the existing conditions and informed by contextual knowledge which includes monitoring data. Furthermore, since services are described semantically, both functional and non-functional properties have clear semantics. This enhances the interpretation of services by humans, and more importantly, it allows data mismatches to be resolved at runtime.

In short, the use of Semantic Web Services provides the following benefits from a process execution perspective [21]:

- Process models are independent of the used partner services.
- Process models are independent of the partner's internal data model. The process model has its own semantically annotated data model. If the partner's data model differs from that, semantic models help to bridge this gap.
- Partner services can be selected based on business aspects.

Non-functional information about cost, quality of service, trust, legal constraints etc. can be taken into account so that the selected service is most suitable from a business perspective.

In execution phase, BPEL process models are deployed to an execution engine which is able to interpret them, consume incoming messages (both XML-based and ontological instances) from services, invoke the partner services (both traditional and semantic), perform data mediation, evaluate logical expressions in control flow conditions and finally emit monitoring events for each step in the execution

## 6 Conclusion and Future Work

In this paper we have proposed an ontology-based framework for the dynamic B2B integration. The approach is motivated by the semantic approach in efforts of bridging business perspective and IT world. It is also trying to solve the challenge in the dynamic B2B integration: formulating collaborative business process with understanding the semantics in a specific domain application.

In this stage of BizKB framework development, we have built prototypes to test functionalities of the framework. The biggest challenge for us is the expertise to provide the knowledge into BizKB. For testing purpose, we are transforming MIT Process Handbook into our BizKB as primitive one.

For the next steps, we plan to use emerging supportive tools to accomplish the goal of the framework in the Process Formulator component. The mapping into the execution level will be focused as well.

## References

1. Bussler, C.: The Application of Workflow Technology in Semantic B2B Integration. Distributed and Parallel Databases 12, 163–191 (2002)
2. Jung, J., Hur, W., Kang, S.H., Kim, H.: Business process choreography for B2B collaboration. Internet Computing, IEEE 8, 37–45 (2004)

3. Leymann, F., Roller, D., Schmidt, M.T.: Web services and business process management. IBM Systems Journal 41, 198–211 (2002)
4. Oh, J.Y., Jung, J.-y., Cho, N.W., Kim, H., Kang, S.-H.: Integrated Process Modeling for Dynamic B2B Collaboration. Knowledge-Based Intelligent Information and Engineering Systems, 602–608 (2005)
5. Rebstock, M., Fengel, J., Paulheim, H.: Ontologies-Based Business Integration. Springer, Heidelberg (2008)
6. Born, M., Drumm, C., Markovic, I., Weber, I.: SUPER - Raising Business Process Management Back to the Business Level. ERCIM News 70, 43 - 44 (2007)
7. Mous, K., Ko, K., Lee, S., Tan, P., Lee, E.: High-level Business Processes for Agile B2B Collaboration. In: Proceedings of International MATADOR Conference, pp. 169–172 (2007)
8. Jenz, D.E.: Ontology-Based Business Process Management: The Vision Statement. Strategic White Paper. Jenz & Partner GmbH, Erlensee, Germany (2003)
9. Dimitrov, M., Simov, A., Stein, S., Konstantinov, M.: A BPMO Based Semantic Business Process Modelling Environment. In: Workshop on Semantic Business Process and Product Lifecycle Management (SBPM 2007), CEUR-WS, Innsbruck, Austria, vol. 251 (2007)
10. Yan, Z., Cimpian, E., Zaremba, M., Mazzara, M.: BPMO: Semantic Business Process Modeling and WSMO Extension. In: IEEE International Conference on Web Services, ICWS 2007, pp. 1185–1186 (2007)
11. Malone, T.W., Crowston, K., Herman, G.A.: Organizing Business Knowledge: The Mit Process Handbook. MIT Press, Cambridge (2003)
12. IFI-UZH: OWL MIT Process Handbook. Dynamic & Distributed Information Systems Group - University of Zürich (2005)
13. Ko, R.K.L.: BPM Standards: A Survey (Technical Report). Nanyang Technological University, Singapore (2007)
14. Koskela, M., Haajanen, J.: Business Process Modeling and Execution: Tools and Technologies Report for the SOAMeS Project. VTT Research Notes 2407. VTT Technical Research Centre of Finland (2007)
15. Mous, K., Ko, K.L., Lee, S.G., Tan, P.S., Lee, E.W.: High-level Business Processes for Agile B2B Collaboration. In: Proceedings of The International Matador Conference, vol. 35, p. 169 (2007)
16. Mayer, R.J., Painter, M.K., de Witte, P.S.: IDEF Family of Methods for Concurrent Engineering and Business Re-engineering Applications. Knowledge Based Systems (1994)
17. UN/CEFACT, OASIS: ebXml, Xmlbpss Version 1.01 Technical Specification (2002)
18. RosettaNet: RosettaNet Implementation Framework: Core Specification (1998)
19. Bosak, J., McGrath, T., Holman, G.K.: Universal Business Language v2.0. OASIS Universal Business Language (UBL) TC (2006)
20. Wombacher, A., Fankhauser, P., Mahleko, B., Neuhold, E.: Matchmaking for business processes. IEEE Comput. Soc., 7–11 (2003)
21. Pedrinaci, C., Domingue, J., Brelage, C., van Lessen, T., Karastoyanova, D., Leymann, F.: Semantic Business Process Management: Scaling Up the Management of Business Processes, pp. 546–553 (2008)
22. Dimitrov, M.S.M.: Semantic Business Process Management (Tutorial). In: Sixth International Semantic Web Conference ISWC 2007 + ASWC 2007, Busan, Korea (2007)

# A Simple Parallel Reasoning System for the $\mathcal{ALC}$ Description Logic

Adam Meissner

Institute of Control and Information Engineering, Poznań University of Technology,
pl. M. Skłodowskiej-Curie 5, 60-965 Poznań, Poland
Adam.Meissner@put.poznan.pl

**Abstract.** In this paper we present a simple, tableau-based, parallel reasoning system for the $\mathcal{ALC}$ description logic. The system is built in relational model in the Oz language and has a form of a short program comprising the implementation of tableau rules. The program can be executed according to various strategies, particularly in parallel on distributed machines. For this purpose, we use a parallel search engine available in the Mozart environment. We describe results of experiments for estimating the speedup obtained by parallel processing.

**Keywords:** parallel reasoning, $\mathcal{ALC}$ description logic, Oz language.

## 1 Introduction

The last fifteen years is a period of a growing interests in Description Logics (DLs) [3]. This term denotes a wide and rather eclectic group of formal systems mainly intended for representing and processing a terminological knowledge. Description Logics can be classified by languages they support. A particular attention is paid to a family of DLs with the $\mathcal{ALC}$ language as a core formalism (so called $\mathcal{ALC}$ DLs). Members of this class show a reasonable expressivity and, in many cases, are decidable. This makes them attractive for knowledge engineers – $\mathcal{ALC}$ DLs are successively applied in various fields, for example in software engineering, object data bases, medical expert systems, control in manufacturing, action planning in robotics and, for last eight years, in Semantic Web [8].

One of the most important issues in DL research is the development of efficient reasoning methods and systems since the majority of inference problems occurring in this area can not be solved in polynomial time. In particular, a time of finding a solution for a given problem can be reduced by parallelizing the inference process. This direction of research in the domain of automated theorem provers (ATPs) has been carrying out since late eighties and it yields many theoretical results [2]. However, bringing up these ideas into practice still causes difficulties. A source of significant problems is often the fact that the method by which the computations in ATP are parallelized is a part of the system execution strategy. The strategy, in turn, is either hardwired in the system if it is implemented by means of imperative programming or it is a fixed part of a runtime platform (as for instance, in the Prolog language). This blurs the system

N.T. Nguyen, R. Kowalczyk, and S.-M. Chen (Eds.): ICCCI 2009, LNAI 5796, pp. 413–424, 2009.

architecture, makes it harder to understand and hence increases the probability of error occurrence. Also, it usually causes problems with the scalability of the environment where the given ATP runs.

In this paper we overcome the problems mentioned above by separating the declarative part of the reasoning system from its execution strategy. More precisely, we take a classical tableau-based inference algorithm for $\mathcal{ALC}$ DL and express it in terms of the *relational model* in the Oz language [5,7]. The reasoning procedure consists of the implementation of inference rules only while the execution strategy is implemented as the *search engine*, to wit a special object which runs the given relational program. This makes possible to perform the computations in various ways practically without any modification of the reasoning procedure. Particularly, the reasoning process can run in parallel on distributed machines. For this purpose, we use a parallel search engine [6] available in the Mozart system [7] being a programming environment for the Oz language. It should be noted, that the presented approach can be regarded as an example of *lean deduction*, which assumes achieving maximal efficiency for minimal means. This idea, including its advantages and limitations, is widely discussed in [1].

The paper is organized as follows. Section 2 contains the principles of the reasoning method, which comprise an outline of $\mathcal{ALC}$ DL tableau calculus and the description of how to express it in the relational model in the Oz programming language. In section 3 we discuss the results of experiments intended for estimating the speedup obtained by parallel computations. Section 4 contains some final remarks.

## 2 Reasoning Method

Firstly, we outline the syntax and the semantics of the $\mathcal{ALC}$ description logic. Then, we briefly present the basic inference problem for $\mathcal{ALC}$ DL, namely testing for concept satisfiability. We also cite the classical tableau-based algorithm, which solves this problem. Finally, we show how the algorithm can be parallelized by defining it in terms of the relational model in the Oz language.

The elementary expressions in $\mathcal{ALC}$ DL are *atomic descriptions* (or synonymously, *names*) of *concepts* and *roles*. A concept is a set of individuals, which are called *instances* of this concept. A role is a binary relation holding between individuals. Any element of a role is called an *instance* of this role. Concepts, besides names, can also be represented by complex descriptions, which are built from simpler descriptions and special symbols called *concept constructors*. We use the letter $A$ to denote a concept name and letters $C$ or $D$ as symbols of any concept descriptions; the letter $R$ stands for a role description. All these symbols can possibly be subscripted. The set of $\mathcal{ALC}$ DL concept constructors comprises five elements, namely *negation* ($\neg C$), *intersection* ($C \sqcap D$), *union* ($C \sqcup D$), *existential quantification* ($\exists R.C$) and *value restrictions* ($\forall R.C$); expressions written in parentheses are schemes of relevant concept descriptions.

In the sequel, if it does not lead to misunderstanding, we often identify descriptions with their meanings (e.g. we say "a concept" instead of "a concept description"). Expressions of the form $C(x)$ and $R(x, y)$ are called *concept assertions* and *role assertions*, respectively. An expression of the first kind states that the individual $x$ is an instance of the concept $C$, while the latter expression declares that the pair of individuals $\langle x, y\rangle$ is an instance of the role $R$. The individual $y$ is called a *filler* of the role $R$ for $x$.

The semantics of concept and role descriptions is defined by means of an interpretation $\mathcal{I}$, which consists of the interpretation domain $\Delta^{\mathcal{I}}$ and the interpretation function $\cdot^{\mathcal{I}}$. The interpretation function assigns a subset of $\Delta^{\mathcal{I}}$ to every concept name and a subset of $\Delta^{\mathcal{I}} \times \Delta^{\mathcal{I}}$ to every role description. The semantics of complex concepts is given as follows:

$$(\neg C)^{\mathcal{I}} = \Delta^{\mathcal{I}} \setminus C^{\mathcal{I}}$$
$$(C \sqcap D)^{\mathcal{I}} = D^{\mathcal{I}} \cap C^{\mathcal{I}}$$
$$(C \sqcup D)^{\mathcal{I}} = D^{\mathcal{I}} \cup C^{\mathcal{I}}$$
$$(\exists R.C)^{\mathcal{I}} = \{x \in \Delta^{\mathcal{I}} \mid (\exists y)\, \langle x, y\rangle \in R^{\mathcal{I}} \wedge y \in C^{\mathcal{I}}\}$$
$$(\forall R.C)^{\mathcal{I}} = \{x \in \Delta^{\mathcal{I}} \mid (\forall y)\, \langle x, y\rangle \in R^{\mathcal{I}} \rightarrow y \in C^{\mathcal{I}}\}$$

Moreover, there are two special concept descriptions, that is to say $\top$ (*top*) and $\bot$ (*bottom*). The first one denotes the most general concept, that is $\top^{\mathcal{I}} = \Delta^{\mathcal{I}}$ while the second represents the least general concept, i.e. $\bot^{\mathcal{I}} = \emptyset$. We say that the interpretation $\mathcal{I}$ *satisfies* the description $C$ if it assigns a nonempty set to it. Such an interpretation is called a *model* of the concept $C$. The concept is *satisfiable* if there exists a model of it, otherwise it is *unsatisfiable*.

The (un)satisfiability of the concept $C_0$ can be checked by a classical tableau-based algorithm [3], which is outlined below. We assume, that the concept $C_0$ (called an *input concept*) is initially converted to *negation normal form* (NNF), where the negation symbol occurs only in front of concept names. The algorithm creates the tree $T$ (called a *tableau*), whose every node is labeled by a set containing, in general, concept and role assertions. For the sake of brevity we often identify a label of the node with the node in the sequel; for example, we say "a formula in the node" instead of "a formula in the label of the node". The label of the root of $T$ is one-element set $\{C_0(x_0)\}$, where $x_0$ is a symbol of some individual. Any other node (symbolized as $\mathcal{A}'$ or $\mathcal{A}''$) can be created from its direct ancestor $\mathcal{A}$ by an application of one of the following *expansion rules*:

$\sqcap$-rule: if (a) $(C_1 \sqcap C_2)(x) \in \mathcal{A}$ and (b) $\{C_1(x), C_2(x)\} \not\subseteq \mathcal{A}$
then $\mathcal{A}' = \mathcal{A} \cup \{C_1(x), C_2(x)\}$

$\sqcup$-rule: if (a) $(C_1 \sqcup C_2)(x) \in \mathcal{A}$ and (b) $C_1(x) \notin \mathcal{A}$ and $C_2(x) \notin \mathcal{A}$
then $\mathcal{A}' = \mathcal{A} \cup \{C_1(x)\}, \mathcal{A}'' = \mathcal{A} \cup \{C_2(x)\}$

$\exists$-rule: if (a) $(\exists R.C)(x) \in \mathcal{A}$ and (b) there is no $y$ that $\{R(x, y), C(y)\} \subseteq \mathcal{A}$
then $\mathcal{A}' = \mathcal{A} \cup \{R(x, z), C(z)\}$ where $z$ does not occur in $\mathcal{A}$

$\forall$-rule: if (a) $(\forall R.C)(x) \in \mathcal{A}$ and (b) $R(x, y) \in \mathcal{A}$ and $C(y) \notin \mathcal{A}$
then $\mathcal{A}' = \mathcal{A} \cup \{C(y)\}$

We say, that the rule is *relevant* to a given concept assertion $D(x)$, if $D(x)$ matches the assertion occurring in the condition (a) of this rule. The application of the rule, however, may be *blocked* (in short: the rule is blocked) if conditions

(b) are not satisfied. A node of the tableau $T$ is dangling if no expansion rule can be applied to it or if it contains a contradiction called a *clash*. In the latter case, the node is called a *clash node*, otherwise it is *clash-free*. The clash can be detected by the following *clash rules*:

*clash*1-rule: if $\bot(x) \in \mathcal{A}$ or $(\neg\top)(x) \in \mathcal{A}$ then mark $\mathcal{A}$ as a clash node
*clash*2-rule: if $A(x) \in \mathcal{A}$ and $(\neg A)(x) \in \mathcal{A}$ then mark $\mathcal{A}$ as a clash node

The *clash*2-rule identifies the presence of *complementary* assertions (i.e. $A(x)$ and $(\neg A)(x)$) in the node. It should be noticed that the application of this rule is restricted to *literals*, namely to atomic concepts (*positive* literals) and their negations (*negative* literals). Expansion rules and clash rules are both called *inference rules* or *tableau rules*.

A branch of the tableau is *closed* if it is ended by a clash node; otherwise, the branch is *open*. A tableau with all branches closed is called a *closed tableau*. A concept $C$ is *unsatisfiable* iff one can construct a closed tableau for it. In the other case, the concept is *satisfiable* and every fully expanded node, which ends an open branch is a straightforward representation of a model of the concept $C$.

In Figure 1 we present an exemplary tableau constructed for the input concept $(D \sqcup \exists R.\neg C) \sqcap \forall R.C$. Every node is additionally labeled by a numerical identifier followed by a colon. A label assigned to an edge, on the other hand, indicates the expansion rule applied to a parent node in order to obtain a child node. For the sake of clarity, we skip already expanded concept assertions in a node label, which do not participate in further inferences. As one can notice, the tableau is open since the node 3 is clash-free. Hence, the input concept is satisfiable and the model for it (extracted from the node 3) is the interpretation with $\Delta^{\mathcal{I}} = \{x_0\}$, $R^{\mathcal{I}} = \emptyset$, $C^{\mathcal{I}} = \emptyset$ and $D^{\mathcal{I}} = \{x_0\}$.

An important property of $\mathcal{ALC}$ tableau calculus is that branches of the tableau can be constructed independently. Roughly speaking, it follows from the fact that there is no information exchange between nodes belonging to two different

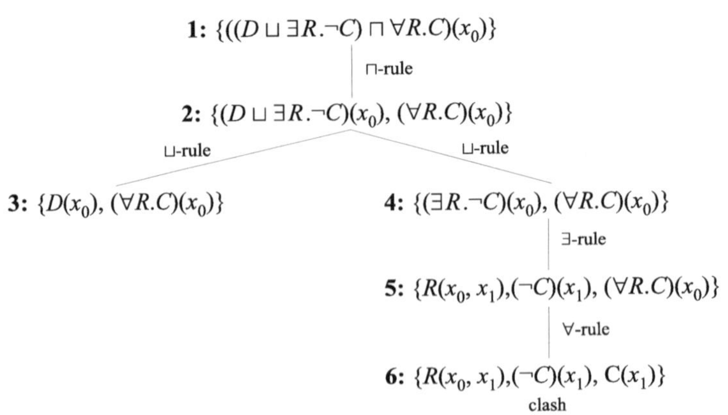


**Fig. 1.** Open tableau for $(D \sqcup \exists R.\neg C) \sqcap \forall R.C$

branches. Hence, a tableau can built in parallel. For this purpose we represent the inference rules in the relational (programming) model in the Oz language. The declarative semantics of relational programs is very similar to normal programs in Prolog, which is a widely known and popular platform for implementing reasoning system. However, unlike in Prolog, the computational strategy is not fixed in the runtime environment but it can be given to a program as a parameter of its execution. Below, we present some basic principles of the relational model of Oz. In this approach, a program is a sequence of statements (in particular, procedure calls) which can either cease normally and produce results (namely, *answers*) or it can terminate by *failure* (i.e. with no answer). A program can also create a *choice point*, which causes a forking of computations into independent paths producing alternative answers.

In order to run a program, one has to store it in a *computation space* [5,6]. A space, among other properties, encapsulates computations so they are separated form the exterior, particularly from processes performed in other spaces. The set of operations defined on spaces comprises the creation of a space for a new process as well as merging, cloning and killing of spaces. These operations are performed by search engines. A space can communicate with an engine by appropriate statements. In particular, the statement `{Space.choose N}`, executed in the given space, tells the engine to create a choice point with `N` alternatives. The engine in the response clones the space `N` times and sends to each copy a numerical identifier ranging from 1 to `N`. The identifier becomes a value of the expression `{Space.choose N}`. The process of the subsequent creation of spaces results in a *search tree*. Every leaf of the fully expanded tree is either a *solved* or a *failed* space. A solved space contains a result of normally terminated computations. It should be noted that computations executed inside spaces determine the shape and the content of the search tree. However, they do not settle the order in which spaces are created and processed – this depends upon the search engine only. In this way the declarative semantics of the program, corresponding to the structure of the tree, can be separated from the operational semantics represented by the search strategy.

We use a search tree as a representation of a tableau. Regarding this, we consider the assumptions given below. The symbol $\mathcal{A}$ denotes any internal tableau node while $S_{\mathcal{A}}$ stands for the space in which the node $\mathcal{A}$ is computed.

1. The root space corresponds to the root of a tableau.
2. If $\mathcal{A}'$ is the only one direct successor of the node $\mathcal{A}$, created by any expansion rule different from the $\sqcup$-rule, then $\mathcal{A}'$ replaces $\mathcal{A}$ in the space $S_{\mathcal{A}}$.
3. If nodes $\mathcal{A}'_1, \ldots, \mathcal{A}'_n$ are direct successors of the node $\mathcal{A}$ obtained by an application of the $\sqcup$-rule, then for every node $\mathcal{A}'_i$ a new space $S_{\mathcal{A}'_i}$ is created, which is a direct successor of the space $S_{\mathcal{A}}$ in the search tree, for $i = 1, \ldots, n$.
4. A clash-free node is mapped to a solved space. The result of computations performed in it represents a model of the input concept.
5. A clash node is represented by a failed space.

The correctness of the second assumption follows from the fact that every node $\mathcal{A}$ of the tableau having the only one successor $\mathcal{A}'$ does not participate in any

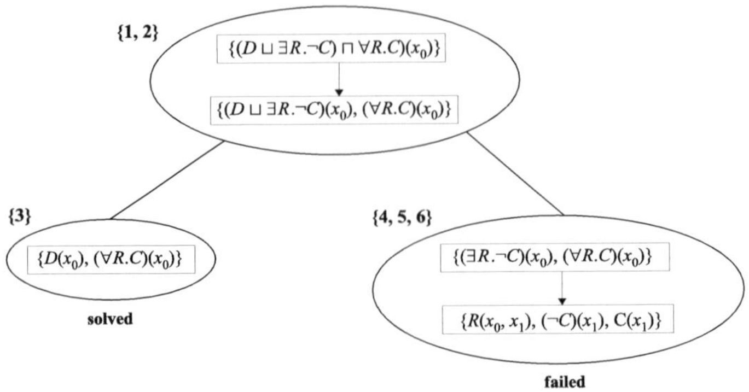


**Fig. 2.** Search tree for the tableau from Figure 1

further inferences after the creation of the node $\mathcal{A}'$, and thus it can be replaced by $\mathcal{A}'$ in the tableau construction process. In Figure 2 we show a search tree for the tableau from Figure 1. The upper and the lower rectangle in a space encloses a label of the starting and, respectively, the resulting node created in this space. If both the nodes are identical then they are depicted by one rectangle. Each space in the tree is additionally labeled by a set of numbers, which identify tableau nodes corresponding to the space.

The Mozart programming system provides various library classes of engines implementing different search strategies. In particular, instances of the class `Search.parallel` are parallel search engines designed to work on distributed machines. The engine can be regarded as a team of concurrent autonomous agents comprising a *manager* and a group of *workers*. The manager controls the computations by finding a work for idle workers and collecting the results whereas the workers construct fragments of the search tree. Members of the team communicate by exchanging messages. The detailed description of this architecture, including the communication protocol, is to be found in [6]. Below, we give an example of the statement creating a new parallel engine.

```
Eng = {New Search.parallel init(w1:2#ssh w2:3#ssh)}
```

It consists of the manager (initiated locally) and five workers, started on remote computers via secure shell (`ssh`) commands – two on the machine `w1` and three on the other machine `w2`. One can tell the engine to execute the given procedure by the following statement.

{`Eng` *ProcedureCall*}

The result of computations becomes a value of the expression written above. It should be remarked that the subexpression *ProcedureCall* is given in pseudocode in order to skip some technical details.

## 3 System Description

The system processes $\mathcal{ALC}$ expressions, which are represented by Oz data structures. Furthermore, we use subsequent numbers starting from 0 to encode individuals. Atomic descriptions (of concepts and roles) are expressed as *atoms* and complex concept descriptions as well as concept and role assertions are denoted by *tuples* [5]. The correspondences between the considered notation and the standard $\mathcal{ALC}$ syntax are given in Table 1. Primed symbols are Oz representations of their unprimed $\mathcal{ALC}$ counterparts – we use this convention also in the sequel. Expressions of the form $C_1 \sqcap \ldots \sqcap C_n$ and $C_1 \sqcup \ldots \sqcup C_n$ stand for

**Table 1.** Oz representation of $\mathcal{ALC}$ expressions

| $\mathcal{ALC}$ syntax | Oz notation |
|---|---|
| $\top$ | `top` |
| $\bot$ | `bot` |
| $\neg A$ | `neg(`$A'$`)` |
| $C_1 \sqcap \ldots \sqcap C_n$ | `and(`$C'_1$ `...` $C'_n$`)` |
| $C_1 \sqcup \ldots \sqcup C_n$ | `alt(`$C'_1$ `...` $C'_n$`)` |
| $\exists R.C$ | `ex(`$R'$ $C'$`)` |
| $\forall R.C$ | `all(`$R'$ $C'$`)` |
| $C(x)$ | $C'$`#`$x'$ |
| $\{R(x, y_1), \ldots, R(x, y_n)\}$ | $R'$`#`$x'$`#[`$y'_1$ `...` $y'_n$`]` |

nested intersections and unions, respectively. It should be noticed that the use of the negation constructor is restricted to atomic concepts, since all concept descriptions are assumed to be in NNF. The representation of role assertions as sets of the form $\{R(x, y_1), \ldots, R(x, y_n)\}$ is motivated by practical reasons, namely by the way the $\forall$-rule is applied and implemented.

The key part of the system is the procedure `Prove` defined below. It is a straightforward realization of the inference method described in section 2. The procedure constructs a tableau for a given concept description and checks whether the description is satisfiable, that is to say if the tableau contains any open branch. If so, the argument `Model` is bound to the list representing a model of the input concept. The list encloses role assertions as well as assertions with literals included in a given clash-free node. Otherwise, to wit when the tableau is closed, the execution of the procedure results in failure. In every subsequent call, the procedure handles one node of the tableau and processes it in the *current space*. For the sake of efficiency, a node label is split into two disjoint subsets containing concept assertions and, respectively, role assertions. The first one is represented by the list `Concs`, while the latter by the list `Roles`. In order to speed up the detection of clashes, assertions with positive and negative literals from the list Concs are additionally stored in two lists, namely PLits and NList, respectively. The subsequent argument of the procedure Prove, i.e. I, is a number standing for an individual, which has been added to the current branch as the last one. This number is successively increased whenever a new concept or role instance is introduced to the tableau. The argument N is used for the detection of a clash-free node. It indicates the current number of assertions from the list Concs to which no expansion rule can be applied.

```
proc {Prove Concs Roles PLits NLits I N Model}                               % 1
   if N == {Length Concs}                                                      % 2
   then                                                                        % 3
      {Flatten [PLits NLits Roles] Model}                                      % 4
   else                                                                        % 5
      Conc = Concs.1 C = Conc.1 X = Conc.2 Op = {Label C} L Inds Roles1 in     % 6
      if C == bot orelse C == neg(top) then fail                               % 7
      elseif {IsAtom C} andthen {Not {Member Conc PLits}} then                 % 8
         if {Member Conc NLits} then fail else                                 % 9
            {Prove {Append Concs.2 [Concs.1]} Roles Conc|PLits NLits I 0 Model}   % 10
         end                                                                   % 11
      elseif Op == neg andthen {Not {Member C.1#X NLits}} then                 % 12
         if {Member C.1#X PLits} then fail else                                % 13
            {Prove {Append Concs.2 [Concs.1]} Roles PLits (C.1#X)|NLits I 0 Model}   % 14
         end                                                                   % 15
      elseif Op == and andthen                                                 % 16
         (L={Filter {Map {Record.toList C} fun {$ E} E#X end}                  % 17
                    fun {$ E} {Not {Member E Concs.2}} end}) \= nil            % 18
      then                                                                     % 19
         {Prove {Append L Concs} Roles PLits NLits I 0 Model}                  % 20
      elseif Op == alt andthen                                                 % 21
            {Not {Some (L={Map {Record.toList C} fun {$ E} E#X end})           % 22
                       fun {$ E} {Member E Concs.2} end}}                      % 23
      then                                                                     % 24
         {Prove {Nth L {Space.choose {Length L}}}|Concs Roles PLits NLits I 0 Model}   % 25
      elseif Op == ex andthen                                                  % 26
            ({Not {SelRol Roles C.1#X#Inds Roles1}} orelse (C.2 \= top andthen   % 27
             {Not {Some Concs fun {$ E} {And (E.1 == C.2) {Member E.2 Inds}} end}}))   % 28
      then                                                                     % 29
         if C.2 == top then L = Concs else L = C.2#(I+1)|Concs end             % 30
         {Prove L C.1#X#(I+1|Inds)|Roles1 PLits NLits I+1 0 Model}             % 31
      elseif Op == all andthen C.2 \= top andthen {SelRol Roles C.1#X#Inds _} andthen   % 32
            (L={Filter Inds fun {$ E} {Not {Member C.2#E Concs.2}} end}) \= nil   % 33
      then                                                                     % 34
         {Prove {Append {Map L fun {$ E} C.2#E end} Concs} Roles PLits NLits I 0 Model}   % 35
      else                                                                     % 36
         {Prove {Append Concs.2 [Concs.1]} Roles PLits NLits I N+1 Model}      % 37
      end                                                                      % 38
   end                                                                         % 39
end                                                                            % 40
```

The reasoning process starts from the execution of the procedure Prove with the following actual arguments

{Prove [$C'$#0] nil nil nil 0 0 Model}

where the symbol $C'$ stands for the input concept and the symbol nil denotes an empty list. In every subsequent call, the procedure takes the first element Conc of the list Concs and tries to apply a relevant tableau rule to it. If it is not possible (because the rule is blocked), the element is moved to the end of the list and the procedure is recursively called with the argument N incremented by 1 (line 37). If the value of this argument is equal to the length of the list Concs, then the current node is regarded as clash-free since no rule can be applied to it. In such case, the argument `Model` is bound to the representation of the model of the input concept (line 4).

Otherwise, namely when `Conc` is a concept assertion to which a tableau rule can possibly be applied, the procedure `Prove` checks at first whether it can be a *clash*1−rule (line 7). The `fail` statement, executed by this rule, causes a failure in the current space. The *clash*2−rule is implemented in two variants. The first one is relevant to an assertion with a positive literal (line 8). More precisely, the assertion `Conc` must contain the description `C`, which denotes an atomic concept. The additional condition for this rule is that `Conc` does not occur in the list `PLits`. In other case, the rule is blocked since it must have been already applied to `Conc` before. The execution of the rule can result in failure if the list `NLits` contains an assertion complementary to `Conc` (line 9). It should be remarked that for the sake of simplicity, negative literals are stored on the list `NLits` without negation symbols (i.e. as apparently positive). If, on the other hand, there is no complementary counterpart for `Conc`, then it is moved to the end of the list `Concs` in the next call of the procedure `Prove` (line 10). Moreover, `Conc` is added to the list `PLits` in order to prevent the subsequent application of the rule to this assertion. The second variant of the *clash*2−rule concerns negative literals, to wit assertions with concepts built from the constructor `neg`. This variant is implemented analogously to the previous one (lines 12–15).

Lines 16–20 implement the ⊓-rule. If the assertion `Conc` is an intersection, then it is decomposed into elements and those of them, which are not members of the list `Concs` (lines 18–19), are added to this list (line 20). If there are no such elements, then the rule is blocked. The implementation of the ⊔-rule is given in lines 21–25. It breaks the assertion `Conc` (containing a union) into components, which are collected on the list `L` (lines 22–23). If non of components is a member of the list `Concs`, then for each of them a clone of the current space is created by the statement `{Space.choose {Length L}}` (line 25). In every clone, this expression evaluates to a distinct number, which is used in turn by the function `Nth` to select the respective element from the list `L`. This component is added to the list `Concs` in the subsequent call of the procedure `Prove`.

Lines 26–31 correspond to the ∃-rule, relevant to an assertion of the form $(\exists R.C)(x)$. In line 27 the function `SelRol` is called, which searches the list `Roles` to find an assertion with the role $R$ (represented by `C.1`) and the list of fillers `Inds` for the individual `X` (standing for $x$). If such the assertion is found, the function returns the value `true` and the variable `Role1` is bound to the result of removing this assertion from the list `Roles`. Otherwise, the value `false` is returned, the variable `Role1` is bound to the list `Role` and the variable `Inds` becomes an empty list. It should be remarked that `SelRol` is the only one non-library function used in the definition of the procedure `Prove`. The ∃-rule is executed if the list `Inds` is empty (line 27) or the concept $C$ (represented by `C.2`) differs from ⊤ and the list `Concs` contains no assertion for $C$ and any individual from the list `Inds` (lines 27–28). The execution of the rule generally consists in inserting appropriate new elements to lists `Concs` and `Roles` in the next call of the procedure `Prove` (line 31). However, when the list `Inds` is nonempty, but the concept $C$ equals $\top$, then no new elements are added to the list `Concs`. This follows from the fact that assertions with the most general concept are unessential for the reasoning process and therefore they can be ignored.

Finally, in lines 32–35 the rule $\forall$-rule is implemented. The execution of this rule takes place if the list `Concs` contains an assertion of the form $(\forall R.C)(x)$ and the list `Roles` includes the assertion `C.1#X#Inds`, where `C.1` and `X` stand for $R$ and $x$, respectively and `Inds` is a nonempty list of fillers of the role $R$ for $x$ (lines 32–33). Moreover, there must exist a nonempty list `L` of assertions `C.2#E` not included in the list `Concs`, where `E` is any element of the list `Inds` and `C.2` represents the concept $C$. If so, the list `L` is appended to the list `Concs` (line 35). Furthermore, the rule is not executed if $C$ is the most general concept (i.e. $\top$), for the same reason as in the case of the $\exists$-rule.

## 4 Experimental Results

The experiments discussed in this section were aimed at estimating the speedup obtained by running the reasoning system in parallel with varying number of workers. The computational environment consisted of the machine, which processed the manager (Pentium-M 760, 2.0 GHz, 1 GB RAM, 1GBit Ethernet, Win. XP HE 2002) and up to five machines (Pentium P4D, 3.4 GHz, 1 GB RAM, 1 GBit Ethernet, Win. 2000 Prof. 5.00) processing one worker each. All the computers were powered by the Mozart system 1.3.2.

The testing data come from the benchmark set T98-sat [4] comprising 18 files. Each of them contains 21 numbered examples of concepts, which are all either satisfiable or unsatisfiable. In the first case, the file name is ended with the letter `n`, while in the latter case, with the letter `p`. In the sequel, we identify the input concept by giving its number in the file it comes from preceded by the file name. For example, the identifier `k_lin_n_6` denotes the 6th problem (i.e. concept example) in the file `k_lin_n`. The problems were initially transformed to NNF and selected under the general criterion that the time of computations performed by one worker should range from 30 to 180 sec. For every test, the computational time is taken from the system clock as an arithmetic mean of five runs.

Results obtained for satisfiable and unsatisfiable concepts differ notably one from another. It follows from both the different character of these two search problems and the indeterministic method, by which the search tree is created. In the first case (i.e. for satisfiable concepts) the computational time is collected in Table 2. The header row contains a number of workers appearing in the given variant of the environment. It should be reminded, that the search engine stops after finding the first open branch in the tree. This means, that for satisfiable concepts the computational time remarkably depends upon the way the search tree is divided into subtrees given to workers. Roughly speaking, every worker explores the tree in a depth-first manner from left to right. If the first open branch is located far from the left-hand side, then the time may seriously decrease if this branch appears as the first path in one of the subtrees. The

**Table 2.** Computational time [s] for selected satisfiable T98-sat concepts

| Problem | 1 worker | 2 workers | 3 workers | 4 workers | 5 workers |
|---|---|---|---|---|---|
| k_lin_n_6 | 142.56 | 142.7 | 145.62 | 147.06 | 150.29 |
| k_path_n_5 | 97.2 | 67.72 | 81.37 | 70.31 | 72.45 |
| k_poly_n_5 | 83.83 | 82.74 | 82.83 | 82.88 | 82.31 |
| k_poly_n_6 | 178.17 | 178.9 | 182.38 | 183.45 | 183.31 |
| k_t4p_n_1 | 142.75 | 25.1 | 25.77 | 25.29 | 26.98 |

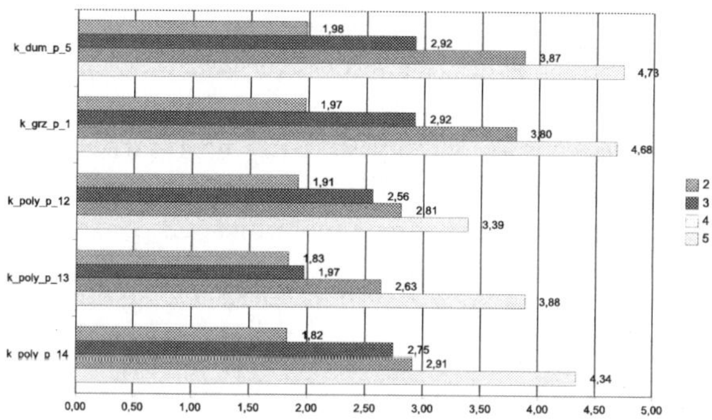


**Fig. 3.** Speedup for selected unsatisfiable T98-sat concepts

computational time may also increase after adding a subsequent worker if the tree is partitioned in a less advantageous manner. This can explain such anomalies as super-linear speedup or slowdown, which occur in case of the problem `k_path_n_5` and `k_t4p_n_1`. On the other hand, if an open branch is close to the left-hand side and it is not considerably longer then the other open branches, then adding new workers can not significantly shorten the computational process (e.g. the problem `k_poly_n_5`). Also, it should be remarked that introducing new machines to the environment increases timing costs of the network communication among parts of the search engine. In certain cases costs can exceed benefits coming from partitioning of the search tree. This may be a reason why the computational time systematically grows after connecting new workers for problems `k_lin_n_6` and `k_poly_n_6`.

The second group of testing problems comprises unsatisfiable concepts only, for which a search tree does not include any open branches and therefore it has to be constructed in whole. Hence, extending the computational environment by new workers always results in a speedup, which is depicted in the bar chart in Figure 3. The speedup is understood as a quotient of time of computations performed in the environment containing one worker, and in the environment with subsequently growing number of workers. Every cluster of bars represents outcomes obtained for the particular problem; a shading of a bar indicates the number of workers in the computational environment as it is given in the legend on the right-hand side of the chart.

The increment of the speedup is positive in all cases. Moreover, it is practically linear for the problem `k_dump_p_6` and `k_grz_p_1` while for the other problems it fluctuates. This may follow from the fact that the distribution of the reasoning process depends upon the structure of the search tree. In the first two cases trees are "nearly balanced", unlike for the other problems, and therefore workers can be loaded uniformly. This issue, however, requires more thorough observations in future.

## 5 Final Remarks

In this paper we present a simple, parallel, tableau-based reasoning system for $\mathcal{ALC}$ description logic. The system is actually independent from the computational strategy thanks to implementing it in the relational model in the Oz language. We use the parallel search engine from the Mozart programming environment to run the system on distributed machines. Experiments show a reasonable speedup achieved by parallelizing the computations for the certain class of problems. However, the comprehensive analysis of the system efficiency requires more tests. The tests have to consider a greater number of workers and also a greater number of exemplary problems selected with regard to the structure of the search tree. These tests are planed for the future.

The work was supported in part by the Polish Ministry of Science and Higher Education under Grant N 516 369536.

## References

1. Beckert, B., Possega, J.: lean$T^AP$: Lean, Tableau-based Deduction. Journal of Automated Reasoning 15(3), 339–358 (1995)
2. Bonacina, M.P.: A taxonomy of parallel strategies for deduction. Annals of Mathematics and Artificial Intelligence 29(1-4), 223–257 (2000)
3. Baader, F., McGuinness, D.L., Nardi, D., Patel-Schneider, P.F. (eds.): The Description Logic Handbook: Theory, implementation, and applications. Cambridge University Press, Cambridge (2003)
4. Horrocks, I., Patel-Schneider, P.F.: DL System Comparison. In: Proc. of the 1998 Int. Workshop on Description Logics (DL 1998), vol. 11 (1998), http://ceur-ws.org/
5. Van Roy, P., Haridi, S.: Concepts, Techniques, and Models of Computer Programming. The MIT Press, Cambridge (2004)
6. Schulte, C.: Programming Constraint Services, PhD thesis. University of Saarlandes (2000)
7. The Mozart Programming System (2008), http://www.mozart-oz.org
8. World Wide Web Consortium, Semantic Web, http://www.w3.org/2001/sw/

# SemCards: A New Representation for Realizing the Semantic Web

Kristinn R. Thórisson[1,2], Nova Spivack[2], and James M. Wissner[2]

[1] Center for Analysis & Design of Intelligent Agents
Reykjavik University, Kringlunni 1, 103 Reykjavik, Iceland
[2] Radar Networks, Inc.
410 Townsend Street, Suite 150, San Francisco, CA 94107 USA
thorisson@ru.is,{nova,jim}@radarnetworks.com

**Abstract.** The Semantic Web promises increased precision in automated information sorting, searching, organizing and summarizing. Realizing this requires significantly more reliable meta-information than is readily available for basic human-readable data types today. Relying solely on hand-crafted ontologies and annotation, or solely on artificial intelligence techniques, seems less likely for success than a combination of the two. How this is best done, however, is far from obvious. We propose an intermediate ontological representational level we call *SemCards* that combines ontological rigour with flexible user interface constructs. SemCards are machine- and human-readable entities that allow non-experts to create and use semantic content with ease, while empowering machines to better assist and participate in the process. We have implemented the SemCard technology on the Semantic Web site *Twine.com*, which to date has a growing 250k subscribers and over 2 million monthly unique visitors. SemCards allow users to quickly create semantically-grounded data that in turn acts as examples for automation processes, creating a positive iterative feedback loop of metadata creation between user and machine. The result is an increasingly larger, more accurate amount of metadata than with either approach alone. The SemCard provides a holistic solution to the Semantic Web, resulting in powerful management of the full lifecycle of data, including its creation, retrieval, classification, sorting and sharing. Here we present the key ideas behind SemCards and describe the initial implementation of the technology on *Twine.com*.

**Keywords:** Semantic Web, Ontologies, Knowledge Management, User Interface, SemCards, Human-Machine Collaboration.

## 1 Introduction

One thing we can all agree on: The world of information workers welcomes any improvement in information processing automation. Intelligent automated retrieval, manipulation, combination, presentation – and organized disposal – of information defines the speed of progress in much of today's high-technology work. Included in many people's vision of the Semantic Web is for machines

N.T. Nguyen, R. Kowalczyk, and S.-M. Chen (Eds.): ICCCI 2009, LNAI 5796, pp. 425–436, 2009.

to have better knowledge of the data they manipulate. This requires metadata – data about the data. Making machines smarter at tasks such as retrieving relevant information at relevant times automatically from the vast collection, even on the typical laptop hard drive, requires much more meta-information than is presently available for such data.

Accurate metadata can only be derived from an understanding of content; classifying photographs according to what they depict, for example, is best done by a recognition of the entities in them, lighting conditions, weather, film stock, lens type used, etc. Hand-crafting metadata for images, to continue with this example, will be an impossible undertaking, even if we limited the metadata to surface phenomena such as the basic objects included in the picture, as the number of photographs generated and shared by people is increasing exponentially. Powertools designed for manual metadata creation would only improve the situation incrementally, not exponentially, as needed.

Although text analysis has come quite a long way and is much further advanced than image analysis, artificial intelligence techniques for analyzing text and images have a long way to go to reliably decipher the complex content of such data. The falling price of computing power could help in this respect, as image analysis is resource-intensive. This will not be sufficient, however, as *general-purpose* image analysis (read: software with commmon sense) is needed to analyze and classify the full range of images produced by people based on content. Relying exclusively on automated creation of metadata seems thus equally doomed as complete reliance on hand-crafting. The question then becomes, What kind of collaborative framework will best address the building of the Semantic Web?

Semantic Cards, or *SemCards*, is a technology that links ontology creation, management and usage to the *user interface*. SemCards provide an intermediate ontological representational level that allows end-users to create rich semantic networks for their information sphere. The technology has three important features: (1) it does not require expertise users, (2) it is tolerant to end-user mistakes, and (3) it provides examples of metadata and semantic relationship links to the underlying machine intelligence. The best known way to create structured metadata is through the use of carefully constructed ontologies. The casual Internet user, however, is not initiated to invest a lot of time in understanding the intricacies of the kinds of advanced ontologies required for this purpose. An important difference between many prior efforts and ours is this separation, done in a way that is tolerant to manual input errors. One of the big problems with automation is low quality of results. While statistics may work reasonably in some cases as a solution to this, for any single user the "average" user is all too often too different on too many dimensions for such an approach to be useful. The SemCard intermediate layer encourages users to create metadata and semantic links, which provides underlying automation with highly specific, user-motivated examples. The net effect is an increase in the possible collaboration between the user and the machine. Semi-intelligent processes can be usefully employed without requiring significant or immediate leaps in AI research. Other important differences between our approach and prior work are an integrated ability to share data between individuals and groups of users over a network, with complex policy control over access and sharing, and the flexible use of SemCards to represent metadata for real-world objects and hypothetical constructs - as "library index cards for digital content, physical things and abstract ideas".

We have built a network portal, Twine.com, for deploying the SemCard technology. Although current enterprise portals are capable of organizing group or team information, they are often inaccessible to the public or to individuals and are expensive and monolithic. Even less utilitarian and intelligent with respect to organizing information are the popular online search engines which are completely unstructured and typically organize information and data by relevance to keywords. From the users' perspective what we have developed is a network portal where they can organize their own information for personal use, publish any of that information to any group – be it "emails" addressed to a single individual or photo albums shared with the world – and manage the information shared with them from others, whether it is documents, books, music, etc. Under the hood are powerful ontology-driven technologies for organizing all categories of data, including access management, relational (semantic) links and display policies, in a way that is relatively transparent to the user. The result is a system that offers improved automation and control over access management, information organization and display features.

Here we describe the ideas behind the approach and give a short overivew of a use-case on the semantic Web site Twine.com.

## 2 Related Work

The full vision of the Semantic Web will require significant amounts of metadata, some of which describes entities themselves, other which describes relationships between entities. Two camps can be seen proposing rather different approaches to this problem. One extreme claims that manual creation of metadata will never work as it is not only slow and error-prone, the level to which it would have to be done would go well beyond the patience of any average user – quite possibly all. To this camp the only real option is automation. The other camp points out that automation is even more error-prone than manual creation, as current efforts to automatic semantic annotation on massive scales produces only moderate results of between 80% and 90% correct, at the very best [1]. They claim that the remaining 10% will always be beyond reach because it requires significant amounts of human-level intelligence to be done correctly. Further, as argued by Etzioni and Gribble [2], metadata augmentation has quite possibly not been done by the general user population because they have seen no benefits in doing so. Lastly, this camp points to the massive amounts of tagging and data entry done on sites such as Wikipedia, Myspace and Facebook as a proof in point that end-users are quite willing to provide (some amount of) metadata. Giving them the right tools might change this. Applications that connect casual end-users with ontologically-driven content and processes are, nevertheless, virtually non-existent.

Many efforts have focused on building digital content management with a focus on the object. Of these, our technology bears perhaps the greatest resemblance to the *Buckets* of Maly et al. (1999) [3] which are "self-contained, intelligent, and aggregative ... objects that are capable of enforcing their own terms and conditions, negotiating access, and displaying their contents". Like SemCards, Buckets are fairly self-contained, with specifications for how they should be displayed. Buckets grew out of Kahn and Wilensky's (1995) [4] proposed infrastructure for digital information services. Key to their proposal was the notion of digital object, composed of essentially the two familar parts, data and metadata. The subsequent work on FEDORA [5] saw the creation of an open-source software framework for the "storage, management, and dissemination of complex objects and the relationships among them" [6]. Buckets represent a focus on storing content in digital libraries, most likely manipulated by experts. In contrast, SemCards aim at enabling casual end-users to create metadata. Buckets are targeted to machime manipulation; SemCards are aimed at machine *manipulation* as well, but more importantly at supporting *automatically generated* meta-information. SemCards also differ from Buckets in that they are especially designed to be sharable between multiple users over mixed-architecture networks.

The separate representation layer provided by SemCards is a key difference between prior efforts and ours. They enable ontologically-driven constructs to be collaboratively built by ontology specialists, algorithms and end-users, encouraging them to provide examples to improve the automation.

## 3 SemCards: Semantic Objects for Collaborative Ontologically-Driven Information Management

*SemCards* form an intermediate separation layer between ontologies and the user interface. By isolating the stochastic nature of end-user activity from underlying semantic networks built with ontological rigour, two important goals are achieved. First, end-users are encouraged to create metadata for their content, as the input methods are familiar and straight-forward. SemCards provide an "isolation layer" that shields the deep ontology from being affected by end-user activity. This does not only help stabilize the system, it also helps the automation processes from having to deal with the "ground shifting from underneath". Second, the automation processes are provided with semantic net examples, created directly and continuously by end-users, that can be used to improve the automatic metadata creation. The net result of this is a significant improvement in automation quality and speed. The result is an improvement in the many tedious details of information management such as data sharing policy maintenance, indexing, sorting – in fact, the of the full data management lifecycle.

A single SemCard can be characterized as an intuitive user interface construct that bridges between a user and an underlying ontology that affords all the benefits of a Semantic Web such as automatic relationship discovery, sorting, data mining, semantic search, etc. Together many SemCards form semantic nets that are in every way the embodiment of what many have envisioned the Semantic Web to be. Instead of being complex, convoluted and non-intuitive as any machine-manipulatable ontology will appear to the uninitiated (c.f. [7]), SemCards provide a powerful and intuitive interface to a unified framework for managing information.

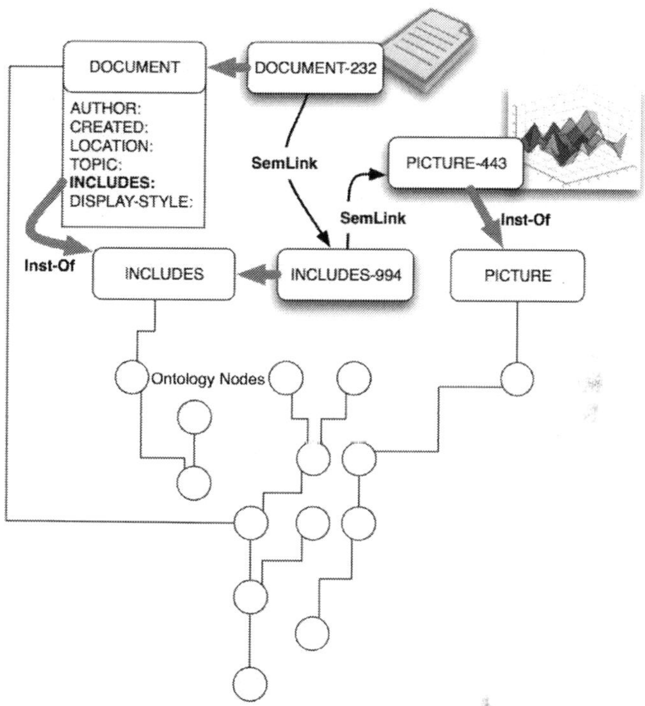


**Fig. 1.** Metadata for entities, digital or physical, is semantically defined by an underlying ontology that appears to the user as (networks of) SemCards

Every SemCard instance has a GUID, timestamps representing time of creation and related temporal aspects such as times of modification, as well as a set of policies. Its author is also represented, and any authors of modifications throughout the SemCard's lifetime. The SemCard's policies allow it to be displayed, shared, copied, etc. in predescribed ways, through the use of rules. In its simplest version a SemCard will appear as a form with fields or slots. A SemCard has one template and one or more instances, which corresponds roughly to the object-oriented programming concepts of object template/class, and object instance, respectively. Under the hood their slots are ontologically defined; however, the end user normally does not see this. To take an example, a SemCard for holding an e-mail message may look exactly like any interface to a regular email program. However, the slots reference an ontology that defines what kinds of data each slot can take, what type of information that is, etc. The e-mail SemCard, when created, will contain information about who authored which part of the content and when. Additionally, the author will not simply be a regular "from" but have a link to the SemCard representing the author of the email SemCard.

SemCards are "dumb" in that they do not carry with them any executable code: We have completely separated the services operating on the SemCards from the SemCards themselves, leaving only a spec for the desired operations to be done on a SemCard in the SemCard itself. A SemCard can thus represent any digital item, like a png image or pdf document, *physical entities* such as a person, building, street, or a kitchen utensil, *as well as immaterial things* like ideas, mythologicical phenomena and intellectual creations.

Any type of digital object or information can be pointed to with a SemCard, e.g. a Web page, a product, a service offer, a data record in a database, a file or other media object, media streams, a link to a remote Web service, etc. Equally importantly, SemCards can represent relationships between SemCards, for example, that a person is the author of an idea. To fill out a SemCard instance, one or more slots are filled with values – these could be semantic links to other SemCards, typed entities or unclassified content. Each SemCard instance, its semantic dimensions and their values, can be stored as an XML (extensible Markup Language) object, using e.g. the RDF (Resource Description Framework) format [9].

*Display rules* dictate how the SemCard itself (as well as its target reference - the thing it represents) should be displayed to the user. These can describe, for example, its owner's preferences or the display device required. As SemCards carry with them their own display specifications their on-screen representation can be customized by their userss; the same SemCard can thus be displayed differently to two different users with different preferences. The rules can specify how metadata and slot values in the SemCard should be organized and what human-readable labels should be used for them, if any, as well as what aspects of the SemCard appear as interactive elements in the interface, and the results of specific interaction with those elements.

For viewing and manipulating SemCards we have developed both client-based editors in the spirit of Haystack [8] and Web-based interfaces. SemCards can be created in many ways; doing so manually from scratch involves selecting a SemCard template type, making an instance of it and customizing its slots using typed entities from an underlying ontology.

As SemCards isolate the user from the related ontologies, classificatory mistakes in their creation does not destroy the underlying ontologies. This results in a kind of graceful degradation; instead of breaking the system such mistakes only make the automated handling of information in the system slightly less accurate. The relationship between SemCards and the unerlying ontology can be likened to non-destructive editing for video: As the creation history (original data, i.e. ontologies) are not changed but rather represented in a separate intermediate layer, the edit history of any SemCard can be traced back and reverted, if need be, with no change to the underlying ontologies.

As explained, a SemCard can represent any digital item, like a png image or pdf document, as well as physical entities, even ideas. SemCards are also be used to represent relationships between SemCards; the type of relationship is then that SemCard's *type*. For example, A SemCard can also represent collections, for example a SemCard representing a group of friends would contain links to the SemCards representing the individuals of that social group.

Behind each SemCard is an ontology that defines the meaning of the SemCard slots, specifies valid values and relations between slots (see Figure 1). An ontology like FOAF (c.f. [10]) or the Dublin Core [11] can be used with SemCards, as each SemCard carries with it a reference to the ontology it is based on. Thus, networks of ontologies can be used with SemCards, whether they use a basic, simple and singleton ontology like the Dublin Core or are definded more deeply in e.g. foundational ontologies such as DOLCE, SUMO [12] [13], or OCHRE [14].

In our current implementation we have created a fairly extensive ontology for important digital data types including Web page, 2-D image, URL, text document, as well as for physical entities such as person, place and organization. This ontology will be made open-source, so as to encourage linking of other ontologies to it, extending its reach and improving its utility.

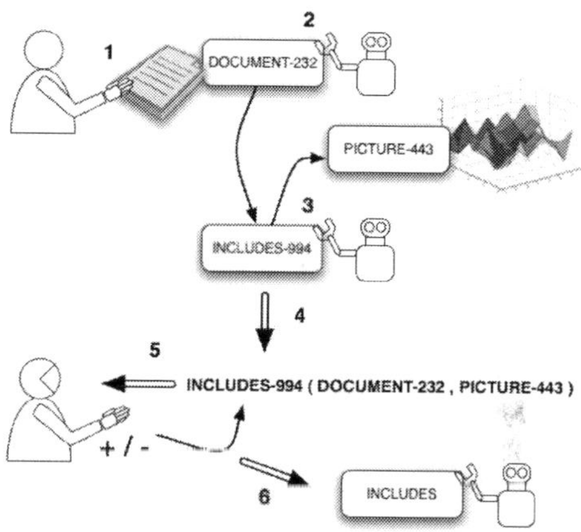


**Fig. 2.** The iterative nature of human-machine metadata annotation. (1) User creates digital document, (2) a SemCard instance is automatically created; the automation infers that a particular image is included in the document and (3) creates a SemCard for it and a SemCard of type *Includes* that links the two; (4) relationship between the SemCards now forms a triplet that the user can inspect (here shown in prepositional form, but is typically graphical); (5) user modifies the results (+/-) from which (6) the automation processes generalizes to improve own performance.

Although SemCards instances ideally derive from a SemCard template that is fully defined by one (or more) ontologies, the case could arise where a user wants to represent an entity for which no template exists. A user can create free-form slots and collect them into a new SemCard (that has no template). As long as the type of the SemCard – or at least one slot in it – has a connection to a known ontology (it will always have its author and date of creation), the automation mechanisms can use this information to base further automatic refinement of the SemCard instance, like linking it to (what are believed to be) related SemCards. Managing such automatic semantic links becomes akin to unstructured database managment; it will of course never be as good as that for fully-specified SemCards, but because these SemCards live in a rich network of other SemCards, this problem is not as large as it may seem.

## 4 Collaborative SemCard Creation by Man and Machine

An important feature of SemCards is that they record significant amounts of metadata about themselves, including their own genesis. This makes automatic creation of SemCards much more flexible as the automation process can make inferences about the quality of the SemCards (based on e.g. edit history). Because the same representational framework - SemCards - can be used for *all* data, including friend networks, author-entity relationships, object-owner, etc., inferencing can use the multiple SemCard relationship types (e.g. not only who created it but also who the creator's friends are) to decide how to perform automatic relationship creation, data-slot filling, automatic correction or deletion. Moreover, as the SemCard stores its edit history, including who/what made the edits, any such changes can be undone with relative ease. Since this history is stored as semantic information, it can be used to sort the SemCards according to their history. This makes managing SemCards over time much more flexible than if they were history-scarce, as e.g the losely-defined metadata of most data on people's hard drives. For example, caching, compressing or any other processe be made history-sensitive to a high level of detail.

**Fig. 3.** The Twine bookmarklet popup enables Web surfers to create SemCards quickly. The system automatically fills in relevant information ("Title", "Description", "icon", etc.).

As an example of collaborative automatic/ manual creation of SemCards, Nova, a SemCard end-user, finds a useful URL and creates a SemCard for it of type "bookmark for a Web-page". He makes personal comments on the Web page's contents by making a "Note" SemCard and linking it to the Webpage SemCard. Nova's automation processes, running on the SemCard hosting site, add two things: They fill the Webpage SemCard with machine-readable metadata from the Web page, and they also link these SemCards to a new SemCard that *it* created, containing further information mined from the Web site. Now Nova shares (a copy) of the SemCard with Jim (it gets saved in his SemCard space), who may add his own comments and links to related SemCards; the fact that the SemCard was shared with Jim by Nova is automatically recorded as part of the SemCard's metadata. Thus, events, data and metadata are created seamlessly and unobtrusively through collaborative paradigm.

As their authorship is automatically recorded in the SemCards, this can be later used to e.g. exclude all SemCards created by particular automation processes, should this be desired. Proactive automatic mining of a user's SemCards can reveal implicit relationships that the system can automatically make explicit, facilitating faster future retrieval through particular relationship chains in the resulting relationship graphs.

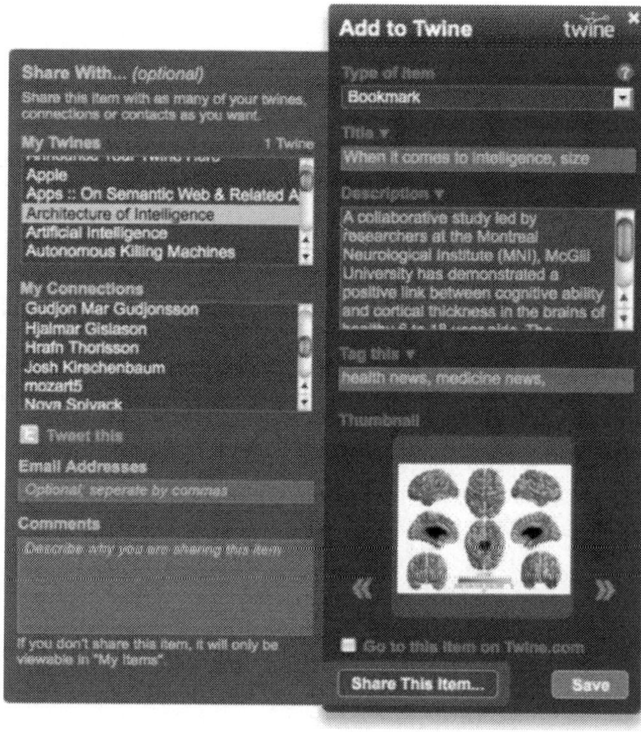


**Fig. 4.** Upon creation of the bookmark, SemCard users can choose to share it (left side of popup) with users via Twines they have created or subscribed to ("My Twines"), or directly with people they have connected with ("My Connections")

A positive feedback loop of iterative improvement on the network is created through the collaboration between a user and his/her data (2): When the initial manual data entry reaches a critical point the automation starts to provide significant enhancements to the user. Increased manual input, especially in the form of additions to automatically generated semantic links, allows the automation system to make inferences about the quality of the data entry, not just for a single user but for many. This enables it to improve the accuracy of its own automation, and suggestions to users about related data will be more relevant and targeted.

As a users's SemCard database grows user-customized automation becomes more relevant; in the long run, as the benefits of automation become increasingly obvious to each user, people will see the benefits of providing a bit of extra meta-information when they create e.g. a word processing document or an image. This will trigger a positive upward spiral where increased use of automation will motivate users to add more pieces of metadata, which will in turn enable better automation.

## 5 Deployment on Twine.com

We have started to implement the SemCard technology and deployed it on the *Twine.com*, an online Semantic Web portal where people can create accounts and use a SemCard-enabled system to manage their online activities and information, including bookmarks, digital files, sharing policies, and more.

At the time of this writing there are well over 4 million SemCards on Twine.com. Twine.com has around 250 thousand registered users and over 2 million unique visits montly. The rate of new SemCard creation has grown to 3K per day, by an estimated 10% of the users. So far, users rarely correct the automatically-generated SemCards, but a subset of the users add extensive additional information to them. We are currently in the process of scaling up the system as well as developing increasingly sophisticated automation and mining processes to provide the user with related items, related users and recommended items. Users' behavior with regards to these items is used for continuous improvement of these processes.

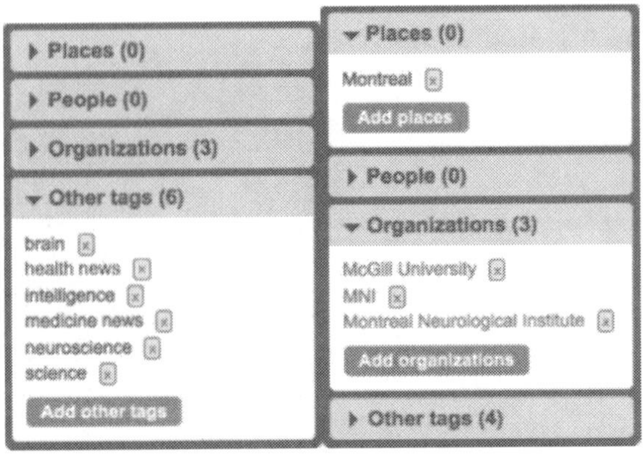


**Fig. 5.** Automatic processing of a bookmarked Web page can detect places, people organizations and various named entities ("tags"). The user can then modify these by deleting (clicking on the [x]) and adding new ones. Two snapshots of the same drop-down box are shown, the left image showing its "other tags" that were auto-recognized, and the right side "organizations", as well as one user-added "place" ("Montreal").

We will now give an actual example of making a SemCard for a Web page, a short article on the Physorg.com Web site[1]. As can be seen in Figure 3, when a user comes to a Web page of interest they can click on the bookmarklet "Twine This", which brings up a simple menu with a few information fields. Parts of the SemCard slots have been filled in; the user can choose to edit these, overwrite them with her own information or to leave them as-is. When the user clicks "save" a SemCard for this Web page is created in their Twine account. The user can choose to share this item with users and/or *twines* (see Figure 4) – a twine is a SemCard that can be described as "a blog with controlled access permissions" – in other words, a SemCard for a set of SemCards with particular visual presentation and adjustable viewing permissions. The twine SemCard shows the dynamic properties of SemCards for specifying dynamic processes, e.g. calling on services from mining, inferencing, etc.

When the "bookmark" SemCard is saved, using the "Save" button on the lower right on the bookmarklet popup, the SemCard is stored on Twine.com. Any sharing selection that the user had made during the creation will make the bookmark SemCard available to the users who have permissions to read those Twines; for example, sharing it with the twine *Architecture of Intelligence* (Figure 4) will enable everyone who has been invited to subscribe to this twine

[1] http://www.physorg.com/news157210821.html

to see it. In their home page on Twine.com this SemCard will now additionally bring forth a lot of information, including auto-tagging (recognized entities, relationships, etc.).

As seen in Figure 5 a cross next to auto-generated tags allow the user to delete the ones that they don't agree with. Further related information is automatically pulled forward, sorted into "places", "people", "organizations", "other tags" and "types of items": The last one is interesting as it is a unique feature of semantic Webs – here one can find related SemCards of type "video", for example, or "product". Vector space representations are used to profile users and their semantic networks and subsequently select related items from other semantic nets. Using a (semantic) drill-down search mechanism a user can keep refining a search for a SemCard, by selecting any combination of type, tags, author, etc. (Figure 6). During such drill-downs, suggestions by the automation of related material become increasingly better.

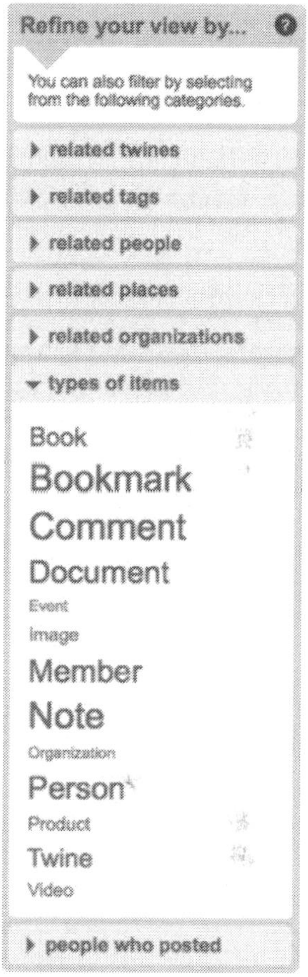


**Fig. 6.** Semantic search box ion *Twine.com*

## 6 Conclusions and Future Work

SemCards are a powerful representation scheme for enabling collaborative human-machine development of Semantic Web information. The technology achieves this by separating hard-core ontologies from the end-user, mediating these via graphical information structures, represented under the hood using RDF and OWL but supplying their own visual representation schemas for on-screen viewing. The SemCard framework allows better sharing, storing, annotating, enhancing and expanding semantic networks, creating true knowledge networks through a collaboration between people and artificial intelligence programs.

We have implemented the SemCard technology on the Semantic Web site *Twine.com*, which has over 2 million monthly unique visitors. In close collaboration with automation processes, these users have created over 4 million SemCards to date. Our results so far show that SemCards can support all of the features described in this paper for hundreds of thousands of users and we have good reason to believe that the technology will scale well. Further, many more features are envisioned, relating to searching, querying and mining. While these have not been described here they have been prototyped in our labs and will be unveiled on Twine.com and in other Radar Networks applications and services in the near future.

**Acknowledgments.** The work was supported by Radar Networks, Inc., by Reykjavik University, Iceland, and by the School of Computer Science, Reykjavik University. The authors would like to thank the Twine team for the implementation of the SemCard technology on Twine.com, as well as the funders of Radar, Vulcan Ventures Inc., Leapfrog Ventures Inc. and the angel investors.

## References

1. Dill, S., Eiron, N., Gibson, D., Gruhl, D., Guha, R., Jhingran, A., Kanungo, T., Rajagopalan, S., Tomkins, A., Tomlin, J.A., Zien, J.Y.: SemTag and Seeker: Bootstrapping the Semantic seb via automated semantic annotation. In: Proceedings of the World Wide Web Conference (2003), DOI: 10.1145/775152.775178
2. Etzioni, O., Gribble, S.: An evolutionary approach to the Semantic Web. In: Poster presentation at the First International Semantic Web Conference (2002)
3. Maly, K., Nelson, M.L., Zubair, M.: Smart objects, dumb archives: A user-centric, layered digital library framework. D-Lib Magazine 5 (1999)
4. Kahn, R., Wilensky, R.: A framework for distributed digital object services. International Journal on Digital Libraries 6, 115–123 (1995)
5. Payette, S., Lagoze, C.: Flexible and extensible digital object and repository architecture (FEDORA). In: Nikolaou, C., Stephanidis, C. (eds.) ECDL 1998. LNCS, vol. 1513, pp. 41–59. Springer, Heidelberg (1998)
6. Lagoze, C., Payette, S., Shin, E., Wilper, C.: An architecture for complex objects and their relationships. Journal of Digital Libraries - Special Issue on Complex Objects (2005), DOI: arXiv:cs/0501012v6
7. Drummond, N., Jupp, S., Moulton, G., Stevens, R.: A practical guide to building OWL ontologies using the Protégé 4 and CO-ODE tools, 1.2 edn. (2009)
8. Huynh, D., Karger, D.R., Quan, D.: Haystack: A platform for creating, organizing and visualizing information using RDF. In: Semantic Web Workshop, WWW2002 (May 2002)
9. Decker, S., Melnik, S., Harmelen, F.V., Fensel, D., Klein, M., Erdmann, M., Horrocks, I.: Knowledge networking in the Semantic Web: The roles of XML and RDF (2000)
10. Ding, L., Zhou, L., Finin, T., Joshi, A.: How the Semantic Web is being used: An analysis of FOAF documents. In: Proceedings of the 38th International Conference on System Sciences, pp. 313.3 (2005), DOI: 10.1109/HICSS.2005.299
11. Sutton, S.A., Mason, J.: The Dublin Core and metadata for educational resources. In: International Conference on Dublin Core and Metadata Applications, pp. 25–31 (2001)
12. Varma, V.: Building large scale ontology networks. In: Language Engineering Conference (LEC 2002), Hyderabad, India, p. 121 (2002)
13. Hong, J.F., Li, X.B., Huang, C.R.: Ontology-based predication of compound relations: A study based on SUMO. In: Proceedings of PACLIC18, pp. 151–160. Waseda University, Japan (2004)
14. Schneider, L.: Designing foundational ontologies: The object-centered highlevel reference ontology OCHRE as a case study. In: Song, I.-Y., Liddle, S.W., Ling, T.-W., Scheuermann, P. (eds.) ER 2003. LNCS, vol. 2813, pp. 91–104. Springer, Heidelberg (2003)

# ExpTime Tableaux for Checking Satisfiability of a Knowledge Base in the Description Logic $\mathcal{ALC}$ *

Linh Anh Nguyen[1] and Andrzej Szałas[1,2]

[1] Institute of Informatics, University of Warsaw
Banacha 2, 02-097 Warsaw, Poland
{nguyen,andsz}@mimuw.edu.pl
[2] Dept. of Computer and Information Science, Linköping University
SE-581 83 Linköping, Sweden

**Abstract.** We give the first ExpTime (optimal) tableau decision procedure for checking satisfiability of a knowledge base in the description logic $\mathcal{ALC}$, not based on transformation that encodes ABoxes by nominals or terminology axioms. Our procedure can be implemented as an extension of the highly optimized tableau prover TGC [12] to obtain an efficient program for the mentioned satisfiability problem.

## 1 Introduction

Description logics (DLs) are a family of knowledge representation languages which can be used to represent the terminological knowledge of an application domain in a structured and formally well-understood way [1]. We can use them, for example, for conceptual modeling or as ontology languages. The logical formalisms of OWL (Web Ontology Language), recommended by W3C, are based on description logics.

Description logics represent the domain of interest in terms of concepts, individuals, and roles. A concept is interpreted as a set of individuals, while a role is interpreted as a binary relation among individuals. A knowledge base in a DL usually has two parts: a TBox consisting of terminology axioms, and an ABox consisting of assertions about individuals. One of the basic inference problems in DLs, which we denote by *Sat*, is to check satisfiability of a knowledge base. Other inference problems in DLs are usually reducible to this problem. For example, the problem of checking consistency of a concept w.r.t. a TBox (further denoted by *Cons*) is linearly reducible to *Sat*.

Both problems, *Sat* and *Cons*, in the basic description logic $\mathcal{ALC}$ are ExpTime-complete. The hardness of *Cons*, and hence also of *Sat*, was proved by Schild in [17]. As shown in [17], $\mathcal{ALC}$ is a sub-logic of propositional dynamic logic (PDL). Therefore, by "internalizing" the TBox, Pratt's ExpTime algorithm [15] for PDL can be used as a decision procedure for *Cons*. In [3], by

* Supported by the grants N N206 3982 33 and N N206 399334 from the Polish Ministry of Science and Higher Education.

N.T. Nguyen, R. Kowalczyk, and S.-M. Chen (Eds.): ICCCI 2009, LNAI 5796, pp. 437–448, 2009.

encoding the ABox by "nominals" and "internalizing" the TBox, De Giacomo showed that the complexity of checking consistency of an ABox w.r.t. a TBox (i.e., checking satisfiability of a knowledge base) in CPDL (PDL with converse) is ExpTime-complete. In [4], using a transformation that encodes the ABox by terminology axioms plus a concept assertion, De Giacomo and Lenzerini showed that the mentioned problem is also ExpTime-complete for the description logic $\mathcal{CIQ}$ (an extension of CPDL).

As it was not clear how to apply optimization techniques to Pratt's algorithm, Donini and Massacci [2] gave an ExpTime tableau algorithm directly for checking consistency of a concept w.r.t. a TBox in $\mathcal{ALC}$ to which different optimization techniques can be applied. Their algorithm, however, permanently caches *"all and only unsatisfiable sets of concepts"*, and temporarily caches visited nodes on the current branch, even though this means that *"many potentially satisfiable sets of concepts are discarded when passing from a branch to another branch"* [2]. In [6], Goré and Nguyen gave an ExpTime tableau decision procedure based on "and-or" graphs with global caching for checking consistency of a concept w.r.t. a TBox in $\mathcal{ALC}$. The idea is similar to the one of Pratt, but their procedure was designed with "on-the-fly" propagation of unsatisfiability, and different optimization techniques can easily be incorporated into it [8].

Despite that the upper-bound ExpTime has been known for the complexity of the satisfiability problem in CPDL (and $\mathcal{ALC}$), implemented tableau provers for description logics like DLP and FaCT [10] usually have non-optimal complexity 2ExpTime. In the overview [1], Baader and Sattler wrote: *"The point in designing these [non-optimal] algorithms was not to prove worst-case complexity results, but ... to obtain 'practical' algorithms ... that are easy to implement and optimise, and which behave well on realistic knowledge bases. Nevertheless, the fact that 'natural' tableau algorithms for such* ExpTime*-complete logics are usually* NExpTime*-algorithms is an unpleasant phenomenon. ... Attempts to design* ExpTime*-tableaux for such logics (De Giacomo et al., 1996; De Giacomo and Massacci, 1996; Donini and Massacci, 1999) usually lead to rather complicated (and thus not easy to implement) algorithms, which (to the best of our knowledge) have not been implemented yet."* [1, page 26].

Recently, the first author has implemented a tableau prover called TGC (Tableaux with Global Caching) [12] for checking consistency of a concept w.r.t. a TBox in $\mathcal{ALC}$. The prover is based on the decision procedure developed by himself and Goré [6] that was mentioned above. He has developed and implemented for TGC a special set of optimizations that co-operates very well with global caching and various search strategies on search spaces of the form of "and-or" graphs. The test results of TGC on the sets T98-sat and T98-kb of DL'98 Systems Comparison are comparable with the test results of the best systems DLP-98 and FaCT-98 that took part in that comparison (see [12]). One can say that the mentioned test sets are not representative for practical applications, but the comparison at least shows that optimization techniques can be applied together with global caching to obtain decision procedures that are both efficient in practice and optimal w.r.t. complexity.

In this paper, we extend the mentioned decision procedure by Goré and Nguyen [6] to give the first ExpTime (optimal) tableau decision procedure for checking satisfiability of a knowledge base in $\mathcal{ALC}$, not based on transformation. Extending TGC [12] with this procedure would result in an efficient program for checking satisfiability of a knowledge base in $\mathcal{ALC}$.

The rest of this paper is structured as follows. In Section 2, we introduce notation and semantics of $\mathcal{ALC}$. In Section 3, we give a tableau calculus for the considered problem. In Section 4, we prove its soundness and completeness. In Section 5, we present our decision procedure, which is based on the calculus. Optimizations for the procedure are discussed in Section 6. Conclusions are given in Section 7.

## 2 Notation and Semantics of $\mathcal{ALC}$

We use $A$ and $B$ to denote *concept names*, $R$ and $S$ for *role names*, and $a$ and $b$ for *individual names*. (We refer to $A$ and $B$ also as *atomic concepts*, to $R$ and $S$ as *roles*, and to $a$ and $b$ as *individuals*.) We use $C$ and $D$ to denote arbitrary concepts.

*Concepts* in $\mathcal{ALC}$ are formed using the following BNF grammar:

$$C, D ::= \top \mid \bot \mid A \mid \neg C \mid C \sqcap D \mid C \sqcup D \mid \forall R.C \mid \exists R.C$$

A *TBox* is a finite set of axioms of the form $C \sqsubseteq D$ or $C \doteq D$. An *ABox* is a finite set of *assertions* of the form $a : C$ (*concept assertion*) or $R(a, b)$ (*role assertion*). A *knowledge base* in $\mathcal{ALC}$ is a pair $(\mathcal{T}, \mathcal{A})$, where $\mathcal{T}$ is a TBox and $\mathcal{A}$ is an ABox. A *formula* is defined to be either a concept or an ABox assertion.

An *interpretation* $\mathcal{I} = \langle \Delta^{\mathcal{I}}, \cdot^{\mathcal{I}} \rangle$ consists of a non-empty set $\Delta^{\mathcal{I}}$, the *domain* of $\mathcal{I}$, and a function $\cdot^{\mathcal{I}}$, the *interpretation function* of $\mathcal{I}$, that maps every concept name $A$ to a subset $A^{\mathcal{I}}$ of $\Delta^{\mathcal{I}}$, maps every role name $R$ to a binary relation $R^{\mathcal{I}}$ on $\Delta^{\mathcal{I}}$, and maps every individual name $a$ to an element $a^{\mathcal{I}} \in \Delta^{\mathcal{I}}$. The interpretation function is extended to complex concepts as follows.

$$\top^{\mathcal{I}} = \Delta^{\mathcal{I}} \qquad \bot^{\mathcal{I}} = \emptyset \qquad (\neg C)^{\mathcal{I}} = \Delta^{\mathcal{I}} \setminus C^{\mathcal{I}}$$
$$(C \sqcap D)^{\mathcal{I}} = C^{\mathcal{I}} \cap D^{\mathcal{I}} \qquad (C \sqcup D)^{\mathcal{I}} = C^{\mathcal{I}} \cup D^{\mathcal{I}}$$
$$(\forall R.C)^{\mathcal{I}} = \{x \in \Delta^{\mathcal{I}} \mid \forall y[(x, y) \in R^{\mathcal{I}} \text{ implies } y \in C^{\mathcal{I}}]\}$$
$$(\exists R.C)^{\mathcal{I}} = \{x \in \Delta^{\mathcal{I}} \mid \exists y[(x, y) \in R^{\mathcal{I}} \text{ and } \ y \in C^{\mathcal{I}}]\}$$

An interpretation $\mathcal{I}$ *satisfies* a concept $C$ if $C^{\mathcal{I}} \neq \emptyset$, and *validates* a concept $C$ if $C^{\mathcal{I}} = \Delta^{\mathcal{I}}$. Clearly, $\mathcal{I}$ *validates* a concept $C$ iff it does not *satisfy* $\neg C$.

An interpretation $\mathcal{I}$ is a *model of a TBox* $\mathcal{T}$ if for every axiom $C \sqsubseteq D$ (resp. $C \doteq D$) of $\mathcal{T}$, we have that $C^{\mathcal{I}} \subseteq D^{\mathcal{I}}$ (resp. $C^{\mathcal{I}} = D^{\mathcal{I}}$).

An interpretation $\mathcal{I}$ is a *model of an ABox* $\mathcal{A}$ if for every assertion $a : C$ (resp. $R(a, b)$) of $\mathcal{A}$, we have that $a^{\mathcal{I}} \in C^{\mathcal{I}}$ (resp. $(a^{\mathcal{I}}, b^{\mathcal{I}}) \in R^{\mathcal{I}}$).

An interpretation $\mathcal{I}$ is a *model of a knowledge base* $(\mathcal{T}, \mathcal{A})$ if $\mathcal{I}$ is a model of both $\mathcal{T}$ and $\mathcal{A}$. A knowledge base $(\mathcal{T}, \mathcal{A})$ is *satisfiable* if it has a model.

We say that an interpretation $\mathcal{I}$ *satisfies* a set $X$ of concepts if there exists $x \in \Delta^{\mathcal{I}}$ such that $x \in C^{\mathcal{I}}$ for all $C \in X$. We say that a set $X$ of concepts is *satisfiable w.r.t. a TBox* $\mathcal{T}$ if there exists a model of $\mathcal{T}$ that satisfies $X$. For $X = Y \cup \mathcal{A}$, where $Y$ is a set of concepts and $\mathcal{A}$ is an ABox, we say that $X$ is *satisfiable w.r.t. a TBox* $\mathcal{T}$ if there exists a model of $\mathcal{T}$ and $\mathcal{A}$ that satisfies $X$.

*Example 2.1.* Consider the domain of web pages. Let

$$\mathcal{T} = \{perfect \sqsubseteq interesting \sqcap \forall link.perfect\}$$
$$\mathcal{A} = \{a\!:\!perfect,\ link(a,b)\}$$

It can be shown that $b$ is an instance of the concept $\forall link.interesting$ w.r.t. the knowledge base $(\mathcal{T}, \mathcal{A})$, i.e., for every model $\mathcal{I}$ of $(\mathcal{T}, \mathcal{A})$, we have that $b^{\mathcal{I}} \in (\forall link.interesting)^{\mathcal{I}}$. To prove this one can show that the knowledge base $(\mathcal{T}, \mathcal{A}')$, where $\mathcal{A}' = \mathcal{A} \cup \{b\!:\!\exists link.\neg interesting\}$, is unsatisfiable. ◁

## 3 A Tableau Calculus for $\mathcal{ALC}$

Let $(\mathcal{T}, \mathcal{A})$ be a knowledge base in $\mathcal{ALC}$. In this section, we present a tableau calculus for the problem of checking satisfiability of $(\mathcal{T}, \mathcal{A})$.

We assume that concepts and ABox assertions are represented in negation normal form (NNF), where $\neg$ occurs only directly before atomic concepts[1]. We denote the NNF of $\neg C$ by $\overline{C}$. For simplicity, we treat axioms of $\mathcal{T}$ as concepts representing global assumptions: an axiom $C \sqsubseteq D$ is treated as $\overline{C} \sqcup D$, while an axiom $C \doteq D$ is treated as $(\overline{C} \sqcup D) \sqcap (\overline{D} \sqcup C)$. That is, we assume that $\mathcal{T}$ consists of concepts in NNF. Thus, an interpretation $\mathcal{I}$ is a model of $\mathcal{T}$ iff $\mathcal{I}$ validates every concept $C \in \mathcal{T}$. As this way of handling the TBox is not efficient in practice, in Section 6 we present an optimization technique called "absorption" for improving the performance of our algorithm.

Tableau rules are written downwards, with a set of formulas above the line as the *premise* and a number of sets of formulas below the line as the *(possible) conclusions*. A $k$-ary tableau rule has $k$ possible conclusions. A tableau rule is either an *"or"-rule* or an *"and"-rule*. Possible conclusions of an "or"-rule are separated by '|', while conclusions of an "and"-rule are separated/specified using '&'. If a rule is a unary rule or an "and"-rule then its conclusions are "firm" and we ignore the word "possible". An "or"-rule has the meaning that, if the premise is satisfiable w.r.t. the TBox $\mathcal{T}$ then some of the possible conclusions is also satisfiable w.r.t. $\mathcal{T}$. On the other hand, an "and"-rule has the meaning that, if the premise is satisfiable w.r.t. $\mathcal{T}$ then all of the conclusions are also satisfiable w.r.t. $\mathcal{T}$.

If $X$ is a set of formulas and $\varphi$ is a formula then we write $X, \varphi$ to denote the set $X \cup \{\varphi\}$. Define tableau calculus $\mathcal{CALC}$ w.r.t. a TBox $\mathcal{T}$ to be the set of the tableau rules given in Table 1, in which $X$ denotes a set of concepts, and $Y$ denotes a set of ABox assertions. Note that, for each of the rules $(\sqcap')$, $(\sqcup')$, $(\forall')$, the premise is a subset of each of the possible conclusions. We assume that the

[1] Every formula can be transformed to an equivalent formula in NNF.

**Table 1.** Rules of the tableau calculus $\mathcal{CALC}$

$$(\bot_0)\ \frac{X,\bot}{\bot} \qquad (\bot)\ \frac{X,A,\neg A}{\bot} \qquad (\sqcap)\ \frac{X,C\sqcap D}{X,C,D} \qquad (\sqcup)\ \frac{X,C\sqcup D}{X,C\mid X,D}$$

$$(trans)\ \frac{X}{\&\{\ (\{C\}\cup\{D \text{ s.t. } \forall R.D\in X\}\cup\mathcal{T}) \text{ s.t. } \exists R.C\in X\ \}}$$

$$(\bot'_0)\ \frac{Y,\ a\!:\!\bot}{\bot} \qquad (\bot')\ \frac{Y,\ a\!:\!A,\ a\!:\!\neg A}{\bot} \qquad (\forall')\ \frac{Y,\ a\!:\!\forall R.C,\ R(a,b)}{Y,\ a\!:\!\forall R.C,\ R(a,b),\ b\!:\!C}$$

$$(\sqcap')\ \frac{Y,\ a\!:\!(C\sqcap D)}{Y,\ a\!:\!(C\sqcap D), a\!:\!C,\ a\!:\!D} \qquad (\sqcup')\ \frac{Y,\ a\!:\!(C\sqcup D)}{Y,\ a\!:\!(C\sqcup D), a\!:\!C\mid Y,\ a\!:\!(C\sqcup D), a\!:\!D}$$

$$(trans')\ \frac{Y}{\&\{\ (\{C\}\cup\{D \text{ s.t. } (a\!:\!\forall R.D)\in Y\}\cup\mathcal{T}) \text{ s.t. } (a\!:\!\exists R.C)\in Y\ \}}$$

rules $(\sqcap')$, $(\sqcup')$, $(\forall')$ are applicable only when the premise is a proper subset of each of the possible conclusions[2].

The rules $(trans)$ and $(trans')$ are the only "and"-rules and the only *transitional rules*. Instantiating, for example, $(trans)$ to $X = \{\exists R.C_1, \exists R.C_2, \exists S.C_3, \forall R.C_4, \forall R.C_5, \forall S.C_6\}$ and $\mathcal{T} = \{C_7\}$ we get three conclusions: $\{C_1, C_4, C_5, C_7\}$, $\{C_2, C_4, C_5, C_7\}$, and $\{C_3, C_6, C_7\}$. The other rules of $\mathcal{CALC}$ are "or"-rules, which are also called *static rules*[3]. Notice that the transitional rules deal with "realization" of formulas of the form $\exists R.C$ or $a : \exists R.C$. Note also that the conclusions of the rule $(trans')$ consist of only concepts (but not ABox assertions). For any static rule of $\mathcal{CALC}$, the distinguished formulas of the premise are called the *principal formulas* of the rule. The principal formulas of the rule $(trans)$ (resp. $(trans')$) are the formulas of the form $\exists R.C$ (resp. $a : \exists R.C$) of the premise.

We assume the following preferences for the rules of $\mathcal{CALC}$: the rules $(\bot_0)$, $(\bot)$, $(\bot'_0)$, and $(\bot')$ have the highest priority; unary static rules have a higher priority than non-unary static rules; all the static rules have a higher priority than the transitional rules.

An *"and-or" graph for* $(\mathcal{T}, \mathcal{A})$, also called a *tableau for* $(\mathcal{T}, \mathcal{A})$, is an "and-or" graph defined as follows. The graph will contain nodes of two kinds: *complex nodes* and *simple nodes*. The label of a complex node consists of ABox assertions, while the label of a simple node consists of concepts. The root of the graph is a complex node with label $\mathcal{A} \cup \{(a\!:\!C) \mid C \in \mathcal{T}$ and $a$ is an individual occurring in $\mathcal{A}\}$. Complex nodes are expanded using the "prime" rules of $\mathcal{CALC}$ (i.e. $(\bot'_0)$, $(\bot')$, $(\sqcap')$, $(\sqcup')$, $(\forall')$, $(trans')$), while simple nodes are expanded using the other

[2] This guarantees that sequences of applications of $(\sqcap')$, $(\sqcup')$, $(\forall')$ are always finite.

[3] As it can be seen later, it does not matter whether a unary static rule is treated as an "and"-rule or as an "or"-rule.

rules of $\mathcal{CALC}$[4]. The graph will never contain edges from a simple node to a complex node. For every node $v$ of the graph, if a $k$-ary rule $\delta$ is applicable to (the label of) $v$ in the sense that an instance of $\delta$ has the label of $v$ as the premise and $Z_1, \ldots, Z_k$ as the possible conclusions, then choose such a rule accordingly to the preference and apply it to $v$ to make $v$ have $k$ successors $w_1, \ldots, w_k$ with labels $Z_1, \ldots, Z_k$, respectively.[5]. If the graph already contains a node $w_i'$ with label $Z_i$ then instead of creating a new node $w_i$ with label $Z_i$ as a successor of $v$ we just connect $v$ to $w_i'$ and assume $w_i = w_i'$. If the applied rule is $(trans)$ or $(trans')$ then we *label* the edge $(v, w_i)$ by the principal formula (of the form $\exists R.C$ or $a\!:\!\exists R.C$) that corresponds to the successor $w_i$. If the rule applied to $v$ is an "or"-rule then $v$ is an *"or"-node*, else $v$ is an *"and"-node*. The information about which rule is applied to $v$ is recorded for later uses. If no rule is applicable to $v$ then $v$ is an *end node*. Note that each node is "expanded" only once (using one rule). Also note that the graph is constructed using *global caching* [15,7,9] and its nodes have unique labels.

A *marking* of an "and-or" graph $G$ is a subgraph $G'$ of $G$ such that:

- the root of $G$ is the root of $G'$.
- if $v$ is a node of $G'$ and is an "or"-node of $G$ then there exists at least one edge $(v, w)$ of $G$ that is an edge of $G'$.
- if $v$ is a node of $G'$ and is an "and"-node of $G$ then every edge $(v, w)$ of $G$ is an edge of $G'$.
- if $(v, w)$ is an edge of $G'$ then $v$ and $w$ are nodes of $G'$.

A marking $G'$ of an "and-or" graph $G$ for $(\mathcal{T}, \mathcal{A})$ is *consistent* if it does not contain any node with label $\{\bot\}$.

**Theorem 3.1 (Soundness and Completeness of $\mathcal{CALC}$).** *Let $(\mathcal{T}, \mathcal{A})$ be a knowledge base in negation normal form in the logic $\mathcal{ALC}$, and let $G$ be an "and-or" graph for $(\mathcal{T}, \mathcal{A})$. Then $(\mathcal{T}, \mathcal{A})$ is satisfiable iff $G$ has a consistent marking.* ◁

The "only if" direction means soundness of $\mathcal{CALC}$, while the "if" direction means completeness of $\mathcal{CALC}$. See Section 4 for the proof of this theorem.

*Example 3.2.* Reconsider Example 2.1. As abbreviations, let $I = interesting$, $P = perfect$, $L = link$, and $\varphi = \neg P \sqcup (I \sqcap \forall L.P)$. We have $\mathcal{A}' = \{a\!:\!P,\ L(a, b),\ b\!:\!\exists L.\neg I\}$, and in NNF, $\mathcal{T} = \{\varphi\}$. In Figure 1, we present an "and-or" graph for $(\mathcal{T}, \mathcal{A}')$. As the graph does not have any consistent marking, by Theorem 3.1, the knowledge base $(\mathcal{T}, \mathcal{A}')$ is unsatisfiable. ◁

## 4 Proofs of Soundness and Completeness of $\mathcal{CALC}$

**Lemma 4.1 (Soundness of $\mathcal{CALC}$).** *Let $(\mathcal{T}, \mathcal{A})$ be a knowledge base in negation normal form in the logic $\mathcal{ALC}$, and let $G$ be an "and-or" graph for $(\mathcal{T}, \mathcal{A})$. Suppose that $(\mathcal{T}, \mathcal{A})$ is satisfiable. Then $G$ has a consistent marking.*

[4] Only the "prime" rules may be applicable to a complex node, and only the remaining rules may be applicable to a simple node.

[5] If there are several applicable rules of the same priority, choose any one of them.

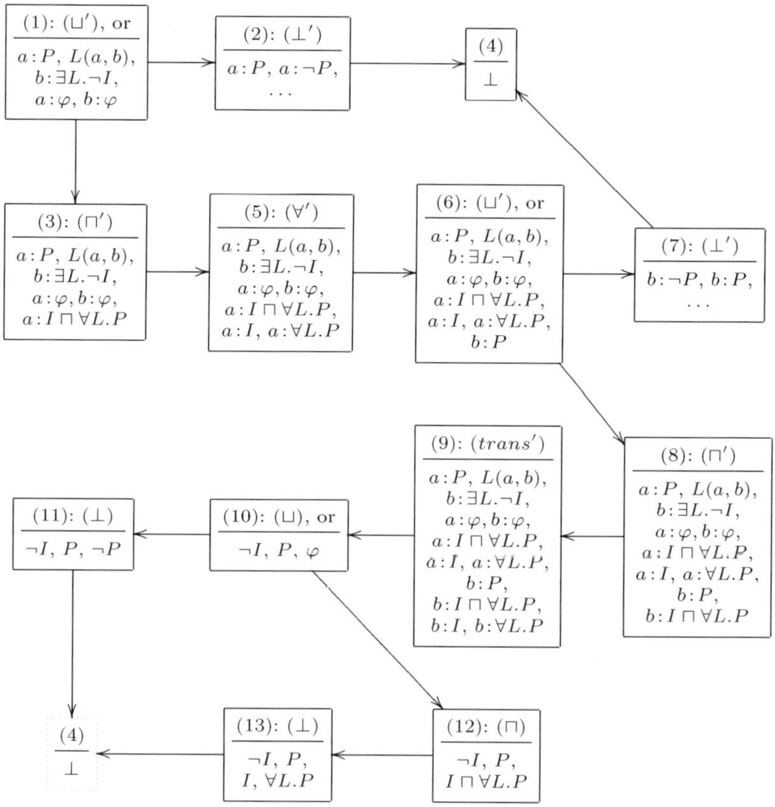


**Fig. 1.** An "and-or" graph for the knowledge base $(\{\varphi\}, \{a : P,\ L(a, b),\ b : \exists L.\neg I\})$, where $\varphi = \neg P \sqcup (I \sqcap \forall L.P)$. The edge $((9), (10))$ is labeled by $b : \exists L.\neg I$.

*Proof.* We construct a consistent marking $G'$ of $G$ as follows. At the beginning, $G'$ contains only the root of $G$. Then, for every node $v$ of $G'$ and for every successor $w$ of $v$ in $G$, if the label of $w$ is satisfiable w.r.t. $\mathcal{T}$, then add $w$ and the edge $(v, w)$ to $G'$. It is easy to check that $G'$ is a consistent marking of $G$. ◁

We prove completeness of $\mathcal{CALC}$ via model graphs. The technique has been used in [16,5,13] for other logics. A *model graph* is a tuple $\langle \Delta, \mathcal{C}, \mathcal{E} \rangle$, where:

- $\Delta$ is a finite set, which contains (amongst others) all individual names (occurring in the considered ABox);
- $\mathcal{C}$ is a function that maps each element of $\Delta$ to a set of concepts;
- $\mathcal{E}$ is a function that maps each role name to a binary relation on $\Delta$.

A model graph $\langle \Delta, \mathcal{C}, \mathcal{E} \rangle$ is *saturated* if every $x \in \Delta$ satisfies:

(1) if $C \sqcap D \in \mathcal{C}(x)$ then $\{C, D\} \subseteq \mathcal{C}(x)$
(2) if $C \sqcup D \in \mathcal{C}(x)$ then $C \in \mathcal{C}(x)$ or $D \in \mathcal{C}(x)$
(3) if $\forall R.C \in \mathcal{C}(x)$ and $(x, y) \in \mathcal{E}(R)$ then $C \in \mathcal{C}(y)$
(4) if $\exists R.C \in \mathcal{C}(x)$ then there exists $y \in \Delta$ s.t. $(x, y) \in \mathcal{E}(R)$ and $C \in \mathcal{C}(y)$.

A saturated model graph $\langle \Delta, \mathcal{C}, \mathcal{E} \rangle$ is *consistent* if no $x \in \Delta$ has $\mathcal{C}(x)$ containing $\bot$ or containing a pair $A$, $\neg A$ for some atomic concept $A$.

Given a model graph $M = \langle \Delta, \mathcal{C}, \mathcal{E} \rangle$, the *interpretation corresponding to* $M$ is the interpretation $\mathcal{I} = \langle \Delta, \cdot^{\mathcal{I}} \rangle$ where $A^{\mathcal{I}} = \{x \in \Delta \mid A \in \mathcal{C}(x)\}$ for every concept name $A$, $R^{\mathcal{I}} = \mathcal{E}(R)$ for every role name $R$, and $a^{\mathcal{I}} = a$ for every individual name $a$.

**Lemma 4.2.** *If $\mathcal{I}$ is the interpretation corresponding to a consistent saturated model graph $\langle \Delta, \mathcal{C}, \mathcal{E} \rangle$, then for every $x \in \Delta$ and $C \in \mathcal{C}(x)$ we have $x \in C^{\mathcal{I}}$.*

*Proof.* By induction on the structure of $C$. $\triangleleft$

Let $G$ be an "and-or" graph for $(\mathcal{T}, \mathcal{A})$ with a consistent marking $G'$ and let $v$ be a node of $G'$. A *saturation path* of $v$ w.r.t. $G'$ is a finite sequence $v_0 = v$, $v_1, \ldots, v_k$ of nodes of $G'$, with $k \geq 0$, such that, for every $0 \leq i < k$, $v_i$ is an "or"-node and $(v_i, v_{i+1})$ is an edge of $G'$, and $v_k$ is an "and"-node. Note that each node $v$ of $G'$ has a saturation path.

*Remark 4.3.* If a sequence $v_0, v_1, \ldots, v_k$ is a saturation path of $v_0$ w.r.t. a consistent marking then:

- if $v_0$ is a complex node then the label of each $v_i$ is a subset of the label of $v_k$,
- if $v_0$ is a simple node then all concepts of the form $\forall R.C$ or $\exists R.C$ of the label of each $v_i$ belong to the label of $v_k$. $\triangleleft$

**Lemma 4.4 (Completeness of $\mathcal{CALC}$).** *Let $(\mathcal{T}, \mathcal{A})$ be a knowledge base in negation normal form in the logic $\mathcal{ALC}$, and let $G$ be an "and-or" graph for $(\mathcal{T}, \mathcal{A})$. Suppose that $G$ has a consistent marking $G'$. Then $(\mathcal{T}, \mathcal{A})$ is satisfiable.*

*Proof.* We construct a model graph $M = \langle \Delta, \mathcal{C}, \mathcal{E} \rangle$ as follows:

1. Let $v_0$ be the root of $G'$ and let $v_0, \ldots, v_k$ be a saturation path of $v_0$ w.r.t. $G'$. Let $\Delta_0$ be the set of all individuals occurring in $\mathcal{A}$ and set $\Delta := \Delta_0$. For each $a \in \Delta_0$, set $\mathcal{C}(a)$ to the set of all concepts $C$ such that $a : C$ belongs to the label of $v_k$, and mark $a$ as *unresolved*. (Each node of $M$ will be marked either as unresolved or as resolved.) For each role name $R$, set $\mathcal{E}(R) := \{(a, b) \mid R(a, b) \in \mathcal{A}\}$.
2. While $\Delta$ contains unresolved nodes, take one unresolved node $x$ and do:
   (a) For every concept $\exists R.C \in \mathcal{C}(x)$ do:
      i. If $x \in \Delta_0$ then:
         - Let $u = v_k$.
         - Let $w_0$ be the node of $G'$ such that the edge $(u, w_0)$ is labeled by $(x{:}\exists R.C)$. (Note that $C$ belongs to the label of $w_0$.)
      ii. Else:
         - Let $u = f(x)$. ($f$ is a constructed mapping that maps each node of $M$ not belonging to $\Delta_0$ to an "and"-node of $G'$. As a maintained property of $f$, $\exists R.C$ belongs to the label of $u$.)
         - Let $w_0$ be the node of $G'$ such that the edge $(u, w_0)$ is labeled by $\exists R.C$. (Note that $C$ belongs to the label of $w_0$.)

iii. Let $w_0, \ldots, w_h$ be a saturation path of $w_0$ w.r.t. $G'$ and let $X$ be the sum of the labels of the nodes $w_0, \ldots, w_h$.
iv. If there does not exist $y \in \Delta$ such that $\mathcal{C}(y) = X$ then: add a new node $y$ to $\Delta$, set $\mathcal{C}(y) = X$, mark $y$ as unresolved, and set $f(y) = w_h$. (One can consider $y$ as the result of sticking together the nodes $w_0, \ldots, w_h$ of a maximal saturation path of $G'$. The above mentioned property of $f$ follows from Remark 4.3.)
v. Add the pair $(x, y)$ to $\mathcal{E}(R)$.

(b) Mark $x$ as resolved.

The above construction terminates and results in a finite model graph because: for every $x, x' \in \Delta \setminus \Delta_0$, $x \neq x'$ implies $\mathcal{C}(x) \neq \mathcal{C}(x')$, and for every $x \in \Delta$, $\mathcal{C}(x)$ is a subset of the set of concepts occurring in $(\mathcal{T}, \mathcal{A})$.

Clearly, $M$ is a consistent model graph.

We show that $M$ satisfies all Conditions (1)-(4) of being a *saturated* model graph. $M$ satisfies Conditions (1) and (2) because at Step 1 (of the construction of $M$), the sequence $v_0, \ldots, v_k$ is a saturation path of $v_0$, and at Step 2a, the sequence $w_0, \ldots, w_h$ is a saturation path of $w_0$. $M$ satisfies Condition (4) because: at Step 2a, $C$ belongs to the label of $w_0$ and hence also to $\mathcal{C}(y)$. For Condition (3), assume $x \in \Delta$, $\forall R.D \in \mathcal{C}(x)$, and $(x, y) \in \mathcal{E}(R)$. We show that $D \in \mathcal{C}(y)$.

Consider the case $x \in \Delta_0$. Since $\forall R.D \in \mathcal{C}(x)$, by Remark 4.3, $(x : \forall R.D)$ belongs to the label of $v_k$. If $y \in \Delta_0$, then $R(x, y) \in \mathcal{A}$ (since $(x, y) \in \mathcal{E}(R)$), and hence $(y : D)$ belongs to the label of $v_k$ (due to the tableau rule $(\forall')$ and Remark 4.3), and hence $D \in \mathcal{C}(y)$. Assume that $y \notin \Delta_0$ and that $y$ is created at Step 2(a)iv. Since $(x : \forall R.D)$ belongs to the label of $u = v_k$, by the tableau rule $(trans')$, $D$ belongs to the label of $w_0$ and hence also to $\mathcal{C}(y)$.

Consider the case $x \notin \Delta_0$ and Step 2a at which the pair $(x, y)$ is added to $\mathcal{E}(R)$. Because $\forall R.D \in \mathcal{C}(x)$ and the label of $x$ is the sum of the labels of nodes of a saturation path that ends at $u$, by Remark 4.3, we have that $\forall R.D$ belongs to the label of $u$. By the tableau rule $(trans)$, it follows that $D$ belongs to the label of $w_0$ and hence also to $\mathcal{C}(y)$.

Therefore $M$ is a consistent saturated model graph.

By the definition of "and-or" graphs for $(\mathcal{T}, \mathcal{A})$ and the construction of $M$: if $(a : C) \in \mathcal{A}$ then $C \in \mathcal{C}(a)$; if $R(a, b) \in \mathcal{A}$ then $(a, b) \in \mathcal{E}(R)$; and $\mathcal{T} \subseteq \mathcal{C}(a)$ for all $a \in \Delta_0$. We also have that $\mathcal{T} \subseteq \mathcal{C}(x)$ for all $x \in \Delta \setminus \Delta_0$. Hence, by Lemma 4.2, the interpretation corresponding to $M$ is a model of $(\mathcal{T}, \mathcal{A})$. ◁

## 5 A Simple ExpTime Decision Procedure for $\mathcal{ALC}$

Algorithm 1 given in Figure 2 is a simple algorithm for checking satisfiability of a knowledge base in $\mathcal{ALC}$. Optimizations for it will be discussed in the next section. Here is our main theorem:

**Theorem 5.1.** *Algorithm 1 is an* ExpTime *(optimal) decision procedure for checking satisfiability of a knowledge base in* $\mathcal{ALC}$.

**Algorithm 1**
Input: a knowledge base $(\mathcal{T}, \mathcal{A})$ in NNF.
Output: *true* if $(\mathcal{T}, \mathcal{A})$ is satisfiable, and *false* otherwise.

1. construct an "and-or" graph $G$ with root $v_0$ for $(\mathcal{T}, \mathcal{A})$;
2. $UnsatNodes := \emptyset$;
3. if $G$ contains a node $v_\bot$ with label $\{\bot\}$ then
   (a) $V := \{v_\bot\}$;
   (b) while $V$ is not empty do:
      i. take out a node $v$ from $V$;
      ii. for every father node $u$ of $v$, if $u \notin UnsatNodes$ and either $u$ is an "and"-node or $u$ is an "or"-node and all the successor nodes of $u$ belong to $UnsatNodes$ then add $u$ to both $UnsatNodes$ and $V$;
4. return *false* if $v_0 \in UnsatNodes$, and *true* otherwise.

**Fig. 2.** Algorithm for checking satisfiability of a knowledge base in $\mathcal{ALC}$

To prove this theorem we need some definitions and a lemma.

For a formula $\varphi$, by $sc(\varphi)$ we denote the set of all subconcepts of $\varphi$, including $\varphi$ if it is a concept. For a set $X$ of formulas, define

$$sc(X) = \{C \mid C \in sc(\varphi) \text{ for some } \varphi \in X\}$$
$$sf(X) = \{C, (a\!:\!C) \mid C \in sc(\varphi) \text{ for some } \varphi \in X, \text{ and } a \text{ occurs in } X\}.$$

Define the *length* of a formula to be the number of its symbols, and the *size* of a finite set of formulas to be the length of the conjunction of its formulas.

**Lemma 5.2.** *Let $(\mathcal{T}, \mathcal{A})$ be a knowledge base in NNF, $n$ be the size of $\mathcal{T} \cup \mathcal{A}$, and $G$ be an "and-or" graph for $(\mathcal{T}, \mathcal{A})$. Then $G$ has $2^{O(n^2)}$ nodes and the label of each node $v$ of $G$ is a subset of $sf(\mathcal{T} \cup \mathcal{A})$, which consists of at most $O(n)$ concepts if $v$ is a simple node, and consists of at most $O(n^2)$ assertions if $v$ is a complex node.*

*Proof.* The label of each simple node of $G$ is a subset of $sc(\mathcal{T} \cup \mathcal{A})$ and therefore consists of at most $O(n)$ concepts. Since the labels of simple nodes are unique, $G$ has $2^{O(n)}$ simple nodes. The label of each complex node of $G$ is a subset of $sf(\mathcal{T} \cup \mathcal{A})$ and therefore consists of at most $O(n^2)$ assertions. It can be seen that each path of complex nodes in $G$ has length of rank $O(n^2)$ (the rank of the cardinality of $sf(\mathcal{T} \cup \mathcal{A})$). Hence $G$ contains $2^{O(n^2)}$ complex nodes. ◁

*Proof (of Theorem 5.1).* It is easy to see that the "and-or" graph $G$ with root $v_0$ for $(\mathcal{T}, \mathcal{A})$ has a consistent marking iff $v_0 \notin UnsatNodes$. Hence, by Theorem 3.1, Algorithm 1 is a decision procedure for checking satisfiability of $(\mathcal{T}, \mathcal{A})$. By Lemma 5.2, $G$ can be constructed in $2^{O(n^2)}$ steps. The set $UnsatNodes$ can be computed in polynomial time in the size of $G$. Hence the algorithm runs in $2^{O(n^2)}$ steps. ◁

## 6 Optimizations

Treating axioms of the TBox as concepts representing global assumptions is not efficient because it generates too many expansions of the kind "or". A solution for this problem is to use *absorption* techniques. A basic kind of absorption is *lazy unfolding* for acyclic TBoxes[6]. For the case the TBox is acyclic and consists of only concept definitions of the form $A \doteq C$, by using lazy unfolding, $A$ is treated as a reference to $C$ and will be "unfolded" only when necessary. Using this technique, TGC runs on the test set DL'98 T98-kb equally well as on the test set DL'98 T98-sat. For the case the TBox is acyclic and contains also concept inclusions of the form $A \sqsubseteq C$, a simple solution can be adopted: treat $A \sqsubseteq C$ as $A \doteq (C \sqcap A')$ for a new atomic concept $A'$. For the case the TBox is cyclic, one can try to divide the TBox into two parts $\mathcal{T}_1$ and $\mathcal{T}_2$, where $\mathcal{T}_1$ is a maximal acyclic sub-TBox "not depending" on the concepts defined in $\mathcal{T}_2$, then one can apply the mentioned "replacing" and "lazy unfolding" techniques for $\mathcal{T}_1$.

Observe that Algorithm 1 first constructs an "and-or" graph and then checks whether the graph contains a consistent marking. To speed up the performance these two tasks can be done concurrently. For this we update the set $UnsatNodes$ mentioned in the algorithm and check the condition $v_0 \in UnsatNodes$ "on-the-fly" during the construction of $G$.

During the construction of the "and-or" graph $G$, if a subgraph of $G$ has been fully expanded (in the sense that all of its nodes and their descendants have been expanded) then each node of the subgraph can be determined to be unsat (unsatisfiable w.r.t. $\mathcal{T}$) or sat (satisfiable w.r.t. $\mathcal{T}$) regardlessly of the rest of $G$: if a node of the subgraph could not have been determined to be unsat then we can set its status to sat.

For further optimizations techniques, we refer the reader to [12,10,2].

## 7 Conclusions

Recall that, in [3] the ABox is encoded by nominals, while in [4] the ABox is encoded by terminology axioms plus a concept assertion. These approaches are not efficient in practice: in the well-known tutorial [11], Horrocks and Sattler wrote *"direct algorithm/implementation instead of encodings"* and *"even simple domain encoding is disastrous with large numbers of roles"*.

In this work we have given the *first* ExpTime (optimal) tableau decision procedure not based on transformation for checking satisfiability of a knowledge base in the description logic $\mathcal{ALC}$. Our procedure can be implemented as an extension of the highly optimized tableau prover TGC [12] to obtain an efficient program for the mentioned satisfiability problem.

As new methods, apart from incorporating individuals into (non-prefixed) tableaux, we defined tableaux directly as "and-or" graphs with global caching,

---

[6] If $A$ is defined by $A \doteq C$ or $A \sqsubseteq C$, and $B$ occurs in $C$, then $A$ *directly depends* on $B$. Let binary relation "depend" be the transitive closure of "directly depend". If in a TBox $\mathcal{T}$ no atomic concept depends on itself, then $\mathcal{T}$ is acyclic. For simplicity, we assume that each atomic concept is defined at most once.

while in [6,7,9] Goré and Nguyen used tree-like tableaux and formulated global caching separately. Consequently, we do not have to prove soundness of global caching when having soundness and completeness of the calculus, while Goré and Nguyen [6,7,9] had to prove soundness of global caching separately after having completeness of their calculi.

We have recently extended our methods for PDL [14]. We intend to extend our methods also for other modal and description logics.

## References

1. Baader, F., Sattler, U.: An overview of tableau algorithms for description logics. Studia Logica 69, 5–40 (2001)
2. Donini, F., Massacci, F.: ExpTime tableaux for $\mathcal{ALC}$. Artificial Intelligence 124, 87–138 (2000)
3. De Giacomo, G.: Decidability of Class-Based Knowledge Representation Formalisms. PhD thesis, Universita' di Roma "La Sapienza" (1995)
4. De Giacomo, G., Lenzerini, M.: TBox and ABox reasoning in expressive description logics. In: Aiello, L.C., Doyle, J., Shapiro, S.C. (eds.) Proceedings of KR 1996, pp. 316–327. Morgan Kaufmann, San Francisco (1996)
5. Goré, R.: Tableau methods for modal and temporal logics. In: D'Agostino, et al. (eds.) Handbook of Tableau Methods, pp. 297–396. Kluwer, Dordrecht (1999)
6. Goré, R., Nguyen, L.A.: ExpTime tableaux for $\mathcal{ALC}$ using sound global caching. In: Calvanese, D., et al. (eds.) Proc. of DL 2007, pp. 299–306 (2007)
7. Goré, R., Nguyen, L.A.: ExpTime tableaux with global caching for description logics with transitive roles, inverse roles and role hierarchies. In: Olivetti, N. (ed.) TABLEAUX 2007. LNCS (LNAI), vol. 4548, pp. 133–148. Springer, Heidelberg (2007)
8. Goré, R., Nguyen, L.A.: Optimised ExpTime tableaux for $\mathcal{ALC}$ using sound global caching, propagation and cutoffs. Manuscript (2007), `http://www.mimuw.edu.pl/~nguyen/papers.html`
9. Goré, R., Nguyen, L.A.: Sound global caching for abstract modal tableaux. In: Burkhard, H.-D., et al. (eds.) Proceedings of CS&P 2008, pp. 157–167 (2008)
10. Horrocks, I., Patel-Schneider, P.F.: Optimizing description logic subsumption. Journal of Logic and Computation 9(3), 267–293 (1999)
11. Horrocks, I., Sattler, U.: Description logics - basics, applications, and more. Tutorial given at ECAI-2002 (2002), `http://www.cs.man.ac.uk/~horrocks/Slides/ecai-handout.pdf`
12. Nguyen, L.A.: An efficient tableau prover using global caching for the description logic $\mathcal{ALC}$. In: Proceedings of CS&P 2008 (2008) (to appear in Fundamenta Informaticae)
13. Nguyen, L.A.: Analytic tableau systems and interpolation for the modal logics KB, KDB, K5, KD5. Studia Logica 69(1), 41–57 (2001)
14. Nguyen, L.A., Szałas, A.: Optimal tableau decision procedures for PDL. CoRR, abs/0904.0721 (2009)
15. Pratt, V.R.: A near-optimal method for reasoning about action. J. Comput. Syst. Sci. 20(2), 231–254 (1980)
16. Rautenberg, W.: Modal tableau calculi and interpolation. JPL 12, 403–423 (1983)
17. Schild, K.: A correspondence theory for terminological logics: Preliminary report. In: Mylopoulos, J., Reiter, R. (eds.) Proceedings of IJCAI 1991, pp. 466–471. Morgan Kaufmann, San Francisco (1991)

# Semantically Enhanced Intellectual Property Protection System - SEIPro2S

Dariusz Ceglarek[1], Konstanty Haniewicz[2], and Wojciech Rutkowski[3]

[1] Wyzsza Szkola Bankowa w Poznaniu, Poland
dariusz.ceglarek@wsb.poznan.pl
[2] Uniwersytet Ekonomiczny w Poznaniu, Poland
konstanty.haniewicz@ue.poznan.pl
[3] Business Consulting Center, Poland
wojciech.rutkowski@bcc.com.pl

**Abstract.** The aim of this work is to present some of the capabilities of a Semantically Intellectual Enhanced Property Protection System. The system has reached a prototype phase where experiments are possible. It uses an extensive semantic net algorithms for Polish language that enable it to detect similarities in two compared documents on a level far beyond simple text matching. SEIPro2S benefits both from using a local document repository and from Web based resources. Main focus of this work is to give a reader overview of architecture and some actual results.

**Keywords:** intellectual property, semantic net, thought matching, natural language processing.

## 1 Introduction

SEIPro2S has been built to be an aid in a situation where some kind of intellectual assets has to be protected from unauthorized usage in terms of copying ideas. An example of this situation is plagiarism which as is commonly known violates ones intellectual property to gain some sort of revenue. One cannot come up with a universally valid definition of plagiarism. As an example, Polish law does not specify a notion of plagiarism, nevertheless it enumerates a list of offences directed at intellectual property.

I. e. :

- claiming of an authorship of some piece of work as a whole or its parts
- reusing someone's work without changes to the narration, examples and the order of arguments
- claiming of an authorship of research project
- claiming of an authorship of invention to be used in evaluation of research achievements that is not protected by intellectual property law or industry property ownership law

SEIPro2S enables its users for a semiautomatic control of publicly available documents and those that are made available to it by other channels in context of using concepts which represent user's intellectual property.

N.T. Nguyen, R. Kowalczyk, and S.-M. Chen (Eds.): ICCCI 2009, LNAI 5796, pp. 449–459, 2009.

This application can be easily transformed into another one, meaning that it lets us monitor a stream of documents for content similar on concept level with base document. One can easily imagine this as a help in automatic building of relatively complete reference base for some research domain.

Working prototype allows to produce a report demonstrating areas copied from other sources along with clear indicator informing user on ratio of copied work to the overall document content.

Thanks to devised algorithms basing on extensive semantic net for Polish language and thesauri prototype is able to detect copied fragments where insincere author tried to hide his deeds by changing of word order, using synonyms, hyponyms or commonly interchangeable phrases when compiling his document.

The structure of the work is as follows: brief discussion of used technologies, description of the system functioning, its architecture, example from the undertaken experiments and summary along with future work.

## 2 Technologies Used by SEIPro2S

The system applies many of well known text mining techniques. Due to fact that text mining is multidisciplinary domain it employs a variety of algorithms coming from various branches of computer science.

Every document submitted to the system has to undergo a set of transformations all commonly defined as a text-refinement. As an output of this text-refinement transformation process the system produces a structure containing ordered concept descriptors coming from the input document.

Following steps are undertaken in the text-refinement process:

- tokenization
- applying stop-list
- concept identification (phrases, order of concepts)
- stemming
- concept generalization with aid of knowledge representation structure (synonyms, hypernyms)
- concept disambiguation
- applying correct set algorithms basing on the type of text-mining task

Further explanation of the knowledge representation structures follows, as this is the crucial element of systems internal workings.

### 2.1 Knowledge Representation

By knowledge representation method it is understood a way in which knowledge of the real world is stored, computed and inferred about by a system. Further, it is strictly defined knowledge description language paired with computation mechanisms.

Every possible entity is mapped to a natural language using a word, collocation or a phrase. This mapping will be referred to from now-on as a concept. A good knowledge representation serves a goal of producing structures capable of being computed by the system and transforming input in a manner that allows it to come up with elements relating to stored system-known concepts.

Classic methods employed by information retrieval systems base on simple knowledge representation where a document is a set of used words. This representation yields well-established methods of querying systems such as Boolean model and vector space model.

A more sophisticated knowledge representation models are: glossary, domain dictionary, taxonomy, thesaurus, semantic net and ontology. In comparison with a bag of words, enlisted methods introduce different, more complicated and of higher information value lexical relations among stored concepts.

We have a special interest in a semantic net as it is employed as a knowledge representation model in the system. It has been chosen due to its superb capability of capturing relationships among concepts. Semantic net is a directed graph with vertices representing concepts and edges representing lexical relationships.

A list of supported types of lexical relationships along with examples follows:

- hypernyms (animal is_more_general_than elephant)
- hyponym (elephant is_less_general than animal)
- meronym (lamb has_part bulb)
- holonym (bulb is_part_of lamp)
- connotation (rose has_trait scent)
- attribute (soft is_specification_of hardness)
- synonym (scent is_synonym_of fragrance)

Every edge can be weighted to introduce a notion of importance of some relation. Inference is realized by edge traversal. One can apply well-known methods of search within graph. Starting with a vertex in question one can reach all relevant vertices by following edges on the way thus inferring of their properties.

Semantic net contains all information concerning concepts semantics, thus its natural application to the domain of interest. Some of benefits one can obtain by applying thesauri and semantic nets were described by [2]. Its greatest advantage is to supply a system with the right meaning of the concept processed based on its contextual usage. What is more, when subjected to be used in classification tasks it levels up quality of classification and categorization.

**Table 1.** Comparison of WordNet and SeNeCa net

| Features | WordNet | SeNeCa net |
|---|---|---|
| Concept count | 155200 | 126800 |
| Polysemic words | 27000 | 18200 |
| Synonyms | 0 | 5100 |
| Homonyms, hypernyms | + | + |
| Antonyms | – | + |
| Connotations | + | + |
| Unnamed relationship | – | + |

Commonly used semantic net in information retrieval systems for processing English is WordNet. Its structure is organized around notion of synset. Every WordNet's synset contains words which are mutually synonyms. Relationships among synsets are hypernyms or hyponyms, when combined with previous data it is easily seen that whole WordNet acts as a thesaurus.

At the current moment, there are following relationships present in WordNet: hypernymy, hyponymy, synonymy, metonymy, homonymy and antonymy. All these make WordNet fully functional semantic net for English.

Facing a challenge of not being able to use a similar structure for Polish, a semantic net produced as an outcome of project SeNeCa was employed in the system. Table 1 demonstrates a comparison between WordNet and SeNeCa net.

## 3 SEIPro2S Functioning

Figure 1 presents a sequence of steps which are undertaken when a document is to be subjected to similarity check against local document repository and documents obtained from the Internet.

When input document $\Delta$ is presented to the system, responsible component starts a procedure of Internet sampling. Samples are obtained by submitting a number of fragments obtained from input document. A list of potentially viable documents is prepared basing on the occurrence of necessarily long identical passages. Documents are downloaded and subjected to text-refining. After completing these steps every downloaded document is indexed. For every document an abstract is created and stored in local repository. A key for the whole process procedure follows. At first, the most relevant of previously sampled, indexed and abstracted documents are selected by comparing the abstracts. Then, input document $\Delta$ is subjected to comparison with every relevant document. As a result of this action, a similarity report is constructed. It conveys following information on the overall similarity of input document $\Delta$ to every checked document:

- length of the longest identical phrase obtained from the documents from the sampling phase
- similarity indicator, which using percent ratio demonstrates how much of the submitted text is identical with documents coming both from the sampling phase and from local repository coming from earlier checks
- checked text along with markup showing which text fragments are identical to those coming from local repository and current Internet sample.

When SEIPro2S acts in monitoring mode some enhancements are introduced. First of all, each analyzed document along with first batch of samples obtained from the Internet is stored in local repository. Thus, when the system is monitoring the Internet it can omit all previously checked documents and focus only on new ones.

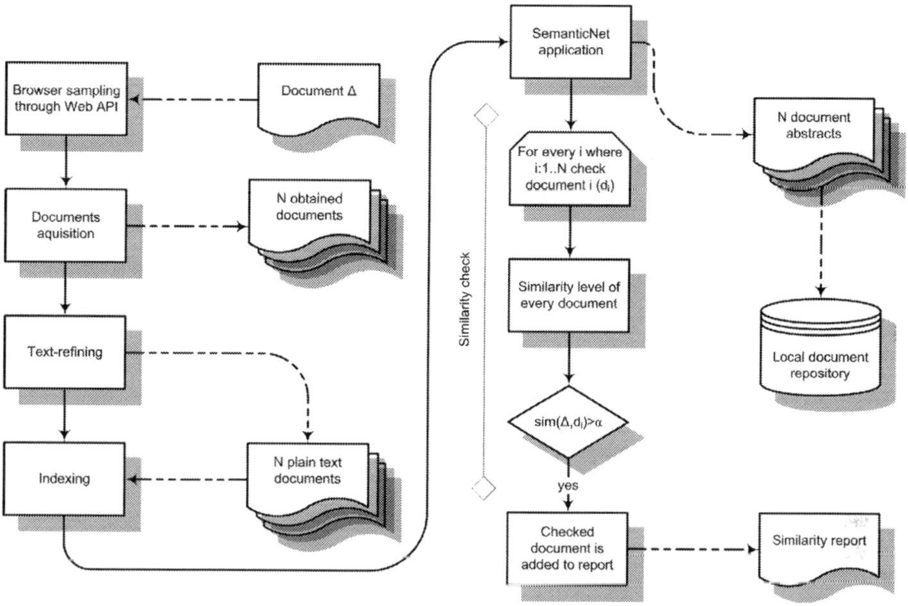


**Fig. 1.** Checking document against a sample of documents obtained from open Internet

# 4 System Architecture

Components constituting SEIPro2S system are illustrated on figure 2. Its main part is a semantic network module, which works as system engine processing semantic relations.

End user communicates with the system through a web interface, defining tasks to be completed by the system. Tasks are characterized by type, subject and such parameters like time limit. After task is done, the user can view a report or have it sent by e-mail.

Other components of SEIPro2S — sampling agent, downloading agent, file format adapters, abstractor, classifier and document corpus — are described below.

## 4.1 Sampling Agent, Downloading Agent

The sampling agent uses internet search engines to find documents relevant to examined text. It uses document abstracts and semantic network module to generate multiple queries. Thanks to heuristics, the agent makes use of domain dictionary to narrow down obtained results to specified domain by determining a frequency of domain-typical words. It can also query the search engine with long text phrases.

The sampling agent outputs a list of URLs and passes it to downloading agent, which downloads relevant documents from the Internet and saves them to local repository. For efficiency matters, the downloading agent works in multiple threads.

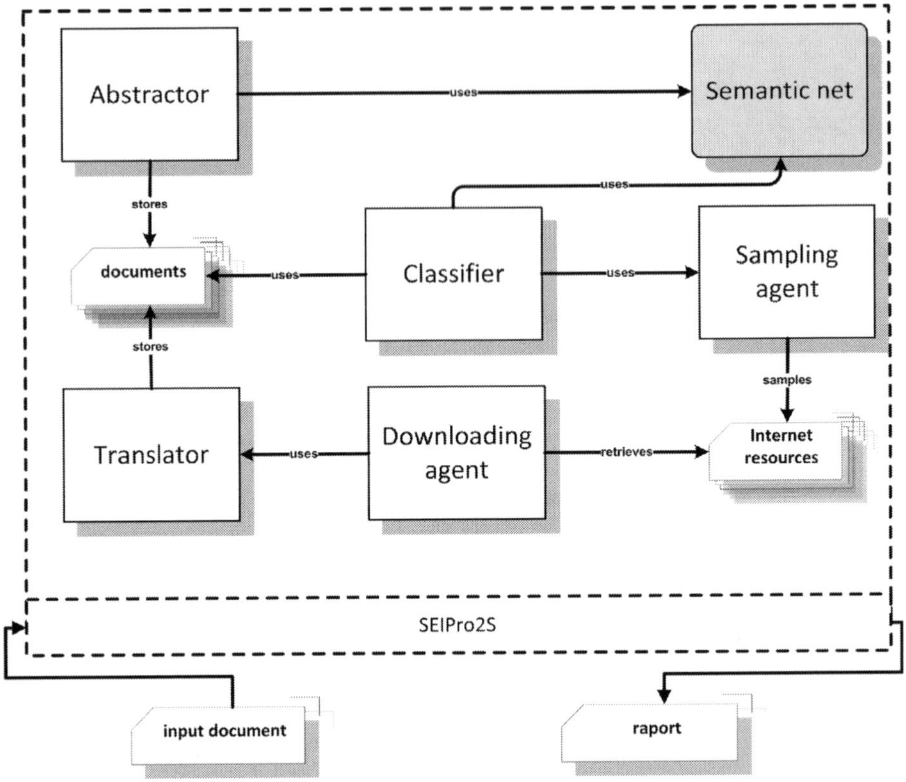


**Fig. 2.** SEIPro2S compontents

## 4.2 File Format Adapters

This module is a strictly functional one, as its sole purpose is to homogenize format of retrieved resources. The module follows well tested approach of flattening any structured file formats into plain text.

## 4.3 Abstractor

In order to conduct similarity analysis, refining of text documents needs to be done before. Text refinement uses multiple information retrieval methods aiming at streamlining resulting quality of the analysis.

First, text units in document need to be identified to allow correct indexing. Following steps are taken by the abstractor:

- lexical analysis,
- concept identification,
- removal of stop-words,
- lemmatization,

– descriptor adding,
– disambiguation.

Lexical analysis is a process of converting an input stream of characters into a stream of words, which can be used for indexing. At this stage, all words need to be split by identifying spaces, punctuation or letter case.

The next step is to drop low significance words, which do not add much information to the context. Such words occur on so called stop-list, which contains mainly prepositions, pronouns or conjunctions.

After stop-words are eliminated, the text is being indexed. At this stage, concepts consisting of two or more words need to be identified to represent one semantic unit and in order to ensure results quality. Then the stream of words is lemmatized to change inflexed forms to main form (subject). It is particularly difficult task for highly flexible languages, such as Polish or French. In Polish, there are multiple noun declination forms and verb conjugation forms. Therefore, for flexible languages word lemmatization is crucial to represent multiple word forms as one concept.

Synonyms need to be represented with concept descriptors using semantic network. It allows correct similarity analysis and also increases classification algorithms efficiency by reducing number of dimensions without loss in comparison quality. Replacing words by their hypernyms [10] is only possible when using semantic network as a form of knowledge representation.

Indexing faces another problem here, which is polysemy. One word can represent multiple meanings, so the apparent similarity need to be eliminated. It is done by concept disambiguation. Disambiguation, which identifies word meaning depending on its context, is important to ensure that no irrelevant documents will be returned in response to a query [14]. A very efficient method of concept disambiguation has been proposed in [7]. It uses semantic network for Polish language and examines word context to determine its meaning, resulting in 82% accuracy. It seems that only linguistic analysis methods can exceed 90% accuracy [16], while human experts are able to recognize correct meaning of 96,8% polysemic words [19].

### 4.4 Classifier

SEIPro2S implements innovative algorithms which ensure, that document comparison methods are insensitive to techniques used to disguise plagiarism, like changes in word order or using synonyms. Classifier module compares documents and searches common phrases which are long enough (phrase length is a configurable parameter set by user). The classifier uses document abstracts which includes concept descriptors, instead of words, to represent multiple synonyms as one descriptor.

The algorithm is not sensitive to word order changes in a few word space. User configurable number of concepts with no match is also ignored, and when a discrepancy occurs inside a longer positively matched text, it can be treated as temporal, and does not eliminate the match.

Metody uwierzytelniania Identyfikator to niepowtarzalna nazwa lub numer nadany obiektowi Uwierzytelnianie to sprawdzenie czy osoba lub obiekt jest tym za kogo się podaje Procedura upoważniania bada czy osoba ta lub obiekt ma prawo do chronionego zasobu Wszystko to jest brane pod uwagę w celu podjęcia decyzji o udzieleniu dostępu Uwierzytelnienie zwykle jest dokonywane jednorazowo ale w instalacjach o dużym stopniu bezpieczeństwa może być wymagana okresowa lub stała weryfikacja Dla uwierzytelniania tożsamości użytkowników komputery używają haseł lub innych metod dialogowych Z punktu widzenia użytkownika znacznie ważniejsze są czynniki takie jak liczba znaków do wprowadzenia wysiłek umysłowy i sposób postępowania w wypadku popełnienia błędu przy wpisywaniu Wyróżnia się następujące informatyczne metody uwierzytelniania 1 Podanie hasła metoda prostych haseł użytkownik wprowadza hasło które może sam sobie wybrać wybrane znaki komputer może zażądać od użytkownika podania pewnych znaków numery znaków mogą być wyliczane na podstawie transformacji zegara wewnętrznego lub generatora hasła jednorazowe użytkownik ma listę N haseł ta sama listę pamięta komputer po użyciu danego hasła użytkownik skreśla je z listy wada jest taka ze użytkownik musi pamiętać lub mięć przy sobie cala listę i znać aktualne hasło oraz w przypadku błędów w transmisji użytkownik nie wie które hasło ma podać 2 Metoda pytań i odpowiedzi system ma zbiór odpowiedzi i zbiór pytań dostarczonych przez użytkownika System zadaje pytania użytkownik odpowiada metoda ta zajmuje dużo pamięci ale jest komunikatywna 3 Uwierzytelnienie tożsamości komputera po uwierzytelnieniu użytkownik komputera ma podąć swoje hasło które wcześniej wprowadził użytkownik 4 Procedura przywitania wykonanie przez użytkownika poprawnie jakiegoś algorytmu metoda ta ma wyższy stopień bezpieczeństwa brak jej jawności ale jest czasochłonna i żmudna dla użytkownika 5 Procedury użytkownika dostarczenie przez użytkownika procedur które są wykonywane przed wejściem do systemu po zakończeniu danej procedury system wywołuje własna kontrole bezpieczeństwa Jeśli system odmówi użytkownikowi dostępu mogą być wykonane następujące działania dziennik systemu i zwłoka czasowa przy źle wprowadzonej odpowiedzi oraz ilość prób wejścia do systemu

**Fig. 3.** Fragment of similarity report for matching document with Internet sources. Positively matched fragments are marked with red bold font, while single discrepancies are ignored (underline only).

The described algorithm has high computational complexity, which is however typical to similar tasks.

One more feature of SEIPro2S classifier is taking into account cases, when text copying is allowed. Citations marked with quotation marks and references are not treated like potential plagiarism, nor are some domain-typical phrases.

### 4.5 End-User Report

The result of SEIPro2S analysis is a similarity report between examined document and text from document corpus (with documents downloaded from the Internet), which is presented to end-user. The report contains following information:

- maximum length of positive matching phrase, which can be used to classify document as violating intellectual property,
- similarity factor — a ratio of document positively matched with document corpus,

*Metody uwierzytelniania*

**Identyfikator** to niepowtarzalna nazwa lub numer nadany obiektowi.

**Uwierzytelnianie** to sprawdzenie, czy osoba lub obiekt jest tym, za kogo się podaje. Procedura upoważniania bada, czy osoba ta lub obiekt ma prawo do chronionego zasobu. Wszystko to jest brane pod uwagę w celu podjęcia decyzji o udzieleniu dostępu.

Uwierzytelnienie zwykle jest dokonywane jednorazowo, ale w instalacjach o dużym stopniu bezpieczeństwa może być wymagana okresowa lub stała weryfikacja. Dla uwierzytelniania tożsamości użytkowników komputery używają haseł lub innych metod dialogowych. Z punktu widzenia użytkownika znacznie ważniejsze są czynniki takie jak: liczba znaków do wprowadzenia, wysiłek umysłowy i sposób postępowania w wypadku popełnienia błędu przy wpisywaniu.

**Wyróżnia się następujące informatyczne metody uwierzytelniania:**

**1.** Podanie hasła:

- metoda prostych haseł - użytkownik wprowadza hasło, które może sam sobie wybrać
- wybrane znaki - komputer może zażądać od użytkownika podania pewnych znaków; numery znaków mogą być wyliczane na podstawie transformacji zegara wewnętrznego lub generatora
- hasła jednorazowe - użytkownik ma listę N haseł, ta sama listę pamięta komputer; po użyciu danego hasła użytkownik skreśla je z listy; wada jest taka, ze użytkownik musi pamiętać lub mięć przy sobie cala listę i znać aktualne hasło oraz w przypadku błędów w transmisji użytkownik nie wie, które hasło ma podać

**2.** Metoda pytań i odpowiedzi - system ma zbiór odpowiedzi i zbiór pytań dostarczonych przez użytkownika. System zadaje pytania, użytkownik odpowiada; metoda ta zajmuje dużo pamięci, ale jest komunikatywna

**3.** Uwierzytelnienie tożsamości komputera - po uwierzytelnieniu użytkownik komputera ma podąć swoje hasło, które wcześniej wprowadził użytkownik

**4.** Procedura przywitania - wykonanie przez użytkownika poprawnie jakiegoś algorytmu; metoda ta ma wyższy stopień bezpieczeństwa, brak jej jawności, ale jest czasochłonna i żmudna dla użytkownika

**5.** Procedury użytkownika - dostarczenie przez użytkownika procedur, które są wykonywane przed wejściem do systemu; po zakończeniu danej procedury system wywołuje własna kontrole bezpieczeństwa

Jeśli system **odmówi użytkownikowi dostępu** mogą być wykonane następujące działania: dziennik systemu i zwłoka czasowa przy źle wprowadzonej odpowiedzi, oraz ilość prób wejścia do systemu.

**Fig. 4.** Original text submitted to the SEIPro2S for analysis. Text formatting has no further significance (and can be easily changed in order to camouflage a plagiarism attempt), hence is cleared in early processing steps.

– examined document with marked text fragments, which have been positively matched (see figure 3), together with found sources.

One can compare the report with the original tested text (see figure 4). In this example systems not only provides user with info which elements were copied but also implies that the tested fragment of text was compiled basing on two different Internet resources.

It contrasts the fragments of the original text with copied passages with the found Internet resources.

What is more, system was able to correctly flatten the structure of Internet resources and cope with Polish diacritics.

## 5 Summary

SEIPro2S has proved to be a successful tool for checking whether submitted content is not an unauthorized copy. It is possible to not only find direct copying but also passages that rephrase the copied content with another set of words thus reproducing the original thought.

This work has presented an example of SEIPro2S working as a authenticity checker. This is only one possible application. Among others, one hast to emphasize it as a tool that can monitor whether important information does not leak out of organization. In addition system can be of great help to researchers checking whether their idea was addressed by their peers granting them with better overview of chosen research field.

SEIPro2S still yields interesting research problems. Above else the system is to be reimplemented so as to allow for performance optimizations, new user interface and introduction of a set of additional algorithms that will enable for further automation.

There are also few topics for further research. Certain phrases are typical for some domains (eg. law, science) and therefore their exact occurrence in different documents cannot be perceived as the act of plagiarism, as well as using public domain documents. An investigation how to recognize allowed proper citations, could also be done.

## References


1. Ah-Hwae, T., Fon-Lin, L.: Text Categorization, Supervised Learning, and Domain Knowledge Integration (2004)
2. Baeza-Yates, R., Ribeiro-Neto, B.: Modern Information Retrieval. ACM Press, Addison-Wesley Longman Publishing Co, New York (1999)
3. Ballinger, K.: .NET Web Services, Architecture and Implementation. Pearson Education, London (2003)
4. Baziz, M.: Towards a Semantic Representation of Documents by Ontology-Document Mapping (2004)
5. Baziz, M., Boughanen, M., Aussenac-Gilles, N.: Semantic Networks for a Conceptual Indexing of Documents in IR (2005)

6. Ben-Ari, M.: Principles of Concurrent and Distributed Programming, 2nd edn. Pearson Education (2005)
7. Ceglarek, D.: Zastosowanie sieci semantycznej do disambiguacji pojec w jezyku naturalnym, red. In: Porebska-Miac T., Sroka H. (eds.). w: Systemy wspomagania organizacji SWO 2006 - Katowice: Wydawnictwo Akademii Ekonomicznej (AE) w Katowicach (2006)
8. Frakes, W.B., Baeza-Yates, R.: Information Retrieval - Data Structures and Algorithms. Prentice Hall, Englewood Cliffs (1992)
9. Gonzalo, J., et al.: Indexing with WordNet Synsets can improve Text Retrieval (1998)
10. Hotho, A., Staab, S., Stumme, S.: Explaining Text Clustering Results using Semantic Structures. In: Lavrač, N., Gamberger, D., Todorovski, L., Blockeel, H. (eds.) PKDD 2003. LNCS (LNAI), vol. 2838, pp. 217–228. Springer, Heidelberg (2003)
11. Labuzek, M.: Wykorzystanie metamodelowania do specyfikacji ontologii znaczenia opisow rzeczywistosci, projekt badawczy KBN (2004)
12. Khan, L., McLeod, D., Hovy, E.: Retrieval effectiveness of an ontology-based model for information selection (2004)
13. Krafzig, D., Banke, K., Slama, D.: Enterprise SOA. Prentice Hall, Englewood Cliffs (2005)
14. Krovetz, R., Croft, W.B.: Lexical Ambiguity and Information Retrieval (1992)
15. Sanderson, M.: Word Sense Disambiguation and Information Retrieval (1997)
16. Sanderson, M.: Retrieving with Good Sense (2000)
17. Stokoe, C., Oakes, M.P., Tait, J.: Word Sense Disambiguation in Information Retrieval Revisited. In: SIGIR 2003 (2003)
18. Van Bakel, B.: Modern Classical Document Indexing. A linguistic contribution to knowledge-based IR (1999)
19. Gale, W., Church, K., Yarowsky, D.: A Method for Disambiguating Word Senses in a Large Corpus. Computers and the Humanities 26, 415–439 (1992)

# Semantic Knowledge Representation in Terrorist Threat Analysis for Crisis Management Systems

Mariusz Chmielewski, Andrzej Gałka, Piotr Jarema, Kamil Krasowski, and Artur Kosiński

Computer Science Department, Cybernetics Faculty, Military University of Technology, Kaliskiego 2, 00-908 Warsaw, Poland

**Abstract.** In recent years problem of identifying terrorist threat has become a priority topic for government and military organizations. We base our ideas on new concepts of indirect association analysis to extract useful information for terrorist threat indication. Method introduces original approach to knowledge representation as a set of ontologies and semantic network, which are then processed by the inference algorithms and structure graph analysis. Described models consist of experience gathered from intelligence experts and several open Internet knowledge systems such as Global Terrorism Database, Profiles in Terror knowledge base. We managed to extract core information from several ontologies and fuse them into one domain model aimed to provide basis for indirect associations identification method.

**Keywords:** counter-terrorism, decision support, ontology, GIS, crisis management, GTD.

## 1 Introduction

Emerging terrorist threats have been extremely difficult in analysis. In recent years this problem has become a priority topic for government and military organizations which often need to use sophisticated methods to detect criminal networks and individuals associated with them. Conclusions received through the multi-analysis allow effectively monitor the activities of criminal organizations, help to identify the phases of actions being prepared or in progress. Opposing irregular groups requires the ability to analyze large data sets that can come from unreliable sources or be incomplete. Researches show that the analysis of activities of terrorist organizations is possible, however, requires the use of interdisciplinary knowledge. In recent years rapid development of information technology, have highlighted the need to introduce methods, algorithms and systems providing precision tools for analysts to support the concept of Semantic Web. This idea is associated with, what is being called, the next stage of development of the Internet grouped around the description languages and ontologies [12] with a strong theoretical basis for inference mechanisms [3].

Semantic net gives one of the most important advantage for terrorist activities data representation – scalability and flexibility of knowledge representation. Presented method of semantic network analysis and association acquiring, aims at:

N.T. Nguyen, R. Kowalczyk, and S.-M. Chen (Eds.): ICCCI 2009, LNAI 5796, pp. 460–471, 2009.

- Providing a tool for operating on large information resources,
- Eliminating the unreliable and unwanted information within the semantic network (essential requirement due to algorithm complexity),
- Selecting significant nodes and relations between them (for the analysis) [8],
- Searching the indirect relations between the nodes based on already stored knowledge in the semantic network (building the new knowledge in the system).

Presented tool is based on a method of identification of hidden semantic associations on the web using the ontology description generation methodology, which provides a dedicated approach to modelling and data processing [11]. The main assumption of presented method is that the generation of semantic network based on registered atomic events, provide the data, in which the proposed algorithms will find indirect links between the vertices of the network. This process can be seen as building of new knowledge [5], thus providing specific inference algorithms for semantic networks. Designed tool gives a scalable architecture able to be developed towards a complex networks reasoning by expanding the number of data sources used and dynamic assigning of semantic graph quantitative characteristics.

Analysis of associations, for the purpose of this work, is understood as seeking through semantic network in order to find the links between vertices, providing modifications to its structure along with addition of new links, new vertices (the facts) or the reclassification of vertices or arcs network [7]. The result of this process is new knowledge acquisition in semantic network, which allows to detect information of potential crises. Developed method make use of a rich set of algorithms (filtering and quantity analysis) in order to eliminate unnecessary information while seeking for new associations.

To extend application of the developed method, process of analysing developed semantic model had been additionally extended towards multi-criteria decision-making [9]. Depending on the tasks analyst may request extended answer taking into account the strength of the connection type. One of the main problems that had been managed in this work is the acquisition of data and its transformation into the semantic model. We base our data on unclassified data sources found on the Internet, both in semantic form (models RDF, OWL - http://profilesinterror.mindswap.org/) and pure relational Global Terrorism Database [13] (models of physical databases and text files obtained from http://209.232.239.37/gtd1). One of the stages of constructing the tool was transformation of relational model into a semantic data and its further integration, in order to provide uniform database of terrorist organizations, incidents, and events merged with spatial services.

One of the main requirements defined based on the method and designed tool is the ability for integration with GIS tools supporting the visualization of geographic features. In addition to ontological mechanisms, two separate sub-libraries based on BBN Openmap™ and NASA WorldWind™ were developed. Chosen GIS tools typically operate on the basis of relational or object-oriented data. Implemented modifications made it possible to extend mechanisms towards semantic data representation and querying subsystem using SPARQL.

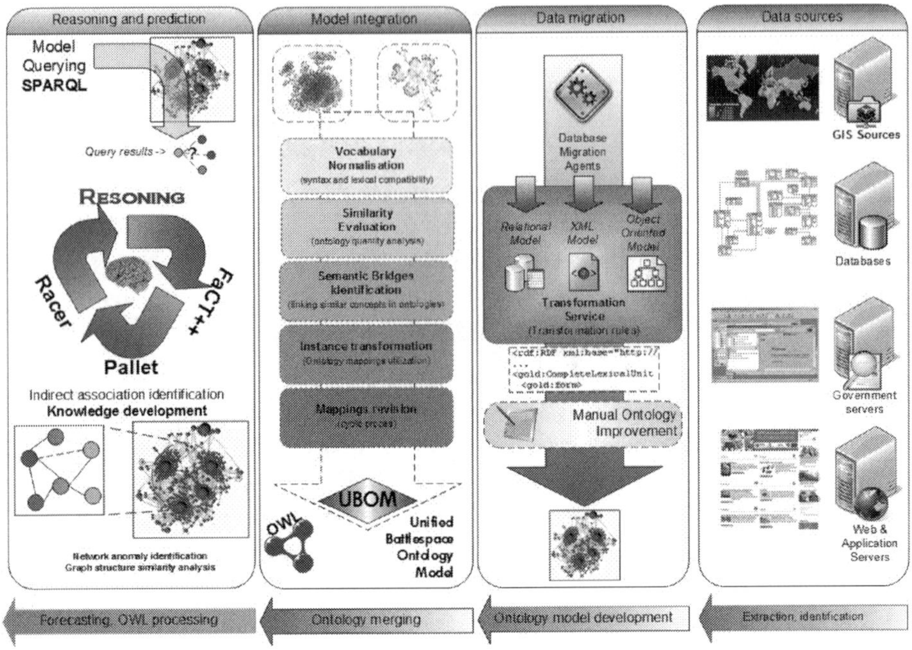


**Fig. 1.** The concept of the collection data in semantic model and its further processing for the analysis of Terrorist Threats

## 2 Spatial Data Subsystem as Important Functionality of Decision Support Systems

Analysis, detection or prevention of terrorist threats is strongly associated with various types of geographic data. Their presentation in the correct form may improve many aspects related to the identification of terrorism. GIS systems may significantly contribute to the identification of risks through [9]:

- accelerating problem-solving and decision-making by presenting the situation on maps
- improving communication - faster and more effective exchange of information
- increasing the effectiveness of learning and training of analysts
- increasing control over the area of operations - visualization of data from different sources at the same time, introduces the phenomenon of synergy in relation to analyzing the data separately
- broadening the horizon of perception of the problem - analyzing the GIS display often allows for a broader look at the problem

To determine the requirements for the terrorism threat identification system we must define what data and dependencies between them can be used. Spatial regularity that may be important in preventing the attacks may have a following backing:

- historical - analysis of historical political geography of the area, ex. non-existing boundaries of administrative divisions
- environment - an analysis of diversity of natural conditions
- cultural heritage - an indication of diversity of cultural regions with their politics
- structural - the level of urbanization, demographic indicators, the level of education of inhabitants, the structure of their employment, and level of income

This information will allow presenting the overall situation and define the likelihood of occurrences of threats in the territories, track tension and aggression between them. Based on these data, it is possible to designate areas potentially at risk and to determine the risk profile. For a full analysis of the problem of terrorism there is a need for more precise data, such that not only will help define areas of danger, but also will help to detect the occurrence of specific attacks - in a particular place and particular time. These should include information about terrorist groups about:

- areas of actions,
- the purpose and type of their distribution,
- method of operation,
- the path of travels,
- used weapons,
- the location of recent activity

An important element is the tracking of communication that occurs between groups of suspects. The geographical representation of the message tracking can be helpful in the analysis of the potential attack site. Increased communication can mean planning the attack, the sudden rupture of the attack that is already close.

To localize locations of probable attacks it is necessary to define the location of strategic sites: the place of mass groupings of people, government institutions, embassies, sports facilities, military units, bridges, airports, stations, etc.

In order to make more detailed proposals of connecting terrorist groups with specific localizations we must define dependencies. For example we may connect different places to be particularly susceptible to attacks of a particular type because of strategic importance for the terrorist group, etc.

## 3 Data Fusion, Analysis and Reasoning

For the purpose of this work, data fusion is defined as integration of data gathered from different sources into one common form, allowing for data management in a uniform manner. The data sources may be not only the new digital maps, satellite imagery but also numerical data as statistical yearbooks and particularly in the case of the identification data for the terrorist attacks all kinds of institutions that can help some way to monitor and detect threats.

Modern standards for developing systems such as SOA (Service Oriented Architecture) head towards the creation of platform-independent interfaces that provide specific services, as for example a standard of exchanging maps such as Web

Map Service (WMS). The problem arises at the stage of the merging data on the target system. One of the most flexible and versatile ways of storing data are semantic models and in consequence ontologies. Such representation allows to create self-describing graph structured data, ready for classifications and reasoning. Modern tools such as the Jena Semantic Web Framework can manage knowledge stored in ontologies not just at level of viewing and filtering, but also allows finding hidden dependencies by reasoners.

An important value of the method is the ability to dynamically accept new information inserted by the analyst. Definition of new spatial elements, shared in the system globally, speed up communication and broadens the horizon of the problem of perception by other users. Since a large problem is verifying completeness of the data provided from external sources system should also allow analysts to verify the incoming data.

## 4 Process of Collecting Facts – Knowledge Base Completion

The easiest way for building a semantic model is provisioning already previously collected data to a model with identical or similar structure to the structure of the data. In practice, however, such situations rarely occur. Models usually have a lot of independent sources. There is no need to do that automatically when a set of data is small, but for large collections of facts, it is almost impossible. Locating thousands of records in the semantic model, requires automation and use of additional software for the analysis of syntactical text. Typically, due to different structures of sets feeding a semantic model, an analysis is performed by which the individual solution is achieved. While designing core ontology authors distinguished main differences between designing ontologies and data models. The key difference is not the language the intended use, but the ability to share the model and its concepts for multiple users and applications. Data models usually live in a relatively small closed world – identified mostly in the system; ontologies are meant for an open, distributed world, systems.

The acquisition of facts about the terrorist incidents, groups and profiles of terrorists for the purpose of ABox is a separate challenge. The study of terrorism engaged in most government institutions, which do not provide data or heavily restrict access to them. Specially constructed for this project tools, the multi-agent system of monitoring RSS feeds, and, found in the network database, containing historical data of terrorist incidents, were used to endow the model. The core of the incidents is a collection of data from GTD and GTD2.

Global Terrorist Database is historical data, set up under the project at the University of Maryland in the U.S. Department of Justice under the supervision of the United States. The set contains a collection of 59 503 incidents of terrorism that took place between 1970 - 1997, with no accompanying data from 1993. Data is based on a database generated by the PGIS (Pinkerton Global Intelligence Services), which contains 67 165 records. Data collection was completed in December 2005. This part of the collection was made public only in 2007 [10]. Due to the nature of the data, the user has access only to a limited number of records. Authors of collections of data made it available to the format of SAS, SPSS, and text format which were the basis for migration mechanisms constructed using XML parsers.

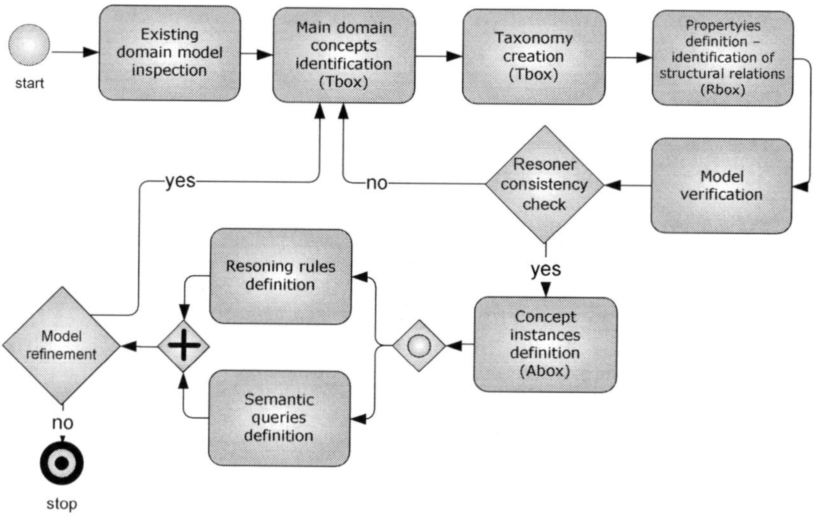


**Fig. 2.** The process of migrating data to the semantic model

GTD2 is continuation of a GTD of the work created at the same university, supplementing incidents registered in the period of 1998 - 2004. Database structure has changed and has been extended with new information. It contains 126 variables of which a large proportion is confidential. GTD and GTD2 provide basic information about the date and place of incident, the terrorist groups involved in the incidents and used weapons. Facts migration to the semantic model from GTD and GTD2 were made in the same way, but the structure of the syntactical analyzer of text had to be revised, because of the differences between sets. The following stages supplying the semantic model:

1. Transforming a set of SAS datasets to XML based format or CSV files;
2. Implementation of appropriate tools built in Java using JDOM package for spatial transformation a intermediate model to the semantic model
3. To manage the semantic model used the open source JENA Semantic Framework library

For the purposes of a semantic model a transitional model in MySQL relational database was created and then loaded into the data posted on the above website. Then the data were loaded, from the database to the semantic model. Until recently, access to the service semantic model was based on the Java Servlet mechanism. Because of small flexibility of presented solution, access to the model is being redesign towards SOA, using WebServices.

## 5 Semantic Model of Terrorist Threat

For the purpouse of this work we define Terrorist Threat Model as:

$$T=\left\langle O_i, I_j, A_k^P, Cr_a, Ev_l \right\rangle \quad (1)$$

where:

$O_i$ - is the i-th terrorist organisation and its description concerning all known leaders, active members, registered events, attacks, statements concerning time, location, media

$I_j$ - is the j-th subject of potential attack (critical infrastructures, people)

$A_k^P$ - is the k-th stored pattern of attack (series of events that are significant for the p-th attack type or one of its phases)

$Cr_a$ - is the a-th crisis situation

$Ev_l$ - l-th vector of significant events that had been or have been registered and are connected or is suspect of the indirect dependency.

A critical infrastructure can an institution, building, facility which is important from the desired point of view connected directly with the domain of attack target.

$$\mathrm{Cr_a} = \left\langle \mathrm{O_i}, \mathrm{I_j}, \mathrm{T_m}, \mathrm{W_a}, state^{\mathrm{p}}, \varphi_{\mathrm{a}} \right\rangle \quad (2)$$

where:

$\mathrm{T_m}$ - is a m-th type of attack, modeled using taxonomy of attacks and the availability of designed attacks scenarios

$\mathrm{W_a}$ - is a collection of weapon types used directly in attack, including conventional and ABC weapons, their classification and their destruction factors

$state^{\mathrm{p}}$ - is the phase of the terrorist act preparation $St = \left\{ state^{\mathrm{p}} \right\}$

$\varphi_{\mathrm{a}}$ - is a vector of quantity factors calculated for the analyzed crisis situation.

Vector consists of two types of factors deterministic and probabilistic defining:

- number of registered suspected events in given time unit,
- probability of given type of attack,
- probability distribution of:

- time to attack,
- human losses,
- critical infrastructure destruction.

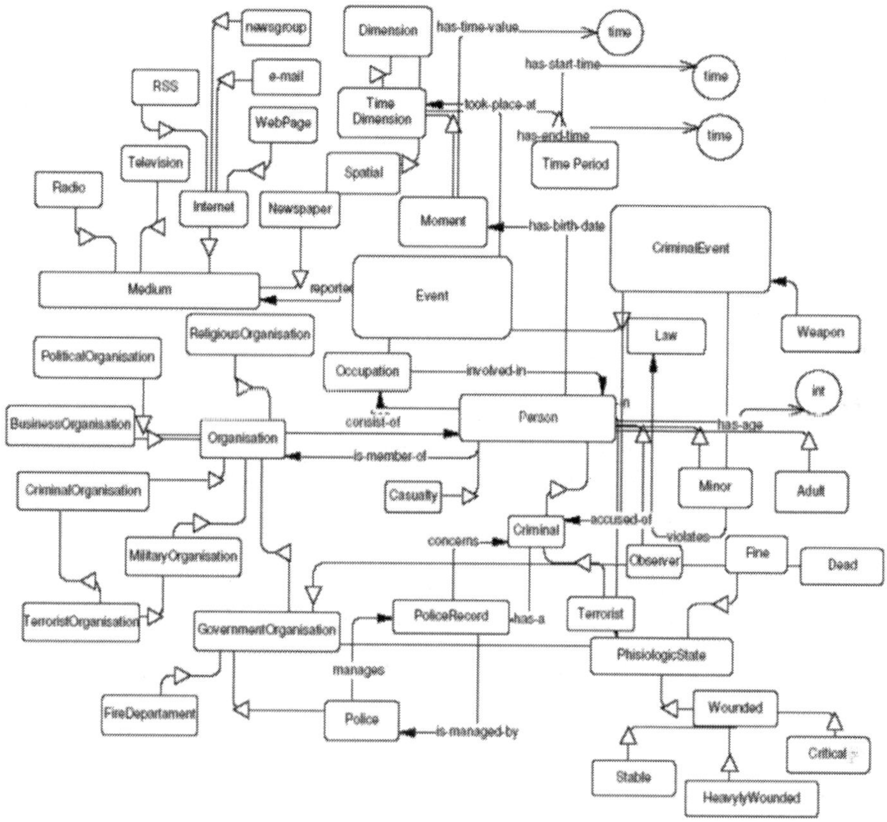


**Fig. 3.** Semantic crisis model presenting core elements using ontology relations

Using model we designed knowledge base of terrorist threat, that additionally had been extended using domain models of:

- Biological weapons (diseases, agents, viruses, bacteria, fungus, and means of their transportation)
- Chemical weapons (chemicals, toxins, venoms, cure, drugs, remedies)
- Conventional warfare (taxonomy, weapons, munitions, explosives, means of deployment)
- Critical infrastructures, urbanised area structures, government and authorities facilities,
- Spatial elements and their characteristics,
- Terrorist organisations and their description including group members,
- Crime taxonomy including criminal registry model

Prepared terminology and model creates a kind of dictionary which aggregates certain data about terrorist organizations, critical infrastructures, attack patterns. Prepared semantic model have been compared in search of missing domain description to available dictionary provided by Taxonomy Warehouse http://www.Taxonomy warehouse.com/

## 6 Semantic Searching – SPARQL Queries

SPARQL Query Language is used to search data in RDF format. RDF (Resource Description Framework) is a metadata specification model, as set out by the W3C, typically implemented in XML. The purpose of RDF is to enable machine processing of abstract descriptions of resources automatically. It can be used both to search for data and tracking information of chosen subject. RDF describes resource by using the expression consisting of three elements: an subject, predicate and object.

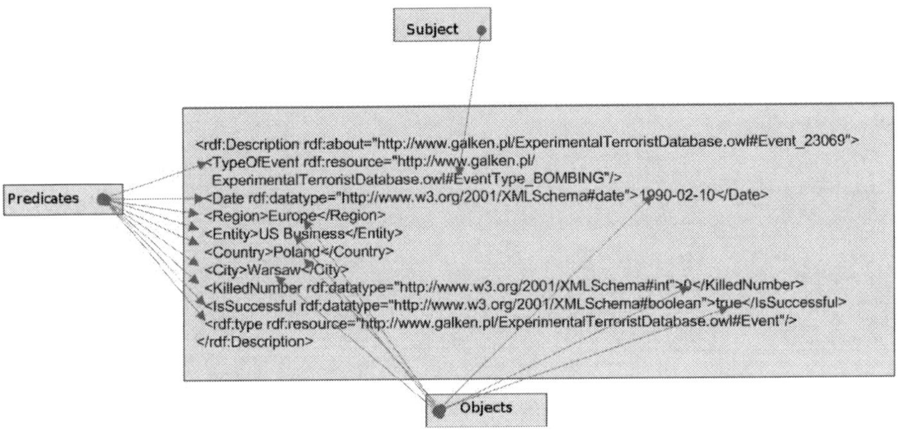


**Fig. 4.** RDF elements in semantic statement presenting the Subject-Predicate-Object triple

The SPARQL query language is executed on the triples collections stored in the RDF format.

Sample SPARQL query:

```
SELECT  ?groupName
WHERE {
    ?y  terrorism:Country  "Poland" .
    ?y  terrorism:Group ?groupName.
}
```

Query compares the three conditions in "where" clause with triples that are contained in the RDF graph. For a given query a predicate and object are given, therefore the pattern will match only values with predicate and object set like in the query. The result will be a collection of subjects of events as defined just after the “SELECT” keyword.

```
SELECT ?suspect
WHERE {
            ?event a terrorism:Event.
            ?event terrorism:Date ?date.
```

```
            ?event terrorism:TypeOfEvent ?eventType.
            ?eventType a terrorism:EventType.

      { ?eventType terrorism:Name "BOMB ATTACK". }
  UNION
      { ?eventType terrorism:Name "ASSASSINATION". }

            ?group terrorism:ResponsibleGroup ?event.
            ?groupMember terrorism:MemberOf ?group.
            ?suspect terrorism:Knows ?groupMember.
            ?suspect a terrorism:Person.
            ?suspect terrorism:Live ?city.
            ?city terrorism:Situated "POLAND".

   FILTER (
            ?date > "2005-01-01"^^xsd:date && ?date <
            "2007-01-01"^^xsd:date
         )
}

  ORDER BY ?city
```

In real conditions more complicated queries are needed. This query shows the syntax of a query with more than one condition and a result of a subgraph. The aim is to find all suspected people living in Poland who had been in contact with terrorist group members involved in bombing and assassination in years 2005 to 2009.

## 7 Acquiring Data for Knowledge Base

Knowledge base of the designed system operates in a distributed environment. The nature of the terrorist incidents and the methods of acquiring information related to them, requires to search suitable sites on the web. These arguments in favour of creating a working system based on agent system - a system consisting of autonomous software components. In contrast to object-oriented projects, in multiagent systems the information may not always be correct and more importantly, may not always reach its target. It is recommended that a single operation should be performed by several agents simultaneously, in order to minimize errors, given that the operations may be performed. Multiagent systems are used in large environments - in computer networks, which can also include the Internet. The information provided in this network may be not fully in line with reality. The devices in the network may be crashed (links, hubs or servers). Agent systems don't need to be precisely synchronized, which is characterized by high tolerance to the delay of communication. Using visualization tool, the analyst can easily determine the location of the selected class instance in the hierarchy. "America" is the subclass of World. It is also higher in the hierarchy from North_America and South_America. Visualization based on interactive graph provides invaluable assistance to the analyst at the time of designing ontologies. Allows him to check the current validity of the defined compounds and present a hierarchy of dependencies between classes. This allows for the most precise representation in the model chosen part of reality. It also provides assistance to the users. User, which is not familiar with the analyst's concepts, looking at the graph representation of the knowledge base, is able to read information about the structure and hierarchy of classes in the model.

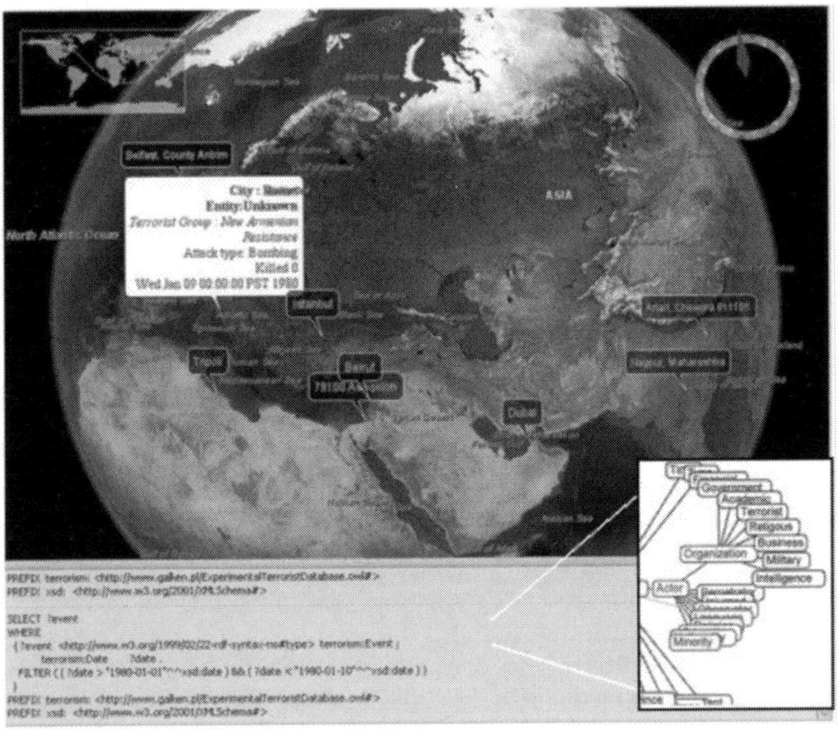

**Fig. 5.** Semantic data visualisation using developed environment for crisis management tools presenting gathered data from GTD and composed knowledge base

# 8 Conclusions

Presented approach for modelling and terrorist threat assessment has been applied in works of NATO Modelling and Simulation Group along with decision support module in Crisis Management System for Warsaw Agglomeration. At this moment we have managed to develop, reliable means of data migration and its instantiation in form of knowledge base, extending available methods, dedicated for ontology reasoning. Conducted research allowed us to verify available mechanism for semantic inferencing and their application in domain of crisis management systems and heterogeneous data sources integration. So far developed method and its implementation has proven that this field of research propose interesting results for huge knowledge bases, especially in form of indirect association analysis which can be further extended towards complex and Bayesian network algorithms indicating quantity characteristics for terrorist threat. Many of presented ideas have already been applied in real world applications, used for demonstration of available decision support methods developed at Cybernetics Faculty in the domain of asymmetric threat management systems. Further research is aimed at extending ontology models in domains of weapons, critical infrastructures and their sensitivity based on the destruction factor and specific environment description.

## References

1. Roman, D., Keller, U., Lausen, H., Bruijn, J., Lara, R., Stollberg, M., Polleres, A., Feier, C., Bussler, C., Fensel, D.: Web Service Modeling Ontology. Applied Ontology (2005)
2. Mohammadian, M.: Intelligent Agents for Data Mining and Information Retrieval. Idea Group Publishing (2004)
3. Davies, J., Fensel, D., Harmelen, F.: Towards the Semantic Web: Ontology-driven Knowledge Management, HPL-2003-173. John Wiley & Sons, Ltd., Chichester (2003)
4. Herrmann, M., Dalferth, O., Aslam, M.A.: Applying Semantics (WSDL, WSDL-S, OWL) in Service Oriented Architectures (SOA). In: 10th Intl. Protégé Conference (2007)
5. Currie, E., Parmelee, M.: Toward s Knowledge-Based Solution for Information Discovery in Complex and Dynamic Domains. In: 7th Intl. Protégé Conference (2004)
6. Krebs, V.: Mapping Networks of Terrorist Cells. Connections 24(3), 43–52 (2005)
7. Mannes, A., Golbeck, J.: Building a Terrorism Ontology. University of Maryland, College Park
8. Barthelemy, M., Chow, E., Eliassi-Rad, T.: Knowledge Representation Issues in Semantic Graphs for Relationship Detection, UCRL-CONF-209845
9. Najgebauer, A.: Systemy wspomagania decyzji w sytuacjach konfliktowych. Modele, metody i środowiska symulacji interaktywnej, Rozprawa habilitacyjna – dodatek do Biuletynu WAT, Warszawa (1999) ISBN 83-908620-6-9
10. Najgebauer, A., et al.: M&S Tool for the Early Warning Identification of Terrorist Activities, Raport finalny w projekcie NATO MSG – 026 – TG – 19 (2007)
11. Najgebauer, A., Antkiewicz, R., Chmielewski, M., Kasprzyk, R.: The prediction of terrorist threat on the basis of semantic association acquisition and complex network evolution. Journal of Telecommunications and Information Technology (February 2008) ISSN 1509-4553
12. http://protege.stanford.edu/publications/ontology_development/ontology101-noy-mcguinness.html
13. National Consortium for the Study of Terrorism and Responses to Terrorism, Global Terrorism Database, http://www.start.umd.edu/start/

# Consensus Choice for Reconciling Social Collaborations on Semantic Wikis

Jason J. Jung[1] and Ngoc Thanh Nguyen[2]

[1] Knowledge Engineering Laboratory
Department of Computer Engineering
Yeungnam University, Korea
j2jung@{intelligent.pe.kr,gmail.com}
[2] Institute of Informatics
Wroclaw University of Technology, Poland
Ngoc-Thanh.Nguyen@pwr.wroc.pl

**Abstract.** Semantic wikis have been regarded as a collaborative knowledge management system which can provide an efficient framework to foster social interactions and collaborations between online people synchronously. However, as such semantic wiki systems allow users to exploit their own semantics for describing their knowledge, there are sometime conflicts between knowledge (or information) published by them. Thereby, the goal of this work is *i*) to automatically detect such conflicts by monitoring the user semantics and *ii*) to reasonably determine consensus choice converged by analyzing social collaborations. In this paper, we want to note major patterns of knowledge dynamics and conflicts through the social interactions on semantic wikis. The consensus choice is effectively selected to be recommended for better understandability about the knowledge conflicts.

**Keywords:** Semantic wiki, Conflict resolution, Consensus theory.

## 1 Introduction

Collective intelligence needs to consider various knowledge processes (e.g., knowledge creation, integration, and so on) in a collaborative manner. Recently, many studies have been focusing on Web 2.0 applications like blogs and wikis to make the collective intelligence implemented [1–3]. These social information spaces based on such Web 2.0 applications can provide an efficient platform to publish many types of knowledge and take various social activities with other users to generate "better" knowledge [4, 5].

However, as knowledge in diverse domains has been "collaboratively and simultaneously" published on the social information space, it may be *inconsistent* and *conflicted* with each other. This problem can be caused by various factors, e.g., simple mistakes and misunderstandings as well as different background knowledge and opinions. Moreover, the problem makes it more difficult to conduct various knowledge processes, e.g., generating knowledge, integrating knowledge, and so on.

N.T. Nguyen, R. Kowalczyk, and S.-M. Chen (Eds.): ICCCI 2009, LNAI 5796, pp. 472–480, 2009.

In this paper, we want to investigate a novel framework to detect inconsistencies and conflicts between knowledge. Particularly, we are focusing on semantic wiki systems, which is a wiki that has an underlying model of the knowledge described in its wiki pages. While regular wikis have simply structured texts and untyped hyperlinks (such as the links in the hypertext documents), the semantic wikis allow the ability to capture or identify further information (e.g., metadata) about resources, webpages and their relations [6]. Thereby, we classify knowledge-based activities within a semantic wiki into four types of knowledge dynamics, as follows;

- Knowledge abstraction (conflicts between generalization and specialization),
- Knowledge refinement (conflicts between accurate and inaccurate),
- Knowledge integration (conflicts between mapping and dividing), and
- Knowledge population (conflicts between instantiation and removal).

As another important issue, by the nature of wikis, the information can be very easily propagated on with any wiki pages. Hence, it is important to employ a certain notification system to inform human experts (or administrators) of the conflicted knowledge as quickly as possible [7]. Instead of asking them to determine which knowledge is better (or more righteous), consensus choices of the "better" knowledge are selected to them.

The outline of this paper is as follows. In the following Sect. 2, we explain the background of consensus theory and how consensus can deal with the conflicts. Sect. 3 addresses a set of knowledge conflicts caused by major patterns of knowledge dynamics through the social interactions on semantic wikis. Sect. 4 explains how to build a consensus decision for a given set of conflicted knowledge. Finally, in Sect. 5, we draw a preliminary conclusion and show our plans in the future.

## 2 The Roles of Consensus in Solving Conflicts

Consensus has usually been understood as a general agreement in situations where some bodies have not been agreed on some matter. What then functions consensus should fulfill in solving conflicts in distributed environments and collaborative workspaces? Before the analysis we should consider what is represented by the conflict content which consists of a number of opinions of the conflict participations. Let's assume that the opinions included in the conflict content represent unknown solution of some problem. The following two cases may take place [8]:

1. This solution is *independent* from the opinions of the conflict participants.
2. This solution is *dependent* on the opinions of the conflict participants.

In the first case the independence means that the proper solution of the problem exists but it is not known for the conflict participants. The reasons of this phenomenon may follow from many aspects, among others, the ignorance of the conflict participations or the random characteristics of the solution which may make the solution impossible to be calculated in a deterministic way. Thus the content of the solution is independent from the conflict content and the conflict participations for some interest have to "guess" it. In this case their solutions have to reflect the proper solution but it is not known if in a valid and complete way.

In the second case this is the opinions of conflict participants which decide about the solution. As the example consider votes at an election. The result of the election is determined only on the basis of these votes.

In both cases there is a need to determine a solution of the problem on the basis of given opinions. This solution should satisfy the following conditions:

- It should best reflect the given opinions, and
- It should possibly to the same degree reflect the opinions given by the conflict participants.

The first condition is rather more suitable to the first case described above because the versions given by the conflict participations reflect the "hidden" and independent solution but it is not known to what degree.

The second condition refers to the second case in which the proper problem solution is dependent on the opinions of the conflict participants. Thus consensus should not only best represent the opinions but also should reflect them in the same degree (with the assumption that each of them is treated in the same way). It should be "acceptable compromise" what means that any of opinions should neither be "harmed" nor "favored". It has been proved that these conditions in general may not be satisfied simultaneously. It has been shown that the choice according to the criterion of minimization of the sum of squared distances between consensus and the profile' elements determines a consensus more uniform than the consensus chosen by minimization of the sum of distances. Therefore, the criterion of the minimal sum of squared distances is also very important. However, the squared distances' minimal sum criterion often generates computationally complex problems (NP-hard problems), which requires working out heuristic algorithms [9]. Fig. 1 below presents the scheme of using consensus methods in the above mentioned cases.

In works [8–11] a methodology for consensus choice and its applications in solving conflicts in distributed systems is presented. It could be partitioned into two parts. In the first part general consensus methods which may effectively serve to solving multi-value conflicts are worked out. For this aim a consensus system, which enables describing multi-value and multi-attribute conflicts, has been defined and analyzed (it is assumed that the attributes of this system are multi-value). Next the structures of tuples representing the contents of conflicts have been defined as distance functions between these tuples. Finally, the consensus and the postulates for its choice have been defined and analyzed. For defined structures algorithms for consensus determination have been worked out. Besides the problems connected with the susceptibility to consensus and the possibility of consensus modification, have also been investigated.

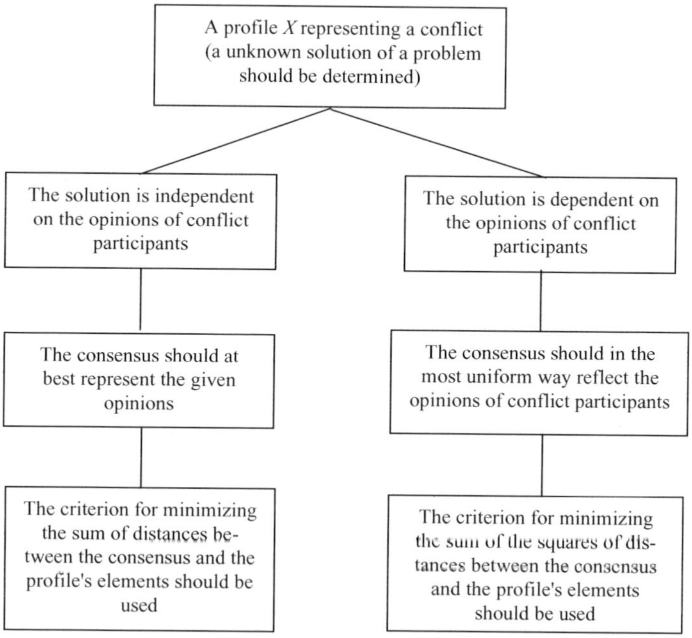


**Fig. 1.** The scheme for using consensus methods

The second part concerns varied applications of consensus methods in solving of different kinds of conflicts which often occur in distributed systems. The following conflict solutions are presented: reconciling inconsistent temporal data; solving conflicts of the states of agents knowledge about the same real world; determining the representation of expert information; creating a uniform version of a faulty situation in a distributed system; resolving the consistency of replicated data and determining optimal interface for user interaction in universal access systems. An additional element of these works is the description of multiagent systems AGWI aiding information retrieval and reconciling in the Web, for which implementation the platform IBM Aglets is used.

## 3 Conflict Profiling between Knowledge on Semantic Wiki

There can be several kinds of social activities, which are collaborative editing, on semantic wikis. During accessing any wiki pages for collaborative editing on semantic wiki, any wiki users can *i*) generate new wiki pages, *ii*) delete the existing wiki pages, and *iii*) modify the existing wiki pages. Through the social activities, we can find out the following knowledge dynamics on semantic wikis, as shown in Table 1.

**Table 1.** Knowledge dynamics on semantic wikis

| Semantics | Knowledge dynamics | Description & Example |
|---|---|---|
| Knowledge Abstraction $\mathcal{A}$ | Generalization | Find a super concept, e.g., "Hominidae is a super concept of Human." |
| | Specialization | Find a subconcept, e.g., "Chimpanzee is a subconcept of Hominidae." |
| Knowledge Refinement $\mathcal{R}$ | More accurate | Modify a fact more correctly, e.g., "evidence indicates that modern humans originated in Africa about 200,000 years ago" |
| | More inaccurate | Modify a fact more incorrectly, e.g., "evidence indicates that modern humans originated in Asia about 2,000 years ago" |
| Knowledge Integration $\mathcal{I}$ | Mapping | Find semantic correspondences between knowledge, e.g., "Chimpanzee is same with Human." |
| | Dividing | Remove semantic correspondences between knowledge, e.g., "Chimpanzee is not same with Human." |
| Knowledge Population $\mathcal{P}$ | Instantiation | Append more instances which are relevant a concept, e.g., "Tiburon is a compact coupe produced by Hyundai." |
| | Removal | Discard more instances which are irrelevant a concept, e.g., "Tiburon is a compact coupe produced by Honda." |

Thus, we have to think of what kind of conflicts can happen among knowledge published on semantic wikis. Four possible cases (i.e., $\mathcal{A}$, $\mathcal{R}$, $\mathcal{I}$, and $\mathcal{P}$) of knowledge inconsistency by social activities on semantic wikis can be considered.

To do so, we want to formulate the knowledge and personal knowledge space on semantic wiki platform.

**Definition 1 (Knowledge).** *Each knowledge t on semantic wiki is represented as*

$$t = \langle k, R, k' \rangle \tag{1}$$

*where k and k′ are ontological entities in the ontologies, and R is a semantic relationship between k and k′. We can easily understand that it is similar to a RDF triple.*

**Definition 2 (Personal knowledge space).** *As i-th user $u_i$ take wiki actions, he can have his own personal knowledge space $\mathcal{T}_i$ with a set of RDF triples.*

### 3.1 Knowledge Abstraction

First case is "knowledge abstraction" ($\mathcal{A}$). As a simple example of Fig. 2, while a wiki user $A$ publishes new knowledge $\langle k, Superclass, k' \rangle$, user $C$ asserts another knowledge $\langle k, Subclass, k'' \rangle$. (The knowledge is simply represented as a

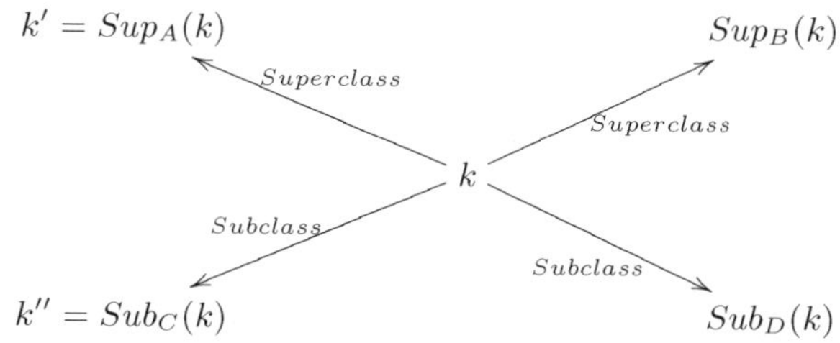


**Fig. 2.** A knowledge conflict on knowledge abstraction

set of RDF triples. Of cause, knowledge representation depends on the semantic wiki systems. We want to skip to discuss this issue in detail.) If $k' = k''$, two knowledge activities are conflicted with each other. Although this example seems rather simple, as the amount of knowledge is getting increased, it is very complex problem. Somehow, we need to exploit ontology reasoners (e.g., Pellet[1] and FacT++[2]) to detect them.

### 3.2 Knowledge Refinement

Second case of knowledge activities that we can be modeled within a semantic wiki is knowledge refinement $\mathcal{R}$. A value $v$ of a certain property $P_k$ of knowledge $k$ can be revised to new value $v'$.

$$\langle k, P_k, v \rangle \longrightarrow \langle k, P_k, v' \rangle \quad (2)$$

This new value becomes either more precise way or more imprecise way by different users' opinions. For example, assume that $k$ and $P_k$ is "World War II" and "StartingDate," respectively, as shown in Fig. 3. Two users $A$ and $B$ can revise $v = 1900$ at time $(t)$ to $v' = 1935$ and $v'' = 1900$ at $(t + 1)$, respectively. In fact, the true value is 1939, so that user $A$'s action is make the knowledge more precise, but $B$'s action does not.

In the similar context, the rest of knowledge dynamics can be considered, but skipped in this paper.

## 4 Consensus Building on Semantic Wikis

Knowledge stored and "revised" within a semantic wiki is conflicted with others over time. Here, we want to introduce a novel approach to select a consensus choice as regarding with a global knowledge structure of semantic wiki. The global structure is acquired by two main factors; $i$) majority voting, and $ii$) semantic relevance.

[1] Pellet, http://pellet.owldl.com/

[2] FacT++, http://owl.man.ac.uk/factplusplus/

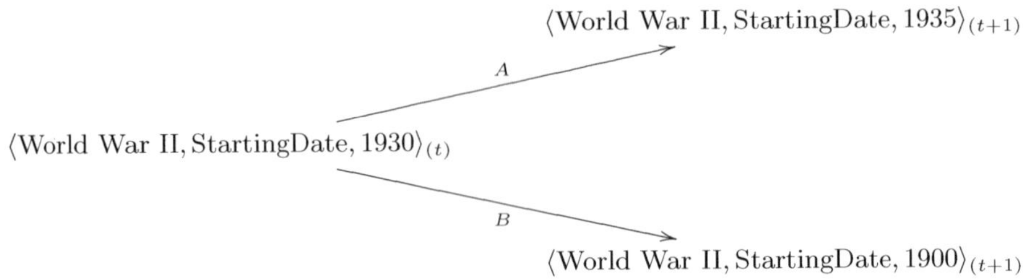


**Fig. 3.** A knowledge conflict on knowledge refinement

### 4.1 Majority Voting

Basic assumption of majority voting is that a decision selected by more individuals in common should be a consensus decision representing the corresponding group [12, 13]. This is still being employed in democratic decision making, e.g., election.

In this context, the voting action can be replaced with social activities (in Table 1) for a certain knowledge dynamics in a semantic wiki. It means that most common social activities resulting in similar knowledge dynamics should be regarded as a major activities for determining the knowledge. For simplicity, we want to mention two main counts of activities, as follows;

1. The number $n_W$ of wiki pages for an identical knowledge dynamics. For example, from the conflicted knowledge, e.g., $\langle k, P, k' \rangle$, $\langle k, P, k'' \rangle$, and $\langle k, P, k''' \rangle$, we can choose $\langle k, P, k' \rangle$ as a consensus choice, when the number of wiki pages for the knowledge is largest, as shown in Equ. 3.

$$n_W(W_{\langle k,P,k' \rangle}) > n_W(W_{\langle k,P,k'' \rangle}) > n_W(W_{\langle k,P,k''' \rangle}) \tag{3}$$

2. The number of wiki users who take actions on the wiki pages for an identical knowledge dynamics. Because the wiki users can publish more than one wiki page, we want to count the number of wiki users.

There are several problems on this majority voting method. One of the problems is that all activities taken by wiki users can make same and identical influence. Thereby, we have to consider weighed voting method by taking into account more additional factors.

### 4.2 Semantic Relevance

To weigh the social activities for knowledge dynamics, we want to measure semantic relevance among the knowledge published on semantic wiki. This issue is important to conduct automated consensus building process. We find out two main information should be exploited to adjust and weigh the number of social activities;

– semantic closeness, and
– expertise of wiki users.

For example, in Equ. 3, assume that knowledge $k''$ is equivalent to $k'''$ with respect to semantic closeness. We can realize that the $k'' \equiv k'''$ should be selected as a consensus decision.

$$n_W(W_{\langle k,P,k''\rangle}) + n_W(W_{\langle k,P,k'''\rangle}) > n_W(W_{\langle k,P,k'\rangle}) \tag{4}$$

This issue is related to some work on ontology mapping methodologies.

## 5 Concluding Remarks and Future Work

Semantic Wiki have been regarded as an important Web 2.0 application for implementing distributed knowledge management systems. In conclusion, we have introduced a new issue to deal with knowledge inconsistency on semantic wikis.

One important limitation of this work is that the consensus choice recommended by the proposed method is only based on the knowledge within an isolated semantic wiki. We have to consider an approach to integrate semantic wiki platforms. There are several issues that you have to take into account in near future, as follows;

- Stupidity of crowd
- Subjective opinions
- Knowledge can change over time.
- Knowledge entailment

As a testing bed, we want to design an semantic wiki authoring system that automatically annotate terms user enters online by referring to background ontology and consensus choices. We will be able to argue that more informative data can be obtained in the case of online annotation, by asking users to give more information. By using that information, the system suggests appropriate words user needs to enter, thereby reducing a change that rubbish knowledge are extracted. Furthermore, with an ontology aligner, semantic inconsistency occurred by collaborative editing is allowed can be more reduced.

We are also planning to implement the system and experiment on it to evaluate whether it operates according to expectation. At the same time, more functionalities are expected to be available on the system as follows. Being equipped with more sophisticated natural language techniques, user is allowed to enter more complex natural language sentence in convenience.

## Acknowledgment

This work was supported by the Korea Science and Engineering Foundation (KOSEF) grant funded by the Korea government (MEST) (No. 2009-0066751).

## References


1. Cayzer, S.: Semantic blogging for bibliographies - lessons learnt. Deliverable 12.1.4, SWAD-Europe (2001)
2. Schaffert, S., Bry, F., Baumeister, J., Kiesel, M.: Semantic wikis. IEEE Software 25(4), 8–11 (2008)
3. Souzis, A.: Building a semantic wiki. IEEE Intelligent Systems 20(5), 87–91 (2005)
4. Goodfellow, T., Graham, S.: The blog as a high-impact institutional communication tool. The Electronic Library 25(4), 395–400 (2007)
5. Jung, J.J.: Social grid platform for collaborative online learning on blogosphere: a case study of eLearning@BlogGrid. Expert Systems with Applications 36(2), 2177–2186 (2009)
6. Noy, N.F., Chugh, A., Alani, H.: The ckc challenge: Exploring tools for collaborative knowledge construction. IEEE Intelligent Systems 23(1), 64–68 (2008)
7. Jung, J.J.: Trustworthy knowledge diffusion model based on risk discovery on peer-to-peer networks. Expert Systems with Applications 36(3), 7123–7128 (2009)
8. Nguyen, N.T., Sobecki, J.: Consensus versus conicts - methodology and applications. In: Wang, G., Liu, Q., Yao, Y., Skowron, A. (eds.) RSFDGrC 2003. LNCS (LNAI), vol. 2639, pp. 565–572. Springer, Heidelberg (2003)
9. Danilowicz, C., Nguyen, N.T.: Consensus methods for solving inconsistency of replicated data in distributed systems. Distributed and Parallel Databases 14(1), 53–69 (2003)
10. Nguyen, N.T.: Using distance functions to solve representation choice problems. Fundamenta Informaticae 48(4), 295–314 (2001)
11. Nguyen, N.T.: Consensus system for solving conicts in distributed systems. Information Sciences 147(1-4), 91–122 (2002)
12. Schapire, R., Freud, Y., Bartlett, P., Lee, W.: Boosting the margin: a new explanation for the effectiveness of voting methods. The Annals of Statistics 26(5), 1651–1686 (1998)
13. Saaty, T.L., Shang, J.S.: Group decision-making: Head-count versus intensity of preference. Socio-Economic Planning Sciences 41(1), 22–37 (2007)

# Ontology Mapping Composition for Query Transformation in Distributed Environment

Jason J. Jung

Knowledge Engineering Laboratory
Department of Computer Engineering
Yeungnam University, Korea
j2jung@{gmail.com,intelligent.pe.kr}

**Abstract.** Semantic heterogeneity should be overcome to support automated information sharing process between information systems in ontology-based distributed environments. To do so, traditional approaches have been based on explicit mapping between ontologies from human experts of the domain. However, the manual tasks are very expensive, so that it is difficult to obtain ontology mappings between all possible pairs of information systems. Thereby, in this paper, we propose a system to make the existing mapping information sharable and exchangeable. It means that the proposed system can collect the existing mapping information and aggregate them. Consequently, we can estimate the ontology mappings in an indirect manner. In particular, this paper focuses on query propagation on the distributed networks. Once we have the indirect mapping between systems, the queries can be efficiently transformed to automatically exchange knowledge between heterogeneous information systems.

**Keywords:** Query transformation; Ontology mapping; Mapping composition.

## 1 Introduction

Information systems in various domains e.g., e-learning [6], telecommunication [5], have been trying to build their own *domain ontologies* for efficiently managing local resources and information. For many purposes, such systems have been interlinked with each other. On this distributed environment, the information systems have to be able to automatically interact with each other. In particular, we are focusing on ontology-based information systems where such interactions can be based on ontology mapping process.

The local ontologies however are generally constructed by domain experts and local database schema. The semantic of each information systems is unique. In [4], each information system tends to include the information which is

- related to the specific and unique topics,
- represented as the consistent linguistic terminologies,

N.T. Nguyen, R. Kowalczyk, and S.-M. Chen (Eds.): ICCCI 2009, LNAI 5796, pp. 481–491, 2009.

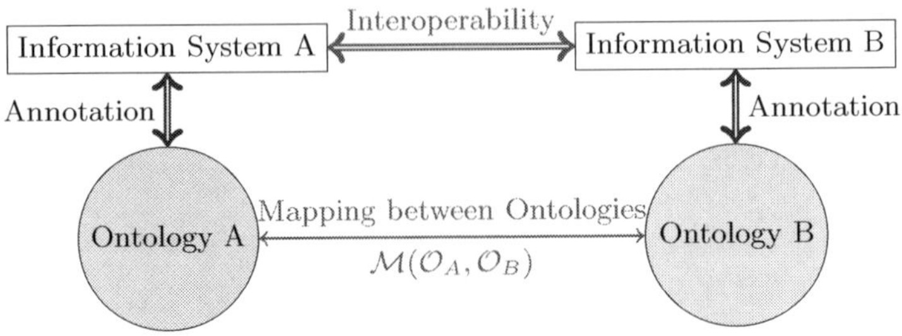


**Fig. 1.** Automatic semantic interoperability between heterogeneous information systems A and B

- organized by local database schema, and
- annotated with local metadata.

It means, in summary, the resources in an information system is semantically encoded by the corresponding local ontologies.

Because of such semantic heterogeneity problems between those local ontologies, it is difficult for the information systems on distributed environment to be semantically interoperable with others. In other words, the systems are supposed to be automatically interact with each other by sharing their resources and knowledge.

In order to solve these problems, the most well-known solution is to find out mapping information between ontologies. As shown Fig. 1, given two ontologies $\mathcal{O}_A$ and $\mathcal{O}_B$, ontology mapping result $\mathcal{M}(\mathcal{O}_A, \mathcal{O}_B)$ is a prerequisite condition for for the interoperability between systems $A$ and $B$.

More importantly, we have already realized that manual ontology mapping by human users are expensive and high cost, because of the lack of domain expertise as well as the complex internal structure of the ontologies (e.g., a large number of concepts and properties). Thus, a number of ontology mapping algorithms for automatically discovering correspondences between ontologies have been proposed [2].

However, most of the mapping algorithms have a scalability problem. While from given two source ontologies they can obtain explicit and direct mapping results, from an increasing number of ontologies on a general distributed environment they have some serious difficulties on scalability.

To deal with this problem, we want to compose an indirect mapping by reusing the mapping information which are already done before. It means that given two ontologies $\mathcal{O}_A$ and $\mathcal{O}_B$, we can make two information systems $\mathcal{S}_A$ and $\mathcal{S}_B$ interoperable by composing two existing mappings $\mathcal{M}(\mathcal{O}_A, \mathcal{O}_C)$ and $\mathcal{M}(\mathcal{O}_C, \mathcal{O}_B)$, instead of finding out direct mapping $\mathcal{M}(\mathcal{O}_A, \mathcal{O}_B)$.

Thereby, In this paper, the ontologies and mapping information should be formalized with some more definitions for the mapping composition. A novel measurement for ontology mapping performance will be introduced. Especially, to evaluate the performance of sharing and composing mapping results, a multi-agent platform has been employed. Any two heterogeneous agents on the multi-agent platform can communicate with each other by query-answering process, so that we can measure how precisely the proposed mapping composition is conducted.

The outline of this paper is as follows. In the following Sect. 2, major components for building ontology-based information systems will be proposed and similarity-based ontology mapping algorithm will be investigated. Sect. 3 will show how to share and compose the existing mapping information. Sect. 4 will exhibit experimental results collected by evaluation. Finally, in Sect. 5 will draw a conclusion of this work.

## 2 Ontology-Based Information Systems

Ontology-based information systems are supposed to be machine processible. In this study, the system is mainly composed of two parts; *i*) ontology $\mathcal{O}$, and *ii*) a mapping set with neighbors.

**Definition 1 (Ontology).** *An ontology $\mathcal{O}$ is represented as*

$$\mathcal{O} := (\mathcal{C}, \mathcal{R}, \mathcal{E}_{\mathcal{R}}, \mathcal{I}_{\mathcal{C}}) \tag{1}$$

*where $\mathcal{C}$ and $\mathcal{R}$ are a set of classes (or concepts), a set of relations (e.g., equivalence, subsumption, disjunction, etc), respectively. $\mathcal{E}_{\mathcal{R}} \subseteq \mathcal{C} \times \mathcal{C}$ is a set of relationships between classes, represented as a set of triples $\{\langle c_i, r, c_j \rangle | c_i, c_j \in \mathcal{C}, r \in \mathcal{R}\}$. $\mathcal{I}_{\mathcal{C}}$ is a power set of instance sets of a class $c_i \in \mathcal{C}$.*

These ontologies are grounded with a set of instances. In terms of description logic, $\mathcal{I}_{\mathcal{C}}$ can be replaced with A-Box.

By any mapping algorithm, each mapping result can be represented as a set of correspondences between ontology entities with confidence value.

**Definition 2 (Correspondences).** *Given two ontologies $\mathcal{O}$ and $\mathcal{O}'$, a set of correspondences are given by*

$$\mathcal{M}(\mathcal{O}, \mathcal{O}') = \{\langle e, e', r_M, CF \rangle | e \in \mathcal{O}, e' \in \mathcal{O}', r_M, CF \in [0, 1]\} \tag{2}$$

*where $e$ and $e'$ are a pair of matched entities. Mapping relationship is $r_M = \{\equiv, \sqsubseteq, \sqsupseteq, \perp\}$. $CF$ indicates a confidence value of the pair.*

**Definition 3 (Confidence).** *Confidence value means the precision between the ontology entities. It can be computed in several different ways. For example, a confidence value can be measured by a string matching function Dist*

$$CF_{\langle e, e' \rangle} = \frac{1 - Dist(L(e), L(e'))}{\max(L(e), L(e'))} \tag{3}$$

*where $L$ is a function for returning a label of ontology entity.*

In this paper, we want to choose similarity-based ontology mapping approach [3]. It defines similarities (e.g., $Sim_C$, $Sim_R$, $Sim_A$) between classes, relationships, attributes, and instances. It is based on the principle that the more features of two entities are similar, the more these entities are similar. Given a pair of classes from two different ontologies, the similarity measure $Sim_C$ is assigned in $[0, 1]$. The similarity ($Sim_C$) between $c$ and $c'$ is defined as

$$Sim_C(c, c') = \sum_{E \in \mathcal{N}(C)} \pi_E^C MSim_Y(E(c), E(c')) \tag{4}$$

where $\mathcal{N}(C) \subseteq \{E^1 \dots E^n\}$ is the set of all relationships in which classes participate (for instance, subclass, instances, or attributes). The weights $\pi_E^C$ are normalized (i.e., $\sum_{E \in \mathcal{N}(C)} \pi_E^C = 1$).

If we restrict ourselves to class labels ($L$) and three relationships in $\mathcal{N}(C)$, which are the superclass ($E^{sup}$), the subclass ($E^{sub}$) and the sibling class ($E^{sib}$), Equ. 4 is rewritten as:

$$\begin{aligned} Sim_C(c, c') = &\ \pi_L^C sim_L(L(A_i), LF(B_j)) \\ &+ \pi_{sup}^C MSim_C(E^{sup}(c), E^{sup}(c')) \\ &+ \pi_{sub}^C MSim_C(E^{sub}(c), E^{sub}(c')) \\ &+ \pi_{sib}^C MSim_C(E^{sib}(c), E^{sib}(c')). \end{aligned} \tag{5}$$

where the set functions $MSim_C$ compute the similarity of two entity collections.

As a matter of fact, a distance between two sets of classes can be established by finding a maximal matching maximizing the summed similarity between the classes:

$$MSim_C(S, S') = \frac{\max(\sum_{\langle c, c' \rangle \in Pairing(S,S')} (Sim_C(c, c'))}{\max(|S|, |S'|)}, \tag{6}$$

in which $Pairing$ provides a matching of the two set of classes. Methods like the Hungarian method allow to find directly the pairing which maximizes similarity. The OLA algorithm is an iterative algorithm that compute this similarity [3]. This measure is normalized because if $Sim_C$ is normalized, the divisor is always greater or equal to the dividend.

A normalized similarity measure can be turned into a distance measure by taking its complement to 1 ($E_C^{dist}(x, y) = 1 - Sim_C(x, y)$). Such a distance introduces a new relation $E_C^{dist}$ in the concept network $\mathcal{C}$.

As a simple example in Fig. 2, once two ontologies $\mathcal{O}_A$ and $\mathcal{O}_B$ are mapped, we can obtain the mapping results (indicated as blue arrows) in Table 1).

At last, given a number of ontology-based systems, we can formulate a distributed ontology-based information system, as follows.

**Definition 4 (Distributed ontology-based information system).** *A distributed ontology-based information system G consists of* $N_G$ *number of ontology-based information systems* $\{S_1, \dots, S_{N_G}\}$. *Some of the information systems are*

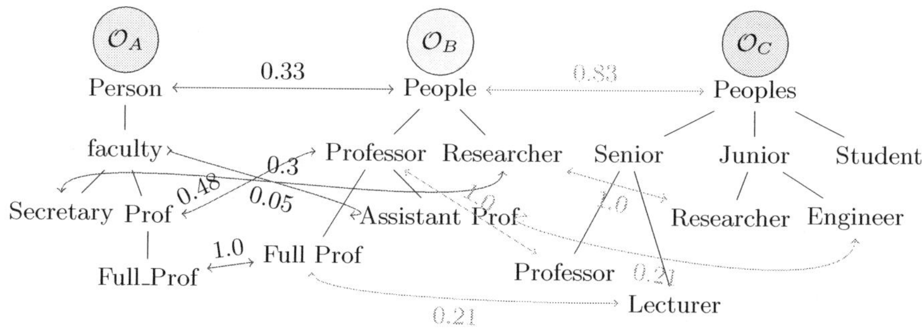


**Fig. 2.** An example of similarity-based ontology mapping and three ontology-based information systems on distributed environment

**Table 1.** Mapping results between ontologies $\mathcal{O}_A$ and $\mathcal{O}_B$

| $\langle e$ | $e'$ | $R$ | $CF\rangle$ |
|---|---|---|---|
| Person | People | $\equiv$ | 0.33 |
| faculty | Assistant Prof | $\equiv$ | 0.05 |
| Secretary | Researcher | $\equiv$ | 0.3 |
| Prof | Professor | $\equiv$ | 0.48 |
| Full_Prof | Full Prof | $\equiv$ | 1.0 |

*interlinked with each other. This linkage between $S_i$ and $S_j$ means the existence of mapping information between the corresponding ontologies $\mathcal{M}((O)_i,(O)_j)$. Thus, a distributed ontology-based information system $G$ is represented as*

$$G = \{\mathcal{M}(\mathcal{O}_i, \mathcal{O}_j)|T_G(S_i, S_j)\} \tag{7}$$

*where function $T_G$ returns topological feature to find out whether $S_i$ and $S_j$ are linked or not.*

*Example 1.* Suppose that a distributed ontology-based information system $G$ is constructed as shown in Fig. 2. There are only two links, which mean mapping results, between $S_A$ and $S_B$ and between $S_B$ and $S_C$.

$$G = \left\{ \mathcal{M}(\mathcal{O}_A, \mathcal{O}_B), \mathcal{M}(\mathcal{O}_B, \mathcal{O}_C)|T_G = \begin{bmatrix} 0 & 1 & 1 \\ 1 & 0 & 0 \\ 1 & 0 & 0 \end{bmatrix} \right\} \tag{8}$$

Queries can be sent for information sharing between $S_A$ and $S_B$. By referring to the direct mapping by $\mathcal{M}(\mathcal{O}_A, \mathcal{O}_B)$, a query $q$ = 'Secretary' which is not understandable in $S_B$ can be rewritten into 'Researcher.'

However, it is still difficult for $S_A$ and $S_C$ to be interoperable, because there is no direct mapping between them.

# 3 Mapping Composition for Query Transformation

In this work, we want to estimate *indirect* mapping between information systems of which ontology mapping result do not exist. To do so, we are focusing on reusing the existing mapping results and properly composing them to reach from the source information system until destination. For example, in Fig. 1, even though there is no direct mapping between $S_A$ and $S_C$ (i.e., a query from system $S_A$ is not understandable in $S_C$), we can compose two mapping results $\mathcal{M}(\mathcal{O}_A, \mathcal{O}_B)$ and $\mathcal{M}(\mathcal{O}_B, \mathcal{O}_C)$.

**Definition 5 (Indirect mapping).** *A indirect mapping $\widetilde{\mathcal{M}}$ in a distributed ontology-based information system $G$ is represented as*

$$\widetilde{\mathcal{M}}(S_{Src}, S_{Dest}) = \Sigma_{S_{Src}}^{S_{Dest}}(\mathcal{M}(\mathcal{O}_i, \mathcal{O}_j)) \tag{9}$$

*where $S_{Src}$ are $S_{Dest}$ the source and destination information sources in $G$. Here, we can find out whether there is a path between them by repeating multiplication of the topology matrix $T_G$.*

Now we want to show a query transformation by using the indirect mapping between ontologies in a distributed environment.

## 3.1 Semantic Query Transformation

A query from a source systems can be transformed to make it understandable to the destination system by referring to the composed mapping results. Thereby, we have to realize a set of query-activated class $C_Q$.

**Definition 6 (Query).** *A query from an ontology-based information system $S_A$ is represented as*

$$q ::= c|\neg q|q \wedge q'|q \vee q' \tag{10}$$

*where $c \in \mathcal{C}_A$.*

**Definition 7 (Query-activated class).** *Given a query traveling to an ontology-based information system $S_k$, a set of query-activated class $C_Q(q)$ can be extracted as*

$$C_Q(q) = \{c|c \in q, c \in \mathcal{C}_A\}. \tag{11}$$

For example, suppose that the following query $q_1$, which is written by SparQL[1], is sent from $S_A$ to $S_C$ in Fig. 2.

```
PREFIX abc: <http://intelligent.pe.kr/TestOntology#>
SELECT ?Secretary ?Full_Prof
WHERE {
  ?Full_Prof abc:Teach abc:Course;
  ?Secretary abc:Assist abc:Prof;
}
```

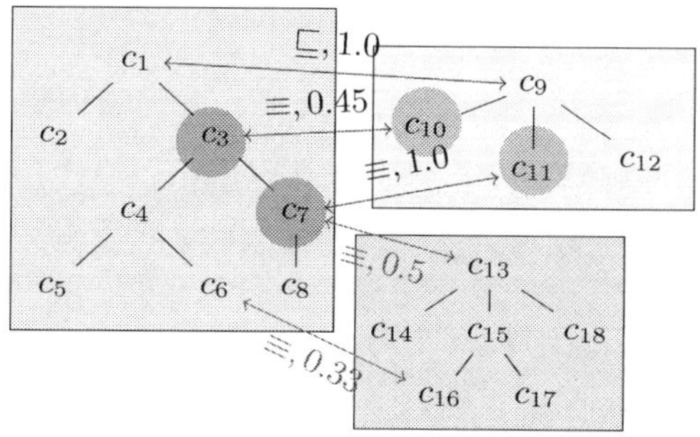


**Fig. 3.** Mapping composition with semantic coverage; $S_A = \{c_1, \ldots, c_8\}$, $S_B = \{c_9, \ldots, c_{12}\}$, and $S_C = \{c_{13}, \ldots, c_{18}\}$. Two sets of query-activated classes $C_Q(q_1) = \{c_{10} = \text{'Secretary'}, c_{11} = \text{'Full\_Prof'}\}$ and $C_Q(q_2) = \{c_3 = \text{'Researcher'}, c_7 = \text{'Full Prof'}\}$.

**Table 2.** Example on query transformation

| Step | Query | Mapping | Query' |
|---|---|---|---|
| 1st | $c_{10} \wedge c_{11}$ | $\langle c_{10}, c_3, \equiv, 0.45\rangle$, $\langle c_{11}, c_7, \equiv, 1.0\rangle$ | $c_3 \wedge c_7$ |
| 2nd | $c_3 \wedge c_7$ | $\langle c_7, c_{13}, \equiv, 0.5\rangle$, $\langle c_6, c_{16}, \equiv, 0.33\rangle$ | $c_{13}$ |

Thus, as shown in Fig. 3, we can extract a set of query-activated class $C_Q(q_1) = \{c_{10} = \text{'Secretary'}, c_{11} = \text{'Full\_Prof'}\}$ but class 'Teach' and 'Course' are not. Consequently, the query $q = c_{10} \wedge c_{11}$ can be transformed through two steps, as shown in Table 2.

Here, we can see *information loss* by mismatching during second step (i.e., $c_3$). This issue is an important issue to discover the optimal state (i.e., minimizing error propagation), and it will be discussed later.

### 3.2 Semantic Coverage

More importantly, we have to take into account more general cases. As shown in Fig. 4, when there are a large number of ontology-based information systems, the network of the distributed ontology-based information system can be more complex. It means that there can be more than one path from arbitrary information systems (i.e., source and destination). The shortest path is not alway the best choice, but the ontology mapping condition and semantics of a given query. For example, if a query should travel from $A$ to $E$, we have to decide the best composition path $\widetilde{\mathcal{M}}(A, E)$ out of the following candidates

[1] SparQL, http://www.w3.org/TR/rdf-sparql-query/

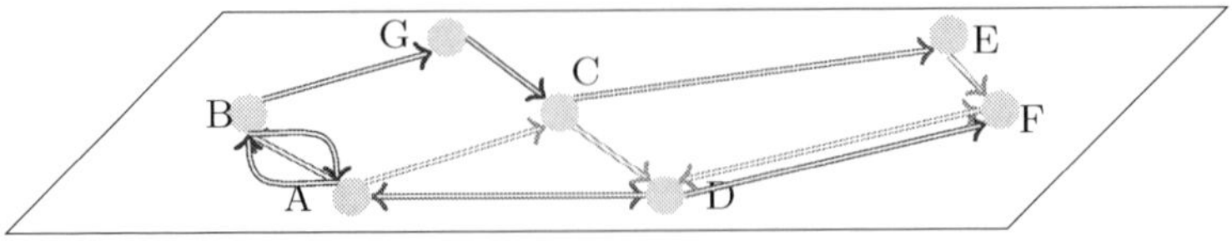


**Fig. 4.** A general case with a large number of information systems

- $\sum(\mathcal{M}(A,C),\mathcal{M}(C,E))$
- $\sum(\mathcal{M}(A,D),\mathcal{M}(D,F),\mathcal{M}(F,E))$
- and more.

Thus, we need to decide which path for composing mapping results will be better than others. In this paper, heuristic approach is exploited, and we want to empirically justify these heuristics. We introduce a semantic coverage ratio for representing two different heuristics ($H_1$ and $H_2$).

**Definition 8 (Semantic coverage ratio).** *A semantic coverage ratio* $\tau_Q$ *means the matching ratio of the size of two correspondence sets to a given query-activated classes. This* $\tau_Q$ *can be defined by the following two heuristics;*

- $H_1$*: As the more correspondences are mapped with query-activated classes, the semantic coverage ratio is increased.*

$$\tau_Q^{H_1}(S_{Src}, S_{Dest}) = \frac{|\{c|c \in e_{Src}, \mathcal{M}(\mathcal{O}_{Src}, \mathcal{O}_{Dest})\}|}{|C_Q|} \tag{12}$$

- $H_2$*: As the confidence values of correspondences are higher, the semantic coverage ratio is increased.*

$$\tau_Q^{H_2}(S_{Src}, S_{Dest}) = \frac{\sum_{e_k \in C_Q} CF_k}{|\{e_k \in C_Q\}|} \tag{13}$$

As an example in Fig. 3, let a query $q_2$ to be sent from $S_A$ though either $S_B$ or $S_C$. To decide the better mapping path, we can measure the semantic coverage ratios by using those two heuristics $H_1$ and $H_2$ in Table 3. It means we can expect that the query should be transformed via $S_B$.

**Table 3.** Measuring semantic coverage ratio by two heuristics

| Heuristics | $\text{Path}_1$ ($S_A \rightarrow S_B$) | $\text{Path}_2$ ($S_A \rightarrow S_C$) |
|---|---|---|
| $H_1$ | $\tau_{q2}^{H_1} = \frac{2}{2} = 1$ | $\tau_{q2}^{H_1} = \frac{0.45+1}{2} = 0.725$ |
| $H_2$ | $\tau_{q2}^{H_2} = \frac{1}{2} = 0.5$ | $\tau_{q2}^{H_2} = \frac{0.5}{1} = 0.5$ |

### 3.3 Transformation Path Selection

The best path for transforming a query is selected by serial aggregation of the semantic coverage ratio. Given two information systems $S_{Src}$ and $S_{Dest}$, the aggregated semantic coverage ratio is computed by

$$\tau_Q(S_{Src}, S_{Dest}) = \max_{Path_k} \prod_{S_{Src}}^{S_{Dest}} \tau_Q(S_i, S_j) \tag{14}$$

where $Path_k$ is a set of all possible paths from $S_{Src}$ to $S_{Dest}$.

## 4 Experimental Results and Discussion

In order to evaluate the proposed distributed ontology-based information system, we have built seven ontology-based information systems (i.e., $S_A$ to $S_G$) with linkages, as shown in Fig. 4. All of the mapping results have been collected by human experts.

We have focused on two evaluation issues (i.e., mapping composition and transformation path selection), and have collected experimental results.

### 4.1 Evaluation on Mapping Composition

By using OLA API [1], we have automatically collected the direct mapping results (i.e., $\mathcal{M}$). The mapping results have been composed in all possible cases

**Table 4.** Recall and precision of mapping composition. $\mathcal{M}(\mathcal{O}_A, \mathcal{O}_B)$ is simply rewritten to $(M)_{AB}$.

| Direct Mapping ($\mathcal{M}$) | Indirect Mapping ($\widetilde{\mathcal{M}}$) | Recall $R$ | Precision $P$ |
|---|---|---|---|
| $\mathcal{M}_{AB}$ | $\mathcal{M}_{AC} \cdot \mathcal{M}_{AB}$ | 0.76 | 0.65 |
| | $\mathcal{M}_{AD} \cdot \mathcal{M}_{DC} \cdot \mathcal{M}_{CB}$ | 0.73 | 0.62 |
| | $\mathcal{M}_{AE} \cdot \mathcal{M}_{ED} \cdot \mathcal{M}_{DC} \cdot \mathcal{M}_{CB}$ | 0.66 | 0.6 |
| | $\mathcal{M}_{AF} \cdot \mathcal{M}_{FE} \cdot \mathcal{M}_{ED} \cdot \mathcal{M}_{DC} \cdot \mathcal{M}_{CB}$ | 0.53 | 0.57 |
| | $\mathcal{M}_{AG} \cdot \mathcal{M}_{GF} \cdot \mathcal{M}_{FE} \cdot \mathcal{M}_{ED} \cdot \mathcal{M}_{DC} \cdot \mathcal{M}_{CB}$ | 0.51 | 0.56 |
| $\mathcal{M}_{BC}$ | $\mathcal{M}_{BD} \cdot \mathcal{M}_{DC}$ | **0.86** | **0.74** |
| | $\mathcal{M}_{BE} \cdot \mathcal{M}_{ED} \cdot \mathcal{M}_{DC}$ | **0.74** | **0.72** |
| | $\mathcal{M}_{BF} \cdot \mathcal{M}_{FE} \cdot \mathcal{M}_{ED} \cdot \mathcal{M}_{DC}$ | **0.72** | **0.65** |
| | $\mathcal{M}_{BG} \cdot \mathcal{M}_{GF} \cdot \mathcal{M}_{FE} \cdot \mathcal{M}_{ED} \cdot \mathcal{M}_{DC}$ | **0.69** | **0.64** |
| | $\mathcal{M}_{BA} \cdot \mathcal{M}_{AG} \cdot \mathcal{M}_{GF} \cdot \mathcal{M}_{FE} \cdot \mathcal{M}_{ED} \cdot \mathcal{M}_{DC}$ | **0.66** | **0.62** |
| $\mathcal{M}_{CD}$ | $\mathcal{M}_{CE} \cdot \mathcal{M}_{ED}$ | 0.73 | 0.66 |
| | $\mathcal{M}_{CF} \cdot \mathcal{M}_{FE} \cdot \mathcal{M}_{ED}$ | 0.67 | 0.63 |
| | $\mathcal{M}_{CG} \cdot \mathcal{M}_{GF} \cdot \mathcal{M}_{FE} \cdot \mathcal{M}_{ED}$ | 0.66 | 0.56 |
| | $\mathcal{M}_{CA} \cdot \mathcal{M}_{AG} \cdot \mathcal{M}_{GF} \cdot \mathcal{M}_{FE} \cdot \mathcal{M}_{ED}$ | 0.54 | 0.52 |
| | $\mathcal{M}_{CB} \cdot \mathcal{M}_{BA} \cdot \mathcal{M}_{AG} \cdot \mathcal{M}_{GF} \cdot \mathcal{M}_{FE} \cdot \mathcal{M}_{ED}$ | 0.51 | 0.52 |

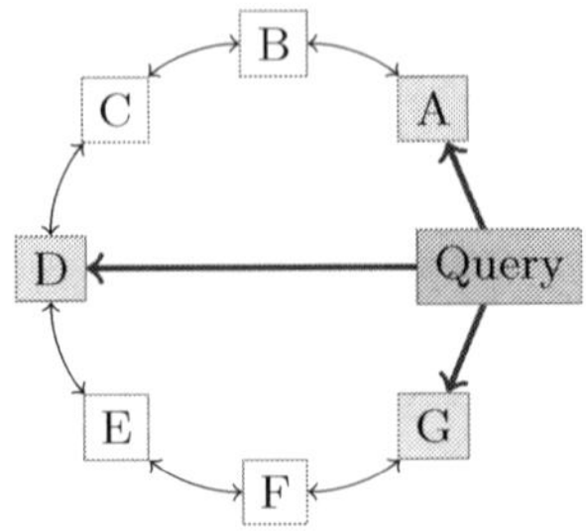


**Fig. 5.** Finding out the best transformation path in a distributed ontology-based information systems

**Table 5.** Performance of transformation path selection for two users

| Heuristics | Users | Information Systems | Recall | Precision |
|---|---|---|---|---|
| $H_1$ | $U_1$ | B | 0.78 | 0.67 |
| | | C | 0.68 | 0.63 |
| | | E | 0.72 | 0.65 |
| | | F | 0.63 | 0.73 |
| | $U_2$ | B | 0.67 | 0.57 |
| | | C | 0.69 | 0.67 |
| | | E | 0.73 | 0.75 |
| | | F | 0.79 | 0.82 |
| $H_2$ | $U_1$ | B | 0.67 | 0.64 |
| | | C | 0.74 | 0.37 |
| | | E | 0.83 | 0.7 |
| | | F | 0.73 | 0.62 |
| | $U_2$ | B | 0.63 | 0.63 |
| | | C | 0.75 | 0.77 |
| | | E | 0.67 | 0.48 |
| | | F | 0.47 | 0.52 |

(i.e., $\widetilde{\mathcal{M}}$). The performance of mapping composition has been tested by precision and recall.

$$Precision = \frac{|\mathcal{M} \cap \widetilde{\mathcal{M}}|}{|\widetilde{\mathcal{M}}|} \text{ and } Recall = \frac{|\mathcal{M} \cap \widetilde{\mathcal{M}}|}{|\mathcal{M}|} \tag{15}$$

The results of three cases (i.e., $\mathcal{M}_{AB}$, $\mathcal{M}_{BC}$, and $\mathcal{M}_{CD}$) are shown in Table 4. In average, we have obtained relatively good results (73% recall and 79% precision). We note that as the mapping results are composed (i.e., the number of mapping composition is increased) in all cases, the recall and precision is getting decreased by nature. This is the information loss cased by mismatching problem of ontology mapping algorithms.

### 4.2 Evaluation on Transformation Path Selection

In second issue, we have tested the performance of transformation path selection resulting from two heuristics (i.e., $H_1$ and $H_2$) by inviting real users. The link topology of the distributed ontology-based information system has been simply built, as shown in Fig. 5. Thirty users were asked to generate 10 queries with SparQL to search for a certain information. These queries have been able to be sent to only three system $S_A$ , $S_D$ , and $S_G$, for considering multiple paths along with the linkages.

Table 5 shows the performance of transformation path selection for two users. In average for all the invited users, heuristic $H_1$ has shown 65.3% recall and 74.2% precision, while $H_2$ has shown 59.5% recall and 68.3% precision. Hence, we found out that $H_1$ outperforms $H_2$ by about 12.4%.

## 5 Concluding Remarks

As the number of ontology-based information systems are getting involved into a global network, we have to somehow establish an efficiently interoperability platform to *semantically* understand the resources from remote and heterogeneous systems. More importantly, this system might count on the scalability of ontology mapping process. In this paper, we have proposed query transformation application on such ontology-based distributed environment.

## Acknowledgements

This work was supported by the Korea Science and Engineering Foundation (KOSEF) grant funded by the Korea government (MEST) (No. 2009-0066751).

## References

1. Euzenat, J.: An API for ontology alignment. In: McIlraith, S.A., Plexousakis, D., van Harmelen, F. (eds.) ISWC 2004. LNCS, vol. 3298, pp. 698–712. Springer, Heidelberg (2004)
2. Euzenat, J., Shvaiko, P.: Ontology matching. Springer, Heidelberg (2007)
3. Euzenat, J., Valtchev, P.: Similarity-based ontology alignment in OWL-Lite. In: de Mántaras, R.L., Saitta, L. (eds.) Proceedings of the 16th European Conference on Artificial Intelligence (ECAI 2004), Valencia, Spain, August 22-27, 2004, pp. 333–337. IOS Press, Amsterdam (2004)
4. Jung, J.J.: Ontological framework based on contextual mediation for collaborative information retrieval. Information Retrieval 10(1), 85–109 (2007)
5. Jung, J.J., Lee, H., Choi, K.S.: Contextualized recommendation based on reality mining from mobile subscribers. Cybernetics and Systems 40(2), 160–175 (2009)
6. Shih, W.-C., Yang, C.-T., Tseng, S.-S.: Ontology-based content organization and retrieval for scorm-compliant teaching materials in data grids. Future Generation Computer Systems 25(6), 687–694 (2009)

# Algebra of Ontology Modules for Semantic Agents

Krzysztof Goczyła, Aleksander Waloszek, and Wojciech Waloszek

Gdańsk University of Technology, Department of Software Engineering,
ul. Gabriela Narutowicza 11/12, 80-233 Gdańsk, Poland
{kris,alwal,wowal}@eti.pg.gda.pl

**Abstract.** In the environment of Semantic Web the problem of precise manipulation of units of knowledge gains importance. In this paper we discuss a possibility of adoption for this purpose the flexible and powerful apparatus of relational algebra. We discuss similarities and differences between knowledge and data bases and their underlying information-processing models. On the basis of the discussion we draw and investigate a special analogy between these two important methods. We use this analogy to introduce the special algebra of ontological modules which may be used to exchange, align and merge the knowledge among agents.

## 1 Introduction

Semantic Web initiative aims at making effective and more intelligent use of the great amount of information hidden in Internet. Semantic description of data is hoped to help answering questions more precisely and accurately. Using Description Logics as a theoretical background and OWL as a practical language knowledge engineers created a large number of ontologies. These ontologies constitute an additional layer which adds semantics to nodes of the Internet network. The layer is utilized by intelligent multi-agent environment which is intended to supply a user with intelligently gathered information.

However, the problem turns out to be complicated, also for the reason that created ontologies are heterogeneous in the way they model the world. Coupling knowledge from different ontologies in a straightforward manner (e.g. by importing one from another) increases complexity and causes "misunderstandings" connected with sometimes very different points of view from which the ontologies describe the domain.

To solve this problems a number of modularization techniques has been proposed ([1], [2], [3]). The techniques allow to manipulate with fragments of knowledge extracted from different knowledge bases in order to build from them a new ontology proper to a given purpose. Moreover, they give methods to express relations between notions defined in joined ontologies.

With use of the proposed techniques of modularization engineers are able to solve the problem. The techniques, however, have also their shortcomings: they propose rather static and persistent methods of insulating and joining modules. A proper technology for highly dynamic multi-agent systems should allow agents to operate on knowledge bases in a very flexible way.

N.T. Nguyen, R. Kowalczyk, and S.-M. Chen (Eds.): ICCCI 2009, LNAI 5796, pp. 492–503, 2009.

In this paper we investigate a possibility of reducing inflexibility of modularization. We are inspired by the relational model which offers a very flexible and powerful mechanism operating on relations: the relational algebra. We found it very desirable to adopt a similar mechanism in the field of DL knowledge bases, and in this paper we show a proposal of a way to accomplishing such adoption.

The next part of the paper is organized as follows. The basic notions of the ontological model are described in Section 2. In Section 3 we analyse the correspondences between ontological model and relational model. We discuss analogies described in literature, and propose a new analogy better suited for our needs. In Section 4 we introduce a semantic algebra of ontological modules which has been created on the basis of the proposed analogy. Section 5 presents examples depicting use of the algebra. Section 6 concludes the paper.

## 2 Ontological Model Basics

Ontological model exploited in the Semantic Web consists of two standards (languages): RDF and OWL. Here we focus on the latter, and especially on its sublanguage OWL DL. OWL-DL is theoretically based on Description Logics [4], the field of study investigating computational properties of various fragments of First Order Logic (FOL).

Every Description Logics ontology describes some selected domain of interest in the terms of *individuals* (individual objects), *concepts* (classes, sets of objects) and *roles* (binary relations between objects). An example of a concept is $Married$ for married people, or $Woman$ for all women. Roles in such domain may be called $hasHusband$ or $hasWife$. The set of used names (terms) forms a *vocabulary*, called a *signature* $\mathbf{S} = \mathbf{C} \uplus \mathbf{R} \uplus \mathbf{I}$ consisting of concept names ($\mathbf{C}$), role names ($\mathbf{R}$), and individual names ($\mathbf{I}$). We assume that there exist sets of acceptable concepts, roles, and individual names, which we denote $\mathcal{N}_\mathrm{C}$, $\mathcal{N}_\mathrm{R}$, $\mathcal{N}_\mathrm{I}$ respectively ($\mathbf{C} \subseteq \mathcal{N}_\mathrm{C}, \mathbf{R} \subseteq \mathcal{N}_\mathrm{R}, \mathbf{I} \subseteq \mathcal{N}_\mathrm{I}$).

Chosen description logics $\mathcal{L}$ determines the set of (possibly complex) concepts, roles and individuals one can build using the *operators* of $\mathcal{L}$ and names from a signature $\mathbf{S}$. We denote these sets with $\mathcal{L}_\mathrm{C}(\mathbf{S})$, $\mathcal{L}_\mathrm{R}(\mathbf{S})$, $\mathcal{L}_\mathrm{I}(\mathbf{S})$ respectively. For instance, if $\mathcal{L}$ is $\mathcal{ALC}$, $\mathcal{L}_\mathrm{C}(\mathbf{S}) ::= A \,|\, \neg C \,|\, C \sqcap D \,|\, C \sqcup D \,|\, \exists R.C \,|\, \forall R.C \,|\, \top \,|\, \bot$ (respectively atomic concept, concept complement, intersection of concepts, union of concepts, existential qualifier, universal qualifier, universal concept) where $A \in \mathbf{C}(\mathbf{S})$, $C, D \in \mathcal{L}_\mathrm{C}(\mathbf{S})$, $R \in \mathcal{L}_\mathrm{R}(\mathbf{S})$. An example of a complex concept is $Married \sqcap Woman$ describing married women, or $\exists hasHusband.\top$ describing individuals who have a husband.

OWL-DL is specifically based on expressive description logic $\mathcal{SHOIN}$, which additionally (to $\mathcal{ALC}$) allows for expressing cardinality constraints, inverse roles, nominals, role hierarchies and transitivity of roles.

Each description logic allows for formulating axioms and assertions (sentences), which describe the interrelationships between terms in the domain of interest. We denote the set of axioms and assertions which can be built from the given signature $\mathbf{S}$ as $\mathcal{L}(\mathbf{S})$. A standard DL ontology is a set of sentences from

$\mathcal{L}(\mathbf{S})$ and consists of two parts: TBox and ABox. Axioms constitute a TBox and introduce relationships primarily between concepts and roles. An example is a concept equivalence axiom $Married \sqcap Woman \equiv \exists hasHusband.\top$ declaring that every married woman has a husband and, conversely, every individual having a husband is a married woman. Assertions constitute an ABox and assign individuals to concepts and individual pairs to roles. An example of an assertion is $hasHusband(Mary, John)$ stating that John is a husband of Mary.

Each ontology is interpreted and the interpretation is understood as a "possible world" which can (but not has to) match the ontological description.

An (base) interpretation $\mathcal{I}$ is a pair $(\Delta^{\mathcal{I}}, \cdot^{\mathcal{I}})$ where $\Delta^{\mathcal{I}}$ is a set called the domain of the interpretation and $\cdot^{\mathcal{I}}$ is an interpretation function assigning each $A \in \mathcal{N}_\mathrm{C}$ a set $A^{\mathcal{I}} \subseteq \Delta$, each $R \in \mathcal{N}_\mathrm{R}$ a binary relation $R^{\mathcal{I}} \subseteq \Delta \times \Delta$, and each $a \in \mathcal{N}_\mathrm{I}$ an element $a^{\mathcal{I}}$ of the domain $\Delta^{\mathcal{I}}$. Every description logic define the conditions under which a specific interpretation $\mathcal{I}$ satisfies a particular sentence. The fact that the interpretation $\mathcal{I}$ satisfies a given sentence (is its model) $\alpha \in \mathcal{L}(\mathbf{S})$ is denoted as $\mathcal{I} \models \alpha$. A model of an ontology has to satisfy every sentence contained in it.

Models and interpretations build the semantic part of the ontology. Solving of inference problems is defined in the terms of models: e.g. checking whether a concept is satisfiable (whether there might exist any individual being a member of this concept) is formally expressed as checking if there exists a model $\mathcal{I}$ in which $C^{\mathcal{I}}$ is non-empty.

## 3 Analogies between Ontological and Relational Models

Our aim is to adopt in the field of DL knowledge bases a similar mechanism like the one offered by the relational algebra. Such adoption is not easy to realize; some of the questions that need to be addressed are: what should be the universe of a new algebra or how to define good operators. To answer these questions it is necessary to establish a proper analogy between the two models.

The problem of drawing relationships between Description Logics and data models used in databases is not new. These comparisons discuss capabilities of modeling some domain of discourse offered by both the technologies. The following differences are often pointed out ([4]):

- Ontologies contain semantic description of data structures, i.e. TBox which allows to treat the relationships between concepts and roles (dictionary terms) as a part of data, while in the relational model this description is rather a part of designing process (E-R diagrams) and is not processed during the runtime.
- The way of interpreting ontologies allows to provide inference, i.e. answering questions about the knowledge which had not been explicitly specified in a model.
- While relational databases concentrate on operating large and relatively simple data, knowledge bases carry out more sophisticated procedures on smaller but much more complex data.

Although semantic description of data structure is not a part of a relational database, we can assume that this role is played by a database schema which is set during logical designing and normalization process. However the schema is not stored in a form readable enough to be rewritten it into the TBox. This is the reason why it is better to look for an analogy between the TBox and E-R model, which we can count as a part of relational model not physically represented in a real database.

There are many correspondences between the way how E-R model and Description Logics are modeling the domain of interest. In [5] a translation of E-R schemas into acyclic $\mathcal{ALN}$ maps entities and relationships into concepts and relationship roles and attributes into ontological roles. In [4] the authors proposed transforming E-R into $\mathcal{DLR}$ where entities were also mapped into concepts but relationships and attributes were mapped into roles (called in $\mathcal{DLR}$ relations) of 2 or more arity. Due to larger expressiveness of DLs it is not possible to make translation in the opposite direction.

In both aforementioned correspondences a database instance, i.e. any consistent state of data, is perceived as the ABox. We may assume that every tuple of a relation, which in fact is an instance of an entity, can be translated into an individual and its attributes.

While discussing this correspondence it is necessary to underline that the most significant difference between the ontological and relational models lays in the way the information is interpreted. Both the models utilize the notion of vocabulary, which contains names (names of E-R entities and relationships in RDBs, or concepts, roles, and individuals in DL). In the case of the relational model, an interpretation of entities and relationships of an E-R schema is the whole set of tuples contained in a database instance. In the case of DL knowledge bases the ABox is not an interpretation. The whole vocabulary (ABox together with TBox) may have many interpretations or, more properly, models.

In addition the relational model interprets names under so-called closed-world assumption (CWA), while in Description Logics open-world semantics is assumed (OWA). In the closed-world semantics the absence of data is interpreted as negative information. For example if there are two tuples containing information about John's children, we can assume that he has only two children. In the open-world semantics the absence of data is treated as lack of knowledge: in the given example the only safe conclusion we may draw is that John has at least two children.

The analysis of properties described above led us to the conclusion that the already described analogies would not be helpful in our task. Although similarities between TBox and database schema and ABox and database contents seems tempting to be utilized, the differences are much more significant. Regarding this, we created our own analogy between the two models.

The centerpiece of our analogy is the notion of interpretation. We draw a correspondence between a tuple in the relational database and a single model of a DL ontology. Although maybe not very intuitive at the first glance, this approach allows us to show that mentioned differences between relational and

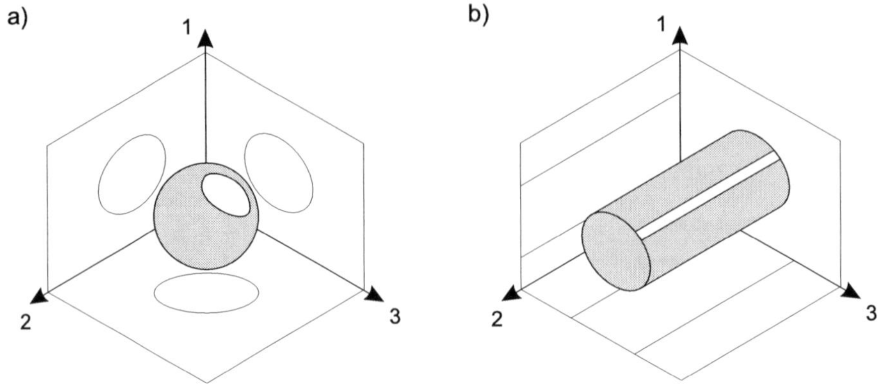


**Fig. 1.** An illustration of the process of cylindrification: sphere before cylindrification (a), and after cylindrification along axis 2 (b)

ontological models are not so severe and the models are more similar than they seem to be. In our considerations we exploit the view on relation database as a special kind of universal relation (see e.g. [6]) described by [7]. This course of work finally led us to creation of algebra, operating on interpretations, analogous to the relational algebra and useful for performing tasks connected with composing, merging and extracting fragments of ontologies.

## 4 Semantic Algebra of Ontological Modules

In the previous section we introduced the idea of drawing the analogy between a tuple in the relational database and a single model of a DL ontology. At the first glance, there are many differences between the two elements of the two models. For instance, the number of tuples in a relation is most often limited to finite values, while the interpretations of a single ontology do not even form a set but rather a proper class. Relational databases rely on the fact that their schema is updated much less often than tuples contained in specific relations. In contrast, in ontologies every new sentence may easily introduce new names, both terminological and individual.

The idea how the analogy may be formalized stems from the work [7] of Imielinski and Lipski. They point out the similarity between relational algebra and cylindric algebra, special algebra introduced by Tarski and his collaborators [8]. The cylindric algebra was originally intended for describing (in algebraic fashion) the universe of predicate calculus sentences with equality, and is basically an extension of a standard Boolean algebra with the operation of *cylindrification*.

A $k$-dimensional cylindric algebra has its geometrical interpretation. In this interpretation the universe of the algebra is the universe of solids (sets of points) in $k$-dimensional space. Boolean operations on solids establish new solids as the intersection, union or negation of the operands. The operation of cylindrification establishes a projection $\mathsf{c}_i$ of a solid on a selected $k-1$-space and enlarges it so that the new set of points contains the whole "cylinder" along a previously "removed" axis $i$ ($i \in [1..k]$, see Fig. 1).

Imielinski and Lipski in [7] approached the problem of unifying the relational model with cylindric world in the following way. They assumed the existence of so-called universal relation with the schema collecting all the attributes of all the relations in the database, but in a slightly different manner than this encountered in the literature under this name. Each attribute (and its domain) constituted a single dimension and all the attributes together established a $k$-th dimensional space of possible relations. The universal relation of Imielinski and Lipski was infinite, and all the relations existing in the database schema were subsets of this universal relation. The operation of projection has been redefined in accordance with previously sketched geometrical approach. Each operation of projection "removes" the unwanted columns but performs it by enlarging the relation so that it assumes every possible value of the "removed" attribute.

Such a view brings very interesting results. For instance, the operation of Carthesian product (in standard approach considered a primitive one) can be simply expressed as intersection of "universalized" relations. The projection operation instead of "contracting" the relation (which is an intuitive perception of this operation on the representations as tables—projection removes columns in that way that the table gets smaller) "enlarges" it.

The last observation is very important from the point of view of bridging the gap between knowledge and data bases. As it has been mentioned in previous discussion, the knowledge bases tend to make no assumptions about facts that have not been explicitly described by the user. We may say that they take into consideration every possibility connected with the non-described part of the world. We can now see a strong similarity between this approach and the way of performing projection in "cylindric" version of relational model. The columns ruled out in the process of projection describe this part of the world which is outside the scope of the newly built relation. In the "cylindric" approach this means that we may assume every value for the removed attributes - which is very akin to the approach employed in knowledge bases.

We should also remind here that the second representation of relation algebra proposed by Tarski and his collaborators was the universe of predicate calculus expressions [8]. Just like Boolean algebra describes propositional statements, the cylindric algebra is suitable for describing formulas containing quantifiers and at most $k$ variables where $k$ is the dimension of the algebra. Conjunctions, disjunctions and negations in these formulas are translated into Boolean intersections, unions and complements. Existential quantifier is translated to projection, while for universal quantifier one can take advantage of the equality $\forall x : \phi(x) = \neg\exists x : \neg\phi(x)$. As pointed out in [7], in the "cylindric" approach to representing relations this makes a straightforward explanation of why relational algebra is equivalent to relational calculus.

In our work we assumed very similar view on describing the space of possibilities. For us the $k$-dimensional space is established by all terms (names of concepts, roles, and individuals) used in the ontology. We cannot, however, constrain ourselves to the names already exploited by the user in sentences building the knowledge base, as the user can at any time create a new sentence with a newly introduce term (especially a new name of an individual object). Therefore in our case $k = \omega$ and our space has an infinite (although countable) number of dimensions.

In this approach every ontology is a projection of a (very large) "universal" ontology involving all the possible names. Every ontology may be represented as a solid in the space, whose points are single interpretations. The solids, thus, represent the sets of valid interpretations of an ontology—the set of their models. Each ontology is in fact a cylinder which allows uninvolved names to have any possible interpretation.

This approach led us to the specific definition of an ontology module and ontology module algebra.

**Definition 1 (s-module).** A s-module $M = (\mathbf{S}, \mathbf{W})$ *is a pair of a signature* $\mathbf{S}$ *(called a signature of the s-module) and a class* $\mathbf{W}$ *of interpretations. The two parts of* $M$ *are denoted respectively as* $\mathbf{S}(M)$ *and* $\mathbf{W}(M)$*. Each interpretation from* $\mathbf{W}$ *is called a model of* $M$*.*

We say that s-module satisfies a particular sentence $\alpha$, which we denote $M \models \alpha$, iff $\forall \mathcal{I} \in \mathbf{W}(M) : \mathcal{I} \models \alpha$. An s-module whose models are all models satisfying a specific $S \subset \mathcal{L}(\mathbf{S})$ can be denoted as $M(S)$. We assume that the module signature consists of all terms used in sentences from $S$, e.g. $M(\{A \sqsubseteq B, A(a)\})$ has the signature $\{A, B, a\}$..

Models of even very simple modules form a proper class because of the fact that any domain set can be used in each interpretation from $\mathbf{W}$. This, however, does not pose much a problem to our formalism. In any case, if we want to constrain ourselves to consider only sets of models we may, without much loss of generality, assume that all the domains of the interpretations in $\mathbf{W}$ are subsets of some given set $\mathbf{\Delta}$.

Boolean operators on modules embrace *union* ($\cup$), *intersection* ($\cap$) and *complement* ($\neg$). These operations in fact manipulate on the set of models.

$$\begin{array}{ll} M_1 \cup M_2 = (\mathbf{S}(M_1), \mathbf{W}(M_1) \cup \mathbf{W}(M_2)) & \mathbf{S}(M_1) = \mathbf{S}(M_2) \\ M_1 \cap M_2 = (\mathbf{S}(M_1), \mathbf{W}(M_1) \cap \mathbf{W}(M_2)) & \mathbf{S}(M_1) = \mathbf{S}(M_2) \\ \neg M = (\mathbf{S}(M), \{\mathcal{I} : \mathcal{I} \notin \mathbf{W}(M)\}) & \end{array}$$

Projection operation changes the signature of a given module. Projection may "remove" some of the names from the signature. According to the "cylindric" analogy this means that we allow any interpretation of the removed terms.

$$\pi_{\mathbf{S}}(M) = (\mathbf{S}, \mathbf{W}(M)|\mathbf{S})$$

Technically, responsible for the "removal" is the operation of signature projection $\mathcal{I}|\mathbf{S}$. The projection is the set of all interpretations interpreting the terms in the remaining signature $\mathbf{S}$ exactly like the interpretation $\mathcal{I}$: $\mathcal{I}|\mathbf{S} = \{\mathcal{J} : \Delta^{\mathcal{I}} = \Delta^{\mathcal{J}} \wedge \forall X \in \mathbf{S} : X^{\mathcal{I}} = X^{\mathcal{J}}\}$. This notion generalizes easily toward projection of sets of interpretations $\mathbf{I}$: $\mathbf{I}|\mathbf{S} = \bigcup_{\mathcal{I}\in\mathbf{I}} \mathcal{I}|\mathbf{S}$. The relationships between original concepts roles and individuals whose names remain in the signature are preserved (e.g. if $M \models \alpha$, $\alpha \in \mathcal{L}(\mathbf{S})$, then $\pi_{\mathbf{S}}(M) \models \alpha$).

Note that projection operation may as well expand the signature of a module. This feature of projection allows us to define with use of projection the generalized versions of union and intersection, generalized union $\cup_g$ can be defined as $M_1 \cup_g M_2 = \pi_{\mathbf{S}(M_1)\cup\mathbf{S}(M_2)}(M_1) \cup \pi_{\mathbf{S}(M_1)\cup\mathbf{S}(M_2)}(M_2)$.

The described operations form the set necessary for creating a cylindric algebra. What is missing from this picture, however, is the operation of selection.

Selection in relational model is an operation which removes certain tuples from a relation and leaves only those which satisfy certain $\theta$-restriction. An example of a simple $\theta$-restriction is $x = y$, where $x$ and $y$ are column names. More complex $\theta$-restrictions may involve arithmetic operations like $x+y < 4\cdot y-x$. We can then say that a $\theta$-restriction is an expression formulated in a language external from the language of relational calculus or algebra, and connected with the specificity of the domain of the column. The counterpart of selection in cylindric algebra is therefore an intersection with a special constant $R_f$, which represent the largest subset of universal relation satisfying $\theta$-restriction $f$. This approach is in line with the original cylindric algebra of formulas which allowed for use of equality of variables with use of special constants called *diagonals*. Each diagonal $\mathsf{d}_{ij}$ in fact represents the restriction $v_i = v_j$.

We took a similar approach in our s-module algebra and treat as the "external" language the specific chosen description logics $\mathcal{L}$ and the sentences in $\mathcal{L}$ as formulas of this language. The selection operation is defined as follows:

$$\sigma_\alpha(M) = (\mathbf{S}(M), \{\mathcal{I} \in \mathbf{W}(M) : \mathcal{I} \models \alpha\}) \qquad \alpha \in \mathcal{L}(\mathbf{S}(M))$$

As mentioned before, the operation may be treated as an intersection with a special module $M_\alpha$ such that $\mathbf{W}(M_\alpha) = \{\mathcal{I} : \mathcal{I} \models \alpha\}$.

The last important operation we took into our consideration was *rename*. Rename is the counterpart to a *substitution* operation in cylindric algebra. Substitution substitute one dimension of another, which in predicate calculus is a counterpart to substituting one variable by another. Similarly rename "substitutes" one concept, role or individual name with another respective one[1].

$$\rho_Y^X(M) = \big(\mathbf{S}(M) - \{X\} \cup \{Y\}, \rho_Y^X(\mathbf{W}(M))\big)$$
$$X, Y \in \mathcal{L}_{\mathrm{C,R\ or I}}(\mathbf{S}(M) \cup Y)$$

By $\rho_Y^X(\mathbf{W}(M))$ we understand a set of interpretations with "substitution" performed: $\rho_Y^X(\mathbf{I}) = \{\mathcal{J} : \exists \mathcal{I} \in \mathbf{I} : \Delta^{\mathcal{I}} = \Delta^{\mathcal{J}} \wedge \forall Z \neq X, Y : Z^{\mathcal{I}} = Z^{\mathcal{J}} \wedge X^{\mathcal{I}} = Y^{\mathcal{J}}\}$.

Such definition of algebra of s-modules gives us the range of operators very similar to those offered by relational algebra. Moreover, it gives us the basis for formulating the following theorem.

[1] Substitution of $j$ by $i$ (we assume $j \neq i$) in cylindric algebra is defined as $\mathsf{s}_i^j x = \mathsf{c}_j(\mathsf{d}_i j \cdot x)$, and similarly we might think about replacing e.g. concept $A$ by $B$ as $\pi_{\mathbf{S}(M)-\{A\}}(\sigma_{A\equiv B}(M))$.

**Theorem 1.** *S-module algebra is homomorphic to a diagonal free cylindric algebra of dimension $\omega$.*

We are proving homomorphism here rather then isomorphism because of the fact that s-modules contain signatures, while cylindric algebra elements are always assumed to contain all the variables. Nevertheless the Theorem 1 is very strong and allows us to transfer all laws of relational (and cylindric) algebra to s-module algebra. The following proposition collects some of them.

**Proposition 1.** *The following laws hold for s-module algebra:*

1. *For union, intersection and negation standard set-theoretic laws hold (e.g. union and intersection are idempotent, associative and commutative).*
2. *For projection we have:*
   a. $\pi_{\mathbf{S}}(M) = \pi_{\mathbf{S}}(\pi_{\mathbf{S}}(M))$,
   b. $\pi_{\mathbf{S}(M)}(M) = M$
   c. $\pi_{\mathbf{S}}(\pi_{\mathbf{S}'}(M)) = \pi_{\mathbf{S}\cap\mathbf{S}'}(M) = \pi_{\mathbf{S}'}(\pi_{\mathbf{S}}(M))$,
   d. $\pi_{\mathbf{S}}(\neg M) = \neg\pi_{\mathbf{S}}(M)$,
   e. $\pi_{\mathbf{S}}(M_1 \cup M_2) = \pi_{\mathbf{S}}(M_1) \cup \pi_{\mathbf{S}}(M_2)$,
   f. $\pi_{\mathbf{S}}(M_1 \cap_g \pi_{\mathbf{S}}(M_2)) = \pi_{\mathbf{S}}(M_1) \cap \pi_{\mathbf{S}}(M_2)$
3. *For selection we have:*
   a. $\sigma_\alpha(M) = \sigma_\alpha(\sigma_\alpha(M))$,
   b. $\sigma_\alpha(\sigma_\beta(M)) = \sigma_\beta(\sigma_\alpha(M))$,
   c. $\sigma_\alpha(M_1 \cup_g M_2) = \sigma_\alpha(M_1) \cup_g \sigma_\alpha(M_2)$,
   d. $\sigma_\alpha(M_1 \cap_g M_2) = \sigma_\alpha(M_1) \cap_g \sigma_\alpha(M_2)$

## 5 Examples of Use

In this section we provide several short examples of how use of the algebra of s-modules may be valuable for the point of view of collaborative agents working in Semantic Web environment. We assume that each modules may be a part of agent's internal knowledge base and can also be published to another agent in order to provide it with information valuable for making a decision. The examples are at high level of abstraction, and, in our opinion, suggest fairly broad range of potential use of the algebra. We assume that the agents and their authors employs some high-level SQL-like language which provides the possibility of building algebraic expressions. The users utilizes it to process s-modules (e.g. extract their fragments, enrich and recombine them) and to produce s-modules as the result.

Here we wanted to stress another similarity between s-module and relational algebra: just like in the case of relational algebra, s-module algebra is used to formulate expression, and the result may be treated as a completely new materialized s-module, which is to be stored by agent in its memory, but also as a perspective through which another modules (operands) are seen or as an *ad-hoc* result which does not need to stored anywhere. For practical use all the approaches may be valuable.

In the following we show some examples of tasks in which use of algebra may prove itself helpful.

*Example 1 (simple import).* We consider two s-modules: $M_1$ describes human resources and $M_2$ a structure of a hospital. The two modules describe internal knowledge bases of two agents ($a_1$ and $a_2$).

$$\begin{aligned} M_1 = M(\{ &\exists isManagerIn.HTBusinessUnit \sqsubseteq Expert, \\ &Expert \sqsubseteq Employee\}) \\ M_2 = M(\{ &leadsDepartment(johnSmith, neurosurgery), \\ &Department(neurosurgery)\}) \end{aligned}$$

To merge the information from the two s-modules in order to infer that $johnSmith$ is an expert we first create an intersection of the s-modules: $M' = M_1 \cap_g M_2$, and then restrict the set of models by introducing additional "bridge" axioms: $M'' = M' \cap_g M(\{leadsDepartment \sqsubseteq isManagerIn, Department \sqsubseteq HTBusinessUnit\})$. The last step can also be accomplished by double use of selection.

*Example 2 (basic ontology alignment).* In the above example we did not encounter any name conflicts. In general, such a conflict may easily occur. In this example we show how to align two s-modules in which the same set of terms has been used to express different meanings.

We consider two s-modules: $M_o$ and $M_l$. Both the modules contain assessment of several rooms for rent. Both the modules utilize the same concepts $\mathbf{C}(\mathbf{S}) = \{HSRoom, ASRoom, LSRoom\}$ where the subsequent concepts describe high, average and low standard rooms. However in $M_o$ and $M_l$ the concepts mean different things as in the case of $M_o$ we were looking for a room for one night and in $M_l$ for a place to stay for longer period. We would like to "import" the assessment from $M_o$ to $M_l$ taking into consideration the necessary translation of classification between the s-modules.

In the first step we have to gather information from both modules, remembering to rename concepts from $M_o$ as they will have different meaning in the new context. To do this we create a new module $M' = \rho^{HSRoom}_{HSRoomON}(\rho^{ASRoom}_{ASRoomON}(\rho^{LSRoom}_{LSRoomON}(M_o))) \cap_g M_l$ ($\mathbf{S}(M') = \{HSRoom, ASRoom, LSRoom, HSRoomON, ASRoomON, LSRoomON\}$).

In the next step we have to establish a translation between "one night" and "longer" concepts. The most natural way to do this is to extend the module by the criteria which were used for the assessment in both cases. In this example we use only one criterion: bathroom. We extend the signature of $M'$: $M'' = \epsilon_{\{RoomWithBathroom\}}(M')$. Afterwards, we have to bind the criteria with the assessment. In "one night" version rooms with bathrooms were automatically considered high standard. In "longer" version no room with bathroom can be considered more than low standard (naturally, the second axiom is valid only if the domain embraces only rooms, which is our assumption):

$$\begin{aligned} M''' = M'' \cap_g M(\{ &RoomWithBathroom \sqsubseteq HSRoomON, \\ &\neg RoomWithBathroom \sqsubseteq LSRoomON\}) \end{aligned}$$

In these steps all the translations possible to perform (considering our awareness of assessment criteria) were done. All average or low standard rooms from $M_o$ were considered low standard in accordance with criteria used in $M_l$.

*Example 3 (different versions and what-if problems).* This example illustrates use of union and negation. Let us consider the s-module $M$:

$$\begin{aligned} M = M(\{ & \top \sqsubseteq \leq 1\, murdered^{-}.\{victim\}, \\ & \exists accuses^{-}.TrustedWitness \sqsubseteq \exists murdered^{-}.\{victim\}, \\ & TrustedWitness \sqsubseteq \exists presentAt.CrimeScene\}) \end{aligned}$$

The s-module describes a "whodunit" world and contains knowledge there is only one murderer and the one who is accused by a trusted witness (who has been present at the crime scene) is certainly the assassin. We want to consider two (mutually exclusive) versions of facts (e.g. collected by two investigating agents):

$$\begin{aligned} M_1 = M \cap_g M(\{ & TrustedWitness(johnShady), \\ & accuses(johnShady, tedInnocent)\}) \\ M_2 = M \cap_g M(\{ & TrustedWitness(henryBrillant), \\ & accuses(henryBrillant, markGuilty)\}) \end{aligned}$$

We can assume that only one of the two version is true and hold the possibilities a single s-module $M' = M_1 \cup_g M_2$. No unambiguous conclusion about the murder can be drawn from $M'$. However the combined module $M'$ is a valuable source of information for the investigator. Many what-if scenarios can be analyzed with use of selection and difference, e.g.:

1. We may assume that *henryBrillant* is not a trusted witness: $M_{p1} = M' \cap_g \neg M(\{TrustedWitness(henryBrillant)\})$. From $M_{p1}$ we can conclude that in this case the murderer is *tedInnocent*.
2. We may assume that *markGuilty* is a murderer, so that $M_{p2} = \sigma_{murdered(markGuilty,victim)}(M')$. In this case we can conclude that *henryBrillant* is a trusted witness (but this does not mean that *johnShady* is not a trusted witness).
3. We may assume that *johnShady* was not present at the crime scene: $M_{p3} = \sigma_{\neg\exists presentAt.CrimeScene(johnShady)}(M')$. In this case we can conclude that *johnShady* is not a trusted witness and *markGuilty* is the murderer.

When some new undeniable evidence is collected (e.g. that *johnShady* was not present at the crime scene) we can add this information to a module permanently and treat the conclusions as the final ones.

## 6 Summary

In the paper we discussed the possibility of adoption of relational algebra to the world of ontologies. We examined similarities and differences between knowledge and data bases and proposed a new analogy between ontological and relational models. Analysis of the analogy allowed us to introduce a semantic algebra of ontological modules, which offers flexibility similar to the relational algebra and calculus. A wide range of operations of the proposed algebra can be used in practical situation, which we illustrate with several examples.

Apart from the possibilities of its practical use we find the proposed algebra interesting for at least two additional reasons. The first reason is the fact that s-module algebra (because of its "completeness" in the meaning of Codd) constitutes the base for describing modularization standards. This branch of research seems to be of undeniable importance for multi-agent systems as without such standards the efficient use of ontology (and their design) is hampered. The first results of our work in this area suggest that with use of the algebra it is possible to devise a theory which allows for creation of new methods of representing knowledge and a framework for the comparison of such methods. For these new representations it will be possible to derive from s-module algebra new specialized algebras for specific purposes.

The second reason is the fact that the algebra is a result of an analogy bridging the two important methods of modeling the universe of discourse - relational and ontological. In effect the more in-depth analysis of similarities and differences between the methods is possible. The analogy may result in enrichment of both models by mutual transfer of algorithms and techniques. Two examples of promising directions are use of relational techniques of normalization for ontological models, and adopting the method of open-world semantics to relational databases.

**Acknowledgements.** This work was supported by Polish Ministry of Research and Higher Education, project No. N N516 4115 33.

## References

1. Borgida, A., Serafini, L.: Distributed description logics: Assimilating information from peer sources. Journal of Data Semantics, 153–184 (2003)
2. Grau, B.C., Parsia, B., Sirin, E.: Tableau Algorithms for $\mathcal{E}$-Connections of Description Logics. UMIACS Research Report, Maryland (2004)
3. Goczyła, K., Waloszek, A., Waloszek, W.: Contextualization of a DL Knowledge Base. In: Proc. of the 20th International Workshop on Description Logics DL 2007, pp. 291–298. Bozen-Bolzano University Press (2007)
4. Baader, F., et al.: The description logic handbook: theory, implementation, and applications. Cambridge University Press, Cambridge (2003)
5. Bergamaschi, S., Sartori, C.: On taxonomic reasoning in conceptual design. ACM Trans. on Database Systems 17(3), 385–422 (1992)
6. Ullman, J.D.: Principles of Database Systems. Computer Science Press (1980)
7. Imielinski, T., Lipski, W.J.: The Relational Model of Data and Cylindric Algebras. J. Comput. Syst. Sci. 28(1), 80–102 (1984)
8. Henkin, L., Monk, J.D., Tarski, A.: Cylindric Algebras pt. 1. Studies in Logic and the Foundations of Mathematics, vol. 64. North-Holland, Amsterdam (1971)

# Grouping Results of Queries to Ontological Knowledge Bases by Conceptual Clustering

Agnieszka Ławrynowicz

Institute of Computing Science, Poznan University of Technology,
ul. Piotrowo 2, 60-965 Poznan, Poland
alawrynowicz@cs.put.poznan.pl

**Abstract.** The paper proposes the framework for clustering results of queries submitted over Semantic Web data. As an instantiation of a framework, an approach is proposed for clustering the results of conjunctive queries submitted to knowledge bases represented in Web Ontology Language (OWL). As components of the approach, a method for the construction of a set of semantic features, as well as a novel method for feature selection are presented. The proposed approach is implemented, and its feasibility is tested empirically.

## 1 Introduction

The Semantic Web [2], commonly named 'Web of data', offers an access to huge data resource. Even though this resource is structured, *data retrieval* from the Web environment face different problems than the retrieval from local databases (e.g. such in which *Relational Database Management System (RDBMS)* is used). In the (Semantic) Web scenarios the user interacts with open and distributed sources, whose structure and content may be not known for him/her apriori.

Instead of finding exact results for queries with Boolean criteria he/she is interested in finding best-matching results. For this purpose the user often performs query results exploration to find out what the data source contains. Therefore, data retrieval in the Semantic Web scenarios may be more of the spirit of *information retrieval* from unstructured text.

The approach of Boolean query conditions as is known from SQL, standard query language for RDBMS, has been as well adopted for *SPARQL*[1], the standard Semantic Web query language. Evaluating exact, boolean criteria may not always be suitable as it may provide incomplete answers due to the overspecification of the query criteria. The exact criteria applied for results aggregation may also lead to huge number of groups. This is due to the standard semantics of SQL clause GROUP BY that creates one row of results for each value of the attribute being in the scope of GROUP BY.

In this work different idea of results grouping is investigated, that is inspired by the functionality of Web search results clustering engines like Clusty[2] or Car-

[1] http://www.w3.org/TR/rdf-sparql-query
[2] http://clusty.com

N.T. Nguyen, R. Kowalczyk, and S.-M. Chen (Eds.): ICCCI 2009, LNAI 5796, pp. 504–515, 2009.

rot2[3] [14,15]. Namely, instead of aggregating the results as GROUP BY clause does, we propose to cluster them. For this purpose we refer to data mining methods. In contrast to the usual offline use of data mining methods, we assume to perform clustering online in query processing time.

In this work we apply our idea to knowledge bases represented in a chosen standard Semantic Web language, *Web Ontology Language (OWL)*[4] (founded formally on description logics [1]). Due to the lack of the standard query language for OWL, SPARQL is commonly used for this purpose. Thus we will present our considerations using the SPARQL syntax.

The rest of the paper is structured as follows. Section 2 provides the motivating scenario, and discusses the requirements for the approach. Section 3 provides the theoretical background. Section 4 introduces the method for extracting semantic features, provided on input to clustering algorithm. Section 5 presents the preliminary evaluation of the approach, and Section 7 concludes the paper.

## 2 Use Case and Requirements

As a motivating scenario of the proposed approach let us consider Example 1.

*Example 1 (Motivating scenario).* Let us suppose that a user may want to search for the touristic offers accross different data sources that are described by one ontology. For example, let us assume that user Anna is looking for a short weekend break offer in East-Central Europe. She would like to have the offers ordered by ascending price (limited to 100 offers), and then grouped by the type of destination (city, resort etc.). She would like also to have the information about the content of each group in terms of intensional descriptions.

From Example 1 follows that the user Anna wants to explore the possibilities offered by the database, and to easier interpret the results of a query have them grouped by some criteria. Let us first consider how the task of data retrieval may be formulated in terms of SQL-like GROUP BY and ORDER BY clauses. Let us write the following GROUP BY clause:

```
GROUP BY destination
ORDER BY price ASCENDING
LIMIT 100
```

As it was mentioned earlier, GROUP BY clause semantics may lead to obtaining too many groups making the results exploration harder for the user. For example, when the names of towns are considered to be the instances of *destination* then SQL GROUP BY would create one row for each name of a town.

Another issues is that this clause is not valid in SQL as the GROUP BY semantics allows results to be ordered only by the attributes present in this clause. What follows, the user when using together GROUP BY and ORDER BY, cannot have at the same time the results grouped and see what belongs to

[3] http://www.carrot2.org
[4] http://www.w3.org/TR/owl-features

the group. Of course this functionality can be achieved by other means, but we concentrate here on the semantics of GROUP BY.

To achieve the behaviour of Vivisimo or Carrot2, the results should be first ordered by some criterion and then *clustered*, not *grouped*:

```
ORDER BY price ASCENDING
LIMIT 100
CLUSTER BY destination
```

In order to be in line with SPARQL syntax the above query needs to be further rewritten to contain variable names instead of table column names:

```
ORDER BY ?y ASCENDING
LIMIT 100
CLUSTER BY ?x
```

Let us summarize the above considerations by presenting the example query.

*Example 2 (Example SPARQL query).* The hypothetical query $Q_1$ that addresses the information needs of the user from Example 1 looks as follows:

```
SELECT ?x ?y ?z
WHERE
?x rdf:type WeekendBreakOffer
?x hasPrice ?y
?x hasDestination ?z
ORDER BY ?y ASCENDING
LIMIT 100
CLUSTER BY ?z
```

To apply such solution we need to define the means for assesing similarity between query tuples to be clustered, and the clustering procedure.

In this work as a language of a knowledge base we assume any description logic language within the subset corresponding to OWL DL sublanguage of OWL, and also assume that the attributes (variables) in a CLUSTER BY clause do not bind to OWL datatypes.

In order to generate intensional descriptions of generated clusters it may be desirable to apply conceptual clustering procedure. For the evaluation of the proposed approach we applied the popular conceptual clustering algorithm COBWEB [9]. COBWEB generates cluster descriptions at clustering time, which is an important feature taking into account the demands of the task.

Before we proceed to the proposed solutions, we present the background on description logics and conceptual clustering algorithm COBWEB.

## 3 Background

### 3.1 Description Logic

*Description logics* (*DLs*) [1] are a family of knowledge representation languages, specifically suited to represent terminological knowledge in a structured and

**Table 1.** Syntax and semantics of example $DL$ constructors

| Constructor | Syntax | Semantics |
|---|---|---|
| Universal concept | $\top$ | $\Delta^{\mathcal{I}}$ |
| Bottom concept | $\bot$ | $\emptyset$ |
| Negation of arbitrary concepts | $(\neg C)$ | $\Delta^{\mathcal{I}} \setminus C^{\mathcal{I}}$ |
| Intersection | $(C \sqcap D)$ | $C^{\mathcal{I}} \cap D^{\mathcal{I}}$ |
| Union | $(C \sqcup D)$ | $C^{\mathcal{I}} \cup D^{\mathcal{I}}$ |
| Value restriction | $(\forall R.C)$ | $\{a \in \Delta^{\mathcal{I}} \mid \forall b.(a,b) \in R^{\mathcal{I}} \rightarrow b \in C^{\mathcal{I}}\}$ |
| Full existential quantification | $(\exists R.C)$ | $\{a \in \Delta^{\mathcal{I}} \mid \exists b.(a,b) \in R^{\mathcal{I}} \wedge b \in C^{\mathcal{I}}\}$ |
| Role domain | $\mathsf{Domain}(R, C)$ | $(a,b) \in R^{\mathcal{I}}$ implies $a \in C^{\mathcal{I}}$ |
| Role range | $\mathsf{Range}(R, C)$ | $(a,b) \in R^{\mathcal{I}}$ implies $b \in C^{\mathcal{I}}$ |

formalized way. Two kinds of atomic symbols are distinguished in any description logic language: *atomic concepts* (denoted by $A$) and *atomic roles* (denoted by $R$ and $S$). Atomic symbols are elementary *descriptions* from which we inductively build *complex descriptions* (denoted by $C$ and $D$) using *concept constructors* and *role constructors*. Description languages (logics) are distinguished by the constructors they provide.

DLs are equipped with logic based, model-theoretic semantics. Semantics is defined by *interpretations* $\mathcal{I}=(\Delta^{\mathcal{I}}, \cdot^{\mathcal{I}})$, where non-empty set $\Delta^{\mathcal{I}}$ is the domain of the interpretation and $\cdot^{\mathcal{I}}$ is an interpretation function. The interpretation function $\cdot^{\mathcal{I}}$ assigns to every atomic concept $A$ a set $A^{\mathcal{I}} \subseteq \Delta^{\mathcal{I}}$, and to every atomic role $R$ a binary relation $R^{\mathcal{I}} \subseteq \Delta^{\mathcal{I}} \times \Delta^{\mathcal{I}}$. The interpretation function is extended to concept descriptions by the inductive definition as the ones presented in column Semantics in Table 1.

A description logic *knowledge base*, $KB$, is typically divided into *intensional* part (*terminological* one, a *TBox*), and *extensional* part (*assertional* one, an *ABox*). A TBox contains axioms dealing with how concepts and roles are related to each other (*terminological axioms*), while an ABox contains assertions about individuals (*assertional axioms*).

A semantics to ABoxes is given by extending interpretations to individual names. The extended interpretation $\mathcal{I}=(\Delta^{\mathcal{I}}, \cdot^{\mathcal{I}})$, additionally to mapping atomic concepts and roles to sets and relations, maps each individual name $a$ to an element $a^{\mathcal{I}} \in \Delta^{\mathcal{I}}$. $\mathcal{I}$ satisfies a set of axioms (e.g. a TBox $\mathcal{T}$ or/and an ABox $\mathcal{A}$) iff $\mathcal{I}$ satisfies each element of this set.

Let by $\mathbf{x}$ and $\mathbf{y}$ denote the sets of distinguished and nondistinguished variables, respectively. A conjunctive query over $KB$, denoted with $Q(\mathbf{x}, \mathbf{y})$, is a finite conjunction of atoms of the form $A(t_1)$ or $R(t_1, t_2)$, for $A$ an atomic concept, $R$ an abstract role, and $t_1$ and $t_2$ individuals from $KB$ or variables from $\mathbf{x}$ or $\mathbf{y}$. An answer to a query $Q(\mathbf{x}, \mathbf{y})$ w.r.t. $KB$ is an assignment $\theta$ of individuals to distinguished variables such that $KB \models \exists \mathbf{y} : Q(\mathbf{x}\theta, \mathbf{y})$.

### 3.2 Cobweb

Cobweb is an incremental conceptual clustering algorithm [9]. As an input it accepts a set of items represented by a set of attribute-value pairs. Given this representation, it generates a clustering in the form of a hierarchical classification tree over items. With each node (class) in the tree the probability is assigned of the occurrence of each attribute-value pair for all the instances of that node.

COBWEB incorporates items into the classification tree incrementally. For each item to classify, the algorithm starts from the root node and descends the tree, applying one of the four operators at each level of the tree: *adding* the item to an existing class, *creating* a new singleton class for the item, *merging* two sibling classes, and *splitting* a class into several ones.

In order to determine proper operation COBWEB uses a heuristic measure named *category utility* [4]. Category utility, as a criterion for evaluating the cluster hierarchy, rewards features traditionally desired in any clustering: high similarity of items within the same cluster, and high dissimilarity of items in different clusters. The operator COBWEB chooses is the one that maximizes the value of a category utility at each level in the tree.

Despite of the known problems from which COBWEB suffers (e.g. exponential complexity in the number of attributes), there is an interest in applying COBWEB in the Semantic Web context, mostly for ontology learning [3,17].

## 4 Construction of the Feature-Vector Representation

Items that are given as input to COBWEB need to be represented by a set of features in the form of attribute-value pairs. In order to cluster items into meaningful clusters, we need to construct the feature sets that would reflect the similarity (or dissimilarity) of the items. Our aim is to use a feature set that does not neglect the availability of the background knowledge base in DL. That is we are interested in defining a set of *semantic* features.

*Feature construction.* The items in our case are vectors of individuals that are retrieved in the query answer tuple. Following the observations made in [6], we assume that semantically similar individuals should behave similarly with regard to the same assertions in a $KB$. Therefore, we propose to measure similarity of individuals in terms of their membership to similar concepts. More precisely, we propose to use a set of concepts from DL knowledge base $KB$ as a set of discriminating features. We formalize that proposition by the following definitions.

**Definition 1 (Feature function).** *A feature function $f$ is a function that maps a pair $(a, C)$, where $a$ denotes an individual, $C$ denotes a concept from DL knowledge base, to values from* $\{\mathbf{false}, \mathbf{true}\}$, $f : (a, C) \rightarrow \{\mathbf{false}, \mathbf{true}\}$.

**Definition 2 (Feature vector construction for individual).** *Let $KB$ be a description logic knowledge base and $\mathbf{C} = \{C_1, C_2, ..., C_m\}$ be a vector of concepts from $KB$. A vector of features for an individual $a$, $\mathbf{F_a}$, is computed by means of feature function $f$ as follows:*

$$f(a, C_i) = \left\{ \begin{array}{ll} \mathbf{true} & KB \models C_i(a) \\ \mathbf{false} & otherwise \end{array} \right\}$$

*where $i \in \{1, ..., m\}$.*

In order to compute a vector of features *instance checking* reasoning service [1] or membership queries are needed. Instance checking reasoning service computes whether an individual $a$ is asserted in $KB$ to belong to the extension of a concept $C$, $C(a)$, or whether $KB$ logically entails so. Membership query (*instance retrieval* service) retrieves the set of individuals being the extension of the given concept. The complexity of the computation performed by the instance checking or instance retrieval services is dependent of the choice of the language from the family of DL languages. Although, some of the computational effort can be saved by the pre-computation.

Example 3 illustrates the definitions introduced in this section.

*Example 3 (Feature vector construction).* Let us consider the knowledge base presented in Table 2, and the query $Q_1$ from Example 2. In query $Q_1$ it is specified that the results are supposed to be clustered by variable $z$. Let us assume that the following concepts (subconcepts of Destination) are taken into account

**Table 2.** Example knowledge base represented in description logic

(a) **Terminology in KB**

| | |
|---|---|
| City $\sqsubseteq$ Destination | *City is a subclass of destination.* |
| Resort $\sqsubseteq$ Destination | *Resort is a subclass of destination.* |
| SeasideResort $\sqsubseteq$ Resort | *Seaside resort is a subclass of resort.* |
| SkiResort $\equiv$ Destination $\sqcap$ $\exists$locatedIn.Mountains$\sqcap$ $\exists$offersActivity.Skiing | *Ski resort is defined as being located in some mountains and offering some skiing activity.* |
| Skiing $\sqsubseteq$ Activity | *Skiing is a subclass of activity.* |
| Swimming $\sqsubseteq$ Activity | *Swimming is a subclass of activity.* |
| Sightseeing $\sqsubseteq$ Activity | *Sightseeing is a subclass of activity.* |
| Skiing $\sqsubseteq$ Sport | *Skiing is a subclass of sport.* |
| Swimming $\sqsubseteq$ Sport | *Swimming is a subclass of sport.* |
| Mountains $\sqsubseteq$ Area | *Mountains are a subclass of area.* |
| $\top \sqsubseteq \forall$hasDestination.Destination | *The range of role* hasDestination *is* Destination. |
| $\top \sqsubseteq \forall$offersActivity.Activity | *The range of role* offersActivity *is* Activity. |

(b) **Assertions in KB**

| | |
|---|---|
| Skiing(DOWNHILLSKIING). | |
| Mountains(TATRA). | |
| Destination(ZAKOPANE) | locatedIn(ZAKOPANE, TATRA) |
| | offersActivity(ZAKOPANE, DOWNHILLSKIING) |
| City(CRACOW) | |
| Resort(SIOFOK) | |
| SeasideResort(JURATA) | |
| SkiResort(HARRACHOV) | |
| WeekendBreakOffer(O1) | hasDestination(O1, CRACOW) |
| WeekendBreakOffer(O2) | hasDestination(O2, JURATA) |
| WeekendBreakOffer(O3) | hasDestination(O3, ZAKOPANE) |
| WeekendBreakOffer(O4) | hasDestination(O4, HARRACHOV) |
| WeekendBreakOffer(O5) | hasDestination(O5, SIOFOK) |

to form the features: City, Resort, SeasideResort, SkiResort. Let us suppose further that we calculate the feature vector for the individual ZAKOPANE, retrieved by $Q_1$. The features are calcutated as follows:

- $f$(ZAKOPANE, City) = **false** (*obvious*),
- $f$(ZAKOPANE, SkiResort) = **true** (*since in the KB it is asserted that* ZAKOPANE *is a destination that is located in mountain area and offers skiing activity, DL reasoner classifies it as* SkiResort)
- $f$(ZAKOPANE, Resort) = **true** (*since* ZAKOPANE *is classified as a ski resort, it is also a resort (superclass of ski resort)*)
- $f$(ZAKOPANE, SeasideResort) = **false** (*obvious*)

For the vector of concepts (City, Resort, SeasideResort, SkiResort), the computation presented above results in the following feature vector for ZAKOPANE: (**false**, **true**, **false**, **true**).

Since an answer to a query consists of a tuple of individuals, feature vectors of the individuals are combined to form one feature vector for one tuple, as presented in Definition 3.

**Definition 3 (Feature vector construction for tuple of individuals).** *Let $KB$ be a DL knowledge base and $\mathbf{F_T} = (a_1, ..., a_n)$ be an answer tuple for a query $Q$ over $KB$. A feature vector for $T$ is computed as: $\mathbf{F_T} = (\mathbf{F_{a_1}}, ...., \mathbf{F_{a_n}})$ where $\mathbf{F_{a_i}}$ denotes the feature vector of the $i^{th}$ individual in the answer tuple.*

*Feature selection* An important concern in any approach based on feature sets is *feature selection*. The choice of a particular subset of features influences the quality of clustering as well as the execution time. From the point of view of the proposed approach the time of execution is crucial.

In the literature, the features are frequently selected by means of search, where the search can be exhaustive, best first, or some specific optimization algorithms can be used founded for example in genetic programming or simulated annealing. For example, in [6] a method based on simulated annealing was presented for feature selection, and in [5] the optimization algorithm founded in genetic programming and simulated annealing.

What is different in case of our approach is that we do not necessarily need to search among all the concepts in $KB$ to find the best set of discriminating features. We have the information, provided by the user in CLUSTER BY clause, indicating the criterion the results are supposed to be clustered by. Therefore, we propose to use this information. For this reason we propose the following procedure, summarized in Algorithm 1. The procedure is executed for each variable appearing in CLUSTER BY clause and consists of two main steps:

1. building a concept description $D$ as an intersection of concepts, and domains and ranges of roles that describe given variable in atoms from the body of a query;

2. selecting as the set of features the subset of concepts from $KB$ that are subsumed w.r.t $KB$ by the created concept description $D$.

**Algorithm 1.** SELECTSETOFDISCRIMINATINGCONCEPTS($KB$, $Q$, $x$)

***input*** *$KB$: knowledge base*
*$Q$: query*
*$x$: a variable by which results are clustered*

***output:***
*set of concepts that serve to form features*

1. $D = \top$ *(initialize concept description D as the universal concept (whole domain))*
2. ***foreach*** *atom* $C_i(x)$, *where* $C$ *is a concept from* $KB$, *and* $C_i(x) \in atomset(Q)$ ***do,***
3. $\quad D \leftarrow D \sqcap C_i$
4. ***foreach*** *atom* $R_i(x,y)$, *where* $R$ *is a role from* $KB$, *and* $R_i(x,y) \in atomset(Q)$ ***do***
5. $\quad D \leftarrow D \sqcap C_i$, *where* Domain$(R_i, C_i)$
6. ***foreach*** *atom* $R_i(y,x)$, *where* $R$ *is a role from* $KB$, *and* $R_i(y,x) \in atomset(Q)$ ***do***
7. $\quad D \leftarrow D \sqcap C_i$, *where* Range$(R_i, C_i)$
8. *retrieve* **C** *a set of all atomic concepts from KB that are subsumed by D*
9. ***return*** **C***;*

The procedure for feature selection is illustrated by means of Example 4.

*Example 4 (Selecting a set of discriminating concepts).* Let us suppose that query $Q_1$ from Example 2 has been refined to the following query $Q_2$:

```
SELECT ?x ?y ?z ?v
WHERE
?x rdf:type WeekendBreakOffer
?x hasPrice ?y
?x hasDestination ?z
?z offersActivity ?v
?v rdf:type Sport
ORDER BY ?y ASCENDING
LIMIT 100
CLUSTER BY ?z, ?v
```

Query $Q_2$ asks for an additional parameter, namely sport activites offered in given destination.

Let us consider creating a vector of discriminating concepts for variable $v$. Following line 1 of Algorithm 1 the description $D$ is created where $D = \top$. After the execution of lines 2-3 of the algorithm $D =$Sport. And finally, after the execution of lines 6-7 of the algorithm $D =$Sport$\sqcap$Activity (since everything which is asserted to be in the range of offersActivity is classified as Activity).

The execution of line 8 of the algorithm retrieves the concepts that are classified by the reasoner to be the subconcepts of $D$: Skiing, Swimming. These concepts are used to create feature vectors for individuals retrieved as bindings of variable $v$ in the answer set of query $Q_2$.

In the future work we may consider also an application of the optimization methods on the feature set obtained by the procedure proposed in this section.

## 5 Preliminary Evaluation

The goal of the preliminary tests was to prove the feasibility of the proposed approach. The proposed method has been implemented in JAVA. The implementation uses KAON2 reasoning engine[5] and Weka data mining software[6].

For tests two datasets have been used (see: Table 3). The first dataset is based on the financial dataset from the PKDD'99 Discovery Challenge (FINANCIAL)[7]. LUBM is a benchmark from the Lehigh University[8], consisting of a university domain ontology and a generator of synthetic data.

**Table 3.** Characteristics of test datasets

| dataset | $DL$ | #concepts | #obj. roles | #individuals |
|---|---|---|---|---|
| FINANCIAL | $\mathcal{ALCIF}$ | 60 | 16 | 17941 |
| LUBM | $\mathcal{SHI}(\mathbf{D})$ | 43 | 25 | 17174 |

For each of the datasets we selected two representational queries that can be submitted over these datasets (see: Table 4). The tests were performed on a laptop computer with a 1.67 GHz Intel processor, 1014 MB of RAM, running Windows Vista.

COBWEB is parameterized with *acuity* (minimum standard deviation) and *cutoff* (minimum category utility). Acuity is only valid for numerical data, which is not our case. Cutoff value influences the number of clusters. A higher cutoff value will promote merging similar clusters into the same cluster. The value of the cuttoff presented in Table 4 was chosen experimentaly to form reasonable (from the point of view of our approach) number of clusters.

The results of the preliminary tests are presented in Table 4. For the clarity of presentation, the queries are presented without the namespaces. Table 4 presents the number of answers for each query, the number of generated discriminating concepts (features), the cutoff, the number of generated clusters, the time of executing query results processing, and the global silhouette index value (*gsi*) [16] for the generated clustering.

It is noteworthy that the systems like Clusty or Carrot2 by default process maximally from about 100 to 500 top ranked results. Hence, we performed the tests for the queries with the order of about 500 results ($Q_1$, $Q_3$). The query processing times for queries $Q_1$ and $Q_3$ are below $1s$. We also tested the method on queries with the number of answers of order of several thousands of results ($Q_2$, $Q_4$). The clustering times for these queries is in order of several seconds. The presented execution times show the feasibility of the proposed approach, and are encouraging for further extensive experimental evaluation, and for an application of the approach in a real-life setting.

---

[5] http://kaon2.semanticweb.org

[6] http://www.cs.waikato.ac.nz/ml/weka

[7] FINANCIAL, http://www.cs.put.poznan.pl/alawrynowicz/financial.owl

[8] LUBM, http://swat.cse.lehigh.edu/projects/lubm/

**Table 4.** The results of the tests

| Ontology | FINANCIAL | FINANCIAL | LUBM | LUBM |
|---|---|---|---|---|
| Query | $Q_1$<br>SELECT ?x ?y ?z ?v<br>WHERE<br>?x rdf:type :Client<br>?x :livesIn ?y<br>?y rdf:type a:Region<br>?x :isOwnerOf ?z<br>?z :hasLoan ?v<br>CLUSTER BY ?y, ?v | $Q_2$<br>SELECT ?x ?y<br>WHERE<br>?x rdf:type :Client<br>?x :livesIn ?y<br>?y rdf:type :Region<br>CLUSTER BY ?y | $Q_3$<br>SELECT ?x ?y<br>WHERE<br>?x rdf:type :Professor<br>?y :member ?x<br>?y rdf:type :Organization<br>CLUSTER BY ?x, ?y | $Q_4$<br>SELECT ?x ?y<br>WHERE<br>?x rdf:type Publication<br>?x :publicationAuthor ?y<br>?y rdf:type :Employee<br>CLUSTER BY ?y |
| answers | 682 | 5369 | 447 | 7325 |
| features | 14 | 8 | 12 | 15 |
| cutoff | 0.235 | 0.2 | 0.05 | 0.2 |
| clusters | 15 | 10 | 6 | 3 |
| time(s) | 0.77 | 4.32 | 0.39 | 17.41 |
| gsi | 0.86 | 1.0 | 1.0 | 0.81 |

It is often indicated that for evaluating the quality of a clustering it is better to use different criterion than the one used during the clustering process. To test the validity of the clustering, Silhouette index measure has been used [16]. Since it does not require any apriori information on classes to which instances belong, silhouette index may be used to judge the quality of any clustering solution. The distance between two instances over a specific feature vector was computed as standard Manhattan distance. The value of Silhouette index varies from -1 to 1, where higher value indicates better clustering result. Hence, according to this measure, the obtained results are of good quality.

## 6 Related Work

To the best of our knowledge this is the first work on clustering the results of conjunctive queries to description logic knowledge bases. Due to this reason we cannot provide direct comparison to other approaches. There have been also no works on clustering SPARQL query results. In standard SPARQL syntax, even grouping is not supported, yet. However, some of the engines (see for example: Virtuoso[9], Jena[10]) implement GROUP BY clause functionality following the approach adopted in SQL.

Clusty, and Carrot2, search results clustering engines, provided the inspirations for this work. There are very few other works on clustering structured query results, and they concern clustering the results of SQL queries [13,18]. The approach proposed in [13], despite of the different language of database, and different language of queries, is designed for numerical data. The approach proposed in [18] is designed for discrete spatial data. However the idea of introducing new constructs into queries whose semantics is to cluster the results may be found already in [13], and [18].

COBWEB has been used in the Semantic Web applications mainly for ontology learning. For instance, in [3] COBWEB was applied to generate hierarchical

[9] http://virtuoso.openlinksw.com
[10] http://jena.sourceforge.net/ARQ/

ontologies of songs. In [17] COBWEB was applied to learn the ontologies used in recommendation systems. In both mentioned approaches, the original dataset was the user-item matrix. In our proposed approach, we operate on the results of queries over description logic knowledge base whose semantics is taken into account during clustering.

The method proposed for computing the feature vector is inspired by that proposed in [6,7,8]. In [6,7,8] for computing the feature vectors, despite the information on instance membership to particular concepts, also its nonmembership derived by a reasoner was taken into account. Due to the open world assumption adopted in DLs it is often not possible to compute nonmembership, especially when there are no disjointness constraints specified in a knowledge base. Computing nonmembership takes additional time which is crucial in our approach. Thus, instead of three possible states (member, nonmember, unknown) we decided to use binary state (member, nonmember/unknown).

Several ways to clustering Semantic Web data were investigated in [10]. However, all of them were applied to RDF graphs, and as such they adopted different language then the one in our proposed approach.

Finally, in our previous work we developed a method for data mining from description logic knowledge bases, but it aimed to solve different tasks, namely frequent pattern discovery, and association rule mining [11,12].

## 7 Conclusions

The contribution of the paper is the proposition of the framework for clustering results of SPARQL queries submitted over Semantic Web data. As an instantiation of a framework a method is proposed for clustering the results of conjunctive queries submitted to knowledge bases represented in OWL (whose underlying formalism is description logic). To the best of our knowledge this is the first proposal for conjunctive query results clustering over such knowledge bases.

To realize the framework, a method for the construction of a set of semantic features is proposed (inspired by literature), as well as a novel method for feature selection. The proposed approach is implemented, and its feasibility is tested empirically.

The plan for the future work assumes an extensive experimental evaluation of the proposed method. In the further versions of the proposed approach it is planned to support OWL datatypes in clustering. Worth considering is also developing a method for clustering that would be dedicated to the task of clustering SPARQL query results. Finally, designing a useful user interface, exploiting the proposed approach, constitutes an interesting task for the future research.

## References

1. Baader, F., Calvanese, D., McGuinness, D., Nardi, D. (eds.): The description logic handbook: Theory, implementation and applications. Cambridge University Press, Cambridge (2003)

2. Berners-Lee, T., Hendler, J., Lassila, O.: The Semantic Web. Scientific American 284(5), 34–43 (2001)
3. Clerkin, P., Cunningham, P., Hayes, C.: Ontology discovery for the Semantic Web using hierarchical clustering. In: Proc. of Semantic Web Mining Workshop (2001)
4. Gluck, M.A., Corter, J.E.: Information, uncertainty, and the utility of categories. In: Proc. of the Seventh Annual Conference on Artificial Intelligence (1985)
5. Fanizzi, N., d'Amato, C., Esposito, F.: Randomized metric induction and evolutionary conceptual clustering for semantic knowledge bases. In: Proc. of the ACM International Conference on Knowledge Management, CIKM 2007 (2007)
6. Fanizzi, N., d'Amato, C., Esposito, F.: Induction of Optimal Semi-distances for Individuals based on Feature Sets. In: Proc. of the DL 2007, Workshop (2007)
7. Fanizzi, N., d'Amato, C., Esposito, F.: Evolutionary conceptual clustering based on induced pseudo-metrics. Int. J. Semantic Web Inf. Syst. 4(3), 44–67 (2008)
8. Fanizzi, N., d'Amato, C., Esposito, F.: Conceptual clustering and its application to concept drift and novelty detection. In: Bechhofer, S., Hauswirth, M., Hoffmann, J., Koubarakis, M. (eds.) ESWC 2008. LNCS, vol. 5021, pp. 318–332. Springer, Heidelberg (2008)
9. Fisher, D.H.: Knowledge acquisition via incremental conceptual clustering. Machine Learning 2, 139–172 (1987)
10. Grimnes, G.A., Edwards, P., Preece, A.D.: Instance Based Clustering of Semantic Web Resources. In: Bechhofer, S., Hauswirth, M., Hoffmann, J., Koubarakis, M. (eds.) ESWC 2008. LNCS, vol. 5021, pp. 303–317. Springer, Heidelberg (2008)
11. Jozefowska, J., Lawrynowicz, A., Lukaszewski, T.: Frequent pattern discovery in owl dlp knowledge bases. In: Staab, S., Svátek, V. (eds.) EKAW 2006. LNCS (LNAI), vol. 4248, pp. 287–302. Springer, Heidelberg (2006)
12. Jozefowska, J., Lawrynowicz, A., Lukaszewski, T.: On reducing redundancy in mining relational association rules from the semantic web. In: Calvanese, D., Lausen, G. (eds.) RR 2008. LNCS, vol. 5341, pp. 205–213. Springer, Heidelberg (2008)
13. Li, C., Wang, M., Lim, L., Wang, H.: Supporting ranking and clustering as generalized order-by and group-by. In: Proc. of SIGMOD 2007, pp. 127–138 (2007)
14. Osiński, S., Weiss, D.: Carrot2: Design of a Flexible and Efficient Web Information Retrieval Framework. In: Szczepaniak, P.S., Kacprzyk, J., Niewiadomski, A. (eds.) AWIC 2005. LNCS (LNAI), vol. 3528, pp. 439–444. Springer, Heidelberg (2005)
15. Osinski, S., Weiss, D.: A Concept-Driven Algorithm for Clustering Search Results. IEEE Intelligent Systems 20, 48–54 (2005)
16. Rousseeuw, P.: Silhouettes: a graphical aid to the interpretation and validation of cluster analysis. J. Comput. Appl. Math. 20, 53–65 (1987)
17. Schickel-Zuber, V., Faltings, B.: Using Hierarchical Clustering for Learning the Ontologies used in Recommendation Systems. In: Proc. of ACM SIGKDD 2007, pp. 599–608 (2007)
18. Zhang, C., Huang, Y.: Cluster By: a new sql extension for spatial data aggregation. In: Proc. of 15th ACM International Symposium on Geographic Information Systems, ACM-GIS 2007, vol. 53 (2007)

# Applying the c.DnS Design Pattern to Obtain an Ontology for Investigation Management System

Jolanta Cybulka

Poznań University of Technology, Institute of Conrol and Information Engineering,
Pl. M. Skłodowskiej-Curie 5,
60-965 Poznań, Poland
Jolanta.Cybulka@put.poznan.pl

**Abstract.** The intelligence of a system aimed at supporting the teamwork of investigators against economy crimes is based on the relevant knowledge base. Such a base implements some view on the state of affairs and should be carefully modeled. We show how it can be done with the help of the *constructive descriptions and situations* (c.DnS) design pattern, which enables to construct an ontology (a conceptual model) in the top-down manner. The modeled state of affairs is constituted by both the domain and the task-based components.

**Keywords:** Constructive descriptions and situations ontology, ontology of criminal and investigation processes, investigation management system.

## 1 Introduction

Our research was motivated by the necessity to have a relevant conceptual model (an ontology) of the information system aimed at supporting the teamwork investigations against the economy crimes [2]. In particular, we focused on the following general issues:

- the way the model supports the implementation of the system's functionality (resulting in the creation of the task-based part of the ontology),
- the scope and the method of the domain knowledge conceptualization, i.e. the method of the creation of the domain-based part of the ontology and
- the ontology design and engineering pattern.

The requirements for the conceptual model are specified in section 2.

In the areas of criminal analysis, crime detection and investigation or crime prevention, the analysts are equipped with software tools, such as, for example, Analyst's Notebook, COPLINK, FraudFocus or FFPoirot. As COPLINK [17] employs the workflow technology in order to support the teamwork it served as the inspiration for the design phase of our system. The FFPoirot [3] is also worth noticing for it uses an ontology. The most significant difference between FFPoirot's approach and ours concerns the strategy of the ontology building. FFpoirot uses the bottom-up method that results in obtaining a layered ontological structure. At first, the universe of 'lexons' is established as the result of the analysis of relevant documents. Lexons are

N.T. Nguyen, R. Kowalczyk, and S.-M. Chen (Eds.): ICCCI 2009, LNAI 5796, pp. 516–527, 2009.

elementary conceptual entities representing (from different perspectives) the relationships between semantic types of some domain. Then, the set of logical 'commitments' is formulated in order to handle some more general semantic entities. The commitments, applied to the lexon base, represent some ontological perspectives of the state of affair. Several such perspectives may be considered leading to a layered architecture comprising application ontologies, topical and domain ontologies and also foundational ontologies (in the case of FFPoirot it is SUMO, [9]), which are the most general. Furthermore, in this approach the knowledge base of an information system is modeled and developed in parallel with the ontology that supports it (AKEM based on RUP, [16]). This approach is quite advantageous however, its essential drawback lies in the necessity of the creation of the lexon's universe with the assumption of not being 'too encoding biased'. In consequence, many of lexons may not be used in any suitable application.

Our approach to ontology construction is quite the opposite - it is a top-down strategy. Similarly to the FFPoirot's method, it enables to create layered and modularized ontologies, but in the first step the top layer of the ontology must be specified. The decision of what top ontology to choose is of great importance. It must provide the modeling means to represent the sufficiently general view on possible states of affairs and also it should be engineered in some formal language. The significant expressiveness of the top level guarantees the possibility to model many different semantic aspects of the states of affairs and, at the same time, to manage the changes and extensions possibly occurring at lower levels in the life cycle of the ontology. After establishing the top ontology its specializations (i.e. more specific layers) may be defined that cover the considered domain or a task-based knowledge. In the proposed solution the lower layers of ontology are gradually created down to the lowest level, which represents the semantics of data in the system's database(s).

In the next section the requirements for the conceptual model are presented. Then the ontology is described (section 3) and in the section 4 the future work is outlined.

## 2 Requirements for the Conceptual Model of the System

The *task-based part* of the ontology is supposed to aid the implementation of the workflow-based investigation management system's functions, which fall into the following categories:

a) data acquisition,
b) data extraction (by means of user queries analysis and answering),
c) data exchange between the system and the environment,
d) the representation of the dynamics of processes within a workflow paradigm.

The system is supplied with data communicated both on-line by the investigation team members and accessed from databases. The on-line methods require the definition of a communication language that should be user friendly and flexible to the most possible extent. The relevantly expressive ontology is a good mean to represent such language semantics. The data residing in relational databases also profit by their semantics specification. Having the meaning of a relational scheme explicitly stated it is possible to automate the acquisition of data from documents (or 'stories' concerning investigated cases) to a database. This may be done through, at first, generation of a textual representation of an interesting part of the ontology (an ideal textual paraphrase; there may exist several such texts with different syntactic variants of semantic ontological entities). Then, syntax-based tools may be used which recognize patterns (data) in real documents that may be further put into the database.

The system's data extraction process may also profit by the semantics - especially when the data are extracted through 'semantic queries' what means queries in which ontological concepts and relations are used.

The ontology has an impact also on the exchange of data (differently formatted in different sources), for the mappings between their sources may be discovered on the basis of semantics attached to data schemas. Having mappings it is possible to use semantics-driven translation to convert data forms.

The system manages the work of an investigation team by using the workflow technology. Here, the conceptual model may facilitate structuring the management of processes (the static part of the workflow conceptualization) and reasoning about the processes in time (the dynamic part). In this angle our solution shows similarity to the COBRA approach (core ontology for business process analysis, [8]).

The *domain part* of the ontology concerns, up to now, the two main specializations of the foundational layer. The first one deals with the three interconnected state of affair types: trade/business situations (as instruments of criminal activities), investigation situations (investigations are carried by the police) and inquiry situations (carried by prosecutors with the help of the police). They are considered in parallel due to the observation that the revealing and the analysis of any crime should be interlaced with the execution of investigative actions. The second specialization concerns both the core conceptualization of law and the legal qualification of the (suspicious) trade/business situations. This specialization needs elaboration and may be supported by the existing ontologies of law, such as CLO [1], LKIF-Core [6], FOLaw [13] or others.

The choice of a foundational ontology that fulfills the listed requirements is very challenging. Fortunately, there are solutions which can be adopted, for example the ontology *of constructive descriptions and situations* [5]. In this conceptual modeling design pattern two components are considered: a basic ontology of 'what really exists' and the ontologically reified means (called here *schematic entities*) to describe the concepts of the basic ontology from the point of view of some perceiving agent. It may be regarded as the epistemological layering or as the "contextualization" of the basic ontology.

The entities of the basic ontology that are classified by relevant schematic entities are regarded as ontologically *grounded.*

The advantages of the c.DnS ontology are as follows.

1. Its expressive power is very high and therefore it may be easily projected on different subdomains.
2. It is well suited to model domains populated with many non-physical entities.
3. It enables to add the contextual or epistemic knowledge to the ontology.
4. The simplified version of the ontology is engineered in the OWL language and can be freely accessed [4].

To model something using the c.DnS paradigm one has to define the relation (1). On this basis some basic binary (or ternary, with the time argument) projections were defined, which specify the relations that hold between the pairs of concepts (their names are given in italics).

$$\text{c.DnS} = <S, E, A, D, C, K, I>. \quad (1)$$

The *S* stands for a *situation* concept that is a kind of a schematic entity and it ontologically reifies the real state of affairs. The situation *S* is constituted (in the c.DnS it is done by means of the *set-for* formal relation) by the *ground entities E* coming from the basic ontology. The situation *S* is located in time by means of a *time interval* Ts (we classify Ts as a kind of *C* concept, see it in the sequel). To be grounded the situation *S* must be described (from some viewpoint) by a *social agent A* (for whom the physical agent, that is to say, some man *acts for*). The agent *A*, in time TD (also classified as a kind of a concept *C*) creates a (potentially many) *description D* of a situation *S*, which is then *assumed* by *A* (and may be also assumed or only *shared* by other agents). Every such description *D* must be *satisfied* by the situation *S*. To make the entities *E* grounded in the description *D* they must be *classified* (named) by the concepts *C*. The agent *A* is a member of some *collective K*, every member of which shares or assumes the description *D* of the situation *S*. In the end, the description *D* is *expressed* by an *information object I* (which may be represented by different physical means).

Recall our core ontology, where there are three main concepts embedded in the foundational layer. They are the situations representing the three main processes represented in our system supporting prosecutor's inquiries and police investigations: a *criminal process*, an *inquiry process* and an *investigation process*. We assume that processes are *goal-oriented situations* and that in addition they are *executions of* special descriptions called *plans*. The processes are constituted (set for) by different real entities, such as persons, objects, real locations, actions, states, social relations, qualities etc. Processes are inspected by agents (*perceiving social agents*) who form their descriptions by means of information objects.

In the phase of formulating a description, social agents who perceive the criminal processes try to recognize and to describe the plan of the criminal activity. In the plan, some concepts classifying the real entities are either used or defined. The main concepts are *tasks* (descriptions of goal-oriented actions), *state descriptions*, *stative activities descriptions* (e.g. being used to sell some kind of goods, to build, to account etc.) and *roles* (that classify detailed functions of different entities which are engaged in actions, states or activities). Besides plans the *goals*, the *norms* and the *constitutive descriptions* are the main types of *descriptions* used in our ontology. We use also different kinds of *social agents* both individual and collective, such as *organizations* and *institutions*. The collections of social agents are associated with them, through the members of which the collective social agents act. The organizations are acted by the members of *intentional collectives* that are *unified by* a common plan of acting. Similarly, the institutions are associated with *normative intentional collectives*, which are unified by *norm and plan bundles* (their plans of acting are restricted by some specific regulative norms).

## 3 An Ontology of Criminal Processes and Investigation Procedures

In the presented ontological view on reality the four main processes are considered. The first of them is a crime. The next two are an inquiry and an investigation - these processes follow directly from the Polish law regulation [10, 12, 14]: an inquiry (in general) carried by a prosecutor and an investigation carried by the police. The fourth process, a workflow one, represents the dynamic aspects of the inquiry/investigation processes. The conceptual scheme composes them and each process is formalized by means of the specialized version of the constructive descriptions and situations relation c.DnS, namely:

- *c.DnSForACriminalProcess*,
- *c.DnSForAnInquiryProcess*,
- *c.DnSForAnInvestigationProcess*,
- *c.DnSForAWorkflowProcess*.

In the sequel we present the c.DnS structures in more detail using the UML[1] graphical notation. The four considered structures are related as shown in Fig. 1. The *classifies* formal ontological relation holds between the *AgentiveInquiryPRole* or the *AgentiveInvestigationRole* (for example, the prosecutor or the investigating policeman) and the *PerceivingSocialAgent* who perceives and re-describes the reality of a criminal process. A *WorkflowProcess* situation representing the dynamics of an *InquiryProcess* situation (or an *InvestigationProcess*) is related to it by the *hasInScope* relation.

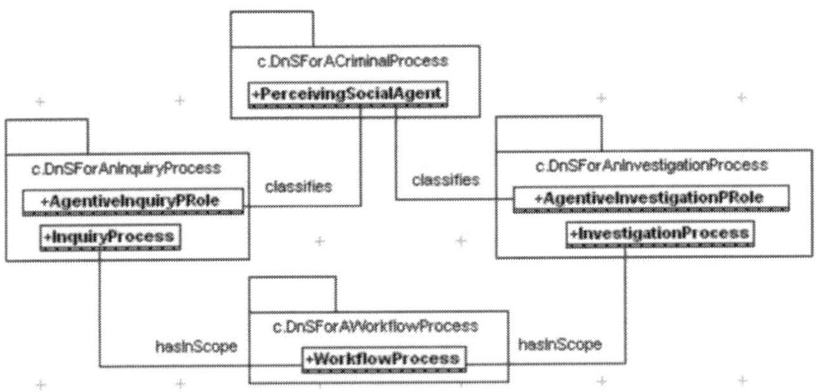


**Fig. 1.** Connections between the four processes

### 3.1 The c.DnS Structure for a Criminal Process

The structure of the main elements of the *c.DnSForACriminalProcess* is depicted in Fig. 2. The criminal process situation consists of the actual criminal process situation (*ActualCPSituation*, a goal-oriented situation) and the inquiry version situation (*InquiryVersion*, a recognized plan realization situation).

There may exist several such versions. The distinction of the two kinds of situations is based on the observation that actual states of affairs of criminal processes (here, situations), though reified in the ontology, are in most cases 'untouchable'

[1] http://www.omg.org/spec/UML

except the *in flagranti* circumstances. The agent who perceives them relies only on their trails trying to recognize the plan of a criminal activity. Thus, in fact, the criminal process reality is represented by a (maybe the set of) special inquiry version situation that is the realization of a recognized plan. Moreover, every criminal process situation is related to some inquiry version situation by means of the *hasInScope* formal ontological relation.

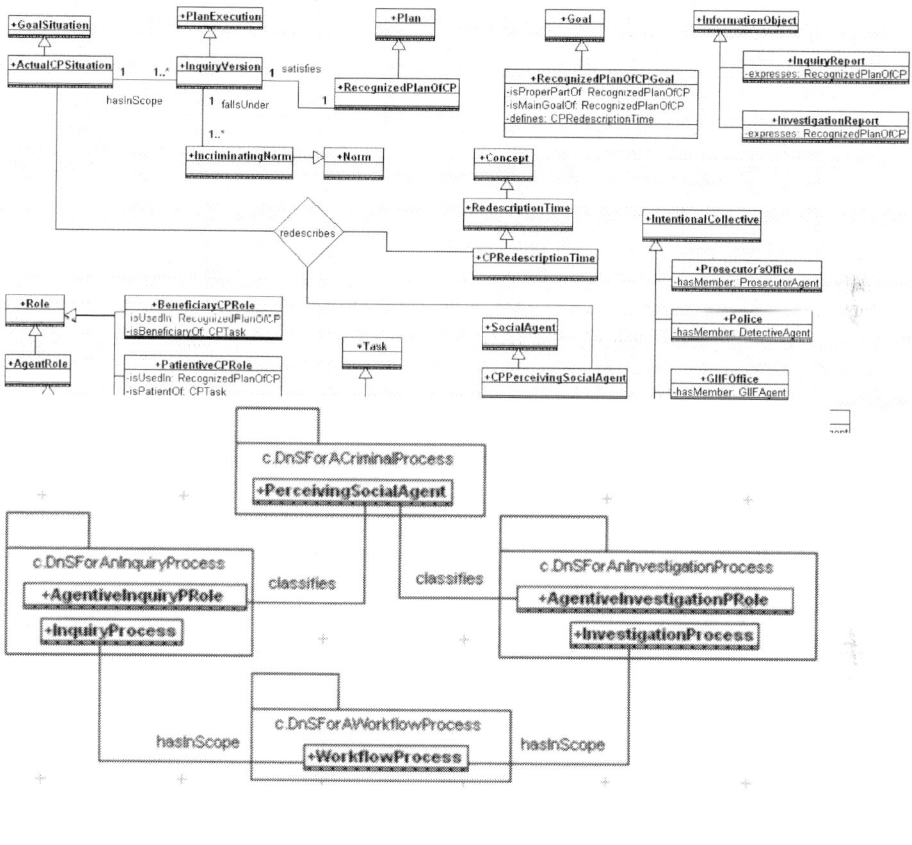


**Fig. 2.** Graphical representation of the main parts of the *c.DnSForACriminalProcess*

The description of a criminal process contains a recognized plan of a criminal process (*RecognizedPlanOfCP*) that is satisfied by the inquiry version situation. Moreover, such a situation is checked for compliance with another description, namely the incriminating norm (*IncriminatingNorm*, formally, the inquiry version situation falls under the incriminating norm). Every plan has as a proper part the goal description (*RecognizedPlanOfCPGoal*) representing its main goal. There are some kinds of descriptions that are not considered yet, for instance the so-called *constitutive descriptions*, which constitute concepts whose existence is declared only by a definition given via such descriptions. For instance, the 'forbidden action' concept exists if it is constituted by a relevant law definition. The same concerns the social agents, such as a prosecutor, a policeman or any institution or an organization - the law provisions or other regulations constitute their existence.

We propose to introduce, among others, the prosecutor agent (*ProsecutorAgent*) and the detective agent (*DetectiveAgent*, an investigating policeman) as the perceiving agents. Both types of agents can perceive (trails of) some actual criminal process situation (*ActualCPSituation*) and then re-describe it in a form of the mentioned recognized plan. There are other social agents that are classified by roles in a recognized plan and in other plans. In particular, we consider a social person, an organization, and an institution. Organizations are collective social agents that are acted by members (being also social agents) of some intentional collectives. Such collective is unified by a common plan. The institution differs from the organization in that it is acted by members of a normative intentional collective, which is unified by a norm-plan bundle. This means that the plans of activity in institutions are constrained by some regulative norms. Among the perceiving agents we also distinguish the institutions like the General Financial Inspector Bureau (*GIIFAgent*, GIIF) and the obliged institution agent (*ObligedInstitutionAgent*). The latter means every financial institution that is legally obliged to inform the GIIF bureau about the suspicious financial activities; bank institutions are among them.

Perceiving agents are members of the mentioned social agent collections (*CPSocialAgentCollection*, in fact, the normative intentional collectives). They share the earlier considered description. Such collectives gather members who act for institutions, e.g. for a prosecutor's office or for a police investigative team. Other collectives (connected with either institutions or organizations) are also defined in our ontology.

As it was mentioned before, the entities are ontologically grounded only when they are classified by concepts. The considered concepts are *CPTask*, *CPStateSpecification*, *CPActivitySpecification* and *CPRole*. *CPTask* is meant for classifying these entities which are real (goal-oriented) actions. Tasks are used in plans where they are related to another concepts called the *criminal process roles*. The skeleton typology of tasks is distinguished on the basis of the observation [15] that, to detect and to legally assess the economic offenses, the actions realizing the five flows should be trailed. These are the flow of business documents (*FlowOfDocumentsCPTask*, e.g. purchase orders, bills of lading, invoices or tax documents), the flow of goods (*FlowOfGoodsCPTask*), the flow of services (*FlowOfServicesCPTask*), the flow of money (*FlowOfMoneyCPTask*, e.g. bank transfer, cash transfer), and the information flow (*FlowOfInformationCPTask*, e.g. phone billings). Thus, we have five generic types of tasks in every *RecognizedPlanOfCP* plan. It is worth noticing that in general each of such tasks can be legal. But the sequences of them performed in special circumstances (for example, with forging or falsifying some documents) may be discovered as symptoms of a crime. For example, let us assume a chain of purchase orders (in one direction) followed by a sequence of invoices (in the opposite direction). If the relevant money transfers (the invoices payment) in the first direction is not accompanied by a shipment of goods (in the opposite direction), then the considered activity is fraudulent. This should be classified as a forbidden action (see the *CPNotion* below) on the basis of some inference rule (programmed in the supporting system). The same concerns criminal roles, which seem to be normal business roles but their engagement in actions being components of criminal sequences enables to classify them as illegal.

*CPStateSpecification* component is used to characterize the entities that are specifications of real states (e.g. being in a relationship or having an attribute of some value). Among such specifications there are *PreConditions*, *TerminatedResults*, and *StartedResults* used to describe the states connected with potential execution of a plan. They also enable to conceptualize the dynamics of a plan execution. The preconditions (*PreConditions*) are descriptions of states that are necessary to start the execution of a plan, the terminated results (*TerminatedResults*) describe states that are terminated through the execution of the plan and the started results (*StartedResults*), describe states, which are started by the plan execution. The state specifications, similarly to tasks, use some roles (namely the subject being in the state and, possibly, some objects concerning the state).

There are also activity specifications (*CPActivityDescription*). They can be useful to describe, for example, the activity category of some business firm. Similarly to tasks and state descriptions, the activity specifications use some roles. The same concerns the event description *CPEventDescription*.

The *CPRoles* are: *AgentiveCPRole*, *PatientiveCPRole*, *BeneficiaryCPRole*, *InstrumentalCPRole*, *ResultativeCPRole*, *MannerCPRole*, *LocativeCPRole*, *ReificationCPRole*, *SubjectiveCPRole*, and *ObjectiveCPRole*. Every role mirrors the semantic roles of arguments of predicates in sentences of a natural language. The reification role serves as a tool to represent tasks concerned with other tasks (the 'higher order tasks' can be defined through this ability). For example, let T1 be the task expressed by the sentence 'The subject S does the job J in the place P'. We have here the social agent S classified by the role, say 'performer', the job J classified by the patient-object role ('job') and the place P described by the locative role 'the place of doing the job'. Then the new task T2 appears expressed by the sentence 'The subject S2 charges the subject S1 with doing the job J in the place P'. Now, the agent S2 is classified by the role 'employer' and the only other argument of the predicate 'to charge with' is the reference to the task T1. Thus we reify T1 using the reification role R-T1 assigned to the task T2.

Among concepts the *CPSituationTime* and the *CPRedescriptionTime* are used to classify real time intervals.

The criminal situation descriptions (in fact, the inquiry versions) are expressed by a document that may be an *InquiryReport* or an *InvestigationReport*. It is an information object, which may have different physical realizations (i.e. paper or electronic ones, which are kinds of not mentioned up to now entities).

The intended meaning of conceptual elements given in Fig. 2 is explicated via the formal relations holding between them.

The being a recognized plan of a criminal process description deserves a special attention. This plan can be a complex structure describing the goal-oriented (and compound) action, so it has the main goal (*RecognizedPlanOfCPGoal*) as a proper part. There may also exist sub-goals that influence the main goal. The plan must use at least one agentive role that is targeted by at least one task. We also assume that plans use state, activity and event specifications (the two latter are not shown in Fig.2.). The plan may define the concepts, and use the state, activity and event descriptions.

## 3.2 The c.DnS Structures for an Inquiry and an Investigation

Similarly to a criminal process description, the *c.DnSForAnInquiryProcess* and *c.DnSForAnInvestigationProcess* structures represent the reality of an inquiry and an investigation. We describe them briefly.

The *InquiryProcess* inquiry process situation is described by mans of the *IPPlan* inquiry process plan that must be constrained by some *RegulativeNorm* regulative norm, in order to be law compliant (the inquiry process situation executes the regulative norm). The plan goal (*IPPlanGoal*) represents the main goal of the inquiry process plan.

There exists the inquiry social agent (*IPSocialAgent*), the member of the social agent collective (*IPSocialAgentCollection*), who perceives the inquiry process. It is mostly a domain of interest of internal intelligence agencies, but among the perceiving agents there may be a prosecutor agent who supervises the inquiry process and in such a position he defines parts of the inquiry process plan.

*IPTask* stands for inquiry process task meant for classifying these entities which are real actions. Tasks are used in plans and are related to another concepts called the *inquiry process roles*. Analogously to *CPRole* the typology of *IPRole* is given. Also, the concepts and documents of *c.DnSForAnInquiryProcess* resemble the elements from *c.DnSForACriminalProcess*.

The typology of the subprocesses of the *InquiryProcess* deserves a special attention. It was established on the basis of [10, 12, 14]. A number of subprocesses were distinguished, comprising two important processes of the generation of inquiry versions and the planning of an inquiry. The latter is conceptually interesting since it is the execution of some plan and its result is also a plan. Let us look closer at them.

The plan of the generation of inquiry versions defines four tasks, and one agentive role (the leader of an investigation team) that targets these tasks.

1. The generation of an inquiry version hypothesis by means of answering the questions about "what had happened?", "where?", "when?", "in what way?", "by means of what?", "why?" and "by whom?".
2. The recognition of gaps which potentially exist in all the inquiry version hypotheses.
3. The checking for verifiability of every inquiry version hypothesis.
4. The ordering of hypotheses according to the degree of their probability.

There are also other roles, and state and activity descriptions not mentioned here. The planning of the inquiry is also a process in which the leader of the investigation team plays the agentive role. The resulting plan is a description of a planning situation that is *assumed* (a kind of formal relation, stronger than *shares*) by a planning agent. The tasks in the inquiry planning process are as follows.

1. The adoption of one of the inquiry versions (ordered within the previously considered process).
2. The definition of the tasks which are connected with the plan describing the adopted inquiry version.
3. The definition of the additional tasks enabling to clarify the gaps in the adopted inquiry version.

4. The definition of the whole plan with tasks defined in 2. and 3.
5. The definition of agentive roles targeting the tasks which are defined in the generated plan.

The other roles and descriptions should also be defined.

The third process situation, namely the *c.DnSForAnInvestigationProcess* investigation process, is defined analogously to the inquiry process. The main differences are in the definition of the perceiving social agent (internal Police agencies) and of the types of the subprocesses. The investigation social agent perceives the investigation process (*InvestigationProcess*). The set of subprocesses resembles the similar set defined for the inquiry process except for some types of tasks, to carry which the Police is not allowed [10, 14].

### 3.3 The c.DnS Structure for a Workflow Process

The fourth considered process (Fig. 3) represents the dynamics of inquiries (and analogously, investigations). It was created on the basis of the COBRA solution [8].

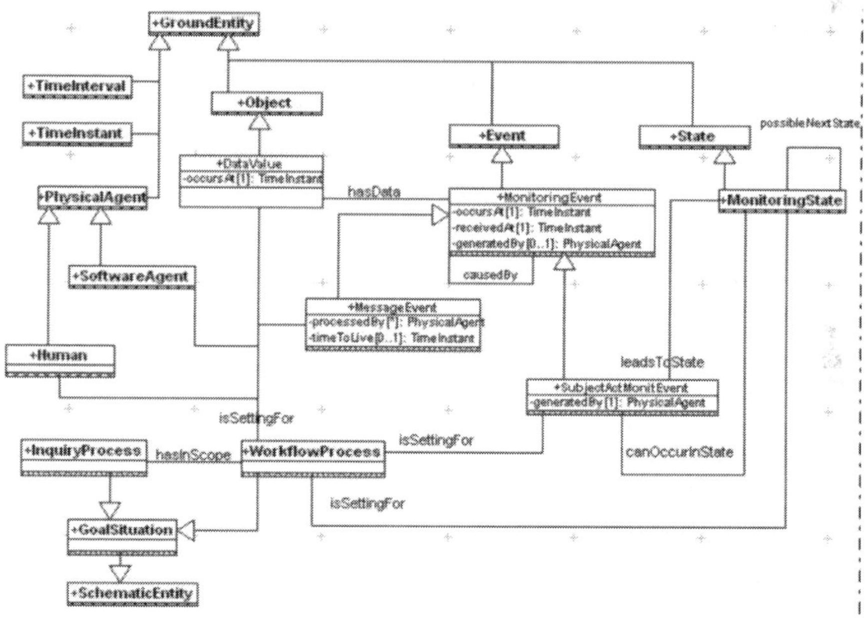


**Fig. 3.** Graphical representation of parts of the *c.DnSForAWorkflowProcess*

This approach requires the modeling of time, so the *TimeInterval* and the *TimeInstant* concepts and the set of temporal relations adopted from Allen and Vilain were introduced (see [8] for references). Also, the temporal entities (that span in *TimeIntervals* or occur at *TimeInstants*) were to be defined - as such the two specializations of the c.DnS *GroundEntity* were used, namely *Event* and *State*. Please note that in Fig. 3 we considered only the situation element (*WorkflowProcess*) of the whole c.DnS structure accompanied by the entities that have setting in this situation. The

*InquiryProcess* (*InvestigationProcess*) situation is also a setting for temporal ground entities, but for the sake of clarity they are not involved in Fig. 3. The temporal entities set in the *InquiryProcess* (*InvestigationProcess*) are subjects to sequences of *MonitoringEvents* of a *WorkflowProcess* that occur at *TimeInstances* and result in *DataValues*. Two types of such monitoring events were distinguished, namely *MessageEvent*, not generated by a *PhysicalAgent* (*Human* or *SoftwareAgent*) but processed by it, and *SubjectActMinitoringEvent* that is generated by a *PhysicalAgent* and leads to some (possible) *MonitoringState*. The following monitoring states are possible [8]: scheduled, assigned, ready (for not started subject actions or activities), running, suspended (for started subject actions or activities), aborted, terminated (for unsuccessfully finished actions or activities) or completed (for successfully finished actions or activities). The monitoring event can concern either an atomic subject action or activity, or a compound one. The compounds may be monitored as follows: instantiated, started, suspended, resumed, terminated, completed or aborted. For example, if the compound subject activity is in the 'ready' monitoring state the 'started' monitoring event leads to the 'running' state. The atomic subject action or activity is monitored by the following types of monitoring events [8]: assigned, started, reassigned, relieved, manually skipped, scheduled, aborted, completed, resumed, withdrawn, suspended or automatically skipped. For example, the atomic activity that is in 'suspended' monitoring state is transferred to the 'running' state after the 'resumed' monitoring event.

## 4 Conclusions

The main reason to build the ontology was to have, at first, a mean to 'handle' the semantics of the system in whole. The presented conceptual scheme is (in the most important parts) formalized in OWL ontology language by using the Protégé 4.0 (with the FaCT++ reasoner plug-in) ontology editor and it is published zipped in [11] (the starting file is PPBW.owl).

This version of the ontology was created on the basis of a description (made by the expert) of several cases of economic crimes that were legally assessed and sentenced a couple years ago. It occurred that there exists the common template by means of which the cases and their investigations can be modeled. We expressed this template using the constructive descriptions and situations paradigm because thinking of the semantics of states of affairs in this way appeared to be both compliant with the commonsense and easy. Additionally, within this framework the whole semantics modeling task can be conveniently divided into subtasks (specialization threads) that are modeled separately but at the same time they form a coherent whole.

Up to now one complete case of fraudulent money laundering was conceptualized and formulated as a specialization layer of the presented ontology. We plan to make more such descriptions to check the relevance of our ontology.

The formalizing of the whole ontology in an ontology language is also an important task. The formalization should be computationally tractable.

The ontology is now applied in the system supporting the prosecutor's inquiries and police investigations. It should be definitely equipped with new specializations such as the ontology of law.

This research is supported by the Polish Ministry of Science and Higher Education, Polish Technological Security Platform grant 0014/R/2/T00/06/02.

## References

1. Core Legal Ontology, http://www.loa-cnr.it/ontologies/CLO/CoreLegal.owl
2. Cybulka, J., Jędrzejek, C., Martinek, J.: Police Investigation Management System Based on the Workflow Technology. In: Legal Knowledge and Information Systems, Frontiers in Artificial Intelligence and Applications, vol. 189, pp. 150–159. IOS Press, Amsterdam (2008)
3. FFPoirot Financial Fraud Prevention Oriented Information Resources using Ontology Technology, http://ffpoirot.org
4. Gangemi, A.: The c.DnS ontology, http://www.loa-cnr.it/ontologies/collInt.owl
5. Gangemi, A., Lehmann, J., Catenacci, C.: Norms and plans as unification criteria for social collectives. In: Boella, G., Torre, L.v.d., Verhagen, H. (eds.) Proceedings of Dagstuhl Seminar 07122, Normative Multi agent Systems, vol. II, pp. 48–87 (2007), http://drops.dagstuhl.de/opus/volltexte/2007/910, ISSN 1862-4405
6. Hoekstra, R., Breuker, J., Bello, M.D., Boer, A.: LKIF core: Principled ontology development for the legal domain. In: Breuker, J., Casanovas, P., Klein, M., Francesconi, E. (eds.) Law, Ontologies and the Semantic Web, Channelling the Legal Information Flood. Frontiers of Artificial Intelligence and Applications, vol. 188. IOS Press, Amsterdam (2008)
7. Masolo, C., et al.: WonderWeb EU Project Deliverable D18. Library of Foundational Ontologies (2004), http://wonderweb.semanticweb.org/deliverables/documents/D18.pdf
8. Pedrinaci, C., Domingue, J., Alves de Medeiros, A.K.: A core ontology for business process analysis. In: Bechhofer, S., Hauswirth, M., Hoffmann, J., Koubarakis, M. (eds.) ESWC 2008. LNCS, vol. 5021, pp. 49–64. Springer, Heidelberg (2008)
9. SUMO Suggested Upper Merged Ontolog, http://www.ontologyportal.org
10. The Code of Penal Procedure (in Polish). Ustawa z 6.06, r. Kodeks postępowania karnego (1997)
11. The OWL version of the ontology of criminal processes and investigation procedures (2008) (in Polish), http://www.man.poznan.pl/~jolac/cDnSPL.zip
12. The Penal Code (in Polish). Ustawa z dnia 6 czerwca 1997 r. Kodeks karny (1997)
13. Valente, A.: Legal Knowledge Engineering: A Modelling Approach. IOS Press, Amsterdam (1995)
14. Widacki, J. (ed.): Criminalistics (in Polish): Kryminalistyka. C.H. Beck, Warszawa (2008)
15. Więckowski, J.: Regional Prosecutors Bureau in Katowice, Poland, Private Communication (February-May 2008)
16. Zhao, G., Gao, Y., Meersman, R.: An Ontology Based Approach to Business Modelling. In: Proc. of the International Conference of Knowledge Engineering and Decision Support (2004)
17. Zhao, J.L., Bi, H.H., et al.: Process-driven Collaboration Support for Intra-agency Crime Analysis. Decision Support Systems 41, 616–663 (2006)

# Ontology Applications for Achieving Situation Awareness in Military Decision Support Systems

Mariusz Chmielewski

Cybernetics Faculty, Military University of Technology, Kaliskiego 2,
00-908 Warsaw, Poland

**Abstract.** Common Operational Picture and in consequence Situation Awareness, had been second class issue treated as a result of decision support procedures. This work concentrates mainly on developing methods for integrating battlefield data using ontologies and presenting such data in distributed GIS environment supported by reasoning methods. Such approach allows to organize and filter data using predefined filters dedicated for various level of command and types of operations. Combining semantic data and spatial querying mechanisms, produce extendable and rapid mechanism for battlespace information presentation which vastly increases the speed of decision process. Published results contain analysis of known military domain models. Based on such study, author proposed, a core ontology for representing current battlefield scenario, filled with geospatial information and tactical data gathered from decision support algorithms and military domain analysis. The concept have been applied in presented designed and developed prototype in Service Oriented Architecture demonstrating, a Network Enabled Capability integrated battlespace picture.

**Keywords:** Military operations, Network Centric Warfare, situation awareness, ontology, semantic models, decision support.

## 1 Introduction

Network Centric Warfare or Network Enabled Capability are the main stream of technology R&D - key to dominating the future battlefield, depending on distributed information system providing filtered data at the right time and place. This article concludes research on the development of core battlespace ontology and its usage as a knowledge base for decision support tools. Current research in this field have aimed at development of methods which can be successfully applied in decision support systems to decrease the decision process innertion allowing maximum flexibility. Presented paper describes also designed layered architecture and a proof-of-concept prototype demonstrating, Network Enabled Capability integrated battlefield picture created using semantic reasoning services. To present real world application of the software have been supplied from external data sources in form of heterogeneous C4I (Command, Control, Communications, Computers, and Intelligence) systems used in Polished Armed Forces.

N.T. Nguyen, R. Kowalczyk, and S.-M. Chen (Eds.): ICCCI 2009, LNAI 5796, pp. 528–539, 2009.

### 1.1 Network Enabled Capabilities definitions

Network Enabled Capabilities [4] is term describing military doctrine or theory of war developed by the United States Department of Defence. The idea is based on information advantage, made available by the information technology, into a competitive combat advantage through the efficient network mechanisms delivered for geographically dispersed forces. Technology, combined with organization, processes, and people — may provide new forms of command (decision makers) behaviour. The theory supports fallowing statements:

- Efficient networked mechanisms can distinctly improve information flow and information sharing abilities;
- Information sharing enhances the quality of transferred information and in result situational awareness of the battlespace;
- Shared situational awareness enables collaboration between sensors, actors and command centers, reducing communication delays which in result speed up the decision process and increase mission effectiveness.

Described theory defines several concepts used as a product of information flow or as a tool for decision support procedures. First of them is Situation Awareness and the other Common Operational Picture. Both terms are highly connected with each other due to their specific domain – battlespace description.

### 1.2 Situation Awareness

Situation awareness (SAW) can be understand as a perception of environmental elements within a volume of time and space, the comprehension of their meaning, and the projection of their status in the near future [3]. SAW bounds perception and environment critical to decision-makers in complex, dynamic processes, in this case - military command and control. Situation awareness as stated before is achieved by the functionalities of C4I systems. Described ontologies and processing tool is aimed at data acquisition, data mining, decision support procedures representing picture of environment in the current combat space. Current research facilitates data gathered from all available battlespace dimensions loaded form C4I systems such as Szafran, Podbiał, Dunaj, MCCIS - Łeba. This process is in fact cumbersome as the complexity of data and described domain differ from each other not only from model point of view but most of all semantics. As an example there is a significant difference in task force representation in naval and ground forces same as with the unit ranks and the details needed for operation scenario. Mapping instances of units and military equipment is the main obstacle in creating COP for Joint Operations.

Imported data is the key to understanding environment, providing battlespace perception and preparing decision variants. Lack of data or having inadequate information and in result limited situation awareness has been identified as one of the primary factors in accidents attributed to human error.

Situation awareness is broader concept compared to Common Operational Picture which could be described as a visualization product used by the C4I systems for Joint Operation Command Centers. The full definition of COP can be provided from several sources which after analysis can be stated as:

*"Common Operational Picture is a single identical display of relevant information shared by more than one command. A common operational picture facilitates collaborative planning and assists all echelons to achieve situational awareness"*

COP's main aim is delivering general pointers and guidelines for command centers in their Area of Responsibility (AOR) and Area of Interest (AOI). Essential requirements for COP has been formulated in [1], in which it can be defined as a service or a set of services providing:

- collection of recognized pictures (identified scenario entrant forces);
- data fusion and correlation of recognized pictures;
- current shared operation picture containing elements from: RAP (Recognised Air Picture), RMP (Recognised Maritime Picture), RGP (Recognised Ground Picture), RLP (Recognised Logistics Picture), Intelligence systems, decision support procedures;
- filtering technology for commander as also if available interoperability link to source C4I system;

Common Operational Picture as a services is supported by Geospatial Information Services based on standardized data sources, supported by graphic and decision support functionalities. Evolving software design concepts influenced also range of decision support tools, providing standardized service decryption, unified registries and in the end semantic support provided by Web Service Modelling Ontology.

## 2 Semantic Models and Ontologies

Semantic data representation is based on simple graph model and is based on RDF concept of triples representing resources and data describing them. Triples identify subject-predicate-object expressions that is subject denotes the resource, predicate denotes traits or aspects of the resource and expresses a relationship between the subject and the object. Such model do not provide enough properties for reasoning except methods designed especially for semantic networks based on graph algorithms. Supporting these features is the ability to reason, which can be applied in field of combat scenarios in form of graph similarity measures representing decision schemas collected in expert system. Ontologies are the direct ancestors of description logic formalisms which were developed as a formal logic methods based on semantic networks and frames knowledge representations.

In case of semantic graphs inference mechanisms are based on Description Logics or specialized reasoner using developed graph based algorithms such as path finding, graph similarity etc. Considering structure of knowledge representation building new knowledge can be described as introducing new nodes or links between nodes to existing graph. The use of semantic metadata is also crucial to integrating information from heterogeneous sources, whether within one organisation or across organisations. Typically, different schemas are used to describe and classify information, and various terminologies are used within the information. Creating mappings between different schemas, allows to develop unified view and to achieve interoperability between the processes which use the information.

Practical approach for ontology definition has been provided by the W3C consortium. An ontology is used as a tool for describing and representing selected knowledge branch that is medicine, finances, battlefield etc. Definitions inside ontology must be defined precisely and unique from the logic point of view. Ontologies also posses the ability to classify knowledge from several domains, allowing it to be accessed by human but also applications and multiagent environments.

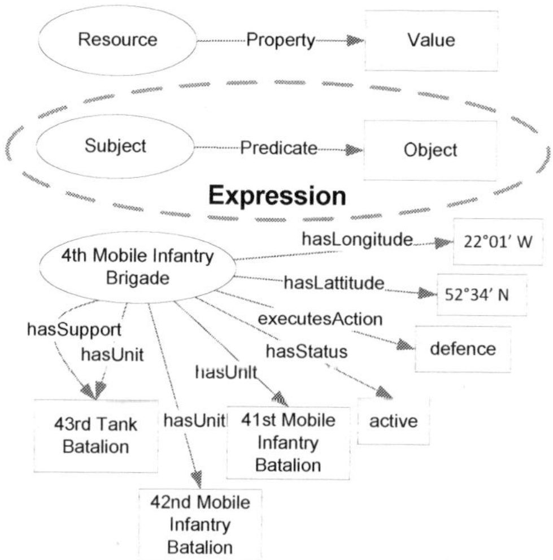


**Fig. 1.** Semantic data representation for simplified combat scenario based on example RDF representation

After analyzing several works describing ontology research, one of most complete definitions of ontology has been presented in [14]. This definition has been extended to provide wider approach for created models.

Ontology is defined as:

$$O = \left\langle L, F, G, \overline{F}, \overline{G}, C, R \right\rangle \tag{1}$$

where :

L is a set called lexicon and defined as follows:

$$L = L_C \cup L_R \tag{2}$$

$L_C$ – set of words describing concepts, called a concepts lexicon

$L_R$ – set of words describing names of relationships, called a relations lexicon

Elements F and G are reference functions defined as:

$$F : L_C \rightarrow 2^C \tag{3}$$

$$G : L_R \rightarrow 2^R \tag{4}$$

Function F assigns words from lexicon $L_C$ concepts which describe them. Analogous function G allows to assign words from relations lexicon $L_R$ to relations between concepts. In other words F and G functions are word interpreters, providing meaning for words. Assigning for each word several concepts is considered as a polyseme definition, that is a word or phrase with multiple meanings.

For modelling purposes definition of ontology has been extended towards definition of functions $\overline{F},\overline{G}$, which provide a tool for identifying words describing specified concepts and relations between them.

$$\overline{F}: C \rightarrow 2^{L_C} \tag{5}$$

$$\overline{G}: R \rightarrow 2^{L_R} \tag{6}$$

Assigning multiple words for separate concept or relationship, describes real world situation in which we call single concept by different words. This feature is represented by a synonym or polyvalent and can be recognized as a thesaurus service.

Mentioned earlier set C identifies the set of concepts used in developed model. It resembles object oriented approach for modelling. A concept is a representation of a group of objects containing shared characteristics set and indentified by the chosen word subset stored in the concepts lexicon $L_C$.

R Element of ontology is defined as follows:

$$R = \{\mathfrak{R} : \mathfrak{R} \subset C \times C\} \tag{7}$$

and is a set of relationships between concepts.

Relationships set is divided on two separate subsets: S identifying structural relationships and subset H containing hierarchical relationships.

$$R = S \cup H \tag{8}$$

It is assumed that hierarchical relationship is antisymmetric relation:

$$\forall a,b \in C \ (a,b) \in H \rightarrow (b,a) \notin H \tag{9}$$

and also transitive relation:

$$\forall a,b,c \in C \ (a,b) \in H \wedge (b,c) \in H \rightarrow (a,c) \in H \tag{10}$$

In object oriented methodology hierarchical relationships define class specialization or generalization. Other relations from set S are used to describe rest of available relationships within the domain.

Using presented ontology definition there can be formed a semantic network model:

$$SN^O = \langle I_C^O, I_R^O \rangle \tag{11}$$

where:

$I_C^O$ - is a set of instance of defined in ontology O concepts;

$I_R^O$ - is a set of instance of defined in ontology O relations;

Additionally we define:

$$I_C^O = \bigcup_{c \in C} Inst_c \tag{12}$$

where: $Inst_c$ - is a set of instances of chosen concept $c \in C$

$$I_R^O = \bigcup_{\Re \in R} Inst_{\Re} \tag{13}$$

where: $$Inst_{\Re} = \left\{(i,j) \in I_C^O \times I_C^O : (V_C^O(i), V_C^O(j)) \in \Re\right\} \tag{14}$$

Previous definition requires explanation of function $V_C^O$:

$$V_C^O : I_C^O \to C \tag{15}$$

which allows to classify an instance to the concept (class).

Such model does not include the instrumental view on the ontology definition which can be considered as graph structure composed with:

1. vertices – representing concepts identified within the modelled domain;
2. edges – representing relationships connecting identified concepts;
3. data records assigned to concepts or relation – representing a set of instances assigned to a particular concepts - directly connected with developed semantic network.

Tools designed for ontology modelling based on dedicated languages such as DAML+OIL and in the end OWL (Ontology Web Language) itself share the same theoretical framework extended towards Description Logics to provide reasoning mechanisms.

## 3 Ontology Modelling Methodology

The core element of the designed tool is usage of an ontology and therefore a semantic model representing the battlespace. Each methodology structural, object oriented, process has well defined language and design approach. Several research projects examined the domain analysis and introduced various views on ontology modelling.

Considering the results and similarities within the methods, the process of ontology development has been updated to cope with automatic model transformation and its refinement and in final stage consist of:

1. Definition of researched domain and boundaries of modelling which was conducted based on domain descriptions of military operation specialist;
2. Existing ontology (domain models) overview and their reference which in this case is mainly connected with the integration of JC3 IEDM model [2],

App6A military symbology model. This phase mainly was connected with the development of transformation rules for the MIRD JC3 model;
3. Definitions and validation of elementary abstractions within the domain – creation of core concept list and term preparation of overlapping terms in source ontologies;
4. Concept taxonomy modelling - class definition and their hierarchy based on main abstracts within battlespace domain;
5. Class property modelling – identifying properties (slots) for classes (domain and range definition);
6. Property restriction definition – object type or datatype specification, in terms of value restrictions and cardinality restrictions which mainly have been extracted from meta model;
7. Properties refinement allowing characterisation of the relation between individuals (specialisation, inverse properties, equivalent properties, special properties)
8. Identification of instances and their classification within the class taxonomy;

Ontology development for the SAW enabled tool required detailed C4I system data sources in form of joint operation scenarios involving all battlespace aspects.

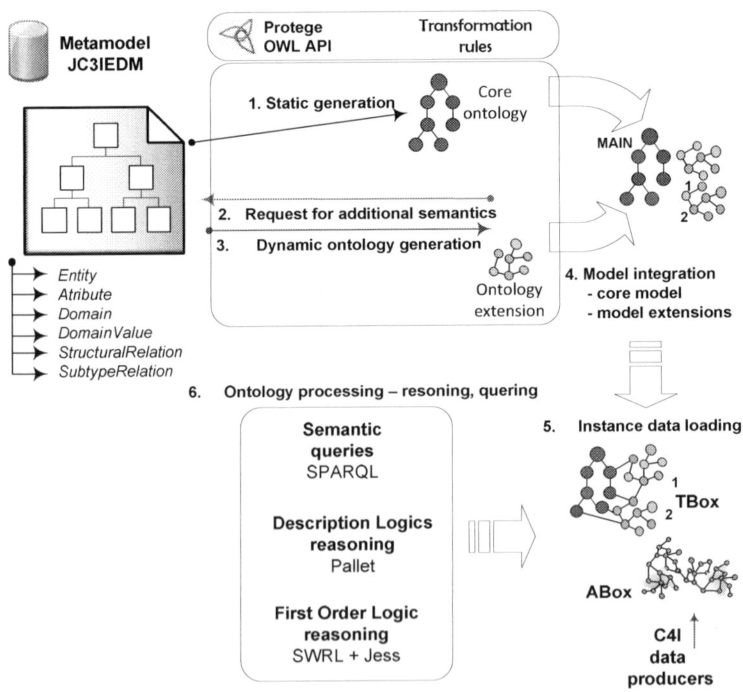


**Fig. 2.** The idea of JC3 model transformation, based on identified rules for full semantic projection of military operations. Transformation rules define detailed concept creation and usage of unions and disjoint clauses along with properties and their refinement in form of value and cardinality restrictions.

Unified Battlespace Ontology Model (UBOM) has been divided into two parts:

- MIP JC3IEDM model based ontology describing domain of military operations, battlespace environment, materiel and equipment;
- Decision process ontology – containing concepts of decision, course of action and the rest of decision process realized on the battlefield;

The basis for the JC3 IEDM ontology was model transformation based on rules that will reflect the source model full semantics. Analysis of JC3 standard identified 290 entities, 1594 attributes, 259 associations, 166 supertype/subtype relations. Transformation of the whole model due to its physical implementation in form of relational database was mainly cumbersome. To overcome filtering of relational model mandatory elements, the method uses JC3 IEDM meta model organized to provide full data of all JC3 contents (entities, attributes, keys, relations, domains). Analysis of all JC3 elements allowed to design and verify ontology equivalent and most of all provide additional semantic meaning for created taxonomy and relations.

Developed ontology as a successor of source JC3 conceptual model has been transformed using generic rules. As a result in several earlier versions it represented relational model stored in OWL rather than ontology itself. Manual refinements of the model helped to extend ontology elements and most of all construct reasoning conditions as class constructors. This directly helped to demonstrate inference abilities of created model. Semantics of JC3 model are mainly stored not in the static structure of the model but in the domain values used as enumeration types and business rules defining valid combinations of domain values (similar to enumerations). This is mainly connected with the semantic statements for the units stored in ABox that define the unit capabilities armor characteristics, firepower and abilities to destroy targets in different battlespace dimensions (air, ground).

## 4 Overcoming Ontology Differences for Data Integration

Building ontology model consisting of separate domain descriptions facilitates overcoming semantic differences between ontologies. Process of reconciliation of these differences is called ontology mediation, which enables reuse of data across C4I systems and, in general, cooperation between different battlespace dimensions. Considering context of semantic knowledge management, ontology mediation is important due to data sharing between heterogeneous knowledge bases and data reuse. Emerging SOA technologies pointed another important application area for ontology mediation, that is Semantic Web Services. In general, such technologies as WSMO [8] (Web Service Modeling Ontology) allow to provide a shared terminology layer for the service requester and service provider which in case of C4I systems is especially important because of separate battlespace characteristics and description in case of Air, Maritime and Ground operations.

Sources describe two ways of providing ontology mediation: ontology mapping and ontology merging. Ontology mapping, the correspondences between two ontologies are stored separately from the ontologies and thus are not part of the ontologies themselves. The correspondences can be used for, querying heterogeneous knowledge bases using a common interface or transforming data between different representations – this is the case proposed in the described data fusion method.

Described method uses five stages of mapping:

1. Importing the content of the ontologies to OWL DL and normalization of defined vocabularies through elimination of lexical and syntactical differences;
2. Similarity evaluation of ontology entities using defined set of parameters (ontology quantity analysis - predefined sets of indicators);
3. Establishing correspondences between similar entities (concept thesaurus), in the form of semantic bridges – linking similar concepts;
4. Utilizing developed mappings for instance transformation
5. Revision of prepared mappings for improvements.

Discovering ontology alignment is the process of discovering similarities between two source ontologies. Te result is specification of similarities between source ontologies based on match operator in this case the base JC3IEDM ontology, App6A ontology and decision model. The input of the operator is a number of ontology and the output is a specification of the correspondences between the ontologies. Sources distinguish several algorithms, that have been applied in this work, which implement the match operator:

- *schema-based matcher* takes different aspects of the concepts and relations in the ontologies and uses some similarity measure to determine correspondence [5],
- *instance-based matcher* takes the instances which belong to the concepts in the different ontologies and compares these to discover similarity between the concepts [7].

## 5 Architecture of Developed Semantic Decision Support Tool

Development of the ontology had been the first step in designing decision support tool, supporting commanders as the model itself is useless without mechanisms for supplying individuals in ABox and providing reasoning and querying algorithms. Developed software is kept as monolith, based on Java technology extended by numerous top rated OpenSource projects. Such design allowed to achieve maximum flexibility of extending and tuning main parts of the project simultaneously accomplishing maximum code transparency. Main obstacle to manage such construction was to pinpoint most extendable framework among the ones that has been chosen – in this case the Protégé based on emerging OSGi framework. The OSGi framework implements dynamic component model, one of the flaws of standalone Java Virtual Machine environments. Applications or components developed in the form of bundles ready for deployment, can be remotely installed, started, stopped, updated and uninstalled without requiring a reboot. Framework itself helps to overcome the JAR versioning problems. Life cycle management is supported by the APIs which allow for remote downloading of management policies. Similar to existing registry technologies CORBA, RMI, UDDI – OSGi defines service registry that allows bundles to detect the addition or removal of services, and adapt accordingly.

The layered architecture links dynamic data streams from many sources from existing command and communication systems using for integration WebServices technology [9], [10], [12]. The whole subsystem has been delivered as set of SOA WebServices, dedicated for different system and device platforms including extension towards mobile devices. This approach has been extensively tested on Java and .NET platform and offers effective migration capabilities for data sources and communication efficiency important in mobile platforms. Data migration solution, uses distributed federation databases dedicated for joint operation planning. Legacy systems in this case incorporate main Polish Armed Forces command and control systems for all battlefield aspects: Kolorado, Szaftan – for the Army, Łeba-MCCIS for the Navy and Dunaj, Podbiał for the Air Forces. Data gathered and stored by the separate federates are often saved in different form, and extracted by different mechanisms. WebServices implementation was based on Apache Axis API. Provided mechanism is used only to collect real data in relational or object oriented form allowing integration of migration server on several levels including transition to ontology models. This solution provide generic way to exchange data through XML data representation inside transmitted SOAP envelopes and moves the data conversion logic to client side which increases the flexibility of solution.

Protégé is treated as an integration platform, providing on one hand a modeling tool for ontology development, processing, and reasoning on the other hand a perfect platform to integrate other sophisticated tools such as:

1. OpenMap - open GIS framework providing core mechanisms for COP visualization and semantic queries;
2. NASA WorldWind – 3d projection visualization toolkit providing a Earth model with a satellite imagery cover and detailed Earth terrain model;
3. GeoServer framework – Web Map Service platform providing spatial information services and conversion tool for raster formats;
4. JENA Semantic Framework – framework dedicated to process semantic models stored in RDF, RDFS, DAML and OWL, supporting SPARQL query language, inference mechanisms and graph persistence;
5. JESS Rules – rule based inference engine which uses an enhanced version of the Rete algorithm to process rules. Tool provides also a extended scripting language for Java allowing to build applications using imperative language;
6. Shrimp Visualization Toolkit – a innovative method of visualization large scale data extended towards ontology models and implemented as a Protégé plug-in Jambalaya;

Each of those tools offer an API provided to use the environment and to extend it towards desired functionalities. Those extensions must be developed with caution especially in case of frameworks providing own execution environment or dynamic management component subsystem.

The environment functionalities have been divided into four main subsystems:

1. Data migration subsystem – responsible for migrating relational or object oriented entities using WebServices and XML representation from chosen C4I systems through JC3 IEDM standard and further data transformation in form of instances to ontology model ;
2. Model processing subsystem – responsible for model transformations and analysis, providing merge procedures, inferencing, querying semantic models;
3. GIS rendering subsystem - Operational Picture visualization using selected GIS frameworks supported by distributed spatial services in form of RMI servers and Web Map Services;
4. Ontology visualization subsystem – responsible for intelligent ontology visualization tool supporting all ontology entities filtering;

During development phase detailed analysis formed requirements for each dimension of battlespace – Ground, Air, Maritime. Those requirements based on the legacy

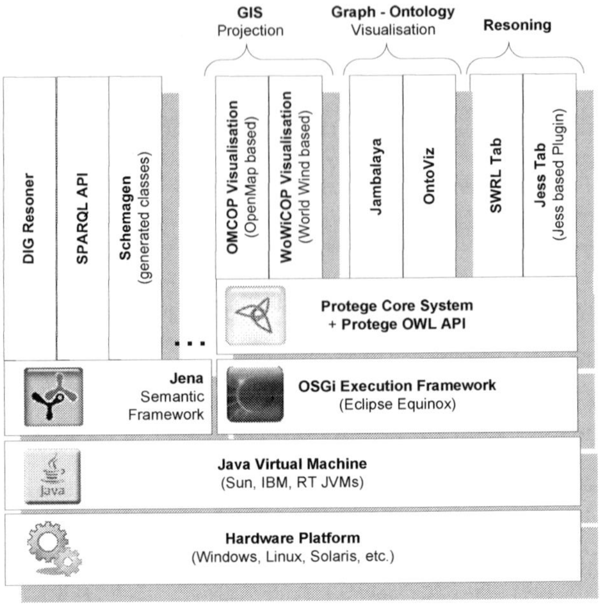


**Fig. 3.** Tool components, based on OSGi Execution Framework including JENA integration layer and Protégé plug-ins

systems visualization engines, contained set of map standards used for rendering and object symbology.

GIS data providing engine has been designed as a distributed environment using Java RMI servers providing mechanisms for CADRG, DTED, ESRI Shape files and NATO App6A based symbology standard. Using such distributed environment can improve system stability and memory utilization due to the fact that each server is maintained in a separate JVM.

Recent development of Web Map Server technology and its deployment especially in satellite imaginary allowed to extend provided layers towards presenting aggregated meteo information or satellite imagery of the battlefield using time filter options. Large spectrum of servers such as MODIS Blue Marble, DEMIS Bathymetry, Onearth Landsat 7 (JPL) extend the variety of delivered Geospatial Information which provides additional information for commanders. OpenMap mechanism allowed also to prepare, a query interface for semantic model which uses the JENA SPARQL engine to filter and merge battlespace information based on the commander responsibilities and preferences. This task could be accomplished due to the extensions of the designed ontology to facilitate geoRSS. Based on such construction of data description (units, objects, infrastructures) we can extend the functionality of tool using Geographic Modeling Language, towards semantic spatial queries.

## 6 Summary

Developed semantic model equipped with ontology tools, demonstrates feasibility study of integrated dynamic battlefield, providing large scale information resources in heterogeneous systems. Extending such system with SOA specific technologies, allowed integration of legacy systems and the ability to easily substitute alternative components to meet specific interoperability requirements. Extensions of battlespace ontology model, developing semantic processing mechanism provide new techniques for accomplishing information superiority. Developed GIS tools present new means of exploring battlespace which can be utilized as a remote reconnaissance tools constructing elements for Situation Awareness in current combat scenario. One of the future development directions are mobile solutions providing CTP or COP solution. SOA based environment and specialized software platform for variety of PDA devices provide means of decision support tools on lower level command and can greatly improve the information flow in decision process.

**Acknowledgments.** This work was partially supported by grant "Research Project No PBZ-MNiSW-DBO-02/I/2007 concerning achieving Network Enabled Capabilities in Polish Armed Forces".

## References

1. NATO, NATO Common Operational Picture (NCOP) - NC3A, 16.11.2006 r
2. NATO MIP, The joint C3 information exchange data model overview
3. Endsley, M.R., Garland, D.J.: Situation Awareness Analysis and Measurement: Analysis and Measurement. Lawrence Erlbaum Associates, Mahwah (2000)
4. Alberts, D., Garstka, J., Stein, F.: Network Centric Warfare: Developing and Leveraging Information Superiority, Library of Congress, 2nd edn. (1999)
5. Noy, N.F., Musen, M.A.: PROMPT: Algorithm and tool for automated ontology merging and alignment. In: Proceedings 17th National Conference on Artificial Intelligence, Austin, Texas, USA (2000)
6. Ehrig, M., Staab, S.: QOM – quick ontology mapping. In: McIlraith, S.A., Plexousakis, D., van Harmelen, F. (eds.) ISWC 2004. LNCS, vol. 3298, pp. 683–697. Springer, Heidelberg (2004)
7. Doan, A., Madhaven, J., Domingos, P., Halevy, A.: Ontology matching: A machine learning approach. In: Handbook on Ontologies in Information Systems. Springer, Heidelberg (2004)
8. Roman, D., Keller, U., Lausen, H., Bruijn, J., Lara, R., Stollberg, M., Polleres, A., Feier, C., Bussler, C., Fensel, D.: Web Service Modeling Ontology. Applied Ontology (2005)
9. Trastour, D., Preist, C., Coleman, D.: Using Semantic Web Technology to Enhance Current Business-to-Business Integration Approaches, HP Labs (2008)
10. Mohammadian, M.: Intelligent Agents for Data Mining and Information Retrieval. Idea Group Publishing, USA (2004)
11. Davies, J., Fensel, D., Harmelen, F.: Towards the Semantic Web: Ontology-driven Knowledge Management. In: HPL-2003-173. John Wiley & Sons, Chichester (2003)
12. Herrmann, M., Dalferth, O., Aslam, M.A.: Applying Semantics (WSDL, WSDL-S, OWL) in Service Oriented Architectures (SOA). In: 10th Intl. Protégé Conference (2007)
13. Currie, E., Parmelee, M.: Towards Knowledge-Based Solution for Information Discovery in Complex and Dynamic Domains. In: 7th Intl. Protégé Conference (2004)
14. Maedche, A., Staab, S.: Comparing Ontologies— Similarity Measures and a Comparison Study, Internal Report No. 408 (March 2001)

# A Collaborative Ontology-Based User Profiles System

Trong Hai Duong, Mohammed Nazim Uddin, Delong Li, and Geun Sik Jo

School of Computer and Information Engineering,
Inha University, Korea
haiduongtrong@gmail.com, nazim@eslab.inha.ac.kr,
lidelong@eslab.inha.ac.kr, gsjo@inha.ac.kr

**Abstract.** The main goal of this research is to investigate the techniques that implicitly build ontology-based *user profiles*. In particular, automatically building profiles based on user's information (blogs, publications, home pages,...) generated from the internet. We proposed a framework to search the user details to build profile automatically. An initial profile is constructed with user interest in hierarchical manner and this profile is learned by assigning user details collected by our search method. Main focused on how quickly we can collect user's information and achieve a profile stability, and how effectively improve profiles. Along with the framework we also consider a new method to extract feature from document. A Wordnet and Lexico-syntactic pattern for hyponyms approach is applied to export the important feature of document to represent the profile ontology. The *user profile* further improved by learning interesting knowledge from similar profiles.

**Keywords:** Ontology, *user profile*, Ontology-based *user profile*, Ontology Integration.

## 1 Introduction

User profiling is a common and important techniques to enhance the personalization in information retrieval. *User profiles* mainly handle the record of user context in various web based application to adapt user needs. Numerous researches are focused on building and managing *user profile*. The *user profile* can be build explicitly by asking the user to provide necessary information about him or implicity, by watching the user's activities on particular application [15].

In personalization, user preference are considered to make the effective search. Most Personalization approach is mainly based on user modeling or user profiling. *user profile* may contains very general information about user such as demographic one, e.g., name, age, address, educational background,etc,and, also contains some specific information such as interest or preference either individual or of a group. Many researches are carried out based on personalization or modeling users for web search and browsing [10], [9] or other purpose such as to enhanced e-commerce activities. Profiles are build according to the different techniques based on personalized system. Nowadays, personalization systems are developed by considering ontology to reduce the limitation of traditional information retrieval such as information overload or cold start problem. So considering ontology to build an accurate profile bring some extra benefit in user modeling. An *user profile* can be presented as a weighted concept hierarchy for searching and browsing [4] in the web. *user profile* can be own

N.T. Nguyen, R. Kowalczyk, and S.-M. Chen (Eds.): ICCCI 2009, LNAI 5796, pp. 540–552, 2009.

created by user with his/her personal information and interest or it can be a reference one. However, profile can be created by manually entering the user's information or automatically by watching the use's activities. In [4] *user profile* is created by analyzing the surfing behavior of user. By analyzing the user's log file information could be collected to create profile. Additionally, user information can be collected from bookmark also [14]. After building the initial profile it can be learned or refined by various methods. Profile can be improved by ranking the concepts as in [15] for searching the web documents. A personal ontology or *user profile* can be described as a reference ontology like ODP [8] and learned by neural network methods to improved [11]. Existing literature, application and specific domain of user context could be taken into account to create a comprehensive *user profile* [5].

In this paper, we investigate the techniques to create an *user profile* automatically using the ontological approach. We use a framework to gather the user information from different search space where user's details could be found. The details include user's general information to specific preferences. We use meta search in user's blog, personal/organization web page, and any other cites to collect information about user. These information are assigned to a pre- structure hierarchy or in a reference ontology to create an initial *user profile*. More clearly, initial profile is learned by the concept/document collected from user's details. In traditional user profiling system feature extraction from document is done by vector space model or considering term frequency, $tf - idf$ methods only. In our case, we considered WordNet and Lexico-Syntactic pattern for hyponyms to extract feature from document. This profile further improved by taking collaborative user methods. Where, we find a group of users with similar interest by taking similarity score among them. After that an ontology matching approach is applied to learn the profile with other similar user which we called improved profile. The paper is organized as following manners. Section 2 describe our framework for building ontology-based *user profile*, in section 3 we build *user profile* according to our framework, improving the user's profile with collaborative matching is explained in section 4. Section 5 includes the experimental explanation and finally, conclude the paper in the end section.

## 2 Ontology-Based User Profiles

### 2.1 Ontologies

An ontology is considered as the specific concepts into a hierarchical manner with relations among them. Ontology can be represented as the structuring information like conceptual graph [6] , where, queries and resources can be presented. It can be comprise of simple taxonomic structure of concept like [8]. The ontologies may differ from simple structural relation to a powerful organization of information like $SHOE$ [7]. Ontology is defined as set of object and the describable relationships among them to form a registrational vocabulary for a knowledge-based [11].

### 2.2 User Profile

*User profile* can static. In static *user profile* user permeant information are included where as in dynamic *user profile* less permanent characteristics like user's current motions, locations are mentioned. We will consider two categories of user's information. First, we will define the demographic information such as name, age, sex, educational background etc. Second, we will consider the user interest which often changes.

**Definition 1 (Interesting Information).** *An user's interesting information is defined as an ontology with a pair:*

$$O = (C, \widehat{\sqsubseteq}) \tag{1}$$

- *$C$ is a set of concepts (interesting topics),*
- *$< C, \widehat{\sqsubseteq} >$ is the taxonomic structure of the concepts from $C$ where $\widehat{\sqsubseteq}$ is the collection of subsumption relationship ($\sqsubseteq$) between any two concepts from C. For two concepts $c_1$, $c_2 \in C, c_2 \sqsubseteq c_1$ if only if any term $(t_2, f_2) \in P_2$ that are the pattern of concept $c_2$ also are the members of the patten $P_1$ of the concept $c_1$, and it is not vice versa*

**Definition 2 (Interesting Topic).** *An interesting topic is defined as the concept of an ontology-based* user profile *with a quintuplet:*

$$\text{concept} = (D^c, S^c, \overrightarrow{S_c}) \tag{2}$$

*where*
- *c is the unique identifier of the concept,*
- *$D^c$ is a collection interesting documents assigned concept c,*
- *$S^c = (t_1, t_2, \ldots, t_n)$ be the collection of all key-objects (terms) in the document collection $D^c$,*
- *$\overrightarrow{S^c} = (w_1, w_2, \ldots, w_n)$ is feature vector of the concept c, where $w_i$ is the weight distribution of term $i$ in $S^c$, $0 \leq w_i \leq 1$.*

## 2.3 System Overview

In this section we will describe our system overview about building ontology for *user profile*. In the figure profile management(PM) is the main component to build and manage the *user profile*. An input is given to Profile Management, input is basically an user name or a person name. PM passes this input to Management Search(MS). MS is basically a meta search where users details are investigated from different sources. the searching details are explained in later sections. After gathering the user's details it's compare with a pre-defined user's information structure. Pre-defined user's information structure is nothing but a reference ontology like ODP (Open directory Project) which we called initial *user profile*. Initial *user profile* is constructed by considering user's interest in a hierarchical manner. The details of this initial profile is stored in database. User information searched by the MS is compered with initial profile. Comparison is based on the similarity measured of feature vector of document. The WordNet is also used to extract the feature vector of the document. Based on similarity between concepts in initial profile and concepts in user's details, the initial profile is learned. After learning the initial profile it is considered as user's personal Profile.

Personal profile can be further improved by taking collaborative approach which is represent as integration part in our system overview. In collaborative approach, number

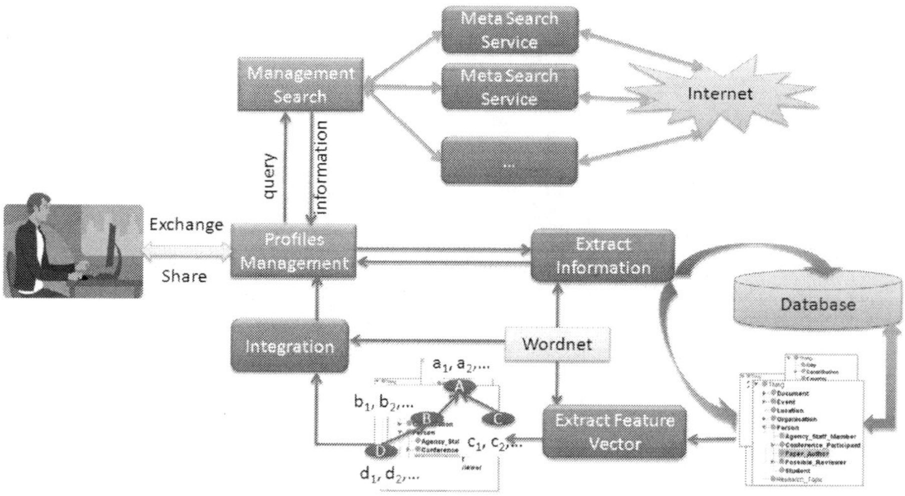


**Fig. 1.** Overview System

of similar personal profiles are considered first and the ontology matching techniques are applied to achieved the improved profile. Onotology matching details explain in section 4.

## 3 Building Ontological User Profile

This section represent the process of building ontological *user profile* based on the techniques described in previous section. Hierarchical user's information in initial *user profile* is treated as an ontology. More clearly, We learned and improved the users profile in the form of ontology.

### 3.1 Information Collection

Information about user(who's profile to be build) collected by searching in the internet by SM as mention in figure 2. Information about user is searched in different spaces like blog, personal web portal or in an organizational web page etc., which is define more clearly in below figure 3.

**Example.** In *Figure* 3 we give an example for a user named Hai. Many results of publications, blogs, webs and home pages talk about Hai. However just some of them tell that Hai has been studied at Inha university and has work for QB University for 6 years. They also tell that Hais birth day: 2009/06/29 Thus they have a common things about Hai. From that MS can identifies Hais information which may be included blogs, home pages, publications, webs.

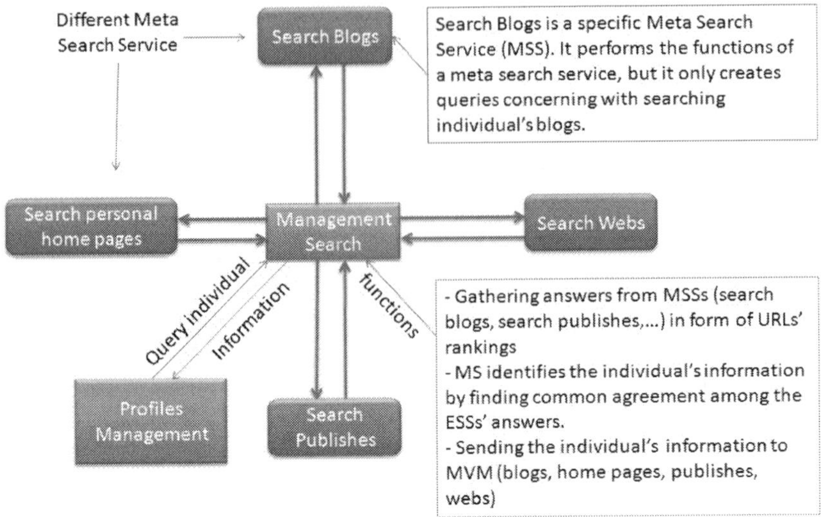


**Fig. 2.** Different Meta Search Service

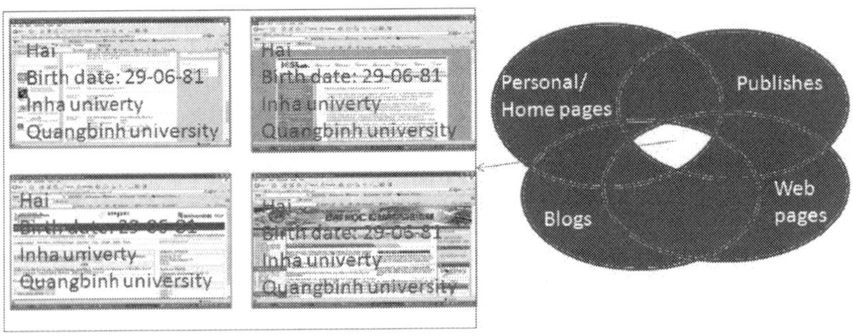


**Fig. 3.** A consensus among different resources

## 3.2 Document Representation

**Feature Extraction.** The $tf–idf$ weight (term frequencyinverse document frequency) is a weight often used in information retrieval and text mining. This weight is a statistical measure used to evaluate how important a word is to a document in a collection or corpus. Here we use traditional vector space model($tf$-$idf$) to define the feature of the documents.

**Definition 3 (Feature Vector of Document).** *Let $T^d = (t_1, t_2, \ldots, t_n)$ be the collection of all of key-words (or terms) of the document d. Term frequency $tf(d, t)$ is defined as the number of occurrences of term t in document d. A set term frequency pairs, $P^d = \{(t, f)|t \in T^d, f > threshold\}$, called the pattern of document. Given a pattern $P^d = \{(t_1, f_1), (t_2, f_2), \ldots, (t_m, f_m)\}$, let $\overrightarrow{d}$ be the feature vector of document d and* *let td be the collection of corresponding terms to the pattern, we have:*

$$\overrightarrow{d} = (w_1, w_2, \ldots, w_m) \tag{3}$$

$$td = (t_1, t_2, ..., t_m) \tag{4}$$

*where*

$$w_i = \frac{f_i}{\sum_{j=1}^{m} f_j} * \log \frac{|D|}{|d : t_i \in d|} \tag{5}$$

**Definition 4 (Feature Vector of Set of Documents).** *Let $P^i = \{(t_1, f_1), (t_2, f_2), \ldots, (t_m, f_m)\}$ be the pattern of the document i belonging to the set of document ds, i=1..n. A set term frequency pairs, $P^c = \sum_{i=1..n} P^i$, called the pattern of the ds. Let $\overrightarrow{ds}$ be the feature vector of the ds and let tds be the collection of corresponding terms to the pattern, we have:*

$$\overrightarrow{ds} = (w_1, w_2, \ldots, w_k) \tag{6}$$

$$tds = (t_1, t_2, ..., t_k) \tag{7}$$

*where*

$$w_i = \frac{f_i}{\sum_{j=1}^{m} f_j} * \log \frac{|D|}{|d : t_i \in d|} \tag{8}$$

- $|D|$*: total number of documents in the corpus*
- $|d : t_i \in d|$*: number of documents where the term $t_i$ appears (that is ). If the term is not in the corpus, this will lead to a division-by-zero. It is therefore common to use 1+ $|d : t_i \in d|$.*

### 3.3 Training

- Bulding a ontology representing common user's profile.
- Assigning documents to the ontology.
- Computing feature vector of each category (concept), including following method:

  - For each leaf concept, the feature vector is calculated as the feature vector of set of documents:

$$\overrightarrow{S^c} = \overrightarrow{ds_c} \tag{9}$$

  - For each none-leaf concept, the feature vector is calculated by taking into consideration contributions from the documents that have been assigned to it ($D^c$), its direct sub concepts ($D^{c'}$, for any $c'$ is direct sub concept of $c$):

$$\overrightarrow{S^c} = \alpha\overrightarrow{ds_c} + \beta\overrightarrow{ds_{c'}} \tag{10}$$

  where $\overrightarrow{ds_c}$ and $\overrightarrow{ds_{c'}}$ is corresponding to the feature vectors of the sets of document $D^c$ and $D^{c'}$, $0 \leq \alpha, \beta \leq 1$ and $\alpha + \beta$=1.

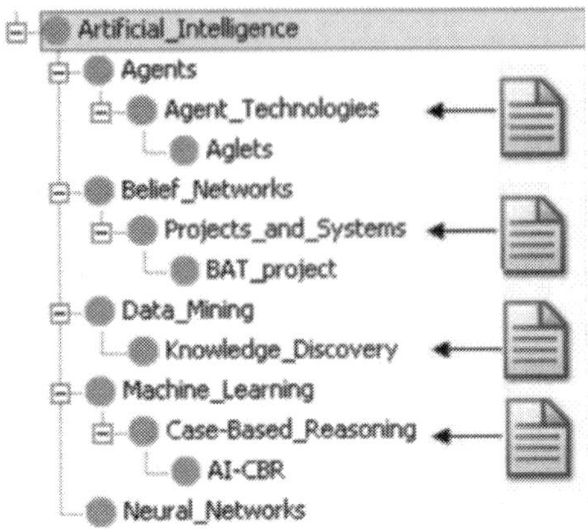


**Fig. 4.** Document Assignment

**input** : Given set of key word (keyWord) concerned with the user and the pre-defined profile (commonProfile) which was trained as above mentioned
**output**: Corresponding user's Profile

/* Collecting use's information */
infoUser ⟵ `CollectingInfoUser` (keyWord);
featureVector ⟵ `ExtractingFeatureVector` (infoUser);
commonProfile ⟵ `AscendingImportantConcept` (commonProfile);
**foreach** *each feature vector d of each document belong to* featureVector **do**
  **foreach** *each concept c belonging to* commonProfile **do**
    /* Similarity between a comcept and document */

$$sim(d,c) = sim(\overrightarrow{d}, \overrightarrow{S^c}) = \frac{\sum_{(T_i^d, T_i^{c2}) \in K} (d_i * S_i^c)}{\sqrt{\sum_{i=1}^{n} (d_i)^2} * \sqrt{\sum_{i=1}^{n} (S_i^c)^2}} \tag{11}$$

    where $K = \{(T_i^d, T_i^{c_2}) | \texttt{Sims}(T_i^d, T_i^{c_2}) = 1\}$
    **if** match ≥ threshold **then**
      **if** *the concept c does not exist in* userProfile **then**
        creating new concept $c$ for the userProfile;
      **end**
      adding document $d$ into the corresponding concept $c$ belonging to userProfile;
      break;
    **end**
  **end**
**end**
Return(userProfile);

**Algorithm 1.** Automatically building user's profile

### 3.4 Automatically Building User's Profile

After collecting the relevant document a number of feature is extracted by above mention equation which are the general candidate concept of personal ontology. With these concepts a general personal ontology can be build automatically as shown algorithm 1.

## 4 Improve User Profile

we can improve the *user profile* by taking collaborative approach. First, a top $N$ user's profile are found by matching Important concept in the ontology by algorithm 2. After form the neighbor missing or dissimilar concepts are added to target user's profile by Algorithm 3.

### 4.1 Ontology Macthing

Ontology Matching is done by applying similarities on selected concept which we called Important concept define in definition below.

**Importance Concepts**

**Definition 5 (Importance Concept)**
*Given a pattern* $P^c = \{(t_1, f_1), (t_2, f_2), \ldots, (t_m, f_m)\}$ *of the concept* $c$

$$support(P^c) = \frac{\sum_{i=1}^{m}(f_i)}{m} \tag{12}$$

*if* $support(P^{c_1}) \geq support(P^{c_2})$ *then the concept* $c_1$ *is more important than the concept* $c_2$

*if* $support(P^{c_1}) = support(P^{c_2})$ *and* $c_1$ *there is more importance sub concepts than* $c_2$ *then the concept* $c_1$ *is more important than the concept* $c_2$

$$support(P^{c_1} \oplus P^{c_2}) = support(P^{c_1}) + support(P^{c_2}) \tag{13}$$

**Matching Algorithm.** In this section, we apply the aforementioned methods to design an effective algorithm for ontology matching. An overview of the algorithm as follows:

### 4.2 Collaboratively Improve Profile

After Matching between ontologies uncommon concepts are added in target ontology to improve the profile. Which can be described as Algorithm 3.

We consider the line 6 in the algorithm. The ImproveProfile($p$, $smatch$) procedure is to improve target profile $p$ from set of simlilar profiles $smatch$. Here we introduce two statements to perform this.

- a new concept $c$ is added to the target, which is the consensus comcept among the ontologies belonging to $smatch$.

- a new concept $c$ is added to the target, which is sub-concept of a strong concept $t$ belonging to $smatch$, where the concept $t$ also belong to the target.

Notice that the strong concept is evaluated by the *importance concept* as above presented.

**input** : Given two ontologies $O$ and $O'$
**output**: Pairs of concepts are equvalent

/* Importance concepts list of each ontology */
dJ ⟵ `ImportanceConcepts` ($O$);
dJ' ⟵ `ImportanceConcepts` ($O'$);
**foreach** *each concept c belonging to* dJ **do**
  **foreach** *each concept c′ belonging to* dJ ′ **do**
    /* Similarity between two concepts */

$$\text{match} = \texttt{Simf}(c,c') = \texttt{Simf}(S^c, S^{c'}) = \frac{\sum_{(T_i^c, T_i^{c'})\in K}(S_i^c * S_i^{c'})}{\sqrt{\sum_{i=1}^{n}(S_i^c)'} * \sqrt{\sum_{i=1}^{n}(S_i^{c'})'}} \tag{14}$$

    where $K = \{(T_i^c, T_i^{c'}) | \texttt{Sims}(T_i^c, T_i^{c'}) = 1\}$;
    **if** match ≥ threshold **then**
      smatch ⟵ $\cup\, c, c'$;
      break;
    **end**
  **end**
**end**
`cConflict` (smatch);
Return(smatch);

**Algorithm 2.** Priority Matching Algorithm

**input** : Given a target profile p and set of profiles resource R
**output**: The target profile p improved by the profiles resource R

**foreach** *each profile* p ′ *belonging to* R **do**
  /* St is set of similar concepts between target p and profile p ′ */
  St =`PriorityMatching` (p, p ′);
  /* Similarity between two profiles will be identified based on St and the following formula */

$$\text{match} = \texttt{Simf}(\text{p}, \text{p}') = \texttt{Simf}(C^c, C^{c'}) = \frac{\sum_{(c\in C, c'\in C')}(\max(\texttt{Simf}(c,c')))}{\sqrt{|C| * |C'|}} \tag{15}$$

  **if** match ≥ threshold **then**
    smatch ⟵ ∪ { p ′ } ;
  **end**
**end**
/* smatch is a set of the profiles resource ⊆ R where search p belong to the set is similar to the target */
`ImproveProfile` (p, smatch);
**return** (p);

**Algorithm 3.** Profile Improving Algorithm

# 5 Experiments

## 5.1 Implements

In this section we present the implementation details of our proposed system and analysis of existing profile building methods and techniques with respect to our system.

**Contructing and Training Common Profile.** We created the common profile which consists of two parts; a demographic information of user and an interesting topic user like. The interesting topic is constructed based on vocabularies and structures from relevant travel sections of the Open Directory Project (ODP) and Yahoo! Category [1]. In particular, for each category we downloaded the first 10 web sites. If there is less than 10 web sites in the category, we downloaded all available information to generate feature vector for each category. If a category has no web site available then its direct sub-concepts are include to the construct its feature vector. Two specific hierarchical relationships are in ODP or Yahoo! Category hierarchy, which are a *is-a* relationship and a *part-of* relationship, counted as *sub-concept* relationship. There exists other nonspecific relationships. Particularly, there are categories *Journals*, *Papers*... look like *part-of* the category *Neural_Networks*. In this case, we consider documents under the categories *Journals*, *Papers*... are direct documents of the *Neural_Networks*.

**Information Collection.** we use meta-search engine to collect user's information. As shown in the $Figure$ 3, the information is generated from consensus among four different kinds of search space included: home/personal pages, blogs, publications, web sites of orgnizations. When a user queries his/her profile, *Management search* creates four different *meta search engines* and assigns each *meta search engine* to collect a specific kind of the information. Each meta search must create itself many queries concerned with assignment and sends each query to a specific search engine sellected by itself based on its *knowledge base*.

**Keyword Extraction.** we applied the traditional Vector space model($tf$-$idf$) with Lexico-syntactic pattern for hyponyms to extract the feature from document. However, in our proposal, we improve the method by analyzing semantic between keywords to improve themselves. For example, we consider the sentence "*A computer is a machine that manipulates data according to a list of instructions such as laptops, notebooks, servers.*". In this case, stead of keywords with pattern $\{(computer, 1), (machine, 1), (data, 1), (laptop, 1), (notebook, 1), (server, 1)\}$ will be extracted by most existing methods, our one is $\{(computer, 4), (machine, 2), (data, 1), (laptop, 1), (notebook, 1), (server, 1)\}$. Because, the hypernym relation from $conputer$ to $laptop$, $notebook$ and $server$, and hypernym relation from $machine$ to $computer$ are counted by our method. This means if a term is hypernym of n other terms, the term occurs on the

[1] The Open Directory Project aims to build the largest human-edited directory of Internet resources and is maintained by community editors who evaluate sites to classify them in the right directory. Yahoo! category is maintained by the Yahoo! directory team for the inclusion of web sites into Yahoo! directory.

text corpus with n+1 times. Notice that it is quickly and effectively to recognise the hypernym relation by the lexico-syntactic patterns mentioned on [2]. Our method, which extracts keywords from text corpus are sketched as follows:

- extracting sentences from text corpus.
- each sentecne, the relations between terms of which are identified via the lexico-syntactic patterns mentioned on [2].
- remove all stop word included: determiner, postdeterminer, modifier, postmodifier.
- filtering out nouns, verbs, adjectives and adverbs. The part-of-speech recognition is done using the WordNet dictionary.
- extracting single keywords (nouns)
- extracting bikeywords based on grammar (compound noun)
- generating the feature vectors based on the definiations above mentioned

**Similarity Measure.** According to above discussed methods, while calculating the similarity between categories, matching among keywords belonging to each category's feature vector is requisite. Thus a similarity measure between keywords (terms) is necessary for the methods to distinguish between categories. Here we applied our previous methods [2][3] to measure the similarity.

### 5.2 Results Analysis

Our system build user's profile automatically based on user's information which are generated from the internet. We evaluate the propose system in two aspects, a comparison of improving $tf - idf$ with original one and results of building user's profiles.

The $Precision$ and $Recall$, two widely used statistical classifications was chosen to comparison between the above mentioned methods. $Precision$ can be seen as a measure of exactness or fidelity, whereas $Recall$ is a measure of completeness. We denote

- $N_{total}$ : the total number of keywords extracted from corpus collection by experts,
- $N_{correct}$ : the number of correct of keywords extracted from corpus collection by the our system, and
- $N_{incorrect}$: the number of incorrect of keywords extracted from corpus collection by the our system.
- *Precision* is used to evaluate the ratio of correctly extracted feature vectors by system:

$$Precision = \frac{N_{correct}}{N_{correct} + N_{incorrect}} \tag{16}$$

- *Recall* is used to evaluate the ratio of completeness extracted feature vectors by the system:

$$Recall = \frac{N_{correct}}{N_{total}} \tag{17}$$

In the $Figure$ 5, we show average results of the comparisons generated from large number of test samples.

Our proposed system Differ from other existing researches [15] [5] in building *user profile*. In our case the information is generated from the internet by user's queries to meta-search engines despite watching the user's activities on the internet or in particular applications [15]. Significantly, we improved the profile by interaction user. We applied ontology integration techniques [2,3,1] to combine the target profile with existing neighbors to exchange interesting knowledge. It is advantage of ontology-based profile. The system built five profiles of research members in the same laboratory as pratical samples. After initially building the profiles, they are excahnged interesting knowledges to enhacne themselves. The result was highly evaluated by experts.

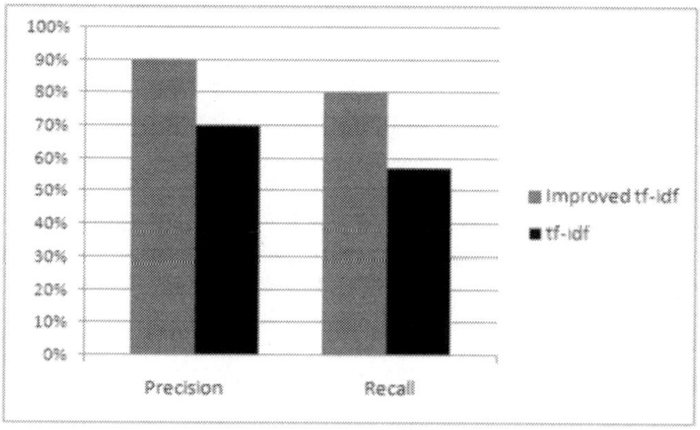


**Fig. 5.** Comparation between tf-idf and our improved one

## 6 Conclusions

We investigated the techniques to create an *user profile* automatically using the ontological approach. A framework to gather the user information from different search space where user's details could be found was proposed. A Wordnet and Lexico-syntactic pattern for hyponyms approach is applied to extort the important feature of document to represent the profile ontology. The *user profile* further improved by learning interesting knowledge from similar profiles. In the future work, we will apply the proposed methods to specific application of user interaction management system.

## References

1. Duong, T.H., Nguyen, N.T., Jo, G.S.: A Method for Integrating Multiple Ontologies. Cybernetics and Systems 40(2), 123–145 (2009)
2. Duong, T.H., Nguyen, N.T., Jo, G.S.: A Method for Integration across Text Corpus and WordNet-based Ontologies. In: IEEE/ACM/WI/IAT 2008 Workshops Proceedings, pp. 1–4. IEEE Computer Society, Los Alamitos (2008)

3. Duong, T.H., Nguyen, N.T., Jo, G.S.: A Method for Integration of WordNet-based Ontologies Using Distance Measures. In: Lovrek, I., Howlett, R.J., Jain, L.C. (eds.) KES 2008, Part I. LNCS (LNAI), vol. 5177, pp. 210–219. Springer, Heidelberg (2008)
4. Gaush, S., Chaffee, J., Pretschner, A.: Ontology-based user profiles for Search and browsing. Journal of Personalization Research Special 1(3-4), 219–234 (2003)
5. Golemati, M., Katifori, A., Vassilakis, C., Lepouras, G., Halatsis, C.: Creating an ontology for the user profile: Methods and Applications. In: Proceedings of the First International Conference on Research Challenges in Information Science, RCIS 2007, pp. 407–412 (2007)
6. Guarino, N., Masolo, C., OntoSeek, G.V.: Content-Based Access to the Web. IEEE Intelligent Systems 14(3), 70–80 (1999)
7. Heflin, J., Hendler, J., Shoe S. L.: A Knowledge Representation Language for Internet Applications; Technical Report CS-TR-4078 (UMIACS TR-99-71), University of Maryland at College Park (1999)
8. `http://www.dmoz.org/`
9. Lieberman, H., Letizia: An Agent That Assists Web Browsing. In: Proceedings of the International Joint Conference on Artificial Intelligence, pp. 924–929. AAAI Press, Menlo Park (1995)
10. Pazzani, M., Muramatsu, B.D.: Syskill Webert Identifying interesting Web sites. In: Proceedings of 13th National Conference of Artificial Intelligence, pp. 54–61. AAAI Press, Menlo Park (1996)
11. Sieg, A., Mobasher, B., Burke, R.: Learning Ontology-Based user profiles: A Semantic Approach to Personalized Web Search. IEEE Intelligent Informatics Bulletin 8(1) (2007)
12. Stermsek, G., Strembeck, M., Neumann, G.: user profile Refinement using Explicit User Interest Modeling. In: Proc. of 37. Jahrestagung der Gesellschaft für Informatik (GI). Lecture Notes in Informatics (LNI), vol. 109, pp. 289–293 (2007)
13. Sugiyama, K., Hatano, K., Yoshikawa, M., Uemura, S.: A method of improving feature vector for web pages reflecting the contents of their out-linked pages. In: Hameurlain, A., Cicchetti, R., Traunmüller, R. (eds.) DEXA 2002. LNCS, vol. 2453, pp. 891–901. Springer, Heidelberg (2002)
14. Thomas, C., Fischer, G.: Using Agents to Personalize the Web. In: Proceedings of the 2nd International Conference on Intelligent User Interfaces, pp. 53–60. ACM, USA (1997)
15. Trajkova, J., Gauch, S.: Improving Ontology-Based user profiles. Presented at RIAO, Vaucluse, France, pp. 380-389 (2004)

# Ontology-Based Intelligent Agent for Grid Resource Management

Kyu Cheol Cho, Chang Hyeon Noh, and Jong Sik Lee

School of Computer Science and Engineering
Inha University
#253, YongHyun-Dong, Nam-Ku,
Incheon 402-751, South Korea
landswell@empal.com, cromirak@hanmail.net, jslee@inha.ac.kr

**Abstract.** The intelligent agent works powerful jobs for handling system complexity and making systems more modular. Especially a reasoning agent is effective on organizing for decision-making process of systems. This paper introduces an Ontology-based Intelligent Agent for a Grid Resource Management System (OIAGRMS), which uses ontology reasoning to select a suitable resource supplier, is proposed. This paper focuses on effective grid resource management and the improvement of resource utilization through transaction management for the OIAGRMS. For performance evaluation with accuracy and reliability, the OIAGRMS is compared with the Prediction-based Agent for Grid Resource Management System(PAGRMS) and the Random-based Agent for Grid Resource Management System(RAGRMS). The OIAGRMS recorded over 90 percents trade success, but the PAGRMS and RAGRMS recorded less than a 90 percents trade success. In comparing of resource utilization rate, maximum deviation, standard deviation, the OIAGRMS were about 9.4 and 9.8 percents but the PAGRMS are about 22.9 and 16.3 percents, the RAGRMS were about 61.6 and 21.7 percents. The empirical results demonstrate the usefulness and improvement utilization with stable performances of the intelligent agent base on ontology reasoning in grid environment.

## 1 Introduction

Intelligent agents[1] offer advanced platforms for composing expert systems to solve complex problems in many sub-fields. Intelligent agents have been developed as next significant breakthroughs to solve complex problems, and the intelligent agent is being used various application, ranged from comparatively small systems. In the grid[2] environment, an end user requires timely grid resources and the intelligent agent is able to interact with others in order to manage grid resources. The intelligent agent distributes resources and data, and controls processes to perform a different set of tasks. Role of the intelligent agent is to answer or recommend a solution and the role is very difficult works to decide.

Recently, the grid middleware maintains grid knowledge based on instances of ontology. Many researchers have been designed and studied to manage grid resources

N.T. Nguyen, R. Kowalczyk, and S.-M. Chen (Eds.): ICCCI 2009, LNAI 5796, pp. 553–564, 2009.

effectively using an grid ontology[3] that have classes and relations. Many researchers designed principal properties, services and relation with knowledge store of resource information by using grid ontology. And prediction of resources usage helps job scheduling, resource discovery and resource information management. The grid ontology and demand prediction induce suppliers to supply resources efficiently and improve the utilization of resources, based on trade evolution trends and ontology reasoning with knowledge of resource information and trade information of suppliers. In addition, good consumption of grid resource is induced by transactions with low prices by competition among suppliers.

This paper is organized as follows: Section 2 describes grid resource management, intelligent gent for grid resource management, and grid ontology and ontology reasoning, which are the background of this research. Section 3 illustrates the intelligent agent for grid resources management. In section 4, the efficiency and effect of the OIAGRMS from this paper is demonstrated, by the analysis of experiment results. Finally, conclusions are in section 5.

## 2 Related Works

### 2.1 Grid Resource Management

Grid resource management is made of complex and various factors like site autonomy, resource heterogeneity and etc, and is a complex tack involving huge scaled resources volume. The Grid Architecture for the Computational Economic (GRACE) model [4] was proposed to control resource distribution with supply and demand of useful resources for continuously huge capacity resources and manages resources with high performance systems. The GRACE structural economic models are organization strategies of resource allocation and system management. And the abstract owner architecture runs ordering and delivery model for resource sharing to manage continuously long time operation[5]. The principal components of the hierarchical architecture[6] are divided into passive and active components. A forecasting system of user demand[7] reduces delay time and offer effective management but the forecasting system did not reflect status of resources and joined grid members. In this paper, we design gird resource management system with knowledge of grid factors as well as user demand forecasting.

### 2.2 Intelligent Agent for Grid Resource Management

List of many issues explain character points of Agent Grid[8]. The Agent Grid offers optimal trades and systemic controls to systems. Agent-based grid management is able to cooperate with other agents to supply, and supply service advertisement and effective applications scheduling[9] that is a centralized broker/agent architecture. The Agent Grid[10] integrates services and resources for building Problem Solving Environments (PSE) in expert fields. The intelligent agents contain behavioral rules and interactions of the rules with other agents. This agent uses the hierarchy structure of homogenous agents that is able to complete remodel agents with other roles in the running time. The agent grid in adaptive negotiation strategies verifies scalability and adaptability with applying intelligent agents to computing resource management and load balancing[11].

### 2.3 Grid Ontology and Ontology Reasoning

The grid ontology is based on the grid knowledge in OIAGRMS. The ontology can explain complex information and improve the quality of information about performance, data and service of grid resources. As the protégé-OWL(Web Ontology Language)[12] enables reasoning, checks consistency and induces class hierarchy, is an extensible OWL ontology. And OWL Query Language(OWL-QL)[13] asks query answering with dialogues that operate automated communication with ontology reasoning using distributed knowledge bases on the semantic web. Ontology reasoning[14] can induce a suitable result among instances of classes. The grid ontology is designed with the terms and classes that include grid entities, services, components, and applications of grid architectures. As OWL[15] can represent the detail meaning of terms and relationships, extends the Resource Description Frame(RDF)[16]. The RDF resource modeling expresses entities and classes easily and extends new entities and classes.

## 3 Intelligent Agent for Grid Resource Management

We propose an Ontology-based Intelligent Agent(OIA) for effective management of grid resources. The intelligent agent solves complex problems and high performance jobs, which should perceive proactive and flexible acts. And the intelligent agent should be able to interact with the other composed components. In this paper, we design the architecture and functions of OIA for grid resource management systems and illustrate the reasoning using the grid ontology in a grid environment.

### 3.1 Use of Ontology and Design of OIA for Grid Resource Management

Grid resources are large-scaled resources and comprise heterogeneous systems that perform different kinds of abilities and services. And, the well composed grid systems induce flexible and effective management with the practical use of grid resources. And, ontology helps advanced intelligences for composing systems with the basic information of resources. Ontology also manages continuously effective and practical resource use. And a multi-agent helps practical use and management of resources through reducing load of jobs and operating multi-management. Moreover ontology is more useful with using resource information and optimal system construction. And intelligent agents have more advantages and works continuous and automatically resource management with ontology.

In the economic grid environment, the OIA manages effective resources and maintains stable performances and controls balanced trades. The OIA takes advantage of workflow solutions to solve complex problems. Figure 1 illustrates a functional structure of the OIA.

The OIA performs interacts among task, knowledge and communication of agent with ideal relations. Tasks of the intelligent agent flexibly process ontology reasoning in the economical grid environment. The ontology reasoning is a frame of knowledge for an intelligent agent. The knowledge has intelligence priorities and trends of user demands and rules with RFQ(Request for Quote) of user demands. The intelligent agent inducts optimal trades with considering demands of users and knowledge of the intelligent agent. The intelligent agent works and answers by reasoning a rule and the reasoning considers the current demand of users. And the intelligent agent communicates to complete trades between intelligent agents and users by information management. In the process of information managements, the intelligent agent gathers available information to induce optimal bid results and filters information of tiny proportion.

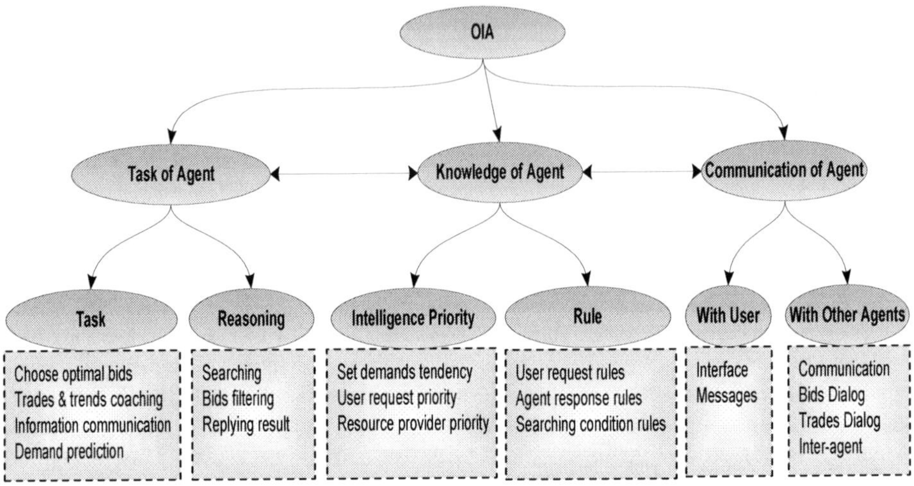


**Fig. 1.** Functional structure of the Ontology-based Intelligent Agent(OIA)

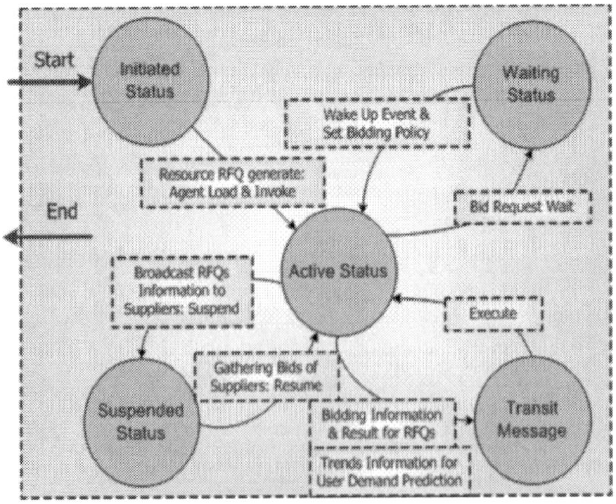


**Fig. 2.** Lifecycle of the OIA

Figure 2 illustrates the lifecycle of OIA for effective grid resources management. The lifecycle has five statuses according to functional abilities. The initialed status is invoked when users generates RFQ with demands of users. The OIA broadcasts RFQ information to supply with RFQ information of user and the OIA gathers bid information of suppliers. The OIA needs waiting time to autonomously and proactively gather and answer information for communications among other agents. The OIA operates entirely to carry optimal trade out. Transit messages do up transport information of intelligent agents, and the messages consisted of by contents with encoded message structure.

### 3.2 Intelligent Agent Using Ontology Reasoning

The grid ontology describes all entities that are useful for applications. And that works inductive reasoning, classification and helps to solve problem in the field of artificial intelligence. Many researcher use web ontology language[12] to implement the grid ontology. The Ontology-based Intelligent Agent for the Grid Resources Management System(OIAGRMS) defines basis relations for resource utilization which is applied to reliability, price and possession volume and so on. The OIAGRMS uses reasoning engine to select the optimal trade condition.

And, the intelligent agent induces active trading that means the grid suppliers supply resources with prediction evolution of user demands. Demand Prediction[7] is set by daily requested demand trend that is fixed resources supply volume by the previous day demand trends and the change of demands compared with previous demands. And the prediction method interacts continually to demands trend and resources, the method can be measured on the basis of RFQ. Both the demands trend and volume of collecting resources are fixed by previous demand trends and volume of collected resources.

The OIA defines relational rules about a grid supplier reasoning and stores timely resource status information. And the grid ontology assists prediction of resource supply for user resource demands. Figure 3 illustrates intelligent agent service ontology for the grid domain ontology. The intelligent agent connects with grid users and grid suppliers, and that operates on resource active demand prediction and usage. Grid users requires prices, due dates, performance of available resources. The grid domain ontology discovers new invoked factors with continuous resource checking, filtering and monitoring. And the intelligent agent composes the grid domain ontology and trends prediction about demands of gird users and strategies of grid suppliers in all grid participation factors. The intelligent agent completes auctions, delivers auction results and saves user trends for the prediction of resources usages.

The grid ontology for OIAGRMS which includes all grid participation factors, implements the gird ontology using the protégé[17]. The protégé provides an independent platform and graphical user interface environment that helps directly usage with creating and modifying classes and properties. The grid ontology attaches the OWL file and is ready for reasoning. The OWL model in OIAGRMS can be used to create, ask and delete resource information of various types. And the model performs processing such as discovery and setting of information and building relationships between resources, grid suppliers and grid users. The OIAGRMS is available to extend

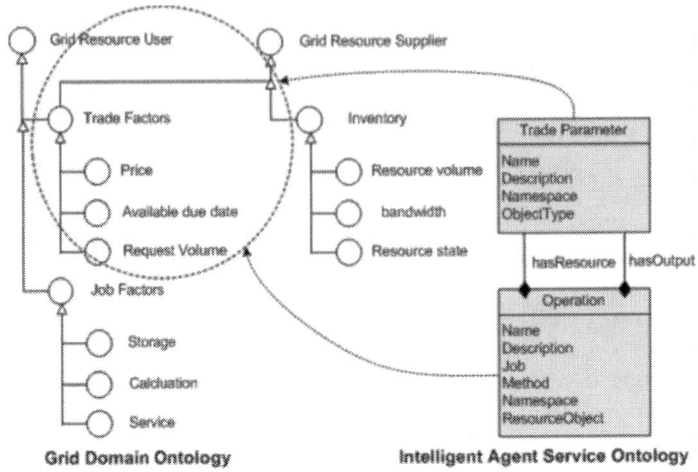


**Fig. 3.** Intelligent agent service ontology for grid domain ontology

for more intelligent knowledge construction using an OWL model and a prediction method for user demand.

The reasoning is based on rules in the OIAGRMS. The reasoning in OIAGRMS allows us to express formal and explicit specification of shared resources. The OIA performs the semantic inference, which can solve the similar problem. The OIA needs grid ontology index services to search collaborative objects, which will take over the job from the former queries and reasoning in the grid ontology repository. The OIA uses the reasoning through trade accomplishment with grid users and grid suppliers. The OIA queries the grid ontology about resource demands and can find the grid supplier with the optimal resource supply strategy.

The grid user generates RFQ to collect resources and make the most of the resources for high performance jobs. The OIA stores information of knowledge about resources and demands evolution which controls the intelligent agent. And the OIA operates reasoning and the result of reasoning induced optimal supply strategies. Generally a query generates several results in database management systems but the reasoning induces an optimal result with several factors, conditions and status in OIAGRMS. Previously the reasoning abstracts all grid suppliers and checks an additional request condition using on prediction of user demand and usage environments. And the reasoning abstracts grid suppliers that have optimal status for supply. If the reasoning has additional conditions, repeats these courses. Finally reasoning results induce appropriate trades. The OIAGRMS accepts the selected trade of grid supplier. The reasoning and prediction method helps operating resource utilization and effective resource management in OIAGRMS.

## 4 Experiment and Results

To measure performance improvement, we implement the Ontology-based Intelligent Agent for Grid Resource Management System(OIAGRMS) based on the discrete event system specification(DEVS) [18] modeling and simulation. We make experiments on transaction success probability, trade rate, and number of transaction and resource utilization rates of the OIAGRMS. And we compare the performance of the OIAGRMS those of the user Prediction-based Agent for Grid Resource Management System(PGARMS) and those of the Random-based Agent for Grid Resource Management System(RAGRMS). The PGARMS uses Demand Prediction[16] by daily requested demand. And the Demand Prediction inducted by the previous demand trends and the change of demand compared with demands. The prediction method connects continually to demand trends and resources can be measured on the basis of demands. The RAGRMS is decided random strategies[19] by supply capacity of suppliers per trade status without considering by the previous demand trends and the change of demand compared with demands.

### 4.1 Experiment 1: Trade Success Probability and Trade Rate

The first experiment measures the trade success probability and trade rate and compares those values of the OIAGRMS with those of the PAGRMS and those of the RAGRMS through resource trading with grid users and grid suppliers during 1000 days. The trade success probability and trade rate are high, that means grid suppliers supply resources in appropriate times with effective management, and the intelligent agent induces appropriate trade with reasoning base on the OIA. Figure 4 presents the result of trade success probability when every 100 day is recorded.

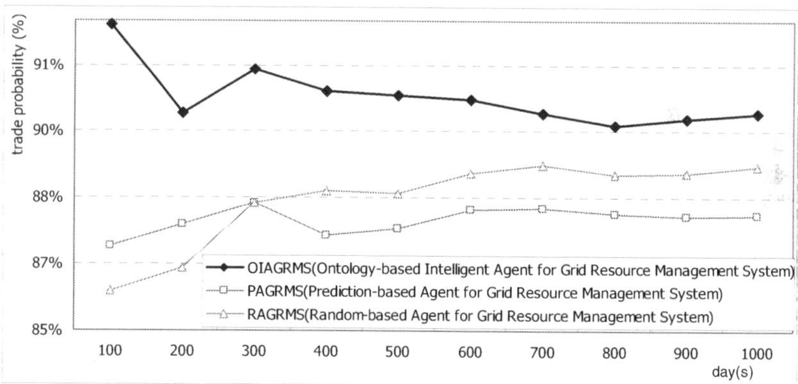


**Fig. 4.** Trade success probability

The trade success probability of OIAGRMS maintains over 90 percent with high probability. The trade success probability of PAGRMS was below 88 percent and that of RAGRMS was below 88.9 percent. This result shows the OIAGRMS supply resources to the grid user by appropriate reasoning and effective resource management with user demand prediction.

Figure 5 presents the result of trade rate when every 100 day is recorded. The result points induce by the equation (1). $Tr_{PAGRMS}$ records 100 percent; this means a standard for comparing performances. $Tr_{OIAGRMS}$ records trade rates from 102.2 to 105.7

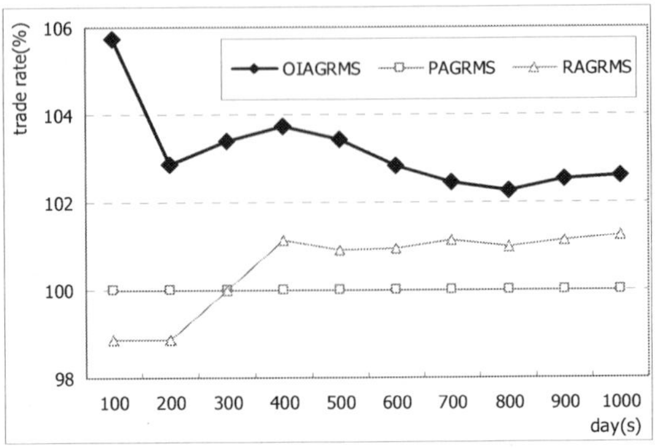


**Fig. 5.** Trade rate on the basis of PAGRMS

$$Tr_{OIAGRMS} = Tv_{OIAGRMS} / Tv_{PAGRMS} \quad 100$$
$$Tr_{RAGRMS} = Tv_{RAGRMS} / Tv_{PAGRMS} \quad 100 \qquad \text{-----} \ (1)$$
$$Tr_{PAGRMS} = Tr_{PAGRMS} / Tr_{PAGRMS} \quad 100 = 100$$

$Tr_{OIAGRMS}$: Trade rate of OIAGRMS

$Tv_{OIAGRMS}$: Trade volume of OIAGRMS

$Tv_{PAGRMS}$: Trade volume of PAGRMS

$Tr_{RAGRMS}$: Trade rate of RAGRMS

$Tv_{RAGRMS}$: Trade volume of RAGRMS

percents and $Tr_{RAGRMS}$ records trade rates from 98.5 to 101.3 percents. The trade rate of OIAGRMS is the highest among all experiment grid resource management system. The OIAGRMS makes more stable and active resource trade with effective resource management. And RAGRMS has low trade rates before 300 days, but has high trade rate end for end after the day. And $Tr_{OIAGRMS}$ record always higher trade rates than RAGRMS yet. This result means that resources trade is accomplished with active and effective resource management with appropriate trade induction by ontology reasoning.

### 4.2 Experiment 2: Transaction of Grid Resource Supplier

This experiment 2 measures transaction of grid suppliers and compares the result of the OIAGRMS with those of the PAGRMS and the RAGRMS. Transaction of grid supplier explains active and harmonious resource usage through steadily effective and balanceable management of resources. Figure 6 presents the result of transaction of grid suppliers when every 100 day is recorded.

Three resource management systems steadily increase transaction in proportion as days. When the day is 1000, the transaction of OIAGRMS recorded as 16799, the

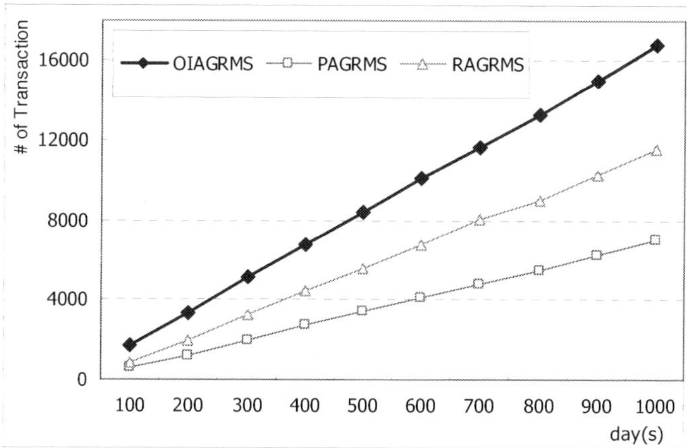


**Fig. 6.** Transaction of grid resource supplier

transaction of RAGRMS record 11538 and the transaction of PAGRMS record 7007. The OIAGRMS accomplishes the highest transaction among the three systems. Especially transaction of the OIAGRMS is twice as much than transaction of the PAGRMS. This result illustrates the OIAGRMS performs active resource trading and manages harmonious resource management.

### 4.3 Experiment 3: Resource Utilization Rate

The third experiment measures resource utilization and compares this result of the OIAGRMS with those of the PAGRMS and the RAGRMS. The resource utilization rate is high means that the resource management system supplies actively resources and induces much resource usage through effective management for available resource. Table 1-(A) illustrates the average resource utilization rate when every 100 day is recorded, and table 1-(B) illustrates deviation of resource utilization averages. And table 1-(C) illustrates the total average and deviation, standard deviation, maximum deviation based on results of table 1-(A) and 1-(B). These results explain the stability and the resource supplying capacity. In table 1-(A), the bold numeric values mean the highest and the lowest value for each of the system. Especially the gap of deviation shows balanced job operation and stable performance. The deviation is low means that resource trade is accomplished with maintaining load balancing of resource demands.

The resource utilization rate of OIAGRMS was recorded from 88 to 95 percent and that of PAGRMS was recorded from 82 to 97 percent. These values are very high resource utilization rate. But that of RAGRMS was recorded from 76.6 to 98 percent with ups and downs. In the table 1-(C), the standard deviation of the OIAGRMS is 0.6 times that of the PAGRMS and is 0.4 times that of the RAGRMS. These results show continuous and stable resource management. Especially the maximum deviation of the OIAGRMS is 0.6 times that of the PAGRMS and is 0.45 times that of the RAGRMS with low ups and downs. These results mean resource trade is continuously accomplished with steady trading and stable resource

**Table 1-(A).** Resource utilization averages (%)

| system \ day(s) | 100 | 200 | 300 | 400 | 500 | 600 | 700 | 800 | 900 | 1000 |
|---|---|---|---|---|---|---|---|---|---|---|
| OIAGRMS | 89.2 | **97.9** | 91.7 | 95.0 | 90.4 | 94.9 | 90.8 | 88.3 | 93.3 | **88.1** |
| PAGRMS | **82.5** | 89.6 | 89.1 | **98.8** | 97.1 | 98.2 | 94.5 | 91.6 | 95.4 | 95.3 |
| RAGRMS | 80.7 | 91.6 | 96.8 | **98.3** | 79.7 | 79.7 | 97.6 | 90.3 | 88.7 | **76.6** |

**Table 1-(B).** Deviation of resource utilization averages (%)

| system \ day(s) | 100 | 200 | 300 | 400 | 500 | 600 | 700 | 800 | 900 | 1000 |
|---|---|---|---|---|---|---|---|---|---|---|
| OIAGRMS | 7.7 | 34.7 | 0.1 | 9.3 | 2.4 | 8.9 | 1.3 | 13.3 | 1.8 | 15.0 |
| PAGRMS | 114.7 | 13.2 | 16.6 | 31.6 | 15.3 | 24.9 | 1.7 | 2.6 | 4.6 | 4.2 |
| RAGRMS | 52.9 | 13.2 | 77.1 | 105.4 | 69.1 | 69.1 | 92.8 | 5.3 | 0.5 | 130.9 |

**Table 1-(C).** Total average, maximum deviation, deviation, standard deviation of resource utilization averages and deviation of those averages

| system | average(%) | maximum deviation | deviation | standard deviation |
|---|---|---|---|---|
| OIAGRMS | **91.96209** | 97.9 - 88.1 = **9.8** | 94.3738411 | **9.71462** |
| PAGRMS | **93.20695** | 98.8 – 82.5 = **16.3** | 229.546606 | **15.15** |
| RAGRMS | **88.01264** | 98.3 – 76.6 = **21.7** | 616.237431 | **24.824** |

management for the available resources using the ontology reasoning with maintaining balance of resource management.

## 5 Conclusion

The Ontology-based Intelligent Agent for Grid Resource Management System(OIAGRMS) uses ontology reasoning to select appropriate resource suppliers and demand prediction with demand trend evolution. Performance evaluation was executed by comparing those of the OIAGRMS the trade success probability, trade rate, transaction and those of the Prediction-based Agent for Grid Resource Management System(PAGRMS) and the Random-based Agent for Grid Resource Management System(RAGRMS).

The OIAGRMS recorded over 90 percent trade success probability with intelligence resource management. In resource management stability for resource utilization and the maximum deviation standard deviation of the OIAGRMS are about 9.4 and

9.8 percents but the PAGRMS are about 22.9 and 16.3 percents, and the RAGRMS are about 61.6 and 21.7 percents. These empirical results demonstrate the usefulness, effectiveness and utilization of the OIAGRMS with ontology-based intelligence reasoning. The ontology-based intelligent agent is quite useful to manage and have more advantages and works automatically stable management using ontology for continuous the large-scaled resources in the grid environment.

## Reference


1. Wooldridge, M., Jennings, N.R.: Intelligent Agents: Theory and Practice. The Knowledge Engineering Review 10(2), 115–152 (1995)
2. Foster, I., Kesselman, C. (eds.): The Grid: Blueprint for a Future Computing Infrastructure. Morgan Kaufmann Publishers, USA (1999)
3. Xing, W., Dikaiakos, M.D., Sakellariou, R., Orlando, S., Laforenza, D.: Design and development of a core grid ontology. In: Gorlatch, Danelutto, M. (eds.) CoreGRID Integration Workshop, Italy, pp. 21–31 (2005)
4. Buyya, R., Chapin, S., DiNucci, D.: Architectural Models for Resource Management in the Grid. Grid Computing GIRD 2000. In: First IEEE/ACM International Workshop, pp. 20–33 (2000)
5. Chapin, S., Clement, M., Snell, Q.: A Grid Resource Management Architecture, Grid Forum Scheduling Working Group (1999)
6. Buyya, R., Abramson, D., Giddy, J.: A Case for Economy Grid Architecture for Service Oriented Grid Computing. In: Proceedings of International Parallel and Distributed Processing Symposium: Heterogeneous Computing Workshop (2001)
7. Cho, K.C., Kim, T.Y., Lee, J.S.: User Demand Prediction-based Resource Management Model in Grid Computing Environment. In: ICHIT 2008, Korea, pp. 627–632 (2008)
8. Thompson, C.: Characterizing the Agent Grid, Technical Report, Object Services and Consulting, Inc., `http://www.objs.com/agility/tech-reports/9812-grid.html`
9. Tierney, B., Johnston, W., Lee, J., Thompson, M.: A data intensive distributed computing architecture for grid applications. Future Generation Computer Systems 16(5), 473–481 (2000)
10. Rana, O.F., Walker, D.W.: The agent grid: agent-based resource integration in PSEs. In: Proceedings of 16th IMACS World Congress on Scientific Computation, Applied Mathematics and Simulation (2000)
11. Shen, W., Li, Y., Ghenniwa, H., Wang, C.: Adaptive negotiation for agent-based grid computing. In: Proceedings of AAMAS Workshop on Agentcities Challenges in Open Agent Environments, pp. 32–36 (2002)
12. Horridge, M., Knublach, H., Rector, A., Stevens, R., Wroe, C.: A practical guide to building OWL ontologies using the Protégé-OWL plugin and CO-ODE tools. University of Manchester (2004)
13. Fikes, R., Hayes, P., Horrocks, I.: Owl-ql: A language for deductive query answering on the semantic web, Technical Report KSL 03-14, Stanford University (2003)
14. Roure, D.D.: A brief history of the semantic grid. In: Semantic Grid: The Convergence of Technologies, number 05271, Dagstuhl Seminar Proceedings. IBFI, Dagstuhl, Germany (2005)

15. McGuinness, D.L., Harmelen, F.V.: OWL Web Ontology Language Overview, W3C Proposed Recommendation (2003), http://www.w3.org/TR/2003/PR-owl-features-20031215/
16. Brickley, D., Guha, R.V.: FDF Vocabulary Description Language 1.0:RDF Scheman, W2C Working Draft (2003)
17. Protégé, http://protege.stanford.edu/
18. Zeigler, B.P., et al.: The DEVS Environment for High-Performance Modeling and Simulation. IEEE C S & E 4(3), 61–71 (1997)
19. Kalpakam, S., Arivarignan, G.: Inventory System with Random Supply Quantity, OR Spektrum, pp. 139–145 (1990)

# The Norm Game on a Model Network: A Critical Line

Marcin Rybak, Antoni Dydejczyk, and Krzysztof Kułakowski

Faculty of Physics and Applied Computer Science, AGH University of Science and Technology, al. Mickiewicza 30, PL-30059 Kraków, Poland
fisher@autocom.pl, dydejczyk@ftj.agh.edu.pl,
kulakowski@novell.ftj.agh.edu.pl

**Abstract.** The norm game (NG) introduced by Robert Axelrod is a convenient frame to disccuss the time evolution of the level of preserving norms in social systems. Recently NG was formulated in terms of a social contagion on a model social network with two stable states: defectors or punishers. Here we calculate the critical line between these states on the plane of parameters, which measure the severities of punishing and of being punished. We show also that the position of this line is more susceptible to the amount of agents who always punish and never defect, than to those who always defect and never punish. The process is discussed in the context of the statistical data on crimes in some European countries close to Wrocław - the place of this Conference - around 1990.

**Keywords:** Social networks, multiagent systems.

## 1 Introduction

The physical boundary between a human being and his or her environment is the skin, but in the space of behaviours the same boundary is less strict. Taking decisions, we are not completely selfish; we are to some extent bound by the social norms. The way of enforcing norms varies between a direct control and a deep internalization; in the latter case we treat our conformity to norms in the same way as our payoff. Norms create the society, where we are formed [1], and norms are modified by the society members. As it was formulated by a leading Polish psychiatrist Antoni Kępiński, to decide where to put limits of our own rebel is one of most difficult problem in human life [2]. Solving this problem in our individual scale emerges in the social scale as a time evolution of norms. To search for laws which rule this process is a worthwhile challenge for the agent-based simulations. A serious advance in this path was done by Robert Axelrod who formulated the norm game: an algorithm to simulate the conditions of persistence and fall of a social norm [3,4]. The simulations done by Axelrod have been questioned [5], but his paper has been cited hundreds times and it triggered a cascade of research; for a recent review of simulations of norms see [6]. Further, the subject of norms overlaps with the theory of cooperation; to cooperate is an example of a social norm. An overview on the latter might

N.T. Nguyen, R. Kowalczyk, and S.-M. Chen (Eds.): ICCCI 2009, LNAI 5796, pp. 565–572, 2009.

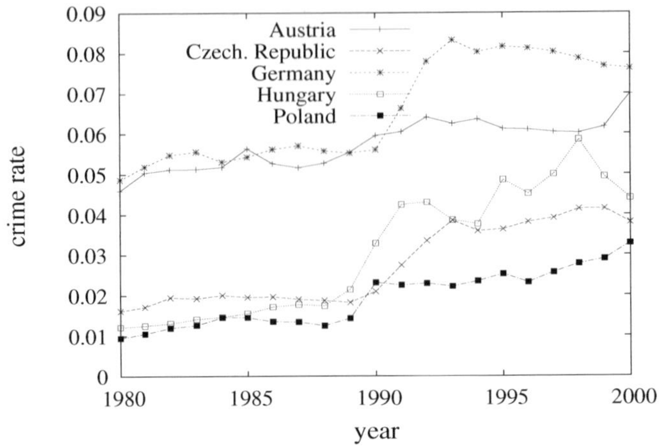


**Fig. 1.** The crime statistics in selected European countries in 1980-2000

provide insight into current trends; still the research in this field seems to be at its intensively rising stage [7,8]. As norms are beliefs, there is also some overlap with the simulations of opinion dynamics; for a review of sociophysical simulations on this matter see [9].

Direct motivation of this research comes from statistical data on norm breaking. Perhaps most striking change we have seen deals with the data on divorces in Portugal during the Carnival Revolution. There, the number of divorces increased from 777 in 1974 to 7773 in 1977 [10]. The plots on crime in countries in Central Europe are more conventional. In Fig. 1 we show the data on Germany, Poland, Czech Republic, Austria and Hungary [10]. In accordance with warnings by Eurostat, our aim is not to compare the amount of crimes in these countries, but rather to show the changes in some of them. Note that opinion shifts were classified into continuous and abrupt by Michard and Bouchaud in 2005 [11] within a theory of imitation. In the Axelrod model, the driving social mechanism is punishment; the interaction inhibits the change rather than releases it.

In NG [3] agents defect a norm with a given probability. Once an agent defects, other agents punish the defector, also with some probability. The defection is gratified with some payoff, but those who are punished lose. Also, those who punish incur some cost. Axelrod considered also a metagame: the possibility of punishment those who do not punish. The overall success of an agent was measured by his income, but the agent himself - represented by his strategy - was not modified. After some number of games, the genetic algoritm was used to select strategies which yielded the best income. As indicated in [6], this kind of modeling has an advantage to deal with the dynamics of the process. We should add that Neumann criticizes the approach of Axelrod for disregarding the functional character of norms [6].

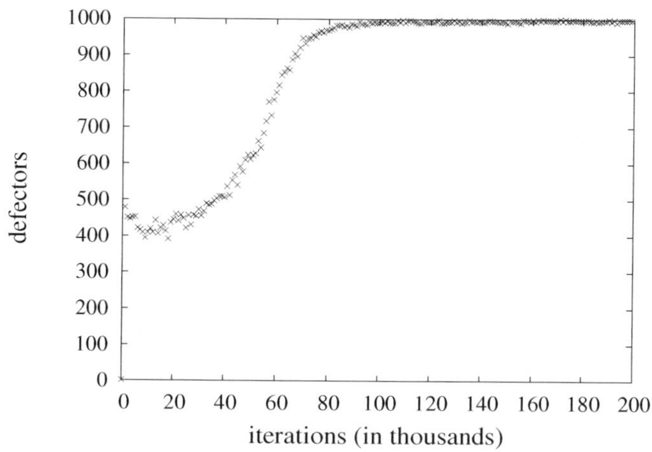


**Fig. 2.** Time evolution of the number of defectors - result of simulation started near the critical line

Recently [12,13] we developed a new realization of the Axelrod model with two new ingredients. First is that the model is freed from the payoff parameters; what is left is just the influence of agents' decisions on decisions of other agents. Further, the probabilities of decisions of individual agents (to defect or not, to punish or not) are not constant, but they are dynamically modified in each game they play. Second modification is less technical: once an agent decides to defect, his ability to punish in his future games vanishes, and his probability to defect (boldness) in future games is kept one until he is punished; then it is multiplied by $(1-\beta)$. Then, the constant $\beta$ describes the severity of the punishment. On the contrary, once an agent punishes, he will never defect the norm, and his probability to punish (vengeance) is set to $(1-\gamma)$, where $\gamma$ is due to the punishment cost. This vengeance can be further reduced if the agent punishes also in his future games. In this way, a kind of social labeling takes place: first decision is irreversible, and the whole process can be seen as a social contagion [9]. As a consequence, a sharp transition of the final boldness as dependent of the initial boldness is found [13]. The threshold value varies with the model parameters $(\beta, \gamma)$. These results depend only quantitatively on the assumed topology of the social network, which determines the probability distribution of the number of punishers, i.e. of the node degree.

The aim of the present paper is twofold. First, we are going to investigate the above mentioned sharp transition in the space of parameters $(\beta, \gamma)$. On the contrary to [13], here we are going to assume a given probability distribution of the initial boldness between agents. This modification makes the calculation closer to a social reality, where different agents present different willingnesses to break the norm and to punish. We should precise that in our model the parameters $(\beta, \gamma)$ represent not the boldness and the vengeance, but rather the modifications of agents' boldness and vengeance due to the decisions of other

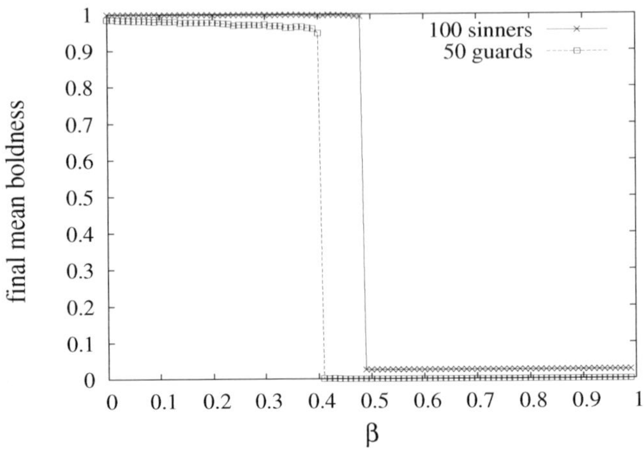


**Fig. 3.** The sharp character of the transition for a small number of guards or sinners

agents. Then, our model parameters describe the interactions between agents and not their actual states. Our goal here is to calculate the critical line on the plane $(\beta, \gamma)$ between the final state 'all defect and nobody is punished' and the final state 'nobody defects'. This critical line is a kind of generalization of the critical concentration, calculated in the problems of directed percolation [14,15]. Our second aim is to investigate the character of the transition influenced by some biased modifications of the structure of the social network. We use two kinds of these modifications: *i)* agents (guards) at some amount of nodes always punish and never defect, *ii)* agents (sinners) at some amount of nodes always defects and never punish.

In our model, agents are placed at nodes of the directed Erdös-Renyi network. This network is selected for its generic topology. Also, the role of hubs would need a separate study; such a study should refer to the homogeneous network. There is much work on the structures of social networks, mostly done by sociologists [16]; for an early list of references see [17]. With the outburst of scale-free networks a common opinion appeared that social networks are scale-free. Some of them can indeed be classified as scale-free, in particular those where direct face-to-face contact is not needed; as citation networks or telephone-calls networks [18,19,20]. Still, the actual structure varies from one social network to another [21,22], and often the network is simply too small to be classified to any type [23]. To end, we have checked [13] that the investigated threshold appears also in the scale-free growing networks, except the case when the direction of links is determined by the sequence of attaching new nodes. Last but not least, here we are not going to discuss the role of hubs, which could complicate the results.

The paper is organized as follows. In the next section the model assumptions and the details of the calculations are listed. In Section 3 we describe the numerical results. These are two: the critical line on the plane of the parameters, and the transition dependence of the amount of guards/sinners. Section 4 is devoted

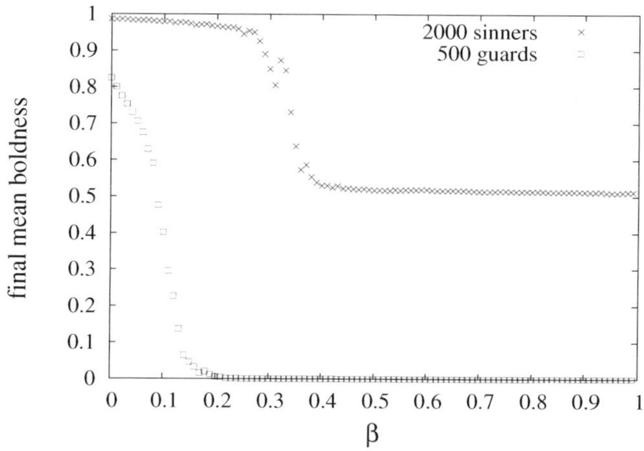


**Fig. 4.** The continuous character of the transition for a large number of guards or sinners

to discussion of the results in the context of a recent classification of contagion processes [24], and of some statistical data on dynamics of crime, presented in Fig. 1.

## 2 Model and Calculations

The network size is $N = 4000$ nodes, the mean number of in-going links (punishers) is $\lambda = 5$, and their distribution is Poissonian. Its construction is as follows: for each node $i$ a random integer number $k_i$ is chosen out of Poissonian distribution. Then, $k_i$ other nodes are selected as potential punishers of $i$, and directed links are drawn from these nodes to $i$-th node.

On the contrary to our former calculations [13], here we assume a homogeneous distribution of the initial probability to defect the norm, i.e. the initial boldness $b(i)$, where $i$ is the node index. As a rule, the initial vengeance $v(i)$, i.e. the initial probability of punishing, is $v(i) = 1 - b(i)$. This condition is not maintained during the simulation; however, $b(i) + v(i) \leq 1$. Third option is to obey the norm and not punish, with the probability $1 - b(i) - v(i)$.

At each time step, an agent $i$ is selected and he breaks the norm with the probability equal to his boldness $b(i)$. If actually he does, his boldness $b(i)$ is set to 1 and his vengeance, i.e. the probability of punishing - is set to zero. Then his neighbours $j(i)$ are asked, one by one, if they punish $i$. If one of them punishes, the boldness of $i$ is multiplied by a factor $1-\beta$, and the vengeance of the punisher $j$ is multiplied by $1 - \gamma$. The defector can be punished only by one neighbour. On the contrary, if a neighbour does not punish, his boldness is set to one and his vengeance is set to zero. In this way, the process is accompanied by kind of social labeling [25]: those who break the norm and those who refrain from punishing cannot punish in their future games, and those who punish cannot break the norm.

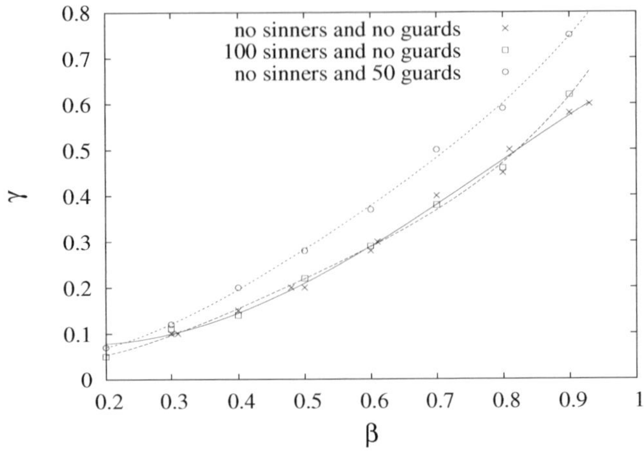


**Fig. 5.** The critical line in the homogeneous system and two lines for small numbers of guards or sinners

As a rule, we calculate the values of the parameters $\beta, \gamma$ where the threshold appears, i.e. the final state changes from the bold state 'all defect' to the vengeant state 'all punish'. Additionally, as a new variant of the game, some amount of sites of the network is selected randomly. Agents at these nodes got special roles of 'guards' or 'sinners'. If an agent is a guard, he always punishes and never defects; sinners do the opposite, i.e. always defect and never punish. The value and character of the threshold is observed against the ratio of the number of those special nodes to the whole population $N$. All these special nodes are either all 'guards' or all 'sinners'.

## 3 Results

An example of the result of simulation is shown in Fig. 2. The horizontal axis is the numerical time, i.e. the number of iterations. At the vertical line we show the amount of those who broke the norm within a given period of time. The simulation is performed with the initial conditions close to the critical line. The character of the evolution indicates that after some transient time, the number of defectors increased and since then they continuously break the norm.

Calculating the final boldness as dependent on $\beta$ and $\gamma$ we observe a sharp change of the result, as in Fig. 3 , at some threshold values of $\gamma$ and $\beta$. This means that we got a critical line $\gamma_c(\beta)$ in the plane of the parameters $(\beta, \gamma)$ - see Fig. 5. This line divides the plane $(\beta, \gamma)$ into two areas; the plane can be treated as a phase diagram, then we can talk about two phases. Above the line, we have a 'Bold' phase, where $\gamma$ is large; there punishment costs too much and is not effective. Below the line, we have a 'Vengeant' phase where the cost of punishment is low. Then, everybody punishes and there is no interest in defection.

Having added a small amount of 'sinners' or 'guards' to the system we observe that the threshold changes differently in these two cases. Basically the character of the threshold remains the same, just the threshold value is more susceptible to the admixture of 'guards' than to the one of 'sinners'. This can be seen at the positions of two additional curves in Fig. 5. While adding of five percent of 'sinners' apparently produces no effect, twice smaller admixture of 'guards' shifts the critical line upwards, reducing the area of the Bold phase. When the number of modified nodes increases more, the character of the plot $b(\gamma;\beta)$ gradually changes from an abrupt to a more continuous one. Examples of these curves are shown in Fig. 4. This change prevents us to investigate the critical line for higher concentration of special nodes; the transition becomes fuzzy.

## 4 Discussion

The crossover from the sharp to the fuzzy character of the transition between the Bold phase and the Vengeant phase, observed at the $(\beta,\gamma)$ plane, fits into a recent classification of contagion processes [24]. According to that scheme, models of contagion processes can be divided into three classes: *I*) independent interaction models, *II*) stochastic threshold models, and *III*) deterministic threshold models. Sharp transitions, like those found in our results, are characteristic for class *III*. Models in class *II* give fuzzy curves, similar to those obtained here for larger amounts of admixtured 'guards' or 'sinners'. These models are also called 'critical mass models'. In our terms, a critical amount of sinners should not be punished to get the transition to the bold phase; this could mean that local concentration of 'guards' should be small. If this is so, our results do fit into the classification proposed in [24]. As we noted in the Introduction, a similar classification was developed in [11]. Note that the basis of [11] was the random-field Ising model, which is far from the picture of contagion.

Coming back to the sociological reality, let us add a few words on the data presented in Fig. 1 in terms of punishment and its cost. We stress that the parameters $\beta$ and $\gamma$ do not mean directly the amount of units which a punished and punishing agent should pay, but just the measure of the decrease of the probability that he will defect and punish again. In these terms, large numbers on the statistics of crimes in a country can be interpreted as an indication that the punishment cost is large or the punishment is weak. Accordingly, an increase of the data can mean that the punishment $\beta$ decreased below some critical value or the punishment cost $\gamma$ increased. Then, new generations faced with an issue, to break a given norm or to preserve it, decide in a collective way, and these decisions are visible in the statistical data. If this point of view is accepted, the data presented in Fig. 1 can be seen as a demonstration, that the punishment cost paid by the policemen in these countries is less after 1993, than before. This cost is not necessarily open and intended; still it can appear as a consequence of burdensome bureaucratic procedures, faulty organization, unclear rights and shifts of political aims.

**Acknowledgements.** The research is partially supported within the FP7 project SOCIONICAL, No. 231288.

## References

1. Dunbar, R.I.M.: Annual Review of Anthropology 32, 163–181 (2003)
2. Kępiński, A.: Self-portrait of a man (in Polish). Wyd. Literackie, Kraków (2003)
3. Axelrod, R.: Amer. Political Sci. Rev. 80, 1095-1111 (1986)
4. Axelrod, R.: Complexity of Cooperation. Player-Based Models of Competition and Collaboration. Princeton UP, Princeton (1997)
5. Galan, M., Izquierdo, L.: JASSS 8(3), 2 (2005)
6. Neumann, M.: JASSS 11(4/6) (2008)
7. Axelrod, R.: Analyse und Kritik 22, 130-151 (2000)
8. Deffuant, G., Moss, S., Jager, W.: JASSS 9(1), 1 (2006)
9. Castellano, C., Fortunato, S., Loreto, V.: Rev. Mod. Phys. 81, 591–646 (2009)
10. Data of Eurostat, `http://epp.eurostat.ec.europa.eu/`
11. Michard, Q., Bouchaud, J.-P.: Eur. Phys. J. B 47, 151-159 (2005)
12. Kułakowski, K.: Int. J. Mod. Phys. C 19, 1105-1111 (2008)
13. Kułakowski, K., Dydejczyk, A., Rybak, M.: The norm game-how a norm fails. In: Allen, G., et al. (eds.) ICCS 2009, Part II. LNCS, vol. 5545, pp. 835–844. Springer, Heidelberg (2009)
14. Stauffer, D., Aharony, A.: Introduction to Percolation Theory. Routledge, London (2003)
15. Bollobás, B., Riordan, O.: Percolation. Cambridge UP, Cambridge (2006)
16. Wasserman, S., Faust, K.: Social Network Analysis: Methods and Applications. Cambridge UP, Cambridge (1994)
17. Hanneman, R.A., Riddle, M.: Introduction to Social Network Methods. University of California, Riverside (2005), `http://faculty.ucr.edu/~hanneman/`
18. Newman, M.E.J.: PNAS 98, 404-409 (2001)
19. González, M.C., Herrmann, H.J., Kertész, J., Vicsek, T.: Physica A 379, 307-316 (2007)
20. Onnela, J.-P., Saramáki, J., Hyvonen, J., Szabó, G., Argollo de Menezes, M., Kaski, K., Barabási, A.-L., Kertész, J.: New J. Phys. 9, 179 (2007)
21. Newman, M.E.J., Watts, D.J., Strogatz, S.H.: PNAS 99, 2566-2572 (2002)
22. Amaral, L.A.N., Scala, A., Barthélémy, M., Stanley, H.E.: PNAS 97, 11149-11152 (2000)
23. Schnegg, M.: Int. J. Mod. Phys. C 17, 1067-1076 (2006)
24. Dodds, P.S., Watts, D.J.: Phys. Rev. Lett. 92, 218701 (2004)
25. Marshall, G. (ed.): The Concise Oxford Dictionary of Sociology. Oxford UP, Oxford (1994)

# Model for Trust Dynamics in Service Oriented Information Systems

Grzegorz Kołaczek

Department of Computer Science and Management,
Wroclaw University of Technology, Wyb.Wyspianskiego 27 str.
Wroclaw, 50-370, Poland
Grzegorz.Kolaczek@pwr.wroc.pl

**Abstract.** The paper presents a novel approach to trust level evaluation for social networks related to service oriented information systems. The proposed framework uses mobile agents to improve the dynamics and accurateness of trust level evaluation and trust evolution modeling. The mobile agents evaluating trust are intelligent entities which use their own experience and records in the global knowledge base to analyze the trust. The agents' classes and relations between elements constituting the framework has been described and discussed as well as the most promising directions for formal expression and calculation of trust level.

**Keywords:** Trust model, Service Oriented Architecture, Service Oriented Knowledge Utility.

## 1 Introduction

One of the most important factors in human interaction and communication is trust. While more and more human activities go to the virtual world, trust is not longer only a typical feature of our usual businesses but it becomes also a crucial element for virtual space societies. Trust evaluation process is performed by all people several times per a day, so it is just an inseparable part of our lives. Most people evaluate the trust for most of typical daily scenarios without any problem. However, there is no simple transformation of this typical human activity to the virtual world. The virtual society differ from the real world one. The much greater level of anonymity and dynamics characteristic for the virtual societies influence significantly the trust relations. This is why we need some new perspective – models and tools to describe and to evaluate trust in virtual societies.

The current direction of the software and information system development is related to the service orientation paradigm. Service Oriented Architecture and Service Oriented Knowledge Utility defines the information systems category where service consumers are not only passive entities which uses the predefined set of services provided by service producers. These architectures enables service consumers to dynamically select the services, services providers and the way how the complex services are composed. The important factor to be taken into account while performing theses

N.T. Nguyen, R. Kowalczyk, and S.-M. Chen (Eds.): ICCCI 2009, LNAI 5796, pp. 573–583, 2009.

actions is the trustworthiness of services. To evaluate the trust of the services, service consumers can collaborate for example to exchange the recommendations and so they constitute in the natural way the social network.

## 2 Related Work

Trust is very important feature of security engineering. In this context it is related to risk management, surveillance, auditing and communication. Extensive knowledge on security engineering has already been collected and analyzed by Taipale [23] and has been studied in the Trusted Systems project, which is a part of Global Information Society Project [22] lead by the World Policy Institute [20]. It investigates systems in which some conditional prediction about the behavior of people or objects within a system that has been determined prior to authorizing access to system resources.

Another research area related to this topic is the concept of "web of trust" systems. This concept emerged in applications related to cryptography and is an element of such technologies like PGP, OpenPGP-compatible (PGP) or public key infrastructure (PKI). They have been defined to support the trust endorsement of the PKI generated certificate authority (CA)-signed certificates.

Finally, the most popular research area in the context of trust is called a trust metric. The aim of it is to propose a measure of how a member of a group is trusted by other members. An exemplary overview of such metrics has been prepared on the Internet community at TrustMetrics Wiki [25] and it presents a brief classification and provides many examples.

There are many different trust related metrics that are diverse in many aspects. For example TrustMail [6]or FilmTrust [5] propose to take advantage of a Semantic Web-based social network. Other different approaches are based on graph walking or use subjective logic to express trust level [8,9,10]. In the PeerTrust Project a decentralized Peer-to-Peer electronic community has been considered and a proposed trust model considers only three factors: the amount of satisfaction established during peer interaction, the number of iterations between peers and a balance factor for trust. At the other hand, the EigenTrust [14] algorithm has been based on similar idea to PageRank [21] but has been used in the context of file-sharing systems. This method computes global trust for peers, where the value is based on the history of uploads. It is dedicated to support the system to choose the peers with a history of reliable downloads and to exclude the malicious peers from the network.

Although there are many trust metrics, the proposed in this paper approach differs from them with regard to several aspects. The proposed trust level evaluation method takes advantage of the capabilities of the social network analysis. The presented method enables the autonomous agent to combine recommendation data obtained from other agents and data related to the agent's context. As the area of possible contexts of agent's interactions is very diverse we will focus on agent's position in agent's society described by its location in the social network.

The rest of the paper has been structured as follows. The subsequent section introduces the concepts of Service Oriented Architecture and Service Knowledge oriented Utility and defines the importance of trust modeling for such environments. The next section presents some basic problems about trust evaluation and modeling and later contains the main contribution – the definition of the mobile agent system for trust evaluation end evolution. Finally, the conclusion and direction of the future work has been presented.

## 3 Service Oriented Architecture and Service Oriented Knowledge Utility

Most organizations deliver their business processes using information technology (IT) applications. Many different software tools are used to capture, transform or report business data. Their role may be for example to structure, define and transform data or to enable or simplify communication. Each such interaction with an IT asset can be defined as a service. The set of delivered from the business processes services provide the incremental building blocks around which business flexibility revolves. In this context, Service Oriented Architecture (SOA) is the application framework that enables organizations to build, deploy and integrate these services independent of the technology systems on which they run [1,4,26,27]. In SOA, applications and infrastructure can be managed as a set of reusable assets and services. The main idea about this architecture was that businesses that use SOA can respond faster to market oppor tunities and get more value from their existing technology assets [2,3,7,27,28].

The vision of service oriented IT systems evolution proposed by [24] and presented in the figure 1. introduces another concept called Service Oriented Knowledge Utility

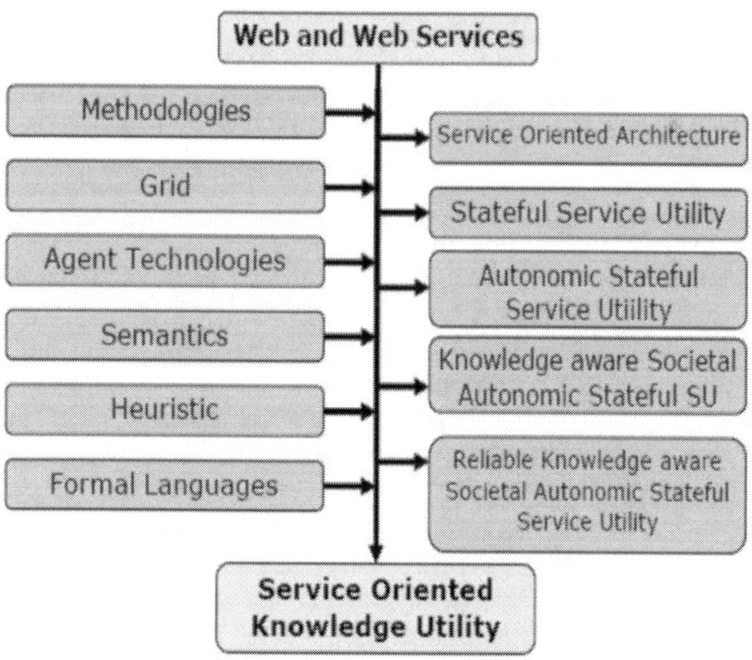


**Fig. 1.** Services Oriented environment evolution diagram

(SOKU). SOKU has been defined there a flexible, powerful and cost-efficient way of building, operating and evolving IT intensive solutions for use by businesses, science and society [24].

The SOKU concept has been established using three fundamental terms:

- Service orientation — what means that architecture comprises services which may be instantiated and assembled dynamically and the structure, behavior and location of software applications are changing at run-time.
- Knowledge — services are knowledge-assisted to facilitate automation and advanced functionality; services delivered by SOKU differs from typical SOA approach that are usually high-level services.
- Utility — an utility is a directly and immediately useable service with established functionality, performance and dependability, illustrating the emphasis on user needs and issues such as trust.

The both architectures for information systems description and development emphasize the role of the trust. It is one of the most top-level layers in these architectures and so the final result of the whole system performance is strictly related to the quality of the trust modeling and evaluation.

## 4 Trust Dynamics Model

There are two main reasons why the trust is so important aspect in contemporary information systems security. The first one is that there is no information system we could believe that is 100% secure [16,18,19]. The consequence of this is that we cannot be completely sure about the subject identity and its true intensions. The second is that SOA or SOKU information systems are a part of open networks. These networks allow subjects to communicate without any prior arrangements like for example organization membership and so this also makes information authenticity difficult to verify.

### 4.1 Trust Evaluation and Evolution Framework

The described in previous sections problems related to the trust modeling and evaluation in Service Oriented Architecture and Service Oriented Knowledge Utility environment can be used to set up basic requirements for trust evaluation and evolution modeling.

The proposed framework will be build on the multi agent system (MAS). Autonomous and intelligent agents of the MAS provides a natural support for the distributed and knowledge aware environment of SOA and SOKU. The crucial requirement related to real life information system environment is not only to evaluate trust, but also to do it effectively. The MAS should support the process of trust evaluation in transparent and nonintrusive way.

The proposed MAS comprises two types of agents: supervising agents (*As*) and testing agents (*At*). Supervising agents are strictly bound to business processes. Their role is to support business processes in evaluation of trust level of available services. This evaluation should support the service selection and complex services composition that are necessary for performing business processes. Testing agents are responsible for evaluation of the registered services reliability, security and so also trustworthiness.

*As* and *At* agents don't interact directly with each other, however each agent can have a real impact on behavior and decision of the other agent. Agents uses the common Knowledge Base as a source of their knowledge about services offered by the information system and about other agents and business processes. They can also contribute to the knowledge development as they record their observations there, assessments, opinions, etc. In this way, all business processes create more or less formal network where each node (business process) interacts with all other nodes. Relations between business processes in such network can influence many dimension of their activity like for example efficiency, accurateness, timeliness, trustworthiness, etc. In this paper we assume that all business processes and related to them *As* and *At* agents constitute the social network (figure 2). The relation between nodes in this social network are analyzed and used to trust level evaluation and dynamics description.

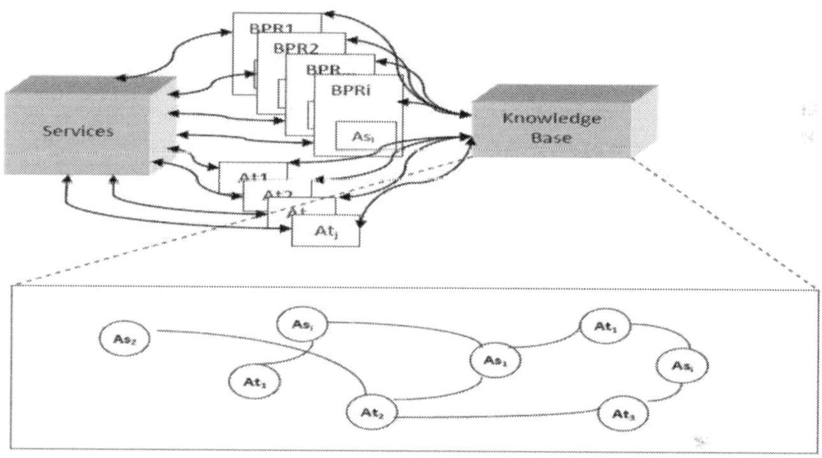


**Fig. 2.** Supervising and testing agents as social network members

More details about the *As* and *At* agents relationships and activities will be presented in the further part of this section. Other elements of the MAS for trust evaluation and evolution modeling are presented in the figure 3

The other important element for the trust modeling framework definition is formalization of the trust notion. The proposed model assumes that the trust level can be evaluated for each service. The final trust level is obtained as a combination of the six following components:

1. Recommendation ($T_r$) – the value of recommendation is calculated from the opinions of the other service consumers
2. Subjective component ($T_s$) – this element allows to reflect the subjective aspect of the trust, what means that the service consumers real life behavior is not always predicable and related only to the crisp trust related parameters (e.g. recommendation)
3. Dynamic component ($T_d$) – the value is set as the result of the agent's observation of the services behavior in the context of the currently performed business process
4. Historical component ($T_h$) – the record of earlier interactions helps agent to set or modify in a suitable way trust value

5. Contextual component ($T_c$) – the execution of the service S1 in task T1 may be related to completely different risks than executing the same service in a different context of e.g. task T2; this component is intended to formally reflect such situations
6. Service validation component ($T_v$) – the testing agents should check the various aspects related to service reliability and security, the result of such evaluation should be also taken into account while setting the trust value of the service

where:

| | | |
|---|---|---|
| BPR | — | Business processes, high level processes that are performed by information system, the BPR are performed as a series of elementary and complex services execution |
| DB | — | Database which is used internally by the information system to collect the data for trust evaluation |
| KB | — | Knowledge Base is a global source of knowledge about the system and its environment, it maintain information about recommendations and the results of security tests performed by testing agents |
| Server registry | — | Typical element of all SOA systems |
| As | — | Supervising agents |
| At | — | Testing agents |
| S | — | Services (both – elementary and complex) |
| SP | — | Service providers |

Mobile agents are responsible for calculation of the correspondent trust components values. In this case, they can get and post records from databases (DB) and knowledge base (these actions are represented by the arrows in the figure 3.)

- Calculate_ $T_d$(as,{bpr})
- Calculate_ $T_h$(as,{db})
- Calculate_ $T_c$(as,s)
- Calculate_ $T_r$(as,{kb})
- Calculate_ $T_v$(at,{kb})

Where: *as* is a member of set of supervising agents, *at* is a member of a set of testing agents, *bpr* is a set of relevant records describing the dynamic parameters of BPR, *db* is a relevant set of records form Database, *kb* is a relevant set of records from Knowledge Databes and *s* i a particular service.

The final trust value will be the product of the data fusion process applied to the six described earlier trust components ($T_r$, $T_s$, $T_d$, $T_h$, $T_c$, $T_v$). The domains of these components can be discrete (e.g. negative, neutral, positive) or continuous (e.g. $x \in <0,1>$). The other important factor to be taken into account while considering the data fusion method selection for trust evaluation is that the values describing trust level are not always crisp. This makes all probabilistic based or fuzzy methods the most interesting approach to be considered while implementing trust models.

The future work will concentrate on the possibility of application of the Subjective Logic formal methods to evaluate the trust components levels[11,12,13]. For example we can consider modification of the algorithm presented in [15,17] for calculation of recommendation factor.

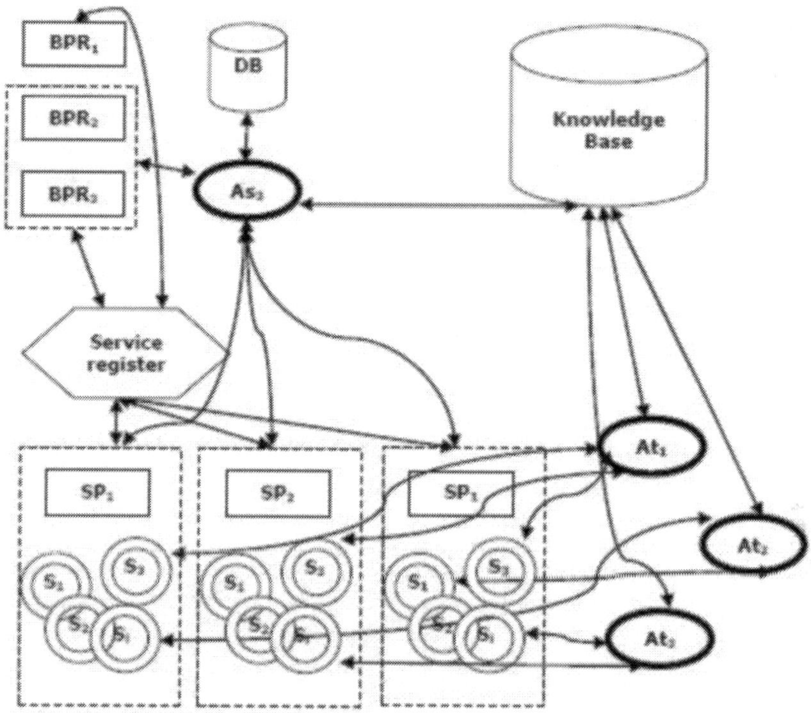


**Fig. 3.** Multi-agent trust dynamic model architecture

**Algorithm 1**

**Given:**

- An agent *A* who calculate unified recommendation about the service *S*,
- KB – includes a set of all *at* agents recommendation about the service *S* (we assume that recommendations have been expressed as the Subjective Logic opinions)

**Result:**

- $T_r = \varpi^{A}_{S}$ - the agent's *A* opinion about the service *S*

BEGIN

1. Find the corresponding recommendations related to the service *S*

$$\left\{ \varpi^{C_1}_{S}, \varpi^{C_3}_{S}, \ldots, \varpi^{C_k}_{S} \right\}$$

2. Calculate the agent *A* opinion about the service *S* as the Subjective Logic consensus of all recommendations found in step 1

$$\varpi^{A}_{S} = \varpi^{C_1}_{S} \oplus \varpi^{C_3}_{S} \oplus \ldots \oplus \varpi^{C_k}_{S} \quad (1)$$

END

Where ⊕ is Subjectiv Logic consensus operator and . $C_1$, $C_2$, ..., $C_k$ – agents that recommended service *S*

We can also consider the situation when the agents exchange opinions not only about the services but also about themselves. It means that the agents can have opinions about other agents trustworthiness. These opinions should be used while calculating some of the earlier defined service trust components (e.g. recommendation).
The modified algorithm of the $T_r$ component evaluation:

**Algorithm 2**

**Given:**

- An agent *A* who calculates its trust level to the service S,
- KB – includes a set of all *at* agents recommendation about the service *S* (we assume that recommendations have been expressed as the Subjective Logic opinions) and the agents' opinion about other agents, so the trust network between all agents can be traced.

**Result:**

$\varpi_S^A$ - the agent's *A* opinion about the service *S*

BEGIN

1. Find the agent $B_l$ that put recommendation record about the service *S* into KB
2. Reconstruct the trust network with agent *A* and agent $B_i$ that gives recommendation to the service *S*
3. Find in the trust network all paths leading from the agent *A* to the agent $B_i$.
4. Let **P**={$P_1$, $P_2$,..., $P_j$} be a set of all paths from *A* to *B* where $P_i$=<$A$,$p_1$, $p_2$,..., $p_z$,$B_i$>
5. Let **D**={$D_1$, $D_2$,..., $D_k$} be a set of all agents that are present in the paths form *A* do $B_i$.
6. Let **C**={$C_1$, $C_2$,..., $C_l$} be a set of all agents that have opinion about $B_i$ and are not the elements of the set **D**.
7. Measure the network context for each agent from the set **C** and present it in a form of Subjective Logic opinion $\varpi_{C_i}^{net} = (b_{netC_i}, d_{netC_i}, u_{netC_i})$. Where $\varpi_{C_i}^{net}$ is a recommendation and it reflects the agent's $C_i$ position in the network and is calculated as follows:

$$b_{net,C_i} = CENTRALITY \tag{2}$$

$$d_{net,C_i} = 1 - b_{C_iB} - u_{C_iB} \tag{3}$$

$$u_{net,C_i} = \min(1 - CENTRALITY, CLIQUE) \tag{4}$$

*CENTRALITY* and *CLIQUE* are normalised values calculated using selected social network metrics.

8. Calculate the modified opinion for each agent from the set **C** using context from Step 5. and Subjective Logic recommendation operation.

$$\varpi'^{C_i}_{B_l} = \varpi^{net}_{C_i} \otimes \varpi^{C_i}_{B_l} \tag{5}$$

Where opinion $\varpi^{C_i}_{B_l}$ is the original opinion of the agent $C_i$ about the agent $B_l$

9. Calculate the opinion for each path from the set **P** using the trust network and Subjective Logic recommendation operation.

$$\varpi'^{P_i}_{B_l} = \varpi^{A}_{p_1} \otimes \varpi^{p_1}_{p_2} \otimes \varpi^{p_2}_{p_3} \ldots \otimes \varpi^{p_z}_{B_l} \tag{6}$$

Where opinion $\varpi^{p_i}_{p_j}$ is the opinion of the agent $p_i$ about the agent $p_j$ which are the elements of the given path leading from $A$ to $B_l$.

10. Calculate the agent $A$ opinion about the agent $B_l$ as the Subjective Logic consensus of all opinions evaluated in steps 8 and 9

$$\varpi^{A}_{B_l} = \varpi'^{C_1}_{B_l} \oplus \varpi'^{C_3}_{B_l} \oplus \ldots \oplus \varpi'^{C_k}_{B_l} \oplus \varpi'^{P_1}_{B_l} \oplus \varpi'^{P_2}_{B_l} \ldots \oplus \varpi'^{P_l}_{B_l} \tag{7}$$

11. Check if there is another agent $B_{l+1}$ that recommended the service $S$
12. If the answer in step 11. is YES, then go to the step 1.
13. If the answer in step 11. is NO, then calculate $T_r$ using the agent $A$ opinions about the agents $B_1, B_2, \ldots, B_z$ and their opinions about the service $S$

$$T_r = \varpi^{A}_{S} = \left(\varpi^{A}_{B_1} \otimes \varpi'^{B_l}_{S}\right) \oplus \ldots \oplus \left(\varpi^{A}_{B_z} \otimes \varpi'^{B_z}_{S}\right)$$

END

The detailed Subjective Logic description and its operators used in the above presented algorithms can be found in [10,12].

## 5 Conclusions

The paper presents a novel approach to trust level evaluation for service oriented information systems The proposed framework uses the multi agent system (MAS) approach. Autonomous and intelligent agents of the MAS provides a natural support for the distributed and knowledge aware environment of SOA and SOKU. The general aim of the proposition is not only to evaluate trust, but also to do it effectively. The MAS should support the process of trust evaluation in transparent and nonintrusive way, however the implementation of all related to the proposed schema mechanisms in an appropriate test bed is required to verify this statement.

Another important problem considered as a next step of the presented research is selection of the formal method for trust level calculation. The most interesting and promising formal methods connected to trust evaluation problem are subjective logic, probabilistic logic or fuzzy logic approach.

## References

1. Chang, E., Dillon, T.S.: Trust and reputation relationships in service-oriented environments. In: Proceedings of the Third International Conference on Information Technology and Applications (ICITA 2005), vol. 1, pp. 4–14 (2005)
2. Chang, E., Hussain, F., Dillon, T.: Trust and Reputation for Service-Oriented Environments: Technologies For Building Business Intelligence and Consumer Confidence. John Wiley and Sons, Chichester (2005)
3. Chou, D.C., Yurov, K.: Security development in web services environment. Computer Standards and Interfaces 27(3) (2005)
4. Damiani, E., Vimercati, D.C., Paraboschi, S., Samarati, P.: Fine grained access control for SOAP E-services. In: WWW 2001: Proceedings of the 10th International Conference on World Wide Web, pp. 504–513. ACM, New York (2001)
5. Golbeck, J.: Trust on the World Wide Web: A Survey. Foundations and Trends in Web Science 1(2), 131–197 (2006)
6. Golbeck, J., Hendler, J.: Reputation Network Analysis for Email Filtering. In: Proceedings of the First Conference on Email and Anti-Spam, Mountain View, California, July 30-31 (2003)
7. Hussain, F.K., Chang, E., Dillon, T.S.: Defining reputation in service oriented environment. In: AICT-ICIW 2006: Proceedings of the Advanced International Conference on Telecommunications and International Conference on Internet and Web Applications and Services. IEEE CS Press, Los Alamitos (2006)
8. Jøsang, A.: A subjective metric of authentication. In: Quisquater, J.-J., Deswarte, Y., Meadows, C., Gollmann, D. (eds.) ESORICS 1998. LNCS, vol. 1485, pp. 329–344. Springer, Heidelberg (1998)
9. Jøsang, A.: An Algebra for Assessing Trust in Certification Chains. In: Kochmar, J. (ed.) Proceedings of the Network and Distributed Systems Security Symposium (NDSS 1999). The Internet Society (1999)
10. Jøsang, A.: Artificial reasoning with subjective logic. In: Nayak, A., Pagnucco, M. (eds.) Proceedings of the 2nd Australian Workshop on Commonsense Reasoning, December. Australian Computer Society, Perth (1997)
11. Jøsang, A.: Trust and Reputation Systems. In: Aldini, A., Gorrieri, R. (eds.) FOSAD 2007. LNCS, vol. 4677, pp. 209–245. Springer, Heidelberg (2007)
12. Jøsang, A., Knapskog, A.: A Metric for Trusted Systems (full paper). In: Proceedings of the 21st National Information Systems Security Conference. NSA (October 1998)
13. Josang, A.: An Algebra for Assessing Trust in Certification Chains. In: J. Kochmar (eds.). In: Proceedings of the Network and Distributed Systems Security Symposium, pp. 124–140 (1999)
14. Kamvar, S.D., Schlosser, M.T., Garcia-Molina, H.: The EigenTrust algorithm for reputation management in P2P networks. In: Proceedings of the Twelfth International World Wide Web Conference, Budapest, May 2003 (2003)
15. Kołaczek, G.: A new concept of trust modeling and management in complex networks. In: Lovrek, I., Howlett, R.J., Jain, L.C. (eds.) KES 2008, Part I. LNCS (LNAI), vol. 5177, pp. 237–244. Springer, Heidelberg (2008)

16. Kołaczek, G.: Dynamika relacji zaufania i jej modelowanie w systemach teleinformatycznych., Internet w społeczeństwie informacyjnym, Zakopane (2004)
17. Kołaczek, G.: Evaluation of trust in multiagent systems using social network analysis metrics. In: Józefczyk, J., Thomas, W., Turowska, M. (eds.) 14th International Congress of Cybernetics and Systems of WOSC. 14th ICCS. Proceedings, September 9-12, pp. 595–603. Oficyna Wydaw. PWroc., Wrocław (2008)
18. Kołaczek, G.: Trust management in E-commerce. W: Electronic commerce. Theory and applications. In: Wachowicz, J. (ed.) Gdańsk: Wydział Zarządzania i Ekonomii PGdań, pp. 53–59 (2005)
19. Kołaczek, G.: Trust networks and cellular automata. W: Information Systems Architecture and Technology ISAT 2004. In: Grzech, A., Wilimowska, Z. (eds.) Proceedings of the 25th international scientific school. Information models, concepts, tools and applications, pp. 272–278. Wrocław Oficyna Wydaw. PWroc. (2004)
20. Michlmayr, A., Rosenberg, F., Platzer, C., Treiber, M., Dustdar, S.: Towards recovering the broken SOA triangle: a software engineering perspective. In: IW-SOSWE 2007: Proceedings of the Second International Workshop on Service Oriented Software Engineering, pp. 22–28. ACM Press, New York (2007)
21. Page, L., Brin, S., Motwani, R., Winograd, T.: The PageRank Citation Ranking: Bringing order to the Web. Technical Report, Computer Science Department, Stanford University (1998)
22. Schillo, M., Funk, P., Rovatsos, M.: Who can you Trust: Dealing with Deception. In: Castelfranchi, C., Tan, Y., Falcone, R., Firozabadi, B.S. (eds.) Proceedings of the Workshop "Deception, Fraud and Trust in Agent Societies" of the Autonomous Agents Conference, pp. 293–320 (1999)
23. Taipale, K.A.: The Trusted Systems Problem: Security Envelopes, Statistical Threat Analysis, and the Presumption of Innocence. IEEE Intelligent Systems 20(5), 80–83 (2005)
24. The Next Generation Expert Group: Future of European Grids: Grids and Service-Oriented Knowledge Utilities (January 2007)
25. Trustlet Wiki, http://www.trustlet.org/wiki
26. Web Services Security: SOAP Message Security 1.1, OASIS WSS TC (2006), http://docs.oasis-open.org/wss/v1.1/
27. Zhengping, L., Xiaoli, W., Guoqing, Y., Min, Z.: A formal framework for trust management of service-oriented systems. In: SOCA 2007: Proceedings of the IEEE International Conference on Service-Oriented Computing and Applications, pp. 241–248. IEEE CS Press, Los Alamitos (2007)
28. Zhu, M., Liu, L., Jin, Z.: A social trust model for services. In: Proceedings of the 11th Australian Workshop on Requirements Engineering, AWRE 2006 (2006)

# Collective Prisoner's Dilemma Model of Artificial Society

Marek Chlebuś, Wojciech Kamiński, and Romuald Kotowski

Polish-Japanese Institute of Information Technology, Koszykowa 86, Warsaw, Poland
marek.chlebus@gmail.com, wkaminski@pjwstk.edu.pl, rkotow@pjwstk.edu.pl

**Abstract.** The Collective Prisoner's Dilemma (CPD) as a tool for modeling the Artificial Society (AS) consisting of the mutually interacting agents is proposed. CPD is an extensive, complex and partially stochastic model constructed as an infinite series of parameterized evolutionary two-agent games conducted in form of the asymmetric Prisoners Dilemma played in intentionally selected pairs of agents. The main ideas and axioms of CPD are presented and discussed. The basic purpose of the model is to examine agents' personal features influence on social relations and society economic evolution. The model can be helpful in analyzing various forms of organization of a state and society. As an example of application, some interesting economic evolution patterns for AS are studied and significant factors of growth identified. CPD can be also used as an engine for computer strategic games.

**Keywords:** agent-based decision-making process, social dynamics, economic growth, evolution of cooperation, prisoner's dilemma, collective prisoner's dilemma, emotions.

## 1 Introduction

In 1950 Merrill Flood and Melvin Dresher proposed the model of cooperation and conflict [4], and Albert Tucker in the same year named it Prisoner's Dilemma (PD). Now it is the most well-known game theoretic paradox. The simplest generalization of PD for two players is to introduce the iterated procedure, when the game is played repeatedly. That can be done either when the number of steps is known in advance or when the number of steps is indefinite. The second approach allows the cooperation to reach equlibiurm.

The proposed here Collective Prisoners Dilemma (CPD) is an extensive generalization of PD, invented and primary formulated by Marek Chlebuś [3], inspired mainly by Robert Axelrod's [1] tournaments for computer algorithms playing the $N$-step PD and Anatol Rapoport's [7] idea of extending Axelrod's approach to population modelling. CPD is a multilevel, complex and partialy stochastic model constructed as an infinite series of parameterized evolutionary multiagent games conducted in form of the asymmetric PD played in intentionally selected pairs of agents.

In this paper the CPD implementation to modelling Artificial Society (AS) as a set of agents, is proposed. The agents are equipped with a number of properties (in this paper seven static and three dynamic individual parameters are

N.T. Nguyen, R. Kowalczyk, and S.-M. Chen (Eds.): ICCCI 2009, LNAI 5796, pp. 584–595, 2009.

proposed) what generates an enormous variety of societies. Each society is given additionally its own collection of properties influencing the types of interactions between agents, and the external circumstances conditioning the final realization of the model. The obtained system demonstrates the living system features in term of the Miller's living systems theory [6]. It means in particular that the system has synergetic, emergent, self-organizing and semi-chaotic properties.

## 2 From Agent to Society

The following assumptions are made in the proposed model. There exist two types of objects only, i.e. agents and societies [2]. Agent represents the single person. Group of agents creates a society. A great number of societies with the diverse types of the social behaviors can be created as the result of the diverse properties of the individuals and additionally imposed external conditions. Each created society is described by the global and individual attributes. In what follows the specific attributes of the agents and societies are described in more details.

### 2.1 The Individuals

Agents are the basic and the most important constituents of the model. The main purpose of the agent is to establish and to conduct the interactions with other agents (these activities will be described more precisely in Section 3). Each agent possesses a collection of attributes with ascribed values characterizing its personality and defining its behavior during interactions with other agents. In the model the agents' features are divided into two subsets, obeying static and dynamic attributes. The static attributes are different for different agents and unchangeable during whole simulation time. The dynamic attributes evolve in time. The existence of the following attributes is assumed:

**Static attributes**

$R$ – rationality, defines the level of influence of the agent's opinion (rationally determined) on the agent's prospective decisions – in opposition to the influence of emotions;
$S$ – sensitivity, defines the level of influence of new experiences on agent's emotions and opinions;
$O$ – openness, defines intention level of interaction with unknown agents;
$T$ – tolerance, defines tolerance threshold in qualifying partners for interaction;
$A$ – activity, defines economic engagement during interaction;
$C$ – conformism, defines the level of influence of public opinion on agents' decisions - in opposition to the influence of its personal opinion;
$M$ – character of morality, defined by following models:

- $E$ – commercial morality, agent values individual income;
- $P$ – utilitarian morality, agent values total income;

- $SO$ – cooperative morality, agent values cooperation.

**Dynamic attributes**

$MEM$ – memory, limited register of past interactions;
$EM$ – emotions, influences agents' interaction decisions, i.e.
- $MOOD$ – mood (agent's emotional state);
- $OP(A_n)$ – opinion about the $A_n$ agent;

$W$ – wealth, agents' economic strength.

### 2.2 $N$-Agents Society

Agents are grouped into $N$-agents societies [5]. Each society is described by the set of attributes corresponding to individual agents' static attributes which define society type. Society attributes values $socAtt \in \{0, 0.5, 1\}$ are assumed and based on the average value of agents individual attributes. Society global parameters are given as follows:

Taxes – society tax rates;
Biological minimum – threshold of agent's existence;
Social minimum – threshold of social aid;
Costs – agents' life costs;
Wealth distribution factor – Gini coefficient[1].

The proposed CPD model [7] is based on the series of multiple semi-deterministic two-agent games conducted in the frames of the multi-parameterized PD.

### 2.3 Social Relations

Agents' personal memory mechanism is given to record every single interaction. Memory time is limited. One prefers interactions with agents already known and remembered. It is assumed in the model that the choosing partners to interact is based upon agents' memory. If the former interactions were evaluated as positive, agent prefers to interact with the same partner again. If the interaction was evaluated as negative, agent refuses to interact again with the same partner. This approach may produce social groups (families, friends) of privileged partners [8]. This aspect will be examined in future papers.

## 3 The Game

### 3.1 Pre-game Preparations

The purpose of each agent is to establish and to conduct the interactions with other agents. The single interaction is conducted in the pairs of agents. The pair

[1] To clarify the model, we use simplified, not exacly the same but very similar redistribution coefficient to Italian statistician Corrado Gini;

is generated according to the following rules. Each agent picks semi-randomly another agent to interact. At the beginning the choice is random (from the group of the already known agents, existing in choosing agents' memory). Next, the tolerance test is conducted to the chosen agent. This test involves the comparison choosing agents' opinion ($OP$) about the chosen agent to choosing agents' tolerance. The more it is tolerant, the less $OP$ value is needed to pass the test. This trial is repeated according to choosing agents' openness ($O$): high value – 3 attempts, mid value – 2 attempts, low value – 1 attempt. If the agent selection fails, the process is repeated with randomly picked agents from the whole society and the tolerance test is conducted again (2 attempts). If this method fails again, the random agent is picked-out.

### 3.2 Prisoner's Dilemma

A PD two-persons game as a model of the interaction of agents is proposed. A pair of agents plays PD game one time per each iteration. Dynamic payoff matrix is used. Each agent contributes a certain part of its wealth as a stake to the game. The stake engaged is an element of dynamic payoff matrix (see Table 1).

In the Table 1, $a_1$ is the agents' $A_1$ contribution (stake) and $a_2$ is an agents' $A_2$ contribution. The variable $a$ defines non-zero sum of the game: $a = (a_1 \cdot a_2)/(a_1 + a_2)$. The contribution $a_i$ is a part of agents' $A_i$ wealth determined by the agents' activity feature ($A$). In the single PD game the optimal individual strategy for the agent $A_1$, irrespectively of the agent $A_2$ strategy, is to defect ($D$). The strategy $D$ puts an agent $A_1$ in situation, when it loses less and wins more (depending on $A_2$ decision). Optimal individual strategy for the agent $A_2$ is $D$ for the same reason. However, optimal strategy for the both agents at the same time, or generally speaking for society, is the both-sided cooperation ($C$, $C$). The last one is the only possibility with the positive payoff sum ($2a$). In other cases the payoff sum equals 0. When interacting more that once optimal strategy changes (see 3.3 Game results). Agents' interaction decision depends on certain factors.

Cooperative interaction probability is given as follows:

$$P = conformism \times [social\ factors] + (1 - conformism) \times [personal\ factors]\ , \quad (1)$$

**Table 1.** The payoff matrix

| $A_1 \setminus A_2$ | C | D |
|---|---|---|
| C | $a$; $a$ | $-a_1$; $a_1$ |
| D | $a_2$; $-a_2$ | 0; 0 |

where composition of $[factors]$ is following:

$$[factors] = rationality \times [experience] + (1 - rationality) \times [emotions] \quad (2)$$

The exact form of the probability equation is:

$$\begin{aligned} P(A_1) = C(A_1) \cdot [R(A_1) \cdot REP(A_2) + (1 - R(A_1) \cdot SM] \\ + (1 - C(A_1)) \cdot [R(A_1) \cdot OP(A_2) + (1 - R(A_1) \cdot MOOD(A_1)] , \end{aligned} \quad (3)$$

where $REP(A_2)$ stands for the agents' $A_2$ reputation calculated as an average opinion of society agents about $A_2$; $SM$ stands for the society mood calculated as an average value of all agents personal mood. As it was shown, the probability of the cooperative behavior depends on the static features (conformism, rationality) and its dynamic features (mood, society mood, opinion, reputation).

### 3.3 Game Results

In each iteration the dynamic properties of the agents and society evolve changing their general characteristics. The main result of the interaction is the change of agents' wealth based on the PD game payoff matrix. A game is also followed by the mutual judgment and mood modification, as a secondary result.

Opinion about game partner and personal mood are calculated as follows:

$$[mood] = sensitivity \times [new\ experience] + (1 - sensitivity) \times [emotional\ state] \quad (4)$$

$$[opinion] = sensitivity \times [opinion] + (1 - sensitivity) \times [oldopinion] \quad (5)$$

The exact forms of the equations are:

$$MOOD = S(A_1) \cdot x + (1 - S(A_1) \cdot MOOD_0 \quad (6)$$

$$OP(A_2) = S(A_1) \cdot x + (1 - S(A_1)) \cdot OP_0(A_2) , \quad (7)$$

where agents' $A_1$ current mood (MOOD) is calculated basing on the values of the previous mood ($MOOD_0$) and the current opinion similarly depends on the previous opinion ($OP_0$). Emotions factor $x$ depends on the current game state and the agent's morality and is defined in Table 2, showing $A_1$ opinion about $A_2$ in accordance to the morality model of $A_1$.

$D$ is the optimal strategy for an individual agent in a single PD game. When playing more than one iteration a possibility to punish the non-cooperative agent occurs. In such a case the optimal individual strategy All-$D$ is no longer profitable. In case of the defection one can be punished during the prospective games, and its chances to win may decrease. Defection causes also the drop down of the reputation, so irrespectively of next agent paired, the possibility of cooperation towards defector decreases.

## 4 After-game Social Mechanisms

In each iteration, when the full set of two-agent games is completed, the following public tasks are implemented: tax collecting, life costs deduction, social aid and budget redistribution.

**Table 2.** Morality models

Commercial morality

| $A_1 \backslash A_2$ | C | D |
|---|---|---|
| C | 2/3 | 1 |
| D | 0 | 1/3 |

Utilitarian morality

| $A_1 \backslash A_2$ | C | D |
|---|---|---|
| C | 1 | 0 |
| D | 1/3 | 2/3 |

Cooperative morality

| $A_1 \backslash A_2$ | C | D |
|---|---|---|
| C | 1 | 0 |
| D | 2/3 | 1/3 |

### 4.1 Tax Collection and Life Costs

When considering an iteration as an elementary time period of an artificial life, each agent has to pay life costs and taxes. The costs of life are equal for every agent. The taxes are individualized. Three tax types are defined:

- turnover tax, where the tax base is PD game stake for each agent;
- income tax, where the tax base is PD game payoff of each agent, if positive;
- wealth tax, where the tax base is the current agents' wealth.

Collected taxes create the temporary public budget used to finance the social aid and the budget redistribution.

### 4.2 Social Aid

Game results, costs and taxes can reduce agents' wealth to a very low value. Wealth social minimum value factor for the given society is defined. The value of the social minimum is calculated at each iteration as a defined portion of current life costs. When agents' wealth goes below the social minimum, society temporary budget is used to compensate agents' poverty. Wealth biological minimum value is parameterized in a similar way. If agents' wealth goes below the biological minimum, agent cannot interact longer, and "dies".

### 4.3 Redistribution

If after social aid the temporary budget surplus remains, it is redistributed among agents using hyperbolic distribution function parameterized by generalized Gini's coefficient $G \in \langle -1; 1 \rangle$.

if $G > 0$ – rich agents obtain more than the poor ones;
if $G < 0$ – poor agents obtain more than rich ones;
if $G = 0$ – budget surplus is divided equally between agents.

## 5 Simulation Results

Computer implementation of the proposed model was provided in order to test the economical behavior of AS. Some interesting economic evolution patterns during time (implemented as number of iterations) were observed.[2] For all examples showed below, the following values of the parameters were set as default:

Rationality – moderate (1/2);
Sensitivity – moderate (1/2);
Openness – moderate (1/2);
Tolerance – moderate (1/2);
Activity – moderate (1/2);
Conformism – moderate (1/2);
Morality model – utilitarian;
Initial wealth distribution – moderate ($G = 0.5$);
Initial mood – moderate (1/2);
Memory time – average (5 iterations);
Income tax – 10%;
Turnover tax – 10%;
Asset tax – 10%;
Life cost – 3% of mean individual wealth;
Social minimum – 20% of life cost;
Biological minimum – 5% of life cost.

Society can evolve differently each time the parameters have changed. It was observed that particular parameters from the listed above have the very significant influence on the economic evolution of AS. Each of the following results shows how single parameter influences economic evolution.

### 5.1 Models of Taxation

The first simulation examines taxation model influence on society overall economy. Four taxation models were considered:

Income tax only;
Turnover tax only;
Asset tax only;
Triple tax model.

This experiment has given interesting (and repeatable) results which have to be examined with details in future.

In a short term scale, asset tax only model is the most favorable for economical growth of AS, and income tax is the less "friendly", but in a long-time period, single income tax achieves surprisingly dominating position over all other solutions (Fig. 1).

[2] Presented results are only exemplifications of AS behavior. The system needs an extensive development before jumping with conclusions beyond a proper area of multi-agent environment.

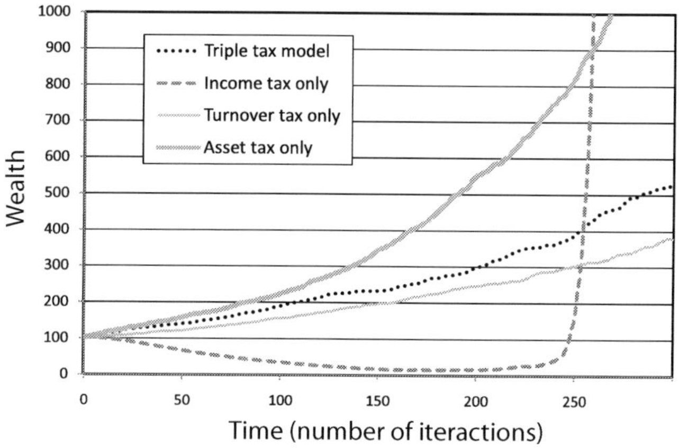


**Fig. 1.** Taxation models

### 5.2 Redistribution Models

Calculations for the whole range of the acceptable values of Gini coefficient were made and the most interesting values were obtained in range $G \in \langle -0.2; 0.2 \rangle$. Three Gini coefficient values are considered basing on observations:

Redistribution = -0.2 (more for poor agents);
Redistribution = 0 (equal redistribution);
Redistribution = 0.2 (more for reach agents).

The best model of redistribution occurs alimenting of poor agents. This model of redistribution decreases effective Gini coefficient of AS, making its wealth distribution more flat. The worst model seems to be alimentation of reach, forcing higher disproportions of individual wealth of AS agents. The correlation seems to be moderate (Fig. 2).

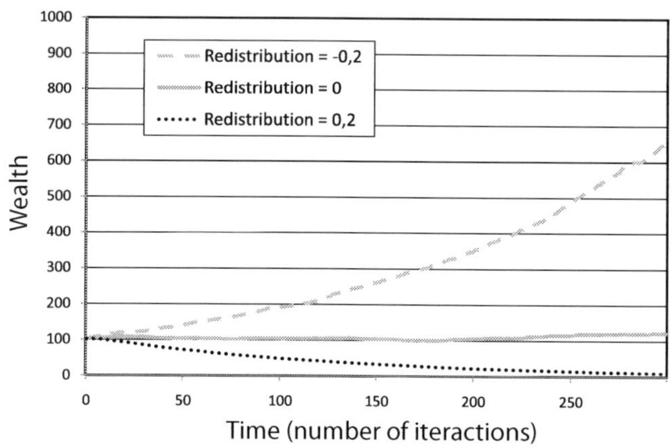


**Fig. 2.** Redistribution models

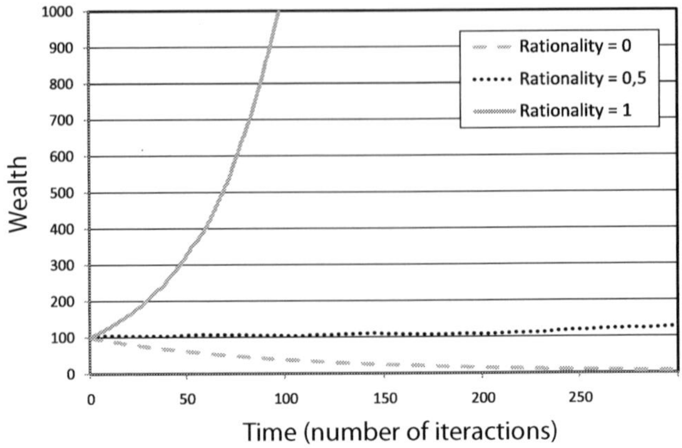


**Fig. 3.** Rationality models

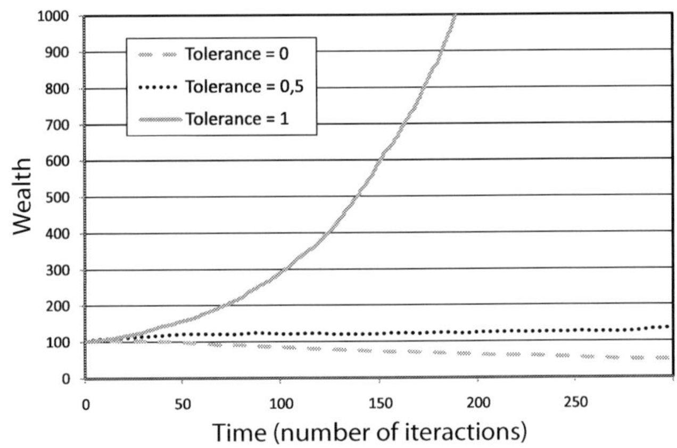


**Fig. 4.** Tolerance models

### 5.3 Rationality Models

Three rationality models were considered:

Rationality = 0 (emotional);
Rationality = 0.5 (neutral);
Rationality = 1 (rational).

High rationality guarantees high economical growths, moderate rationality - stabilization, and low rationality (high emotionality) correlates with economic depression of AS. The correlation seems to be very strong (Fig. 3).

### 5.4 Tolerance Models

Three toleration models were considered:

Toleration = 0 (low);
Toleration = 0.5 (average);
Toleration = 1 (high).

The influence of tolerance on economy looks very similar to the influence of rationality: the higher value of tolerance, the higher growth of AS wealth. The correlation seems to be strong (Fig. 4).

### 5.5 Memory Models

Three settings were considered:

Lack of memory (agent doesn't remember single interaction; memory = 0);
Average memory (agent remembers few past interactions; memory = 5);
Long memory (agent remembers many past interactions; memory = 10).

The situation with memory is similar to these with rationality and tolerance. High values of memory feature (agent remembers more past iterations) seem to guarantee economic growth, moderate values - stabilization, and low values - depression. The correlation seems to be moderate (Fig. 5).

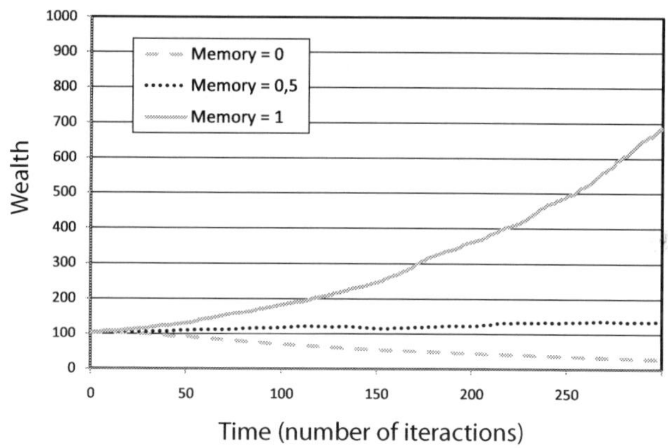


**Fig. 5.** Memory models

### 5.6 Morality Models

Morality models tested:

Commercial morality – individual success on top of hierarchy of values;
Utilitarian morality – common success on top of hierarchy of values;
Cooperative morality – cooperativeness on top of hierarchy of values.

While keeping all remaining parameters values default, only utilitarian morality model correlates with economic growth of AS. Other models are depressive. The correlation seems to be comparatively weak (Fig. 6).

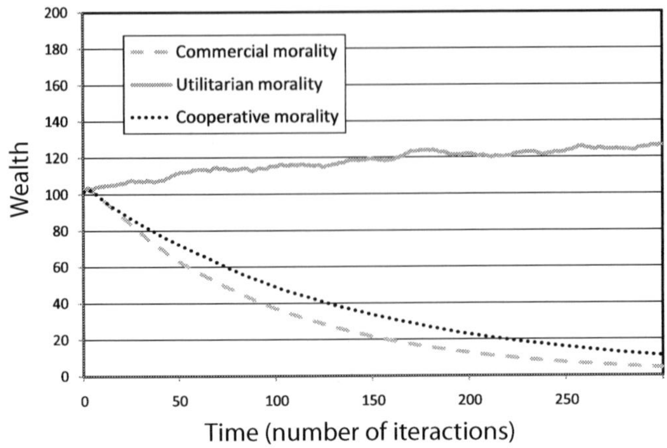


**Fig. 6.** Morality models

## 6 Conclusions and Future Development

The model of AS based on the modified CPD game was proposed. The model gave an overview of the structures of various societies following from the properties of individual agents. The presented plots inform about dominating atributes of the society caused by the changing of atributes of the agents.

It has occured that our model gave a realistic description of real societies. Nevertheless the further studies are needed. Future model development plans are:

1. Model implemented as a real-time simulation;
2. Demographic mechanisms – an agent can be born due to certain rules; new agents' personal character features may depend on "parents" features and real-time simulation society settings;
3. Possibility to make an external influence on society as a e.g. crisis, disaster, economic acceleration;
4. Spatial model – each agent located in a cell of cellular automaton; territorially bounded groups of agents can form a kind of "countries";
5. Interactive on-line game – spatial model can be used as a simulation game between $N$ on-line players reigning their own "countries".

## References

1. Axelrod, R.: The Evolution of Cooperation. Basic Books, New York (1984)
2. Bonabeau, E.: Agent-based modeling: methods and techniques for simulating human systems. Proc. National Academy of Sciences 99(3), 7280–7287 (2002)
3. Chlebuś, M.: Series of lectures presented at the Polish-Japanese Institute of Information Technology during the seminar Application of IT in social research (2008)
4. Dresher, M.: The Mathematics of Games of Strategy: Theory and Applications. Dover Publ., New York (1981)

5. Epstein, J.M., Axtell, R.: Growing Artificial Societies: Social Science From the Bottom Up. MIT Press/Brookings Institution (1996)
6. Miller, J.G.: Living systems. McGraw-Hill, New York (1978)
7. Rapoport, A., Chammah, A.M.: Prisoner's Dilemma, 1965 e CRAS Paris 287, pp. 1013–1015. The University of Michigan Press (1978)
8. Takahashi, I., Okada, I.: Sympathy and Interaction Frequency in the Prisoner's Dilemma. In: Proceedings of the Proceedings of the 38th Annual Hawaii International Conference on System Sciences (HICSS 2005), vol. 03 (2005)

# DES Control Synthesis and Cooperation of Agents

František Čapkovič*

Institute of Informatics, Slovak Academy of Sciences
Dúbravská cesta 9, 845 07 Bratislava, Slovak Republic
Frantisek.Capkovic@savba.sk
http://www.ui.sav.sk/home/capkovic/capkhome.htm

**Abstract.** A procedure of control synthesis for DES (discrete-event systems) modelled by the special class of place/transition Petri nets (P/T PNs), the so called state machines (SMs), is introduced here, generalized and extended for DES modelled by P/T PNs with general structure. Its applicability to the automatic DES control synthesis is presented as well as the possibility of its utilizing at supervision of the agent cooperation in multi-agent systems (MAS). For modelling the agents P/T PNs are utilized too. The usefulness of the approach is demonstrated by examples.

**Keywords:** Agents, bipartite graphs, control, cooperation, discrete-event systems, Petri nets, supervisor, synthesis.

## 1 Introduction

Discrete-event systems (DES) are widely used in human practice - e.g. flexible manufacturing systems, communication systems of different kinds, transport systems, etc. They are systems driven by discrete events. Thus, the DES behaviour is discrete in nature. The behaviour of agents cooperating each other can also be understood to be a kind of DES. The causality in the DES behaviour was described in [1]. DES are frequently modelled and analysed by means of Petri nets (PNs), especially P/T PNs (place/transition PNs). The control synthesis for DES modelled by PNs can be performed also in virtue of P/T PN-based methods. In [1] the method of such a kind was introduced. It is based on ordinary reachability graphs (RGs) and yields the space of feasible state trajectories from a given initial state to a desired terminal one. This paper immediately goes on to solve the DES control synthesis problem in order to automate this process as soon as possible, even to achieve the automated control synthesis. The earlier results presented in [2],[3] are also utilized in this paper. In the denotation from system theory, the P/T PN-based model of DES can be formally expressed as the linear discrete system constrained by the inequality as follows

$$\mathbf{x}_{k+1} = \mathbf{x}_k + \mathbf{B}.\mathbf{u}_k \quad , \quad k = 0, ..., N \tag{1}$$

* Partially supported by the Slovak Grant Agency for Science (VEGA) under grant # 2/0075/09 and by the Slovak Grant Agency APVV under grant LPP-0231-06.

N.T. Nguyen, R. Kowalczyk, and S.-M. Chen (Eds.): ICCCI 2009, LNAI 5796, pp. 596–607, 2009.

$$\mathbf{B} = \mathbf{G}^T - \mathbf{F} \tag{2}$$

$$\mathbf{F}.\mathbf{u}_k \leq \mathbf{x}_k \tag{3}$$

where $k$ is the discrete step of the dynamics development; $\mathbf{x}_k = (\sigma_{p_1}^k, ..., \sigma_{p_n}^k)^T$ is the $n$-dimensional state vector; $\sigma_{p_i}^k \in \{0, 1, ..., c_{p_i}\}$, $i = 1, ..., n$ express the states of the elementary subprocesses or operations by 0 (passivity) or by $0 < \sigma_{p_i} \leq c_{p_i}$ (activity); $c_{p_i}$ is the capacity of the subprocess $p_i$ as to its activities; $\mathbf{u}_k = (\gamma_{t_1}^k, ..., \gamma_{t_m}^k)^T$ is the $m$-dimensional control vector; its components $\gamma_{t_j}^k \in \{0, 1\}$, $j = 1, ..., m$ represent occurring of elementary discrete events (e.g. starting or ending the elementary subprocesses or their activities, failures, etc.) by 1 (presence of the corresponding discrete event) or by 0 (absence of the event); $\mathbf{B}$, $\mathbf{F}$, $\mathbf{G}$ are matrices of integers; $\mathbf{F} = \{f_{ij}\}_{n \times m}$, $f_{ij} \in \{0, M_{f_{ij}}\}$, $i = 1, ..., n$ , $j = 1, ..., m$ expresses the causal relations among the states (as causes) and the discrete events occurring during the DES operation (as consequences) by 0 (nonexistence of the relation) or by $M_{f_{ij}} > 0$ (existence and multiplicity of the relation); $\mathbf{G} = \{g_{ij}\}_{m \times n}$, $g_{ij} \in \{0, M_{g_{ij}}\}$, $i = 1, ..., m$, $j = 1, ..., n$ expresses very analogically the causal relations among the discrete events (as causes) and the DES states (as consequences); the matrix $\mathbf{B}$ is given by means of the arcs incidence matrices $\mathbf{F}$ and $\mathbf{G}$ according to (2); $(.)^T$ symbolizes the matrix or vector transposition.

## 2 DES Control Problem

Considering the introduced system (1)-(3) to be the model of DES, the problem of control is the problem of finding a suitable sequence of control vectors $\{\mathbf{u}_0, \mathbf{u}_1, ..., \mathbf{u}_{N-1}\}$ transferring the system form a given initial state $\mathbf{x}_0$ to a prescribed terminal state $\mathbf{x}_t$. In any step $k$, i.e. in any state $\mathbf{x}_k$ of the system, it is necessary to find possible alternatives of the further course. In the PN terminology, it is necessary to find PN transitions which are enabled in the step $k$. In any step $k$ it is possible to proceed on the base of the simple logical consideration: "if we are not able to detect directly which transitions are enabled in the step $k$, but we are able to detect only which transitions are disabled in this step, we can eliminate the disabled transitions". This can be mathematically described as follows

$$\mathbf{u}_k = \text{neg}\,(\mathbf{F}^T.\text{neg}\,(\mathbf{x}_k)) \tag{4}$$

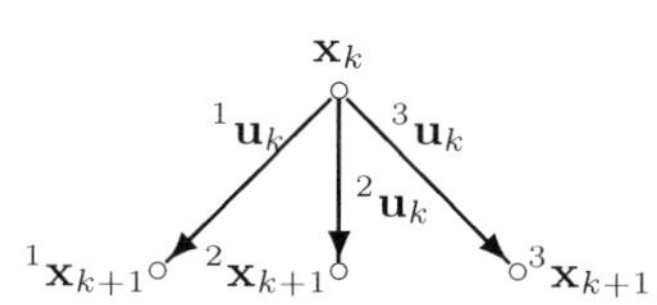


**Fig. 1.** The example of alternative courses of the DES states development

where neg (.) formally represents the operator of negation. The applying of this operator for a vector $\mathbf{v} = (v_1, v_2, ..., v_n)^T$ of integers yields the vector $\text{neg}(\mathbf{v}) = \mathbf{w} = (w_1, w_2, ..., w_n)^T$. If $v_i = 0$ then $w_i = 1$, otherwise $w_i = 0$. But simultaneously, the condition (3) has to be satisfied. However, because the development of the PN-based model is consecutive (only one transition can be fired in the step $k$), in case when several transitions are enabled in the step $k$, only one of them can be chosen to be fired. When the vector $\mathbf{u}_k$ contains more than one non-zero entry, it often does not met (3). But when such $\mathbf{u}_k$ is decomposed into a set of vectors ${}^i\mathbf{u}_k$, $i = 1, ..., r$, each with the different single non-zero entry, any single vector has to satisfy (3) in order to be used as the control vector. It means that there exists only one way how to proceed from the existing state $\mathbf{x}_k$ to another one $\mathbf{x}_{k+1}$ - see illustration in Fig. 1. Every ${}^i\mathbf{u}_k$, $i = 1, 2, 3$ in Fig. 1 has to contain only one non-zero entry. Hence, if $\mathbf{u}_k$ contains in general $r$ non-zero entries (i.e. when $r$ transitions are enabled in the step $k$), theoretically there are $r$ possibilities of the further development of the system dynamics. Thus, $\mathbf{u}_k$ has to be decomposed into $r$ control vectors ${}^i\mathbf{u}_k$, $i = 1, ..., r$ with single non-zero entry in such a way that $\sum_{i=1}^{r} {}^i\mathbf{u}_k = \mathbf{u}_k$. Thus, the branching makes such a procedure too complicated. Moreover, we do not know which way is a part of trajectory leading to the prescribed terminal state. In general, without additional "backtracking" information we are not able to find solution of the control problem by means of such a "blind" procedure. Therefore, it is better (and simpler in general) to compute the reachability tree (RT) corresponding to P/T PN-based model. The RT yields information about the branching process in the whole. However, even the RT does not give us directly any solution of the control problem. Namely, we have to find the trajectory (or trajectories) between the concrete states from the pair $(\mathbf{x}_0, \mathbf{x}_t)$ and to "extract" the corresponding trajectory (or trajectories) from the global RT. In case of large scale RTs it is not any simple problem. Namely, in general the terminal state can be a multiple leaf of the tree, i.e. there exist many trajectories from $\mathbf{x}_0$ to $\mathbf{x}_t$. Consequently, this paper is motivated by the endeavour to find the automatic procedure yielding control trajectories. Namely, the state trajectories represent only the system responses on the control ones.

## 3 Forming the Procedure of the Automated Control Synthesis

Starting from results presented in [1]-[3] we are able to compute the functional $(N_{RT} \times N_{RT})$-dimensional adjacency matrix $\mathbf{A}_k$ of the RT as well as the set of feasible state vectors $\{\mathbb{X}_1, ..., \mathbb{X}_{N_{RT}}\}$ reachable from the given initial state $\mathbf{x}_0 \equiv \mathbb{X}_1$ of the PN-based model (icluding this state too). Such a space of feasible states is represented by the matrix $\mathbf{X}_{reach}$, where the vectors $\mathbb{X}_i$, $i = 1, ..., N_{RT}$ create its columns. The Matlab procedure generating on its output the matrices $\mathbf{A}_k$, $\mathbf{X}_{reach}$ (after entering $\mathbf{F}$, $\mathbf{G}$, $\mathbf{x}_0$ on its input) is introduced in [3]. The method for DES control synthesis based on the intersection of both the straight-lined RT and the backward RT was presented in [1]. Here, the new approach based on bipartite graphs will be presented, because it yields the control trajectories.

### 3.1 The Procedure for State Machines

A PT/PN is a state machine (SM), if every transition has exactly one input place and one output place. There can not be concurrence in SMs, but conflicts can occur there. Namely, the incidence matrices **F**, **G** have a special structure - any column of **F** and any row of **G** contains only one non-zero entry. As it is known [7], from the structural point of view PNs are bipartite directed graphs (BDGs) with two kinds of nodes (places and transitions) and two kinds of edges (the arcs directed from places to transitions and the arcs directed contrariwise). Thus, the SM is given as $\langle P, T, F, G\rangle$, where $P$ is the set of PN places, $T$ is the set of PN transitions, $F$ is the set of the edges directed from places to transitions and $G$ is the set of edges directed from transitions to places. The sets $F$, $G$ can be expressed by the PN incidence matrices matrices **F**, **G**. Thus, we can proceed as follows. Let $S = \{P, T\}$ is the set of BDG nodes. Let $D$ is, formally, the set $S \times S$ of BDG edges. Thus, the occurrence of the edges can be expressed by the $((n+m) \times (n+m))$ BDG incidence matrix $\mathbf{A}_{BDG}$ and its transpose $\mathbf{D}$ (as we will see below, it will be very useful)

$$\mathbf{A}_{BDG} = \begin{pmatrix} \emptyset_{n\times n} & \mathbf{F} \\ \mathbf{G} & \emptyset_{m\times m} \end{pmatrix}; \; \mathbf{D} = \mathbf{A}^T_{BDG} = \begin{pmatrix} \emptyset_{n\times n} & \mathbf{G}^T \\ \mathbf{F}^T & \emptyset_{m\times m} \end{pmatrix} \tag{5}$$

where $\emptyset_{i\times j}$ in general is the $(i \times j)$ zero matrix. The matrices **F** and **G** are the same as the matrices in the PN-based model (1)-(3). In general, being in a state $\mathbf{x}_k$ the system can develop its dynamic behaviour either in the straight-lined direction or (formally) also in the backward one. The former development is performed by means of the matrix **D** given in (5) used in the state equation (6) while the latter one by means of the transpose $\mathbf{D}^T$.

To synthesize the DES control from a given initial state $\mathbf{x}_0$ to a desired terminal state $\mathbf{x}_t = \mathbf{x}_N$ the straight-lined development of the system dynamic can be computed - in analogy with [1] - by means of the following state equation

$$\{\mathbf{s}_{k+1}\} = \mathbf{D}.\{\mathbf{s}_k\}\ ,\ k = 0, 1, \ldots, 2N-1 \tag{6}$$

with $\mathbf{s}_k$ being the augmented $(n+m)$-dimensional vector defined as follows

$$\{\mathbf{s}_k\} = \begin{cases} (\{\mathbf{x}_{k/2}\}^T, \emptyset^T_m)^T & \text{if } k = 0, 2, 4, ..., 2N-2 \\ (\emptyset^T_n, \{\mathbf{u}_{(k-1)/2}\}^T)^T & \text{if } k = 1, 3, 5, ..., 2N-1 \end{cases} \tag{7}$$

where $\emptyset_j$ in general is the j-dimensional zero vector; $\mathbf{x}_{k/2} = \mathbf{G}^T.\mathbf{u}_{(k-2)/2}$, $k = 2, 4, ..., 2N-2$; $\mathbf{u}_{(k-1)/2} = \mathbf{F}^T.\mathbf{x}_{(k-1)/2}$, $k = 1, 3, 5, ..., 2N-1$. To be sure that during the straight-lined development the prescribed terminal state will be met, the backward development of the system dynamic can be computed by means of the following state equation

$$\{\mathbf{s}_{k-1}\} = \mathbf{D}^T.\{\mathbf{s}_k\} = \mathbf{A}_{BDG}.\{\mathbf{s}_k\}\ ,\ k = K, K-1, \ldots, 1 \tag{8}$$

However, because of the special block form of both the matrix **D** and the vector $\mathbf{s}_k$ we can alternate step-by-step two procedures - the straight-lined procedure and the backward one. The former procedure is the following

$$
\begin{aligned}
&{}^{1}\mathbf{X} = (\mathbf{x}_0, \emptyset_{n\times N});\ {}^{1}\{\mathbf{u}_0\} = \mathbf{F}^T.\mathbf{x}_0;\\
&{}^{1}\mathbf{U} = ({}^{1}\{\mathbf{u}_0\}, \emptyset_{m\times(N-1)});\ {}^{1}\{\mathbf{x}_1\} = \mathbf{G}^T.{}^{1}\{\mathbf{u}_0\}\\
&{}^{1}\mathbf{X} = (\mathbf{x}_0, {}^{1}\{\mathbf{x}_1\}, \emptyset_{n\times(N-1)});\ {}^{1}\{\mathbf{u}_1\} = \mathbf{F}^T.{}^{1}\mathbf{x}_1\\
&{}^{1}\mathbf{U} = ({}^{1}\{\mathbf{u}_0\}, {}^{1}\{\mathbf{u}_1\}, \emptyset_{m\times(N-2)})\\
&{}^{1}\mathbf{X} = (\mathbf{x}_0, {}^{1}\{\mathbf{x}_1\}, {}^{1}\{\mathbf{x}_2\}, \emptyset_{n\times(N-2)});\ \dots\ \dots {}^{1}\{\mathbf{u}_{N-1}\} = \mathbf{F}^T.{}^{1}\mathbf{x}_{N-1};\\
&{}^{1}\mathbf{U} = ({}^{1}\{\mathbf{u}_0\}, {}^{1}\{\mathbf{u}_1\}, \dots, {}^{1}\{\mathbf{u}_{N-1}\});\ {}^{1}\{\mathbf{x}_N\} = \mathbf{G}^T.{}^{1}\{\mathbf{u}_{N-1}\}\\
&{}^{1}\mathbf{X} = (\mathbf{x}_0, {}^{1}\{\mathbf{x}_1\}, {}^{1}\{\mathbf{x}_2\}, \dots, {}^{1}\{\mathbf{x}_N\})
\end{aligned}
$$

where ${}^{1}\mathbf{U}$ and ${}^{1}\mathbf{X}$ are, respectively, $(m \times N)$ and $(n \times (N+1))$ matrices. The left upper index ${}^{1}(.)$ points out performing the straight-lined procedure. The backtracking procedure runs as follows

$$
\begin{aligned}
&{}^{2}\mathbf{X} = (\emptyset_{n\times N}, \mathbf{x}_N);\ {}^{2}\{\mathbf{u}_{N-1}\} = \mathbf{G}.\mathbf{x}_N\\
&{}^{2}\mathbf{U} = (\emptyset_{m\times(N-1)}, {}^{2}\{\mathbf{u}_{N-1}\});\ {}^{2}\{\mathbf{x}_{N-1}\} = \mathbf{F}.{}^{2}\{\mathbf{u}_{N-1}\}\\
&{}^{2}\mathbf{X} = (\emptyset_{n\times(N-1)}, {}^{2}\{\mathbf{x}_{N-1}\}, \mathbf{x}_N);\ {}^{2}\{\mathbf{u}_{N-2}\} = \mathbf{G}.{}^{2}\{\mathbf{x}_{N-1}\}\\
&{}^{2}\mathbf{U} = (\emptyset_{m\times(N-1)}, {}^{2}\{\mathbf{u}_{N-2}\}, {}^{2}\{\mathbf{u}_{N-1}\});\ {}^{2}\{\mathbf{x}_{N-2}\} = \mathbf{F}.{}^{2}\{\mathbf{u}_{N-1}\}\\
&{}^{2}\mathbf{X} = (\emptyset_{n\times(N-2)}, {}^{2}\{\mathbf{x}_{N-2}\}, {}^{2}\{\mathbf{x}_{N-1}\}, \mathbf{x}_N);\ \dots\ \dots {}^{2}\{\mathbf{u}_1\} = \mathbf{G}.{}^{2}\{\mathbf{x}_2\}\\
&{}^{2}\mathbf{U} = (\emptyset_{n\times 1}, {}^{2}\{\mathbf{u}_1\}, \dots, {}^{2}\{\mathbf{u}_{N-1}\});\ {}^{2}\{\mathbf{x}_1\} = \mathbf{F}.{}^{2}\{\mathbf{u}_1\}\\
&{}^{2}\mathbf{X} = (\emptyset_{n\times 1}, {}^{2}\{\mathbf{x}_1\}, {}^{2}\{\mathbf{x}_2\}, \dots, \mathbf{x}_N);\ {}^{2}\{\mathbf{u}_0\} = \mathbf{G}.{}^{2}\{\mathbf{x}_1\}\\
&{}^{2}\mathbf{U} = ({}^{2}\{\mathbf{u}_0\}, {}^{2}\{\mathbf{u}_1\}, \dots, {}^{2}\{\mathbf{u}_{N-1}\});\ {}^{2}\{\mathbf{x}_0\} = \mathbf{F}.{}^{2}\{\mathbf{x}_0\}\\
&{}^{2}\mathbf{X} = ({}^{2}\{\mathbf{x}_0\}, {}^{2}\{\mathbf{x}_1\}, {}^{2}\{\mathbf{x}_2\}, \dots, \mathbf{x}_N)
\end{aligned}
$$

where ${}^{2}\mathbf{U}$ is $(m \times N)$ matrix and ${}^{2}\mathbf{X}$ is $(n \times (N+1))$ matrix. The left upper index ${}^{2}(.)$ points out performing the backtracking procedure. The final phase of the control problem solving consists in the special intersection. Namely, $\{\mathbf{x}_i\} = \max(\{{}^{1}\mathbf{x}_i\}, \{{}^{2}\mathbf{x}_i\}), i = 0, ..., N;\ \{\mathbf{u}_j\} = \max(\{{}^{1}\mathbf{u}_j\}, \{{}^{2}\mathbf{u}_j\}), j = 0, ..., N-1$. Hence, we have at disposal following both kinds of system trajectories - the control trajectories and the corresponding state ones

$$\mathbf{U} = {}^{1}\mathbf{U} \cap {}^{2}\mathbf{U};\quad \mathbf{U} = (\{\mathbf{u}_0\}, \{\mathbf{u}_1\}, \dots, \{\mathbf{u}_{N-1}\}) \tag{9}$$

$$\mathbf{X} = {}^{1}\mathbf{X} \cap {}^{2}\mathbf{X};\quad \mathbf{X} = (\mathbf{x}_0, \{\mathbf{x}_1\}, \dots, \{\mathbf{x}_{N-1}\}, \mathbf{x}_N) \tag{10}$$

Handling the zero blocks is eliminated on this way as well. Although the described approach seems to be very hopeful as to the automated synthesis of the DES control, its usage is limited. In such a form it is suitable only for SMs. In order to utilize such an approach also for P/T PNs that do not fulfil this restriction, we have to work with the RT adjacency matrix appertaining to such PNs.

### 3.2 The Generalization of the Procedure

Although SMs create the important class of P/T PNs, in comparison with the diversity of P/T PN structure SMs represent only a very small part of P/T PNs in the whole. Consequently, more general methods have to be found in order to deal with general P/T PNs. Here, the generalization of the method introduced above for SMs will be performed in order to be suitable also for P/T PNs with general structure. The main idea emerges from the fact that the P/T PN reachability graph (RG) is state machine in the sense of the SM definition. However, the adjacency matrix of the RG does not contain explicitly the incidence matrices $\mathbf{F}$, $\mathbf{G}$. Namely, in contrast to SMs, where transitions assigned to edges have only one input place and only one output place (and consequently any ambiguity, as to transitions, cannot occur) in P/T PN with general structure transitions can have more input places and more output ones. Hence, some ambiguities can occur. Therefore, in order to use the previous method here, we have to deal with this problem. Fortunately, the RG adjacency matrix $\mathbf{A}_{RG}$ can be formally decomposed into the fictive incidence matrices $\mathbf{F}_{RG}$, $\mathbf{G}_{RG}$, provided that the original transitions were renamed before in order to remove any ambiguity. Namely, any column of $\mathbf{F}_{RG}$ and any row of $\mathbf{G}_{RG}$ must not contain more than one non-zero entry. Thus, a fictive SM corresponding to P/T PN arises.

**Computing the RG Parameters.** The RG adjacency matrix $\mathbf{A}_{RG}$, more precisely the so called $k$ variant adjacency matrix $\mathbf{A}_k = a_{ij}^k$, $i = 1, ..., N_{RT}$; $j = 1, ..., N_{RT}$ [1], where $N_{RT}$ is the number of feasible states $\mathbb{X}_k$, $k = 1, ..., N_{RT}$ reachable from the initial state $\mathbf{x}_0 \equiv \mathbb{X}_1$ (including $\mathbf{x}_0$), can be computed by means the procedure presented in [3]. The $\mathbf{A}_k$ entries $a_{ij}^k$ are the transition functions $\gamma_{t_{\mathbb{X}_i \to \mathbb{X}_j}}^k$. Such an entry concerns the transition fixed to the RG edge connecting two RG nodes $\mathbb{X}_i$, $\mathbb{X}_j$. This matrix is given on the procedure output in the form of *quasi* $k$ variant adjacency matrix $\mathbf{A}_{Qk}$. Its entries are given as integers corresponding to indices of the PN transitions. For example, when $t_{\mathbb{X}_i \to \mathbb{X}_j} = t_q$ - i.e. when it has the index $q$ - the integer $q$ will represent the entry $a_{ij}^{Qk}$.

**Decomposition of the RG Adjacency Matrix.** To find the fictive matrices $\mathbf{F}_{RG}$, $\mathbf{G}_{RG}$ we have to disassemble the matrix $\mathbf{A}_{Qk}$. Namely, some of its entries may not be mutually different. They may acquire the same value because the original P/T PN transitions may occur more then once among the entries. Consequently, a confusion could occur during the computational process and decline it. To avoid these difficulties, it is necessary to rename the original P/T PN transitions in order to obtain fictive transitions that occur only once. The number of them is $T_r$, being the global number of the non-zero elements of $\mathbf{A}_{Qk}$. The renaming is performed raw-by-raw so that the non-zero elements are replaced by integers - ordinal numbers starting from 1 and finishing at $T_r$. Thus, the auxiliary (fictive) adjacency matrix $\mathbf{A}_{T_r}$ is obtained. The disassembling of the matrix $\mathbf{A}_{T_r}$ into the incidence matrices $\mathbf{F}_{RG}$ and $\mathbf{G}_{RG}$ is accomplished as follows. The elements of these matrices for $i = 1, ..., N_{RT}$, $j = 1, ...N_{RT}$ are

$$\mathbf{T}_{tr}(\mathbf{A}_{Qk}(i,j), \mathbf{A}_{T_r}(i,j)) = \begin{cases} 1 & \text{if } \mathbf{A}_{Qk}(i,j) \neq 0 \;\&\; \mathbf{A}_{T_r}(i,j) \neq 0 \\ 0 & \text{otherwise} \end{cases} \tag{11}$$

$$\mathbf{F}_{RG}(i, \mathbf{A}_{T_r}(i,j)) = \begin{cases} 1 & \text{if } \mathbf{A}_{Qk}(i,j) \neq 0 \;\&\; \mathbf{A}_{T_r}(i,j) \neq 0 \\ 0 & \text{otherwise} \end{cases} \tag{12}$$

$$\mathbf{G}_{RG}(\mathbf{A}_{T_r}(i,j), j) = \begin{cases} 1 & \text{if } \mathbf{A}_{Qk}(i,j) \neq 0 \;\&\; \mathbf{A}_{T_r}(i,j) \neq 0 \\ 0 & \text{otherwise} \end{cases} \tag{13}$$

Here, $\mathbf{T}_{tr}$ is the transformation matrix between the original set of transitions and the fictive ones. Hence, actual $\mathbf{U} = \mathbf{T}_{tr}.\mathbf{U}^*$ where $\mathbf{U}^*$ yields the fictive control strategies given by (9), computed by means of the set of the fictive transitions.

**Importance of the Generalized Approach.** The generalized approach (i) makes possible the automated control synthesis of the P/T PNs with the general structure - i.e. it extends the applicability of the method originally proposed for SMs; (ii) enables us to analyse the dynamic behaviour of arbitrary structure of agents modelled by P/T PNs. Namely, agents often (even when they are modelled by SMs and supervised by a supervisor) do not represent together any SM but rather P/T PN with a general structure (because of the interconnections between the elementary agents and the supervisor). The supervisor is usually used in order to bring DES subsystems or agents in MAS (e.g. robots, automatically guided vehicles, etc.) to a (better) cooperation or forbid them to be selfish at making use of common sources (like energy, row materials, yard, roads, etc.).

**Example 1.** Consider an agent modelled by the P/T PN with more general structure (i.e. different from SM) given in Fig. 2 on the left. It represents the behaviour of the autonomous agent. Let us demonstrate the applicability of the approach introduced in the previous subsections directly on this agent. The corresponding RG is given in Fig. 2 on the right. As we can see $n = 4$, $m = 3$, $\mathbf{x}_0 \equiv \mathbb{X}_1 = (2, 0, 1, 0)^T$, $N_{RT} = 7$. The PN incidence matrices and the RG parameters are the following

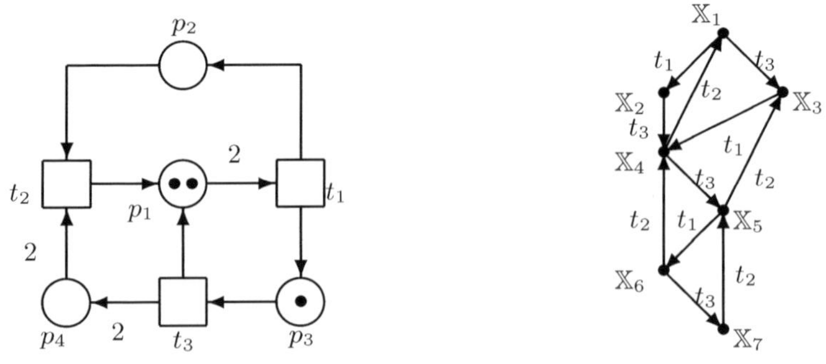


**Fig. 2.** The PN-based model of the agent (on the left) and the corresponding RG (on the right)

$$\mathbf{F}=\begin{pmatrix}2\,0\,0\\0\,1\,0\\0\,0\,1\\0\,2\,0\end{pmatrix}\ \mathbf{G}=\begin{pmatrix}0\,1\,1\,0\\1\,0\,0\,0\\1\,0\,0\,2\end{pmatrix}\ \mathbf{A}_{Qk}=\begin{pmatrix}0\,1\,3\,0\,0\,0\,0\\0\,0\,0\,3\,0\,0\,0\\0\,0\,0\,1\,0\,0\,0\\2\,0\,0\,0\,3\,0\,0\\0\,0\,2\,0\,0\,1\,0\\0\,0\,0\,2\,0\,0\,3\\0\,0\,0\,0\,2\,0\,0\end{pmatrix}\ \mathbf{X}_{reach}=\begin{pmatrix}2\,0\,3\,1\,2\,0\,1\\0\,1\,0\,1\,1\,2\,2\\1\,2\,0\,1\,0\,1\,0\\0\,0\,2\,2\,4\,4\,6\end{pmatrix}$$

After renaming the transitions, decomposing the matrix $\mathbf{A}_{T_r}$ and applying the proposed method we can synthesize the control $\mathbf{U}$ from the initial state $\mathbf{x}_0 \equiv \mathbb{X}_1 = (2,\ 0,\ 1,\ 0)^T$ to a desired terminal state - e.g. to $\mathbf{x}_t = \mathbb{X}_7 = (1,\ 2,\ 0,\ 6)^T$

$$\mathbf{A}_{T_r}=\begin{pmatrix}0\,1\,2\,0 & 0\,0 & 0\\0\,0\,0\,3 & 0\,0 & 0\\0\,0\,0\,4 & 0\,0 & 0\\5\,0\,0\,0 & 6\,0 & 0\\0\,0\,7\,0 & 0\,8 & 0\\0\,0\,0\,9 & 0\,0 & 10\\0\,0\,0\,0 & 11\,0 & 0\end{pmatrix}\ \mathbf{T}_{tr}^T=\begin{pmatrix}1\,0\,0\\0\,0\,1\\0\,0\,1\\1\,0\,0\\0\,1\,0\\0\,0\,1\\0\,1\,0\\1\,0\,0\\0\,1\,0\\0\,0\,1\\0\,1\,0\end{pmatrix}\ \mathbf{U}^*=\begin{pmatrix}1\,0\,0\,0\,0\\1\,0\,0\,0\,0\\0\,1\,0\,0\,0\\0\,1\,0\,0\,0\\0\,0\,0\,0\,0\\0\,0\,1\,0\,0\\0\,0\,0\,0\,0\\0\,0\,0\,1\,0\\0\,0\,0\,0\,0\\0\,0\,0\,0\,1\\0\,0\,0\,0\,0\end{pmatrix}$$

$$\mathbf{X}=\begin{pmatrix}1\,0\,0\,0\,0\,0\\0\,1\,0\,0\,0\,0\\0\,1\,0\,0\,0\,0\\0\,0\,1\,0\,0\,0\\0\,0\,0\,1\,0\,0\\0\,0\,0\,0\,1\,0\\0\,0\,0\,0\,0\,1\end{pmatrix}\ \mathbf{U}=\mathbf{T}_{tr}.\mathbf{U}^*=\begin{pmatrix}1\,1\,0\,1\,0\\0\,0\,0\,0\,0\\1\,1\,1\,0\,1\end{pmatrix}$$

Its graphical expression is on the left in Fig. 3. To obtain the actual matrix $\mathbf{U}$ the fictive matrix $\mathbf{U}^*$ (obtained together with $\mathbf{X}$ at using the proposed procedure) had to be transformed by means of the matrix $\mathbf{T}_{tr}$.

## 4 Synthesis of Agents Cooperation

Using the method presented in [4] we can synthesize the structure of the agent supervisor. The supervisor can be suitable (i) for the avoidance of the egoistic effort of autonomous agents by means of prohibition some states of the global system describing the MAS, e.g. like the so called mutex (mutual exclusion) - see examples introduced in [4]; (ii) in order to synthesize the desired cooperation of agents in MAS. The principle of the method is based on the PN place invariants

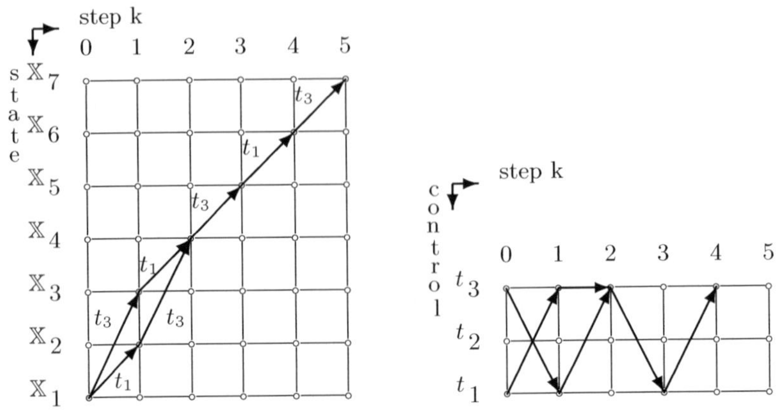


**Fig. 3.** The state trajectories from $\mathbf{x}_0 \equiv \mathbb{X}_1$ to $\mathbf{x}_t \equiv \mathbb{X}_7$ (on the left) as the responses on the synthesized control trajectories given on the right

(P-invariants) [7,5,6]. P-invariants are vectors, $\mathbf{v}$, with the property that multiplication of these vectors with any state vector $\mathbf{x}$ reachable from a given initial state vector $\mathbf{x}_0$ yields the same result (the relation of the state conservation)

$$\mathbf{v}^T.\mathbf{x} = \mathbf{v}^T.\mathbf{x}_0, \qquad \text{i.e.} \qquad \mathbf{v}^T.\mathbf{x} = \mathbf{v}^T.\mathbb{X}_1 \tag{14}$$

Hence, taking into account (1) and the consecutiveness of states, for column $col_t(\mathbf{B})$ of $\mathbf{B}$ corresponding to a transition $t$

$$\mathbf{v}^T.col_t(\mathbf{B}) = 0, \qquad \text{i.e.} \qquad \mathbf{v}^T.\mathbf{B} = \boldsymbol{\emptyset} \tag{15}$$

where $\mathbf{v}$ is $n$-dimensional vector ($n$ is the number of PN places) and $\boldsymbol{\emptyset}$ is $m$-dimensional zero vector. This equation is usually presented as the definition of the P-invariant $\mathbf{v}$ of PN.

The P-invariants can be utilized at the supervisor synthesis [10,8,9,5,6]. In case of several invariants the set of the P-invariants is created by the columns of the $(n \times n_x)$-dimensional matrix $\mathbf{V}$ ($n_x$ expresses the number of invariants) being the solution of the equation

$$\mathbf{V}^T.\mathbf{B} = \boldsymbol{\emptyset} \tag{16}$$

Just this equation represents the base for the supervisor synthesis method. Some additional PN places (slacks) can be added to the PN-model in question. The slacks create the places of the supervisor. Hence, (16) can be rewritten as

$$[\mathbf{L}, \mathbf{I}_s].\begin{bmatrix} \mathbf{B} \\ \mathbf{B}_s \end{bmatrix} = \boldsymbol{\emptyset}$$

where $\mathbf{I}_s$ is $(n_s \times n_s)$-dimensional identity matrix with $n_s \leq n_x$ being the number of slacks, $(n_s \times n)$-dimensional matrix $\mathbf{L}$ of integers represents (in a suitable form) the conditions $\mathbf{L}.\mathbf{x} \leq \mathbf{b}$ ($\mathbf{b}$ is the vector of integers), imposed on marking of the original PN and the $(n_s \times m)$-dimensional matrix $\mathbf{B}_s$ yields (after its finding by computing) the structure of the PN-based model of the supervisor. Consequently,

$$\mathbf{L}.\mathbf{B} + \mathbf{B}_s = \emptyset; \quad \mathbf{B}_s = -\mathbf{L}.\mathbf{B}; \quad \mathbf{B}_s = \mathbf{G}_s^T - \mathbf{F}_s \tag{17}$$

where the actual structure of the matrix $\mathbf{L}$ has to be respected. The augmented state vector (i.e. the state vector of the original PN together with the supervisor) and the augmented matrices are as follows

$$\mathbf{x}_a = \begin{bmatrix} \mathbf{x} \\ \mathbf{x}_s \end{bmatrix}; \; \mathbf{F}_a = \begin{pmatrix} \mathbf{F} \\ \mathbf{F}_s \end{pmatrix}; \; \mathbf{G}_a^T = \begin{pmatrix} \mathbf{G}^T \\ \mathbf{G}_s^T \end{pmatrix}$$

where the submatrices $\mathbf{F}_s$ and $\mathbf{G}_s^T$ correspond to the interconnections of the incorporated slacks with the actual PN structure. Because of the prescribed conditions we have

$$[\mathbf{L} \,|\, \mathbf{I}_s].\begin{bmatrix} \mathbf{x}_0 \\ {}^s\mathbf{x}_0 \end{bmatrix} = \mathbf{b} \quad \text{i.e. the supervisor initial state is:} \qquad {}^s\mathbf{x}_0 = \mathbf{b} - \mathbf{L}.\mathbf{x}_0$$

where $\mathbf{b}$ is the vector of the corresponding dimensionality (i.e. $n_s$) with integer entries representing the limits for number of common tokens - i.e. the maximum numbers of tokens that the corresponding places can possess altogether (i.e. share). Many times $\mathbf{b} = \mathbf{1}$. Here $\mathbf{1}$ is the vector with all its entries equal to 1.

**Example 2.** Consider two agents with the same structure given in Example 1. Distinguish their models by the left upper index in the vectors and matrices. When the agents are autonomous, their structure can be written as follows

$$\mathbf{F} = \begin{pmatrix} {}^1\mathbf{F} & \emptyset \\ \emptyset & {}^2\mathbf{F} \end{pmatrix}; \; \mathbf{G} = \begin{pmatrix} {}^1\mathbf{G} & \emptyset \\ \emptyset & {}^2\mathbf{G} \end{pmatrix}; \; \mathbf{B} = \mathbf{G}^T - \mathbf{F}; \; \mathbf{x}_0 = \begin{pmatrix} {}^1\mathbf{x}_0 \\ {}^2\mathbf{x}_0 \end{pmatrix}$$

Consider their initial states ${}^1\mathbf{x}_0 = (2,\, 0,\, 1,\, 0)^T$, ${}^2\mathbf{x}_0 = (0,\, 1,\, 1,\, 0)^T$. Let us request that $p_1$ and $p_5$ may share (keep together) maximally two tokens, i.e.

$$p_1 + p_5 \leq 2, \text{ after introducing the slack } p_9\text{:} \quad p_1 + p_5 + p_9 = 2; \; s = 1; \; \mathbf{I}_s = 1$$
$$\mathbf{L} = \begin{pmatrix} 1\,0\,0\,0\,1\,0\,0\,0 \end{pmatrix}; \; \mathbf{b} = 2; \; \mathbf{B}_s = -\mathbf{L}.\mathbf{B} = \begin{pmatrix} 2, & -1, & -1, & 2, & -1, & -1 \end{pmatrix}$$
$$\mathbf{F}_s = \begin{pmatrix} 0\,1\,1\,0\,1\,1 \end{pmatrix}; \; \mathbf{G}_s^T = \begin{pmatrix} 2\,0\,0\,2\,0\,0 \end{pmatrix}; \; {}^s\mathbf{x}_0 = \mathbf{b} - \mathbf{L}.\mathbf{x}_0 = 2 - 2 = 0$$

The closed loop structure of the agents together with their supervisor is in Fig.4 on the left. The corresponding RG is given on the right. The supervisor structure (i.e. $p_9$ together with the interaction edges to and from both agents) and its initial state ensure that the pair of agents will behave in the prescribed fashion (the prescribed condition will never be disturbed). However, in order to know the course of both the the control trajectories and the responses on them (i.e. the state trajectories) we can use the proposed method of the automatic control synthesis. Starting from the initial state $\mathbf{x}_0 \equiv \mathbb{X}_1 = (2,\, 0,\, 1,\, 0 \,|\, 0,\, 1,\, 1,\, 0 \,|\, 0)^T$ being the

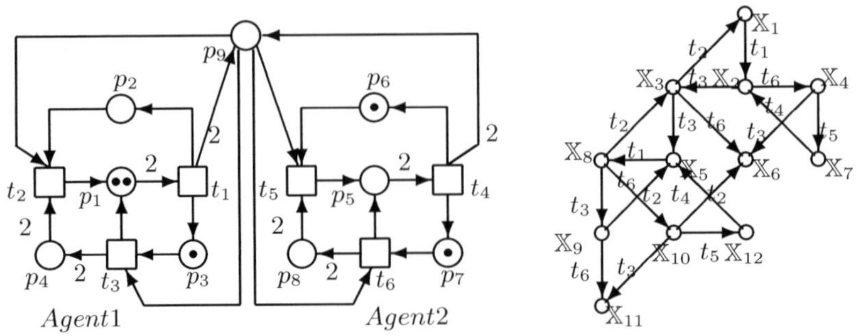


**Fig. 4.** The PN-based model of the supervised agent cooperation (on the left) and the corresponding RG (on the right)

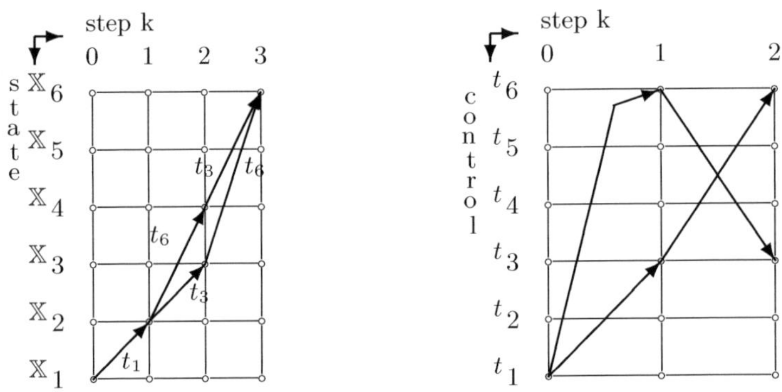


**Fig. 5.** The state trajectories from $\mathbf{x}_0 \equiv \mathbb{X}_1$ to $\mathbf{x}_t \equiv \mathbb{X}_6$ (on the left) as the responses on the synthesized control trajectories (on the right)

first column of $\mathbf{X}_{reach}$ to the terminal state $\mathbf{x}_t \equiv \mathbb{X}_6 = (1, 1, 1, 2| 1, 1, 0, 2 | 0)^T$ being the sixth column of $\mathbf{X}_{reach}$ we have at disposal below the synthesized trajectories in the form of matrices and in the graphical form given in Fig.5.

$$\mathbf{X}_{reach} = \left( \begin{array}{cccccccccccc} 2&0&1&0&2&1&0&0&1&0&1&0 \\ 0&1&1&1&1&1&1&2&2&2&2&2 \\ 1&2&1&2&0&1&2&1&0&1&0&1 \\ 0&0&2&0&4&2&0&4&6&4&6&4 \\ \hline 0&0&0&1&0&1&2&0&0&1&1&2 \\ 1&1&1&1&1&1&0&1&1&1&1&0 \\ 1&1&1&0&1&0&0&1&1&0&0&0 \\ 0&0&0&2&0&2&0&0&0&2&2&0 \\ \hline 0&2&1&1&0&0&0&2&1&1&0&0 \end{array} \right) \begin{array}{l} \left.\begin{array}{c} \\ \\ \\ \\ \end{array}\right\} \text{Agent 1} \\ \\ \left.\begin{array}{c} \\ \\ \\ \\ \end{array}\right\} \text{Agent 2} \\ \\ \} \text{Supervisor} \end{array}$$

$$\mathbf{X}^T = \begin{pmatrix} 1\,0\,0\,0\,0\,0\,0\,0\,0\,0\,0\,0\,0 \\ 0\,1\,0\,0\,0\,0\,0\,0\,0\,0\,0\,0\,0 \\ 0\,0\,1\,1\,0\,0\,0\,0\,0\,0\,0\,0\,0 \\ 0\,0\,0\,0\,0\,1\,0\,0\,0\,0\,0\,0\,0 \end{pmatrix} ; \ \mathbf{U} = \begin{pmatrix} 1\,0\,0 \\ 0\,0\,0 \\ 0\,1\,1 \\ 0\,0\,0 \\ 0\,0\,0 \\ 0\,1\,1 \end{pmatrix}$$

## 5 Conclusion

The procedure of the automatic DES control synthesis was proposed and tested. The approach suitable for DES modelled by the specific class of P/T PNs (by SMs) was presented and extended to the procedure suitable also for DES modelled by P/T PNs with general structure. It utilizes the fact that RG of P/T PN is SM. The control synthesis is performed automatically. Such an approach can be utilized also for the supervised agent cooperation. Although the supervisor meets the constraints imposed on its structure, jointly with agents it represents only a PN structure. Both the control and state trajectories are not explicitly at disposal and they have to be computed. The illustrative examples were introduced in order to document the soundness and applicability of the approach.

## References

1. Čapkovič, F.: Modelling, Analysing and Control of Interactions among Agents in MAS. Computing and Informatics 26(5), 507–541 (2007)
2. Čapkovič, F.: Control Synthesis of a Class of DEDS. Kybernetes 31(9/10), 1274–1281 (2002)
3. Čapkovič, F.: The Generalised Method for Solving Problems of DEDS Control Synthesis. In: Chung, P.W.H., Hinde, C., Ali, M. (eds.) IEA/AIE 2003. LNCS, vol. 2718, pp. 702–711. Springer, Heidelberg (2003)
4. Čapkovič, F.: Supervisory Control of Agents Cooperation. In: Proceedings of 4th IEEE International Conference on Intelligent Systems, vol. 1, pp. 6_8–6_13. IEEE Press, Piscataway (2008)
5. Iordache, M.V., Antsaklis, P.J.: Supervision Based on Place Invariants: A Survey. Discrete Event Dynamic Systems 16(4), 451–492 (2006)
6. Iordache, M.V., Antsaklis, P.J.: Supervisory Control of Concurrent Systems: A Petri Net Structural Approach. Birkhauser, Boston (2006)
7. Murata, T.: Petri Nets: Properties, Analysis and Applications. Proceedings of the IEEE 77(4), 541–580 (1989)
8. Ramadge, P., Wonham, W.: Supervisory Control of a Class of Discrete Event Processes. SIAM J. on Control and Optimization 25(1), 206–230 (1989)
9. Moody, J.O., Antsaklis, P.J.: Supervisory Control of Discrete Event Systems Using Petri Nets. Kluwer Academic Publishers, Norwell (1998)
10. Yamalidou, E., Moody, J.O., Antsaklis, P.J., Lemmon, M.D.: Feedback Control of Petri Nets Based on Place Invariants. Automatica 32(1), 1–28 (1996)

# Parameter Tuning for the Artificial Bee Colony Algorithm

Bahriye Akay and Dervis Karaboga

Erciyes University, The Dept. of Computer Engineering, 38039,
Melikgazi, Kayseri, Turkiye
bahriye@erciyes.edu.tr, karaboga@erciyes.edu.tr

**Abstract.** While solving a problem by an optimization algorithm, adjusting algorithm parameters have significant importance on the performance of the algorithm. A fine tuning of control parameters is required for most of the algorithms to obtain desired solutions. In this study, performance of the Artificial Bee Colony (ABC) algorithm, which simulates the foraging behaviour of a honey bee swarm, was investigated by analyzing the effect of control parameters.

## 1 Introduction

Algorithms dealing with solving optimization problems can be classified into groups upon their characteristics such as stochastic, deterministic, iterative, population-based, nature-based, etc. Performance of an algorithm is related with the manner of the algorithm coming from the phenomena it mimics. Operators indicate this behavioral manner of the algorithm and dictate the search efficiency. Each procedure in operators makes some modifications or decisions on the individuals in the population. While making them, the algorithm requires determining a set of control parameters of the algorithm such as the population size, the amount of reproduction and mutation, etc. However, while a parameter set produces optimal solutions on a particular landscape, the same set can fail for another one. Another issue is that an appropriate parameter set at the beginning of the search may become inappropriate in the next evaluations of the algorithm. Therefore, a parameter set is required to be optimal in general, to be robust, to produce good performance [1]. An analysis is desired to determine how a parameter change affects the performance of the algorithm and it should be examined how parameters can be dynamically and self-adaptively set considering landscape characteristics.

Artificial Bee Colony (ABC) algorithm is a swarm intelligence based algorithm simulating the foraging behavior of a honey bee colony [2–5]. It also has several control parameters dictating the search behaviour. In this paper, an analysis on the parameter tuning for ABC algorithm is conducted. In the experiments, performance of the ABC algorithm is examined for different colony sizes, problem dimensions, initialization ranges and limit values on nine well-known benchmark problems in relevant literature.

N.T. Nguyen, R. Kowalczyk, and S.-M. Chen (Eds.): ICCCI 2009, LNAI 5796, pp. 608–619, 2009.

Paper is organized as follows: In the second section ABC algorithm is described, in the third section parameter tuning of the ABC algorithm is conducted and results are presented, finally it is concluded.

## 2 Artificial Bee Colony Algorithm

Tereshko developed a minimal model of foraging behaviour of a honeybee colony based on the reaction-diffusion equations [6]. This model leads to the emergence of collective intelligence of honey bee swarms. The model consists of three essential components: food sources, employed foragers, and unemployed foragers, and defines two leading modes of the honey bee colony behaviour: recruitment to a food source and abandonment of a source [6]. In order to select a food source, a forager bee evaluates several properties related with the food source such as its closeness to the hive, richness of the energy, taste of its nectar, and the ease or difficulty of extracting this energy. An employed forager is employed at a specific food source which she is currently exploiting. She carries information about this specific source and shares it with other bees waiting in the hive. The information includes the distance, the direction and the profitability of the food source. Unemployed forager: A forager bee who looks for a food source to exploit is called unemployed. It can be either a scout who searches the environment randomly or an onlooker who tries to find a food source by means of the information given by employed bee.

Artificial Bee Colony algorithm is a recently proposed optimization algorithm that simulates this minimal model of honey bee swarm [2].The main steps of the algorithm are as given in Algorithm 1:

**Algorithm 1.** Main steps of the ABC algorithm

1: cycle=1
2: Initialize the food source positions $\boldsymbol{x}_i, i = 1 \dots SN$
3: Evaluate the nectar amount of food sources
4: **repeat**
5: Employed Bees' Phase
6: Onlooker Bees' Phase
7: Scout Bee Phase
8: Memorize the best solution achieved so far
9: **until** cycle=Maximum Cycle Number

In ABC algorithm, the position of a food source represents a possible solution to the optimization problem and the nectar amount of the food source corresponds to the quality (fitness) of the associated solution. The number of the employed bees or the onlooker bees is equal to the number of solutions in the population. At the first step, the ABC generates a randomly distributed initial food source positions of $SN$ solutions, where $SN$ denotes the size of employed bees or onlooker bees. Each solution $\boldsymbol{x}_i$ $(i = 1, 2, ..., SN)$ is a $D$-dimensional vector. Here, D is the number of optimization parameters. After initialization, the population of the positions (solutions) repeat the search cycles of the employed bees, the onlooker bees and scout bees.

In the employed bees' phase, each employed bee finds a new food source $\upsilon_i$ in the neighbourhood of its current source $\boldsymbol{x}_i$ by (1) and compares the new one against the current solution. Employed bee memorizes the better one by means of a greedy selection mechanism.

$$\upsilon_{ij} = x_{ij} + \phi_{ij}(x_{ij} - x_{kj}) \tag{1}$$

where $k \in \{1,2,...,SN\}$ and $j \in \{1,2,...,D\}$ are randomly chosen indexes. Although $k$ is determined randomly, it has to be different from $i$. $\phi_{i,j}$ is a random number between [-1, 1]. It controls the production of a neighbour food source around $x_{i,j}$.

After employed bees complete their searches, they share information about the quality of their sources. An onlooker bee chooses a food source depending on the nectar amount (fitness) of a food source shared by employed bees. Probability of a food source to be selected by an onlooker is calculated by (2):

$$p_i = \frac{fit_i}{\sum\limits_{n=1}^{SN} fit_n} \tag{2}$$

After onlooker bees choose a food source, each of them finds a new solution in the neighbourhood of the food source chosen by (1) and a greedy selection is applied between the new solution and the food source chosen in order to improve the solution at hand.

If a food source can not be improved through a predetermined cycles, called "limit", it is removed from the population and the employed bee of that food source becomes scout. The scout bee finds a new random food source position by (3):

$$x_i^j = x_{\min}^j + \mathrm{rand}[0,1](x_{\max}^j - x_{\min}^j) \tag{3}$$

where $x_{\min}^j$ and $x_{\max}^j$ are lower and upper bounds of parameter $j$, respectively.

These steps are repeated through a predetermined number of cycles, called Maximum Number of Cycle ($MCN$), or until a termination criterion is satisfied.

From the explanation given above, it is clear that there are three control parameters of basic ABC: the number of food sources which is equal to the number of employed or onlooker bees ($SN$), the value of *limit* and the maximum cycle number ($MCN$).

## 3 Parameter Tuning for the ABC Algorithm

As mentioned before, ABC algorithm employs three control parameters which are the number of food sources which is equal to the number of employed or onlooker bees ($SN$), the value of *limit* and the maximum cycle number ($MCN$).

In this work, we investigated the parameter tuning for the ABC algorithm via some experiments. In all experiments, maximum cycle number was 10000. Effect of initialization range on the performance of ABC algorithm was also examined. Experiments are conducted on the well-known benchmark functions given on Table 1. In tables reporting the experimental results, mean values and standard deviations of 30 runs are presented. Beside comparing the results of ABC algorithm obtained by the different parameter values, the results were also compared to the results of differential evolution (DE) and particle swarm optimization (PSO) algorithms in order to see which algorithm is more sensitive to the parameter values. Although DE and PSO algorithms have also control parameters that should be tuned, analyzing the best parameter settings of these algorithms is out of our purpose and we employed the recommended values in the literature [7, 8].

### 3.1 Effect of Dimension

In order to analyze how colony size (population size) affects the performance of the ABC algorithm or in order to decide which value of colony size is large enough for finding the minima, we made empirical experiments. In the first experiment, we kept the colony size constant and changed the dimension of the problem as 10, 100, and 1000. Results of this experiment are reported on Table 2. From the results in Table 2, for a fixed population size, performance of the PSO algorithm deteriorated as the dimension of the problem was increased (10, 100, 1000). DE algorithm had superior performance on Rosenbrock function in case of dimension was 10 while ABC algorithm produced the best results on the other eight functions (Sphere, Rastrigin, Griewank, Schwefel, Ackley, Step, Penalized, Dixon-Price). When the dimension was incremented, for a fixed number of population, ABC algorithm produced reasonable results even if for 1000 dimension.

### 3.2 Effect of Colony Size

In the second experiment, we analyzed which size of the population is large enough for a fixed dimension. This experiment was also repeated for the functions on Table 1 and a comparison was conducted between PSO, DE and ABC algorithms in Table 3. Results in Table 3 show that ABC algorithm is able to preserve its robustness for different colony sizes considered in this table. Hence, ABC algorithm does not need a fine tuning for the colony size in order to obtain satisfactory good results.

### 3.3 Effect of Region Scaling

Initialization range of the solutions also acts on the performance of an algorithm. If an algorithm is poor in exploration of new areas, it fails in finding optima when it is started in the range that does not include the optimal parameter values.

It is said to be not robust against initialization conditions and region scaling. When global optimum is located in the center of search space or initialization is symmetric, algorithms can easily find the optimum. For this reason, in order to investigate the performance of the ABC algorithm under different initialization conditions, its initial solutions were generated in the different subspaces of all feasible space and its performance was again compared to DE and PSO algorithms. For this examination, we divided the entire search space into sub-regions including left quarter and right half of the search space and the solutions were initialized in these sub-regions. Results of the PSO, DE and ABC algorithms in case of initialization in these sub-regions and entire space are presented on Table 4 for test functions in Table 1. Results show that ABC algorithm is less sensitive to the initialization range respect to DE and PSO algorithms since ABC algorithm has a powerful exploration capability.

### 3.4 Effect of Limit

*Limit* is another control parameter of the ABC algorithm that determines the food source to be abandoned. For analyzing the effect of this control parameter we used different values for *limit* and gave the results obtained on Tables 5-7. In Table 5, dimension of the problems was 100, and the colony size was 100 in each instance of runs. Results in this table show that very low values of the limit worsen the performance of the algorithm, but when the limit values were gradually increased, performance of the ABC algorithm did not change significantly. Reason of this fact is that diversity of the individuals in the population is sufficient and the solutions proceed with evolution since the number of solutions is large enough. Therefore, the algorithm did not need scouts to explore new regions in this case.

In order to see the effect of *limit* parameter, we reduced the colony size from 100 to 10. Results produced when colony size is 10 are presented in Table 6. From the results on this table, on some functions, ABC algorithm produced better results with the value of *limit* about 500. This result is reasonable because basic ABC algorithm articulated in Section II, modifies only one parameter of a solution in each cycle and at least D cycle is needed to change all parameters of one solution. Therefore, a value around $SN*D$ is generally appropriate for *limit*. But, on some functions which have relatively flat surfaces (Sphere, Ackley, Step, Dixon-Price), the performance of the algorithm was not highly influenced by the increment in the value of the *limit*.

By taking a step further, in order to make the search process more difficult for these functions and in order to see the effect of the scout production mechanism, initialization was conducted in a subregion of the parameter ranges that was left quarter of the entire range. Results of the algorithm when dimension was 100, colony size was equal to 10 and initial population was generated in a subregion are given on Table 7 and Figures 1(a)-1(i). Computational results in Table7 indicate that lower limit values cause more scouts to be produced than needed while higher limit values cause scouts not to occur often and this

**Table 1.** Test Functions used in Experiments

| | Function | Range | Formulation | Minimum |
|---|---|---|---|---|
| 1 | Sphere | $[-100, 100]^n$ | $f(x) = \sum_{i=1}^{n} x_i^2$ | 0 |
| 2 | Rosenbrock | $[-100, 100]^n$ | $f(x) = \sum_{i=1}^{n-1} [100(x_{i+1} - x_i^2)^2 + (x_i - 1)^2]$ | 0 |
| 3 | Rastrigin | $[-5.12, 5.12]^n$ | $f(x) = \sum_{i=1}^{n} [x_i^2 - 10\cos(2\pi x_i) + 10]$ | 0 |
| 4 | Griewank | $[-600, 600]^n$ | $f(x) = \frac{1}{4000} \sum_{i=1}^{n} x_i^2 - \prod_{i=1}^{n} \cos\left(\frac{x_i}{\sqrt{i}}\right) + 1$ | 0 |
| 5 | Schwefel | $[-500, 500]^n$ | $f(x) = \sum_{=1}^{n} -x_i \sin(\sqrt{\|x_i\|})$ | -418.983*n |
| 6 | Ackley | $[-32, 32]^n$ | $f(x) = -20\exp\left(-0.2\sqrt{\frac{1}{n}\sum_{i=1}^{n} x_i^2}\right) - \exp\left(\frac{1}{n}\sum_{i=1}^{n}\cos(2\pi x_i)\right) + 20 + e$ | 0 |
| 7 | Step | $[-100, 100]^n$ | $f(x) = \sum_{i=1}^{n} (\lfloor x_i + 0.5 \rfloor)^2$ | 0 |
| 8 | Penalized | $[-50, 50]^n$ | $f(x) = \frac{\pi}{n}\left\{10\sin^2(\pi y_1) + \sum_{i=1}^{n-1}(y_i - 1)^2[1 + 10\sin^2(\pi y_{i+1})] + (y_n - 1)^2\right\} + \sum_{i=1}^{n} u(x_i, 10, 100, 4)$<br>$yi = 1 + \frac{1}{4}(x_i + 1)$<br>$u(x_i, a, k, m) = \begin{cases} k(x_i - a)^m, & x_i > a \\ 0, & -a \le x_i \le a \\ k(-x_i - a)^m, & x_i < -a \end{cases}$ | 0 |
| 9 | Dixon-Price | $[-10, 10]^n$ | $f(x) = (x_1 - 1)^2 + \sum_{i=2}^{n} i(2x_i^2 - x_{i-1})^2$ | 0 |

**Table 2.** Experimental Results of PSO, DE and ABC Algorithms under varying Dimensions, Population Size/Colony Size was fixed to 100, SD:Standard Deviation

| | | PSO | | | DE | | | ABC | | |
|---|---|---|---|---|---|---|---|---|---|---|
| | D | 10 | 100 | 1000 | 10 | 100 | 1000 | 10 | 100 | 1000 |
| Sphere | Mean | 4.13E-17 | 5.14E-16 | 9723.034942 | 4.41E-17 | 8.84E-17 | 329214.6744 | 4.88E-17 | 1.08E-15 | 0.058275686 |
| | SD | 7.71E-18 | 3.12E-16 | 3920.944041 | 8.09E-18 | 4.29E-17 | 917847.3604 | 5.21E-18 | 1.04E-16 | 0.021093306 |
| Rosenbrock | Mean | 0.425645397 | 113.143751 | 1679629.019 | 4.22E-17 | 132.3488752 | 14373397912 | 0.013107593 | 0.054865327 | 2603.968539 |
| | SD | 1.187984439 | 48.99432331 | 648462.4744 | 1.09E-17 | 41.72265261 | 361340776.6 | 0.008658828 | 0.045566135 | 599.4022496 |
| Rastrigin | Mean | 7.362692992 | 148.2486456 | 2722.799729 | 0.099495906 | 133.1138439 | 1674.782779 | 4.76E-17 | 1.08E-15 | 735.8480014 |
| | SD | 2.485404145 | 17.76489083 | 83.14754621 | 0.298487717 | 106.6728854 | 96.86409615 | 4.40E-18 | 8.99E-17 | 24.75231998 |
| Griewank | Mean | 0.059270504 | 0.048643996 | 86.03568115 | 0.008127282 | 0.000739604 | 266.1639753 | 5.10E-19 | 4.92E-17 | 0.10290266 |
| | SD | 0.03371245 | 0.063166266 | 29.1502045 | 0.009476456 | 0.002218812 | 335.3504904 | 1.93E-19 | 4.25E-18 | 0.068217103 |
| Schwefel | Mean | -2654.033431 | -20100.36156 | -187704.1438 | -4166.141206 | -31182.49983 | -252854.5198 | -4189.828873 | -41898.28873 | -350890.8062 |
| | SD | 246.5263242 | 1763.156655 | 11097.95553 | 47.37533385 | 2078.47339 | 17724.02042 | 9.09E-13 | 3.30E-10 | 2279.801625 |
| Ackley | Mean | 4.67E-17 | 0.732022399 | 8.741445965 | 4.86E-17 | 2.14E-16 | 17.47129372 | 1.71E-16 | 4.21E-15 | 3.200412604 |
| | SD | 8.06E-18 | 0.755456829 | 0.784830594 | 6.55E-18 | 4.53E-17 | 3.815946124 | 3.57E-17 | 3.09E-16 | 0.133628837 |
| Step | Mean | 0 | 1.7 | 13538.3 | 0 | 0 | 47653.1 | 0 | 0 | 0 |
| | SD | 0 | 2.60959767 | 3112.434162 | 0 | 0 | 27093.57104 | 0 | 0 | 0 |
| Penalized | Mean | 4.13E-17 | 0.184850201 | 9706.265611 | 3.65E-17 | 0.01243899 | 35667997246 | 4.57E-17 | 1.07E-15 | 0.000332724 |
| | SD | 9.11E-18 | 0.314058054 | 6077.475023 | 9.90E-18 | 0.028500601 | 891866689.6 | 7.28E-18 | 6.72E-17 | 0.00093282 |
| Dixon Price | Mean | 0.666666667 | 2.07647225 | 615801.4332 | 0.666666667 | 0.666666667 | 3246360171 | 4.07E-16 | 1.26E-06 | 2704.747482 |
| | SD | 1.67E-14 | 4.229416739 | 642782.3094 | 4.97E-17 | 3.51E-17 | 1082157842 | 1.22E-16 | 7.21E-07 | 360.341568 |

**Table 3.** Experimental Results of PSO, DE and ABC Algorithms under varying Population Sizes/Colony Sizes, Dimensions of the functions were 100, SD:Standard Deviation

| | | PSO | | | DE | | | ABC | | |
|---|---|---|---|---|---|---|---|---|---|---|
| | Pop | 50 | 100 | 200 | 50 | 100 | 200 | 50 | 100 | 200 |
| Sphere | Mean | 6.16E-16 | 5.14E-16 | 3.11E-16 | 1.54E-16 | 8.84E-17 | 1.11E-16 | 1.12E-15 | 1.08E-15 | 1.02E-15 |
| | SD | 2.86E-16 | 3.12E-16 | 5.02E-17 | 1.08E-17 | 4.29E-17 | 4.87E-17 | 8.78E-17 | 1.04E-16 | 7.56E-17 |
| Rosenbrock | Mean | 152.6300403 | 113.143751 | 129.8648065 | 244.4621408 | 132.3488752 | 65.57549295 | 0.334214887 | 0.054865327 | 0.033840879 |
| | SD | 55.22881247 | 48.99432331 | 72.04063248 | 167.3699563 | 41.72265261 | 2.539552946 | 0.579373464 | 0.045566135 | 0.058330247 |
| Rastrigin | Mean | 161.6805719 | 148.2486456 | 159.3921715 | 90.740202 | 133.1138439 | 666.5000853 | 1.14E-15 | 1.08E-15 | 1.04E-15 |
| | SD | 30.0214669 | 17.76489083 | 25.03970003 | 11.60140207 | 106.6728854 | 53.08918163 | 6.26E-17 | 8.99E-17 | 1.13E-16 |
| Griewank | Mean | 0.066057787 | 0.048643996 | 0.00320331 | 0.034770873 | 0.000739604 | 2.52E-17 | 8.33E-17 | 4.92E-17 | 4.93E-17 |
| | SD | 0.073721225 | 0.063166266 | 0.00505551 | 0.038642093 | 0.002218812 | 2.86E-18 | 9.96E-17 | 4.25E-18 | 2.90E-18 |
| Schwefel | Mean | -18931.71298 | -20100.36156 | -21200.01235 | -30783.55835 | -31182.49983 | -16666.82437 | -41898.21377 | -41898.28873 | -41898.28873 |
| | SD | 1453.154666 | 1763.156655 | 2155.846443 | 930.3749272 | 2078.47339 | 1663.709053 | 0.223900743 | 3.30E-10 | 3.99E-12 |
| Ackley | Mean | 1.857181245 | 0.732022399 | 6.80E-16 | 2.666357404 | 2.14E-16 | 1.74E-14 | 4.48E-15 | 4.21E-15 | 4.11E-15 |
| | SD | 0.739376943 | 0.755456829 | 9.64E-17 | 0.607483624 | 4.53E-17 | 5.61E-15 | 2.14E-16 | 3.09E-16 | 1.59E-16 |
| Step | Mean | 38.5 | 1.7 | 0.2 | 0 | 0 | 0 | 0 | 0 | 0 |
| | SD | 56.90210892 | 2.60959767 | 0.4 | 0 | 0 | 0 | 0 | 0 | 0 |
| Penalized | Mean | 0.196111794 | 0.184850201 | 0.049804791 | 7000.171624 | 0.01243899 | 0.003110071 | 1.17E-15 | 1.07E-15 | 1.00E-15 |
| | SD | 0.54742191 | 0.314058054 | 0.077774663 | 12773.11331 | 0.028500601 | 0.009330212 | 9.12E-17 | 6.72E-17 | 6.86E-17 |
| Dixon Price | Mean | 1.697096319 | 2.07647225 | 0.666666667 | 12.31123978 | 0.666666667 | 0.666666667 | 1.64E-06 | 1.26E-06 | 6.23E-07 |
| | SD | 3.090995081 | 4.229416739 | 2.81E-16 | 8.511926958 | 3.51E-17 | 3.51E-17 | 1.25E-06 | 7.21E-07 | 4.45E-07 |

**Table 4.** Experimental Results of PSO, DE and ABC Algorithms under varying initialization ranges, Dimensions of the functions were 100, Population Sizes/ Colony Sizes=100, SD:Standard Deviation, E:Entire Range, RH:Right Half, LQ:Left Quarter

| | | PSO | | | DE | | | ABC | | |
|---|---|---|---|---|---|---|---|---|---|---|
| | | E | RH | LQ | E | RH | LQ | E | RH | LQ |
| Sphere | Mean | 5.14E-16 | 7000 | 110000 | 8.84E-17 | 1.36E-16 | 1.22E-16 | 1.08E-15 | 1.09E-15 | 1.09E-15 |
| | SD | 3.12E-16 | 7810.249676 | 31304.95168 | 4.29E-17 | 4.19E-17 | 4.65E-17 | 1.04E-16 | 1.11E-16 | 8.80E-17 |
| Rosenbrock | Mean | 113.143751 | 18142.03013 | 315476109.5 | 132.3488752 | 93.66374186 | 92.95788738 | 0.054865327 | 71.39586669 | 37.53711903 |
| | SD | 48.99432331 | 35980.68932 | 105760465.7 | 41.72265261 | 22.08092455 | 22.88409265 | 0.045566135 | 23.41091877 | 34.92147863 |
| Rastrigin | Mean | 148.2486456 | 532.9408297 | 1030.111121 | 133.1138439 | 109.1917708 | 121.9071695 | 1.08E-15 | 1.08E-15 | 1.09E-15 |
| | SD | 17.76489083 | 49.12656056 | 85.72984831 | 106.6728854 | 75.71280927 | 96.40118191 | 8.99E-17 | 1.27E-16 | 1.14E-16 |
| Griewank | Mean | 0.048643996 | 18.17940342 | 873.9824102 | 0.000739604 | 2.57E-17 | 0.004184622 | 4.92E-17 | 4.81E-17 | 7.21E-17 |
| | SD | 0.063166266 | 36.23688903 | 306.6712183 | 0.002218812 | 3.48E-18 | 0.008386537 | 4.25E-18 | 3.19E-18 | 6.68E-17 |
| Schwefel | Mean | -20100.36156 | -24547.79772 | -24611.33859 | -31182.49983 | -28521.95094 | -27250.87531 | -41898.28873 | -40048.20134 | -38637.4862 |
| | SD | 1763.156655 | 546.1026552 | 1232.055624 | 2078.47339 | 1612.081026 | 1412.287108 | 3.30E-10 | 285.5155238 | 478.6030682 |
| Ackley | Mean | 0.732022399 | 5.165491095 | 19.82115921 | 2.14E-16 | 2.09E-16 | 19.57582237 | 4.21E-15 | 4.42E-15 | 8.22E-15 |
| | SD | 0.755456829 | 4.71352016 | 0.018115853 | 4.53E-17 | 4.92E-17 | 1.113679862 | 3.09E-16 | 1.72E-16 | 7.98E-16 |
| Step | Mean | 1.7 | 8825.2 | 109771.5 | 0 | 0 | 0 | 0 | 0 | 0 |
| | SD | 2.60959767 | 6866.461968 | 19004.7616 | 0 | 0 | 0 | 0 | 0 | 0 |
| Penalized | Mean | 0.184850201 | 0.193568778 | 256000005 | 0.01243899 | 0.028107741 | 496.7207752 | 1.07E-15 | 1.01E-15 | 1.15E-15 |
| | SD | 0.314058054 | 0.349715544 | 114486682.7 | 0.028500601 | 0.084323224 | 1490.100125 | 6.72E-17 | 9.09E-17 | 5.50E-17 |
| Dixon Price | Mean | 2.07647225 | 69492.9502 | 10338445.49 | 0.666666667 | 0.666666667 | 0.666666667 | 1.26E-06 | 0.839076595 | 1.596308148 |
| | SD | 4.229416739 | 127273.8011 | 2073835.779 | 3.51E-17 | 6.08E-17 | 0 | 7.21E-07 | 0.376101069 | 1.847784176 |

**Table 5.** Experimental Results of ABC Algorithm under varying limit values, Dimensions were 100, Colony Size was 100

| Function | | Limit | | | | | | |
|---|---|---|---|---|---|---|---|---|
| | | 5 | 50 | 100 | 500 | 1000 | 5000 | 10000 |
| Sphere | Mean | 40172.619 | 1.30E-14 | 2.21E-15 | 1.27E-15 | 1.11E-15 | 1.09E-15 | 1.05E-15 |
| | SD | 4636.9765 | 1.28E-14 | 3.23E-16 | 1.05E-16 | 8.94E-17 | 1.00E-16 | 9.55E-17 |
| Rosenbrock | Mean | 75479633 | 0.3687739 | 0.1066848 | 0.0577759 | 0.0305455 | 0.0400428 | 0.0903778 |
| | SD | 16643440 | 0.2485734 | 0.1231579 | 0.0384831 | 0.0295326 | 0.048434 | 0.1663324 |
| Rastrigin | Mean | 722.41549 | 4.13E-08 | 5.27E-11 | 1.24E-15 | 1.12E-15 | 1.09E-15 | 1.07E-15 |
| | SD | 30.956511 | 8.44E-08 | 1.55E-10 | 1.95E-16 | 1.14E-16 | 1.04E-16 | 1.04E-16 |
| Griewank | Mean | 387.05656 | 2.36E-13 | 6.56E-15 | 5.94E-17 | 4.98E-17 | 4.76E-17 | 4.92E-17 |
| | SD | 44.678379 | 3.23E-13 | 4.18E-15 | 3.33E-17 | 4.13E-18 | 4.93E-18 | 3.69E-18 |
| Schwefel | Mean | -20848.434 | -41347.734 | -41612.546 | -41898.289 | -41898.289 | -41898.289 | -41898.289 |
| | SD | 626.53929 | 106.94234 | 106.48415 | 6.75E-08 | 0.0002771 | 3.88E-08 | 2.97E-06 |
| Ackley | Mean | 18.376048 | 2.62E-13 | 1.01E-14 | 4.75E-15 | 4.35E-15 | 4.31E-15 | 4.29E-15 |
| | SD | 0.390422 | 1.56E-13 | 1.57E-15 | 2.12E-16 | 2.27E-16 | 1.72E-16 | 2.68E-16 |
| Step | Mean | 46155.1 | 13.5 | 0.7 | 0 | 0 | 0 | 0 |
| | SD | 8465.072 | 5.4267854 | 0.9 | 0 | 0 | 0 | 0 |
| Penalized | Mean | 86851655 | 1.64E-14 | 2.48E-15 | 1.15E-15 | 1.10E-15 | 1.06E-15 | 1.05E-15 |
| | SD | 42828908 | 5.48E-15 | 5.41E-16 | 8.45E-17 | 9.15E-17 | 1.12E-16 | 8.59E-17 |
| Dixon Price | Mean | 1565358.2 | 0.0031539 | 1.07E-06 | 1.47E-06 | 1.96E-06 | 9.17E-07 | 1.01E-06 |
| | SD | 674594.38 | 0.0017079 | 5.76E-07 | 7.28E-07 | 2.26E-06 | 3.82E-07 | 4.45E-07 |

**Table 6.** Experimental Results of ABC Algorithm under varying limit values, Dimensions were 100, Colony Size was 10

| Function | | Limit | | | | | | |
|---|---|---|---|---|---|---|---|---|
| | | 5 | 50 | 100 | 500 | 1000 | 5000 | 10000 |
| Sphere | Mean | 98095.07116 | 3.37E-12 | 1.72E-13 | 1.77E-15 | 1.36E-15 | 1.46E-15 | 1.36E-15 |
| | Std | 12959.09176 | 1.02E-11 | 3.58E-13 | 3.05E-15 | 1.64E-16 | 6.73E-16 | 1.90E-16 |
| Rosenbrock | Mean | 267422928 | 59.2079779 | 15.18515019 | 10.23990112 | 27.60392216 | 17.75898437 | 36.92995963 |
| | Std | 51812195.52 | 218.0286192 | 43.30865642 | 23.98853007 | 56.06136362 | 47.47330859 | 114.2961123 |
| Rastrigin | Mean | 942.1587016 | 8.565987682 | 8.172342827 | 7.604827352 | 10.70040148 | 9.658915175 | 9.883292388 |
| | Std | 59.80687384 | 3.194695616 | 4.648093126 | 3.697719518 | 6.798092714 | 4.932444925 | 4.839869249 |
| Griewank | Mean | 869.2096345 | 0.003233202 | 0.002041184 | 0.002975775 | 0.028078594 | 0.00810774 | 0.010963982 |
| | Std | 134.3235844 | 0.010931443 | 0.005640278 | 0.008799799 | 0.085084827 | 0.017752733 | 0.028803579 |
| Schwefel | Mean | -17790.24013 | -39738.06939 | -40051.94037 | -40339.09933 | -40219.75069 | -40230.36359 | -40297.86110 |
| | Std | 1015.674426 | 336.6597611 | 503.2839908 | 407.1759947 | 422.4181709 | 461.630346 | 367.496268 |
| Ackley | Mean | 19.52803902 | 0.212870212 | 0.255519831 | 0.255407435 | 0.40063048 | 0.319323308 | 0.412635114 |
| | Std | 0.22340354 | 0.296340225 | 0.340478681 | 0.381650479 | 0.511608274 | 0.483124854 | 0.454188474 |
| Step | Mean | 106992.4 | 55.66666667 | 7.866666667 | 0 | 0 | 0 | 0 |
| | Std | 14232.35432 | 22.35073949 | 4.030991055 | 0 | 0 | 0 | 0 |
| Penalized | Mean | 576542253.4 | 9.44E-10 | 6.10E-14 | 3.02E-14 | 1.55E-12 | 1.61E-06 | 0.00207338 |
| | Std | 156824154.4 | 4.68E-09 | 1.11E-13 | 8.81E-14 | 8.36E-12 | 8.67E-06 | 0.007757879 |
| Dixon Price | Mean | 6197348.721 | 1.919382041 | 3.109303975 | 2.467438828 | 1.862482692 | 1.686539242 | 3.110083529 |
| | Std | 1661787.691 | 4.344332443 | 5.509126939 | 5.304917011 | 4.863661272 | 3.869273525 | 5.447932166 |

**Table 7.** Experimental Results of ABC Algorithm under varying limit values when population initialized in the left quarter of the entire search space, Dimensions were 100, Colony Size was 10

| Function | | Limit | | | | | | |
|---|---|---|---|---|---|---|---|---|
| | | 5 | 50 | 100 | 500 | 1000 | 5000 | 10000 |
| Sphere | Mean | 95105.3328 | 1.07E-10 | 1.67E-08 | 5.54E-12 | 5.11E-11 | 1.09E-15 | 1.33E-15 |
| | Std | 14661.92645 | 4.28E-10 | 8.28E-08 | 2.89E-11 | 2.75E-10 | 1.92E-16 | 5.25E-16 |
| Rosenbrock | Mean | 269505615.3 | 190.3554476 | 1199.876802 | 838.9745817 | 363.3571176 | 624.7432321 | 1121.62295 |
| | Std | 68373898.46 | 322.9181696 | 4998.402971 | 2472.735357 | 571.2721707 | 849.2494919 | 4397.815502 |
| Rastrigin | Mean | 950.5869637 | 14.94996831 | 43.07122725 | 247.1658087 | 253.5954414 | 259.1623807 | 253.0372257 |
| | Std | 56.02863864 | 9.560372871 | 22.61337802 | 43.24912076 | 36.28455454 | 42.23473145 | 39.77043485 |
| Griewank | Mean | 859.7045978 | 0.008499821 | 0.004393589 | 0.032915998 | 0.043504983 | 1.4181218 | 0.566366173 |
| | Std | 141.3025005 | 0.018106107 | 0.014055128 | 0.048141907 | 0.061808438 | 4.657248824 | 1.952489982 |
| Schwefel | Mean | -17107.32873 | -39327.94192 | -39491.03497 | -36680.47542 | -35317.9738 | -32379.39516 | -29995.05183 |
| | Std | 805.7904857 | 474.3960215 | 344.835107 | 1299.968374 | 1191.019816 | 2034.384921 | 128.6856694 |
| Ackley | Mean | 19.61754314 | 0.392724383 | 9.043197628 | 10.16441257 | 10.27369945 | 10.71427402 | 10.27599214 |
| | Std | 0.225559909 | 0.463505056 | 3.231658614 | 2.571042729 | 2.953414239 | 1.46314876 | 2.926421499 |
| Step | Mean | 101259.8667 | 61.6 | 9.466666667 | 0 | 0 | 5.1 | 182.4666667 |
| | Std | 15550.22787 | 24.55415783 | 5.155795014 | 0 | 0 | 27.46434052 | 601.9868622 |
| Penalized | Mean | 587043270.1 | 7.38E-06 | 3.66E-08 | 0.014249608 | 0.004113517 | 0.001008872 | 0.001617202 |
| | Std | 160656052.9 | 3.98E-05 | 1.13E-07 | 0.055924082 | 0.022151949 | 0.005432943 | 0.006300558 |
| Dixon Price | Mean | 6185260.626 | 19.0649427 | 34.04251776 | 21.89179569 | 20.64629783 | 20.92190235 | 34.0333849 |
| | Std | 1514959.996 | 15.58836441 | 57.80617854 | 13.03460008 | 20.90865202 | 10.89836716 | 59.58264989 |

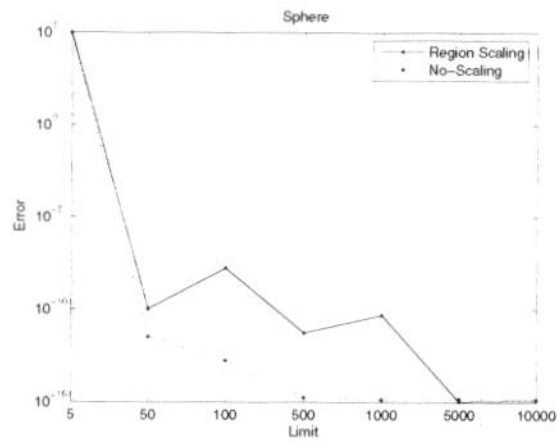


(a) Sphere

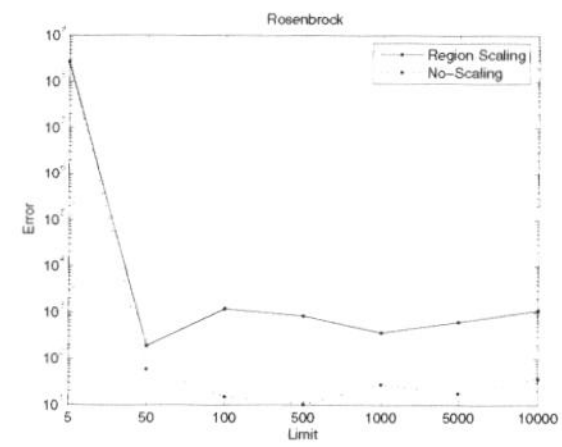


(b) Rosenbrock

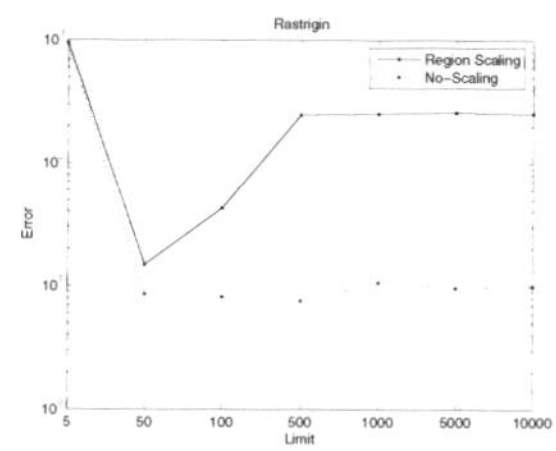


(c) Rastrigin

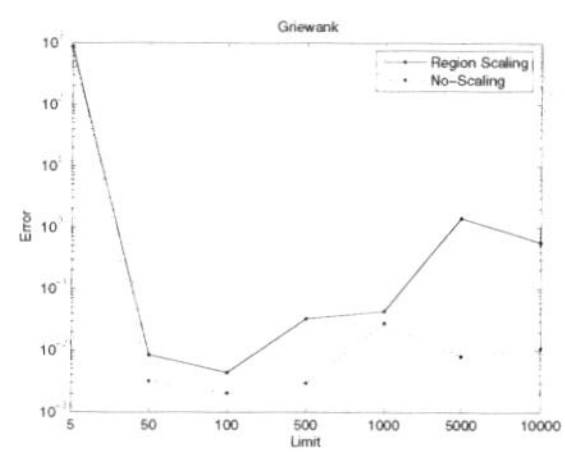


(d) Griewank

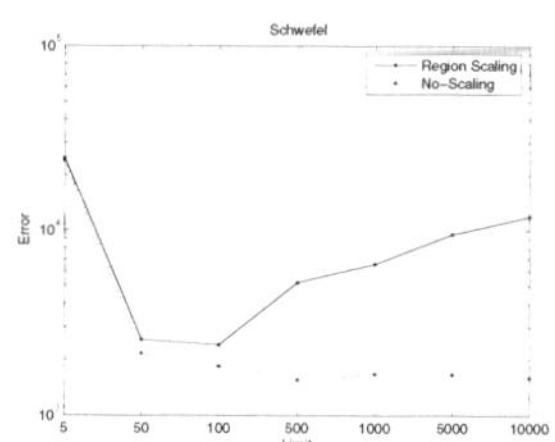


(e) Schwefel

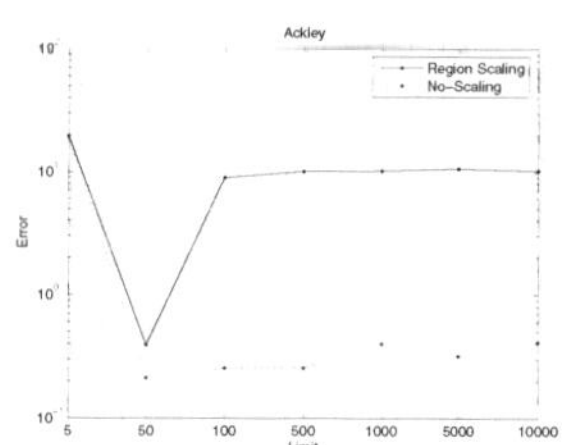


(f) Ackley

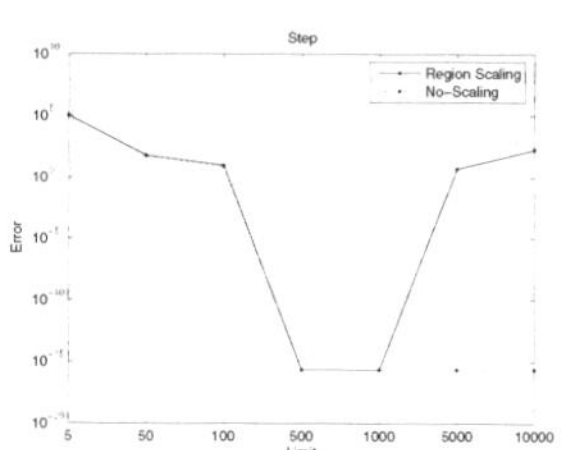


(g) Step

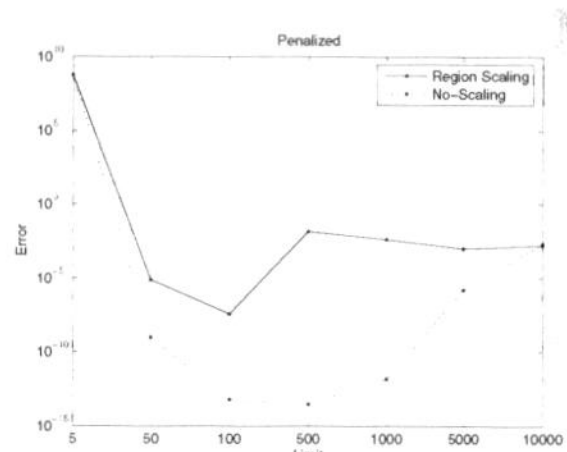


(h) Penalized

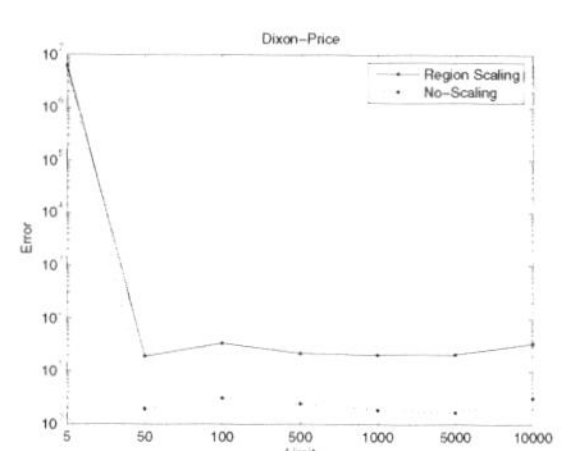


(i) Dixon-Price

**Fig. 1.** Effect of limit parameter on Functions when the initial populations are generated in the entire region and the subregion. Dimension was 100, Colony size was 10.

attenuates the exploration capability of the ABC algorithm. However, when the search process gets more difficult, the value of limit parameter shifts to lower values from the value that is around ($SN*D$) as mentioned earlier. Figures 1(a)-1(i) clearly demonstrate the results in Table 7.

## 4 Conclusion

ABC algorithm is a relatively new nature inspired algorithm simulating the foraging behaviour of a honey bee swarm. Like most of the optimization algorithms, ABC algorithm also has control parameters to be set before running the algorithm. Values of these parameters affect the performance of the algorithm. In this paper, we examined the effect of different parameter values of Artificial Bee Colony algorithm in the optimization of well-known benchmark functions and compared the performance of Artificial Bee Colony algorithm against those of Differential Evolution and Particle Swarm Optimization algorithms. Different problem dimensions, colony sizes, initialization ranges and limit values are tested on the performance of ABC algorithm.

Since ABC algorithm uses exploitative process efficiently to converge minima and explorative process to provide sufficient diversity in the population for a given colony size, ABC algorithm does not need big number of colony size to solve optimization problems with high dimensions.

Control parameter *limit* is the core parameter of the algorithm dictating the occurrence of scout bees that are responsible for providing the diversity in the population. For relatively smaller populations, control parameter *limit* should not be so low in order to avoid the population to be comprised of so many random solutions. However, since large populations have sufficient diversity and improvement in their progress, *limit* parameter does not stand out in these situations.

Another conclusion is that ABC algorithm is not very sensitive to initialization ranges compared to other algorithms considered in the paper. Initialization in the region that does not contain the optimum solution surely influences the search efficiency since the exploited information is poor. However, the scout unit of the ABC algorithm avoids the population to get stuck a region and allows the individuals of population to spread other regions of the space.

While producing new individuals of the population, ABC algorithm uses random step size in the range [-1,1]. By parameter control, this step size can be adaptively determined in the run time of the algorithm using some statistics of the population such as variance, which is our future work.

## Acknowledgement

This paper is a part of the project contract numbered FBA-06-22 of Erciyes University, Scientific Research Department.

## References

1. DeJong, K.: Parameter setting in eas: a 30 year perspective. Parameter Setting in Evolutionary Algorithms, 1–18 (2007)
2. Karaboga, D.: An idea based on honey bee swarm for numerical optimization. Technical Report TR06, Erciyes University, Engineering Faculty, Computer Engineering Department (2005)
3. Basturk, B., Karaboga, D.: An artificial bee colony (abc) algorithm for numeric function optimization. In: IEEE Swarm Intelligence Symposium 2006, Indianapolis, Indiana, USA (May 2006)
4. Karaboga, D., Basturk, B.: A powerful and efficient algorithm for numerical function optimization: Artificial bee colony (abc) algorithm. Journal of Global Optimization 39(3), 459–471 (2007)
5. Karaboga, D., Basturk, B.: On the performance of artificial bee colony (abc) algorithm. Applied Soft Computing 8(1), 687–697 (2008)
6. Tereshko, V.: Reaction–diffusion model of a honeybee colony's foraging behaviour. In: Deb, K., Rudolph, G., Lutton, E., Merelo, J.J., Schoenauer, M., Schwefel, H.-P., Yao, X. (eds.) PPSN 2000. LNCS, vol. 1917, pp. 807–816. Springer, Heidelberg (2000)
7. Corne, D., Dorigo, M., Glover, F.: New Ideas in Optimization. McGraw-Hill, New York (1999)
8. Vesterstrom, J., Thomsen, R.: A comparative study of differential evolution particle swarm optimization and evolutionary algorithms on numerical benchmark problems. In: IEEE Congress on Evolutionary Computation (CEC 2004), Piscataway, New Jersey, June 2004, vol. 3, pp. 1980–1987 (2004)

# A Modified Ant-Based Approach to Edge Detection

Doğan Aydın

Department of Computer Engineering, Ege University, Izmir, 35100 Turkey
dogan.aydin@ege.edu.tr

**Abstract.** A modified ant-based edge detector is presented. It is based on the distribution of ants on an image. Ants try to find possible edges by using a heuristic function based on 3x3 ideal edge structures. Visual comparisons show that the modified algorithm gives finer details and thinner edges at lesser computational times when compared to earlier ant-based approaches. When compared to standard edge detection methods, it provides finer details than others except Canny's, with slightly more occurrence of broadening.

**Keywords:** Edge detection, ant-based algorithm, ideal edge structures.

## 1 Introduction

Edges separate two different regions and characterize object boundaries in images. Therefore, detecting of edges is an important problem in pattern recognition, computer vision and image processing. Many edge detection methods have been proposed in the past 40 years. In recent years, ant-based approaches are also proposed as a new approach to detect edges.

Ant-based algorithms are inspired from Ant Colony Optimization algorithms proposed by Dorigo as a metaheuristic method [1]. ACO has been applied to a wide range of different discrete optimization problems successfully. Besides, ACO algorithms have been used to solve many image processing and computer vision problems such as multilevel thresholding [2], and image segmentation [3, 4] in recent years. To apply ACO to solve image processing problems, two factors are needed, optimization function model to represent the image features and ACO algorithm to solve the model. However, Ant-based approaches to solve edge detection problem don't need to formulize the problem as an optimization function model. Instead, with some modifications to ACO algorithms, they can as well be adopted as a new approach to edge detection. Some Ant-based approaches [5,6] used with other edge detection methods give good results if they used to solve the problem of linking disjointed edges produced by other edge detector. But, pure Ant-based approaches to edge detection problems [7,8,9] are not yet satisfactory. These works aim to show that Ant-based approaches can be used as an edge detector, which suffer from detailed detection and increasing time complexity. Besides, they clearly indicate that more research is needed to improve detection quality and decrease time complexity [7,8]. In this paper, we propose a fast and modified ant-based approach to edge detection that provides both detail improvement and error minimization.

N.T. Nguyen, R. Kowalczyk, and S.-M. Chen (Eds.): ICCCI 2009, LNAI 5796, pp. 620–628, 2009.

## 2 The Proposed Ant Algorithm

### 2.1 Ant Colony Optimization

Swarm Intelligence (SI), inspired from the biological examples by swarming, flocking and herding phenomena in vertebrates, is an innovative intelligent solution method for optimization problems [10]. Ant Colony Optimization (ACO) is another interesting swarm intelligence technique derived from real ant colonies.

Many ant colonies have trail-laying trail-following behavior while foraging. Ants deposit a chemical substance called pheromone on the trace from where they find a food source to their nest. The amount of pheromone deposited depends on the quality and amount of the food source. They forage while wandering in the vicinity of pheromone trails. This feedback (tracing) mechanism that reinforces finding good solutions directly or indirectly is the basic idea underlying all ant-based (ACO) algorithms.

ACO algorithms are not interested in simulation of real ant behaviors but uses artificial ant colonies as an optimization technique. In ACO, artificial ants act randomly and deposit pheromones according to their solution quality as the real ants do, but they have some differences as follows:

- artificial ants have some memory,
- they are not completely blind,
- they live in an environment where time is discrete.

In ACO algorithms, each virtual ant tries to find good solutions simultaneously and individually in a problem space. At each step of the construction of solution, they randomly choose their direction where the probability distribution associated in choosing any direction is specified. The first implementation of ACO algorithm, Ant System (AS)[11], defined the transition probability from state $i$ to $j$ by

$$p_{ij} = \begin{cases} \dfrac{[\tau_{ij}]^{\alpha}[\eta_{ij}]^{\beta}}{\sum_{h\in\Omega}[\tau_{ih}]^{\alpha}[\eta_{ih}]^{\beta}} & if \quad j\in\Omega \\ 0 & otherwise \end{cases} \tag{1}$$

where $\tau_{ij}$ is trail level of artificial pheromone between state $i$ and $j$, $\eta_{ij}$ is a heuristic function, $\Omega$ is the set of unvisited states, also $\alpha$ and $\beta$ are two adjustable tuning parameters. Ant Colony System (ACS) [12] uses a different transition probability rule, where the rule 1 is modified to allow explicitly for exploration as follows:

$$j = \begin{cases} \arg \max\limits_{j\in\Omega} \left\{ [\tau(i,j)]\cdot[\eta(i,j)]^{\beta} \right\} & if\ q \le q_0 \\ J & otherwise \end{cases} \tag{2}$$

where $q$ is a random number, $q_0$ is a parameter $(0 \le q_0 \le 1)$ and $J$ is a random variable selected according to the previous probability distribution function given in eq. (1). At each step, an ant in state $i$ chooses a state $j$ to go, according to the value of $q$ that is randomly generated. If $q \le q_0$ then the best edge is chosen (exploitation), otherwise an arbitrary edge is chosen according to eq. (1)(Biased exploration) [12].

Once all tour constructions (at the end of the each iteration) have been completed, the best-so-far solution is recorded/updated. Meanwhile, colony members deposit pheromone on their trail path according to the solution qualities. Pheromone depositing provides indirect communication in colony as a positive feedback. Therefore, probability of finding new good solutions is reinforced. In addition, a negative feedback is applied through the pheromone evaporation on paths. In this way, pheromone levels on previous bad paths are decreased thus preventing algorithm to stop at a local optima. Generally, pheromone updating rule (for both depositing and evaporation) is defined as:

$$\tau_{new}(i, j) = (1-\rho)\tau_{old}(i, j) + \sum_{k=1}^{m} \Delta\tau^k(i, j) \tag{3}$$

where $\rho$ is pheromone decay value, $m$ is the number of ant and

$$\Delta\tau^k(i, j) = \begin{cases} f_k / Q & (i, j) \in T_k \\ 0 & otherwise \end{cases} \tag{4}$$

where $f_k$ is the fitness function is the fitness function (i.e., $f_k$ is tour length for traveling salesman problem), $Q$ is a constant parameter usually set as 1 and $T_k$ is the route of the $k^{th}$ ant.

## 2.2 The Proposed Ant-Based Approach to Edge Detection

First, a gray level image is considered as a two-dimensional graph where the pixels are assumed as vertices. At the beginning of the each iteration, ants are located on pixels randomly. Then, they wander on the graph pixel by pixel in fixed number of steps (ant memory) and deposit pheromone according to the path quality. The path quality indicates how closely that path belongs to an edge field. Unlike original ant-based approaches, each ant try to find a part of edge fields (a part of a solution, not a solution itself). After they complete their trips, they turn back to their initial states and start their tours again. The algorithm is terminated after a specific number of iterations.

The heuristic function, $\eta(j)$, plays the most critical role for ants in selecting suitable pixels to visit around the position they are at. Since the pixel $j$ to be visited is determined independent of the current position, $\eta$ becomes only a function of $j$. In order to minimize occurrence of speckles and thick edges, we determined $\eta(j)$ using the structures of 3x3 ideal edge images described in [11,12]. We used four structures of 3x3 ideal edge images, and each had two possible variants shown in Fig. 1.

In each pattern, edges divide the pixels into two sets, giving $s_1 = \{p_1, p_4, p_6\}$, $s_2 = \{p_3, p_5, p_8\}$, $s_3 = \{p_4, p_6, p_7\}$, $s_4 = \{p_2, p_3, p_5\}$, $s_5 = \{p_1, p_2, p_3\}$, $s_6 = \{p_6, p_7, p_8\}$, $s_7 = \{p_1, p_2, p_4\}$ and $s_8 = \{p_5, p_7, p_8\}$, where $p_j$ is the 8-bit quantized intensity of the $j^{th}$ pixel (see in Figure 2). If a pixel $j$ belongs to an ideal edge region, total pixel intensity variation between two possibly divided regions should be

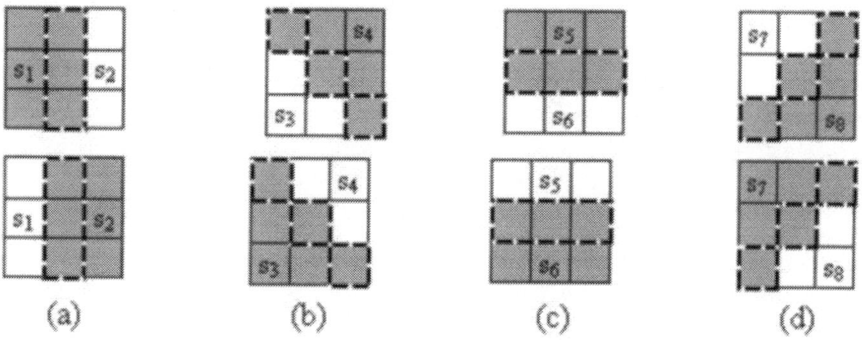


**Fig. 1.** 3x3 ideal edge structures

bigger. The more the variation, the higher the probability of being an edge, and the more the pheromone. By using this assumption, we define $\eta(j)$ ($j$ is any pixel at the center of a 3x3 image region) as

$$\eta(j) = \frac{\arg\max\{|s_1 - s_2|, |s_3 - s_4|, |s_5 - s_6|, |s_7 - s_8|\}}{p_{\max}} \tag{5}$$

where $p_{max}$ is the maximum gray valued pixel in the entire image. Consequently, ants move towards pixels that belong to ideal edges on the image. Furthermore, if $\eta(j)$ is smaller than intensity threshold T for all possible neighbors, it indicates that there are no suitable pixels to visit. In this situation, the ant does not move to any new pixel and dies. Otherwise, an ant at pixel $i$ can select the pixel $j$ with respect to the following equation based on the ACS rule:

$$j = \begin{cases} \arg \max_{j \in \Omega} \{[\tau(j)] \cdot [\eta(j)]^{\beta}\} & \mathit{if}\ q \le q_0 \\ J & \mathit{otherwise} \end{cases} \tag{6}$$

where $\tau(j)$ is pheromone intensity on pixel $j$ and is the set of unvisited pixels. Eq.(6) indicates that only the unvisited pixels are allowed to be visited. Fig. 2 shows an example where an ant can not move back to $p$ but can move to any of other seven neighbors according to their probabilities.

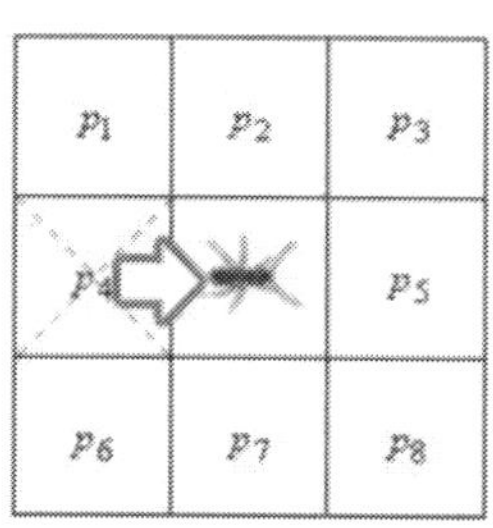


**Fig. 2.** Suitable and unsuitable neighbors to traverse

Unlike the original ant algorithms, each ant deposits some amount of pheromone after it moves from one pixel to another. Evaporation mechanism assists to discard mistakenly traced irrelevant pixels as edges. After one step of iteration, the pheromone intensity becomes

$$\tau_{new}(j) = (1-\rho)\tau_{old}(j) + \sum_{k=1}^{v} \Delta\tau^k(j) \tag{7}$$

where v is the number of ants that has visited to pixel $j$, and $\Delta\tau^k(j) = \eta(j)$.

After the iterations are completed, pheromone levels on pixels imply the likelihood of the pixels to belong to an edge. To determine the edge points, thresholding is applied on pheromone levels over the value of $\Psi$ and a binary image indicating the presence of edges is constructed.

## 3 Experimental Results

The proposed algorithm has many parameters to set and it is hard to assign them the best values. Therefore, like other ant-based methods, the convenient ant algorithm parameters are determined by the experiments. On the other hand, there is not yet any general performance index to judge the performance of the edge detection methods [11]. So, a suitable parameter set (Table 1) is arrived at after over 100 trials on the "Cameraman" image.

**Table 1.** Suitable parameters of the proposed algorithm

| *Parameter* | *Value* |
|---|---|
| $\rho$ | 0.85 |
| $\beta$ | 1 |
| $q_0$ | 0.95 |
| T | 30 |
| $\Psi$ (pheromone level threshold value) | 4 |
| **number of iterations** | 10 |

**Fig. 3.** Parameter effects on ant-based edge-detection of a 512x512 "Cameraman" image. (a)125 ants with 20-pixel memory (b)500 ants with 50-pixel memory (c)1000 ants with 100-pixel memory

Other ant algorithm parameters, the number of ants and ant memory, affecting the performance depend on the varying image size. For small sizes, too many ants with large memories lead to edge-broadening and longer computational times. For large sizes, too few ants with small memories would not produce expected synergy, giving bad results. Such parameters effects are shown in Fig. 3, for a 512x512 image.

We applied our proposed method on 256x256 and 512x512 sized standard images. We observed that 125(500) ants each with 20(50)-pixel memory are adequate to give suitable results for 256x256(512x512) sized images, respectively. We conclude that the memory and the number of ants should be set according to the image size. Fig. 4 shows the edge images of "Lena," "Home," and "Peppers" for the sizes of 256x256 and 512x512, obtained using the parameters of Table 1.

We as well compared our new technique with other two popular ant-based edge detection approaches. We implemented them using both the Matlab programming language and C#. All experiments are run on a PC with a Intel CoreTM DUO 2.6 GHz

**Fig. 4.** Quantitative results obtained by using the new ant-based edge detection with the parameters of Table 1. First row show the original images, second and third are the 256x256 and 512x512 sized edge-images, respectively.

**Fig. 5** Comparison of ant-based edge detection algorithms. The results of Nezamabadi-pour et al. method with thinning operation is at first row, second row shows the results of Tian et al. method without thinning, and last row shows the results of our approach.

CPU and a 2 GB RAM on 512x512 sized standard images. Nezamabadi-pour et al. [7] applied ant algorithms to detect edges for the first time. They used the transition probability function in Eq.1 that caused ants to wander excessively, resulting in edge-broadening. Therefore, they had to use a morphological thinning algorithm in order to thin the binary output image. Their heuristic function operates on single pixels rather than on a region around a given pixel, making their algorithm vulnerable to noise so much. The first row in Fig.5 shows its best results, where 1500 ants with random memory lengths between 15 and 30 were used. In contrast, the third row in Fig. 5 shows our results obtained with 500 ants each with memory length of 50 with no thinning operation. The new approach provides finer details, and yet almost 3-times as fast. (Our method takes about 3 and 20 seconds, but Nezamabadi-pour method takes about 10 and 60 seconds at C# and Matlab respectively.). Furthermore, Tian et al. [9] method gives the worst results in comparison. We used the parameters as the authors suggested, but its execution time is about 5 hours and it gives worst results that are shown at second row in Fig. 5 according to the detailed detection.

Finally, we compared our method with Canny, Sobel, Laplacian of the Gaussian (LoG) and Prewitt methods that are available in MATLAB. All experiments are done using default MATLAB parameters and results are shown in Fig. 6. Our method gives slightly more occurrences of edge-broadening and needs more execution time, but provides finer details than others except Canny's.

(a) Original (b) Canny (c) Sobel

(d) LoG (e) Prewitt (f) Proposed

**Fig. 6.** Comparison with other edge detection methods for 512x512 "Cameraman" image

## 4 Conclusion

In conclusion, this paper presents a modified and improved ant-based edge detector. It is based on the distribution of ants on an image. We define a new heuristic function based on 3x3 ideal edge structures and a transition probability rule, with which ants try to find edges. The new function attracts ants towards to edge fields. It avoids excessive wandering and increases sensitivity. Visual comparisons show that the new algorithm gives finer details and thinner edges at lesser computational times when compared to earlier ant-based approaches. When compared to conventional edge-detectors, it provides finer details than others except Canny's, with slightly more occurrences of broadening.

For another point of view, the algorithm is amenable to parallelization, which can improve the performance. The parallelization of the algorithm can be implemented by distributing ants on different processors. By exchanging some number of ants and pheromone values asynchronously, edges can be detected on each processor. For the future, besides working on parallelization of the proposed approach, we try to increase details detection ratio with minimizing the errors. We will work on a new heuristic functions that can clearly detect edges in digital images corrupted by noises as well.

## References

1. Dorigo, M., Di Caro, G.: Ant colony optimization: A new metaheuristic. In: Proc. of the Congress on Evolutionary Computation, vol. 2, pp. 1470–1477 (1999)
2. Liang, Y.-C., Chen, A.H.-L., Chyu, C.-C.: Application of a hybrid ant colony optimization for the multilevel thresholding in image processing. In: King, I., Wang, J., Chan, L.-W., Wang, D. (eds.) ICONIP 2006. LNCS, vol. 4233, pp. 1183–1192. Springer, Heidelberg (2006)
3. Tao, W., Jin, H., Liu, L.: Object segmentation using ant colony optimization algorithm and fuzzy entropy. Pattern Recognition Letters 28(7), 788–796 (2007)
4. Zhuang, X.: Image segmentation with ant swarm - A case study of digital signal processing with biological mechanism of intelligence. In: 11th Digital Sig. Proc. Workshop 2004, pp. 143–146 (2004)
5. Wong, Y.-P., Soh, V.C.-M., Ban, K.-W., Bau, Y.-T.: Improved Canny Edges Using Ant Colony Optimization. In: 5th Int. Conf. on Computer Graphics, Imaging and Visualisation, pp. 197–202 (2008)
6. Lu, D.-S., Chen, C.-C.: Edge detection improvement by ant colony optimization. Pattern Recognition Letters 29(4), 416–425 (2008)
7. Nezamabadi-pour, H., Saryazdi, S., Rashedi, E.: Edge detection using ant algorithms. Soft Computing 10, 623–628 (2006)
8. Rezaee, A.: Extracting edge of images with ant colony. Journal of Electrical Engineering 59(1), 57–59 (2008)
9. Tian, J., Yu, W., Xie, S.: An Ant Colony Optimization Algorithm For Image Edge Detection. In: IEEE Congress on Evolutionary Computation 2008 (CEC 2008), pp. 751–756 (2008)
10. Abraham, A., Guo, H., Liu, H.: Swarm intelligence: foundations, perspectives and applications, Swarm Intelligent System, N. Nedjah, L.d.M. Mourelle (eds.), pp. 1-25 (2006)
11. Dorigo, M., Maniezzo, V., Colorni, A.: The ant system: Optimization by a colony of cooperating agents. IEEE Trans. on Systems. Man. and Cyber. Part-B 26, 29–41 (1996)
12. Dorigo, M., Gambardella, L.M.: Ant colonies for the traveling salesman problem. Bio Systems 43, 73–81 (1997)
13. Kanga, C.-C., Wang, W.-J.: A novel edge detection method based on the maximizing objective function. Pattern Recognition 40(2), 609–618 (2007)
14. Kim, D., Lee, W., Kweon, I.: Automatic edge detection using 3 x 3 ideal binary pixel patterns and fuzzy-based edge Thresholding. Patt. Recog. Letters 25(1), 101–106 (2004)

# A Hybrid Evolutionary Approach for the Protein Classification Problem*

Denise F. Tsunoda[1,2], Heitor S. Lopes[2], and Alex A. Freitas[3]

[1] Bioinformatics Laboratory, Federal University of Technology - Paraná, Av. 7 de setembro, 3165 80230-901, Curitiba (PR), Brazil
[2] Department of Information Sciences, Federal University of Paraná, Brazil
[3] Computing Laboratory, University of Kent at Canterbury, UK
dtsunoda@ufpr.br, hslopes@utfpr.edu.br, A.A.Freitas@ukc.ac.uk

**Abstract.** This paper proposes a hybrid algorithm that combines characteristics of both Genetic Programming (GP) and Genetic Algorithms (GAs), for discovering motifs in proteins and predicting their functional classes, based on the discovered motifs. In this algorithm, individuals are represented as IF-THEN classification rules. The rule antecedent consists of a combination of motifs automatically extracted from protein sequences. The rule consequent consists of the functional class predicted for a protein whose sequence satisfies the combination of motifs in the rule antecedent. The system can be used in two different ways. First, as a stand-alone classification system, where the evolved classification rules are directly used to predict the functional classes of proteins. Second, the system can be used just as an "attribute construction" method, discovering motifs that are given, as predictor attributes, to another classification algorithm. In this usage of the system, a classical decision tree induction algorithm was used as the classifier. The proposed system was evaluated in these two scenarios and compared with another Genetic Algorithm designed specifically for the discovery of motifs – and therefore used only as an attribute construction algorithm. This comparison was performed by mining an enzyme data set extracted from the Protein Data Bank. The best results were obtained when using the proposed hybrid GP/GA as an attribute construction algorithm and performing the classification (using the constructed attributes) with the decision tree induction algorithm.

**Keywords:** Bioinformatics, Protein classification problem, evolutionary computation, genetic programming, genetic algorithm.

## 1 Introduction

Bioinformatics has become a major field in scientific research due to two main reasons. First, the life sciences community is overwhelmed by the huge amount of data available in public databases. Indeed, the size of biological databases is

* This work was partially supported by the Brazilian National Research Council – CNPq, under research grant no. 309262/2007-0 to H.S. Lopes.

N.T. Nguyen, R. Kowalczyk, and S.-M. Chen (Eds.): ICCCI 2009, LNAI 5796, pp. 629–640, 2009.

growing at an exponential rate as a consequence of advancements in genome sequencing technology. Second, there has been a significant increase in the amount of available computational power. As a result, some complex problems such as protein folding simulation can now be handled within a reasonable time.

Proteins have several functions within an organism and there are thousands of different types of proteins. They are composed of amino acids linked in linear chains through peptide connections. Proteins are grouped into families according to their biological function, and the classification of proteins is an important task for the molecular biologist.

There are several protein databases freely available in the Internet, and this work is based on the PDB (Protein Data Bank)[3]. This database contains information about the primary, secondary and tertiary structures of more than 56600 proteins (as in march/2009).

The protein classification problem (PCP) is a very important research area in bioinformatics. Basically, the PCP is the discovery of the functional class of an unknown-function protein, by means of analyzing its structure. Most proteins share similar structures (in particular, considering the primary structure), since many of them have a common evolutionary origin [15]. Common structures may be characteristic of a given family of proteins but, on the other hand, unrelated families can also share common structures. This twofold characteristic makes protein classification a difficult problem.

Computer science researchers have been using many different methods to find possible solutions for the PCP, for instance: neural networks [20], clustering algorithms [12], particle swarm optimization [8], genetic algorithms [18] and other data mining algorithms [11],[9].

This paper reports the development and application of a hybrid Genetic Programming (GP)/Genetic Algorithm (GA), specially devised for the automatic discovery of motifs using as input the primary structure information of proteins. The system generates rules based on discovered sequences of amino acids (motifs). These rules cover most proteins of a given class (family) without covering many proteins in other classes. Further, these discovered rules can be used for the characterization of families of proteins as well as for the automatic classification of unknown-function proteins.

## 2 A Hybrid GP/GA System for Discovering Protein Motifs

The hybrid Genetic Programming/Genetic Algorithm system is detailed here. This system was named HEADMOP (*H*ybrid *E*volutionary *A*lgorithm for the *D*iscovery of *MO*tifs in *P*roteins) and was used for discovering motifs and classification of protein functions.

### 2.1 Individual Representation

In Genetic Programming (GP) [10], each individual corresponds to a candidate solution to the problem. In HEADMOP, an individual corresponds to a classification rule of the form: *IF (a certain combination of motifs) THEN (predict a certain functional class)*.

The genetic material of an individual consists of a tree containing two kinds of nodes: internal and leaf nodes. Each internal node contains one of the following logical operations: AND, OR, NOT. Each leaf node contains a sequence of amino acids (considering the 20 standard amino acids), also known as a feature or motif. Leaf nodes can contain motifs of different lengths. An example of an individual is shown in Figure 1.

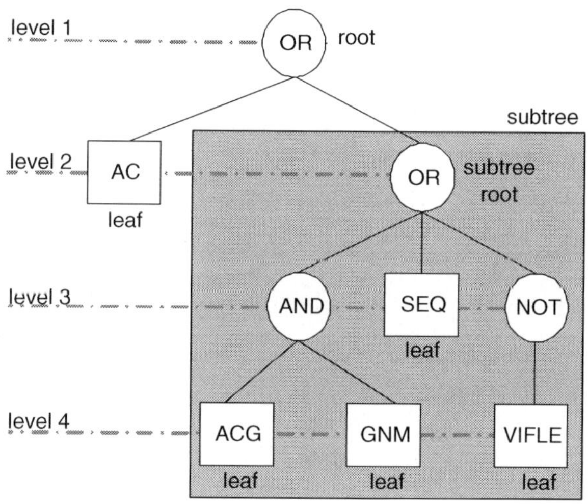


**Fig. 1.** Example of an individual in the proposed hybrid GP/GA system

Two categories of genetic operators were defined in HEADMOP: structural and leaf operators. Structural operators work on internal nodes of the tree, modifying the structure of the rule represented by the individual. Leaf operators modify only the sequences of amino acids (the motifs) in leaf nodes. These operators will be discussed in detail in section 2.2.

It should be noted that an individual consists of logical conditions and motifs specifying only the antecedent (the "IF part") of a classification rule. The class predicted by the consequent (the "THEN part") of the rule is not represented in the individual. The class associated with a rule is computed by using a deterministic procedure that assigns the best possible class to the rule, as will be explained later.

Actually, the proposed system can be regarded as a hybrid GP/GA, in the following sense. The individuals' genetic material has internal and leaf nodes (like in GP), but these leafs are encoded as linear genomes (like in GA), where different genomes can have different lengths of amino acid sequences (corresponding to chromosomes with a variable number of genes).

### 2.2 Selection Method and Genetic Operators

HEADMOP uses stochastic tournament selection, which works as follows. First, $k$ individuals are randomly drawn from the current population, with replacement, where $k$ is a user-specified parameter determined as a percentage of the population size. In this work, $k$ was set to 3% of the population size. Then, these k individuals are put to "play a tournament", in which the probability of an individual to win the tournament is proportional to its fitness value. A copy of the winner of the tournament is then set apart to further undergo the action of the genetic operators.

**A) Structural Operators.** The proposed system uses the classical sub-tree crossover, where a crossover point is randomly selected in each of the two parent individuals, and then they swap their genetic material rooted at their corresponding crossover points.

Mutation is the other structural operator. It is an asexual operation, involving only one parent individual. This operation consists of randomly selecting one node of the individual and applying a mutation to that point, producing a new child individual. In our system two different structural mutation operations were defined:

1. Sub-tree mutation: this operator replaces the sub-tree rooted at the selected mutation point by a new randomly-generated sub-tree
2. Point mutation: this operator simply replaces the logical function at the selected mutation point (OR, AND or NOT) by another randomly-generated logical function. If an OR or an AND is changed to a NOT, only one of the two sub-trees below the former OR/AND node is maintained (that one with better evaluation) – the other is simply deleted to maintain the individual's integrity. If a NOT is changed to an OR or an AND a sub-tree is randomly-generated and inserted as the other sub-tree below the new OR/AND node, also to maintain the individual's integrity.

Both structural crossover and mutation operators can be used as "hill-climbing operators", in the following sense. In the case of the mutation operator, immediately after the creation of a new child individual, its fitness is computed. If the child's fitness is lower than the parent's fitness, the parent (rather than the child) is copied to the next generation. In the case of the crossover operator, the same holds. The individual with best fitness among parents and children is copied to the next generation. The hill-climbing behavior of the structural operators is probabilistic, controlled by a user-defined parameter named "probability of hill climbing" (see section 3.2).

**B) Leaf Operators.** The conventional one-point crossover in genetic algorithms was originally designed for a fixed-length chromosome representation [6]. Hence, it cannot be applied directly to the leaf nodes of our approach, where a motif is represented by a variable-length sequence of amino acids. Therefore, we adapted the conventional one-point crossover to a variable-length representation, as follows. The crossover point (which is still randomly generated) indicates the percentage of the amino acid sequence of the leaf node in each parent where the swapping of amino acids ("genes") starts. The percentage (relative position) is the same for both leaf nodes (i.e., one leaf node of each parent GP individual), but the actual (absolute) position where the amino acid swapping starts can be different, since the two leaf nodes can have different numbers of amino acids.

This work introduces four different mutation operators devised for dealing with the variable-length sequence of amino acids contained in a leaf node of an individual, as follows.

1. Addition to the Left (AL): a letter (representing an amino acid) is randomly generated and inserted into the leftmost end of the sequence of amino acids;
2. Addition to the Right (AR): analogous to AL, except that the new amino acid is inserted into the rightmost end of the sequence of amino acids;
3. Multiple Mutations (MM): all the amino acids starting from a randomly-generated position up to the end of the sequence are replaced by other randomly-generated amino acids. The starting position can be any in the sequence, except the first and the last ones.
4. Removal (RM): the amino acid in a randomly chosen position is removed from the sequence. The application of this operator is subject to the constraint that after removal of an amino acid the remaining sequence still has at least three amino acids. If this condition is not met then this operator is not applied, and another mutation operator is applied instead.

### 2.3 Fitness Function

Since the goal is to maximize classification accuracy, the quality of a rule is determined by its ability in discriminating proteins of different classes. That is, ideally a rule should cover all the proteins of a given class and none of all other classes. The fitness function was designed to take this basic principle into account. First, for each class $i$ ($i = 1, \ldots, n$) we compute a measure of coverage of the rule for that class, called $F_i$. This measure is defined in the range [0..1] and corresponds to the number of proteins of the $i$-th class covered by the rule, divided by the total number of proteins belonging to the $i$-th class. A protein is said to be covered by a rule if and only if the protein's primary sequence of amino acids satisfies the conditions of the rule antecedent. Next, for each class $i$, a measure of the ability of the rule to discriminate between class $i$ and the other classes is computed. This measure is denoted $Disc_i$ and it is defined by equation 1:

$$Disc_i = F_i \cdot \left[ 1 - \sum_{j=1}^{n} \left( \frac{F_{j,j \neq i}}{n-1} \right) \right] \quad (1)$$

### 2.4 Result Designation

As explained earlier, each individual represents a rule that is associated with a given class of proteins. Therefore, it is not enough to return as a solution only the best rule found throughout the evolutionary process – as usual in conventional evolutionary algorithms. Indeed, it is necessary to return a set of rules, since different rules can predict different classes. Hence, the result provided by our system consists of the best $M$ rules for each class found throughout the evolutionary process, where $M$ is a user-defined parameter. This was implemented by elitism, in such a way that, at every generation, the $M$ best individuals of each class are preserved in the population. Recall that the quality of a rule (Equation 1 is based on its coverage.

The set of rules returned by HEADMOP was used for classification in two ways. First, each rule (individual) was used as a complete classification rule. In this case, when a new protein is to be classified, the system counts how many rules associated with each class cover that protein. The protein is assigned to the class with the highest value of that count, i.e., the class with the largest number of rules covering the protein. This is effectively a majority voting strategy, where each class has as many votes as the number of rules that predict that class and cover the protein.

In the second approach, each returned rule is interpreted as a binary predictor attribute, which takes on the value true if a protein satisfies the antecedent of the rule and takes on the value false otherwise. Note that, in this case, the rule consequent is effectively ignored, i.e., the new predictor attribute is constructed based only on the information derived from the rule antecedent - the motifs in the leaf nodes and the logical conditions in the internal nodes. The motivation for this approach – which is more complex than the previously-discussed first approach – is that this new, higher-level data set can be given to any classification algorithm. In this case the system is effectively acting as a sophisticated "attribute construction" (feature discovery) algorithm and it delegates the task of actually building the classifier to a another classification algorithm.

In order to implement this second approach, the chosen classification algorithm was J48, which is a Java implementation of the very well-known, and widely used, C4.5 algorithm [16]. J48 is available as part of a data mining tool named WEKA [22], which has the advantage of being freely available and widely used[1]. The choice of J48 was further motivated by the fact that this algorithm produces a decision tree, a classification model that tends to be comprehensible to the user, allowing him/her to interpret discovered knowledge. This is important not only in data mining in general, but also in Bioinformatics applications [14], [4], [17], where the goal is to give the user some new insight about the relationships that hold in the data.

## 3 Computational Experiments and Results

### 3.1 Data Set Used in the Experiments

The data set used in the experiments was extracted from the PDB. First, records which had an EC number were set apart and, from these, a random subset was

[1] http://www.cs.waikato.ac.nz/~ml/weka/

**Table 1.** Characteristics of the enzyme dataset used in the experiments

| Enzyme class level | Number of enzymes | Number of classes |
|---|---|---|
| EC.X | 11,493 | 6 |
| EC.X.X | 11,455 | 56 |
| EC.X.X.X | 10,094 | 178 |
| EC.X.X.X.X | 7,725 | 1036 |

extracted. Proteins having an EC number are enzymes and the code is provided by IUBMB (International Union of Biochemistry and Molecular Biology). From a data mining viewpoint, each EC number corresponds to a class, i.e., a specific protein function. More precisely, the EC number consists of four digits, where each pair of adjacent digits is separated by a dot ("."), and it specifies the chemical reaction catalyzed by the corresponding enzyme. For instance, the enzyme Alcohol dehydrogenase has the number EC.1.1.1.1.

Note that this is a four-level hierarchical classification, so that the first digit represents the most general class and the last digit the most specific subclass. In this work we report results of enzyme classification at the four levels, one-level-at-a-time. That is, first an experiment is carried out to classify enzymes at the first level (i.e. predict the first EC code digit); then a separate, independent experiment is carried out to classify enzymes at the second level; and so on, until experiment four that predicts the fourth EC code digit. Ultimately, these experiments aim at predicting the functional class of enzymes.

It should be noted that these experiments are more extensive than other experiments with enzyme classification reported in the literature, where sometimes just the first EC code digit is predicted – see e.g. [20], [2] and [1]. A direct comparison between different methods is not possible due to the differences in the data sets used and in the methodology for computing results. Also, presumably, the main reason why the literature sometimes focuses on just the first-level classification is that, as we consider deeper levels of classification, the number of examples per class becomes smaller and smaller, so that the classification problem becomes harder and harder. Indeed, in the enzyme data set, many classes at the third and fourth level have so few examples (less than 10) that they can hardly be predicted with a reasonable accuracy. This is not a limitation associated with any classification algorithm by itself, it is simply a limitation associated with the data being mined. Therefore, as part of our data preparation procedure, we have retrieved from PDB only the enzymes belonging to subclasses with at least 10 elements. After this simple filtering, the total number of enzymes retrieved from PDB and the corresponding number of classes are shown in Table 1 for all four levels of classification.

### 3.2 Running Parameters

Preliminary experiments were done to adjust the parameters of the algorithm. These experiments were done evolving motifs with the primary structure of proteins. As a result of the preliminary experiments, parameters were adjusted as follows: population size = 500; number of generations = 20; tournament size = 3% (of the population size); probability of structural crossover = 60%; probability of structural mutation = 60%; probability of leaf crossover = 20%; probability of leaf mutation = 80%; probability of hill climbing = 40%; number of discovered rules per class ($M$) = 10; stopping criterion = maximum number of generations.

From now on these parameter values will be referred to as default values, and they were used in all other experiments reported in this paper. It should be stressed that these preliminary experiments were not exhaustive and it is possible that other set of running parameters could lead to better performance. Actually, in general, adjusting running parameters in evolutionary computation systems is still an open issue. As a consequence of the lack of a widely accepted methodology for adjusting parameters, current research is pointing to self-adapting strategies [13]. Possibly, this issue will be addressed in future developments.

### 3.3 Comparative Results for Motif Discovery

In order to evaluate the performance of HEADMOP, we have compared it with a Genetic Algorithm (GA) that was also designed for motif discovery. This GA was previously described in the literature, as GAMDI [18], and an improved version as GAMBIT [19] The main differences between GAMDI and HEADMOP, are as follows. In GAMDI each individual corresponds to a single motif, whereas in HEADMOP each individual corresponds to a rule consisting of logical operators that combine several motifs (one motif for each leaf node of the individual). Also, the motifs discovered by GAMDI have to be given, as predictor attributes, to another classification algorithm which, actually, will discover classification rules. By contrast, since each individual in HEADMOP already corresponds to a potentially complex rule, each rule it discovers can be used on its own, without the need for a classification algorithm. Notwithstanding, it is also possible to use each of the individuals returned by this system as a single predictor attribute to be given to a classification algorithm. This approach is particularly interesting to implement a fair comparison between the two evolutionary algorithms, since the motifs discovered by both are used by the same classification algorithm. Hence, we report two types of results for HEADMOP:

1. Results using the system as a stand-alone data mining algorithm – in which the system both discovers motif-based rules and uses those rules directly for enzyme classification.
2. Results when the system is used to discover motifs that are then used by another separate classification algorithm as predictor attributes.

Results of both experiments are compared with the results of GAMDI discovering motifs that are then used, as predictor attributes, by a separate classification algorithm. In order to make the comparison between HEADMOP and GAMDI as fair as possible, both the previously-mentioned experiment (2) and the experiments with GAMDI used the same classification algorithm, namely J48 [22], as explained earlier.

**Table 2.** Classification accuracy rates (%) on the test set

| Enzyme class level | GAMDI | HEADMOP | HEADMOP+J48 |
|---|---|---|---|
| EC.X | 83.32±1.34 | 78.25±0.65 | 87.82±1.31 |
| EC.X.X | 79.64±0.43 | 77.19±0.98 | 83.33±0.54 |
| EC.X.X.X | 72.54±1.41 | 78.79±0.74 | 79.10±0.56 |
| EC.X.X.X.X | 75.85±5.26 | 79.19±0.89 | 83.13±0.49 |

Many works in current literature deals with a two-classes problem, rather than a multiclass problem (such as the one approached in this work). Viewing a multi-class problem as a two-classes one, the cases of a given class could be considered as "positive" and the cases of all remaining classes together are considered "negative". Table 2 shows the accuracy rates obtained by the algorithms for each class level of the EC code hierarchical classification. In this table, numbers after the "±" symbol denote the corresponding standard deviations. All accuracy rates in this paper refer to the average predictive accuracy on a test set unseen during training, as measured by a 5-fold cross-validation procedure [22]. The test set was randomly chosen with 1/3 of the cases of the data set shown in Table 1, maintaining the proportion of classes. We choose to present only the accuracy rate as a measure of quality in order to make possible the comparison with other published work.

As shown in Table 2, using the HEADMOP system as a stand-alone data mining algorithm mixed results were obtained. That is, HEADMOP achieved classification accuracies lower than the GAMDI algorithm at the first and second class levels, but higher than GAMDI at the third and fourth class levels. The differences in accuracies are significant at the first, second and third class levels, considering that the corresponding standard deviation intervals do not overlap. The difference in accuracy between GAMDI and HEADMOP alone is not significant, however, at the fourth level, where the standard deviation intervals overlap.

A better result was obtained when HEADMOP was used only to produce the predictor attributes used by J48, as explained before. In this case HEADMOP+J48 outperformed both GAMDI and the use of HEADMOP alone for all class levels. Additionally, the differences in accuracy between HEADMOP+J48 and GAMDI were significant at all class levels. These results suggest that HEADMOP is actually discovering good motifs, encoded in the leaf nodes of individuals. However, the logical combination of the motifs represented in the leaf nodes – performed by the logical operations in the internal nodes of individuals – still leaves room for improvement. Indeed, the use of J48 to combine the motifs represented by HEADMOP leaf nodes seems a good way to implement such an improvement, since the accuracies of HEADMOP+J48 are clearly higher than the accuracies of HEADMOP alone in general. This result is consistent with the fact that J48 is the product of several decades of research in machine learning, as mentioned earlier. Anyway, HEADMOP still has the important merit of discovering high-quality motifs, without which J48 would be helpless – since, of course, J48 cannot discover protein motifs.

### 3.4 Comparative Results for Processing Time

It is also interesting to compare the computational time taken by GAMDI and HEADMOP. The result of this comparison is presented in Table 3, in the format *hours* : *minutes*. All experiments were run in PC desktops with AMD Athlon XP 2.4 processors.

**Table 3.** Total running time of the algorithms(hr:min)

| Enzyme class level | GAMDI | HEADMOP |
|---|---|---|
| EC.X | 03:42 | 0:39 |
| EC.X.X | 10:04 | 0:30 |
| EC.X.X.X | 16:06 | 1:04 |
| EC.X.X.X.X | 19:34 | 0:57 |

As can be observed in Table 3, HEADMOP is considerably faster than GAMDI. In the case of the enzyme data set used in this project, the differences in the processing time of the algorithms is not crucial, since the longest running time, taken by GAMDI, was shorter than one day. However, in much larger data sets the long processing time taken by GAMDI would probably be a serious limitation to the use of that algorithm, whilst the hybrid GP/GA system proposed in this paper seems much more scalable to larger data sets. Further research will also focus on the parallelization of these algorithms.

## 4 Conclusions

We have proposed a hybrid Genetic Programming/Genetic Algorithm system for rule discovery, aiming at the automatic functional classification of proteins with unknown function. The system, named HEADMOP, was evaluated using an enzyme data set extracted from the Protein Data Bank. The solution returned by the proposed system consists of a set of rules (individuals), each of them using logical operators to combine a set of motifs and predicting a certain class for all enzymes satisfying the rule represented by the individual.

More precisely, each solution returned by HEADMOP consists of two major components evolved by the algorithm, namely, a set of protein motifs in the leaf nodes and a set of logical operators in the internal nodes of the individual. Hence, it is desirable to do controlled experiments evaluating the effectiveness of each these two solution components separately. In this spirit, we evaluated the results of the system when it was used in two different ways. In the first approach HEADMOP was used as a stand-alone classification algorithm, so that both kinds of evolved solution components (the set of motifs and the logical operators combining the motifs) were used to predict the class of an enzyme. In the second approach, HEADMOP was used only as an "attribute construction" algorithm, in the sense that only the motifs discovered by the system (and not the logical operators) were used. In this case the discovered motifs were used as attributes by J48, a standard classification algorithm based on decision trees.

The results of both approaches for using the solution returned by HEADMOP were compared with the results of a previous work using GA (GAMDI) that was also especially designed for the discovery of protein motifs. Unlike HEADMOP, GAMDI evolves only motifs, and not logical operators combining motifs. Hence, the motifs discovered by GAMDI have to be given (as attributes) to a classification algorithm. Again, J48 was used, to make the comparison between HEADMOP and GAMDI as fair as possible.

The performance measure used in this comparison was classification accuracy on a test set separated from the training set, as usual in the data mining literature. The two main findings from these experiments are as follows. First, there was no clear winner in the comparison between HEADMOP used as a stand-alone classification algorithm and J48 using motifs discovered by the GAMDI. Second, the results of J48 using the motifs discovered by HEADMOP were clearly better than the results of J48 using the motifs discovered by the former GA-based system. Hence, the general conclusion from these results is that the hybrid GP/GA system is evolving good sets of motifs in the leaf nodes of the individuals, but the combination of those motifs via the logical operators in the internal nodes is not fully effective – suggesting more research in this topic in the future.

It is intended to develop a more sophisticated version of HEADMOP where the motifs use information about the secondary structure of proteins. We believe that incorporating more information in the motifs can improve the classification accuracy of the system, especially covering atypical proteins, considered as small disjuncts [21] in the classification task. Authors intend to make such version freely available in the internet so as to foster further research in the area.

Results, in general, encourages the continuity of the research. Future work will include the use of HEADMOP with other categories of proteins, such as the globins, or else, the use the obtained classification rules to determine the functional class of recently discovered proteins, not yet classified. Overall, we believe that this work can be useful not only for biologists, but also for those working in this important field of Bioinformatics.

## References

1. Arakaki, A.K., Zhang, Y., Skolnick, J.: Large-scale assessment of the utility of low-resolution protein structures for biochemical function assignment. Bioinformatics 20, 1087–1096 (2004)
2. Ben-Hur, A., Brutlag, D.: Remote homology detection: a motif based approach. Bioinformatics 19, i26–i33 (2003)
3. Berman, H.M., Westbrook, J., Feng, Z., Gilliland, G., Bhat, T.N., Weissig, H., Shindyalov, I.N., Bourne, P.E.: The protein data bank. Nucleic Acids Research 28, 235–242 (2000)
4. Clare, A., King, R.D.: Machine learning of functional class from phenotype data. Bioinformatics 18, 160–166 (2002)

5. Fayyad, U.M., Piatetsky-Shapiro, G., Smyth, P.: From data mining to knowledge discovery: an overview. In: Fayyad, U.M., et al. (eds.) Advances in Knowledge Discovery and Data Mining, pp. 1–34 (1996)
6. Goldberg, D.E.: Genetic Algorithms in Search, Optimization & Machine Learning. Addison-Wesley, Reading (1989)
7. Han, J., Kamber, M.: Data Mining: Concepts and Techniques. Morgan Kaufmann, San Francisco (2001)
8. Holden, N., Freitas, A.A.: Hierarchical classification of protein function with ensembles of rules and particle swarm optimisation. Soft Computing 13, 259–272 (2009)
9. King, R.D., Karwath, A., Clare, A., Dehaspe, L.: The Utility of different representations of protein sequence for predicting functional class. Bioinformatics 17, 445–454 (2001)
10. Koza, J.R.: Genetic Programming: on the Programming of Computers by Means of Natural Selection. MIT Press, Cambridge (1992)
11. Kretschmann, K., Fleischmann, W., Apweiler, R.: Automatic rule generation for protein annotation with the C4.5 data mining algorithm applied on Swiss-Prot. Bioinformatics 17, 920–926 (2001)
12. Manning, A.M., Brass, A., Goble, C.A., Keane, J.A.: Clustering techniques in biological sequence analysis. In: Proceedings of the 1st European Symposium on Principles of Data Mining and Knowledge Discovery, pp. 315–322 (1997)
13. Maruo, M.H., Lopes, H.S., Delgado, M.R.B.S.: Self-adapting evolutionary parameters: encoding aspects for combinatorial optimization problems. In: Raidl, G.R., Gottlieb, J. (eds.) EvoCOP 2005. LNCS, vol. 3448, pp. 154–165. Springer, Heidelberg (2005)
14. Mirkin, B., Ritter, O.: A Feature-based approach to discrimination and prediction of protein folding groups. In: Suhai, S. (ed.) Genomics and Proteomics: Functional and Computational Aspects, pp. 157–177. Kluwer, Dordrecht (2000)
15. Murzin, A.G., Brenner, S.E., Hubbard, T., Chothia, C.: A structural classification of proteins database for the investigation of sequences and structures. Journal of Molecular Biology 247, 536–540 (1995)
16. Quinlan, J.R.: C4.5: Programs for Machine Learning. Morgan Kaufmann, San Francisco (1993)
17. Sebban, M., Mokrousov, I., Rastogi, N., Sola, C.: A data mining approach to spacer oligonucleotide typing of mycobacterium tuberculosis. Bioinformatics 18, 235–243 (2002)
18. Tsunoda, D.F., Lopes, H.S.: Automatic motif discovery in an enzyme database using a genetic algorithm-based approach. Soft Computing 10, 325–330 (2006)
19. Tsunoda, D.F., Lopes, H.S., Freitas, A.A.: An evolutionary approach for motif discovery and transmembrane protein classification. In: Rothlauf, F., Branke, J., Cagnoni, S., Corne, D.W., Drechsler, R., Jin, Y., Machado, P., Marchiori, E., Romero, J., Smith, G.D., Squillero, G. (eds.) EvoWorkshops 2005. LNCS, vol. 3449, pp. 105–114. Springer, Heidelberg (2005)
20. Weinert, W.R., Lopes, H.S.: Neural networks for protein classification. Applied Bioinformatics 3, 41–48 (2004)
21. Weiss, G.M.: Learning with rare cases and small disjuncts. In: Proc. of Twelfth International Conference on Machine Learning, pp. 558–565 (1995)
22. Witten, I.H., Frank, E.: Data Mining: Practical Machine Learning Tools and Techniques, 2nd edn. Morgan Kaufmann, San Francisco (2005)

# A Family of GEP-Induced Ensemble Classifiers

Joanna Jędrzejowicz[1] and Piotr Jędrzejowicz[2]

[1] Institute of Informatics, Gdańsk University,
Wita Stwosza 57, 80-952 Gdańsk, Poland
jj@inf.univ.gda.pl

[2] Department of Information Systems, Gdynia Maritime University,
Morska 83, 81-225 Gdynia, Poland
pj@am.gdynia.pl

**Abstract.** The paper proposes applying Gene Expression Programming (GEP) to induce ensemble classifiers. Four algorithms inducing such classifiers are proposed. The first one, denoted GEPA, based on the Adaboost method, is the two-class specific. The second, denoted MV is based on majority voting learning. Third one, denoted MVI, assumes incremental learning where for some classes more genes may be needed than for other ones. Finally, the last one denoted MVC involves partitioning of the training dataset into clusters prior to expression trees induction. The proposed algorithms were validated experimentally using several datasets.

**Keywords:** Gene expression programming, classification.

## 1 Introduction

Gene expression programming introduced by Ferreira [3] is an automatic programming approach. In GEP computer programs are represented as linear character strings of fixed-length called chromosomes which, in the subsequent fitness evaluation, can be expressed as expression trees of different sizes and shapes. The approach has flexibility and power to explore the entire search space, which comes from the separation of genotype and phenotype. As it has been observed by Ferreira [4] GEP can be used to design decision trees, with the advantage that all the decisions concerning the growth of the tree are made by the algorithm itself without any human input, that is, the growth of the tree is totally determined and refined by evolution.

The ability of GEP to generate decision trees makes it a natural tool for solving classification problems. Ferreira [4] showed several example applications of GEP including classification. Weinert and Lopes [15] apply GEP to the data mining task of classification by inducing rules. The authors proposed a new method for rule encoding and genetic operators that preserve rule integrity. They also implemented a system, named GEPCLASS which allows for the automatic discovery of flexible rules, better fitted to data. Duan with co-authors [2] claimed to improve efficiency of GEP used as a classification tool. Their contribution includes proposing new strategies for generating the classification threshold dynamically and designing a new approach called Distance Guided Evolution Algorithm.

N.T. Nguyen, R. Kowalczyk, and S.-M. Chen (Eds.): ICCCI 2009, LNAI 5796, pp. 641–652, 2009.

Zeng with co-authors [16] proposed a novel Immune Gene Expression Programming as a tool for rule mining. Another approach to evolving classification rules with gene expression programming was proposed in [17]. A different example of GEP application to classification problems was proposed by Li and co-authors [12]. They proposed a new representation scheme based on prefix notation which brings some advantages as compared with the traditional approach. Wang and co-authors [14] proposed a GEP decision tree system. The system can construct decision tree for classification without prior knowledge about the distribution of data. Karakasis and Stafylopatis [9] proposed a hybrid evolutionary technique for data mining tasks, which combines the Clonal Selection Principle with Gene Expression Programming. The authors claim that their approach outperforms GEP in terms of convergence rate and computational efficiency.

In this paper we propose a family of GEP-induced ensemble classifiers. Ensemble methods first solve a classification problem by creating multiple learners each attempting to solve the task independently, then use a procedure specified by the particular ensemble method for selecting or integrating the individual learners. In section 2 four ensemble classifiers induced by gene expression programming are proposed and discussed. In section 3 the example set of rules derived using one of the proposed ensembles is shown with a view to present possibility of practical interpretation of the approach. In section 4 the results of validating experiment are shown. Finally, section 5 contains conclusions.

## 2 Using Gene Expression Programming to Induce Ensemble Classifiers

In [8] GEP was used for two-class classification of data with numeric and categorical attributes. The idea was to induce expression trees which correspond to rules. Data that satisfied the rule were classified as the first class and those for which the rule did not work - as the second class. In this paper the general data classification problem is considered with no restriction to two classes. In what follows $C$ is the set of categorical classes which are denoted $1, \dots, |C|$. We assume that the learning algorithm is provided with the training set $TD = \{< d, c > \mid d \in D, c \in C\} \subset D \times C$, where $D$ is the space of attribute vectors $d = (w_1^d, \dots, w_n^d)$ with $w_i^d$ being symbolic or numeric values. The learning algorithm is used to find the best possible approximation $\bar{f}$ of the unknown function $f$ such that $f(d) = c$. Then $\bar{f}$ can be used to find the class $\bar{c} = \bar{f}(\bar{d})$ for any $\bar{d} \in D - TD|D$.

As usual when applying GEP methodology, the algorithm uses a population of chromosomes, selects them according to fitness and introduces genetic variation using several genetic operators. Each chromosome is composed of a single gene (chromosome and gene mean the same in what follows) divided into two parts as in the original head-tail method [3]. The size of the head ($h$) is determined by the user with the suggested size not less than the number of attributes in the dataset. The size of the tail ($t$) is computed as $t = h(n-1)+1$ where $n$ is the largest arity found in the function set. In the computational experiments the functions are: logical AND, OR, XOR, NOR and NOT. Thus $n = 2$ and the size of the gene is $h + t = 2h + 1$. The terminal set contains triples $(op, attrib, const)$ where $op$ is one of relational operators $<$, $\leq$, $>$, $\geq$, $=$, $\neq$, $attrib$ is the attribute number, and finally $const$ is a value belonging to the domain of the attribute $attrib$. As usual in GEP, the tail part of a gene always contains terminals and head can have both, terminals and functions. Observe that in this model each gene is syntactically correct and corresponds to a valid expression. Each attribute can appear once, many times or not at all. This allows to define flexible characteristics like for example $(attribute1 > 0.57)$ AND $(attribute1 < 0.80)$. On the other hand, it can also introduce inconsistencies like for example $(attribute1 > 0.57)$ AND $(attribute1 < 0.40)$. This does not cause problems since a decision subtree corresponding to such a subexpression would evaluate it to $false$. Besides, when studying the structure of the best classifiers in our experiments the above inconsistencies did not appear.

The expression of genes is done in exactly the same manner as in all GEP systems. Consider the chromosome C with head=6, defined below.

| 0 | 1 | 2 | 3 | 4 | 5 | 6 | 7 | 8 | 9 |
|---|---|---|---|---|---|---|---|---|---|
| OR | AND | AND | $(>, 1, 0)$ | $(=, 2, 5)$ | $(>, 1, 2)$ | $(<, 1, 10)$ | $(>, 3, 0)$ | $(<, 3, 5)$ | $\cdots$ |

The start position (position 0) in the chromosome corresponds to the root of the expression tree (OR, in the example). Then, below each function branches are attached and there are as many of them as the arity of the function - 2 in our case. The following symbols in the chromosome are attached to the branches on a given level. The process is complete when each branch is completed with a terminal. The number of symbols from the chromosome to form the expression tree is denoted as the termination point. For the discussed example, the termination point is 7.

The algorithm for learning the best classifier using GEP works as follows. Suppose that a training dataset is given and each row in the dataset has a correct label representing the class. In the initial step the minimal and maximal value of each attribute is calculated and a random population of chromosomes is generated. For each gene the symbols in the head part are randomly selected from the set of functions AND, OR, NOT, XOR, NOR and the set of terminals of type $(op, attrib, const)$, where the value of $const$ is in the range of $attrib$. The symbols in the tail part are all terminals. To introduce variation in the population the following genetic operators are used:

- mutation,
- transposition of insertion sequence elements (IS transposition),
- root transposition (RIS transposition),
- one-point recombination,
- two-point recombination.

Mutation can occur anywhere in the chromosome. We consider one-point mutation which means that with a probability, called mutation rate, one symbol in a chromosome is changed. In case of a functional symbol it is replaced by another randomly selected function, otherwise for $g = (op, attrib, const)$ a random relational operator $op'$, an attribute $attrib'$ and a constant $const'$ in the range of $attrib'$ are selected. Note that mutation can change the respective expression tree since a function of one argument may be mutated into a function of two arguments, or vice versa.

Transposition stands for moving part of a chromosome to another location. Here we consider two kinds of transposable elements. In case of transposition of insertion sequence (IS) three values are randomly chosen: a position in the chromosome (start of IS), the length of the sequence and the target site in the **head** - a bond between two positions. For example consider the chromosome C defined above. Suppose that IS is defined as: start position=6, length=2, target=0. Then a cut is made in the bond defined by the target site (in the example between symbol 0 and 1), and the insertion sequence (the symbols from positions 6 and 7) is copied into the site of the insertion. The sequence downstream from the copied IS element loses, at the end of the head, as many symbols as the length of the transposon. The resulting chromosome is shown below:

| 0 | 1 | 2 | 3 | 4 | 5 | 6 | 7 | 8 | 9 |
|---|---|---|---|---|---|---|---|---|---|
| OR | $(<,1,10)$ | $(>,3,0)$ | AND | AND | $(>,1,0)$ ‖ | $(<,1,10)$ | $(>,3,0)$ | $(<,3,5)$ | $\cdots$ |

Observe that since the target site is in the head, the newly created individual is always syntactically correct though it can reshape the tree quite dramatically as in the above case. The termination point is 3 for the new chromosome.
In case of root transposition, a position in the head is randomly selected, the first function following this position is chosen - it is the start of the RIS element. If no function is found then no change is performed. The length of the insertion sequence is chosen. The insertion sequence is copied at the root position and at the same time the last symbols of the head (as many as RIS length) are deleted. For the chromosome C defined before and RIS sequence starting with the function AND at the position 2, of length 2 the resulting chromosome is defined as:

| 0 | 1 | 2 | 3 | 4 | 5 | 6 | 7 | 8 | 9 |
|---|---|---|---|---|---|---|---|---|---|
| AND | $(>,1,0)$ | OR | AND | AND | $(>,1,0)$ ‖ | $(<,1,10)$ | $(>,3,0)$ | $(<,3,5)$ | $\cdots$ |

Again the change has quite an effect since the termination point is now 9.

For both kinds of recombination two parent chromosomes $P_1$, $P_2$ are randomly chosen and two new child chromosomes $C_1$, $C_2$ are formed. In case of one-point recombination one position is randomly generated and both parent chromosomes are split by this position into two parts. Child chromosomes $C_1$ (respectively, $C_2$) is formed as containing the first part from $P_1$ (respectively, $P_2$) and the second part from $P_2$ (and $P_1$). In two-point recombination two positions are randomly chosen and the symbols between recombination positions are exchanged between two parent chromosomes forming two new child chromosomes. Observe that again, in both cases, the newly formed chromosomes are syntactically correct no matter whether the recombination positions were taken from the head or tail.

During GEP learning, the individuals are selected and copied into the next generation based on their fitness and the roulette wheel sampling with elitism which guarantees the survival and cloning of the best gene in the next generation. More details of applied GEP learning can be found in [8].

To find ensemble classifiers four algorithms were used and compared. They differ in two aspects:

1. the learning process, that is the method for creating the population of genes,
2. the combination method used to output the final class depending on the values of genes from the population.

The learning stage is class-specific:

- the fitness function is class dependent: for class *cl* and gene $g$ the value of $fitness^{cl}(g)$ is defined as the difference between:
  the number of rows from class *cl* for which $g$ is 'true', and
  the number of rows from classes $\neq cl$ for which $g$ is 'true',
- learning takes place separately for each class $cl \in C$ and the first step of each algorithm is data preprocessing: the training data is divided into $|C|$ sets $TD_1, \ldots, TD_{|C|}$ containing separately data from different classes,
- all the algorithms make use of the algorithm GEP defined below.

For each of the algorithms the following parameters were used (on top of some other ones): P-the number of supporting genes (usually much smaller than the initial size of the population), NG - number of iterations; pm,pris,pis,pr1,pr2 - standing for the probability of mutation, RIS transposition, IS transposition, one-point and two-point recombination.
*Algorithm GEP class-oriented*
*Input: class cl, training data with correct labels representing $|C|$ classes,*
*Output: gene g (best fitting class cl)*

```
create genes of the initial population,
repeat NG  times the following:
1. express genes as expression trees,
   calculate fitness of each gene with respect to class cl,
2. keep best chromosome,
3. select chromosomes,
4. mutation,
5. transposition of insertion sequence elements (IS transposition),
6. root transposition (RIS transposition),
7. one-point recombination,
8. two-point recombination.
calculate fitness and keep the best gene g.
```

The first algorithm, denoted GEPA, is two-class specific and it is the ensemble learning using Adaboost of [8]. The other algorithms are MV, MVI and MVC.

*Algorithm MV (majority voting - learning)*
*Input: training data with correct labels representing $|C|$ classes divided into $|C|$ subsets, two parameters: supp_cl (support in class), supp_ncl (support outside class),*
*Output: population of $P$ genes for each class*

```
for each class i = 1,...,|C| and training data TD_i do
repeat
call algorithm GEP to find one gene g satisfying:
g is 'true' for at least supp_cl % of rows from class i
g is 'true' for not more than supp_ncl % of rows from classes other
than i
add g to the population for class i
until the population for class i contains at least P elements
end for i
```

*Algorithm MV - testing*
Given a new instance $d$ the decision profile is calculated:

$$DP(d) = \begin{pmatrix} g_1^1(d) & \dots & g_P^1(d) \\ \dots & \dots & \dots \\ g_1^i(d) & \dots & g_P^i(d) \\ \dots & \dots & \dots \\ g_1^{|C|}(d) & \dots & g_P^{|C|}(d) \end{pmatrix}$$

where the row $(g_1^i(d), \dots, g_P^i(d))$ is the support for class $i$. And $g_j^i(d) = 1$ if the expression tree for gene $g_j^i$ is 'true' for instance $d$, otherwise $g_j^i(d) = 0$. Observe that after an ideal learning, if $i$ is the right class for data $d$ then in matrix $DP(d)$ all the elements in row $i$ are 1 and 0 everywhere else.

To assign a class label $j$ to instance $d$, the entire decision profile $DP(d)$ is considered and

$$j = arg \max_{1 \le j \le |C|} (\sum_{k=1}^{P} g_k^j(d))$$

that is, the majority voting is performed among genes which evaluate to 'true'.

*Algorithm MVI (incremental - learning)*
*Input: training data with correct labels representing $|C|$ classes divided into $|C|$ subsets,*
*Output: population of at most $P$ genes for each class*

```
for each class i = 1,...,|C| and training data TD_i do
while TD_i ≠ ∅ and (population size does not exceed P) do
call algorithm GEP to find one best gene g for class i,
add g to the population
delete from data set TD_i those data for which g evaluates to 'true'
end while
end for i
```

In this case the decision profile matrix has a different number of meaningful elements in rows since in the process of learning for some classes more genes may be needed than for other ones. The procedure for assigning a class to data is as for Algorithm MV, except that the number of genes is not constant for classes.

*Algorithm MVC (cluster - learning)*
*Input: training data with correct labels representing $|C|$ classes divided into $|C|$ subsets, number of clusters k,*
*Output: population of P genes for each class and each cluster*

```
for each class i = 1,...,|C| do
use k-means algorithm to partition the data set TD_i into k clusters
CL^i_1,...,CL^i_k with the centroid CNT_i
for each cluster j = 1,...,k do
repeat P times
call algorithm GEP to find one gene g best fitting the class i with
data set CL^i_j
end for j
end for i
```

The k-means algorithm [6] is used for each class $i$ to partition the training data set $TD_i$ so that the resulting intercluster similarity is high but the intracluster similarity is low. Cluster similarity is measured in regard to the mean value of the objects in a cluster. Observe that in this case the population is a three dimensional structure since for each class $i$ the gene matrix contains $k \times P$ genes denoted $g^i_j$.

*Algorithm MVC - testing.*
For instance $d$ and each class $i$, $1 \leq i \leq |C|$ the nearest cluster $C_i^{j_i}$ is found, i.e

$$dist(d, C_i^{j_i}) = \min_{1 \leq l \leq k} (dist(d, C_i^l))$$

Then for each class $i$, genes generated for cluster $j_i$ make up for the decision profile:

$$DP(d) = \begin{pmatrix} g_1^{j_1}(d) & \dots & g_P^{j_1}(d) \\ \dots & \dots & \dots \\ g_1^{j_i}(d) & \dots & g_P^{j_i}(d) \\ \dots & \dots & \dots \\ g_1^{j_{|C|}}(d) & \dots & g_P^{j_{|C|}}(d) \end{pmatrix}$$

## 3 Practical Interpretation of GEP-Induced Ensemble Classifiers

Each GEP-induced ensemble classifier can be seen as a set of expression trees, which, in turn, can be easily transformed into classification rules. The following example of the set of rules was induced by the MVI ensemble in one of the

**Table 1.** Attributes of the Contraceptive Method Choice dataset

| Attribute name | Attribute type | Attribute value |
|---|---|---|
| 1. wife's age | numerical | |
| 2. wife's education | categorical | 1=low, 2, 3, 4=high |
| 3. husband's education | categorical | 1=low, 2, 3, 4=high |
| 4. no. of children ever born | numerical | |
| 5. wife's religion | binary | 0=non-Islam, 1=Islam |
| 6. wife's now working | binary | 0=yes, 1= no |
| 7. husband's occupation | categorical | 1, 2, 3, 4 |
| 8. standard-of-living index | categorical | 1=low, 2, 3, 4=high |
| 9. media exposure | binary | 0=low, 1=high |
| 10. contraceptive method | class attribute | 1=no-use, 2=long-term, 3=short-term |

experiment runs described in Section 4. These rules can be used to classify the Contraceptive Method Choice cases and were induced from the respective dataset. In Table 1 the dataset attribute names and possible values are shown while in Figure 1 the resulting set of rules is depicted.

## 4 Computational Experiment Results

The performance of the proposed family of GEP-induced ensemble classifiers was evaluated experimentally. The experiment involved the following medical datasets taken from the UCI Machine Learning Repository [1]: Wisconsin Breast Cancer (WBC), Cleveland Heart Disease (Heart), Lung Cancer (Lung), Contraceptive Method Choice (CMC), Ecoli, Pima Indians Diabetes (Pima). Basic characteristics of these datasets are shown in Table 2.

The experiment plan was based on ten independent repetitions of the ten-fold cross validation experiment for each benchmark dataset and each ensemble classifier (GEPA, MV, MVC, MVI). Performance criteria were percentage of correct classifications and its standard deviation averaged over all 100 runs. The experiment was carried out on a standard PC. Parameter settings for the considered GEP ensembles are shown in Table 3, the other were: probability of mutation pm=0.1,root transposition pris, transposition of insertion sequence elements pis, probability of one-point recombination pr1 and two-point recombination pr2 with

**Table 2.** Characteristics of the benchmark datasets

| dataset | no. cases | no. attributes | no. class |
|---|---|---|---|
| WBC | 699 | 9 | 2 |
| Heart | 303 | 14 | 2 |
| Lung | 32 | 56 | 2 |
| CMC | 1473 | 9 | 3 |
| Ecoli | 336 | 8 | 8 |
| PIMA | 768 | 8 | 2 |

**Class 1 No-use**
Gene 1:
husband's education $< 3$
Gene 2:
wife's age $\geq 44$
**Class 2 Long-term**
Gene 1:
husband's occupation $\geq 3$ AND husband's education $< 2$
Gene 2:
(wife's religion $= 1$ AND (media exposure $= 1$ AND (standard of living $\geq 3$ AND no of children $> 6$) OR (standard of living $<3$ AND no of children $\leq 6$)))
OR
(media exposure $= 0$ AND wife's education $< 3$ OR (husband's education $<3$ XOR wife's age $\geq 48$))
**Class 3 Short-term**
Gene 1:
(standard of living $< 3$ AND (husband's education $\leq 2$ OR media exposure $= 1$))
OR
(wife's age $\neq 31$ AND husband's education $\leq 2$ AND media exposure $= 0$)
OR (wife's education $= 4$ AND wife's working $= 0$)
Gene 2:
husband's education $\leq 2$ AND ( wife's age $\neq 46$ OR
(husband's occupation $\geq 3$ AND wife's age $\geq 47$))

**Fig. 1.** Set of rules induced by the MVI ensemble based on the Contraceptive Method Choice dataset

**Table 3.** GEP ensembles parameter setting

| ensemble type | dataset | population size/no.iter | values of other param. |
|---|---|---|---|
| GEPA | WBC | 100/400 | no.iter AdaBoost =40 |
| | Heart | 100/300 | |
| MV | Ecoli | 100/300 | supp_cl=0.8, supp_ncl=0.2, P=50 |
| MVC | Lung | 200/400 | P=20, no_clusters=3 |
| | PIMA | 100/500 | P=20, no_clusters=5 |
| MVI | CMC | 50/100 | P=10, no_repetitions=3 |

pris=pis=pr1=pr2=0.2. In Table 4 the performance of the ensemble classifiers is shown.

Table 5 contains comparison of the best GEP induced ensemble classifiers with several literature-reported solutions. It can be easily seen that the proposed ensembles as a rule perform very well. There are, of course some drawbacks. Each of the proposed algorithms requires setting values of several parameters including population size, number of iterations and probabilities of genetic operations. This requires some computational effort at the fine-tuning phase. On the other

**Table 4.** Performance of the proposed GEP ensembles

| Dataset | GEPA | | MV | | MVC | | MVI | |
|---|---|---|---|---|---|---|---|---|
| | % correct | st.dev | % correct | st.dev | % correct | st.dev | % correct | st.dev |
| WBC | **98.57** | 1.79 | 97.80 | 2.78 | 96.70 | 2.49 | 95.60 | 4.97 |
| Heart | **88.00** | 6.12 | 86.50 | 4.56 | 84.40 | 5.17 | 82.75 | 2.39 |
| Lung | - | - | 80.10 | 23.30 | **90.10** | 15.94 | 86.70 | 23.32 |
| CMC | - | - | 53.35 | 3.35 | 50.49 | 3.20 | **56.20** | 3.05 |
| Ecoli | - | - | **89.77** | 4.24 | 85.90 | 6.90 | 87.60 | 4.30 |
| PIMA | - | - | 75.20 | 4,32 | **76.70** | 4.99 | 74.90 | 4.28 |

**Table 5.** Best GEP ensemble results (% correct classifications) versus some literature reported results

| Dataset | The best GEP-induced ensemble | Type of the GEP-induced ensemble | Literature reported results | Method & source of the literature |
|---|---|---|---|---|
| WBC | 98.57 | GEPA | 98.50 | GEPCLASS (5CV) [15] |
| | | | 96.20 | GEP rules (5CV) [17] |
| | | | 97.00 | AdaBoost with LGG [5] |
| Heart | 88.00 | GEPA | 100.00 | GEP rules (5CV) [17] |
| | | | 85.10 | 28-NN [13] |
| | | | 77.50 | C4.5 [9] |
| Lung | 90.10 | MVC | 77.00 | KNN [7] |
| | | | 57.62 | GEP rules (5CV) [17] |
| CMC | 56.20 | MVI | 55.70 | GDT-MA [11] |
| | | | 55.29 | ADTree [5] |
| | | | 97.00 | C4.5 [5] |
| Ecoli | 89.77 | MV | 84.4' | ADTree [5] |
| | | | 84.11 | C4.5 [5] |
| PIMA | 76.70 | MVC | 75.40 | GDT [10] |
| | | | 75.10 | GEP rules (5CV) [17] |
| | | | 74.60 | C4.5 [10] |

hand, we have observed that apart from the population size and the number of iterations the remaining parameters do not play a decisive role from the point of view of the efficiency of computations and classifier performance.

## 5 Conclusions

The presented research allows to draw the following conclusions:

- Gene expression programming is a versatile and useful tool to automatically induce expression trees.
- Using GEP-induced expression trees allows for construction of a high quality ensemble classifiers competitive, in terms of classification accuracy, to many other recently published solutions.

- High quality of the ensemble classifier performance can be attributed to the expression trees induced by GEP.
- Expression trees induced GEP can be easily converted into sets of rules easy to understand and interpret.

Future research should focus on extending the approach through integration with the data reduction algorithms to achieve better performance and reduce computation time required.

## References

1. Asuncion, A., Newman, D.J.: UCI Machine Learning Repository University of California, School of Information and Computer Science (2007), `http://www.ics.uci.edu/~mlearn/MLRepository.html`
2. Duan, L., Tang, C., Zhang, T., Wei, D., Zhang, H.: Distance guided classification with gene expression programming. In: Li, X., Zaïane, O.R., Li, Z.-h. (eds.) ADMA 2006. LNCS (LNAI), vol. 4093, pp. 239–246. Springer, Heidelberg (2006)
3. Ferreira, C.: Gene Expression Programming: a New Adaptive Algorithm for Solving Problems. Complex Systems 13(2), 87–129 (2001)
4. Ferreira, C.: Gene Expression Programming. Studies in Computational Intelligence 21, 337–380 (2006)
5. FT Datasets (2008), `http://www.grappa.univ-lille3.fr/torre/Recherche/Experiments/Results/tables.php` (accessed December 2008)
6. Hartigan, J.A., Wong, M.A.: A k-means clustering algorithm. Applied Statistics 28(1), 100–108 (1979)
7. Hong, Z.Q., Yang, J.Y.: Optimal Discriminant Plane for a Small Number of Samples and Design Method of Classifier on the Plane. Pattern Recognition 24(4), 317–324 (1991)
8. Jedrzejowicz, J., Jedrzejowicz, P.: GEP-induced expression trees as weak classifiers. In: Perner, P. (ed.) ICDM 2008. LNCS (LNAI), vol. 5077, pp. 129–141. Springer, Heidelberg (2008)
9. Karakasis, V.K., Stafylopatis, A.: Data Mining based on Gene Expression Programming and Clonal Selection. In: Proc. IEEE Congress on Evolutionary Computation, pp. 514–521 (2006)
10. Kretowski, M., Popczynski, P.: Global induction of decision trees: From parallel implementation to distributed evolution. In: Rutkowski, L., Tadeusiewicz, R., Zadeh, L.A., Zurada, J.M. (eds.) ICAISC 2008. LNCS (LNAI), vol. 5097, pp. 426–437. Springer, Heidelberg (2008)
11. Kretowski, M.: A memetic algorithm for global induction of decision trees. In: Geffert, V., Karhumäki, J., Bertoni, A., Preneel, B., Návrat, P., Bieliková, M. (eds.) SOFSEM 2008. LNCS, vol. 4910, pp. 531–540. Springer, Heidelberg (2008)
12. Li, X., Zhou, C., Xiao, W., Nelson, P.C.: Prefix Gene Expression Programming. In: Proc. Genetic and Evolutionary Computation Conference, Washington, pp. 25–31 (2005)
13. Statlog Datasets: comparison of results (2008), `http://www.is.umk.pl/projects/datasets.html#Cleveland` (accessed on December 2008)

14. Wang, W., Li, Q., Han, S., Lin, H.: A Preliminary Study on Constructing Decision Tree with Gene Expression Programming. In: Proc. First International Conference on Innovative Computing, Information and Control, vol. 1, pp. 222–225 (2006)
15. Weinert, W.R., Lopes, H.S.: GEPCLASS: A classification rule discovery tool using gene expression programming. In: Li, X., Zaïane, O.R., Li, Z.-h. (eds.) ADMA 2006. LNCS (LNAI), vol. 4093, pp. 871–880. Springer, Heidelberg (2006)
16. Zeng, T., Xiang, Y., Chen, P., Liu, Y.: A Model of Immune Gene Expression Programming for Rule Mining. Journal of Universal Computer Science 13(7), 1239–1252 (2007)
17. Zhou, C., Xiao, W., Tirpak, T.M., Nelson, P.C.: Evolving Accurate and Compact Classification Rules with Gene Expression Programming. IEEE Transactions on Evolutionary Computation 7(6), 519–531 (2003)

# Handling Dynamic Networks Using Ant Colony Optimization on a Distributed Architecture

Sorin Ilie and Costin Badica

Software Engineering Department, University of Craiova, Romania
{ilie_sorin,badica_costin}@software.ucv.ro

**Abstract.** Nowadays organizations are willing to share and cooperate in building better services and products. A distributed framework is needed to support these current trends. An ant colony metaphor is a great source of inspiration to build such a framework. This paper proposes a study of Ant Colony Optimization on handling dynamic networks. The novelty of our work consists in using a multi-agent architecture to model the dynamic network and artificial intelligence to decide on the type of ants needed. Our approach allows greater flexibility in adapting to network changes.

**Keywords:** Artificial Intelligence, Ant Colony Optimization, Handling Dynamic Networks, Multi-Agent Systems.

## 1 Introduction

This paper tackles the problem of finding *the shortest path in dynamic networks* using asynchronous message passing and local knowledge. A dynamic network contains a set of interconnected entities (nodes) that can change its configuration over time. These changes can be: (i) updates of the cost of traveling between adjacent nodes, and (ii) creation and/or destruction of nodes and their interconnections with the rest of the network. The nodes need to interact locally with each other following simple rules. Out of these simple decentralized interactions, a collective organized behavior will emerge. This is a type of artificial intelligence called *swarm intelligence*.

In this paper we introduce ACODA – *Ant Colony Optimization on a Distributed Architecture*. With ACODA the *nodes of the network are represented by software agents* and the communication between them will involve exchanges of messages following the Ant Colony Optimization (ACO) algorithm. Ants use pheromones to direct each other to their findings with no real physical contact. We propose to model artificial ants as software objects that are managed by agents of the network. Ant migration will involve messages exchanged between the agents. Messages will contain the information needed to simulate the traveling of ants between nodes, i.e. no real ant code mobility is necessary. Decisions and actions of ants are determined according to the values of their attributes, following the rules of ACO algorithms.

In order to simulate instinctual choices, we use a randomness factor such that the ants will be spread out evenly when searching. All ants will eventually return to the node they originated from (the *anthill*). Ants that reach the destination (*food source*) node as

N.T. Nguyen, R. Kowalczyk, and S.-M. Chen (Eds.): ICCCI 2009, LNAI 5796, pp. 653–664, 2009.

well as unsuccessful ants will be re-configured in order to achieve better results. Overall in our solution we propose a relatively simple design and implementation of a software agent that can be re-used to create a dynamic network of software agents that are able to collectively solve complex search problems using a distributed approach.

## 2 Background

### 2.1 Ant Colony Behavior

Natural ants have a very organized collective life. When they go searching for food, they mark the way back to the anthill with pheromones such that themselves and other ants as well will find the food source again. Other members of the colony will be attracted to the marked paths. The shortest routes are quickly refreshed; but the pheromones are volatile and evaporation erases long paths as well as those that are not of interest anymore. This process results in *shorter paths being favored over longer paths* as shown in the example from Fig.1.

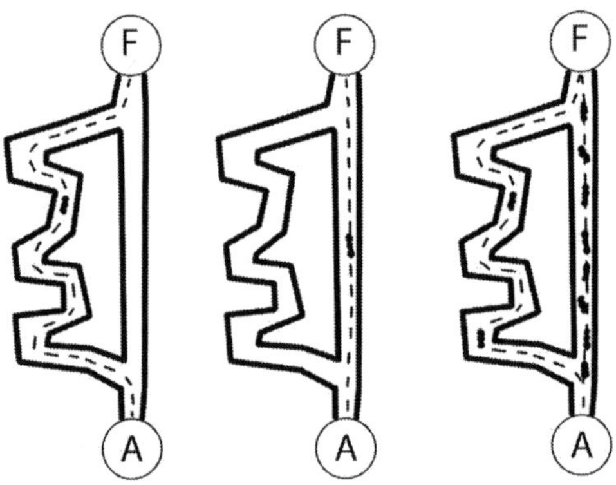


**Fig. 1.** Ants choosing the shortest path

Ants will deposit pheromones only on their way back to the anthill (A). In Fig.1 the ants that follow a pheromone trail to food (F) using the longer path on the left will take longer to get back to the anthill, thus allowing the deposited pheromone more time to evaporate before refreshing the pheromone concentration by returning home. On the contrary, the ants that randomly pick the shorter route on the right will leave less time for the evaporation. The more pheromone there is on one path the more it will be preferred. Intuitively, *ants will converge towards the most efficient path* because that path gets the strongest concentration of pheromone. As soon as they exhaust the food source, the ants will resume their instinct guided search for another source, so the colony is *adaptable to changes in the environment*.

We can identify positive feedback in the form of the pheromone and negative feedback in the form of evaporation. Ants use the environment (i.e. the network of software agents in our solution) as a medium of communication.

Note that this ACO approach does not guarantee the computation of the best solution; however, in most situations the computed solution will be in the close vicinity of the best solution. From a practical point of view, that is good enough, as proven by the thriving ant populations.

## 2.2 The ACO Algorithm

Ant colony algorithms are part of the more general concept of swarm intelligence. They can be used to solve problems that can be reduced to finding paths between two nodes of a weighted graph. Artificial ants are programmed to mimic the behavior of real ants while searching for food. Generally random solutions are found that are improved and combined in order to produce more viable ones. The shortest path problem fits well within ACO's capabilities.

Existing implementations invest all decisional abilities in the ants. These are programmed to move through the graph and, when reaching the destination, to mark with pheromones the path on their way back to the anthill. The intensity of the pheromones needs to be inversely proportional to the path length; that way, the shorter paths will receive more pheromone. Ants need to remember not only the way back to the anthill, but the costs of the edges on that way as well. We have addressed this problem by adding a *lifespan attribute* to our ants. Every cost unit in the path decreases the ant's life by one unit. Knowing the ant's age we also know the path's length. The lifespan can also be used to favor shorter paths at the beginning of the search.

Existing implementations of the shortest path problem with ACO simulate evaporation by letting the ants decrease the amount of pheromone on the edges they are about to mark. In our approach we propose *a more realistic form of evaporation*: the pheromone deposited on the edges is decreased with the help of a separate ''evaporation behavior'' that is defined by every software agent of the network. This behavior is run in parallel with the other behaviors of our agents and it is periodically activated with a given frequency.

Another problem with implementing the artificial intelligence in the ants is the pheromone update in the case of fine-grained parallel implementations (one process per ant), as the ants end up spending most of their time communicating the changes made. *Our implementation is already parallel by nature*. When a returning ant is received by a software agent representing a network node, it will also mark with the ant's pheromone the neighbor node which sent it. Therefore no communication overhead is needed. Note that our nodes are software agents and therefore they can be distributed on an arbitrary number of machines, while multiple agents are still allowed to run on the same machine, depending on the network size and on the number of available machines.

In order to favor the flexibility required by the shortest path problem in dynamic networks some authors (see [4]) introduced evolution and lifespan in ACO, achieving superior results in avoidance of local optima. In our approach we introduce an adjustable randomness factor that, depending on its value, guides the ant either to more rationally guided paths or to more random paths. Keeping a minimum value to this factor we ensure that some ants will always have the chance to look for better solutions.

Note that our approach has several advantages over existing proposals. These advantages can be summarized as follows: more natural distributed representation of edge costs rather than usual centralized approach, less overhead than parallel implementations and more realistic evaporation and adaptation to changes. Adding the fact that our implementation respects typical ant behavior patterns makes our architecture to worth a closer look and motivates our work.

# 3 The Ant Colony Optimization on a Distributed Architecture (ACODA) Algorithm

## 3.1 Theoretical Background

We now present the mathematical engine behind our algorithm. The function that determines the most lucrative hop is the same as in other ACO implementations. A move from node $i$ to node $j$ has the probability:

$$p_{i,j} = \frac{(\tau_{i,j}^{\alpha})(\eta_{i,j}^{\beta})}{\sum(\tau_{i,j}^{\alpha})(\eta_{i,j}^{\beta})} \tag{1}$$

where:

$\tau_{i,j}$ is the amount of pheromone on edge $(i, j)$

$\alpha$ is a parameter to control the influence of $\tau_{i,j}$

$\eta_{i,j}$ is the desirability of edge $(i, j)$

$\beta$ is a parameter to control the influence of $\eta_{i,j}$

We also introduced a random number that allows the temperament tweaking of the ant who can be more of a gatherer (sticking to the $p$ calculated with (1) ) or more of a scout (searches randomly for the destination). This is particularly handy when trying to free oneself from local optima or getting an even spread at the beginning.

$$P_{i,j}=(1-r_f)p_{i,j}+r_f*R \tag{2}$$

where:

$P_{i,j}$ is the final probability of an ant moving from node $i$ to node $j$

$r_f$ is the randomness factor

$p_{i,j}$ is the conservative probability defined at (1)

$R$ is a random number

Pheromone deposits need to be recalculated locally for each node-agent to simulate evaporation.

$$\tau_{i,j}(t) = (1 - \rho)\tau_{i,j}(t - 1), \tag{3}$$

where:

$\tau_{i,j}(t)$ is the amount of pheromone on edge $(i, j)$ at time $t$

$\rho$ is the evaporation rate $0 \leq \rho < 1$

The pheromone deposited on an edge is inversely proportional on the length of the path found.

$$\Delta\tau_{i,j}^{k} = \begin{cases} 1/L_k & \text{if ant } k \text{ travels on edge } i,j \\ 0 & \text{otherwise} \end{cases} \quad (4)$$

where:

$L_k$ is the cost of the $k$-th ant's path. As our ants have the lifespan attribute and they age with every cost unit of their current path, $L_k$ is already readily available.

### 3.2 Overview of the Algorithm

The nodes of the network are implemented as software agents. There are many definitions of the agent concept [10]. However, for the purpose of our project we understand by agent an autonomous software entity that (i) has its own thread of control and can decide autonomously if and when to perform a given action; (ii) communicates with other agents by asynchronous message passing. Each agent is referenced using its name, also known as agent id.

The actual program performed by a given agent is given as a set of behaviors. A behavior is defined as a sequence of primitive actions. Behaviors are executed in parallel using interleaving of actions on the agent's thread with the help of a non-preemptive scheduler, internal to the agent [11].

For each agent we defined three behaviors that are run while the agent is active: *receiveAnt*(), *moveAnt*() and *evaporate*(). An agent also manages a list of agents representing neighboring nodes and a list *antQueue* of the currently ''visiting'' ants. For each neighbor $N$ in the list we record its agent id together with the cost and pheromone deposit of the arc connecting the current node with $N$.

The ants are modeled as software objects with the following attributes: *lifespan*, *pheromone strength*, *randomness factor*, *returning flag*, *goal reached flag* and a list of agent ids representing the path the ant followed to reach its current location. Note that ants do not have to migrate between nodes; transporting their knowledge and attributes is sufficient.

The *antQueue* is accessible to both *receiveAnt*() and *moveAnt*() behaviors (see Fig.2 for visual aid). When an ant arrives in a node it is added to the *antQueue* with the help of *receiveAnt*() behavior. However, beforehand this behavior checks if the current node is the destination of the ant. If yes then the behavior *receiveAnt*() tweaks the ant, i.e. it updates its pheromone strength, sets the goal reached flag and sets the ant to returning mode. Otherwise, if the ant is already returning and the goal has been reached then its pheromone is deposited on the arc connecting the sender node with the current node. Note that no message exchanges are needed to update pheromone deposits. The update is done locally using the list of neighboring nodes. Once the ant arrives at the destination or runs out of life it is set to return to the anthill by following the memorized path in its local memory.

The behavior *moveAnt*() determines the id of the agent representing the network node where the ant on the top of the queue of the current node must be transported. This behavior utilizes the ants' attributes and also either consults: (i) the list of neighbors of the current node if the ant has not yet reached its destination or (ii) the ant's path stored in its local memory if the ant was set to returning mode. If the ant is not returning then *moveAnt*() behavior adds the agent id of the current node to the

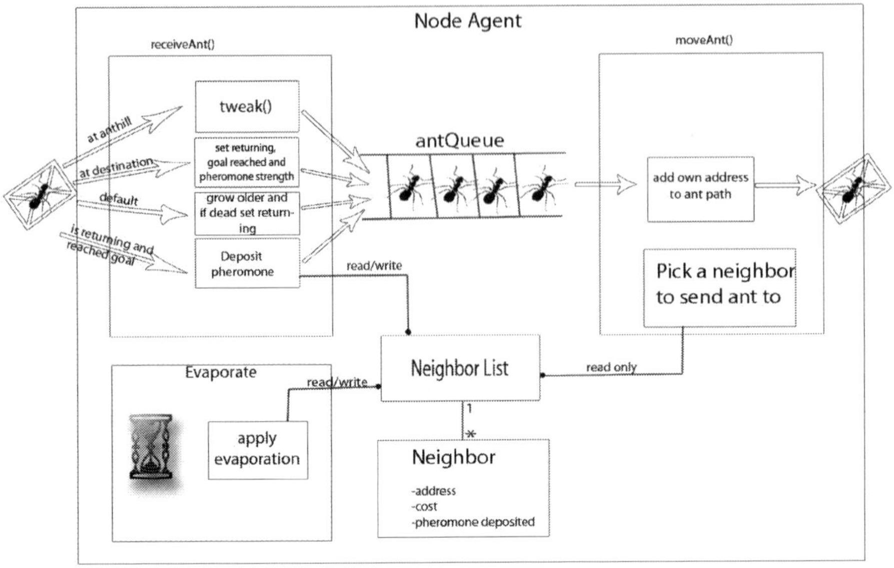


**Fig. 2.** Structure of an agent representing a network node

path that is currently stored by the ant. The transportation of the ant to the next node in the network is simulated by sending a message to the agent representing that node as well as by deleting the ant from the queue of the current node. The message will be received and processed by the *receiveAnt*() behavior of the neighbor node. The behavior *moveAnt*() does not stop until the ant queue becomes empty. Therefore the agent representing the anthill node can easily introduce new ants or re-introduce returned (tweaked) ants by adding them to its internal queue. Note that the *moveAnt*() behavior is set to wait an amount of time proportional to the cost of the edge (distance between current node and next node) before migrating the ant to simulate physical traversal of this distance; this is vital for the evaporation process to be effective.

One of the network nodes plays the role of the anthill. So it creates the ant population deciding on ants' initial attribute values and uses the *tweak*() method to modify returning ants. Ants that did not reach the destination get their randomness factor increased while successful ants receive a pheromone bonus and get their randomness factor decreased.

The other special node in the network is the destination node which calculates pheromone strength for the ants according to the path length (see formula 4) and sets the goal reached flag. All other nodes just pass ants between themselves depositing pheromone when needed, decreasing ant lives and setting the return flag of dead ants.

Evaporation is implemented by a separate ticker behavior [11]. This behavior is activated periodically to decrease the amount of pheromone that was deposited on each arc adjacent out from the current node by updating its value in the list of neighbor nodes.

For further clarification we have included sketches for the behaviors in the form of Algorithms 1, 2, 3 and 4 (see below).

```
behavior moveAnt
  if (!antQueue.isEmpty())
              if (antQueue.getFirst().isReturning())
                sendTo(antQueue.getFirst().retraceSteps());
              else
                sendTo(advantageousNeighbor(antQueue.getFirst()));
  antQueue.removeFirst();
```

**Algorithm 1.** The node agent's *moveAnt*() behavior sketch

Algorithm 1 is using a series of functions as follows: (i) *isReturning*() asks the ant if it is going back to the anthill or it is searching for food and returns true if the ant is returning to the anthill, (ii) *retraceSteps*() returns the next node back towards the anthill, (iii) *sendTo*(*node*) moves the ant to *node* by sending the ant's attributes and knowledge to the agent *node*, (iv) *advantageousNeighbor*(*ant*) returns the address of the neighbor with the highest $P_{i,j}$ value according to formula (2).

```
behavior receiveAnt
  Ant a=receiveAttributesAndKnowledge();
  if (atAnthill())
              tweak();
  else {
              if (a.isAlive())
                if (atDestination()) {
                  a.setGoalReached();
                  a.setReturning();
                  a.rememberTrack(getOwnAdress());
                } else
                  if (a.isReturning()&& a.goalReached())
                    applyPheromoneToSender(a);
                  else a.growOlder();
              else a.setReturning();
              antQueue.addLast(new Ant(a));
  }
```

**Algorithm 2.** The node agent's *receiveAnt*() behavior sketch

Algorithm 2 is using a series of functions as follows: (i) *receiveAttributesAndKnowledge*() returns the ant received from a neighbor node, (ii) *a.setGoalReached*() tells to ant *a* that it was successful, i.e. goal node was reached, (iii) *a.rememberTrack*(*getOwnAdress*()) makes the ant *a* to remember its current traveling path by recording the current node, (iv) *applyPheromoneToSender(a)* applies the ant's pheromone on the arc connecting the current node with the sender node, (v) *a.growOlder()* tells the ant to age with one unit, (vi) *a.setReturning()* tells the ant to go back to the anthill, and (vii) *atAnthill()* returns true if the ant arrived back to the anthill.

```
function tweak(Ant a)
  returnedAnts++;
  if (a.getGoalReached()) {
          ant.increaseFeromoneStrength();
          ant.decreaseRandomnessFactor();
          goalReached=true;
  } else
          ant.increaseRandomnessFactor();
  ant.clearMemory();
  antQueue.addFirst(ant);
  if (returnedAnts==totalAntPopulation && !goalReached) {
          antLifespan++;
          returnedAnts=0;
  }
```

**Algorithm 3.** Sketch of anthill's *tweak(Ant ant)* method

Algorithm 3 is using a series of functions as follows: (i) *ant.clearMemory()* resets the ant's lifespan, path queue and all flags such including "retuning" and "goalReached", and (ii) *antQueue.addFirst(ant)* the *ant* is added to the queue and it is prepared to be released again.

```
behavior aplyEvaporation()
  doWait(period);
  for (Neighbor n: neighborList)
          applyEvaporation(n);
```

**Algorithm 4**. The node agent's *applyEvaporation()* behavior sketch. It decreases neighbors' quantity of deposited pheromone at every *period* of milliseconds.

Summarizing, we have designed and implemented a society of node agents that collaborate by exchanging ant information according to the ACO algorithm. The anthill agent receives only the result of the search as well as the returning ants which are then modified and released again in order to achieve better solutions. This relatively simple code is designed to generate viable results by acting as a society.

## 4 Experiments

We have implemented the system using JADE agent platform [11]. All agents use the same code base, excepting the setup code which differs for the anthill as it must create and initialize the ant population.

The initialization of the ants is a very important step as their attributes and the population size will influence the results. A too small randomness factor will bunch the ants together at the beginning possibly missing good solutions altogether. A very high lifespan might guide the ants towards an inefficient solution. However these situations should be recoverable. Differently, too low (or too high) default pheromone strength and too low ant population size are more problematic, as they can render the evaporation useless.

We tested the system on the graph shown in Fig.3 that has also been used for experiments in paper [4].

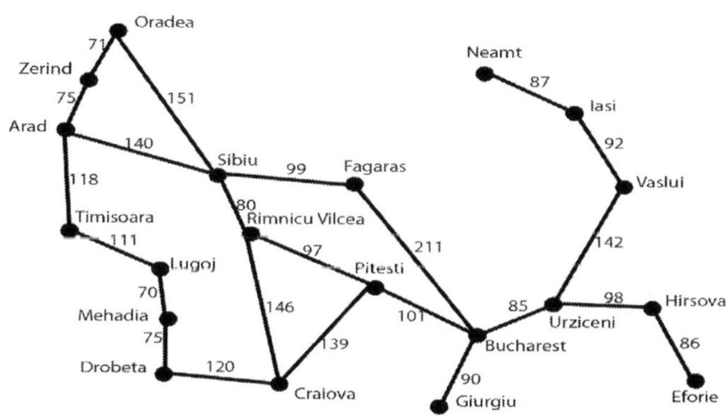


**Fig. 3.** Graph with map of Romania

For the experiment setup, all the necessary agents were created manually from a batch file containing the following command:

*java jade.Boot Oradea:node.Node Zerind:node.Node Arad:node.Node Timisoara:node.Node Lugoj:node.Node Mehadia:node.Node Drobeta:node.Node Craiova:node.Node RimnicuVilcea:node.Node Sibiu:node.Node Fagaras:node.Node Pitesti:node.Node Bucuresti:node.Node Giurgiu:node.Node Urziceni:node.Node Hirsova:node.Node Vaslui:node.Node Iasi:node.Node Neamt:node.Node*

where *Node* is the class describing the node software agent. JADE software agents have a *setup*() method that is run only once during their startup. We have used this method to initialize the agents' neighbor lists with values read from a configuration file and to create and initialize the agents' behaviors.

We have used *log4j* library to create log files for each software agent in the system where we have captured experimental data. Probably the most interesting log file is that corresponding to the destination agent that contains information about all the successful ants and their solution paths. Other nodes' logs contain the ants that die in their custody. We also let the anthill agent to count the total number *NMAX* of ant migrations originating at the anthill to get a sense of the effort needed to reach a solution.

Given the randomness of their movements, usually the ants do not really agree on following the same path, while, however, most of them will be eventually carried over the same path as the value of *NMAX* is increasing. In our experiments we set *NMAX* to 500. The criterion we use to identify the "convergence" of ants to the solution path was the detection of the most frequent path from the last $k = 20$ solutions found. This criterion was applied by analyzing the log file of the agent representing the destination node.

In our experiments we set the anthill to Arad and the destination node to Bucharest. In this scenario the shortest path would be: Arad, Sibiu, Rimnicu Vilcea, Pitesti, Bucharest. The minimum value of the randomness factor $r_f$ was set to 5% in order to facilitate ants to discover new routes. We ran a total number of 50 tests on a single computer in which the anthill generated 100 ants with the lifespan of 500, allowing the anthill to execute *NMAX* = 500 ant migrations before stopping each test. Evaporation was set to execute every 50 milliseconds. The algorithm found the best path for 80% of the found solutions. The second best path (Arad, Sibiu, Fagaras, Bucharest) was also found by the rest of the tests.

Using the same parameters (i.e. under the same experimental conditions) we allowed the anthill to execute *NMAX* = 500 ant migrations, then we changed the cost of the edge linking Fagaras to Bucharest from 211 to 50, and finally we let the anthill perform other *NMAX* = 500 ant migrations. Note that after this update of the network, the new best path was Arad, Sibiu, Fagaras, Bucharest. We ran 50 tests and found that for 70% of the situations the algorithm was able to recover itself from local optima. Note that in order to get correct percentages of actually recovering itself we repeated the tests in which initially the local optimum was the updated best path: Arad, Sibiu, Fagaras, Bucharest (as it was no point to count recovering from this trivial situation).

We believe that similar results can be reached if nodes together with their adjacent edges are allowed to be dynamically added on or removed from the map. Note however that experimenting with such tests is left as future work. We also expect that efficiency and scalability would improve as the number of machines on which agents are run will be increased.

## 5 Related Work

The ACO algorithm proposed in this paper is based on the Ant System (AS) presented in Dorigo's seminal work [3] and the technique for handling dynamic networks that was proposed by Roach and Menezes in [4].

AS was improved by Dorigo and Gambardella by using a pseudorandom probability rule, resulting Ant Colony System (ACS) [9]. ACS also introduced an offline pheromone update that was performed only by the best ant. However, in our context it is impractical to have the whole distributed system of agents waiting for one agent to finish its updates; so this was not an option for us.

Stützle and Hoos proposed the MAX-MIN ant system (MMAS) [6] in which only the best ant marks its path. In a distributed architecture one cannot easily determine which ant has found the best path without centralizing all data.

In [1], Botee and Bonabeu attempted to use genetic algorithms (GA) to tune the ACO algorithm. In [5], White, Pagurek and Oppacher used ants with different sensitivities for cost and pheromone, while the best ants were used to update the least successful ones when the GA was applied. However, both these algorithms had a long execution time. Therefore, Roach and Menezes introduced in [4] a faster GA approach that is able to react promptly when a new shorter path dynamically appears in the network. The GA is activated whenever the colony becomes incomplete and new ants are created based on the most successful ones. However, this approach requires the system to know how many ants are still alive to replenish the population if needed, so it has a centralized flavor.

Attempts at a parallel implementation of ACO assumed the use of a single process per ant and this approach incurs a significant overhead as pheromone updates were accomplished through messaging [8]. Other attempts of parallelization used multiple colonies acting on separated identical maps that exchange information about their progress [12]. The interchange of results strays from real ant colony behavior and adds to the overhead of the implementation. However, we believe the problems encountered by them can be addressed by using nodes as software agents and ants as messages exchanged between those agents.

Summarizing, the main advantage of our approach over existing approaches is that it tries to envision a fully distributed Swarm Intelligence framework that does not require the existence of a central entity with global knowledge about the system. In particular: (i) the number of agents needed is given by the size of the network, and the addition/removal of network nodes can be easily and naturally mapped to creation and destruction of agents, (ii) agents only need to have local knowledge of the network, so no global map is needed, (iii) ants do not require additional processor overhead in terms of additional number of threads, as their management only involves messages exchanged between agents to simulate their migration, and (iv) pheromones updates do not require additional exchange of messages.

## 5 Conclusions and Future Work

In this paper we introduced the new approach ACODA that proposes a natural mapping of well-known ACO algorithms on a multi-agent distributed architecture. We have motivated our solution, presented the core design and implementation of this proposal and also provided an initial experimental evaluation that shows that our approach is able to handle shortest path problems in dynamic networks in a quite efficient way.

As future work we plan: (i) to extend the design for handling dynamic addition and deletion of nodes to the network; (ii) to evaluate the system by configuring and running it on several machines; (iii) to extend the design for handling other types of network optimization problems. We shall report on our progress in subsequent papers.

## References

1. Botee, H.M., Bonabeau, E.: Evolving ant colony optimization. Advanced Complex Systems 1, 149–159 (1998)
2. Dorigo, M., Gambardella, L.: Ant colonies for the traveling salesman problem. BioSystems 43, 73–81 (1997)
3. Dorigo, M., Stützle, T.: Ant Colony Optimization. MIT Press, Cambridge (2004)

4. Roach, C., Menezes, R.: Handling dynamic networks using evolution in ant-colony optimization. In: Nguyen, N.T., Borzemski, L., Grzech, A., Ali, M. (eds.) IEA/AIE 2008. LNCS (LNAI), vol. 5027, pp. 795–804. Springer, Heidelberg (2008)
5. White, T., Pagurek, B., Oppacher, F.: ASGA: Improving the ant system by integration with genetic algorithms. In: Genetic Programming 1998: Proceedings of the Third Annual Conference, pp. 610–617. Morgan Kaufmann, San Francisco (1998)
6. Stützle, T., Hoos, H.H.: MAX–MIN Ant System. Future Generation Computer Systems 16(8), 889–914 (2000)
7. Dorigo, M.: ANTS 1998, From Ant Colonies to Artificial Ants: First International Workshop on Ant Colony Optimization, ANTS 1998, Bruxelles, Belgique (October 1998)
8. Bullnheimer, B., Kotsis, G., Strauss, C.: Parallelization strategies for the Ant System. In: Leone, R.D., Murli, A., Pardalos, P., Toraldo, G. (eds.) High Performance Algorithms and Software in Nonlinear Optimization. Kluwer Series of Applied Optmization, vol. 24, pp. 87–100. Kluwer Academic Publishers, Dordrecht (1998)
9. Dorigo, M., Gambardella, L.M.: Ant Colony System: A cooperative learning approach to the traveling salesman problem. IEEE Transactions on Evolutionary Computation 1(1), 53–66 (1997)
10. Wooldridge, M.: An Introduction to MultiAgent Systems. John Wiley & Sons, Chichester (2002)
11. Bellifemine, F.L., Caire, G., Greenwood, D.: Developing Multi-Agent Systems with JADE. John Wiley & Sons, Chichester (2007)
12. Middendorf, M., Reischle, F., Schmeck, H.: Multi colony ant algorithms. Journal of Heuristics 8(3), 305–320 (2002)

# Modelling Shortest Path Search Techniques by Colonies of Cooperating Agents

Dariusz Król[1] and Łukasz Popiela[2]

[1] Institute of Informatics, Wrocław University of Technology
Wybrzeże S. Wyspiańskiego 27, 50-370 Wrocław, Poland
dariusz.krol@pwr.wroc.pl
[2] Faculty of Computer Science and Management, Wrocław University of Technology
Łukasiewicza 5, 50-371 Wrocław, Poland
160384@student.pwr.wroc.pl

**Abstract.** In this paper, we propose a nature-inspired agent coordination model for comparing ants', bees' and beents' behaviour in finding the shortest path when searching for food. The model is based on Jadex. Simulations demonstrate that the system can be trained to solve a wide range of problems efficiently. The model also provides means for estimating and interpreting common errors of these multi-agents simulations.

**Keywords:** swarm intelligence, nature-inspired networking, multi-agent system.

## 1 Introduction

In connection with the growing size of the modern networks, independently of whether they are wired or wireless, the problem of shortest path finding and efficient routing mechanisms remains the same [1]. In general terms, the shortest (optimal) path between sender and receiver needs to be found, so that the network can operate efficiently and provide the largest benefits for its users. In this research we take the inspiration from natural biological systems that are faced with similar problem [2]. Watching behaviours of organisms living in large swarms and collaborating to achieve planned purposes, we can draw a conclusion and adopt observed solutions to our needs [3], [4], [5]. In this paper mechanisms used by ants and bees, adopted by mimicking social insects' behaviour in multi-agent system will be presented.

Among all social insects, the ants and bees dominate the environment [6]. The individuals are naturally equipped with the ability to conduct complicated voting mechanisms and also include interaction and self organization for solving real world problems [7], [8]. When searching for food, insects initially explore the area surrounding their nest in a random manner. After some time they are ready to guide other insects to the food source. The indirect or direct communication enables them to find quite efficiently shortest paths between home and food sources [9], [10], [11].

For the purpose of presenting the social insects' behaviour simulations based on Jadex [12] will be introduced. The Jadex reasoning engine follows the Belief Desire

N.T. Nguyen, R. Kowalczyk, and S.-M. Chen (Eds.): ICCCI 2009, LNAI 5796, pp. 665–675, 2009.

Intention (BDI) model and facilitates simple intelligent agent construction. Thanks to this framework, the simulations will allow to examine potential benefits and limitations resulting from applied food searching algorithms. Action taken by agents will be divided into two crucial stages. The first stage of finding the food, and the second stage of transporting the food home (the hive or the nest, depending on the kind of the agent) [13]. The number of steps (referred as 'distance') carried out during each iteration will change together with the extent of the population of insects. This number will be the base for the measure, thus the assessment of the agent cooperation will be available [14].

Although many studies on swarm intelligence have been presented (provide some references at this point), there are no general approaches how to evaluate a swarm intelligent system's performance in terms of errors, or deviations from the ideal path. [15], [16] and [17] tried to make such an analysis for ants. However, there is a growing need for more analytical studies not only for ants, but also for other swarm-based creatures, e.g. combination of bees and ants for example.

The paper is organized as follows. Section 2 characterises the environment and the configuration of simulations. In Section 3, we present simulation results We compared swarm agent based systems with respect to average number of steps (lifts and drops). Finally, we conclude the paper in Section 4.

## 2 Test-Bed and the Configuration of Simulations

Tasks to be performed by individual agents are divided in two groups: seeking the location of the food out and raising it (lift), moving back the raised food to the nest or the hive (drop).

Bees are able to communicate between themselves with waggle dance informing about food location. Ants, similarly to bees, are able to communicate between themselves. However, ants do not communicate directly, but rather through the mechanisms based on pheromones. When the ant agent found the food source, during the return to the nest will leave pheromones who can be used through different ant. However, ants don't have the global knowledge about surroundings. In order to find the food ant must be guided by a trace of pheromones left in the local environment. Beents like bees are able to remember locations as well as to communicate directly. However, this insect kind doesn't require the closeness of the hive (beents can communicate in the any place on the map). Additionally, beents also use (like ants) the mechanism based on deposits of pheromones. Thanks to this beents can communicate in non-direct fashion.

The direct or indirect communication between the insects enables them to find shortest paths between their nest/hive and food sources as illustrated in Fig. 1. The legend is described in Table 1. As shown on the diagram, there are three food sources. The shortest paths between home (node 11) and food locations are the following: 26 steps for the distance between home and first food location (node 1), 28 steps for the distance between home and second food location (node 7), and finally 32 steps for the distance between home and third food location (node 14).

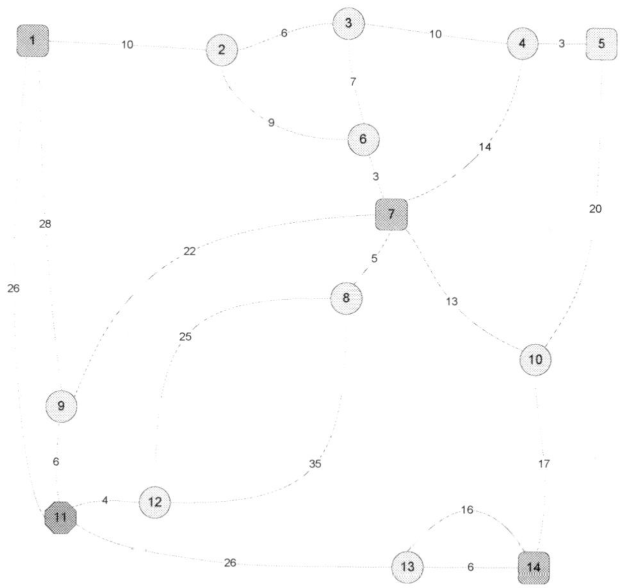


**Fig. 1.** An example of .connections used for the experiments. Agents explore possible routes.

**Table 1.** List of graphical symbols.

| Nest or hive | Food source | Empty node (3x3) | Empty node (1x1) | Path and the path length |
|---|---|---|---|---|
|  |  |  |  | 10 |

All our simulations were carried out on the following computer: Centrino Duo 2.0 GHz, 1 GB RAM. The speed of the computer or the size of the main memory does not influence the decisions of individual agents. We use three kinds of creatures: ants, bees and new type emerged from connecting of ants and bees features - beents. The number of agents are changing from 1 till 10. We have examined 9 simulations. The simulations are the base to the assessment of the productivity of individual solutions applied by agent system for solving the problem of finding the optimal path in the graph.

There is a small difference in the way how received findings should be interpreted. They differ in the speed of bringing up the food and the size of charge to raise back to home. For example, ants are moving twice as slow as others, but cargo can be twice as big; bees and beents are moving quickly, but can transport only single food value. Hence, a number of operations in logs made is different. For all kinds operations lift is remembered in the identical way (raising charge twice as large by ants was written as two single charges). However operations bustard (drop) for ants has to be doubled (in cases when in one iteration double value of the food was dropped).

Results of experiments are presented in two categories. In the first, we are focussed on the kind of agents (ants, bees and beents) and what's more they will be compared between themselves. In the second, we are focused on a number of the population that has been used (1, 5 and 10 agents).

## 3 Simulations and the Results

During the presentation of the results, two attempts of interpretation will be provided. In the first, both kinds of events (the lift and the drop) will be treated as equivalent In the second, they will be treated separately.

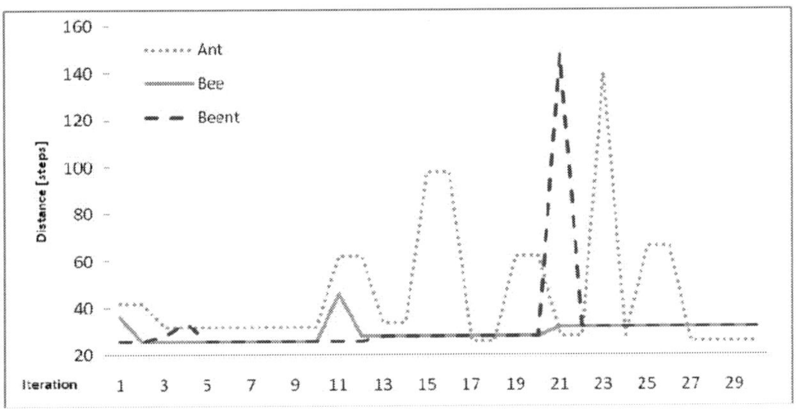


**Fig. 2.** Route discovery in lifts and drops with 1 agent

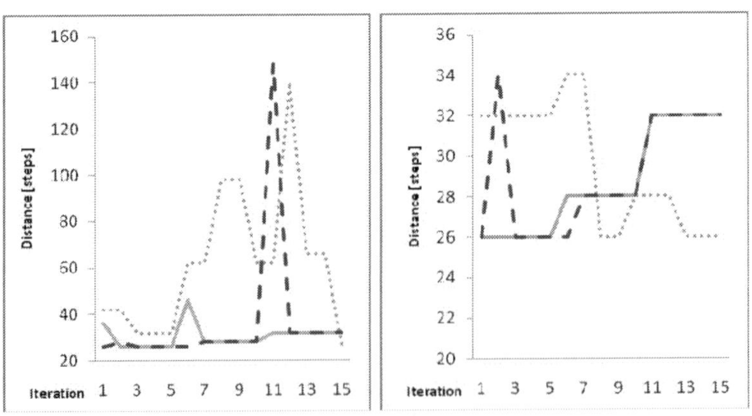


**Fig. 3.** Route discovery in lifts (left) and drops (right) with 1 agent

All sudden growths of the distance (Fig. 3, 5 and 7) that agents made during their task realization could be explained in two ways. First, in the predominating number of cases it was caused by too slow reaction to the changing conditions of the environment and the surroundings. The scale of the problem depends on pheromones left, the number of times the given path was used where the pheromones were deposited and longer they were kept on the map. Second, more rarely appearing situation, when agents needed to find the new food source when the existing source has been exhausted. Of course, bees have to also contend with this problem, however, they have often already had the knowledge about earlier found food locations and they do not have to search for a new one.

Bees were performing the best because they have the global knowledge about world surrounding them, and they are able to communicate between themselves in the more direct way passing on the information about the food location. Therefore, the number of errors made by them is practically zero.

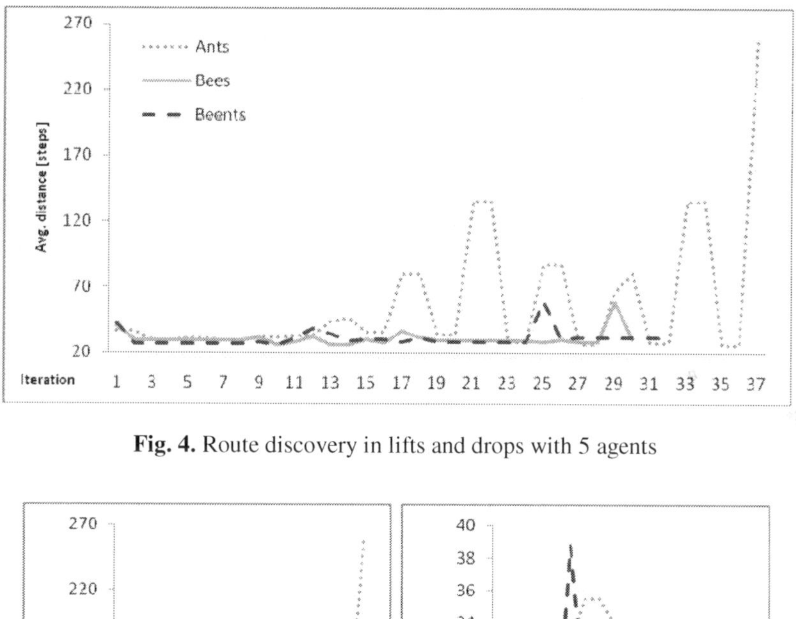


**Fig. 4.** Route discovery in lifts and drops with 5 agents

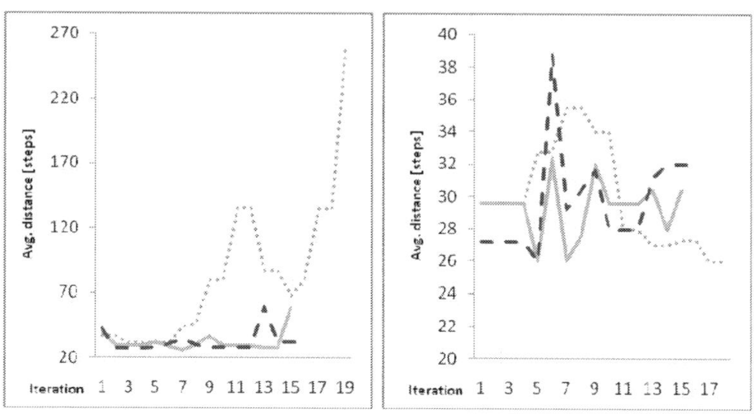


**Fig. 5.** Route discovery in lifts (left) and drops (right) with 5 agents

Beents too, share those features with bees. Differences consist in the fact that for the direct communication beents do not need to be close to home and as ants they use the non-direct communication thanks to the deposited pheromones. Those pheromone deposits are just a reason for the rare errors committed by beents.

Fig. 2, 4 and 6 show that ants' results came out poorly. To explain these results we should remember: (i) limited abilities (ii) lack of the global view on world, (iii) implementation difficulties in selection of correct value of the pheromones weakening process. First from mentioned features one should however treat not only as limitation, but also as their main advantage. In this case, agents are less complex, they require minimum amount of memory for correct action.

Analysing graphs for ants, given in Fig. 8, we notice significant differences in behaviour of agents (depending on the number of their population). Fig. 9 illustrates that the increasing number of agents negatively influences the scale of errors. Large number of agents in a shorter time cause longer persistence on the map of undesirable pheromones. This is a cause for a large number of steps (almost 300 steps).

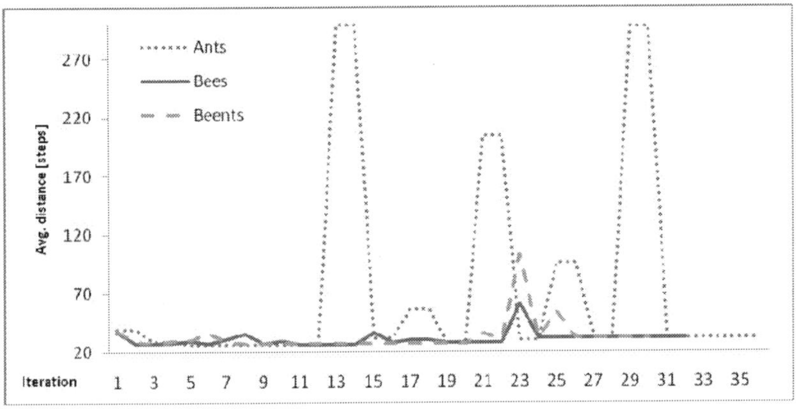


**Fig. 6.** Route discovery in lifts and drops with 10 agents

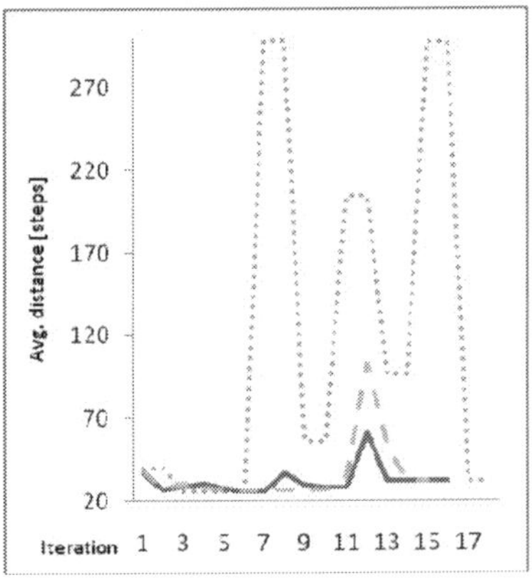


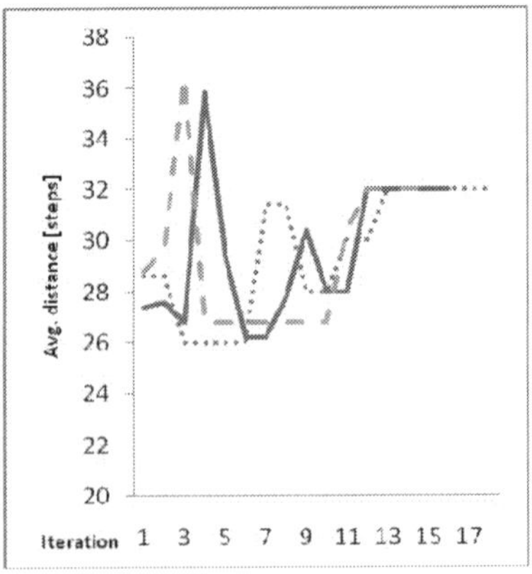


**Fig. 7.** Route discovery in lifts (left) and drops (right) with 10 agents

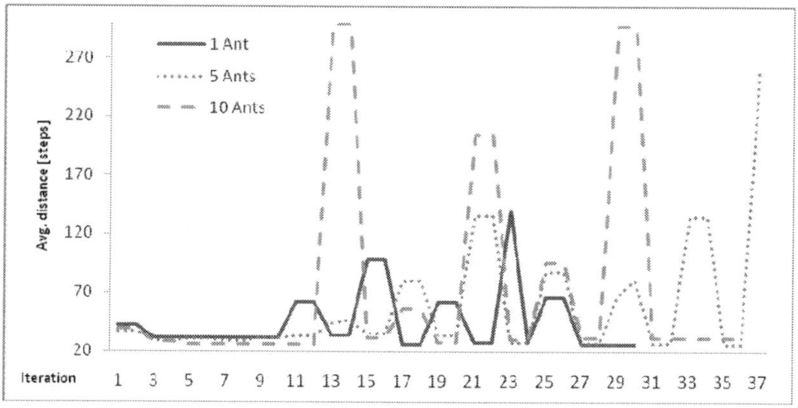


**Fig. 8.** Ant's route discovery in lifts and drops with 1-5-10 agent(s)

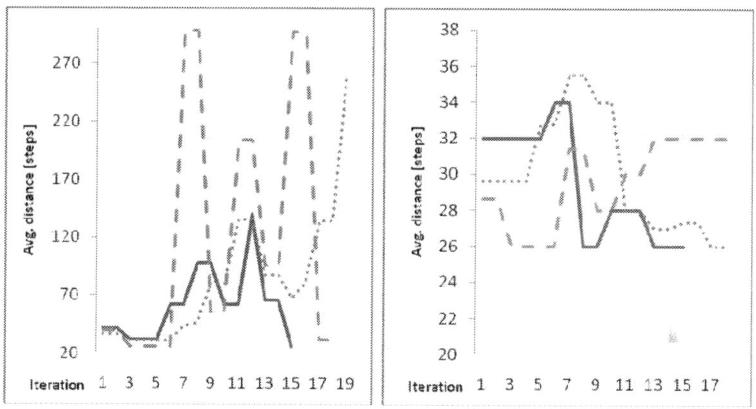


**Fig. 9.** Ant's route discovery in lifts (left) and drops (right) with 1-5-10 agent(s)

For bees (Fig. 10 and 11) we can also see occurring anomalies, however their number and scale are much lower. For every population we are observing one each more serious jump of the average distance required for lift. It was only one reason. Agent needs to seek the new food location, because the previous one is exhausted. These situations should not be perceived as errors.

For experiments with beents (Fig. 12 and 13), similarly to bees we observed also one serious rise in the average number of steps for every population. However their scale is much greater (e.g. the number in the 11th iteration for one agent reached close 150 steps). The general presented trend on graphs is optimal and observed single errors result from too slow evaporation of pheromones (similarly as for ants). The difference lies in that beents have abilities of more complex communication system and have the global view on the environment. Therefore the scale of this problem is lower and appears less frequently.

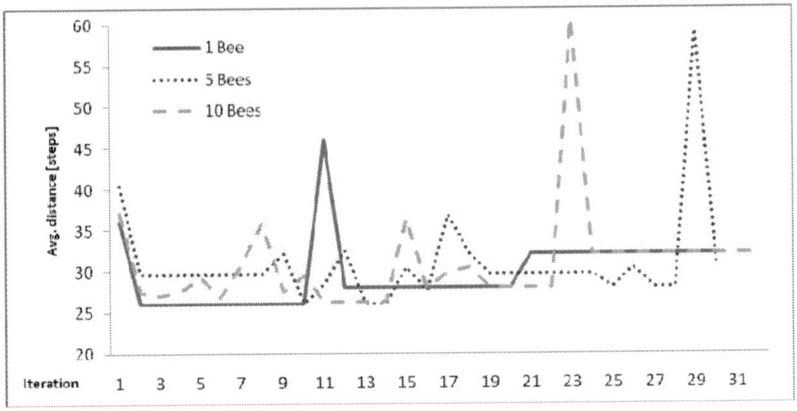


**Fig. 10.** Bee's route discovery in lifts and drops with 1-5-10 agent(s)

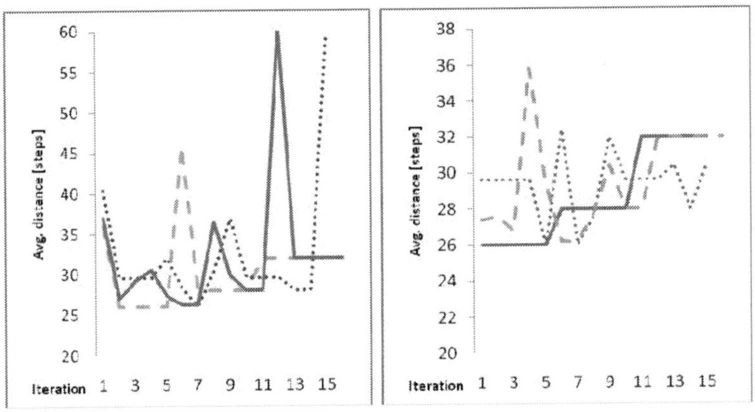


**Fig. 11.** Bee's route discovery in lifts (left) and drops (right) with 1-5-10 agent(s)

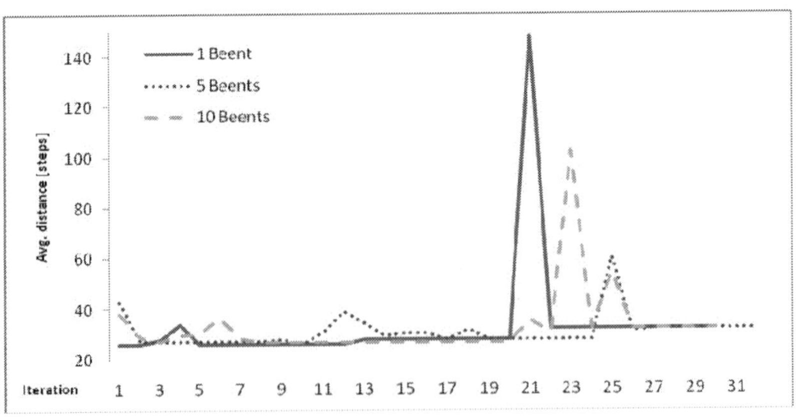


**Fig. 12.** Beent's route discovery in lifts and drops with 1-5-10 agent(s)

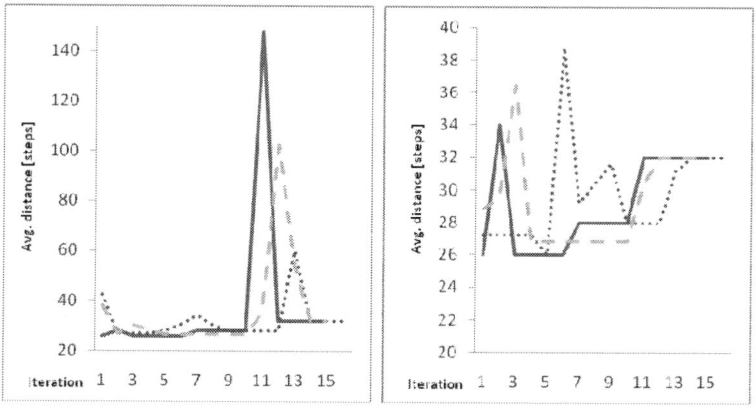


**Fig. 13.** Beent's route discovery in lifts (left) and drops (right) with 1-5-10 agent(s)

## 4 Discussion and the Summary

All experiments presented in the paper aim at proving that multi-agent system based on different behaviours appearing in the natural environment will advise themselves with the task of finding the shortest path in the environment. During the implementation also a new kind of agents is used – beents. Beents combine features of both, bees and ants.

The simulations were divided into two groups. In the first, see Table 2, the best in type respect are bees, the worst – ants. Having the global view on environments and more advanced communication skills are more important than carrying (cargo) capacity. Apart from that, we should remember, that ants administering lower communication skills and the shorter demand for the memory are more simpler agents under construction as well as for action what isn't also meaningless in the case of complex systems without a doubt real systems are. We should also recall why beents having the greatest abilities and who theoretically should win rivalry didn't make it. A mechanism taken from ants - pheromones which allows the non-direct transport, in practice slow evaporation leads to the too long persistence out of date already of paths on the map what inserted agents into mistake. In the second where we focus on comparing the behaviour of the system depending on the number of the agent population, differences are low-order. How it could be foreseen a large number of agents has a positive effect to the decisions taken by the system as the whole, however not in every case. The divergence is possible mainly for ants. The more numerous population increases the problem with long evaporation time of pheromones.

From the comparative experimental results, it is confirmed that we can additionally detail at least three kinds of changes which have potentially a positive effect on behaviour of the system based on a mechanism of pheromones, for ants and beents: (i) the parameter deciding how fast pheromones may evaporate, (ii) a way of the incrementing the strength of pheromones modified from linear to logarithmic, (iii) agent who lifted last value of the food in the given location during the return home

**Table 2.** Various results of shortest path finding with 1-5-10 agent(s). The value is a triple: distance for lifts and drops, distance for lifts, distance for drops.

| Agents-Distance | Ants | Bees | Beents |
|---|---|---|---|
| 1-Min | 26 (26/26) | 26 (26/26) | 26 (26/26) |
| 1-Max | 140 (140/34) | 46 (46/32) | 148 (148/34) |
| 1-Average | 45,47 (61,47/29,47) | 29,60 (30,53/28,67) | 32,73 (36,40/29,07) |
| 5-Min | 24 (24/26) | 26 (26/26) | 26 (26/26) |
| 5-Max | 262 (262/42) | 80 (80/42) | 80 (80/42) |
| 5-Average | 49,84 (68,56/30,33) | 30,89 (32,45/29,33) | 30,93 (32,25/29,60) |
| 10-Min | 26 (26/26) | 26 (26/26) | 26 (26/26) |
| 10-Max | 426 (426/34) | 126 (126/42) | 126 (126/42) |
| 10-Average | 72,45 (115,84/29,07) | 30,8 (32,16/29,44) | 32,71 (36,03/29,39) |

instead of leaving more pheromones will remove them. The entry results are in line with our expectations.

## References


1. Di Caro, G., Ducatelle, F., Gambardella, L.M.: AntHocNet: An Adaptive Nature-Inspired Algorithm for Routing in Mobile Ad Hoc Networks. European Transactions on Telecommunications 16, 443–455 (2005)
2. Dorigo, M., Maniezzo, V., Colorni, A.: Ant System: Optimization by a Colony of Cooperating Agents. IEEE Transactions on Systems, Man, and Cybernetics-Part B Cybernetics 26(1), 29–41 (1996)
3. Wedde, H.F., Farooq, M.: A comprehensive review of nature inspired routing algorithms for fixed telecommunication networks. Journal of Systems Architecture 52, 461–484 (2006)
4. Zieliński, C., Trojanek, P.: Stigmergic cooperation of autonomous robots. Mechanism and Machine Theory 44, 656–670 (2009)
5. Teodorovic, D.: Swarm intelligence systems for transportation engineering: Principles and applications. Transportation Research Part C 16, 651–667 (2008)
6. Garnier, S., Gautrais, J., Theraulaz, G.: The biological principles of swarm intelligence. Swarm Intell 1, 3–31 (2007)
7. Xiang, W., Lee, H.P.: Ant colony intelligence in multi-agent dynamic manufacturing scheduling. Engineering Applications of Artificial Intelligence 21, 73–85 (2008)
8. Lee, U., et al.: Bio-inspired multi-agent data harvesting in a proactive urban monitoring environment. Ad Hoc Networks 7, 725–741 (2009)
9. Grosan, C., Ajith, A.: Stigmergic Optimization: Inspiration, Technologies and Perspectives. Studies in Computational Intelligence, vol. 31, pp. 1–24. Springer, Heidelberg (2006)
10. White, T.: Expert Assessment of Stigmergy: A Report for the Department of National Defence. Technical report, Carleton University (2005)
11. Hinchey, M.G., Sterritt, R., Rouff, C.: Swarms and Swarm Intelligence. Computer 40(4), 111–113 (2007)
12. Jadex BDI Agent System - Overview, `http://jadex.informatik.uni-hamburg.de`

13. Hadeli, K., et al.: Multi-agent coordination and control using stigmergy. Computers in Industry 53, 75–96 (2004)
14. Murthy, V.K., Krishnamurthy, E.V.: Can Artificial Life Emerge in a Network of Interacting Agents? In: IEEE International Conference on Fuzzy Systems, pp. 106–113. IEEE Press, New York (2006)
15. Dhillon, S.S., Van Mieghem, P.: Performance analysis of the AntNet algorithm. Computer Networks 51, 2104–2125 (2007)
16. Mullen, R.J., et al.: A review of ant algorithms. Expert Systems with Applications 36, 9609–9617 (2009)
17. Yu, B., et al.: An improved ant colony optimization for vehicle routing problem. European Journal of Operational Research 196, 171–176 (2009)

# Natural Scene Retrieval Based on Graph Semantic Similarity for Adaptive Scene Classification

Nuraini Jamil and Sanggil Kang

Department of Computer and Information Engineering, Inha University,
253 Yonghyun-dong, Incheon, Korea 402-751
Tel.: +82-32-860 8377, Fax: +82-32-860 8377
nuraini@inhaian.net and sgkang@inha.ac.kr

**Abstract.** In this paper, we introduce our method for image retrieval to access and measuring the similarity of natural scenes by using graph semantic similarity. The proposed method is motivated by continuing effort from our previous work in adaptive image classification based on semantic concepts and edge detection. The method will learn the image information by concept occurrence vector of semantic concepts such as water, grass, sky and foliage. We constructed the graph using this information and illustrate the similarity with connecting edges. The empirical results demonstrated promising performance in terms of accuracy.

**Keywords:** Image retrieval, Graph, Semantic similarity, Semantic concepts.

## 1 Introduction

Natural scene retrieval is a challenging task due to various image information or contents. Thus, the efficient browsing and querying become increasingly important. In general most approaches concisely convey information by collections of low-level features such as color, texture and edge position. These features are used to measure the similarity among the images. This has long been recognized as a problem in Semantic Web for natural scene retrieval because this matrix of these features is insufficient for image retrieval. Therefore, image information needs to be linked and communicate to each other.

From our previous work [1], we have a benchmark for find a similarity between two images. We used matrix based on pairwise distance linear calculation between images. In order to solve the problem, this paper presents a novel natural scene retrieval method by using graph semantic similarity for adaptive image classification. We propose a segmentation of image sub-regions into semantic concept classes such as sky, water, sand and rock, where by dividing an image into sub-regions based on edge detection. Sub-region descriptions are combined to a global image representation by tabulates the frequency of occurrence of each semantic concept. Our graph semantic similarity will model the relationship between semantic concept and image and value of occurrence of each semantic concept will be an edge to this graph.

N.T. Nguyen, R. Kowalczyk, and S.-M. Chen (Eds.): ICCCI 2009, LNAI 5796, pp. 676–684, 2009.

The rest of the paper is organized as follows. Related works are reviewed in next section. In Section 3, we describe our adaptive segmentation approach generally. Then in Section 4, we proposed our graph semantic similarity model for natural scene retrieval. We implement our method and present performance evaluation in Section 5 and the concluding remarks are given in Section 6.

## 2 Related Works

### 2.1 Image Classification

Nowadays, many natural scene image categorization systems have been created, mainly focused on modeling a scene using the global statistical information of an image against the local details. Szummer et al. [2] proposed an algorithm for indoor/outdoor classification based on the k-NN classifier and three types of features: color, texture and frequency information. Particularly, the proposed algorithm has been based on the extraction of only low-level features. When classifying an image, it is assumed that color and texture are the most important and invariant visual features. Color structure and homogeneous texture descriptors of MPEG-7 standard [3] are selected to represent these features.

Murphy et al. [4] built four graphical models to relate features of image blocks to objects, and to perform joint scene and object recognition. Barnard and Forsyth [5] applied a hierarchical statistic model to generate keywords for classification based on a sequence of semantically meaningful regions. Fan et. al. [6] provided simplified geometric structures of the image regions using their color-edge detector. Later some methods were proposed to classify images into multiple categories by Olivia and Torralba [7]. In general, these approaches aim at learning the correspondence between global annotations and image or image regions. However, the region labels are usually not combined to a global image representation that can be employed in scene classification. Vogel and Schiele [8] employed a novel image representation enable to classify natural scenes by semantic concepts description. These "semantic concepts" are objects or object patches in the scene. Semantic concept for each sub-region is much easier to obtain ground-truth than that for entire image. The method used a fixed regular grid of 10 x 10 sub-regions. First, this sub-region might annotate double concepts. Second, if the sub-regions include two concepts then these sub-regions are not used for training or testing of the concept classifiers.

We introduce an adaptive segmentation method instead of fixed regular grid based segmentation method. We use edge detection technique to split an image into sub-regions which eliminates the double annotation problem. Moreover, our adaptive segmentation method promises the use of all sub-regions for testing and training.

### 2.2 Image Retrieval

Many applications dealing with image retrieval require knowledge and information of semantic similarity between images. Some popular technique to construct semantic content is based on user's relevance feedback [9]. A relevance feedback system allows the user to indicate to the system, which of these instances are desirable, or relevant. Relevance feedback technique uses decision trees to learn a common thread among instances marked relevant. However, this system learns a nonlinear embedding that maps the cluster image into a hidden space of semantic attributes and has various limitations like time consuming because it is a sizeable number of examples.

Text based image retrieval or image annotation [10] assumes that all images are labeled with keywords. Aslandogan et al.[11] proposed thesaurus-based techniques by taking into account the relationship between keywords in the keyword matching process. However, this is not an ideal descriptor of semantics because it is usually done by humans and single keyword might have several meanings or different keywords might share the same meaning.

Most content-based image retrieval (CBIR) approaches [12, 13] have difficulty in bridging the gap between low-level feature representation and high-level semantic interpretation. The semantic of images cannot be easily captured by even a combination of low-level features and can only retrieve visually similar rather than semantically relevant images to query image. Furthermore, CBIR assumes each image as an entire semantic unit. In a single image, at least have two different categories, foreground and background. In order to retrieve those images containing the content of interest, each category should be treated as an individual semantic category during the image retrieval process.

Earth Mover's Distance (EMD) [14], which has a rigorous probabilistic interpretation and has been successfully applied to image retrieval. However, these metrics are based on pairwise distance calculation and oversimplify the relationship among all the images in the database. Therefore, their effectiveness is quite limited. Here we propose graph semantic similarity method that represents the problem in terms of a graph where each node corresponds to semantic concepts in the image, and the edges connect certain nodes. A weight is associated with each edge based on frequency of occurrence for semantic concepts.

## 3 Adaptive Scene Classification

### 3.1 Overview

An image is constructed with regular 2x2 sub-regions. And then we compute edges for each sub-region based on grey-level information. Edge is defined to split a sub-region into more regular 2x2 sub-regions. We use the Canny edge detector [15] to detect edges in sub-regions. The input parameters were selected to optimize the quality of edges for the purpose of higher-level perceptual organization. The optimal parameter set for our images is (1.0, 0.15, 0.35). Feature vectors are extracted by using color histogram and texture for each sub-region. By dividing an image into sub-regions and storing local histograms for each of these areas, the signature for each image becomes increasingly robust. By observing psychological studies in the human visual perception, Tamura [16] explored the texture representation using computational approximations to the three main texture features of coarseness, contrast, and directionality. We label the sub-regions based on semantic concepts. Then, we classify the image into the scene category using k-Nearest Neighbor (k-NN) classifier [17] to classify all the images into the semantic categories (*beach, forest, mountain* and *field).*

### 3.2 Concept Occurrence Vector (COV)

The sub-regions information of the semantic concepts is combined to determine the frequency with which sub-regions of each concept occur in the image. *Concept-occurrence vector (COV)* refers to a semantic concepts of sub-regions developed to observe the normalization histogram of the concept occurrences in an image. The COV can be formulized as follows:

$$\boldsymbol{COV} = \sum_{k=1}^{n} \frac{N}{2^{2k}} \tag{1}$$

where $N$ is known as semantic concepts, and $n$ is the number of sub-regions level.

The additional information that can be used based on COV is that these can also reveal interface information for each scene category. For example, beach is characterized through a large amount of water. In contrast, forest can be differentiated when a large amount of foliage is detected.

## 4 Graph Semantic Similarity Model

In this section, we describe how to construct a graph semantic similarity for natural scene retrieval. Our primary purpose is to retrieve similar images with query image based on semantic concepts that appear in an image. In the case of natural scene retrieval, we are mapping the image with semantic concepts that appear in image based on graph design and the example is illustrated in Fig.1.

We begin with some definitions. Let $G = (V, E)$ be a graph semantic similarity with nodes $v_i \in V$, the elements in $V$ are semantic concepts, and edges $(v_i, v_j, \ldots, v_m) \in E$ corresponding to connection between nodes and image. Each edge has a corresponding weight $w_{vi}$, which is a non-negative value of frequency of occurrence for semantic concepts. This weight is usually normalized to prevent a bias towards many edges to give a measure of the importance of the nodes within the particular image $p_y$. The exact formula for weight can be computed as follow:

$$\boldsymbol{wvi_y} = vi_y \times \frac{x}{E} \tag{2}$$

where $x$ is the number of edges connecting the considered vertex in image $p_x$, and the denominator is the sum of image.

Each semantic concepts of query image $q$ corresponds to a connected nodes in a graph $G' = (V,E')$, where $E' \subseteq E$. In other words, any semantic concept is induced by a subset of the edges in $E$. This means, we assume, the image and query image have a similarity if both of them connected with edges to a same node or nodes. Fig.2 illustrates the structure of graph semantic similarity.

We defined a similarity measure in terms of a matching process. This measure produces a similarity value which is in agreement to an information-theoretic definition of similarity [18]. The similarity we used does a comparison of the edges of the sets of same nodes. For all possible combinations of two nodes we compute a value that is proportional to the degree of relation of the two nodes. This means the

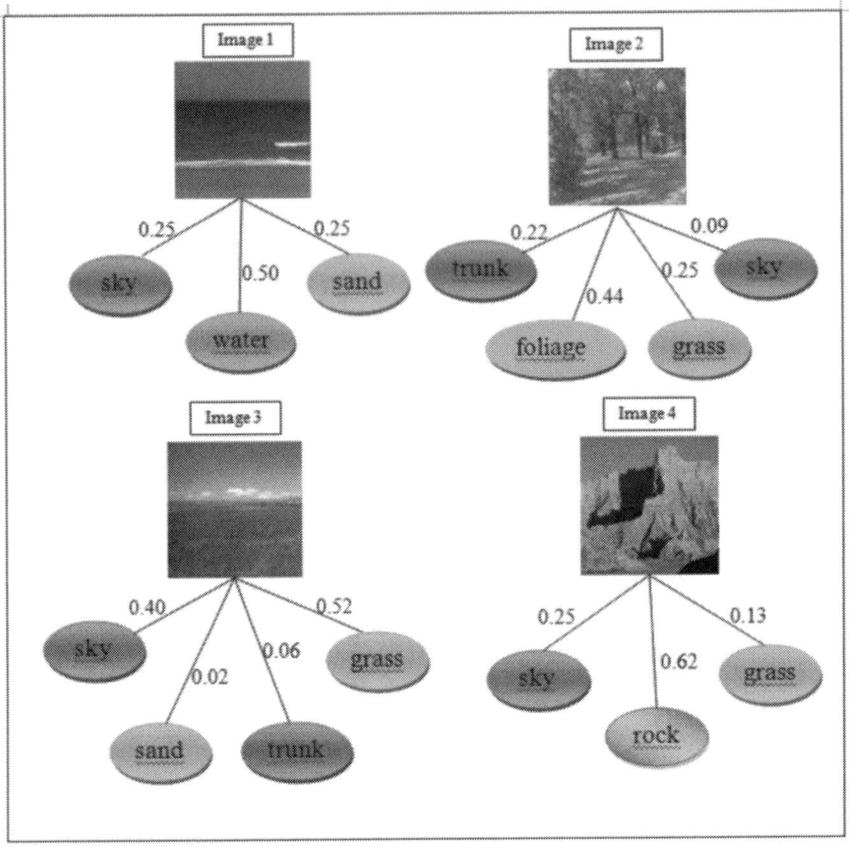


**Fig.1**. Image example represented by graph. Each image has its own vertices corresponding to semantic concept, and the edges represent the COV value.

similarity between two images is based on the similarity between corresponding node vectors. The inner vector product for similarity between image $p$ and query image $q$ is computed as:

$$\boldsymbol{sim}(\boldsymbol{p},\boldsymbol{q}) = \sum_i (\boldsymbol{wvi_y}\, p_i)(\boldsymbol{wvi_y}\, q_i) \tag{3}$$

where $\boldsymbol{wvi_y}\, p_i$ and $\boldsymbol{wvi_y}\, q_i$ are the weights in the two vector representation. Given a query, all images are retrieved according to their similarity with the query. The lack of common nodes in two images does not necessarily mean that the images are unrelated. Semantic similarity may not be expressed the small values of edges.

## 5 Experiment and Result

The experiments results were measured running on Windows XP platform with 3.2 GHz Intel Pentium and 1.00 GB RAM. A k-Nearest Neighbor is trained for each scene category. The input to the k-NN is the Concept Occurrence Vector (COV) of the relevant images. For the experiments, the MATLABArsenal[1] package [19] was employed.

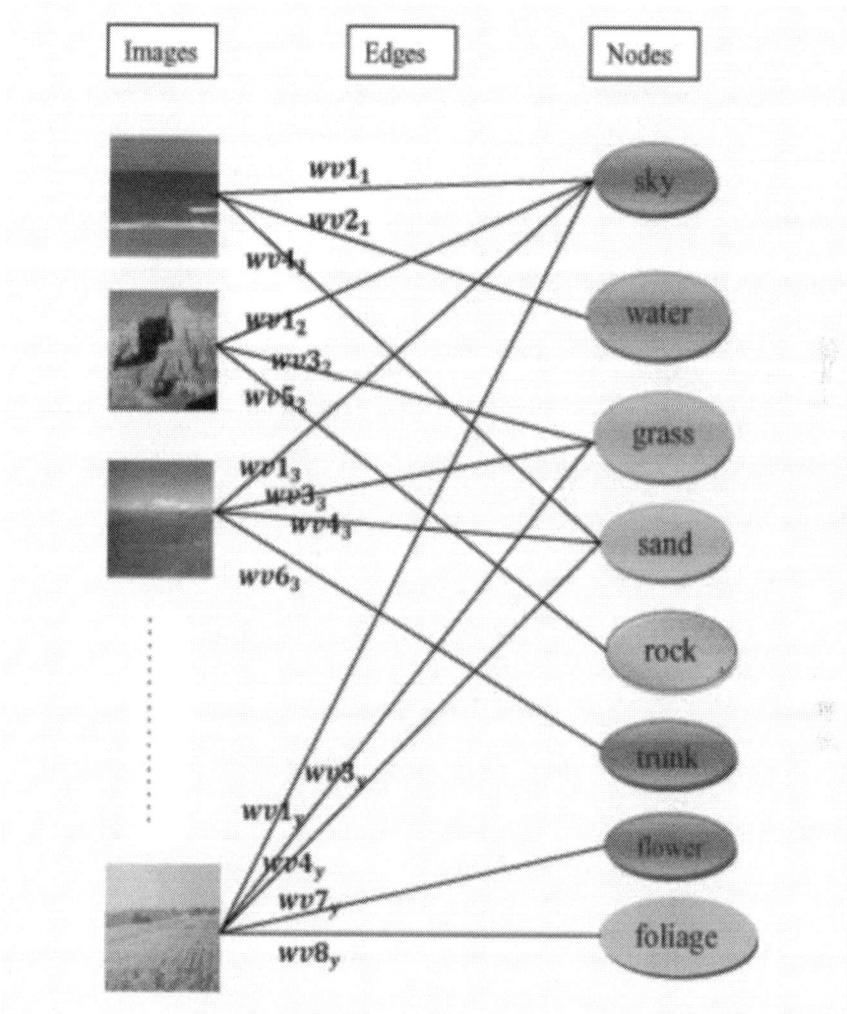


**Fig. 2.** Graph semantic similarity structure

[1] This software is free only for non-commercial use. It must not be modified and distributed without prior permission of the author. Software available at http://www.informedia.cs.cmu.edu/yanrong/MATLABArsenal/MATLABArsenal.htm

## 5.1 Image and Dataset

We used the Oliva and Torralba [7] dataset. It consists of few hundred images of scenes belonging to the same semantic category. It is the diverse data containing images of scene categories including beach/coast (360 images), forest (328 images), mountain (374 images), and field (210 images).

## 5.2 Performance Using Graph Semantic Similarity

In order to measure retrieval effectiveness for an image retrieval system, the accuracy was measured using F-score, defined as

$$\boldsymbol{Fscore} = \frac{2.P.R}{P+R} \tag{4}$$

where $P$ is precision and $R$ is recall. Although this effectiveness measure is useful for testing a retrieval system objectively, it cannot exclude subjectivity inherent in image retrieval system. It's because the category itself was classified by humans.

Fig.3 shows retrieval results for several images. As one can see, query 1, 2, and 4 from top rows performs very well. The remaining row shows, the returned results are suboptimal, for the last row because the mountain has a snow for semantic concept. The problem our semantic concepts do not include the snow concept. Table 1 summarizes the average F-score from those 4 scene categories. Graph semantic similarity method achieved high accuracy for three scene categories *Beach, Forest,*

**Fig.3** Retrieval results obtained by our graph semantic similarity. The left most images in each row shows the query image, the five images to the right show the most relevant images retrieved from left to right.

**Table 1.** Accuracy of each category for image retrieval based on precision and recall

| | Graph Semantic Similarity | Our Previous method |
|---|---|---|
| **Beach** | 0.8451 | 0.6167 |
| **Forest** | 0.8253 | 0.6898 |
| **Mount** | 0.4224 | 0.5330 |
| **Field** | 0.7992 | 0.6287 |

and *Field.* Graph semantic similarity method performs better than our previous method with a 22.84% difference in the overall accuracy for beach category. This method also improves the accuracy for other categories, forest and field by 13.55% and 17.05% respectively over the previous method. For *Mountain* category, post hoc image examination reveals that the confusion between 'forest' and 'mountain' comes mostly from the co-occurrence of trees. Significantly, these images are also ambiguous for human to classify into forest or mountain.

## 6 Conclusion

In this paper, we presented an effective image retrieval method by using graph semantic similarity. From the experimental result, we can conclude that our method can improve the accuracy of retrieval for natural scene images, compared to our existing method. However, the graph semantic similarity only used semantic concepts as a content to find a similarity between images. For the future method, we are trying to construct multiple graphs based on different content, and then incorporate information from these graphs to perform retrieval, which will hopefully make better use of this content and give even higher precision.

## Acknowledgments

This research was supported by the Ministry of Knowledge Economy, Korea, under the Information Technology Research Center support program supervised by the Institute of Information Technology Advancement. (Grant number IITA-2008-C1090-0801-0031).

## References

1. Nuraini, J., Shohel, A., Kangseok, K., Sanggil, K.: Adaptive Scene Classification based on Semantic Concept and Edge Detection. Journal of Korea Industrial Information Systems Society (KIISS) (will be published)

2. Szummer, M., Picard, R.: Indoor-outdoor image classification. In: International Workshop on Content-based Access of Image and Video Databases, Bombay, India (1998)
3. Manjunath, B.S., Salembier, P., Sikora, T.: Introduction to MPEG-7 Multimedia Content Description Interface. John Wiley & Sons Ltd., England (2002)
4. Murphy, K., Torralba, A., Freeman, W.: Using the Forest to See the Trees: A Graphical Model Relating Features, Objects, and Scenes. In: Advances in Neural Information Processing Systems, vol. 16 (2004)
5. Barnard, K., Forsyth, D.: Learning the Semantics of Words and Pictures. In: Proceeding of International Conference on Computer Vision, pp. 408–415 (2001)
6. Fan, J., David, K., Yau, Y., Ahmed, K., Walid, G.: Automatic Image Segmentation by Integrating Color-Edge Extraction and Seeded Region Growing. IEEE Transactions On Image Processing 10, 1454–1466 (2001)
7. Oliva, A., Torralba, A.: Modeling the shape of the scene: A holistic representation of the spatial envelope. International Journal of Computer Vision 42(3), 145–175 (2001)
8. Vogel, J., Schiele, B.: Natural Scene Retrieval based on a Semantic Modeling Step. In: Conference on Image and Video Retrieval, Dublin, Ireland (2004)
9. MacArthur, S.D., Brodley, C.E., Kak, A.C.: Interactive content-based image retrieval using relevance feedback. Computer Vision and Image Understanding 88, 55–75 (2002)
10. Mori, J., Tsujishita, T., Matsuo, Y., Ishizuka, M.: Extracting Relations in Social Networks from theWeb Using Similarity Between Collective Contexts. In: Proceeding. 20th IJCAI (2006)
11. Aslandogan, Y.A., Thier, C., Yu, T.C., Zou, J., Rishe, S.: Using Semantic Contents and WordNet in Image Retrieval. In: Proceeding 20th International ACM SIGIR Conference on Research and Development in Information Retrieval, Philadelphia, US, pp. 286–295 (1997)
12. Cox, I.J., et al.: PicHunter: Bayesian relevance feedback for image retrieval. In: Proceeding 13th International Conference on Pattern Recognition, Vienna, Austria, pp. 361–369 (1996)
13. Smith, J.R., Chang, S.F.: VisualSEEk: a fully automated content-based image query system. Proceeding ACM International Conference Multimedia (1995)
14. Rubner, Y., Tomasi, C., Guibas, L.: A metric for distributions with applications to image databases. In: Proceeding of IEEE International Conference Computer Vision, pp. 59–66 (1998)
15. Canny, J.: A Computational Approach To Edge Detection. IEEE Transactions on Pattern Analysis and Machine Intelligence 8, 679–698 (1986)
16. Tamura, H., Mori, S., Yamawaki, T.: Texture features corresponding to visual perception. IEEE Transactions System Man, and Cybernetics 8(6) (1978)
17. Duda, R.O., Hart, P.E., Stork, D.G.: Pattern Classification, 2nd edn. A Wiley-Interscience Publication, Hoboken (2000)
18. Lin, D.: An Information-Theoretic Definition of Similarity. In: Proceeding 15th International Conference on Machine Learning, pp. 296–304 (1998)
19. Yan, R.: MATLABArsenal; a library for k-nearest neighbor, Software (2006), `http://www.informedia.cs.cmu.edu/yanrong/MATLAMATLABAr/MATLABArsenal.htm`

# A Hybrid Architecture for E-Procurement

Giner Alor-Hernandez[1], Alberto Aguilar-Lasserre[1], Ulises Juarez-Martinez[1], Ruben Posada-Gomez[2], Guillermo Cortes-Robles[1], Mario Alberto Garcia Martinez[1], Juan Miguel Gomez[2], Myriam Mencke[2], and Alejandro Rodriguez Gonzalez[2]

[1] Division of Research and Postgraduate Studies, Instituto Tecnológico de Orizaba.
[2] Computer Science Department, Universidad Carlos III de Madrid
{galor,aaguilar,ujuarez,rposada,gcortes,
magarcia}@itorizaba.edu.mx,
{juanmiguel.gomez,myriam.mencke,alejandro.rodriguez}@uc3m.es

**Abstract.** In this paper, we propose a Web service-based system that offers a brokering service for the procurement of products in a Supply Chain Management (SCM) scenario. As salient contributions, our system provides a hybrid architecture combining features of both SOA and EDA and a set of mechanisms for business processes pattern management, monitoring based on UML sequence diagrams, Web services-based management, event publish/subscription and reliable messaging service.

**Keywords:** EDA, SOA, Supply Chain Management, Web services.

## 1 Introduction

Commonly, the main goal of a supply chain is satisfy the customer's requests as soon as they appear. However, to satisfy these requests, it is necessary to consider notification services and event management as required mechanisms due to continuously changing data. Service-Oriented Architecture (SOA) is an architectural paradigm for creating and managing "business services" that can access these functions, assets, and pieces of information with a common interface regardless of the location or technical makeup of the function or piece of data [1]. SOA uses Web services. A Web service is a software component that is accessible by means of messages sent using standard web protocols, notations and naming conventions, including the XML protocol [2]. The notorious success that the application of the Web service technology has achieved in B2B e-Commerce has also lead it to be viewed as a promising technology for designing and building effective business collaboration in supply chains. Deploying Web services reduces the integration costs and brings in the required infrastructure for business automation, obtaining a quality of service that could not be achieved otherwise [3], [4]. However, an SOA infrastructure does not address all the capabilities needed in a typical SCM scenario. It does not have the ability to monitor, filter, analyze, correlate, and respond in real time to events. These limitations are addressed with an Event-Driven Architecture (EDA). An EDA combined with SOA, provides that ability to create a SCM architecture that enables business. An EDA is an architectural paradigm based on

N.T. Nguyen, R. Kowalczyk, and S.-M. Chen (Eds.): ICCCI 2009, LNAI 5796, pp. 685–696, 2009.

using events that initiate the immediate delivery of a message that informs to numerous recipients about the event so they can take appropriate action [5].
Based on this understanding, in this paper we propose a Web service-based system that offers a brokering service to facilitate the business processes integration in supply chains. Our brokering service is an extended part of a complex system named BPIMS-WS which provides a virtual marketplace where people, agents and trading partners can collaborate by using current Web services technology in a flexible and automated manner [6, 7]. Our brokering service provides the following contributions:(1) A hybrid architecture that borrows features from service-oriented and event-driven architectures to provide support for B2B and SCM collaborations, (2)A business processes pattern management component for the orchestrations of business processes, (3)A mechanism based on UML sequence diagrams to monitor the interactions involved in business collaborations, and (4) An execution event publish/subscription mechanism to incorporate information into business processes and decisions through event publication.

## 2 Hybrid Architecture for e-Procurement

The hybrid architecture has a layered design. Furthermore, our proposal presents a component-based and hybrid architecture, borrowing features from SOA and EDA. In an SOA context, our approach acts as a Business Process Management (BPM) platform based on the SOA paradigm, facilitating the creation and execution of highly transparent and modular process-oriented applications and enterprise workflows In an EDA context, our approach provides a software infrastructure designed to support a more real-time method of integrating event-driven application processes that occur throughout existing applications, and are largely defined by their meaning to the business and their granularity. Regardless of the event's granularity, our proposal focuses on ensuring that interested parties, usually other applications, are notified immediately when an event happens. These features are performed by our brokering service. Its general architecture is shown in Fig. 1. Each component has a function explained as follows:

**SOAP Message Analyzer** determines the structure and content of the documents exchanged in business processes involved in SCM collaborations. Since our proposal is based on Web services, this component determines the information involved in the incoming SOAP messages by means of XML parsers and tools. A DOM API is used to generate the tree structure of the SOAP messages, whereas SAX is used to determine the application logic for every node in the SOAP messages.

**Service Registry** is the mechanism for registering and publishing information about business processes, products and services among supply chain partners, and to update and adapt to SCM scenarios. We used a UDDI node which is an industry initiative to create a platform-independent, open framework for describing services, discovering businesses, and integrating business services. In our UDDI node, commercial enterprises, services and products both are classified and registered. For the classification of business processes, products and services in the registry, we use broadly accepted ontologies/taxonomies like NAICS, UNSPSC and RosettaNet.

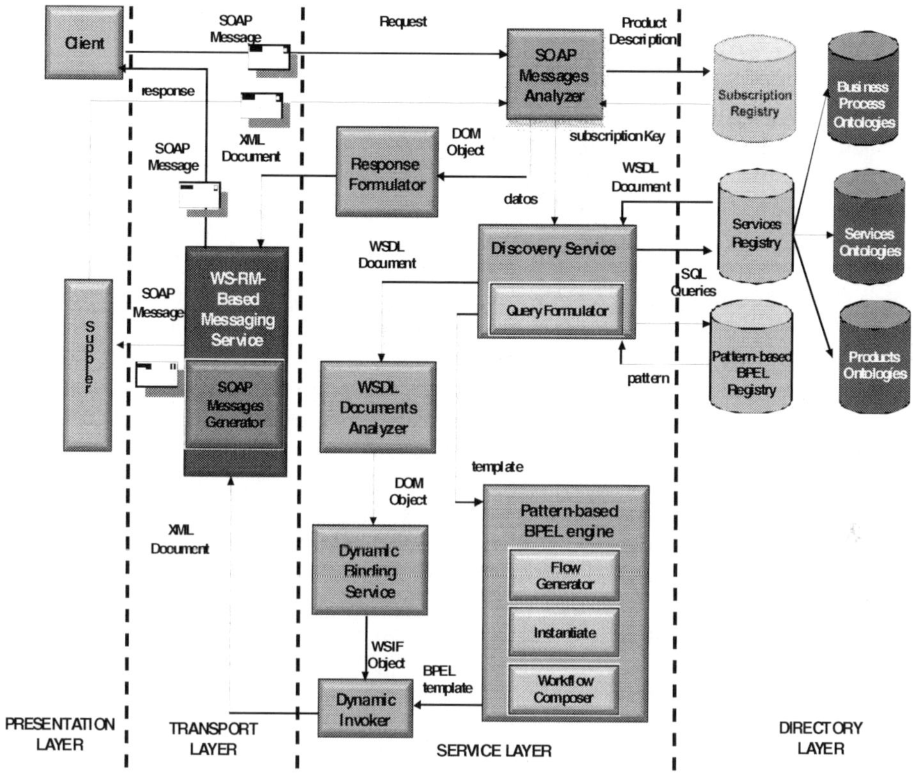


**Fig. 1.** General architecture of the Web services-based brokering service

**Subscription Registry** is the mechanism for registering interactions in which systems publish information about an event to the network so that other systems, which have subscribed and authorized to receive such messages, can receive that information and act on it appropriately. According to the cause of an event, knowledge often referred to as event causality in this work, we have considered both vertical and horizontal causality meaning that the event's source and cause reside both on different and on the same conceptual layers in an architectural stack, respectively. From vertical causality, this registry has support for storing execution events that represent runtime occurrences such as service or component invocations under a vertical causality. Lifecycle (such as stopping or starting a business process), management (when thresholds have exceeded defined limits or ranges) events, which are also part of a vertical causality, are considered as future work. From horizontal causality, this registry has support for Platform-layer, Component-layer and Business-layer events. Platform-layer events signify platform-level activities, such as the modification of a data source or the addition of a new service. Component-layer events signify component-level activities, such as the transformation of a view or a state-machine transition. Finally, Business-layer events signify business-level activities, such as the creation of a new user or the removal of an account.

**Discovery Service** is a component used to discover business processes implementations. Given the dynamic environment in SCM, the power of being able to find business processes on the fly to create new business processes is highly desirable. A key step in achieving this capability is the automated discovery of business processes described as Web services. In this sense, this component discovers Web services like authentication, payments, and shipping at run time from a SCM scenario. These Web services can be obtained from suitable service providers and can be combined into innovative and attractive product offerings to customers. When there is more than one service provider of the same function, it can be used to choose one service based on the client's requirements. Inside the discovery service, there is a query formulator which builds queries based on the domain ontology that will be sent to the registry service. This module retrieves a set of suitable services selected from the previous step and creates feasible/compatible sets of services ready for binding. The discovery service uses sophisticated techniques to dynamically discover web services and to formulate queries to UDDI nodes.

**Dynamic Binding Service** is a component that binds compatible business processes described as Web services. The binding of a Web Service refers to how strong the degree of coupling with other Web Services is. For instance, the technology of one Web service provider might be incompatible with that of another, even though the capabilities of both of them match with some requirements. In this sense, the module acts as an API wrapper that maps the interface source or target business process to a common interface supported by our proposal.

**Dynamic Invoker** transforms data from one format to another. This component can be seen as a data transfer object which contains the data (i.e., request or response) flowing between the requester to the provider applications of Web services.

**WSDL Document Analyzer** validates WSDL documents that describe business processes by their interfaces which are provided and used by supply chain partners. WSDL documents employ XML Schema for the specification of information items either product technical information or business processes operations. In this context, this component reports the business processes operations, input and output parameters, and their data types in a XML DOM tree.

**WS-RM-based Messaging Service** is the communication mechanism for the collaboration among the parties involved along the whole chain. One of the most critical aspects in SCM is to maintain its continuous operation as long as possible. In order to bring effective communication mechanisms along the chain, information technologies are considered to be the ideal solution for solving the problems related to reliability. Reliability of Web services is impacted by several factors including but not limited to, the reliability of the Web service end-points; the performance and fault-tolerance characteristics and the extent to which Web services can handle concurrent client access, among others.

Our architecture uses the Web Services Reliable Messaging (WS-RM) which is a protocol that provides a standard, interoperable way to guarantee message delivery to applications or Web services. In this sense, our proposal provides a guaranteed delivery and processing that allows in a reliable way, delivery of messages between distributed applications in the presence of software components, systems, or network failures through WS-RM.

**Response Formulator** receives the responses from the suppliers about a requested product. This module retrieves useful information from the responses and builds a XML document with information coming from the service registry and the invocations' responses. This XML document is presented in HTML format using the Extensible Stylesheet Language (XSL). The answer contains information pertaining to the product (according to the requested product) and the electronic address of the enterprise that offers that product.

**Workflow Engine** coordinates Web services by using a BPEL-based business process language. It consists of building a fully instantiated workflow description at design time, where business partners are dynamically defined at execution time. In supply chain management, workflows cannot be determined since business partners are not known beforehand, and because they are continuously changing their client-provider roles through collaboration. For this reason, we have designed and implemented a repository of generic BPEL workflow definitions which describe increasingly complex forms of recurring situations abstracted from the various stages from SCM. Its design is presented in [8]. This repository contains workflow patterns of interactions involved in an e-procurement scenario. These workflows patterns describe the types of interactions behind each business process, and the types of messages that are exchanged in each interaction.

According to the emphasis on automation, our hybrid architecture can be accessed in two modes of interaction, either as a proxy server or as an Internet portal. In the first mode, the brokering service can interoperate with other systems or software agents. In the second mode, our hybrid architecture acts as an Internet portal that provides to the users a range of options among the Web services available through the brokering service.

## 3 Case of Study: An e-procurement scenario

The case of study describes how our brokering service facilitates the discovery of Web services that are offered by different enterprises that sell electronic components.

Suppose the following scenario:

1. The enterprises sell on-line electronic components. The enterprises have registered their products and their business processes as Web services in the UDDI node of our brokering service.
2. A potential client (enterprise) starts a supply chain to procure products by requesting a purchase order by means of Web services.

In this scenario, we approach the fundamental problem of determining how a client can discover and invoke the Web services available to carry out e-procurement?

On the one hand, we assume that the business processes of the registered enterprises in our architecture are based on the commercial behaviors described in RosettaNet PIPs. On the other hand, we do not assume, however, that all suppliers use the same data fields for product descriptions and follow the same protocol for their interactions with customers. The whole e-procurement business process based on the fourth kind of response (detailed below) is shown in figure 2. In this scenario four types of responses can be observed:

**(1)** The request is not understood, i.e. the hybrid architecture cannot process the client's request because it is ill-formed. Some factors, such as lack of parameters or invoking a Web service with a wrong operation name, produce an ill-formed request. In this case, an error message indicating that the request cannot be processed is returned.

**(2)** No provider is found for the requested product, i.e. the hybrid architecture returns an empty list indicating that no provider can supply the requested product. This situation is derived from two factors: (a) the requested product is not registered in our architecture and therefore providers are not available, and (b) the requested product is registered but is not associated with a provider.

**(3)** A provider for the requested product is found, i.e. the hybrid architecture returns a list of the enterprises that appear as the product providers. This is the better case because the requested product is registered and is associated with a provider. In this case, a list of providers that offer the requested product in their stocks is displayed.

**(4)** A provider is found but the product is not available at that time. In this case, our hybrid architecture may engage in obtaining such a product from the registered providers.

In what follows, we address only the fourth case observed, when some providers are found but none of them have enough number of products at the time to fulfill the request. The proposed solution is based on the use of business processes patterns described as BPEL workflows, after an event-based subscription pattern that occurs in the fourth kind of response has been identified. The event-based subscription pattern is used to obtain a list of suppliers offering the requested product in sufficient quantities.

The event-based subscription pattern is explained as follows: Firstly, the hybrid architecture creates new instances for both service and subscription registries (Steps 1-4 in Fig. 2). This step is an initialization phase in which our architecture is ready for listening client's requests. The client submits a request by sending a SOAP message containing the product description (Step 5 in Fig. 2). This request corresponds to the RosettaNet PIP 2A5 (Query Technical Information). A screenshot of products selection is shown in Fig. 3a. The SOAP message analyzer extracts the product description to be sent to the discovery service. The query formulator builds a query to determine the qualified suppliers. The query formulator transforms the product description into SQL sentences and builds a query to the service registry.

The query is executed and the response is a set of WSDL documents of candidate suppliers (Steps 6-9 in Fig. 2). Each WSDL document represents the RosettaNet PIP 2A5 from available suppliers. For each WSDL document, the access point, operations, messages and parameters are retrieved through the WSDL document analyzer. Then, the binding service identifies the type of communication protocol specified in each WSDL document and invokes the Web service of each candidate supplier (Step 10 and 11 in Fig. 2). The hybrid architecture receives the supplier's response (Steps 12 and 13 in Fig. 2). Then, the SOAP message analyzer determines which of the four types of responses mentioned before was included in the message. Whenever the response corresponds to the fourth case, an execution event is triggered by asking the client if it is willing to subscribe for a product supplier when it becomes available

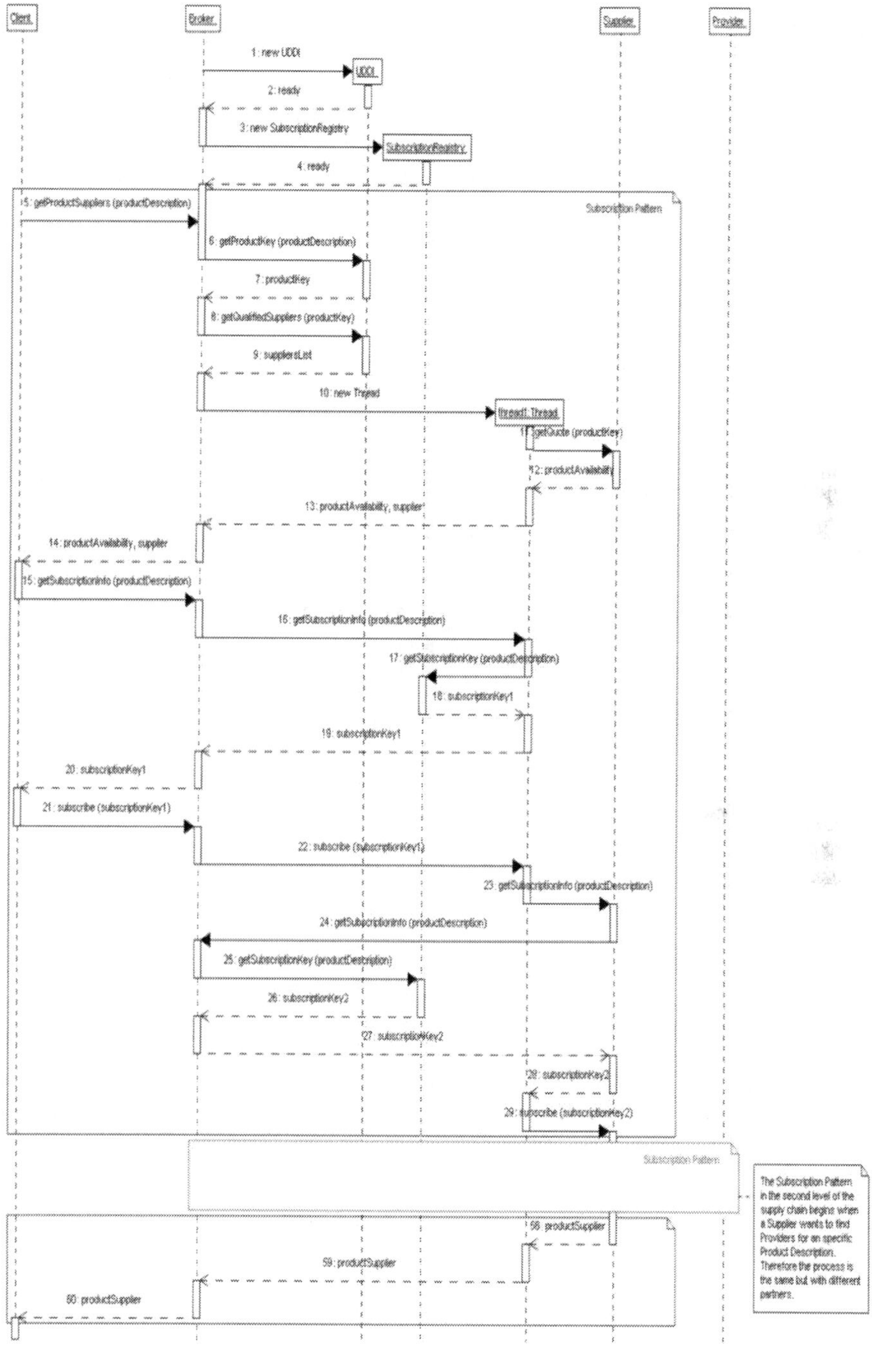


**Fig. 2.** Interactions involved in e-procurement business process

(Step 14 in Fig. 2). If the client accepts, the hybrid architecture uses a publish/subscription mechanism within the subscription registry, publishing the client's requested product description and generating a subscriptionKey that will be returned to her (Steps 15-20 in Fig. 2). Then the client invokes the subscription service by sending a SOAP message containing the subscriptionKey. In this case, she will be waiting for an answer (Step 21 in Fig. 2). During this wait, the architecture requests a corresponding subscriptionKey from the supplier by sending the client's requested product description (Steps 22 and 23 in Fig. 2). The supplier stores the product description and generates a subscriptionKey as a response to the request (Steps 24-26 in Fig. 2). Next, another execution event is triggered by invoking the supplier's subscribe service and waits for a supplier's answer (Steps 27-29 in Fig. 2). During this wait, the event-based subscription pattern in the second level of the supply chain management begins when a supplier wants to find providers for a product description. Therefore, the process is the same but the partners take different roles, i.e., the supplier acts as a client and the provider acts as a supplier of our brokering service. When the provider's response is returned the second level of the SCM, the hybrid architecture analyzes it and retrieves the useful information about the requested product. With this information, the architecture is able to send a response to the supplier's subscribe service request. Here the second level of the supply chain management is completed.

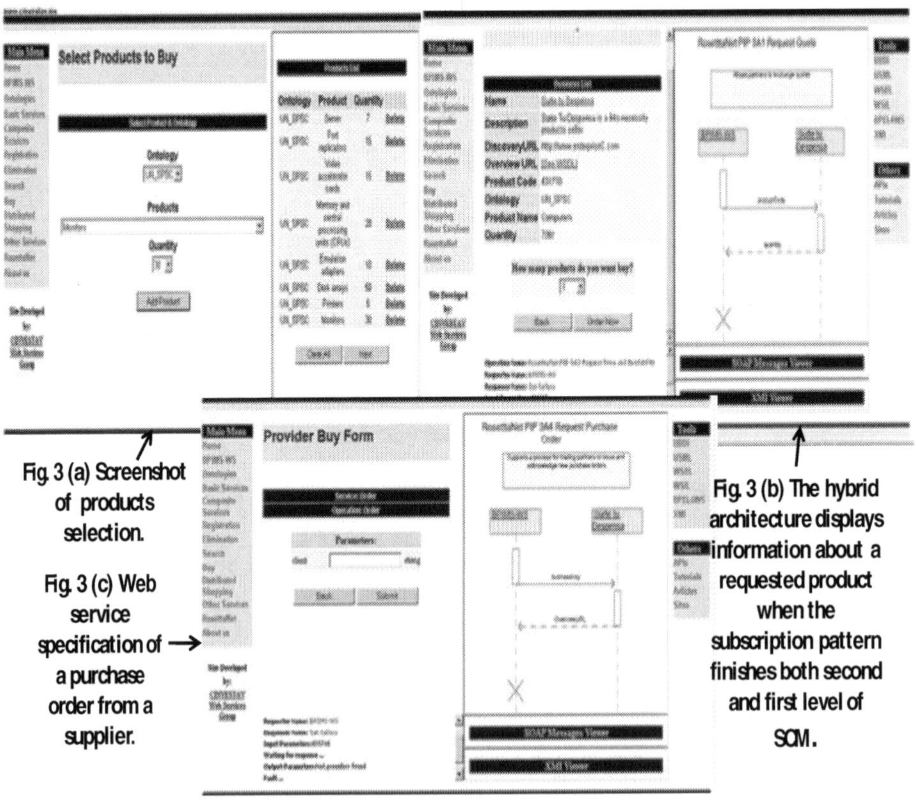


Fig. 3 (a) Screenshot of products selection.

Fig. 3 (b) The hybrid architecture displays information about a requested product when the subscription pattern finishes both second and first level of SCM.

Fig. 3 (c) Web service specification of a purchase order from a supplier.

In a similar way, the supplier will answer to the subscribe service request and its wait will finish (Step 58-59 in Fig. 2). Finally, the hybrid architecture is able to answer the client's subscribe service request and as a consequence, the client's wait will finish and the client is getting the information about her requested product (Step 60 in Fig. 2). The information about a requested product by the client is shown in Fig. 3b. At this point, the client is already able to select the quantity of the requested product. Once selected, the architecture makes a query to the service registry to locate the URL where the RosettaNet PIP 3A4 (Request Purchase Order) is located to obtain and analyze the Web service specification. Next, a graphic user interface of the Web service specification is displayed, enabling the visualization of the activities involved in the purchasing order process. The client is then asked to provide the information required to complete the purchase. This graphic interface is shown in Fig 3c. Upon completion, the Web service corresponding to the RosettaNet PIP 3A5 is invoked (Query Order Status) to verify that the purchase was successful. Finally, the results to the user are displayed.

## 4 Related Work and Discussion

In [9] a system named eXFlow for business processes integration on EAI and B2B e-commerce is presented. However, eXFlow only provides support for Web services discovery, invocation, orchestration and monitoring. Web services management is not considered and since eXFlow is based on SOA architecture, asynchronous messaging is not provided. In [10] another system named THROWS is proposed, an architecture for highly available distributed execution of Web services compositions. In the THROWS architecture, the execution control is hierarchically delegated among dynamically discovered engines. However, THROWS is in the design phase, thus it is still under development. In [11] an architecture for semi-automatic Web services composition is provided, combining both centralized model and peer-to-peer approaches. This proposal only has support for Web services discovery, invocation, orchestration and monitoring. Web services management is not considered, and the architecture is being developed. In [12] a system which acts as an Integration Broker for Heterogeneous Information Sources (IBHIS) was developed. IBHIS is already implemented but process activity monitoring is not included. In [13], a system named IRIS (Interoperability and Reusability of Internet Services) for Web services composition through a set of graphic interfaces. In IRIS, Web services discovery and orchestration are provided by an ontology-based registry approach. However, IRIS is focused on the simulation of Web services composition, therefore Web services execution is not included. In [14] is proposed a framework named KDSWS (Knowledge-based Dynamic Semantic Web Services) which addresses in an integrated end-to-end manner, the life-cycle of activities involved in brokering and managing of Semantic Web Services. However, agent monitoring is not considered and the architecture is subject to ongoing work. In [15] an architecture for Web services discovery is proposed by using a goal-oriented approach. Web services discovery is carried out by means of services chains satisfying certain constraints. This architecture only provides support for Web services management and monitoring, and it is in the design phase. In [16] is proposed a framework for Dynamic Web Service Invocation. This framework is based on SOA architecture. Publication/subscription and notification mechanisms are used in Web services discovery in UDDI nodes. However, an experimental prototype is provided, which does not consider Web services orchestration, monitoring and management. In [17] was developed a system named METEOR-S (Managing End-To-End OpeRations for Semantic Web services) which is a constraint driven Web Service composition tool. In the METEOR-S architecture, web services management is not considered. Nevertheless, METEOR-S has been implemented and is working well. More recent works have proposed others approaches for e-procurement in SCM. In [18] is presented an overview and a classification of B2B approaches. Business- and implementation-related specifications are classified in terms of the Open-EDI reference model. Furthermore, the path from a business model down to deployments artifacts for SOA environments is outlined. In [19] an agent-mediated coordination approach is proposed to dynamic supply chain integration. A prototype has been implemented with simulated experiments highlighting the effectiveness of the approach. In [20] is proposed a service oriented knowledge-based architecture to support supply chain application. The proposed architecture has the capability to meet the on-demand requirements of dynamic supply chains. Business requirements and potential benefits associated are discussed. In [21] an approach that improves requirements engineering as part of the UN/CEFACT's Modeling Methodology for specifying B2B systems is presented. This approach helps to avoid inconsistencies between worksheets capturing the business experts' knowledge and the modeling artifacts. This is realized by integrating a worksheet editor into a UMM modeling tool. In [22] is established a two-period replenishment model of (s, S) strategy to analyze the operation cost efficiency of the traditional procurement in strategic partnership, e-procurement and the mixed procurement strategy respectively. The model quantifies the effect of the Internet on procurement management. Finally, in [23] is proposed a new solving scheme for modern e-commerce system frameworks. The solution of sharing function and data for each node of a modern business flow is designed including data reuse, elastic cart, order of automatic collection, trusteeship calculating of logistics cost request and automatic chamberlain based on Web Services.

## 5 Conclusions

SCM is an important yet difficult problem to be addressed in its full complexity. However, we believe that a hybrid architecture, borrowing features from SOA and EDA, may provide the fundamental structure in which the solutions to the diverse problems that SCM conveys can be accommodated. In this paper, we have described a hybrid architecture we have developed so far that provides a comprehensive framework for developing business integration, collaboration and monitoring in SCM scenarios. Among the applications we envisioned for our proposal, the orchestration of long-term supply chains involving operation research methods to minimize costs, reduce delivery times and maximize quality of service along with artificial intelligence methods to provide semantic matching and to define business partners profile management is now under consideration.

## Acknowledgements

This work is supported by the General Council of Superior Technological Education of Mexico (DGEST). Additionally, this work is sponsored by the National Council of Science and Technology (CONACYT) and the Public Education Secretary (SEP) through PROMEP.

## References

1. Papazoglou, M.P.: Service-Oriented Computing: Concepts, Characteristics and Directions. In: Proceedings of the Fourth International Conference on Web Information Systems Engineering (WISE 2003), IEEE Press, Los Alamitos (2003)
2. Vinoski, S.: Integration with Web Services. IEEE Internet Computing, 75–77 (November-December 2003)
3. Adams, H., Gisolfi, D., Snell, J., Varadan, R.: Custom Extended Enterprise Exposed Business Services Application Pattern Scenario (January 1, 2003), http://www-106.ibm.com/developerworks/webservices/library/ws-best5/
4. Samtani, G., Sadhwani, D.: Enterprise Application Integration and Web Services. In: Fletcher, P., Waterhouse, M. (eds.) Web Services Business Strategies and Architectures, pp. 39–54. Expert Press, Ltd., Birmingham (2002a)
5. Sriraman, B., Radhakrishnan, R.: Event-Driven Architecture augmenting Service-Oriented Architecture. Sun Microsystems (January 2005)
6. Alor Hernandez, G., Olmedo Aguirre, J.O.: BPIMS-WS: Service-Oriented Architecture for Business Processes Management. Advances in Artificial Intelligence, Computing Science and Computer Engineering. Journal of Research on Computing Science 10, 255–264 (2005)
7. Alor Hernandez, G., Sandoval Hernandez, C., Olmedo Aguirre, J.O.: BPIMS-WS: Brokering Architecture for Business Processes Integration in B2B E-commerce. In: Proceedings of the XV International Conference on Electronics, Communications, and Computers, CONIELECOMP 2005. IEEE Press, Los Alamitos (2005)
8. Sandoval Hernandez, C., Alor-Hernandez, G., Olmedo, J.O.: Dynamic generation of organizational BPEL4WS workflows. In: Proceedings of the 1st International Conference on Electrical and Electronics Engineering and X Conference on Electrical Engineering (ICEEE-CIE 2004), pp. 130–135. IEEE Press, Los Alamitos (2004a)
9. Chung Nin, N.C., Wen-Shih, H., Tse-Ming, Seng-cho, T.: eXFlow:A Web Services-Compliant System to Support B2B Process Integration. In: Proceedings of the 37th Hawaii International Conference on System Sciences 2004 (2004)
10. Lakhal, N.B., Kobayashi, T., Yokota, H.: THROWS: An Architecture for Highly Available Distributed Execution of Web Services Compositions. In: Proceedings of the 14th International Workshop on Research Issues on Data Engineering: Web Services for E-Commerce and E-Government Applications. IEEE Press, Los Alamitos
11. Budak, A.I., Aleman-Meza, B., Zhang, R., Maduko, A.: Ontology-Driven Web Services Composition Platform. In: Proceedings of the IEEE International Conference on E-Commerce Technology. IEEE Press, Los Alamitos

12. Turner, M., Zhu, F., Kotsiopoulos, I., Russell, M., Budgen, D., Bennett, K., Brereton, P., Keane, J., Layzell, P., Rigby, M.: Using Web Service Technologies to create an Information Broker: An Experience Report. In: Proceedings of the 26th International Conference on Software Engineering (ICSE 2004). IEEE Press, Los Alamitos (2004)
13. Radetzki, U., Cremers, A.B.: IRIS: A Framework for Mediator-Based Composition of Service-Oriented Software. In: Proceedings of the IEEE International Conference on Web Services (ICWS 2004). IEEE Press, Los Alamitos (2004)
14. Howard, R., Kerschberg, L.: A Knowledge-based Framework for Dynamic Semantic Web Services Brokering and Management. In: Proceedings of the 15th International Workshop on Database and Expert Systems Applications, DEXA 2004 (2004)
15. Srinivasan, A., Sundaram, D.: Web Services for Enterprise Collaboration: A Framework and a Prototype. In: Proceedings of the 30th EUROMICRO Conference (EUROMICRO 2004), IEEE Press, Los Alamitos (2004)
16. Jian-Jun, Y., Zhou, G.: Dynamic Web Service Invocation. In: Proceedings of the IEEE International Conference on E-Commerce Technology for Dynamic E-Business (CEC-East 2004). IEEE Press, Los Alamitos (2004)
17. Aggarwal, R., Verma, K., Miller, J., Milnor, W.: Constraint Driven Web Service Composition in METEOR-S. In: Proceedings of the 2004 IEEE International Conference on Services Computing (SCC 2004). IEEE Press, Los Alamitos (2004)
18. Jurgen, D., Christoph, G., Hannes, W., Marco, Z.: A Survey of B2B Methodologies and Technologies: From Business Models towards Deployment Artifacts. In: Proceedings of the 40th Hawaii International Conference on System Sciences – 2007. IEEE, Los Alamitos (2007)
19. Minhong, W., Huaiqing, W., Jiming, L.: Dynamic Supply Chain Integration through Intelligent Agents. In: Proceedings of the 40th Hawaii International Conference on System Sciences – 2007, pp. 1–10. IEEE Press, Los Alamitos (2007)
20. Wei, D., Moynihan, P., Juanqiong, G., Ping, Z., Xi, Y., Tiedong, C., Xin, W.: Services Oriented Knowledge-based Supply Chain Application. In: Proceedings of the 2007 IEEE International Conference on Services Computing (SCC 2007). IEEE Press, Los Alamitos (2007)
21. Christian, H., Marco, Z., Philipp, L., Rainer, S.: Worksheet-Driven UMM Modeling of B2B Services. In: Proceedings of the IEEE International Conference on e-Business Engineering, pp. 30–38. IEEE Press, Los Alamitos
22. Guangshu, C.: The Analysis of Supply Chain Procurement Strategy in E-Commerce. In: Proceedings of the International Symposium on Electronic Commerce and Security 2008 (2008)
23. Ying, H., Cheng, L.V.: Research on the Solving Scheme of E-commerce System ase don Web Services. In: Proceedings of the International Conference on Management of e-Commerce and e-Government, pp. 230–233. IEEE Press, Los Alamitos

# Localization by Wireless Technologies for Managing of Large Scale Data Artifacts on Mobile Devices

Ondrej Krejcar

VSB Technical University of Ostrava, Center for Applied Cybernetics,
Department of measurement and control, 17. Listopadu 15,
70833 Ostrava Poruba, Czech Republic
Ondrej.Krejcar@remoteworld.net

**Abstract.** The ability to locate a mobile device by a wireless network is well known possibilities. Except traditional WiFi the BT or GSM network can be used. Utility of position info is in many of current areas. New kind of mobile devices are equipped with high capacity of hardware like RAM, ROM, SD Cards etc. However the memory bus and CPU are not able to process large amount of data which leads to slow response in case of mobile software application with large files. To allow an adequate work with such kind of applications with same comfort as on desktop devices the prebuffering techniques can be used to solve it. Main area of interest is in a use of locating and tracking users of a mobile information system to prebuffer possible large amount data to before usage. All large data files are stored as artifacts along with its position information in building or larger area environment. The accessing of prebuffered data on mobile device can highly improve response time needed to view large multimedia data. This fact can help with design of new full scale applications for mobile devices.

**Keywords:** Mobile Device, Localization, Prebuffering, Response Time.

## 1 Introduction

The usage of various mobile wireless technologies and mobile embedded devices has been increasing dramatically every year. This will lead to the rise of new application domains in network-connected PDAs (Personal Digital Assistants) that provide more or less the same functionality as their desktop application equivalents. The idea of full scale applications pursuable on mobile devices is based on current hi-tech devices with large scale display, large memory capabilities, and wide spectrum of network standards plus embedded GPS module (e.g. HTC Touch HD).

Users of these portable devices use them all time in context of their life. Context is relevant to the mobile user, because in a mobile environment the context is often very dynamic and the user interacts differently with the applications on his mobile device when the context is different [1].

On localization basis, a special framework called PDPT (Predictive Data Push Technology) was created to improve a usage of large data artifacts (image, audio or video stream, technology plans, etc.) of mobile devices [3]. The continual user

N.T. Nguyen, R. Kowalczyk, and S.-M. Chen (Eds.): ICCCI 2009, LNAI 5796, pp. 697–708, 2009.

position information is used to determine a predictive user position. The data artifacts linked to user predicted position are prebuffered to user mobile device. When user arrives to position which was correctly determined by PDPT Core, the data artifacts are in local memory of PDA. The time to display the artifacts from local memory is much shorter than in case of remotely requested artifact.

The WiFi is not only one wireless network to use for localization of user device. WiFi has advantage in speed in indoor positioning therefore the GSM/UMTS can be used in outdoor. The GPS sensor is also embedded in several types of current mobile devices, or it can be plugged by SDIO or BT interface.

### 1.1 The Need of Predictive Data Push Technology

The created PDPT framework is based on a model of location-aware enhancement which is applied in framework to increase the real dataflow from wireless access point – AP (server side) to PDA (client side). The real download speed for WiFi network is about 160 kB/s for actual PDA devices. More details about the facts of slow transfer speed on mobile devices can be found in chapter 2.5 [5].

Primary dataflow is enlarged by data prebuffering. Selecting of data objects to be buffered to mobile device cache is made on the base of position of user's device and objects with relevant position for such user's position. PDPT Core pushes the data from SQL database (WLA database [Fig. 5]) to clients PDA to be helpful when user comes at final location which was expected by PDPT Core. The benefit of PDPT consists in time delay reducing needed to display desired artifacts requested by a user from PDA. First of all the maximum response time of an application (PDPT Client) for user must be specified. Nielsen [6] specified this time delay to 10 seconds [7]. During this time the user was focused on the application and was willing to wait for an answer. The book is over 20 years old (published in 1994) but it is a basic literature for this phenomena. Galletta, Henry, McCoy and Polak (2002) findings suggest that, 'decreases in performance and behavioral intentions begin to flatten when the delays extend to 4 seconds or longer, and attitudes flatten when the delays extend to 8 seconds or longer'. The time period of 10 seconds is used to calculate the maximum possible data size of a file transferred from server to client (during this period). For transfers speed 160 kB/s the result file size is 1600 kB.

The next step was an average artifact size definition. The network architecture building plan is used as a sample database, which contained 100 files of average size of 470 kB. The client application can download during the 10 second period from 2 to 3 artifacts. The problem is the long time delay in displaying of artifacts in some original file types (e.g. Autocad in case of vector graphic or MS Office in general case). It is needed to use only basic data formats, which can be displayed by PDA natively (bmp, jpg, wav, mpg, etc.) without any additional striking time consumption. If other file types are used, the delay for presentation of file must be included.

The final result of several real tests and consequential calculations is definition of artifact size to average value of 500 kB. The buffer size may differ from 50 to 100 MB in case of 100 to 200 artifacts.

The position obtaining from wireless networks background will be described in the next chapter to give a reader more information about these themes.

## 2 Obtaining the Position of a User from Wireless Networks

If the mobile device knows the position of the stationary device (transmitter), it also knows that its own position within a range of this location provider. The typical range wary from 30 to 100 m in WiFi case, respectively 50 m in BT case or 30 km for GSM. Granularity of location can be improved by triangulation of two or more visible APs (Access Points). The PDA client currently supports the application in automatically retrieving location information from nearby WiFi, BT and GSM location providers, and in interacting with the PDPT server. The application (locator) is implemented in C# language using the MS Visual Studio .NET with .NET Compact Framework and a special OpenNETCF library enhancement.

A first key step of the localization is a data collection phase. The information about the radio signals is recorded as a function of a user's location. The signal information is used to construct and validate models for signal propagation. Among other information, the signal strength (SS) is available in case of WiFi, BT and GSM networks.

To get a user position with more accuracy, the triangulation is currently used in PDPT framework. Other localization techniques like Monte Carlo localization can be used to get a better position if it is needed, but PDPT framework provides good results only with triangulation techniques on basic level of localization.

In model case [Fig. 1] the mobile client gets the SS info of three BSs (Base Stations) with some inaccuracy. Inaccuracy is caused by SS value from mobile device wireless module, where only SS in present. Circles around the BSs (in real 3D space the sphere is used around the BSs representing SS value) are crossed in red points on figure. The intersection red point (centre of three) is the best computed location of mobile user. The user track is also computed from these locations and it is stored in database for later use. This idea is applicable in case of WiFi as well as BT and GSM networks.

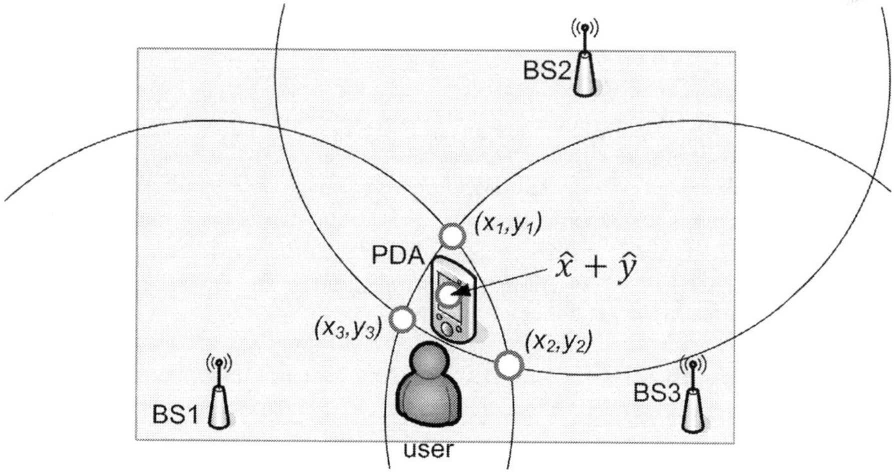


**Fig. 1.** Localization principle – triangulation

## 2.1 WiFi Localization

In real case of indoor localization by WiFi networks the several types of environment is used like open spaces, walls and mixed spaces. The Cisco APs (Cisco Aironet 1121 and 1131) are used in the test environment at Campus of Technical University of Ostrava.

The measurements on three selected (representing three types of environment) APs of all APs have been performed to get signal strength (SS) characteristics. Three test batches with 60 iterations in each selected distance were performed on each of these APs [Tab. 2]. Every batch measurement contains 60 iterations where the time delay is 5 seconds. Between each individual batch the one week delay was dropped. Distribution of a SS values is nearly the same at each batch.

Important computed values of batches are in table [Tab. 1]. The Mode values are joined for each selected distance of each AP. The resulted values are in [Tab. 2].

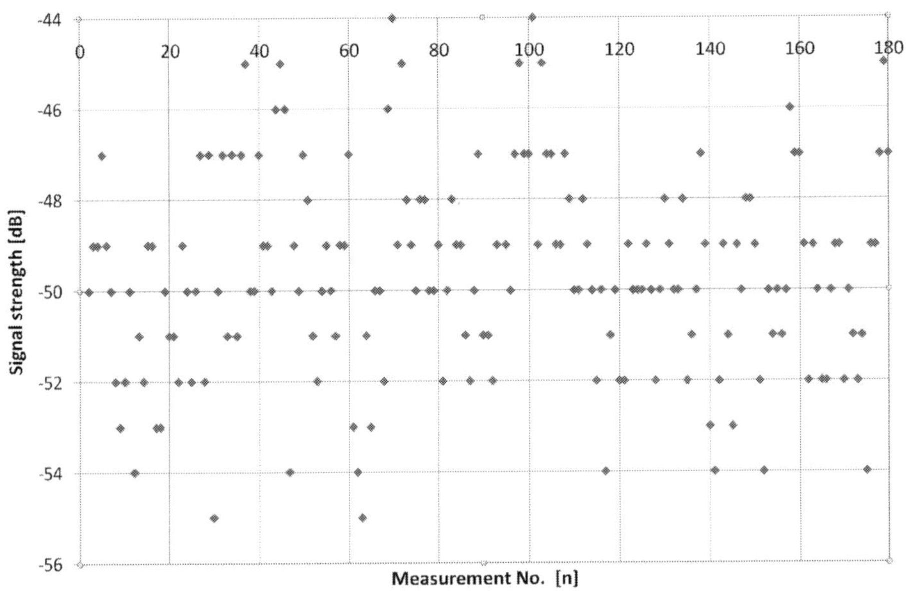


**Fig. 2.** Test measurement of WiFi SS of open space AP in 3 m distance from AP (3 batches integration)

**Table 1.** Average, Mode, Median and the Range computed values for measured batches of WiFi SS of open space AP in 3 m distance from AP

| Batch | Average [dB] | Mode [dB] | Median [dB] | Range [dB] | Distance [m] |
|---|---|---|---|---|---|
| 1 | -49.8833 | -50 | -50 | 12 | 3 |
| 2 | -49.2333 | -50 | -49 | 12 | 3 |
| 3 | -49.9833 | -50 | -50 | 9 | 3 |

**Table 2.** WiFi SS [dB] on selected AP´s in chosen distances. AP1 represent space with walls, AP2 is mixed space and AP3 is open space. All WiFi SS value is negative (-34 dB e.g.)!

| AP | Distance [m] | | | | | | | | | | | | | | | |
|---|---|---|---|---|---|---|---|---|---|---|---|---|---|---|---|---|
| | **1** | **2** | **3** | **4** | **5** | **6** | **7** | **8** | **10** | **12** | **14** | **17** | **20** | **24** | **28** | **32** |
| 1 | 34 | 35 | 45 | 47 | 46 | 47 | 57 | 50 | 54 | 63 | 69 | 64 | 63 | 69 | | |
| 2 | 34 | 35 | 34 | 43 | 41 | 42 | 51 | 57 | 63 | 60 | 63 | 70 | 74 | 75 | 75 | |
| 3 | 34 | 45 | 52 | 58 | 57 | 59 | 67 | 63 | 70 | 70 | 70 | 69 | 72 | 75 | 71 | 71 |

The characteristics from [Tab. 2] were combined to get a one characteristic called “Super-Ideal characteristic” [Fig. 3]. The computed equation for Super-Ideal characteristic is taken as basic equation for PDPT Core to compute the real distance from WiFi SS. The equation has a sufficient coefficient of determination $R^2 = 80\ \%$ ($R^2$= ssreg/sstotal).

From this Super-Ideal characteristic [Fig. 3] it is also evident the signal strength is present only in 30 meters range from the base station. This small range is caused by using of Cisco APs. These APs has only 2 dB WiFi omnidirectional internal antenna.

In the case of in-door location the damping effect of walls especially when the number of BS´s is small could hamper the positioning. However the precise positioning is not needed in all cases. When the granularity of object areas to be prebuffered into the mobile device cache is in level less than tens of meters, the localization by one or two visible BS´s is possible with high level of success. Maximal location error for static localization is 25 meters for 1 visible WiFi AP, 7

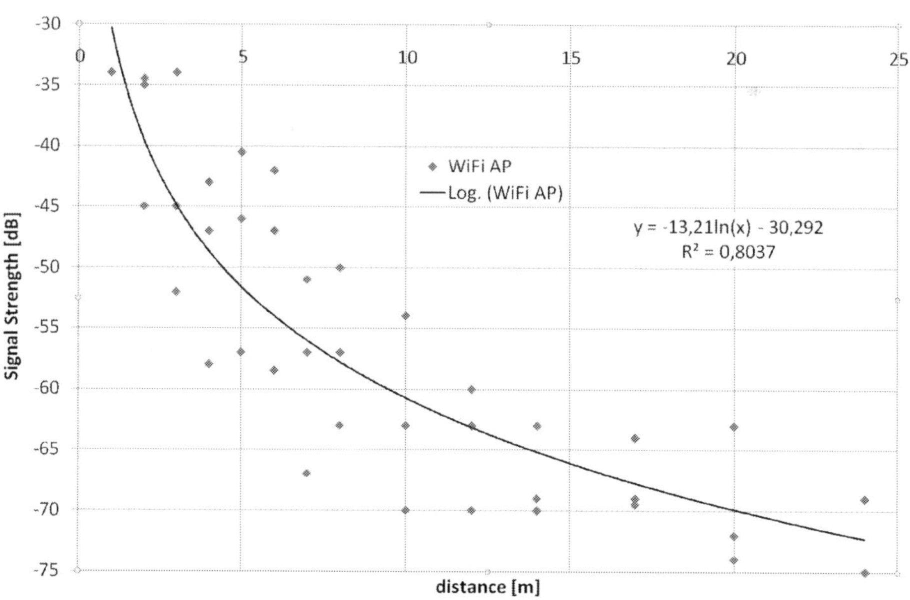


**Fig. 3.** Super-Ideal characteristic of WiFi SS propagation in mixed environment (three Cisco APs with 2dB internal antennas)

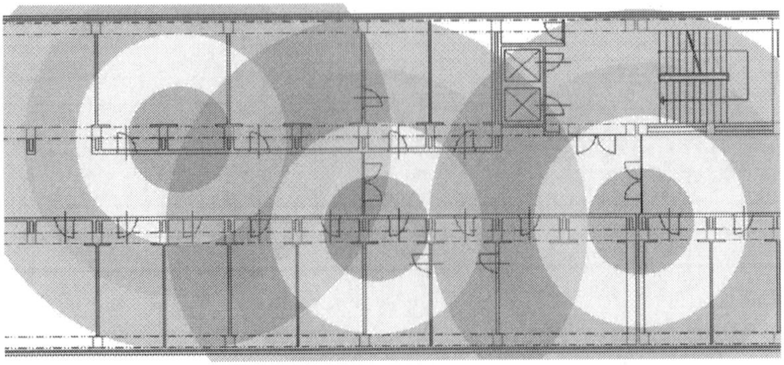

**Fig. 4.** Simple WiFi SS propagation model on real test environment. Colors represent levels of SS where red is -40 dBm, yellow is -57 dBm, green is -63 dBm and blue is -67 dBm.

meters for 2 APs and less than 5 meters in case of 3 or more visible WiFi APs (mentioned Cisco models). This localization error can be rapidly reduced by use of dynamic localization in a sense of user movement trajectory computation. Naturally, this localization principle can be applied to other wireless technologies like Bluetooth, GSM or WiMAX.

To let a mobile device determine its own position is needed to have a WiFi adapter still powered on. This fact provides a small limitation of use of mobile devices. The complex test with several types of battery is described in article [4] in chapter (3). The test results with a possibly to use a PDA with turned on WiFi adapter for a period of about 5 hours.

In BT network case, the position of BT APs must be known to allow the position determination. To manage with BT hardware of mobile device another library InTheHand 32Feet.NET is used.

### 2.2 GSM Localization

GSM network is not local network but a cellular network. The first problem is in position information of GSM BTSs (Base Transceiver Stations). However the network operator doesn't provide any of such information. One of possible solutions is based on unofficial BTSs lists which can be found on internet. The lists are typically available in HTML, TXT or CSV file format. The medium rate for BTs with GPS position information is about 90 % of all BTs in European countries. In case of PDPT Framework, the list must be converted to PDPT server database – GSM_BTS table [Fig. 5].

In all three described cases of nearby BSs scanning, the data are saved to Locator Table in PDPT server DB [Fig. 5]. Data are processed from Locator Table through the PDPT Core to Position Table. The processing techniques depend on selected wireless network. WiFi and BT network provide all visible APs nearby the user. From list of these APs, the actual position is computed (by triangulation [Fig. 1]).

Retrieving the BTS info in mobile devices with windows mobile is a problem, because operating system does not provide any GSM info to .NET Compact Framework. Even any special framework as in previous two cases does not exist. The

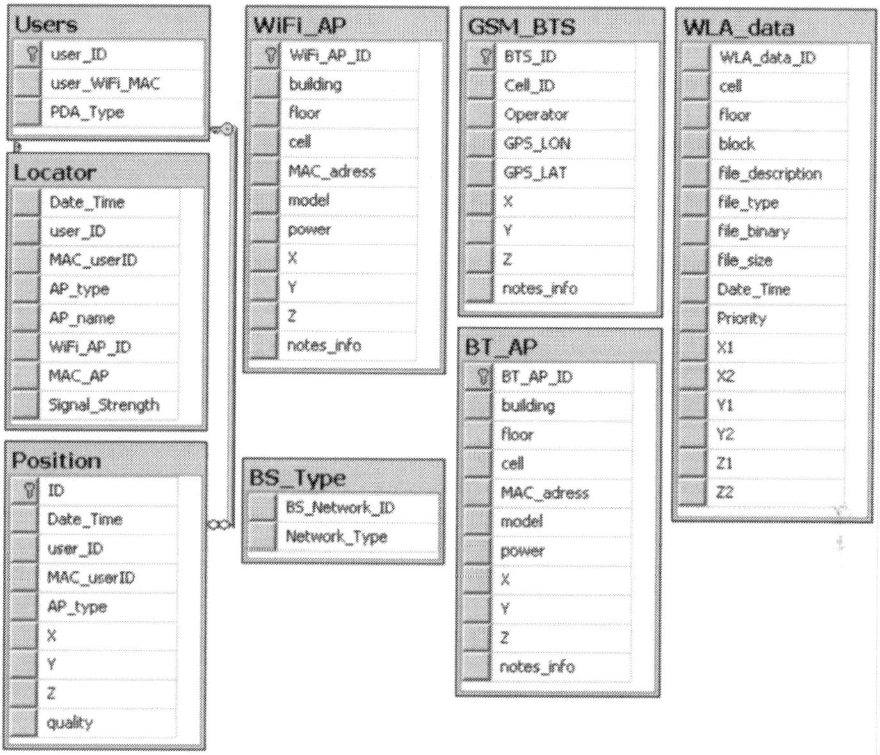


**Fig. 5.** PDPT server DB (Data Base) - New database architecture

only one possibility is in use of RIL (Radio Interface Layer) library. When a module calls the RIL to get the signal strength, the function call immediately returns a response identifier. The RIL uses the function response callback to convey signal strength information to the module. The GSM network provide only one BTS info in each search cycle. The more BTSs info is collected by a several iteration cycles. During 10 cycles (per 10 seconds) the 4 BTS info is obtained in average.

The important info from BTSs is Signal Strength and Time Advance (TA). SS is refreshed every several seconds (in every scan) whereas TA is provided only during some type of communication with selected BTS (e.g. request to talk, move to another area - Location Area Code (LAC)). The list of these BTSs with info is further processed as in previous case for WiFi and BT networks. Only change is in usage of TA if it is accessible.

## 3 The PDPT Framework

The PDPT framework server is developed as a web service to act as a bridge between SQL Server (contain WLA database) and PDPT PDA Clients (windows mobile application) [Fig. 6]. Client PDA application contain a location sensor component to scan nearby for WiFi APs. WiFi SS info is continuously transferred to the PDPT Framework Core. This component computes the user's location information from WiFi SS. In next step the PDPT Core makes a decision to which part of WLA Server database needs to be replicated to client's SQL Server CE database [9][10]. The PDPT Core decisions constitute the most important part of PDPT framework, because the kernel must continually compute the position of the user and track, and predict his future movement. After doing this prediction the appropriate data are prebuffered to client's database for the future possible requirements. This data represent artifacts list of PDA buffer imaginary image [Fig. 9].

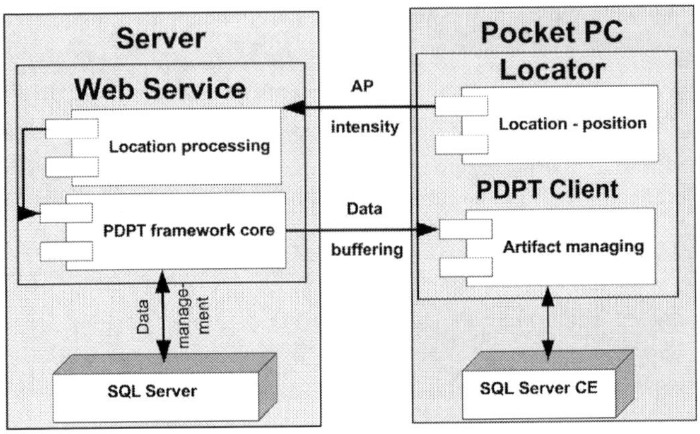


**Fig. 6.** PDPT architecture – UML design

### 3.1 Data Artifact Creating

Artifact represent an object in WLA SQL server database with image, audio, video or other file types. Every artifact must have associated with position coordinates in 3D environment (S-JTSK format is used). Open source software Quantum GIS [Fig. 7] is used to manage all data in 3D spaces like building map basis, APs location and artifacts location. To manage and work with locations of artifacts, firstly the building floor map is needed to obtain. In most cases the scanned version is adequate. The obtained map needs to be converted to Tagged Image File Format (TIFF). Location coordinates for such file must be created in TFW separate file. TFW file contains coordinates that describe the location, scale, and rotation of a map formatted as a raster TIFF image.

To get an AP location the creating of new layer in GIS project is the possible solution. The screenshot of QGIS [Fig. 7] contain a building floor map and two WiFi APs (violet colored points). Visible layers of project are marked by cross (left part of screenshot). All obtained position info must be stored in PDPT Core web service. Artifacts with position coordinates are stored in WLA database [Fig. 5] by "WLA Database Artifact Manager" [Fig. 8].

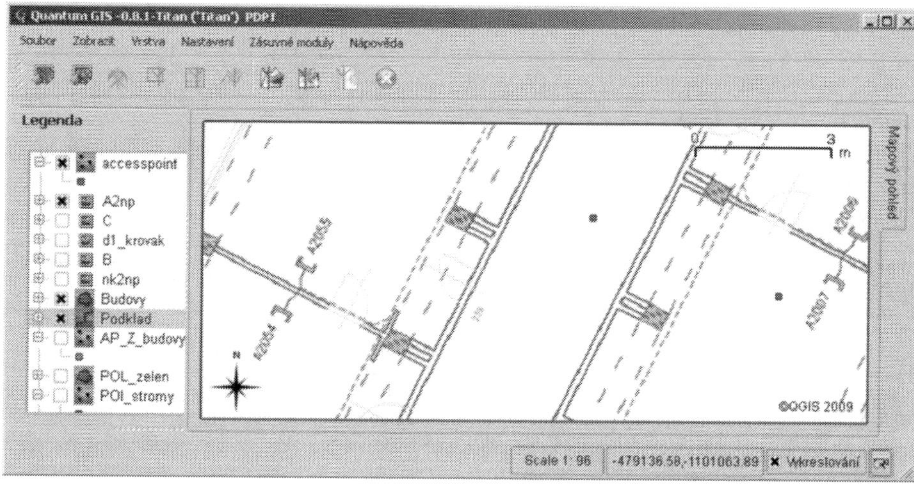


**Fig. 7.** 3D environment definition for artifacts. Every artifact must have allocated an area in which is valid

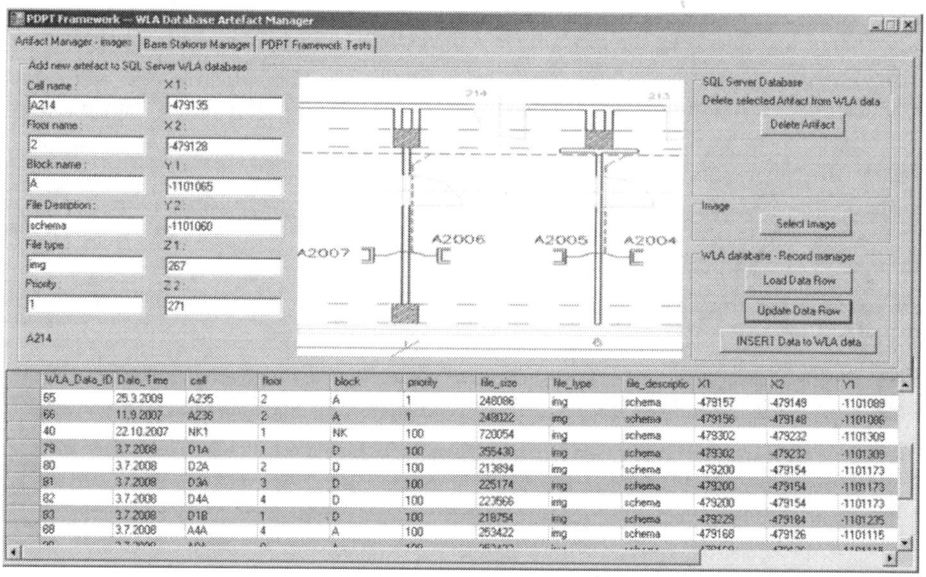


**Fig. 8.** PDPT Framework Data Artifact Manager

## 3.2 Data Artifact Managing

The WLA server database manages the artifacts in the context of their location in building environment. The PDPT Core selecting the data to be copied from PDPT server to PDA client by context information (position info). Each database artifacts must be saved in database with the information about area to which it is belong.

A software application called "Data Artifacts Manager" was created to manage the artifacts in WLA database. User can set the priority, location, and other metadata of the artifact [Fig 8]. The Manager allows creating a new artifact from multimedia file source (image, video, audio, etc.), and work with existing artifacts [5].

The needs of interface to operate with APs info arose out of the developing process of PDPT Framework. The enhancement of Artifact manager was created on that ground. Now the Artifact Manager contains a new tab "Base Stations Manager" to operate with APs or BSs of selected networks [Fig. 8]. This manager is connected directly to PDPT Server database, to tables WiFi_AP, BT_AP, GSM_BTS [Fig. 5].

### 3.3 PDPT Core - Area Definition for selecting Artifacts to Buffering

The PDPT buffering and predictive PDPT buffering principle consists of several following steps. Firstly the client must activate the PDPT on PDPT Client. This client creates a list of artifacts (PDA buffer image), which are contained in his mobile SQL Server CE database. Server create own list of artifacts (imaginary image of PDA buffer) based on area definition for actual user position and compare it with real PDA buffer image. The area is defined as an object where the user position is in the center of object. The cuboid form is used in present time for initial PDPT buffering. This cuboid has a predefined area with a size of 10 x 10 x 3 (high) meters. The PDPT Core continues in next step with comparing of both images. In case of some difference, the rest artifacts are prebuffered to PDA buffer. When all artifacts for current user position are in PDA buffer, there is no difference between images. In such case the PDPT Core is going to make a predicted user position. On base of this new predicted user position it makes a new predictive enlarged imaginary image of PDA buffer. The size of this new cuboid is predefined area of size 20 x 20 x 6 meters. The new cuboid has a center in direction of predicted user moving and includes a cuboid area for current user position. The PDPT Core compares the both new images (imaginary and real PDA buffer) and it will continue with buffering of rest artifacts until they are same. Creation of an algorithm for dynamic area definition is better in real case of usage to adapt a system to user needs more flexible in real time [11].

### 3.4 Accessing the Artifacts in PDPT Client Application

The PDPT Client application realizes thick client and PDPT and Locator modules extension. The screenshot [Fig. 9a] shows the Locator module with selected GSM scanning. The info text box "Locator AP ret." provides info about last founded GSM BTSs and number of recognized BTSs (BTSs with GPS position). In current case the 6 BTSs was founded and 5 of them was recognized by PDPT Framework. Figure [Fig. 9b] shows the data artifact presentation from MS SQL CE database to user (Ethernet plan of the current area). The PDPT tab [Fig. 9c] presents a way to tune the settings of PDPT Framework. The middle section shows the log info about the prebuffering process. The right side means measure the time of one artifact loading ("part time") and full time of prebuffering in millisecond resolution.

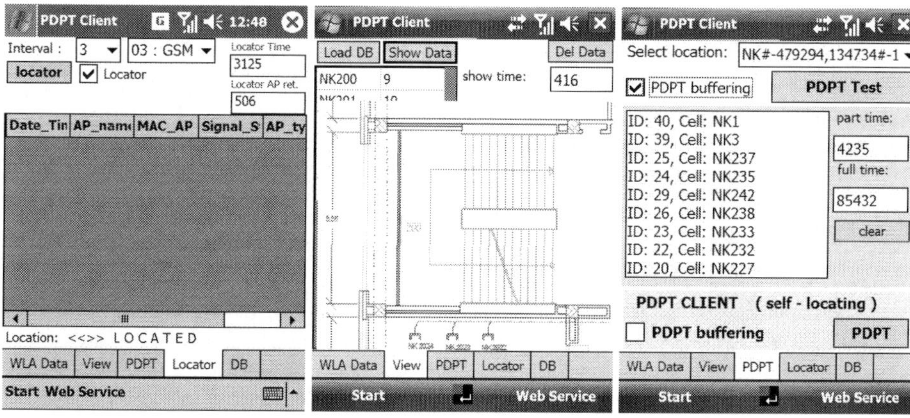


**Fig. 9.** PDPT Client – Left one figure 9a – Locator component with GSM scanning. Middle one figure 9b – View of classical client data representation. Right one figure 9c - The PDPT options screen allow to start and control the PDPT buffering

### 3.5 Test Results of WiFi PDPT Buffering

Developed PDPT Framework solution was tested with WiFi locator module. For testing purpose, five mobile devices were selected with different hardware and software capabilities. Six types of tests batches were executed in test environment. Every test was between two points with 132 meter distance. Every even test was in reversed direction. Five iterations (five devices used) were made during one batch. Results [Tab. 3] provide a good level of usability when user is moving slowly (less than 0,5 m/s). This fact is caused by low number of visible WiFi APs in test environment, where 60 % of time only 1 AP was visible, 20 % for 2 visible and 5 % for 3 or more visible WiFi APs. 15 % of time represents a time without any WiFi connections. Reached values of prebuffering quality in such case are very good.

**Table 3.** Average, Mode, Median and the Range computed values for measured batches of WiFi SS of open space AP in 3 m distance from AP

| | Test 1 | Test 2 | Test 3 | Test 4 | Test 5 | Test 6 |
|---|---|---|---|---|---|---|
| Execution Time [min:s] | 1:40 | 1:48 | 3:37 | 3:24 | 5:12 | 5:26 |
| Move speed [m/s] | 1,32 | 1,22 | 0,61 | 0,65 | 0,42 | 0,40 |
| Quality of prebuffering [%] | 25 | 18,18 | 75 | 36,36 | 97 | 54,55 |

## 4 Conclusions

The focus was targeted on the real usage of the developed PDPT Framework on a wide range of wireless mobile devices and its main issue at increased data transfer. The high success rate (varies from 37 to 97 %) surpassed our expectations. The PDPT prebuffering techniques can improve the using of medium or large artifacts on wireless mobile devices connected to information systems. The localization part of

PDPT framework is currently used in another project of biotelemetrical system for home care named Guardian to make a patient's life safer [12].

**Acknowledgement.** This work was supported by the Ministry of Education of the Czech Republic under Project 1M0567.

## References

1. Abowd, G., Dey, A., Brown, P., et al.: Towards a better understanding of context and context-awareness. In: Gellersen, H.-W. (ed.) HUC 1999. LNCS, vol. 1707, p. 304. Springer, Heidelberg (1999)
2. Krejcar, O.: User Localization for Intelligent Crisis Management. In: AIAI 2006, 3rd IFIP Conference on Artificial Intelligence Applications and Innovation, pp. 221–227 (2006)
3. Krejcar, O., Cernohorsky, J.: Database Prebuffering as a Way to Create a Mobile Control and Information System with Better Response Time. In: Bubak, M., van Albada, G.D., Dongarra, J., Sloot, P.M.A. (eds.) ICCS 2008, Part I. LNCS, vol. 5101, pp. 489–498. Springer, Heidelberg (2008)
4. Krejcar, O.: PDPT Framework - Building Information System with Wireless Connected Mobile Devices. In: ICINCO 2006, 3rd International Conference on Informatics in Control, Automation and Robotics, Setubal, Portugal, August 01-05, pp. 162–167 (2006)
5. Krejcar, O., Cernohorsky, J.: New Possibilities of Intelligent Crisis Management by Large Multimedia Artifacts Prebuffering. In: Ulieru, M., Palensky, P., Doursat, R. (eds.) I.T. Revolutions 2008, Venice, Italy, December 17- 19. LNICST, vol. 11, pp. 44–59. Springer, Heidelberg (2008)
6. Nielsen, J.: Usability Engineering. Morgan Kaufmann, San Francisco (1994)
7. Haklay, M., Zafiri, A.: Usability engineering for GIS: learning from a screenshot. The Cartographic Journal 45(2), 87–97 (2008)
8. Evennou, F., Marx, F.: Advanced integration of WiFi and inertial navigation systems for indoor mobile positioning. Eurasip journal on applied signal processing (2006)
9. Arikan, E., Jenq, J.: Microsoft SQL Server interface for mobile devices. In: Proceedings of the 4th International Conference on Cybernetics and Information Technologies, Systems and Applications/5th Int Conf on Computing, Communications and Control Technologies, Orlando, FL, USA, July 12-15 (2007)
10. Jewett, M., Lasker, S., Swigart, S.: SQL server everywhere: Just another database? Developer focused from start to finish. DR DOBBS Journal 31(12) (2006)
11. Krejcar, O.: Utilization Possibilities of Area Definition in User Space for User-Centric Pervasive-Adaptive Systems. In: Hesselman, C., Giannelli, C. (eds.) Mobilware 2009 Workshops. LNICST, vol. 12, pp. 124–130. Springer, Heidelberg (2009)
12. Krejcar, O., Janckulik, D., Motalova, L., Kufel, J.: Mobile Monitoring Stations and Web Visualization of Biotelemetric System - Guardian II. In: EuropeComm 2009. LNICST, vol. 16. Springer, Heidelberg (to appear)

# Avoiding Threats Using Multi Agent System Planning for Web Based Systems

Punam Bedi, Vandana Gandotra, Archana Singhal, Vandita Vats, and Neha Mishra

Department of Computer Science, Univ. Of Delhi, Delhi, India
pbedi@cs.du.ac.in,
{vandanarla,vandita_11,neha_1_11}@yahoo.co.in,
singhal_archana@yahoo.com

**Abstract.** We have seen security challenges are increasing due to extensive use of modern technologies with expanding connectivity during last few years. This has created a high risk threat environment which needs effective steps to counter these threats. Multi agent system planning for threat avoidance (MASPTA) is a proposed system that uses goal oriented action planning (GOAP) strategy with threat modeling process in a multi agent environment. In multi agent system planning, agents will work in unison to avoid the threats with their action plans executed by individual agents as per the schedule generated. The main goal of this approach is to avoid threats to a web based system. Firstly, it identifies the threats in the system using the concepts of threat modeling and create attack tree for each identified threat using Hierarchal Task Network (HTN) technique Thereafter action plans are generated and executed as planned by using Goal Oriented Action Planning (GOAP) to avoid the threat. Case study explains the multi agent system planning that helps in avoiding the threats to the online banking system.

**Keywords:** Threat modeling, Multi agent system, HTN, GOAP, MASPTA.

## 1 Introduction

With growing connectivity and diversity of software systems, technical challenges especially in the security zone of web based systems have largely increased in last few years. Software security is the idea of engineering software to make it to function correctly under malicious environment. This has led us to work on how we can avoid these attacks or threats to a web based system and can protect our assets from malicious users. As we know purpose of planning is achievement of goal and here our goal is to protect our assets from attacker. In our proposed approach we have combined different techniques to generate scheduling plans for agents to avoid the threats to a web based system. The agents in the system will execute their action plans according to the generated schedules in multi agent environment. These multiple agents will coordinate and communicate with each other to achieve their goal. This proposed approach named as MASPTA is executed in four phases. We try to understand and analyze the system to be protected against attacks in first phase. The second phase deals with identification of threats to the system. Possible attacks for each threat are

N.T. Nguyen, R. Kowalczyk, and S.-M. Chen (Eds.): ICCCI 2009, LNAI 5796, pp. 709–719, 2009.

enumerated to create an attack tree in the next phase. In the final phase action plans are generated for each agent who will work in unison to execute these plans to avoid the threats in multi agent environment. Graphical representation of MASPTA is given in Fig. 1.

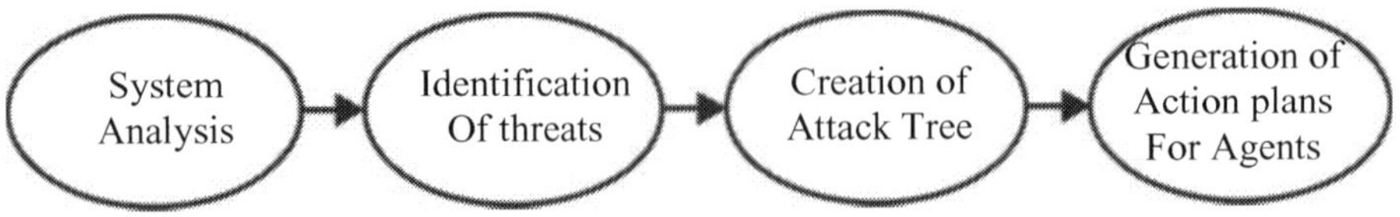


**Fig. 1.** Graphical Representation of MASPTA

Our work is organized as follows. Section 2 reviews the related works in this area. Concepts used in our proposed approach have been discussed in Section 3. Section 4 deals with our proposed approach MASPTA. Case study is given in Section 5 with conclusion and future work in Section 6.

## 2 Related Works

Glenn Wissing [1] in his work has discussed the concept of HTN and how it can be integrated with GOAP to reach the goal state efficiently. Axel van Lamsweerde in his paper [2] has elaborated the generation of security requirement through construction of intentional Anti models. Philiparning Bjarnolf [3] has presented how GOAP can be used in a multi agent environment for game planning. Jeff Orkin [4] has also worked on use of goal oriented action planning in a multi agent gaming environment. Suvda Myagmar, Adam J.Lee, William Yurcik [5] has used threat modeling process for identification of threats. Agent planning and its types have been discussed in detail by Runan Moa [6]. Xin Li and Leen-Kiat Soh have proposed an innovative two-step learning approach to planning-instantiation for multi-agent coalition formation in dynamic, uncertain, real-time, and noisy environments [8].

In the present work, we are integrating threat modeling process with GOAP strategy in a multi agent environment to avoid threat to a web based system. Action plans are executed by individual agents against each attack according to a schedule planned to avoid the threat taking place. We mean to provide a comprehensive system design to guard against the threat by integrating various useful techniques mentioned above.

## 3 Background

This section briefly describes the concepts of Threat Modeling Process, HTN Technique, GOAP Technique, and Multi agent system planning.

### 3.1 Threat Modeling Process

Threat modeling process involves understanding an adversary's goal in attacking a system based on system's assets of interest [7]. It is important to be systematic during the threat modeling process to ensure that as many as possible threats and vulnerabilities are discovered by developers not by attackers [5]. Threat modeling process consists of characterizing the security of the system, identifying assets and access points and determining threats [7]. Characterizing the security of the system involves complete understanding of the system. This includes gathering information about different components and their interconnections, usage scenarios and identifying assumptions and dependencies. Understanding the system completely will help us to identify vulnerable assets that need to be protected from malicious users. Then we identify the threats to a system using the information gathered so far. Threats can be identified by going through each of these security critical entities and creating threat hypotheses that violate confidentiality, integrity, or availability of the entity [5].

### 3.2 HTN Technique

HTN is hierarchical task network planning technique which consists of a set of operators and a set of methods. The input to HTN planner is set of operators and set of methods and it produces hierarchical task network with non-primitive task as the root and primitive tasks as the leaf nodes as output [1]. There is an initial state and a goal to be achieved. Here the goal is not set of states but it is in the form of task(s), which is referred to as goal task. Methods are used to break down a non-primitive task (compound task) into primitive (actions) and non-primitive tasks. The operators are then used to execute the primitive tasks.

### 3.3 GOAP Technique

GOAP goal oriented action planning technique that deals with only actions, i.e. a set of preconditions and a set of effects. It offers dynamic problem solving, which can understand an agent's environment and react in a logical manner. We view agents as pieces of code that execute action plans to accomplish a given task. It is a technique that produces a sequence of actions (called a plan) to achieve a desired goal state. The primary objects in GOAP are Goals and Actions. A Goal is described by relevant world state properties. An Action is described in terms of its effects on the world state and its world state. A plan of action (actions) is formulated by a component called a planner. GOAP is a regressive approach that searches until every goal state is satisfied. It starts its search at the goal state and then works its way towards the current state by finding actions with appropriate preconditions and effects. The search does not find all possible plans, but instead finds a valid plan i.e. cheapest and most promising. In GOAP, each action has a static cost value which helps the planner to determine what actions are more favorable than others, i.e. the planner considers more specified actions as the priority demands.

### 3.4 Multi agent System Planning

The term "multi-agent system planning" is an approach to a planning problem with complex goals that splits the problem into manageable pieces, and lets each agent deal with such a sub-problem. In this approach, the solutions to the sub-problems have to be combined afterwards to achieve a coherent, feasible solution to the original problem. It is a planning in dynamic environment with multiple autonomous agents working together to solve a common problem by coordinating their actions. In this planning each agent works out its plan more or less independently but there is also a need to coordinate these plans.

## 4 Proposed Approach

MASPTA as the name suggests is a system that works in a multi-agent environment. Multiple agents will work together, cooperating and communicating with each other in order to avoid the threat to an asset which is an attacker's goal. Corresponding to each threat an attack tree is built, in which the root node will be the threat which needs to be mitigated and leaves will correspond to all possible attacks that an attacker can make to achieve the threat. In MASPTA, sequence of action plans will be generated and then each agent will be assigned one action plan to guard against each attack. Goal in HTN will be an anti-goal or attack in MASPTA. Using the Situation Awareness (SA) of GOAP that is being aware of "what is going on" in the environment in which the agent is working. The agent will be able to determine how big a threat an asset poses. Multiple agents will work in unison communicating with each other thereby avoiding threat to the system. MASPTA strategy is given below.

### 4.1 System Analysis

This phase performs complete system analysis of the web-based system for which threats are to be avoided. It includes understanding and defining the system, analyzing its characteristics and determining assets, value of each asset, entry points and trust levels associated with each asset. It includes defining the system in totality that is giving advantages, disadvantages, applications etc. all about the system. This step details the system properly so that the architecture and working of system becomes clear to the user.

### 4.2 Identification and Analysis of Threats

This phase deals with identification of all possible threats associated with each asset of the system using the concept of threat modeling. The identified threats are then analyzed to see how harmful a threat is by associating a threat level with each threat. Generally the DREAD algorithm is used to compute a risk value of each threat, which is an average of all five categories (D+R+E+A+D/5). The calculation always produces a number between 0 and 10; the higher the number, the more serious is the risk [11]. This step will help us to know whether it is a case of risk avoidance, risk transfer, risk reduction and risk retention. High risks threats need to be treated accordingly.

### 4.3 Identification of Attacks and Creation of Attack Tree

Once a threat is identified, all possible attacks that can accomplish the identified threat are determined and an attack tree is created. We adopted the HTN technique to create an attack tree by refining the composite attack into simpler attacks following AND/OR refinement that can be executed by the attacker. The threat that is to be avoided is at the root level of the attack tree. Goal in HTN will be represented as anti-goal (attack) in attack tree. This attack is now required to be mitigated. Root node of the tree is the threat, each leaf node is an attack to accomplish the threat and the path from leaf to root is the way an attacker can achieve the threat. AND refinement means all attacks should succeed for the corresponding threat to occur while OR refinement means presence of at least one attack can cause the threat to occur as shown in Fig. 2.

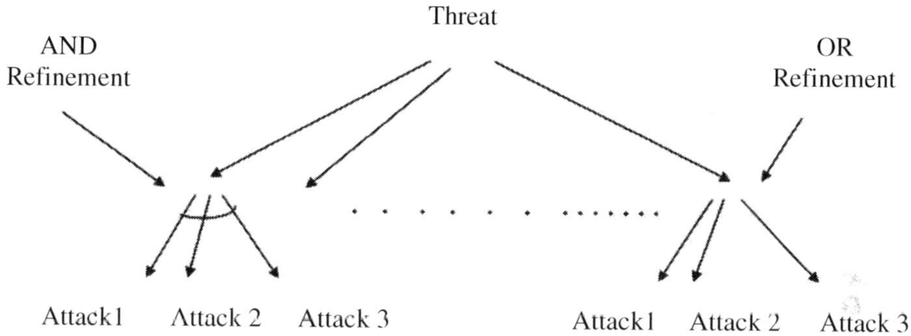


**Fig. 2.** View of an Attack Tree

### 4.4 Creation of Action Plans and Schedules for Agents

The attack tree that is created using HTN technique has threat at the root node and attacks at leaf nodes. These attacks are now required to be mitigated. So the agents will prune the attack branches from the attack tree to avoid the threat. To prune the attack branches, we followed GOAP technique, which will work from leaf to root. In HTN technique operators are used to execute the actions and reach the goal state; in contrast with our approach in which agents will be given action plans to avoid the threat which is the goal of the attacker. These action plans will prune the attack branches at lower level only so that the threat is not accomplished.

In this phase, an action plan is created corresponding to each attack. A plan generated is a sequence of actions of the agents that are executed to avoid the threat at root level. Thus multiple agents will work in parallel to avoid the threat. The generation of plans is explained using the example of the attack tree shown in Fig. 3. In this attack tree the Threat is at the root level. The threat and each attack are refined into simpler attacks at each level. As shown in Fig. 3 Attack 7 and Attack 8 will be responsible for Attack 3. Similarly Attack 3 and Attack 4 will give birth to Attack 1. Attack 5 and Attack 6 will result in Attack 2. Realization of Attack 1 and Attack 2 will cause the threat to be avoided at the root level. The attacks that are at the leaf nodes can not be refined further, and are executed by the attacker to accomplish the threat. Action plans are generated against these attacks and are assigned to agents. There are five leaf nodes that indicate attacks and five agents: A1, A2, A3, A4, and A5 to work against them. There is an AND arc between Attack 7 and Attack 8, so only one of the agents A1 or A2 needs to execute their action plans to avoid Attack 3. Similarly there is an AND arc between Attack 3 and Attack 4, so either (A1 or A2) or A3 needs to be executed to

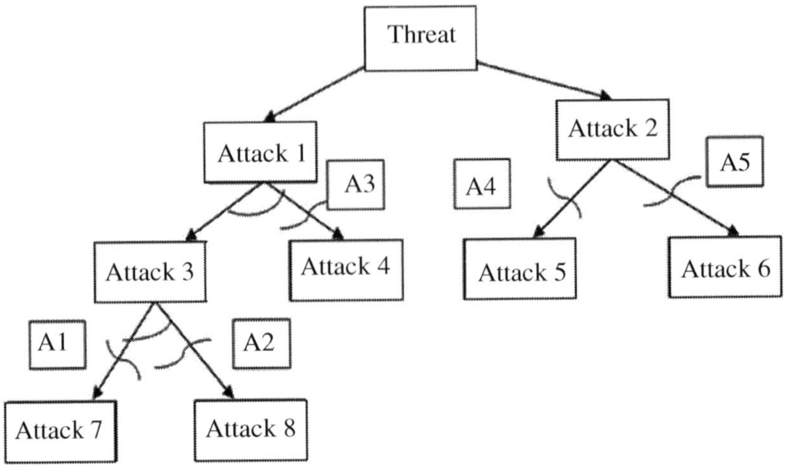


**Fig. 3.** Illustration of an Attack Tree with Agents

avoid Attack1. In order to avoid Attack 2 both the agents A4 and A5 will execute their action plans. Avoidance of Attack 1 and Attack 2 will result in elimination of the threat to the system.

Now as only one of the agents A1 or A2 need to execute their action plan to avoid the Attack 3, then one of them who finishes its task first immediately communicate to the other agent to stop its task as the attack has already been successfully blocked by the previous agent. This is how agent communication will help in planning against the attack. Agents that are dependent on each other carry out communication among themselves, and the agents that are independent work in parallel, with a common aim to avoid the threat. Thus the possible generated scheduling plans to avoid the Threat to the system are given as follows:

1. A1 A4 A5
2. A2 A4 A5
3. A3 A4 A5

Every agent's action plan is executed on the basis of certain preconditions, and has corresponding actions, that will avoid the threat. The agents whose preconditions are true at the same time will work in parallel.

## 5 Case Study

Our case study is on 'Online Banking' system. Online Banking involves use of Internet for delivery of banking products & services. Access to one's accounts at anytime and from any location via the World Wide Web is convenient. Through online banking the customers can access their account just by a click of mouse. But in case user's username and password is hacked by an attacker, he can gain access to the user's bank account and can carry out illegal money transactions through his account. Thus we have considered security of online banking as a subject. We are applying our proposed system MASPTA to online banking system and trying to avoid threats to that system to preserve its assets. Four phases of the system and experimental study is given below.

## 5.1 System Analysis

Online Banking involves use of Internet for delivery of banking products & services. Assets to Online Banking system are: User's bank account information comprising of Username and Password that have to be protected. Illegitimate links, illegitimate e-mail and TCP connection are the weak points through which the attacker can enter in the system.

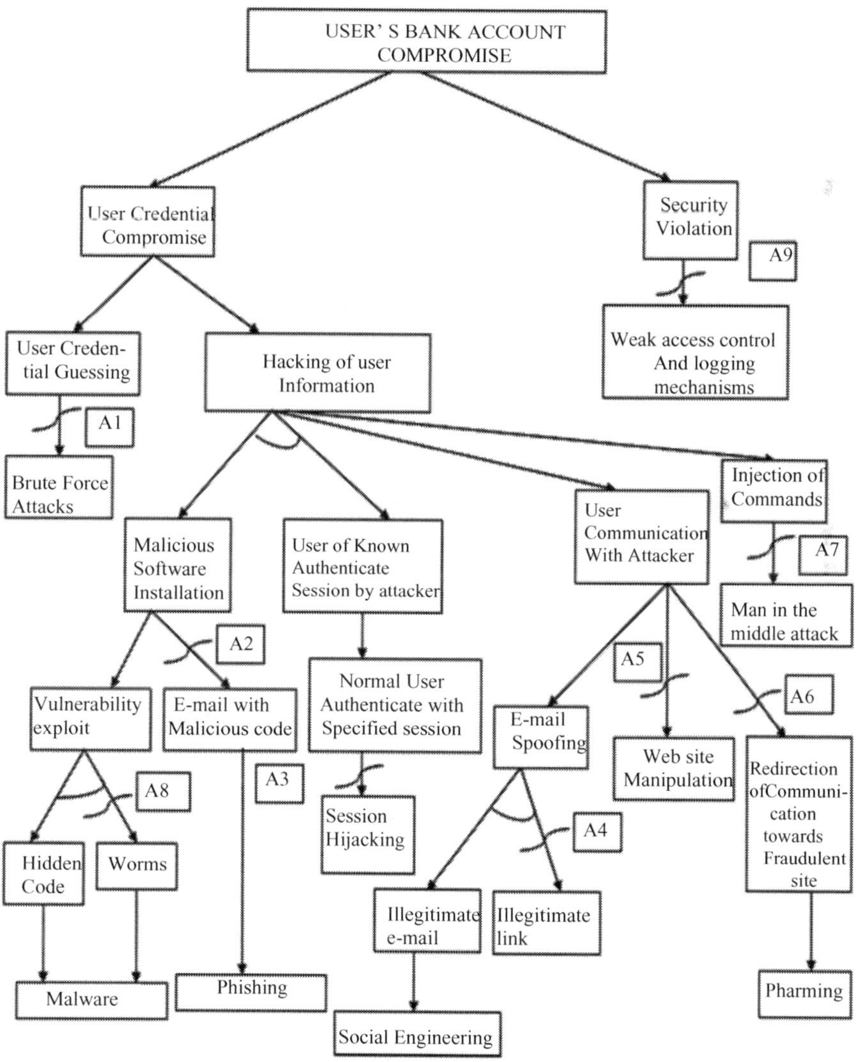


**Fig. 4.** Attack Tree with Agents for avoidance of threat to Online Banking System

### 5.2 Identification and Analysis of Threats

Most common threat to online banking system is the 'compromise of a user's bank account' by deceiving user's login data. Since this information has to be exclusively used by the user, its leak can be considered as one of the biggest threat in this system.

### 5.3 Identification of Attacks and Creation of Attack Tree

In this phase we will first determine all possible attacks that can maliciously compromise user bank account. We will then create an attack tree corresponding to the identified attacks. The threat of user's bank account compromised can be accomplished by mainly two attacks that are user credentials compromise or some security violation in the system.

These attacks can be further refined, resulting in the creation of the attack tree shown in Fig. 4, that is used by our system. Various modes through which the attacker can get access to the secret data are shown in the attack tree. For example, as indicated in the diagram, through the brute force attack the attacker can reach the user's credentials maliciously resulting in the realization of the threat.

### 5.4 Creation of Action Plans and Schedules for Agents

The agents, action plan of each agent with reference to Fig. 4, pre-conditions and effects of each action plan are detailed below:

Agent 1 (A1) : Agent against brute force attack
This agent would check for the strength of the password that user have selected for his account and would warn the user in case of weak password.
Pre-condition: User registers an account and enters a password.
Effect/Action: Prompts the user if password is weak.

Agent 2 (A2) : Agent against phishing
This agent would check for https:// instead of http:// in the URL using parser.
Pre-condition : When a URL is clicked.
Effect /Action : Block the access to site, if the site is insecure.
Agent 3 (A3) : Agent against session hijacking
This agent would delete the cookies whenever a session of the user is completed, and the user logs off from his account, so that no attacker can extract those cookies that may contain valuable user specific information regarding his session.
Pre-condition: The size of cookies folder becomes greater than two.
Effect /Action : All temporary files of the cookies folder are deleted.

Agent 4 (A4) : Agent against illegitimate link
This agent would check if a link is legitimate or not. It will do it by parsing the given link, so that if it is found to be unauthorized, then the user gets a pop up menu, or a warning signifying it to be suspicious.
Pre-condition: When ever a link is clicked.
Effect /Action : Prompt a pop-up to the user if the link is non-bank site.

Agent 5 (A5) : Agent against web-site manipulation

This agent would check if a web-site is authorized or not, by performing check over the values of certain parameters of the site.
Pre-condition: If certain parameters of the website are changed.
Effect /Action: Prompt to the user that the site is manipulated and block its access

Agent 6 (A6) : Agent against Pharming
This agent would perform check over SSL certification, so that the user does not get connected to a fraudulent site and upload his data on it. Thus it allows a site to be browsed by the user if it is legitimate.
Pre-condition: When a URL is clicked
Effect /Action: Block the access to site if the site is insecure

Agent 7 (A7) : Agent against Man-in the middle attack
This agent would send the data out on the TCP connection, after encrypting it with a good encryption key using private key, so that even if the attacker traces the user's private data, he can not decrypt the user specific information out of it.
Pre-condition: User logs-in and enters his password.
Effect /Action: Decrypted password is send to the server.

Agent 8 (A8) : Agent against worms
This agent would keep a check on the daily updates of the anti-viruses installed on the system, and would prompt an alert to the user in case the software need to be updated.
Pre-condition: When ever anti-virus requires updation.
Effect /Action: Updates the antivirus.

Agent 9 (A9) : Agent against security violation
This agent would keep a check on firewall and anti-virus update. And will prompt a pop up window to the user in case of notification for the updates to take place.
Pre-condition: When ever anti-virus requires updation
Effect /Action: Updates the antivirus.

The above mentioned agents will execute when their pre-condition are true, to block the corresponding attack and thus all will work together communicating with each other, in parallel, to avoid the threat of compromise to user's bank account of the online banking system. The possible scheduling plans thereby generated to avoid the threat to the system are given as follows:

1. A6 A2 A8 A1 A7 A5 A4 A9
2. A6 A1 A7 A5 A4 A9 A3

Communication will be carried out between agents A2, A8 and A3, as they are working against same attack, under an AND refinement. Other agents will work in parallel. The agents will be executing their action plans whenever their corresponding preconditions are true. Thus it will be seen that induction of agents will work as watchdogs and help them to block attacker's path .

## 5.5 Experimental Results

To give strength to our case study, we have observed experimental results to watch the efficiency of the steps taken for security of online banking system. As a part of our implementation, we have first implemented a site prototype for online banking in java server programming (JSP). Thereafter we have implemented our nine agents in Java Agent Development Environment (JADE) that are meant to block the attacks that can accomplish the threat of user's bank account compromise. Snapshot of the website of online banking is given in Fig. 5.

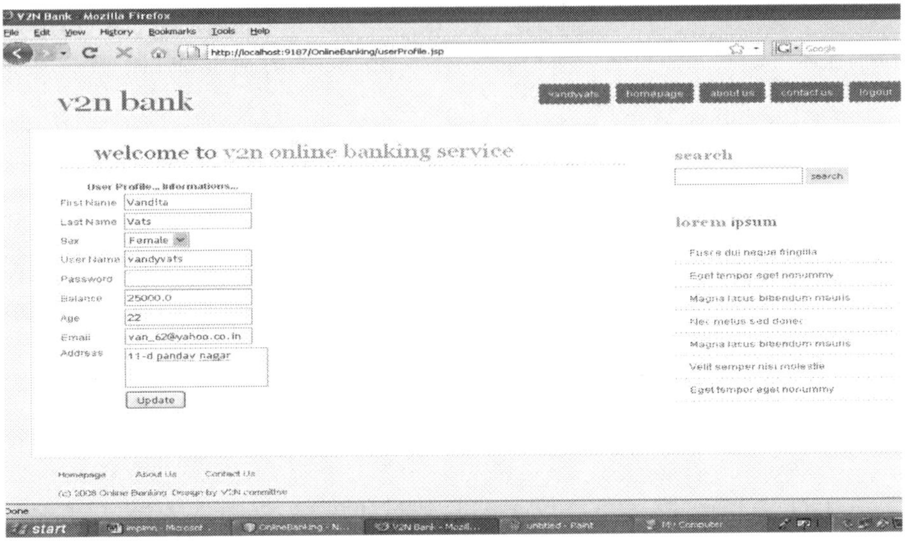


**Fig. 5.** Snapshot of the website of Online Banking

Working of our agent A3 against the attack of session hijacking is shown in Fig. 6. This will delete the session of the user once the user has logged out.

**Fig. 6.** Snapshot of Agent 3

## 6 Conclusion and Future Work

Vulnerable software is one of the major challenges faced by IT industry today. The facts show that that more than 60 percent security vulnerabilities discovered in last two years are related to Web applications. This challenge can be met by developing better and more secure web applications with thorough understanding of the security aspects. The aim of this paper is to present a system named as MASPTA to defend against security attacks to any web based system. The main emphasis of this approach is identification and avoidance of threats from malicious users or hackers in such cases. MASPTA has emerged by integrating the concepts of threat modeling, HTN and GOAP technique. This will aid in producing scheduling plans of agents working in a multi agent environment to avoid threats to the system. As a future work a global monitoring system will be designed for overall view of the system including the performance of the agents regarding execution of their specified action plans to avoid the threat. We can further simplify the proposed approach by dividing the attacks into two groups. First group will correspond to the attacks which are general in nature like malicious software installation and illegitimate mail etc. which are common to most of the web based systems. The second group of attacks will be tailored to the specific needs of the application. This arrangement will also help simplify creation of attack trees. We may have to enhance this area of research to create an authenticated approach which can be made applicable to any web based system.

## References


1. Glenn, W.: Multi-agent planning using HTN and GOAP. Lulea university of Technology, Department of Skelleftea campus, Division of Leaisure and Entertainment (2007:16HIP)
2. van Axel, L.: Elaborating Security Requirements by Construction of Intentional Anti-Models. In: 26th international Conference on Software Engineering (ICSE 2004), pp. 148–157. United Kingdom (2004)
3. Philiparning, B.: Threat Analysis using Goal Oriented Action Planning. University of Skovde (May 2008)
4. Jeff, O.: Applying Goal Oriented Action Planning to games. In: AI Game Programming Wisdom, vol. 2. Charles River Media (2003)
5. Myagmar, S., Lee, J.A., Yurcik, W.: Threat Modeling as a basis for Security Requirements. National Centre for Supercomputing Applications, University of Illinois at Urbana-Champaign (2005)
6. Moa Runan: Planning with agents. Matrikelnummer: 1008277 (2001)
7. Swiderski, F., Synder, W.: Threat Modeling. Microsoft Press (2005)
8. Li, X., Soh Leen, K.: Learning how to plan and instantiate a Plan in Multi-Agent Coalition Formation. In: International Conference on Agent Technology, Proceedings, pp. 479–482. IEEE/ WIC / ACM (2004)
9. Dimitriadis, K.C.: Analyzing the security of Internet Banking Authentication Mechanism. Information Systems Control Journal 3 (2007)
10. Tally, G., Thomas, R., Vleck Van, T.: Anti-Phishing: best Practices for Institutions and consumers. McAfee Research, Technical Report # 04-004 (2004)
11. Howard, M., LeBlanc, D.: Writing Secure Code. Microsoft Press (2003)

# Using WordNet to Measure the Similarity of Link Texts

Andrzej Siemiński

Wrocław University of Technology, Institute of Informatics,
Wybrzeże Wyspiańskiego 27, 50-370 Wroclaw, Poland
Andrzej.Sieminski@pwr.wroc.pl

**Abstract.** The primary aim of the study is to evaluate the extend to which the introduction of word similarity defined by the WordNet database could improve the link texts similarity assessment. The proper assessment is crucial for focused crawlers. The crawlers need it to select which links are to be followed. The proposed WordNet based semantic similarity algorithm has increased the recall of the link selection process. To mitigate the co-occurring loss of precision an adaptive algorithm for modifying the initial word similarity levels is introduced and evaluated. The proposed algorithms are verified by an experiment.

**Keywords:** WordNet, focused browsers, word semantic similarity, link text selection.

## 1 Introduction

Focused crawler is a Web crawler that tries do download only pages that are relevant to a given topic of interest [1]. Their main advantage over the general search engines that use general crawlers is that they are better suited to monitor Web sites where content changes quickly. General crawlers require enormous, storage and bandwidth resources. The limited recourses consumption of focused crawlers make them an ideal tool for performing comprehensive searches of web-related materials even on PC based implementations [2]. Focused crawlers rely heavily on the link text. The main aim of the paper is to increase the usefulness by using WordNet based similarity of words.

Each implementation of a focused crawler must provide solutions for two main problems. The first one is the selection of links that point to potentially interesting pages. The number of links on a page could well exceed one hundred which clearly indicates how selective the algorithm must be. Selecting too many of links diminishes the one of the main advantages of focused crawlers – sparse usage of resources. Rejecting too many links inevitably results in the loss of valuable data. The second problem is deciding whether an already downloaded page should be used as starting point for further search or should it be discarded. A focused crawler must be able to accept non relevant pages, hoping that they will lead to pages of interest. The short-term gains should are not be pursued at the expense of less-obvious crawl paths that ultimately yield larger sets of valuable pages. The early focused crawlers imitated the

N.T. Nguyen, R. Kowalczyk, and S.-M. Chen (Eds.): ICCCI 2009, LNAI 5796, pp. 720–731, 2009.

behavior of a school of fish or a shark [3] to deal with the problem. More recent solutions are presented in [4]. The paper addresses only the first problem.

The rest of the paper is organized as follows: Section 2 presents briefly the selected problems of focused crawlers that are pertinent to the papers subject area. The traditional text similarity assessment is presented in the Section 3. Section 4 introduces the WordNet based word similarity algorithm. The Section 5 presents the extension of the standard methods of measuring the text similarity by incorporating the WordNet based term semantic similarity. Preliminary results of an experiment are given in Section 6. The paper concludes and discusses possible extensions in Section 7.

## 2 Focused Crawlers

A focused crawler needs a precise selection links algorithm. Ideally crawler should reproduce the behavior of a human being. This is only possible in the environment of a Semantic Web in which the pages not only define the way text should be displayed but also the semantics of information. Sophisticated schemas involving multiple crawlers discovering semantic instances from annotated resources and then integrating the knowledge by a centralized metacrawler are described in [5]. The idea is promising but its practical feasibility is questionable. The origins of Semantic Web date back to the ideas firstly introduced by Berners-Lee almost ten years ago. The evolution from the existing web to its semantic version is yet to occur. In its very nature the Web is decentralized. The far less complex transition from the protocol HTTP 1.0 to HTTP 1.1 has not been fully completed yet. Therefore it is doubtful that web masters at large would take the trouble and annotate their web sites by additional semantic information, especially when we take into account the complexity of the proposed formats. One should also not forget the problem of the credibility of the semantic information. The possibility of providing misleading data considerably hinders the relatively simple task of filtering pages with illegal and objectionable content as described in [6]. The various tricks used by web masters to cheat the search engines and thus influence the ranking of pages is also to be remembered. Therefore, we believe that solutions based on Semantic Web could be used for specialized web sites and not for the Web at large.

Over the years the link texts have proven to be extremely useful for the focused crawlers. The texts are short, they describe very well the content of the target page and are the last but not least are ready available – there is no need to download the target page. The link text must confirm to the page content otherwise the users would surely avoid pages with text and links that confuse them. Any intentionally misleading action is counterproductive. The principal challenge is to identify the links that most probably lead to relevant pages.

The paper studies to what extent the full texts of pages the texts of links are pointing to are similar. To go beyond the traditional text similarity assessment methods one can use the semantics of the text being displayed to the user. The deep text analysis is not yet feasible. On the other hand, the shallow text analysis combined with a content extraction mechanism could produce interesting results, as is recently shown by the European Media Monitor [7]. Such solutions, although useful, are restricted to selected subject areas such as disaster or threat monitoring. The removal of this limitation is possible only when we use a semantic description of the unrestricted natural language. Such a description is available in the WordNet database.

## 3 Measuring the Text Similarity

Almost all systems do not take advantage of the syntactic nor semantic properties of natural texts and treat them just as bags of words. A text similarity measure should take into account two basic text properties:

- texts vary considerably in their lengths;
- words are more popular then the others.

The classical method of compensating the different texts lengths is the cosine measure. A text t is represented be a vector of numbers: t= ($t_1$,..,$t_m$). Each position in the vector is mapped to a word and the respective number represents the word frequency (number of occurrences) of that word in the text. The frequency of i-th term in the text t is defined as:

$$t_i = \frac{f_{ti}}{\max_{j=1..m}(f_{tj})} \tag{1}$$

where $f_{tj}$ is the number of occurrences of term j in the text t and m is the total number of terms.

The cosine measure of similarity of the texts represented by vectors s and w is defined as follows:

$$sim_{\cos}(s,w) = \frac{s \circ w}{|s||w|} \tag{2}$$

where:

$$|s| = \sqrt{s_1^2 + \ \ldots \ + s_m^2}$$

and $s \circ w = \sum_{i=1}^{m} s_i w_i$ is the inner product of vectors s and w.

In order to take into account the number of texts in which a term appears (the second property) the term frequencies are weighted (multiplied) by the idf factor. The idf (inverse document frequency) weight schema for word i is defined as:

$$idf(i) = \log_2(n/n_i) + 1 \quad | n_i > 0 \tag{3}$$

where n is the total number texts and of $n_i$ is the number of texts in which the term i occurs at least one time.

Practical experience has demonstrated that the standard idf (inverse document frequency) weighting scheme performs very well in a wide variety of circumstances [8] and in what follows it is referred to as a standard cosine similarity.

## 4 Using WordNet to Measure Word Similarity

A term mentioned in the above section represents all inflectional forms of a word. Identifying all forms of a word is not trivial even for English is not a highly inflectional language. The regular expressions that is so popular in the database world (LIKE clause of the SQL language) are clearly not satisfactory. To remedy the problem the stemming algorithms are used. The six stage Porter algorithm is probably the most popular [9]. It takes into account several properties of the English language but it is unable to cope with irregular words so that words like children or woman are not related to relation to child or woman respectively.

The semantic similarity of words has been subject of many research projects on in area of Artificial Intelligence for some 40 years. The traditional approaches rely heavily statistical data on the occurrence of words. To measure the similarity of words first the Information Content (ic) of a word must be calculated by the simple formula:

$$ic(w) = -\log(p(w)) \tag{4}$$

where p(c) is the probability of encountering c in a given corpus. The assumption is, that the more frequent the word is the less meaning it has. This is an extrinsic measure and it requires having a database on the frequencies of terms from a large language corpora.

To measure the similarity of two concepts c1 and c2 one has to find the Most Specific Common Abstraction (MSCA) that subsumes c1 and c2. The similarity is thus defined by the following formula:

$$sim(c1, c2) = \max_{c \in S(c1,c2)} ic(c) \tag{5}$$

where S(c1, c2) are the set of concepts that subsume c1 and c2.

An alternative way to process words is to use the WordNet - a comprehensive dictionary and lexical data base for the English language. It was created and is being maintained at the Cognitive Science Laboratory of Princeton University [10]. The WordNet can be interpreted and used as a lexical ontology in the computer science sense. One of its functions converts any word to its basic form. The algorithm removes not only the standard word endings but also handles properly the irregular forms. The relationships between words in WordNet describe the properties of a natural language and they remain relatively stable. To improve performance the similarity between words is calculated using a local version of the databaseThe WordNet taxonomic structure could be used to assess the semantic similarity of words. The structure was used to propose an intrinsic word similarity measure [11]. The frequency of word occurrences in the formula 4 is replaced by the number of hyponyms that a given word has. In this approach the hypothetical word at the top of word hierarchy is so general that its Information Content is minimal, the words occupying leaves of the hierarchy are most specified so they carry maximal information content. Below is the revised formula for calculating the Information Content of a word.

$$ic(w) = 1 - \frac{\log(hypo(w)+1)}{\log(\max c)} \tag{6}$$

where hypo is a function that returns the number of hyponyms of a given concept w and max c is a constant that is set to the maximum number of concepts that exist in the taxonomy. In what follows semantic similarity between terms t and calculated using the above formula is denoted by wns(t,w).

The empirical studies described in [11] compare the values of the wns function with human judgment. The results indicate that the similarity measure outperforms traditional algorithms.

## 5 Link Text Similarity

The standard cosine measure is based on statistical properties of texts and performs well while comparing two texts or a query with a text. The link texts are very short, their average length barely exceeds 3 words. Links leading to similar pages with similar content contain often semantically related but different terms. To increase the recall the word semantic similarity has to incorporate into the similarity measure.

Let best(w,α) denote the term from the string α that is most similar to the word w, that is: $y \in \alpha, \quad \forall x \in \alpha \quad wns(w,y) \geq wns(w,x)$.

The next function sBest extends the notion of the best function on a the string β by the string α. The function sBest produces a sequence of terms defined in the following way:

$$sBest(\beta,\alpha) = \bigcup_{t \in st(\beta,\alpha)} w \mid \quad w = best(t,\beta) \tag{7}$$

where *st(β, α)* is the sequence of all terms in alphabetical order taken from α or β without repetitions.

The SSWS (Semantic Similarity Weighting Schema) assigns the weights to terms of in sequence β in the context of the sequence α by calculating the wns similarity for terms on the same positions in the sequences st(β,α) and sBest(β,α).

The cosine similarity function that uses the SSWS is called the semantic cosine measure and is denoted by $SemSim_{cos}(\alpha, \beta)$.

Example 1.

α: "french basketball team" , β: "english football team" ;

st(α, β)= "basketball english football french team"

Let us further suppose that wns(English, French)= 0.80, wns(foolball, basketball)= 0.73 and for all other non identical terms the wns function has value of 0.

st(β,α)= (basketball, english, football, french, team)
sBest(α, β) = (basketball, french, basketball, french, team)
sBest(β, α)= (football, english, football, english, team)
SSWS(α, β) = (1.0, 0.8, 0.73, 1.0, 1.0)
SSWS(β, α,) = (0,73, 1.0, 1.0, 0.8, 1.0)

The standard similarity $sim_{cos}(\alpha, \beta) = 0,333$ whereas $SemSim_{cos}(\alpha, \beta)=0.97$ a value which intuitively better reflects the relationship that exists between the two texts. The usefulness of using the word similarity is even more evident when we consider another set of texts: "French football team" and "soccer player group". The original cosine measure does not detect any similarity (not a single word appears in both texts) whereas the $SimSem_{cos}$ measure produces a value of 0,93 which is undoubtedly more appropriate.

$SSWS_{\lambda}$ is the λ-level SWS that is used to diminish the influence of words that do not have sufficiently similar counterparts. Instead of the standard wns(α,w) function it uses its λ-level version defined as follows:

$$wns_{\lambda}(\alpha, w) = \begin{cases} wns(\alpha, w) & if \quad wns(\alpha, w) \geq \lambda \\ 0 & otherwize \end{cases} \tag{8}$$

The usefulness of all of the introduced above functions was evaluated during the experiment.

## 6 Experiment

During the experiment data from a news site CNN were downloaded over o period of several days and then parsed. The similarity matrix of the texts of all downloaded pages was calculated and it was later compared with the similarity matrix of link texts.

### 6.1 Data Extraction

The CNN site is a typical news site and its home page consists almost entirely of link texts. The link texts were the primary target of investigation. For 7 consecutive days the sites' home page was downloaded and then analyzed. The links located in the main news sections were extracted and the pages to which they were pointing to were also downloaded. The total number of such pages was around 500.

The engine of Lynx text browser was used to parse the obtained HTML code [12]. The browser successfully handled many HTML syntactic errors that are present on the pages of the CNN server [13]. In the process the text of the pages was extracted together with the link addresses and link texts. The number of links was greater than the number of texts as it was possible that more then one link pointed to the same target page. The text extracted from a page has always undergone some rudimentary normalization:

- Common abbreviations were replaced by their textual equivalents. The abbreviation dictionary was not extensive. It contained about 30 replacement rules like U.S. ➔ United_States or a.m. ➔ ante_meridiem.
- The text was cut to strings, any character non-letter character was treated as a separator and is omitted.
- The text was changed to lower case.

The resulting text is referred to as raw text. The next step involved converting the raw text to base text. It was a 4 step process:

- Elimination of common words which are found in the English stop list. The stop list used had a moderate size and was used by the Google search engine [14].
- Removal of all words shorter than 3 characters.
- Replacing words by their root form using the WordNet database.
- If the resulting text was empty then it was replaced by string "_none_"

In what follows the RT(u) and BT(u) represent respectively the raw and base text extracted from a page with identified by the URL u. The RL(u) and BL(u) denote the sets of raw and base texts that are extracted from links that point to the target page with the URL u.

### 6.2 Similarity Matrixes

During the experiment the similarity matrixes between pages described by full texts and link texts were calculated. In what fallows the reference matrix refers to the similarity matrix of RT(u) texts that was calculated using the standard cosine measure with the idf weighting schema. The primary aim of the experiment is to study the influence of using WordNet and therefore the Raw Text and not the Base page text was chosen as a reference. In order to evaluate the proposed algorithms a number of other matrixes was calculated. The link texts were compared using both the standard and semantic cosine measure with three threshold values: 0.0, 0.5 and 0.75. The link text yielding maximum similarity was selected for the matrix. The codes of the resulting similarity matrixes are given in the table 1.

**Table 1.** Codes for the similarity matrixes

| input text tape | similarity measure | code |
|---|---|---|
| Page raw text | standard cosine | RT |
| Page base text | standard cosine | BT |
| Raw link text | standard cosine | RL |
| Base Link text | standard cosine | BL |
| Base link text | semantic cosine with α threshold | $BL_\alpha$ |

The next Table 2 indicates that the Raw Text to Base Text conversion does not cause significant differences of the resulting similarity matrixes for the standard cosine measure. The table contains the root mean square deviation (RMSD) of the matrixes A and B.

The values of the RMSD for the first two rows are negligible. This illustrates the efficiency of the idf weighting schema. Significantly higher values could be expected for derivational languages. The raw text ➔ base text replacement has far greater importance for the semantic measure. Some of the raw word forms are not present in the WordNet database and therefore their semantic similarity is equal to 0.0. As expected, increasing the minimal value of accepted similarity decreases the similarity of words and so diminishes the value of RMSD.

**Table 2.** The RMSD values for matrixes of raw and base texts

| Matrix A | Matrix B | RMSD |
|---|---|---|
| RT | BT | 0.032 |
| RL | BL | 0.024 |
| $RL_{0.0}$ | $BL_{0.0}$ | 0.230 |
| $RL_{0.5}$ | $BL_{0.5}$ | 0.177 |
| $RL_{0.75}$ | $BL_{0.75}$ | 0.094 |

### 6.3 Case Studies

The main goal of the study is to answer the following question: what would happen if the full texts were replaced by link texts. To answer the question the analysis of RMSD between matrixes is not is much relevant and we need measures that are more akin to the way of work of a focused crawler.

Firstly we try to assess the extent to which the recall of crawling is effected. The approximation of the crawler recall starts with the creation of a reference set. The set consists of all elements from the reference matrix that have similarity value >=α. The tested set includes all elements from the reference set that have similarity value >=β in the tested similarity matrix. The quotient of the cardinality of the two sets is referred to the matrix recall. The resulting values are presented in the table 3, the second raw contains the value of β.

**Table 3.** The recall of Link similarity matrixes

| α | RL | | | $BL_{0.0}$ | | | $BL_{0.5}$ | | | $BL_{0.75}$ | | |
|---|---|---|---|---|---|---|---|---|---|---|---|---|
| | 0.50 | 0,75 | 0,9 | 0,5 | 0,75 | 0,9 | 0,5 | 0,75 | 0,9 | 0,5 | 0,75 | 0,9 |
| 0,6 | 0,86 | 0,86 | 0,80 | 0,93 | 0,82 | 0,82 | 0,90 | 0,81 | 0,80 | 0.87 | 0,80 | 0.80 |
| 0,7 | 0,88 | 0,88 | 0,85 | 0,93 | 0,86 | 0,85 | 0,92 | 0,85 | 0.85 | 0.89 | 0,85 | 0,85 |
| 0,8 | 0,92 | 0,92 | 0,89 | 0,96 | 0,90 | 0,89 | 0,95 | 0,89 | 0,89 | 0,92 | 0,89 | 0,89 |
| 0,9 | 0.93 | 0,93 | 0,92 | 0,97 | 0,92 | 0.92 | 0,96 | 0.92 | 0,92 | 0,94 | 0,92 | 0,92 |

The RL recall values are astonishingly high. e.g. if the we are interested in achieving α>=0.9 then the accepting RL similarity >= 0,75 would retrieve 93% of all relevant pages. The recall for semantic similarity is even higher. For the $BL_{0.0}$ matrix and α>=0.9 and β>0.5 the recall is equal to 0.97 - almost all relevant pages would be retrieved. The performance of $BL_{0.0}$ and $BL_{0.5}$ matrixes does not differ very much but the $BL_{0.75}$ matrix lags behind especially for the lower values of α. The values for the RL and $BL_{0.75}$ matrixes are almost identical.

The increased matrix recall is a positive sign but its cost is also to be estimated. As usually increasing recall results in the loss of precision. To asses the phenomena all elements from the tested matrix with values >= 0.9 were selected and then the proportion elements with similarity level >= α in the reference matrix was calculated. The resulting values are presented in the table 4.

The table shows once more again that raw link texts perform surprisingly well. Even for α=0.8 more than half of all accepted pairs is correct. On the other hand the precision of semantic similarity matrixes is significantly lower. The reduced precision

**Table 4.** The precision of text matrixes

| α | RL | $BL_{0.0}$ | $BL_{0.5}$ | $BL_{0.75}$ |
|---|---|---|---|---|
| 0.6 | 0.63 | 0.26 | 0.35 | 0.44 |
| 0.7 | 0.57 | 0.23 | 0.31 | 0.40 |
| 0.8 | 0.53 | 0.22 | 0.29 | 0.37 |
| 0.9 | 0.46 | 0.19 | 0.25 | 0.32 |

would drain the resources of a crawler and ways of increasing the precision are to be found.

The search for solution starts with presentation of a few examples that illustrate the cases in which the semantic similarity has performed better and cases in which it was worse then the standard similarity measure.

The table 5 contains three pairs of link texts for which the reference similarity exceeds 0.6 and the respective values from the RL and BL0.0 matrixes. In the first case the better performance of BL.0 is purely accidental, the semantic similarity of home and top exists in WordNet but could not be inferred form the link texts alone. In the second and the third case the semantic similarity really shines, due to relatively high values of similarity for word pairs (offcial, london), (light. impact) or (report, fight).

**Table 5.** Semantic similarity successes

| No | Link Text A | Link Text B | RL | BL.0 |
|---|---|---|---|---|
| 1 | show top | home | 0 | 0.7 |
| 2 | reality star jade goody lose cancer fight | report celebrity jade goody die | 0,41 | 0.77 |
| 3 | official blunt impact kill richardson | broadway london dim light richardson | 0.19 | 0.71 |

The next table contains three link text pairs for which the semantic similarity has levels well over 0.90 but this time both the reference and RL similarity and was less than 0.25.

**Table 6.** Semantic similarity failures

| No | Link Text1 | Link Text2 | RL | BL.0 |
|---|---|---|---|---|
| 1 | africa | europe | 0.00 | 0.97 |
| 2 | transcript | transcripts | 0.00 | 1.00 |
| 3 | unite state | unite face porto champion league | 0.07 | 0.95 |

The case number 1 is somewhat justified. It could be expected from a focused browser that in the case or geographically oriented queries it will follow links labeled by the text Africa or Europe. In the second case, the proper value from the RL matrix is due to its simplicity – the semantic closeness of terms transcript and transcripts is not recognized. Naturally, the semantic similarity of such two terms does not imply a subsequent similarity of the target pages. In a real world crawler the link text should be treated in a manner described for the texts: “see more” or “next”, see Section 7 for details. The third case is more difficult to eliminate. The two links point obviously to pages with different content: the former one is of political nature and the latter is a page devoted to the world of soccer. The surprisingly high level of semantic similarity is due the fact that their terms have general meaning e.g. the similarity of state and league is equal to 0.82 and of face and state is 0,67. In that case a more refined text preprocessing could be helpful. The abbreviations and word sequences having the same meaning should be replaced by the same code.

### 6.4 Adoptive Semantic Similarity

To improve the precision semantic similarity an adaptive algorithm is proposed. A word similarity database is maintained for each similarity matrix. At start it contains the values obtained from the WordNet database. The values are later modified. The modification algorithm starts with the comparison of the similarity value of an element from the reference matrix and the corresponding value taken from one of the link text matrixes. If the reference value is greater than the link value the similarity level of terms used in calculation of the semantic similarity is increased. The set of such word pairs is defined by the Formula 7 in the Section 5. In the opposite case the similarity of the terms is reduced. The exact value by which the similarity is increased or decreased is calculated using two formulas. The first one, denoted by Ab(s), is referenced to as the absolute approach with step s. It uses the following formulas for changing the term similarity level:

For increasing the silimarity level: newSim = prevSim + (1-prevSim)*Step
For decreasing the silimarity level: newSim= prevSim - prevSim*Step;

where prevSim and newSim represent the values of similarity level of words before and after the change respectively.

The second way of changing the word similarity level, denoted by Re(s), is a relative approach with step s. To calculate the gain or loss one takes into account the similarity level from the reference matrix in the following way:

newSim = prevSim + (refSim-prevSim)*Step;
where the refSim is the level of similarity of texts taken from the reference matrix.

To asses the usefulness of these approaches an experiment was conducted. A number of randomly selected pairs of link texts were selected, similarity obtained from the reference and the current matrix were compared and then the term similarity levels were changed using the absolute or the relative approach. The new value of term similarity was stored in the modified word similarity matrix. At the same time the root mean square deviation of all compared so far text similarity values was calculated. The values of the RMSD for steps equal to 0.25 and 0.5 and the number of selected pairs ranging from 1000 up to 100 000 are presented on the Figure 1.

As can be easily seen the modification of term similarity brings about considerable improvement in performance. The best results were obtained for the relative approach with step set to 0.5. Increasing further the step size has produced any significant improvement.

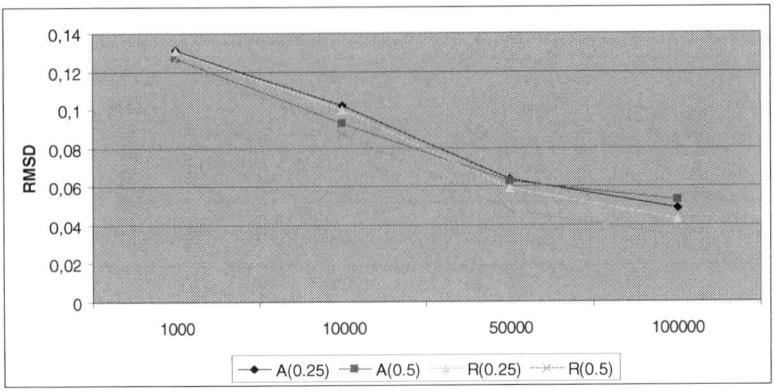


**Fig. 1.** The RDMS of absolute and relative approach to modifying word similarity

## 7 Conclusions

The link text have proved once more again their ability to describe very efficiently the content of pages the links are pointing to. A focused crawler analyzing the link text harvests the collaborative effort of all the web site content authors. A page with not meaningful or inappropriate link texts would be rejected by the community of users and has either to improve or to perish. This objective evaluation process makes the links texts superior to the additionally added semantic tags which do not undergo such a verification procedure.

Extending the link texts comparing process by incorporating the WordNet based semantic similarity of terms increases the recall but at the cost of a falling precision. The proposed adaptive modification of term similarity level helps us to mitigate the disadvantage.

The experiment has covered only one news site. We have a strong premonition that the WordNet could be even more useful in the case of pages originating from different sites. The WordNet describes general properties or words that are common to language as a whole and thus the individual styles of texts could be unified.

Contrary to the classical weighting schemas used for the cosine measure the proposed algorithm does not take into account the frequencies of term occurrences. The decision was taken by premeditation not by omission. A link text resembles more a keyword section of a paper than a regular text in natural language. The keywords based text retrieval usually does not weigh the keywords by their frequency distribution but rather by their importance for a user. This the way which is followed in the proposed algorithm.

The process of calculating and modifying words similarity is not a simple one and in real life application it may slow down the crawling. The word frequency distribution is extremely uneven as described the Zipf law. One can easily speed up the similarity assessment by cashing the word similarity of most frequent word pairs.

Attaching surrounding text to the link texts is a possibility definitely worth considering. The surrounding text is especially important for link text that that have only contextual meaning such as "for more" or "see also". In this case a text window encompassing the link text should probably replace the link text altogether.

The next stage of the study is to extend the database so it would encompass more sites both in English and in other languages. The WordNet databases for other languages are already available and they share the same database structure. This facilitates the porting to other languages. The study of the Polish and the Russian sites is planed for the near future. With their rich inflectional structure the languages have strikingly different character than English what makes them an additionally interesting study area.

The transformation from original text to raw text needs to be further extended. The proper handling abbreviations and multi term concepts improves performance on areas that are not covered by the WordNet term similarity algorithm.

## References

1. De Bra, P.M.E., Post, R.: Information retrieval in the World-Wide Web: making client-based searching feasible. Comput Netw. ISDN Systems 27, 183–192 (1994)
2. Novak, B.: A Survey Of Focused Web Crawling Algorithms. In: Proceedings of SIGKDD (2004)
3. Jacovi, M., Maarek, Y., Pelleg, Y., Shtalhaim, Y., Ur, S.: The Shark-Search Algorithm - An Application: Tailored Web Site Mapping. In: Proceedings of the Seventh International World Wide Web Conference, Brisbane, Australia (April 1998)
4. Diligenti, M., Coetzee, F.M., Lawrence, S., Giles, C.L., Gori, M.: Focused Crawling Using Context Graphs. In: 26th International Conference on Very Large Databases VLDB 2000, pp. 527–534 (2000)
5. Jung, J.J.: Using evolution strategy for cooperative focused crawling on semantic web. Neural Comput. & Applic. 18, 213–221 (2009)
6. Nol, M.G., Meinel, C.: Web Page Classification: An Exploratory Study of the Usage of Internet Content Rating Systems. In: HACK 2005 Conference (2005)
7. Piskorski, J., Tanev, H., Atkinson, M.: Cluster-Centric Approach to News Event Extraction. In: New Trends in Multimedia and Network Information Systems, pp. 277–290. IOS Press, Amsterdam (2008)
8. Manning, C., Raghavan, P., Schütze, H.: An Introduction to Information Retrieval. Cambridge University Press, Cambridge (2008)
9. Porter, M.F.: An algorithm for suffix stripping. Program 4(3), 130–137 (1980)
10. Fellbaum, C.: WordNet: an electronic lexical database. MIT Press, Cambridge (1998)
11. Seco, N., Veale, T., Hayes, J.: An Intrinsic Information Content Metric for Semantic Similarity in WordNet. In: Proceedings of the European Conference of Artificial Intelligence (2004)
12. `http://www.vordweb.co.uk/standards/lynx_text_browser.htm`
13. `http://validator.w3.org/`
14. `http://www.ranks.nl/resources/stopwords.html`

# Mining Frequent Purchase Behavior Patterns for Commercial Websites

Li-Fu Hsu[1], Chuin-Chieh Hsu[2], and Yi-Chen Ku[2]

[1] Hwa-Hsia institute of Technology, Department of Information Management, 23554 Taipei County, Taiwan
[2] National Taiwan University of Science and Technology, Department of Information Management, 69042 Taipei, Taiwan
Lhhsu@cc.hwh.edu.tw, {CChsu,M9209004}@mail.ntust.edu.tw

**Abstract.** Due to the rapid growth in the field of electronic commerce (EC), a huge amount of data has been gathered in many EC sites since their inception. Although many studies have focused on the mining of an EC site's frequent traversal paths and frequent purchase items, an efficient combination of the two types of mining, however, is still not available up to date. To resolve this problem, we first combine both types of data, i.e. the traversal paths and the purchase records, and then mine the combined data for the frequent purchase behavior pattern. In this study, we propose an effective algorithm named Mining Frequent Purchase Behavior (MFPB), which will dig for all frequent path patterns and all frequent purchase records, with a pattern growth concept for an efficient and complete pattern mining, within the projected transactions.

**Keywords:** Mining, frequent purchase, behavior patterns.

## 1 Introduction

In today's widespread internets, how to effectively provide the users customized website guidance and rapidly respond the users' requests has represented a key success factor to a website. Many studies related to traversal path prediction and web prefetching thus emerge one after another [1],[2]; the mining techniques for frequent path traversal patterns [1] also become these studies' main subject, too.

To effectively dig for useful information within the huge amount of data, many studies have devoted to various kinds of data mining techniques. The data mining techniques, according to various application needs and purposes, can be divided into: Association rules, sequence pattern, time series prediction and categorization and classification. In this study, we will combine electronic-commerce (EC) website customers' path traversal pattern and frequent purchase items. Only a handful studies have effectively and completely combined these two [2],[11],[12] although quite many traditional studies have separately explored association rules and frequent path traversal patterns [1],[8],[9],[10]. We will combine the web log and the transaction records to extract the more valuable frequent purchase behavior patterns.

N.T. Nguyen, R. Kowalczyk, and S.-M. Chen (Eds.): ICCCI 2009, LNAI 5796, pp. 732–742, 2009.

Within the frequent purchase behavior patterns, the traversal paths and the purchase records all must be of frequent patterns. We propose a mining frequent purchase behavior (MFPB) algorithm in this study to, first, mining frequent purchase behavior patterns using simulation approach to generate web transversals and transaction records and, then, extracting all frequent purchase behavior patterns with MFPB algorithm.

## 2 Related Work

The EC website purchase behavior pattern concept was first proposed by Yun and Chen in 2000 [11],[12] which expanded the traversal paths initiated by Chen in 1996 [1]. The portion related to frequent purchase items in the concept mainly follows the Apriori-like algorithm [3],[13].

Association rules were first proposed by Agrawal and Srikant [3] to, mainly, extract the items, out of a data bank, that frequently occur jointly in transactions. Hence, the rules are often used in a market basket analysis. For example, the items within purchase transactions are the purchased items in a market's sales data bank, and a pattern extracted by association rules will be the items often jointly purchased. Agarwal et al. proposed an Apriori feature: To a frequent itemset, all its non-empty subsets must also be frequent [3]. The Lattice-based method Zaki proposed in 2000 [5] adopts a transaction identifier (TID) list data structure and may effectively improve the Apriori compulsory multiple-retrieving data bank problem. The TID List may be taken as an inverted-transaction structure. A transaction within a conventional data bank will record what items the transaction contains; an inverted transaction will, on the other hand, record that by what transactions an item is contained. After adding the TID list, we only need to AND the two patterns' TID lists, in the process of joining two frequent patterns to generate a new pattern, to extract the transactions jointly contain the two patterns without needing to re-retrieve the data bank to determine if each transaction contains a new pattern.

The Apriori algorithm will generate a big volume of candidate itemsets in the Join portion, while using a frequent pattern tree (FP tree) will not. The tree adopts pattern growth style to recursively establish even longer patterns. Since Apriori-like algorithm requires multiple scanning a data bank, Han and Pei et al. proposed a pattern growth concept in 2000 [17],[18] and initiated a free span [18] and a prefix span [16],[17] in a pattern growth algorithm to mine sequence patterns.

Chen et al. initiated a path traversal pattern concept in 1998 [1]. The mining problem of frequent path traversal patterns was later divided into two types, i.e. simple traversal sequences and non-simple traversal sequences. If we permit repeated webpages to occur within traversal sequences and within path traversal patterns extracted from said sequences, we will term this nature as a non-simple traversal sequence problem; otherwise, a simple traversal sequence problem.

The web transaction pattern concept was first raised by Yun and Chen in 2000 [11],[12]; this pattern combines EC website customers' simple path traversal patterns and frequent purchase patterns.

## 3 Mining Purchase Behavior Patterns

The method for mining purchase behavior patterns will use the inverted transaction concept and improve previous TID-List data structure [3] to meet the Join needs of a purchase behavior pattern, which is different from the association-rule pattern. The mining method will first retrieve a data bank once, covert the bank into two inverted transactions, according to the traversal paths and purchased items, and use the two inverted transaction tables and one FB-tree data structure to extract all frequent purchase behaviors.

### 3.1 The Definition of Purchase Behavior Patterns

In an EC website, N represents the site's webpage set, i, the sale item set, and P, the purchase behavior pattern, will be: $P = < s_1, s_2,..., s_p ; s'_1(i_1), s'_2(i_2),..., s'_m(i_m) >$ ,where $S = < s_1, s_2,..., s_p >$ , a traversal path which is a sequence in order; p is the length of the traversal path, and $s_j \in N$, $1 \le j \le p$. The purchase record, $\{s'_1(i_1), s'_2(i_2),..., s'_m(i_m)\}$, represents that the consumers purchase Item $i_k$, where $i_k \in i$, $1 \le k \le m$ and $\{s'_k \mid 1 \le k \le m\} \in \{s_j \mid 1 \le j \le p\}$. For example, P equals to $< A, B, C, D ; A(i_1), D(i_3) >$. Since the length of the traversal path for Pattern P, $P = < s_1, s_2,..., s_p ; s'_1(i_1), s'_2(i_2),..., s'_m(i_m) >$, is p, the length of the pattern is also p.

### 3.2 The Data Bank Conversion

The web transaction database that combines the website browsers' traversal paths and purchase behaviors is shown in Table 1. Since we are mining purchase behavior patterns, the databank only preserves the sessions with purchase behaviors and records their transaction IDs (TIDs) and the browsed webpages (stored in the Path column in order ) along with the purchase records; e.g., $A(i_1)$ represents a purchase of Item $i_1$ on Webpage A.

**Table 1.** The web transaction database (DB)

| TID | Path | Purchase |
|---|---|---|
| $T_1$ | H, A, C, D, A, B, F | $A(i_1), D(i_2), F(i_3)$ |
| $T_2$ | A, D, H, B, E, F, G | $D(i_2), B(i_4), E(i_3)$ |
| $T_3$ | C, D, A, H, B, F, E | $C(i_5), D(i_2), B(i_4), F(i_3)$ |
| $T_4$ | H, B, E, B, F, A, D | $E(i_3), B(i_4), D(i_2)$ |

Retrieving the data bank once, in the TID order, will render two inverted transaction data banks, where one is an inverted page traversal file, as shown in Table 2, and the other an inverted purchase file, as shown in Table 3. The data bank uses the TID as the primary key while the two inverted transaction data banks use a traversal path, with a length of one, and a purchase-record pattern as the primary key to separately record the TIDs of that pattern's transaction behavior(s). Since the browsed webpages in a traversal path are sequential, the inverted page traversal file's TID column not only records that pattern's TIDs but also that pattern's location appearing in that transaction.

**Table 2.** The inverted page traversal file

| Page | TID | Count |
|---|---|---|
| A | $T_1(1:1)$, $T_2(3:3)$, $T_3(4:4)$, $T_4(1:1)$ | 4 |
| B | $T_1(2:2)$, $T_1(5:5)$, $T_2(1:1)$, $T_3(3:3)$, $T_4(6:6)$ | 4 |
| C | $T_1(6:6)$, $T_2(4:4)$, $T_3(5:5)$, $T_4(2:2)$, $T_4(4:4)$ | 4 |
| D | $T_1(3:3)$, $T_3(1:1)$ | 2 |
| E | $T_1(4:4)$, $T_2(2:2)$, $T_3(2:2)$, $T_4(7:7)$ | 4 |
| F | $T_2(5:5)$, $T_3(7:7)$, $T_4(3:3)$ | 3 |
| G | $T_1(7:7)$, $T_2(6:6)$, $T_3(6:6)$, $T_4(5:5)$ | 4 |

**Table 3.** The inverted purchase file

| Purchase Behavior | TID | Count |
|---|---|---|
| $A(i_1)$ | $T_1$ | 1 |
| $B(i_4)$ | $T_2$, $T_3$, $T_4$ | 3 |
| $C(i_5)$ | $T_3$ | 1 |
| $D(i_2)$ | $T_2$, $T_3$, $T_4$ | 3 |
| $D(i_3)$ | $T_1$ | 1 |
| $E(i_3)$ | $T_2$, $T_4$ | 2 |
| $F(i_3)$ | $T_1$, $T_3$ | 2 |

**Definition 1.** If Transaction $T_i$ equals to $< t_1, t_2,..., t_q \,;\, t'_1(i'_1), t'_2(i'_2),..., t'_r(i'_r) >$, Ti(s:e) represents the path formed by the s-th to the e-th browsed webpages, namely $Ti(s{:}e) = < t_s, t_{s+1},\ldots, t_e >$.

### 3.2.1 The TID-List Data Structure

Each pattern possesses a TID-List data structure to record what transactions that pattern contains; for a path traversal pattern, the structure will record the location(s) that pattern appears in the transactions. For example, a path traversal pattern's, P's, TID equals to {Ti(s:e)}; that indicates that Pattern P appears in Transaction Ti, and the path formed by Ti's s-th to e-th webpages just minimally contains such Pattern P. This represents that the path formed by the s-th to the e-th webpages may contain Pattern P, but the two paths formed by the (s+1)-th to the e-th webpages and by the s-th to the e-1-th webpages both don't contain Pattern P (see Definition 2).

**Definition 2.** A traversal path with a length of m, T, with $T = < N_1,\ldots, N_m >$, minimally contains a path traversal pattern, P, if and only if T contains P, and T's continuous sub-paths all don't contain P.

For example: For a certain path traversal pattern, P, with $P = < A, F >$, we can learn from the web transaction database, as shown in Table 1, that Transaction $T1 = <$ H, A, C, D, A, B, F ; $A(i_1)$, $D(i_2)$, $F(i_3) >$, then $T_1(5:7)$, with $T_1(5:7) = < A, B, F >$, minimally contains Pattern P, but $T_1(2:7)$, with $T_1(2:7) = < A, C, D, A, B, F >$, although also containing P, does not minimally contain P.

A path traversal pattern's TID List possesses more information, such as that pattern's beginning and ending locations, and so on, than a purchase-record pattern's TID List. For example: If $<E>$'s TID = $<T_2(5{:}5), T_3(7{:}7), T_4(3{:}3)>$, then $<E>$'s TID-List structure is as shown in Fig. 1; if $<B(i_4)>$'s TID = $<T_2, T_3, T_4>$, then its TID-List structure is as shown in Fig. 2.

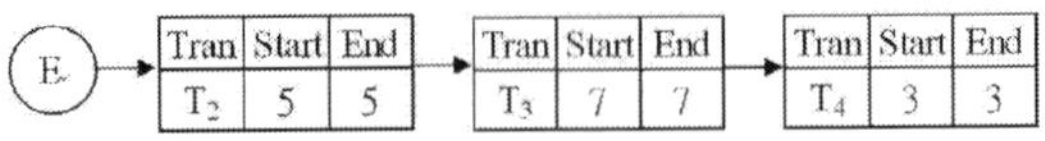


**Fig. 1.** The path traversal pattern's TID List

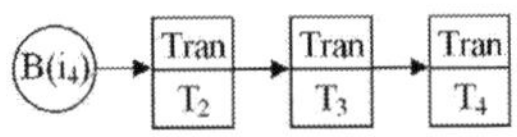


**Fig. 2.** The purchase-record pattern's TID List

### 3.2.2 TID Join

The concept of TID Join is: When we are to make sure if Frequent Pattern A is still qualified to be frequent under Frequent Pattern B's conditions, we only need to use the Join treatment on the two patterns' TID Lists once with a complexity of O(max(a,b)), where a and b are the length of Patterns A's and B's TID List, respectively.

In view of the sequential feature a path traversal pattern possesses, a pattern growth will be divided into backward-wise pattern growth and forward-wise pattern growth. For example: That $<A, B, C>$ Join $<B, C, D>$ growing into $<A, B, C, D>$ pattern represents $<A, B, C>$ pattern grows backward while $<B, C, D>$ pattern grows forward. Due to the sequential feature, if a frequent path traversal pattern, P, with P = $<p_1, p_2,\ldots, p_k>$, is of k-length, while its TID List is $\{T_i(s{:}e)\}$, and that P is to grow backward by 1 in length to become a new path, i.e. $<p_1, p_2,\ldots, p_k, p_{k+1}>$, then the condition projection will be the webpage after Transaction Ti's e-th webpage.

A purchase record pattern does not possess a sequential feature; hence, the TID List only needs to record the transaction ID containing that pattern's transactions. In addition, when two patterns are making Join, we only need to extract the transaction IDs concurrently containing two patterns to form a new TID List without considering the patterns' backward or forward sequence feature, nor considering the difference of growing backward or forward. Hence, making Join with two purchase record patterns will grow into a new purchase record pattern.

## 3.3 The Path Traversal Pattern Growth

The MFPB algorithm uses path traversal pattern growth method to grow all k-length frequent purchase traversal paths into (k+1)-length frequent traversal paths and, then, determine if there exist frequent purchase records under such path conditions. If yes, the frequent traversal path(s) will belong to frequent purchase traversal path(s). The combination of the path(s) and that frequent purchase record(s) will be the frequent purchase behavior. To speed up the searching of frequent purchase path traversal

```
Algorithm : MFPB (Mining Frequent Purchase Behavior)
Input : Web Transaction W
Output : Frequent Pattern P
{
  FNS = Transform W into Inverted Page Traversal File //Frequent Node Set
  FPS = Transform W into Inverted Purchase File //Frequent Purchase Set
  FB tree = ∅
  Add FPS into FB tree as Purchase Node //Purchase Node represents a purchase node
  k=1
  while(the k-level in the FB tree ≠ ∅)
  {
    FB tree =TPG(FB tree, FNS)
    //(k+1)-Traversal Path Generating, and add the Frequent (k+1)-Path into the FB tree
    k++
  }
  return the Frequent Purchase Behavior in the FB tree
}
```

**Fig. 3.** The MFPB Algorithm

patterns suitable for Join-making during a pattern growth, we establish a frequent behavior tree (FB tree) to rapidly extract frequent-purchase path-traversal patterns. The MFPB algorithm will, within an FB tree of k-height, extract k-length frequent purchase traversal paths suitable for growth, grow such patterns into (k+1)-length frequent traversal paths, and then, if such paths are also frequent purchase traversal paths, add them into the FB tree.

### 3.3.1 The FB Tree

The MFPB algorithm will add all extracted frequent purchase traversal paths into an FB tree. The method for adding a frequent purchase traversal path, P, with $P = < p_0, p_1, \ldots, p_k >$, is to search for a child node, $p_0$, from the root; if there exists no $p_0$, add in $p_0$. Again, search for the child node $p_1$ of $p_0$; if $p_1$ does not exist, add in $p_1$; until Node pk of k-height is added in, label $p_k$ as a purchase node and add the purchase records, which become frequent under that path conditions, onto said purchase node.

The nodes within an FB tree can be divided into purchase nodes and regular nodes; each leaf node must be a purchase node. Each k-height purchase node represents a k-length frequent purchase traversal path, i.e. the path formed from the root of the FB tree to said node is k-FPTP. For example: The shaded nodes in the FB tree as shown in Fig. 4 stand for purchase nodes. This indicates that the path formed from the root to that node represents a frequent purchase traversal path; the frequent purchase records under such path conditions are linked by dotted lines. Walking through the HBK path from the root node {} to Purchase Node F, with a height of 3, stands for a 3-length frequent purchase traversal path < HBF >; the frequent purchase records on that path include the 1-length $F(i_3)$ and the 2-length $\{F(i_3), H(i_1)\}$, namely < H, B, F ; $F(i_3)$ >, < H, B, F ; $H(i_1)$ >, < H, B, F ; $F(i_3)$, $H(i_1)$ >, as the frequent purchase behavior patterns.

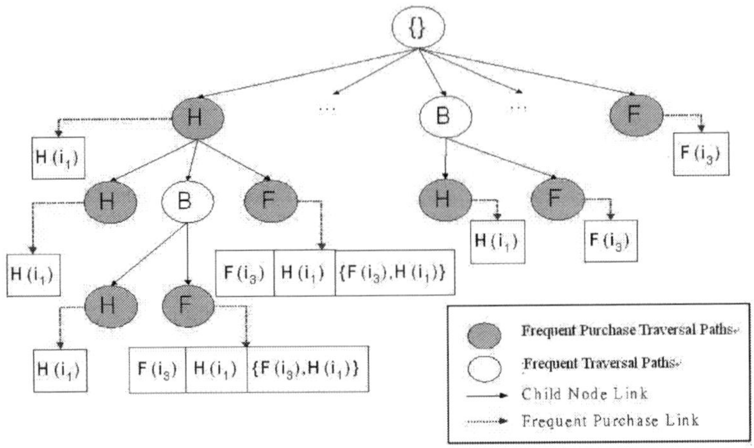


**Fig. 4.** An FB Tree

### 3.3.2 The TPG Algorithm

Use the TID List generated through TID Join to determine if the newly born paths are frequent traversal paths. If they are, determine which frequent purchase records are also frequent under the newly grown frequent traversal paths so as to decide if the newly grown paths are frequent purchase traversal paths. If yes, add them into the FB tree.For example: Pattern < B > 's TID equals to < $T_1(6:6)$, $T_2(4:4)$, $T_3(5:5)$, $T_4(2:2)$, $T_4(4:4)$ >; carrying out forward pattern growth, we can find three frequent purchase traversal paths, i.e. < D, B >, < H, B >, and < A, B >. Carrying out backward pattern growth, we arrive two frequent purchase traversal paths, i.e. < B, E > and < B, F >. We may add them in the FB tree, as shown in Fig. 5.

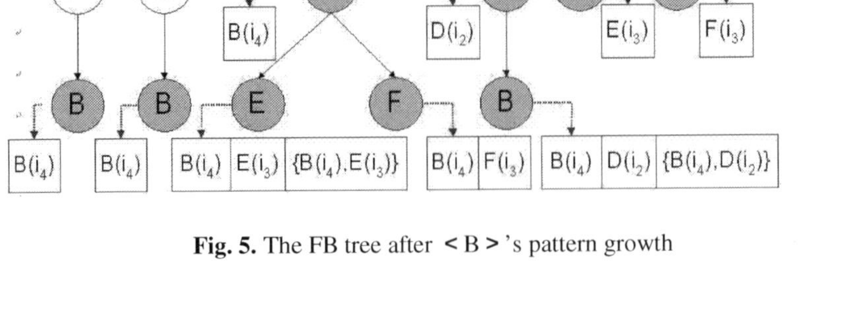


**Fig. 5.** The FB tree after < B > 's pattern growth

# 4 The Experiment Setup and Analysis

Referring to the generation method of Quest Dataset Simulation Generator provided by IBM, our experiment setup adds in the web tree structure and the website browsers' stochastic browsing purchase behaviors [1],[2],[12] to simulate the users' behaviors on EC websites. We will next introduce the dataset's simulation method and, then, the MFPB algorithm's implementation and result analysis.

## 4.1 The Simulation Method for Experiment Datasets

We use a simulation method to build a web tree [1],[2],[12], the parameters of which are shown in Table 4, to simulate an EC website's webpage structure.

**Table 4.** The web tree-generation parameters

| Title | Explanation |
|---|---|
| N | The number of the web-tree nodes, namely the number of the site's webpages, as a simulation program parameter |
| M | The number of kinds of items sold on the website, a parameter of the simulation program |
| $fanout_i$ | The fanout of Node i on the web tree; determined by Uniform (fmin, fmax) |
| fmin | A program parameter for fanout simulation |
| fman | ditto |
| $N_I$ | The number of sold-item nodes on the web tree; a parameter of the simulation program |
| $N_J$ | The number of exit nodes on the web tree; a parameter of the simulation program |
| $N_D$ | The number of destination nodes on the web tree; a parameter of the simulation program |
| $M_i$ | The number of the kinds of items sold on Sold-item Node I; determined by Poisson (mi) |
| $m_i$ | A program parameter for Mi simulation |
| $PB_i$ | The probability of purchasing items on Sold-item Node i; randomly determined |
| $PB_{im}$ | The purchasing probability of buying Item m on a sold-item node that is deemed to purchase items |
| $PG_0$ | Except the root node, each node has a PG0, representing the probability walking toward a parent node; randomly determined |
| $PG_i$ | Except the leaf node, each node has a probability, PGi, walking toward the i-th child node; randomly determined |
| $PG_J$ | Each exit node has a PGJ, representing the probability walking toward a destination node; randomly determined |

Normalize the browsing probability for each web-tree node to make the total of all the nodes' probabilities equal to 1, see equation (1), where NC is the number of child nodes.

$$\sum_{i=0}^{NC} PG_i + PG_J = 1 \quad (1)$$

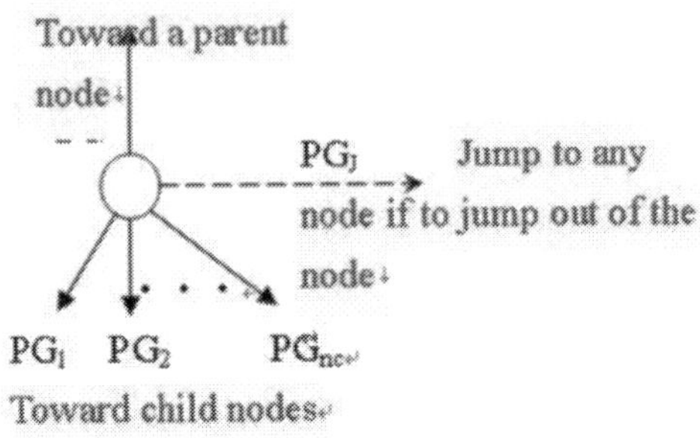


**Fig. 6.** Illustration for simulating browsing probabilities

## 4.2 Implementation and Analysis of the Website Behavior-Pattern Mining Method

In this experiment, we adopt a 500MHz Pentium III computer, with an environment of 384MB main memory and 576MB virtue memory, and use Borland C++ Builder 5.0 to develop codes used from simulating the generation of datasets to implementing

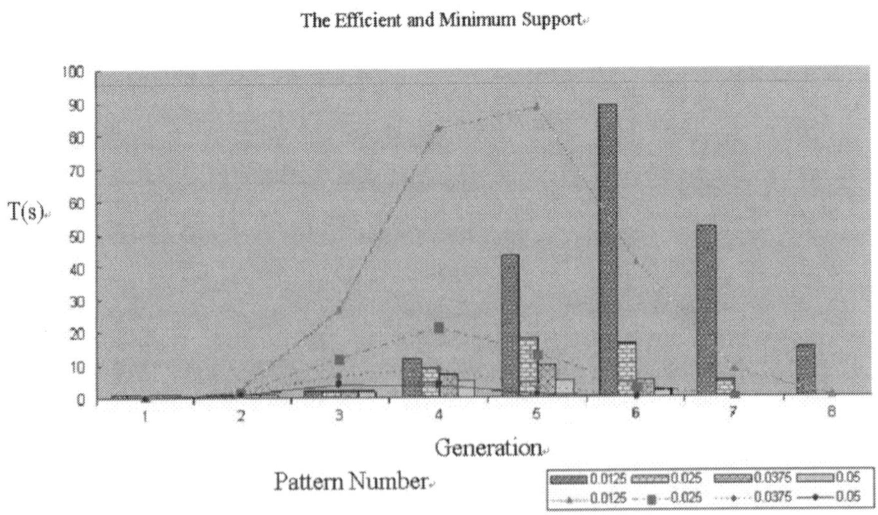


**Fig. 7.** The minimal support degrees and the execution performance

the MFPB algorithm. We adopt the simulation method to generate two purchase behavior datasets, N50M30 and N20M10, and use the N50M30 dataset to implement the MFPB algorithm. Via this experiment, we observe that the relationship between the number of datasets and the executing time is linear. We then use Dataset N20M10, with various minimal support degrees, to evaluate the correlations between the minimal support degrees and the number of the mined frequent patterns with the execution time, as shown in Fig. 7. From this experiment, we observe that the less the support degrees are, resulting in a rapid increase on the number of the mined frequent patterns by the middle, multiple pattern growth, the longer the relative execution time will, consequently, be.

## 5 Conclusion and Prospective Outlook

A conventional website-transaction pattern cannot extract a complete transaction pattern which is formed by simple path-traversal patterns; neither can it extract any transaction pattern formed by non-simple path-traversal patterns. Even if we can establish the sequence-pattern algorithm for the non-simple path-traversal patterns, we still cannot link it with an EC website's consumer purchase information.

The proposed frequent-purchase behavior pattern not only takes the EC website customers' traversal path information into consideration but also adds in the purchased item information. We also adopt the mining frequent-purchase behavior (MFPB) algorithm to avoid multiple-retrieving a data bank; in addition, we use the TID-List structure to reduce a large demand in memory by general pattern-growth algorithms; hence, we can rapidly extract consumer purchase behavior patterns in an EC website. Responding to the purchase-behavior records' continuous increase, the MFPB we adopt uses TID-List data structure to record a pattern's projected transactions so that we can rapidly update the TID List when new transactions are added.

We also use the FP-tree structure to record the pattern growth relationship among frequent purchase behavior patterns; hence, we can execute an incremental mining for the added purchase-behavior recorded patterns to a database and partially update the FP trees. We suggest adding in the purchased item's time information in a purchase-behavior pattern, in the future, so that an EC website can effectively predict a consumer's re-consuming time.

## References

1. Chen, M.S., Park, J.S., Yu, P.S.: Efficient Data Mining for Path Traversal Patterns. IEEE Transactions on Knowledge and Data Engineering, 209–220 (1998)
2. Vallamkondu, S., Gruenwald, L.: Integrating Purchase Patterns and Traversal Patterns to Predict HTTP Requests in E-Commerce Sites. In: Proc. of IEEE Int. Conf. on E-Commerce (2003)
3. Agrawal, R., Srikant, R.: Fast algorithms for mining association rules. In: Proc. 1994 Int. Conf. Very Large Data Bases (VLDB 1994), Santiago, Chile, pp. 487–499 (1994)
4. Yen, S.J., Chen, A.L.P.: An Efficient Approach to Discovering Knowledge from Large Databases. In: Int. Conf. on Parallel and Distributed Information Systems (1996)

5. Zaki, M.J.: Scalable Algorithms for Association Mining. IEEE Transactions on Knowledge and Data Engineering 12(3), 372–390 (2000)
6. Han, J., Pei, J., Yin, Y.: Mining Frequent Patterns without Candidate Generation. In: ACM SIGMOD (2000)
7. Yen, S.J., Chen, A.L.P.: A Graph-based Approach for Discovering various types of Association Rule. IEEE Transactions on Knowledge and Data Engineering 13(5) (2001)
8. Cooley, R., Mobasher, B., Srivastava, J.: Data Preparation for Mining World Wide Web Browsing Patterns. Knowledge and Information System, 5–32 (1999)
9. Lin, I.Y., Huang, X.M., Chen, M.S.: Capturing User Access Patterns in the Web for Data Mining. In: Proc. of the 11th IEEE Int. Conf. on Tools with Artificial Intelligence, pp. 345–348 (1999)
10. Pei, J., Han, J., Mortazavi-Asl, B.: Mining Access Patterns Efficiently from Web log. In: Terano, T., Chen, A.L.P. (eds.) PAKDD 2000. LNCS, vol. 1805. Springer, Heidelberg (2000)
11. Chen, M.S., Yun, C.H.: Mining Web Transaction Patterns in an Electronic Commerce Environment. In: Proc. of the 4th Pacific-Asia Conf. on Knowledge Discovery and Data Mining, pp. 216–219 (2000)
12. Yun, C.H., Chen, M.S.: Using Pattern-Join and Purchase-Combination for Mining Web Transaction Patterns in an Electronic Commerce Environment. In: Int. Conf. on Computer Software and Applications (2000)
13. Park, J.S., Chen, M.S., Yu, P.S.: An Effective Hash Based Algorithm for Mining Association Rules. In: ACM SIGMOD Int. Conf. on Management of Data (1995)
14. Srikant, R., Agrawal, R.: Mining Sequential Patterns: Generalizations and Performance Improvements. In: Apers, P.M.G., Bouzeghoub, M., Gardarin, G. (eds.) EDBT 1996. LNCS, vol. 1057, pp. 3–17. Springer, Heidelberg (1996)
15. Agrawal, R., Srikant, R.: Mining Sequential Patterns. In: Proc. 1995 Int. Conf. Data Engineering (ICDE 1995), Taipei, Taiwan, pp. 3–14 (1995)
16. Pei, J., Han, J., Mortazavi-Asl, B., Pinto, H., Chen, Q., Dayal, U., Hsu, M.C.: PrefixSpan: Mining Sequential Patterns Efficiently by Prefix-Projected Pattern Growth. In: Proc. 2001 Int. Conf. on Data Engineering (ICDE 2001), Heidelberg, Germany, pp. 215–224 (2001)
17. Han, J., Pei, J.: Mining Frequent Patterns by Pattern-Growth: Methodology and Implications, ACM SIGKDD Explorations (Special Issue on Scalable Data Mining Algorithms) (2000)
18. Han, J., Pei, J., Mortazavi-Asl, B.: FreeSpan: Frequent Pattern-Projected Sequential Pattern Mining. In: Proc. Int. Conf. on Knowledge Discovery and Data Ming (2000)

# Block Map Technique for the Usability Evaluation of a Website

Kazimierz Choros[1] and Monika Muskala[2]

Institute of Informatics, Wroclaw University of Technology,
kazimierz.choros@pwr.wroc.pl
Wybrzeze Wyspianskiego 27, 50-370 Wroclaw, Poland
monika.muskala@gmail.com

**Abstract.** Usability refers to the extent to which a product can be used by specified users to achieve specified goals with effectiveness, efficiency and satisfaction in a specified context of user. In the Web world usability is defined as set of layout, structure, arrangements, typography, and many other properties that makes website simple and easy to use. There are many techniques of the evaluation of the websites. The analysis of click and heat maps leads to the recognition of these parts of the website which are not used or of these parts of the websites in which the user is intuitively expecting a link to the next part of the site visited. The proposed block map technique used for the usability evaluation of the website is a technique which encourages to restructure the layout of the examined website and to improve the user satisfaction.

**Keywords:** block maps, click maps, heat maps, website pages, Web designing, usability evaluation, usability in practice.

## 1 Introduction

The satisfaction of the client/user is a priority not only in business but also in the Web. It has been determined that there is a great influence of the usability on the user's loyalty to websites that they already visited. The results of the empirical analysis confirmed [1] that the trust of the user increases when the user perceived that the system was usable and that there was a consequent increase in the degree of website loyalty. Greater usability has a positive influence on user satisfaction and also generates greater website loyalty. It was observed [2] that website success is significantly associated with website download delay (speed of access and display rate within the Web site), navigation (organization, arrangement, layout, and sequencing), content (amount and variety of product information), interactivity (customization and interactivity), and responsiveness (feedback options and FAQs).

Various measures for evaluating user satisfaction have been developed [3-6]. Although these measures use different definitions of user satisfaction, all of them are based on an evaluative response provided by a user.

Nowadays, the usability of many specific websites is evaluated. For example, academic library websites [7,8], e-learning systems [9], hypermedia systems [10],

N.T. Nguyen, R. Kowalczyk, and S.-M. Chen (Eds.): ICCCI 2009, LNAI 5796, pp. 743–751, 2009.

commercial websites [11,12], or museum websites [13] have been examined. We expect that the analysis and the evaluation of the usability of websites lead to the improvement of the user satisfaction at website navigation time.

The paper is organized as follows. The next section describes the usability evaluation methods, mainly the most popular in practice: testing, inspection, and inquiry. The third section discusses the clicks and heat maps showing the website areas which are most frequently clicked on by the users. Moreover, these two kinds of maps are presented for the tested website. In the sections 4 the proposed block maps technique is described and the experimental results for the tested website are reported. Section 5 presents the suggested improvements for the tested website which have been formulated after the analysis of click, heat and block maps. The final conclusion and the future research work areas are discussed in the last 6th section.

## 2 Usability in Practice

In general, usability is the measure of a product's potential to accomplish the goals of the user. It refers to the extent to which a product can be used by specified users to achieve specified goals with effectiveness, efficiency and satisfaction in a specified context of a user. In the Web world usability is defined as set of layout, structure, arrangements, typography, and many other properties that makes website simple and easy to use. Usability reflects the ease with which visitors can employ a particular website in order to achieve their particular goal. Usability is not a single, one-dimensional property of a user interface. Usability is a combination of factors including [14]:

- ease of learning – how fast can a user who has never seen the user interface before learn it sufficiently well to accomplish basic tasks?
- efficiency of use – once an experienced user has learned to use the system, how fast can he or she accomplish tasks?
- memorability – if a user has used the system before, can he or she remember enough to use it effectively the next time or does the user have to start once more the learning everything?
- error frequency and severity – how often do users make errors while using the system, how serious are these errors, and how do users recover from these errors?
- subjective satisfaction – how much does the user like using the system?

There are many usability evaluation methods leading to the best website designing. Three types are the most popular in practice: testing, inspection, and inquiry [15,16].

In usability testing approach, representative users work on typical tasks using the system and the evaluators use the results to see how the user interface supports the users to do their tasks. Testing methods include the following: coaching method, co-discovery learning, performance measurement, question-asking protocol, remote testing, retrospective testing, shadowing method, teaching method, and thinking aloud protocol.

In usability inspection approach, usability specialists, sometimes software developers, users and other professionals, examine usability-related aspects of a user interface. Commonly used inspection methods are: cognitive walkthroughs, feature inspection, heuristic evaluation, pluralistic walkthrough, and perspective-based inspection.

In inquiry methods, usability evaluators demand information about users' likes, dislikes, needs, and understanding of the system by talking to them, observing them using the system in real work (not for the purpose of usability testing), or letting them answer questions verbally or in written form. Inquiry methods include: field observation, focus groups, interviews, logging actual use, proactive field study, and questionnaires.

Relatively new techniques of the evaluation of websites are click maps as well as heat maps.

## 3 Click and Heat Maps

Maps show the website areas which are most frequently clicked on by the users. The data may be displayed in two ways: click map and heat map [17]. The map of clicks is

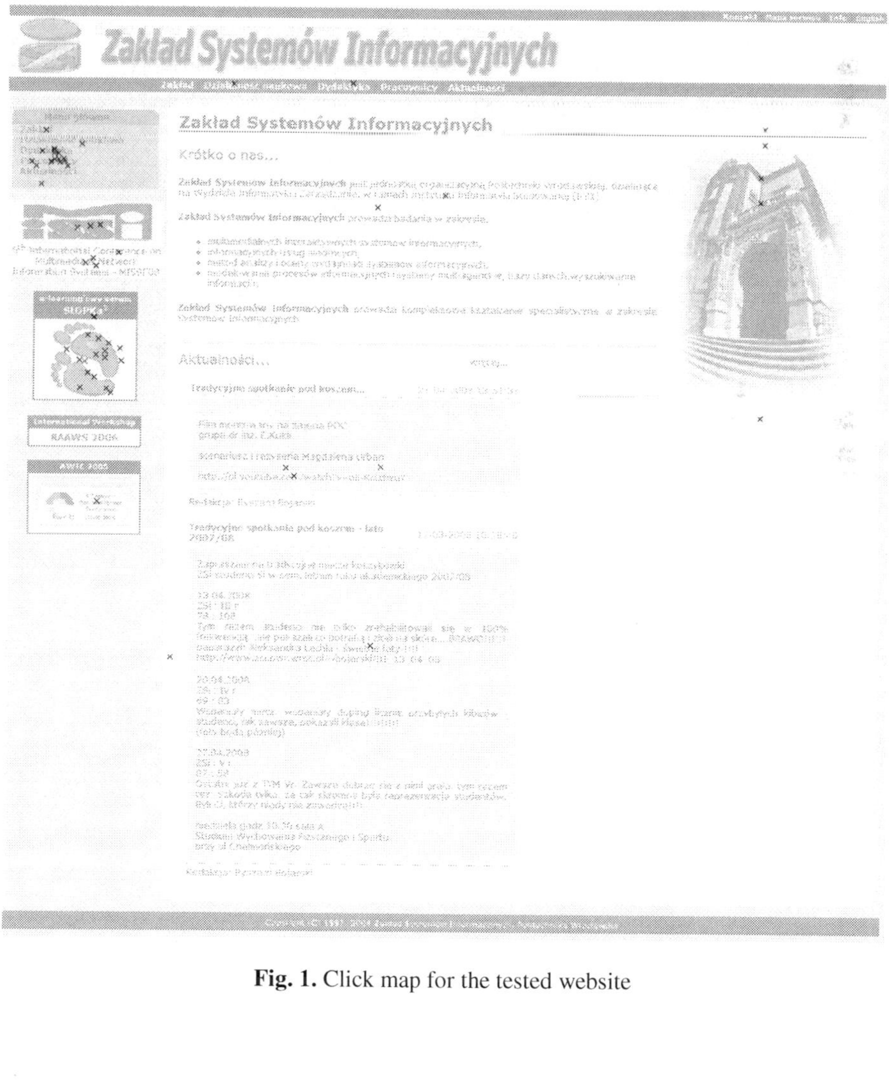


**Fig. 1.** Click map for the tested website

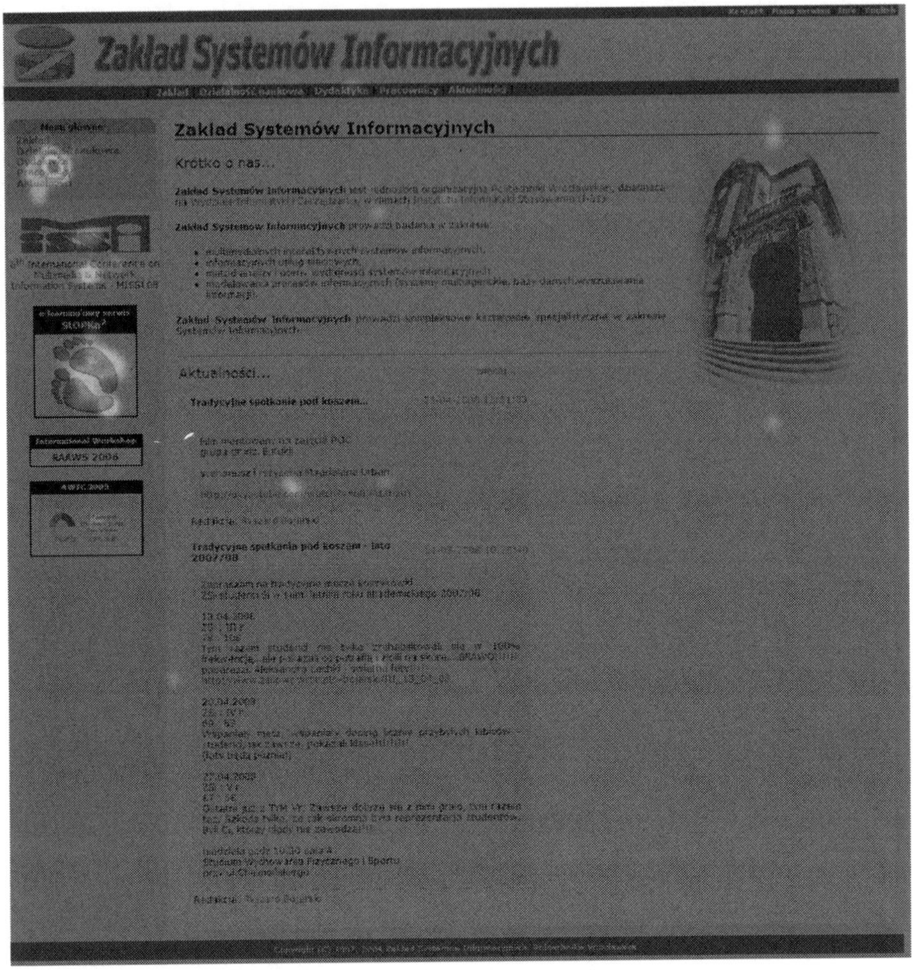


**Fig. 2.** Heat map for the tested website

represented by crosses, whereas the hot map is represented by circles. Click map shows the position of every click on a page, creating a virtual map of visitor activity. Many clicks on the same point are represented by one cross. The number of crosses on the map is equal to the number of the different points on the website clicking by the user. Whereas, on the heat maps the areas that are clicked most often appear in red (these are so-called hot spot), while the areas clicked least often appear in blue or with no color at all. Therefore, a heat map is a visualization of data which uses color to represent data values in a two-dimensional image. Data are gathered during a normal functioning of the website. Monitoring program works in background and is totally invisible and transparent for the user.

Click and heat maps are [18]:

- tools which help us to identify the most popular areas of the website,
- tools which allow us to enhance the usefulness of the website,
- knowledge of how most efficiently the content and the advertising components should be organized on the website,
- information on the users' behavior during website visits,
- information on the website presented directly on Web pages,
- help in e-marketing activity,
- verification of changes on the website,
- assessment of advantages and disadvantages of the website.

Click maps as well as heat maps are nowadays used for the evaluation of the website. Therefore, we have used these techniques to evaluate the website of our scientific department (Fig. 1-2).

The data have been gathered during one week. In this period the page tested has been not changed. During this period 339 clicks have been registered in the Internet Explorer browser. The users have been not anticipated and did not know that their clicks have been registered. So, their behavior was natural and not influenced by anyone nor anything.

Naturally, the question arises why people are clicking where they shouldn't. Probably, they expect there a specific link leading to the information they are seeking.

## 4 Block Maps

However, we have found that the data we received could be better understood if some part of the screen would be analyzed together. Therefore, we have defined some blocks and then we have generated maps for these blocks. In consequence, we have received the block map of the website, see Figure 3.

The blocks have been defined as such a part of area which contains several links, placed every one near another one, forming logical entity:

- top-right menu – block with four general links: contact, site map, info, and language;
- top menu – block with the link under the main heading of the site;
- left menu – block with the navigation link on the left side of the page;
- MISSI (conference) – block with the MISSI Conference link ("Multimedia & Network Information Systems");
- StOPKa – block with the link to the e-learning system;
- RAAWS (conference) – block with two links to the conferences: "RAAWS" and "AWIC".

The most attractive areas of the website are the block with the link to the e-learning system and the left menu. The other parts of the website are significantly less important (Tab. 1).

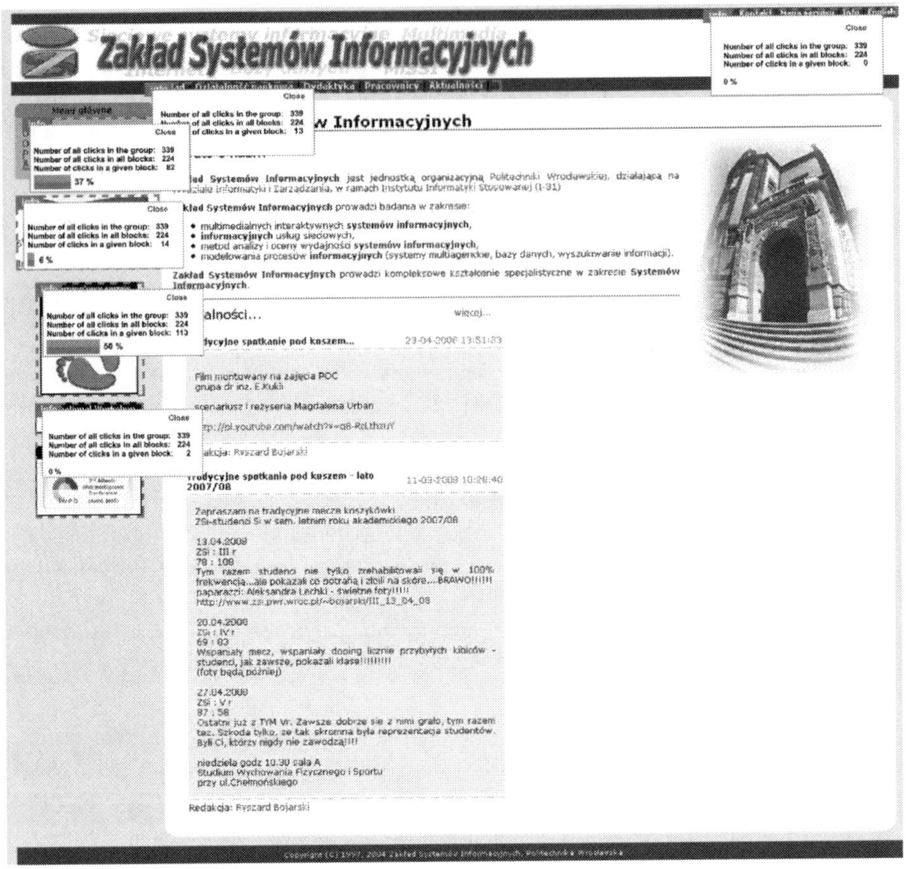

**Fig. 3.** Block map for the tested website

**Table 1.** Number of clicks in the defined blocks

| Block | Number of clicks in a block |
|---|---|
| top-right menu | 0 |
| top menu | 13 |
| left menu | 82 |
| MISSI (conference) | 14 |
| StOPKa (e-learning) | 113 |
| RAAAWS (conference) | 2 |

## 5 Website Improvements

The main goal of a map creation is to decide how we can improve the website. Table 2 presents the suggested improvements for the tested website which have been formulated after the analysis of click, heat and block maps. What is the benefit of

**Table 2.** Conclusions formulated after the analysis of click, heat and block maps leading to some improvements of the tested website

| Problem observed | Source of information | Proposed suggestions |
|---|---|---|
| 1. The top-right menu is totally omitted by the users. Although, the links in this menu are rather fundamentals, e.g. link to the contact data or the website structure and they should be well seen by the user. | Block map: number of clicks in this block is equal to zero. | Removal of this top-right menu under the main heading. |
| 2. Left menu is duplicated under the main header. The links from the left menu are more frequently chosen than those from the menu under the heading. | Click map and heat map. | Deletion of redundant links under the main heading. |
| 3. Little use of links: "MISSI", "RAAWS" and "AWIC". | Block map. | Enlargement of this part of data with some kind of comments (explanation). |
| 4. Insufficiently highlighted the link leading to the news. | Click map and heat map. | Changing of font or changing of term "news" for example to "all news". |
| 5. Right part of the page totally not used. | Click map and heat map. | Reorganization of the site layout or introduction of a new data or links. |

such an approach? These maps identify usability and information architecture issues on the website. Click density maps are not another statistics or Web analytics tool. Namely, they allow us instead to analyze the user experience and identify usability and information architecture issues.

The maps are useful for many reasons, some of which include:

- knowing exactly how Web viewers use a specific page,
- seeing parts of a page that are most frequently used, are the most attractive,
- seeing which parts of a page are totally unused,
- recognizing such part of the page that are not linked but where the users are clicking, so, identifying areas where a link is expected,
- understanding typical patterns of use on the site as a whole,
- predicting how people will use the website in the future.

The visualizations in the form of click, heat, and block maps make them ideal for presentation to non-technical audiences. The results of any change made on the page could be easily observed by comparing a map before and after change. We can exactly see where the users are clicking. But we can also see where people are clicking even if they should not. For example, some blue text on a page is mistaken for a link. An unlinked logo or any other graphical element may also occur that users think they can click on. It seems that Web analysts cannot easily detect any of these situation, but with click, heat, or block maps it immediately becomes obvious.

## 6 Final Conclusion and Further Studies

The click and heat maps are very useful tools for usability evaluation. The click and heat maps are visual maps of clicks on a given Web page, showing attractive, frequently used zones where the user is expecting and in general finds links as well as unused areas. Click and heat maps record the exact x and y coordinates of every click. The map techniques are totally invisible and transparent for the users. Their behavior is natural and not influenced by anyone nor anything.

The statistical data are more impressive if we group together user clicks registered in logically associated parts of the webpage, called click blocks. In the test performed all three maps have been done: click, heat and block map. These maps are very useful for those who search usability improvements.

The block map should be – beside click maps and heat maps – a third indispensable tool for each webmaster, administrator or those who care for effectiveness of their websites and increasing their websites' usability. A block is an area of a Web page which contains several links, placed next to one another, forming logical entity. Such a solution allow us to aggregate the statistical results obtained for click or heat maps.

The aggregation of click statistics in a form of a block map should facilitate to change the Web page layout, i.e. to improve website usability. The problem arises how to define automatically or semi-automatically a block and then how to suggest the optimal size of a surface where a given block of links is planed to be placed. A further research should also help us to determine where the most important block of links should be placed because this place is the most frequently clicked.

And then, it seems obvious that such a technique could be also applied in a personalization process for websites. We are going to examine if the maps have a personal specificity.

We are also going to study the usefulness of click, heat, and block maps in a dynamically developing Web2.0 environment.

## References


1. Flavian, C., Guinaliu, M., Gurrea, R.: The role played by perceived usability, satisfaction and consumer trust on website loyalty. Information & Management 43, 1–14 (2006)
2. Palmer, J.W.: Web Site Usability, Design, and Performance Metrics. Information Systems Research 13, 151–167 (2002)
3. Muylle, S., Moenaert, R., Despontin, M.: The conceptualization and empirical validation of web site user satisfaction. Information & Management 41, 543–560 (2004)
4. Brinck, T., Gergle, D., Wood, S.D.: Usability for the Web: designing Web sites that work. Morgan Kaufmann, San Francisco (2002)

5. Nielsen, J.: Designing Web Usability: The Practice of Simplicity. New Riders Publishing, Indianapolis (2000)
6. Nielsen, J., Loranger, H.: Prioritizing Web Usability. New Riders Press, Berkeley (2006)
7. Battleson, B., Booth, A., Weintrop, J.: Usability testing of an academic library Web site: a case study. The Journal of Academic Librarianship 27, 188–198 (2009)
8. Hilyer, L.A.: User-Centred Library Websites: Usability Evaluation Methods. The Journal of Academic Librarianship 35, 101–102 (2009)
9. Triacca, L., Bolchini, D., Botturi, L., Inversini, A.: MiLE: Systematic Usability Evaluation for E-learning Web Applications. In: Cantoni, L., McLoughlin, C. (eds.) Proceedings of World Conference on Educational Multimedia, Hypermedia and Telecommunications, pp. 4398–4405 (2004)
10. Matera, M., Costabile, M.F., Garzotto, F., Paolini, P.: SUE Inspection: An Effective Method for Systematic Usability Evaluation of Hypermedia. IEEE Transaction on Systems, Man and Cybernetics 32, 93–103 (2002)
11. Benbunan-Fich, R.: Using protocol analysis to evaluate the usability of a commercial web site. Information and Management 39, 151–163 (2001)
12. Zviran, M., Glezer, C., Avni, I.: User satisfaction from commercial web sites: The effect of design and use. Information & Management 43, 157–178 (2006)
13. Cunliffe, D., Kritou, E., Tudhope, D.: Usability evaluation for museum web sites. Museum Management and Curatorship 19, 229–252 (2001)
14. Usability gov. Your Guide for Developing Usable & Useful Web Sites, `http://www.usability.gov/basics/whatusa.html/`
15. Folmer, E., Bosch, J.: Architecting for usability: a survey. Journal of Systems and Software 70, 61–78 (2004)
16. Usability Evaluation, `http://www.usabilityhome.com/`
17. Brown, S.: Click Survey Analysis & Heatmap. Interpretations and ramifications from a web design point of view (2006), `http://modernl.com/article/click-survey-heatmap-analysis`
18. Graphical Map of Clicks on a Website, `http://stat24.com/en/statistics/clickmap/`

# Global Distribution of HTTP Requests Using the Fuzzy-Neural Decision-Making Mechanism

Leszek Borzemski[1], Anna Zatwarnicka[2], and Krzysztof Zatwarnicki[2]

[1] Institute of Informatics, Wroclaw University of Technology, Wroclaw, Poland
leszek.borzemski@pwr.wroc.pl
[2] Faculty of Electrical Engineering, Institute of Automatic Control and Computer Science, Opole University of Technology, Opole, Poland
a.zatwarnicka@po.opole.pl, k.zatwarnicki@gmail.com

**Abstract.** This paper presents the HTTP request global distribution using the fuzzy-neural decision making mechanism. Two efficient algorithms GARDiB and GARDiB2 are proposed to support HTTP request routing to the Websites. The algorithms use the fuzzy-neural decision-making method to assign each incoming request to the Website with the least expected response time. The response time includes the transmission time over the network; as well as the time elapsed on the responding Website server. Simulation experiments showed that GARDiB performed slightly better than GARDiB2 and both proposed algorithms outperformed other competitive distribution algorithms in all simulated workload scenarios.

**Keywords:** HTTP request distribution algorithms, Fuzzy-neural decision-making, Web service quality, Web content delivery.

## 1 Introduction

Web traffic redirection plays an important role in assuring the availability and quality of services (QoS) appropriate for every category of end user. To enable QoS based redirection the specific designs of local Web sites and network infrastructure are needed. They can be realized by the replacement of a single Web server with a few number of Web servers, forming so called locally distributed Web cluster (if Web servers are in the same geographical localization) or globally distributed Web cluster (if Web servers are localized on different geographical localizations) [10, 11, 17]. Other designs assume inclusion into the network infrastructure some additional servers, called intermediary servers, to support the redirection on the wide area network.

This paper presents the design where both techniques are applied and where the distribution decision making mechanism of Web (HTTP) requests is based on the fuzzy-neural model. Our aim is to redirect HTTP requests as to complete them in time-effective manner from the point of view of the end users issuing these requests in a global content delivery system consisting of a set of local services (LSs) equipped with clusters of Web servers possessing the same content and in the equal manner having possibility to serve every request [4, 17]. The system includes the intermediary

N.T. Nguyen, R. Kowalczyk, and S.-M. Chen (Eds.): ICCCI 2009, LNAI 5796, pp. 752–763, 2009.

servers called brokers for request redirection using request global distribution algorithms GARDiB (Global Adaptive Request Distribution with Broker) and GARDiB2 (Global Adaptive Request Distribution with Broker version 2) with built-in fuzzy-neural decision-making mechanism. The global level concerns the distribution to the LSs at different wide network locations, whereas at the local level the specific local distribution algorithms are used. GARDiB and GARDiB2 estimate the expected response time of request based on the class of the request, the load of the wide area network and the load of LS. The classification of requests and determination of the network load take place in the same manner for both algorithms. The difference between them is the way they provide and use information about local services. The aim of this work is to present a performance comparison of these algorithms.

The paper is structured in the following way. First we discuss related work. Next we present our global HTTP request distribution system. After that we introduce the simulation model and present the results of simulations.

## 2 Related Work

Numerous HTTP requests distribution algorithms were proposed as theoretical proposals and practical solutions. They can be divided in two major sub classes, namely static and dynamic. Static algorithms do not take into account any system state information. Dynamic algorithms utilize state information to make distribution decisions. A comprehensive research on different distribution approaches and algorithms is presented in [10] where various local and global, static and dynamic, content-blind and content-aware, client-aware and server-aware HTTP request distribution methods and algorithms are discussed. Among many proposals the most often used in the comparative research analysis distribution algorithms are: Round Robin (RR) [10], Weighted Round Robin (WRR) [10], Locality Aware Request Distribution (LARD) [18], and Client Aware Policy (CAP) [8, 13].

Recent research perspectives in the field, especially in the area of content-aware Web switch architectures are now focused on dynamic and adaptive policies, combining effectively client and server information. To meet this challenge we proposed to use a soft computing approach based on fuzzy logic and neural networks. Neural and fuzzy systems have many significant applications in intelligent computer systems [e.g. 15, 16, 19]. Fuzzy logic approach may be especially profitable when a small amount of uncertain and imprecise information about system under consideration is provided whereas neural networks have learning and adaptive capabilities. Mathematical models do not exist for HTTP request distribution problems and environment is uncertain, and information about clients, servers, and network is incomplete, and can be wrong and contradictory, therefore fuzzy logic can help. Combining fuzzy and neuro approaches we are able to deploy a learning machine that is able to tune (adapt) parameters of a fuzzy-based request distribution system by exploiting approximation techniques from neural networks [14].

Our first work on the application of fuzzy logic and neural networks in Web request distribution research was presented in 2003 [7]. It was the first algorithm in the literature that used fuzzy logic and neural network for response time estimation in content-aware HTTP distribution, especially in local request switching. It was also an important novelty because our algorithm was frankly developed to minimize of the response time of each issued request whereas other algorithms were aimed rather at load balancing and load sharing. An IRD (Intelligence Request Dispatcher) scheduling policy followed our approach [19]. Our recent fuzzy-neural based research in the field of local request distribution include Fuzzy Neural Request Distribution (FNRD) algorithm [5, 6]. In global request distribution we developed two algorithms GARDiB (Global Adaptive Request Distribution with Broker) and GARDiB2 (Global Adaptive Request Distribution with Broker version 2) [2, 3, 4]. The contribution of this work is to present a performance comparison of these algorithms.

## 3 Global Request Distribution System

In this paper we consider the global request distribution system as shown in Fig.1. Requests are HTTP (Hypertext Transfer Protocol) requests issued by Web browsers or similar software to get Web content from Web servers.

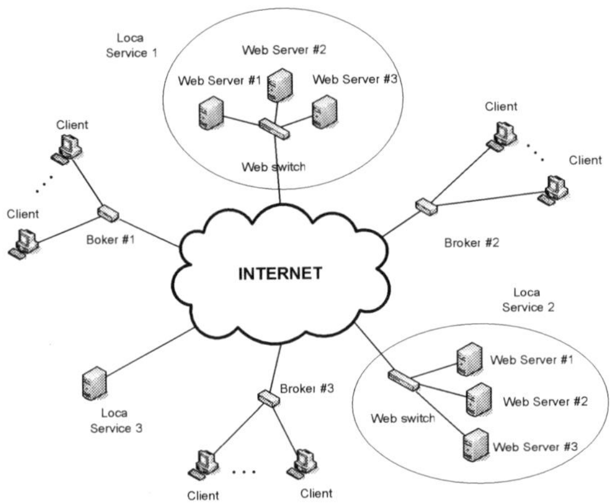


**Fig. 1.** Global request distribution system

The system under consideration is aimed at the delivering to the clients some content from Local Services distributed over the Internet. Local Service (LS) is constructed on the basis of Web server cluster where each server can service every request. The system utilizes the *global distribution algorithm* which is used to decide which LS is to be used to answer to the request. This algorithm is used by intermediary systems called *brokers*. Each domain of clients is using the broker which is in the closest distance (in the network sense) to them. Clients send their requests for Web resources via their broker. Broker supports every request individually and looks for LS that could offer the minimum *request response time*. The response time is evaluated as perceived by the broker. It is estimated by means of fuzzy-neural mechanism. After determining best LS, the broker redirects given request to that LS, where the request is locally redirected to one of Web servers using the *local distribution algorithm*.

The brokers are the key elements of the system. There can be any number of brokers in the system, and they should be installed in different geographical localizations as close as possible to the client domains. The request response time is estimated from the point of view of each broker individually and includes time needed to transfer the request and its response via Internet as well as the time needed to service request at LS. More precisely, it is the time between the start of the transmission of the first byte of the request by the broker, and the end of transmission of the last byte of the response received by the broker.

### 3.1 Architecture of the Broker

Fig. 2 shows the broker's architecture. The broker consists of the following main components: the mechanism of the classification (*Classification module*), models of local services (*Local Services modules*), the module of the decision making (*Decision making module*), the executive module (*Execution module*) and the measuring module (*Measurement module*).

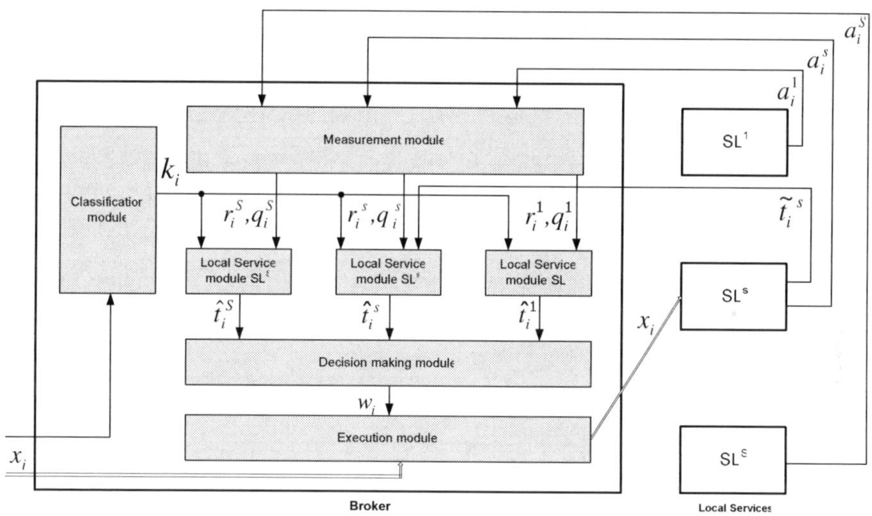


**Fig. 2.** Broker's architecture (GARDiB case)

Every broker can serve $S$ Local Services $LS^1$, $LS^2$,..., $LS^s$,..., $LS^S$. All requests incoming to the broker are initially classified to one of classes by the classification module, $k_i$ where $k$=1, 2,...,$K$. Every model of the local service estimates the request response time $\hat{t}_i^s$ as expected by the $s$-th LS for answering for $x_i$ - th request. The estimation of the request response time is based on: $k_i$ - the class of the request, $q_i^s$ - the network load evaluated by the transfer time of the probing object send from LS to the broker and $r_i^s$ - the load of the local service. The transfer times of probing object are collected by the broker's measurement module every 5 seconds. GARDiB evaluates the load of local service by the number of serviced requests $a_i^s$ whereas GARDiB2 evaluates the load of local service by the expected request completion times offered by the local services in processing the request of the $i$-th class. The decision making module chooses a local service $w_i$, $w_i \in \{1,...,S\}$ characterizing by the shortest estimated request response time. Then the executive module sends the $x_i$ – th request to that local service. After request completion the real value of the request response time $\tilde{t}_i^s$ is measured. $\tilde{t}_i^s$ is used by the particular $s$-th local service model to update the knowledge about the processing possibilities of $s$-th local service.

## 3.2 Architecture of Local Service Module

The model of the local service is shown in Fig. 3. The model of the local service consists of the *Estimation mechanism*, *Adaptation mechanism* and the *Load State Database*. Further in this paper we omit the index $s$ for the simplicity resons.

The estimation and adaptation mechanisms are realized with the use of fuzzy-neural network approach based on the classical fuzzy model including fuzzification, rule based inference mechanism and defuzzification [15, 16]. The network inputs are: $q_i$ - the load of the Internet network, $r_i$ - the load of the local service; the class $k$ of required resource $\{k=1,\ldots,K\}$ and $U_{ki}$ from the Load State Database $U_i = [U_{1i},...,U_{ki},...,U_{Ki}]^T$ which includes information about the LS state for $k$-th class of requests at the $i$-th moment. On the basis of this data the decision making module estimates the response time $\hat{t}_i^s$ for request $x_i$. Based on the measured real values of the request response time $\tilde{t}_i^s$, the adaptation mechanism updates $U_i$ database in such a way that the state $U_{ki}$ for the $k_i$ - the class is updated to $U_{k(i+1)}$.

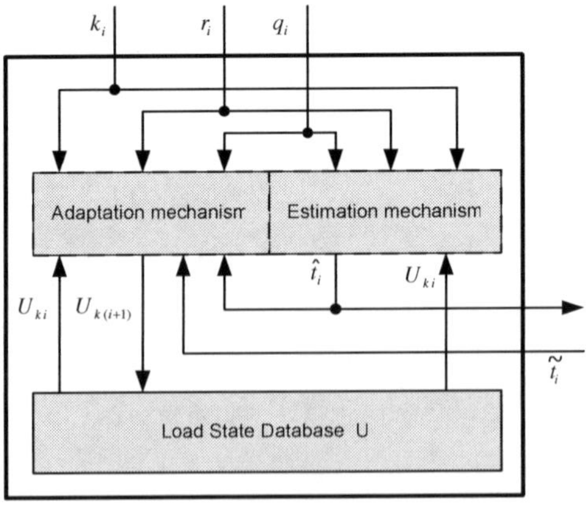


**Fig. 3.** Local Service's architecture

For every input to the fuzzy-neural network, as well as for the outputs, there were defined appropriate fuzzy sets. Values of all parameters of fuzzy sets are stored in $U_i = [U_{1i},...,U_{ki},...,U_{Ki}]^T$. These parameters are used as weights of neurons and are tuned during the adaptation process. After the completion of the service of the request, the value $\tilde{t}_i^s$ is used to tune parameters of fuzzy sets. The weights are tuned with the use of the backward error propagation method every time, when the request completion occurs.

## 4 GARDiB and GARDiB2 Algorithms

Both request distribution algorithms estimate the response time of request based on the class of the request, the load of the Internet network and the load of local services. The classification of requests and determination of the network load take place in the same manner for both algorithms. The load of the network is determined based on the measurement of the transfer time of the probing object requested from local services by the broker. The transfer time of that object is counted by each broker individually. The measurement module of the broker sends the HTTP request to every local service. In reply, the local services send back an object of the size of 20 KB. This object's size was chosen because we wanted to use at least a dozen of IP datagrams to send it. However it was important not to generate to much heavy traffic and obtain transfer time not longer than the permissible measurement interval.

The difference between the GARDiB and GARDiB2 algorithms lies in the manner they provide and use information about the Local Services load state. This information is sent every 5 seconds by Local Services to the brokers. Algorithm GARDiB determines the LS load taking into account the number of HTTP requests $r_i$ simultaneously served by the LS at $i$-th moment. In GARDiB2 the LS's load state is determined using the vector $P_i = [p_{1i},..., p_{ki},..., p_{Ki}]$ including the service times offered by LS for $k$-th class request at the $i$-th moment. However these times can be obtained from Local Services only if they use the FNRD (Fuzzy Neural Request Distribution) local distribution algorithm [5, 6] which estimates the completion time of the request in a local cluster of Web servers using fuzzy-neural approach.

## 5 Simulation Environment

We built a simulation environment using C programming language and CSIM simulation package. CSIM is a simulation package of C libraries for discrete systems simulation. It is well-known and approved in the domain of HTTP request distribution simulation [7, 8, 9, 12]. Fig. 4 shows a general simulation model used in our research. The simulation program consists of the following modules: Clients, Internet, Broker and Local Service (LS) modules.

As data for simulation of the network load in the Internet module we used real-life measurements collected in data transmission experiments performed between Opole, Poland and eleven Web sites localized in different network distances from Opole in various regions of the World including the Netherlands, Australia and USA. We

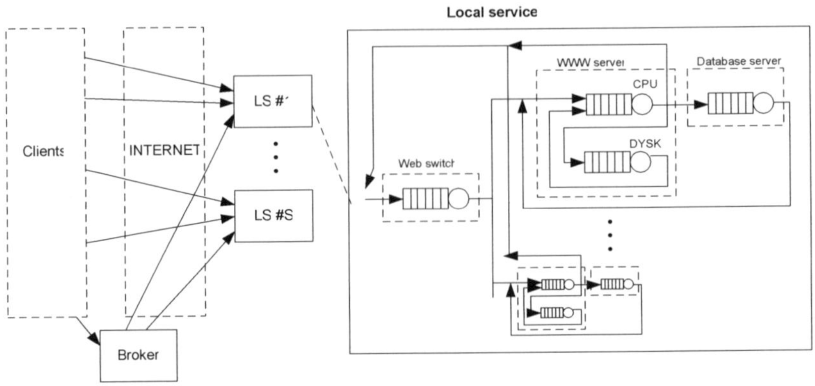


**Fig. 4.** A simulation model

downloaded the RFC 1832 document every 10 seconds over a sample period of 48 hours starting at 7 p.m. on 17 June 2006. We used wget, pcap as well as our own software. We measured the TCP connection time as the Round-Trip Time and transfer time of requested document to evaluate the effective throughput of theese Internet paths.

The request generator (Clients module) is similar to traffic generators which have been designed in other our projects. Generated HTTP requests are sent either to the broker or to the LSs directly as the background traffic. The traffic includes 80% of static requests which are serviced directly by the WWW servers and 20% of dynamic requests which are processed by back-end database servers. Dynamic requests are prepared for low, medium and intensive workload sizes.

In Broker module we implemented GARDiB and GARDiB2 algorithms, as well as three algorithms used in the comparative analysis, that is RR (Round Robin), WRR_L (Weighted Round Robin based on the load state of the LS) and WRR_T (Weighted Round Robin based on the transfer time of probing object).

LS module has several functional modules: the module of the Web switch, modules of WWWservers and modules of database servers.

In simulations we made the similar assumptions about the processing costs as in [18] but we decreased all time parameters by the factor of ten in order to make our assumptions up-to-date ([18] determins the processing costs for Pentium II 300 MHz PC with FreeBSD 2.2.5). Thus the connection establishment and teardown operation cost were set to 14.5 μs of CPU time each, the transmit processing incurred 4.0 μs per 512 bytes. Disc costs were the following: reading a file from disk had a latency of 2.8 ms, the disk transfer time was 41.0 μs per 4 KByte. For files larger than 44 KBytes, an additional 1.4 ms (seek plus rotational latency) was charged for every 44 Kbytes of file length in excess of 44 KBytes. The Least Recently Used cache replacement policy was used; but files with a size of more than 500 KB were never cached. The total memory size used for caching was 85 MB. The database servers are modelled as the single queues [8]. Times of the service of dynamic requests on the database server are modelled according to hyperexponential distribution [9]. Each LS has the same Web content, i.e. Web pages, images and others Web objects. In local Web switches we implemented local distribution algorithms, namely: Round Robin (RR), Locality Aware Request Distribution (LARD), Client Aware Policy (CAP), and Fuzzy Neural Request Distribution (FNRD). All of them can work with GARDiB and GARDiB2 global algorithms used by the broker, but only FNRD algorithm is able to run together with GARDiB2.

## 6 Simulation Results

We simulated a system consistiong of one broker, three local services, each composed of three WWW servers and three database servers. The broker has been localized in Opole, Poland. According to real-life network measurements we were able to study three different situations we wanted to study: short network distance with good effective throughput (the Netherlands, NL), long network distance with high effective throughput (USA) and long network distance with low effective throughput (Australia, AU. Such selection gives us the choice to observe how our algorithms perform.

### 6.1 Workload Scenario #1

Here we present experiments in which the whole generated traffic is divided so that 25% of clients access the content via broker whereas the rest 75% of clients has been evenly sent to LSs as a background load.

In Fig. 5 we show the performance results for five studied distribution algorithms: RR, WRR_L, WRR_T, GARDiB and GARDiB2. A dotted straight line in Fig. 5 shows a level of 8 seconds and indicates the time, which an average user is ready to wait on gaining of Internet site [1]. This 90-percentile graph of the response time for the whole Web page shows the best performance for GARDiB and GARDiB2 distribution algorithms. Algorithm GARDiB seems to be a slightly better. The system with WRR_T algorithm also achieved quite good results, however only for loads not bigger than 490 clients/sec. Above that load the WRR_T algorithm overloaded local service in the Netherlands and reqests are not serviced as quickly as at lower loads.

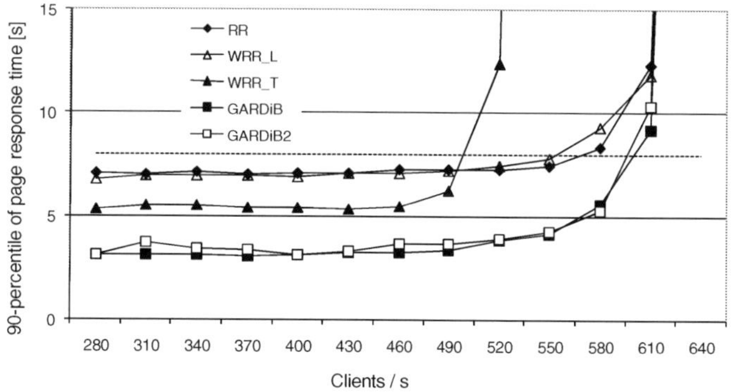


**Fig. 5.** 90-percentile of response time vs. number of clients for workload scenario #1

Algorithms RR and WRR_L for load 550 clients/sec balance on the edge of the time that user is ready to wait. For the load more than 580 clients/sec the system supplies Web access service on the accepted level only when equipped with GARDiB or GARDiB2 algorithms.

Other results concerning the workload scenario #1 are presented in next two figures. In Fig. 6 our proposed algorithms show the best performance taking into account the mean request's response time. In Fig. 7 we display a 95-percentile of the response time for the whole Web page. The results are similar to that obatained in more rigorous 90-percentile analysis (Fig. 5) but in this case only GARDiB and GARDiB2 meet expectations of 8s for the load up to 550 clients/sec.

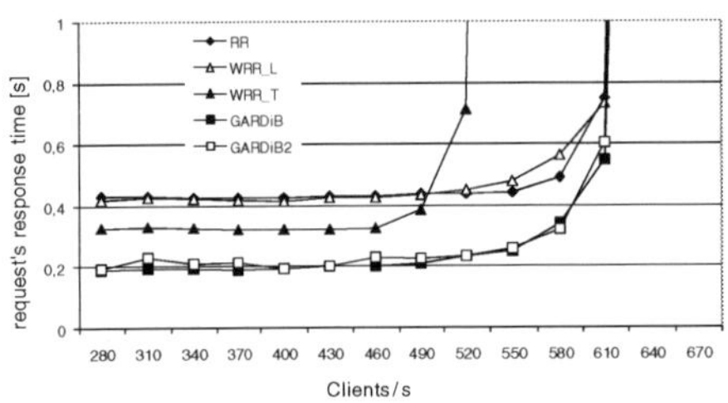


**Fig. 6.** Mean response time vs. number of clients in workload scenario #1

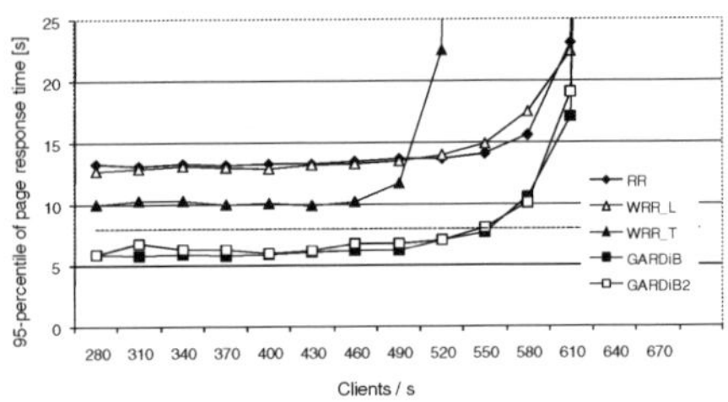


**Fig. 7.** 95-percentile of response time vs. number of clients for workload scenario #1

### 6.2 Workload scenario #2

Workload scenarios #2 and #3 have differentiated background trafiic what may reflect World's time zones effect. The load distribution in #2 scenario has bigger background load in the Netherlands (40%), lower background load in USA (20%), and small backround load in Australia (10%). 30% of traffic is distributed via broker.

In Fig. 8 we present the results of 90-percentile of the response time in this experiment. RR and WRR_T algorithms perform badly. For WRR_L algorithm the system is moderately effective however with page response time slightly greater than expected 8 s. For GARDiB and GARDiB2 algorithms the page response times are good even for large loads. Let us show how GARDiB and GARDiB2 distribute the requests according to its type (static or dynamic) and service localization. This is displayed in Fig. 9.We can see that for the GARDiB algorithm (Fig. 9a) almost all static requests were sent to the nearest local service that is to the Netherlands. At light loads the broker distributed the most of dynamic requests to that nearest service, too. At the greater load the broker engaged USA site for dynamic requests, and decreased the number of dynamic requests sent to the Netherlands. At the heavy load the broker also sent the certain part of dynamic request to Australia in spite of the low effective throughput on this direction. It happened so, because the local service in Australia was very lightly loaded, and it was effectively to use that site for the service of time-consuming dynamic requests.

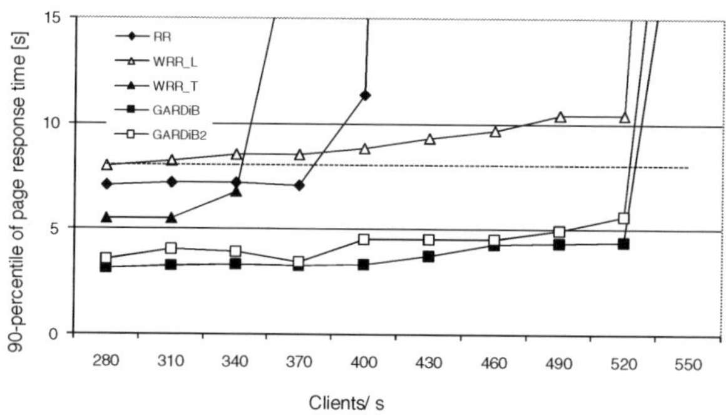


**Fig. 8.** 90-percentile of response time vs. number of clients for workload scenario#2

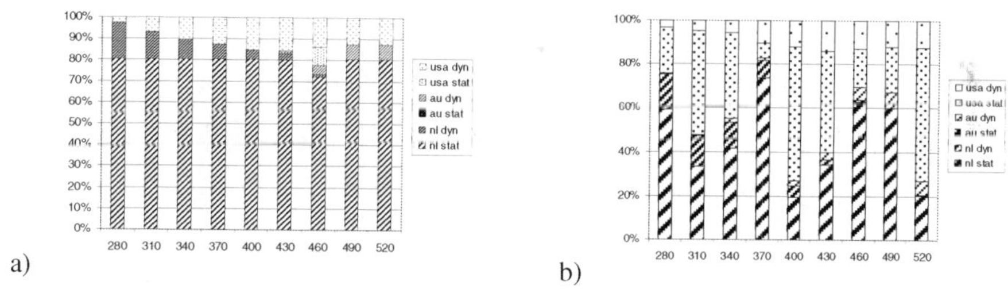


**Fig. 9.** Percentage shares of local services in request processing for workload scenario #2 for a) GARDiB, b) GARDiB2

GARDiB2 performed in a different way. It redirected large number of static and dynamic requests to the local service in USA. At the greater load the service in the Netherlands stopped to serve dynamic requests effectively, so these requests had to be serviced in Australia.

### 6.3 Workload scenario #3

Let us consider the case, when the most loaded LSs are services in the Netherlands (35%) and in Australia (30%). USA has only 10% of background load and the broker supports 25% of clients.

Fig. 10 shows the page response times for 90% of total number of pages in this workload scenario. The results show that the times of the service of total pages are the shortest for the GARDiB algorithm and slightly longer for the GARDiB2 algorithm. For all remaining algorithms, at the load above 460 clients/s, they are greater than 8 s. Fig. 11 shows that GARDiB, like in scenario#2, redirected static requests to the nearest service (in the Netherlands), however at the greater load level the dynamic requests were sent to the unloaded service in USA. Only at the very large load the algorithm sent somewhat less than of 1% dynamic requests to the local service in Australia. GARDiB2 sent the large number of requests to the non-overloaded service in USA.

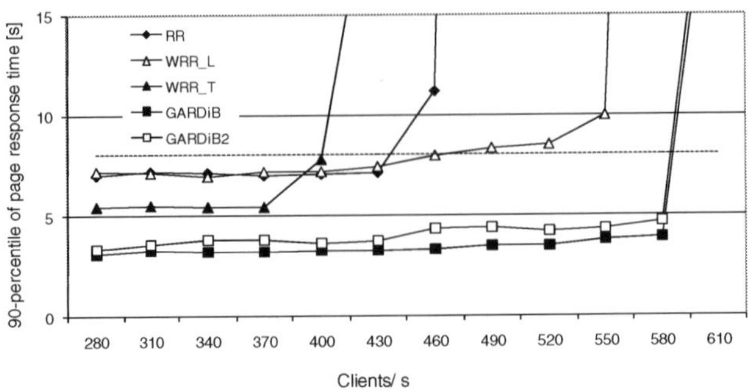


**Fig. 10.** 90-percentile of response time vs. number of clients for workload scenario #3

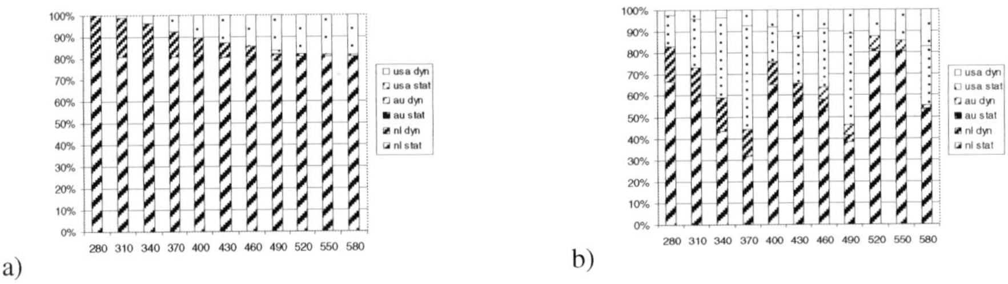


**Fig. 11.** Percentage shares of local services in request processing for workload scenario #3 for a) GARDiB, b) GARDiB2

## 7 Conclusions

In this paper we presented the global distribution of HTTP requests based on the fuzzy-neural algorithms called GARDiB and GARDiB2. GARDiB performed slightly better than GARDiB2 and both proposed algorithms outperformed other competitive distribution algorithms in all simulated workload scenarios. The algorithms exhibit the $O(S)$ linear computational complexity, which makes the proposed methodology of the fuzzy-neural HTTP request global distribution very competitive in practical applications.

## Acknowledgements

This work was supported by the Polish Ministry of Science and Higher Education under Grant No. N516 032 31/3359 (2006-2009).

## References

1. Bhatti, N., Bouch, A., Kuchinsky, A.: Integrating User-Perceived Quality into Web Server Design. In: Proc. of the 9th Int. World-Wide Web Conference, pp. 1–16 (2000)
2. Borzemski, L., Zatwarnicki, K.: CDNs with global adaptive request distribution. In: Lovrek, I., Howlett, R.J., Jain, L.C. (eds.) KES 2008, Part II. LNCS (LNAI), vol. 5178, pp. 117–124. Springer, Heidelberg (2008)

3. Borzemski, L., Zatwarnicki, K., Zatwarnicka, A.: Adaptive and Intelligent Request Distribution for Content Delivery Networks. Cybernetics and Systems 38(8), 837–857 (2007)
4. Borzemski, L., Zatwarnicki, K., Zatwarnicka, A.: Global Adaptive Request Distribution with Broker. In: Apolloni, B., Howlett, R.J., Jain, L. (eds.) KES 2007, Part II. LNCS (LNAI), vol. 4693, pp. 271–278. Springer, Heidelberg (2007)
5. Borzemski, L., Zatwarnicki, K.: Fuzzy-Neural Web Switch Supporting Differentiated Service. In: Gabryś, B., Howlett, R.J., Jain, L.C. (eds.) KES 2006. LNCS (LNAI), vol. 4252, pp. 195–203. Springer, Heidelberg (2006)
6. Borzemski, L., Zatwarnicki, K.: Using Adaptive Fuzzy-Neural Control to Minimize Response Time in Cluster-Based Web Systems. In: Szczepaniak, P.S., Kacprzyk, J., Niewiadomski, A. (eds.) AWIC 2005. LNCS (LNAI), vol. 3528, pp. 63–68. Springer, Heidelberg (2005)
7. Borzemski, L., Zatwarnicki, K.: A Fuzzy Adaptive Request Distribution Algorithm for Cluster-Based Web Systems. In: Proc. 11th Euromicro Conference on Parallel, Distributed and Network-Based Processing, pp. 119–126 (2003)
8. Casalicchio, E., Tucci, S.: Static and Dynamic Scheduling Algorithm for Scalable Web Server Farm. In: Proc. of 9th Euromicro Workshop on PDP, pp. 369–376 (2001)
9. Cardellini, V., Casalicchio, E., Colajanni, M., Mambelli, M.: Web Switch Support for Differentiated Services. ACM Perf. Eval. Rev. 29(2), 14–19 (2001)
10. Cardellini, V., Casalicchio, E., Colajanni, M., Yu, P.S.: The State of the Art in Locally Distributed Web-Server Systems. ACM Comp. Surv. 34(2), 263–311 (2002)
11. Cardellini, V., Colojanni, M., Yu, P.S.: Geographic Load Balancing for Scalable Distributed Web Systems. In: Proc. of 8-th Symposium on Modeling, analysis and Symulation of Computer and Telecommunicaton Systems (MASCOS 2000), pp. 20–27 (2000)
12. Cardellini, V., Colajanni, M., Yu, P.S.: DNS Dispatching Algorithms with State Estimators for Scalable Web-Server Clusters. World Wide Web Journal 2(3), 117–124 (1999)
13. Casalicchio, E., Colajanni, M.: A Client-Aware Dispatching Algorithm for Web Clusters Providing Multiple Services. In: Proc.10th International Conference on WWW, pp. 535–544 (2001)
14. Horikawa, S., Furahashi, T., Uchikawa, Y.: On Fuzzy Modeling Using Fuzzy Neural Networks with Back Propagation Algorithm. IEEE Trans. Neural Networks 3, 801–806 (1992)
15. Lee, K.-M., Kwak, D.-H., Leekwang, H.: Tuning of Fuzzy Models by Fuzzy Neural Networks. Fuzzy Sets and Systems 76(1), 47–61 (1995)
16. Mamdami, E.H.: Application of Fuzzy Logic to Approximate Reasoning Using Linguistic Synthesis. IEEE Trans. on Computers C-26(12), 1182–1191 (1977)
17. Markus, H., Beaumont, L.: Content Networking: Architecture, Protocols, and Practice. Morgan Kaufmann Publishers, USA (2005)
18. Pai, V.S., Aron, M., Banga, G., Svendsen, M., Druschel, P., Zwaenpoel, W., Nahum, E.: Locality-Aware Request Distribution in Cluster-Based Network Servers. In: 8th ACM Conf. on Arch. Support for Progr. Lang. and Oper. Syst., pp. 205–216 (1998)
19. Sharifian, S., Akbari, M.K., Motamedi, S.A.: An Intelligence Layer-7 Switch for Web Server Clusters. In: 3rd International Conference: Sciences of Electronic, Technologies of Information and Telecommunications SETIT 2005, p. 8 (2005)

# Deterministic Processing of WWW Pages by the Web Service*

Krzysztof Zatwarnicki

Faculty of Electrical Engineering, Automatic Control and Computer Science,
Opole University of Technology, Opole, Poland
KZatwarnicki@po.opole.pl

**Abstract.** The article presents a new method for scheduling HTTP requests at the front of the Internet service, in a way that the service time of the whole WWW pages wouldn't exceed the service time imposed by the service provider. The presented way of scheduling is a novelty in the way how the service is perceived, not only as a provider of individual HTTP objects, but also whole WWW pages.

**Keywords:** HTTP request scheduling, deterministic Web page processing, Web service.

## 1 Introduction

Over the last few years the Internet became one of the basic sources of information and enabled running varied services of informative and business character. It has evolved from a medium for only privileged users into the medium we can't imagine living without. It is the medium with the use of which many institutions earn a lot of money. Therefore, being the basis for the business services, it is required to provide a high level of the information technology services. For many years, there have been many researches conducted on the improvement in the quality of the Internet services. The clients, for the benefit of whom the services were provided, signed the SLA documents (Service Level Agreement) with the Internet services providers, in which both a level and quality of the services were stated. However, the protocols, mainly HTTP and means of serving the Internet requests, do not guarantee an appropriate level of the service. These are the best–effort services, which means that the centres accomplishing specific services do their best, but there is no guarantee that the quality of the service will be satisfactory or whether the service would be accomplished at all. The same concerns the process of data transmission on the Internet. There is a need of conducting researches, not only on improving the quality of the Internet services, but also on providing determined quality of them, to make them deterministic.

In the further part of the chapter, the method and algorithm WEDF enabling the service of the requests on the specified level will be described, as well as the research environment and the results obtained.

---

* This work was supported by the Polish Ministry of Science and Higher Education under Grant No. N516 032 31/3359 (2006-2009).

N.T. Nguyen, R. Kowalczyk, and S.-M. Chen (Eds.): ICCCI 2009, LNAI 5796, pp. 764–775, 2009.

## 2 Related Works

Many of the papers on the HTTP request services were dedicated to the quality of the services. The first collective papers concerned mainly improving the quality of the services are [5,7,12,3,4]. The following review paper mentioned guarantee of the service quality is [13]. The more significant papers on the creation of the Internet service infrastructure, whose behavior is deterministic within some range include the following [8,9,11]. The researches conducted up till now refer to the service of an individual HTTP request, an improvement of its quality and the guarantee of its accomplishing within fixed time. However, there are no papers devoted to the guarantee of the quality of the services of downloading by the clients complete WWW pages.

The innovative method, for scheduling the HTTP requests at the front of Web service, described in the following paper will enable processing complete WWW pages in more deterministic way, than the methods used currently.

## 3 Description of WEDF Method

The WEDF method (Web Page EDF) is designed for the Web services which are serviced mainly by the single WWW servers. In the system, working in accordance with the method described, a broker and a service can be singled out. The broker receives the HTTP requests sent by the clients, schedules them in a specified way and sent to the service processing the requests. Fig. 1 presents the diagram of the junction between the broker and the service.

The broker's task is to schedule the HTTP requests in a way that the response time required to process the whole WWW page was equal or shorter than the time imposed. The maximum time of serving the whole page $t_{max}$ should be imposed by the client entrusting the management of the Internet service. At the same time, it is assumed that the response time of the page is the total of individual response times of each HTTP request generated by one client in order to download a frame of Web page and embedded objects. The response time of the HTTP request is measured from the moment of time then the request arrived to the broker to the moment when the broker started to send the HTTP response.

Current publications do not refer to any methods enabling serving the whole Web pages within the imposed and fixed, deterministic page response time. The solution presented herewith is entirely pioneer. It comes as a result of an innovatory attitude to processing methods accomplished in WWW servers: servers should be perceived as the whole WWW pages providers, instead of single HTTP objects.

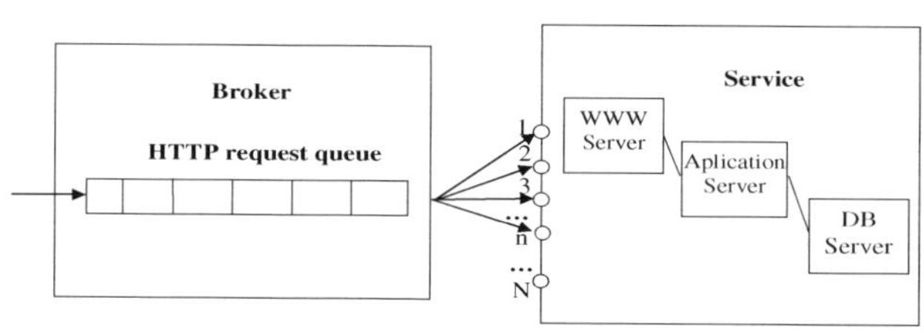


**Fig. 1.** The broker and the service

The broker in the described method can be both, an individual device relaying the clients' requests to the WWW server and software applied on the computer of the WWW server, or it can also be a part of the software of the WWW server itself.

The service can comprise a WWW server, an application server and a database server. It can service many HTTP requests in a concurrent or a parallel way. It is assumed, that the maximum number of requests $N$, serviced by the service should be equal to the minimal number of the requests, for which the average throughput of the system $X(n)$ has the maximum value, where $n$ applies to the number of the requests serviced simultaneously by the service and the throughput of the system is expressed by the number of HTTP requests serviced within the time unit.

Owing to this, the efficiency of the Web service will not decrease after WEDF application. $N$ should have the lowest acceptable value, so that the requests form a queue on the broker, but at the same time the value should be high enough to make the WWW server achieve the efficiency approximate to the maximum value.

Fig. 2 presents a diagram of dependence of a throughput $X(n)$ upon a number $n$ of simultaneously serviced requests together with a number $N$ appointed appropriately.

The logical scheme of the broker is presented in Fig. 3. The broker consists of a service module, a queue module and an execution module. The service module determines the deadline $d_i$ pointing the time limit, within which the request should be completed.

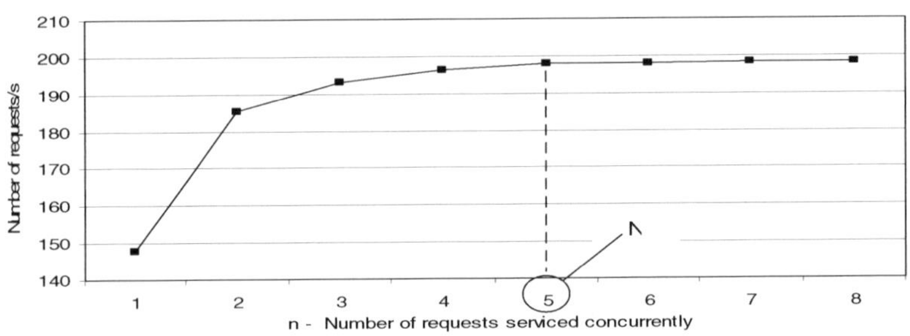


**Fig. 2.** A diagram of dependence of a throughput vs. number of simultaneously serviced requests

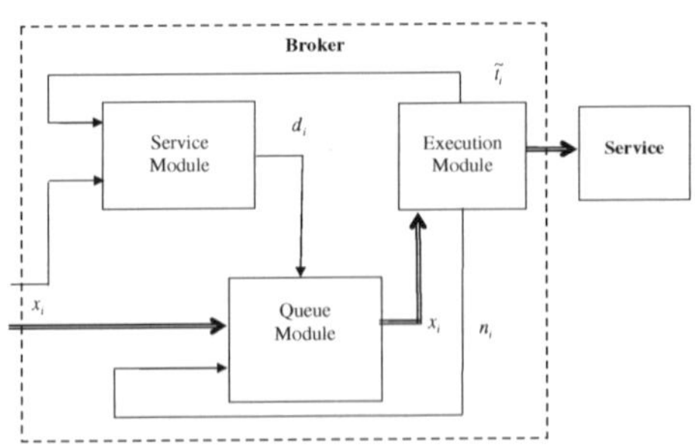


**Fig. 3.** The broker scheme

The functions and work of respective modules of the broker are presented in the following sub-sections.

### 3.1 Execution Module

An execution module sends the request obtained from the queue module to the service. The object obtained in the response to the HTTP request is sent by the execution module back to the client. At the same time, the execution module can supervise $N$ requests serviced by the service.

The execution module measures the time $\widetilde{t}_i$ of serving request $x_i$ and after its completion passes the time to the service module.

### 3.2 Queue Module

Queue module contains the queue of requests and is responsible for placing the HTTP requests according to a scheduling algorithm. The queue module obtains information, on the number of requests $n_i$ being serviced simultaneously by the service, from the execution module (index $i$ points that the information is up-to-date in the moment of the $x_i$ request arrival). It is assumed, that if the number of requests $n_i$ serviced simultaneously by the service is equal to or lower than $N$, then a new HTTP request $x_i$ is not queued, but passed directly onto the execution module and further to the execution by the service. If $n_i > N$, then the request is placed in the queue arranged according to deadline $d_i$, determined by the service module for each of the request queued. The requests with the earliest deadlines are being queued first and the ones with the latest deadlines are queued last in accordance with the EDF algorithm (Earliest Deadline First). The request is taken from the queue and passed onto the execution module only when the service for any kind of request is completed by the service. The scheduling system discussed, does not apply a dispossession, as it would require interference in the service work.

### 3.3 Service Module

The basic role of the service module is to determine the moment of the time, in which the request should be completed. In its construction 4 sub-modules can be named: a classification service model, an estimation mechanism, a load state database and an adaptation mechanism. Fig. 4 presents logical scheme of the service module.

The classification service model stores information on the HTTP objects accessible in the service, their size, type (whether they are dynamic or static) and the affiliation of individual objects to the whole pages.

The information can change in time and is updated in the classification service model, especially in case of pages, whose frame is generated dynamically. At the entrance, the service model receives information on the client's request $x_i$. The information includes:

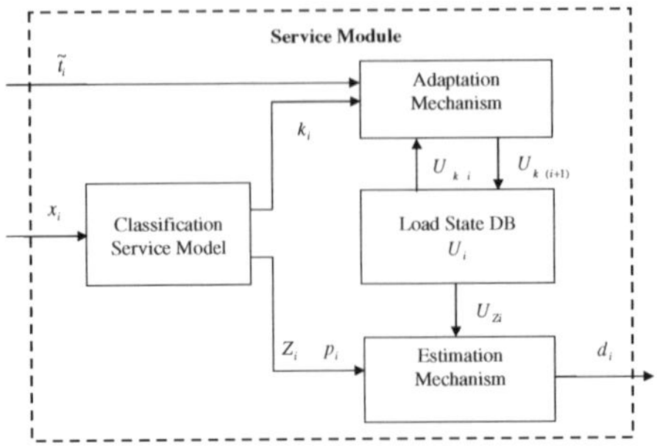


**Fig. 4.** The scheme of the service module

- the address of requested object,
- the client identifier,
- the identifier of the page, within which the object is requested.

The address of requested object is always in the HTTP request sent by the client to the service. The client identifier can be given to the client when entering the service for the first time within the session and passed to the Internet browser. Next, included in the HTTP request, it can be passed to the service in the information included in the cookie. The identifier of the page can be passed onto the client's browser when downloading the frame of the page, and next as above, from the browser to the service in the HTTP request in the information included in the cookie.

The classification service model stores information on the objects downloaded by the client within a single WWW page.

The classification service model passes the information about the class $k_i$ of requested object, to the module of the adaptation mechanism, where $k_i = 1,2,3,..., K$. The class of the object is determined on the basis of the object's size, in case of static objects. The objects of similar size belong to one class. Every dynamic object has a different class.

The classification service model passes onto the module of the estimation mechanism a vector $Z_i = \left[k_i^0, k_i^1, k_i^2,..., k_i^l,..., k_i^{L-1}\right]$ of the object classes belonging to the page, not the ones downloaded within one page by the client.

Class $k_i^0$ is the class of the requested object ($k_i^0$ is also further indicated as $k_i$). Additionally the time $p_i$ of the first request arrival concerning the downloaded page is also passed on.

The module of the load state database stores information about the times of services referring to the objects belonging to the individual classes. The load state database was indicated as $U_i = \left[\hat{t}_{0i}, \hat{t}_{1i}, \hat{t}_{2i},..., \hat{t}_{ki},..., \hat{t}_{Ki}\right]$, $i$ index points that the information in the load state is up-to-date in the moment of the request $x_i$ arrival. $\hat{t}_{ki}$ is the estimated time of the $k$ class request serviced by the service.

The load state database is updated by the adaptation mechanism in the process when the service module adapts to changeable conditions of processing.

The load state database passes the service times $U_{Zi} = \left[\hat{t}_{k^0{}_i}, \hat{t}_{k^1{}_i}, \ldots, \hat{t}_{k^l{}_i}, \ldots, \hat{t}_{k^L{}_i}\right]$ of the objects belonging to the page, not the ones downloaded within one page by the client, onto the estimation mechanism. The objects of these classes were indicated in $Z_i$.

The estimation mechanism determines the deadline $d_i$, pointing the time in which request should be served. In order to specify it, the maximum time $t_{\max}$ of response for the whole page is required. The value of $t_{\max}$ can be equal for every page and every client.

The deadline $d_i$ is determined in accordance with the formula $d_i = \Delta d_i + c_i$ where $c_i$ is the moment in time of the $x_i$ request arrival. $\Delta d_i$ is calculated according to formula 1:

$$\Delta d_i = \hat{t}_{k^0{}_i} \frac{(p_i + t_{\max} - c_i)}{\sum_{l=0}^{L-1} \hat{t}_{k^l{}_i}}. \tag{1}$$

The time range $\Delta d_i$ is proportional to the time $\hat{t}_{k^0{}_i}$ of the request service in the service. It may occur, that the value of $\Delta d_i$ is negative, if the sum of service times for the individual objects $\sum_{l=0}^{L-1} \hat{t}_{k^l{}_i}$ would be higher than the value of $t_{\max}$.

The adaptation mechanism updates the service times of the requests of the individual classes in the load state database $U_i$. The adaptation is proceeded after service completion of the request $x_i$ belonging to class $k_i$. The load state database is updated only when the request was placed in the queue prior to serving.

The service time $\tilde{t}_i$ is measured by the execution module starting from the moment when the first byte is sent to the service by the executor module till the moment the last byte with the response is received by the executor.

The estimated time of the service $\hat{t}_{k\ i}$ is updated according to formula 2:

$$\hat{t}_{k\ (i+1)} = \hat{t}_{k\ i} + \eta(\tilde{t}_i - \hat{t}_{k\ i}), \tag{2}$$

where $\eta$ is the adaptation coefficient with the value in the range between $(0,1)$.

### 3.4 Additional Remarks

To close this point it is important to mention that the time $t_{\max}$ imposed by the client should refer to the achievable response time of a page. Therefore for each page $t_{\max}$ should be bigger than the total of service times of individual requests of a page at the service, whose load (number of requests served at the same time) is equal to $N - 1$.

It is also to be stated that using WEDF method by the web service will neither improve its performance nor move the limit of the maximum number of requests being served. Using this method while the load is heavy (but not exceeding the service capability) will result in similar service times of both small simple pages as well as complex ones which need loading data from databases. Simultaneously, these times will not exceed the determined time limit $t_{max}$. Therefore, while the load is heavy, there will not be situations when some clients are served very quickly and others, needing complex services, have to wait for a long time.

# 4 Simulation Model and Experiment Results

In order to determine properties, functioning ranges and usability of proposed solutions, simulation studies have been conducted. A simulation environment built in an appropriate way can afford possibilities for conducting research without necessity of simplifying the adopted model (as in case of analytic methods) or bearing high costs as in case of conducting research with servers and Web infrastructure.

The package CSIM19 [10] has been selected as a simulation environment. It gives possibilities of creating process-oriented simulation models which use discrete events. This well-known package has been tested and is one of the most popular in the realm of Web environments simulations [3,4,5].

## 4.1 Simulation Environment

The simulator has included the following functional modules: requests generator module, broker module, service module. The general scheme of simulator construction is shown in Fig. 5.

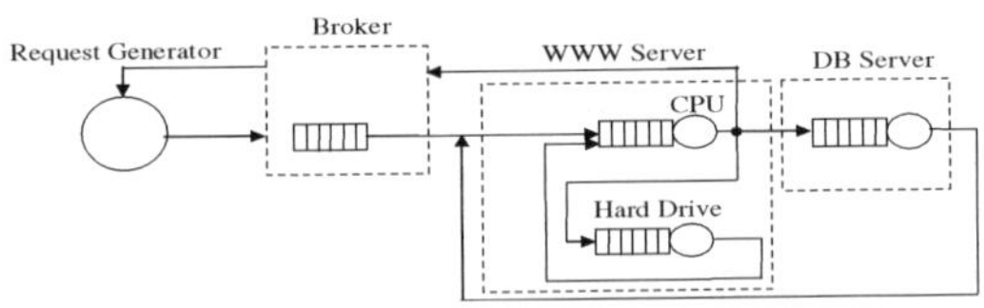


**Fig. 5.** A simulation model

The requests generator module is to create clients sending HTTP requests to the broker. During the simulation studies, a particular stress has been laid on the correct modeling of the clients' functioning so that the character of WWW requests' generated load would be compatible with the one observed in the Internet [2], which means to be distinguished by: variability, self-similarity and heavy-tailed distributions. Such a load can be generated with use of the heavy-tailed distributions like Pareto and lognormal distribution.

Table 1 shows distributions and parameters used while constructing a client model.

**Table 1.** Workload model parameters

| Category | Distribution | Parameters |
|---|---|---|
| Requests per session | Inverse Gaussian | $\mu$=3.86, $\lambda$=9.46 |
| User think time | Pareto | $\alpha$=1.4, k=1 |
| Objects per request | Pareto | $\alpha$=1.33, k=2 |
| HTML object size | Lognormal | $\mu$=7.630, $\sigma$=1.001 |
| | Pareto | $\alpha$=1, k=10240 |
| Embedded object size | Lognormal | $\mu$=8.215, $\sigma$=1.46 |

It has been assumed that the size of the whole site downloaded by the users is about 400 MB. The client can generate requests referring to static resources (files located in the server files system) as well as to dynamic resources (created on demand by WWW and database server). During the research it has been assumed that 20% of the requests refers to dynamic resources, other requests refer to static resources. The dynamic resources downloaded by a client have been divided into three classes [5]:

- highly intensive, loading a database in a highly intensive way,
- medium intensive, loading a database in a medium intensive way,
- low intensive, loading a database in a low intensive way.

Table 2 shows service times of dynamic requests on a database server for the respective classes of the objects.

**Table 2.** Workload model parameters of dynamic objects

| Type | Mean service time | Frequency |
|---|---|---|
| High intensive | 20 ms | 0.85 |
| Medium intensive | 10 ms | 0.14 |
| Low intensive | 5 ms | 0.01 |

The broker module obtains the requests generated by users. In order to compare the quality of the broker performance, working in accordance with WEDF method, with other methods of scheduling requests used nowadays, four algorithms of scheduling requests have been implemented:

- WEDF;
- FIFO – with this method of scheduling, HTTP requests have been served in accordance with their sequence of appearing, successive requests have been taken from the queue if the number of the requests served by the service was lower than $N$ ;
- SJF – requests have been scheduled in the queue in accordance with the algorithm Shortest Job First;

- DS – requests have not been placed in the queue and directly after its arrival, requests has been transferred to the service regardless to the number of requestes being served by the service.

The WWW server module in the simulator has contained the following elements [1,11]: processor, disc and cache functioning in accordance with the LRU (Last Recently Used) algorithm.

The following times of serving requests on the processor have been applied:

- time of establishing a connection – 0,00010097s,
- time needed by WWW demon to prepare a response to a static request - 0,00014533s,
- time needed by WWW demon to prepare a response to a dynamic request - 0,029s,
- data transfer time - $0.0000042914 * w$, where $w$ is the object size expressed in KB.

The time of serving requests on the hard drive can by described as:

$$S_d(x) = \begin{cases} 0.0045 \; for \; w \in < 0KB, 128KB > \\ 0.00003813125 \cdot w - 0.0003808 \; for \; w \in (128KB, \infty) \end{cases}. \quad (3)$$

The times of serving requests described above have been obtained after numerous studies performed for the Apache WWW server version 2.2.4, PHP 5.1 functioning under control of the Linux operating system Linux (Fedora Core 6) functioning on the computer with the processor Intel Pentium 4, 2 GHz, with the hard disk drive Seagate ST340810A, 80GB, IDE with the network interface 1000 Mbps.

The service has also contained a database server which has been modelled as a single queue [7]. The times of serving requests on the database server post have been modeled in accordance with the hyperexpotential distribution.

### 4.2 Evaluation of the Solutions

An average value of satisfaction has been adopted as a measure of the quality of the service working in accordance with the WEDF method. Satisfaction is often used to evaluate the quality of the service of soft Real Time Systems [9]. Taking into consideration our needs, it is assumed that satisfaction is the level of user's satisfaction from the page response time. The user's satisfaction depends on the response time of the whole page according to the formula 4:

$$Satisfaction(\bar{t}) = \begin{cases} 1 \text{ when } \bar{t} < t_{\max}^{s} \\ 0 \text{ when } \bar{t} > t_{\max}^{h} \\ 1 - \dfrac{t_{\max}^{h} - \bar{t}}{t_{\max}^{h} - t_{\max}^{s}} \text{ in other cases} \end{cases}, \quad (4)$$

where $\bar{t}$ is the response time of the page, $t^s_{\max}$ is the time that can be exceeded, although it is advised to avoid such situation, while $t^h_{\max}$ is the deadline of the service, after which the user gives up watching the WWW page. The times $t^s_{\max}$ and $t^h_{\max}$ should be imposed by the customer, ordering the service. Fig. 6 picture of the satisfaction function.

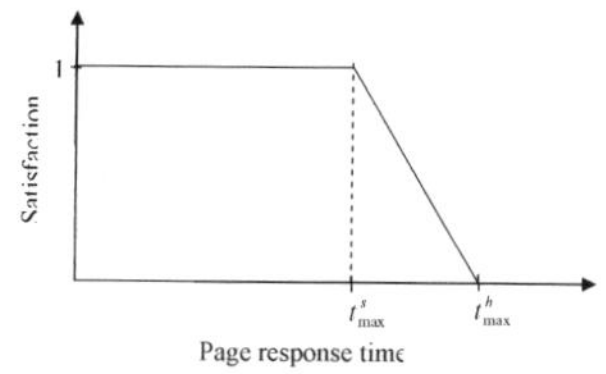


**Fig. 6.** Satisfaction function

The adopted method has not been used yet to evaluate the quality of Web services because the Web services are best-effort type. By changing the service character, it seems appropriate to indicate the adequate method of evaluating the proposed solutions.

### 4.3 Research Results

The research has been conduced on the testbed described above. For the same Web site and conditions of processing request, there have been different maximum times of serving WWW pages adopted. The Fig. 7a shows the research results for the assumed $t^s_{\max} = 300ms$ and $t^h_{\max} = 600ms$, the Fig. 7b shows the results for $t^s_{\max} = 400ms$ and $t^h_{\max} = 800ms$, the Fig. 7c shows the results for $t^s_{\max} = 500ms$ and $t^h_{\max} = 1000ms$. For each of the assumed times, a research has been conducted at a different changing load of the Web service. In each experiment the value of $t_{\max}$ was equal to $t^s_{\max}$.

The results of the conducted research clearly show that the highest level of satisfaction is obtained independently of the load for the WEDF method. The best results have been obtained for the heavy loads, when the maximum times of serving requests have been quite long. The results obtained when the broker has been working in accordance with the algorithm DS and transferring HTTP requests to be attended directly on the service seem to be surprising. The satisfaction obtained for this option is relatively low. At present almost all Web services do not queue requests at the entrance but serve them directly.

The Fig. 7c shows the values of the average HTTP request response time in the load function with the assumed times $t^s_{\max} = 300ms$ and $t^h_{\max} = 600ms$. As it can be seen the service working under control of WEDF has obtained rather weak results in comparison to other solutions. In spite of this, as it results from Fig. 7, the satisfaction obtained for WEDF has been the highest which indicates that obtaining minimum response times for single requests do not mean that the client will receive the whole WWW page in the satisfying time.

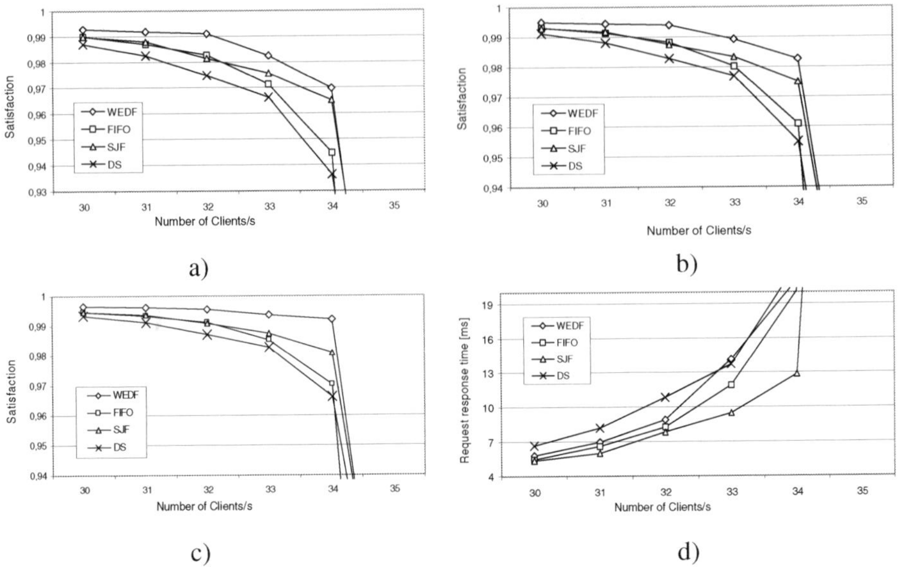


**Fig. 7.** Satisfaction vs. number of clients per second for: a) $t^{s}_{\max} = 300ms$ and $t^{h}_{\max} = 600ms$, b) $t^{s}_{\max} = 400ms$ and $t^{h}_{\max} = 800ms$, c) $t^{s}_{\max} = 500ms$ and $t^{h}_{\max} = 1000ms$; c) Mean HTTP request response time vs. number of new clients per second for $t^{s}_{\max} = 300ms$ and $t^{h}_{\max} = 600ms$

The obtained results confirm that it is worth scheduling the HTTP requests in the front of Web service, improving at the same time the Web services quality and making WWW pages to be served in a deterministic way.

## 5 Summary

The article presents a new WEDF method for scheduling HTTP request in a Web service. According to the method described, requests are scheduled at the front of the service in a way that the response times of the whole WWW pages do not exceed the time imposed by the customer. Thanks to this new method the users using a loaded service will obtain similar times of serving WWW pages, for both small simple pages as well as complex ones requiring transferring data from databases. At the same time, response times of the whole WWW pages should not exceed the determined time limit.

## References

1. Aron, M., Druschel, P., Zwaenepoel, W.: Efficient Support for P-HTTP in Cluster Based Web Servers. In: Proc. Usenix Ann. Techn. Conf., Monterey, CA (1999)
2. Barford, P., Crovella, M.E.: A Performance Evaluation of Hyper Text Transfer Protocols. In: Proc. ACM SIGMETRICS 1999, Atlanta, pp. 188–197 (1999)

3. Borzemski, L., Zatwarnicki, K.: A Fuzzy Adaptive Request Distribution Algorithm for Cluster-Based Web Systems. In: Proc. of 11th Conf. on Parallel, Distributed and Network-based Processing, pp. 119–126. IEEE CS Press, Los Alamitos (2003)
4. Borzemski, L., Zatwarnicki, K.: Using adaptive fuzzy-neural control to minimize response time in cluster-based web systems. In: Szczepaniak, P.S., Kacprzyk, J., Niewiadomski, A. (eds.) AWIC 2005. LNCS (LNAI), vol. 3528, pp. 63–68. Springer, Heidelberg (2005)
5. Cardellini, V., Casalicchio, E., Colajanni, M., Yu, P.S.: The State of the Art in Locally Distributed Web-Server Systems. ACM Comp. Surv. 34(2), 263–311 (2002)
6. Cardellini, V., Casalicchio, E., Colajanni, M., Mambelli, M.: Web Switch Support for Differentiated Services. ACM Perf. Eval. Rev. 29(2), 14–19 (2001)
7. Colajanni, M., Yu, P.S., Cardellini, V.: Scalable Web-Server systems: architectures, models and load balancing algorithms. In: SIGMETRICS (2000)
8. Elnikety, S., Nahum, E., Tracey, J., Zwaenepoel, W.: A method for transparent admission control and request scheduling in e-commerce web sites. In: Proceedings of the 13th international conference on World Wide Web, New York, USA (2004)
9. Lee, S.C.M., Lui, J.C.S., Yau, D.K.Y.: A Proportional-Delay DiffServ-Enabled Web Server: Admission Control and Dynamic Adaptation. IEEE Transactions on Parallel and Distributed Systems 15(5), 385–400 (2004)
10. Mesquite Software, CSIM 19 library, `http://www.mesquite.com`
11. Pai, V.S., Aront, M., Banga, G., Svendsen, M., Druschel, P., Zwaenpoel, W., Nahum, E.: Locality-Aware Request Distribution in Cluster-Based Network Servers. In: Proc. of 8th ACM Conf. on Arch. Support for Progr. Languages (1998)
12. Pathan, A.-M.K., Buyya, R.: A Taxonomy and Survey of Content Delivery Networks, Technical Report, GRIDS-TR-2007-4, Grid Computing and Distributed Systems Laboratory, The University of Melbourne, Australia (2007)
13. Zhou, X., Wei, J., Xu, C.-Z.: Quality-of-service differentiation on the Internet: a taxonomy. Journal of Network and Computer Applications 30(1), 354–383 (2007)

# Comparative Analysis of Regression Tree Models for Premises Valuation Using Statistica Data Miner

Tadeusz Lasota[1], Piotr Sachnowski[2], and Bogdan Trawiński[2]

[1] Wrocław University of Environmental and Life Sciences, Dept. of Spatial Management
Ul. Norwida 25/27, 50-375 Wroclaw, Poland
[2] Wrocław University of Technology, Institute of Informatics,
Wybrzeże Wyspiańskiego 27, 50-370 Wrocław, Poland
tadeusz.lasota@wp.pl, sachnowski@wp.pl,
bogdan.trawinski@pwr.wroc.pl

**Abstract.** Several dozen of experiments were conducted with Statistica Data Miner in order to assess the suitability of different machine learning algorithms for an Internet expert system to assist with real estate appraisal. The investigations concentrated first of all on regression trees and ensemble tree models. Moreover, decision tree approaches were compared with commonly used algorithms as KNN, SVM and a multilayer perceptron neural network. The results provided by the collection of twelve predictive accuracy measures were also analyzed. The study proved the usefulness of majority of algorithms to build the real estate valuation models.

**Keywords:** regression trees, machine learning, real estate appraisal, Statistica.

## 1 Introduction

The application of trees in discrimination and regression analysis was introduced by Breiman et al. in 1984 [1]. Shortcut of the title of their book (CART) gave the name to the whole family of methods and algorithms. New methods and algorithms referring to trees i.e. CHAID, M5, boosted trees, MAR Splines, forests and others were developed in the course of time. Integrating regression trees with other regression approaches is also alternative use of trees which overcome the limitations of individual methods both in terms of accuracy as well in computational efficiency [2]. Trees found their application in many areas of life, e.g. in economy [3], industry [4], mechanics [5], computer science [6], mathematics [7], medicine [8], robotics [9], real estate appraisal [10]. The results presented in tree-form are intuitive, aid decision making and are easy to use. Interesting subsets of models can be indentified along with their characteristics by following the path from the subset to the root of the tree. The technique is fairly robust to the presence of outliers, is stable with highly uncorrelated data, and can handle missing values [11]. Regression trees can be learned from large datasets, and they perform better on larger data sets than on smaller ones [12], [13].

In our pervious works [14], [15] we investigated different machine learning algorithms, among others genetic fuzzy systems and artificial neural networks, devoted to

N.T. Nguyen, R. Kowalczyk, and S.-M. Chen (Eds.): ICCCI 2009, LNAI 5796, pp. 776–787, 2009.

build data driven models to assist with real estate appraisals using MATLAB and KEEL. In this paper, which is the continuation of our research, we report the results of experiments conducted with Statistica Data Miner aimed at the comparison of several decision tree methods using actual data taken from cadastral system in order to assess their appropriateness to an Internet expert system assisting appraisers' work.

## 2 Cadastral Systems as the Source Base for Model Generation

The concept of a data driven models for premises valuation, presented in the paper, was developed basing on sales comparison method. It was assumed that whole appraisal area, that means the area of a city or a district, is split into sections (e.g. clusters) of comparable property attributes. The architecture of the proposed system is shown in Fig. 1. The appraiser accesses the system through the internet and chooses an appropriate section and input the values of the attributes of the premises being evaluated into the system, which calculates the output using a given model. The final result as a suggested value of the property is sent back to the appraiser.

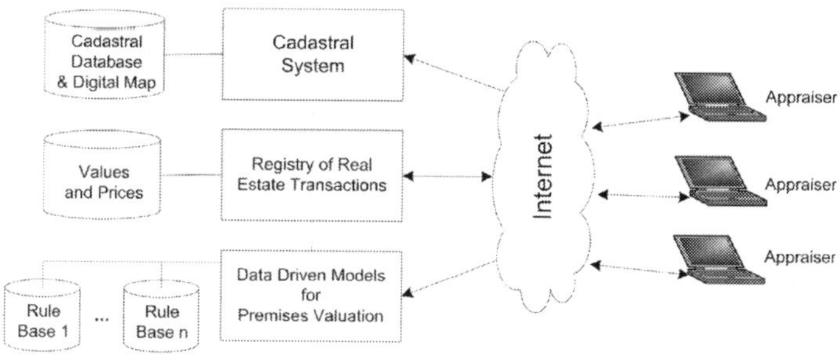


**Fig. 1.** Information systems to assist with real estate appraisals

Actual data used to generate and learn appraisal models came from the cadastral system and the registry of real estate transactions referring to residential premises sold in one of big Polish cities at market prices within two years 2001 and 2002. They constituted original data set of 1098 cases of sales/purchase transactions. Four attributes were pointed out as price drivers: usable area of premises, floor on which premises were located, year of building construction, number of storeys in the building, in turn, price of premises was the output variable.

## 3 Regression Algorithms Used in Experiments

In our study we used nine machine learning algorithms implemented in Statistica, one of the most popular statistic and analytics software developed by StatSoft (www.statsoft.com). Statistica Data Miner can be characterized by some useful features including a comprehensive selection of data mining methods, e.g., clustering techniques, neural networks architectures, classification/regression trees, multivariate modeling, and many other predictive techniques; a large selection of graphics and visualization procedures. Moreover, it provides complete data mining projects, ready to run and set up to competitively evaluate alternative models, using bagging, boosting, stacking, meta-learning, etc., and to produce presentation-quality summary reports [16]. The algorithms are listed in Table 1, they comprise four algorithms to construct single regression trees (CRT, CHI, ICR, IEC), two to build ensemble tree models (BST, FOR), and three other popular machine learning algorithms (KNN, ANN, and SVM).

**Table 1.** Statistica machine learning algorithms used in study

| Code | Description |
|---|---|
| CRT | C&RT algorithm - Classification and Regression Trees |
| CHI | CHAID tree - Chi-squared Automatic Interaction Detector |
| ICR | Interactive C&RT algorithm |
| IEC | Interactive Exhaustive CHAID algorithm |
| BST | Boosting trees |
| FOR | Random forests |
| KNN | K-Nearest Neighbors with Euclidean distance measure |
| ANN | Artificial Neural Networks with MLP architecture |
| SVM | Support Vector Machines with RBF core |

*CRT*. C&RT algorithm is based on binary recursive partitioning. At each node, which represents a group - subsets of the training data, there is applied binary rule which produces two branches. Such process, which divides input spaces into regions, is repeated till the given threshold is reached or when the algorithm finds out that no further gain is made.

*CHI*. CHAID algorithm builds non-binary trees, i.e., trees where more than two branches can attach to a single root or node. Construct trees, where each non-terminal node identifies a split condition, to yield optimum prediction of continuous dependent variables. It is well suited for the analysis of larger datasets.

*ICR*. Applying this algorithm user can add and remove nodes and whole levels of trees. User have full control in process of building tree. User can in few seconds make a tree and judge its quality.

*IEC*. This modification to the basic CHAID algorithm performs a more thorough merging and testing of predictor variables, specifically, the merging of categories continues until only two categories remain for each predictor. The algorithm then proceeds the split variable step, and selects among the predictors the one that yields the most significant split.

*BST*. The algorithm evolved from the application of boosting methods to regression trees. The general idea is to compute a sequence of simple trees, where each successive tree is built for the prediction residuals of the preceding tree. This method builds binary trees, i.e., partition the data into two samples at each split node.

*FOR*. It is an implementation of the Random Forest algorithm developed by Breiman [17]. A Random Forest consists of an ensemble of simple tree classifiers, each capable of producing a response when presented with a set of predictor values.

*KNN*. It is a memory-based method that requires no training. Predictions are based on a set of prototype examples that are used to predict new (i.e., unseen) data based on averaging over a set of k-nearest prototypes. In this algorithm Euclidean measure was employed for measuring the distance between the query point and cases from the examples sample.

*ANN*. During network training, the weights and thresholds are first initialized to small, random values. This implies that the units' response surfaces are each aligned randomly with low slope: they are effectively uncommitted. As training progresses, the units' response surfaces are rotated and shifted into appropriate positions, and the magnitudes of the weights grow as they commit to modeling particular parts of the target response surface.

*SVM*. This method performs regression tasks by constructing nonlinear decision boundaries. It assumes that the relationship between the independent and dependent variables is given by a deterministic function f plus the addition of some additive noise. The task is then to find a functional form for f that can correctly predict new cases that the SVM has not been presented with before. In this algorithm as kernel the radial basis function (RBF) was used.

## 4 Plan of Experiments

The goal of the research was to carry out comparative analysis of several regression tree algorithms implemented in the Statistica Data Miner and a few the most popular machine learning regression algorithms. Some preliminary tests were accomplished in order to determine the best parameters of respective algorithms using trial and error method.

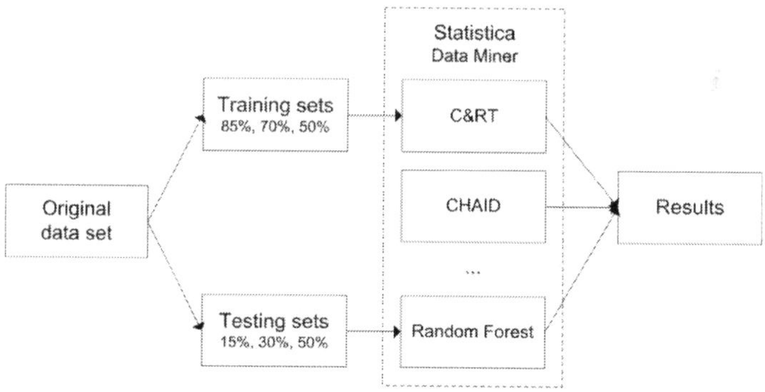


**Fig. 2.** Schema of the experiments with Statistica Data Miner

Schema of the experiments comprising algorithms with preselected parameters is depicted in Fig. 2. Three different random splits of the original data set into training and testing sets were performed. The latter contained 15%, 30% and 50% of cases included in the original data set, that means 152, 337, and 582 cases respectively. In order to avoid the impact of an unfavourable drawing, the random splits were repeated 5 times for each case and as averaging function median was applied. Normalization of data was accomplished using the min-max approach. As fitness function the mean absolute percentage error (MAPE) was applied and cross validation available in

Statistica was employed. SVM, ANN and KNN were run using 25-fold cross validation, in turn, trees except BST and FOR were run using 15-fold cross validation. A dozen of commonly used performance measures [18], [19] was applied to evaluate models built by respective algorithms used in our study. These measures are listed in Table 2 and expressed in the form of following formulas below, where $y_i$ denotes actual price and $\hat{y}_i$ – predicted price of i-th case, *avg(v), var(v), std(v)* – average, variance, and standard deviation of variables $v_1, v_2, \ldots, v_N$, respectively and $N$ – number of cases in the testing set.

**Table 2.** Performance measures used in study

| Denot. | Description | Dimen-sion | Min value | Max value | Desirable outcome | No. of form. |
|---|---|---|---|---|---|---|
| *MSE* | Mean squared error | $d^2$ | 0 | ∞ | min | 1 |
| *RMSE* | Root mean squared error | d | 0 | ∞ | min | 2 |
| *RSE* | Relative squared error | no | 0 | ∞ | min | 3 |
| *RRSE* | Root relative squared error | no | 0 | ∞ | min | 4 |
| *MAE* | Mean absolute error | d | 0 | ∞ | min | 5 |
| *RAE* | Relative absolute error | no | 0 | ∞ | min | 6 |
| *MAPE* | Mean absolute percent. error | % | 0 | ∞ | min | 7 |
| *NDEI* | Non-dimensional error index | no | 0 | ∞ | min | 8 |
| *r* | Linear correlation coefficient | no | -1 | 1 | close to 1 | 9 |
| $R^2$ | Coefficient of determination | % | 0 | ∞ | close to 100% | 10 |
| *var(AE)* | Variance of absolute errors | $d^2$ | 0 | ∞ | min | 11 |
| *var(APE)* | Variance of absolute percentage errors | no | 0 | ∞ | min | 12 |

$$MSE = \frac{1}{N}\sum_{i=1}^{N}(y_i - \hat{y}_i)^2 \tag{1}$$

$$RMSE = \sqrt{\frac{1}{N}\sum_{i=1}^{N}(y_i - \hat{y}_i)^2} \tag{2}$$

$$RSE = \frac{\sum_{i=1}^{N}(y_i - \hat{y}_i)^2}{\sum_{i=1}^{N}(y_i - avg(y))^2} \tag{3}$$

$$RRSE = \sqrt{\frac{\sum_{i=1}^{N}(y_i - \hat{y}_i)^2}{\sum_{i=1}^{N}(y_i - avg(y))^2}} \tag{4}$$

$$MAE = \frac{1}{N}\sum_{i=1}^{N}|y_i - \hat{y}_i| \tag{5}$$

$$RAE = \frac{\sum_{i=1}^{N}|y_i - \hat{y}_i|}{\sum_{i=1}^{N}|y_i - avg(y)|} \tag{6}$$

$$MAPE = \frac{1}{N}\sum_{i=1}^{N} \frac{|y_i - \hat{y}_i|}{y_i} * 100\% \quad (7)$$

$$NDEI = \frac{RMSE}{std(y)} \quad (8)$$

$$r = \frac{\sum_{i=1}^{N}(y_i - avg(y))\,(\hat{y}_i - avg(\hat{y}))}{\sqrt{\sum_{i=1}^{N}(y_i - avg(y))^2}\,\sqrt{\sum_{i=1}^{N}(\hat{y}_i - avg(\hat{y}))^2}} \quad (9)$$

$$R^2 = \frac{\sum_{i=1}^{N}(\hat{y}_i - avg(y))^2}{\sum_{i=1}^{N}(y_i - avg(y))^2} * 100\% \quad (10)$$

$$var(AE) = var(|y - \hat{y}|) \quad (11)$$

$$var(APE) = var(\frac{|y - \hat{y}|}{y}) \quad (12)$$

## 5 Results of Experiments

The performance of the models built by all nine algorithms for respective measures and for 15%, 30%, and 50% splits was presented in Fig. 3-14. For clarity, all measures were calculated for normalized values of output variables except for MAPE, where in order to avoid the division by zero, actual and predicted prices had to be denormalized. Almost in each instance the 50% split led to worse performance than the 15% split, what may be explained that the latter provided more data to learn models. In turn the 30% split gave various results, difficult to explain: they were once better and another time worse.

Fig. 11 shows there is high correlation, ranging from 61.1% to 83.7%, between actual and predicted prices for each model. In turn, Fig.12 illustrating the coefficients of determination indicates that from 50.7% to 82.6% of total variation in the dependent variable (prices) is accounted for by the models. This can be explained that data derived from the cadastral system and the register of property values and prices could cover only some part of potential price drivers.

The nonparametric Wilcoxon signed-rank tests were carried out for three measures: MSE, MAE, and MAPE and for each repeated split separately. The results are shown in Tables 3, 4, and 5 in the form of symmetric matrices, where a triple, eg. <035>, in each cell reflects the outcome for a given pair of models for 15%, 30% and 50% splits respectively. The numbers denote how many times there were statistically significant differences in mean values of respective errors within five repeated splits. Thus the triple <555> would denote that there were evidences against the H0 hypothesis in each instance, with the 0.05 significance level. However that result and <545> and <455> occurred in Tables 3, 4, and 5 only few times, majority of them for FOR models which outperformed CHI, ICA, and IEC models.

Due to the non-decisive results of majority of Wilcoxon tests, rank positions of individual algorithms were determined for each measure and each split (see Table 6). Observing median, average, minimal and maximal ranks it can be noticed that highest rank positions gained ANN, FOR, SVN, CRT, and KNN algorithms and the lowest IEC, ICA and CHI ones.

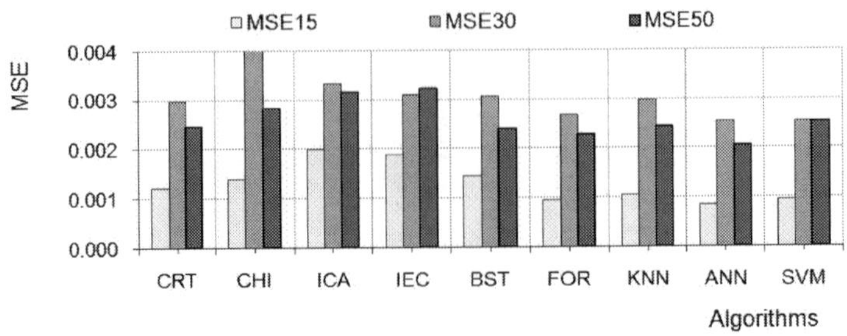


**Fig. 3.** Comparison of MSE values

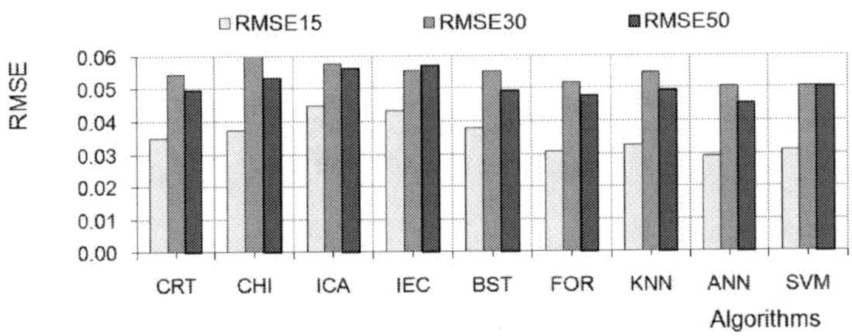


**Fig. 4.** Comparison of RMSE values

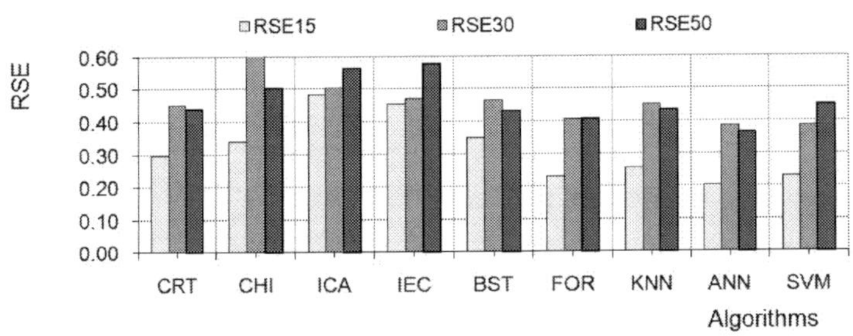


**Fig. 5.** Comparison of RSE values

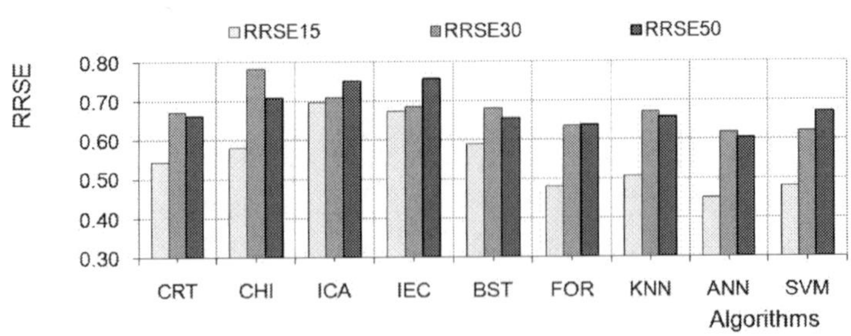


**Fig. 6.** Comparison of RRSE values

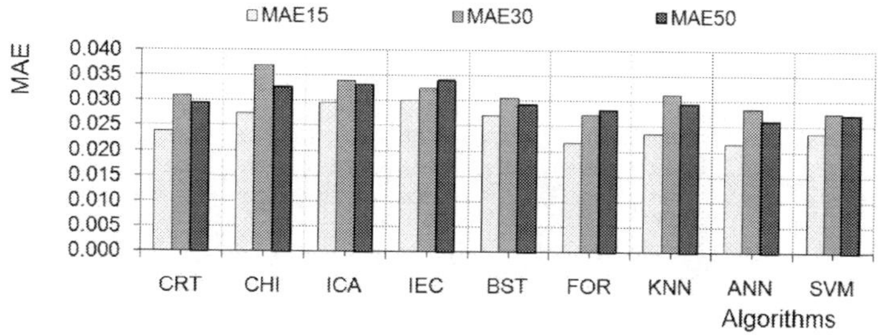


**Fig. 7.** Comparison of MAE values

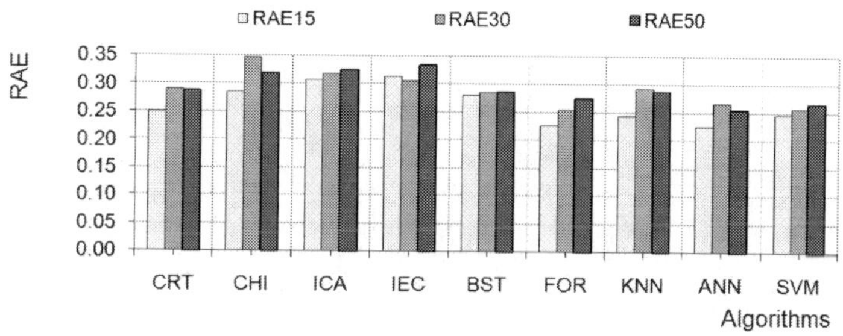


**Fig. 8.** Comparison of RAE values

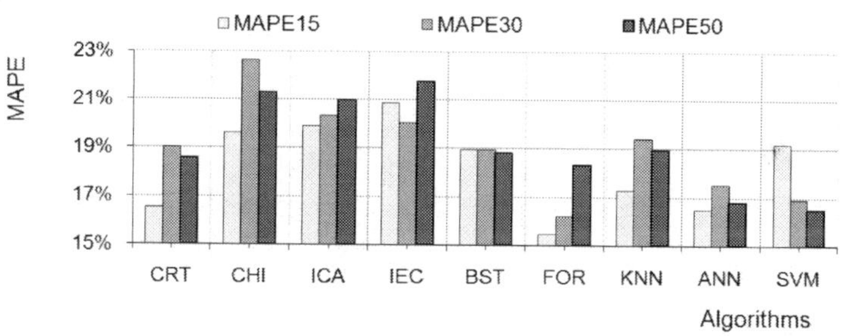


**Fig. 9.** Comparison of MAPE values

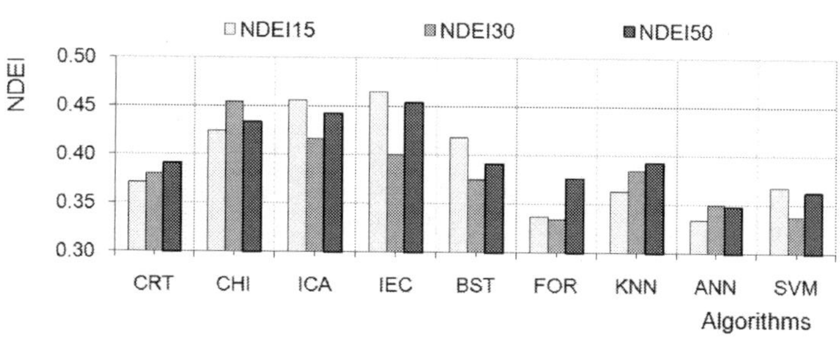


**Fig. 10.** Comparison of NDEI values

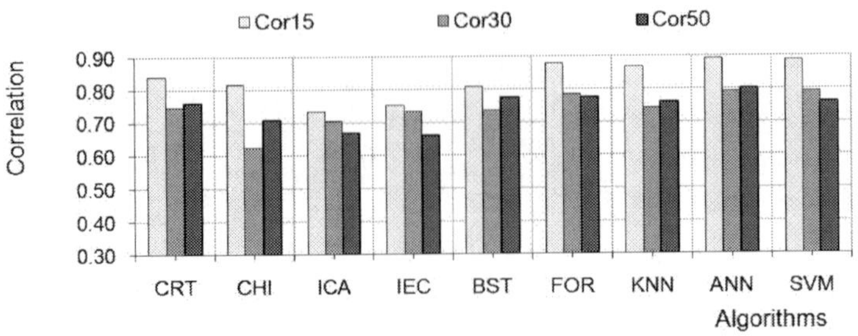


**Fig. 11.** Correlation between predicted and actual prices

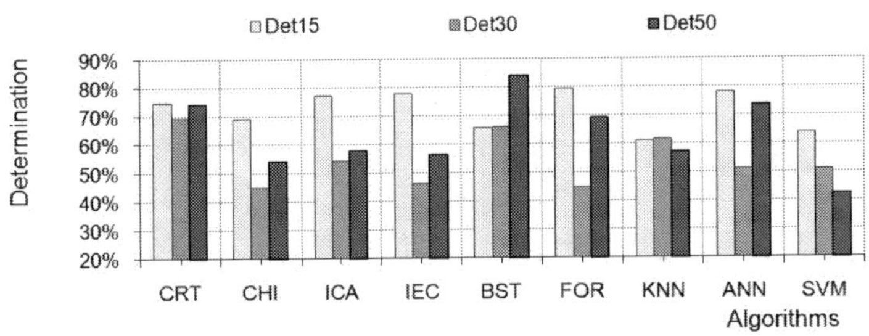


**Fig. 12.** Comparison of determination coefficient values

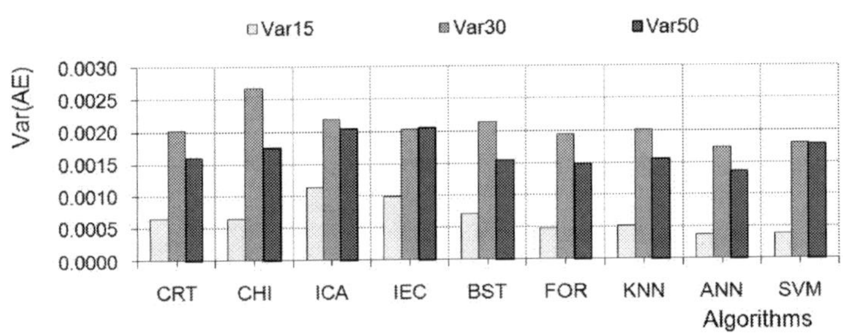


**Fig. 13.** Variance of absolute errors - Var(AE)

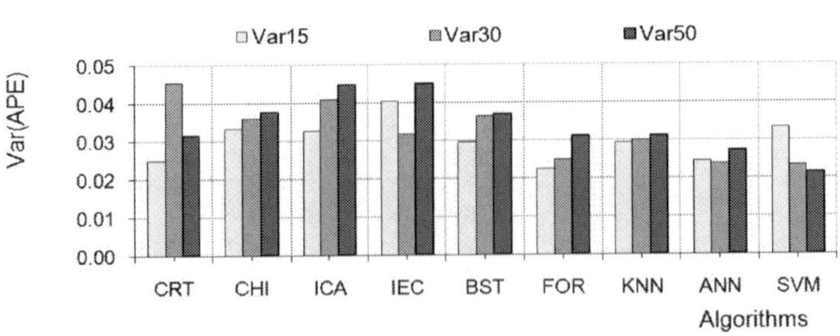


**Fig. 14.** Variance of absolute percentage errors - Var(APE)

**Table 3.** Results of Wilcoxon signed-rank test for squared errors comprised by MSE

| | CRT | CHI | ICA | IEC | BST | FOR | KNN | ANN | SVM |
|---|---|---|---|---|---|---|---|---|---|
| **CRT** | - | 145 | 232 | 332 | 300 | 224 | 103 | 023 | 012 |
| **CHI** | 145 | - | 151 | 144 | 345 | 545 | 243 | 555 | 344 |
| **ICA** | 232 | 151 | - | 011 | 334 | 555 | 110 | 344 | 133 |
| **IEC** | 332 | 144 | 011 | - | 243 | 555 | 121 | 344 | 123 |
| **BST** | 300 | 345 | 334 | 243 | - | 222 | 214 | 132 | 110 |
| **FOR** | 224 | 545 | 555 | 555 | 222 | - | 325 | 102 | 321 |
| **KNN** | 103 | 243 | 110 | 121 | 214 | 325 | - | 125 | 124 |
| **ANN** | 023 | 555 | 344 | 344 | 132 | 102 | 125 | - | 133 |
| **SVM** | 012 | 344 | 133 | 123 | 110 | 321 | 124 | 133 | - |

**Table 4.** Results of Wilcoxon test for absolute percentage errors comprised by MAPE

| | CRT | CHI | ICA | IEC | BST | FOR | KNN | ANN | SVM |
|---|---|---|---|---|---|---|---|---|---|
| **CRT** | - | 345 | 133 | 332 | 310 | 223 | 003 | 022 | 202 |
| **CHI** | 345 | - | 143 | 234 | 355 | 545 | 243 | 555 | 244 |
| **ICA** | 133 | 143 | - | 032 | 334 | 555 | 120 | 244 | 033 |
| **IEC** | 332 | 234 | 032 | - | 243 | 555 | 131 | 354 | 133 |
| **BST** | 310 | 355 | 334 | 243 | - | 221 | 214 | 202 | 132 |
| **FOR** | 223 | 545 | 555 | 555 | 221 | - | 335 | 211 | 303 |
| **KNN** | 003 | 243 | 120 | 131 | 214 | 335 | - | 115 | 334 |
| **ANN** | 022 | 555 | 244 | 354 | 202 | 211 | 115 | - | 332 |
| **SVM** | 202 | 244 | 033 | 133 | 132 | 303 | 334 | 332 | - |

**Table 5.** Results of Wilcoxon signed-rank test for absolute errors comprised by MAE

| | CRT | CHI | ICA | IEC | BST | FOR | KNN | ANN | SVM |
|---|---|---|---|---|---|---|---|---|---|
| **CRT** | - | 345 | 122 | 332 | 320 | 223 | 103 | 123 | 002 |
| **CHI** | 345 | - | 151 | 234 | 455 | 545 | 243 | 555 | 344 |
| **ICA** | 122 | 151 | - | 022 | 334 | 555 | 120 | 344 | 033 |
| **IEC** | 332 | 234 | 022 | - | 243 | 555 | 131 | 354 | 143 |
| **BST** | 320 | 455 | 334 | 243 | - | 222 | 214 | 102 | 130 |
| **FOR** | 223 | 545 | 555 | 555 | 222 | - | 235 | 202 | 301 |
| **KNN** | 103 | 243 | 120 | 131 | 214 | 235 | - | 115 | 324 |
| **ANN** | 123 | 555 | 344 | 354 | 102 | 202 | 115 | - | 233 |
| **SVM** | 002 | 344 | 033 | 143 | 130 | 301 | 324 | 233 | - |

**Table 6.** Rank positions of algorithms with respect to performance measures (1 means the best)

| | Split | CRT | CHI | ICA | IEC | BST | FOR | KNN | ANN | SVM |
|---|---|---|---|---|---|---|---|---|---|---|
| MSE | 15% | 5 | 7 | 8 | 9 | 4 | 1 | 6 | 2 | 3 |
| MSE | 30% | 4 | 5 | 9 | 7 | 8 | 1 | 3 | 6 | 2 |
| MSE | 50% | 5 | 7 | 9 | 8 | 3 | 2 | 6 | 4 | 1 |
| RMSE | 15% | 5 | 7 | 8 | 9 | 4 | 1 | 6 | 2 | 3 |
| RMSE | 30% | 4 | 5 | 9 | 7 | 8 | 1 | 3 | 6 | 2 |
| RMSE | 50% | 5 | 7 | 9 | 8 | 3 | 2 | 6 | 4 | 1 |
| RSE | 15% | 4 | 7 | 9 | 8 | 5 | 1 | 6 | 3 | 2 |
| RSE | 30% | 5 | 3 | 9 | 7 | 8 | 2 | 6 | 4 | 1 |
| RSE | 50% | 7 | 5 | 8 | 9 | 4 | 3 | 6 | 2 | 1 |
| RRSE | 15% | 4 | 7 | 9 | 8 | 5 | 1 | 6 | 3 | 2 |
| RRSE | 30% | 5 | 3 | 9 | 7 | 8 | 2 | 6 | 4 | 1 |
| RRSE | 50% | 7 | 5 | 8 | 9 | 4 | 3 | 6 | 2 | 1 |

| | | | | | | | | | | |
|---|---|---|---|---|---|---|---|---|---|---|
| MAE | 15% | 5 | 9 | 6 | 8 | 4 | 1 | 7 | 2 | 3 |
| MAE | 30% | 6 | 5 | 9 | 7 | 8 | 1 | 4 | 2 | 3 |
| MAE | 50% | 5 | 7 | 8 | 9 | 4 | 2 | 6 | 3 | 1 |
| RAE | 15% | 4 | 8 | 7 | 9 | 5 | 1 | 6 | 2 | 3 |
| RAE | 30% | 6 | 5 | 9 | 7 | 8 | 3 | 4 | 2 | 1 |
| RAE | 50% | 4 | 8 | 7 | 9 | 5 | 2 | 6 | 3 | 1 |
| MAPE | 15% | 5 | 9 | 7 | 8 | 3 | 1 | 4 | 2 | 6 |
| MAPE | 30% | 6 | 5 | 9 | 7 | 8 | 2 | 4 | 1 | 3 |
| MAPE | 50% | 4 | 8 | 7 | 9 | 5 | 3 | 6 | 2 | 1 |
| NDEI | 15% | 5 | 7 | 8 | 9 | 4 | 1 | 6 | 2 | 3 |
| NDEI | 30% | 4 | 5 | 9 | 7 | 8 | 1 | 3 | 6 | 2 |
| NDEI | 50% | 5 | 7 | 9 | 8 | 3 | 2 | 6 | 4 | 1 |
| r | 15% | 4 | 8 | 9 | 7 | 5 | 2 | 6 | 1 | 3 |
| r | 30% | 3 | 6 | 9 | 7 | 8 | 1 | 5 | 4 | 2 |
| r | 50% | 7 | 6 | 8 | 9 | 4 | 3 | 5 | 2 | 1 |
| $R^2$ | 15% | 3 | 9 | 7 | 5 | 2 | 6 | 4 | 1 | 8 |
| $R^2$ | 30% | 1 | 5 | 9 | 6 | 7 | 2 | 3 | 8 | 4 |
| $R^2$ | 50% | 3 | 9 | 8 | 9 | 4 | 1 | 7 | 6 | 2 |
| Var(AE) | 15% | 6 | 7 | 8 | 9 | 4 | 3 | 5 | 2 | 1 |
| Var(AE) | 30% | 3 | 5 | 9 | 7 | 8 | 1 | 4 | 6 | 2 |
| Var(AE) | 50% | 5 | 8 | 9 | 6 | 3 | 2 | 4 | 7 | 1 |
| Var(APE) | 15% | 4 | 6 | 8 | 9 | 3 | 5 | 7 | 1 | 2 |
| Var(APE) | 30% | 6 | 2 | 9 | 7 | 9 | 3 | 5 | 4 | 1 |
| Var(APE) | 50% | 5 | 8 | 7 | 9 | 6 | 3 | 4 | 1 | 2 |
| median | | 5.00 | 7.00 | 9.00 | 8.00 | 5.00 | 2.00 | 6.00 | 2.50 | 2.00 |
| average | | 4.69 | 6.39 | 8.31 | 7.86 | 5.33 | 2.00 | 5.19 | 3.22 | 2.14 |
| min | | 1 | 2 | 6 | 5 | 2 | 1 | 3 | 1 | 1 |
| max | | 7 | 9 | 9 | 9 | 9 | 6 | 7 | 8 | 8 |

## 6 Conclusions and Future Work

The experiments aimed to compare machine learning algorithms to create models for the valuation of residential premises which were conducted using the Statistica Data Miner. Nine methods comprising four algorithms to construct single regression trees: CRT, CHI, ICA, and IEC, two to build ensemble tree models: FOR and BST, and three others KNN, ANN, and SVM were applied to actual data sets derived from the cadastral system and the registry of real estate transactions. The overall conclusion is that ANN, FOR, SVN, CRT, KNN , and BST algorithms performed comparably well and the IEC, ICA and CHI algorithms should be rather discarded. Processing times of generating appraisal models using six best above mentioned algorithms, were not prohibitively long, therefore they can be incorporated into proposed internet system for real estate appraisals.

MAPE obtained in all tests ranged from 16.2% do 24.1%. This can be explained that data derived from the cadastral system and the register of property values and prices can cover only some part of potential price drivers. Physical condition of the premises and their building, their equipment and facilities, the neighbourhood of the building, the location in a given part of a city should also be taken into account, moreover, overall subjective assessment after inspection in site should be done. Therefore, we intend to test data obtained from public registers and then supplemented by experts conducting on-site inspections and evaluating more aspects of properties being appraised. What is more, further investigations of multiple models comprising ensembles of different regression trees using bagging and boosting techniques is planned.

## References

1. Breiman, L., Friedman, J., Olshen, R., Stone, C.: Classification and Regression Trees, Wadsworth Int. Group, Belmont (1984)
2. Torgo, L.: Functional Models for Regression Tree Leaves. In: Proc. 14th International Conference on Machine Learning, pp. 385–393. Morgan Kaufmann, San Francisco (1997)
3. Badulescu, L.A., Nicula, A.: Data Mining Decision Trees in Economy. Analele Universitatii din Oradea, Stiinte Economice XVI II, 723–727 (2007)
4. Giudici, P., Figini, S.: Applied Data Mining for Business and Industry. Wiley, Chichester (2009)
5. Trana, V.T., Yanga, B.S., Oha, M.S., Tanb, A.C.C.: Machine condition prognosis based on regression trees and one-step-ahead prediction, Mech. Systems and Signal Process. 22(5), 1179–1193 (2008)
6. Khoshgoftaar, T.M., Seliya, N.: Tree-based software quality estimation models for fault prediction. In: Proc. 8th IEEE Symp. on Software Metrics, pp. 203–214 (2002)
7. Wang, X., Dietterich, T.G.: Efficient Value Function Approximation Using Regression Trees. In: Proceedings of the IJCAI Workshop on Statistical Machine Learning for Large-Scale Optimization (1999)
8. Podgorelec, V., Kokol, P., Stiglic, B., Rozman, I.: Decision trees: an overview and their use in medicine. Journal of Medical Systems 26, 445–463 (2002)
9. Balac, N., Gaines, D.M., Fisher, D.: Using Regression Trees to Learn Action Models. IEEE Int. Conference on Man and Cybernetics 5, 3378–3383 (2000)
10. Acciani, C., Fucilli, V., Sardaro, R.: Model Tree: An Application in Real Estate Appraisal, 109th EAAE Seminar, no. 44853, Viterbo (2008)
11. Gokhale, S.S., Lyu, M.R.: Regression tree modeling for the prediction of software quality. In: Proc. of ISSAT 1997, pp. 31–36 (1997)
12. Perlich, C., Provost, F., Simonoff, J.S.: Tree induction vs. logistic regression: a learning-curve analysis. J. of Machine Learning Res. 4, 211–255 (2003)
13. Quinlan, J.R.: Learning with Continuous Classes. In: 5th Australian Joint Conference on Artificial Intelligence (AI 1992), Singapore, pp. 343–348 (1992)
14. Król, D., Lasota, T., Trawiński, B., Trawiński, K.: Investigation of evolutionary optimization methods of TSK fuzzy model for real estate appraisal. International Journal of Hybrid Intelligent Systems 5(3), 111–128 (2008)
15. Lasota, T., Pronobis, E., Trawiński, B., Trawiński, K.: Exploration of Soft Computing Models for the Valuation of Residential Premises using the KEEL Tool. In: Nguyen, N.T., et al. (eds.) 1st Asian Conference on Intelligent Information and Database Systems (ACIIDS 2009), pp. 253–258. IEEE, Los Alamitos (2009)
16. Hill, T., Lewicki, P.: Statistics: Methods and Applications. StatSoft, Tulsa (2007)
17. Breiman, L.: Random Forests. Machine Learning 45, 5–32 (2001)
18. Hagquist, C., Stenbeck, M.: Goodness of Fit in Regression Analysis – $R^2$ and $G^2$ Reconsidered. Quality & Quantity 32, 229–245 (1998)
19. Witten, I.H., Frank, E.: Data Mining: Practical machine learning tools and techniques, 2nd edn. Morgan Kaufmann, San Francisco (2005)

# Electronic Trading on Electricity Markets within a Multi-agent Framework*

Mariusz Kaleta, Piotr Pałka, Eugeniusz Toczyłowski, and Tomasz Traczyk

Warsaw University of Technology, Nowowiejska 15/19, 00-665 Warsaw, Poland
M.Kaleta@ia.pw.edu.pl
http://www.ia.pw.edu.pl

**Abstract.** Specific features of trade on electrical energy markets make the automation of certain negotiation and contracting processes attractive and desirable. Multi-agent systems are a natural solution for automating processes in a distributed environment, where individual market entities posses their own objectives. In the paper we consider a concept of a multi-agent framework for an electronic market trade. Since both energy markets and trade mechanisms are complex and permanently evolving, we require the highest expressiveness from this concept. Development of such a system involves solving several design issues related to embedding the agents in the environment, agent communications schemas, language, and offers modeling, especially in the case of the distributed, multilateral trade. We present a solution for the most important design issues. Our concept, including new information technologies supplementing multi-agent systems and Web services, are capable to support enterprises in the decision processes, to facilitate the buy/sell offers preparation, to select parties and enter into business relations with entities, and to support contract negotiations.

## 1 Introduction

Negotiations and contracting on electrical energy markets are usually supported by computer aided systems. Typically, the software tools available to the energy market entities provide forecasting modules and some methods for evaluation of trade offers and energy delivery schedules, with estimation of the risk measures. Only few of them are capable to perform the optimization process of selecting the best trade offers under some conditions. A decision maker (DM) tries to achieve the best trade decisions through the process of his/her interactions with computer-aided system. DM controls the process of taking the decisions by using expert knowledge, intelligence and intuition. Nevertheless, taking into consideration the specific energy market trade requirements (i.e., real time and resource-constrained requirements), the needs for more sophisticated computer support and wider automation of trade processes becomes more evident. Based on some strategy, automation may comprise some decision processes related to

* Partially supported by MNiSW grant no N N516 375736.

N.T. Nguyen, R. Kowalczyk, and S.-M. Chen (Eds.): ICCCI 2009, LNAI 5796, pp. 788–799, 2009.

trade up to a certain stage, after which DM may enter into the process. In typical interactions between DM and commercial computer-aided systems, this strategy is usually only a domain of DM and is not represented in software.

The need for automating decision processes for trade on energy markets results from specific requirements on the intra-day and real-time (security-constrained) balancing markets and repeatability of some processes. For instance, the detailed schedules of energy delivery are negotiated repeatedly within the framework resulting from the long-term contracts.

Multi-agent systems (MAS) approach is a natural solution for process automation in distributed environment since it imitates well the human being trade methods. There are several application areas related to electrical energy markets, including modeling and simulation of power systems, distributed control and monitoring and diagnostics. A comprehensive picture of state of the art is presented in [8] and [9]. However, in the literature, there is very little pertaining to electronic energy trading. In [12] a MAS system is used for support negotiate-based decision making, but it is used only as a simulation tool without trade automation.

In the paper we consider a multi-agent model of a local electrical energy market, where agents are software components, which independently take actions tending to make agreements at the most attractive conditions. We use this case to prove, that multi-agent systems are proper solution for process automation in distributed environment. The power system (at the distribution network level) is being balanced in a complex, multi-stage process. At each stage of this process, the trade mechanisms can take different forms: from bilateral distributed trade, to formalized and centralized trade platforms, like power exchanges. The MAS-based system for an electronic energy trading should be capable to support as much of the trade mechanisms from this wide range, as possible. Thus, there are strong expectations for achieving high expressiveness of the system, due to changing environment, e.g. emerging new technologies for generation, dissemination of dispersed generation, micro grids and new equipments for demand side management.

In the literature, only a few aspects of MAS applications as the market simulation tools are relatively close to applications considered in this paper. Most of the current simulation agent-based platforms are limited and targeted to some specific market mechanisms. For instance, PowerWeb is limited to a single uniform auction with fixed demand [2], the Auction Agents for the Electric Power implements Dutch auction [3], work [6] concerns model of market in England and Wales, [13] is targeted at hydroelectric power stations, MASCEM is quite interesting, but is also limited to a given market structure [11].

A proper implementation of MAS-based electronic trading requires solving several design issues. In this paper we present the solutions of the most important ones: the issues related to the environment organization, agent communication mechanisms, comprehensible data interchange mechanism for diversified negotiating entities, and modeling of offers. In the case of the distributed trade, also the issues pertaining to localization of the potential contractors and their offers are addressed.

Our concept, with the new information technologies supplementing multi-agent systems and Web services, is capable to support enterprises in the decision processes, to facilitate the buy/sell offers preparation, to select parties and enter into business relations with entities, and to support contract negotiations.

# 2 Trade Models

## 2.1 Distributed Markets

In the distributed market, participants (or agents) are directly engaged in exchange of goods, by negotiating the best, from their point of view, contracts. In such markets the various agents reach bilateral or multilateral agreements, usually after some, often complex negotiations. The simplest type of agreement is a bilateral contract. Note, however, that in the case of multicommodity trade (e.g. simultaneous energy and ancillary services trade), contracts may affect not only many goods, but also a number of suppliers and customers. Such agreements will be called multilateral.

**Morris column concept.** Agents, to be able to negotiate, need to obtain some information about other agents, and particularly on the current negotiation process. We can certainly imagine situation, where each agent broadcasts the messages about initiating negotiations to all other agents. However, in large systems, such a solution would involve a massive communication effort. Moreover, agents would obtain a huge amount of unnecessary information. In the centralized systems it is possible to create a repository, which collects the data about the agents, and existing and planned negotiation processes.

As the distributed systems do not have (and often cannot have) any central repository that could collect such information, there is a need to share this information in a different way. For this purpose we define a special agent called *Morris column* [7]. The task of such agent is to offer some public location, where other agents may "hang a notice" to report certain information about their negotiations. In addition, the Morris column should provide functionality of searching and removing the information. In a broader context, agents may leave a notice on initiating various types of auction processes with some market-specific characteristics.

Both in distributed and in centralized market systems, there can be more than one agent who plays the *Morris Column* role. However, in distributed market system, there is a strong need for multiple instances of such agents. Every single *Morris Column* stores only a part of global information, and particular pair of such agents shares some parts of information.

## 2.2 Exchange-Like Markets

In this class of the markets there are central agents (e.g. the market operators), which collect offers, bids and asks sent by other agents involved in the trade.

Such agents balance the demand and supply on the market, according to some established rules (generally in the best possible way). The operator distributes the balance results to all agents involved. The operator of balancing market may further ensure compliance with various restrictions, e.g. including physical constraints associated with the production and transmission of electricity.

The simplest type of such mechanism is the stock market, where there are no physical limitations for delivery. The bids can be ordered by price and most favorable offers are accepted, while others are rejected. In the simplest case, the equilibrium price and total volume are determined in the intersection of supply and demand curves. Such rules are used e.g. in the wholesale electric power exchanges. However, there is often a need to reflect the above-mentioned physical limitations in the balancing processes.

## 3 Key Elements of the Framework

To design MAS in a context of electronic trade, several requirements must be met. Below, we discuss the most important ones.

### 3.1 Communication Models in Multi-agent Environment

Agents, implemented as software components, must be embedded in some environment. FIPA (Foundation of Intelligent Physical Agents) standards form a framework for agents operations [1]. FIPA has developed standards pertaining to inter-agents communications, and detailed communicative acts and certain schemas for inter-agent communication.

Mechanisms defined by FIPA are not satisfactory for modeling communication in a case of multilateral negotiations and distributed environment (without central entity). Nevertheless, FIPA standards are a good basis for developing communication schemas meeting the requirements. Necessary modifications of FIPA standards are described later.

### 3.2 Communication Language

Expressiveness of the system manifests, among the other things, in rich possibilities of definition various commodities, offers and other information elements required for negotiation process. Information necessary for negotiation process is passed as a content of a message. A language of this content must be defined by MAS developer. An interpretation of message content is determined by two elements: language and ontology. Language defines the syntactic of message content and ontology can provide semantics for elements used in message content and expressed by defined language. FIPA has four languages, however none of them has a potential for modeling data in a manner sufficiently flexible for our needs. To our best knowledge, there exists only one standard proposition, which can meet strong requirements for our system. It is the Multi-commodity Market Model $M^3$ which is an open general data model for trading many commodities simultaneously [14].

### 3.3 Modeling of Offers

One of the most challenging issues is a method for expressing the entity preferences during the bidding process. In both bilateral negotiations and wide spectrum of organized trade mechanisms, bids and offers can take diversified forms. Moreover, these forms evolve during market development, mechanisms and software development, and market entities knowledge and requirements growth. Thus, it is extremely important to have ability to express a wide range of possible offer forms. This target can be reached due to open data model of $M^3$. $M^3$ standard generalizes the offer objects expressed by volume and offered price, including possibilities for offering bundles of commodities (e.g. bundling energy in the consecutive hours to offer a complex block of energy), and including complex bindings (e.g. flexible offer for storing energy in dispenser that considers buying energy for charging, then losses and efficiency depended on selling moment). Till now, there are no examples of specific important requirements on the electrical energy market with market data that could not be modeled by $M^3$.

## 4 Agent Model

### 4.1 Agents and Their Roles

Multi-agent platform should allow efficient exchange of goods between participants represented by some software agents. Consistency and unambiguous communication must be ensured. The platform should allow agents to carry out multilateral negotiations.

The main type of agent is a Dealing Agent (DA). It intermediates between the actual decision entity (e.g. marketing company, power plants etc.) and the multi-agent system. Its mission is to establish the most advantageous (from the point of view of company which it represents) agreements, preceded by the process of negotiations. Negotiations are conducted in accordance with implemented strategies. We assume that the role of dealing agent can be extended to the role of a broker (intermediary trade), which also buys and sells goods, drawing profits from such turnover.

To provide a means of communication between the agents, we have to ensure that there are some agents, which will act as the repositories that provide information about the location of DA. Repository includes, in the simplest case, the URI address of the individual agents. Agent acting for this role is called Bilateral Transaction Broker (BTB). In the $M^3$ model, the roles of repositories in the distributed trade may be played by Morris columns. Thus, Bilateral Transaction Brokers can be implemented with the features of Morris column.

### 4.2 Information Flow

To assure trade, agents have to exchange some information. Specified agent, which wants to start negotiations, has to declare willingness to negotiate by sending a message to BTB agent. Other agents could send queries concerning

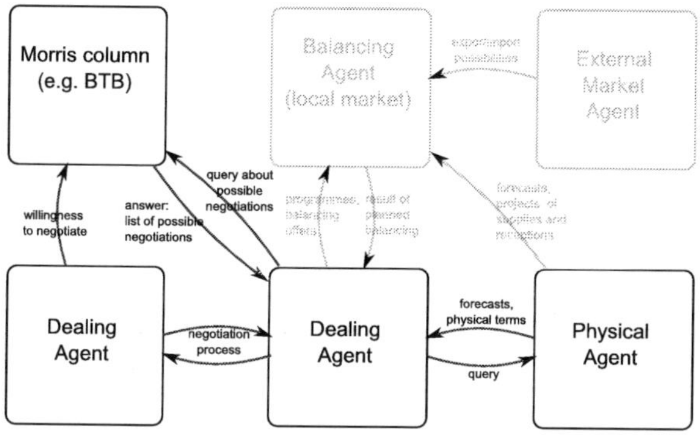


**Fig. 1.** Information flow between agents

ongoing negotiations. Other information being exchanged between agents is related to the estimates of demand and supply, or physical conditions.

Figure 1 presents basic types of agents and acts of communication. Agents marked with grey colour exist on additional market segments located closer to the real-time realisation of the contract.

## 5 Agents' Interactions

Agents interact using some typical messaging schemas. FIPA systematized these schemas and called them 'protocols'. From our point of view, two of them are interesting: Contract Net Interaction Protocol and Iterated Contract Net Interaction Protocol.

Contract Net Interaction Protocol defines messages interchange for centralized trade (in auction or exchange). Agent, which submits an offer using this protocol, has no possibility to negotiate.

Iterated Contract Net Interaction Protocol is an extension of the preceding one. This protocol assumes that offers may be negotiated, so it can be used on markets, where negotiations are necessary, e.g. for bilateral or multilateral trading.

In the following sections our extension of these two protocols is presented, which enables them to be used for multi-commodity trade.

### 5.1 Centralized Markets

On the centralized market there is a system operator that is responsible, among other things, for creating and maintaining a platform, where market participants can submit their sell and buy offers. The operator specifies the market model (under some regulation rules), determines the range of traded commodities, their characteristics, the rules for transaction making and clearing, infrastructural (e.g. network) constraints, other system requirements and constraints, etc. The operator usually defines all necessary dictionaries (sometimes defined in the form of ontologies): lists of commodity kinds, market entities, etc. The operator defines also a calendar, which determines time periods in which commodities can be offered.

Agents willing to trade on the centralized market must know where such platform can be found and how it can be used. If such a platform is not commonly known, it must be somehow announced to interested agents. It is of course impossible to notify all potentially interested agents directly, because they are numerous, and much of them are probably not known to the operator. A Morris column can be used to publish necessary information. An operator submits announcements to one or more Morris columns, where they become available to all interested agents.

A sequence diagram in Figure 2 presents a modified Contract Net Interaction Protocol. The operator submits a cfp (*call for proposal*) message to Morris column(s), which informs that the platform is available and waits for offers. Agents scans commonly known Morris columns, looking for interesting trade opportunities. They use *query-ref* messages that contain patterns specifying, which information is interesting to the sender. A Morris column responds to these queries sending *inform-result* messages, which contain only information fitting the patterns.

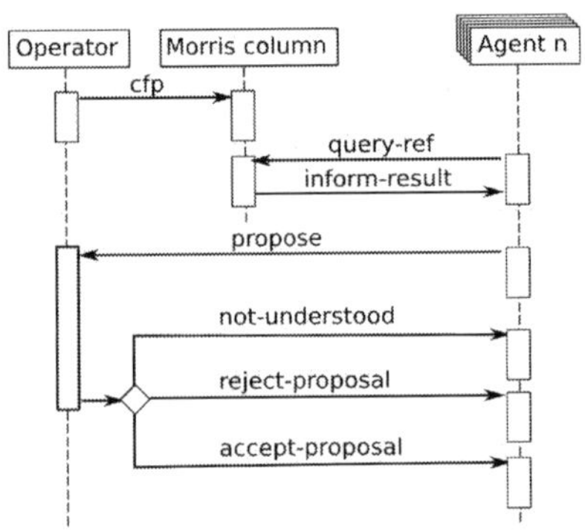


**Fig. 2.** Modified Contract Net Interaction Protocol

Agents interested in trade on the particular platform submit their offers, using *propose* message, to the platform announced on the Morris column. Depending on market model, offers should be submitted in specific time window, or can be submitted at any time. If the offer is not correct (e.g. it can not be understood or is incomplete), a *not-understood* message is sent to the offeror. The operator joins the offers using algorithms specific to the market model. Next, *accept-proposal* messages are send to the offerors, whose offers succeeded, and *reject-proposal* are send to the agents, whose offers failed. In this kind of market an offer is binding – an agent, which submits the offer, is obliged to fulfil the commitment.

## 5.2 Decentralized Markets and Negotiations

Decentralized markets lack of a central institution, where offers can be placed, and which determines the rules and common dictionaries for the market. Such markets are based on bilateral or multilateral trade, but before any trade can take place, potential participants must find each other. One of possible methods is the use of Morris columns, on which agents place their announcements.

Figure 3 shows proposed scenario of such bilateral interaction.

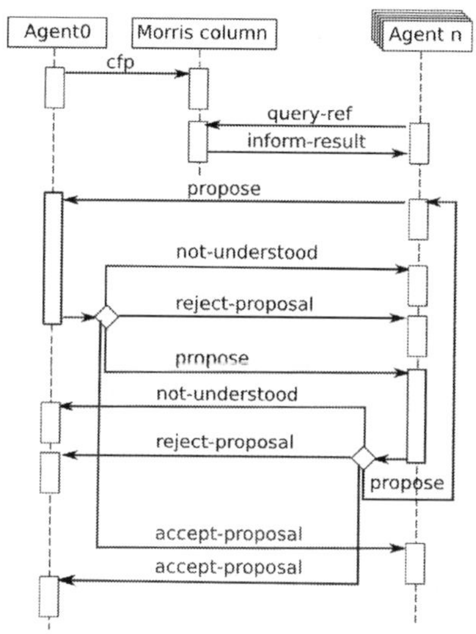


**Fig. 3.** Modified Iterated Contract Net Interaction Protocol

Agent submits a *call-for-proposal* message to one or more Morris columns. The message contains detailed description of the offer: commodities, quantities, time windows, etc. Other agents looks for interesting offers, sending *query-ref* messages to appropriate Morris columns. As a result of the query they obtain an *inform-result* messages containing lists of offers that fit the query; these lists contain also URL addresses of agents-offerors. Next, agents enter into direct bilateral or multilateral negotiations. They send *propose* messages to each other in one ore more iterations, negotiating the terms of trade. If negotiations succeed, one of the agents sends *accept-proposal* message the other one.

Unfortunately, this quite simple schema does not work if any of the parties continues searching for better offers simultaneously to negotiations. Offers on this type of market usually are not binding, and each party can break off the negotiations at any time, whenever it finds a more profitable solution, even when the negotiations are almost finished. So, submitting *accept-proposal* message does not confirm the transaction, it must be additionally committed using special protocol, which ensures that all the parties involved in the transaction accept the result of the negotiations.

### 5.3 Two-Phase Commit

An adaptation of two-phase commit protocol (2PC), well known in distributed OLTP (On-line Transaction Processing) systems, can be used as a method for accepting transactions in decentralized markets. It guaranties safe completion of transactions, ensuring that all involved parties ultimately accept the contract.

Figure 4 presents a two phase commit protocol.

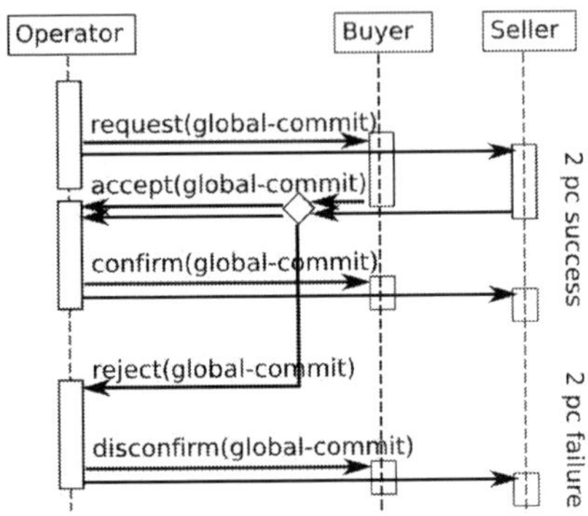


**Fig. 4.** Two Phase Commit Protocol

Unfortunately, FIPA standards do not directly define this protocol, but it can be quite easily defined on base of communication acts existing in the standards. In [10] an appropriate communication schema is proposed for bilateral transactions, but the 2PC protocol can also be used for multilateral ones.

Agent, which initiates completion of the transaction, sends *request(global-commit)* message to the other participants of the transaction. The parties should answer using *accept(global-commit)* message if they accept the transaction, or *reject(global-commit)* otherwise. If all of the parties accept the transaction, an initiating agent sends *confirm(global-commit)* message and the transaction is irrevocably accepted. If any of the parties rejects the transaction, a *disconfirm (global-commit)* is send and the transaction fails. If any of the parties does not answer in a given time, the transaction is timed-out and it also fails.

## 6 Communication Acts

Single communicative act contains some portion of control data, e.g. sender's and receiver's identifiers, type of the message (including semantic meaning), etc. This is defined in FIPA specifications, so we do not consider it in this paper. The body of the message contains the main content. This content can be understood differently, depending on used language and ontology [1]. In this section, the proposals for the content-language of particular messages are presented.

*Call For Proposal* This communicative act is used to report willingness to negotiate, or to the declaration that some kind of market/auction begins. Such communicative act should contain elements like, e.g., unique identifier of market/negotiations, kind of market/negotiations, time window on offers/proposal submitting, list of commodities, which are available for trade, etc.

We propose the following notation of these elements, using the XML notation (due to the fact that some items, as the commodities or the time windows, are defined by means of XML-based language M3-XML). An exemplary structure of the message is as follows:

```
(cfp
 :sender ex:Agent0
 :receiver ex:MorrisColumn01
 :content (
      <market id="ex:001" type="stock" quotation="continuous">
         <m3:calendar>
            <m3:CalendarPeriod id="ex:Y2009" periodType="ex:one-year"
                                 startTime="2009-01-01T00:00:00"
                                 endTime="2010-01-01T00:00:00"/>
         </m3:calendar>
         <m3:commodities>
            <m3:Commodity dref="op:ElectricEnergy" id="ex:eY2009">
              <m3:description>Energy on Copper Plate</m3:description>
              <m3:availableAt ref="op:CPNode" />
              <m3:CalendarScheduledCommodity ref="ex:Y2009" />
            </m3:Commodity>
         </m3:commodities>
      </market> )
 :language xml+m3
 :ontology EnergyMarket )
```

*Query-ref* Such communicative act is used by agents to submit a query to Morris column. Agent, which wants to obtain a list of advertisements, sends a query to the Morris column. Since the advertisements are XML documents, the most appropriate tool is XPath language [4] or XQuery language [5].

Let us assume that agent wants to query the Morris column about all advertisements, which comprise selling electric energy in negotiations process. An exemplary message, containing a query written in XPath language, is as follows:

```
(query-ref
 :sender ex:Agent2
 :receiver ex:MorrisColumn01
 :content ([/announces/announce/market[@type = 'negotiations' \
           and @offerType = 'ask']/m3:commodities/ \
           m3:Commodity[@dref = 'op:ElectricEnergy']] )
 :language xpath
 :ontology EnergyMarket
 :reply-with xpath:query02 )
```

A similar query in XQuery language is as follows:

```
for $x in doc("MorrisColumnAnnounces.xml")
  where $x/announces/market/@type = 'negotiations'
     and $x/announces/market/@offerType = 'ask'
     and $x/announces/market/m3:commodities/m3:Commodity/@dref =
                                                'op:ElectricEnergy'
return $x
```

*Inform-result* The content of this communicative act is a list of the announcements. As the advertisements are formed in the XML language, the answer contains a list of inform-results; each result is related to particular XML advertisement.

*Propose* This act contains an ask/bid (proposal in the negotiation process may also be regarded as a type of a deal). Thus, for this communicative act we propose to use M3-XML as a language for defining offers. An example of a message is presented below.

```
(propose
 :sender ex:Buyer
 :receiver ex:Operator
 :content (
      <m3:Offer id="ex:o12345-67" offeredPrice="-140.00">
        <m3:description>Buy offer</m3:description>
        <m3:offeredBy ref="ex:Buyer"/>
        <m3:volumeRange minValue="0" maxValue="100"/>
        <m3:ElementaryOffer>
         <m3:offeredCommodity shareFactor="-1"
                        ref="ex:energyOn2007-04-09T08:00:00"/>
        </m3:ElementaryOffer>
      </m3:Offer>)
 :language m3
 :ontology EnergyMarket
 :reply-with ex:o12345-67)
```

## 7 Summary

The presented MAS framework enables distributed trade activities by autonomous agents and, in the result, increases the efficiency of the market processes. To make it possible, several conceptual and technical issues must be resolved. In the paper the solutions of the most important issues are presented, that result in a flexible open system, ready for applications in a wide range of possible market processes, ranging from distributed, multilateral trade up to more centralized approaches. Our concept is based on general acknowledged FIPA standards with some minimal extensions required for achieving a desirable level of flexibility. We have also pointed out some open solutions and technical standards, developed in scientific groups, that are ready for transfer into some important practical areas. In further research works the proposed solution should be verified by a simulation.

## References


1. Foundation for Intelligent Physical Agents, `http://fipa.org/`
2. PowerWeb, `http://stealth.ee.cornell.edu/powerweb`
3. The Auction Agents for the Electric Power Industry, `http://www.agentbuilder.com/Documentation/EPRI`
4. XML Path Language (XPath), `http://www.w3.org/TR/xpath`
5. XQuery 1.0: An XML Query Language, `http://www.w3.org/TR/xquery/`
6. Bunn, D., Oliveira, F.S.: Agent-Based Simulation – An Application to the New Electricity Trading Arrangements of England and Wales. IEE Trans. Evolutionary Computation 5(5), 493–503 (2001)
7. Kacprzak, P., Kaleta, M., Pałka, P., Smolira, K., Toczyłowski, E., Traczyk., T.: Modeling distributed multilateral markets using Multi-commodity Market Model. In: Information Systems Architecture and Technology: Decision Making Models, pp. 15–22. Oficyna Wydawnicza Politechniki Wrocławskiej, Wrocław (2007)
8. McArthur, S., Davidson, E., Catterson, V., Dimeas, A., Hatziagyriou, N., Ponci, F., Funabashi, T.: Multi-Agent Systems for Power Engineering Applications - Part I: Concepts, Approaches and technical Challenges. IEEE Transactions on Power Systems 22(4), 1743–1752 (2007)
9. McArthur, S., Davidson, E., Catterson, V., Dimeas, A., Hatziagyriou, N., Ponci, F., Funabashi, T.: Multi-Agent Systems for Power Engineering Applications - Part II: Technologies, Standards and Tools for Building Multi-agent Systems. IEEE Transactions on Power Systems 22(4), 1753–1759 (2007)
10. Nimis, J., Lockemann, P.C.: Robust Multi-Agent Systems The Transactional Conversation Approach. In: SASEMAS 2004 – Third International Conference on Autonomous Agents and Multi-Agent Systems, New York (2004)
11. Praça, I., Ramos, C., Vale, Z., Cordeiro, M.: MASCEM: A Multiagent System that Simulates Competitive Electricity Markets. IEEE Inteligent Systems 6(18), 54–60 (2003)
12. Praça, I., Viamonte, M.J., Ramos, C., Vale, Z.: A Multi-Agent Market Simulator to Support Negotiation Decision Making (2008)
13. Villar, J., Rudnick, H.: Hydrothermal Market Simulator Using Game Theory: Assessment of Market Power. IEEE Trans. on Power Sys. 1(18), 91–98 (2003)
14. Multi-commodity Market Model ($M^3$) Project, `http://www.openM3.org`

# Comparative Analysis of Premises Valuation Models Using KEEL, RapidMiner, and WEKA

Magdalena Graczyk[1], Tadeusz Lasota[2], and Bogdan Trawiński[1]

[1] Wrocław University of Technology, Institute of Informatics,
Wybrzeże Wyspiańskiego 27, 50-370 Wrocław, Poland
[2] Wrocław University of Environmental and Life Sciences, Dept. of Spatial Management
Ul. Norwida 25/27, 50-375 Wroclaw, Poland
mag.graczyk@gmail.com, tadeusz.lasota@wp.pl,
bogdan.trawinski@pwr.wroc.pl

**Abstract.** The experiments aimed to compare machine learning algorithms to create models for the valuation of residential premises, implemented in popular data mining systems KEEL, RapidMiner and WEKA, were carried out. Six common methods comprising two neural network algorithms, two decision trees for regression, and linear regression and support vector machine were applied to actual data sets derived from the cadastral system and the registry of real estate transactions. A dozen of commonly used performance measures was applied to evaluate models built by respective algorithms. Some differences between models were observed.

**Keywords:** machine learning, property valuation, KEEL, RapidMiner, WEKA.

## 1 Introduction

Sales comparison approach is the most popular way of determining the market value of a property. When applying this method it is necessary to have transaction prices of the properties sold which attributes are similar to the one being appraised. If good comparable sales/purchase transactions are available, then it is possible to obtain reliable estimates. Prior to the evaluation the appraiser must conduct a thorough study of the appraised property using available sources of information such as cadastral systems, transaction registers, performing market analyses, accomplishing on-site inspection. His estimations are usually subjective and are based on his experience and intuition. Automated valuation models (AVMs), devoted to support appraisers' work, are based primarily on multiple regression analysis [8], [11], soft computing and geographic information systems (GIS) [14]. Many intelligent methods have been developed to support appraisers' works: neural networks [13], fuzzy systems [2], case-based reasoning [10], data mining [9] and hybrid approaches [6].

So far the authors have investigated several methods to construct models to assist with real estate appraisal: evolutionary fuzzy systems, neural networks and statistical algorithms using MATLAB and KEEL [4], [5].

N.T. Nguyen, R. Kowalczyk, and S.-M. Chen (Eds.): ICCCI 2009, LNAI 5796, pp. 800–812, 2009.

Three non-commercial data mining tools, developed in Java, KEEL [1], RapidMiner [7], and WEKA [12] were chosen to conduct tests. A few common machine learning algorithms including neural networks, decision trees, linear regression methods and support vector machines were included in these tools. Thus we decided to test whether these common algorithms were implemented similarly and allow to generate appraisal models with comparable prediction accuracy. Actual data applied to the experiments with these popular data mining systems came from the cadastral system and the registry of real estate transactions.

## 2 Cadastral Systems as the Source Base for Model Generation

The concept of data driven models for premises valuation, presented in the paper, was developed on the basis of sales comparison method. It was assumed that whole appraisal area, that means the area of a city or a district, is split into sections (e.g. clusters) of comparable property attributes. The architecture of the proposed system is shown in Fig. 1. The appraiser accesses the system through the internet and chooses an appropriate section and input the values of the attributes of the premises being evaluated into the system, which calculates the output using a given model. The final result as a suggested value of the property is sent back to the appraiser.

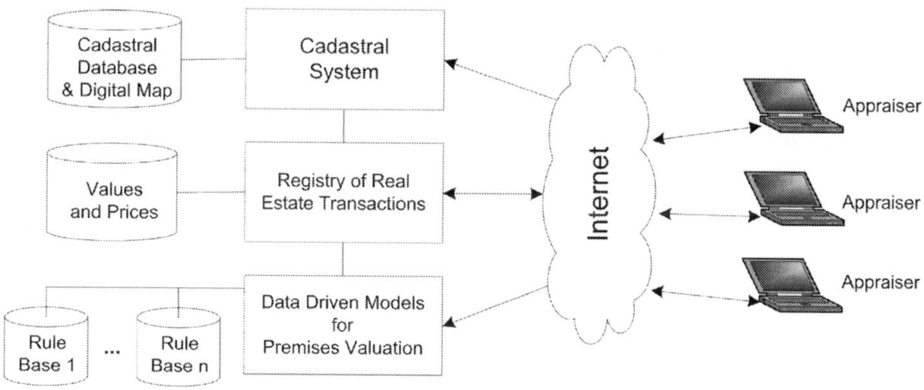


**Fig. 1.** Information systems to assist with real estate appraisals

Actual data used to generate and learn appraisal models came from the cadastral system and the registry of real estate transactions referring to residential premises sold in one of the big Polish cities at market prices within two years 2001 and 2002. The data set comprised 1098 cases of sales/purchase transactions. Four attributes were pointed out as price drivers: usable area of premises, floor on which premises were located, year of building construction, number of storeys in the building, in turn, price of premises was the output variable.

## 3 Data Mining Systems Used in Experiments

*KEEL (Knowledge Extraction based on Evolutionary Learning)* is a software tool to assess evolutionary algorithms for data mining problems including regression, classification, clustering, pattern mining, unsupervised learning, etc. [1]. It comprises evolutionary learning algorithms based on different approaches: Pittsburgh, Michigan, IRL (iterative rule learning), and GCCL (genetic cooperative-competitive learning), as well as the integration of evolutionary learning methods with different pre-processing techniques, allowing it to perform a complete analysis of any learning model.

*RapidMiner (RM)* is an environment for machine learning and data mining processes [7]. It is open-source, free project implemented in Java. It represents a new approach to design even very complicated problems - a modular operator concept which allows the design of complex nested operator chains for a huge number of learning problems. RM uses XML to describe the operator trees modeling knowledge discovery (KD) processes. RM has flexible operators for data input and output in different file formats. It contains more than 100 learning schemes for regression, classification and clustering tasks.

*WEKA (Waikato Environment for Knowledge Analysis)* is a non-commercial and open-source project [12]. WEKA contains tools for data pre-processing, classification, regression, clustering, association rules, and visualization. It is also well-suited for developing new machine learning schemes.

## 4 Regression Algorithms Used in Experiments

In this paper common algorithms for KEEL, RM, and WEKA were chosen. The algorithms represent the same approach to build regression models, but sometimes they have different parameters. Following KEEL, RM, and WEKA algorithms for building, learning and optimizing models were employed to carry out the experiments.

*MLP - MultiLayerPerceptron.* Algorithm is performed on networks consisting of multiple layers, usually interconnected in a feed-forward way, where each neuron on layer has directed connections to the neurons of the subsequent layer.

*RBF - Radial Basis Function Neural Network for Regression Problems.* The algorithm is based on feed-forward neural networks with radial activation function on every hidden layer. The output layer represents a weighted sum of hidden neurons signals.

*M5P.* The algorithm is based on decision trees, however, instead of having values at tree's nodes, it contains a multivariate linear regression model at each node. The input space is divided into cells using training data and their outcomes, then a regression model is built in each cell as a leaf of the tree.

*M5R - M5Rules.* The algorithm divides the parameter space into areas (subspaces) and builds in each of them a linear regression model. It is based on M5 algorithm. In each iteration a M5 Tree is generated and its best rule is extracted according to a given heuristic. The algorithm terminates when all the examples are covered.

*LRM - Linear Regression Model.* Algorithm is a standard statistical approach to build a linear model predicting a value of the variable while knowing the values of the other variables. It uses the least mean square method in order to adjust the parameters of the linear model/function.

*SVM - NU-Support Vector Machine.* Algorithm constructs support vectors in high-dimensional feature space. Then, hyperplane with the maximal margin is constructed. Kernel function is used to transform the data, which augments the dimensionality of the data. This augmentation provokes that the data can be separated with an hyperplane with much higher probability, and establish a minimal prediction probability error measure.

## 5 Plan of Experiments

The main goal of our study was to compare six algorithms for regression, which are common for KEEL, RM and WEKA. There were: multilayer perceptron, radial-basis-function networks, two types of model trees, linear regression, and support vector machine, and they are listed in Table 1. They were arranged in 4 groups: neural networks for regression (NNR), decision tree for regression (DTR), statistical regression model (SRM), and support vector machine (SVM).

**Table 1.** Machine learning algorithms used in study

| Type | Code | KEEL name | RapidMiner name | WEKA name |
|---|---|---|---|---|
| NNR | MLP | Regr-MLPerceptronConj-Grad | W-MultilayerPerceptron | MultilayerPerceptron |
| | RBF | Regr-RBFN | W-RBFNetwork | RBFNetwork |
| DTR | M5P | Regr-M5 | W-M5P | M5P |
| | M5R | Regr-M5Rules | W-M5Rules | M5Rules |
| SRM | LRM | Regr-LinearLMS | LinearRegression | LinearRegression |
| SVM | SVM | Regr-NU_SVR | LibSVM | LibSVM |

Optimal parameters were selected for every algorithm to get the best result for the dataset by means of trial and error method. Having determined the best parameters of respective algorithms, final experiments were carried out in order to compare predictive accuracy of models created using all six selected algorithms in KEEL, RM, and WEKA. Schema of the experiments is depicted in Figure 2. All the experiments were run for our set of data using 10-fold cross validation. In order to obtain comparable results, the normalization of data was performed using the min-max approach. As fitness function the mean square error (MSE) programmed in KEEL, and root mean square error (RMSE) implemented in WEKA and RM were used ($MSE = RMSE^2$). Nonparametric Wilcoxon signed-rank test was employed to evaluate the outcome. A dozen of commonly used performance measures [3], [12] were applied to evaluate models built by respective algorithms used in our study. These measures are listed in Table 2 and expressed in the form of following formulas below, where $y_i$ denotes actual price and $\hat{y}_i$ – predicted price of i-th case, *avg(v), var(v), std(v)* – average, variance, and standard deviation of variables $v_1, v_2, \ldots, v_N$, respectively and $N$ – number of cases in the testing set.

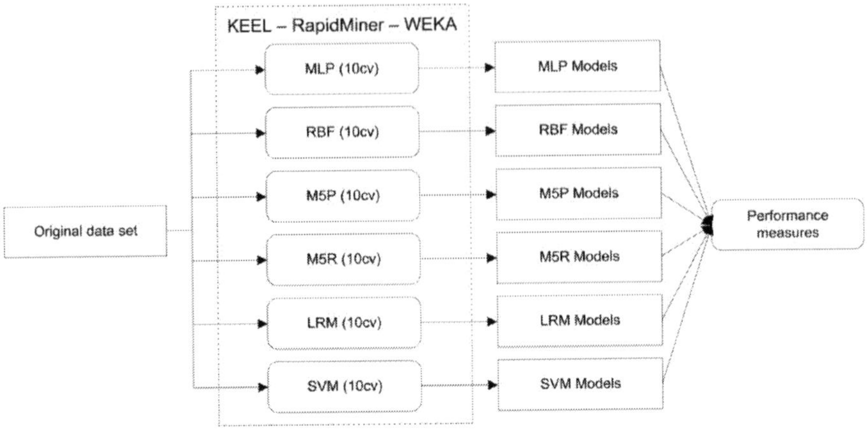


**Fig. 2.** Schema of the experiments with KEEL, RapidMiner, and WEKA

**Table 2.** Performance measures used in study

| Denot. | Description | Dimen-sion | Min value | Max value | Desirable outcome | No. of form. |
|---|---|---|---|---|---|---|
| *MSE* | Mean squared error | $d^2$ | 0 | ∞ | min | 1 |
| *RMSE* | Root mean squared error | d | 0 | ∞ | min | 2 |
| *RSE* | Relative squared error | no | 0 | ∞ | min | 3 |
| *RRSE* | Root relative squared error | no | 0 | ∞ | min | 4 |
| *MAE* | Mean absolute error | d | 0 | ∞ | min | 5 |
| *RAE* | Relative absolute error | no | 0 | ∞ | min | 6 |
| *MAPE* | Mean absolute percent. error | % | 0 | ∞ | min | 7 |
| *NDEI* | Non-dimensional error index | no | 0 | ∞ | min | 8 |
| *r* | Linear correlation coefficient | no | -1 | 1 | close to 1 | 9 |
| $R^2$ | Coefficient of determination | % | 0 | ∞ | close to 100% | 10 |
| *var(AE)* | Variance of absolute errors | $d^2$ | 0 | ∞ | min | 11 |
| *var(APE)* | Variance of absolute percentage errors | no | 0 | ∞ | min | 12 |

$$MSE = \frac{1}{N}\sum_{i=1}^{N}(y_i - \hat{y}_i)^2 \tag{1}$$

$$RMSE = \sqrt{\frac{1}{N}\sum_{i=1}^{N}(y_i - \hat{y}_i)^2} \tag{2}$$

$$RSE = \frac{\sum_{i=1}^{N}(y_i - \hat{y}_i)^2}{\sum_{i=1}^{N}(y_i - avg(y))^2} \tag{3}$$

$$RRSE = \sqrt{\frac{\sum_{i=1}^{N}(y_i - \hat{y}_i)^2}{\sum_{i=1}^{N}(y_i - avg(y))^2}} \quad (4)$$

$$MAE = \frac{1}{N}\sum_{i=1}^{N}|y_i - \hat{y}_i| \quad (5)$$

$$RAE = \frac{\sum_{i=1}^{N}|y_i - \hat{y}_i|}{\sum_{i=1}^{N}|y_i - avg(y)|} \quad (6)$$

$$MAPE = \frac{1}{N}\sum_{i=1}^{N}\frac{|y_i - \hat{y}_i|}{y_i} * 100\% \quad (7)$$

$$NDEI = \frac{RMSE}{std(y)} \quad (8)$$

$$r = \frac{\sum_{i=1}^{N}(y_i - avg(y))\,(\hat{y}_i - avg(\hat{y}))}{\sqrt{\sum_{i=1}^{N}(y_i - avg(y))^2}\,\sqrt{\sum_{i=1}^{N}(\hat{y}_i - avg(\hat{y}))^2}} \quad (9)$$

$$R^2 = \frac{\sum_{i=1}^{N}(\hat{y}_i - avg(y))^2}{\sum_{i=1}^{N}(y_i - avg(y))^2} * 100\% \quad (10)$$

$$var(AE) = var(|y - \hat{y}|) \quad (11)$$

$$var(APE) = var(\frac{|y - \hat{y}|}{y}) \quad (12)$$

## 6 Results of Experiments

The performance of the models built by all six algorithms for respective measures and for KEEL, RM, and WEKA systems was presented in Fig. 3-14. For clarity, all measures were calculated for normalized values of output variables except for MAPE, where in order to avoid the division by zero, actual and predicted prices had to be denormalized. It can be observed that, some measures, especially those based on square errors reveal similar relationships between model performance. Most of the models provided similar values of error measures, besides the one created by MLP algorithm implemented in RapidMiner. Its worst performance can be seen particularly in Figures 8, 9, 11, and 13.

Fig. 9 depicts that the values of MAPE range from 16.2% to 19.3%, except for MLP in RapidMiner with 25.3%, what is a fairly good result, especially when you take into account, that no all price drivers were available in our sources of experimental data.

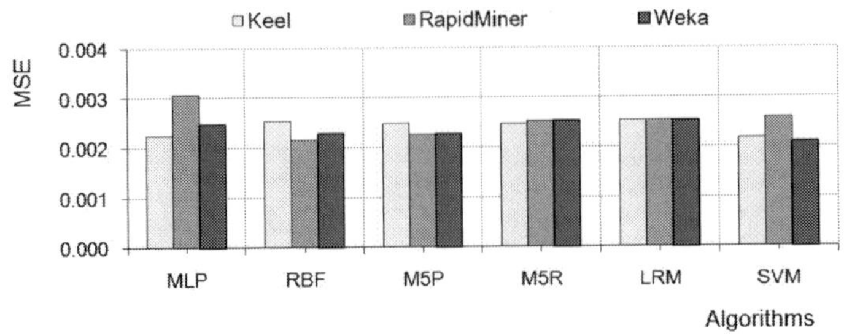


**Fig. 3.** Comparison of MSE values

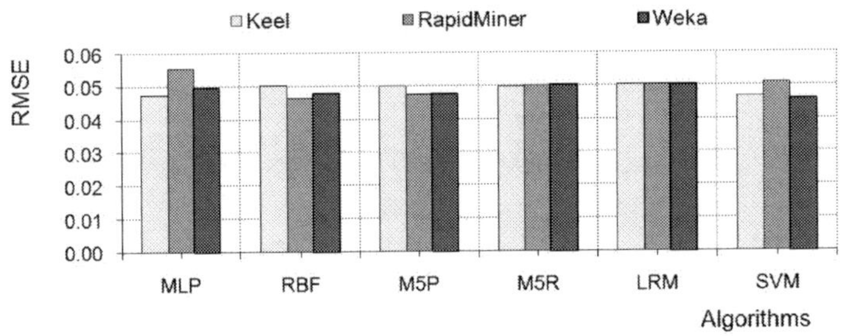


**Fig. 4.** Comparison of RMSE values

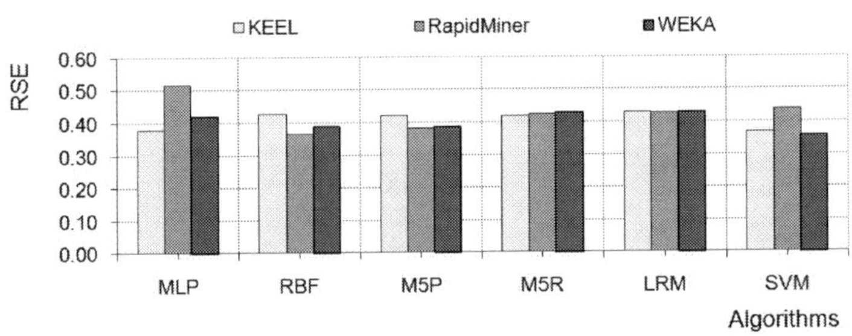


**Fig. 5.** Comparison of RSE values

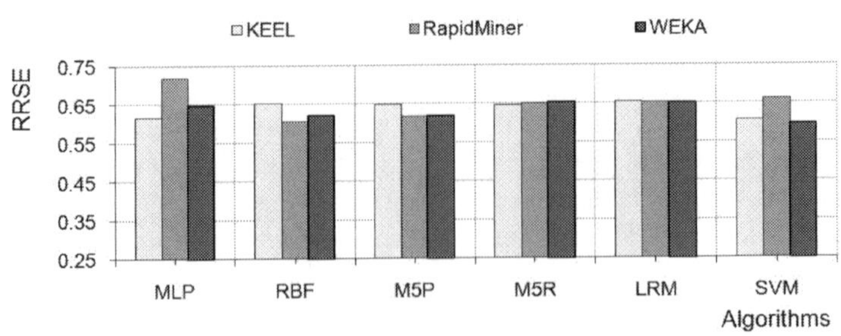


**Fig. 6.** Comparison of RRSE values

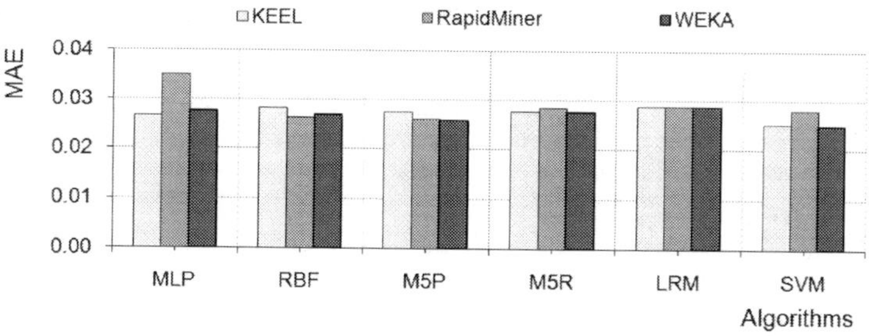


**Fig. 7.** Comparison of MAE values

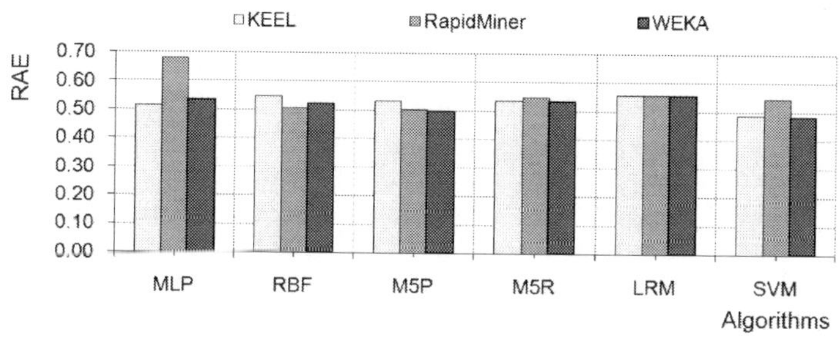


**Fig. 8.** Comparison of RAE values

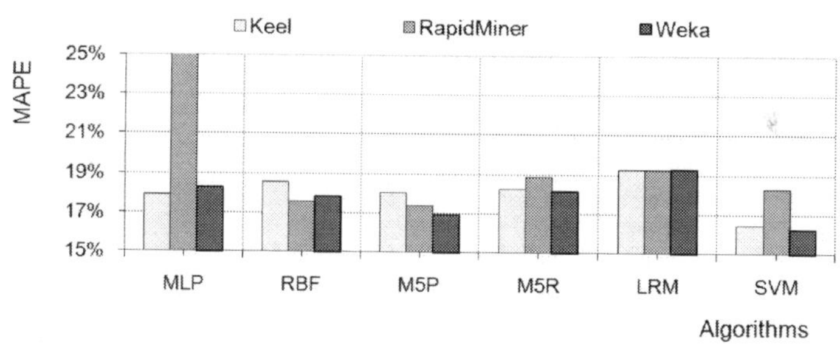


**Fig. 9.** Comparison of MAPE values

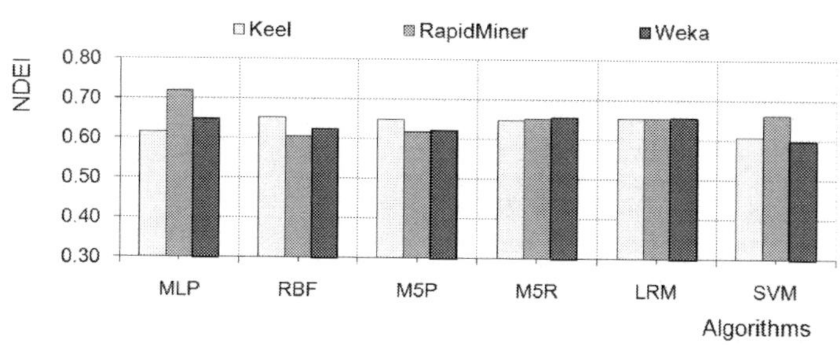


**Fig. 10.** Comparison of NDEI values

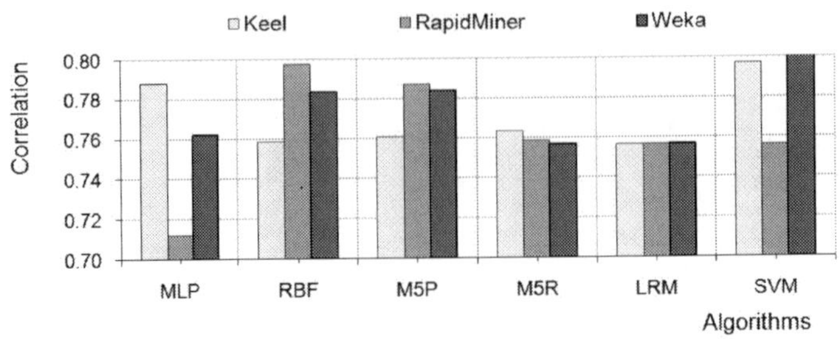


**Fig. 11.** Correlation between predicted and actual prices

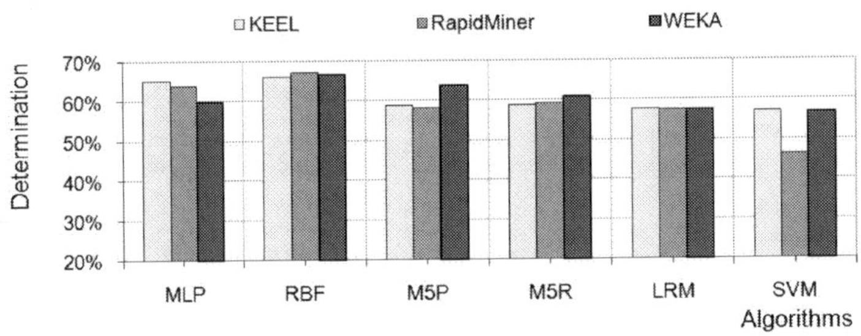


**Fig. 12.** Comparison of $R^2$ - determination coefficient values

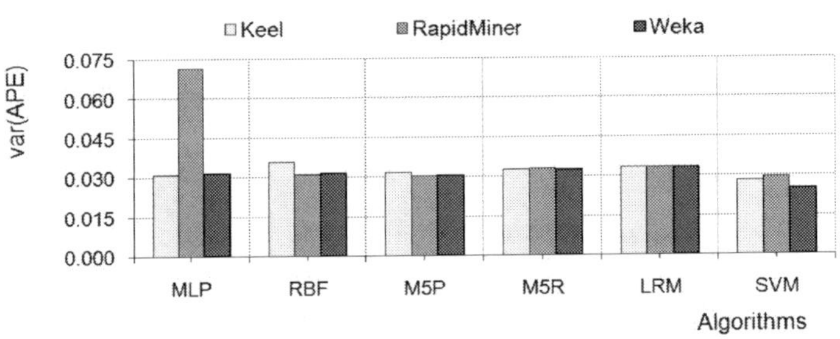


**Fig. 13.** Variance of absolute percentage errors - Var(APE)

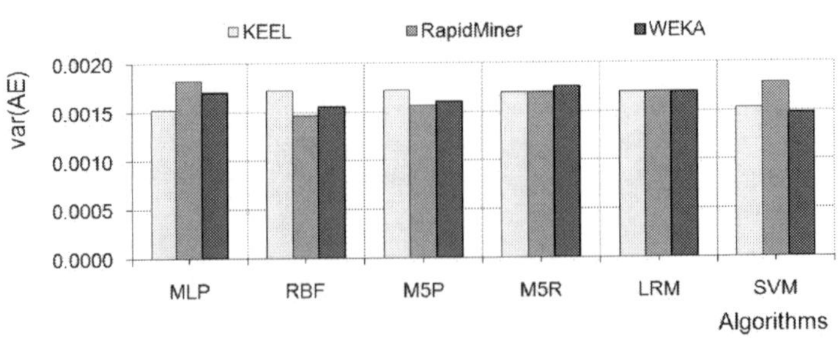


**Fig. 14.** Variance of absolute errors - Var(AE)

High correlation between actual and predicted prices for each model, ranging from 71.2% to 80.4%, was shown in Fig. 11. In turn, the coefficients of determination, presented in Fig. 12, indicate that from 46.2% to 67.2% of total variation in the dependent variable (prices) is accounted for by the models. This can be explained that data derived from the cadastral system and the register of property values and prices cover only some part of potential price drivers.

The nonparametric Wilcoxon signed-rank tests were carried out for two commonly used measures: MSE and MAPE. The results are shown in Tables 3 and 4, where a triple in each cell, eg <NNN>, reflects the outcome for a given pair of models and for KEEL, RM, and WEKA respectively. N - denotes that there are no differences in mean values of respective errors, and Y - indicates that there are statistically significant differences between particular performance measures. For clarity Tables 3 and 4 were presented in form of symmetric matrices.

Almost in all cases but one SVM revealed significantly better performance than other algorithms, whereas LRM turned out to be worse.

**Table 3.** Results of Wilcoxon signed-rank test for squared errors comprised by MSE

| Model | MLP | RBF | M5P | M5R | LRM | SVM |
|---|---|---|---|---|---|---|
| MLP | - | YYY | YYY | YYN | YYY | YYY |
| RBF | YYY | - | NNY | NYN | YYY | YYY |
| M5P | YYY | NNY | - | NYY | YYY | YYN |
| M5R | YYN | NYN | NYY | - | YNY | YYY |
| LRM | YYY | YYY | YYY | YNY | - | YYY |
| SVM | YYY | YYY | YYN | YYY | YYY | - |

**Table 4.** Results of Wilcoxon test for absolute percentage errors comprised by MAPE

| Model | MLP | RBF | M5P | M5R | LRM | SVM |
|---|---|---|---|---|---|---|
| MLP | - | NYY | NYY | NYN | YYY | YYY |
| RBF | NYY | - | NNY | NYN | YYY | YYY |
| M5P | NYY | NNY | - | NYY | YYY | YYN |
| M5R | NYN | NYN | NYY | - | YNY | YYY |
| LRM | YYY | YYY | YYY | YNY | - | YYY |
| SVM | YYY | YYY | YYN | YYY | YYY | - |

Due to the non-decisive results of statistical tests for other algorithms, rank positions of individual algorithms were determined for each measure (see Table 5). It can be noticed that highest rank positions gained SVM, RBF, and M5P algorithms and the lowest LRM, M5R, and MLP. There are differences in rankings made on the basis of the performance of algorithms within respective data mining systems.

In order to find out to what extent the three data mining systems produce uniform models for the same algorithms the nonparametric Wilcoxon signed-rank tests were carried out for three measures: MSE, MAE and MAPE. The results are shown in Table 6, where a triple in each cell, eg <NNN>, reflects outcome for each pair of models created by KEEL-RM, KEEL-WEKA, and RM-WEKA respectively. N - denotes that there are no differences in mean values of respective errors, and Y - indicates that there are statistically significant differences between particular performance measures. Some general conclusions can be drawn when analysing Table 6. For LRM models there is

**Table 5.** Rank positions of algorithms with respect to performance measures (1 means the best)

| Measure | Tool | MLP | RBF | M5P | M5R | LRM | SVM |
|---|---|---|---|---|---|---|---|
| MSE | KEEL | 2 | 5 | 4 | 3 | 6 | 1 |
| MSE | RM | 6 | 1 | 2 | 3 | 4 | 5 |
| MSE | WEKA | 4 | 3 | 2 | 6 | 5 | 1 |
| RMSE | KEEL | 2 | 5 | 4 | 3 | 6 | 1 |
| RMSE | RM | 6 | 1 | 2 | 3 | 4 | 5 |
| RMSE | WEKA | 4 | 3 | 2 | 6 | 5 | 1 |
| RSE | KEEL | 3 | 1 | 5 | 4 | 6 | 2 |
| RSE | RM | 6 | 4 | 1 | 2 | 3 | 5 |
| RSE | WEKA | 4 | 3 | 2 | 6 | 5 | 1 |
| RRSE | KEEL | 3 | 1 | 5 | 4 | 6 | 2 |
| RRSE | RM | 6 | 4 | 1 | 2 | 3 | 5 |
| RRSE | WEKA | 4 | 3 | 2 | 6 | 5 | 1 |
| MAE | KEEL | 3 | 2 | 4 | 5 | 6 | 1 |
| MAE | RM | 6 | 3 | 1 | 4 | 5 | 2 |
| MAE | WEKA | 5 | 3 | 2 | 4 | 6 | 1 |
| RAE | KEEL | 3 | 2 | 4 | 5 | 6 | 1 |
| RAE | RM | 6 | 3 | 1 | 4 | 5 | 2 |
| RAE | WEKA | 5 | 3 | 2 | 4 | 6 | 1 |
| r | KEEL | 3 | 1 | 5 | 4 | 6 | 2 |
| r | RM | 6 | 2 | 1 | 3 | 4 | 5 |
| r | WEKA | 4 | 3 | 2 | 5 | 6 | 1 |
| MAPE | KEEL | 3 | 2 | 4 | 5 | 6 | 1 |
| MAPE | RM | 6 | 3 | 1 | 4 | 5 | 2 |
| MAPE | WEKA | 5 | 3 | 2 | 4 | 6 | 1 |
| NDEI | KEEL | 2 | 5 | 4 | 3 | 6 | 1 |
| NDEI | RM | 6 | 1 | 2 | 3 | 4 | 5 |
| NDEI | WEKA | 4 | 3 | 2 | 6 | 5 | 1 |
| r | KEEL | 3 | 1 | 5 | 4 | 6 | 2 |
| r | RM | 6 | 2 | 1 | 3 | 4 | 5 |
| r | WEKA | 4 | 3 | 2 | 5 | 6 | 1 |
| $R^2$ | KEEL | 5 | 6 | 3 | 4 | 2 | 1 |
| $R^2$ | RM | 6 | 5 | 3 | 4 | 2 | 1 |
| $R^2$ | WEKA | 3 | 6 | 5 | 4 | 2 | 1 |
| Var(AE) | KEEL | 2 | 1 | 6 | 4 | 5 | 3 |
| Var(AE) | RM | 6 | 4 | 1 | 3 | 2 | 5 |
| Var(AE) | WEKA | 5 | 2 | 3 | 6 | 4 | 1 |
| Var(APE) | KEEL | 3 | 2 | 4 | 5 | 6 | 1 |
| Var(APE) | RM | 6 | 5 | 2 | 3 | 4 | 1 |
| Var(APE) | WEKA | 4 | 3 | 2 | 5 | 6 | 1 |
| median | | 4.00 | 3.00 | 2.00 | 4.00 | 6.00 | 1.00 |
| average | | 4.32 | 2.58 | 2.63 | 4.16 | 5.42 | 1.89 |
| min | | 2 | 1 | 1 | 3 | 4 | 1 |
| max | | 6 | 5 | 5 | 6 | 6 | 5 |

**Table 6.** Results of Wilcoxon test for common algorithms in KEEL, RM, and WEKA

| Measure | MLP | RBF | M5P | M5R | LRM | SVM |
|---|---|---|---|---|---|---|
| MSE | YNY | YNN | YNN | YNN | NNN | YNY |
| MAE | YNY | YNN | YNN | YNN | NNN | YNY |
| MAPE | YNY | YNN | YNN | YNN | NNN | YNY |

no significant difference in performance, the same applies to all models built by KEEL and WEKA. For each model created by means of KEEL and RapidMiner there are significant differences in prediction accuracy.

## 7 Conclusions and Future Work

The experiments aimed to compare machine learning algorithms to create models for the valuation of residential premises, implemented in popular data mining systems KEEL, RapidMiner and WEKA, were carried out. Six common methods were applied to actual data sets derived from the cadastral systems, therefore it was naturally to expect that the same algorithms implemented in respective systems will produce similar results. However, this was true only for KEEL and WEKA systems, and for linear regression method. For each algorithm there were significant differences between KEEL and RapidMiner.

Some performance measures provide the same distinction abilities of respective models, thus it can be concluded that in order to compare a number of models it is not necessary to employ all measures, but the representatives of different groups.

MAPE obtained in all tests ranged from 16% do 25%. This can be explained that data derived from the cadastral system and the register of property values and prices can cover only some part of potential price drivers. Physical condition of the premises and their building, their equipment and facilities, the neighbourhood of the building, the location in a given part of a city should also be taken into account, moreover overall subjective assessment after inspection in site should be done. Therefore we intend to test data obtained from public registers and then supplemented by experts conducting on-site inspections and evaluating more aspects of properties being appraised. Moreover further investigations of multiple models comprising ensembles using bagging and boosting techniques is planned.

## References

1. Alcalá-Fdez, J., Sánchez, L., García, S., del Jesus, M.J., Ventura, S., Garrell, J.M., Otero, J., Romero, C., Bacardit, J., Rivas, V.M., Fernández, J.C., Herrera., F.: KEEL: A Software Tool to Assess Evolutionary Algorithms to Data Mining Problems. Soft Computing 13(3), 307–318 (2009)
2. González, M.A.S., Formoso, C.T.: Mass appraisal with genetic fuzzy rule-based systems. Property Management 24(1), 20–30 (2006)
3. Hagquist, C., Stenbeck, M.: Goodness of Fit in Regression Analysis – $R^2$ and $G^2$ Reconsidered. Quality & Quantity 32, 229–245 (1998)
4. Król, D., Lasota, T., Trawiński, B., Trawiński, K.: Investigation of evolutionary optimization methods of TSK fuzzy model for real estate appraisal. International Journal of Hybrid Intelligent Systems 5(3), 111–128 (2008)
5. Lasota, T., Pronobis, E., Trawiński, B., Trawiński, K.: Exploration of Soft Computing Models for the Valuation of Residential Premises using the KEEL Tool. In: Nguyen, N.T., et al. (eds.) 1st Asian Conference on Intelligent Information and Database Systems (ACIIDS 2009), pp. 253–258. IEEE, Los Alamitos (2009)

6. McCluskey, W.J., Anand, S.: The application of intelligent hybrid techniques for the mass appraisal of residential properties. Journal of Property Investment and Finance 17(3), 218–239 (1999)
7. Mierswa, I., Wurst, M., Klinkenberg, R., Scholz, M., Euler, T.: YALE: Rapid Prototyping for Complex Data Mining Tasks. In: Proceedings of the 12th ACM SIGKDD International Conference on Knowledge Discovery and Data Mining (KDD-2006), pp. 935–940 (2006)
8. Nguyen, N., Cripps, A.: Predicting housing value: A comparison of multiple regression analysis and artificial neural networks. J. of Real Estate Res. 22(3), 3131–3336 (2001)
9. Soibelman, W.L., González, M.A.S.: A Knowledge Discovery in Databases Framework for Property Valuation. J. of Property Tax Assessment and Admin. 7(2), 77–106 (2002)
10. Taffese, W.Z.: Case-based reasoning and neural networks for real state valuation. In: Proceedings of the 25th IASTED International Multi-Conference: Artificial Intelligence and Applications, Innsbruck, Austria (2007)
11. Waller, B.D., Greer, T.H., Riley, N.F.: An Appraisal Tool for the 21st Century: Automated Valuation Models. Australian Property Journal 36(7), 636–641 (2001)
12. Witten, I.H., Frank, E.: Data Mining: Practical machine learning tools and techniques, 2nd edn. Morgan Kaufmann, San Francisco (2005)
13. Worzala, E., Lenk, M., Silva, A.: An Exploration of Neural Networks and Its Application to Real Estate Valuation. J. of Real Estate Res. 10(2), 185–201 (1995)
14. Wyatt, P.: The development of a GIS-based property information system for real estate valuation. Int. J. Geographical Information Science 111(5), 435–450 (1997)

# A Multi-agent System to Assist with Real Estate Appraisals Using Bagging Ensembles

Tadeusz Lasota[1], Zbigniew Telec[2], Bogdan Trawiński[2], and Krzysztof Trawiński[3]

[1] Wrocław University of Environmental and Life Sciences, Dept. of Spatial Management
Ul. Norwida 25/27, 50-375 Wroclaw, Poland
[2] Wrocław University of Technology, Institute of Informatics,
Wybrzeże Wyspiańskiego 27, 50-370 Wrocław, Poland
[3] European Centre for Soft Computing, Edificio Científico-Tecnológico, 3ª Planta,
C. Gonzalo Gutiérrez Quirós S/N, 33600 Mieres, Asturias, Spain

**Abstract.** The multi-agent system for real estate appraisals MAREA was extended to include aggregating agents, which could create ensemble models applying the bagging approach, was presented in the paper. The major part of the study was devoted to investigate to what extent bagging approach could lead to the improvement of the accuracy machine learning regression models. Four algorithms implemented in the KEEL tool, including linear regression, decision trees for regression, support vector machines, and artificial neural network of MLP type, were used in the experiments. The results showed that bagging ensembles ensured higher prediction accuracy than single models.

## 1 Introduction

In the last ten years the ensemble learning systems gained a large attention of the researchers. This technique combine the output of machine learning algorithms, called "weak learners", in order to get smaller prediction errors (in regression) or lower error rates (in classification). The individual estimator must provide different patterns of generalization, thus in the training process diversity is employed. Otherwise, the ensemble would be composed of the same predictors and would provide as good accuracy as the single one. It has been proved that the ensemble performs better when each individual machine learning system is accurate and makes errors on different examples [19], [20], [22], [27].

Although, there are many taxonomies for that, there is one recognized group the so-called data resampling, which generates different training sets to obtain unique regressor or classificator. To this group we may include bagging [5], [6] and boosting [10], [11], [15], [16], [30], [31]. In boostrap aggregation (bagging), each machine learning system is independently learned on resampled training set, which is randomly picked from the original samples of the training set. Hence, bagging is devoted to the unstable algorithms, where the small changes in the training set, result in large changes in the output of that system. Training of each predictor could be in parallel, in due to the independence of training of each machine. In turn, boosting provides sequential learning of the predictors. The first one is learned on the whole

N.T. Nguyen, R. Kowalczyk, and S.-M. Chen (Eds.): ICCCI 2009, LNAI 5796, pp. 813–824, 2009.

data set, while the following are learned on the training set based on the performance of the previous one. In other words, the examples which were predicted improperly are noted. Then, such examples are more probable to appear in the training set of the next predictor. It results with different machines being specialized in predicting some areas of the dataset.

Although the ensemble learning systems are more common in the classification problems, they found some applications in the regression field of machine learning. Bagging has been applied to a large number of model structures, i.e. decision trees [8], regression with variable selection [33], and neural networks [18], [35]. In the literature several types of bagging have been offered: "nice" bagging [32], subbagging the so-called subsample aggregating [9], and iterated bagging [7]. In regression, to combine generated predictors in the ensemble, weighted average [13], [19], [34] is usually used Apart from that, studies on weighted median [2], [3], [9], [12], [13], [21] were also carried out. Traditional average or median are the simplest approaches to do so.

In our previous work [24] we outlined the concept of a multi-agent system for real estate appraisals, called MAREA, employing six different machine learning algorithms. The main purpose of this paper is to extend the design of the MAREA system to include aggregating agents, which create ensemble models applying the bagging procedure to single ones. The next goal was to investigate to what extent bagging approach could lead to the improvement of the accuracy machine learning regression models devoted to assist with real estate appraisals.

## 2 Multi-agent System for Real Estate Appraisal

The general architecture of the extended multi-agent system for real estate appraisals (MAREA) is presented in Figure 1. Source databases are heterogeneous and distributed, they comprise a cadastral database, cadastral digital map and registry of real estate sales/purchase transactions. The source databases are continuously updated by an information centre located at a self-government office. Data agents retrieve data necessary to build appraisal models, cleanse and integrate them and save into an appraisal data warehouse. Learning agents create data driven appraisal models from scratch or update existing ones basing on the analysis of data included in the warehouse. Interface, query and result agents serve the appraisers.

*Interface Agents.* Interface agents interact with appraisers, i.e. system users, gather parameters of properties being appraised and present suggested prices of properties exposed to valuation.

*Query Agents.* Query agents are generated at each demand of a user. They apply to available models assisting with real estate appraisals. These models are used to generate the suggested prices of properties being appraised.

*Result Agents.* Result agents process the results provided by the models assisting with valuation and prepare suggested prices to be presented to the users.

*Data Agents.* Data agents process data gathered in source databases. They extract, cleanse and integrate data preparing them to be useful to learn real estate appraisal models. They create and maintain a certain kind of appraisal data warehouse.

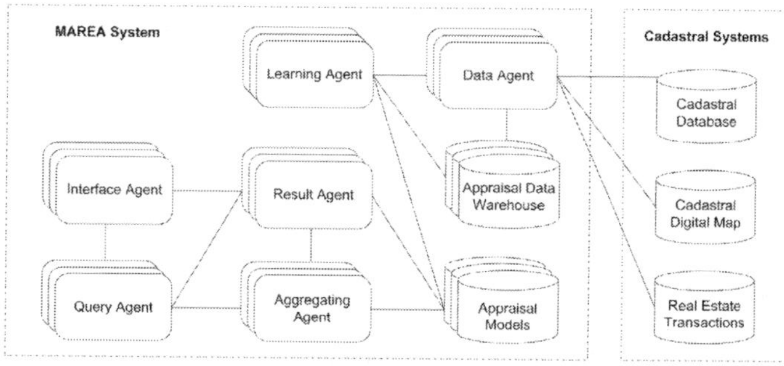


**Fig. 1.** General architecture of the MAREA system

*Learning Agents.* Learning agents perform overall and supplementary learning, creating in result different data driven models for real estate appraisal.

*Aggregating Agents.* Aggregating agents create ensemble models.

The JADE platform supplies mechanisms ready for communication between agents and their collaboration and mobility [4]. JADE's advantage is also possibility to create distributed systems working on many computers at once. Architecture of the MAREA system in the JADE platform is depicted in Figure 2. A MAREA system construction is brought to proper implementation of programming agents in the JADE environment. Each agent is attributed with different role during creation of MAREA system on the JADE platform: A1 – Interface Agents, A2 – Query Agents, A3 – Result Agents, A4 – Learning Agents using linear regression algorithm, A5 – Learning Agents using decision trees for regression, A6 – Learning Agents using support vector machines for regression, A7 – Learning Agents using multilayer perceptron for modelling, A8 – Aggregating Agents, and A9 – Data Agents.

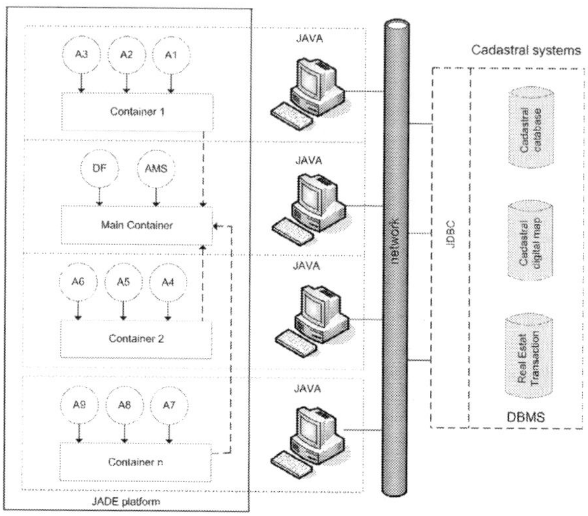


**Fig. 2.** Architecture of the MAREA system in the JADE platform

## 3 Machine Learning Algorithms Used in Experiments

KEEL is a non-commercial Java software tool designed to assess various algorithms for regression, classification, clustering, pattern mining problems [1]. KEEL contains several dozen of algorithms for data pre-processing, designing and conducting the experiments, data post-processing and evaluating and visualizing the results obtained, which have been bound into one flexible and user friendly system. The great advantage of KEEL is the possibility to create different experimental sets of data automatically and to perform cross-validation of learned models within the same process, what substantially decreases time needed to prepare and accomplish experiments and allows to avoid or diminish the threat of model overfitting. Previous investigations by the authors of this paper [23] proved the appropriateness of KEEL to build and evaluate data driven models for real estate appraisal. Four algorithms implemented in KEEL, including linear regression, decision trees for regression, support vector machines, and artificial neural network of MLP type, were applied to carry out our experiments. They are listed in Table 1:

**Table 1.** KEEL algorithms used in study

| Code | KEEL name | Description |
|---|---|---|
| LRM | Regr-LinearLMS | Linear regression models |
| DTR | Regr-M5 | Decision trees for regression |
| SVM | Regr-NU_SVR | Support vector machines for regression |
| ANN | Regr-MLPerceptronConj-Grad | Multilayer perceptron for modeling |

*Regr-LinearLMS (LRM).* Linear regression method is a standard statistical approach to build a linear model predicting a value of the variable while knowing the values of the other variables. It uses least mean squares in order to adjust the parameters of the linear model/function [29].

*Regr-M5 (DTR).* The M5 model tree is a system solving regression problems. It is based on decision tree approach which can learn efficiently being capable to solve tasks with high dimensionality. As it is in decision trees, it builds tree based model, however instead of having values at their nodes it contains multivariate linear regression models at each node. The main advantage of M5 approach over traditional regression trees is that model trees are much smaller than regression trees [28].

*Regr-NU_SVR (SVM).* The SVM (Support Vector Machine) model uses the sequential minimal optimization training algorithm and treats a given problem in terms of solving a quadratic optimization problem. The NU-SVR, called also v-SVM, for regression problems is an extension of the traditional SVM and it aims to build a loss function [14].

*MLPerceptronConj-Grad (ANN).* Proposed by Moller [25] conjugate gradient based algorithm is an approach for supervised learning of the neural nets avoiding a time consuming line-search. Algorithm is performed on networks consisting of multiple layers, where each neuron on layer has directed connections to the neurons of the subsequent layer.

## 3 Results of Experiments

The goal of research reported was to investigate to what extent bagging approach led to the improvement of the accuracy machine learning regression models devoted to assist with real estate appraisals. Actual data used to generate and learn appraisal models came from the cadastral system and the registry of real estate transactions referring to residential premises sold in one of the big Polish cities at market prices within two years 2001 and 2002. They constituted original data set of 1098 cases of sales/purchase transactions. Four attributes were pointed out as price drivers: usable area of premises, floor on which premises are located, year of construction, number of storeys in the building, in turn, price of premises established the output variable.

Schema of the experiments is depicted in Figure 3. On the basis of the original data set 28 bootstrap replicates, called also bags, were created by drawing with replacement the elements from original data set. When determining the number of bags we followed Breiman [5], who stated that about 25 replicates seemed reasonable to obtain satisfactory ensemble. These replicates were used to generate models employing LMR, DTR, SVM and ANN regression algorithms implemented in KEEL. Normalization of data was performed using the min-max approach. As fitness function the mean square error (MSE) was applied and 10-fold cross validation (10cv) was accomplished. In consequence 112 different models were created. As aggregation functions simple averages and weighted averages of mean squared error obtained for test sets were used.

Descriptive statistics of the results obtained for individual bootstrap sets were presented in Table 2. The lowest median values of MSE were provided by SVM and ANN models and they were smaller than MSE achieved by the models constructed using original data set. By contrast, the LRM models revealed the highest value of MSE median. On the other hand the DTR models were characterized by the highest variance of MSE. In turn, the performance of models created by LRM, DTR, SVM and ANN algorithms for individual bootstrap sets against the MSE obtained for the model built over original data set are shown in Figures 4, 5, 6 and 7 respectively.

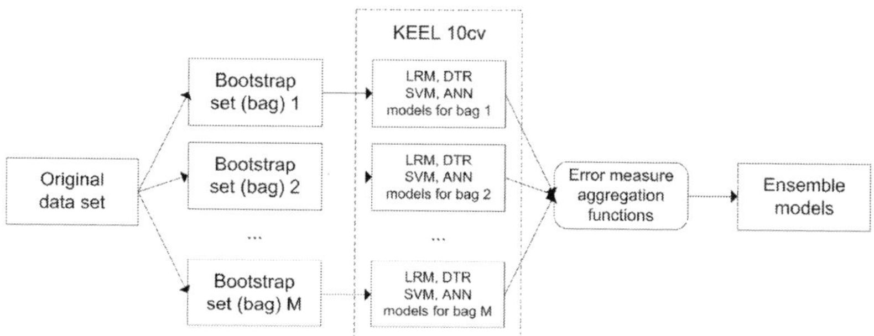


**Fig. 3.** Schema of bagging ensemble model development

**Table 2.** Descriptive statistics of performance (MSE) of individual bootstrap replicates

| Bootstrap sets | LRM | DTR | SVM | ANN |
|---|---|---|---|---|
| min | 0.00376 | 0.00385 | 0.00349 | 0.00346 |
| max | 0.00135 | 0.00117 | 0.00105 | 0.00112 |
| median | 0.00232 | 0.00209 | 0.00186 | 0.00198 |
| avg | 0.00243 | 0.00226 | 0.00203 | 0.00215 |
| var* | 0.44607 | 0.60825 | 0.41737 | 0.44531 |
| Original set | **0.00253** | **0.00252** | **0.00216** | **0.00232** |

* *For clarity of presentation actual values of variance were multiplied by $10^6$*

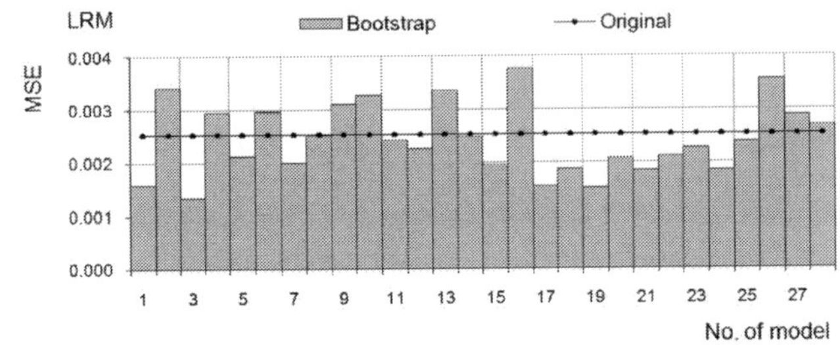


**Fig. 4.** Performance of individual bootstrap models created by LRM

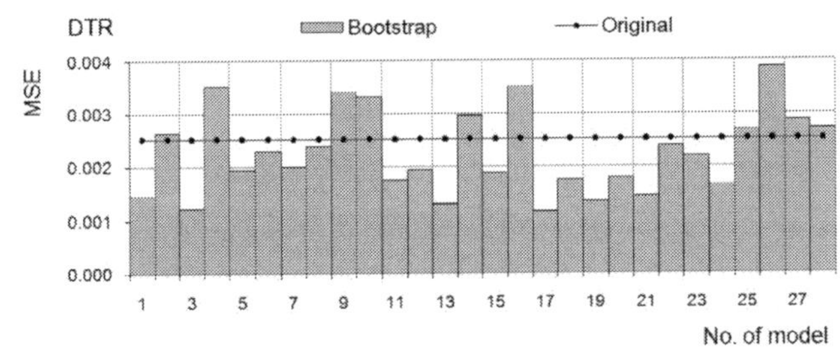


**Fig. 5.** Performance of individual bootstrap models created by DTR

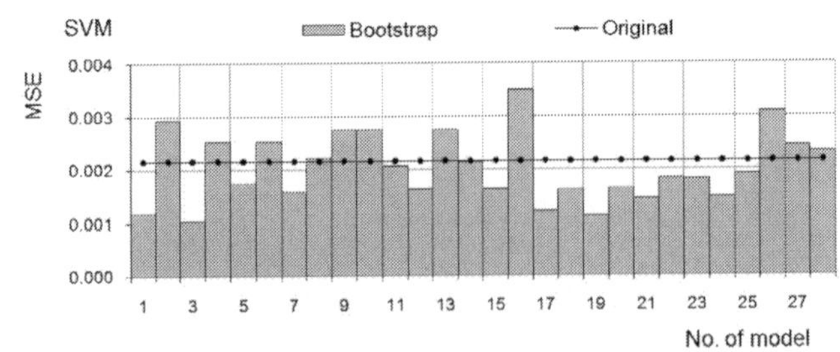


**Fig. 6.** Performance of individual bootstrap models created by SVM

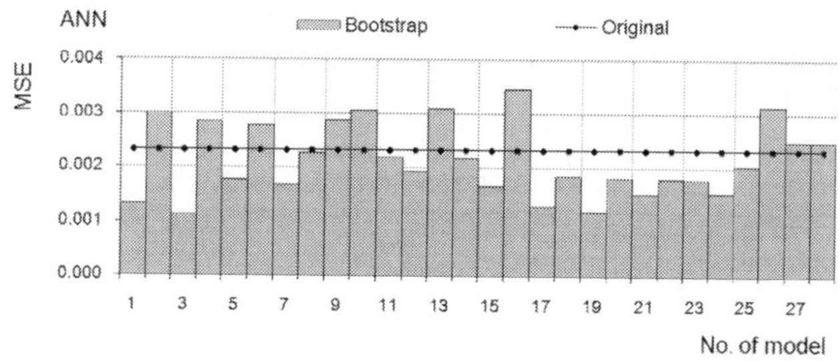


**Fig. 7.** Performance of individual bootstrap models created by ANN

First series of bagging models were created for individual algorithms using simple averages of prediction errors expressed in terms of MSE. The performance of bagging ensembles comprising from 1 to 28 bootstrap models was presented in Table 3 and illustrated in Figures 8, 9, 10, and 11. Table 3 contains also percentage reduction of MSE of respective ensembles compared to MSE provided by the original model. It can be observed that, except for some first ensembles, which should be passed over in order to avoid the effect of favourable drawing, the bagging models including 24 and

**Table 3.** Performance of bagging ensembles for individual algorithms

| No. of models | LRM | Error reduc. | DTR | Error reduc. | SVM | Error reduc. | ANN | Error reduc. |
|---|---|---|---|---|---|---|---|---|
| 1 | 0.00160 | 36.5% | 0.00146 | 41.9% | 0.00118 | 45.5% | 0.00132 | 42.8% |
| 2 | 0.00251 | 0.6% | 0.00205 | 18.6% | 0.00205 | 4.9% | 0.00217 | 6.4% |
| 3 | 0.00212 | 16.0% | 0.00177 | 29.6% | 0.00172 | 20.4% | 0.00182 | 21.5% |
| 4 | 0.00233 | 7.8% | 0.00221 | 12.2% | 0.00192 | 11.1% | 0.00207 | 10.5% |
| 5 | 0.00229 | 9.5% | 0.00216 | 14.3% | 0.00188 | 12.7% | 0.00201 | 13.1% |
| 6 | 0.00240 | 5.1% | 0.00218 | 13.4% | 0.00199 | 7.7% | 0.00214 | 7.6% |
| 7 | 0.00234 | 7.4% | 0.00215 | 14.4% | 0.00193 | 10.5% | 0.00207 | 10.4% |
| 8 | 0.00236 | 6.5% | 0.00218 | 13.3% | 0.00197 | 8.8% | 0.00210 | 9.4% |
| 9 | 0.00244 | 3.2% | 0.00232 | 7.9% | 0.00205 | 4.8% | 0.00218 | 5.7% |
| 10 | 0.00253 | 0.0% | 0.00242 | 3.9% | 0.00212 | 1.6% | 0.00227 | 2.0% |
| 11 | 0.00252 | 0.3% | 0.00236 | 6.3% | 0.00212 | 1.9% | 0.00226 | 2.3% |
| 12 | 0.00250 | 1.2% | 0.00232 | 7.7% | 0.00208 | 3.8% | 0.00223 | 3.5% |
| 13 | 0.00256 | -1.4% | 0.00224 | 10.9% | 0.00213 | 1.4% | 0.00230 | 0.7% |
| 14 | 0.00256 | -1.2% | 0.00229 | 8.8% | 0.00213 | 1.4% | 0.00229 | 1.0% |
| 15 | 0.00252 | 0.4% | 0.00227 | 9.9% | 0.00209 | 2.9% | 0.00225 | 2.9% |
| 16 | 0.00259 | -2.7% | 0.00234 | 6.8% | 0.00218 | -1.1% | 0.00232 | -0.4% |
| 17 | 0.00253 | -0.3% | 0.00227 | 9.6% | 0.00212 | 1.5% | 0.00226 | 2.2% |
| 18 | 0.00250 | 1.2% | 0.00225 | 10.7% | 0.00210 | 2.9% | 0.00224 | 3.2% |
| 19 | 0.00244 | 3.2% | 0.00220 | 12.6% | 0.00205 | 5.2% | 0.00219 | 5.6% |
| 20 | 0.00243 | 3.9% | 0.00218 | 13.4% | 0.00202 | 6.2% | 0.00217 | 6.4% |
| 21 | 0.00240 | 5.0% | 0.00214 | 14.8% | 0.00200 | 7.5% | 0.00214 | 7.8% |
| 22 | 0.00239 | 5.5% | 0.00215 | 14.4% | 0.00199 | 7.8% | 0.00212 | 8.4% |
| 23 | 0.00238 | 5.8% | 0.00216 | 14.3% | 0.00198 | 8.2% | 0.00211 | 9.0% |
| **24** | **0.00236** | **6.6%** | **0.00213** | **15.2%** | **0.00196** | **9.2%** | **0.00208** | **10.0%** |
| **25** | **0.00236** | **6.6%** | **0.00216** | **14.3%** | **0.00196** | **9.3%** | **0.00208** | **10.1%** |
| 26 | 0.00240 | 4.8% | 0.00222 | 11.7% | 0.00200 | 7.3% | 0.00212 | 8.4% |
| 27 | 0.00242 | 4.1% | 0.00225 | 10.8% | 0.00202 | 6.5% | 0.00213 | 7.8% |
| 28 | 0.00243 | 3.8% | 0.00226 | 10.1% | 0.00203 | 6.0% | 0.00215 | 7.3% |

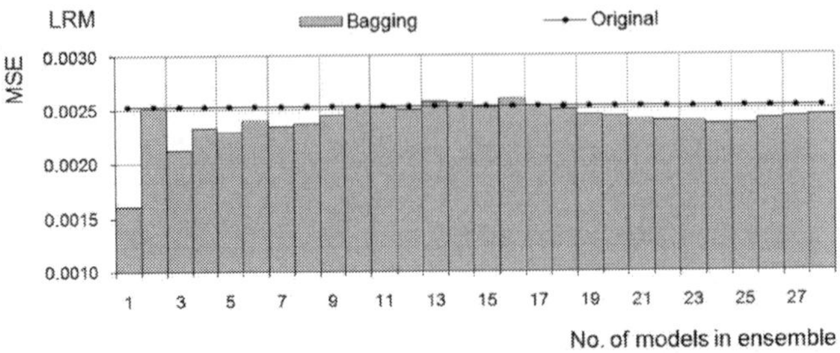


**Fig. 8.** Performance of bagging ensembles compared with original model created by LRM

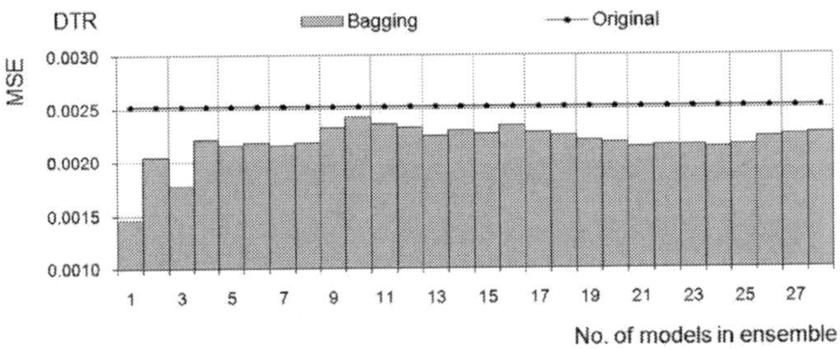


**Fig. 9.** Performance of bagging ensembles compared with original model created by DTR

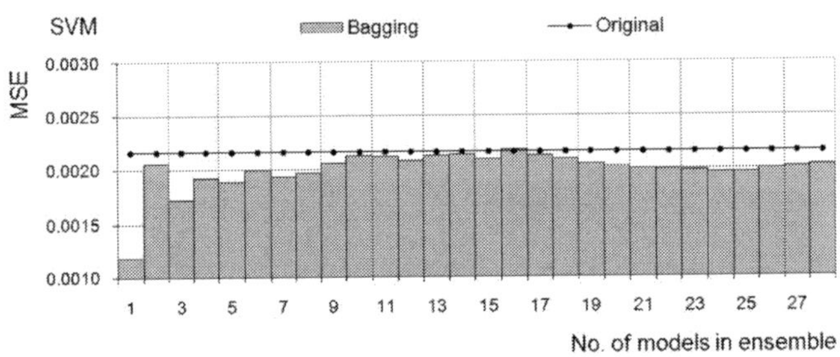


**Fig. 10.** Performance of bagging ensembles compared with original model created by SVM

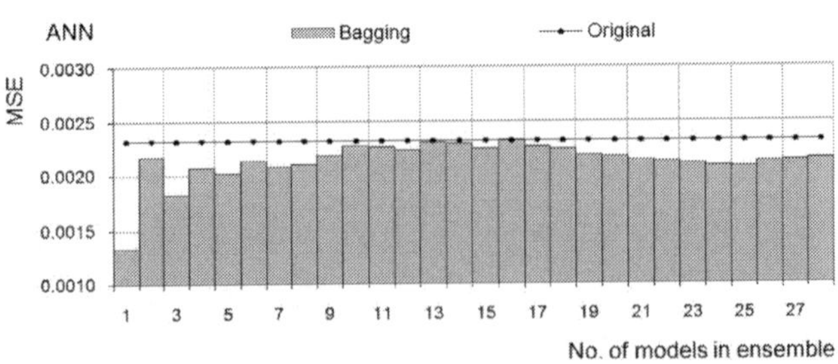


**Fig. 11.** Performance of bagging ensembles compared with original model created by ANN

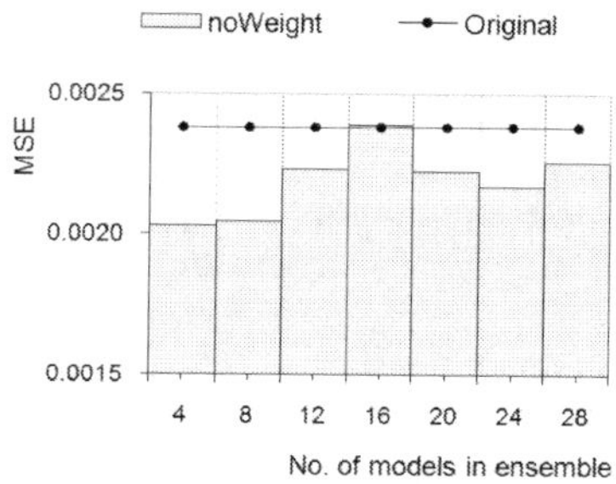


**Fig. 12.** Performance of ensembles using unweighted averages

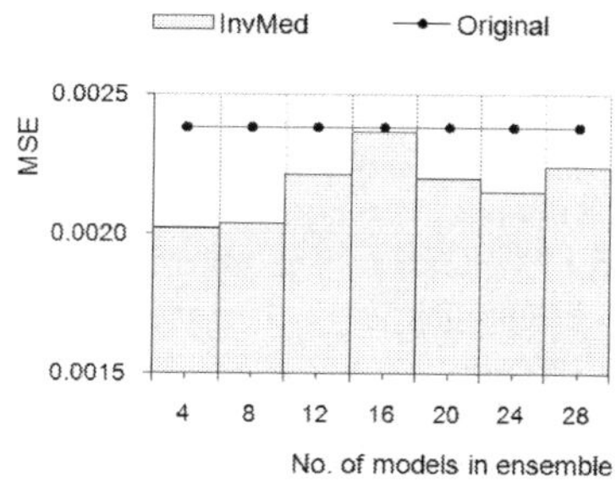


**Fig. 13.** Performance of ensembles using InvMed weighted averages

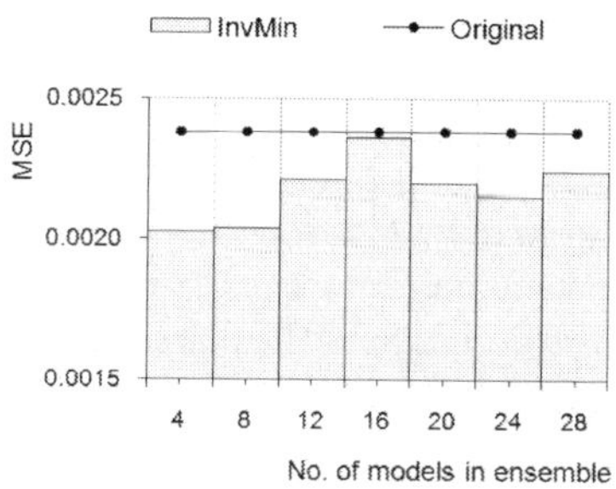


**Fig. 14.** Performance of ensembles using InvMin weighted averages

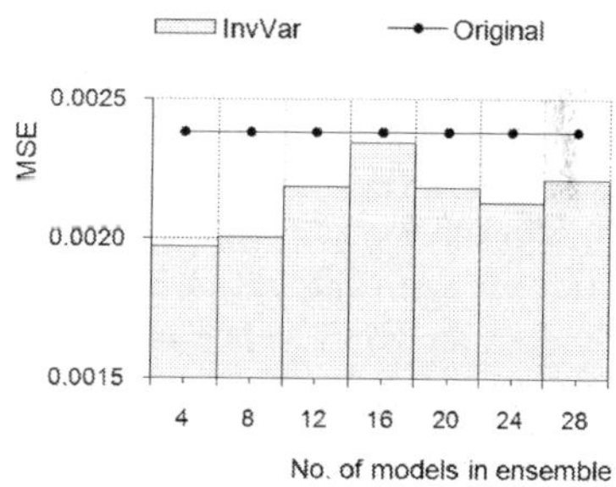


**Fig. 15.** Performance of ensembles using InvVar weighted averages

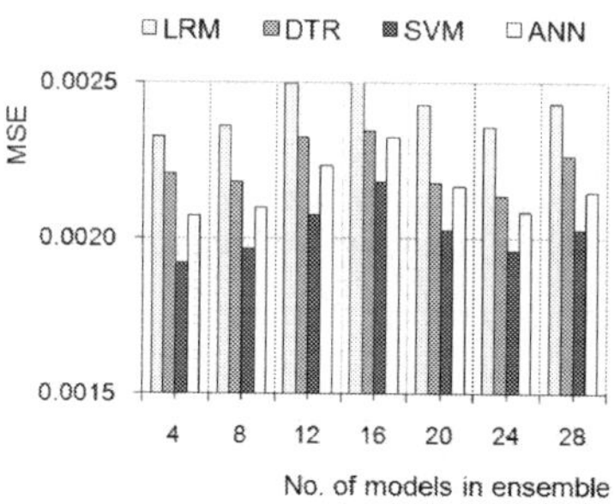


**Fig. 16.** Performance comparison of bagging ensembles created by individual algorithms

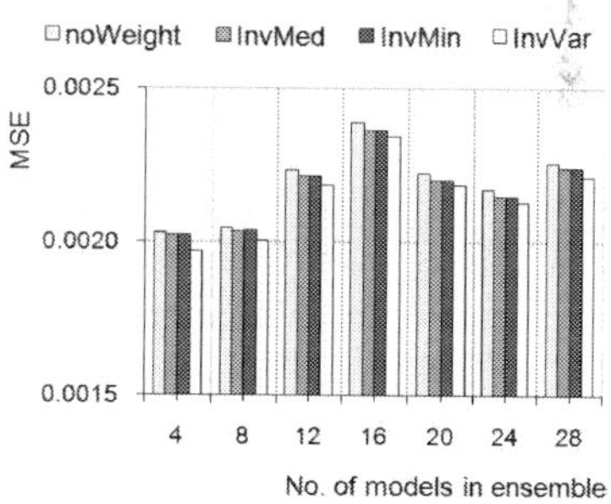


**Fig. 17.** Comparison of ensembles using unweighted and weighted averages

25 bootstrap replicates achieved the best prediction performance. They outperformed single models based on original data set created by LRM, DTR, SVM, and ANN algorithms, reducing average MSE by 6.6%, 15.2%, 9.3%, and 10.1% respectively. It should be also noted that at the combination of 16 models there was a point where the bagged predictors had larger prediction error than the unbagged. But this did not occurred in the case of DTR models, which were characterized by the greatest

variance of MSE. The figures confirmed the superior prediction accuracy of the bagging method to single models based on the original data set. Moreover, the results conformed to the intuitive Breiman's statement [5] that 25 bootstrap replicates could be reasonable to find the ensemble with the lowest prediction error.

Next series of bagging models encompassed the combinations of bootstrap models generated by different algorithms using different bootstrap replicates. This time, simple averages as well as weighted averages of prediction errors expressed in terms of MSE were applied. Bagging ensembles contained from 1 to 7 groups of four models obtained employing LMR, DTR, SVM and ANN to successive bootstrap sets (the results are presented in Table 4). Weights assigned to prediction errors were calculated on the basis of descriptive statistics presented in Table 3, and they were proportional to inverse median (InvMed), inverse minimum (InvMin), and inverse variance (InvVar). The performance of bagging ensembles comprising from 4 to 28 bootstrap models was illustrated in Figures 12, 13, 14, and 15. Similar observations to those referring to the first series could be done, namely except two first ensembles, the bagging models including 24 bootstrap replicates achieved the best prediction performance, and they outperformed single models based on original data set created by LRM, DTR, SVM, and ANN algorithms. The average of MSE averages calculated for all algorithms, which was equal to 0.00238, was reduced by 8.9%, 9.7%, 9.7%, and 10.5% for noWeight, InvMed, InvMin, and InvVar ensembles respectively. At the combination of 16 models only the unweighted bagged predictor had larger prediction error than the unbagged. The figures also confirmed the higher prediction accuracy of the bagging method to average of models based on the original data set.

**Table 4.** Performance of unweighted and weighted bagging ensembles

| No. of models | noWeight | Error reduc. | InvMed | Error reduc. | InvMin | Error reduc. | InvVar | Error reduc. |
|---|---|---|---|---|---|---|---|---|
| 4 | 0.00203 | 14.7% | 0.00202 | 15.0% | 0.00202 | 15.0% | 0.00197 | 17.2% |
| 8 | 0.00205 | 14.0% | 0.00204 | 14.4% | 0.00204 | 14.4% | 0.00200 | 15.8% |
| 12 | 0.00223 | 6.2% | 0.00221 | 7.0% | 0.00221 | 7.0% | 0.00218 | 8.2% |
| 16 | 0.00239 | -0.3% | 0.00236 | 0.7% | 0.00236 | 0.8% | 0.00234 | 1.5% |
| 20 | 0.00222 | 6.6% | 0.00220 | 7.5% | 0.00220 | 7.6% | 0.00218 | 8.2% |
| **24** | **0.00217** | **8.9%** | **0.00215** | **9.7%** | **0.00215** | **9.7%** | **0.00213** | **10.5%** |
| 28 | 0.00226 | 5.2% | 0.00224 | 5.9% | 0.00224 | 5.9% | 0.00221 | 7.0% |

## 4 Conclusions and Future Work

An extension of the MAREA multi-agent system for real estate appraisals to include aggregating agents, which created ensemble models applying the bagging approach, was presented in the paper. The major part of the study was devoted to investigate to what extent bagging approach could lead to the improvement of the accuracy machine learning regression models. Four algorithms implemented in KEEL including linear regression, decision trees for regression, support vector machines, and artificial neural network of MLP type, were used in the experiments. 112 bootstrap models and 140 bagging ensembles combining the former by means of unweighted and weighted averages were created and evaluated in respect of prediction accuracy. The smallest values of MSE ensured bagging ensembles encompassing 24 and 25 bootstrap SVM and ANN models. However the largest percentage reduction of prediction error achieved the ensemble of DTR models, which were characterized by the greatest variance of MSE.

Moreover, bagging models comprising the combinations of bootstrap models generated by all algorithms using different bootstrap replicates were investigated. Similar observations could be done, the bagging models including 24 bootstrap replicates achieved the best prediction performance, and they outperformed single models based on original data set created by individual algorithms. The largest percentage reduction of prediction error achieved the ensemble with weights proportional to inverse variance of MSE.

Further research on ensemble models devoted to real estate appraisal is planned in order to find components assuring error diversity and to extend the algorithm range to evolutionary fuzzy systems. Moreover the impact of different model aggregation methods ranging from algebraic expressions, such as minimum, maximum, sum, mean, product, median, etc. to more sophisticated measures will be investigated.

## References


1. Alcalá-Fdez, J., et al.: KEEL: A software tool to assess evolutionary algorithms for data mining problems. Soft Computing 13(3), 307–318 (2009)
2. Avnimelech, R., Intrator, N.: Boosting regression estimators. Neural Computation 11, 491–513 (1999)
3. Bertoni, A., Campadelli, P., Parodi, M.: A boosting algorithm for regression. In: Proc. Int. Conference on Artificial Neural Networks, pp. 343–348 (1997)
4. Bellifemine, F., Caire, G., Poggi, A., Rimassa, G.: JADE. A White Paper. EXP 3(3), 6–19 (2003)
5. Breiman, L.: Bagging Predictors. Machine Learning 24(2), 123–140 (1996)
6. Breiman, L.: Stacked Regressions. Machine Learning 24(1), 49–64 (1996)
7. Breiman, L.: Using iterated bagging to debias regressions. Machine Learning 45, 261–277 (2001)
8. Büchlmann, P., Yu, B.: Analyzing bagging. Annals of Statistics 30, 927–961 (2002)
9. Büchlmann, P.: Bagging, subagging and bragging for improving some prediction algorithms. In: Akritas, M.G., Politis, D.N. (eds.) Recent Advances and Trends in Nonparametric Statistics, pp. 19–34. Elsevier, Amsterdam (2003)
10. Drucker, H., Cortes, C.: Boosting decision trees. In: Touretzky, D.S., Mozer, M.C., Hasselmo, M.E. (eds.) Advances in Neural Information Processing Systems, vol. 8, pp. 479–485. Morgan Kaufmann, San Francisco (1996)
11. Drucker, H., Schapire, R.E., Simard, P.: Boosting performance in neural networks. Int. J. of Pattern Recogn. and Artificial Intel. 7(4), 705–719 (1993)
12. Drucker, H.: Improving regressors using boosting techniques. In: Proc. 14th Int. Conf. on Machine Learning, pp. 107–115. Morgan Kaufmann, San Francisco (1997)
13. Duffy, N., Helmbold, D.P.: Leveraging for regression. In: Proceedings of the 13th Conference on Computational Learning Theory, pp. 208–219 (2000)
14. Fan, R.E., Chen, P.H., Lin, C.J.: Working set selection using the second order information for training SVM. J. of Mach. Learning Res. 6, 1889–1918 (2005)
15. Freund, Y., Schapire, R.E.: A decision-theoretic generalization of on-line learning and an application to boosting, J. of Comp 55(1), 119–139 (1997)

16. Freund, Y., Schapire, R.E.: Experiments with a New Boosting Algorithm. In: Proc. of the Thirteenth Int. Conf. on Machine Learning, pp. 148–156 (1996)
17. Friedman, J.: Greedy function approximation: a gradient boosting machine. Technical report, Dept. of Statistics, Stanford University (1999)
18. Gencay, R., Qi, M.: Pricing and hedging derivative securities with neural networks: Bayesian regularization, early stopping, and bagging. IEEE Transactions on Neural Networks 12, 726–734 (2001)
19. Hansen, L., Salamon, P.: Neural network ensembles. IEEE Transactions on Pattern Analysis and Machine Intelligence 12(10), 993–1001 (1990)
20. Hashem, S.: Optimal linear combinations of neural networks. Neural Networks 10(4), 599–614 (1997)
21. Kégl, B.: Robust regression by boosting the median. In: Proc. of the 16th Conference on Computational Learning Theory, pp. 258–272 (2003)
22. Krogh, A., Vedelsby, J.: Neural network ensembles, cross validation, and active learning. In: Advances in Neural Inf. Proc. Systems, pp. 231–238. MIT Press, Cambridge (1995)
23. Lasota, T., Mazurkiewicz, J., Trawiński, B., Trawiński, K.: Comparison of Data Driven Models for the Validation of Residential Premises using KEEL. International Journal of Hybrid Intelligent Systems (in press, 2009)
24. Lasota, T., Telec, Z., Trawiński, B., Trawiński, K.: Concept of a Multi-agent System for Assisting in Real Estate Appraisals. In: Håkansson, A., et al. (eds.) KES-AMSTA 2009. LNCS (LNAI), vol. 5559, pp. 50–59. Springer, Heidelberg (2009)
25. Moller, F.: A scaled conjugate gradient algorithm for fast supervised learning. Neural Networks 6, 525–533 (1990)
26. Optiz, D., Maclin, R.: Popular Ensemble Methods: An Empirical Study. Journal of Artificial Intelligence Research 11, 169–198 (1999)
27. Opitz, D., Shavlik, J.W.: Actively searching for an effective neural network ensemble. Connection Science 8(3-4), 337–353 (1996)
28. Quinlan, J.R.: Learning with Continuous Classes. In: Proc. 5th Australian Joint Conference on Artificial Intelligence (AI 1992), Singapore, pp. 343–348 (1992)
29. Rustagi, J.S.: Optimization Techniques in Statistics. Academic Press, London (1994)
30. Schapire, R.E.: The Boosting approach to machine learning: An overview. In: Denison, D.D., et al. (eds.) Nonlinear Estimation and Classification. Springer, Heidelberg (2003)
31. Schapire, R.E.: The strength of weak learnability. Machine Learning 5(2), 197–227 (1990)
32. Skurichina, M., Duin, R.P.W.: Bagging for linear classifiers. Pattern Recognition 31, 909–930 (1998)
33. Triadaphillou, S., et al.: Fermentation process tracking through enhanced spectral calibration modelling. Biotechnology and Bioengineering 97, 554–567 (2007)
34. Zemel, R.S., Pitassi, T.: A gradient based boosting algorithm for regression problems. In: Adv. in Neural Inf. Processing Systems, vol. 13, pp. 696–702 (2001)
35. Zhang, J.: Inferential estimation of polymer quality using bootstrap aggregated neural networks. Neural Networks 12, 927–938 (1999)

# Reputation Tracking Procurement Auctions

Ana Petric and Gordan Jezic

University of Zagreb
Faculty of Electrical Engineering and Computing
Unska 3, HR-10000, Zagreb, Croatia
{ana.petric,gordan.jezic}@fer.hr

**Abstract.** The introduction of e-markets has created great challenges for both buyers and suppliers. Buyers have to decide how to take advantage of the possibilities offered on e-markets (i.e., higher savings, expanding the supplier base) but at the same time they should preserve values that are associated with their long-term relationships. As the use of reverse auctions in the industry grows, involved parties are increasingly concerned with how these auctions influence their previously established business relationships. Due to those existing relationships, reverse auctions should not be analyzed as stand-alone auctions. If the relationship is specified as a value-generating asset in the procurement process, then neither business party should expect relationships to be harmed. We propose a reputation tracking reverse auction model that exploits the advantages reverse auctions bring to buyers, while decreasing the disadvantages they bring to sellers. Several experiments were conducted and the analysis was focused on auctions that have different outcomes depending on whether they took reputation into account or not.

**Keywords:** reputation, reverse auctions, procurement, multi-attribute auctions, B2B e-market.

## 1 Introduction

In the past, available markets and their associated choices were much more limited than today. Consequently, the volatility of supply and demand functions was much more inert. Under such market conditions, companies based their business transactions on long-term partnerships. However, the advent of the Internet and the accelerated economic globalization trend in the past decade is leading us closer to the existence of just one market - the global one. On this global market, electronic commerce (e-commerce) has emerged as the new way of conducting business transactions (i.e., including buying, selling and/or exchanging products, services and information), communicating primarily via the Internet [1]. In this new regime, most attention is focused on Business-to-Business (B2B) and Business-to-Consumer (B2C) e-commerce. B2B e-commerce is considered to be the more lucrative of the two, due to its larger trading volumes and the fact that businesses are less fickle than consumers [1].

N.T. Nguyen, R. Kowalczyk, and S.-M. Chen (Eds.): ICCCI 2009, LNAI 5796, pp. 825–837, 2009.

Electronic markets (e-markets) can be seen as Internet-based business transaction platforms that serve as digital intermediaries that create value by bringing consumers and providers together. In most industries, e-markets provide opportunities for increasing supply chain management efficiency by changing competition mechanisms [2]. By reducing transaction costs and supporting the exchange of information, they create transactional immediacy and supply liquidity.

In this paper, we focus on the B2B e-market since it is widely believed that it will become the primary way of doing business [3]. Transactions conducted on B2B e-markets are a lot more that just one-shot deals. The relationships between the parties involved in the transactions are ongoing, systematic exchange structures and can affect behavior of all the parties involved over the course of the transaction [4]. Special intention is paid to the negotiation phase of the transaction since the outcome (i.e. financial efficiency) is still the premier performance measure for most businesses [5,6].

The paper is organized as follows. Section 2 gives an overview of research in the area of reverse auctions. A reputation tracking reverse auction model is proposed in Section 3. Section 4 presents the conducted experiments and the obtained results, while Section 5 concludes the paper and gives an outline for future work.

## 2 Related Work

Due to their well defined protocols, auctions are suitable enablers of negotiations in e-markets. Auctions are defined as a market institution that acts in pursuit of a set of predefined rules in order to compute the desired economic outcome (i.e., high allocation efficiency) of social interactions [7]. Based on bids and asks placed by market participants, resource allocation and prices are determined.

According to [8], buy-side (i.e., reverse, procurement) auctions are one out of three commonly used auction types on B2B e-markets. In a reverse auction we have one buyer that negotiates the purchase of a particular item (i.e., good, service) with multiple sellers. As the use of reverse auctions in the industry grows, involved parties are increasingly concerned with how these auctions influence their previously established business relationships. Due to those existing relationships, reverse auctions should not be analyzed as stand-alone ones.

Various studies concerning buyers' and sellers' points of view on reverse auctions have been conducted. Most buyers feel that such auctions will enhance their productivity and improve their relationships with suppliers [9]. Buyers also believe that suppliers will embrace participating in reverse auctions since they open new opportunities to increase their sales and penetrate new markets [10]. On the other hand, one of the most common sellers' beliefs includes suspicions of buyers' opportunistic behavior. Suspicions grow as the number of sellers participating in an auction decreases. Furthermore, such suspicions are higher in open-bid auctions than in seal-bid auctions [11]. However, as the economic stakes of the auction increase, the suspicions of buyer's opportunistic behavior decrease [12]. Suppliers' willingness to make dedicated investments towards a buyer increases as the price visibility in an auction decreases. Looking at suppliers' trading history with buyers, this willingness is higher with current and incumbent suppliers than with new suppliers [12]. Willingness to make specific investments shows suppliers' willingness to develop long-term relationships with buyers, but also results with suppliers' decreased bid activity (i.e., submitting fewer bids, bidding less often) and a decrease in making price concessions [4].

Relationships on B2B markets gain on importance if we consider that the number of participants in a certain industry does not have large oscillations. Exceptions are always possible, but they are not common. From time to time a new participant appears on the market or an old one goes out of business, but this does not have a significant influence on other existing relationships. Participants on the market must be aware that their actions (e.g., avoiding commitments, late deliveries, respecting the negotiated terms) influence their reputation and relationships with other participants. B2B trading can be viewed as an ongoing process that connects conducted and ongoing business transactions into a complex arranged structure that can affect future behavior [4]. If the relationship is specified as a value-generating asset in the procurement process, then neither business party should expect relationships to be harmed [13].

## 3 A Reputation Tracking Reverse Auction Model

In this section, we give a formal description of our reputation tracking reverse auction model and present the environment the model is placed in.

Item characteristics (i.e., attributes) represent an important factor in deciding which auction should be used. Negotiation on commodities focuses mainly on the price of the item. These items are mostly sold in conventional single-attribute auctions. On the other hand, complex items often require negotiation of several attributes, and not just the price [14]. Such items are sold in multi-attribute auctions [15]. Multi-attribute auctions have been attracting more and more attention in B2B markets, since the price is not the only attribute considered in the decision making process[1]. Most reverse auctions are so-called buyer-determined auctions in which the buyer chooses the winner by integrating non-monetary attributes into the winner determination process [4].

The first step in a multi-attribute auction is for the buyer to specify his preferences regarding the item he wishes to purchase. Preferences are usually defined in the form of a scoring function based on the buyer's utility function [14]. In order to familiarize sellers with buyer's valuations of relevant attributes, the buyer can publicly announce his scoring function. Sellers are not obligated to disclose their private values of an item. The winner of the multi-attribute auction is the seller that provided the highest overall utility for the buyer.

Our model is based on reverse auctions and takes into account the price, as well as other non-monetary attributes of the purchased items. Furthermore, exogenous attributes (e.g., the quality of suppliers' offers in previous auctions and the orderly fulfilment of suppliers' obligations) are also considered.

[1] http://www.cindywaxer.com/viewArticle.aspx?artID=149

### 3.1 Motivation for Introducing Reputation Tracking

The introduction of e-markets has created great challenges for both buyers and suppliers. Buyers have to decide how to take advantage of the possibilities offered on e-markets (i.e., higher savings, expanding the supplier base) but at the same time they should preserve values that are associated with their long-term relationships [16]. Buyers usually granted various privileges (e.g., volume discounts, higher service levels, favorable credit terms) to strategically important suppliers. Those privileges do not exist on e-markets were all suppliers are treated as equals. This can result in suppliers' withdrawal from participating on the e-market. If a significant number of suppliers refuse to participate, adequate liquidity cannot be achieved and the likelihood of market manipulation increases.

The reputation tracking reverse auction was developed primarily for maintaining good business relationships between buyers and their strategically important suppliers, as well as those suppliers who deliver their items in an orderly manner. For this purpose, we developed an auction model that takes into account two important facts. First, that traded items usually require negotiation on different attributes, and not just the price. Second, that with the advent of the Internet, a global market is forming and market conditions are changing. Consequently, long-term relationships between business partners are slowly being replaced by short-term arrangements with new and sometimes unknown suppliers that offer the most favorable deal.

Most of the work in the area of reputation auctions assumes that purchased items are delivered on time and does not take into account possible delays in delivery. Unfortunately, late deliveries are not very unusual and can cause significant financial damage considering the buyer is often just a link in the supply chain and not the end user of the purchased item. Namely, the buyer commonly uses the purchased good to manufacture new goods or combines purchased services into new value added services. These items are then sold further to other business entities on the B2B market or to consumers on the B2C market.

The buyer most often forms arrangements with his business partners in accordance with his planed production schedules and on-time delivery assessments. Due to other arrangements and limited resources (e.g., production plant capacities, available man power, resources used in ongoing projects), project schedules are not subject to significant changes. The damages caused by disturbances in the delivery of items that are used as resources in further production are twofold. First of all, the value of the item will most probably fall compared with the value it would have had on the agreed delivery date. Second of all, late delivery of the resources will probably cause a delay in the production of the buyer's items. Consequently, the delivery of the items will also be late and will cause financial damage and/or a tainted reputation to other buyer's business partners.

Similar problems are modeled in the Trading Agent Competition Supply Chain Management Game (TAC SCM) [17,18]. In TAC SCM on the B2B market participants trade by using sell-side auctions and suppliers can deliver the purchased item late. Due to limitations of the game (i.e., small number of suppliers) the buyer cannot efficiently apply penalties nor change his supplier base.

### 3.2 The Formal Model Description

The reputation tracking auction is designed for businesses with high purchasing frequencies on a market with a limited number of suppliers. In our simulated environment, the buyer uses a sealed-bid multi-attribute reverse auction to determine the winning seller. The basic assumption is that a single buyer uses an auction to purchase a non-commodity item from one out of $n_s$ sellers. The buyer sends a request for quotes (RFQ) to $n_s$ sellers that produce the item he is looking to buy. The sellers decide to participate in the auction depending on their current production capacity. Based on his current schedule, the seller calculates his on-time delivery probability for the item. If this probability corresponds with his risk policy, he sends an offer. After the auction closes, the buyer evaluates received offers and determines the winner by taking into account the total utility of the offered item and sellers' reputations.Research has shown that participants of reverse auctions consider that single-attribute auctions, where the bidding evolves only around the price, are not constructive in the development of long-term relationships between buyers and suppliers [19].

In our work, we make the following assumptions; the number of sellers invited to participate in an auction is known to all sellers while the number of sellers that decide to participate in an auction in not known to other sellers. Sellers are also allowed to drop out of a current auction and join another auction later.

A reverse auction can be defined as a tuple $< b, S, t >$, where

- $b$ is the buyer agent;
- $S$ (of size $s$) denotes the set of all seller agents that are invited to participate in buyer $b$'s reverse auction; while $S' \subseteq S$, where $S'$ (of size $s'$) denotes the set of seller agents that decide to participate in the auction;
- $t : \mathbb{R}^{s'} \rightarrow \mathbb{R}$ is the winner determination function.

The reputation tracking auction consists of two descriptions and three functions: an *item evaluation model* that contains a description of all the relevant attributes of an item that is being sold in an auction; a *reputation tracking model* containing a description of sellers' reputation for all sellers that participate in reverse auctions; a function that assigns values to the item; a function that assigns values to sellers' reputations; and a function that combines the two previous ones and determines the auction outcome.

The *item evaluation model* is represented with a tuple $< x, w_I, I >$, where

- $x = (x_1, \ldots, x_j, \ldots, x_n)$ is the set of attributes used to describe an item; each attribute $j$ has a reserve and aspiration value, denoted as $x_j^r$ and $x_j^a$, respectively, determined by the buyer;
- $w_I = \{w_{I1}, \ldots, w_{Ij}, \ldots, w_{In}\}$ is a set of weights that determines the importance of each attribute from $x$ for the buyer, where $w_{Ij}$ is the weight of attribute $j$;

- $O \in \mathbb{R}^{s' \times n}$, is the offer matrix which describes the attribute values of offered items, where $O_{ij}$ denotes the value that seller $i$ places for attribute $j$;
- $I : \mathbb{R}^{s' \times n} \times \mathbb{R}^n \rightarrow \mathbb{R}^{s'}$ is a utility function that calculates the buyer's utility of sellers' offers.

Utility function $I(O_i)$ takes as input an offer placed by seller $i$ (i.e., row $i$ of the offer matrix denoted as $O_i$) and, together with the set of weights $w_I$ maps it to a real value. Function $I(O_i)$ can be defined as an additive scoring function that assumes the existence of mutual *preferential independence* between attributes [20]. In order to calculate the utility of the multi-attribute item, reserve values and weights for each attribute need to be considered [21]. Function $I(O_i)$ is defined as follows:

$$I(O_i) = \sum_{j=1}^{n} w_{Ij} I(O_{ij}), where \sum_{j=1}^{n} w_{Ij} = 1 \tag{1}$$

Existing models of multi-attribute auctions use different approaches to calculate the buyer's utility of an item (e.g., by defining utility functions [15,20], by using fuzzy multi-attribute decision making algorithms [22], by introducing pricing functions and preference relations for determining acceptable offers [23], by defining reserved and aspiration levels of attributes and distinguishing negotiable and non-negotiable attributes [24]).

We distinguish between two types of attributes: ascending and descending ones. Ascending attributes are evaluated according to positive criteria, where higher values of the attribute increase its utility (e.g., Quality of Service). On the other hand, the utility of descending attributes increases if the value of the attribute decreases (e.g., price). In our model, $I(O_{ij})$ depends on the reserve and aspiration values, $x_j^r$ and $x_j^a$, respectively, that the buyer defines for each attribute $j$. Reserve value $x_j^r$ marks the lowest value of ascending attribute $j$ that is acceptable to the buyer. The aspiration value $x_j^a$ is the highest value of ascending attribute $j$ that the buyer is interested in. Values offered higher than the aspiration value are acceptable, but they do not cause a further increase in the buyer's utility for that attribute. This prevents the seller from increasing the total utility of his offer by assigning unnecessarily high values to some (less important) attributes and unreasonably low values to other important ones. Analogously, for descending attributes, $x_j^r$ marks the highest acceptable value, i.e. any value higher than $x_j^r$ is not acceptable, while $x_j^a$ denotes the lowest value the buyer is interested in after which lower values do not increase the utility for that attribute.

Attribute $j = 1$ marks the price of an item, where $x_1^r$ is the highest price the buyer is willing to pay. The buyer's aspiration price is $x_1^a = 0$ money units, so it is not stated in equation (2). Value N.A. in (2) marks a non-acceptable value for an attribute, i.e., it is worse than the reserve value. An offer is rejected if the utility of at least one attribute is N.A. The utility of an offered value higher than the aspiration value cannot be higher than 1.

$$I(O_{ij}) = \begin{cases} 1 - \frac{O_{ij}}{x_j^r}, & j = 1 \text{ } and \text{ } O_{ij} \leq x_j^r \\ N.A., & j = 1 \text{ } and \text{ } O_{ij} > x_j^r \\ \frac{O_{ij} - x_j^r}{x_j^a - x_j^r}, & j \geq 2 \text{ } and \text{ } x_j^r \neq x_j^a \text{ } and \text{ } x_j^r \leq O_{ij} \leq x_j^a \\ N.A., & j \geq 2 \text{ } and \text{ } O_{ij} < x_j^r \\ 1, & j \geq 2 \text{ } and \text{ } O_{ij} \geq x_j^a \end{cases} \quad (2)$$

The *reputation tracking model* is described by a tuple $< y, w_R, V, R >$, where

- $y = (y_1, y_2, y_3)$ is a set of attributes used to calculate sellers' reputations;
- $w_R = \{w_{R1}, w_{R2}, w_{R3}\}$ is a set of weights that determines the importance of each attribute from $y$ for the buyer;
- $V$ is the buyer's sliding window which is used to store information regarding the last $v$ auctions the buyer participated in;
- $R : \mathbb{R} \rightarrow \mathbb{R}$ is a utility function that calculates the reputation of all sellers that participated in a least one auction in the sliding window.

The reputation of seller $i$ is defined as follows:

$$R(i) = \sum_{l=1}^{3} w_{Rl} y_l, \text{ } where \sum_{l=1}^{3} w_{Rl} = 1 \quad (3)$$

The buyer calculates sellers' reputations from information gathered in the last $v$ auctions. There are three attributes we consider relevant when calculating sellers' reputations:

- $y1$ – the average utility of the items the seller offers. We compare the item utility of the seller's offer with the item utility of the winning offer in that auction;
- $y2$ – the share of total transactions carried out by the seller. We calculate the value of all the items the seller sold to the buyer and compare it with the value of all transactions conducted in the observed period of time and stored in the sliding window;
- $y3$ – the financial damage inflicted on the buyer due to the seller's late delivery of purchased items. We keep track of all the damage that was caused by the seller's late delivery and compare it with the overall damage caused by all the sellers. The advantage is given to sellers that deliver items on time or with smaller delays.

The exact formulas and explanations of the attributes follow:

$$\begin{aligned} R(i) = w_{R1} \tfrac{1}{a_i} \textstyle\sum_{k=1}^{v} \frac{a_i^k S(x^{i,k})}{S(x^{winner,k})} + w_{R2} \frac{\sum_{k=1}^{v} c_i^k x_1^{i,k}}{\sum_{k=1}^{v} x_1^{winner,k}} + \\ w_{R3}(1 - \frac{\sum_{k=1}^{v} c_i^k d_i^k}{\sum_{m=1}^{s} \sum_{k=0}^{v} c_m^k d_i^k}) \end{aligned} \quad (4)$$

$$a_i = \sum_{k=1}^{v} a_i^k, \text{ } a_i^k = \begin{cases} 1, & if \text{ } seller \text{ } i \text{ } participated \text{ } in \text{ } auction \text{ } k \\ 0, & otherwise \end{cases} \quad (5)$$

$$c_i^k = \begin{cases} 1, & if\ seller\ i\ is\ the\ winner\ of\ auction\ k \\ 0, & otherwise \end{cases} \quad (6)$$

$$d_i^k = x_1^i(t_{onTime}) - x_1^i(t_{late}^i) \quad (7)$$

In (4), $a_i^k$ (5) denotes seller $i$'s participation in auction $k$; $c_i^k$ (6) denotes whether seller $i$ won auction $k$; and $d_i^k$ (7) denotes the decline in the value of the purchased item due to the seller's late delivery.

The *reputation tracking auction model* is defined as a tuple $< x, w_I, I, y, w_R, V, R, b, S, w_T, T, t >$, where

- $< x, w_I, I >$ is an *item evaluation model*;
- $< y, w_R, V, R >$ is a *reputation tracking model*;
- $w_T = \{w_{T1}, w_{T2}\}$ is a pair of weights where $w_{T1}$ denotes the weight of the offered item considered in the winner determination function, while $w_{T2}$ denotes the weight of the seller's reputation;
- $T$ is the total score function.

The total score function $T(i)$ takes as input the value of an item offered by seller $i$ and seller $i$'s reputation and maps it to a total score for the offer. $T(i)$ is defined as follows:

$$T(i) = w_{T1} I(x^i) + w_{T2} R(i),\ where\ w_{T1} + w_{T2} = 1 \quad (8)$$

The winner determination function takes as input the total scores of all received offers and maps it to a real value in order to determine the winning offer.

$$t = \max_i\ T(i) \quad (9)$$

## 4 Experiments and Results

In this section, we first give a short description of how sellers determine their on-time delivery probability and how the buyer calculates the amount of damage caused by late deliveries. It is followed by the presentation of the experimental method and obtained results of simulations.

### 4.1 Supplier Settings

The probability that seller $i$ will deliver the purchased item on time $P_i(t)$ is calculated according to the expression shown in (10) while $a_i^k$ (11) denotes seller $i$'s participation in auction $k$ and $c_i^k$ (12) denotes whether seller $i$ won auction $k$. Since suppliers try to maintain good business relationships with their buyers, they increase the probability of on-time delivery for their more significant ones. However, this causes them to increase the risk of late deliveries to their less significant buyers. Each supplier also has a sliding window keeping track of the $v$ previous transactions conducted with each buyer. He increases his on-time delivery probability proportionally to the ratio of the total monetary value of all transactions won by the seller, compared to the total values of all the auctions the seller participated in. This additional increase has an upper limit of $D_{max}$. Depending on his preferences and willingness to accept risks, the supplier participates in an auction if his $P_i(t)$ is higher than a certain threshold.

$$P_i(t) = P_i(t-1) + random[-0.05, 0.05] + \frac{\sum_{k=1}^{v} c_i^k a_i^k x_1^k}{\sum_{k=1}^{v} a_i^k x_1^k} D_{max} \tag{10}$$

$$a_i^k = \begin{cases} 1, & if\ seller\ i\ participated\ in\ auction\ k \\ 0, & otherwise \end{cases} \tag{11}$$

$$c_i^k = \begin{cases} 1, & if\ seller\ i\ is\ the\ winner\ of\ auction\ k \\ 0, & otherwise \end{cases} \tag{12}$$

### 4.2 Buyer Damages

The buyer negotiates the delivery date of each purchased item. If the item is delivered after this date, its value is usually lower than it would have been on the negotiated delivery date. In our experiments, we consider only the direct financial damage (i.e., the decrease in the value of the purchased item). However, it is important to note that indirect financial damage (i.e. penalties for late delivery to buyer's business partners, a decrease in the buyer's reputation and the possible loss of new business projects) can be even several times higher. In our experiments, we modeled the decline in value as:

$$D_i = x_1^i (1 - e^{-(1-P_i(t))}) \tag{13}$$

where $D_i$ is the decline in the value of the purchased item in auction $i$, $x_1^i$ is the price the buyer paid for the item, and $P_i(t)$ is the seller's on-time delivery probability for the purchased item.

### 4.3 Experiment Settings

We conducted three sets of experiments as follows. In each set we changed the values of reputation attribute weights (i.e., $w_{R1}, w_{R2}, w_{R3}$) and conducted 10 experiments per set with different total attribute weights (i.e., $w_{T1}$, $w_{T2}$). One experiment includes 2700 seal-bid multi-attribute reverse auctions held sequentially one after the other. The size of the buyer's sliding window $v$ is 200, so the first 200 out of 2700 auctions were conducted without reputation tracking but they were used to calculate sellers' reputations later. The remaining 2500 auctions took reputation into account when determining the winner of the auction. The items sold in auctions within one experiment all had different weights, reserve and aspiration values, but the auctions with the same ordinal number in different experiments sold the same item (e.g., in auction 361 with $w_{R1} = 0.35, w_{R2} = 0.35, w_{R3} = 0.3$ and $w_{T1} = 0.85, w_{T2} = 0.15$ the same item was sold as in auction 361 with $w_{R1} = 0.5, w_{R2} = 0.3, w_{R3} = 0.2$ and $w_{T1} = 0.75, w_{T2} = 0.25$). While the reserve and aspiration values for non-monetary attributes were chosen randomly from [0, 1], the reserve prices of items were chosen randomly from [100 000, 1 000 000] in monetary units.

For each experiment, we compared auctions in which we tracked sellers' reputation with auctions in which reputation was not considered. The remainder of the analysis in the paper is conducted only on auctions with different outcomes for these two cases. At the end of the experiment, we calculated the amount of money saved by awarding contracts to suppliers with higher reputations. Figure 1. shows the total savings obtained for all three sets of experiments. In each set the values of reputation attribute weights are different. We can see that the savings are quite close for different values of reputation attribute weights for auctions where the influence of the reputation in the total score (i.e., $w_{T2}$) was lower than 35%. It is clear that the influence of reputation in the decision making process should not be too high since this can result in significant financial damage.

The total value of all the items the buyer bought in auctions with different outcomes can significantly vary in different experiments. Namely an increase in reputation impact in the decision making process is followed by a higher number of auctions with different outcomes. The value of all items bought in such auctions rises accordingly. Figure 2. shows the ratio of the total savings achieved in auctions with different outcomes in each experiment (shown in Figure 1.) and the value of all items bought in these auctions. As in Figure 1, we can see that reputation tracking only achieves savings in auctions if the impact of the reputation in the total score is lower than 35%.

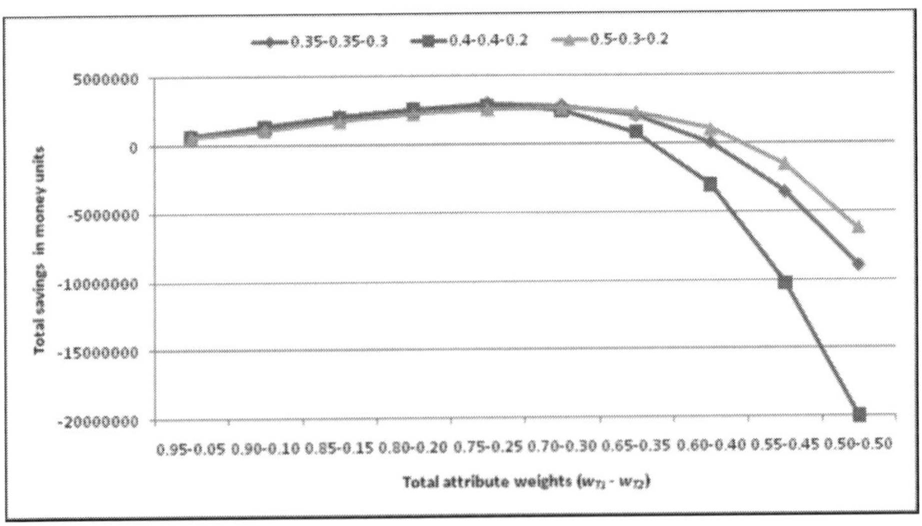


**Fig. 1.** Total savings in reputation tracking auctions

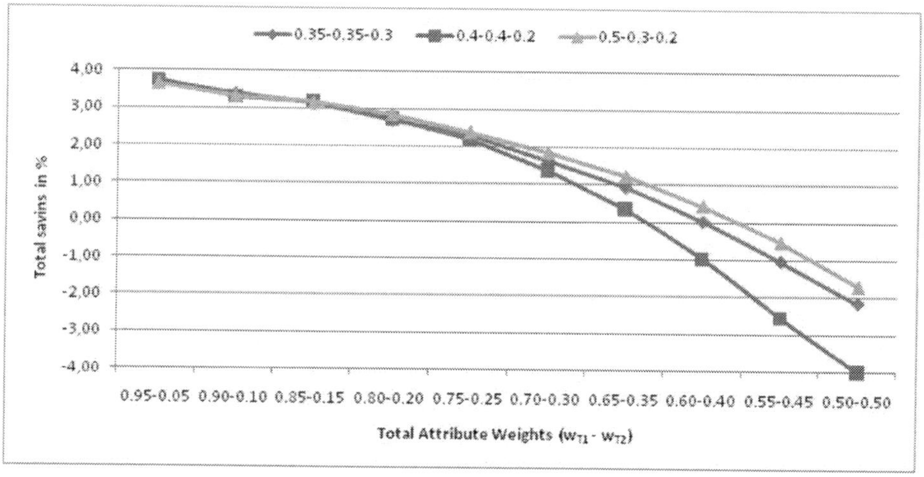


**Fig. 2.** Total savings compared with the value of all auctions

## 5 Conclusion and Future Work

In this paper, we presented a reputation tracking model developed for reverse auctions on business-to-business e-markets. As the use of reverse auctions in the industry grows, business partners are concerned with how they will influence previously established business relationships. We propose a model that exploits the advantages auctions bring to buyers, while decreasing the disadvantages they bring to sellers. The buyer calculates seller's reputations from a combination of different attributes, such as the quality of items the supplier offers, the volume of all transactions conducted with the supplier, and the financial damages caused by the supplier's late delivery of purchased items. We conducted several experiments and focused our analysis on auctions that have different outcomes depending on whether they took reputation into account or not. From those experiments, we can see that the reputation tracking reverse auction model achieves savings for the buyer only when the weight of seller reputation in the decision making process is less than 35%.

For future work, we plan the upgrade our supplier model with attributes modeling a supplier's willingness to make dedicated investments towards the buyer, as well as improving the way sellers make offers to their significant buyers. Furthermore, we plan to model the indirect financial damage caused to the buyer by late delivery of purchased items.

**Acknowledgments.** The work presented in this paper was carried out within research projects 036-0362027-1639 "Content Delivery and Mobility of Users and Services in New Generation Networks", supported by the Ministry of Science, Education and Sports of the Republic of Croatia, and "Agent-based Service & Telecom Operations Management", supported by Ericsson Nikola Tesla, Croatia.

## References


1. Dou, W., Chou, D.C.: A structural analysis of business-to-business digital markets. Industrial Marketing Management 31(2), 165–176 (2002)
2. Pressey, A.D., Ashton, J.K.: The antitrust implications of electronic business-to-business marketplaces. Industrial Marketing Management (in press, 2008) (Corrected Proof)
3. Shaw, M.: Electronic commerce: State of the art. In: Shaw, M., Blanning, R., Strader, T., Whinston, A. (eds.) Handbook on Electronic Commerce, pp. 3–24. Springer-Verlag New York, Inc., Secaucus (2000)
4. Jap, S.D., Haruvy, E.: Interorganizational relationships and bidding behavior in industrial online reverse auctions. Journal of Marketing Research 45(5), 550–561 (2008)
5. He, S., Cattelan, R.G., Kirovski, D.: Modeling viral economies for digital media. In: Sventek, J.S., Hand, S. (eds.) EuroSys, pp. 149–162. ACM, New York (2008)
6. Maes, P., Guttman, R.H., Moukas, A.: Agents that buy and sell. Communications of the ACM 42(3), 81–91 (1999)
7. Wurman, P.R., Wellman, M.P., Walsh, W.E.: Specifying rules for electronic auctions. AI Magazine 23(3), 15–24 (2002)
8. He, M., Jennings, N.R., Leung, H.F.: On agent-mediated electronic commerce. IEEE Transactions on Knowledge and Data Engineering 15(4), 985–1003 (2003)
9. Carter, C.R., Kaufmann, L., Beall, S., Carter, P.L., Hendrick, T.E., Petersen, K.J.: Reverse auctions–grounded theory from the buyer and supplier perspective. Transportation Research Part E: Logistics and Transportation Review 40(3), 229–254 (2004)
10. Smeltzer, L.R., Carr, A.S.: Electronic reverse auctions: Promises, risks and conditions for success. Industrial Marketing Management 32(6), 481–488 (2003)
11. Jap, S.D.: The impact of online reverse auction design on buyer-supplier relationships. Journal of Marketing 71(1), 146–159 (2007)
12. Jap, S.D.: An exploratory study of the introduction of online reverse auctions. Journal of Marketing 67(3), 96–107 (2003)
13. Engelbrecht-Wiggans, R., Haruvy, E., Katok, E.: A comparison of buyer-determined and price-based multiattribute mechanisms. Marketing Science 26(5), 629–641 (2007)
14. Do, V.T., Halatchev, M., Neumann, D.: A context-based approach to support virtual enterprises. Hawaii International Conference on System Sciences 6, 6005 (2000)
15. Bichler, M., Kalagnanam, J.: Configurable offers and winner determination in multi-attribute auctions. European Journal of Operational Research 160(2), 380–394 (2005)
16. Grey, W., Olavson, T., Shi, D.: The role of e-marketplaces in relationship-based supply chains: A survey. IBM Systems Journal 44(1), 109–124 (2005)
17. Arunachalam, R., Sadeh, N.M.: The supply chain trading agent competition. Electronic Commerce Research and Applications 4(1), 66–84 (2005)
18. Podobnik, V., Petric, A., Jezic, G.: An agent-based solution for dynamic supply chain management. Journal of Universal Computer Science 14(7), 1080–1104 (2008)
19. Tassabehji, R., Taylor, A., Beach, R., Wood, A.: Reverse e-auctions and supplier-buyer relationships: an exploratory study. International Journal of Operations & Production Management 26(2), 166–184 (2006)
20. Bichler, M.: An experimental analysis of multi-attribute auctions. Decision Support Systems 29(3), 249–268 (2000)

21. Vulkan, N., Jennings, N.R.: Efficient mechanisms for the supply of services in multi-agent environments. Decision Support Systems 28(1-2), 5–19 (2000)
22. Tong, H., Zhang, S.: A fuzzy multi-attribute decision making algorithm for web services selection based on qos. In: APSCC 2006: Proceedings of the 2006 IEEE Asia-Pacific Conference on Services Computing, Washington, DC, USA, pp. 51–57. IEEE Computer Society, Los Alamitos (2006)
23. Bellosta, M.J., Kornman, S., Vanderpooten, D.: An agent-based mechanism for autonomous multiple criteria auctions. In: IAT, pp. 587–594. IEEE Computer Society, Los Alamitos (2006)
24. Bui, T., Yen, J., Hu, J., Sankaran, S.: A multi-attribute negotiation support system with market signaling for electronic markets. Group Decision and Negotiation 10(6), 515–537 (2001)

# Comparative Analysis of Evolutionary Fuzzy Models for Premises Valuation Using KEEL

Marek Krzystanek[1], Tadeusz Lasota[2], and Bogdan Trawiński[1]

[1] Wrocław University of Technology, Institute of Informatics,
Wybrzeże Wyspiańskiego 27, 50-370 Wrocław, Poland
[2] Wrocław University of Environmental and Life Sciences, Dept. of Spatial Management
Ul. Norwida 25/27, 50-375 Wroclaw, Poland
marek.krzystanek@gmail.com, tadeusz.lasota@wp.pl,
bogdan.trawinski@pwr.wroc.pl

**Abstract.** The experiments aimed to compare evolutionary fuzzy algorithms to create models for the valuation of residential premises were conducted using KEEL. Out of 20 algorithms divided into 5 groups to final comparison five best were selected. All models were applied to actual data sets derived from the cadastral system and the registry of real estate transactions. A dozen of predictive accuracy measures were employed. Although statistical tests were not decisive, final evaluation of models could be done on the basis of the measures used.

**Keywords:** genetic fuzzy systems, real estate appraisal, KEEL.

## 1 Introduction

Applying the most popular approach to determining the market value of a property, i.e. sales comparison approach, it is necessary to have transaction prices of the properties sold with attributes similar to the one being appraised. If good comparable transactions are available, then it is possible to obtain reliable estimates. Prior to the evaluation the appraiser must conduct a thorough study of the appraised property using available sources of information such as cadastral systems, transaction registers, performing market analyses, accomplishing on-site inspection. His estimations are usually subjective and are based on his experience and intuition.

Automated valuation models (AVMs), devoted to support appraisers' work, are based primarily on multiple regression analysis [17], [26], soft computing and geographic information systems (GIS) [28]. Many intelligent methods have been developed to support appraisers' works: neural networks [27], fuzzy systems [10], case-based reasoning [22], data mining [21] and hybrid approaches [16].

In our previous works [14], [15] we investigated different machine learning algorithms, among others genetic fuzzy systems devoted to build data driven models to assist with real estate appraisals using MATLAB and KEEL tools. In this paper we report the results of experiments conducted with KEEL aimed at the comparison of several evolutionary fuzzy algorithms with respect to a dozen performance measures,

N.T. Nguyen, R. Kowalczyk, and S.-M. Chen (Eds.): ICCCI 2009, LNAI 5796, pp. 838–849, 2009.

using actual data taken from cadastral system in order to assess their appropriateness to an internet expert system assisting appraisers' work. Compared to our earlier works we tried to test all appropriate genetic fuzzy algorithms for regression included in KEEL and our preliminary experiment comprised 20 of them.

## 2 Cadastral Systems as the Source Base for Model Generation

The concept of a data driven models for premises valuation, presented in the paper, was developed based on the sales comparison method. It was assumed that whole appraisal area, that means the area of a city or a district, is split into sections (e.g. clusters) of comparable property attributes. The architecture of the proposed system is shown in Fig. 1. The appraiser accesses the system through the internet and chooses an appropriate section and input the values of the attributes of the premises being evaluated into the system, which calculates the output using a given model. The final result as a suggested value of the property is sent back to the appraiser.

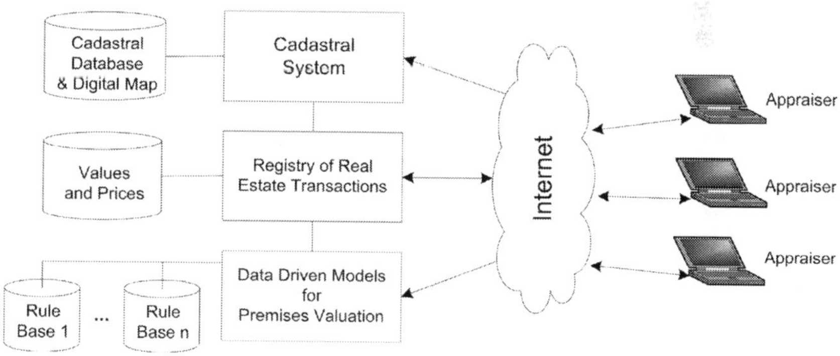


**Fig. 1.** Information systems to assist with real estate appraisals

Actual data used to generate and learn appraisal models came from the cadastral system and the registry of real estate transactions referring to residential premises sold in one of the big Polish cities at market prices within two years 2001 and 2002. They constituted original data set of 1098 cases of sales/purchase transactions. Four attributes were pointed out as price drivers: usable area of premises, floor on which premises were located, year of building construction, number of storeys in the building, in turn, price of premises was the output variable.

## 3 KEEL as the Tool for Data Driven Model Exploration

KEEL is a non-commercial Java software tool [1] designed to assess evolutionary algorithms for data mining problems. It enables problem solving through regression, classification, clustering, and data mining. Genetic fuzzy algorithms based on different approaches such as Pittsburgh, Michigan, IRL (iterative rule learning), and GCCL (genetic cooperative-competitive learning), are encapsulated into one system. KEEL is designed for different users with different expectations and provides three main functionalities: *Data Management*, which is used to set up new data, data import and export, data edition and visualization, apply data transformations and partitioning etc.; *Experiments*, which is used to design and evaluate experiments with use of selected data and provided parameters; *Education*, which is used to design an experiment and run it step-by-step in order to display learning process.

KEEL algorithms for building, learning and tuning fuzzy models employed to carry out the experiments are listed in Table 1, where references to source articles are shown. Details of the algorithms can also be found on KEEL web site: www.keel.es.

**Table 1.** Evolutionary fuzzy algorithms used in study

| Group | Code | KEEL name | Description |
|---|---|---|---|
| A | MOG1 | Regr-Fuzzy-MOGUL-IRL | MOGUL: Iterative Rule Learning of Descriptive Mamdani Rules [6] |
| | MOG2 | Regr-Fuzzy-MOGUL-IRLHC | MOGUL: Iterative Rule Learning of Mamdani Rules - High Constrained Approach [8] |
| | MOG3 | Regr-Fuzzy-MOGUL-IRLSC | MOGUL: Iterative Rule Learning of Mamdani Rules - Small Constrained Approach [6] |
| | MOG4 | Regr-Fuzzy-MOGUL-TSK | Local Evolutionary Learning of TSK fuzzy rule-based system (LEL-TSK) [2] |
| B | WMT1 | Regr-Fuzzy-WM & Post-G-G-Tuning-FRBSs | Wang-Mendel algorithm tuned using Global Genetic Tuning of the Fuzzy Partition of Linguistic FRBSs [6], [25] |
| | WMT2 | Regr-Fuzzy-WM & Post-A-G-Tuning-FRBSs | Wang-Mendel algorithm tuned using Approximative Genetic Tuning of FRBSs [9], [25] |
| | WMT3 | Regr-Fuzzy-WM & Post-G-T-Weights-FRBSs | Wang-Mendel Algorithm tuned using Genetic Tuning of FRBSs Weights [3], [25] |
| | WMT4 | Regr-Fuzzy-WM &Post-G-S-Weight-RRBS | Wang-Mendel Algorithm tuned using Genetic Selection of rules and rule weight tuning [3] |
| C | EFR1 | Regr-COR_GA | Genetic Fuzzy Rule Learning, COR algorithm inducing cooperation among rules [5] |
| | EFR2 | Regr-Thrift | Genetic Fuzzy Rule Learning, Thrift Algorithm [23] |
| | EFR3 | Regr-GFS-RB-MF | Genetic-Based Fuzzy Rule Base Construction and Membership Functions Tuning [6], [12] |
| | EFR4 | Regr-Fuzzy-P_FCS1 | Pittsburgh Fuzzy Classifier System #1 [4] |
| D | SYM1 | Regr-Fuzzy-GAP-RegSym | Symbolic Regression for fuzzy-Valued Data, Grammar-based GAP Algorithm [19] |
| | SYM2 | Regr-Fuzzy-SAP-RegSym | Symbolic Regression for fuzzy-valued data, Grammar-GP based operators and Simulated Annealing-based algorithm [19], [20] |
| | SYM3 | Regr-SAP | Symbolic Regression, Grammar-GP based operators and Simulated Annealing-based algorithm [19], [20] |
| | SYM4 | Regr-GAP | Symbolic Regression, Grammar-based GAP Algorithm [19] |
| E | FUZ1 | Regr-Fuzzy-SAP | Fuzzy Rule Learning, Grammar-GP based operators and Simulated Annealing-based algorithm [20] |
| | FUZ2 | Regr-Fuzzy-SEFC | SEFC: Symbiotic-Evolution-based Fuzzy Controller design method [13] |
| | FUZ3 | Regr-FRSBM | Fuzzy and Random Sets based Modeling [18] |
| | FUZ4 | Regr-Fuzzy-TSK-IRL | Iterative Rule Learning of TSK Rules [7] |

All the algorithms were divided into five groups for four methods: group A contains all kinds of MOGUL algorithms, group B - the Wang-Mendel algorithm for fuzzy rule learning tuned by means of evolutionary post-processing algorithms, group C - simple evolutionary fuzzy rule based systems for regression, group D - evolutionary symbolic regression algorithms, and other fuzzy systems were assigned to group E.

## 4 Experiment Description

The main goal of our investigations was to carry out the comparative analysis of 20 evolutionary fuzzy algorithms for regression implemented in KEEL, which task was to create and learn data driven models for premises property valuation. Our study consisted of two stages: the group one and the final comparison one (like in tournament). The first one aimed at selecting from each group one algorithm, which produces the best evolutionary fuzzy model. The second one was final contest between groups winners. All experiments were run for data described in section 2 using 10-fold cross validation from KEEL. In order to obtain comparable results, the normalization of data was accomplished using the min-max approach. As fitness function the mean square error (MSE) implemented in KEEL was applied. Within each group preliminary tuning was performed using empirical trial and error method in order to choose the values of parameters of individual algorithms providing the best prediction accuracy in terms of MSE.

At final stage, a dozen of commonly used performance measures [11], [26] was applied to evaluate models built by respective algorithms. The values of the measures were calculated using actual and predicted prices, obtained for testing sets, and which were saved as the results of experiments in one of KEEL experiment folders. These measures are listed in Table 2 and expressed in the form of following formulas below, where $y_i$ denotes actual price and $\hat{y}_i$ – predicted price of i-th case, *avg(v), var(v), std(v)* – average, variance, and standard deviation of variables $v_1, v_2, \ldots, v_N$, respectively and $N$ – number of cases in the testing set.

**Table 2.** Performance measures used in study

| Denot. | Description | Dimen-sion | Min value | Max value | Desirable outcome | No. of form. |
|---|---|---|---|---|---|---|
| *MSE* | Mean squared error | $d^2$ | 0 | ∞ | min | 1 |
| *RMSE* | Root mean squared error | d | 0 | ∞ | min | 2 |
| *RSE* | Relative squared error | no | 0 | ∞ | min | 3 |
| *RRSE* | Root relative squared error | no | 0 | ∞ | min | 4 |
| *MAE* | Mean absolute error | d | 0 | ∞ | min | 5 |
| *RAE* | Relative absolute error | no | 0 | ∞ | min | 6 |
| *MAPE* | Mean absolute percentage error | % | 0 | ∞ | min | 7 |
| *NDEI* | Non-dimensional error index | no | 0 | ∞ | min | 8 |
| *r* | Linear correlation coefficient | no | -1 | 1 | close to 1 | 9 |
| $R^2$ | Coefficient of determination | % | 0 | ∞ | close to 100% | 10 |
| *var(AE)* | Variance of absolute errors | $d^2$ | 0 | ∞ | min | 11 |
| *var(APE)* | Variance of absolute percentage errors | no | 0 | ∞ | min | 12 |

$$MSE = \frac{1}{N}\sum_{i=1}^{N}(y_i - \hat{y}_i)^2 \tag{1}$$

$$RMSE = \sqrt{\frac{1}{N}\sum_{i=1}^{N}(y_i - \hat{y}_i)^2} \tag{2}$$

$$RSE = \frac{\sum_{i=1}^{N}(y_i - \hat{y}_i)^2}{\sum_{i=1}^{N}(y_i - avg(y))^2} \tag{3}$$

$$RRSE = \sqrt{\frac{\sum_{i=1}^{N}(y_i - \hat{y}_i)^2}{\sum_{i=1}^{N}(y_i - avg(y))^2}} \tag{4}$$

$$MAE = \frac{1}{N}\sum_{i=1}^{N}|y_i - \hat{y}_i| \tag{5}$$

$$RAE = \frac{\sum_{i=1}^{N}|y_i - \hat{y}_i|}{\sum_{i=1}^{N}|y_i - avg(y)|} \tag{6}$$

$$MAPE = \frac{1}{N}\sum_{i=1}^{N}\frac{|y_i - \hat{y}_i|}{y_i} * 100\% \tag{7}$$

$$NDEI = \frac{RMSE}{std(y)} \tag{8}$$

$$r = \frac{\sum_{i=1}^{N}(y_i - avg(y))\,(\hat{y}_i - avg(\hat{y}))}{\sqrt{\sum_{i=1}^{N}(y_i - avg(y))^2}\,\sqrt{\sum_{i=1}^{N}(\hat{y}_i - avg(\hat{y}))^2}} \tag{9}$$

$$R^2 = \frac{\sum_{i=1}^{N}(\hat{y}_i - avg(y))^2}{\sum_{i=1}^{N}(y_i - avg(y))^2} * 100\% \tag{10}$$

$$var(AE) = var(|y - \hat{y}|) \tag{11}$$

$$var(APE) = var(\frac{|y - \hat{y}|}{y}) \tag{12}$$

## 5 Results of Experiments

Results of preliminary stage are shown in Fig. 2-4. It can be observed that MSE and MAPE differentiate the algorithms similarly. The algorithms which created models with the lowest MSE within individual groups are listed in Table 3.

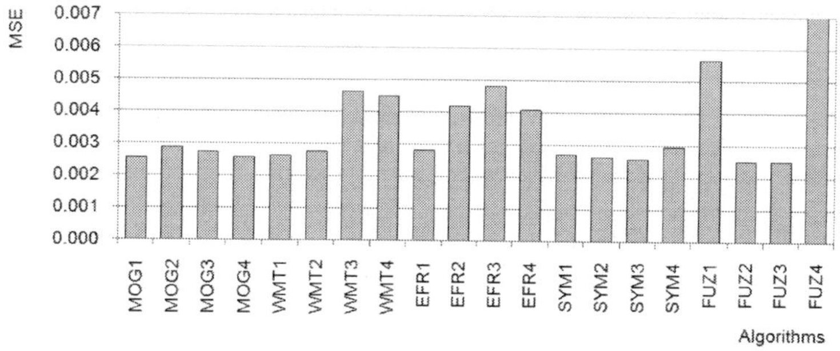


**Fig. 2.** Comparison of MSE values at preliminary stage

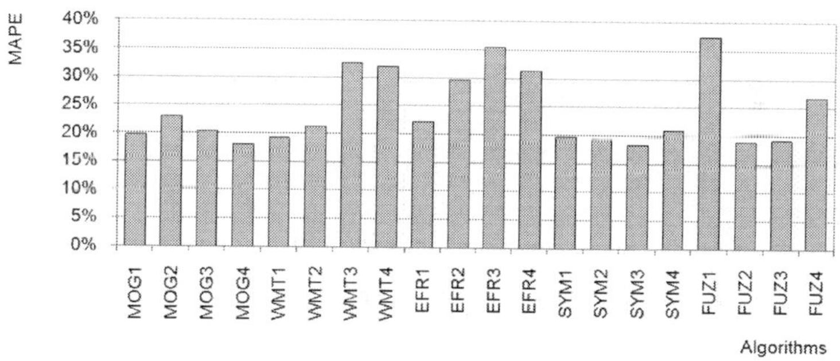


**Fig. 3.** Comparison of MAPE values at preliminary stage

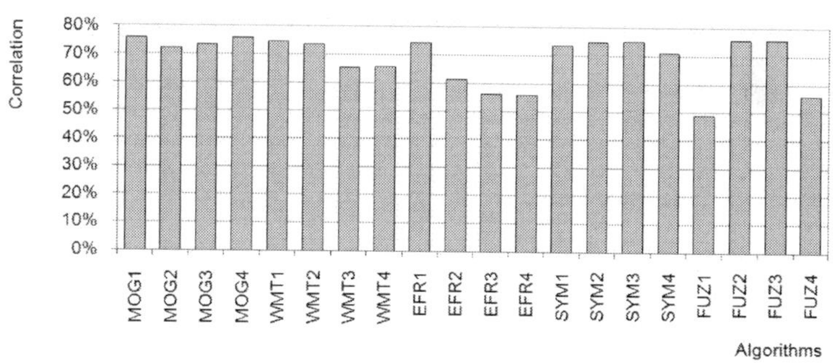


**Fig. 4.** Comparison of correlation between actual and predicted prices at preliminary stage

**Table 3.** Best algorithms within respective groups of algorithms

| Group | Final stage | Group winner | KEEL name |
|---|---|---|---|
| A | MOG | MOG1 | Regr-Fuzzy-MOGUL-IRL |
| B | WMT | WMT1 | Regr-Fuzzy-WM & Post-G-G-Tuning-FRBSs |
| C | EFR | EFR1 | Regr-COR_GA |
| D | SYM | SYM3 | Regr-SAP |
| E | FUZ | FUZ2 | Regr-Fuzzy-SEFC |

Final stage of study contained comparison of algorithms listed in Table 3, using all 12 performance measures enumerated in pervious section. The results of respective measures for all models are shown in Fig. 5-16, it can be easily noticed that relationship among individual models are very similar for some groups of measures.

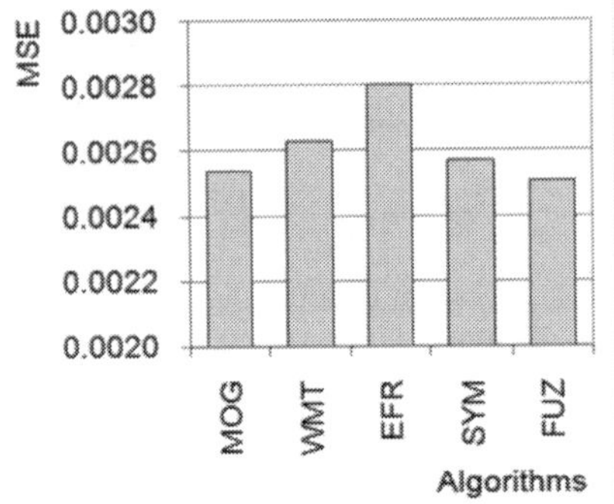


Fig. 5. Comparison of MSE values

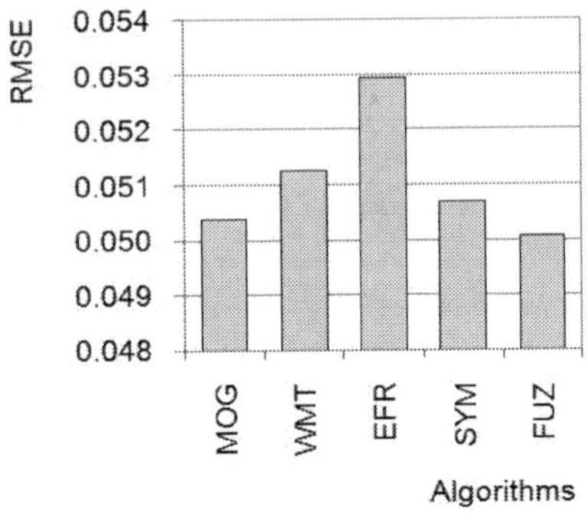


Fig. 6. Comparison of RMSE values

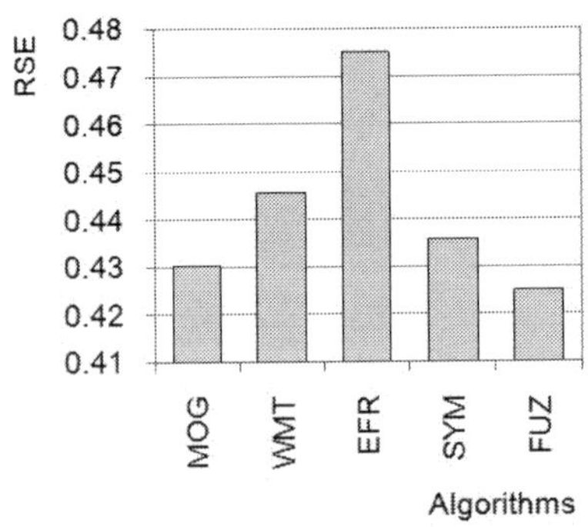


Fig. 7. Comparison of RSE values

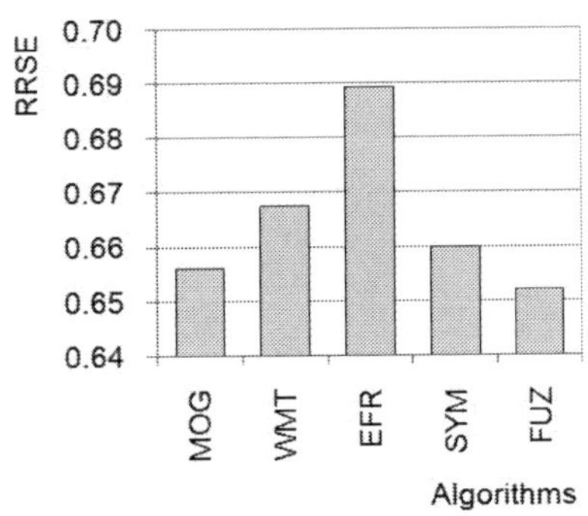


Fig. 8. Comparison of RRSE values

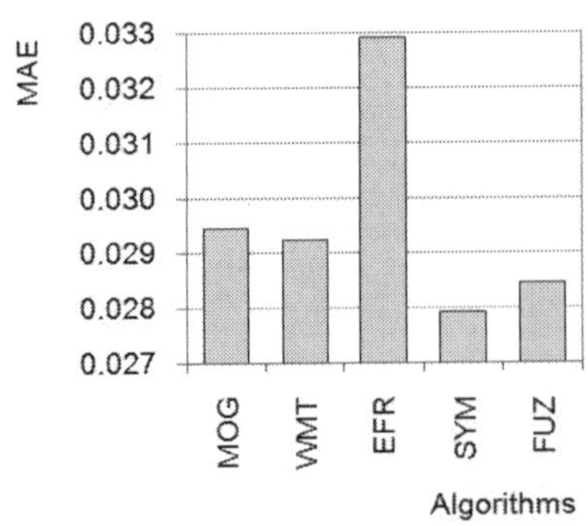


Fig. 9. Comparison of MAE values

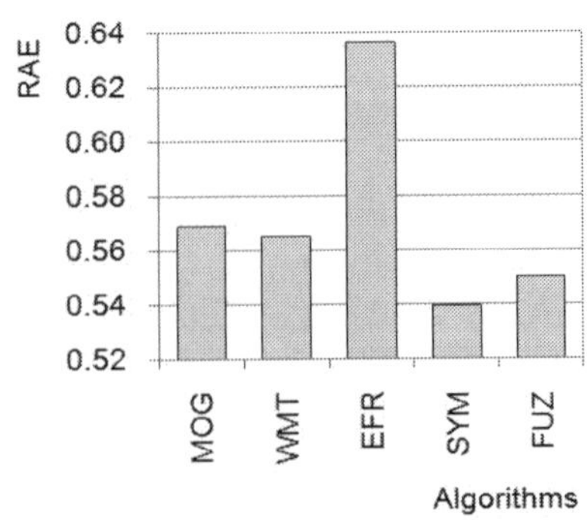


Fig. 10. Comparison of RAE values

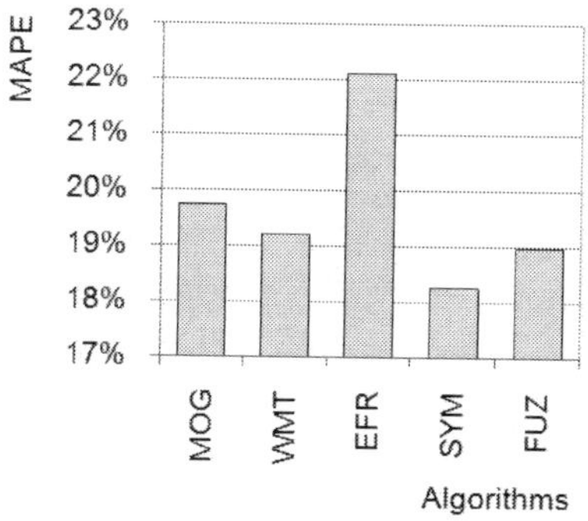


**Fig. 11.** Comparison of MAPE values

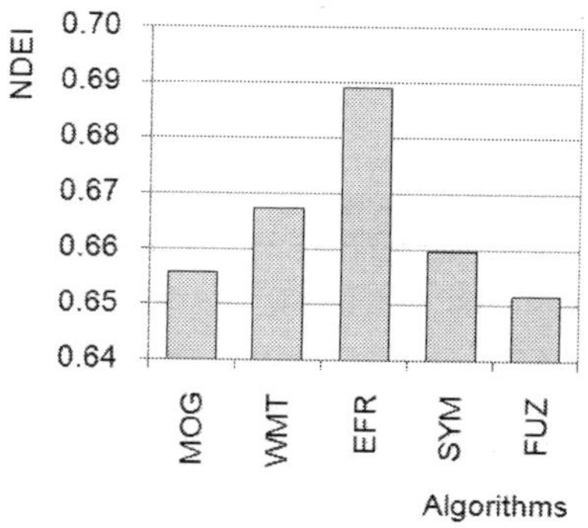


**Fig. 12.** Comparison of NDEI values

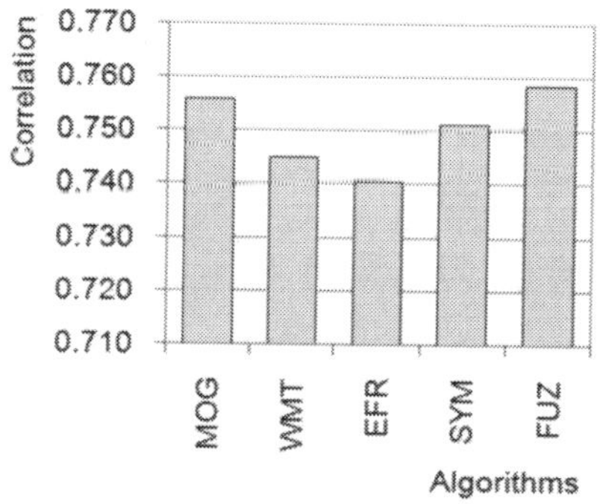


**Fig. 13.** Comparison of correlation coefficient (r) values

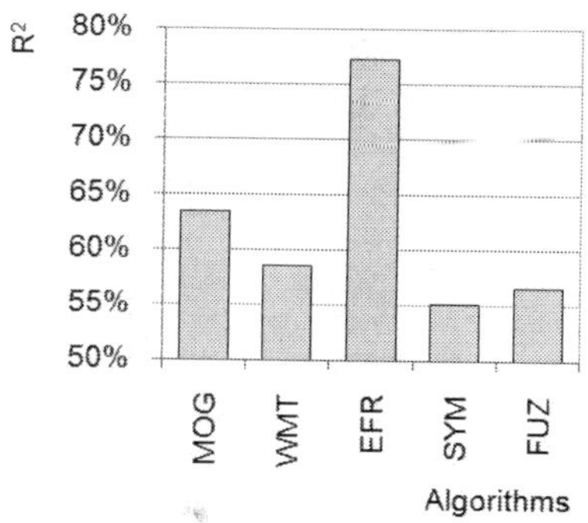


**Fig. 14.** Comparison of determination coefficient (R2) values

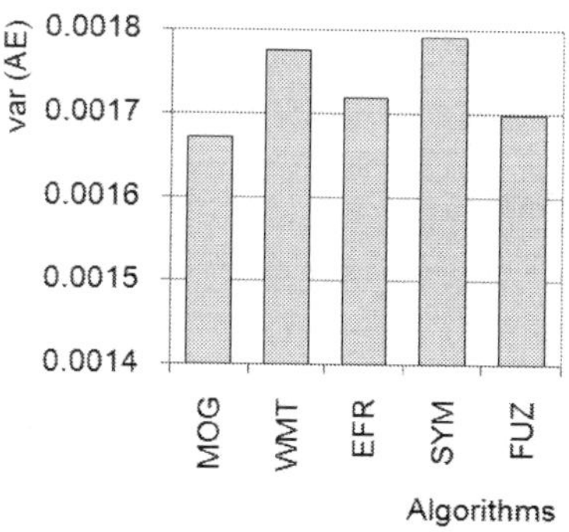


**Fig. 15.** Comparison of var(AE) values

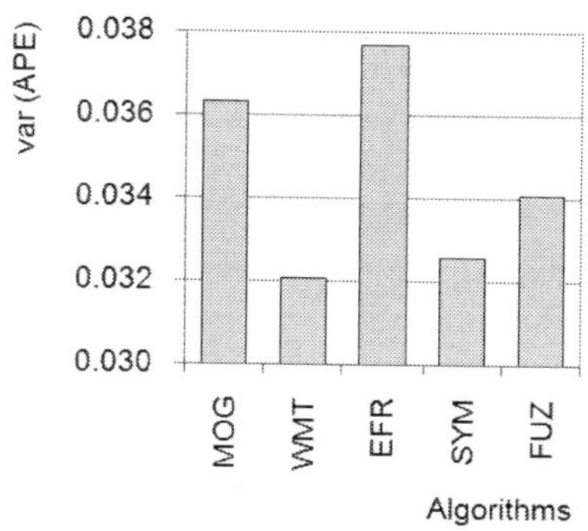


**Fig. 16.** Comparison of var(APE) values

Fig. 11 depicts that the values of MAPE range from 18.2% to 22.1%, what can be regarded as fairly good, especially when you take into account that not all price drivers were available in our sources of experimental data.

Fig. 13 shows there is high correlation, i.e. about 0.75, between actual and predicted prices for each model. In turn, Fig.14 illustrating the coefficients of determination indicates that from 55.2% to 77.2% of total variation in the dependent variable (prices) is accounted for by the models. This can be explained that data derived from the cadastral system and the register of property values and prices could cover only some part of potential price drivers.

The nonparametric Wilcoxon signed-rank tests were carried out for three measures: MSE, MAPE, and MAE. The results are shown in Tables 4, 5, and 6. In each cell results for a given pair of models were placed, in upper halves of the tables – p-values, and in lower halves - final outcome, where N denotes that there are no differences in mean values of respective errors, and Y indicates that there are statistically significant differences between particular performance measures. For EFR algorithm all tests resulted in significant differences, thus it can be stated this algorithm provides the worst results. However, for other algorithms the Wilcoxon signed-rank tests did not provide any decisive result.

**Table 4.** Results of Wilcoxon signed-rank test for squared errors comprised by MSE

| MSE | MOG | WMT | EFR | SYM | FUZ |
|---|---|---|---|---|---|
| **MOG** | | 0.374 | 0.000 | 0.048 | 0.438 |
| **WMT** | N | | 0.001 | 0.161 | 0.885 |
| **EFR** | Y | Y | | 0.000 | 0.000 |
| **SYM** | Y | N | Y | | 0.077 |
| **FUZ** | N | N | Y | N | |

**Table 5.** Results of Wilcoxon test for absolute percentage errors comprised by MAPE

| MAPE | MOG | WMT | EFR | SYM | FUZ |
|---|---|---|---|---|---|
| **MOG** | | 0.247 | 0.000 | 0.000 | 0.208 |
| **WMT** | N | | 0.000 | 0.039 | 0.982 |
| **EFR** | Y | Y | | 0.000 | 0.000 |
| **SYM** | Y | Y | Y | | 0.010 |
| **FUZ** | N | N | Y | Y | |

**Table 6.** Results of Wilcoxon signed-rank test for absolute errors comprised by MAE

| MAE | MOG | WMT | EFR | SYM | FUZ |
|---|---|---|---|---|---|
| **MOG** | | 0.362 | 0.000 | 0.011 | 0.308 |
| **WMT** | N | | 0.000 | 0.071 | 0.727 |
| **EFR** | Y | Y | | 0.000 | 0.000 |
| **SYM** | Y | N | Y | | 0.047 |
| **FUZ** | N | N | Y | Y | |

Due to the non-decisive results of majority of statistical tests, rank positions of individual algorithms were determined for each measure (see Table 7). Observing median, average, minimal and maximal ranks it can be noticed that highest rank positions gained FUZ, MOG, SYM algorithms and the lowest WMT and EFR. Table 7 indicates also that some performance measures provide the same rank positions, and

**Table 7.** Rank positions of algorithms with respect to performance measures (1 means the best)

| | MOG | WMT | EFR | SYM | FUZ |
|---|---|---|---|---|---|
| MSE | 2 | 4 | 5 | 3 | 1 |
| RMSE | 2 | 4 | 5 | 3 | 1 |
| RSE | 2 | 4 | 5 | 3 | 1 |
| RRSE | 2 | 4 | 5 | 3 | 1 |
| MAE | 4 | 3 | 5 | 1 | 2 |
| RAE | 4 | 3 | 5 | 1 | 2 |
| MAPE | 4 | 3 | 5 | 1 | 2 |
| NDEI | 2 | 4 | 5 | 3 | 1 |
| r | 2 | 4 | 5 | 3 | 1 |
| $R^2$ | 2 | 3 | 1 | 5 | 4 |
| var(AE) | 1 | 4 | 3 | 5 | 2 |
| var(APE) | 4 | 1 | 5 | 2 | 3 |
| median | 2.00 | 4.00 | 5.00 | 3.00 | 1.50 |
| average | 2.58 | 3.42 | 4.50 | 2.75 | 1.75 |
| min | 1 | 1 | 1 | 1 | 1 |
| max | 4 | 4 | 5 | 5 | 4 |

two groups of those measures can be distinguished. First one based on mean square errors contains MSE, RMSE, RSE, RRSE, NDEI, and the second one based on mean absolute errors comprises MAE, RAE, and MAPE.

# 6 Conclusions and Future Work

The goal of experiments was to compare several evolutionary fuzzy algorithms to create models for the valuation of residential premises, implemented in KEEL. Out of 20 algorithms divided into 5 groups to final experimental comparison five best were selected. A dozen of commonly used performance measures were applied to models generated using actual data set derived from cadastral system and the registry of real estate transactions. Although nonparametric Wilcoxon signed-rank tests were not decisive, the models could be ranked according to decreasing accuracy in following way: FUZ, MOG, SYM, WMT, EFR.

Some performance measures provide the same distinction abilities of respective models, thus it can be concluded that in order to compare a number of models it is not necessary to employ all measures, but the representatives of different groups. Of course the measures within groups differ in their interpretation, because some are non-dimensional as well as in their sensitivity understood as the ability to show the differences between algorithms more or less distinctly.

High correlation between actual and predicted prices was observed for each model and the coefficients of determination ranged from 55% to 77% .

MAPE obtained in all tests ranged from 18% do 22%. This can be explained that data derived from the cadastral system and the register of property values and prices can cover only some part of potential price drivers. Physical condition of the premises and their building, their equipment and facilities, the neighbourhood of the building, the location in a given part of a city should also be taken into account, moreover overall subjective assessment after inspection in site should be done. Therefore we intend to test data obtained from public registers and then supplemented by experts conducting on-site inspections and evaluating more aspects of properties being appraised. This kind of data will be very interesting from the point of view of evolutionary fuzzy systems since most of them will be imprecise linguistic information. Moreover further investigations of multiple models comprising ensembles of different genetic fuzzy systems using bagging and boosting techniques is planned.

## References


1. Alcalá-Fdez, J., Sánchez, L., García, S., del Jesus, M.J., Ventura, S., Garrell, J.M., Otero, J., Romero, C., Bacardit, J., Rivas, V.M., Fernández, J.C., Herrera., F.: KEEL: A Software Tool to Assess Evolutionary Algorithms to Data Mining Problems. Soft Computing 13(3), 307–318 (2009)
2. Alcalá, R., Alcalá-Fdez, J., Casillas, J., Cordón, O., Herrera, F.: Local identification of prototypes for genetic learning of accurate TSK fuzzy rule-based systems. International Journal of Intelligent Systems 22(9), 909–941 (2007)
3. Alcala, R., Cordón, O., Herrera, F.: Combining Rule Weight Learning and Rule Selection to Obtain Simpler and More Accurate Linguistic Fuzzy Models. In: Lawry, J., G. Shanahan, J., L. Ralescu, A. (eds.) Modelling with Words. LNCS, vol. 2873, pp. 44–63. Springer, Heidelberg (2003)
4. Carse, B., Fogarty, T.C., Munro, A.: Evolving fuzzy rule based controllers using genetic algorithms. Fuzzy Sets and Systems 80(3), 273–293 (1996)
5. Casillas, J., Cordón, O., Herrera, F.: COR: A methodology to improve ad hoc data-driven linguistic rule learning methods by inducing cooperation among rules. IEEE Trans. on System, Man and Cybernetics, Part B: Cybernetics 32(4), 526–537 (2002)
6. Cordón, O., Herrera, F.: A three-stage evolutionary process for learning descriptive and approximate fuzzy logic controller knowledge bases from examples. International Journal of Approximate Reasoning 17(4), 369–407 (1997)
7. Cordón, O., Herrera, F.: A Two-Stage Evolutionary Process for Designing TSK Fuzzy Rule-Based Systems. IEEE Trans. on Systems, Man and Cybernetics, Part B: Cybernetics 29(6), 703–715 (1999)
8. Cordón, O., Herrera, F.: Hybridizing genetic algorithms with sharing scheme and evolution strategies for designing approximate fuzzy rule-based systems. Fuzzy Sets and Systems 118(2), 235–255 (2001)
9. Herrera, F., Lozano, M., Verdegay, J.L.: Tuning Fuzzy Logic Controllers by Genetic Algorithms. International Journal of Approximate Reasoning 12, 299–315 (1995)
10. González, M.A.S., Formoso, C.T.: Mass appraisal with genetic fuzzy rule-based systems. Property Management 24(1), 20–30 (2006)
11. Hagquist, C., Stenbeck, M.: Goodness of Fit in Regression Analysis – R2 and G2 Reconsidered. Quality & Quantity 32, 229–245 (1998)
12. Homaifar, A., McCormick., A.E.: Simultaneous Design of Membership Functions and Rule Sets for Fuzzy Controllers Using Genetic Algorithms. IEEE Trans. on Fuzzy Systems 3(2), 129–139 (1995)
13. Juang, C.-F., Lin, J.-Y., Lin, C.-T.: Genetic reinforcement learning through symbiotic evolution for fuzzy controller design. IEEE Trans. on Systems, Man, and Cybernetics, Part B: Cybernetics 30(2), 290–302 (2000)

14. Król, D., Lasota, T., Trawiński, B., Trawiński, K.: Investigation of evolutionary optimization methods of TSK fuzzy model for real estate appraisal. International Journal of Hybrid Intelligent Systems 5(3), 111–128 (2008)
15. Lasota, T., Pronobis, E., Trawiński, B., Trawiński, K.: Exploration of Soft Computing Models for the Valuation of Residential Premises using the KEEL Tool. In: Hoi, D., Vietnam, Nguyen, N.T., et al. (eds.) 1st Asian Conference on Intelligent Information and Database Systems (ACIIDS 2009), pp. 253–258. IEEE, Los Alamitos (2009)
16. McCluskey, W.J., Anand, S.: The application of intelligent hybrid techniques for the mass appraisal of residential properties. Journal of Property Investment and Finance 17(3), 218–239 (1999)
17. Nguyen, N., Cripps, A.: Predicting housing value: A comparison of multiple regression analysis and artificial neural networks. J. of Real Estate Res. 22(3), 3131–3336 (2001)
18. Sánchez., L.: A random sets-based method for identifying fuzzy models. Fuzzy Sets and Systems 98(3), 343–354 (1998)
19. Sánchez, L., Couso., I.: Fuzzy random variables-based modeling with GA-P Algorithms. In: Yager, R., Bouchon-Menier, B., Zadeh, L. (eds.) Information, Uncertainty and Fusion, pp. 245–256. Kluwer, Dordrecht (2000)
20. Sánchez, L., Couso., I.: Combining GP operators with SA search to evolve fuzzy rule based classifiers. Information Sciences 136, 175–192 (2001)
21. Soibelman, W.L., González, M.A.S.: A Knowledge Discovery in Databases Framework for Property Valuation. Journal of Property Tax Assessment and Admin. 7(2), 77–106 (2002)
22. Taffese, W.Z.: Case-based reasoning and neural networks for real state valuation. In: Proceedings of the 25th IASTED International Multi-Conference: Artificial Intelligence and Applications, Innsbruck, Austria (2007)
23. Thrift, P.: Fuzzy logic synthesis with genetic algorithms. In: Proceedings of the Fourth Int. Conference on Genetic Algorithms (ICGA 1991), San Diego, pp. 509–513 (1991)
24. Waller, B.D., Greer, T.H., Riley, N.F.: An Appraisal Tool for the 21st Century: Automated Valuation Models. Australian Property Journal 36(7), 636–641 (2001)
25. Wang, L.X., Mendel, J.M.: Generating Fuzzy Rules by Learning from Examples. IEEE Trans. on Systems, Man and Cybernetics 22(6), 1414–1427 (1992)
26. Witten, I.H., Frank, E.: Data Mining: Practical machine learning tools and techniques, 2nd edn. Morgan Kaufmann, San Francisco (2005)
27. Worzala, E., Lenk, M., Silva, A.: An Exploration of Neural Networks and Its Application to Real Estate Valuation. J. of Real Estate Res. 10(2), 185–201 (1995)
28. Wyatt, P.: The development of a GIS-based property information system for real estate valuation. Int. J. Geographical Inf. Sci. 111(5), 435–450 (1997)

# Hybrid Repayment Prediction for Debt Portfolio

Tomasz Kajdanowicz and Przemysław Kazienko

Wrocław University of Technology, Wyb. Wyspiańskiego 27, 50-370 Wrocław, Poland
{tomasz.kajdanowicz,kazienko}@pwr.wroc.pl

**Abstract.** The concept of the new hybrid method for debt portfolio repayment prediction has been presented and examined. The method provides functionality for repayment value prediction over time that describes the recovery profile of the debt portfolio. Experimental studies on hybrid combination of various data mining methods like clustering and decision trees into one complex process revealed usefulness of the proposed method for claim appraisals.

**Keywords:** hybrid information system, prediction, repayment prediction, claim appraisal, competence regions modeling.

## 1 Introduction

There exists a wide variety of studies on prediction and classification in the literature e.g. [8, 11]. Overall, existing data mining methods usually provide better classification and prediction accuracy than techniques based only on common statistical techniques such as regression [12, 13]. A better precision of the prediction may be obtained by combination of several existing methods into one hybrid solution [1, 2, 3]. In general, hybridization could be achieved either by application of additional external mechanisms into existing prediction models (low level), e.g. neuro-fuzzy systems [7] or by combination of different methods on high level, e.g. multiple classifier systems, where separate classifiers are treated more likewise 'black boxes' [4, 5].

Hybrid prediction methods have been successfully used in a number of domains such as medicine, engineering and industry. Other application areas of these methods are economy and finance, where hybrid systems provide specialized knowledge in order to support business decisions [10].

The paper is focused on the description of a new hybrid method for debt portfolio appraisal. The correct prediction of target repayment value in debt recovery is of great practical importance, because it reveals the level of possible expected benefit and chances to collect receivables. The crucial concept of this method is clustering of the reference set and application of multiple classifiers based on their region of competence. Additionally, a sequence of classifiers is built to obtain predictions over periods. Apart from the general idea, the proposed hybrid prediction method has been examined on real data. According to the findings achieved, the method appears to return more precise results in comparison with some regular approaches.

N.T. Nguyen, R. Kowalczyk, and S.-M. Chen (Eds.): ICCCI 2009, LNAI 5796, pp. 850–857, 2009.

## 2 Business Process of Debt Portfolio Recovery

Each company whose activity is strictly connected with consumer market and which relays on the cash flow from receivables is very interested in clients' debt to be retained on the lowest possible level. Possession of creditors for a long time is very ineffective, especially when debtors are not able to repay theirs arrears in short term. Under some circumstances, it is better for companies to sell the liabilities to a specialist debt collection company and to obtain at least a part of their nominal value. In the process of selling the debt portfolio, the transaction price is usually estimated based on the possible repayment to be obtained in the long term. Overall, it is expected that the method of debt portfolio value appraisal will bring values matching with the reality.

The process of debt portfolio value prediction starts when the first company offers a package of debts and expects a purchase proposal from the second one, see Fig. 1. The second company is usually a specialist debt recovery entity. Based on historical data of debt recovery available for the second company, a prediction model is prepared. The model provides estimation of possible return from the package. The bid is supplemented by additional cost of repayment procedures and cash flow abilities as far as risk and final purchase price is proposed to the first company. The most significant part of the process is the repayment value prediction for debt portfolio as there is a strong business need for the method to be designed for efficient and accurate prediction with the time factor.

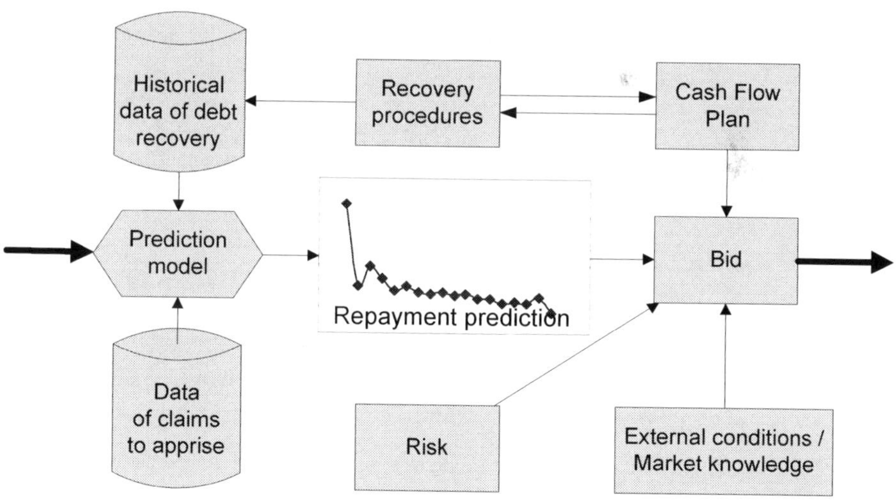


**Fig. 1.** The business proces of purchasing a debt portfolio based on repayment prediction

## 3 Repayment Prediction over Time

Having the data of historical claim cases together with their repayment profiles over time, a debt collection company can build a model to predict receivables for the new claim set invited for bids, see Fig. 2. However, in order to be able to evaluate cash flows in the following periods (usually months), the company needs to have possibly precise distribution of the receivables collection. It helps to estimate the final upper value for the considered input debt portfolio. Hence, not only the total aggregated value of the receivables is useful for bidding but also their probable timing, period by period.

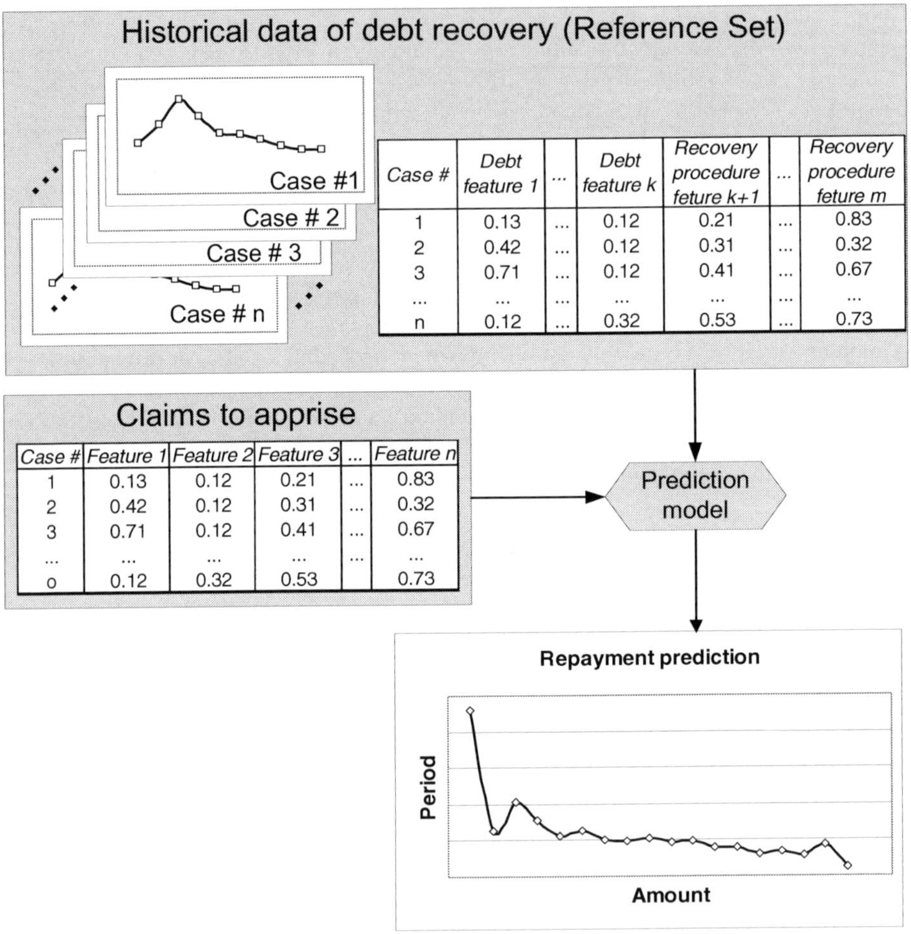


**Fig. 2.** The concept of repayment prediction based on the reference data set

### 3.1 General Concept of the Hybrid Prediction Method

The idea of the hybrid method for debt recovery value prediction consists of many data flows that are executed separately for each period (*M* times), see Fig. 3. First, the prepared historical data is clustered into groups of similar debt cases. Next, a set of models is created (learnt) separately for each cluster using the fixed set of common, predefined models. The best one is selected for each cluster and becomes the cluster's

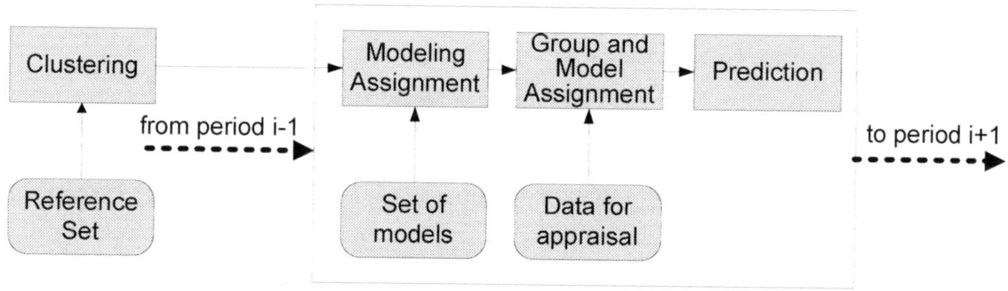


**Fig. 3.** Prediction of repayment for the single *i*th period

predictive model. This assignment is done based on minimization of the average square error. This is the main learning phase followed by the final prediction for the debt portfolio. For each of debt cases, the closest cluster of historical data is determined and the prediction for this case is done based on the model assigned and learnt on that cluster.

The important characteristic of the method is that the predicted value of return on debt in period $i$ is taken as the input variable for $i$+1th period prediction as an additional feature in the model, see Fig 4. Symbol "G & C" represents the group together with its classifier (model) assignment.

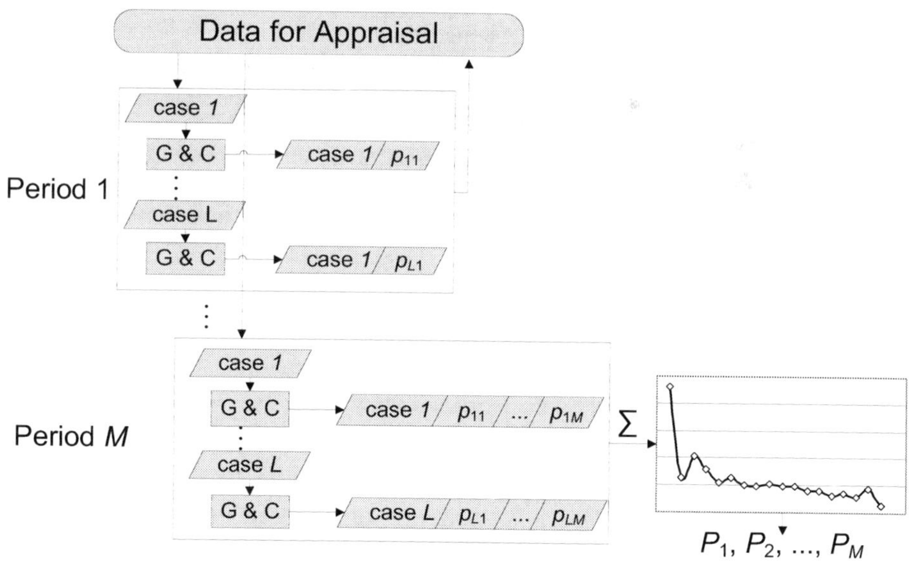


**Fig. 4.** Modeling prediction of repayment for the sequence of periods

### 3.2 Features Processed

During the data preparation procedure, some of the most meaningful features are selected and some new ones are created. This is valid both for the reference data set and the input set (claims to appraise). Finally, we obtain a fixed set of $N^f$ features. In practice, this can be quite complex process supported by expert knowledge. However, it is out of scope of this paper.

The prediction is performed separately for each period $i$ and the predictive model for each period provides a new, predictive variable $p_i$. This variable, in turn, is used as an additional input, descriptive feature for all following period $i$+1,$i$+2, $i$+3, …, see Fig. 4. As a result, a separate number of features $N^f+i-1$ is used in the prediction for period $i$.

### 3.3 Clustering of the Reference Set

Due to different origin, all historical cases can contain more or less information about repayments, i.e. for smaller or greater number of periods. In general, the longer period the less historical data is available.

Historical, reference cases are clustered into $N^G$ groups using any partitioning method and the best prediction model is separately assigned to each group and each period $i$. Only features directly available within the input data set or new ones derived from them are used in the clustering process. Besides, clustering is performed for the whole reference set, i.e. for cases being after at least one period of the recovery procedure (period 1). For the following periods, e.g. for period $i$, cases with to short history (being recovered shorter than $i$ periods), are just removed from their clusters without re-clustering. As a result, the quantity of each cluster $G_{ij}$ may vary depending on the period $i$ and it is smaller for larger $i$. For the $j$th group $G_{ij}$ and the $i$th period, we have: $card(G_{ij}) \geq card(G_{(i+1)j})$. In consequence, there are the same reference groups for all periods but their content decreases for the following periods. This is obvious, because the debt collection company possesses many pending recovery cases, which can be used as references in prediction only for the beginning periods.

Each group $G_{ij}$ possesses its own representation and the common similarity function is used to evaluate closeness between group $G_{ij}$ and each input case $x$ just being predicted. Next, the single closest group, or more precise the assigned model, is applied to the input case $x$.

## 4 Experiments

In the experiment, the debt recovery value prediction has been conducted for five consecutive periods. Two separate data sets were used: one reference set containing several debt portfolios and one debt portfolio as the input set. The latter had the predictive variable set up (real output values) but it was used only to assess the accuracy of the method. Prediction is realized in three different ways, according to separate scenarios, which have been finally compared. Only different decision trees were used as prediction models but in each scenario, the output of period $i$ is used as the input variable for the following periods.

The first scenario assumes simple prediction to be carried out on the single model, which has been learnt on all historical data. The learning is executed separately for each period. In the second scenario, clustering of historical reference data was applied. Next, the clusters were used to learn a single type of model with fixed

**Table 1.** The results of debt recovery value prediction for three different scenarios

| Period | Appraisal data (real output values) | Prediction scenario 1 | Average square error | Prediction scenario 2 | Average square error | Prediction scenario 3 | Average square error |
|---|---|---|---|---|---|---|---|
| 1 | 643 115 | 980 955 | 0.00681 | 811 408 | 0.00621 | 761 947 | 0.00515 |
| 2 | 396 759 | 200 814 | 0.00154 | 267 037 | 0.00147 | 319 547 | 0.00138 |
| 3 | 250 042 | 129 803 | 0.00179 | 185 267 | 0.00167 | 189 545 | 0.00163 |
| 4 | 173 877 | 190 690 | 0.00253 | 183 841 | 0.00183 | 183 615 | 0.00183 |
| 5 | 154 497 | 109 362 | 0.00142 | 125 170 | 0.00164 | 166 196 | 0.00116 |
| Average | | | 0.00282 | | 0.00256 | | 0.00223 |
| Std. dev. | | | 0.00227 | | 0.00204 | | 0.00165 |
| Sum | 1 618 290 | 1 611 623 | | 1 572 723 | | 1 620 849 | |

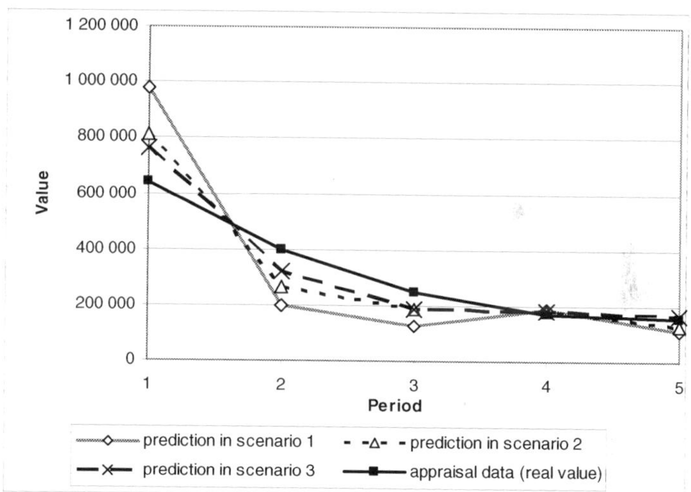


**Fig. 5.** Debt recovery value prediction in three different scenarios

parameters, one model for each cluster. Input cases were assigned to the closest cluster and processed by the assigned predictor. The full hybrid process, see Fig. 3 and 4, is performed in the third scenario. It is the extension of the second model, i.e. only one most competent model, from within a set of models, is assigned to each cluster. Similarly to the second scenario, if an appraisal case is close to the certain cluster, the return value would be predicted by the model assigned to that cluster.

The study is based on historical data that contained the repayment history of about 15,000 debt cases. The precise number of cases must have been hidden due to security policy of the debt collection company. It also refers to the values of appropriate features, however the error rates remained real.

The initial set of models was composed of 10 distinct in parameterization regression trees, i.e. confidence factor in range of 0.1 to 0.5, minimum number of instances in a leaf from 5 to 30 and maximal depth of 3 to 10.

The results of experiments are presented in Tab. 1.

Average square error in the results of three distinct prediction scenarios revealed that the method in the third scenario performs better by 26% than in the first one and by 14% better than the in second. The absolute error on total predicted value was reduced 2.6 times.

The sum of predicted values between scenarios cannot be compared unambiguously, because it is composed of prediction for periods that overestimate (period 1) and underestimate (periods 2, 3, 4, 5). However, comparison of figures predicted for particular periods points at smaller absolute error in following scenarios that can be observed on the Fig. 5 and Fig. 6.

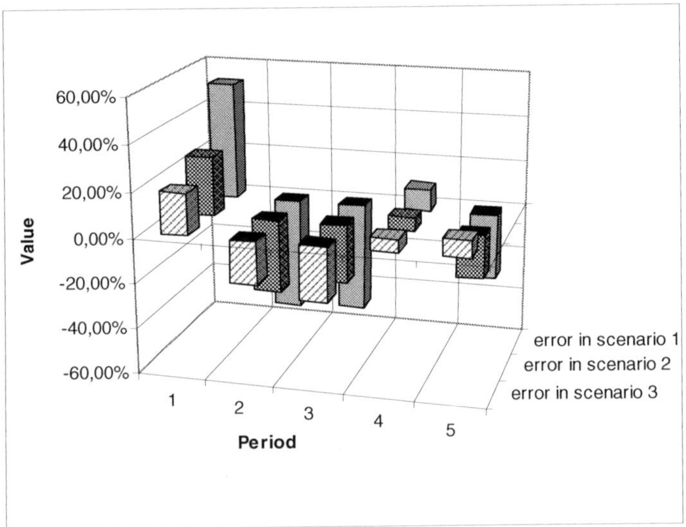


**Fig. 6.** Average square error in debt recovery value prediction for three different scenarios

## 6 Conclusions and Future Work

In order to predict debt portfolio value, the hybrid method has been suggested. Application of this method in the process of claim appraisal reduces error and makes the prediction better match the reality. Results obtained from the examined experiments support the conclusion that hybrid prediction solutions may be efficiently applied to debt recovery value prediction.

In the future studies, many further aspects improving and popularizing the method will be considered, in particular: combination of distinct types of classifiers, models' tuning using genetic based optimization [9] and adaptive clustering.

The application of the similar hybrid concept is also considered to be applied to social-based recommender system [6].

**Acknowledgments.** The work was supported by The Polish Ministry of Science and Higher Education, grant no. N516 037 31/3708.

## References

1. Aburto, L., Weber, R.: A Sequential Hybrid Forecasting System for Demand Prediction. In: Perner, P. (ed.) MLDM 2007. LNCS (LNAI), vol. 4571, pp. 518–532. Springer, Heidelberg (2007)
2. Ali, S., Smith, K.: On learning algorithm selection for classification. Applied Soft Computing 6(2), 119–138 (2006)
3. Chou, C.-H., Lin, C.-C., Liu, Y.-H., Chang, F.: A prototype classification method and its use in a hybrid solution for multiclass pattern recognition. Pattern Recognition 39(4), 624–634 (2006)
4. Eastwood, M., Gabrys, B.: Building Combined Classifiers. In: Nguyen, N.T., Kolaczek, G., Gabrys, B. (eds.) A book chapter in Knowledge Processing and Reasoning for Information Society, pp. 139–163. EXIT Publishing House, Warsaw (2008)
5. Gabrys, B., Ruta, D.: Genetic algorithms in classifier fusion. Applied Soft Computing 6(4), 337–347 (2006)
6. Kazienko, P., Musiał, K., Kajdanowicz, T.: Multidimensional Social Network and Its Application to the Social Recommender System. IEEE Transactions on Systems, Man, and Cybernetics - Part A: Systems and Humans (2009) (accepted)
7. Keles, A., Kolcak, M., Keles, A.I.: The adaptive neuro-fuzzy model for forecasting the domestic debt. Knowledge-Based Systems 21(8), 951–957 (2008)
8. Kuncheva, L.: Combining Pattern Classifiers. In: Methods and Algorithms. John Wiley & Sons, Inc., Chichester (2004)
9. Lin, P.-C., Chen, J.-S.: A genetic-based hybrid approach to corporate failure prediction. International Journal of Electronic Finance 2(2), 241–255 (2008)
10. Ravi, V., Kurniawan, H., Nwee, P., Kumar, R.: Soft computing system for bank performance prediction. Applied Soft Computing 8(1), 305–315 (2008)
11. Rud, O.: Data Mining Cookbook. In: Modeling Data for Marketing, Risk, and Customer Relationship Management, John Wiley & Sons, Inc., Chichester (2001)
12. Swanson, N.R., White, H.: A model selection approach to assessing the information in the term structure using linear model and the artificial neural network. Journal of Business and Economics Statistics 13, 265–275 (1995)
13. Zurada, J., Lonial, S.: Comparison Of The Performance Of Several Data Mining Methods For Bad Debt Recovery In The Healthcare Industry. The Journal of Applied Business Research 21(2), 37–53 (2005)

# Author Index


Aguilar-Lasserre, Alberto 685
Akay, Bahriye 608
Alor-Hernandez, Giner 685
Anutariya, Chutiporn 75
Ayanegui-Santiago, Huberto 163
Aydın, Doğan 620

Babič, František 365
Badica, Costin 653
Bagić Babac, Marina 217
Barla, Michal 309
Bedi, Punam 709
Bednár, Peter 365
Bieliková, Mária 309
Bogunović, Nikola 377
Borzemski, Leszek 752

Čapkovič, František 596
Ceglarek, Dariusz 449
Chen, Da-Ren 152
Chiu, Sheng-Yung 152
Chlebuś, Marek 584
Chmielewski, Mariusz 460, 528
Cho, Kyu Cheol 553
Chohra, Amine 252
Choros, Kazimierz 743
Cortes-Robles, Guillermo 685
Cybulka, Jolanta 516

Díaz-Agudo, Belén 321
Duong, Trong Hai 540
Dydejczyk, Antoni 565

Freitas, Alex A. 629
Furdík, Karol 365

Galetic, Vedran 193
Gandotra, Vandana 709
Gałka, Andrzej 460
Garcia Martinez, Mario Alberto 685
Gautam, Divya 240
Gledec, Gordan 377
Goczyła, Krzysztof 492
Gomez, Juan Miguel 685
González, Carlos 266
González-Sanz, Sergio 321
Gorawski, Marcin 107
Graczyk, Magdalena 800
Gupta, Ashok K. 240

Haniewicz, Konstanty 449
Herrero, Pilar 118
Hoang, Hanh Huu 401
Hofstede, Gert Jan 205
Horvat, Marko 377
Hsu, Chuin-Chieh 152, 732
Hsu, Li-Fu 732

Ilie, Sorin 653

Jamil, Nuraini 676
Jang, Kyung-Soo 183
Jankowska, Beata 128
Jarema, Piotr 460
Jędrzejowicz, Joanna 641
Jędrzejowicz, Piotr 36, 641
Jezic, Gordan 193, 825
Jiménez, Luis 266
Jo, Geun Sik 540
Jonker, Catholijn M. 205
Juarez-Martinez, Ulises 685
Jung, Jason J. 472, 481
Juszczyszyn, Krzysztof 357

Kajdanowicz, Tomasz 850
Kaleta, Mariusz 788
Kamiński, Wojciech 584
Kang, Sanggil 676
Kanzari, Dalel 252
Karaboga, Dervis 608
Karwowski, Kamil 389
Kasianiuk, Krzysztof 229
Katarzyniak, Radosław 301
Kazienko, Przemysław 850
Keinänen, Helena 51
Kołaczek, Grzegorz 573
Kolbusz, Rolisław J. 229
Kosiński, Artur 460
Kotowski, Romuald 229, 584
Kozierkiewicz-Hetmańska, Adrianna 87

Krasowski, Kamil 460
Krejcar, Ondrej 697
Król, Dariusz 665
Krzystanek, Marek 838
Ku, Yi-Chen 732
Kułakowski, Konrad 294
Kułakowski, Krzysztof 565
Kunštić, Marijan 217

Lasota, Tadeusz 776, 800, 813, 838
Ławrynowicz, Agnieszka 504
Le, Thanh Manh 401
Lee, Jong Sik 553
Lee, Seung-Hyun 183
Lévy, Pierre 22
Li, Delong 540
Lin, Tsai-Duan 152
Lopes, Heitor S. 629
Luca, Longo 63
Łukasik, Szymon 97

Madani, Kurosh 252
Magdalena, Ciszczyk 333
Maleewong, Krissada 75
Maries, Iulia 140
Meissner, Adam 413
Memon, Zulfiqar A. 279
Mencke, Myriam 685
Mishra, Neha 709
Monekosso, Dorothy N. 266
Musiał, Katarzyna 357
Muskala, Monika 743

Nalepa, Grzegorz J. 345
Nguyen, Linh Anh 437
Nguyen, Ngoc Thanh 472
Noh, Chang Hyeon 553

Paletta, Mauricio 118
Pałka, Piotr 788
Paralič, Ján 365
Petric, Ana 193, 825
Pierpaolo, Dondio 63
Płuciennik-Psota, Ewa 107
Podobnik, Vedran 193
Popiela, Łukasz 665
Posada-Gomez, Ruben 685
Przemysław, Różewski 333

Recio-García, Juan A. 321
Remagnino, Paolo 266
Rodriguez Gonzalez, Alejandro 685
Rutkowski, Wojciech 449
Rybak, Marcin 565
Ryu, Sang-Hwan 183

Sachnowski, Piotr 776
Scarlat, Emil 140
Shin, Dong-Ryeol 183
Shin, Ho-Jin 183
Siemiński, Andrzej 720
Singh, Rishi Raj 240
Singh, Vivek Kumar 240
Singhal, Archana 709
Skorupa, Grzegorz 301
Słowiński, Roman 1
Spivack, Nova 425
Stephen, Barrett 63
Szałas, Andrzej 437
Szymkowiak, Magdalena 128

Telec, Zbigniew 813
Thanh, Hoang Chi 174
Thórisson, Kristinn R. 425
Toczyłowski, Eugeniusz 788
Traczyk, Tomasz 788
Trawiński, Bogdan 776, 800, 813, 838
Trawiński, Krzysztof 813
Treur, Jan 279
Trzec, Krunoslav 193
Tsunoda, Denise F. 629

Uddin, Mohammed Nazim 540

Vallejo, David 266
Vats, Vandita 709
Verwaart, Tim 205

Wagner, Jozef 365
Waloszek, Aleksander 492
Waloszek, Wojciech 492
Wąs, Jarosław 294
Wissner, James M. 425
Wuwongse, Vilas 75
Wysota, Witold 389
Wytrębowicz, Jacek 389

Żak, Sławomir 97
Zatwarnicka, Anna 752
Zatwarnicki, Krzysztof 752, 764

FSC
www.fsc.org
MIX
Papier | Fördert gute Waldnutzung
FSC® C083411

Kolibri 360 GmbH
Ferdinand-Jühlke-Straße 7
99095 Erfurt, Deutschland
produktsicherheit@kolibri360.de